Principles and Processes of Chemical Reactions

Principles and Processes of Chemical Reactions

Editor:

Pam Wang, Ph.D.

Delve Publishing
www.delvepublishing.com

Principles and Processes of Chemical Reactions

Editor: Pam Wang, Ph.D.

708 3rd Avenue, 6th Floor
New York, NY 10017
United States of America
www.delvepublishing.com

ISBN: 978-1-68095-902-4

Library of Congress Control Number: 2016951760

About the Editor

Pam Wang, Ph.D.

Pam Wang is a chemical biologist by training who has research experience in a variety of interdisciplinary fields. She received her PhD in 2014 studying artificial proteins and is now an industrial postdoc working on bacterial genome engineering. Her interests lie primarily in the application of biophysical, biochemical, and molecular engineering techniques to biotechnology and the pharmaceutical sciences.

Contents

List of Contributors

Maryam Salehi
Department of Medical Biophysics, University of Toronto, Toronto, ON, Canada

Theodore J. Perkins
Regenerative Medicine Program, Ottawa Hospital Research Institute, Ottawa, ON, Canada
Department of Biochemistry, Microbiology and Immunology, University of Ottawa, Ottawa, ON, Canada

Masaki Nakagawa
Department of Mathematical and Life Sciences, Graduate School of Science, Hiroshima University, Higashi-Hiroshima, Japan

Yuichi Togashi
Department of Mathematical and Life Sciences, Graduate School of Science, Hiroshima University, Higashi-Hiroshima, Japan
Research Center for the Mathematics on Chromatin Live Dynamics, Hiroshima University, Higashi-Hiroshima, Japan

Dhiman Maitra
Department of Obstetrics and Gynecology, The C.S. Mott Center for Human Growth and Development, Wayne State University School of Medicine, Detroit, MI, United States of America

Iyad Ali
Department of Obstetrics and Gynecology, The C.S. Mott Center for Human Growth and Development, Wayne State University School of Medicine, Detroit, MI, United States of America
Department of Biochemistry and Genetics, Faculty of Medicine and Health Sciences, An-Najah National University, Nablus, Palestine

Rasha M. Abdulridha
Department of Obstetrics and Gynecology, The C.S. Mott Center for Human Growth and Development, Wayne State University School of Medicine, Detroit, MI, United States of America

Faten Shaeib
Department of Obstetrics and Gynecology, The C.S. Mott Center for Human Growth and Development, Wayne State University School of Medicine, Detroit, MI, United States of America

Sana N. Khan
Department of Obstetrics and Gynecology, The C.S. Mott Center for Human Growth and Development, Wayne State University School of Medicine, Detroit, MI, United States of America

Ghassan M. Saed
Department of Obstetrics and Gynecology, The C.S. Mott Center for Human Growth and Development, Wayne State University School of Medicine, Detroit, MI, United States of America

Subramaniam Pennathur
Division of Nephrology, Department of Internal Medicine, University of Michigan Medical School, Ann Arbor, MI, United States of America

Husam M. Abu-Soud
Department of Obstetrics and Gynecology, The C.S. Mott Center for Human Growth and Development, Wayne State University School of Medicine, Detroit, MI, United States of America
Department of Biochemistry and Molecular Biology, The C.S. Mott Center for Human Growth and Development, Wayne State University School of Medicine, Detroit, MI, United States of America

Jakob Fischer
Bio Systems Analysis Group, Institute of Computer Science, Jena Centre for Bioinformatics and Friedrich Schiller University, Jena, Germany
Max-Planck-Institute for Biogeochemistry, Jena, Germany
International Max Planck Research School for Global Biogeochemical Cycles, Jena, Germany

Axel Kleidon
Max-Planck-Institute for Biogeochemistry, Jena, Germany

Peter Dittrich
Bio Systems Analysis Group, Institute of Computer Science, Jena Centre for Bioinformatics and Friedrich Schiller University, Jena, Germany

Ian J. Keyte
Division of Environmental Health & Risk Management, School of Geography, Earth & Environmental Sciences, University of Birmingham, Edgbaston, UK

Roy M. Harrison
Division of Environmental Health & Risk Management, School of Geography, Earth & Environmental Sciences, University of Birmingham, Edgbaston, UK

Gerhard Lammel
Max Planck Institute for Chemistry, Hahn-Meitner-Weg 1, 55128 Mainz, Germany
Masaryk University, Research Centre for Toxic Compounds in the Environment, Kamenice 5, 62500 Brno, Czech Republic

Bart Danon
Energy Technology, Delft University of Technology, Leeghwaterstraat 44, 2628CA, Delft, The Netherlands

Gianluca Marcotullio
SEA Servizi Energia Ambiente srl, Via Miraflores 2, 67100, L'Aquila, Italy

Wiebren de Jong
Energy Technology, Delft University of Technology, Leeghwaterstraat 44, 2628CA, Delft, The Netherlands

Carsten Hennig
Institute of Physical Chemistry, University of Göttingen, Tammannstr. 6, D-37077 Göttingen, Germany

Stefan Schmatz
Institute of Physical Chemistry, University of Göttingen, Tammannstr. 6, D-37077 Göttingen, Germany

Moritz P. Haag
Laboratory of Physical Chemistry, ETH Zurich, Vladimir Prelog route 2, CH-8093 Zurich, Switzerland

Markus Reiher
Laboratory of Physical Chemistry, ETH Zurich, Vladimir Prelog route 2, CH-8093 Zurich, Switzerland

Matthias Peter
Department of Chemical Physics, Fritz-Haber-Institut der Max-Planck-Gesellschaft, Faradayweg 4-6, 14195 Berlin, Germany

Sergey Adamovsky
Department of Chemical Physics, Fritz-Haber-Institut der Max-Planck-Gesellschaft, Faradayweg 4-6, 14195 Berlin, Germany

Jose Manuel Flores Camacho
Department of Chemical Physics, Fritz-Haber-Institut der Max-Planck-Gesellschaft, Faradayweg 4-6, 14195 Berlin, Germany

Swetlana Schauermann
Department of Chemical Physics, Fritz-Haber-Institut der Max-Planck-Gesellschaft, Faradayweg 4-6, 14195 Berlin, Germany

Hua Guo
Department of Chemistry and Chemical Biology, University of New Mexico, Albuquerque, New Mexico 87131, USA

Kopin Liu
Institute of Atomic and Molecular Sciences, Academia Sinica, Taipei 10617, Taiwan
Department of Physics, National Taiwan University, Taipei 10617, Taiwan

P. Kovaříček
Institute of Science and Supramolecular Engineering, University of Strasbourg, 8 avenue Gaspard Monge, 67000 Strasbourg, France

A. C. Meister
Institute of Science and Supramolecular Engineering, University of Strasbourg, 8 avenue Gaspard Monge, 67000 Strasbourg, France

K. Flídrová
Institute of Science and Supramolecular Engineering, University of Strasbourg, 8 avenue Gaspard Monge, 67000 Strasbourg, France

R. Cabot
Institute of Science and Supramolecular Engineering, University of Strasbourg, 8 avenue Gaspard Monge, 67000 Strasbourg, France

K. Kovaříčková
Institute of Science and Supramolecular Engineering, University of Strasbourg, 8 avenue Gaspard Monge, 67000 Strasbourg, France

J.-M. Lehn
Institute of Science and Supramolecular Engineering, University of Strasbourg, 8 avenue Gaspard Monge, 67000 Strasbourg, France

Martin E. Maier
Institute of Organic Chemistry, University of Tübingen, Auf der Morgenstelle 18, 72076 Tübingen, Germany

Matthew T. Clough
Department of Chemistry, Imperial College London, London, SW7 2AZ, UK

Karolin Geyer
BASF SE, Ludwigshafen, Germany

Patricia A. Hunt
Department of Chemistry, Imperial College London, London, SW7 2AZ, UK

Jürgen Mertes
BASF SE, Ludwigshafen, Germany

Tom Welton
Department of Chemistry, Imperial College London, London, SW7 2AZ, UK

Satish Balasaheb Nimse
Institute for Applied Chemistry, Department of Chemistry, Hallym University, Chuncheon, 200-702, Korea

Dilipkumar Pal
Institute of Pharmaceutical Sciences, Guru Ghasidas Vishwavidyalaya (A Central University), Koni, Bilaspur, Chhattisgarh-495009, India

Tobias Müller
Biokatalyse & Organische Chemie, Gebouw voor Scheikunde, Afdeling Biotechnologie, Technische Universiteit Delft, Julianalaan 136, 2628BL Delft, The Netherlands

Kristina Djanashvili
Biokatalyse & Organische Chemie, Gebouw voor Scheikunde, Afdeling Biotechnologie, Technische Universiteit Delft, Julianalaan 136, 2628BL Delft, The Netherlands

Isabel W. C. E. Arends
Biokatalyse & Organische Chemie, Gebouw voor Scheikunde, Afdeling Biotechnologie, Technische Universiteit Delft, Julianalaan 136, 2628BL Delft, The Netherlands

Joop A. Peters
Biokatalyse & Organische Chemie, Gebouw voor Scheikunde, Afdeling Biotechnologie, Technische Universiteit Delft, Julianalaan 136, 2628BL Delft, The Netherlands

Ulf Hanefeld
Biokatalyse & Organische Chemie, Gebouw voor Scheikunde, Afdeling Biotechnologie, Technische Universiteit Delft, Julianalaan 136, 2628BL Delft, The Netherlands

Atanu Modak
Department of Chemistry, Indian Institute of Technology, Powai, Mumbai, 400076, India

Debabrata Maiti
Department of Chemistry, Indian Institute of Technology, Powai, Mumbai, 400076, India

T. Gensch
Westfälische Wilhelms-Universität Münster, Organisch-Chemisches Institut, Corrensstrasse 40, 48149 Münster, Germany

M. N. Hopkinson
Westfälische Wilhelms-Universität Münster, Organisch-Chemisches Institut, Corrensstrasse 40, 48149 Münster, Germany

F. Glorius
Westfälische Wilhelms-Universität Münster, Organisch-Chemisches Institut, Corrensstrasse 40, 48149 Münster, Germany

J. Wencel-Delord
Laboratoire de Chimie Moléculaire (UMR CNRS 7509), Université de Strasbourg, ECPM 25 Rue Becquerel, 67087 Strasbourg, France

Nicole M. G. Franssen
Van't Hoff Institute for Molecular Sciences (HIMS), Universiteit van Amsterdam, Science Park 904, 1098 XH Amsterdam, The Netherlands.
Dutch Polymer Institute DPI, 5600 AX Eindhoven, The Netherlands

Joost N. H. Reek
Van't Hoff Institute for Molecular Sciences (HIMS), Universiteit van Amsterdam, Science Park 904, 1098 XH Amsterdam, The Netherlands

Bas de Bruin
Van't Hoff Institute for Molecular Sciences (HIMS), Universiteit van Amsterdam, Science Park 904, 1098 XH Amsterdam, The Netherlands.

Mark D. Greenhalgh
School of Chemistry, University of Edinburgh, Joseph Black Building, West Mains Road, Edinburgh, EH9 3JJ, UK

Stephen P. Thomas
School of Chemistry, University of Edinburgh, Joseph Black Building, West Mains Road, Edinburgh, EH9 3JJ, UK

Kenichi Horisawa
Division of Organogenesis and Regeneration, Department of Molecular and Cellular Biology, Medical Institute of Bioregulation, Kyushu University, Fukuoka, Japan

Preface

Chemical reactions are processes that transform chemical substances to create new substances with different identities. Traditionally, these processes involve solely the movement of electrons in forming and breaking bonds between atoms; processes that change the composition of nuclei are nuclear reactions. Equations are often used to represent the path from starting materials to end products, and provide details such as reaction conditions, stoichiometry, and intermediates. In a chemical equation, the number of atoms for each species must be equal on either side of the arrow, which indicates the direction in which the reaction proceeds. Reversible reactions are denoted by double arrows, which means the forward and reverse reactions will compete with each other until the system reaches equilibrium. The simplest form of a chemical reaction—in which reactants are converted into products in a single step with a single transition state—are known as elementary reactions. Although more commonly, reactions take place in a single phase, in certain cases, they occur at the interface between two different states (e.g. solid/gas interface).

Depending on the specific conditions, a chemical reaction can proceed under thermodynamic or kinetic control. A thermodynamically favored reaction is one where the products are more stable than the reactants (i.e. energy state of products is lower than that of reactants), resulting in the release of energy or heat (exothermic). On the other hand, a reaction is kinetically favored when the activation energy is low (i.e. low energy transition state), leading to faster reaction rates. Catalysts can lower the intrinsic activation energy of a reaction and thus speed it up. Generally speaking, thermodynamic and kinetic variables both need to be taken into account in determining whether or not a chemical reaction is favorable. There are four basic reaction types: synthesis, decomposition, single replacement, and double replacement. Synthesis refers to bond-forming reactions in which two or more simple substances combine into a more complex substance. Decomposition is the reverse of a synthesis reaction, involving the breakage of bonds in a complex substance to form simpler ones. Single replacement reactions involve the exchange of a single, uncombined element for another within a compound, while double replacement reactions involve a switch between two elements that are both within compounds. Reactions that lead to a change in oxidation states of atoms are classified as either oxidation or reduction reactions ("redox"). Photochemical reactions are ones that are catalyzed by light energy. Acid-base reactions involve the transfer of protons from the acidic species to the basic species. In organic chemistry, reactions are generally classified according to their mechanism (movement of electrons). In substitution reactions, a functional group is replaced by another group (the "nucleophile") in a compound. Elimination reactions lead to the loss of a functional group from a compound via the formation of a C=C double bond. By contrast, rearrangement reactions give rise to changes in the carbon skeleton, often resulting in structural isomers. Other organic reactions include radical reactions, oxidations/reductions, condensations, hydrolysis, polymerizations, among many others. Many molecules will have more than one reactive site – much research effort in synthetic chemistry has been put into developing conditions and catalysts that will selectively or specifically functionalize one site over another. Chemical reactions lie at the core of biological, industrial, and many other important processes. This book aims to provide a general overview of the type, mechanism, and application of chemical reactions.

Editor
Pam Wang, Ph.D.

INTRODUCTION

Chemical reaction a process in which one or more substances, the reactants, are converted to one or more different substances, the products. Substances are either chemical elements or compounds. A chemical reaction rearranges the constituent atoms of the reactants to create different substances as products.

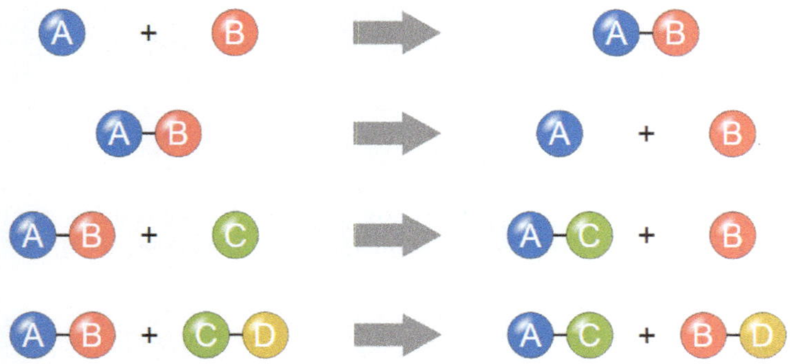

Chemical reactions are an integral part of technology, of culture, and indeed of life itself. Burning fuels, smelting iron, making glass and pottery, brewing beer, and making wine and cheese are among many examples of activities incorporating chemical reactions that have been known and used for thousands of years. Chemical reactions abound in the geology of Earth, in the atmosphere and oceans, and in a vast array of complicated processes that occur in all living systems.

Chemical reactions must be distinguished from physical changes. Physical changes include changes of state, such as ice melting to water and water evaporating to vapour. If a physical change occurs, the physical properties of a substance will change, but its chemical identity will remain the same. No matter what its physical state, water (H_2O) is the same compound, with each molecule composed of two atoms of hydrogen and one atom of oxygen. However, if water, as ice, liquid, or vapour, encounters sodium metal (Na), the atoms will be redistributed to give the new substances molecular hydrogen (H_2) and sodium hydroxide (NaOH). By this, we know that a chemical change or reaction has occurred.

Historical Overview

The concept of a chemical reaction dates back about 250 years. It had its origins in early experiments that classified substances as elements and compounds and in theories that explained these processes. Development of the concept of a chemical reaction had a primary role in defining the science of chemistry as it is known today.

The first substantive studies in this area were on gases. The identification of oxygen in the 18th century by Swedish chemist Carl Wilhelm Scheele and English clergyman Joseph Priestley had particular significance. The influence of French chemist Antoine-Laurent Lavoisier was especially notable, in that his insights confirmed the importance of quantitative measurements of chemical processes. The concept of chemical reactions involving the combination of elements clearly emerged from his writing, and his approach led others to pursue experimental chemistry as a quantitative science.

The other occurrence of historical significance concerning chemical reactions was the development of atomic theory. For this, much credit goes to English chemist John Dalton, who postulated his atomic theory early in the 19th century. Dalton maintained that matter is composed of small, indivisible particles, that the particles, or atoms, of each element were unique, and that chemical reactions were involved in rearranging atoms to form new substances. This view of chemical reactions accurately defines the current subject. Dalton's theory provided a basis for understanding the results of earlier experimentalists, including the law of conservation of matter (matter is neither created nor destroyed) and the law of constant composition (all samples of a substance have identical elemental compositions).

Thus, experiment and theory, the two cornerstones of chemical science in the modern world, together defined the concept of chemical reactions. Today experimental chemistry provides innumerable examples, and theoretical chemistry allows an understanding of their meaning.

Synthesis

When making a new substance from other substances, chemists say either that they carry out a synthesis or that they synthesize the new material. Reactants are converted to products, and the process is symbolized by a chemical equation. For example, iron (Fe) and sulfur (S) combine to form iron sulfide (FeS).

$$Fe(s) + S(s) \rightarrow FeS(s)$$

The plus sign indicates that iron reacts with sulfur. The arrow signifies that the reaction "forms" or "yields" iron sulfide, the product. The state of matter of reactants and products is designated with the symbols (s) for solids, (l) for liquids, and (g) for gases.

The Conservation of Matter

In reactions under normal laboratory conditions, matter is neither created nor destroyed, and elements are not transformed into other elements. Therefore, equations depicting reactions must be balanced; that is, the same number of atoms of each kind must appear on opposite sides of the equation. The balanced equation for the iron-sulfur reaction shows that one iron atom can react with one sulfur atom to give one formula unit of iron sulfide.

Chemists ordinarily work with weighable quantities of elements and compounds. For example, in the iron-sulfur equation the symbol Fe represents 55.845 grams of iron, S represents 32.066 grams of sulfur, and FeS represents 87.911 grams of iron sulfide. Because matter is not created or destroyed in a chemical reaction, the total mass of reactants is the same as the total mass of products. If some other amount of iron is used, say, one-tenth as much (5.585 grams), only one-tenth as much sulfur can be consumed (3.207 grams), and only one-tenth as much iron sulfide is produced (8.791 grams). If 32.066 grams of sulfur were initially present with 5.585 grams of iron, then 28.859 grams of sulfur would be left over when the reaction was complete.

The reaction of methane (CH_4, a major component of natural gas) with molecular oxygen (O_2) to produce carbon dioxide (CO_2) and water can be depicted by the chemical equation

$$CH_4(g) + 2O_2(g) \rightarrow CO_2(g) + 2H_2O(l)$$

Here another feature of chemical equations appears. The number 2 preceding O_2 and H_2O is a stoichiometric factor. (The number 1 preceding CH_4 and CO_2 is implied.) This indicates that one molecule of methane reacts with two molecules of oxygen to produce one molecule of carbon dioxide and two molecules of water. The equation is balanced because the same number of atoms of each element appears on both sides of the equation (here one carbon, four hydrogen, and four oxygen atoms). Analogously with the iron-sulfur example, we can say that 16 grams of methane and 64 grams of oxygen will produce 44 grams of carbon dioxide and 36 grams of water. That is, 80 grams of reactants will lead to 80 grams of products.

The ratio of reactants and products in a chemical reaction is called chemical stoichiometry. Stoichiometry depends on the fact that matter is conserved in chemical processes, and calculations giving mass relationships are based on the concept of the mole. One mole of any element or compound contains the same number of atoms or molecules, respectively, as one mole of any other element or compound. By international agreement, one mole of the most common isotope of carbon (carbon-12) has a mass of exactly 12 grams (this is called the molar mass) and represents $6.022140857 \times 10^{23}$ atoms (Avogadro's number). One mole of iron contains 55.847 grams; one mole of methane contains 16.043 grams; one mole of molecular oxygen is equivalent to 31.999 grams; and one mole of water is 18.015 grams. Each of these masses represents $6.022140857 \times 10^{23}$ molecules.

Energy Considerations

Energy plays a key role in chemical processes. According to the modern view of chemical reactions, bonds between atoms in the reactants must be broken, and the atoms or pieces of molecules are reassembled into products by forming new bonds. Energy is absorbed to break bonds, and energy is evolved as bonds are made. In some reactions the energy required to break bonds is larger than the energy evolved on making new bonds, and the net result is the absorption of energy. Such a reaction is said to be endothermic if the energy is in the form of heat. The opposite of endothermic is exothermic; in an exothermic reaction, energy as heat is evolved. The more general terms exoergic (energy evolved) and endoergic (energy required) are used when forms of energy other than heat are involved.

A great many common reactions are exothermic. The formation of compounds from the constituent elements is almost always exothermic. Formation of water from molecular hydrogen and oxygen and the formation of a metal oxide such as calcium oxide (CaO) from calcium metal and oxygen gas are examples. Among widely recognizable exothermic reactions is the combustion of fuels.

The formation of slaked lime (calcium hydroxide, $Ca(OH)_2$) when water is added to lime (CaO) is exothermic.

$$CaO(s) + H_2O\ (l) \rightarrow Ca(OH)_2(s)$$

This reaction occurs when water is added to dry portland cement to make concrete, and heat evolution of energy as heat is evident because the mixture becomes warm.

Not all reactions are exothermic (or exoergic). A few compounds, such as nitric oxide (NO) and hydrazine (N_2H_4), require energy input when they are formed from the elements. The decomposition of limestone ($CaCO_3$) to make lime (CaO) is also an endothermic process; it is necessary to heat limestone to a high temperature for this reaction to occur.

$$CaCO_3(s) \rightarrow CaO(s) + CO_2(g)$$

The decomposition of water into its elements by the process of electrolysis is another endoergic process. Electrical energy is used rather than heat energy to carry out this reaction.

$$2\ H_2O(g) \rightarrow 2\ H_2(g) + O_2(g)$$

Generally, evolution of heat in a reaction favors the conversion of reactants to products. However, entropy is important in determining the favorability of a reaction. Entropy is a measure of the number of ways in which energy can be distributed in any system. Entropy accounts for the fact that not all energy available in a process can be manipulated to do work.

A chemical reaction will favor the formation of products if the sum of the changes in entropy for the reaction system and its surroundings is positive. An example is burning wood. Wood has a low entropy. When wood burns, it produces ash as well as the high-entropy substances carbon dioxide gas and water vapor. The entropy of the reacting system increases during combustion. Just as important, the heat energy transferred by the combustion to its surroundings increases the entropy in the surroundings. The total of entropy changes for the substances in the reaction and the surroundings is positive, and the reaction is product-favored.

When hydrogen and oxygen react to form water, the entropy of the products is less than that of the reactants. Offsetting this decrease in entropy, however, is the increase in entropy of the surroundings owing to the heat transferred to it by the exothermic reaction. Again because of the overall increase in entropy, the combustion of hydrogen is product-favored.

Kinetic Considerations

Chemical reactions commonly need an initial input of energy to begin the process. Although the combustion of wood, paper, or methane is an exothermic process, a burning match or a spark is needed to initiate this reaction. The energy supplied by a match arises from an exothermic chemical reaction that is itself initiated by the frictional heat generated by rubbing the match on a suitable surface.

In some reactions, the energy to initiate a reaction can be provided by light. Numerous reactions in Earth's atmosphere are photochemical, or light-driven, reactions initiated by solar radiation. One example is the transformation of ozone (O_3) into oxygen (O_2) in the troposphere. The absorption of ultraviolet light (hv) from the Sun to initiate this reaction prevents potentially harmful high-energy radiation from reaching Earth's surface.

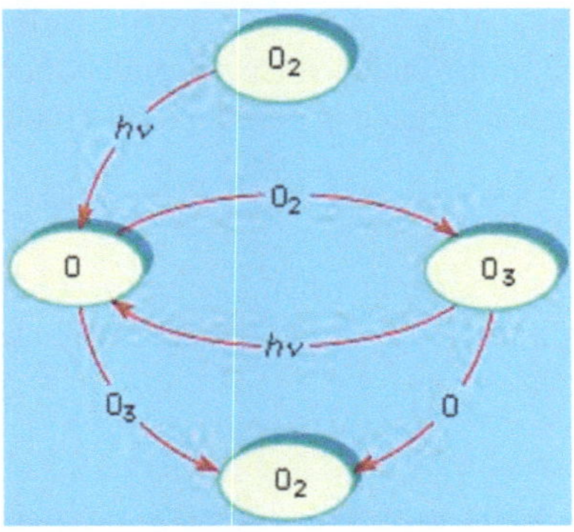

For a reaction to occur, it is not sufficient that it be energetically product-favored. The reaction must also occur at an observable rate. Several factors influence reaction rates, including the concentrations of reactants, the temperature, and the presence of catalysts. The concentration affects the rate at which reacting molecules collide, a prerequisite for any reaction. Temperature is influential because reactions occur only if collisions between reactant molecules are sufficiently energetic. The proportion of molecules with sufficient energy to react is related to the temperature. Catalysts affect rates by providing a lower energy pathway by which a reaction can occur. Among common catalysts are precious metal compounds used in automotive exhaust systems that accelerate the breakdown of pollutants such as nitrogen dioxide into harmless nitrogen and oxygen. A wide array of biochemical catalysts are also known, including chlorophyll in plants (which facilitates the reaction by which atmospheric carbon dioxide is converted to complex organic molecules such as glucose) and many biochemical catalysts called enzymes. The enzyme pepsin, for example, assists in the breakup of large protein molecules during digestion.

Classifying Chemical Reactions

Chemists classify reactions in a number of ways: (a) by the type of product, (b) by the types of reactants, (c) by reaction outcome, and (d) by reaction mechanism. Often, a given reaction can be placed in two or even three categories.

Classification by Type of Product

Gas-forming Reactions

Many reactions produce a gas such as carbon dioxide, hydrogen sulfide (H_2S), ammonia (NH_3), or sulfur dioxide (SO2). An example of a gas-forming reaction is that which occurs when a metal carbonate such as calcium carbonate ($CaCO_3$, the chief component of limestone, seashells, and marble) is mixed with hydrochloric acid (HCl) to produce carbon dioxide.

$$CaCO_3(s) + 2\ HCl(aq) \rightarrow CaCl_2(aq) + CO_2(g) + H_2O\ (l)$$

In this equation, the symbol (aq) signifies that a compound is in an aqueous, or water, solution.

Cake-batter rising is caused by a gas-forming reaction between an acid and baking soda, sodium hydrogen carbonate (sodium bicarbonate, $NaHCO_3$). Tartaric acid ($C_4H_6O_6$), an acid found in many foods, is often the acidic reactant.

$$C_4H_6O_6(aq) + NaHCO_3(aq) \rightarrow NaC_4H_5O_6(aq) + H_2O\ (l) + CO_2(g)$$

In this equation, $NaC_4H_5O_6$ is sodium tartrate.

Most baking powders contain both tartaric acid and sodium hydrogen carbonate, which are kept apart by using starch as a filler. When baking powder is mixed into the moist batter, the acid and sodium hydrogen carbonate dissolve slightly, which allows them to come into contact and react. Carbon dioxide is produced, and the batter rises.

Precipitation Reactions

Formation of an insoluble compound will sometimes occur when a solution containing a particular cation (a positively charged ion) is mixed with another solution containing a particular anion (a negatively charged ion). The solid that separates is called a precipitate.

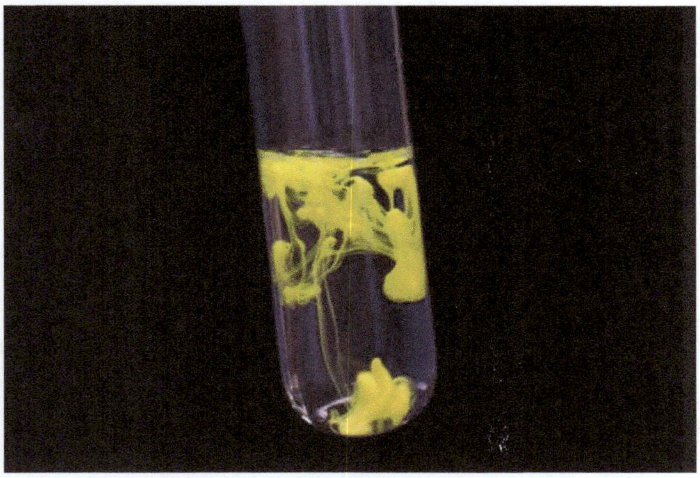

Compounds having anions such as sulfide (S^{2-}), hydroxide (OH^-), carbonate (CO_3^{2-}), and phosphate (PO_4^{3-}) are often insoluble in water. A precipitate will form if a solution containing one of these anions is added to a solution containing a metal cation such as Fe^{2+}, Cu^{2+}, or Al^{3+}.

$Fe^{2+}(aq) + 2\ OH^-(aq) \rightarrow Fe(OH)2(s)$

$Al^{3+}(aq) + PO_4^{3-}(aq) \rightarrow AlPO^4(s)$

Minerals are water-insoluble compounds. Precipitation reactions in nature can account for mineral formation in many cases, such as with undersea vents called "black smokers" that form metal sulfides.

Classification by Types of Reactants

Two types of reactions involve transfer of a charged species. Oxidation-reduction reactions occur with electron transfer between reagents. In contrast, reactions of acids with bases in water involve proton (H^+) transfer from an acid to a base.

Oxidation-reduction Reactions

Oxidation-reduction (redox) reactions involve the transfer of one or more electrons from a reducing agent to an oxidizing agent. This has the effect of reducing the real or apparent electric charge on an atom in the substance being reduced and of increasing the real or apparent electric charge on an atom in the substance being oxidized. Simple redox reactions include the reactions of an element with oxygen. For example, magnesium burns in oxygen to form magnesium oxide (MgO). The product is an ionic compound, made up of Mg^{2+} and O^{2-} ions. The reaction occurs with each magnesium atom giving up two electrons and being oxidized and each oxygen atom accepting two electrons and being reduced.

Another common redox reaction is one step in the rusting of iron in damp air.

$$2Fe(s) + 2H_2O(l) + O_2(g) \rightarrow 2Fe(OH)_2(s)$$

Here iron metal is oxidized to iron dihydroxide ($Fe(OH)_2$); elemental oxygen (O_2) is the oxidizing agent.

Redox reactions are the source of the energy of batteries. The electric current generated by a battery arises because electrons are transferred from a reducing agent to an oxidizing agent through the external circuitry. In a common dry cell and in alkaline batteries, two electrons per zinc atom are transferred to the oxidizing agent, thereby converting zinc metal to the Zn^{2+} ion. In dry-cell batteries, which are often used in flashlights, the electrons given up by zinc are taken up by ammonium ions (NH_4^+) present in the battery as ammonium chloride (NH_4Cl). In alkaline batteries, which are used in calculators and watches, the electrons are transferred to a metal oxide such as silver oxide (AgO), which is reduced to silver metal in the process.

The Arrhenius Theory

The Arrhenius theory, named after Swedish physicist Svante August Arrhenius, views an acid as a substance that increases the concentration of the hydronium ion (H_3O^+) in an aqueous solution and a base as a substance that increases the hydroxide ion (OH^-) concentration. Well-known acids include hydrochloric acid (HCl), sulfuric acid (H_2SO_4), nitric acid (HNO_3), and acetic acid (CH3COOH). Bases includes such common substances as caustic soda (sodium hydroxide, NaOH) and slaked lime (calcium hydroxide, $Ca(OH)_2$). Another common base is ammonia (NH_3), which reacts with water to give a basic solution according to the following balanced equation.

$$NH_3(aq) + H_2O(l) \rightarrow NH_4^+(aq) + OH^-(aq)$$

(This reaction occurs to a very small extent; the hydroxide ion concentration is small but measurable.)

A large number of natural bases are known, including morphine, cocaine, nicotine, and caffeine; many synthetic drugs are also bases. All of these contain a nitrogen atom bonded to three other groups, and all behave similarly to ammonia in that they can react with water to give a solution containing the hydroxide ion.

Amino acids, a very important class of compounds, are able to function both as acids and as bases. Amino acidmolecules contain both acidic (–COOH) and basic ($-NH_2$) sites. In an aqueous solution, amino acids exist in both the molecular form and the so-called "zwitterionic" form, $H_3N + CH_2CO_2^-$. In this structure the nitrogen atom bears a positive charge, and the oxygen atom of the acid group bears a negative charge.

According to the Arrhenius theory, acid-base reactions involve the combination of the hydrogen ion (H^+) and the hydroxide ion to form water. An example is the reaction of aqueous solutions of sodium hydroxide and hydrochloric acid.

$$HCl(aq) + NaOH(aq) \rightarrow NaCl(aq) + H_2O\ (l)$$

The Brønsted-Lowry Theory

A somewhat more general acid-base theory, the Brønsted-Lowry theory, named after Danish chemist Johannes Nicolaus Brønsted and English chemist Thomas Martin Lowry, defines an acid as a proton donor and a base as a proton acceptor. In this theory, the reaction of an acid and base is represented as an equilibrium reaction.

$$\text{acid (1)} + \text{base (2)} \rightleftharpoons \text{base (1)} + \text{acid (2)}$$

(The double arrows, $\rightleftharpoons$, indicate that the products can re-form the reactants in a dynamic process.)

Acid (1) and base (1) are called a conjugate acid-base pair, as are acid (2) and base (2). The advantage of this theory is its predictive capacity. Whether the equilibrium lies toward the reactants (reactant-favoured) or the products (product-favoured) is determined by the relative strengths of the acids and bases.

The Brønsted-Lowry theory is often closely associated with the solvent water. Dissolving an acid in water to form the hydronium ion and the anion of the acid is an acid-base reaction. Acids are classified as strong or weak, depending on whether the equilibrium favours the reactants or products. Hydrochloric acid, a strong acid, ionizes completely in water to form the hydronium and chlorine (Cl^-) ions in a product-favoured reaction.

$$HCl(aq) + H_2O\ (l) \rightarrow H_3O^+(aq) + Cl^-(aq)$$

Using the Brønsted-Lowry theory, the reaction of ammonia and hydrochloric acid in water is represented by the following equation:

$$NH_3(aq) + HCl(aq) \rightarrow NH_4^+(aq) + Cl^-(aq)$$

Hydrochloric acid and the chlorine ion are one conjugate acid-base pair, and the ammonium ion and ammonia are the other. The acid-base reaction is the transfer of the hydrogen ion from the acid (HCl) to the base (NH_3). The equilibrium favours the weaker acid and base, in this case the products. Note that the hydroxide ion does not appear in this equation, a point differentiating the Arrhenius and Brønsted-Lowry theories.

The Lewis theory

A still broader acid and base theory was proposed by American physical chemist Gilbert Newton Lewis. In the Lewis theory, bases are defined as electron-pair donors and acids as electron-pair acceptors. Acid-base reactions involve the combination of the Lewis acid and base through sharing of the base's electron pair.

Ammonia is an example of a Lewis base. A pair of electrons located on the nitrogen atom may be used to form a chemical bond to a Lewis acid such as boron trifluoride (BF_3). (In the following equation, the colon represents an electron pair.)

$$H_3N: + BF_3 \rightarrow H_3N\text{–}BF_3$$

Ammonia, water, and many other Lewis bases react with metal ions to form a group of species known as coordination compounds. The reaction to form these species is another example of a Lewis acid-base reaction. For example, the light blue color of a solution of Cu^{2+} ions in water is due to the $[Cu(H_2O)_6]^{2+}$ ion. If ammonia is added to this solution, the water molecules attached to copper are replaced by ammonia molecules, and the beautiful deep blue ion $[Cu(NH_3)_4]^{2+}$ is formed.

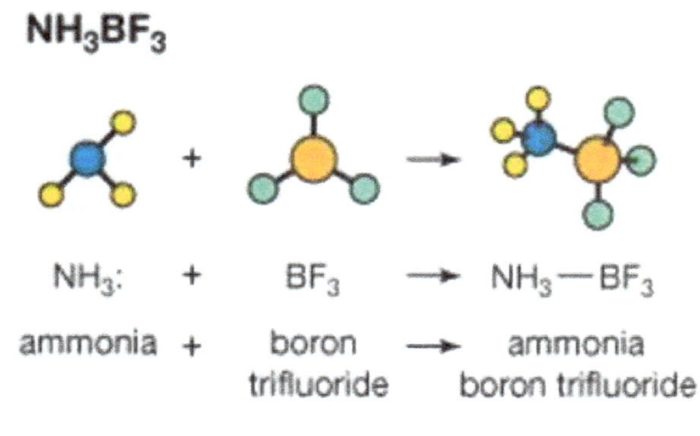

CHAPTER 1

MAXIMUM PROBABILITY REACTION SEQUENCES IN STOCHASTIC CHEMICAL KINETIC SYSTEMS

Maryam Salehi and Theodore J. Perkins

Keywords: Stochastic chemical kinetics, trajectory inference, peptide folding, Trp-cage, ant colony optimization, population-based incremental learning

INTRODUCTION

Increasingly, we have detailed knowledge about the chemical interactions or transformations in which various biomolecules participate. For example, high-throughput assays such as ChIP-chip or ChIP-seq are rapidly identifying the DNA binding sites of transcription factors (Horak and Snyder, 2002; Valouev et al., 2008). Quantitative tandem mass spectrometry allows us to not only identify protein complexes, but also to study how they form (Gingras et al., 2007; Link et al., 1999). Protein folding simulations identify intra-molecular binding events that define different folding paths (Snow et al., 2002). Real-time fluorescence microscopy allows us to see single molecules moving and interacting (Nie et al., 1995; Sekar and Periasamy, 2003).

When studying such interactions at single-cell or even single-molecule levels, mass-action chemical kinetics can be either misleading or simply inapplicable. In such cases, dynamics are more accurately represented by stochastic chemical kinetic models (SCKMs) (Gardiner, 2004; Van Kampen, 2008). These models define a chemical system in terms of the types of molecules or molecular configurations that are possible, the types of chemical interactions or transformations that may occur, and the state-dependent probabilities with which they occur. The state of the chemical system is given by the number of molecules of each type that are present, hence is discrete, and the state evolves stochastically in continuous time, as described in greater detail below. SCKMs are often used, for example, to model stochastic thermodynamic switching between different configurations of a protein (Marinelli et al., 2009), to model the opening and closing of ion channels (Ball and Rice, 1992), and to study the sources of noise in gene expression (Swain et al., 2002; Thattai and van Oudenaarden, 2001).

Considerable effort has been devoted to various computational problems surrounding SCKMs. For example, computing stationary distributions and computing time-dependent state probabilities given an initial state (usually via the chemical master equation) are two classical and well-studied problems (Van Kampen, 2008). The development of correct and efficient methods for simulating the dynamics of such models is another major area of research (Gibson and Bruck, 2000; Gillespie, 1977). More recently, there have been efforts to learn the parameters or even structure of SCKMs based on time series data (Henderson et al., 2010; Tian et al., 2007). SCKMs, which we define carefully below, can also be viewed as a means for describing continuous-time Markov chains (Anderson, 1991). Thus, many theoretical results and methods from the Markov chain literature also apply to SCKMs, although historically the two communities developed largely separately and have focused on different applications.

We focus on a problem that has received very little attention in the SCKM or continuous-time Markov chain communities: finding most-probable reaction sequences connecting temporally-separated observations of the system. Such problems can arise very naturally – for example, as a means of estimating the behavior of a molecular system between experimental observations, or for extracting "prototypical" behaviors connecting different endpoints. For discrete-time stochastic models, such as Markov chains or hidden Markov models, similar trajectory inference problems are of enormous importance, with applications in areas such as path planning, speech recognition, error correction, robot navigation, DNA sequence analysis, user categorization, etc. Moreover, efficient dynamic programming algorithms can be used to find most-probable system trajectories (Bertsekas, 1995; Rabiner, 1989). For SCKMs, there has been work on

Citation: Salehi M and Perkins TJ (2010) Maximum probability reaction sequences in stochastic chemical kinetic systems. Front. Physio. 1:170. doi: 10.3389/fphys.2010.00170.

finding most-probable trajectories in the case of large system sizes, where the SCKM dynamics can be approximated by a stochastic differential equation (Dykman et al., 1994; Liu, 2008). However, we know of no previous work on finding most-probable trajectories in which the continuous-time and discrete-state nature of the SCKM is retained – a distinction that is crucial in settings where molecule counts are very small, including binary presence/absence situations such as the binding state of a gene's promoter region or the folding state of an individual protein molecule.

We expect that solutions to trajectory inference problems for SCKMs will have widespread utility. For example, towards the end of this paper, we demonstrate this possibility by analyzing an SCKM describing the stochastic folding of the Trp-cage peptide. Other potential applications include finding most-probable sequences of transcription factor binding and interaction leading to gene activation, sequences of protein complex assembly, or sequences of nucleic acid polymer folding. Maximum probability reaction sequences are, of course, only one aspect of the stochastic dynamics of a chemical system. For systems in which the most-probable trajectory is much more probable than the alternatives, or when other probable trajectories are "similar" to the most-probable one, then the most-probable trajectory is representative of the dynamics as a whole. However, if there are many alternative trajectories with similar probability, including possibly quite different end states, then looking at only the single most-probable trajectory may be misleading. There are well-established methods for computing transient and stationary probability distributions over the state space (Van Kampen, 2008), and for performing stochastic simulations of chemical systems (Gillespie, 1977). These provide alternative views of system behavior which are not sensitive to issues of the uniqueness or representativeness of maximum probability trajectories. However, transient or stationary distributions do not include trajectory-based information, such as which states follow which other ones. Stochastic simulations, although they include trajectory-based information, are based on random sampling rather than exact calculations. Thus, maximum probability trajectories provide a novel and potentially useful form of information for understanding stochastic chemical systems, and one that is complementary to other computational approaches.

It turns out that trajectory inference for SCKMs is not so readily or elegantly solvable as it is for discrete-time models (Markov chains or hidden Markov models) or for stochastic differential equation models. We show that the probability of a particular trajectory can be evaluated in polynomial time. However, results from the study of Petri nets tell us that finding most-probable trajectories is at least EXPSPACE-hard in general. Nonetheless, we develop a correct exhaustive search method that is sufficient to answer questions in a peptide folding domain. We also propose and evaluate several heuristic optimization strategies on a hard optimization instance from the peptide domain. Our work thus establishes baseline expectations and algorithms for what we believe will turn out to be an important computational problem in analyzing SCKMs and in making inferences based on observational data.

RESULTS AND DISCUSSION

Stochastic Chemical Kinetic Models and Problem Statement

An SCKM is a stochastic, continuous-time, discrete-state model of the dynamics of a chemical system. It is specified in terms of a finite set of chemical reactions that may take place between a finite set of chemical species, or types of molecules. In the standard interpretation, one assumes that the molecules are contained in a bounded, well-mixed, fixed-volume, constant-temperature space (Van Kampen, 2008). Extensions to modeling spatial inhomogeneity or temporal variation in parameters are possible (Andrews and Bray, 2004; Gibson and Bruck, 2000; Shahrezaei et al., 2008), but we do not consider them here.

Let there be N chemical species and M chemical reactions that may take place between them. Let $\mathbb{N} = \{0, 1, 2,\ldots\}$. The state of the system at any time is given by the number of molecules of each species, $\mathbf{X} = (X_1,\ldots,X_N) \in \mathbb{N}^N$. Reaction i consumes C_{ij} molecules of species j (the *reactants*) and produces D_{ij} molecules of species j (the *products*). Reaction i can occur in state $\mathbf{X}$ only if there are sufficient reactants to match those consumed, in which case we say the reaction is *valid* in that state. If it occurs, then the resulting state is

$$\mathbf{X}' = \left(X_1 - C_{i1} + D_{i1},\ldots,X_N - C_{iN} + D_{iN}\right).$$

While the effects of a reaction occurring are deterministic, which reactions occur and when they occur are both stochastic. The *propensity* of reaction i to occur in state $\mathbf{X}$ is

$$\lambda_i(\mathbf{X}) = r_i \prod_{j=1}^{N} \binom{X_j}{C_{ij}}, \qquad (1)$$

where r_i is the kinetic rate constant associated to reaction i. Let $\lambda(\mathbf{X}) = \Sigma_i \lambda_i(\mathbf{X})$ be the total propensity of the reactions in state $\mathbf{X}$. Then, the time until the next reaction is exponentially distributed with parameter $\lambda(\mathbf{X})$, and, independently of the reaction time, reaction i is the next one to occur with probability $\lambda_i(\mathbf{X})/\lambda(\mathbf{X})$. The only exception to these rules is when $\lambda(\mathbf{X}) = 0$, in which case no reactions can occur, and the system stays in state $\mathbf{X}$ forever.

In a given amount of time t, there is no strict limit on the number of reactions that may occur. However, each possible sequence of reactions occurs with some probability. The computational problem we study is: given an initial state $\mathbf{X}_0$, a final time $t > 0$, and a non-empty target set of states $\mathbb{X}$, find the/a most-probable sequence of valid reactions that brings the system from $\mathbf{X}_0$ to any state in $\mathbb{X}$ in time t. In the special case that $\mathbb{X} = \mathbb{N}^N$, one is simply asking for the most-probable reaction sequence to occur over time t, starting from $\mathbf{X}_0$.

The Probability of a Reaction Sequence is Computable in Polynomial Time

Before we address the problem of finding most-probable reaction sequences, we consider how to compute the probability of a particular reaction sequence. This problem is more involved than one might think, though it can be solved in polynomial time[1]. Consider a reaction sequence $R_1,\ldots,R_k$ which, from initial state $\mathbf{X}_0$ produces successive states $\mathbf{X}_1,\ldots,\mathbf{X}_K$. Let $t_0,\ldots,t_K$ be the random waiting times spent in each state before the next reaction occurs. The probability of the reaction sequence $R_1,\ldots,R_K$ occurring in time t from state $\mathbf{X}_0$ depends on three events: (i) each reaction must be selected among the alternatives, (ii) the reactions must complete by time t, and (iii) no reaction can occur in state $\mathbf{X}_K$ until after time t. We can write this as

$$P(R_1,\ldots,R_K \mid \mathbf{X}_0,t) = \left(\prod_{i=0}^{K-1}\frac{\lambda_{R_{i+1}}(\mathbf{X}_i)}{\lambda(\mathbf{X}_i)}\right)P(t_0+\cdots+t_{K-1}\le t \text{ and } t_0+\cdots+t_K>t) \quad (2)$$

The first term is easy to compute. The second term concerns sums of exponentially distributed random variables – that is, hypoexponential distributions. In two special cases, there are simple analytical formulae for the densities of such sums. When all the $\lambda(\mathbf{X}_i)$ are equal to a common λ, then the sum of the first K times follows the Erlang distribution.

$$l(t_0+\cdots+t_{K-1}=t)=\frac{\lambda^K t^{K-1}e^{-\lambda t}}{(K-1)!} \quad (3)$$

If all the $\lambda(\mathbf{X}_i)$ are distinct, then the sum of the first K times follows a different distribution.

$$l(t_0+\cdots+t_{K-1}=t)=\left(\prod_{i=0}^{K-1}\lambda(\mathbf{X}_i)\right)\left(\sum_{i=0}^{K-1}\frac{e^{-\lambda(\mathbf{X}_i)t}}{\prod_{j\ne i}\lambda(\mathbf{X}_j)-\lambda(\mathbf{X}_i)}\right) \quad (4)$$

In either case, the desired probability can then be obtained by computing the convolution:

$$P(t_0+\cdots+t_{K-1}\le t \text{ and } t_0+\cdots+t_K>t) = \int_{\tau=0}^{t} l(t_0+\cdots,t_{K-1}=\tau)P(t_K>t-\tau)d\tau$$

where $P(t_K>t-\tau)=e^{-\lambda(\mathbf{X}_k)(t-\tau)}$. However, as we will see, we commonly encounter situations in which some of the $\lambda(\mathbf{X}_i)$ are the same and some are different, in which case there is no convenient, explicit formula.

We propose a general, polynomial-time approach to computing the probability based on standard transient-distribution calculations from the theory of continuous-time Markov chains. As shown in Figure 1, our strategy is to imagine a continuous-time Markov chain in which each step along the reaction path is one state, and the exponential waiting times in each state are the same as in the corresponding states of the SCKM. In essence, we strip away any alternative reaction paths that might be followed. If we assume that the chain starts in state $\mathbf{X}_0$, then the probability that the chain is in state $\mathbf{X}_K$ at time t is precisely the probability that all K reactions of the SCKM complete in time t, but no other reactions occur afterward. Let $\rho(t)$ denote the probability distribution over states of the continuous-time chain at time t. With Q being the matrix shown in Figure 1C, the forward Chapman–Kolmogorov equation states that $\rho(t)$ obeys the linear differential equation $(d/dt)\rho = Q\rho$. The solution to this equation is $\rho(t) = e^{Qt}\rho(0)$. The matrix exponential, e^{Qt}, can be computed in a polynomial number of operations (e.g., by converting Qt to Jordan normal form, in which matrix exponentiation is easy). Evaluating the next-to-last component of $\rho(t)$ gives us the second term in Eq. 2, and thus the entire probability of the reaction sequence, in polynomial time. Such a direct approach to computing transient distributions is not generally followed in the SCKM literature because an SCKM can in general have a very large or even countably infinite number of states, leading to an intractably large matrix $\mathbf{Q}$. In the present use, however, we are concerned with only the states along a particular reaction path, so that the computation is tractable. See Materials and Methods for details of our implementation.

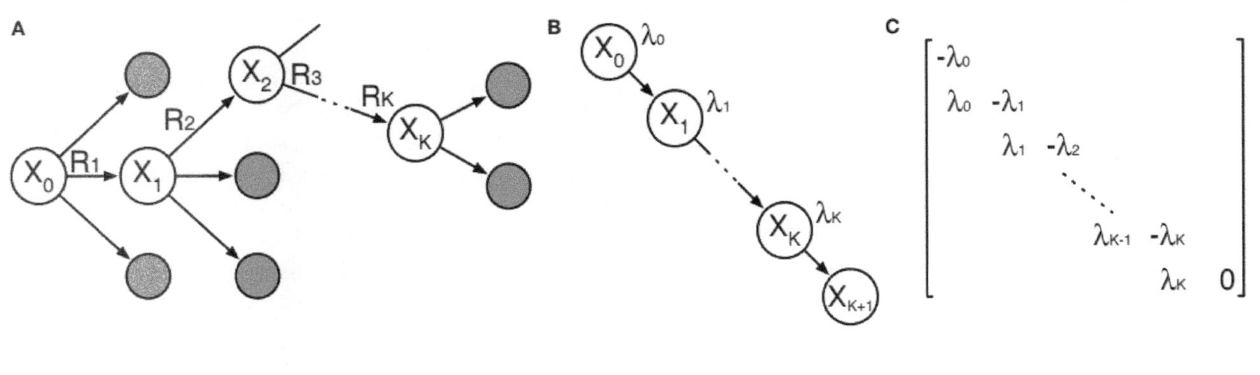

Figure 1. Outline of proposed strategy for computing the probability that a reaction sequence completes in a given amount of time. (A) The reaction sequence picks out a sequence of states starting from X_0. (B) A continuous-time Markov chain describing the timing of transitions along that sequence of states. The waiting time in state $\mathbf{X}_i$ is exponentially distributed with parameter $\lambda_i = \lambda(\mathbf{X}_i)$. **(C)** The rate matrix for the chain shown in **(B)**.

Finding Most-Probable Reaction Sequences is Computationally Hard

Because the state space, $\mathbb{N}^N$, is countably infinite, it should be no surprise that it is difficult to find a most-probable reaction sequence leading to $\mathbb{X}$. Indeed, results regarding Petri nets tell us that even establishing the reachability of $\mathbb{X}$ from $\mathbf{X}_0$ is computationally difficult. Although they have come into much more general use, Petri nets originated as a mean for describing chemical systems (Petri and Reisig, 2008). They have states (species) with markings (molecule counts) and transformations (reactions) which subtract markings from some states and add markings to other states (consume reactants, and produce products). An SCKM is a particular kind of stochastic Petri net, by virtue of its special probabilistic rules for the timing and selection of reactions. The reachability question for general Petri nets, which means finding a sequence of transformations leading from one marking of the states to a different marking of the states, is decidable and EXPSPACE-hard (Esparza and Nielsen, 1994). Various restrictions on Petri nets can reduce the complexity of reachability questions. For example, if markings are restricted to being binary (each molecular species is either present or absent), then the problem is PSPACE-complete (see Esparza and Nielsen, 1994 for more). These results immediately imply similar hardness results for our SCKM problem, because finding a most-probable reaction sequence from $\mathbf{X}_0$ to $\mathbb{X}$ requires determining whether there is any such reaction sequence at all. More formally, for any Petri net reachability problem, we can pose the decision problem: consider an SCKM with species and reactions corresponding to the states and transitions of the Petri net, let all reaction rates be $r_i = 1$, let $\mathbf{X}_0$ correspond to the initial marking of the Petri net, and let $\mathbb{X}$ be a singleton set corresponding to the final marking of the Petri net; does there exist a sequence of valid reactions bringing the SCKM from $\mathbf{X}_0$ to $\mathbb{X}$ with probability ≥ 0. Clearly, reachability is true for the Petri net problem if and only if the desired reaction sequence exists for the SCKM (regardless of its probability, as long as the sequence is valid).

Correct Exhaustive Search for the Most-Probable Reaction Sequence

Having established that efficient identification of the most-probable reaction sequence is impossible in general, we next propose an exhaustive approach based on best-first search. Our procedure is "correct" in the sense that if it terminates and returns a reaction sequence, then that sequence is guaranteed to be a maximum probability sequence. Moreover, we establish conditions, which turn out to be readily satisfied in practice, that ensure termination of the search procedure. Thus, when these conditions are met, the algorithm is guaranteed to terminate and return a maximum probability reaction sequence.

Recall that the best-first search procedure maintains a priority queue of partial solutions to the optimization problem (the *search frontier*), which it repeatedly expands until identifying the optimal solution (Russell and Norvig, 1995). In the present context, a partial solution is any valid reaction sequence beginning in state $\mathbf{X}_0$, whether or not a state in the target set $\mathbb{X}$ is the result of the reaction sequence. We propose to prioritize a reaction sequence $\mathcal{R} = (R_1,\ldots,R_K)$ resulting in state sequence $\mathbf{X}_0,\ldots,\mathbf{X}_K$ by

$$pri(\mathcal{R}) = \left(\prod_{i=0}^{K-1} \frac{\lambda_{R_{i+1}}(\mathbf{X}_i)}{\lambda(\mathbf{X}_i)} \right) P\left(t_0 + \cdots + t_{K-1} \leq t\right) \quad (6)$$

where the t_i are the random waiting times in the states $\mathbf{X}_i$. Comparing to Eq. 2, we have dropped the requirement that no other reactions happen before time t. Indeed, Eq. 6 is the probability that $\mathcal{R}$ or any extension of that sequence occur in time t. As a consequence, the priority assigned to the sequence is an upper bound on the actual probability of $\mathcal{R}$ occurring (Eq. 2), as well as on the actual probability of any extension of the sequence.

Now, the best-first search operates as follows. At all times, the procedure keeps track of the most-probable reaction sequence $\mathcal{R}^*$ from $\mathbf{X}_0$ to $\mathbb{X}$ found so far, if any. Let $P(\mathcal{R}^*)$ denote the probability of that sequence. If no such solution has been found so far, then we write $\mathcal{R}^* =$ **undef**. The search is initialized by placing the empty reaction sequence, $\mathcal{R} = \emptyset$, on the priority queue with priority 1. At each iteration of the search, the highest-priority reaction sequence $\mathcal{R}$ is removed from the queue, its priority $pri(\mathcal{R})$ defined by Eq. 6. $\mathcal{R}$ is processed according to the following steps. (See Materials and Methods for more detailed pseudocode).

- If $\mathcal{R}^* \neq$ undef and $pri(\mathcal{R}) \leq P(\mathcal{R}^*)$, then terminate the search, returning $\mathcal{R}^*$ as the solution.
- If $\mathcal{R}$ results in a state in $\mathbb{X}$, then compute the exact probability of that reaction sequence, $P(\mathcal{R})$. If $P(\mathcal{R}) > P(\mathcal{R}^*)$, then set $\mathcal{R}^*$ equal to $\mathcal{R}$.
- Add every possible single-reaction extension of $\mathcal{R}$ to the priority queue.

If the algorithm terminates in line 1, then the following reasoning establishes the correctness of the solution returned. First, if the search finds a solution of probability $P(\mathcal{R}^*)$, and the highest-priority partial solution in the queue has priority $pri(\mathcal{R}) < P(\mathcal{R}^*)$, then neither that partial solution nor any extension of it can have probability higher than $P(\mathcal{R}^*)$. This is true because of how we have chosen the priority function. Moreover, all other partial solutions in the queue have priority no higher than $P(\mathcal{R})$, and so neither they nor any extensions of them can have probability higher than $P(\mathcal{R}^*)$. This establishes the correctness of the algorithm.

Whether or not the algorithm terminates at all is another question. If there are infinitely many possible reaction sequences, but none leads to the target set $\mathbb{X}$, then the algorithm does not terminate. If $\mathbb{X}$ is reachable, then the only way the algorithm can fail to terminate is if the search explores arbitrarily long reaction sequences without ever triggering the termination condition. From Eq. 6, for this to happen requires an infinite reaction sequence for which $\lambda_{R_{i+1}}(\mathbf{X}_i)/\lambda(\mathbf{X}_i)$ converges rapidly to one and for which $P(t_0 + t_1 + t_2 + \ldots \leq t) > 0$ is not too small. The latter condition implies that $\lambda(\mathbf{X}_i) \rightarrow +\infty$, which, considering Eq. 1, implies that the number of molecules of at least one of the species must escape to infinity in finite time. One can write down SCKMs with this property. For example, the SCKM with a single species A, and a single-reaction rule $2A \rightarrow 3A$, has finite probability of generating infinitely many A's over any finite time interval. Such SCKMs are obviously physically impossible, let alone biologically realistic, so we do not consider this a great concern. As long as reachability and probability one finiteness of all molecular counts over time t can be established, both of which are often straightforward in practice, then the search will terminate.

An Analysis of Trp-Cage Folding

To demonstrate the potential utility of the problem we have formulated, we consider a recently-proposed model of Trp-cage folding (Marinelli et al., 2009). Trp-cage is a synthetic peptide of 20 amino acids. Marinelli et al. (2009) used molecular dynamics computations to estimate the major configurations in which the peptide can reside, and the transitions between those configurations. They proposed the five-state model shown in

Figure 2A. This model can be analyzed within the framework of SCKMs by equating each possible configuration with a distinct chemical species and the transitions between them as different chemical reactions.

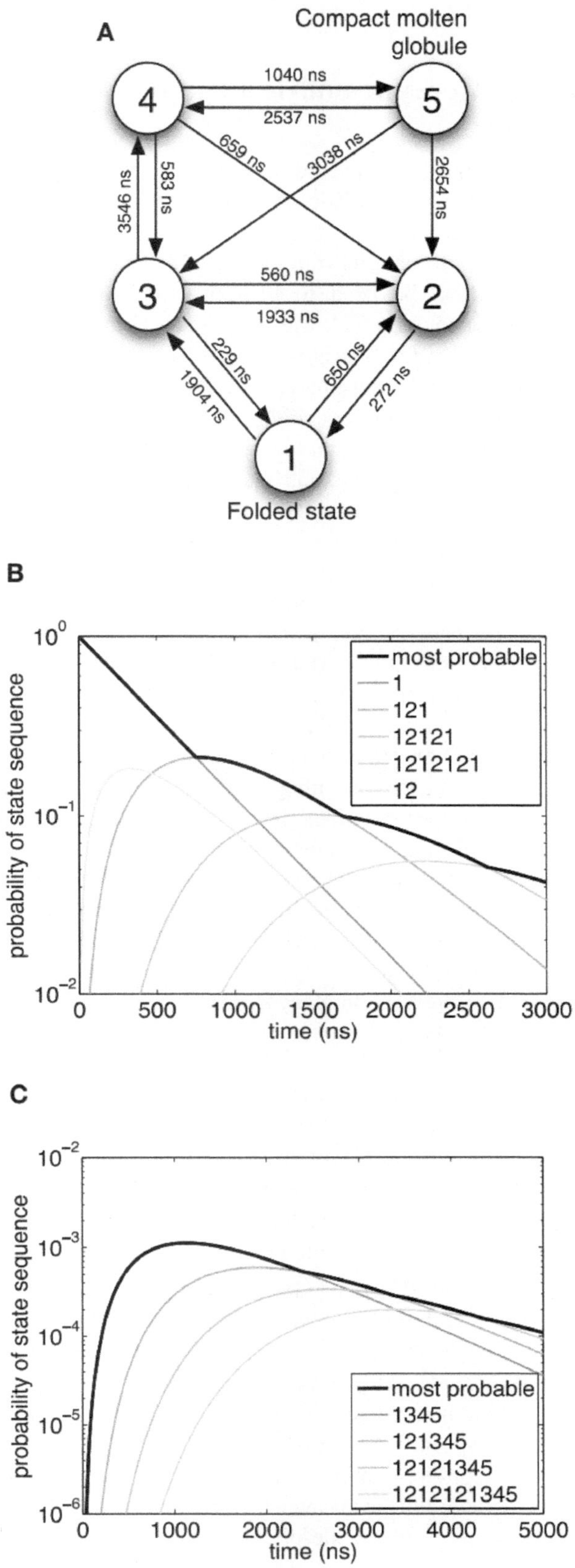

Figure 2. Most probable reaction sequences for a model of Trp-cage folding. (A) High-level model of Trp-cage stochastic folding dynamics from Marinelli et al. (2009). Circles correspond to major configurations, with "1" being the most stable, folded configuration, and "5" being an un/mis-folded configuration that is rarely visited but difficult to escape from. Arcs represent possible transitions and are labeled with the expected time for a transition to occur – the inverse of the kinetic rate constant r_i for that reaction. **(B)** The probability of the most-probable state sequence, and of several specific state sequences, as a function of time, assuming the peptide begins in state 1 and can end in any state. **(C)** The probability of the most-probable state sequence, and of several specific state sequences, as a function of time, assuming the peptide begins in state 1 and ends in state 5.

We used best-first search to compute answers to two questions. For the first question, we consider the problem in which the peptide is initially in the folded state, 1, and we ask what is the most-probable sequence of reactions (or equivalently, configurations of the peptide) for varying end times t, allowing for any possible end state. The results are shown in Figure 2B. For short end times, the peptide stays in the folded state the whole time. However, with more time, it becomes more likely for the peptide to switch to state 2 and then back to the folded state. Interestingly, the intermediate possibility of moving to state 2 and still being there at time t is not a most-probable outcome for any t. With even more time, the peptide is most likely to switch back and forth two or more times between state 1 and state 2. Figure 2B also shows the probabilities of these individual state sequences as a function of time.

Figure 2C shows the results of our second analysis, in which we assumed that the peptide starts in the folded state and ends in state 5. At the smallest times, the shortest possible state path, $1 \rightarrow 3 \rightarrow 4 \rightarrow 5$, is most-probable. Unlike the previous example, however, this most-probable path has probability approaching zero as $t \rightarrow 0$, because it is very unlikely for the peptide to go from the folded state to the molten globule state in so little time. For larger amounts of time, the most-probable path has the peptide flipping between states 1 and 2 before following the $3 \rightarrow 4 \rightarrow 5$ route to the molten globule state.

As one would expect for exhaustive search, the time complexity increases approximately exponentially with increasing end time and/or optimal solution length. For example, for the second problem, at $t = 1000, 2000, 3000, 4000$, and 5000 ns, the search either evaluates a candidate solution or evaluates the lower bound (Eq. 6) 210, 787, 2492, 6633, and 15,286 times respectively. The evaluation of solutions or lower bounds is the dominant computational cost in the search.

Probabilistically Correct Search by Stochastic Sampling

Although best-first search can identify optimal solutions, its exponential time complexity encouraged us to look for simpler, heuristic approaches. A simple, non-exhaustive method to search for high-probability reaction sequences is to generate sequences randomly, computing the probability of all those that result in a state in $\mathbb{X}$, and keeping track of the best solution found. In particular, we propose to generate reaction sequences using Gillespie's stochastic simulation algorithm (Gillespie, 1977), which generates random realizations of the stochastic dynamics of an SCKM. From start state $\mathbf{X}_0$, Gillespie's algorithm repeatedly generates a random waiting time until the next reaction and a random next reaction, until reaching the final time t. While simple, this approach is not entirely naive. Of all possible reaction sequences that result in a state in $\mathbb{X}$, the one with the highest true probability has the greatest chance of being generated. If the highest probability reaction sequence has probability p, then the chance that this approach fails to find it in K random realizations of the dynamics is $(1 - p)^K$. In the limit as $K \rightarrow +\infty$, the optimal reaction sequence is found with probability one, and in particular, the probability of not finding the optimal solution decreases exponentially in the number of attempts made. However, the expected number of random samples until the optimal sequence is found is $1/p$, which can be quite large when the most-probable path has low probability.

We tested the Gillespie generate-and-test approach on the Trp-Cage domain, using the problem with initial state 1, final state 5, and final time t = 5000 ns. From the best-first search results, we knew the optimal reaction sequence, and we knew its probability $p \approx 1.09 \times 10^{-4}$. We allowed the Gillespie approach to generate K = 15,286 trajectories, equal to the number of trajectories scored or lower-bounded by the best-first search. We repeated this procedure 100 independent times. The results are shown in Figure 3. Figure 3A shows that as more and more trajectories are generated, an increasing fraction of the 100 runs found the optimal solution. By the end, 78 of the runs had done so, consistent with the predicted fraction of $1 - (1 - p)^K \approx 0.81$. Figure 3B shows the geometric average of the probabilities of the best solutions found by the 100 runs as more trajectories are generated. Although it is not obvious from the figure, many runs find alternative solutions before identifying the optimal one – some only a little worse, and some startlingly worse, with probabilities of 10^{-10} or smaller. After 15,286 trajectories, the worst of the 100 runs had found a best state sequence of 1212121212121345, which has probability $\approx 4.61 \times 10^{-5}$.

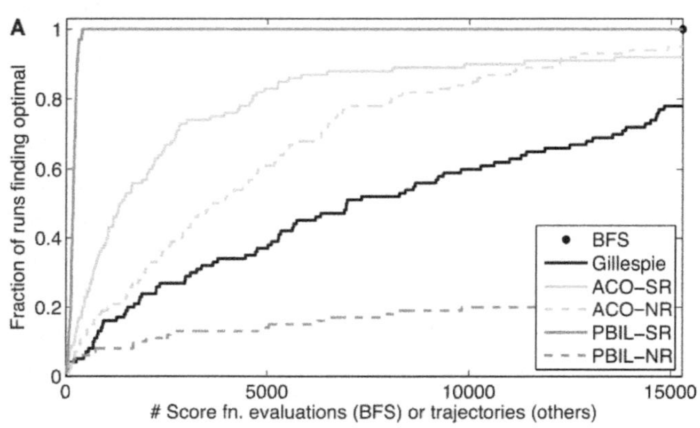

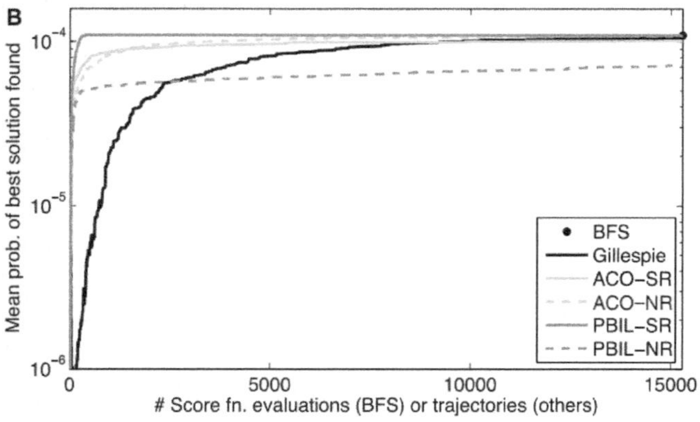

Figure 3. Heuristic optimization results on the peptide problem with initial state 1, final state 5, and final time t = **5000 ns.** "BFS" refers to best-first search, "Gillespie" to the generate-and-test approach using the Gillespie algorithm, "ACO" to ant colony optimization, and "PBIL" to population-based incremental learning. The suffix "-SR" denotes the SR matrix representation for ACO or PBIL, whereas "-NR" denotes the number-of-reactions/reaction matrix representation. **(A)** The fraction of 100 independent runs discovering the optimal solution as a function of the number of sample trajectories. **(B)** The geometric mean, across the 100 runs, of the probabilities of the best solutions found as a function of the number of sample trajectories.

Adaptive Generate-and-Test Optimization Approaches

Given the partial success of the Gillespie approach, we decided to try more general heuristic optimization approaches, while staying within the overall generate-and-test scheme. We supposed that information from previously-evaluated reaction sequences could be used to bias the random generation towards better solutions. We tried two approaches, Ant Colony Optimization (ACO) (Dorigo et al., 2006) and Population-Based Incremental Learning (PBIL) (Baluja, 1994). ACO has been applied to protein folding before (Shmygelska and Hoos, 2005), albeit in a rather different problem formulation. Both approaches employ a matrix to summarize the results of previous solution attempts, and to guide the generation of future solutions. We tried two different matrix representations. In our SR representation, the matrix is indexed by states of the system (1–5) and "actions" (1–14, corresponding to the 13 reactions and an "end here" action which ends a reaction sequence at the current state; this action was allowed only in state 5, and of course, other reactions were only allowed from their source states). Intuitively, the entries of the matrix represent an estimate of how good it is to take the action (extending the reaction sequence, or ending it) in the given state. In our number-of-reactions/reactions (NR) representation, the matrix is indexed by the number of reactions so far (0–20) and actions (1–14). Here, the entries represent an estimate of how good it is to choose a next reaction or end the reaction sequence depending on how long the reaction sequence already is. In this representation, if 20 reactions are generated without reaching the target state, the reaction sequence is discarded. ACO updates its matrix after every attempt at generating a solution, whereas PBIL makes a set of attempts and updates its matrix based on the best (and optionally worst) solution(s) found in that set. We used standard implementations and default parameter settings for these algorithms; see Section "Materials and Methods" for details. Our goal was not necessarily to obtain the best possible performance, but rather to test how well off-the-shelf heuristic approaches would perform without any tuning.

The results of 100 independent runs of these two approaches and two representations are shown in Figure 3. With the SR representation, both ACO and PBIL significantly outperform the Gillespie approach, both in terms of how many of the runs discovered the optimal solution and in mean solution quality. PBIL was especially successful. All 100 runs discovered the optimal solution, taking just 183.14 sampled trajectories on average, and 399 for the worst of the 100 runs. Although ACO-NR outperformed Gillespie, PBIL-NR did rather poorly. Only 22 runs of the latter found the optimal solution, although the mean solution score rose quickly early on. These results show that, at least on this problem, heuristic approaches can improve on the performance of both Gillespie and best-first search, by using information from earlier trajectories to guide the generation of new candidate solutions.

MATERIALS AND METHODS

Computing Reaction Sequence Probabilities

As described above, given an SCKM, an initial state $\mathbf{X}_0$, a reaction sequence $R_1,\ldots,R_K$, and a final time t, we compute the probability of that reaction sequence as follows. Let $\mathbf{X}_1,\ldots,\mathbf{X}_K$ be the states that result from the corresponding reactions. First, we compute the propensities of every reaction i in every state $\mathbf{X}_j$, $0 \leq j \leq K$, along the reaction path, $\lambda_i(\mathbf{X}_j)$. By summing over reactions, this gives us the total propensities in each state, $\lambda(\mathbf{X}_j)$. From these propensities, we straightforwardly compute the first term of Eq. 2. For the second term, we form the matrix Q, as described in Figure 1. We multiply this by t element-wise, to obtain Qt. We then compute the matrix exponential of this using the expm function of matlab. The entry in the first column and next-to-last row of this matrix is the probability that all reactions complete in time t, but no reaction occurs in state $\mathbf{X}_K$.

Best-First Search for a Most-Probable Reaction Sequence

Suppose we are given an SCKM, an initial state $\mathbf{X}_0$, and a target set $\mathbb{X}$. For our best-first search, we assume access to several helper functions. RSP computes the probability of a particular reaction sequence (Eq. 2).RSPUB computes Eq. 6, which is an upper bound on the probability of a reaction sequence as well as any extension of that reaction sequence. EndsInTargetSet tell us whether a given reaction sequence results in a state in $\mathbb{X}$. Extensions returns a list of all valid one-reaction extensions of a given reaction sequence. We also use a priority queue PQ, which supports enqueue, dequeue, and not-empty functions.

Ant Colony Optimization and Population-Based Incremental Learning Details

Both ant colony optimization (ACO) and population-based incremental learning (PBIL) use a non-negative matrix, M, to summarize the successes and failures of previous attempts at generating solutions to the problem. In the state/reaction (SR) representation, M has 5 rows (corresponding to the five possible states of the system) and 14 columns. The columns correspond to extending a partial solution by one of the 13 possible reactions, though in any given state only a few of these are actually possible, or choosing to end a reaction sequence, which is allowed only if the previous reactions have brought us to the target state. In the number-of-steps/reaction representation (NR), M has 21 rows (corresponding to 0 to 20 reactions having been chosen so far), and 14 columns (which have the same meaning as in the SR representation). Both ACO and PBIL use M to generate candidate solutions in the same way, though the exact method depends on whether the SR or NR representation is used, as shown below. Note that the NRSeq method does not always succeed in generating a reaction path ending at the target state, whereas the SRSeq method keeps going until it chooses "end here" at the target state.

Algorithm 1 Best-first search for optimal reaction sequence

```
// Initialization
RSeq ← ∅
PQ.enqueue(1,RSeq)
BestRSeq ← undef
BestProb ← 0
// Continue search as long as nodes on priority queue
while PQ.notempty do
  (Priority,RSeq) ← PQ.dequeue
  // If highest-priority node has worse score than a
     solution already found, we're done
  if BestProb ≥ Priority then
     return BestRSeq
  end if
  // If reaction sequence brings us to 𝕏, compute
     probability and record, if appropriate
  if EndsInTargetSet(RSeq) then
     Prob ← RSP(RSeq)
     if Prob > BestProb then
        BestProb ← Prob
        BestRSeq ← RSeq
     end if

  // Consider all possible extensions of the reaction
     sequence
  for all RSeqExt in Extensions(RSeq) do
    Priority ← RSPUB(RSeqExt)
    PQ.enqueue(Priority,RSeqExt)
  end for
end while
return BestRSeq
```

Algorithm 2 SRSeq: SR method for generating random sample reaction sequences

```
RSeq ← ∅
KeepGoing ← true
while KeepGoing do
  i ← the state resulting from RSeq
  J ← the set of reactions possible from state i, and
     the "end here" choice, if i is the target set
  Choose j ∈ J with probability proportional to M_ij
  if j = "end here" then
     KeepGoing ← false
  else
     Add j to the end of RSeq
  end if
end while
return RSeq
```

Algorithm 3 NRSeq: NR method for generating random sample reaction sequences

```
RSeq ← ∅
KeepGoing ← true
while KeepGoing do
  i ← the state resulting from RSeq
  J ← the set of reactions possible from state i, and
      the "end here" choice, if i is the target set
  Choose j ∈ J with probability proportional to M_ij
  if j = "end here" or length(Rseq)=20 then
     KeepGoing ← false
  else
     Add j to the end of RSeq
  end if
end while
return RSeq
```

Ant colony optimization updates its matrix M after every target-reaching reaction sequence is generated and its probability computed. The intuition is that choices leading to higher-probability solutions get bigger increments in the M matrix. Pseudocode for the algorithm is given below. The outermost loop can be done a fixed number of times, or until some other criterion is satisfied. We stopped each run when it had either generated 15,286 candidate reaction paths or it had found the optimal solution, whichever came first. The matrix update depends on a parameter δ, which we take to be 0.1.

```
Algorithm 4 Ant Colony Optimization (ACO)
  BestRSeq ← undef
  BestProb ← 0
  while not <Termination Condition> do
   // Generate a candidate solution
   RSeq ← SRSeq (or NRSeq)
   // If it's a valid solution, compare with best solution
       and update M
   if RSeq results in a state in 𝕏 then
       Prob ← RSP(RSeq)
       if Prob > BestProb then
          BestProb ← Prob
          BestRSeq ← RSeq
       end if
       for all State- (or step-) reaction pairs (i,j)
            specified by RSeq do
          M_{i,j} ← (1 − δ) M_{i,j} − 1/log(Prob)
       end for
   end if
  end while
  return BestRSeq
Algorithm 5 Population Based Incremental Learning (PBIL)
  BestRSeq ← undef
  BestProb ← 0
  while not <Termination Condition> do
    Generate PopSize candidate solutions by SRSeq (or
    NRSeq)
    Compute the probabilities of any valid solutions
    generated
    if at least one valid solution was generated then
       RSeq ← the highest-probability valid solution
       Prob ← RSP(RSeq)
       // Compare against best solution so far
       if Prob > BestProb then
         BestProb ← Prob
         BestRSeq ← RSeq
       end if

    // Update matrix M
    for all State- (or step-) reaction pairs (i,j)
       specified by RSeq do
       M_{i,j} ← (1 - δ) M_{i,j} + δ
       M_{i,j'}  ← (1 - δ)M_{i,j'}δ for all j' ≠ j
    end for
  end if
  // "Mutate" the matrix M
  for all State- (or step-) reaction pairs (i,j) do
     With probability ρ, perform the following update,
     where r is a uniform random variable on [0, 1]:
     M_{i,j} ← M_{i,j} + μ × r
  end for
  Normalize rows of M so they sum to one
end while
```

Population-based incremental learning takes a more complicated approPBIL takes a more complicated approach to updating M. It repeatedly generates a batch of PopSize = 10 candidate solutions using either the SR or NR method. It checks whether each solution is valid in the sense of leading to the target set. It finds the best of those, and increases the probabilities of the choices that led to it, decreasing the probabilities of the other choices. Then, it randomizes the matrix M by a small amount. The rows of the matrix M are maintained to sum to one, so that they can be interpreted as probabilities. The algorithm depends on a learning rate parameter $\delta = 0.1$, a mutation probability $\rho = 0.02$, and a mutation shift $\mu = 0.05$, all recommended default values (Baluja, 1994).

CONCLUSION

We have proposed a novel computational problem, that of finding most-probable reaction sequences for SCKMs, and studied its solution. We believe that this problem is relevant to molecular biology now, and will grow in relevance as we obtain better characterizations of the stochastic chemical behavior of biomolecules and biomolecular networks. We have demonstrated the potential utility of the problem formulation by analyzing a stochastic model of peptide folding. Many other applications can be envisaged. For example, one might, in a similar fashion, study most-probable sequences of transcription factor-DNA binding events that lead to gene activation or most-probable sequences of protein-protein interaction events in protein complex assembly.

We showed that computing the probability of an individual reaction sequence can be done in polynomial time, but that finding most-probable reaction sequences is computationally intractable in general. In essence, this is because the reachability question itself is hard. It is an open question whether our SCKM problem is any harder than the reachability problem. Conversely, if we can establish reachability by other means, as is often possible in practical applications, does the optimization problem become any easier? Even in the case that the reachable state space is polynomial in the size of the problem description, we have not been able to derive an efficient algorithm for finding optimal reaction sequences. The complex, non-linear form of the reaction sequence probability (see Eq. 4 especially), defies convenient decomposition. The probability of a particular reaction sequence is non-monotone in t, while, for fixed t, adding extra reactions to the end of a sequence may either increase or decrease its probability. We conjecture that finding most-probable reaction sequences is intractable even if reachability of $\mathbb{X}$ is guaranteed and the reachable state space is polynomial in size, but this remains to be determined definitively.

We proposed a correct exhaustive search method for finding most-probable reaction sequences, and demonstrated its use in analyzing folding behavior in a high-level model of Trp-Cage molecular dynamics. Although the focus of this paper was on establishing basic hardness results and a first correct algorithm, we also evaluated several heuristic approaches to solving the problem. The results on the peptide folding problem indicate that heuristics have the potential to outperform exhaustive search, with one form of PBIL doing particularly well. However, no general conclusions can be drawn without a more thorough empirical study. We have no doubt that improved heuristic solution methods can be developed, and we are actively working on doing so. Another important avenue for further research is to look at more sophisticated types of queries. For example, one may be interested in most-probable folding paths of a protein that avoid certain states or interactions – such as misfolded states that may be associated with diseases (Selkoe, 2004) or interactions that may be the target of pharmaceutical interventions. More generally, one can imagine marrying temporal logics with SCKMs (similar to Chabrier-Rivier et al., 2004) to ask a host of important questions about the stochastic behavior of biomolecular systems.

ACKNOWLEDGMENTS

This work was supported in part by grants from the National Sciences and Engineering Research Council of Canada and The Ottawa Hospital Research Institute.

REFERENCES

1. Anderson, W. J. (1991). *Continuous-Time Markov Chains: An Applications-Oriented Approach*. New York: Springer-Verlag.
2. Andrews, S. S., and Bray, D. (2004). Stochastic simulation of chemical reactions with spatial resolution and single molecule detail. *Phys. Biol.* 1, 137–151.
3. Ball, F. G., and Rice, J. A. (1992). Stochastic models for ion channels: introduction and bibliography. *Math. Biosci.* 112, 189.
4. Baluja, S. (1994). *Population-Based Incremental Learning*. Pittsburgh: School of Computer Science, Carnegie-Mellon University. Technical Report Number CMU-CS-94-163.
5. Bertsekas, D. P. (1995). *Dynamic Programming and Optimal Control*. Athena, MA: Scientific Belmont.
6. Chabrier-Rivier, N., Chiaverini, M., Danos, V., Fages, F., and Schächter, V. (2004). Modeling and querying biomolecular interaction networks.*Theor. Comp. Sci.* 325, 25–44.
7. Dorigo, M., Birattari, M., and Stutzle, T. (2006). Ant colony optimization. *IEEE Comput. Intell. Mag.* 1, 28–39.
8. Dykman, M. I., Mori, E., Ross, J., and Hunt, P. M. (1994). Large fluctuations and optimal paths in chemical kinetics. *J. Chem. Phys.* 100, 5735–5750.
9. Esparza, J., and Nielsen, M. (1994). Decidability issues for petri nets – a survey. *J. Inf. Process. Cybern.* 30, 143–160.
10. Gardiner, C. W. (2004). *Handbook of Stochastic Methods: For Physics, Chemistry and the Natural Sciences*. New York: Springer-Verlag.
11. Gibson, M. A., and Bruck, J. (2000). Efficient exact stochastic simulation of chemical systems with many species and many channels. *J. Phys. Chem. A* 104, 1876–1889.
12. Gillespie, D. T. (1977). Exact stochastic simulation of coupled chemical reactions. *J. Phys. Chem.* 81, 2340–2361.
13. Gingras, A. C., Gstaiger, M., Raught, B., and Aebersold, R. (2007). Analysis of protein complexes using mass spectrometry. *Nat. Rev. Mol. Cell Biol.* 8, 645–654.
14. Henderson, D. A., Boys, R. J., and Wilkinson, D. J. (2010). Bayesian calibration of a stochastic kinetic computer model using multiple data sources. *Biometrics* 66, 249–256.
15. Horak, C. E., and Snyder, M. (2002). ChIP-chip: a genomic approach for identifying transcription factor binding sites. *Meth. Enzymol.* 350, 469–483.
16. Link, A. J., Eng, J., Schieltz, D. M., Carmack, E., Mize, G. J., Morris, D. R., Garvik, B. M., and Yates, J. R. III. (1999). Direct analysis of protein complexes using mass spectrometry. *Nat. Biotechnol.* 17, 676–682.
17. Liu, D. (2008). A numerical scheme for optimal transition paths of stochastic chemical kinetic systems. *J. Comput. Phys.* 227, 8672–8684.
18. Macintyre, A., and Wilkie, A. J. (1996). *On the decidability of the real exponential field*. Kreiseliana: about and around Georg Kreisel. Wellesley: Peters, 441–467.
19. Marinelli, F., Pietrucci, F., Laio, A., and Piana, S. (2009). A kinetic model of trp-cage folding from multiple biased molecular dynamics simulations. *PLoS Comput. Biol.* 5, e1000452. doi: 10.1371/journal.pcbi.1000452.
20. Nie, S., Chiu, D. T., and Zare, R. N. (1995). Real-time detection of single molecules in solution by confocal fluorescence microscopy. *Anal. Chem.*67, 2849–2857.
21. Petri, C. A., and Reisig, W. (2008). Petri net. *Scholarpedia* 3, 6477.
22. Rabiner, L. R. (1989). A tutorial on hidden Markov models and selected applications in speech recognition. *Proc. IEEE* 77, 257–286.
23. Russell, S., and Norvig, P. (1995). *Artificial Intelligence: A Modern Approach*. Upper Saddle River, NJ: Prentice Hall.
24. Sekar, R. B., and Periasamy, A. (2003). Fluorescence resonance energy transfer (FRET) microscopy imaging of live cell protein localizations. *J. Cell Biol.* 160, 629–633.
25. Selkoe, D. J. (2004). Cell biology of protein misfolding: the examples of Alzheimer's and Parkinson's diseases. *Nat. Cell Biol.* 6, 1054–1061.
26. Shahrezaei, V., Ollivier, J. F., and Swain, P. S. (2008). Colored extrinsic fluctuations and stochastic gene expression. *Mol. Syst. Biol.* 4, 196.
27. Shmygelska, A., and Hoos, H. H. (2005). An ant colony optimisation algorithm for the 2D and 3D hydrophobic polar protein folding problem.*BMC Bioinformatics* 6, 30. doi: 10.1186/1471-2105-6-30.
28. Snow, C. D., Nguyen, H., Pande, V. S., and Gruebele, M. (2002). Absolute comparison of simulated and experimental protein-folding dynamics.*Nature* 420, 102–106.
29. Swain, P. S., Elowitz, M. B., and Siggia, E. D. (2002). Intrinsic and extrinsic contributions to stochasticity in gene expression. *Proc. Natl. Acad. Sci. U.S.A.* 99, 12795–12800.

30. Thattai, M., and van Oudenaarden, A. (2001). Intrinsic noise in gene regulatory networks. *Proc. Natl. Acad. Sci. U.S.A.* 98, 8614–8619.

31. Tian, T., Xu, S., Gao, J., and Burrage, K. (2007). Simulated maximum likelihood method for estimating kinetic rates in gene expression.*Bioinformatics* 23, 84.

32. Valouev, A., Johnson, D. S., Sundquist, A., Medina, C., Anton, E., Batzoglou, S., Myers, R. M., and Sidow, A. (2008). Genome-wide analysis of transcription factor binding sites based on ChIP-Seq data. *Nat. Methods* 5, 829–834.

33. Van Kampen, N. G. (2008). *Stochastic Processes in Physics and Chemistry*. New York: Elsevier.

34. Weispfenning, V. (2000). "Deciding linear-transcendental problems," in *Computer Algebra in Scientific Computing: CASC 2000: Proceedings of the Third Workshop on Computer Algebra in Scientific Computing*, October 5–9 (Samarkand: Springer Verlag), 423.

CHAPTER 2 CATALYTIC REACTION NETWORKS

Masaki Nakagawa and Yuichi Togashi

Keywords: Small-number issues, catalytic reaction networks, chemical master equations, probability generating functions, systems biology, mathematical biology, population dynamics

INTRODUCTION

Biochemical systems consist of a variety of chemicals, including proteins, nucleic acids, and also small metabolites. Enzymatic reactions, which play an important role in catalyzing many biological reactions, are particularly important to maintain the structure and activity of these systems. Hence, biochemical systems are often modeled as catalytic reaction networks.

These networks are typically analyzed by using deterministic ordinary differential equations with respect to concentrations of chemical species, so-called reaction rate equations [or partial differential equations (reaction-diffusion equations) for spatially distributed, non-uniform cases]; i.e., the concentrations are represented by continuous variables. However, because each chemical actually consists of molecules, the concentration of each species should be a discrete variable. The effects of such discreteness as well as finite-size fluctuations in stochastic reactions become non-negligible if the number of molecules in the system is small. In theory, situations such as these can result in phenomena that cannot be described by rate equations (as well as those equations with additional noise; Togashi and Kaneko, 2001, 2007; Awazu and Kaneko, 2007, 2009).

In contrast, gene regulations are often modeled as a combination of binary (on/off) switches, typically represented by Boolean network models (Kauffman, 1969). However, this approach does not consider the quantities of chemicals such as DNA, mRNA, and proteins. Even for the seemingly digital expression of genes, a stoichiometry involving DNA cannot be ignored, as seen in X-chromosome inactivation (2 to 1) and the trisomy syndrome (2 to 3). Thus, the use of a binary (on/off) representation would also be inappropriate. We therefore need to consider the number of molecules.

Recent experiments have shown the existence in the cell of proteins that only consist of a few molecules each (Taniguchi et al., 2010), and these potential biological phenomena (sometimes referred to as the *small-number issue*) have been gradually recognized by biologists. General theories that would enable predictions in small number situations would be helpful to every biological scientist seeking to understand biochemical processes on any level. Of course, the distributional behavior of such discrete stochastic systems is described by the chemical master equation. However, it is generally difficult to obtain its solution; hence, most efforts have been devoted to the development of approximation and simulation methods and their application (Gillespie, 1976,1977; Munsky and Khammash, 2006; Lee et al., 2009; Kim and Lee, 2012).

The effects of small numbers in particular systems, e.g., small autocatalytic systems have been mathematically analyzed (Ohkubo et al., 2008; Biancalani et al., 2012, 2014; Houchmandzadeh and Vallade, 2015; Saito and Kaneko, 2015). In addition, a parameter representing the degree of discreteness in the number of molecules has been introduced (Haruna, 2010, 2015). In this work, we pursue analytical frameworks for studying the effects of small numbers in catalytic reaction networks by following an approach that is as general as possible. Our aim is to obtain the steady-state analytical solution for the chemical master equation without specifying parameters, rather than developing numerical schemes to solve it as was previously done (Kim and Lee, 2012). Furthermore, we try to theoretically describe the long-term behavior of the system by only using information about relationships between elements, thereby implying that we aim to produce results that could be applied to studies in any field.

Our framework provides good operability because our formulas have a specific and satisfyingly simple form, and enables us to obtain the steady state for a wide class of

Citation: Nakagawa M and Togashi Y (2016), Catalytic Reaction Networks. Front. Physiol. 7:89. doi: 10.3389/fphys.2016.00089.

catalytic reaction networks because our framework never specifies any parameters for the networks. We use a probability generating function approach. The probability generating function approach to stochastic chemical kinetics itself has been proposed a long time ago (e.g., Krieger and Gans, 1960; McQuarrie, 1967), after which it spread to biological stochastic kinetics studies (e.g., stochastic gene expression Thattai and van Oudenaarden, 2001; Shahrezaei and Swain, 2008); thus, physicists as well as chemists and mathematical biologists are familiar with the approach. Our main contributions in this paper relate to the efficient usage of the probability generating function (as in Gadgil et al., 2005 regarding first-order reaction networks). Therefore, our approach for obtaining the steady state is understandably easier than that followed in a previous study (Anderson et al., 2010) while, as a consequence, our results are consistent with that study (see Theorem 4.1 and 4.2 in Anderson et al., 2010). Furthermore, since our method uses a procedure based on analytical calculations, it can be easily converted to a computer algorithm.

The present paper is organized as follows. In Section 2.1, we define the catalytic reaction network considered in this paper. The chemical master equation (CME) is provided in Section 2.2. We introduce the probability generating function (PGF) and derive the generating function equation (GFE) in Section 2.3. In subsequent Sections (2.3.1–2.3.3), we show that the GFE introduces the time evolution equation of the first-order and second-order moments of concentrations (we refer to the first-order moment time evolution equation as the pre-rate equation, PRE), and the second-order moment expression of time-averaged concentrations (SME). Section 2.4 is devoted to obtaining the steady-state solutions of the GFE. To simplify the GFE, we neglect the non-catalytic reactions considered as perturbation for the catalytic reaction network if the system is "entirely ergodic." In Section 2.4.2.3 as the main result, we obtain the probability generating function without winner-takes-all states (PGFwoWTAS) including the solutions of the corresponding rate equation. In Section 3, we describe applications of these results: the rank conservation law, the connecting state to the winner-takes-all state, analysis of 2-molecules M-species systems, and non-autocatalyzation of autocatalytic reaction networks. The prospects of our theory in terms of the small-number issue are briefly discussed in Section 4.

METHODS AND RESULTS

Catalytic Reaction Networks

Consider an abstract catalytic reaction network consisting of M chemical species and N molecules in a well-stirred reactor of volume V, as in Awazu and Kaneko (2007). Each species is labeled by each of the integers between 1 and M. The total number of molecules N is always conserved in reaction processes. This chemical reaction system involves both catalytic reactions and non-catalytic reactions, to prevent catalytic reactions from stopping:

(i) Catalytic reactions (two-body catalysis):

$$i \xrightarrow{j} k, \tag{1}$$

where the species $i, j, k \in [1, M]$ represent a substrate, a catalyst, and a product. The reaction rate constant is represented by $R_{ijk} > 0$. If this catalytic reaction does not exist, we specify $R_{ijk} = 0$. Therefore, the catalytic reaction networks are determined by R_{ijk}. In this paper, we impose the following conditions for the catalytic reaction networks;

A) $R_{IIK} = 0$; SUBSTRATE ≠ CATALYST.

B) $R_{IJI} = 0$; SUBSTRATE ≠ PRODUCT.

C) $R_{IKK} = 0$; AUTOCATALYTIC REACTIONS ARE NOT INCLUDED.

D) $\#\{K : R_{IJK} > 0\ (\forall I, J)\} = 1$;

One product against a substrate and a catalyst.

(ii) Non-catalytic reactions (one-body reactions):

$$i \xrightarrow{\text{Prob. } 1/M} j. \tag{2}$$

This reaction exists for all combinations between each species, but a product $j \in [1, M]$ is uniformly-randomly selected from all species 1 to M. The rate constant is set $\varepsilon > 0$ in common, where ε is very small, i.e., $\varepsilon \ll \min\{R_{ijk} > 0\}$.

The state of this catalytic reaction network is specified by the combination of M natural numbers $n = (n_1, n_2, \cdots, n_M)$, where $n_i \in [0, N]$ is the number of molecules of the ith-species. For later convenience, we introduce the following notations: the *state space of the catalytic reaction network* $W_{M,N}$ (abbr. W) is represented by

$$W := \{\boldsymbol{n} \in [0, N]^M \mid n_1 + \cdots + n_M = N\}, \tag{3}$$

which consists of $\left(\!\binom{M}{N}\!\right) := \frac{(N+M-1)!}{(M-1)!N!}$ points; a *collection consisting of (one species) winner-takes-all states* $I_{M,N}$ (abbr. I) is represented by

$$I := \{\boldsymbol{n} \in W_{M,N} \mid \exists i \in [1, M] \text{ s.t. } n_i = N\}, \tag{4}$$

which consists of M points. Of course, the winner*s*-take-all states of more than one species can be considered, but we focus on the winner-takes-all states of one species by supposing the system satisfies a certain condition, *entire ergodicity*.

In the present paper, we are interested in the N-dependence of the concentration of each species $x_i = n_i/V$. We basically consider a situation in which the total density of molecules $\rho = N/V$ is conserved, even if the total number of molecules N is changed.

The Chemical Master Equation

The rate constant R_{ijk} in the catalytic reaction is defined as the number of reactions per unit volume, unit concentration, and unit time. Therefore, the number of reactions per unit time in the catalytic reaction $i \xrightarrow{j} k$, such that the concentrations are x_i and x_j, is

$$R_{ijk} x_i x_j V = R_{ijk} \frac{n_i n_j}{V} = R_{ijk} \frac{n_i n_j}{N} \rho, \tag{5}$$

where $\rho = N/V$ is the total density of molecules in the vessel. On the other hand, the number of reactions per unit time in the non-catalytic reaction $i \to j$ with probability $1/M$ is

$$\varepsilon x_i V \times \frac{1}{M} = \frac{\varepsilon}{M} n_i. \tag{6}$$

Then, the time-evolution of the probability $P(\boldsymbol{n}, t)$, with which the system is in the state $\boldsymbol{n}$ at time t, obeys the chemical master equation (CME):

$$\frac{dP(\boldsymbol{n}, t)}{dt} = \frac{\rho}{N} \sum_{i,j,k} R_{ijk}(E_i^{+1} E_k^{-1} - 1) n_i n_j P(\boldsymbol{n}, t) + \frac{\varepsilon}{M} \sum_{i,j} (E_i^{+1} E_j^{-1} - 1) n_i P(\boldsymbol{n}, t), \tag{7}$$

where $E_i^{\pm m}$ are step operators, i.e.,

$$E_i^{\pm m} f(n_1, \cdots, n_i, \cdots, n_M) := f(n_1, \cdots, n_i \pm m, \cdots, n_M). \tag{8}$$

Of course, the state space on which the probability $P(\cdot, t)$ is supported, is the previously described $W_{M,N}$.

Probability Generating Function Method

The probability generating function (PGF) is useful to analyze the CME:

$$\phi(\boldsymbol{z}, t) := \sum_{n_1, \cdots, n_M} P(\boldsymbol{n}, t)\, {z_1}^{n_1} \cdots {z_M}^{n_M}. \tag{9}$$

Note that the following expressions are translated to differential forms of the PGF:

$$n_i P(\boldsymbol{n}, t) \longmapsto z_i \frac{\partial}{\partial z_i} \phi(\boldsymbol{z}, t), \tag{10a}$$

$$n_i n_j P(\boldsymbol{n}, t) \longmapsto z_i z_j \frac{\partial}{\partial z_i} \frac{\partial}{\partial z_j} \phi(\boldsymbol{z}, t) \qquad (i \neq j), \tag{10b}$$

$$E_i^{+1} E_j^{-1} n_i P(\boldsymbol{n}, t) \longmapsto z_j \frac{\partial}{\partial z_i} \phi(\boldsymbol{z}, t) \qquad (i \neq j), \tag{10c}$$

$$E_i^{+1} E_k^{-1} n_i n_j P(\boldsymbol{n}, t) \longmapsto z_j z_k \frac{\partial}{\partial z_i} \frac{\partial}{\partial z_j} \phi(\boldsymbol{z}, t) \qquad (i \neq j \neq k \neq i). \tag{10d}$$

Therefore, rewriting the CME to the equation governing the PGF enables the generating function equation (GFE) to be obtained:

$$\frac{\partial \phi(\boldsymbol{z}, t)}{\partial t} = \frac{\rho}{N} \sum_{i,j,k} R_{ijk}(z_k - z_i)\, z_j \frac{\partial}{\partial z_i} \frac{\partial}{\partial z_j} \phi(\boldsymbol{z}, t) + \frac{\varepsilon}{M} \sum_{i,j} (z_i - z_j) \frac{\partial}{\partial z_j} \phi(\boldsymbol{z}, t). \tag{11}$$

The GFE consists of continuous variables, unlike the CME, which consists of discrete variables.

Once we obtain the PGF ϕ as a solution to the GFE, we can derive all statistics of the catalytic reaction network; for example, the ensemble averages (first-order moments) and second-order moments become

$$\langle n_i \rangle (t) = \left. \frac{\partial}{\partial z_i} \phi(\boldsymbol{z}, t) \right|_{\boldsymbol{z}=\boldsymbol{1}}, \tag{12a}$$

$$\langle n_i n_j \rangle (t) = \left. \frac{\partial^2}{\partial z_i \partial z_j} \phi(\boldsymbol{z}, t) \right|_{\boldsymbol{z}=\boldsymbol{1}} \qquad (i \neq j), \tag{12b}$$

$$\langle n_l (n_l - 1) \rangle (t) = \left. \frac{\partial^2}{\partial z_l^2} \phi(\boldsymbol{z}, t) \right|_{\boldsymbol{z}=\boldsymbol{1}}, \tag{12c}$$

and the marginal distributions become

$$p_i(n, t) = \frac{1}{n!} \left. \frac{\partial^n}{\partial z_i^n} \phi(\boldsymbol{z}, t) \right|_{z_i=0;\ z_j=1 (j \neq i)} \qquad (0 \leq n \leq N), \tag{13}$$

where $p_i(n, t)$ is defined by

$$p_i(n, t) := \sum_{\substack{n_1, \cdots, n_M \\ (\text{except } n_i)}} \left. P(\boldsymbol{n}, t) \right|_{n_i = n}. \tag{14}$$

The Pre-rate Equation

Rate equations are differential equations for the concentrations x_i of chemical species $i = 1, \cdots, M$ when $N \to \infty$. We use the GFE to derive a formula, which is reminiscent of the rate equation.

Differentiating both sides of Equation (11) by z_i, substituting $\boldsymbol{z} = \boldsymbol{1}$, and writing $x_i = n_i/V$, leads to the following pre-rate equation (PRE);

$$\frac{d}{dt} \langle x_i \rangle = \sum_{j,k} \left(R_{jki} \langle x_j x_k \rangle - R_{ikj} \langle x_i x_k \rangle \right) + \varepsilon \left(\frac{\rho}{M} - \langle x_i \rangle \right). \tag{15}$$

Note that, as we are considering a system of which the total number of molecules N is conserved, the concentration conservation law (CCL) must be satisfied;

$$\sum_i \langle x_i \rangle (t) = \rho \qquad (\forall t), \tag{16}$$

where the PRE (Equation 15) is consistent with the CCL (Equation 16).

If the independence $\langle x_i x_j \rangle = \langle x_i \rangle \langle x_j \rangle$ $(i \neq j)$ holds, the PRE (Equation 15) becomes the same expression as the rate equation. On the other hand, the PRE does not explicitly include extensive variables, such as the total number of molecules N. Therefore, the formula (Equation 15) always holds for an arbitrary molecular number N unless the total molecular density ρ is changed.

Second-Order Moment Expression for Time-Averaged Concentrations

We suppose the following ergodicity to replace ensemble-averages with time-averages; $\overline{n_i} = \langle n_i \rangle_*$ i.e.,

$$\lim_{T\to\infty}\frac{1}{T}\int_0^T n_i(t)dt = \sum_{n_1,\cdots,n_M} n_i\, P_*(\boldsymbol{n}), \tag{17}$$

where $P_*(\boldsymbol{n})$ is the steady-state solution of the CME corresponding to almost all initial conditions. Considering the steady state in the PRE [that is, taking $t \to \infty$ on both sides of Equation (15)], one obtains the following second-order moment expression (SME) for time-averaged concentrations:

$$\overline{x_i} = \frac{\rho}{M} + \frac{1}{\varepsilon}\sum_{j,k}\left(R_{jki}\overline{x_j x_k} - R_{ikj}\overline{x_i x_k}\right). \tag{18}$$

The second term on the right hand side of Equation (18) represents the difference from the uniform concentration ρ/*M*.

If the independence $\overline{x_i x_j} = \overline{x_i}\,\overline{x_j}$ holds, Equation (18) becomes an equation that only includes the concentrations $\overline{x_i}$. Therefore, in combination with the time-averaged version of the CCL (Equation 16), the concentrations $\overline{x_i}$ can be determined without including unknown quantities. However, the actual concentrations $\overline{x_i}$ depend on their second-order moments $\overline{x_i x_j}$.

Time-Evolution of Second-Order Moments

Determining the concentrations $\overline{x_i}$ from the SME (Equation 18) requires us to know their second-order moments $\overline{x_i x_j}$.

Differentiating both sides of the GFE (Equation 11) by z_l and z_m ($l \neq m$), substituting $\boldsymbol{z} = \boldsymbol{1}$, and writing $x_i = n_i/V$, then, after simplification, the following time-evolution equation of second-order moments (TESM) is obtained:

$$\begin{aligned}\frac{d}{dt}\langle x_l x_m\rangle = &-\frac{\rho}{N}\sum_i\left(R_{lim}\langle x_i x_l\rangle + R_{mil}\langle x_i x_m\rangle\right)\\ &+\sum_{i,j}\{R_{ijl}\langle x_i x_j x_m\rangle + R_{ijm}\langle x_i x_j x_l\rangle - (R_{lij}+R_{mij})\\ &\langle x_i x_l x_m\rangle\}\\ &+\frac{\varepsilon}{M}\left\{\frac{N-1}{N}\rho(\langle x_l\rangle + \langle x_m\rangle) - 2M\langle x_l x_m\rangle\right\}\\ &(1 \le l < m \le M).\end{aligned} \tag{19a}$$

On the other hand, twice differentiating both sides of the GFE (Equation 11) by z_l, substituting $\boldsymbol{z} = \boldsymbol{1}$, and writing $x_i = n_i/V$, then after some simplification, the following alternative TESM is obtained:

$$\begin{aligned}\frac{d}{dt}\langle x_l(x_l - \tfrac{\rho}{N})\rangle = &\,2\sum_{i,j}\left(R_{ijl}\langle x_i x_j x_l\rangle - R_{lij}\langle x_i x_l(x_l - \tfrac{\rho}{N})\rangle\right)\\ &+\frac{2\varepsilon}{M}\left\{\frac{N+M-1}{N}\rho\langle x_l\rangle - M\langle x_l^2\rangle\right\}\\ &(1 \le l \le M).\end{aligned} \tag{19b}$$

The expression of TESMs (Equation 19) consists of $M(M+1)/2$ equations containing the third-order moments $\langle x_i x_j x_k\rangle(t)$. Therefore, the TESMs are not effective unless the systems are restricted such that the third-order moments vanish (e.g., 2-molecules systems). If $N = 2$, because $\partial_{z_i}\partial_{z_j}\partial_{z_k}\phi = 0$ (ϕ is a second-order polynomial), one can calculate the 2-molecules version of the TESMs (2mTESMs);

$$\begin{aligned}\frac{d}{dt}\langle x_l x_m\rangle = &\,\frac{\rho}{2}\sum_i\{R_{iml}\langle x_i x_m\rangle + R_{ilm}\langle x_i x_l\rangle\\ &-(R_{lmi}+R_{mli})\langle x_l x_m\rangle\}\\ &+\frac{\rho\varepsilon}{2M}(\langle x_l\rangle + \langle x_m\rangle) - 2\varepsilon\langle x_l x_m\rangle\\ &(1 \le l < m \le M),\end{aligned} \tag{20a}$$

$$\frac{d}{dt}\langle x_l(x_l - \tfrac{\rho}{N})\rangle = \frac{2\varepsilon}{M}\left(\frac{M+1}{2}\rho\langle x_l\rangle - M\langle x_l^2\rangle\right) \quad (1 \le l \le M). \tag{20b}$$

Treatment of the 2mTESMs (Equation 20) to demonstrate their effectiveness appears in a later section.

Steady-State Solutions of GFE

If the GFE (Equation 11) can be solved, this would enable us to obtain all the statistics of the catalytic reaction networks. However, it is generally difficult to solve. Here, we focus on the steady-state solutions of the GFE and consider the case that the ε-term in the GFE can be ignored. Through the following discussion, we see that the approximation is effective only if the system is ergodic.

The Case of Non-catalytic Reactions Only

First, we consider the steady-state solutions of non-catalytic reactions only as an introduction. The PGF $\phi_*^{\mathrm{nc}}(\boldsymbol{z})$, corresponding to the steady state in the case of non-catalytic reactions only, should satisfy the following equation:

$$0 = \sum_{i,j}(z_i - z_j)\frac{\partial}{\partial z_j}\phi_*^{\mathrm{nc}}(\boldsymbol{z}). \tag{21}$$

By exchanging the subscripts *i* and *j* in the second term, one obtains

$$0 = \sum_{i,j} z_i\left(\frac{\partial\phi_*^{\mathrm{nc}}}{\partial z_j} - \frac{\partial\phi_*^{\mathrm{nc}}}{\partial z_i}\right). \tag{22}$$

Because the coefficients of variables z_i must be zero, ϕ_*^{nc} must satisfy the following equations:

$$0 = \sum_j\left(\frac{\partial\phi_*^{\mathrm{nc}}}{\partial z_j} - \frac{\partial\phi_*^{\mathrm{nc}}}{\partial z_i}\right) \quad (\forall i), \tag{23}$$

equivalently,

$$\frac{\partial\phi_*^{\mathrm{nc}}}{\partial z_i} = \frac{1}{M}\sum_{j=1}^{M}\frac{\partial\phi_*^{\mathrm{nc}}}{\partial z_j} \quad (\forall i). \tag{24}$$

Equation (24) implies that all of $\partial_{z_i}\phi_*^{\mathrm{nc}}$ $(i = 1,\cdots,M)$ are equal to each other, i.e.,

$$\frac{\partial\phi_*^{\mathrm{nc}}}{\partial z_i} = \frac{\partial\phi_*^{\mathrm{nc}}}{\partial z_j} \quad (1 \le i < j \le M). \tag{25}$$

Considering that the PGF ϕ_*^{nc} (z) is an *N*th order polynomial of z_i and must satisfy the condition ϕ_*^{nc} (1)=1 by definition, the following solution of Equation (25) can be found:

$$\phi_*^{\mathrm{nc}}(z) = \left(\frac{z_1 + z_2 + \cdots + z_M}{M}\right)^N. \tag{26}$$

Therefore, we obtain the following stationary distribution in the case of non-catalytic reactions only:

$$P_*^{\mathrm{nc}}(\boldsymbol{n}) = \frac{1}{M^N}\binom{N}{n_1, n_2, \cdots, n_M} \quad (\boldsymbol{n} \in W), \tag{27}$$

where $\binom{N}{n_1,n_2,\cdots,n_M} := \frac{N!}{n_1!n_2!\cdots n_M!}$ are multinomial coefficients. Furthermore, Equation (13) can also be used to derive the marginal distribution of the i-th species:

$$p_{i*}^{\mathrm{nc}}(n) = \binom{N}{n}\left(\tfrac{1}{M}\right)^n\left(1 - \tfrac{1}{M}\right)^{N-n} \quad (0 \le n \le N), \tag{28}$$

which is the binomial distribution with parameters N and $1/M$.

If we suppose the ergodicity $\langle n_i \rangle = \overline{n}_i$, the following statistics can be calculated from Equation (12):

$$\overline{x}_i = \frac{\rho}{M} \quad (i = 1, \cdots, M), \tag{29a}$$

$$\overline{x_i x_j} = \left(\frac{\rho}{M}\right)^2\left(1 - \frac{1}{N}\right) \quad (i \neq j), \tag{29b}$$

$$\mathrm{Var}[x_i] = \left(\frac{\rho}{M}\right)^2\frac{M-1}{N} \quad (i = 1, \cdots, M), \tag{29c}$$

where $\mathrm{Var}[x_i] := \overline{x_i^2} - \overline{x_i}^2$ is the variance of the concentration $x_i = n_i/V$.

The Case of Catalytic Reactions Only

Next, we consider the steady-state solutions of catalytic reactions only, assuming that the ε-term in the GFE (Equation 11) can be ignored. The steady-state solutions are assumed to have a form similar to Equation (26), including undetermined coefficients (λ_i) deriving from the network structure (R_{ijk}).

A Condition for Finding the Steady State: λ-Condition

The PGF ϕ_*^{nc} (z), corresponding to the steady state in the case of catalytic reactions only, should satisfy the following equation;

$$0 = \sum_{i,j,k} R_{ijk}(z_k - z_i)z_j\frac{\partial}{\partial z_i}\frac{\partial}{\partial z_j}\phi_*^{\mathrm{c}}(z). \tag{30}$$

Construction of the steady-state solution requires us to find a particular solution of Equation (30) as bases of linear space consisting of N-th order polynomials. Accordingly, let us assume the following extended form of Equation (26) by introducing parameters $\{\lambda_i\}_{i=1}^M$, which eventually correspond to the concentrations (per total density ρ) of each species in the continuous limit $N \to \infty$ as shown in Section 2.4.2.3:

$$\phi_*^{\mathrm{c}}(z) = (\lambda_1 z_1 + \lambda_2 z_2 + \cdots + \lambda_M z_M)^N, \tag{31}$$

where

$$\sum_{i=1}^M \lambda_i = 1 \quad (0 \le \lambda_1, \lambda_2 \cdots, \lambda_M \le 1), \tag{32}$$

because $\phi_*(1) = 1$. Substituting Equation (31) into Equation (31) and setting the coefficients of variables $z_i z_j$ as zero, gives the following condition for $\{\lambda_i\lambda_j\}$:

$$\sum_{k=1}^M \left\{R_{kij}\lambda_k\lambda_i + R_{kji}\lambda_k\lambda_j - \left(R_{ijk} + R_{jik}\right)\lambda_i\lambda_j\right\} = 0 \quad (1 \le i < j \le M). \tag{33}$$

The λ-condition (Equation 33) represents $M(M-1)/2$ homogeneous equations for $\lambda_i\lambda_j$; therefore, λ_i can be calculated by combining the λ-condition (Equation 33) with Equation (32). The λ-condition has trivial solutions:

$$(\lambda_1, \cdots, \lambda_i, \cdots, \lambda_M) = (0, \cdots, 0, 1, 0, \cdots, 0) \quad \exists i \in [1, M]. \tag{34}$$

These solutions represent the states for which the i-th species take all molecules (the *winner-takes-all state*). On the other hand, there is also a non-trivial solution. In the following paragraph, we treat three-species systems ($M = 3$) to demonstrate that non-trivial solutions do exist. The following procedures can easily be extended to those for arbitrary M species systems.

Demonstration for Non-Trivial Solutions of the λ -Condition

The λ-condition (Equation 33) can be rewritten in matrix form; i.e., in the case of $M = 3$ for example,

$$\begin{bmatrix} -R_{123} - R_{213} & R_{312} & R_{321} \\ R_{213} & -R_{132} - R_{312} & R_{231} \\ R_{123} & R_{132} & -R_{231} - R_{321} \end{bmatrix}\begin{bmatrix} \lambda_1\lambda_2 \\ \lambda_1\lambda_3 \\ \lambda_2\lambda_3 \end{bmatrix} = 0. \tag{35}$$

The 3 × 3 matrix (A) on the left-hand side can be rewritten in terms of its column vectors:

$$A = \left[-R_{123}\boldsymbol{e}_1 - R_{213}\boldsymbol{e}_2,\ R_{312}\boldsymbol{e}_2 - R_{132}\boldsymbol{e}_3,\ R_{321}\boldsymbol{e}_1 + R_{231}\boldsymbol{e}_3\right], \tag{36}$$

where

$\boldsymbol{e}_1{}^T = (1, 0, -1)$, $\boldsymbol{e}_2{}^T = (1, -1, 0)$, and $\boldsymbol{e}_3{}^T = (0, 1, -1)$. According to the determinant property, i.e., det[***a, b, c***] = − det[***b, a, c***], one has

$$\det(A) = 0 \tag{37}$$

Therefore, the non-trivial solution of Equation (35), i.e., the eigenvector corresponding to the eigenvalue 0 of the matrix A, certainly exists and it has the following expression:

$$\begin{bmatrix} \lambda_1\lambda_2 \\ \lambda_1\lambda_3 \\ \lambda_2\lambda_3 \end{bmatrix} \propto \begin{bmatrix} R_{312}R_{231} + R_{321}R_{132} + R_{312}R_{321} \\ R_{213}R_{321} + R_{231}R_{123} + R_{213}R_{231} \\ R_{123}R_{312} + R_{132}R_{213} + R_{123}R_{132} \end{bmatrix} \equiv \begin{bmatrix} \Lambda_3 \\ \Lambda_2 \\ \Lambda_1 \end{bmatrix}. \tag{38}$$

The proportionality constant (> 0) can be determined by the condition (Equation 32), and thus the desired non-trivial solution of the λ-condition (Equation 33) for $M = 3$ is obtained:

$$\lambda_1 = \frac{\Lambda_2\Lambda_3}{\Lambda_1\Lambda_2 + \Lambda_1\Lambda_3 + \Lambda_2\Lambda_3}, \tag{39a}$$

$$\lambda_2 = \frac{\Lambda_1\Lambda_3}{\Lambda_1\Lambda_2 + \Lambda_1\Lambda_3 + \Lambda_2\Lambda_3}, \tag{39b}$$

$$\lambda_3 = \frac{\Lambda_1\Lambda_2}{\Lambda_1\Lambda_2 + \Lambda_1\Lambda_3 + \Lambda_2\Lambda_3}. \tag{39c}$$

Note that it is not guaranteed that the above solution always represents a non-trivial solution of the λ-condition (Equation 33). For example, the case of $\Lambda_1 = 0$ (where Λ_2 and Λ_3 are not zero) implies the existence of a trivial solution $(\lambda_1, \lambda_2, \lambda_3) = (1, 0, 0)$. In another example, the case of $\Lambda_1 = \Lambda_2 = 0$ implies that the denominator becomes zero; thus, the expression becomes indefinite.

PGF without Winner-Takes-All States

We are interested in those states in which any species does not take all molecules, because the actual simulations are performed by using the initial states excluding the winner-takes-all states. The PGF without the winner-takes-all states is represented by a linear summation of winner-takes-all states z_i^N and Equation (31) if the λ-condition has a non-trivial solution like Equation (39);

$$\phi_*^c(z) = \sum_{i=1}^{M} b_i z_i^N + b_{M+1}\left(\sum_{i=1}^{M} \lambda_i z_i\right)^N, \tag{40}$$

where $b_1 + \cdots + b_{M+1} = 1$. The *take-all exclusion conditions* are that the coefficients of z_i^N are zero;

$$b_i + b_{M+1}\lambda_i^N = 0 \qquad (i = 1, 2, \cdots, M). \tag{41}$$

Therefore, we obtain the desired PGF without the winner-takes-all states (PGFwoWTAS);

$$\phi_*^c(z) = \frac{\left(\sum_{i=1}^M \lambda_i z_i\right)^N - \sum_{i=1}^M (\lambda_i z_i)^N}{1 - \sum_{i=1}^M \lambda_i^N}, \qquad \sum_{i=1}^{M} \lambda_i = 1,\ 0 \le \lambda_i < 1, \tag{42}$$

which immediately implies the following stationary distribution in the case of catalytic reactions only;

$$P_*^c(\boldsymbol{n}) = \begin{cases} \binom{N}{n_1, n_2, \cdots, n_M} \dfrac{\prod_{i=1}^M \lambda_i^{n_i}}{1 - \sum_{i=1}^M \lambda_i^N} & (\boldsymbol{n} \in W \setminus I), \\ 0 & (\boldsymbol{n} \in I), \end{cases} \tag{43}$$

where $\binom{N}{n_1, n_2, \cdots, n_M} := \frac{N!}{n_1! n_2! \cdots n_M!}$ are multinomial coefficients. Furthermore, we can calculate the marginal distribution of the i-th species from Equation (13);

$$p_{i*}^c(n) = \begin{cases} \dfrac{(1-\lambda_i)^N + \lambda_i^N - \sum_{l=1}^M \lambda_l^N}{1 - \sum_{l=1}^M \lambda_l^N} & (n = 0), \\ \binom{N}{n} \dfrac{\lambda_i^n (1-\lambda_i)^{N-n}}{1 - \sum_{l=1}^M \lambda_l^N} & (1 \le n \le N-1), \\ 0 & (n = N) \end{cases} \tag{44}$$

If we suppose the ergodicity $\langle n_i \rangle = \overline{n}_i$, the time-averaged concentrations can be derived from Equation (12);

$$\overline{x}_i = \frac{\lambda_i - \lambda_i^N}{1 - \sum_{i=1}^M \lambda_i^N}\rho \quad (i = 1, \cdots, M). \tag{45a}$$

The above equations indicate that λ_i means the concentration per total density ρ in the continuous limit $N \to \infty$, that is, λ_i should be the solution of the classical rate equation. We can also calculate the second-order moments $\overline{x_i x_j}$ and the variances $\mathrm{Var}[x_i] := \overline{x_i^2} - \overline{x}_i^2$ from Equation (12);

$$\overline{x_i x_j} = \frac{\lambda_i \lambda_j}{1 - \sum_{k=1}^M \lambda_k^N}\left(1 - \frac{1}{N}\right)\rho^2 \quad (i \ne j). \tag{45b}$$

$$\mathrm{Var}[x_i] = \left[\frac{\lambda_i^2 - \lambda_i^N - \frac{1}{N}\lambda_i(\lambda_i - 1)}{1 - \sum_{i=1}^M \lambda_i^N} - \left(\frac{\lambda_i - \lambda_i^N}{1 - \sum_{i=1}^M \lambda_i^N}\right)^2\right]\rho^2 \quad (i = 1, \cdots, M), \tag{45c}$$

In the continuous limit $N \to \infty$, the above Equations (45b) and (45c) become $\lambda_i\lambda_j$ and 0, respectively, that is, the concentrations become mutually independent without fluctuating variables.

We compare our formulas with simulation results that are obtained by applying the Gillespie algorithm (Gillespie, 1977) to the following three-species system (Figure 1A) as an example:

$$R_{123} = R_{132} = R_{213} = R_{231} = R_{312} = 1, \quad R_{321} = 0; \quad \rho = 1, \tag{46}$$

which implies the following by Equation (39):

$$\lambda_1 = 2/11, \quad \lambda_2 = 3/11, \quad \lambda_3 = 6/11. \tag{47}$$

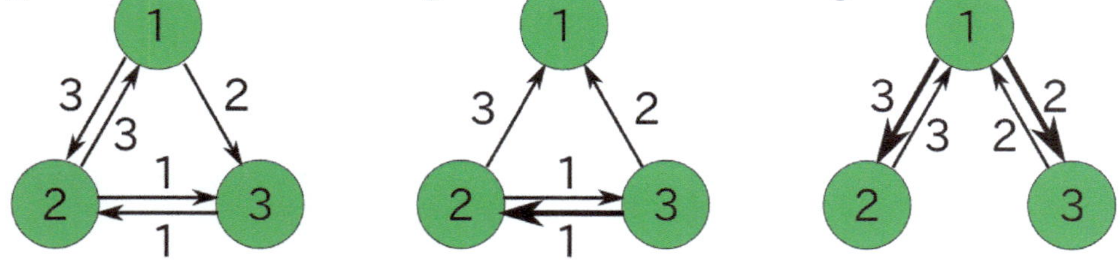

Figure 1. Examples of catalytic reaction networks consisting of 3 species. The rate constants are(A) $R_{123} = 1$, $R_{132} = 1$, $R_{213} = 1$, $R_{231} = 1$, $R_{312} = 1$, $R_{321} = 0$, **(B)** $R_{123} = 0$, $R_{132} = 0$, $R_{213} = 1$, $R_{231} = 1$, $R_{312} = 2$, $R_{321} = 1$, and **(C)** $R_{123} = 1997/3$, $R_{132} = 1000/3$, $R_{231} = 1$, $R_{321} = 1$, $R_{213} = 0$, $R_{312} = 0$.

The marginal distributions are shown in Figure 2. Our formula Equation (44) is entirely in agreement with the simulation results. When the total number of molecules N is large, the marginal distributions can be approximated by normal distributions. Similarly, the formulas of the concentrations and the variances, Equations (45a) and (45c), are completely in agreement with the simulation results as shown in Figure 3. One can see that the variances monotonically decrease as N becomes larger.

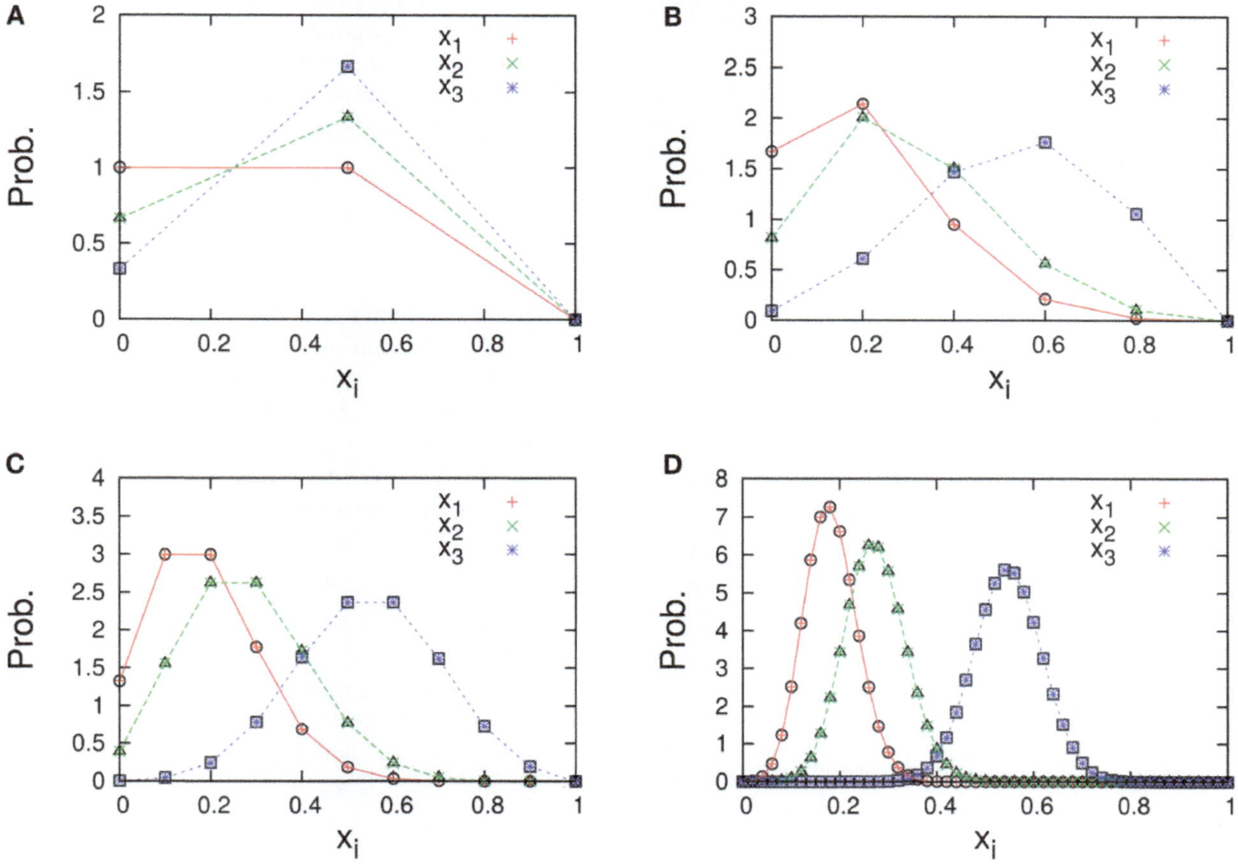

Figure 2. Marginal distributions of $x_i = n_i/N$ ($i = 1, 2, 3$) in the three-species system of Figure1A. The total number of molecules in each figure is (A) $N = 2$, (B) $N = 5$, (C) $N = 10$, and (D) $N = 50$. Cross symbols (red, green, blue) represent the simulation results obtained numerically using the Gillespie algorithm (Gillespie, 1977), which is performed under the following condition; the total number of reactions: 10^8, the number of reactions for transient exclusion: 10^7, and the initial value ($n_1(0)$, $n_2(0)$, $n_3(0)$) is randomly selected from $W \setminus I$ such that the average per one-species is N/M. Empty symbols (circle, triangle, square) represent the theoretical expression $Np^c_{i*}(Nx)$ [Equation (44)] for $\lambda_1 = 2/11$, $\lambda_2 = 3/11$, and $\lambda_3 = 6/11$.

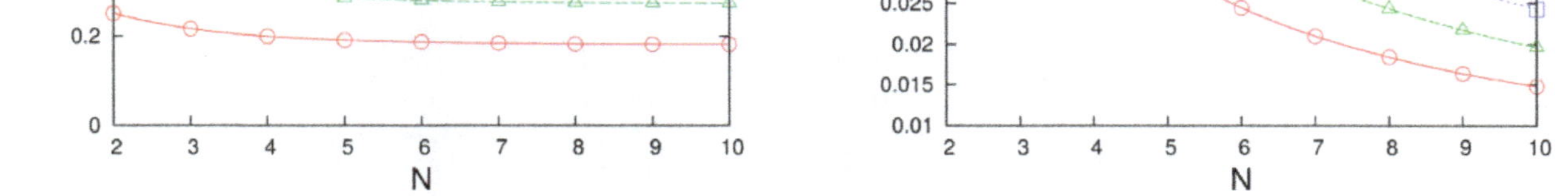

Figure 3. N-dependence of (A) the time-averaged concentrations $\bar{x}_i = \bar{n}_i/N$ and (B) the variances of concentrations $\mathrm{Var}[x_i] = (\overline{n_i^2} - \bar{n}_i^2)/N^2$ in the three-species system of Figure 1A. Empty symbols are numerically obtained using the Gillespie algorithm, where the time averages are performed over a long time series $n_i(t)$ such that the total number of reactions reaches 10^8. The initial values are randomly selected from $W \setminus I$ such that the average per one-species is N/M, and the number of reactions for transient exclusion is 10^7. Lines in each figure represent the theoretical expressions, Equations (45a) and (45c), for $\lambda_1 = 2/11$, $\lambda_2 = 3/11$, and $\lambda_3 = 6/11$. One can see that the rank of concentrations is conserved but the rank of variances is exchanged between $N = 2$ and 4.

Note that if the λ-condition (Equation 33) does not have a non-trivial solution like Equation (39), the expression for the PGF (Equation 42) cannot be applied. Such special cases are treated in the following section.

Ergodicity as a Sufficient Condition for Applying Our PGF

The PGFwoWTAS (Equation 42) would be applicable if the catalytic reaction network was "entirely ergodic," which means the following in this paper (it is reminiscent of the ω-limit set):

$$\omega(\boldsymbol{n}_0;\xi) = W \setminus I \quad (\forall \xi), \tag{48}$$

where

$$\omega(\boldsymbol{n}_0;\xi) := \bigcap_{t \geq 0} \{\boldsymbol{n}(\tau;\xi) : \tau \geq t\}, \tag{49}$$

and ξ represents one of the trials of the stochastic process $\boldsymbol{n}(t)$ according to the catalytic reaction network.

We use a specific three-species system to intuitively illustrate what Equation (48) means. In the case of the three-species system of Figure 1A, the ergodic condition (Equation 48) can be rewritten in terms of simplified conditions. As previously described, the state space of catalytic reaction networks consists of $\left(\binom{M}{N}\right) = \frac{(M+N-1)!}{(M-N)!N!}$ points. Therefore, in the case of $M = 3$ for example, there are $(N+2)!/(2N!)$ points. The state space of the three-species system forms a regular triangle in three-dimensional Euclidean space (see Figure 4A). The possible motion from each state point is shown in Figure 4B, where there potentially exist six possible directions, but the network structure forbids the direction of R_{321} in the case of the network of Figure1A. Note that the state point on the boundary cannot move in a direction parallel to the boundary. In this case obviously, a trajectory starting from an initial point $\boldsymbol{n}_0 \in W \setminus I$ visits every point in $W \setminus I$ as shown by the circles in Figure 4A (the set I is equivalent to the vertexes of a triangle). Therefore, as seen in Figure 4B, the necessary and sufficient condition to hold the ergodic condition (Equation 48) for $M = 3$ is:

$$\sum_{i,j} R_{ijk} > 0 \quad \& \quad \sum_{i,j} R_{kij} > 0 \qquad (\forall k), \tag{50}$$

where the first condition represents that at least one direction for moving away from the boundary $n_k = 0$ is allowed, and the second condition represents that at least one direction for approaching the boundary $n_k = 0$ is allowed. Note that the condition (Equation 50) is no longer a sufficient condition for entire ergodicity in the case of four-species systems. In fact, in the following four-species system (Figure 5A), the condition (Equation 50) does not imply entire ergodicity;

$$R_{132} > 0,\quad R_{143} > 0,\quad R_{231} > 0,\quad R_{243} > 0,\quad R_{314} > 0,\quad R_{413} > 0,\quad \text{(others 0)}. \tag{51}$$

This four-species system obviously satisfies the condition (Equation 50), but most initial states [in particular, $n_k(0) > 0 (\forall k)$] eventually fall into any of 2-species winners-take-all states, i.e., $\{n_1 = n_2 = 0, n_3 + n_4 = N\}$.

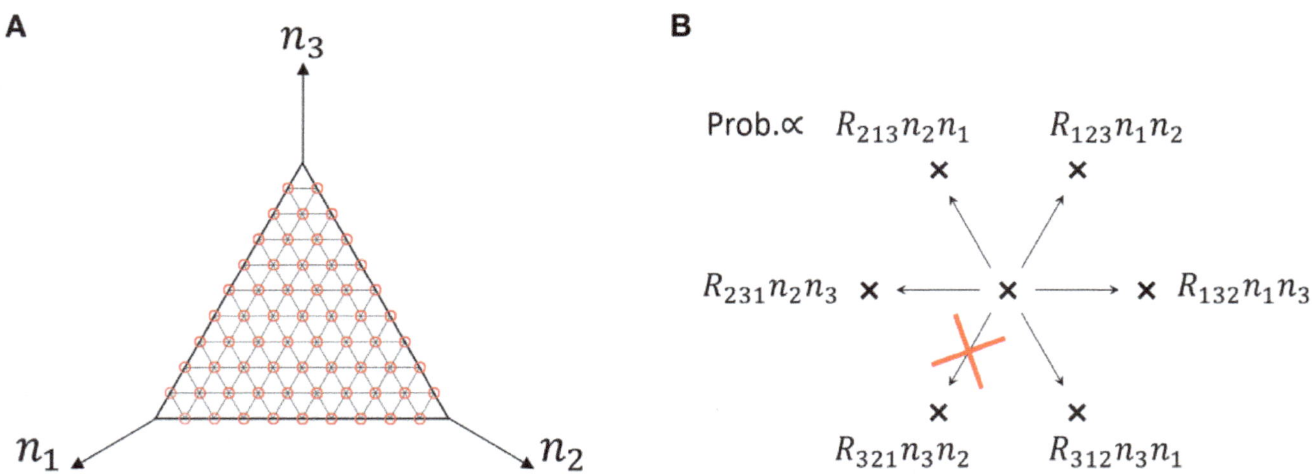

Figure 4. (A) State space of the three-species system of Figure 1A in the case of $N = 10$. Circles (red) represent the states that a single trajectory can visit, which means that the system is "entirely ergodic." **(B)**Motion of the state point when one reaction occurs in the three-species system of Figure 1A. There are 6 possible directions in which to move from one state point. Each direction is randomly selected in proportion to the specified probability. In the case of Figure 1A, the state point cannot move in the direction of R_{321} since $R_{321} = 0$. Note that the state point on the boundary (e.g., $n_1 = 0$) cannot move parallel to the boundary (in this case, the directions of R_{213} and R_{312}).

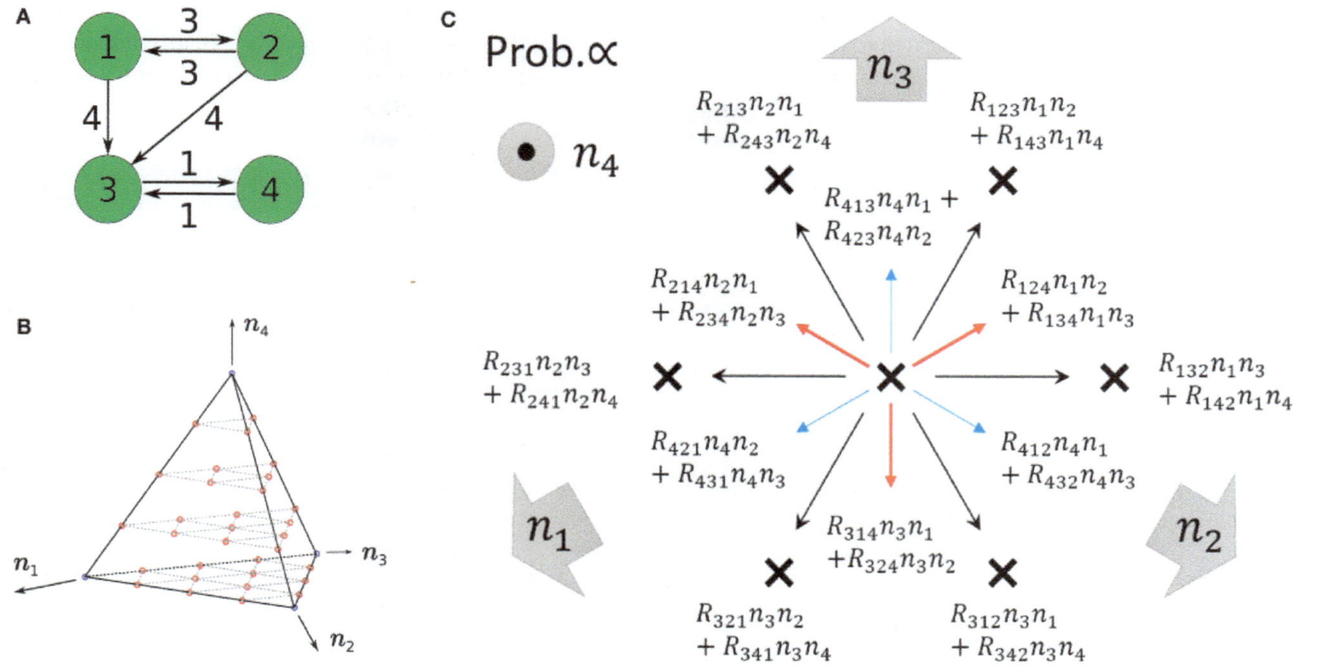

Figure 5. (A) Example of non-ergodic catalytic networks consisting of four species; the rate constants are$R_{132} > 0$, $R_{143} > 0$, $R_{231} > 0$, $R_{243} > 0$,$R_{314} > 0$, $R_{413} > 0$, and others 0. **(B)** State space of four-species systems in the case of $N = 4$. Red circles represent the states a single trajectory can visit in the case that the system is entirely ergodic. Blue circles represent winner-takes-all states, i.e., $n_k = 4$ ($\exists k$). **(C)** Motion of the state point when one reaction occurs in four-species systems (viewed from the top of **B**). Red and blue arrows represent upward and downward arrows, respectively. There are 12 possible directions in which to move from one state point. Each direction is randomly selected in proportion to the specified probability. Note that the state point on each I(2)(i1,i2)I(2)(i1,i2) cannot move parallel to itself (e.g., on $I^{(2)}(1, 2)$, the directions of R_{231}, R_{241} as well as R_{132}, R_{142} are not allowed).

In the case of M-species systems, a necessary and sufficient condition for entire ergodicity is difficult to derive, although we may consider a sufficient condition instead. We discuss the sufficient condition for entire ergodicity by introducing the collection of all *l-species winners-take-all states* ($l = 1, 2, \cdots, M$):

$$I_{M,N}^{(l)} := \bigcup_{1 \le i_1 < \cdots < i_l \le M} I_{M,N}^{(l)}(i_1, i_2, \cdots, i_l), \tag{52}$$

where

$$I_{M,N}^{(l)}(i_1, i_2, \cdots, i_l) := \{\boldsymbol{n} \in W_{M,N} \mid n_{i_1} + n_{i_2} + \cdots + n_{i_l} = N \;\&\; n_{i_k} > 0\ (\forall k)\}. \tag{53}$$

We omit the subscripts M and N provided there is no concern for confusion. The above collections are not empty sets if $l \le N$. Note that $I = I^{(1)}$ and $W = \bigcup_{l=1}^{M} I^{(l)}$. One can see that the following diagram holds if the catalytic network is fully connected [i.e., $R_{ijk} > 0$ for all i, j, k ($i \neq j \neq k \neq i$)];

$$I^{(1)} \not\leftrightarrows I^{(2)} \leftrightarrows I^{(3)} \leftrightarrows \cdots \leftrightarrows I^{(M-1)} \leftrightarrows I^{(M)} \tag{54}$$
$$\circlearrowleft \quad \circlearrowleft \quad \circlearrowleft \quad \circlearrowleft$$

where each arrow represents the direction in which a state point can move by one reaction. Each I(l)(i1,⋯,il)I(l)(i1,⋯,il) is a $(l-1)$-dimensional object in the whole $(M-1)$-dimensional state space W. For example, the state space of four-species systems forms a regular tetrahedron $W_{M=4,N}$ (see Figure 5B); $I^{(4)}(1, 2, 3, 4)$ is the interior of the regular tetrahedron; $I^{(3)}(1, 2, 3)$, $I^{(3)}(1, 2, 4)$, $I^{(3)}(1, 3, 4)$, and $I^{(3)}(2, 3, 4)$ are regular triangles that form the boundaries of $I^{(4)}$; and $I^{(2)}(1, 2)$, $I^{(2)}(1, 3)$, $I^{(2)}(1, 4)$, $I^{(2)}(2, 3)$, $I^{(2)}(2, 4)$, and $I^{(2)}(3, 4)$ are line segments that form the boundaries of $I^{(3)}$. Note that I(2)(i1,i2)→I(2)(j1,j2) I(2)(i1,i2)→I(2)(j1,j2) for $(i_1, i_2) \neq (j_1, j_2)$ is possible but I(2)(i1,i2)→I(2)(i1,i2)I(2)(i1,i2)→I(2)(i1,i2) is always impossible as seen from Figure 5C. By considering the diagram (Equation 54) and keeping in mind Figures 5B,C, we can expect the following fact: *if entire ergodicity holds in the restricted system on $I^{(l)}$ for $l \ge 3$, then entire ergodicity holds in the whole state space W.* In other words, in systems consisting of M-species, the sufficient condition under which the ergodic condition (Equation 48) holds can be geometrically described as follows: *on each* I(l)(i1,⋯,il)I(l)(i1,⋯,il) *for $l \ge 3$, at least one direction for moving away from each boundary* I(l−1)(j1,⋯,jl−1)I(l-1)(j1,⋯,jl-1) *should be allowed, and simultaneously at least one direction for approaching each boundary* I(l−1)(j1,⋯,jl−1)I(l-1)(j1,⋯,jl-1) *should also be allowed, where* $\{j_1, \cdots, j_{l-1}\} \subset \{i_1, \cdots, i_l\}$.

The Case of Catalytic-Noncatalytic Mixed Reactions

We could not obtain the steady-state solution of the general GFE (Equation 11) in the case of catalytic-noncatalytic mixed reactions ($\varepsilon > 0$), but we expect our PGF (Equation 42) to be a good approximation for mixed-reaction systems if ε is sufficiently small ($\varepsilon \ll \min\{R_{ijk} > 0\}$). More specifically, we expect the PGF (Equation 42) to be robust against non-catalytic reactions if the catalytic reaction system constituting the mixed reaction system has an ergodic component spread across the entire state space (*entire ergodicity*). Otherwise, if the catalytic reaction system has several ergodic components in the entire state space, the non-catalytic reactions may imply that a certain highly stable ergodic component attracts all possible trajectories, which obviously means that our PGF is not applicable to such a mixed reaction system. Although we were unable to prove this mathematically, we used numerical simulations to determine whether our PGF is a good approximation in the case of "entirely ergodic."

Figure 6 represents the non-catalytic reaction rate constant ε dependencies of time-averaged concentrations x_i for the three-species system of Figures 1A,B. (As shown later, the system shown in Figure 1B is not entirely ergodic, which is the reason why the systems shown in Figures 1A,B are compared here.) It can be seen that the differences of x_i in Figure 6A between ε = 0 and ε = 0.1 are smaller than those in Figure 6B. We consider the robustness in the case of Figure 6A to originate from the entire ergodicity of Figure 1A, which means that the ergodic component spreads on the entire state space except for winner-takes-all states (see Figure 4A).

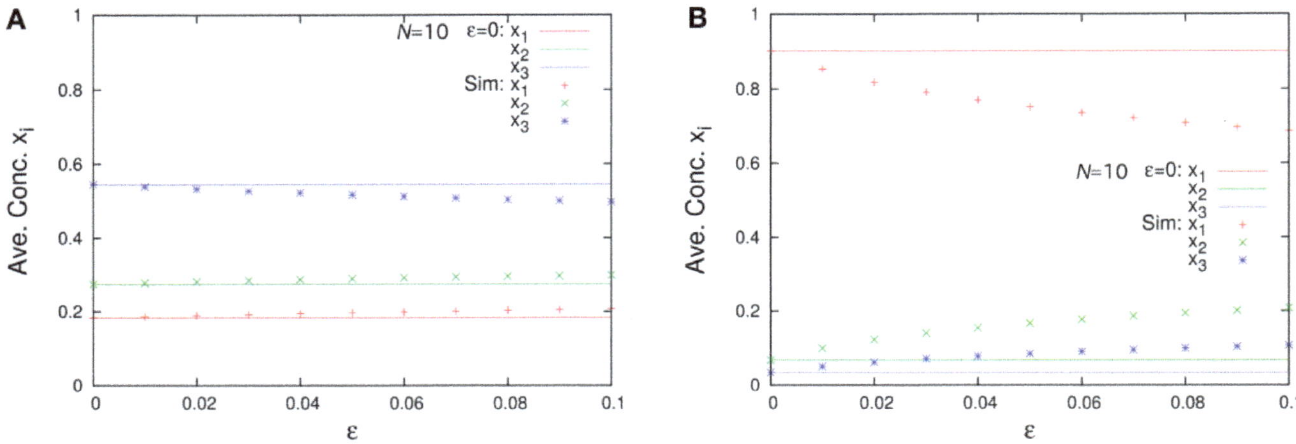

Figure 6. ε-dependence of the time-averaged concentrations x_i (N = 10, ρ = 1) in the case of (A) the three-species system of Figure 1A with non-catalytic reactions and (B) that of Figure1B with non-catalytic reactions. Points in the figures are numerically obtained using the Gillespie algorithm in the same way as Figure 3. The catalytic reaction network of Figure 1A is entirely ergodic; thus, the non-catalytic reactions may be treated as perturbation. On the other hand, that of Figure 1B is non-ergodic; thus, the non-catalytic reactions introduce relatively large differences compared with the case of ε = 0.

APPLICATIONS

The starting point of our analysis is the GFE (Equation 11), from which several useful formulas are derived, namely the PRE (Equation 15), SME (Equation 18), TESMs (Equation 19) and (Equation 20), the λ-condition (Equation 33), and the PGFwoWTAS (Equation 42). In this section, we reveal the effectiveness of these formulas by showing important applications for several catalytic reaction networks.

Rank Conservation Law for Concentrations

We show that the rank of concentrations is conserved even if the total number of molecules changes in catalytic reaction networks (excluding non-catalytic and auto-catalytic reactions).

Suppose the concentration of the ith-species is expressed by Equation (45a) in the state corresponding to the PGFwoWTAS. When determining the rank conservation, it suffices to confirm that the relation between the amount of two arbitrary species is unchanged if the total number of molecules is changed. Let λ_1, λ_2 be the concentrations per total density in the continuous limit such that $\lambda_1 < \lambda_2$. Because $\sum_{i=1}^{M} \lambda_i = 1$, the inequality $\lambda_1 + \lambda_2 < 1$ must be satisfied. Therefore, the following evaluation holds:

$$\begin{aligned} \bar{x}_2 - \bar{x}_1 &= \frac{(\lambda_2 - \lambda_2^N) - (\lambda_1 - \lambda_1^N)}{1 - \sum_{i=1}^{M} \lambda_i^N}\rho \\ &= \frac{(\lambda_2 - \lambda_1)\rho}{1 - \sum_{i=1}^{M} \lambda_i^N}\left[1 - (\lambda_2 + \lambda_1)^{N-1} \right. \\ &\quad \left. + \sum_{k=1}^{N-2}\left\{\binom{N-1}{k} - 1\right\}\lambda_2^{N-1-k}\lambda_1^k\right] \\ &> 0 \qquad (\forall N \in \mathbb{N}_{\geq 2}). \end{aligned} \tag{55}$$

This is the rank conservation law of concentrations.

Note that the rank of the variances of concentrations is generally not conserved when the total number of molecules changes. For example, let us consider the following three-species system (Figure 1C);

$$R_{123} = 1997/3,\ R_{132} = 1000/3,\ R_{231} = R_{321} = 1,$$
$$R_{213} = R_{312} = 0;\ \rho = 1. \tag{56}$$

As shown in Figure 7, this system depends on the total number of molecules for the time-averaged concentrations $\bar{x}_i$ and the variances of concentrations Var[x_i], represented by Equations (45a) and (45c) with λ_1 = 1/1000, λ_2 = 1/3, and λ_3 = 1997/3000. The rank of time-averaged concentrations is always conserved, but the rank of variances is exchanged at certain N (see Figure 8).

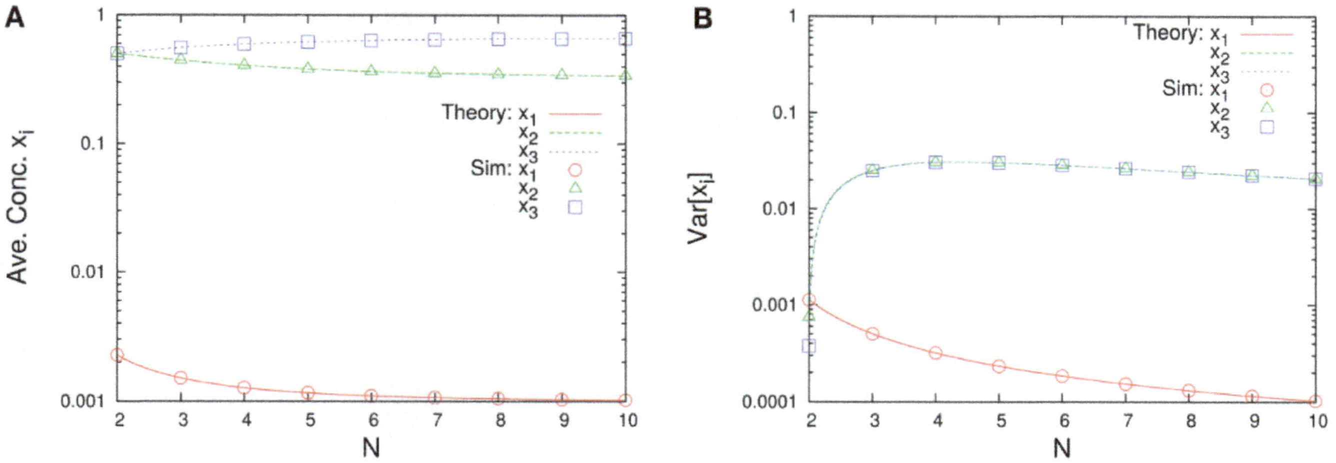

Figure 7. N-dependence of (A) the time-averaged concentrations $\bar{x}_i = \bar{n}_i/N$ and (B) the variances of concentrations $\mathrm{Var}[x_i] = (\overline{n_i^2} - \bar{n}_i^2)/N^2$ in the three-species system of Figure 1C. Empty symbols are numerically obtained using the Gillespie algorithm in the same way as Figure 3. Lines in each figure represent the theoretical expressions, Equations (45a) and (45c), for $\lambda_1 = 1/1000$, $\lambda_2 = 1/3$, and $\lambda_3 = 1997/3000$. It is clear that the rank of time-averaged concentrations is conserved.

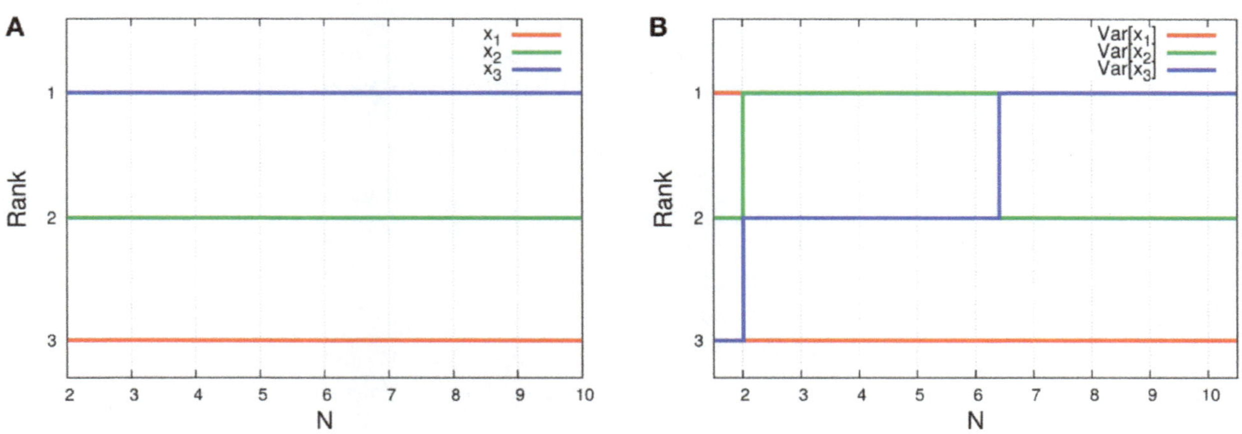

Figure 8. N-dependence of (A) the rank of time-averaged concentrations and (B) the rank of its variances in the three-species system of Figure 1C. The ranks (lines) in the figures are depicted using the formulas (Equations 45a and 45c). Clearly, the rank of concentrations is conserved, although the rank of variances is exchanged between $N = 2$ and 3, and also $N = 6$ and 7.

A Special Case: The Connecting State to the Winner-Takes-All State

There exists an N-molecular state, which connects to the winner-takes-all state in the continuous limit $N \to \infty$. As is later shown, such a special case is not the "weakly reversible" case (Anderson et al., 2010).

We show this by taking the following limit in the PGFwoWTAS (Equation 42):

$$\lambda_1 \longrightarrow 1, \qquad \lambda_2, \cdots, \lambda_M \longrightarrow 0, \tag{57}$$

where we suppose the following constraints are always satisfied:

$$\frac{\lambda_i}{\lambda_2 + \lambda_3 + \cdots + \lambda_M} = \kappa_i \qquad (i = 2, 3, \cdots, M), \tag{58}$$

where $\{\kappa_i\}$ are positive constants, which are determined from the network structure $\{R_{ijk}\}$ (it is explained later). Evidently, the following holds;

$$\sum_{i=2}^{M} \kappa_i = 1. \tag{59}$$

The PGFwoWTAS (Equation 42) has the following limiting expression:

$$\phi_*^{c}(\boldsymbol{z}) = \frac{\left(\frac{\lambda_1}{1-\lambda_1} z_1 + \sum_{i=2}^{M} \kappa_i z_i\right)^N - \left(\frac{\lambda_1}{1-\lambda_1} z_1\right)^N - \sum_{i=2}^{M} (\kappa_i z_i)^N}{\left(\frac{1}{1-\lambda_1}\right)^N - \left(\frac{\lambda_1}{1-\lambda_1}\right)^N - \sum_{i=2}^{M} \kappa_i^N},$$

$$\xrightarrow[\lambda_1 \to 1]{} \quad z_1^{N-1}(\kappa_2 z_2 + \cdots + \kappa_M z_M) \equiv \phi_*^{cs}(\boldsymbol{z}). \tag{60}$$

The state corresponding to the PGF (Equation 60) is the connecting state to the winner-takes-all state (CStoWTAS). The stationary distribution corresponding to the CStoWTAS is immediately obtained:

$$P_*^{cs}(\boldsymbol{n}) = \begin{cases} \kappa_i & (\boldsymbol{n} \in J), \\ 0 & (\boldsymbol{n} \in W \setminus J), \end{cases} \tag{61}$$

where $J_{M,N}$ (abbr. J) is defined as

$$J = \{\boldsymbol{n} \in W_{M,N} \mid n_1 = N-1 \,\&\, \exists i \in [2, M] \text{ s.t. } n_i = 1\}. \tag{62}$$

Furthermore, the marginal distributions of the i-th species can be derived by Equation (13):

$$p_{i*}^{\mathrm{cs}}(n) = \begin{cases} \delta_{n,N-1} & \text{for} \quad i = 1, \\ (1-\kappa_i)\delta_{n,0} + \kappa_i \delta_{n,1} & \text{for} \quad i = 2, \cdots, M, \end{cases} \quad (0 \le n \le N), \tag{63}$$

where δ_{ij} is the Kronecker delta, $\delta_{ij} = 0$ (if $i \neq j$) or 1 (if $i = j$). The time-averaged concentrations calculated by Equation (12) are

$$\bar{x}_1 = \left(1 - \frac{1}{N}\right)\rho, \tag{64a}$$

$$\bar{x}_i = \frac{\kappa_i}{N}\rho \qquad (i = 2, \cdots, M), \tag{64b}$$

and the variances of the time-averaged concentrations calculated by Equation (12) are

$$\mathrm{Var}[x_1] = 0, \tag{65a}$$

$$\mathrm{Var}[x_i] = \frac{\kappa_i(1-\kappa_i)}{N^2}\rho^2 \qquad (i = 2, \cdots, M). \tag{65b}$$

Next, we derive the relation between the positive constants $\{\kappa_i\}$ and the network structure $\{R_{ijk}\}$. The λ-condition (Equation 33) can be converged to conditions for $\{\kappa_i\}$ (the κ-condition) by taking the limit $\lambda_1 \to 1$ as follows. Divide the λ-condition (Equation 33) into two groups, of which one group is the case of $2 \le i < j \le M$,

$$R_{1ij}\lambda_1\kappa_i + R_{1ji}\lambda_1\kappa_j - \left(R_{ij1} + R_{ji1}\right)\kappa_i\kappa_j(1-\lambda_1) + (1-\lambda_1)\sum_{k=2}^{M}\left\{R_{kij}\kappa_k\kappa_i + R_{kji}\kappa_k\kappa_j - \left(R_{ijk} + R_{jik}\right)\kappa_i\kappa_j\right\} = 0, \tag{66a}$$

and the other group is the case of $i = 1$, $2 \le j \le M$,

$$\sum_{k=2}^{M}\left\{R_{k1j}\kappa_k\lambda_1 + R_{kj1}\kappa_k\kappa_j(1-\lambda_1) - \left(R_{1jk} + R_{j1k}\right)\lambda_1\kappa_j\right\} = 0, \tag{66b}$$

where we substituted $\lambda_i = \kappa_i(1-\lambda_1)$, $i \ge 2$ and divided by $1-\lambda_1$. Then, taking the limit $\lambda_1 \to 1$ in Equation (66a),

$$\kappa_i R_{1ij} + \kappa_j R_{1ji} = 0 \qquad \text{i.e.,} \quad R_{1ij} = 0 \quad (\forall i, j), \tag{67a}$$

and taking that in Equation (66b),

$$\sum_{k=2}^{M}\left(\kappa_k R_{k1j} - \kappa_j R_{j1k}\right) = 0 \qquad (2 \le j \le M). \tag{67b}$$

The condition (Equation 67a) expresses that the 1st-species cannot be a substrate, and the condition (Equation 67b) represents the desired κ-condition. Note that the CStoWTAS must be a limiting state corresponding to the PGFwoWTAS [see Equation (60)]. Therefore, the network structure $\{R_{ijk}\}$ must have definite $\{\lambda_i\}$.

For example, let us consider the following three-species system (Figure 1B):

$$R_{123} = R_{132} = 0, R_{213} = R_{231} = R_{321} = 1, R_{312} = 2; \rho = 1, \tag{68}$$

which implies $\kappa_2 = 2/3$ and $\kappa_3 = 1/3$. Obviously, this system is not weakly reversible, which means that the theorems (Theorem 4.1 and 4.2) in the previous study (Anderson et al., 2010) cannot be applied. This system should have a CStoWTAS because λ_1, λ_2, and λ_3 are definite from Equation (39) with $\Lambda_1 = 0$, $\Lambda_2 = 2$, and $\Lambda_3 = 4$. As shown in Figure 9, this system certainly has the time-averaged concentrations $\bar{x}_i$ represented by Equation (64) and the variances $Var[x_i]$ represented by Equation (65). In this case also, the rank conservation law of concentrations holds.

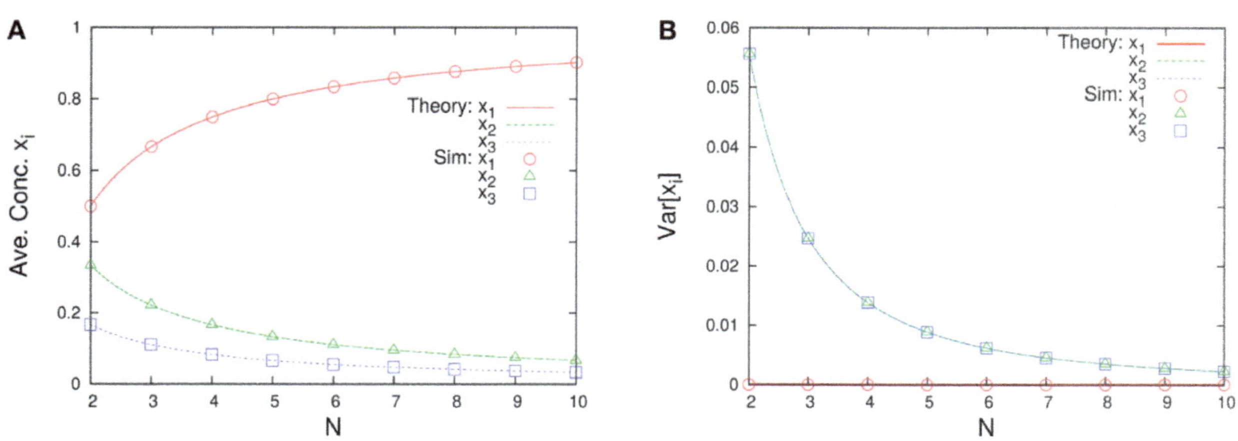

Figure 9. N-dependence of (A) time-averaged concentrations $\bar{x}_i = \bar{n}_i/N$ and (B) the variances of concentrations $\mathrm{Var}[x_i] = (\overline{n_i^2} - \bar{n}_i^2)/N^2$ in the three-species system of Figure 1B. Empty symbols were numerically obtained using the Gillespie algorithm in the same way as Figure 3. Lines in each figure represent the theoretical expressions, Equations (64) and (65), for $\kappa_2 = 2/3$ and $\kappa_3 = 1/3$.

M-Species 2-Molecules Systems with Non-catalytic Reactions

The 2mTESM (Equation 20) becomes the closed equation of the second-order moments $\langle x_l x_m \rangle$ if the first-order moments $\langle x_i \rangle$ are substituted by the second-order moments according to the PRE (Equation 20). In this subsection, we consider catalytic-non-catalytic mixed reaction systems of M species, consisting of only 2 molecules in total.

We first focus on the second formula in the 2mTESMs (Equation 20). It can be seen that each variance of time-averaged concentration $\mathrm{Var}[x_i] = \overline{x_i^2} - \overline{x_i}^2$ in the steady state depends solely on its concentration:

$$\mathrm{Var}[x_i] = \left(\frac{M+1}{2M}\rho - \overline{x}_i\right)\overline{x}_i, \tag{69}$$

where we supposed the ergodicity $\langle x_i \rangle = \overline{x}_i$. The above equation expresses that the fluctuation of the concentration x_i becomes larger as its time-averaged concentration approaches

$$\overline{x} = \frac{M+1}{4M}\rho, \tag{70}$$

at which the fluctuation takes the largest value $\left(\frac{M+1}{4M}\rho\right)^2$. Furthermore, the time-averaged concentrations potentially have the maximum value because the variance must be positive;

$$\overline{x}_i \leq \frac{M+1}{2M}\rho \qquad (\forall i \in [1, M]). \tag{71}$$

Next, we focus on the first formula in the 2mTESMs (Equation 20). Let us consider the steady state of Equation (20a) and suppose the ergodicity $\langle x_i \rangle = \overline{x}_i$. Eliminating the first-order moments in Equation (20a) by using the SME (Equation 18) gives the determination equation of second-order moments in the 2 molecules system (2mDESM);

$$\frac{4\varepsilon}{\rho}\overline{x_l x_m} + \sum_i \{(R_{lmi} + R_{mli})\,\overline{x_l x_m} - R_{ilm}\overline{x_i x_l} - R_{iml}\overline{x_i x_m}\}$$
$$- \frac{1}{M}\sum_{ij}\left\{(R_{jil} + R_{jim})\,\overline{x_i x_j} - R_{lij}\overline{x_i x_l} - R_{mij}\overline{x_i x_m}\right\}$$
$$= \frac{2\rho\varepsilon}{M^2} \quad (1 \leq l < m \leq M). \tag{72}$$

The above 2mDESM can be rewritten in the $M(M-1)/2 \times M(M-1)/2$ matrix form. We demonstrate the procedure for processing the 2mDESM by considering the case of $M = 3$. In this case, the 2mDESM (Equation 72) has a matrix of the following form:

$$\begin{bmatrix} \frac{4}{3}(R_{123}+R_{213})+\frac{4\varepsilon}{\rho} & -\frac{4}{3}R_{312} & -\frac{4}{3}R_{321} \\ -\frac{4}{3}R_{213} & \frac{4}{3}(R_{132}+R_{312})+\frac{4\varepsilon}{\rho} & -\frac{4}{3}R_{231} \\ -\frac{4}{3}R_{123} & -\frac{4}{3}R_{132} & \frac{4}{3}(R_{231}+R_{321})+\frac{4\varepsilon}{\rho} \end{bmatrix} \begin{bmatrix} \overline{x_1x_2} \\ \overline{x_1x_3} \\ \overline{x_2x_3} \end{bmatrix} = \frac{2\rho\varepsilon}{9}\begin{bmatrix} 1 \\ 1 \\ 1 \end{bmatrix}. \tag{73}$$

Moreover, we consider the specific three-species system of Figure 1A (ρ = 1) including non-catalytic reactions. Solving Equation (73) as ε > 0 and ρ = 1 in the case of Figure 1A, then

$$\begin{bmatrix} \overline{x_1x_2} \\ \overline{x_1x_3} \\ \overline{x_2x_3} \end{bmatrix} = \begin{bmatrix} \frac{1+3\varepsilon}{18(2+3\varepsilon)} \\ \frac{1}{18} \\ \frac{1+\varepsilon}{6(2+3\varepsilon)} \end{bmatrix}, \tag{74}$$

which is consistent with the result obtained when solving the CME (Equation 7) directly [recall $\langle x_1x_2\rangle = \frac{1}{4}P(n_1 = 1, n_2 = 1, n_3 = 0)$, $\langle x_1x_3\rangle = \cdots$ and so on]. By the SME (Equation 18), the concentrations become

$$\overline{x}_1 = \frac{1+2\varepsilon}{2(2+3\varepsilon)}, \quad \overline{x}_2 = \frac{1}{3}, \quad \overline{x}_3 = \frac{5+6\varepsilon}{6(2+3\varepsilon)}. \tag{75}$$

Figure 10 shows the second-order moments $\overline{x_ix_j}$ and the time-averaged concentrations $\overline{x}_i$ as functions of ε. Note that in the case of ε = 0 (catalytic reactions only), we need to derive the second-order moments $\overline{x_ix_j}$ from Equation (45b) corresponding to $\lambda_1 = 2/11$, $\lambda_2 = 3/11$, and $\lambda_3 = 6/11$. The non-catalytic reaction rate constant ε seems to be a singular perturbation against the second-order moments (not the concentrations).

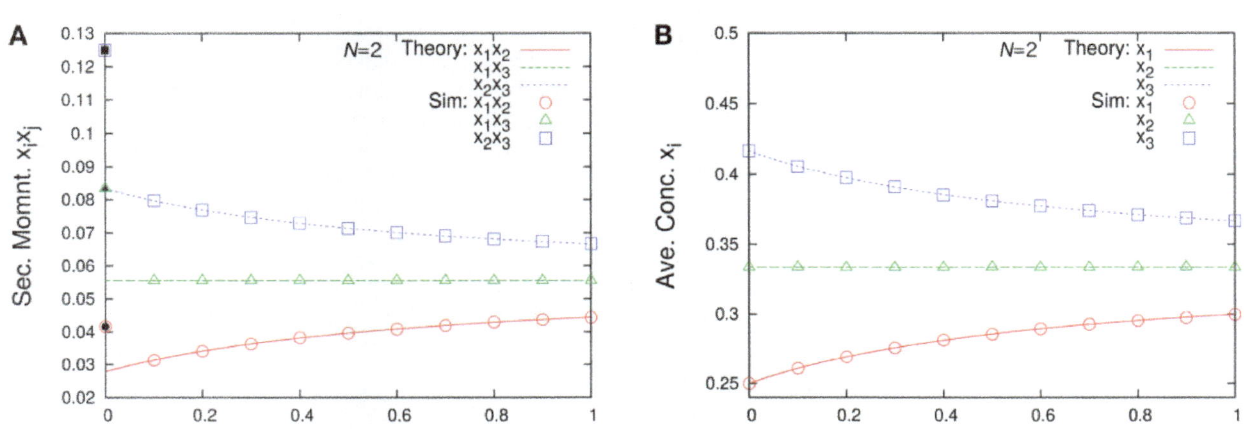

Figure 10. ε-dependence of (A) the second-order moments $\overline{x_ix_j} = \overline{n_in_j}/N^2$ and (B) the time-averaged concentrations $\overline{x}_i = \overline{n}_i/N$ in the three-species system of Figure 1A with non-catalytic reactions (N=2, ρ = 1). Empty symbols are numerically obtained using the Gillespie algorithm in the same way as Figure 3. Lines in each figure represent the theoretical expressions, Equations (74) and

(75). Note that there exist differences between the limit $\varepsilon \to 0$ and the case $\varepsilon = 0$ for the second-order moments (solid black points) because the SME (Equation 18) breaks down at $\varepsilon = 0$. The solid black points are calculated from Equation (45b) as $\lambda_1 = 2/11$, $\lambda_2 = 3/11$, $\lambda_3 = 6/11$, $N = 2$, and $\rho = 1$.

Non-Autocatalyzation of Autocatalytic Reaction Networks

The framework we developed in this paper applies to non-autocatalytic reaction networks. However, our framework may be applicable to autocatalytic reaction networks if it were possible to convert autocatalytic to non-autocatalytic networks. Here, we show several examples of such conversions using a minimal autocatalytic reaction network "2TK model" (Ohkubo et al., 2008; Saito and Kaneko, 2015), which is a well-studied model in the context of discreteness-induced phenomena.

The 2TK model consists of only two species, and includes both autocatalytic reactions (rate const. r) and non-catalytic reactions (rate const. $\varepsilon \ll r$) (see Figure 11A);

$$A \underset{A}{\overset{B}{\rightleftarrows}} B \quad \text{(autocat. react.)}, \tag{76}$$

$$A \rightleftarrows B \quad \text{(non-cat. react.)}. \tag{77}$$

If we suppose that each of the species A and B consists of two further species, we can convert autocatalytic reactions Equation (76) of the 2TK model to non-autocatalytic reactions as shown in Figure 11B. The catalysis of the species 5 between the species 1 and 3 (or, 2 and 4) is to establish an equilibrium between the concentrations of the species 1 and 3 (or, 2 and 4), that would make it possible to regard these as one species. The non-autocatalytic reactions of Figure 11B plus non-catalytic reactions $i \xrightarrow{\text{Prob. }1/5} j$ is the desired five-component model, of which the CME is simply described by Equation (7) with

$$R_{142} = R_{153} = R_{231} = R_{254} = R_{351} = R_{452} = 1 \quad \text{(others 0)}. \tag{78}$$

As shown in Figure 12, the behavior of the variables $n_A = n_1+n_3+n_5/2$ and $n_B = n_2+n_4+n_5/2$ in the five-component model is similar to that of the 2TK model (compare with Figures 1A,B in Saito and Kaneko, 2015). The λ-condition (Equation 33) of catalytic reactions in the five-component model can be written in matrix form,

$$\begin{bmatrix} 0&0&0&0&0&0&0&0&0&0 \\ 0&0&0&0&1&0&0&0&0&0 \\ 0&0&-1&0&0&0&0&0&0&0 \\ 0&0&0&-1&0&0&0&0&1&0 \\ 0&0&0&0&-1&0&0&0&0&0 \\ 0&0&1&0&0&0&0&0&0&0 \\ 0&0&0&0&0&0&-1&0&0&1 \\ 0&0&0&0&0&0&0&0&0&0 \\ 0&0&0&1&0&0&0&0&-1&0 \\ 0&0&0&0&0&0&1&0&0&-1 \end{bmatrix} \begin{bmatrix} \lambda_1\lambda_2 \\ \lambda_1\lambda_3 \\ \lambda_1\lambda_4 \\ \lambda_1\lambda_5 \\ \lambda_2\lambda_3 \\ \lambda_2\lambda_4 \\ \lambda_2\lambda_5 \\ \lambda_3\lambda_4 \\ \lambda_3\lambda_5 \\ \lambda_4\lambda_5 \end{bmatrix} = 0, \tag{79}$$

in which there exist six non-trivial eigenvectors corresponding to the 0-eigenvalue; hence, the general eigenvector corresponding to the 0-eigenvalue can be expressed as a linear summation of those six non-trivial eigenvectors:

$$\begin{bmatrix} \lambda_1\lambda_2 \\ \lambda_1\lambda_3 \\ \lambda_1\lambda_4 \\ \lambda_1\lambda_5 \\ \lambda_2\lambda_3 \\ \lambda_2\lambda_4 \\ \lambda_2\lambda_5 \\ \lambda_3\lambda_4 \\ \lambda_3\lambda_5 \\ \lambda_4\lambda_5 \end{bmatrix} = c_1 \begin{bmatrix} 1\\0\\0\\0\\0\\0\\0\\0\\0\\0 \end{bmatrix} + c_2 \begin{bmatrix} 0\\1\\0\\0\\0\\0\\0\\0\\0\\0 \end{bmatrix} + c_3 \begin{bmatrix} 0\\0\\0\\0\\0\\1\\0\\0\\0\\0 \end{bmatrix} + c_4 \begin{bmatrix} 0\\0\\0\\0\\0\\0\\0\\1\\0\\0 \end{bmatrix} + c_5 \begin{bmatrix} 0\\0\\0\\1\\0\\0\\0\\0\\1\\0 \end{bmatrix} + c_6 \begin{bmatrix} 0\\0\\0\\0\\0\\0\\1\\0\\0\\1 \end{bmatrix}. \tag{80}$$

The above equation can be used to derive six types of non-trivial solutions $(\lambda_1, \lambda_2, \cdots, \lambda_5)$, which immediately correspond to the stationary states of catalytic reactions in the five-component model by using Equation (43):

$$(\lambda_1, \lambda_2, \lambda_3, \lambda_4, \lambda_5) = \begin{cases} \text{(i)} & (0, c, 0, c, 1-2c), \\ \text{(ii)} & (c, 0, c, 0, 1-2c), \\ \text{(iii)} & (2c, 1-2c, 0, 0, 0), \\ \text{(iv)} & (2c, 0, 1-2c, 0, 0), \\ \text{(v)} & (0, 2c, 0, 1-2c, 0), \\ \text{(vi)} & (0, 0, 2c, 1-2c, 0), \end{cases} \quad \forall c \in (0, 1/2), \tag{81}$$

where each case (i-vi) corresponds to (i) $c_3 = c^2$, $c_6 = c(1-2c)$ (others 0), (ii) $c_2 = c^2$, $c_5 = c(1-2c)$ (others 0), (iii) $c_1 = 2c(1-2c)$ (others 0), (iv) $c_2 = 2c(1-2c)$ (others 0), (v) $c_3 = 2c(1-2c)$ (others 0), and (vi) $c_4 = 2c(1-2c)$ (others 0). The switching behavior of the five-component model may be explained in terms of transition processes between the above six types of steady states and the five trivial steady states $\lambda_i = 1$ (others 0) of catalytic reactions in the five-component model, which are sometimes caused by non-catalytic reactions. Furthermore, the marginal distribution of the species A, $p_A(n)$, can be derived by calculating the convolution of the marginal distributions $p_1(n)$, $p_3(n)$, and $p_5(n)$, if $p_1(n)$, $p_3(n)$, and $p_5(n)$ were obtained for the case including non-catalytic reactions.

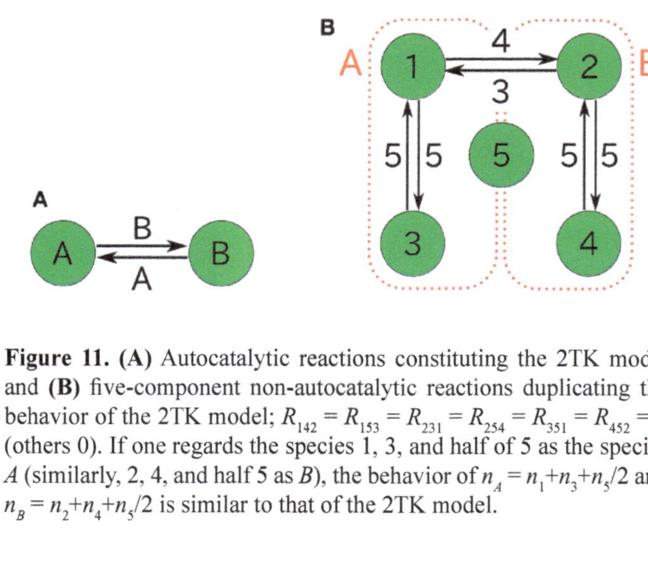

Figure 11. (A) Autocatalytic reactions constituting the 2TK model and **(B)** five-component non-autocatalytic reactions duplicating the behavior of the 2TK model; $R_{142} = R_{153} = R_{231} = R_{254} = R_{351} = R_{452} = 1$ (others 0). If one regards the species 1, 3, and half of 5 as the species A (similarly, 2, 4, and half 5 as B), the behavior of $n_A = n_1+n_3+n_5/2$ and $n_B = n_2+n_4+n_5/2$ is similar to that of the 2TK model.

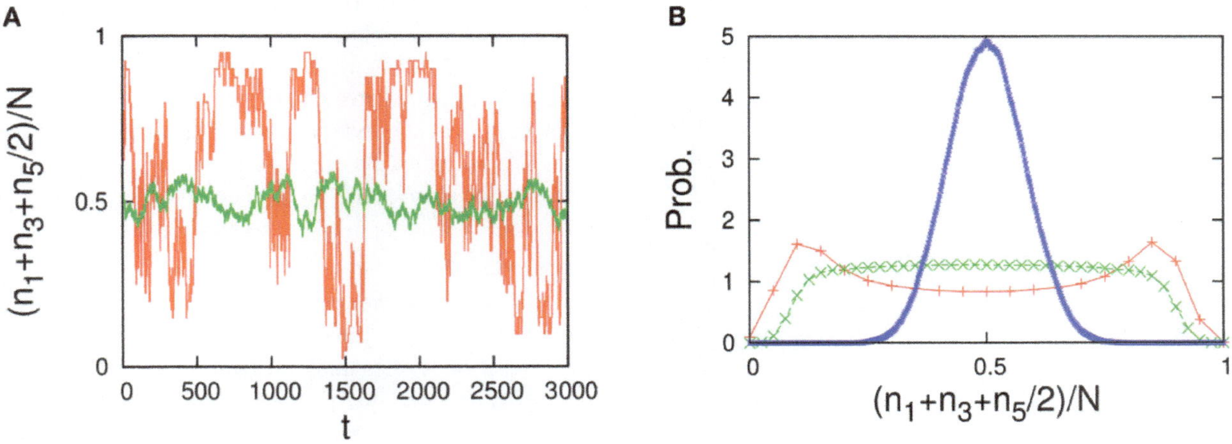

Figure 12. Behavior of the five-component model [Figure 11B plus non-catalytic reactions ($\varepsilon = 0.01$, $\rho = 1$)]. The behavior of the five-component model is reminiscent of the behavior of the 2TK model (compare with Figures 1A,B in Saito and Kaneko, 2015). **(A)** Time series of the total concentration of the species 1, 3, and half of 5 for $N = 20$ (red line) and $N = 2000$ (green line). **(B)** Stationary distributions of $(n_1+n_3+n_5/2)/N$, obtained numerically from a long time series $n_i(t)$ using the Gillespie algorithm whose simulation conditions are the total number of reactions: 10^8, the number of reactions for transient exclusion: 10^7, and the initial value is randomly selected from $W \setminus I$ such that the average per one-species is N/M. The unimodal distribution (blue), the flat distribution (green), and the bimodal distribution (red) indicate the stationary distribution for $N = 500$, 40, and 20, respectively.

Other non-autocatalytic reaction networks duplicating the 2TK model are shown in Figure 13A, and consist of 4 components only. Although readers would think that the four-component models do not function properly because the transitions between the species group (1, 3) [or (2, 4)] depend on the other species group (2, 4) [or (1, 3)], the four-component models actually generate much the same behavior with the 2TK model (see Figure14). The switching behavior of the four-component model may also be explained as a transition process between the following three types of steady states

$$(\lambda_1, \lambda_2, \lambda_3, \lambda_4) = \begin{cases} (0, 0, c, 1-c), \\ (0, c, 0, 1-c), \\ (c, 0, 1-c, 0), \end{cases} \quad \forall c \in (0, 1), \tag{82}$$

and the four trivial steady states $\lambda_i = 1$ (others 0) of catalytic reactions in the four-component model, which are sometimes caused by non-catalytic reactions. We also confirmed that the results are not changed even in the other four-component model of Figure 13B, of which the catalysis role of each the species 1 and 2 is exchanged by that of 3 and 4, respectively.

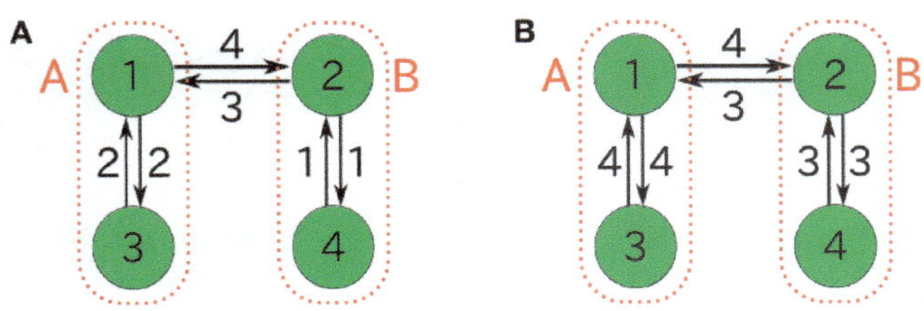

Figure 13. Four-component non-autocatalytic reactions duplicating behaviors of the 2TK model; (A) $R_{123} = R_{142} = R_{214} = R_{231} = R_{321} = R_{412} = 1$ (others 0), and (B) $R_{142} = R_{143} = R_{231} = R_{234} = R_{341} = R_{432} = 1$ (others 0). If one regards the species 1 and 3 as the species A (similarly, 2 and 4 as$B$), the behavior of $n_A = n_1+n_3$ and $n_B = n_2+n_4$ is almost equivalent to that of the 2TK model.

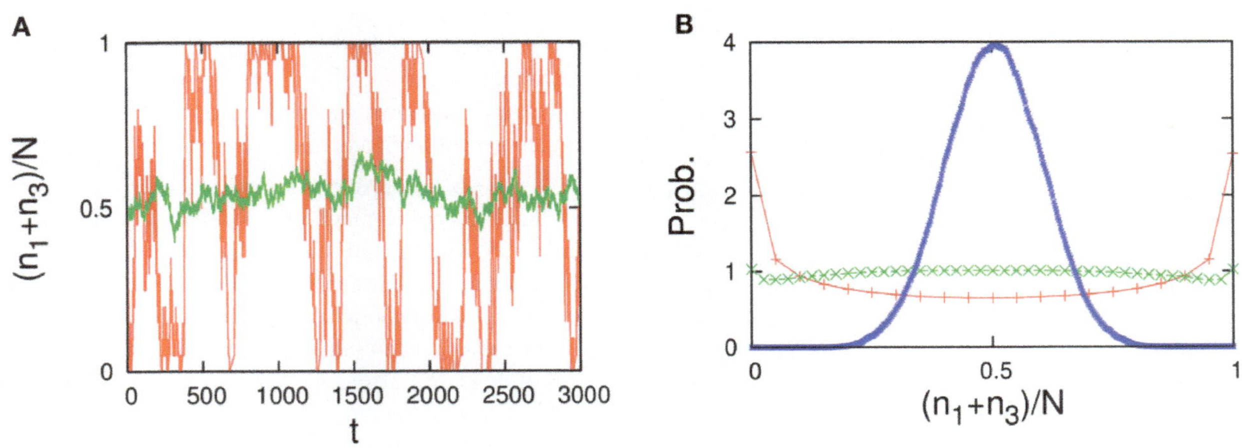

Figure 14. Behavior of the four-component model [Figure 13A plus non-catalytic reactions ($\varepsilon = 0.01$, $\rho = 1$)]. The four-component model reproduces the behavior of the 2TK model (compare with Figures 1A,B in Saito and Kaneko, 2015). (A) Time series of the total concentration of the species 1 and 3 for $N = 20$ (red line) and $N = 2000$ (green line). (B) Stationary distributions of $(n_1+n_3)/N$, obtained numerically from a long time series $n_i(t)$ using the Gillespie algorithm in the same way as Figure 12. The unimodal distribution (blue), the uniform distribution (green), and the bimodal distribution (red) indicate the stationary distribution for $N = 500$, 40, and 20, respectively.

SUMMARY AND DISCUSSIONS

The framework we presented in this paper facilitates the prediction of the effect of the small-number issue on the concentration of each species in catalytic reaction networks. This can be described in an extreme manner by comparing the concentrations between the continuous limit ($N \to \infty$) and the case of 2 molecules ($N = 2$). If the reaction network does not include non-catalytic reactions (or, includes negligible non-catalytic reactions), we can use the formula (Equation 45a) to compare them. On the other hand, if the reaction network includes (non-negligible) non-catalytic reactions, we need to apply the formula, 2mDESM (Equation 72), with the SME (Equation 18) to obtain the concentrations in the case of 2 molecules. Although our theory has a presupposition referred to as *entire ergodicity*, the presupposition is intuitively verifiable if the system is specified, as in Section 2.4.2.4. We also demonstrated three examples for *non-autocatalyzation conversions of autocatalytic reaction networks* in Section 3.4. We consider this type of conversion to be generalized relatively easily such that our analytical framework can be applied to more general catalytic reaction networks including autocatalytic reactions.

One might think that the analysis presented in this paper can be straightforwardly extended to the case including autocatalytic reactions [in fact, the CME (Equation 7) and GFE (Equation 11) themselves hold even in the case including autocatalytic reactions]. However, if autocatalytic reactions are included (i.e., the case $R_{ikk} > 0$ is allowed), we cannot consider catalytic reactions and non-catalytic reactions to be completely separate. The reason is that, in the case including autocatalytic reactions, the absence of non-catalytic reactions generally implies winner-takes-all steady states. Generally, solving the CME (or GFE) of catalytic-noncatalytic mixed reactions systems is more advanced and a more difficult task than that of catalytic reactions only. The proposed strategy, i.e., non-autocatalyzation conversions, is one of our ideas to address the problem.

The formulas obtained in the present work are specific and satisfactorily simple. Therefore, our theory has the capabilities to be developed into a general theory for catalytic reaction networks. On the other hand, there exists a mathematical theory for a certain class of catalytic reaction networks that are "weakly reversible" and "deficiency zero" (Anderson et al., 2010). Our formulas (Equations 27 and 43) are consistent with the main theorems (theorem 4.1 and 4.2 in Anderson et al., 2010) in the above-mentioned mathematical theory. One of the advantages of our theory compared to the above mathematical theory is understandably the probability generating function (PGF) approach, because the PGF is a major analytical tool for physicists, chemists, and mathematical biologists. Therefore, our theory is easily verifiable, and one can design a computer algorithm to calculate our analytical formulas. We also showed the extensibility of our theory by using applications (in Section 3), especially because of the CStoWTAS (Equation 61), which was not suggested by the above mathematical theory.

Actual biochemical pathways in the cell involve thousands of chemical species, and their chemical properties vary. Our theoretical framework is general and extensible to such complex reaction networks, if they can be represented by CMEs such as Equation (7). As our current model consists of simple two-body catalytic reactions, it is difficult to point out examples in actual biological systems that correspond exactly to our model. Biochemical reactions in reality may involve a number of intermediates. There are also autocatalytic processes such as autophosphorylation, and replication of templates such as DNA, in which the catalyst or template species is also a substrate or a product. Our framework is applicable to many such cases involving network conversion, as shown for simple autocatalytic cases.

Nevertheless, the reaction kinetics of each enzyme is not always simple. Enzymes are complex macromolecules and their reaction cycles may depend on their conformational states. Therefore, the prediction of biological phenomena caused by small-number effects in real biochemical reactions, would entail further analytical challenges for catalytic reaction networks including arbitrary higher-order mixed reactions (rather than first- and second-order reactions only) or internal dynamics of the enzymes (as modeled and analyzed in Togashi and Casagrande, 2015) as important issues.

Throughout this work, our primary intention is to approach small-number issues in biological systems. One might wonder how general these small-number issues appear, and how important they are, in living cells. Recently, absolute quantification of various proteins and mRNAs in the cell has become possible, and the integration of experimental results (e.g., the construction of a database Milo et al., 2010) is also underway. Taniguchi et al. investigated the copy number distribution for more than a thousand protein species in bacteria (Taniguchi et al., 2010). Li et al. further discussed the relationship between the copy number and synthesis rate, and also the role, of proteins (Li et al., 2014). According to the result, some transcription factors, particularly activators, are rare, of the order of 0.1 to 10 molecules per genome equivalent. Although stochastic gene expression has been intensively discussed for years (McAdams and Arkin, 1997; Thattai and vanOudenaarden, 2001; Elowitz et al., 2002; Raj and van Oudenaarden, 2008; Shahrezaei and Swain, 2008), the discrete small-number nature of transcription factors has often been ignored; hence, the finding may urge us to reconsider the issue. Synthetic approaches are also becoming popular to confirm small-number effects. Ma et al. reported that an additional stable state in a genetic bistable toggle switch attributable to the small-number effect, which was predicted by stochastic simulations, was indeed observed in bacteria containing the genetically engineered switch (Ma et al., 2012). These results suggest that such rare proteins, of the order of one molecule per cell, are common and affect regulatory function in bacteria.

Although eukaryotic cells are much larger than bacteria, they have complex membrane structures and cytoskeletons inside, and the small-number issues can be particularly significant in compartments or bottlenecks (e.g., if we consider the volume of a synaptic vesicle represented by a sphere 40 nm in diameter, then 1 molecule corresponds to ca. 50 μmol/L). Rare proteins are also involved in physiologically important signaling pathways in eukaryotes. In the Wnt signaling pathway, for example, the concentration of axin is reported to be 20 pmol/L in Xenopus eggs (Lee et al., 2003) (though suggested to be higher in mammalian cells Tan et al., 2012). Another example is the MAP-kinase cascade, where proteins of the order of merely 10^2molecules (e.g., Ste5) exist in a yeast cell (Thomson et al., 2011). For scaffold proteins such as axin and Ste5, specifically, localization (locally high concentrations) of other chemical species around the scaffold may drastically change the reaction behavior, as spatial discreteness of the scaffolds becomes significant (Shnerb et al., 2000; Togashi and Kaneko, 2004). Further studies in which spatial structures (cf. reaction-diffusion equations) are considered are also expected to be important.

In the presented framework, we mainly focused on the steady-state solutions of GFE. Of course, temporal courses are biologically crucial in some cases. A well-studied example is oscillatory behavior in circadian clocks (Bell-Pedersen et al., 2005). In such oscillations, if a chemical factor is depleted down to a small number in a certain phase, then, the period can be susceptible to stochastic reactions involving the factor in that phase; on the other hand, sequestration of a factor may contribute to regular oscillations (Jolley et al., 2012). Again, the internal dynamics of enzymes can also be relevant in some systems. Although stochastic simulations are indeed powerful and many attempts are currently underway, further theoretical understanding as well as the experimental quantitative observation of rare factors would be required.

Note that a chemical "species" here can also be interpreted as a specific state of a molecule; e.g., we can consider proteins or genes, with and without modification, as separate species. Moreover, a similar interpretation is also applicable to ecology and ethology (Biancalani et al., 2014), if the laws governing the system are analogous to reactions. We remain hopeful that theoretical frameworks including ours will facilitate the exploration of small-number issues at equally higher levels of biological systems in future.

AUTHOR CONTRIBUTIONS

MN and YT conceived the research. MN performed the analysis and simulations. Both authors discussed the results and wrote the paper.

ACKNOWLEDGMENTS

We would like to thank Dr. Nen Saito for valuable discussion and useful comments.

REFERENCES

1. Anderson, D. F., Craciun, G., and Kurtz, T. G. (2010). Product-form stationary distributions for deficiency zero chemical reaction networks. *Bull. Math. Biol.* 72, 1947–1970. doi: 10.1007/s11538-010-9517-4
2. Awazu, A., and Kaneko, K. (2007). Discreteness-induced transition in catalytic reaction networks. *Phys. Rev. E* 76:041915. doi: 10.1103/PhysRevE.76.041915
3. Awazu, A., and Kaneko, K. (2009). Ubiquitous "glassy" relaxation in catalytic reaction networks. *Phys. Rev. E* 80:041931. doi: 10.1103/PhysRevE.80.041931
4. Bell-Pedersen, D., Cassone, V. M., Earnest, D. J., Golden, S. S., Hardin, P. E., Thomas, T. L., et al. (2005). Circadian rhythms from multiple oscillators: lessons from diverse organisms. *Nat. Rev. Genet.* 6, 544–556. doi: 10.1038/nrg1633
5. Biancalani, T., Dyson, L., and McKane, A. J. (2014). Noise-induced bistable states and their mean switching time in foraging colonies. *Phys. Rev. Lett.* 112:038101. doi: 10.1103/PhysRevLett.112.038101
6. Biancalani, T., Rogers, T., and McKane, A. J. (2012). Noise-induced metastability in biochemical networks. *Phys. Rev. E* 86:010106. doi: 10.1103/PhysRevE.86.010106
7. Elowitz, M. B., Levine, A. J., Siggia, E. D., and Swain, P. S. (2002). Stochastic gene expression in a single cell. *Science* 297, 1183–1186. doi: 10.1126/science.1070919
8. Gadgil, C., Lee, C. H., and Othmer, H. G. (2005). A stochastic analysis of first-order reaction networks. *Bull. Math. Biol.* 67, 901–946. doi: 10.1016/j.bulm.2004.09.009
9. Gillespie, D. T. (1976). A general method for numerically simulating the stochastic time evolution of coupled chemical reactions. *J. Comput. Phys.* 22, 403–434. doi: 10.1016/0021-9991(76)90041-3
10. Gillespie, D. T. (1977). Exact stochastic simulation of coupled chemical reactions. *J. Phys. Chem.* 81, 2340–2361. doi: 10.1021/j100540a008
11. Haruna, T. (2010). Investigating the gap between discrete and continuous models of chemically reacting systems. *J. Comput. Chem. Jpn.* 9, 135–142. doi: 10.2477/jccj.H2110
12. Haruna, T. (2015). Distinguishing between discreteness effects in stochastic reaction processes. *Phys. Rev. E* 91:052814. doi: 10.1103/PhysRevE.91.052814
13. Houchmandzadeh, B., and Vallade, M. (2015). Exact results for a noise-induced bistable system. *Phys. Rev. E* 91:022115. doi: 10.1103/PhysRevE.91.022115
14. Jolley, C. C., Ode, K. L., and Ueda, H. R. (2012). A design principle for a posttranslational biochemical oscillator. *Cell Rep.* 2, 938–950. doi: 10.1016/j.celrep.2012.09.006
15. Kauffman, S. A. (1969). Metabolic stability and epigenesis in randomly constructed genetic nets. *J. Theor. Biol.* 22, 437–467. doi: 10.1016/0022-5193(69)90015-0
16. Kim, P., and Lee, C. H. (2012). A probability generating function method for stochastic reaction networks. *J. Chem. Phys.* 136, 234108. doi: 10.1063/1.4729374
17. Krieger, I. M., and Gans, P. J. (1960). First-order stochastic processes. *J. Chem. Phys.* 32, 247–250. doi: 10.1063/1.1700909
18. Lee, C. H., Kim, K. H., and Kim, P. (2009). A moment closure method for stochastic reaction networks. *J. Chem. Phys.* 130, 134107. doi: 10.1063/1.3103264
19. Lee, E., Salic, A., Krüger, R., Heinrich, R., and Kirschner, M. W. (2003). The roles of APC and axin derived from experimental and theoretical analysis of the Wnt pathway. *PLoS Biol.* 1:e10. doi: 10.1371/journal.pbio.0000010
20. Li, G.-W., Burkhardt, D., Gross, C., and Weissman, J. S. (2014). Quantifying absolute protein synthesis rates reveals principles underlying allocation of cellular resources. *Cell* 157, 624–635. doi: 10.1016/j.cell.2014.02.033
21. Ma, R., Wang, J., Hou, Z., and Liu, H. (2012). Small-number effects: a third stable state in a genetic bistable toggle switch. *Phys. Rev. Lett.* 109, 248107. doi: 10.1103/

PhysRevLett.109.248107

22. McAdams, H. H., and Arkin, A. (1997). Stochastic mechanisms in gene expression. *Proc. Natl. Acad. Sci. U.S.A.* 94, 814–819. doi: 10.1103/PhysRevLett.109.248107
23. McQuarrie, D. A. (1967). Stochastic approach to chemical kinetics. *J. Appl. Prob.* 4, 413–478. doi: 10.2307/3212214
24. Milo, R., Jorgensen, P., Moran, U., Weber, G., and Springer, M. (2010). BioNumbers — the database of key numbers in molecular and cell biology. *Nucl. Acids Res.* 38(Suppl. 1), D750–D753. doi: 10.1093/nar/gkp889
25. Munsky, B., and Khammash, M. (2006). The finite state projection algorithm for the solution of the chemical master equation. *J. Chem. Phys.*124:044104. doi: 10.1063/1.2145882
26. Ohkubo, J., Shnerb, N., and Kessler, D. A. (2008). Transition phenomena induced by internal noise and quasi-absorbing state. *J. Phys. Soc. Jpn.*77:044002. doi: 10.1143/JPSJ.77.044002
27. Raj, A., and van Oudenaarden, A. (2008). Nature, nurture, or chance: stochastic gene expression and its consequences. *Cell* 135, 216–226. doi: 10.1016/j.cell.2008.09.050
28. Saito, N., and Kaneko, K. (2015). Theoretical analysis of discreteness-induced transition in autocatalytic reaction dynamics. *Phys. Rev. E*91:022707. doi: 10.1103/PhysRevE.91.022707
29. Shahrezaei, V., and Swain, P. S. (2008). Analytical distributions for stochastic gene expression. *Proc. Natl. Acad. Sci. U.S.A.* 105, 17256–17261. doi: 10.1073/pnas.0803850105
30. Shnerb, N. M., Louzoun, Y., Bettelheim, E., and Solomon, S. (2000). The importance of being discrete: life always wins on the surface. *Proc. Natl. Acad. Sci. U.S.A.* 97, 10322–10324. doi: 10.1073/pnas.180263697
31. Tan, C. W., Gardiner, B. S., Hirokawa, Y., Layton, M. J., Smith, D. W., and Burgess, A. W. (2012). Wnt signalling pathway parameters for mammalian cells. *PLoS ONE* 7:e31882. doi: 10.1371/journal.pone.0031882
32. Taniguchi, Y., Choi, P. J., Li, G.-W., Chen, H., Babu, M., Hearn, J., et al. (2010). Quantifying *E.* coli proteome and transcriptome with single-molecule sensitivity in single cell. *Science* 329, 533–538. doi: 10.1126/science.1188308
33. Thattai, M., and van Oudenaarden, A. (2001). Intrinsic noise in gene regulatory networks. *Proc. Natl. Acad. Sci. U.S.A.* 98, 8614–8619. doi: 10.1073/pnas.151588598
34. Thomson, T. M., Benjamin, K. R., Bush, A., Love, T., Pincus, D., Resnekov, O., et al. (2011). Scaffold number in yeast signaling system sets tradeoff between system output and dynamic range. *Proc. Natl. Acad. Sci. U.S.A.* 108, 20265–20270. doi: 10.1073/pnas.1004042108
35. Togashi, Y., and Casagrande, V. (2015). Spatiotemporal patterns enhanced by intra- and inter-molecular fluctuations in arrays of allosterically regulated enzymes. *New J. Phys.* 17:033024. doi: 10.1088/1367-2630/17/3/033024
36. Togashi, Y., and Kaneko, K. (2001). Transitions induced by the discreteness of molecules in a small autocatalytic system. *Phys. Rev. Lett.* 86, 2459–2462. doi: 10.1103/PhysRevLett.86.2459
37. Togashi, Y., and Kaneko, K. (2004). Molecular discreteness in reaction-diffusion systems yields steady states not seen in the continuum limit.*Phys. Rev. E* 70:020901. doi: 10.1103/PhysRevE.70.020901
38. Togashi, Y., and Kaneko, K. (2007). Switching dynamics in reaction networks induced by molecular discreteness. *J. Phys. Condens. Matter*19:065150. doi: 10.1088/0953-8984/19/6/065150

CHAPTER 3

KINETIC STUDIES ON THE REACTION BETWEEN DICYANOCOBINAMIDE AND HYPOCHLOROUS ACID

Dhiman Maitra, Iyad Ali, Rasha M. Abdulridha, Faten Shaeib, Sana N. Khan, Ghassan M. Saed, Subramaniam Pennathur, Husam M. Abu-Soud

INTRODUCTION

Cyanogen chloride (CNCl, CAS 506-77-4) is a disinfectant byproduct found in drinking water treated with free chlorine and chloramines at concentrations ranging from 0.45–0.8 μg/L [1], [2]. CNCl can be hydrolyzed to form hydrogen cyanide (HCN), and hypochlorous acid (HOCl) [3]. Cyanogen chloride and its reduced final product cyanide (CN^-) are highly toxic, and their damage potential depends mainly on the concentration and duration of exposure [4]. Any exposure to higher concentrations causes immediate injury to multiple organs in the central nervous, cardiovascular, and pulmonary systems [5]. Prolonged exposure can cause permanent brain damage, muscle paralysis, coma, and death [4]. The ability of these molecules to react with sulfhydryl compounds such as protein thiols and reduced glutathione (GSH) causes the toxicity elicited in biological systems [6]. They are also known to block the electron transport chain by inhibiting mitochondrial cytochrome C oxidase, initiating a fatal series of events by decreased oxidative metabolism and oxygen utilization [7]. Recently, we have shown that significant amounts of CNCl/ CN^- are liberated by mixing cyanocobalamin (Cbl), the most common supplemental form of vitamin B_{12}, with HOCl through a mechanism that involves disruption of axial coordination of the Co atom and cleavage of the corrin ring [8].

The structure of the Cbl and dicyanocobinamide ($(CN)_2$-Cbi), a naturally occurring intermediate of vitamin B_{12} synthesis, are based on a corrin ring. In the corrin ring four of the six coordination sites of the cobalt (Co) atom are the pyrrole nitrogen atoms, which are provided by the corrin ring (Figure 1). In Cbl, the fifth position of the Co atom (the lower or α- axial ligand) is taken by one of the nitrogens of the 5,6-dimethylbenzimidazole group [9]. The other nitrogen of the 5,6-dimethylbenzimidazole is connected to a five-carbon sugar, which in turn links to a phosphate group, and then back to the corrin ring via one of the seven-amide groups at the periphery of the corrin ring. Finally, the sixth position (the upper/β-axial ligand, also called the site of reactivity, is occupied by a cyano group (-CN) (Figure 1A). Cobinamides (Cbi) are vitamin B_{12} derivatives that lack the dimethylbenzimidazole group at the α-axial ligand. The solubility, stability, and the CN^- binding ability of cobalamin derivatives depend on the type of the β-axial ligand and the presence or lack of 5,6-dimethylbenzimidazole at the α-axial ligand[10]–[16] The presence of this group in Cbi causes the increased binding ability to CN^- (100 times greater than Cbl), explaining the difference of Cbi as compared to Cbl in CN^-detoxification of massive CN^- poisoning [17], [18]. Dicyanocobinamide plays an important role, both in vitro and in intact cells, as a soluble guanylate cyclase (sGC) co-activator by targeting the N-terminal regulatory regions, an action which resembles the effect of forskolin on adenylyl cyclases [19]. It increases intracellular cGMP levels and displays vasorelaxant activity in phenylephrine-constricted rat aortic rings in an endothelium-independent manner. Both effects are synergistically potentiated by BAY41-2272, (an NO-independent soluble guanylate cyclase stimulator consisting of structurally diverse benzylindazole/pyrazolopyridine and acrylamide derivatives) [19]. Recently, it has been shown that $(CN)_2$-Cbi and related vitamin B_{12}derivatives may serve not only as nitric oxide (NO) scavengers [20], but also as NO synthase (NOS) inhibitors [21]. They have also been shown to serve as potent inhibitors of HIV-1 integrase and may prove useful as anti-viral treatments [22].

Citation: Maitra D, Ali I, Abdulridha RM, Shaeib F, Khan SN, Saed GM, et al. (2014) Kinetic Studies on the Reaction between Dicyanocobinamide and Hypochlorous Acid. PLoS ONE 9(11): e110595. doi:10.1371/journal.pone.0110595.

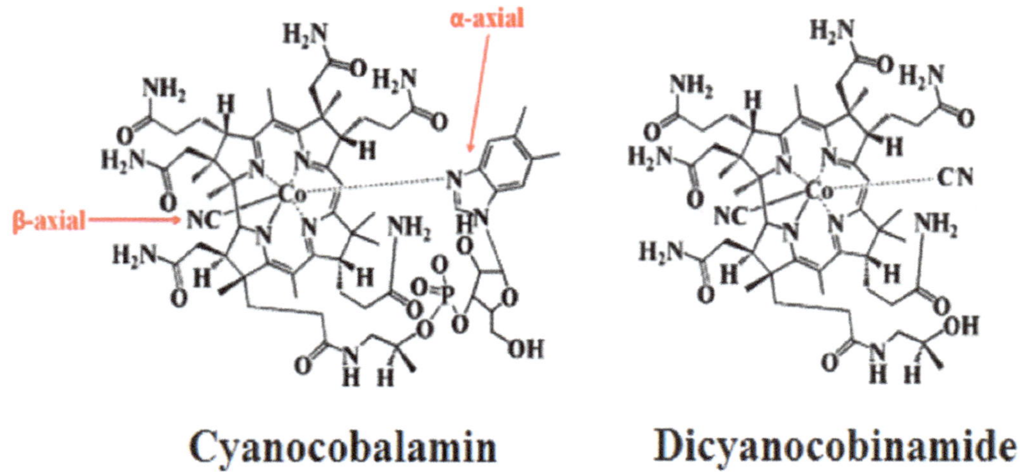

Figure 1. Structures of cyanocobalamin (Cbl) (left panel) and dicyanocobinamide ((CN)$_2$-Cbi)) (right panel).

Hypochlorous acide, which is produced by heme peroxidase and myeloperoxidase (MPO) using hydrogen peroxide (H_2O_2) and chloride (Cl^-) as co-substrates, is a vital component of the innate immune system, which protects the host through the oxidative destruction of invading pathogens and microbes [23], [24]. Activated neutrophils have been previously found to generate around 150–425 μM of HOCl per hour [24]–[27], and it has been estimated that HOCl levels can reach up to 5 mM at sites of inflammation [28]. Furthermore, the progression of a number of pathological conditions including atherosclerosis, pulmonary fibrosis, vascular and diabetic complications, glomerulonephritis, cancer and even oocyte aging have been linked to high levels of reactive oxygen species (ROS) such as HOCl [29]–[33].

With the use of a combination of optical absorbance and rapid kinetic approaches, we now show that HOCl mediates (CN)$_2$-Cbi corrin destruction through a complex multistep process. This sequence includes the generation of at least two kinetically and spectrophotometrically distinct intermediates prior to corrin ring destruction and the generation of CNCl. Collectively, these kinetic behaviors distinguish HOCl from other ROS, and provide a foundation to advance our understanding of the role of HOCl plays in destroying the tetrapyrrole-based compounds as well as the associated toxicity.

MATERIALS AND METHODS

Materials

All the materials used were of highest purity grade and used without further purification. Sodium hypochlorite (NaOCl), pyridine, 1,3 dimethyl barbituric acid, Cbl, (CN)$_2$-Cbi, L-methionine, methanol- HPLC grade, were obtained from Sigma Aldrich (St. Louis, MO, USA).

Absorbance Measurements

A Cary 100 Bio UV–visible spectrophotometer was used to record the absorbance spectra, at 25°C, pH 7.0. Experiments were performed in a 1-ml phosphate buffer solution (200 mM, pH 7.4), and then supplemented with fixed amount of (CN)$_2$-Cbi (10 μM) followed by increasing concentrations of HOCl. Reaction completion was assured after 2 hours of incubation, and methionine (5-fold of the final HOCl concentration) was added to eliminate excess HOCl, and absorbance changes were recorded from 300 to 700 nm.

Rapid Kinetic Measurements

A dual syringe stopped-flow instrument from Hi-Tech, Ltd. (Model SF-61) was used to perform kinetic measurements of HOCl mediated (CN)$_2$-Cbi destruction. Measurements were carried out in an aerobic atmosphere at 25°C and 37°C following rapid mixing of equal volumes of a buffer solution containing a fixed amount of (CN)$_2$-Cbi (20 μM; initial), and a buffer solution containing increasing concentration of HOCl. The time course of the absorbance change was fitted to a single-exponential (Eq. 1), or a double-exponential (Eq. 2) function as indicated. Signal-to-noise ratios for all kinetic analyses were improved by averaging at least six to eight individual traces. In some experiments, the stopped-flow instrument was attached to a rapid scanning diode array device (Hi-Tech). This instrument was designed to collect multiple complete spectra (200–800 nm) for specific time ranges. The detector was automatically calibrated relative to a holmium oxide filter, as it had spectral peaks at 360.8, 418.5, 446.0, 453.4, 460.4, 536.4, and 637.5 nm, which were used by the software to correctly align pixel positions with wavelength.

$$Y = 1 - e^{-kt} + C \quad \textbf{(Eq. 1)}$$

$$Y = Ae^{-k_1 t} + Be^{-k_2 t} + C \quad \textbf{(Eq. 2)}$$

Colorimetric detection of CNCl

CNCl was detected colorimetrically using the pyridine-1,3 dimethyl barbituric acid reagent as previously reported [34]. The amount of CNCl was determined from the extinction coefficient of 1.03×10^5 $M^{-1}cm^{-1}$ for the violet colored complex [35].

pH Measurements

An experiment on the effect of pH on HOCl mediated (CN)$_2$-Cbi corrin destruction was carried out using 50 mM sodium acetate buffer (pH 4.5 and 5.7), and 50 mM phosphate buffer (pH 6–9). The different spectra were collected at 25°C.

Solution Preparation

HOCl Preparation

HOCl was prepared using a slight modification of a previously published method [36]. Briefly, a stock solution of HOCl was prepared by adding 1 mL NaOCl solution to 40 mL of 154 mM NaCl and the pH was adjusted to approximately 3 by adding HCl. The concentration of active total chlorine species in solution expressed as $[HOCl]_T$ (where $[HOCl]_T = [HOCl] + [Cl_2] + [Cl_3^-] + [OCl^-]$) in 154 mM NaCl was determined by converting all the active chlorine species to OCl^- by adding a single bolus of 40 µL 5 M NaOH and measuring the concentration of OCl^-. The concentration of OCl^- was determined spectrophotometrically at 292 nm (ε=362 M^{-1} cm^{-1}). Secondary to the fact that HOCl is unstable, the stock solution was freshly prepared daily, stored on ice, and used within one hour of preparation. For further experiments, dilutions were made from the stock solution using 200 mM phosphate buffer pH 7, to give working solutions of lower HOCl concentrations.

$(CN)_2$-Cbi preparation

Dicyanocobinamide stock solution was prepared by dissolving an exact weight of $(CN)_2$-Cbi in distilled water.

RESULTS

Stability of $(CN)_2$-Cbi and the Distribution of Chlorine Species as a Function of pH

We first investigated the stability of $(CN)_2$-Cbi as a function of pH. Figure 2 shows the UV-visible absorption spectra of $(CN)_2$-Cbi prepared at different pHs ranging from 4.5–9.0. At pH range 6.5–9.0, the spectra exhibited corrin maxima centered at 366 with two resolved visible peaks at 538 and 578 nm. At pH 4.5, the spectra exhibited corrin maxima centered at 354 with visible bands centered at 497, and 526 nm, indicating the replacement of one of the CN molecules with H_2O and the formation of an H_2O-Cbi-CN complex [37]. At pHs between 5.7 and 6 the spectra demonstrated an equilibrium mixture of $(CN)_2$-Cbi and (CN)-Cbi-(H_2O). These results are consistent with those previously reported by Ye et al. in which the instability of Cbi/Cbl at lower pHs was described [38]. The pH profiles of Cl_2, OCl^-, and HOCl populations have been previously reported by Kettle et al. [39]. The HOCl population shows a bell-shaped pattern, with optimum pH ranging from 3.0 to 7.0 with a sharp drop to ~50% at pH 2 (Cl_2 and HOCl are about equal) and to ~50% at pH 7.4 (HOCl and OCl^- are about equal). At pH greater than 7, OCl^- starts to form and becomes the predominant species at pH 9 and above. Collectively, these observations have been used as a guide to investigate the reaction of HOCl against $(CN)_2$-Cbi utilizing rapid kinetic measurements.

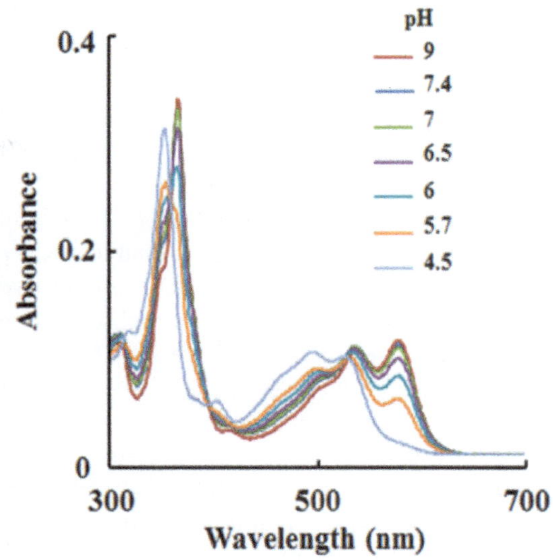

Figure 2. Stability of $(CN)_2$-Cbi as a function of pH. UV-visible difference measurements recorded after dissolving equal amounts of $(CN)_2$-Cbi (14 µM, final) at different pH values as described under "Materials and Methods."

Spectrophotometric and Rapid Kinetic Characterization of the Interaction between HOCl and $(CN)_2$-Cbi

We next utilized a diode array stopped-flow spectrophotometer to investigate the kinetics of the reaction of $(CN)_2$-Cbi with increasing concentrations of HOCl. Reactions were run in the dark (vitamin B_{12} derivatives have been considered light sensitive analytes) under aerobic conditions, at 25°C, pH 6.4–9.0. At physiologic pH, rapid mixing of a solution of 20 µM $(CN)_2$-Cbi with 80-fold molar excess of HOCl resulted in the rapid formation of a transient intermediate that displayed a decrease in the corrin absorbance peak at 366 nm and broad visible bands centered at 493 and 613 nm, typical of a six-coordinate complex Figure 3 (upper Panel). This spectrum differs from that of $(CN)_2$-Cbi, whose corrin maxima are centered at 366 with two resolved visible peaks at 538 and 578 nm [37]. This spectrum also differs from that of both α-cyano, β-aqua-cobinamide (UV/Vis 353, 496, and 525 nm) and α-aqua, β-cyano-cobinamide (UV/Vis 354, 497, and 527 nm) [37]. The spectrum for $(CN)_2$-Cbi reported by Zhou and Zelder that was also supported by ^{1}H NMR [37] matches our spectrum indicating that $(CN)_2$-Cbi is the starting compound. Furthermore, there is no alteration in the $(CN)_2$-Cbi upon the addition of excess CN^- to the reaction mixture solidifying the evidence in favor of $(CN)_2$-Cbi as the starting compound. The spectrum of the intermediate that initially formed following addition of HOCl/OCl^- to $(CN)_2$-Cbi is consistent with the replacement of one of the CN^- groups with hypochlorite ion (OCl^-). This Cbi intermediate formed within 2.4 s after mixing, but was unstable and rapidly converted into a more stable intermediate within 12 s, as judged by further decrease in the absorbance at 366 nm and a time-dependent shift in visible absorbance peak from 613 to 493 nm. These spectral changes were attributed to the replacement of the second CN^- group with OCl^-, as shown in Figure 3 (middle Panel). The second transient intermediate that formed within 12 s of initiating the reaction was also unstable and this was followed by spontaneous corrin ring destruction, within minutes of initiating the reaction (Figure 3, lower Panel). This conclusion was made by the decrease and flattening in the absorbance spectra indicating oxidative destruction of the corrin ring. Spectral transitions between each intermediate formed revealed distinct and well-defined isosbestic points (Figure 3). Thus sequential formation and decay of Cbi intermediates occur at sufficiently different rates to enable each process to be studied by conventional (i.e., single mixing) stopped-flow methods. Similar intermediates were observed when the experiments were repeated at pH 6.4 and 9.0 (where HOCl and OCl^-, respectively, predominated); however the rate constants of the transition from one intermediate to another were slower (see below). Collectively, these results indicate that HOCl binds $(CN)_2$-Cbi generating two unstable transient intermediates before corrin ring fragmentation. The generation of the transient octahedral intermediate, (OCl)(CN)-Cbi, and its conversion to $(OCl)_2$-Cbi was characterized by one set of isosbestic points at 520 and 602 nm, as well as differing responses to varying HOCl concentrations.

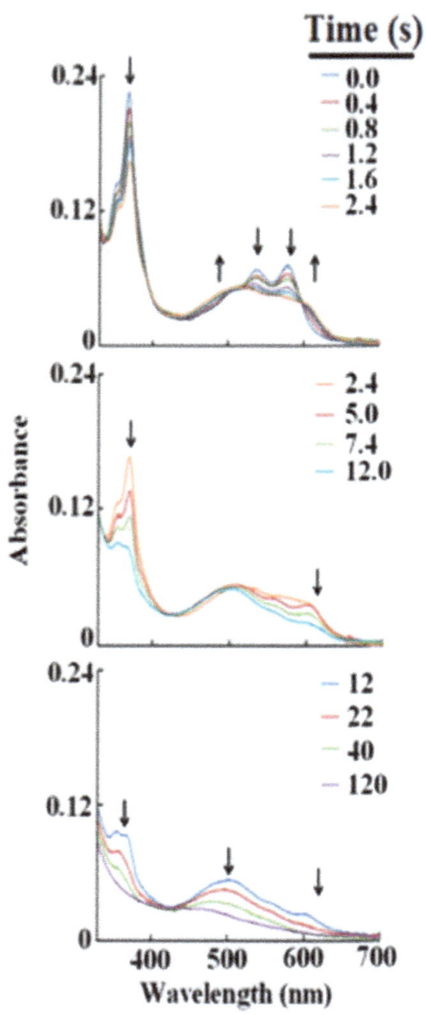

Figure 3. Diode array rapid scanning spectra for the intermediates and corrin ring destruction by reacting $(CN)_2$-Cbi with HOCl at three sequential time frames. Panel A, spectra traces collected at 0.0, 0.4, 0.8, 1.2, 1.6, and 2.4 s and was attributed to the replacement of the first molecule of CN^- with OCl in $(CN)_2$-Cbi. Panel B, spectra traces collected at 2.4, 5.0, 7.4, and 12.0 and were attributed to the replacement of the second molecule of CN^- with OCl in $(CN)_2$-Cbi. Panel C, spectra collected at 12.0, 22.0, 40.0, and 120.0 s and was attributed to corrin ring destruction. Experiments were carried out by rapid mixing a phosphate buffer solution (200 mM, pH 7.0), at 25°C, supplemented with 20 μM $(CN)_2$-Cbi with a same volume of a buffer solution supplemented with 80-fold excess of HOCl. Arrows indicate the direction of spectral change over time as each intermediate advanced to the next. These data are representative of three independent experiments.

We next utilized single wavelength stopped-flow to elucidate the mechanism by which HOCl mediates ligand replacement and corrin ring destruction. At pH 7.4 when the reaction was monitored at 613 nm, there was a rapid increase in absorbance which reached a maximum intensity in less than 10 s, and then decayed over a period of approximately 100 s when a solution of 10 μM $(CN)_2$-Cbi was rapidly mixed with 200 μM HOCl (Figure 4). The time course for this reaction was fitted to Eq. 2 with values of k_1 and k_2 of 0.30 and 0.026 s^{-1}, respectively. The increase in absorbance at the start of the reaction can be attributed to axial ligand replacement of one of the CN^- groups while the subsequent decrease in absorbance that took place over the next 100 s can be attributed to the replacement of the second CN^- group. These experiments were conducted with fixed amounts of $(CN)_2$-Cbi with increasing concentrations of HOCl. The addition of increasing concentrations of HOCl to $(CN)_2$-Cbi resulted in dramatic effects on the rates of the buildup, duration, and decay as assessed by stopped-flow spectroscopy (Figure 4). HOCl was readily used as a ligand of $(CN)_2$-Cbi, as indicated by the enhanced rate of this intermediate formation (Figures 3 and 4). As shown in Figure 5A, the rate of both phases increased linearly when plotted as a function of HOCl concentration, yielding second-order rate constants of 0.002 μM^{-1} s^{-1} and 0.0002 μM^{-1} s^{-1} for the first and second phases, respectively. Figure 3 shows the decrease in absorbance at 613 nm that took place during the next 100 s of the reaction when increasing concentrations of HOCl were rapidly mixed with a fixed concentration of $(CN)_2$-Cbi (10 μM). This decrease and flattening in the spectra associated with a lack of isosbestic points was attributed to Cbi destruction that could be due to the loss of hyperconjugation in the molecule, Co atom release, and/or Cbi fragmentation. Plots of HOCl concentration versus observed rates of corrin destruction demonstrated linear kinetics and yielded a second-order rate constant (k_3) of 4.0×10^{-5} μM^{-1} s^{-1} (Figure 5B). In all cases, the plots intersected the axes near the origin indicating that the transient intermediates formed and subsequent Cbi corrin destruction are essentially irreversible (Figure 5).

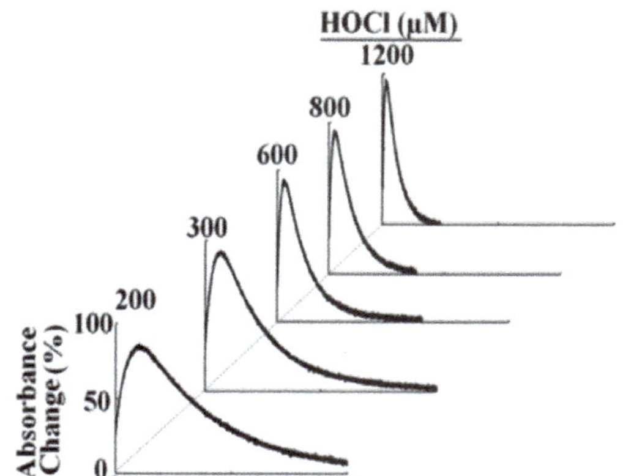

Figure 4. The effect of HOCl concentration on the formation, duration of (OCl) (CN)-Cbi and its conversion to $(OCl)_2$-Cbi. A solution containing sodium phosphate buffer (200 mM, pH 7.0) supplemented

with 5 µm (final) dicyanocobiamide was rapidly mixed with an equal volume of buffer containing increasing concentrations of HOCl (200, 300, 600, 800, and 1200 µM, final) at 25°C. Replacement of the first CN^- molecule by OCl^-, duration, and its decay to $(OCl)_2$-Cbl were monitored as a function of time by observing spectral changes at 613 nm. The final concentration of HOCl in mixtures is indicated.

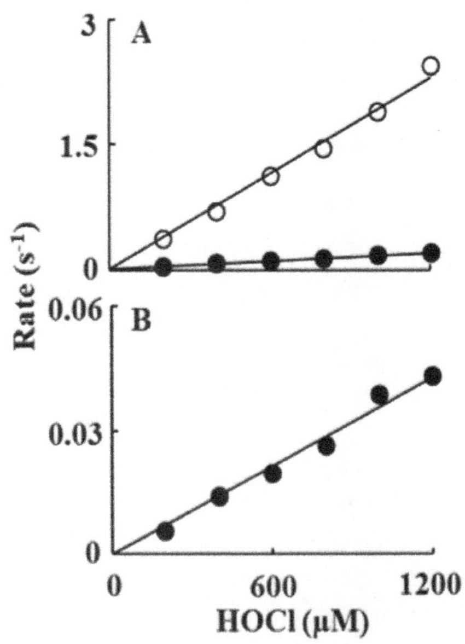

Figure 5. Rate constants of the axial ligands replacement and corrin ring destruction of dicyanocobinamide as a function of HOCl concentration. Upper panel, the observed rate constants of (OCl)(CN)-Cbi complex formation (open circles) and its conversion to $(OCl)_2$-Cbi (monitored at 613 nm) (closed circles) observed in Fig. 3 were plotted as a function of HOCl concentration. R^2 values for the first and second phases were 0.992 and 0.995 respectively. Data represent the mean of triplicate determinations from an experiment performed three times. Lower panel, the rate constants for the corrin ring destruction, for the same reaction, monitored at 493 nm as a function of HOCl concentration. R^2 value was found to be 0.987. These data are representative of three independent experiments and the standard error for each individual rate constant has been estimated to be less than 8%.

At pH 6.5 where HOCl is the predominant species in the solution there is little or no effect on the rate constants or the intermediates that formed upon mixing HOCl with $(CN)_2$-Cbi compared to pH 7.4, as judged by single wavelength stopped-flow measurements (Figure 7). Rapid kinetic measurements performed at pH 9, where OCl^- is the predominant species in the solution, revealed that the rate constants of ligand replacement and corrin ring destruction were 3–4 fold lower than those observed at a lower pH range (6.5–7.4) (Figure 7), indicating the stability of $(CN)_2$-Cbi at this pH range.

Corrin ring destruction can also occur at a lower pH (4–5) where one of the CN^- molecules is replaced with H_2O (data not shown). Importantly, replacement of one of the CN^- molecules with H_2O did not show any sign of protection against HOCl destruction. The characteristic spectrum of (CN)-Cbi-(H_2O) was also observed when $(CN)_2$-Cbi was mixed with low concentration of HOCl (eg., 1:1, or 1:2 mole equivalent of $(CN)_2$Cbi:HOCl, respectively). The (CN)-Cbi-(H_2O) complexes that formed at low HOCl concentrations were identified by liquid chromatography mass spectrometry (LC-MS) analysis and rapid kinetic measurements (data not shown).

In increase in the temperature from 25 to 37°C nearly doubles the rate of the reactions, but the kinetic behavior and the sequence of the reaction stays the same. We selected to show the results at 25°C to slow the reaction for clarity purposes.

Liberation of CNCl and Co from $(CN)_2$-Cbi after HOCl Treatment

We next investigated whether the CN^- released from the $(CN)_2$-Cbi molecule reacts with the excess HOCl in the reaction mixture to form cyanogen chloride (CNCl). Dicyanocobinamide (110 µM) was treated with 50-fold molar excess of HOCl and the accumulation of CNCl was measured using the pyridine-1,3 dimethyl barbituric acid colorimetric assay as previously reported. As shown in Figure 6, the amount of CNCl generated from the reaction of $(CN)_2$-Cbi with HOCl was approximately 1.8 fold the amount generated from the same reaction of Cbl with HOCl. Thus, the destruction of $(CN)_2$-Cbi mediated by HOCl is more toxic than that of Cbl.

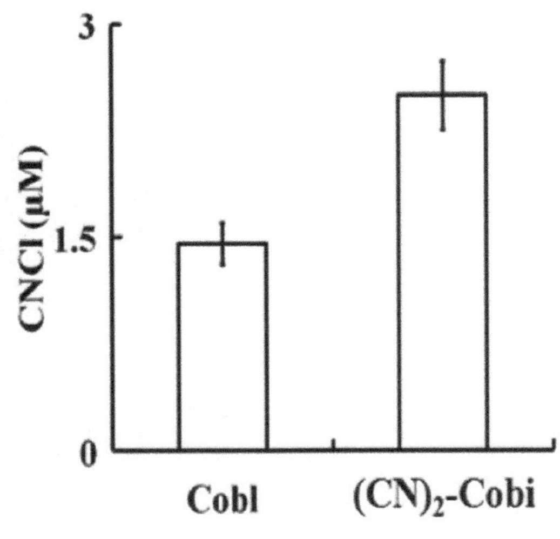

Figure 6. Cyanocobalamin and $(CN)_2$-Cbi destruction mediated by HOCl causes the liberation of CNCl. Equal concentrations of Cbl and $(CN)_2$-Cbi (110 µM) were treated with 50-fold molar excess of HOCl and CNCl generation were assayed colorimetrically as detailed underMaterials and methods. The data are representative of three independent experiments with the error bars representing the standard error measurements.

$$(CN)_2\text{-Cobi} \xrightarrow{k_1} (OCl)\text{-}(CN)\text{-Cobi} \xrightarrow{k_2} (OCl)_2\text{-Cobi} \xrightarrow{k_3} \text{Corrin destruction formation of CNCl}$$

pH	k_1 ($\mu M^{-1}s^{-1}$)	k_2 ($\mu M^{-1}s^{-1}$)	k_3 ($\mu M^{-1}s^{-1}$)	Starting compound (%)
6.5	1.60×10^{-3}	2.50×10^{-4}	2.50×10^{-5}	HOCl (~100)
7.4	2.00×10^{-3}	2.00×10^{-4}	4.00×10^{-5}	HOCl:OCl^- (50:50)
9.0	0.38×10^{-3}	1.00×10^{-4}	1.13×10^{-5}	OCl^- (~100)

Figure 7. Rate constants for the formation and decay of the intermediates that formed upon mixing $(CN)_2$-Cbi with HOCl. $(CN)_2$-Cbi was rapid-mixed with buffer containing HOCl at various concentrations. Rates of complex formation and decay were determined at three different pHs (6.5, 7.4, and 9.0) by following absorbance change at 613 or 493 nm, at 10°C, and the rate constants determined as described in the text. These data are representative of three independent experiments and the standard error for each individual rate constant has been estimated to be less than 3%.

In contrast to what was observed for the reaction of hemoproteins and free heme with HOCl where we detected Fe^{2+} release [40]–[43], we were unable to determine the oxidation state of the Co ion and the nature of the axial ligation to the Co after corrin ring destruction by HOCl. This is due to several different smaller fragments of the corrin ring generated that might coordinate to Co. Colorless Cobalt (II)/(III) complexes have previously reported [44]. The exact nature of the Co-complex is currently under investigation in our laboratory. Collectively, HOCl not only mediated $(CN)_2$-Cbi ring destruction, but also mediated the generation of CN-/CNCl.

DISCUSSION

In this work, we showed that HOCl mediates $(CN)_2$-Cbi corrin ring destruction and subsequent liberation of toxic CNCl. Corrin ring destruction occurs through a mechanism that initially involved the sequential replacement of the CN^- groups by OCl^- molecules. Changes in the corrin ring geometry either from the alteration in oxidation state of the Co atom or by the replacement of the axial ligands might make the ring more susceptible to HOCl-mediated oxidation and destruction. Thus, secondary to the liberation of toxic CNCl/HCN/CN^- molecules, cyanocobalamin derivatives may display harmful effects in chronic inflammatory states where HOCl is elevated.

In general, the characteristic spectroscopic features of vitamin B_{12} derivatives are extremely sensitive to the alteration of both the axial ligands, the modification of the corrin ring, as well as the oxidation state of the Co atom [45], [46]. These events may have various effects on the geometric parameters of Cbi including the modulation of the Co atom›s electronically charged microenvironment by changing a variety of factors. These include the location of the Co atom (in the plane or out of the plane of the corrin ring), the distance between the Co atom and the axial ligands, the distance between the Co atom and the four Ns of the pyrrole rings, the corrin ring fold angle, and the Co-ligand bond angles and their ability to make Van der Waals contact with the corrin ring [47]. There is a considerable amount of evidence from several X-ray crystallographic studies that indicate that these parameters are interconnected and alteration of one modulates the others [13]. In $(CN)_2$-Cbi, the coordination around the central Co atom is a distorted octahedron [47]. Although the Co-C and C-N bond lengths of both axial ligands along with the Co-C-CN angles are almost equal (within the experimental error), the X-ray analysis indicated that the two CN^- molecules located at both axial ligands exhibit differences in conformation [47]. These differences resulted, namely, from the orientation of the two CN^-molecules in two different directions. Indeed, one of the N-C-Co linkages is consistently bent away from the bond linking the two-pyrrole groups directly while the other linkage can take up a variety of orientations. This flexibility allows for the interaction of the corrin edge to adjacent atoms of the side chains with CN^-, and the occurrence and effects of specific CN-H bonding[47]. These events, accompanied by changes in the corrin ring fold angle, may cause a steric effect that masks one site and minimally restricts access of the ligand replacement to the other site. This behavior is an exceptional case among other tetrapyrrole compounds and explains why the replacement of CN^- molecules at both axial sites happened sequentially rather than simultaneously when a fixed amount of $(CN)_2$-Cbi was rapidly mixed with increasing concentrations of HOCl.

The shift and alteration in the UV-visible spectra indicate that the ligand replacement of the first CN^- molecule by OCl^- and the formation of the (OCl)(CN)-Cbl complex displays distinct effects on the corrin Co microenvironment. This is expected, since the OCl^- molecule has different physical and chemical properties such as ion size, electronegativity, and affinity toward Co. Given the radius, polarizability, flexibility, and charge of OCl^- compared to CN^-, OCl^- displays the potential capacity to adopt a bent geometry and bind $(CN)_2$-Cbi to form a six-coordinate complex. The interactions of (OCl)(CN)-Cbi with the second OCl^- molecule suggest that the bond of OCl^- has electronic influences on the Co-bound CN. Hypochlorite, like other bulky ligands, may display a longer cobalt-ligand (Co-L) bond and a large corrin ring fold angle due to a large area of contact (Van der Waals) with the corrin ring which shortens the trans Co-L bond[48]. Therefore, replacement of the first CN^- with OCl^- in $(CN)_2$-Cbi is likely accompanied by a strengthening of the remaining Co-CN linkage which may allow the Co ion to move away from the corrin ring center and closer to the CN^- molecule. Such a process may increase the affinity of the corrin ring towards CN^-, which ceases the pyrrole ring movement. Thus, replacement of the second CN^- molecule with OCl^- should be more difficult than that observed for the first CN^-molecule. This notion is supported by the slower rate constant of OCl^- binding to the second site.

The perturbation in the electronic structure of the corrin ring mediated by HOCl modification and oxidation may also play an important role in deactivating the axial coordination site toward ligand substitution. This assumption is built on a previous study by Brown et al., which showed that the substitution of the C10 by an electron-withdrawing group such as NO deactivates the axial coordination site toward ligand substitution [48]. Chemaly et al., have also demonstrated that modifications of C10 and C5, in which the H atoms are replaced by Cl and OH, respectively, depend strongly on the polarizability of the axial ligand [49]. The hypochlorite ion is a more stable and less active form of chlorine. It is still capable of replacing both CN^- molecules of $(CN)_2$-Cbi and mediating corrin ring destruction but at slower rates. Indeed, our rapid kinetic measurements reveal that the rate constants of ligand replacement and corrin ring destruction are 3–4 fold lower than those observed at a lower pH where HOCl is the predominant species (Figure 7). A mechanism that describes the involvement of HOCl in tetrapyrrole macrocyclic compound destruction (e.g. corrin ring, heme, and porphyrin) that could apply to $(CN)_2$-Cbi corrin ring damage was described by Maitra et al., and Abu-Soud et al., [8], [40]. In this model, we showed that following ligand replacement, HOCl oxidatively cleaved the corrin ring leading to the generation of a dicorrinic derivative. The attack of HOCl to the carbon–methyne bridges between the adjacent corrin rings led to the formation of chlorinated adducts, which by releasing chloride, formed an epoxide or aminal. The epoxide gave rise to a hydroxylated compound with the hydroxyl group being attached to the carbon-methyne bridge of the tetrapyrrole moiety, where the initial attack by HOCl occurred.

Comparison of the kinetic data obtained for the reaction of $(CN)_2$-Cbi with HOCl showed inFigures 3–5 with recently reported reactions of Cbl with HOCl showed the involvement of several kinetically and spectrophotometrically distinguishable marks. Interaction of HOCl with Cbl consists of a three-step mechanism: a lag phase interval parallel to the weakening and release of the α-axial ligand; α-axial ligand replacement leading to the formation of a new hexa-coordinated intermediate; and corrin destruction with subsequent CNCl formation [8]. The dependence of the pseudo-first-order rate constants of the two initial phases on the concentration of HOCl are noticeably curved indicating that these reactions are second order with respect to HOCl and follow the rate expression ($k_{obs}=k_1[HOCl]^2$) with second order rate constants for the first and second phases of 1×10^{-7} $\mu M^{-2}s^{-1}$ and 3×10^{-7} $\mu M^{-2}s^{-1}$, respectively [8]. It was concluded that the chlorinated intermediate was generated through two consecutive steps based on the similar trends and close proximity of the values of the rate constants for the first and second steps. The rate constant of Cbl destruction is relatively slow with a second order rate constant of (2×10^{-5} $\mu M^{-1}s^{-1}$). The second order rate constant of HOCl mediated Cbl corrin ring destruction is comparable to the rate constants of HOCl with $(CN)_2$-Cbi and other biomolecules such as C=C in lipid molecules, adenine mononucleotide phosphate, and backbone amides in protein chains (for a detailed review see [23], [50]). Multiple fragments were identified utilizing HPLC and mass spectrometry when Cbl was treated with HOCl in phosphate buffer associated with a buildup of CNCl. Similar products have been obtained when the reaction was carried out in plasma, indicating the biological relevance of our finding [40]. A biphasic reaction has also been reported for CN^- with (OH)(H_2O)-Cbi to form $(CN)_2$-Cbi [2], [45]. In this reaction, the replacement of the H_2O molecule by CN^- has been suggested as an initial step due to favorable thermodynamic and kinetic reasons, subsequently (OH)(CN)-Cbi was attacked by the second CN^- molecule to form $(CN)_2$-Cbi. In addition, Baldwin et al. [51] have reported that cyanoaquacobinamide reacts with CN^- much faster than either $(H_2O)_2$-Cbi or (H_2O)(OH)-Cbi due presumably to the stronger *trans*-labializing effect of CN^- compared to OH^- [2], [45].

Sharina et al., showed that $(CN)_2$-Cbi acts as a sGC coactivator both in vitro and in intact cells[19]. Heme depletion or heme oxidation does not affect $(CN)_2$-Cbi-dependent activation. In addition to their antimalarial activity, a recent study by Weinberg et al. has shown that cobalamins and cobinamides inhibit the three NOS isoforms to various degrees [21]. The authors suggested that these compounds could be administered acutely, subacutely, or chronically in certain diseases in which NO acts in a deleterious fashion (e.g., inflammatory diseases) with very few side effects. Our current finding opposes this implication. While cobalamins are widely used to treat human cyanide poisoning because of their efficacy and high affinity for CN^-, exposure of $(CN)_2$-Cbi to HOCl can undermine these beneficial effects and generate proinflammatory reaction products, such as CNCl. Hypochlorous acid mediates corrin ring destruction and liberates toxic molecular products such as CN^-/HCN/CNCl, all potent blood agents. More recently, we have shown that HOCl can promote heme destruction in free heme, hemoproteins, and red blood cells leading to iron release and protein aggregation [40]. Our work, thus, provides some exciting evidence to support the potential relationship between elevated levels of free metals and elevated activity of MPO. This deleterious relationship, during inflammation, could manifest in lipoprotein oxidation in vivo, the initiation of immunologic reactions, and the production of damaging ROS such as hydroxyl radicals (•OH) via a Fenton-like reaction when reducing agents are available (Cl^- can reduce Co^{3+} to Co^{2+} [33], [52]. The higher toxicity of $(CN)_2$-Cbi destruction mediated by HOCl compared to Cbl destruction is reflected by the liberation of almost twice the amount of CNCl. The overall kinetic mechanism that describes our current finding is shown in Figure 7. Thus, supplementation with HOCl scavenger may provide a beneficial effect through prevention of the formation of CNCl. Indeed, previous studies by Matthews et al. have shown that treatment of HL60 cells (cells expressing MPO) by Cbl led to cytotoxicity, which could be prevented by adding methionine, a potent scavenger of HOCl, to the medium [53].

ACKNOWLEDGMENTS

This work was supported by the National Institutes of Health grant RO1HL066367 and R01HL094230, the Children's Hospital of Michigan, and the Molecular Phenotyping Core of the Michigan Nutrition and Obesity Research Center (DK089503). Dr. Iyad. A. Ali is a Fulbright Senior Research Scholar in Dr. H. M. A-S laboratory.

AUTHOR CONTRIBUTIONS

Conceived and designed the experiments: DM IA HAS SP. Performed the experiments: DM IA RA FS. Analyzed the data: DM IA RA FS. Contributed reagents/materials/analysis tools: DM IA RA FS. Wrote the paper: HAS GS SP SK.

REFERENCES

1. Na CZ, Olson TM (2006) Mechanism and kinetics of cyanogen chloride formation from the chlorination of glycine. Environmental Science & Technology 40: 1469–1477. doi: 10.1021/es0512273
2. Krasner SW, McGuire MJ, Jacangelo JG, Patania NL, Reagan KM, et al. (1989) The occurrence of disinfection by-products in US drinking water. J Am Water Works Assoc 81: 41–53.
3. Ma J, Dasgupta PK (2010) Recent developments in cyanide detection: a review. Anal Chim Acta 673: 117–125. doi: 10.1016/j.aca.2010.05.042
4. Brennan RJ, Waeckerle JF, Sharp TW, Lillibridge SR (1999) Chemical warfare agents: emergency medical and emergency public health issues. Ann Emerg Med 34: 191–204. doi: 10.1016/s0196-0644(99)70229-5
5. Szinicz L (2005) History of chemical and biological warfare agents. Toxicology 214: 167–181. doi: 10.1016/j.tox.2005.06.011
6. Aldridge WN (1951) The conversion of cyanogen chloride to cyanide in the presence of blood proteins and sulphydryl compounds. Biochem J 48: 271–276.
7. Turrens JF, Alexandre A, Lehninger AL (1985) Ubisemiquinone is the electron donor for superoxide formation by complex III of heart mitochondria. Arch Biochem Biophys 237: 408–414. doi: 10.1016/0003-9861(85)90293-0
8. Abu-Soud HM, Maitra D, Byun J, Souza CE, Banerjee J, et al. (2012) The reaction of HOCl and cyanocobalamin: corrin destruction and the liberation of cyanogen chloride. Free Radic Biol Med 52: 616–625. doi: 10.1016/j.freeradbiomed.2011.10.496
9. Banerjee R, Ragsdale SW (2003) The many faces of vitamin B12: catalysis by cobalamin-dependent enzymes. Annu Rev Biochem 72: 209–247.
10. Hodgkin DG, Pickworth J, Robertson JH, Trueblood KN, Prosen RJ, et al. (1955) The crystal structure of the hexacarboxylic acid derived from B12 and the molecular structure of the vitamin. Nature 176: 325–328. doi: 10.1038/176325a0
11. Gruber K, Puffer B, Krautler B (2011) Vitamin B12-derivatives-enzyme cofactors and ligands of proteins and nucleic acids. Chem Soc Rev 40: 4346–4363. doi: 10.1039/c1cs15118e
12. Brown KL, Cheng S, Zou X, Zubkowski JD, Valente EJ, et al. (1997) Cis Effects in the Cobalt Corrins. 1. Crystal Structures of 10-Chloroaquacobalamin Perchlorate, 10-Chlorocyanocobalamin, and 10-Chloromethylcobalamin. Inorg Chem 36: 3666–3675. doi: 10.1021/ic9615077
13. Brescianipahor N, Calligaris M, Randaccio L, Toscano PJ (1982) Structural Aspects of a Vitamin-B12 Model - Crystal and Molecular-Structure of Trans-Cyanobis(Dimethylglyoximato)(Trimethyl Phosphite)-Cobalt(Iii). Journal of the Chemical Society-Dalton Transactions: 1009–1011.
14. Randaccio L, Furlan M, Geremia S, Slouf M (1998) Preparation and X-ray analysis of crystals of azido- and chlorocobalamin containing LiCl: A structural model for the interactions of the corrin ring with ionic species. Inorg Chem 37: 5390–5393. doi: 10.1021/ic980181u
15. Kamiya K, Kennard O (1982) The X-Ray Structure-Analysis of Anhydrous Heptamethyl Dicyanocobyrinate (Cobester). Journal of the Chemical Society-Perkin Transactions 1: 2279–2288. doi: 10.1039/p19820002279
16. Krautler B, Konrat R, Stupperich E, Farber G, Gruber K, et al. (1994) Direct Evidence for the Conformational Deformation of the Corrin Ring by the Nucleotide Base in Vitamin-B12 - Synthesis and Solution Spectroscopic and Crystal-Structure Analysis of Co-Beta-Cyanoimidazolylcobamide. Inorg Chem 33: 4128–4139. doi: 10.1021/ic00096a043
17. Broderick KE, Balasubramanian M, Chan A, Potluri P, Feala J, et al. (2007) The cobalamin precursor cobinamide detoxifies nitroprusside-generated cyanide. Exp Biol Med (Maywood) 232: 789–798.
18. Broderick KE, Potluri P, Zhuang S, Scheffler IE, Sharma VS, et al. (2006) Cyanide detoxification by the cobalamin precursor cobinamide. Exp Biol Med (Maywood) 231: 641–649.
19. Sharina I, Sobolevsky M, Doursout MF, Gryko D, Martin E (2012) Cobinamides are novel coactivators of nitric oxide receptor that target soluble guanylyl cyclase catalytic domain. J Pharmacol Exp Ther 340: 723–732. doi: 10.1124/jpet.111.186957
20. Broderick KE, Singh V, Zhuang S, Kambo A, Chen JC, et al. (2005) Nitric oxide scavenging by the cobalamin precursor cobinamide. J Biol Chem 280: 8678–8685. doi: 10.1074/jbc.m410498200
21. Weinberg JB, Chen Y, Jiang N, Beasley BE, Salerno JC, et al. (2009) Inhibition of nitric oxide synthase by cobalamins and cobinamides. Free Radic Biol Med 46: 1626–1632. doi: 10.1016/j.freeradbiomed.2011.07.010
22. Weinberg JB, Shugars DC, Sherman PA, Sauls DL, Fyfe JA (1998) Cobalamin inhibition of HIV-1 integrase and integration of HIV-1 DNA into cellular DNA. Biochem Biophys Res Commun 246: 393–397. doi: 10.1006/bbrc.1998.8629
23. Davies MJ, Hawkins CL, Pattison DI, Rees MD (2008) Mammalian heme peroxidases: from molecular mechanisms to health implications. Antioxid Redox Signal 10: 1199–1234. doi: 10.1089/ars.2007.1927
24. Kettle AJ, Winterbourn CC (1994) Assays for the chlorination activity of myeloperoxidase. Methods Enzymol 233: 502–512. doi: 10.1016/s0076-6879(94)33056-5
25. Malech HL, Gallin JI (1987) Current concepts: immunology. Neutrophils in human diseases. N Engl J Med 317: 687–694. doi: 10.1056/nejm198709103171107
26. Pullar JM, Vissers MC, Winterbourn CC (2000) Living with a killer: the effects of hypochlorous acid

on mammalian cells. IUBMB Life 50: 259–266. doi: 10.1080/713803731

27. Weiss SJ, Klein R, Slivka A, Wei M (1982) Chlorination of taurine by human neutrophils. Evidence for hypochlorous acid generation. J Clin Invest 70: 598–607. doi: 10.1172/jci110652
28. Hampton MB, Kettle AJ, Winterbourn CC (1998) Inside the neutrophil phagosome: oxidants, myeloperoxidase, and bacterial killing. Blood 92: 3007–3017.
29. Goud AP, Goud PT, Diamond MP, Gonik B, Abu-Soud HM (2008) Reactive oxygen species and oocyte aging: role of superoxide, hydrogen peroxide, and hypochlorous acid. Free Radic Biol Med 44: 1295–1304. doi: 10.1016/j.freeradbiomed.2007.11.014
30. Vivekanadan-Giri A, Wang JH, Byun J, Pennathur S (2008) Mass spectrometric quantification of amino acid oxidation products identifies oxidative mechanisms of diabetic end-organ damage. Rev Endocr Metab Disord 9: 275–287. doi: 10.1007/s11154-008-9093-1
31. Pennathur S, Heinecke JW (2007) Mechanisms for oxidative stress in diabetic cardiovascular disease. Antioxid Redox Signal 9: 955–969. doi: 10.1089/ars.2007.1595
32. Fletcher NM, Jiang Z, Ali-Fehmi R, Levin NK, Belotte J, et al. (2011) Myeloperoxidase and free iron levels: potential biomarkers for early detection and prognosis of ovarian cancer. Cancer Biomark 10: 267–275. doi: 10.1158/1538-7445.am2011-3192
33. Nicholls SJ, Hazen SL (2009) Myeloperoxidase, modified lipoproteins, and atherogenesis. J Lipid Res 50 Suppl: S346–351 doi: 10.1194/jlr.r800086-jlr200
34. Lundquist P, Rosling H, Sorbo B (1985) Determination of cyanide in whole blood, erythrocytes, and plasma. Clin Chem 31: 591–595.
35. Gumus G, Demirata B, Apak R (2000) Simultaneous spectrophotometric determination of cyanide and thiocyanate after separation on a melamine-formaldehyde resin. Talanta 53: 305–315. doi: 10.1016/s0039-9140(00)00485-9
36. Wang L, Bassiri M, Najafi R, Najafi K, Yang J, et al. (2007) Hypochlorous acid as a potential wound care agent: part I. Stabilized hypochlorous acid: a component of the inorganic armamentarium of innate immunity. J Burns Wounds 6: e5.
37. Zhou K, Zelder F (2011) Identification of Diastereomeric Cyano–Aqua Cobinamides with a Backbone-Modified Vitamin B12 Derivative and with 1H NMR Spectroscopy. European Journal of Inorganic Chemistry 2011: 53–57. doi: 10.1002/ejic.201001146
38. Eitenmiller RR, Ye L, Landen WO (2008) Vitamin analysis for the health and food sciences. Boca Raton: CRC Press. xxi, 637 p.
39. Kettle AJ, Albrett AM, Chapman AL, Dickerhof N, Forbes LV, et al. (2014) Measuring chlorine bleach in biology and medicine. Biochim Biophys Acta 1840: 781–793. doi: 10.1016/j.bbagen.2013.07.004
40. Maitra D, Byun J, Andreana PR, Abdulhamid I, Diamond MP, et al. (2011) Reaction of hemoglobin with HOCl: mechanism of heme destruction and free iron release. Free Radic Biol Med 51: 374–386. doi: 10.1016/j.freeradbiomed.2011.04.011
41. Maitra D, Shaeib F, Abdulhamid I, Abdulridha RM, Saed GM, et al. (2013) Myeloperoxidase acts as a source of free iron during steady-state catalysis by a feedback inhibitory pathway. Free Radic Biol Med 63: 90–98. doi: 10.1016/j.freeradbiomed.2013.04.009
42. Souza CE, Maitra D, Saed GM, Diamond MP, Moura AA, et al. (2011) Hypochlorous acid-induced heme degradation from lactoperoxidase as a novel mechanism of free iron release and tissue injury in inflammatory diseases. PLoS One 6: e27641. doi: 10.1371/journal.pone.0027641
43. Maitra D, Byun J, Andreana PR, Abdulhamid I, Saed GM, et al. (2011) Mechanism of hypochlorous acid-mediated heme destruction and free iron release. Free Radic Biol Med 51: 364–373. doi: 10.1016/j.freeradbiomed.2011.03.040
44. Huttemann TJ, Foxman BM, Sperati CR, Verkade JG (1965) Transition Metal Complexes of a Constrained Phosphite Ester. IV. Compounds of Cobalt(I), Cobalt(III), Nickel(II), and Nickel(0). Inorganic Chemistry 4: 950–953. doi: 10.1021/ic50029a005
45. Ma J, Dasgupta PK, Zelder FH, Boss GR (2012) Cobinamide chemistries for photometric cyanide determination. A merging zone liquid core waveguide cyanide analyzer using cyanoaquacobinamide. Anal Chim Acta 736: 78–84. doi: 10.1016/j.aca.2012.05.028
46. Schneider Z, Stroinski A (1987) Comprehensive B12: chemistry, biochemistry, nutrition, ecology, medicine: de Gruyter.
47. Markwell AJ, Pratt JM, Shaikjee MS, Toerien JG (1987) The chemistry of vitamin B12. Part 28. Crystal structure of dicyanocobyrinic acid heptamethyl ester and its interaction with alcohols: the effects of hydrogen bonding to co-ordinated cyanide. J Chem Soc, Dalton Trans: 1349–1357.
48. Brown KL (2005) Chemistry and enzymology of vitamin B12. Chem Rev 105: 2075–2149. doi: 10.1021/cr030720z
49. Chemaly SM, Brown KL, Fernandes MA, Munro OQ, Grimmer C, et al. (2011) Probing the nature of the Co(III) ion in corrins: the structural and electronic properties of dicyano- and aquacyanocobyrinic acid heptamethyl ester and a stable yellow dicyano- and aquacyanocobyrinic acid heptamethyl ester. Inorg Chem 50: 8700–8718. doi: 10.1021/ic200285k
50. Pattison DI, Davies MJ (2006) Reactions of myeloperoxidase-derived oxidants with biological substrates: Gaining chemical insight into human inflammatory diseases. Current Medicinal Chemistry 13: 3271–3290. doi: 10.2174/092986706778773095
51. Baldwin DA, Betterton EA, Pratt JM (1983) The chemistry of vitamin B12. Part 20. Diaquocobinamide: pK values and evidence for conformational isomers. J Chem Soc, Dalton Trans: 217–223.
52. Saini Y, Greenwood KK, Merrill C, Kim KY, Patial S, et al. (2010) Acute cobalt-induced lung injury and the

role of hypoxia-inducible factor 1alpha in modulating inflammation. Toxicol Sci 116: 673–681. doi: 10.1093/toxsci/kfq155

53. Matthews JH (1997) Cyanocobalamin [c-lactam] inhibits vitamin B12 and causes cytotoxicity in HL60 cells: methionine protects cells completely. Blood 89: 4600–4607.

CHAPTER 4

THERMODYNAMICS OF RANDOM REACTION NETWORKS

Jakob Fischer, Axel Kleidon, Peter Dittrich

INTRODUCTION

Connecting network theory with thermodynamics was an idea already present more than 40 years ago under the term network thermodynamics [1]. Despite the fact that the terms were used in combination, the theory was merely a graphical representation of conservation equations and did not make any statements about complex networks, as they are known today. In 2006 Cantú and Nicolis [2] studied thermodynamic properties of linear networks, but limited themselves to small networks, which they were able to handle analytically. Here, we extend their study by generating big random linear and nonlinear reaction networks and simulating them to a thermodynamically constrained steady state. This might contribute to a framework that allows to test methods for reconstructing thermodynamic data of reaction networks [3, 4] and lead to a better thermodynamic understanding of reaction networks in general. Possible applications of this approach include the thermodynamic investigation of reaction models in biology [3–5], origin of life [6] and also Earth system and planetary science [7, 8].

We look at reaction networks as thermodynamic systems that transforms two chemical species into one another [2]. The environment is driving the network to thermodynamic disequilibrium by keeping the concentration of two species constant. In the following, we will call the chemical species that are kept constant 'boundary species', because they are the species to which the boundary conditions are applied to.

Our basic assumption is that the network is able to transform the two boundary species into each other. This is not always possible in real reaction networks where the transformations are constrained by stoichiometry of chemical constituents. For example, any chemically sound reaction model will implicitly forbid pathways that transform N_2O into H_2O. Even if the artificial networks we create are comparable in density, they are not created with this constraint. This is due to the implications this constraint would have on the complexity of the boundary conditions. Omitting it leads to the existence of many transformation pathways between most pairs of randomly chosen boundary species, otherwise almost all pairs of boundary species would just have a steady state flow of zero between them.

We study different quantitative properties of the networks at steady state. In particular, because cycles have been reported to have important functions in networks [9–11], we look at the cycles that appear in the flow pattern. These cycles depend on the direction of the flow of each reaction, which in turn depends on the strength of the thermodynamic disequilibrium caused by the boundary condition.

In the next section we describe our method for generating reaction networks so they resemble different complex network models and how we simulate them to find their non-equilibrium steady state. We then present our results concerning the flow through the networks, the distribution of entropy production of individual reactions, and the dependency of cycle number from flow through the nonlinear networks.

METHODS

Reaction Networks

Reaction networks [12] consist of a set of species $\mathcal{M}$ combined with a set of reactions $\mathcal{R}$. They contain information on the connection of chemical species through reactions and include the stoichiometric constraints given by the reactions. Mathematically, a reaction network can be described by two

Citation: Fischer J, Kleidon A, Dittrich P (2015) Thermodynamics of Random Reaction Networks. PLoS ONE 10(2): e0117312. doi:10.1371/journal.pone.0117312.

stoichiometric matrices **L** and **R**. L_{ij} is the coefficient of the i-th species on the left side of the j-th reaction and R_{ij} is the coefficient of the i-th species on the right side of the j-th reaction. Combining both matrices gives the stoichiometric matrix $\mathbf{N} = \mathbf{R}-\mathbf{L}$, for which the element N_{ij} in i-th row and j-th column gives the effective change of species i by reaction j. Given a relation $\mathbf{v} = \mathbf{v}(\mathbf{x})$ between reaction rates **v** and species concentrations **x**, one can associate the reaction network with the dynamics of an ordinary differential equation (ODE):

$$\frac{d\mathbf{x}}{dt} = \mathbf{N} \cdot \mathbf{v}(\mathbf{x}).$$

In complex network science instead of looking at a bipartite graph, where reactions and species are represented by different types of nodes which are connected by edges, often the substrate graph is used. In this simplified view the nodes represent the species and an edge between two species is present if and only if there is a reaction having those two species on different sides of the reaction equation (Fig. 1, (B')) [13, 14].

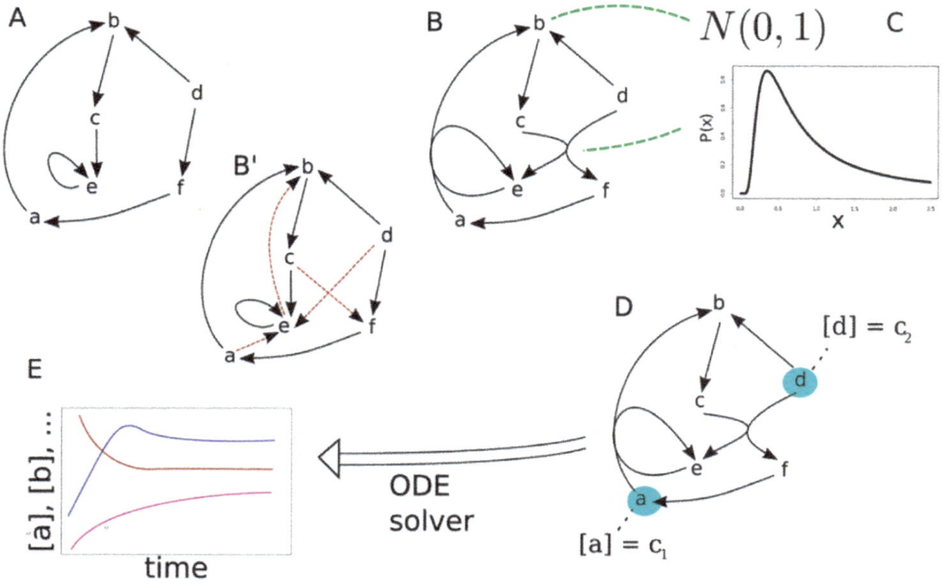

Figure 1. Illustration of realized simulations. (A) Linear reaction networks are generated from existing complex network models. (Arrows represent reactions, chemical species are indicated by lowercase letters.) (B)Pairs of linear reactions are combined to form nonlinear reactions. (B') Substrate graph that should maintain its characteristic properties while coupling. Edges invoked by coupling are depicted with dotted arrows. **(C)** Gibbs energies of formation are assigned to species from a normal distribution, activation energies to reactions from a Planck-like distribution (Eq. 8). **(D)** Two boundary species whose concentrations are kept constant are selected while the others are initialized randomly. **(E)** Reaction equation is solved numerically and final rates are taken as steady state rates.

Network Construction

Our artificial reaction networks are generated in three steps. We first generate a simple directed network (graph) consisting out of a set V of N nodes and a multiset $E \subset V \times V$ of M edges. These networks are generated following the models of Erdős-Rényi [15], Barabási-Albert [16] (scale-free), Watts-Strogatz [17] (small-world, clustering) and Pan-Sinha [18, 19] (hierarchically-modular). We are always using variants of these network models that allow formation of self loops and multiple edges between the same nodes. Also, we generate networks with a fixed number of edges. From these complex networks the reaction network is constructed.

Simple reaction networks are created by translating each edge into a reaction of the form X⇋Ywith X being the first and Y being the second node. In the rate equation of mass action kinetics this leads to a linear dependence of the reaction rates from the concentrations and thus we are calling these networks "linear" reaction networks.

Nonlinear reaction networks are generated out of directed networks by combining pairs of edges to second order reactions of the form X+Z⇋Y+W. The selection of pairs is done with a probability distribution that maintains the characteristic properties of the substrate graph as much as possible. This is done by considering the probability of newly introduced edges in the substrate graph in the originally used network model. For example, consider the combination of the reactions A⇋B and C⇋D to create the reaction A+C⇋B+D. This leads to two new edges in the substrate graph between A and D as well as between C and B. The probabilities of these two edges in the original network model are then used to calculate the probability of the combined reactions.

Finally, the thermodynamic data is generated and assigned to species and reactions. In the following, the generation process of nonlinear networks specific to the different network models is explained before the generation of thermodynamic data is specified in detail.

Erdős-Rényi (ER)

In the Erdős-Rényi network model [15] all possible edges have the same probability. We create these networks by simply drawing the nodes of every edge from the set of all nodes with uniform probability. For the construction of nonlinear reaction networks, second order reaction equations are then chosen from the set of pairs of linear equations with uniform distribution. Note that linear equations that are used as part of a nonlinear equation are not returned leading to the probability of all pairs of linear equations containing it being set to zero for subsequent couplings.

Barabási-Albert (BA)

For generating scale free networks, the Barabási-Albert models are used [16]. In this model nodes are added consecutively. Newly added nodes are connected to the network by introducing edges between it and already existing nodes. The selection of nodes to attach to is done with probability scaling with their node degree (preferential attachment).

The coupling probability of linear reactions is calculated from the product of the node degrees of the chemical products. In principle, other functional dependencies are possible, but for simplicity we choose this one and check that it maintains the power law distribution of the node degree in the associated substrate graph (Fig. 2 (A)).

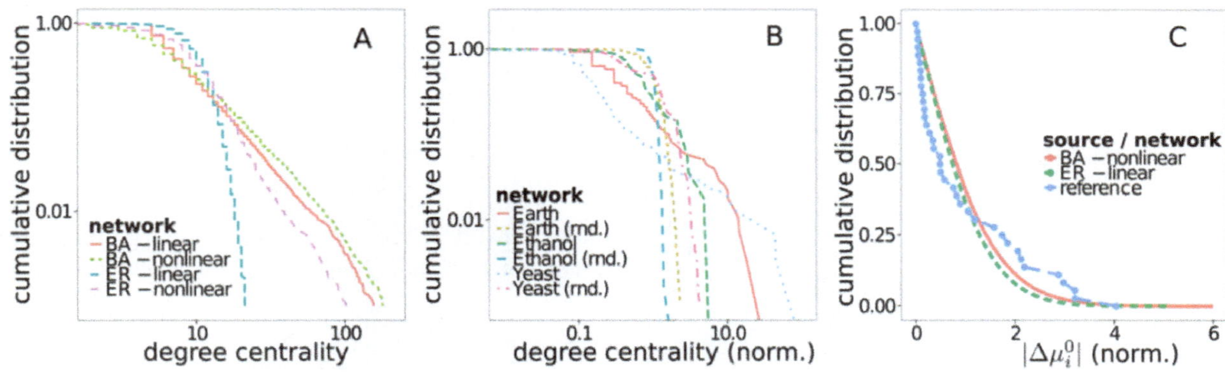

Figure 2. Comparison of artificial and real networks. (A) Nonlinear BA networks maintain scale-free degree distribution. (B) Cumulative degree scaling of real network's substrate graphs shows pronounced scale free property in comparison with their null models (randomized counterparts) for Earth's photochemistry [22] and a kinetik model of Yeast's metabolism [25]. (C) Cumulative distribution of the standard change of Gibbs energy of formation $(\Delta\mu_i^0 = \sum_i N_{ij}\mu_i^0)$ for artificial networks and respective thermodynamic reference data for glycolysis (see [28], Table 4, $\Delta_r G'^0$). Distributions were (linearly) rescaled to have a mean of one.

Watts-Strogatz (WS)

Networks having a comparable average path length to Erdős-Rényi but with a higher clustering are generated with the Watts-Strogatz model [17]. From a circular lattice like structure, a fraction α is randomly reordered. For the size of our networks we choose a value of α = 0.1 (see Table 1).

Table 1. Network properties. Properties of the substrate graphs of artificially generated networks as well as of examples of real networks. Table contains the number of vertices (|V|) and edges (|E|), the mean shortest path length (< L >) and the clustering coefficient (< C >) of the respective undirected network. The modularity is calculated using the walktrap community finding algorithm [38]. Data for real networks is taken from a database for Earth's photochemical reactions [22], models for the combustion of Ethanol [23] and Dimethyl ether [24] and a kinetic model of Yeast's metabolism [25]. For the artificial networks and the randomizations of the real networks mean values and standard deviations are calculated from 10 samples

network	$\|V\|$	$\|E\|$	$< L >$	$< C >$	modularity	1-cycles	2-cycles	4-cycles
ER (linear)	1000	5000	4.5 ± 0.02	0.0098 ± 5e-04	0.234 ± 0.0037	4.6 ± 2.4	10 ± 2.3	151 ± 14
ER (nonlinear)	1000	5000	4.5 ± 0.03	0.0141 ± 0.002	0.296 ± 0.00921	4.1 ± 2	1634 ± 274	2964 ± 939
BA (linear)	1000	5000	3.9 ± 0.04	0.0277 ± 0.001	0.178 ± 0.00412	5.2 ± 2.4	47 ± 9.3	2138 ± 245
BA (nonlinear)	1000	5000	3.8 ± 0.04	0.0386 ± 0.002	0.266 ± 0.00664	9 ± 3.9	1254 ± 273	9321 ± 5306
WS (linear)	1000	5000	6.4 ± 0.1	0.484 ± 0.009	0.805 ± 0.00569	0.3 ± 0.48	2.4 ± 1.8	4188 ± 158
WS (nonlinear)	1000	5000	6.7 ± 0.1	0.255 ± 0.008	0.748 ± 0.00733	90 ± 5.1	1002 ± 58	2201 ± 161
PS (linear)	1000	5000	5 ± 0.03	0.0414 ± 0.003	0.51 ± 0.0124	469 ± 17	208 ± 24	316 ± 31
PS (nonlinear)	1000	5000	4.6 ± 0.06	0.0297 ± 0.001	0.314 ± 0.00817	285 ± 23	1782 ± 553	3758 ± 1745
Earth's atm.	280	1846	2.9	0.147	0.301	3	1337	48503
(randomized)	280	1846	3.3 ± 0.03	0.0513 ± 0.003	0.202 ± 0.011	6.2 ± 1.6	21 ± 5	480 ± 42
Ethanol	57	2902	1.7	0.4977	0.171	18	14264	1.59e+07
(randomized)	57	2902	1.4 ± 0.005	0.831 ± 0.006	0.051 ± 0.014	56 ± 7.4	1265 ± 22	1.49e+6 ± 19103
Dimethyl ether	79	2492	1.9	0.416	0.285	10	9995	3.45e+06
(randomized)	79	2492	1.7 ± 0.005	0.551 ± 0.007	0.073 ± 0.016	30 ± 4.9	492 ± 23	226961 ± 5134
Yeast Metab.	295	16954	2.6	0.1005	0.0175	27	355713	3.34e+09
(randomized)	295	16954	2 ± 0.01	0.51 ± 0.02	0.025 ± 0.045	59 ± 6.1	1715 ± 208	2.77e+06 ± 6.5e+05

For the creation of nonlinear networks we only form those couplings between linear reactions which lead to two new close edges in the substrate graph. Here "close" means that their distance in terms of the circular lattice is not larger than the largest distance of non reordered edges in it. It would have be possible to use a more sophisticated approach and use the parameter α as the probability of introducing a far edge in the substrate graph while coupling. But because even our simple method does not achieve a clustering coefficient as high as equivalent linear networks (Table 1) we use this simple method.

Pan-Sinha (PS)

Hierarchically-modular networks are generated starting with uniformly partitioning the nodes into 2^h elementary modules, with h being the number of hierarchical levels of the network. On the first level two pairs of modules on the elementary level are joined to form a new module, leading to $2^{(h-1)}$ modules on the first level. Analogous, for all other levels modules of the level below are joined pairwise, up to the h-th level where there is just one module consisting out of the entire network. When edges are added to the network this happens with a probability proportional to the lowest level l in which the two nodes to be connected share a module. Two nodes that share an elementary module are connected with the probability p_0 whose value is given by normalization. Nodes whose lowest common level is l are connected with probability $p_0 p^l$. For our networks we choose p = 0.5 and h = 8.

When creating nonlinear reactions we assign each possible coupling a probability proportional to the product of the probability of the two newly introduced edges in the original model. Assuming a coupling leads to new edges in the substrate graph between nodes with lowest common module on levels l_1 and l_2, then the probability of choosing this coupling is scaled with $p^{l1}p^{l2}$.

Parameters

For network construction we generate linear reaction networks with N = 1000 species and M = 5000 first-order reactions. Nonlinear networks are build by generating a linear network with M = 3000 reactions and connecting C = 1000 of them to second-order reactions. To compare linear and nonlinear networks directly we also generate linear networks from the substrate graph of the nonlinear networks. This comparison is not possible with the generated linear reaction networks because their substrate graph is not as clustered. An overview of all generated networks is shown in Table 1.

Thermodynamics of Reaction Networks

Thermodynamic properties of reaction networks can be described by non-equilibrium thermodynamics [20, 21]. For simplicity and due to the artificial nature of our simulations we use unitless equations with the Boltzmann constant k_B and the temperature T set to one in this work (we do not consider variations in T).

The change of entropy dS can be separated into the exchange of the system with the environment $d_e S$ and change through processes in the system $d_i S$:

$$dS = d_e S + d_i S. \tag{1}$$

The entropy exchange with the environment (with constant temperature and pressure) is given by

$$d_e S = dU + pdV - \sum_k \mu_k d_e x_k, \tag{2}$$

with $d_e x_k$ being the change of concentration due to interaction with the environment and μ_k being the chemical potential. The entropy change through internal processes $d_i S$ is given accordingly:

$$d_i S = \sum_k \mu_k d_i x_k. \tag{3}$$

From this equation the rate of entropy production can be calculated. By rewriting we see that the entropy production of the network $\sigma_{tot} = \frac{d_i S}{dt}$ is merely the sum of the entropy production σ_i of all individual reactions with $\sigma_i = \sum_m \mu_m N_{mi} v_i$:

$$\frac{d_i S}{dt} = \sum_k \mu_k \frac{d_i x_k}{dt} = \sum_l \sum_m \mu_m N_{ml} v_l = \sum_l \sigma_l \tag{4}$$

In steady state the entropy production of individual reaction i can be also written as function of forward and backward reaction rates $v_{i,+}$ and $v_{i,-}$ [21]:

$$\sigma_i = (v_{i,+} - v_{i,t-}) \ln \frac{v_{i,+}}{v_{i,-}}. \tag{5}$$

This relation can be applied to calculate the total entropy production rate $\sigma_{tot} = \sum_i \sigma_i$ of a reaction network acting between two boundary species b_1 and b_2 kept at concentrations c_1 and c_2. As the entropy production rate in steady state only depends on the boundary conditions ($c_1, c_2, v = v_+ - v_-$) we can replace the entire network with one imaginary linear reaction $b_1 \leftrightharpoons b_2$. If we assume the Gibbs energies of formation of boundary species to be zero, the forward and backward rate coefficients are equal and we obtain the equation

$$\sigma_{tot} = v \ln \frac{c_1}{c_2}. \tag{6}$$

Alternatively one could also get this result by calculating the boundary species entropy exchange with the environment, because in steady state $0 = dS = d_e S + d_i S$.

Generating Thermodynamic Data

The Gibbs energies of formation of the species μ0iμi0 are drawn from a normal distribution N(0,1). Reaction rates are calculated using the Arrhenius equation (with the prefactor A set to 1):

$$k = Ae^{-E_a} = e^{-E_a}. \tag{7}$$

Here, E_a is the activation energy which is sampled from the distribution

$$P(x) = \frac{6}{\pi^2} \frac{1}{x^3(\exp(1/x) - 1)} \tag{8}$$

for every reaction. We have chosen this distribution, which resembles the Planck-distribution, because it has an effective non-zero lower bound while still having a large tail to the right

(Fig. 1). We simulate all reactions reversibly. Forward and backward reaction are energetically constrained by the Gibbs energies of the species. Thus, we sample E_a just once for every reaction and assign it to that reaction direction which respective products have a higher Gibbs energy of formation, either $E_{e,i} = \sum_{j,N_{ij}>0} \mu_i^0 N_{ij}$ (forward direction) or $E_{p,i} = \sum_{j,N_{ij}<0} \mu_i^0 N_{ij}$ (backward direction). The second reaction directions activation energy is then given by the constraint $E_e' = E_e + |\sum_j \mu_i^0 N_{ij}|$. This expression is a reflection of the fact that in equilibrium, forward and backward reaction rates need to balance.

Network Simulation

As we are interested in the steady state of the network under thermodynamic boundary conditions we solve the reaction equation while keeping the concentration of two selected chemical species b_1, b_2 at fixed concentration c_1, c_2. To remove the effects of the energy difference between the boundary species on the flow, we set their Gibbs energy of formation μ_i^0 to zero and recalculate reaction rates before the simulation. For solving the reactions' ODE, the integrator of C++'s boost library is used. The selected algorithm is "Dormand-Prince 5". Concentrations are initialized normally distributed with $\frac{c_1+c_2}{2}$ taken as mean and $|c_1-c_2|$ as standard deviation. Dynamics are simulated up to a time t of 50000 or up to the time when the mean square change of concentration (per species and time-step size) is smaller than 10^{-20}.

We assume that the greatest topological factor influencing flow through the reaction network is the shortest path distance between the boundary species. Because we cannot perform simulation and analysis for one million pairs of boundary species, we sample 50 pairs of boundary species for all values of the shortest path occurring in the network.

A simple investigation of network flow and entropy distribution is done with boundary species concentrations set to $c_1 = 0.1$ and $c_2 = 1$. To get an error estimate, we generate 10 independent samples of every network type.

To investigate the response of the nonlinear networks to an increase in thermodynamic disequilibrium we vary the boundary conditions accordingly. For this we keep c_1 at 0.1 while varying c_2 from 0.2 up to 60. With higher values of boundary concentrations we notice an extreme increase of computational time needed to solve the individual ODEs. Thus, we are only able to simulate one network sample of every type for this setup.

The software for generating (https://github.com/jakob-fischer/jrnf_tools), running (https://github.com/jakob-fischer/jrnf_int), and analyzing (https://github.com/jakob-fischer/jrnf_R_tools) the simulations is developed in R and C++ and freely available through the platform github.

RESULTS

For our study we generate various random networks for each network model (Table 1). We simulate these networks for different boundary conditions and analyze the resulting steady state. In the following, we first compare the artificial networks with real networks and then show in detail how the flow and energy dissipation depend on network structure and boundary condition.

Network Structure

We compare the topological features of our artificially generated networks with real world networks (Table 1). For this we use a compilation of chemical reactions in Earth's atmosphere [22] and models for the combustion of Methane [23] and Dimethyl ether [24]. Also a kinetic model of the metabolic network of Yeast [25], available through the BioModels Database [26], is investigated. To avoid that the representation of networks as substrate graphs biases our results [27] we compare each network with a randomized version of itself. When randomizing an artificial network we would obtain an Erdős-Rényi network with the same density and the same types of reactions. Thus, rows for randomized BA, WS, and PS networks are omitted in Table 1.

The power law scaling for Earth's atmospheric reaction network and the metabolic network of Yeast are clearly pronounced in comparison with their respective null models (Fig. 2 (B)). This is not true for the Ethanol combustion chemistry whose size of 57 species (nodes) does not allow to unambiguously decide on the scale-free property. The substrate graphs of the two combustion chemistries show the properties of small world networks, they have a small mean shortest path length and a high clustering coefficient. As their null models show the same properties, this can be attributed to their high density. All reaction networks have more cycles than their randomized counterparts. With the exception of the network from Yeast's metabolism all real networks also have a higher clustering coefficient or a higher value for modularity.

For a comparison of the artificial reaction networks with real thermodynamic data we use a table of reaction free energies ($\Delta_r G^0$) of reactions in glycolysis [28]. In our networks this corresponds to $\Delta\mu_j^0 = \sum_j N_{ij}\mu_i^0$. Because there is no way to assign a unique reaction direction to the reactions in the artificial networks, we are only comparing the distributions of absolute values $|\mu_j^0|$. The normalized (mean set to one) cumulative distributions show a more localized distribution with a wider tail for the data from glycolysis. The distribution for the artificial networks is over all more regular. The bimodal distribution for the data from glycolysis might be related to the fact that it describes two distinct processes, the tricarboxylic acid cycle and the pentose phosphate pathway.

Distance Dependency of Flow

To characterize the strength of the steady state flow for different network types, we start with the intuitive assumption that the main factor determining the flow is the distance between the two boundary species in the reaction network, measured by shortest path length d in the substrate graph. The dependency of the mean flow on shortest path length is shown in Fig. 3.

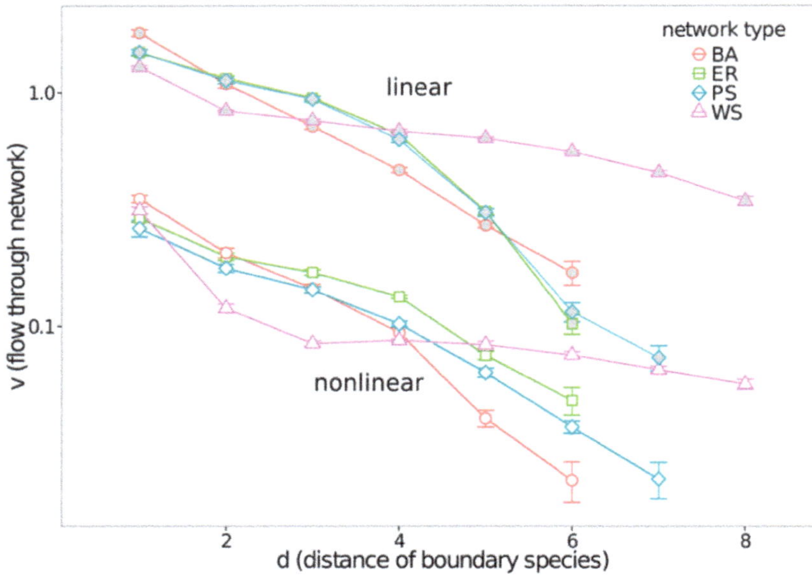

Figure 3. The flow v **through the network depending on boundary species distance**d. All networks are simulated with a boundary concentration difference of $|c_1-c_2| = 0.9$ and a base concentration of $\min(c_1,c_2) = 0.1$. Filled (grey) symbols represent linear networks, empty (white) the nonlinear ones. Error bars show the standard error of the mean.

The flow through reaction networks created with small-world and clustering topology (Watts-Strogatz model) shows to be especially weakly dependent on boundary species distance d. In the linear as well as the nonlinear case these networks have a lower mean flow for small d (≤ 4) while for larger values of d, they have generally a larger flow than the other networks. We hypothesize that the flow for boundary points whose distance is close to the diameter is limited by the sparse connection of those boundary species to the network. The high clustering of Watts-Strogatz networks (cf. Table 1) apparently leads to their exceptional high flow for boundary points with a large distance d. This also agrees with the low sensitivity to boundary species distance that the Watts-Strogatz networks show.

The linear networks generated out of the Erdős-Rényi model and those generated with the Pan-Sinha model show a strikingly similar behavior. This may be due to their similar degree distribution (not shown).

Varying Flow through Nonlinear Networks

Unlike in linear networks, the flow and dissipation distribution in nonlinear networks depend on the absolute concentrations of the boundary species. For the variation of boundary concentration flow dependency of the concentration difference is in an intermediate regime (Fig. 4 (A)) and the slope in log-log plot takes a value between 1 and 2. This is plausible since the network consists of a mix of linear reactions and nonlinear reactions with at best quadratic behavior. Theoretically a stronger than quadratic dependency of flow from concentration difference would be possible for a specific boundary condition and a specific concentration range, but this possibility seems not to influence the mean behavior.

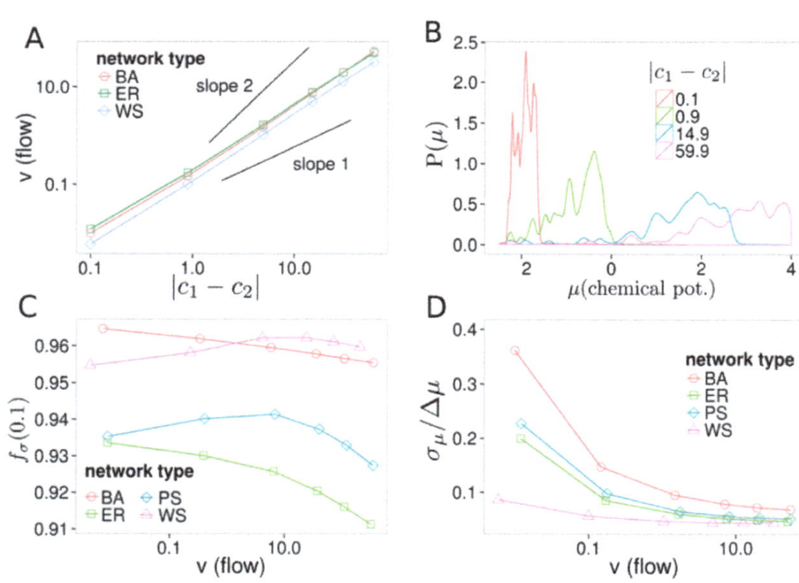

Figure 4. Varying flow through nonlinear networks. Each data point is the average of all simulations with specific boundary species concentration ($c_1 = 0.1$ $c_2 = 0.2\ldots60$) and a shortest path between boundary species of 3. **(A)** Dependency of flow from concentration difference. Pan-Sinha results are not shown as they overlap with the Erdős-Rényi ones. **(B)** Distribution of

species chemical potential μ_i for different boundary condition strengths of BarabsiAlbert (BA) networks.**(C)** The fraction of dissipation in the network explained by the most dissipating 10 percent of reactions, $f_\sigma(0.1)$. **(D)** Standard deviation of chemical potentials σ_μ normalized by difference between boundary species' potentials $\Delta\mu = |\mu_{b2}-\mu_{b1}|$ shows a more localized distribution of chemical potentials for larger flows.

We look at the distribution of chemical potentials $\mu = \mu_0+\ln(x_i)$ inside the reaction network for different strengths of the boundary condition. In Fig. 4 (B) the distributions $P(\mu)$ are shown for the simulated Barabási-Albert (BA) reaction networks with boundary species distance d of 3. The distributions in general are localized between the chemical potentials of the boundary species $\mu_{b1} = \ln(x_{b1})$ and $\mu_{b2} = \ln(x_{b2})$ (remember that μ_0 for boundary species is set to zero). While the distributions are almost uniform in this range for low flows, at higher flows the distributions are more shifted towards the upper part. Normalizing the standard deviation σ_μ by $\Delta\mu = |\mu_{b2}-\mu_{b1}|$ confirms this finding (Fig. 4 (D)) and shows a narrower distribution relative to the chemical potentials of the boundary species.

The distributions of dissipation values of the reactions are to noisy to find out if they also get narrower for higher flows. Thus, we calculate the fraction of the dissipation explained by the 10% of reactions with the highest dissipation, $f_\sigma(0.1)$. We see that with higher flows the fraction of dissipation explained by these 10 percent of the network decreases (Fig. 4 (C)). The networks generated from the Watts-Strogatz (WS) and the Pan-Sinha (PS) networks show an increase of $f_\sigma(0.1)$ for lower values, but above a flow of around 5 they also decrease. Explained differently, for higher flows one needs a larger part of the network to explain a given fraction of its dissipation. Together with the narrower distribution of chemical potentials we interpret this as the thermodynamic disequilibrium leading to a tighter coupling of the reaction network. This coupling leads to the chemical potential of different species to be closer and to the dissipation being more evenly distributed among reactions.

Flow Dependency of Cycle Number in Nonlinear Networks

There are many indicators that cycles have an important function in networks [9–11]. Cycles function as feedback mechanisms and stabilize the dynamics of the system against perturbations. Also cyclicity has been related to thermodynamic efficiency [29]. To check if there is a dependency of the number of cycles on the flow through the networks we count the number of small cycles (2- and 4-cycles) in the directed substrate graph for different values of v. Note, that even if the simulated reactions do not change, a change in the effective flow of a reaction can imply a change of direction and by this a change in the directed substrate graph.

The number of small cycles is dependent on local topological properties of the network models. Thus, for evaluation we subtract the number of cycles found in networks with randomly chosen reaction directions (Table 1). For all network types we find a clear increase in the number of cycles with increasing flow (Fig. 5). This formation of additional cycles can be understood as the network self-organizes in thermodynamic disequilibrium to increase its flow and dynamic stability. Note that this supports the idea of the previous section of a closer coupling of the network with higher degree of disequilibrium.

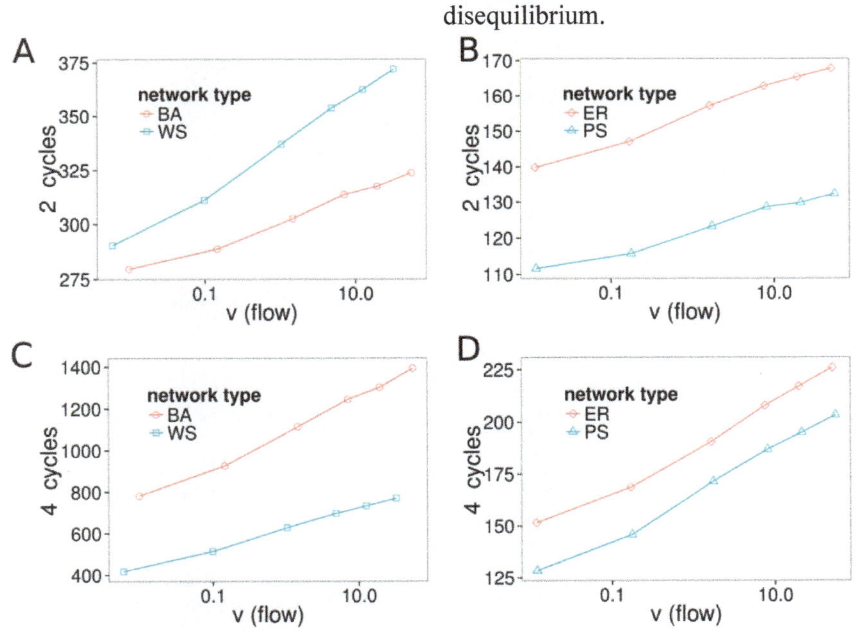

Figure 5. Number of 2- and 4-cycles in the (directed) substrate graphs of the nonlinear reaction networks. The plots show the number of additional cycles depending on the flow through the network in comparison to the same network with random reaction directions (Table 1). Each data point is the average of all simulations with boundary points distance of 3 and fixed boundary concentrations ($c_1 = 0.1$ $c_2 = 0.2\ldots60$).

Distribution of Entropy Production Rates

To see how dissipation is distributed inside of the networks, we calculate the entropy production rate for the individual reactions σ (Eq. 5) and look at their distribution for specific network topologies and boundary conditions. To better see the power law dependency, we plot the cumulative distribution $1 - \int_{-\infty}^{\sigma} P(\sigma')d\sigma'$, which describes the probability of the entropy production rate being higher than σ[30] instead of P(σ).

The distributions show no large qualitative differences between the different network models (Fig. 6). The power law in the intermediate regime is differently pronounced in its extent for different network types but the greatest difference is clearly seen between the slopes of linear and nonlinear networks. Assuming that P(σ) follows a power law, we get an exponent of about −1.5 for linear networks and of −1.66 for nonlinear networks. The steeper slope of the nonlinear networks can be interpreted as an effect of their reactions being better coupled. This can be seen by the fact that nonlinear (A+B⇋C+D) reactions are not depleting a potential between two species directly but there is always the probability that they increase the potential between two other species. The coupling implies a stronger connection of the flow between individual reactions and by this a stronger connection with the magnitude of dissipation.

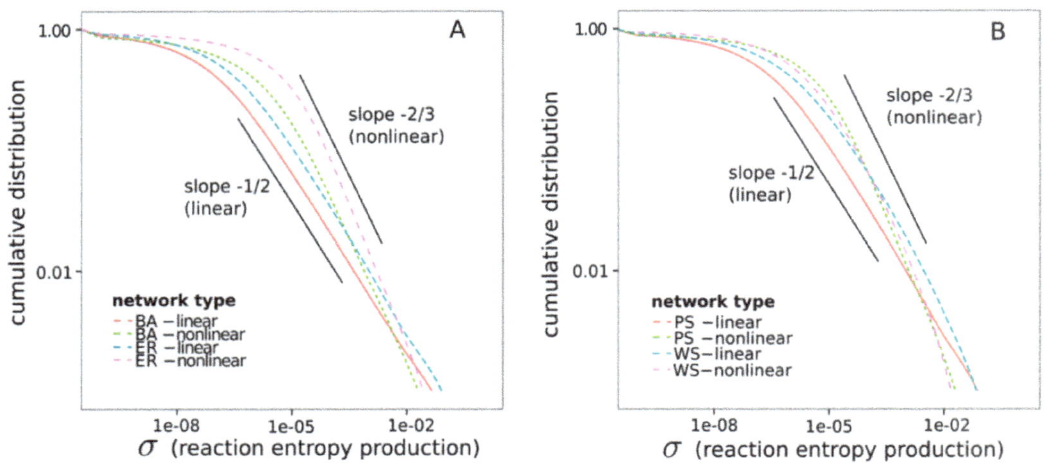

Figure 6. Cumulative distribution of the entropy production of the reactions. All simulations are performed with boundary concentration values of $c_1 = 0.1$, $c_2 = 1.0$ and a shortest path between boundary species of length 4. **(A)** Distributions for Barabási-Albert (BA) and Erdős-Rényi (ER) networks. **(B)** Distributions for Watts-Strogatz (WS) and Pan-Sinha (PS) networks.

Connectivity Dependence of Dissipation

To evaluate how the dissipation of a reaction depends on the connectivity of the involved species, for every species we calculate the mean dissipation of all reactions connected to it. Plotting the mean dissipation depending on the degree centrality of the species (in the substrate graph) shows a relatively high dissipation for reactions adjacent to lowly connected species (Fig. 7). This effect is more pronounced for nonlinear networks. When looking for reactions with high dissipation we should search in the vicinity of lowly connected species. This can be explained by the stronger connection between reactions generating and consuming the species. When the rate of a reaction that produces a species is increased, the additional flow has to be distributed over the consuming reactions. If there are many consuming reactions, there are more potential pathways to forward the flow while keeping the mean dissipation rate low.

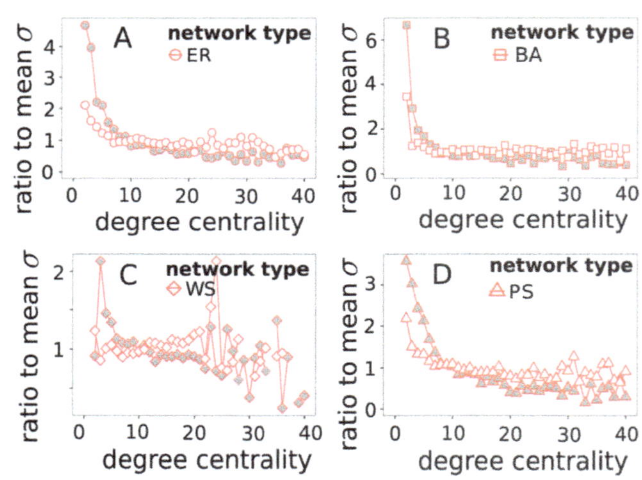

Figure 7. Mean entropy production ***σ*** **associated with nodes of degree** f. Values are normalized by mean entropy production in the sample network. Grey filled points show nonlinear networks, white filled points show linear networks. Data was taken from all simulation runs of the specified network type with $\min(c_1,c_2) = 0.1$, $|c_1-c_2| = 0.9$ and shortest path d = 4.

CONCLUSIONS

We have simulated random reaction networks under thermodynamic constraints in order to provide insight into how energy is dissipated in complex reaction networks in thermodynamic disequilibrium. We observe a clear difference between linear and nonlinear networks. The power law has an exponent of ≈ -1.5 for linear and a slightly lower exponent of ≈ -1.66 for nonlinear networks. However, there are no qualitative differences between the distributions of entropy production rate for different complex network models like Erdős-Rényi, Barabási-Albert, etc. (Fig. 6). The differences between thermodynamic reaction networks of different topologies are more pronounced in the flow (Fig. 3) than in the other properties investigated.

We found that a greater disequilibrium in nonlinear reaction networks is associated with a more tightly coupled network. For a greater flow, the network self-organizes and maintains a greater number of cycles (Fig. 5). A greater flow also leads to a narrower distribution of chemical potentials (Fig. 4 (B), (D)). This is associated with results that suggest that for higher flows, a larger fraction of the network is necessary to explain a given fraction of its dissipation (Fig. 4 (C)). We interpret this as an increase in the system's complexity that comes along with a higher thermodynamic disequilibrium.

Finally, we found that reactions involving lowly connected species tend to dissipate more energy, which is more pronounced in nonlinear networks, but is also found in linear networks (Fig. 7). This might help to identify reactions that play central roles in the energy dissipation of a complex reaction network.

We also showed how our artificial networks share topological properties with real reaction networks. The artificial networks are toplogically more similar to Earth's atmospheric chemistry and Yeast's metabolism than to the two investigated combustion chemistries. The main discriminating factor here is the high density of those two combustion chemistries. The distribution of thermodynamic parameters in the artificial networks only roughly matches data from reactions of glycolysis (Fig. 2 (B)). Obviously, the amount of thermodynamic data (37 reactions) is quite limited. Current progress in bioinformatic methods to reconstruct thermodynamic data [31–33] may improve the availability of such data in future and allow a better analysis.

Nevertheless, a fundamental problem of such a comparison remains. It is the way the data of reaction networks is obtained. In networks from chemical models, experimentalists and modelers have made a decision on which reactions are relevant. Experimentalists only find reactions that are occurring and are measurable in the systems they investigate. Also the modelers might just decide to exclude reactions with low reaction rates from their models. Hence, the reaction network taken from a model is already biased with respect to the model's intention. Our approach with artificial networks, however, assumes the artificial network is a set of (hypothetically) possible reactions; which reactions become important emerges from the dynamics and can be different depending on the boundary conditions.

Thus, we suggest to investigate such emergent phenomena in the future. This is possible by taking smaller artificial reaction networks and then looking at their reaction pathways using elementary flux modes [34]. This would also allow to test the relationship between the rate of an elementary mode and its entropy production [35, 36]. The thermodynamics of reaction networks and of cycling processes therein may also provide insight into the origins of life. Revealing how thermodynamics constraints the behavior of complex reaction networks will be an important ingredient in understanding the role of thermodynamics in domains like prebiotic chemistry [11, 37], biogeochemistry, and cellular systems.

AUTHOR CONTRIBUTIONS

Conceived and designed the experiments: JF AK PD. Performed the experiments: JF. Analyzed the data: JF AK PD. Contributed reagents/materials/analysis tools: JF. Wrote the paper: JF AK PD.

REFERENCES

1. Oster G, Perelson A, Katchalsky A (1971) Network thermodynamics. Nature 234: 393–399. doi: 10.1038/234393a0.
2. Cantú AG, Nicolis G (2006) Toward a Thermodynamic Characterization of Chemical Reaction Networks. J Non-Equil Thermodyn 31: 23–46. doi: 10.1515/jnetdy.2006.003
3. Soh KC, Hatzimanikatis V (2010) Network thermodynamics in the post-genomic era. Curr Opin Microbiol 13: 350–357. doi: 10.1016/j.mib.2010.03.001. pmid:20378394
4. Soh KC, Miskovic L, Hatzimanikatis V (2012) From network models to network responses: integration of thermodynamic and kinetic properties of yeast genome-scale metabolic networks. FEMS Yeast Res 12: 129–143. doi: 10.1111/j.1567-1364.2011.00771.x. pmid:22129227
5. Martínez VS, Quek LE, Nielsen LK (2014) Network thermodynamic curation of human and yeast genome-scale metabolic models. Biophys J 107: 493–503. doi: 10.1016/j.bpj.2014.05.029. pmid:25028891
6. Hordijk W, Steel M (2012) Autocatalytic sets extended: Dynamics, inhibition, and a generalization. J Syst Chem 3: 1–12. doi: 10.1186/1759-2208-3-5.
7. Kasting JF, Holland HD, Pinto JP (1985) Oxidant abundances in rainwater and the evolution of atmospheric oxygen. J Geophys Res-Atmos 90: 497–510. doi: 10.1029/JD090iD06p10497.
8. Saunders SM, Jenkin ME, Derwent RG, Pilling MJ (2003) Protocol for the development of the master chemical mechanism, mcm v3 (part a): tropospheric degradation of non-aromatic volatile organic compounds. Atmos Chem Phys 3: 161–180. doi: 10.5194/acp-3-161-2003.
9. Gleiss PM, Stadler PF, Wagner A, Fell DA (2001) Relevant cycles in chemical reaction networks. Adv Complex Syst 4: 207–226. doi: 10.1142/S0219525901000140.
10. Kreyssig P, Escuela G, Reynaert B, Veloz T, Ibrahim B, et al. (2012) Cycles and the qualitative evolution of chemical systems. PLOS ONE 7: e45772. doi: 10.1371/journal.pone.0045772. pmid:23071525
11. Pascal R, Pross A, Sutherland JD (2013) Towards an evolutionary theory of the origin of life based on kinetics and thermodynamics. Open Biol 3: 130–156.

doi: 10.1098/rsob.130156.

12. Clarke BL (1988) Stoichiometric network analysis. Cell Biochem Biophys 12: 237–253. doi: 10.1007/bf02918360
13. Solé RV, Munteanu A (2004) The large-scale organization of chemical reaction networks in astrophysics. Europhys Lett 3: 1–7. doi: 10.1209/epl/i2004-10241-3
14. Wagner A, Fell DA (2001) The small world inside large metabolic networks. P Roy Soc Lond B Bio 268: 1803–1810. doi: 10.1098/rspb.2001.1711.
15. Erdős P, Rényi A (1959) On random graphs i. Publ Math Debrecen 6: 290–297.
16. Albert R, Barabási AL (2002) Statistical mechanics of complex networks. Rev Mod Phys 74: 47–97. doi: 10.1103/RevModPhys.74.47.
17. Watts DJ, Strogatz SH (1998) Collective dynamics of 'small-world' networks. Nature 393: 440–442. doi: 10.1038/30918. pmid:9623998
18. Pan RK, Sinha S (2009) Modular networks with hierarchical organization: The dynamical implications of complex structure. Pramana 71: 331–340. doi: 10.1007/s12043-008-0166-1.
19. Pan RK, Sinha S (2007) Modular networks emerge from multiconstraint optimization. Phys Rev E 76: 1–4. doi: 10.1103/PhysRevE.76.045103.
20. Kschischo M (2010) A gentle introduction to the thermodynamics of biochemical stoichiometric networks in steady state. Eur Phys J Special Topics 187: 225–274. doi: 10.1140/epjst/e2010-01290-3.
21. Kondepudi D, Prigogine I (1998) Modern thermodynamics / from heat engines to dissipative structures. Chichester; Weinheim: Wiley.
22. Yung YL, DeMore WB (1998) Photochemistry of Planetary Atmospheres. New York: Oxford University Press.
23. Marinov NM (1999) A detailed chemical kinetic model for high temperature ethanol oxidation. Int J Chem Kinet 31: 183–220. doi: 10.1002/(SICI)1097-4601(1999)31:3%3C183::AID-KIN3%3E3.0.CO;2-X.
24. Kaiser EW, Wallington TJ, Hurley MD, Platz J, Curran HJ, et al. (2000) Experimental and modeling study of premixed atmospheric-pressure dimethyl ether-air flames. J Phys Chem A 104: 8194–8206. doi: 10.1021/jp994074c.
25. Stanford NJ, Lubitz T, Smallbone K, Klipp E, Mendes P, et al. (2013) Systematic construction of kinetic models from genome-scale metabolic networks. PLOS ONE 8: e79195. doi: 10.1371/journal.pone.0079195. pmid:24324546
26. Li C, Donizelli M, Rodriguez N, Dharuri H, Endler L, et al. (2010) Biomodels database: An enhanced, curated and annotated resource for published quantitative kinetic models. BMC Syst Biol 4: 92. doi: 10.1186/1752-0509-4-92. pmid:20587024
27. Zhou W, Nakhleh L (2011) Properties of metabolic graphs: biological organization or representation artifacts? BMC Bioinformatics 12: 132. doi: 10.1186/1471-2105-12-132. pmid:21542923
28. Li X, Wu F, Qi F, Beard DA (2011) A database of thermodynamic properties of the reactions of glycolysis, the tricarboxylic acid cycle, and the pentose phosphate pathway. Database 2011: bar005.
29. Layton A, Reap J, Bras B, Weissburg M (2012) Correlation between thermodynamic efficiency and ecological cyclicity for thermodynamic power cycles. PLOS ONE 7: e51841. doi: 10.1371/journal.pone.0051841. pmid:23251638
30. Newman MEJ (2005) Power laws, pareto distributions and zipf's law. Contemp Phys 46: 323–351. doi: 10.1080/00107510500052444.
31. Cannon WR (2014) Simulating metabolism with statistical thermodynamics. PLOS ONE 9: e103582. doi: 10.1371/journal.pone.0103582. pmid:25089525
32. Chakrabarti A, Miskovic L, Soh KC, Hatzimanikatis V (2013) Towards kinetic modeling of genome-scale metabolic networks without sacrificing stoichiometric, thermodynamic and physiological constraints. Biotechnol J 8: 1043'1057. doi: 10.1002/biot.201300091. pmid:23868566
33. De Martino D (2013) Thermodynamics of biochemical networks and duality theorems. Phys Rev E 87: 052108. doi: 10.1103/PhysRevE.87.052108.
34. Schuster R, Schuster S (1993) Refined algorithm and computer program for calculating all non–negative fluxes admissible in steady states of biochemical reaction systems with or without some flux rates fixed. Comput Appl Biosci 9: 79–85. pmid:8435772 doi: 10.1093/bioinformatics/9.1.79
35. Srienc F, Unrean P (2010) A statistical thermodynamical interpretation of metabolism. Entropy 12: 1921–1935. doi: 10.3390/e12081921.
36. Unrean P, Srienc F (2011) Metabolic networks evolve towards states of maximum entropy production. Metab Eng 13: 666–673. doi: 10.1016/j.ymben.2011.08.003. pmid:21903175
37. Bagley RJ, Farmer JD (1992) Spontaneous emergence of a metabolism. In: Langton CG, Taylor C, Farmer JD, Rasmussen S, editors, Artificial Life II. Redwood City, CA: Addison-Wesley, pp. 93–140.
38. Pons P, Latapy M (2006) Computing communities in large networks using random walks. J Graph Algorithms Appl 10: 191–218. doi: 10.7155/jgaa.00124.

CHAPTER 5

CHEMICAL REACTIVITY AND LONG-RANGE TRANSPORT POTENTIAL OF POLYCYCLIC AROMATIC HYDROCARBONS

Ian J. Keyte, Roy M. Harrison and Gerhard Lammel

INTRODUCTION

Polycyclic aromatic hydrocarbons (PAHs) are a group of compounds with proven carcinogenicity. Although the carcinogenic potency varies greatly between congeners,[1,2] they are generally emitted to the atmosphere as a mixture which presents a substantial toxic hazard. Levels in the atmosphere are almost everywhere dominated by combustion sources, while aluminium and iron and steel production as well as petrogenic sources are significant in some environments.[3] The World Health Organisation recommends guidelines in terms of a carcinogenic slope factor[4] and the European Union indicative limit value is set at 1 ng m^{-3} of benzo[a]pyrene.[5] The United Kingdom has set an air quality standard of 0.25 ng m^{-3} benzo[a]pyrene.[6] Because of the complexity of PAH mixtures, the standards and guidelines are set in terms of the concentration of benzo[a]pyrene as this compound typically represents a substantial proportion of the total carcinogenic potential of at least the parent PAH mixture.[2] In turn, the carcinogenic potential of PAH frequently makes a substantial contribution to the carcinogenicity of urban air.[7–10]

Although often referred to as persistent organic pollutants (POPs), PAHs are reactive in the atmosphere. In fact, such reactivity presents problems for receptor modelling methods in source attribution studies. Oxidized products are formed, the most notable being oxy-derivatives (mostly quinones) and nitrated compounds. Some such compounds are also present in primary emissions. PAH derivatives have attracted interest because some are very potent mutagens and carcinogens.[11–13]

A further complexity in studying PAH and their derivatives arises from their semi-volatility. Compounds partition between the particle and vapor phases and hence undergo chemical reactions in both phases. While for the lower molecular weight compounds, reactions in the gas phase are likely to dominate, for compounds with four aromatic rings and higher, reactions in both phases need to be considered. Furthermore, if contact with the ground does not imply immediate reaction, semi-volatility implies the potential of re-volatilization.

Observations of PAHs at remote sites, i.e. far from industrial and transport sources, indicate long-range transport. This is a concern, because of the related hazard for human health and ecosystems. Some PAHs, benzo[a]pyrene, benzo[b]fluoranthene, benzo[k]fluoranthene, and indeno[1,2,3-cd]pyrene beside others, are bio-accumulative. Therefore, and because of resistance to degradation PAHs are considered persistent organic pollutants (POPs) by the Convention on Long-range Transboundary Air Pollution and are listed by conventions for UNECE[14] and OSPAR.[15] Monitoring in the atmospheric environment mandated by convention processes is on-going in Europe and North America.[14,16,17]

Much of the scientific literature is focused upon the 16 USEPA priority PAHs i.e., acenaphthene (Ace), acenaphthylene (Acy), fluorene (Fln), naphthalene (Nap), anthracene (Ant), fluoranthene (Flt), phenanthrene (Phe), benzo[a]anthracene (BaA), benzo[b]fluoranthene (BbF), benzo[k]fluoranthene (BkF), chrysene (Chr), pyrene (Pyr), benzo[ghi]perylene (BgP), benzo[a]pyrene (BaP), dibenzo[a,h]anthracene (DBahA), and indeno[1,2,3-cd]pyrene (IPy) (Table 1). This list is of largely historic significance and was developed when knowledge of relative toxicity of PAH congeners was more limited than at present. It does not include some of the more potent carcinogens, but focuses upon those compounds typically present at higher concentrations which are hence measurable in many environmental samples.

Citation: Ian J. Keyte, Roy M. Harrison and Gerhard Lammel, "Chemical reactivity and long-range transport potential of polycyclic aromatic hydrocarbons," DOI: 10.1039/C3CS60147A.

Table 1: Physico-chemical properties for 298 K, i.e. vapour pressure of the supercooled liquid p_L (Pa; semivolatility is corresponding to $p_L = 10^{-6}$–10^{-2}), Henry coefficient K^H ($M\ atm^{-1}$), octanol–air partition coefficient K_{oa}, rate coefficients of the gaseous substance with the hydroxyl radical and ozone, and total photochemical residence time τ_{total} of 17 parent PAHs (USEPA 16 priority PAHs + BeP). Estimated values (EPI-WIN; USEPA)[18] in lack of experimental data are given in brackets. τ_{total} refers to oxidation by O_3 and OH in the gaseous and particulate phases but neglects the possible particle matrix effect

	p_L (Pa)	K^H ($M\ atm^{-1}$)	$\log K_{oa}{}^{c}$	$k_{g\,OH}$ ($10^{-12}\ cm^3\ molec^{-1}\ s^{-1}$)	$k_{g\,O3}$ ($10^{-18}\ cm^3\ molec^{-1}\ s^{-1}$)	τ_{total} (h)
Naphthalene (Nap) 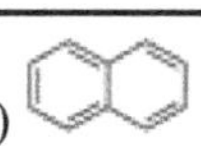	38	2.2	5.19	22	<0.3	13
Acenaphthylene (Acy) 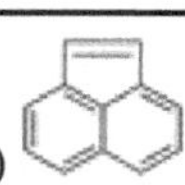	2.6	10.5	6.46	110	550	1
Acenaphthene (Ace) 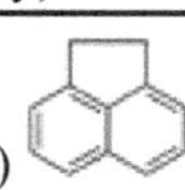	1.7	7.1	6.44	58	<0.5	5
Fluorene (Fln) 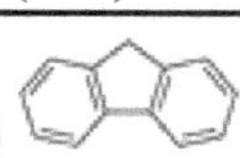	0.54	11.3	6.85	13		21
Phenanthrene (Phe) 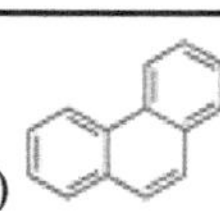	0.10	23.5	7.64	31	0.40	9
Anthracene (Ant) 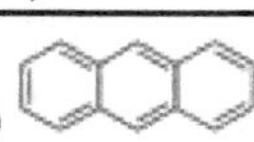	5.9×10^{-2}	20.0	7.70	130		2
Fluoranthene (Flt) 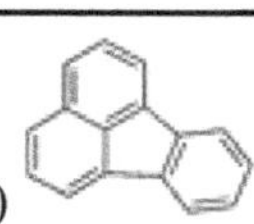	6.8×10^{-3}	76	8.81	11		14–25
Pyrene (Pyr) 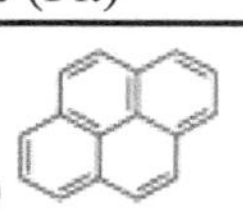	4.2×10^{-3}	76	8.86	50		5–6
Benzo(a)anthracene (BaA) 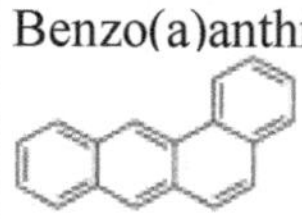	3.8×10^{-4}	159	10.28			3–11
Chrysene (Chr) 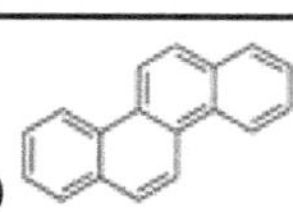	1.3×10^{-4}	270	10.30	(50)		9–27
Benzo(b)fluoranthene (BbF) 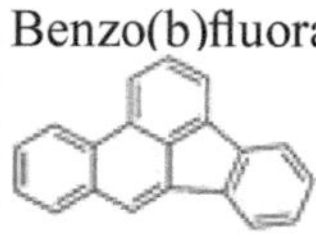	1.0×10^{-5}	1550	11.34	(16)		34 to >330
Benzo(k)fluoranthene (BkF) 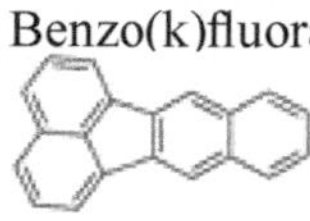	7.8×10^{-6}	1790	11.37	(54)		8–21
Benzo(e)pyrene (BeP) 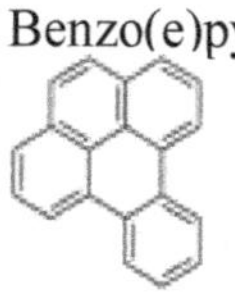	$1.8 \times 10^{-5\ e}$	276[e] (1240)	(11.35)	(50)		10–15
Benzo(a)pyrene (BaP) 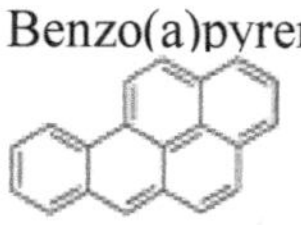	7.9×10^{-6}	1320	11.48	(50)		2–5
Indeno(1,2,3-c,d)pyrene (IPy) 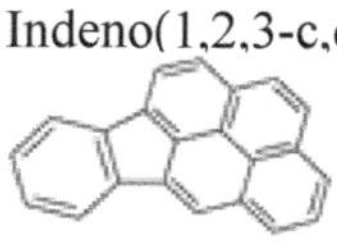	6.6×10^{-7}	2050	12.43	150		6–9

Dibenz(a,h)anthracene (DBA) 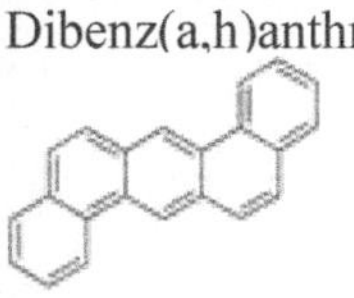	9.5 × 10^{-8} [e]	120[e] (2045)	12.59[f]	(50)		34 to >330
Benzo(g,h,i)perylene (BPe) 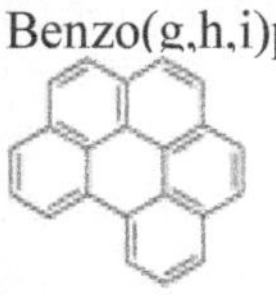	4.6 × 10^{-7}	2410	12.55	5.9		5

[a] Ma et al.,[19] for temperature dependent data see Tenhulscher et al.[20]

[b] Ma et al.,[19] for temperature dependent data see Lei et al.[21]

[c] Ma et al.,[19] for temperature dependent data see Odabasi et al.[22]

[d] Based on oxidant levels characteristic of the continental background in mid latitudes (10^6 OH per cm^3, 50 ppbV O_3) and assuming model aerosols being representative for ambient aerosols (silica for unspecific particulate mass, diesel and flame soot for BC). Ranges given reflect uncertainties of kinetic data (see Tables 3, 5 and 6), particulate mass fraction, $\boldsymbol{\theta}$ (adopted from a range of field observations during various seasons and in various climates, i.e. Simcik et al.;[23] Mandalakis et al.;[24] Tsapakis and Stephanou;[25] Ding et al.;[26] He and Balasubramanian;[27] Lammel et al.;[10] Demircioglu et al.;[28] besides others), and of BC content (adopted from Spindler et al.[29]).

[e] Paasivirta et al.[30]

[f] Odabasiet al.[22]

In this article, we review current knowledge of PAH and nitro-PAH with a focus upon chemical reactivity in both the gaseous and condensed phases, and where possible elucidate rates and mechanisms. Reactivity is put in the context of exposure of remote atmospheric environments, i.e. long-range transport (LRT) potential of these pollutants.

SOURCES

Primary Emissions

Polycyclic aromatic hydrocarbons (PAHs) are by-products of all types of combustion, in particular incomplete. Major anthropogenic emission sources to air are, domestic burning of coal and wood, fossil fuel burning power plants, road transport, wood burning, cooking, and others. For global sources in 2004, Zhang and Tao[31] reported biofuel (i.e. straw, firewood and animal dung, 56.7%) and wildfire (i.e. forest and savanna fires, 17.0%) as the main contributions, while traffic, domestic coal combustion and burning of agricultural waste contribute only 4.8%, 3.7% and 2.7%, respectively (referring to the sum of the 16 USEPA priority PAHs). In open fires of agricultural debris, forest or savannah grass, PAHs are not only emitted into air but also into soil. Additionally, petrogenic and biological[32] sources may contribute, but anthropogenic sources prevail by far.

The countries ranking highest in emissions of the 16 USEPA priority PAHs in 2004 were China (114 kt $year^{-1}$), India (90 kt $year^{-1}$) and the USA (32 kt $year^{-1}$).[31] Dominated by the use of biofuels, the PAH pattern may differ considerably among countries. For example, health risk, expressed in BaPeq/sum of 16 PAHs, ranges 0.18–3.58% among countries with highest values expected for countries with high usage of non-anthracite coal or wheat straw.[31] High spatial resolution emission inventories are available for a number of countries (usually only BaP or few substances, namely BaP, BbF, BkF and InP; e.g. EEA;[33] Xu et al.[34]). Emission estimates have also been compiled for whole regions (e.g., Gusev et al.[35]) as well as globally[31,36] (Table 2). Global anthropogenic emissions of the 16 USEPA priority PAHs in 2004 were close to 4 kg km^{-2} $year^{-1}$ with biofuel (56.7%), wildfire (17.0%) and consumer product usage (6.9%) contributing most.[31] Power plants, open biomass burning, road transport (mostly diesel), industrial processes, air and sea transport also contribute. For BaP 1990 global anthropogenic emissions, it was estimated that 13% is due to biomass burning and 87% due to fossil fuel combustion.[37]

Table 2: Regional and global emission estimates of (a) the 16 USEPA priority PAHs and (b) benzo[a]pyrene (t $year^{-1}$)

Europe and Russia	USA	China	World	Year	
(a)					
	26 500			1990	USEPA[39]
		25 300		2003	Xu et al.[34]
51 000	32 000	114 000	520 000	2004	Zhang and Tao[31]
(b)					
1047			22 436	1990	Münch and Axenfeld[37]
590				1995	Pacyna et al.[40]
264	88	239	3700	1996	Lammel et al.[36]

		540		2003	Xu et al.[34]
907				2000	Denier van der Gon[41]

Emission inventories are based on emission factors from the various combustion technologies and fuel consumption and transformation data and usually apply for one particular year. Emission inventory determination is very uncertain as substance-specific emission factors (ng g^{-1} fuel burnt or μg km^{-1} for transport) determined experimentally are very sensitive to even minor differences in combustion technology, engine operation, fuel composition, etc. and, hence, may vary by several orders of magnitude for the same type of emission. Most uncertain are emission factors of open fires (e.g. forest fires), because of difficulties in measuring under realistic and representative conditions. Data gaps have been filled assuming a constant ratio of emission factors of PAH and another trace substance (e.g. black carbon[36]). To distribute on a grid, country-based data are often scaled according to population density (under the assumption that emission per capita is constant) and annual data are temporally distributed evenly neglecting seasonal variation of major sources, such as residential heating related fossil fuel burning. More realistic temporal functions have been successfully used in regional PAH modelling.[35,38]

Known substance patterns of emission sources partly combined with known emission fluxes have been used to quantify the contribution of individual sources to pollution levels at receptor sites using various types of source-receptor modelling.[42,43] The most simple approach in this context is the comparison of ratios of pairs individual PAH's concentrations at the receptor and source sites. Pairs of PAHs of similar degradability in air have been suggested for this purpose (so-called diagnostic ratios[44–46]). This approach is not successful due to the variability of sink (degradation, deposition) conditions and incomplete knowledge of chemical kinetic data in both the gaseous and particulate phases (Section 3[47,48]). The success of multivariate statistical approaches, making use of substance patterns of emission sources which encompass many PAHs (so-called fingerprint methods), is often limited by the similarity of sources in terms of the PAH pattern.[49] The inclusion of other substances, such as the anhydrous sugar levoglucosan as a marker for biomass combustion[50] and hopanes as markers for motor vehicle exhausts[51] and coal combustion,[52] can solve the problem. In the future, this challenge may consistently be covered using inverse modelling (adjoint model). In this context, the ratio of the concentrations of relatively fast degrading benzo[a]pyrene over relatively slow degrading benzo[e]pyrene (BeP) has been suggested as a measure for age of air mass.[53]

Re-Volatilization from Ground Compartments

Most of the 3–4 ring PAHs are semivolatile (i.e., their saturation vapor pressure at 298 K is in the range $p_{sat} = 10^{-6}$–10^{-2} Pa; Table 1). Therefore, these substances are subject to re-volatilization upon condensation on surfaces and to gas exchange with vegetation,[54–57] soils[28,58–62] and water surfaces.[63–65] Persistence of PAHs in soils is on the order of decades.[61] Therefore, LRT of PAHs may be enhanced by re-evaporation.[36] Secondary emissions due to re-volatilization from surfaces need to be included in modelling of semivolatiles, in particular if the substance had been accumulated historically in the ground compartments. For example, at the central European background observatory Košetice, Czech Republic, the burden of fluoranthene and pyrene in soil, 121 and 95 ng g^{-1} (mean of 2006–2008, depth sampled 10 cm) exceeds the burden in air by 4 orders of magnitude (under the assumption of a bulk soil density of 1.5 g cm^{-3} and that the concentrations in ground-level air, 1.83 and 1.19 ng m^{-3}, respectively, were representative for the air column within the planetary boundary layer of 1000 m depth) and concentrations in soil and air have been approaching equilibrium during the last decade.[66] Much of the soil PAH in temperate regions is historic and concentrations in air and soil are relaxing to equilibrium on a multi-year time scale.

The volatilization flux is described in models based on an empirical parameterization or as given by diffusion. However, not all contributions to effective diffusion are represented in models and additional sources to the atmosphere exist. For example, the possible enhancement of soil–air exchange by co-evaporation with soil water vapour and of air–sea exchange by turbulence in surface water (breaking wave phenomena, bubbles[67,68]) is usually neglected. Apart from volatilization additional secondary emission paths are possibly relevant, e.g. the formation of aerosol particles from drying sea spray droplets. This is a relevant source as PAHs, like other hydrophobic organics, were found enriched in the sea surface microlayer by up to a factor of 10, sometimes even up to a factor of 100 as compared to the subsurface water.[69,70] The fraction of the PAH in surface seawater which is partitioning to suspended or sinking particulate matter can volatilize following dissolution in water in relaxation to phase equilibria – or be removed from the surface layer carried by sedimenting particles or by deep water formation. Similarly, the PAH burden stored in soils may only partly be mobile and subject to re-volatilization: in soil, through chemical interaction with fulvic acid or other organic matter[71–75] and black carbon[76–78] components PAH may in the long-term turn immobile ('ageing') and more persistent. Sorption strength is dependent on the type of soil organic matter (OM) and sorption history.[79–81]

Degradation in ground compartments is in models generally assumed to obey first order chemical kinetics without spatial variability in soils or in seawater. However, there is evidence for significant spatial heterogeneity of PAH biodegradability at least in soils.[82] The rates used for seawater are often merely extrapolated from experimental data obtained in freshwater using a default factor to account for a lower level of bioactivity. The temperature dependence of the degradation reaction coefficient, if not neglected, is usually fixed to some default function, e.g. doubling per 10 K temperature increase.

SINKS

Gas–Particle Partitioning and Deposition

Gas–Particle Partitioning

The phase partitioning of a semivolatile organic compound, i, in aerosols can be quantified by the gas–particle partitioning coefficient K_p ($m^3\ \mu g^{-1}$):

$$K_p = c_{ip}/(c_{ig} c_m) \quad (1)$$

with the concentrations in the gaseous and particulate phases, c_{ig} and c_{ip}, and particulate matter mass concentration c_m. Note that K_p is operationally defined, which means that data are often biased by sampling artefacts.[83–89] The most relevant contributions under many ambient and sampling conditions are due to oxidative or evaporative losses from particulate matter deposited on filter membranes, which leads to an underestimation of the particulate phase mass fraction.[86,89]

Adsorption:

If adsorption is dominating the process, then for the particulate phase mass fraction of semivolatile substances, $\theta = c_{ip}/(c_{ig} + c_{ip}) = [1 + 1/(K_p c_{TSP})]^{-1}$, it holds:[90,91]

$$\theta = c_J S/[c_J S + p^o_L] \quad (2)$$

with constant c_J, aerosol surface S (cm^{-1}), and saturation vapour pressure of the super-cooled liquid p^o_L (Pa). This relationship is empirically based. c_J should be substance specific, but was hardly ever specified, but rather generally adopted as 17.2 Pa cm. K_p could be expressed as:[91]

$$K_p = N_s A_m T e^{(Q_l - Q_v)/RT}/(16 p^o_L) \quad (3)$$

with surface concentration of sorption sites N_s, specific surface area of particulate matter $A_m = 10^6$ S cm^{-1} ($cm^2\ \mu g^{-1}$), the enthalpy of desorption from the surface Q_l and the enthalpy of vapourisation of the super-cooled liquid Q_v. For equilibrium partitioning the slope, m, of plots of the form:

$$\log K_p = m \log p^o_L + b \quad (4)$$

should be equal to −1 and the intercept, b, for constant temperature should be determined by N_s and A_m. Deviation of m from −1, however, is inconclusive, as various possible reasons exist which usually cannot be addressed, such as non-equilibrium, temperature change during sampling, weaker interaction between the gas and solid surface, and absorption into OM.[92,93] Non-equilibrium can be caused by kinetic constraints (introduction of fresh particles or clean air) or by the presence of non-exchangeable compounds on or in the particles.

As $\log p^o_L \sim 1/T$ (Clausius–Clapeyron equation) eqn (4) is physically equivalent to plots of the form

$$\log K_p = A/T + B \quad (5)$$

which was derived from Langmuir adsorption theory.[94] For PAH absorption in primary or secondary organic particulate matter K_p would be given as:[95]

$$K_p = K_{OA} f_{OM} MW_{ocr} \zeta_{ocr}/(\rho_{ocr} MW_{OM} \zeta_{OM} 10^{12}) \quad (6)$$

with K_{OA} being the octanol–air partition coefficient (values in Table 1), f_{OM} the mass fraction of OM in particulate matter, MW_{OCT}, and MW_{OM} the mean molecular weights of octanol and the OM phase (g mol^{-1}), respectively, ρ_{OCT} the density of octanol (0.820 kg L^{-1}), ζ_{OCT} the activity coefficient of the absorbing compound in octanol, ζ_{OM} the activity coefficient of the compound in the OM phase. With the assumptions that $\zeta_{OCT} = \zeta_{OM}$ and $MW_{OCT}/MW_{OM} = 1$, eqn (6) can be simplified as:

$$\log K_p = m' \log K_{OA} + b' \quad (7)$$

then, and when phase equilibrium is established the slope, m′, of plots of eqn (7) should be equal to +1 and the intercept, b′, equals $\log f_{OM} - 11.91$.

$\log K_p$ of PAHs based on differentiating sampling of the gaseous and particulate fractions was found to be correlated with 1/T (or $\log p^o_L$) in a number of studies.[92,96–98] Therefore, as a general trend a doubling of θ per ≈5 K temperature decrease had been found,[95] θ being significantly higher in winter,[10,25,99,100] similar to other semivolatile aerosol constituents. This temperature trend is in accordance with both adsorptive and absorptive mechanisms.[95] However, K_{OA} was suggested to be a better descriptor for PAH gas–particle partitioning than p^o_L.[101] Measurement of f_{OM} has partly been included, however. Over-prediction of θ by the surface adsorption model had been suggested by other studies at both source and receptor sites.[65] In these studies, however, S has been estimated and assumed to be constant, rather than measured. Gas–particle partitioning of PAHs is presently understood as having adsorptive and absorptive contributions. In particular, a combination of absorption into OM and adsorption onto soot seems to have a high potential to explain the observations.[102,103]

With, again, the assumptions $\zeta_{OCT} = \zeta_{OM}$ and $MW_{OCT}/MW_{OM} = 1$ adsorptive and absorptive terms can be combined to eqn (8) (Lohmann and Lammel;[103] similar equation suggested by Dachs and Eisenreich[102]).

$$K_p = 10^{-12}[f_{OM}/\rho_{oct} K_{oa} + f_{BC}/\rho_{BC}(S_{BC}/S_{soot}) K_{soot-air}] \quad (8)$$

with f_{BC} being the mass fraction of soot (black carbon) in particulate matter, $K_{soot-air}$ the soot–air partitioning coefficient, ρ_{oct} and ρ_{BC} the densities of octanol and soot, respectively, and S_{BC} and S_{soot} available specific surface areas. Diesel soot is taken as representative for soot. A particularly high affinity to particulate matter and to soot was found for the semivolatile PAHs Ant and BaA.[96,104]

While f_{OM}, likely to influence absorption, and soot, f_{BC}, likely to influence adsorption, had been covered in field studies,[65,88,105,106] the aerosol surface, S, has only rarely been covered so far,[10,107] and never together with the other parameters, i.e. f_{OM}, f_{BC}, S_{BC} and S_{soot}. Moreover, a more differentiated specification of carbonaceous fractions of PM would be needed in order to advance the understanding of PAH's gas–particle partitioning.

Matrix specific adsorption to and absorption of PAH in particulate matter components can in principle be described accounting for all types of chemical interaction at the molecular level (so-called poly-parameter linear free energy relationships).[108] The substance specific[109,110] and many relevant aerosol matrix parameters are available to quantify PAH's electron donor and acceptor, and van der Waals interactions (for adsorption) and a cavity formation term accounting for solvation (absorption into the particulate matter organic phase), including their humidity dependence.[111,112] However, there are still limitations due to data gaps in the experimentally determined substance and matrix parameters.

PAH Mass Size Distribution

Upon emission in combustion processes and close to the sources mass size distributions of PAHs are peaking in the ultrafine and accumulation modes (0.01–0.1 and 0.1–1 μm aerodynamic diameter size ranges, respectively). One or two mass size modes are found at urban and rural sites, with mass median diameters, D_m, almost exclusively in the accumulation mode (0.1–1.0 μm of size).[113–118] This should be seen in the context of limitations by particle size resolution (number of impactor stages) and analytical sensitivity. During atmospheric transport re-distribution in the aerosol occurs, at least of the semivolatile PAHs, which undergo gas–particle partitioning. Their mass size distribution is determined by condensation to surfaces, which leads to a maximum in the accumulation mode[113,119] and the number median diameter is in the Aitken mode. This was also confirmed for PAHs.[113,119–121]

Dry and Wet Deposition Contributions

Wet scavenging of hydrophobic organic substances (low Henry coefficients, K^H, listed in Table 1) is expectedly of low efficiency, unless the substance is particle-associated.[122] Atmospheric lifetimes with respect to removal by dry and wet deposition at a Central European site were ranging between <2 weeks and >2 weeks, respectively.[121] Correspondingly, for temperate climates dry deposition of PAHs may dominate over wet deposition, although this has not been substantiated yet. Atmospheric lifetime of PAHs towards removal by dry deposition is dominated by particle deposition.[122,123] However, for individual PAHs the opposite may apply, as was found for anthracene, fluoranthene and pyrene at a suburban site,[28] and is obviously related to air–soil exchange and, hence, to the concentration of PAH in soil subject to volatilization.

Particulate dry deposition fluxes of individual PAHs were measured and mean particle dry deposition velocities ranged between 0.4–10 cm s^{-1}.[28,64,124,125] In samples of deposited particulate matter 3-ring PAHs were found enriched as compared to air samples, i.e. had higher dry deposition velocities than 4- and 5-ring PAHs.[28]

When there is no exchange of material between the particulate and dissolved phases in rain, the total scavenging ratio, W_t, i.e. apparent volume of air washed out by volume of rainwater of a given compound can be expressed by:

$$W_t = 10^3 c_r/c_a = W_g \times (1 - \theta) + W_p \times \theta \qquad (9)$$

where c_r and c_a are the rainwater and near-ground air concentrations in units of ng L^{-1} and ng m^{-3}, respectively.

The gaseous PAH scavenging ratio, W_g, i.e. the apparent volume of air washed out by volume of rainwater, ranges from 5×10^2–5×10^5 in precipitation events.[63,121,126,127] W_g was found to be in good agreement with predictions,[126] suggesting that phase equilibria were established.[128,129] Besides dissolution, adsorption of gaseous molecules onto the surface of rain drops or snow flakes could contribute to the phase transfer process.[128] The particle scavenging ratio W_p is dependent on particle size and chemical surface properties, rain intensity, raindrop diameter and collision efficiency,[129] as it is the carrier's (particle) washout efficiency which is determining the process. Typical values of W_p of PAH are mostly in the range $W_p = 5 \times 10^2$–10^5,[63,121,124,130,131] and even higher in a tropical climate.[127] In conclusion, the particle scavenging contribution to the total scavenging exceeds the gas scavenging contribution. The particle scavenging contribution, $W_p \times \theta/W_t$, accounted for 14–100% in a temperate (Central Europe[121]) and 86–100% in a tropical environment (Singapore[65]). More efficient scavenging for semivolatile PAHs than for non-volatile PAHs found in various studies was attributed to a different mass size distribution.[130]

Snow is considered to be a more efficient scavenger for nonpolar organics in air than rain.[124,132] Efficiency is negatively correlated with the vapour pressure of a substance, indicating that adsorption onto the air–ice interface is the process responsible for vapour scavenging.[133]

Recommendations for Future Work

- Artefact-free (e.g., in situ) method to determine PAH in the gas and particulate phases at remote sites
- Studies of PAH air–soil exchange
- Improved characterisation of the carbonaceous fraction of PM in field and laboratory studies in order to better account for specific sorption of PAHs to these fractions and, hence, better understand PAH's gas–particle partitioning
- Extension of sorption studies to fill the data gaps describing chemical interaction of PAH at the molecular level with regard to relevant aerosol matrix parameters and their humidity dependence (for poly-parameter linear free energy relationships).

Chemical Reactions

Gas Phase PAH Reaction Chemistry

In the troposphere, lower molecular weight (LMW) PAHs exist predominantly in the gas phase due to their relatively high vapour pressure.[134–137] As discussed in Section 3.1.2, wet and dry deposition of gas phase PAHs is unlikely to be a significant atmospheric loss process for these compounds.[121,122,138,139] Additionally, no evidence has been observed for direct photolysis as a sink for gas phase PAHs.[135]

The key sink for PAHs in the atmosphere is chemical reactions with atmospheric oxidants, i.e. OH, NO_3 and O_3.[135,140–142] While it has been suggested that reaction with Cl atoms could also be an important loss process for gas phase PAHs in the marine boundary layer and coastal environments,[143] there is, as yet, no conclusive evidence of this process being significant in the wider environment. Direct reaction with NO_2 or HNO_3 is also not an important atmospheric loss process.[144]

A significant amount of pioneering work investigating the kinetics of atmospheric reactions of PAHs over the past 30 years has been conducted by Atkinson, Arey and co-workers at the University of California, Riverside, USA. Through this research, much insight has been gained into the reaction chemistry (rate coefficients, mechanisms and products) of atmospheric reactions of gas phase 2–4 ring PAHs with OH, NO_3 and O_3 and their relative importance in the removal of PAHs from the atmosphere and conversion to derivative compounds.

Here, we review the reactions of PAHs with OH radicals, NO_3 radicals and O_3 and their transformation into nitro- and oxy-derivative compounds.

Calculated rate coefficients, proposed reaction mechanisms and observed products are discussed. Rate coefficient values and reaction products for the reactions of LWM PAHs with OH, NO_3 and O_3 have previously been reviewed by Atkinson and Arey.[135] Since this review was published, additional research has been undertaken to validate or enhance our understanding of the reaction kinetics of these processes. We therefore provide an updated overview and discussion for specific 2–4 ring PAHs.

Gas Phase Reactions of PAHs with OH Radicals

- Overview: The hydroxyl radical, OH, plays an essential role in atmospheric behavior and transformations of organic compounds.[145–147] For a wide range of organic compounds, reaction with OH represents an important removal mechanism from the atmosphere and is also a potential pathway for the transformation into more toxic compounds.[148]

  In the troposphere, OH radicals can be formed from a series of reactions initiated by the photolysis of ozone in the presence of water vapour:[147,149]

  $O_3 + h\nu\ (\lambda < 319\ nm) \rightarrow O(^1D) + O_2(^1\Delta_g)$

  $O(^1D) + H_2O \rightarrow 2OH$

  $O(^1D) + M \rightarrow O(^3P) + M\ (M = N_2, O_2, CO_2)$

  Other mechanisms involve the reaction of RO_2 with NO,[145] and photolysis of nitrous acid and formaldehyde.[150]

  The average 24 h OH concentration has been estimated to be 1.0×10^6 molecules cm^{-3}, but is generally reported as 2.0×10^6 molecules cm^{-3}, a daytime average adjusted for a 12 h average daytime period.[151–153]

  Due to the nature of its formation and reaction chemistry, OH radical concentrations are dependent on temporal and meteorological factors including time of day, season, latitude, cloud cover, and relative humidity[151,152] as well as concentration of O_3 (a key source compound) and NO_2 (a key sink reactant). For example, the highest OH concentrations are expected in the tropics where humidity and UV light intensity levels will be highest.[152]

- Reaction Mechanisms: Despite the considerable amount of research into the reactions of PAHs with OH and calculations of rate coefficient values for these reactions, the precise mechanisms by which these processes occur are not fully understood. Experimental studies have a limited capability to assess these reaction routes due to the lack of analytical methods to detect the radical intermediate species believed to be involved in these processes.[154] Therefore a combination of experimental and theoretical approaches is required.

The potential mechanism(s) for the reaction of OH radicals with aromatic hydrocarbons have been discussed by Atkinson[147] and Atkinson and Arey[135] and more recently by Atkinson and Arey.[151]

The reaction of PAHs with OH radicals can be initiated via two possible reaction pathways. These are illustrated for the acenaphthene and acenaphthylene in Scheme 1. The proposed pathways involve: (1) OH radical interaction with substituent groups either through H-atom abstraction from C–H groups (Scheme 1a), or, in the e of acenaphthylene which contains an unsaturated cyclopentafused ring, addition to the >C=C< bond of this substituent (Scheme 1b);[135,155] or (2) OH addition to the aromatic ring, forming an initially energy-rich hydroxycyclohexadienyl-type radical intermediate (hereafter referred to as the PAH–OH adduct), which then further reacts with NO_2 or O_2, to form products or thermally decompose back to reactants[135,155] (Scheme 1c).

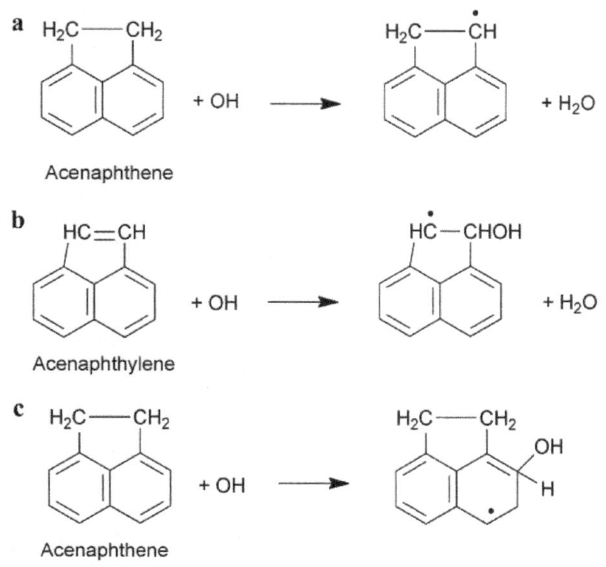

Scheme 1: Mechanism for the reaction of gas-phase PAHs with OH radicals; (a) H-atom abstraction; (b) OH addition to substituent groups; (c) OH addition to the aromatic ring.[135,147]

The relative importance of the two proposed reaction pathways is dependent on the reaction temperature and pressure conditions. While the OH-addition mechanism is expected to dominate at room temperature, at elevated temperatures, H-atom abstraction from the C–H bonds will become of increasing importance as the PAH–OH adduct will be too thermally unstable and will decompose back to the original reactants.[147,151] This shift in the reaction mechanism is indicated by the observation of non-exponential OH radical decay at higher temperature ranges.[147,151]

The dominance of the OH-addition pathway at room temperature has been demonstrated for naphthalene,[156] anthracene[157,158] and phenanthrene.[159,160] These studies have generally reported a negative dependence of the rate constant on temperature. Rate coefficient expressions to fit these experimental data, derived from these studies, are shown in Table 3. A lack of a significant 'isotope effect' i.e. differences in rate coefficients between unsubstituted PAHs and their deuterated equivalent, as noted by Ananthula et al.[157,160] and Lee et al.[159] also indicates a minor contribution of the H-abstraction mechanisms over the temperature ranges studied.

Table 3: Second-order rate coefficients $k^{(2)}$ for gas-phase reactions of PAH with OH radicals

	k_{OH} (cm^{-3} molecules^{-1} s^{-1})	Ref.	*T* (K)	Notes
Nap	2.4×10^{-11}	Phousongphouang and Arey[171]	298 ± 2	RR relative to k(1,2,3-trimethylbenzene) = 3.27×10^{-11} cm^{-3} moleculues^{-1} s^{-1}
	2.2×10^{-11}	Atkinson[147]	298	Recommended value based on previous data, overall uncertainty of ±30%
	2.3×10^{-11}	Brubaker and Hites[140]	298	Measured over the temperature range 306–366 K
	2.7×10^{-11}	Klamt[172]	n/a	Theoretical calculation based on a new molecular orbital based estimation method
	2.4×10^{-11}	Biermann et al.[169]	298 ± 1	RR, relative to k(propene) = 2.63×10^{-11} cm^{-3} molecules^{-1} s^{-1}
	1.9×10^{-11}	Lorenz and Zellner[156]	300	Absolute rate, temperature range 300–873 K, extrapolated using Arrhenius parameter
	2.2×10^{-11}	Klöpffer et al.[170]	300	RR, relative to k(ethene) = 8.44×10^{-12} cm^{-3} molecules^{-1} s^{-1}
	2.4×10^{-11}	Atkinson et al.[168]	294 ± 1	RR, relative to k(n-nonane) = 1.07×10^{-11} cm^{-3} molecules^{-1} s^{-1}
	2.6×10^{-11}	Atkinson and Aschmann[144]	295 ± 1	RR, relative to k(2-methyl-1,3-butadiene) = 1.02×10^{-10} cm^{-3} molecules^{-1} s^{-1}
1M-Nap	4.1×10^{-11}	Phousongphouang and Arey[171]	298 ± 2	RR, relative to k(naphthalene) = 2.39×10^{-11} cm^{-3} molecules^{-1} s^{-1}, derived from the same work
	5.3×10^{-11}	Atkinson and Aschmann[173]	298 ± 2	RR, 2-methyl-1,3-butadiene used as reference compound, T = 298 ± 2
	6.0×10^{-11}	Klamt[172]	n/a	Theoretical calculation based on a new molecular orbital based estimation method
2M-Nap	4.9×10^{-11}	Phousongphouang and Arey[171]	298 ± 2	RR, relative to k(naphthalene) = 2.39×10^{-11} cm^{-3} molecules^{-1} s^{-1}, derived from the same work
	5.2×10^{-11}	Atkinson and Aschmann[144]	295 ± 1	RR, relative to k(2-methyl-1,3-butadiene) = 1.02×10^{-10} cm^{-3} molecules^{-1} s^{-1}
	5.7×10^{-11}	Klamt[172]	n/a	Theoretical calculation based on a new molecular orbital based estimation method
1E-Nap	3.6×10^{-11}	Phousongphouang and Arey[171]	298 ± 2	RR, relative to k(naphthalene) = 2.39×10^{-11} cm^{-3} molecules^{-1} s^{-1}, derived from the same work
2E-Nap	4.0×10^{-11}	Phousongphouang and Arey[171]	298 ± 2	RR, relative to k(naphthalene) = 2.39×10^{-11} cm^{-3} molecules^{-1} s^{-1}, derived from the same work
1,2DM-Nap	6.0×10^{-11}	Phousongphouang and Arey[171]	298 ± 2	RR, relative to k(naphthalene) = 2.39×10^{-11} cm^{-3} molecules^{-1} s^{-1}, derived from the same work
1,3DM-Nap	2.2×10^{-11}	Banceu et al.[175]	295	RR [relative to k(naphthalene) = 2.39×10^{-11} cm^{-3} molecules^{-1} s^{-1}]
	7.5×10^{-11}	Phousongphouang and Arey[171]	298 ± 2	RR, relative to k(naphthalene) = 2.39×10^{-11} cm^{-3} molecules^{-1} s^{-1}, derived from the same work
1,4DM-Nap	5.8×10^{-12}	Klamt[172]	n/a	Theoretical calculation based on a new molecular orbital based estimation method
	5.8×10^{-11}	Phousongphouang and Arey[171]	298 ± 2	RR, relative to k(naphthalene) = 2.39×10^{-11} cm^{-3} molecules^{-1} s^{-1}, derived from the same work
1,5DM-Nap	6.0×10^{-11}	Phousongphouang and Arey[171]	298 ± 2	RR, relative to k(naphthalene) = 2.39×10^{-11} cm^{-3} molecules^{-1} s^{-1}, derived from the same work

1,6DM-Nap	6.3×10^{-11}	Phousongphouang and Arey[171]	298 ± 2	RR, relative to k(naphthalene) = 2.39×10^{-11} cm^{-3} $molecules^{-1}$ s^{-1}, derived from the same work
1,7DM-Nap	6.8×10^{-11}	Phousongphouang and Arey[171]	298 ± 2	RR, relative to k(naphthalene) = 2.39×10^{-11} cm^{-3} $molecules^{-1}$ s^{-1}, derived from the same work
1,8DM-Nap	6.3×10^{-11}	Phousongphouang and Arey[171]	298 ± 2	RR, relative to k(naphthalene) = 2.39×10^{-11} cm^{-3} $molecules^{-1}$ s^{-1}, derived from the same work
2,3DM-Nap	6.2×10^{-11}	Phousongphouang and Arey[171]	298 ± 2	RR, relative to k(naphthalene) = 2.39×10^{-11} cm^{-3} $molecules^{-1}$ s^{-1}, derived from the same work
	1.0×10^{-10}	Klamt[172]	n/a	Theoretical calculation based on a new molecular orbital based estimation method
	7.7×10^{-11}	Atkinson and Aschmann[144]	295 ± 1	RR, relative to k(2-methyl-1,3-butadiene) = 1.02×10^{-10} cm^{-3} $molecules^{-1}$ s^{-1}
2,6DM-Nap	6.7×10^{-11}	Phousongphouang and Arey[171]	298 ± 2	RR, relative to k(naphthalene) = 2.39×10^{-11} cm^{-3} $molecules^{-1}$ s^{-1}, derived from the same work
2,7DM-Nap	6.9×10^{-11}	Phousongphouang and Arey[171]	298 ± 2	RR, relative to k(naphthalene) = 2.39×10^{-11} cm^{-3} $molecules^{-1}$ s^{-1}, derived from the same work
Ace	8.0×10^{-11}	Reisen and Arey[155]	296	RR [relative to k(trans-2-butene) = 6.48×10^{-11} cm^{-3} $molecules^{-1}$ s^{-1}]
	5.8×10^{-11}	Brubaker and Hites[140]	298	Measured over the temperature range 325–365 K
	1.0×10^{-10}	Atkinson and Aschmann[173]	296 ± 1	RR [relative to k(2,3-dimethyl-2-butene) = 1.11×10^{-10} cm^{-3} $molecules^{-1}$ s^{-1}]
	5.8×10^{-11}	Klöpffer et al.[170]	300	RR [relative to k(ethene) = 10^{-12} cm^{-3} $molecules^{-1}$ s^{-1}]
	6.4×10^{-11}	Banceu et al.[175]	295	RR [relative to k(naphthalene) = 2.2×10^{-11} cm^{-3} $molecules^{-1}$ s^{-1}]
	8.0×10^{-11}	Klamt[172]	n/a	Theoretical calculation based on a new molecular orbital based estimation method
Acy	1.2×10^{-10}	Reisen and Arey[155]	296	RR [relative to k(trans-2-butene) = 6.48×10^{-11} cm^{-3} $molecules^{-1}$ s^{-1}]
	1.3×10^{-10}	Banceu et al.[175]	295	RR [relative to k(naphthalene) = 2.2×10^{-11} cm^{-3} $molecules^{1}$ s^{-1}]
	1.1×10^{-10}	Atkinson and Aschmann[173]	296 ± 1	RR [relative to k(2,3-dimethyl-2-butene) = 1.11×10^{-10} cm^{-3} $molecules^{-1}$ s^{-1}]
Fln	1.6×10^{-11}	Kwok et al.[176]	297	Placed on an absolute basis by using k_2(cyclohexane) = 7.47×10^{-11} cm^{-3} $molecules^{-1}$ s^{-1}
	1.3×10^{-11}	Brubaker and Hites[140]	298	Measured over the temperature range 326–366 K
	9.9×10^{-12}	Klamt[172]	n/a	Theoretical calculation based on a new molecular orbital based estimation method
	1.3×10^{-11}	Klöpffer et al.[170]	300	RR [relative to k(ethene) = 7.47×10^{-12} cm^{-3} $molecules^{-1}$ s^{-1}]
Phe	3.4×10^{-11}	Biermann et al.[169]	298 ± 1	RR [relative to k(propene) = $4.85 \times 10^{-12} e^{504/T}$ cm^{-3} $molecules^{-1}$ s^{-1}]
	3.1×10^{-11}	Atkinson[147]	298	Recommended value based on previous data, overall uncertainty of ±30%
	2.6×10^{-11}	Klamt[172]	n/a	Theoretical calculation based on a new molecular orbital based estimation method
	1.6×10^{-11}	Lorenz and Zellner[156]	338	Absolute rate study, measured over a temperature range 338–748 K

	1.3×10^{-11}	Kwok et al.[176]	296	RR [relative to k(propene) = 2.66×10^{-11} cm^{-3} $molecules^{-1}$ s^{-1}]
	2.7×10^{-11}	Brubaker and Hites[140]	298	Measured over the temperature range 346–386 K, extrapolated using Arrhenius parameters
	3.2×10^{-11}	Lee et al.[159]	298	Measured over the temperature range 298–386 K, extrapolated using Arrhenius parameters
	$4.98 \pm 2.96 \times 10^{-6}\ T^{-1.97\pm0.10}$	Ananthula et al.[160]	373–1000 K	Two-parameter expression to best fit experimental data
1M-Phe	2.9×10^{-11}	Lee et al.[159]	298	Measured over the temperature range 363–403 K, extrapolated using Arrhenius parameters
2M-Phe	6.5×10^{-11}	Lee et al.[159]	298	Measured over the temperature range 338–398 K, extrapolated using Arrhenius parameters
3M-Phe	6.6×10^{-11}	Lee et al.[159]	298	Measured over the temperature range 353–388 K, extrapolated using Arrhenius parameters
9M-Phe	7.6×10^{-11}	Lee et al.[159]	298	Measured over the temperature range 333–373 K, extrapolated using Arrhenius parameters
Ant	1.1×10^{-10}	Biermann et al.[169]	325 ± 1	RR [relative to k(propene) = 2.29×10^{-11} cm^{-3} $molecules^{-1}$ s^{-1}]
	1.9×10^{-10}	Brubaker and Hites[140]	298	Measured over the temperature range 346–365 K
	1.3×10^{-11}	Kwok et al.[176]	296	Based on a derived k(anthracene)/k(phenanthrene) value of 1.0 ± 0.5
	2.0×10^{-10}	Klamt[172]	n/a	Theoretical calculation based on a new molecular orbital based estimation method
	1.3×10^{-10}	Atkinson;[147] Biermann et al.[169]	298	Recommended value based on previous data, overall uncertainty of ±30%
	$1.12 \times 10^{-10}\ (T/298)^{-0.46}$	Goulay et al.[158]	58–470	Two-parameter expression to best fit experimental data
	$8.17 \times 10^{-14}\ T^{-8.3}\ e^{(-3171.71/T)}$	Ananthula et al.[157]	373–923	Modified Arrhenius equation to best fit experimental data
	$2.18 \times 10^{-11}\ e^{(-1734.11/T)}$	Ananthula et al.[157]	999–1200	Modified Arrhenius equation to best fit experimental data
Flt	1.1×10^{-11}	Brubaker and Hites[140]	298	Measured over the temperature range 346–366 K
Pyr	5.0×10^{-11}	Atkinson et al.[162]	296 ± 2	RR Relative to k(naphthalene) = 3.6×10^{-28} cm^{-3} $molecules^{-1}$ s^{-1}
1N-Nap	5.4×10^{-11}	Atkinson[147]	298	Recommended value
2N-Nap	5.6×10^{-11}	Atkinson[147]	298	Recommended value

Studies indicate that the OH-addition mechanism is the dominant reaction route up to 773 K for anthracene,[157] 525 K for naphthalene[156] and ~380 K for monocyclic aromatic compounds.[161] It is therefore suggested that an increase in molecular size (number of aromatic rings) could have a positive effect on the OH-addition rate and/or the thermal stability of the PAH–OH adduct, with 3-ring PAHs being more stable at higher temperatures than 2-ring structures. Furthermore, Brubaker and Hites[140] investigated the kinetics of the 4-ring PAH fluoranthene and indicated that this compound would form a more stable OH-adduct than 2- or 3-ring structures.

Ananthula et al.[160] noted from assessment of previous kinetic studies that the rate of H-atom abstraction at elevated temperatures appears to be largely unaffected by the molecular size. This is presumably due to the dependence of reactivity on the number of possible reaction sites, which does not vary significantly between 1-, 2- and 3-ring species. The reason for the apparent dependence of the OH reaction rate upon molecular size at lower (<773 K) temperatures is not currently known.

- Reactions of the PAH–OH Adducts: Reactions of the PAH–OH adduct with NO_2 are expected based on the observation of nitro-PAH compounds in a wide number of ambient measurements with relative concentrations and isomer profiles consistent with their formation from gas phase reactions (as discussed in detail in Section 5).

Indeed, initial investigations into the mechanism(s) of PAH reactions suggested that the PAH–OH adduct will react with NO_2 to form nitro-derivative products[162–164] and would not react with O_2 to a significant degree.

However, kinetic and product data from experimental studies now suggest that for OH–PAH adducts, the reaction with O_2 dominates over NO_2 reactions under atmospheric conditions.[151,165,166] It is suggested that the NO_2 and O_2 reactions with the naphthalene–OH adduct may be of equal importance for NO_2 mixing ratios down to 60 ppbV.[166]

In addition to laboratory studies, the reactions of naphthalene with OH radicals and subsequent reactions of the Nap–OH adduct have been investigated theoretically in studies by Ricca and Bauschlicher[167] and Qu et al.[154] Their investigations of the bond energies of the OH–naphthalene adduct indicate that OH will react with naphthalene at either C_1 or C_2 without an activation energy barrier above the reactants. It was inferred that, after formation of the OH–Nap adduct, decomposition back to reactants is the most favorable pathway.

Qu et al.[154] also carried out theoretical study of the mechanism of the subsequent reactions of the Nap–OH adduct with NO_2 and/or O_2 using molecular orbital calculations. The derived pathways for these reactions are summarized in Scheme 2.

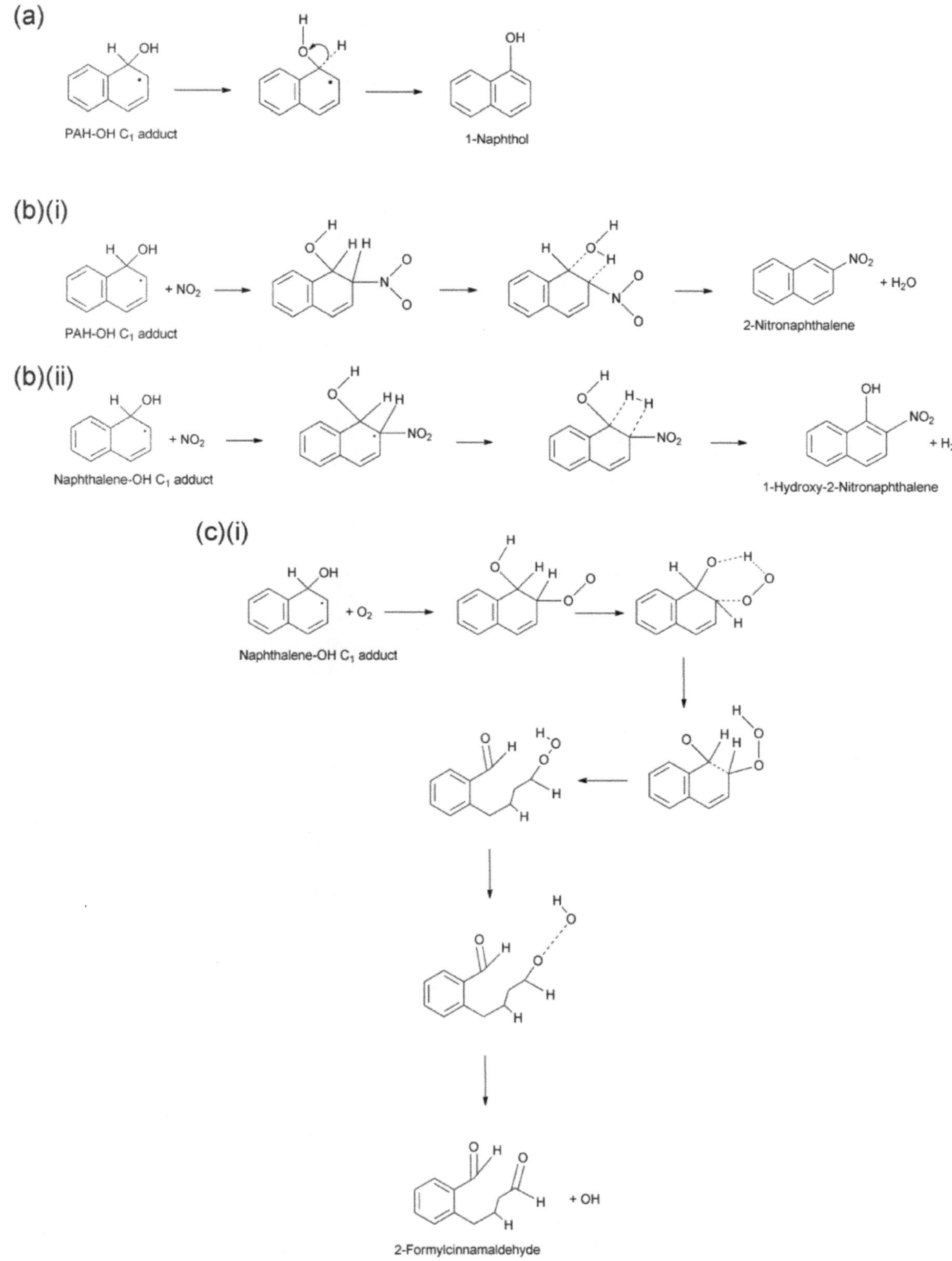

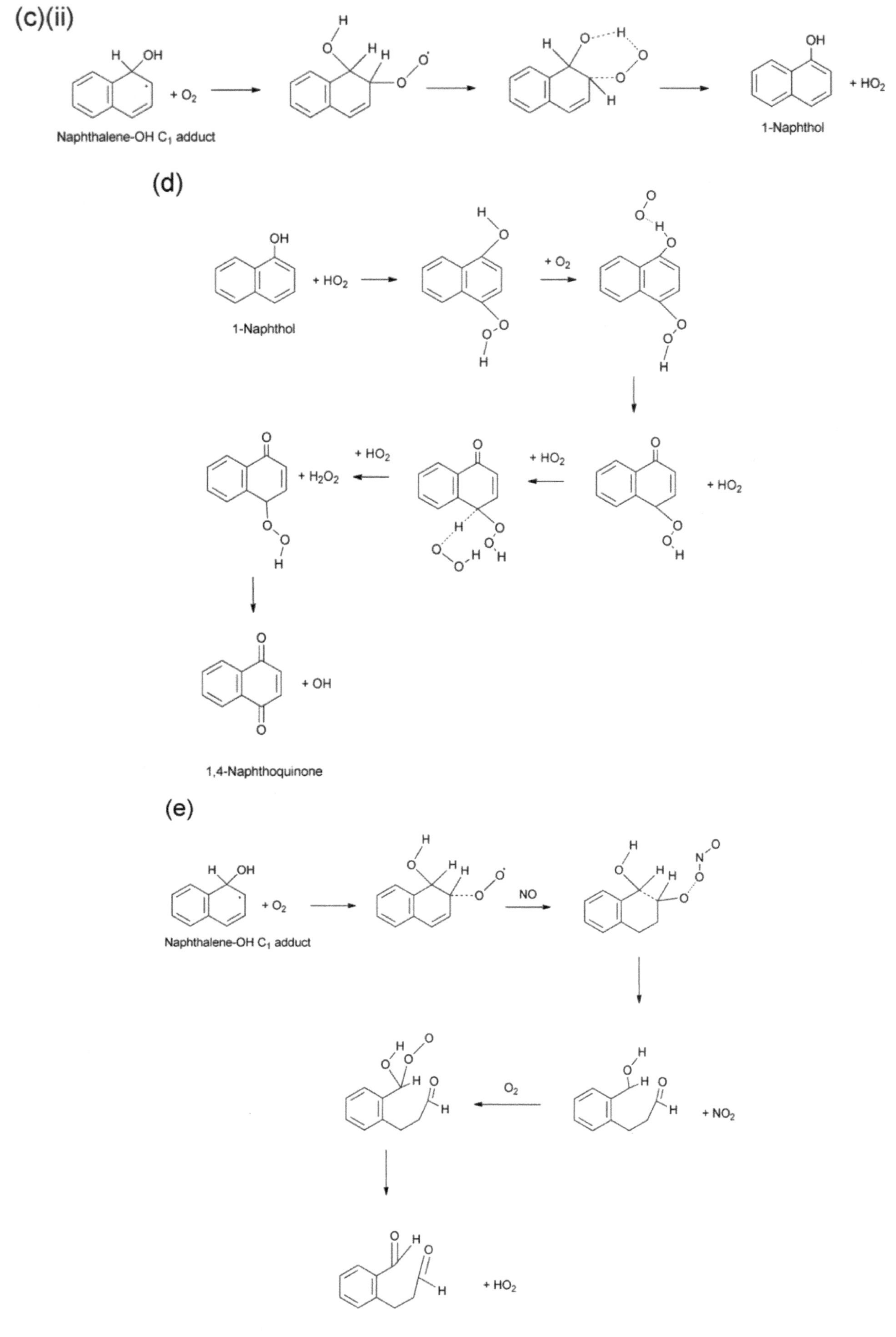

Scheme 2: Proposed mechanisms for the further reaction of the PAH–OH adduct: (a) H-atom abstraction; (b) reactions with NO_2; (c) reaction with O_2; (d) further reaction of 1-naphthol; (e) reaction with NO/O_2.[154]

It was suggested that H-atom abstraction from the OH–naphthalene adduct would primarily yield the closed-ring 1- and 2-napthol isomers, from C_1 and C_2 interaction of OH respectively, which would be more stable than open-shell intermediate species[154,167] (Scheme 2a). It was indicated that the C_1-isomer will be more stable than the C_2-isomer for these hydroxyl products with a ratio of 2 : 1.[167]

It was expected that the C_1 and C_2 naphthalene–OH adduct would undergo a competitive decomposition with O_2 and/or NO_x. The N atom of NO_2 can attack the OH–Nap adduct at either the trans- or cis-position of OH. Attack at the trans-position is expected to yield 1- and 2-nitronaphthalene isomers for the reactions of the C_2 and C_1 adducts respectively (Scheme 2bi), while attack at the cis-position would yield 2-hydroxy-1-nitronaphthalene and 1-hydroxy-2-nitronaphthalene (Scheme 2bii). It was indicated that the nitronaphthalene isomers will be the dominant products formed.

O_2 is expected to attack the C_1- and C_2-adducts forming energy-rich intermediates, which will decompose via different pathways to yield either 2-formylcinnamaldehyde (Scheme 2ci) or 1- or 2-naphthol isomers (Scheme 2cii). It was also indicated that naphthols can further react with HO_2 which yields 1,4-naphthoquinone (1,4-NQu) (Scheme 2d). It was suggested that 1,4-NQu can also be formed from 1-naphthol by further addition of OH to yield a diol intermediate which can rearrange to form 1,4-diketone followed by loss of H_2.[167] Furthermore, it is indicated that the C_1 and C_2 adducts can react with an NO–O_2 mixture to form an additional route to 2-formylcinnamaldehyde (FCA) (Scheme 2e).

- Rate coefficients: Experimental studies have been carried out to derive second-order rate coefficients for the gas phase reaction of OH with a number of individual 2–4 ring PAHs, including naphthalene, alkylnaphthalene derivatives, acenaphthene, acenaphthylene, anthracene, phenanthrene, fluorene, fluoranthene and pyrene. Values of the calculated second-order rate coefficient, $k^{(2)}$, derived in these studies and the experimental conditions used are detailed in Table 3.

While for most studies experimental investigation of $k^{(2)}$ values for PAH reaction with OH has been conducted over relatively narrow temperature ranges, Lorenz and Zellner[156]investigated the rate coefficients for the reactions of OH with naphthalene and phenanthrene over a wide range of elevated temperatures (~300–873 K). Similarly Brubaker and Hites[140]measured rate coefficients for a range of temperatures above ambient levels and then extrapolated to environmentally relevant temperatures using the Arrhenius equation.

Atkinson[147] recommended a $k^{(2)}$ value for the reaction of naphthalene with OH based on a unit-weighted least squares analysis of data from previous studies by Atkinson et al.;[168]Atkinson and Aschmann;[144] Lorenz and Zellner[156] and Biermann et al.[169] The value reported by Klöpffer et al.[170] was not used in this calculation. Subsequent experimental studies by Brubaker and Hites[140] and Phousongphouang and Arey,[171] and a theoretical value derived by Klamt et al.,[172] using molecular orbital calculations, are shown to be in good agreement with this recommended $k^{(2)}$ value.

$k^{(2)}$ values for a range of alkylnaphthalene derivatives (MNs, DMNs and ENs) with OH have been studied experimentally.[144,171,173] The $k^{(2)}$ values for 1M-Nap, 2M-Nap and 2,3DM-Nap calculated by Phousongphouang and Arey[171] and Atkinson and Aschmann[144,173] are shown to be in reasonable agreement, the small difference in measured rate coefficient values being attributed to differences in product quantification methods.[171] These $k^{(2)}$ values are also within a factor 2 of the theoretical value derived by Klamt.[172]

$k^{(2)}$ values derived in different experimental studies for the reaction of acenaphthene with OH are clearly not in agreement. Differences between $k^{(2)}$ reported by Reisen and Arey[155] and Atkinson and Aschmann[174] could be attributed to the use of capillary GC columns in the former and packed column GC techniques in the latter, leading to a more accurate measurement in the later study.[155] The $k^{(2)}$ value reported by Klöpffer et al.[170] may be erroneously low as the reference compound, ethene, reacts substantially slower than acenaphthene.[155] The $k^{(2)}$ value reported by Reisen and Arey[155] is very similar to the theoretical value derived by Klamt.[172] There is generally good agreement between the three $k^{(2)}$ values derived for the OH reaction with acenaphthylene.[155,174,175]

For reactions of anthracene, there is reasonable agreement, between $k^{(2)}$ values calculated theoretically by Klamt[172] and experimentally by Brubaker and Hites.[140]$k^{(2)}$ values measured at 325 K by Biermann et al.[169] and Goulay et al.[158] were in agreement but both lower than that reported by Brubaker and Hites.[140] Kwok et al.[176] reported a much lower $k^{(2)}$ based on ak(Ant)/k(Phe) value of 1.0 ± 0.5 derived from samples taken during the reaction of phenanthrene with OH. It has therefore been suggested that the $k^{(2)}$ value measured by Kwok et al.[176] may be erroneously low.

For the reaction of phenanthrene with OH, Atkinson[147] recommended a $k^{(2)}$ value based on previous studies by Lorenz and Zellner[156] and Biermann et al.[169] This $k^{(2)}$ value is shown to be in good agreement with the subsequent studies by Brubaker and Hites[140] and Lee et al.,[159] both extrapolated to 298 K using the Arrhenius relationship, and slightly higher than the theoretical value derived by Klamt.[172] The $k^{(2)}$ value derived by Kwok et al.[176] is a factor of two lower than that of the other rate coefficient values in Table 3 for a similar temperature.

The $k^{(2)}$ values for the reaction of fluorene with OH from three experimental studies[140,170,177] and a theoretical study[172] do agree well. To our knowledge, the only $k^{(2)}$ measurements for the gas phase reaction of fluoranthene and pyrene with OH have been made by Brubaker and Hites[140] and Atkinson et al.,[163] respectively.

It is clear that $k^{(2)}$ values for gas-phase OH reactions depend strongly on the specific PAH molecular structure. For example, as noted by Brubaker and Hites,[140] Ace has a higher reaction rate than Nap. Both compounds consist of two fused aromatic rings but it is suggested that the reactivity of Ace towards OH is enhanced by H-atom abstraction occurring at the cyclopentafused ring.[155] Similarly, Acy is shown to be more reactive than Ace. This has been attributed to the higher reactivity of the unsaturated cyclopentafused ring of Acy with a more rapid OH addition to the $>C=C<$ than H-atom abstraction from the saturated cyclopenta-fused ring.[155,173,174]

The $k^{(2)}$ values for 1M-Nap and 2M-Nap are shown to be essentially identical suggesting that the mechanism of this reaction with OH will be the same for the two isomers. The presence of alkyl groups has been observed to have an 'activating effect' on OH reactivity for Nap[171] and Phe[159]

(Table 4). In contrast, Banceu et al.[175] reported $k^{(2)}$ values for 1,3DM-Nap and 2,3DM-Nap, which are lower than that of Nap. Furthermore, enhanced reactivity is not observed for 1-methylphenanthrene (1M-Phe). As yet, it is not known why 1M-Phe is not similarly activated.

Table 4: Rate coefficient ratios for the reaction of methylnaphthalenes and methylphenanthrenes with OH radicals, relative to the reactions of their parent compound

Compound	Relative reactivity^a ($k_{1(obs)}/k_{nap(obs)}$)
Nap	1.00
1M-Nap	1.71
2M-Nap	2.02
1E-Nap	1.52
2E-Nap	1.68
1,2DM-Nap	2.49
1,3DM-Nap	3.13
1,4DM-Nap	2.42
1,5DM-Nap	2.51
1,6DM-Nap	2.65
1,7DM-Nap	2.84
1,8DM-Nap	2.62
2,3DM-Nap	2.57
2,6DM-Nap	2.78
2,7DM-Nap	2.87
Compound	Relative reactivity^b ($k_{1(obs)}/k_{phen(obs)}$)
Phe	1.00
1M-Phe	0.90
2M-Phe	2.02
3M-Phe	2.08
9M-Phe	2.36

^{a}From Phousongphouang and Arey.[171] b From Lee et al.[159]

It is indicated that the reactivity of Ant towards OH is much faster than that of other 3-ring PAHs such as Phe and Fln.

The difference between Ant and Fln reactivity could possibly be explained by differences in the reaction mechanism at C_9 which is the most likely position for OH attack for these compounds.[178,179] It is suggested, based on reaction product studies, that the H-abstraction mechanism will dominate at the C_9 position for Fln,[178] while the OH addition mechanism will dominate for reactions of Ant,[140] which could explain the enhanced reactivity of Ant relative to Fln.

Ant and Phe have essentially the same 3-ring structure, only differing by the relative position of their aromatic rings. However, Ant appears to be significantly more reactive.

The most reactive sites for Phe are expected to be at the C_{11} and C_{12} positions, where net charge is greatest. However, product studies show that reaction occurs predominantly at the $C_9$and C_{10} positions.[179–182] This suggests that the reactivity at the C_{11}- and C_{12}-positions could be sterically hindered, resulting in lower overall reactivity. Conversely, Ant is expected to have highest net charge at C_9 and C_{10}, which are sterically unhindered. This is confirmed by product studies which indicate reactions occurring at these sites,[158,180] which possibly explain the relatively enhanced reactivity.

- Reaction Products: Products from the reactions of LMW PAHs with OH radicals in the presence of O_2 and NO_x have been investigated in a number of laboratory studies. More attention in the literature has been paid to the reactions of naphthalene but product investigations have also been carried out for alkylnaphthalenes, acenaphthene, acenaphylene, phenanthrene, anthracene, fluoranthene and pyrene.
- Naphthalene Reaction Products: The gas phase reaction of Nap with OH radicals has been shown to yield both ring-retaining and ring-opened products. Key ring-retaining products identified include naphthols, nitronaphthalenes and naphthoquinones.

1- and 2-naphthol isomers have been identified in product studies[162,181–183] in yields of approximately 3–7%. The formation of naphthol isomers is expected to result from OH–Nap adduct rearomatization (Scheme 2a) or reaction with O_2 and decomposition of the peroxy radical (Scheme 2cii).

1- and 2N-Nap isomers are also formed in these reactions.[162,166,181,183,184] This would result from the pathway shown in Scheme 2b. Nishino et al.[166] measured the formation yields of 1- and 2N-Nap to be 0.35% and 0.60% respectively. The higher yield observed for reaction occurring at C_1 compared to C_2 was also noted by Wang et al.[185] who indicated that 68% of OH addition occurs at C_1. This is in agreement with the theoretical calculations of Qu et al.,[154] who calculated that the C_1 adduct lies 10 kcal mol^{-1} lower in energy than the C_2 adduct.

1,2- and 1,4-NQu isomers have also been observed as reaction products in yields of 1–6%.[181,183,184] This could be expected to occur via the reaction in Scheme 2d. An alternative pathway is suggested by Lee and Lane[181] and Kautzman et al.[184]via 1-hydroxynaphthalene-2-one and 1-hydroxynaphthalene-4-one to yield 1,2-NQu and 1,4-NQu respectively (Scheme 3). It has also been suggested that 1,4-NQu can be formed from the photolysis of 1-nitronaphthalene.[162] Other ring retaining products identified from the reaction of Nap with OH include 2,3-epoxy-naphthoquinone, oxygenated indenes such as inden-1-one and 1,3-indene-dione, benzopyrones and nitro-naphthol isomers.[181]

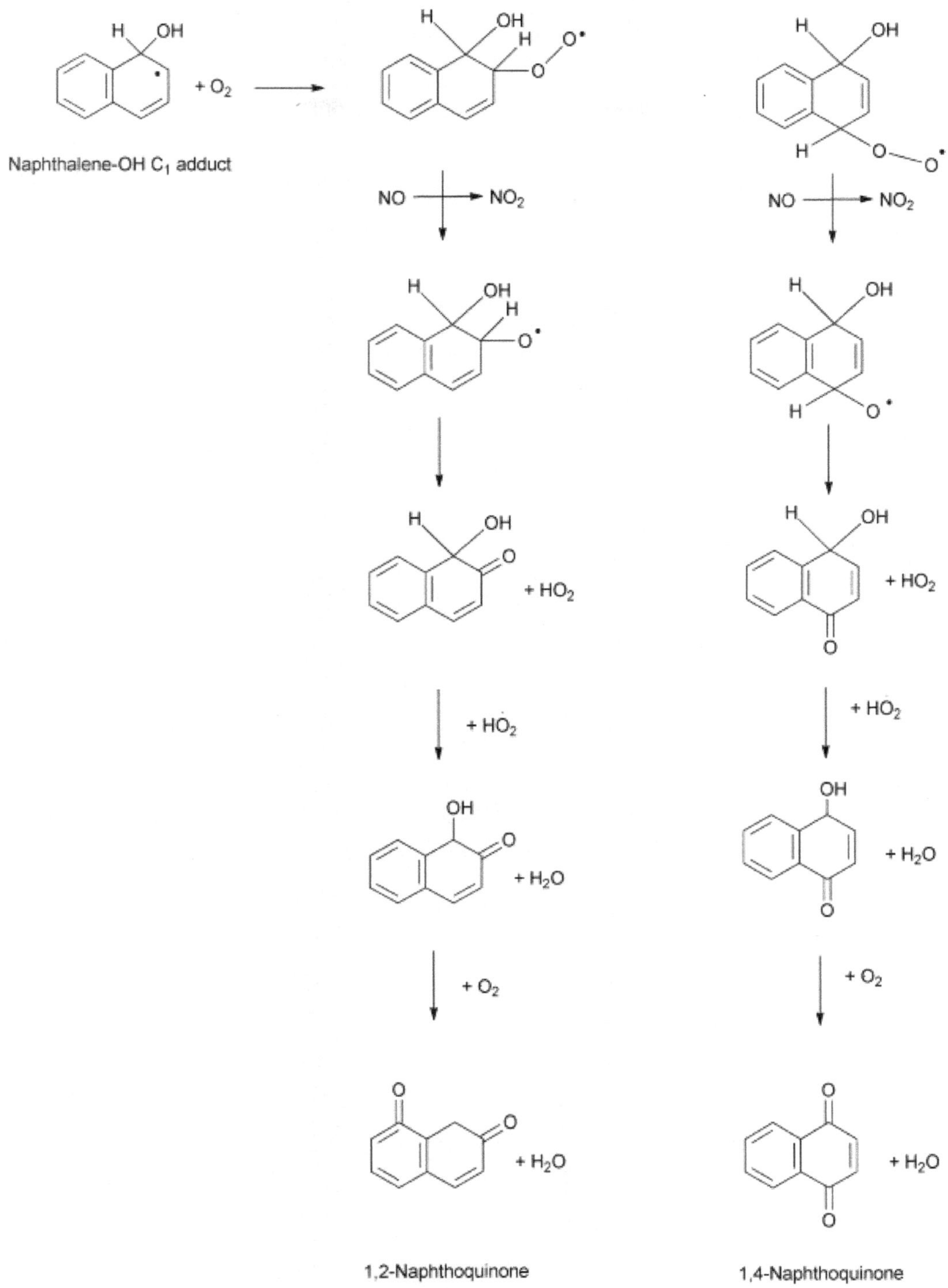

Scheme 3: A proposed alternative reaction pathway for the formation of naphthoquinone isomers.[183]

Sasaki et al.[182] reported that ring-retaining products represented only ~30% of the reaction products from this process. Therefore understanding the ring-opening mechanism is important to fully understand these reactions. In particular, much interest has been paid to the formation and reactivity of the ring-opened species 2-FCA, identified as the main reaction product of naphthalene with OH.[166,181–186]

The product yield of 2-FCA from this reaction was estimated to be 46–71% by Nishino et al.[186] Other studies have reported lower formation yields[182,183] but these did not take the rapid loss of 2-FCA due to photolysis and reactivity with OH into account.[186] The formation of FCA can be attributed to the mechanism shown Scheme 2ci, involving Nap–OH adduct reaction with O_2 to form the peroxy radical intermediate or 2e involving reaction with O_2 and NO, as predicted by Qu et al.[154]

Other ring opened products identified include phthalic acid, phthaldialdehyde and phthalic anhydride.[181,182,184] Kautzman et al.[184] suggested that these were formed from the further reaction of 2-FCA with OH. An alternative reaction pathway was also suggested involving reaction with O_2 to yield C_7 and C_9 compounds such as benzoic acid.

Kautzman et al.[184] investigated the products from the naphthalene reaction with OH under both high (~80 ppbV) and low NO_x conditions. Two reaction pathways were proposed to yield 2-FCA: (i) a series of reactions initiated by the reaction of the Nap–OH adduct with O_2 to yield a peroxy radical intermediate; (ii) reaction of the Nap–OH adduct involving O_2 and NO. These are summarized in Scheme 4. These mechanisms are in agreement with other product studies[182,183] and theoretical calculations.[154] It is not currently possible to establish the relative dominance of either of these proposed routes.

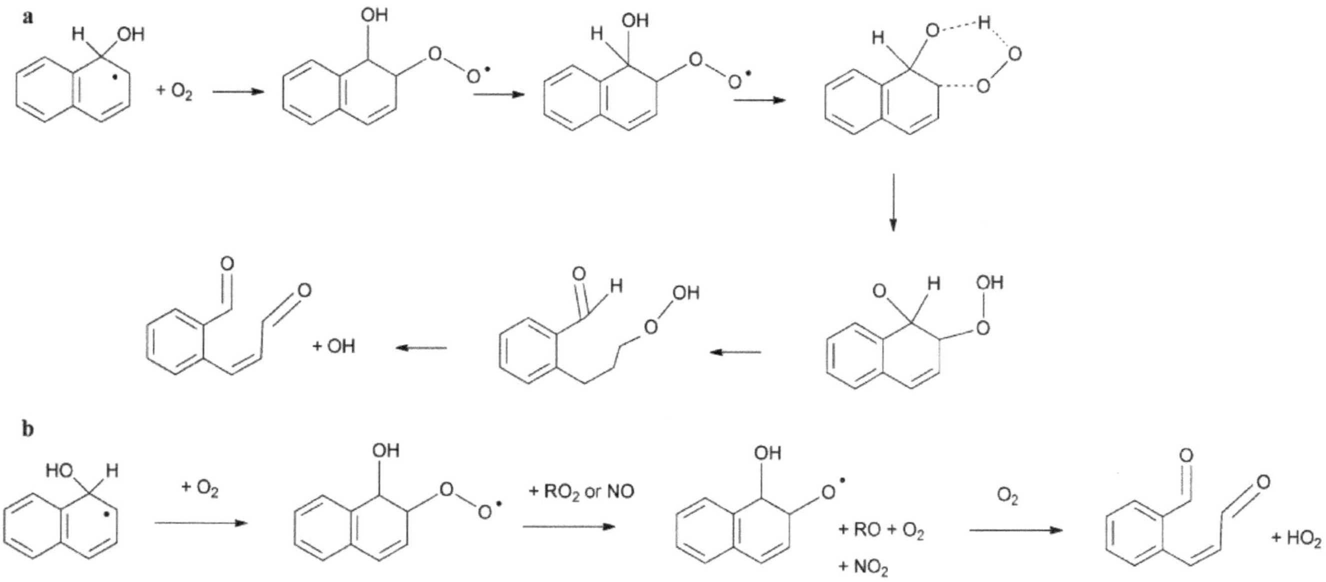

Scheme 4: Summary of pathways to 2-formylcinnamaldehyde from the naphthalene–OH adduct: (a) via reaction with O_2, (b) via reaction with O_2/NO/RO_2.

The presence of organic peroxides in the product study by Kautzman et al.[184] was attributed to an alternative reaction mechanism, initiated similarly by the reaction of the Nap–OH adduct with O_2, but proceeding via the formation of a bicyclic peroxy radical (Scheme 5). This mechanism is suggested to lead to the formation of both observed ring-opened species as well as peroxide or epoxide compounds.

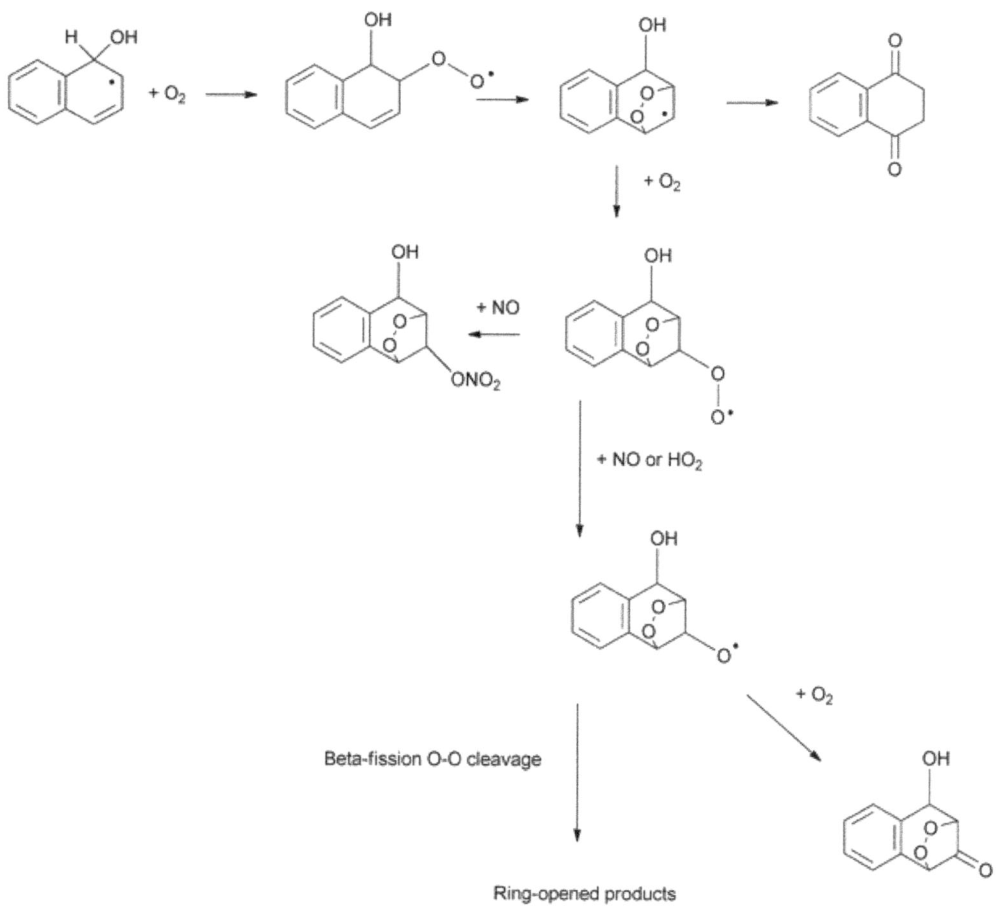

Scheme 5: An alternative reaction scheme for the reaction of naphthalene–PAH adduct proceeding via a bicyclic peroxy radical.[184]

Under low NO_x conditions, Kautzman et al.[184] indicated that the dominant mechanism for the formation of products from naphthalene will differ from that at higher NO_x levels. It was suggested that the reaction is dominated by the alkoxy radical reaction predominantly with HO_2 to form ring-opened products such as 2-FCA. Under these low NO_x conditions, 1,4-NQu is suggested to result from the reaction pathway detailed in Scheme 2d involving the addition of HO_2 to 1-naphthol followed by reaction with O_2 and two further additions of HO_2.[184]

The identified products are broadly consistent with the suggested reaction mechanisms of Qu et al.[154] (as detailed in Scheme 2). The product studies detailed above have allowed the identification of >90% of the products formed from the reaction of naphthalene with OH radicals. This has allowed relatively detailed reaction pathways for these processes to be suggested, despite some discrepancies in the literature. However, obtaining a full mass balance for this reaction is difficult due to the secondary reactions of the PAH reaction products.[183] More work is clearly needed to fully understand these precise reaction mechanisms.

- Reactions of other PAHs: Products identified from the reactions of Phe have included ring-opened species such as 2,2′-diformylbiphenyl, 9-fluorenone (9-Flr), 1-,2-,3-,4-, and 9-phenanthrols, 2-,3-,4- and 9-nitrophenanthenes (N-Phe), 1,4- and 9,10-phenanthraquinones (PQu), dibenzopyranone and 2- and 4-nitrodibenzopyranones, and ring-opened products such as 1,2-naphthalenedicarboxaldehyde and 1,2-naphthalic anhydride.[180,185,187]

Corresponding to the OH addition mechanism, it was suggested that the Phe–OH adduct would be formed via addition of OH to C_1.[182,187] This can then form the OH–phenanthrene–O_2peroxy radical, analogous to the reaction of naphthalene (Scheme 2), from which stable products can be formed through reaction with O_2. Phenanthrols could also be formed through H-atom abstraction from the Phe–OH adduct.[187] It was suggested that ring-opened products such as 1,2-naphthalenedicarboxaldehyde and 1,2-naphthalic anhydride are formed through cleavage of the central ring between C_9–C_{10}, the most active site for electrophilic attack for Phe.[178,187]

Wang et al.[185] suggested that 2,2′-diformylbiphenyl is a primary product in this reaction, which then reacts further to produce 9-fluorenone and dibenzopyranone.[178,179,185] The suggested pathway leading to PQu involved initial formation of a keto alcohol at the C_9–C_{10} bond.[179,185] The calculated formation yield of 9,10-PQu was ~3%.[185] Combining the measured formation yield with the rate coefficient,[159] the authors calculated a PQu formation rate of 80 pg m^{-3} h^{-1}.

Helmig et al.[178] identified products from the reaction of Fln with OH in the presence of NO_x, including hydroxyfluorenes, nitrofluorenes (N-Fln), 9-fluorenone (9-Flr) and nitrofluorenones. The formation of N-Fln isomers is attributed to the OH-addition mechanism analogous to that discussed for Nap and Phen (Scheme 1c and 2bi). The formation of 9-Flr is expected to proceed via H-atom abstraction of the $-CH_2-$ group followed by reaction with O_2.[178] The relatively high fluorenone yield (~9%) compared to nitro-fluorenes (~1%) may indicate a higher importance of the H-atom abstraction mechanism for Fln.

Nitro-PAH isomers have been identified from the gas-phase reactions of Ant, Flth and Pyr with OH.[163,180,188] As with other studies, nitro-PAH product yields are low: 1-nitroAnt (0.2%), 2-nitroAnt (0.2%), 2-nitroPyr (0.5%) and 4-nitroPyr (0.06%) 2-nitroFlt (3%) 7-nitroFlt (1%) and 8-nitroFlt (0.3%).

It is clear that the formation of the observed products from the gas phase reactions of PAHs with OH radicals is initiated by the formation of the PAH–OH adduct. The nature and extent of the subsequent reaction pathway will then dictate the precise products formed. Different pathways have been suggested leading to the formation of both ring-retaining and ring-opened products. The relatively low yield of nitro-PAH products from these processes clearly suggests that the reaction of PAH–OH adducts with NO_2 does not dominate and other mechanisms such as peroxy radical formation and reaction/decomposition are more significant.

In general, formation yields of nitro-PAHs are low, ranging from 0.2 to 5%. It should be noted that, for most PAHs, despite this considerable work, the majority of OH-radical reaction products remain unidentified.[135] While it can be envisaged that reactions of 3- and 4-ring PAHs will proceed in ways analogous to naphthalene, to fully understand the mechanisms of these processes, the products from these reactions need to be investigated in more detail.

Gas Phase Reactions of PAHs with NO_3 Radicals

- Overview: The nitrate radical, NO_3, plays an important role in tropospheric chemistry and is thought to be a key contributor to the oxidizing capacity towards VOCs in the lower atmosphere.[189–191]

This has been demonstrated through investigations of NO_3 reactions with alkanes,[192–194] organosulphur compounds,[192] aldehydes[192,195–197] and hydroxy-substituted aromatics.[198] These reactions may have further impacts on tropospheric reactions through the generation of peroxy radicals.[192,199,200] A detailed overview of NO_3 radical chemistry is provided by Geyer et al.[191,201] but a brief discussion is provided here.

The levels of NO_3 in the troposphere are controlled by the complex interplay of the reactions that lead to its formation or removal. In the troposphere, NO_3 radicals are formed via the sequential reactions of NO and NO_2 with O_3:[202–205]

$$NO + O_3 \rightarrow NO_2 + O_2$$

$$NO_2 + O_3 \rightarrow NO_3 + O_2$$

NO_3 can be removed from the troposphere via photolysis by solar radiation.[204–206]

$$NO_3 + h\nu \rightarrow NO + O_2$$

$$NO_3 + h\nu \rightarrow NO_2 + O(^3P)$$

NO_3 reaction with NO will also lead to its removal from the troposphere:[207]

$$NO_3 + NO \rightarrow 2NO_2$$

NO_3 radicals can also be removed from the troposphere via reaction with NO_2 to form N_2O_5, which can subsequently thermally decompose to NO_3:[202,208]

$$NO_3 + NO_2 + M \leftrightarrow N_2O_5 + M \ (M = N_2, O_2)$$

The reaction rates for this process are shown to be the same for M = O_2 and M = N_2.[209] It is also suggested that N_2O_5 can be removed by hydrolysis to form HNO_3, which leads to the removal of NO_3 from the atmosphere:[189]

$$N_2O_5 + H_2O \rightarrow 2HNO_3$$

Because of the rapid photolysis of NO_3 and reaction with NO and O_3 (as well as the reaction of NO with O_3) the concentrations of the NO_3 radical in the ambient troposphere will be low during daytime hours.[210] Concentrations of NO_3 will therefore be present in the troposphere at significant levels only during evening and night time hours when photolysis is absent and NO levels are low,[135,151,203] except close to major sources of emissions.

Levels of NO_3 in the troposphere will therefore be subject to substantial temporal and spatial variability,[151] ranging from $<5 \times 10^7$ molecules cm^{-3} to 1×10^{10} molecules cm^{-3}.[204,210] Atkinson[210] has suggested an average 12 h night time NO_3 radical concentration in the lower troposphere of $\sim 5 \times 10^8$ molecules cm^{-3} (~20 pptV) over continental areas. This value is expected to be significantly lower in marine environments, with mixing ratios of ~0.25 pptV measured at 3 km altitude in Mauna Loa, Hawaii,[211] due to lower NO_2 concentrations.[194,211]

- NO_3 reaction mechanism: N_2O_5, NO_2 and NO_3 are effectively in equilibrium in the lower troposphere.[203]

Initial kinetic studies in mixtures of N_2O_5–NO_3–NO_2 suggested that these reactions are kinetically equivalent to the reaction with N_2O_5 alone.[162,173,174,212] However, in a study by Atkinson et al.,[203] 2,3-dimethyl-2-butene was added to the reaction mixture of N_2O_5–NO_3–NO_2–N_2–O_2 and Nap to scavenge NO_3 radicals. The removal of NO_3 was shown to reduce the amount of naphthalene reacted by a factor of up to >100 indicating that Nap does not undergo elementary reaction with N_2O_5 to a significant degree.

It is therefore suggested that the reaction of PAHs in N_2O_5–NO_3–NO_2 mixtures involves the addition of NO_3, formed from the thermal decomposition of N_2O_5, to the aromatic ring, in a reaction analogous to the reaction of OH radicals with PAH. The reaction scheme for this process is represented in Scheme 6. The reaction proceeds via initial addition of NO_3 to the aromatic ring to form a nitratocyclohexadienyl-type radical (**6a**) (the PAH–NO_3 adduct), which, similarly to the PAH–OH adduct can decompose back to reactants (**6b**), reacts with NO_2(**6c**) or O_2 (**6d**) to form products[135,151,210] or can undergo unimolecular decomposition (**6e**) that would lead to the formation of hydroxy-PAH products and further reaction to form nitro-hydroxy-PAHs.[213]

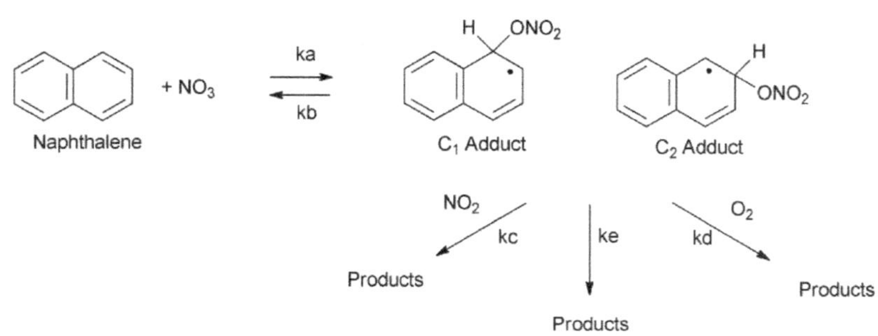

Scheme 6: Potential pathways for the reaction of PAHs with NO_3.[151]

NO_3 radicals can also interact with the substituent groups of compounds such as Ace, Acy and M-Nap[210,214] (Scheme 7). For Ace, this involves H-atom abstraction from the C–H bond of the cyclopenta-fused ring and the formation of HNO_3.[174,180] For Acy, NO_3 addition to the >C=C< bond of the cyclopenta-fused ring is considered to be the dominant reaction pathway.[174,180] However, for unsubstituted PAHs, the NO_3-addition pathway is expected to dominate.[210] It is thought that, for PAHs containing substituent groups, reaction will not result in nitro-PAH formation.[180,210]

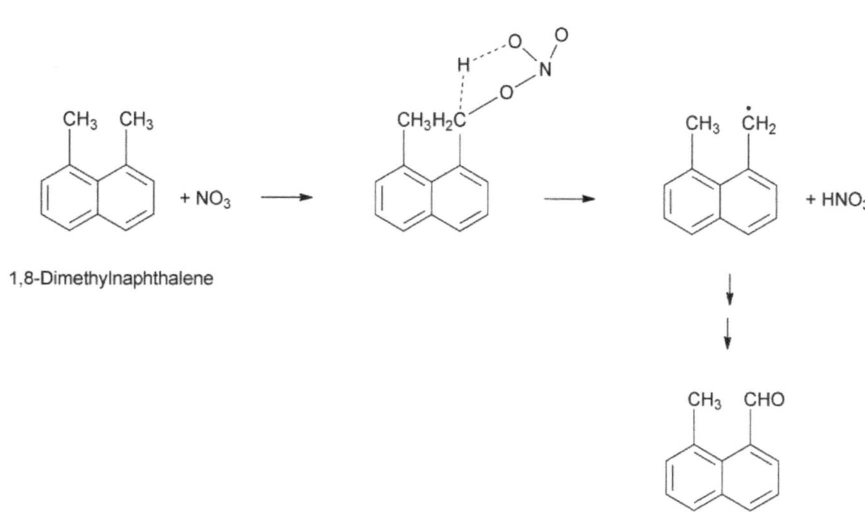

Scheme 7: H-atom abstraction mechanism for the reaction of a methyl-substituted PAH with NO_3.[214]

The measured rate coefficient for this reaction will therefore be:

$$k_{obs} = k_a(k_c[NO_2] + k_d[O_2] + k_e)/(k_b + k_c[NO_2] + k_d[O_2] + k_e)$$

Assuming, as indicated by experimental data,[162,203,210,212–215] under conditions used in reaction studies and in the ambient troposphere:

$$k_b \gg (k_c[NO_2]) \text{ and } k_c[NO_2] > (k_d[O_2] + k_e)$$

this equation simplifies to:

$$k_{obs} = k_a(k_c[NO_2])/k_b$$

The observed rate coefficient will therefore be proportional to NO_2 concentration. The decomposition rate coefficient of the Nap–OH adduct was calculated to be 5×10^5 s^{-1}. The PAH–NO_3 adduct is therefore more stable towards decomposition compared to the monocyclic-NO_3 adduct (decomposition rate coefficient of 3×10^8 s^{-1} at 298 K). However, the NO_3 adduct is estimated to be 46 kJ mol^{-1} less stable towards decomposition than the OH-adduct.[210]

It is also suggested from kinetic data that reaction of the NO_3–PAH adduct with NO_2 will dominate over the reaction with O_2 under atmospherically relevant conditions. Atkinson et al.[213] measured the rate coefficient of the NO_3 reaction with Nap as a function of NO_2 and O_2 concentration. An upper limit of $k_d/k_c < 4 \times 10^{-7}$ was obtained at 298 ± 2 K. Therefore the reaction of the Nap–NO_3 adduct with NO_2 is expected to dominate relative to the reaction with O_2 down to a NO_2 mixing ratio of at least 80 ppbV (2×10^{12} molecules cm^{-3}) and possibly much lower.[151]

Qu et al.[216] used high-level molecular orbital theory to predict NO_3-initiated PAH reactions theoretically. It was shown that the Nap–NO_3 adduct is formed by the attack of the O atom of NO_3 at either the C_1 or C_2 positions. The proposed further reactions of the PAH–NO_3 adduct are illustrated in Scheme 8. Two energetically favorable primary reaction pathways are proposed for the Nap–NO_3 adduct(s) formed In the first pathway, loss of NO_2 from the adduct leads to the formation of 1,2-epoxynaphthalene or 2,3-epoxynaphthalene for C_1 and C_2adducts respectively (Scheme 8a). In the second pathway, the isomerization of the NO_3–naphthalene adduct followed by unimolecular decomposition leads to the formation of 2-FCA for the C_1 adduct or 2-acetyl-cinnamaldehyde for the C_2 adduct (Scheme 8b).

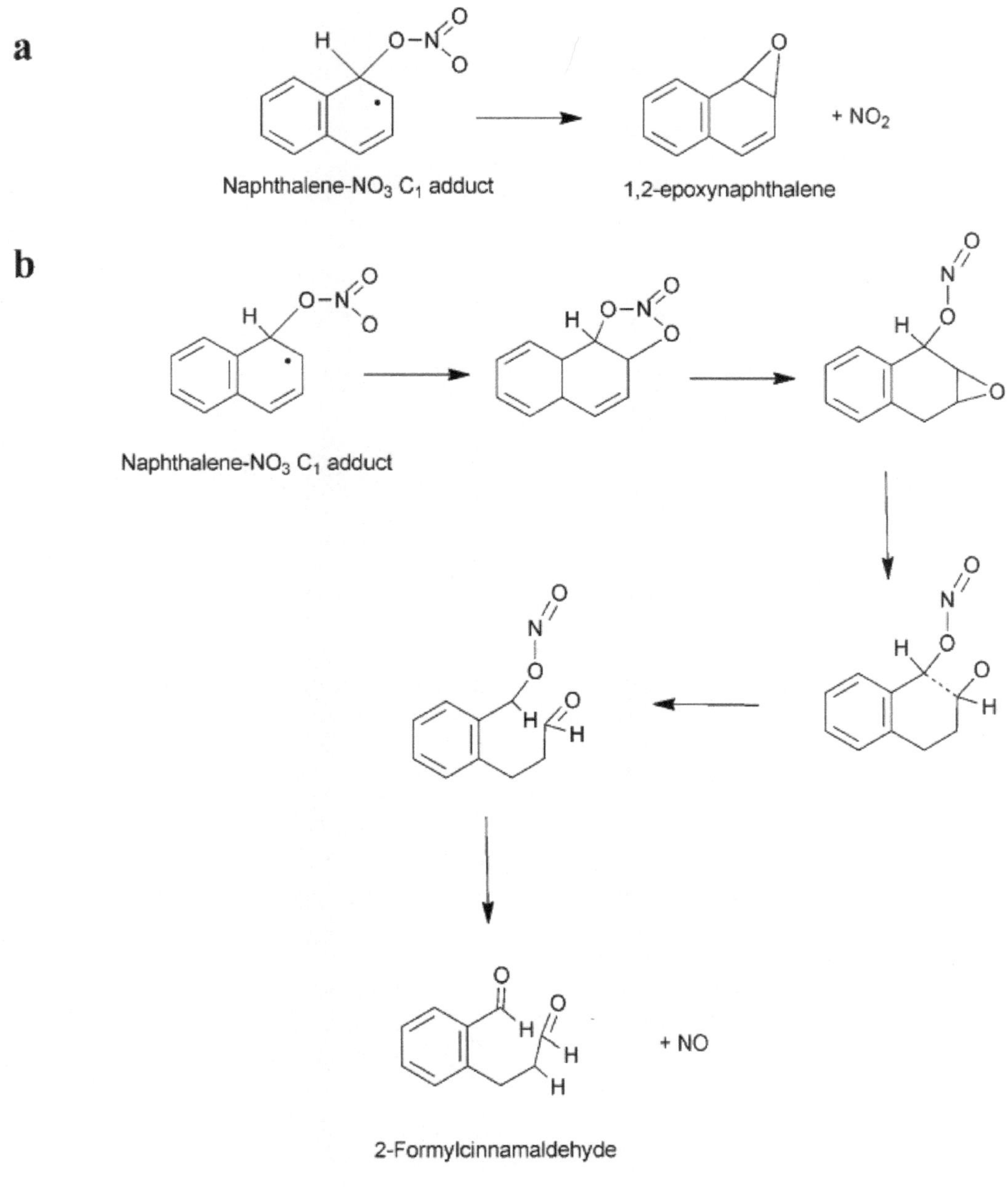

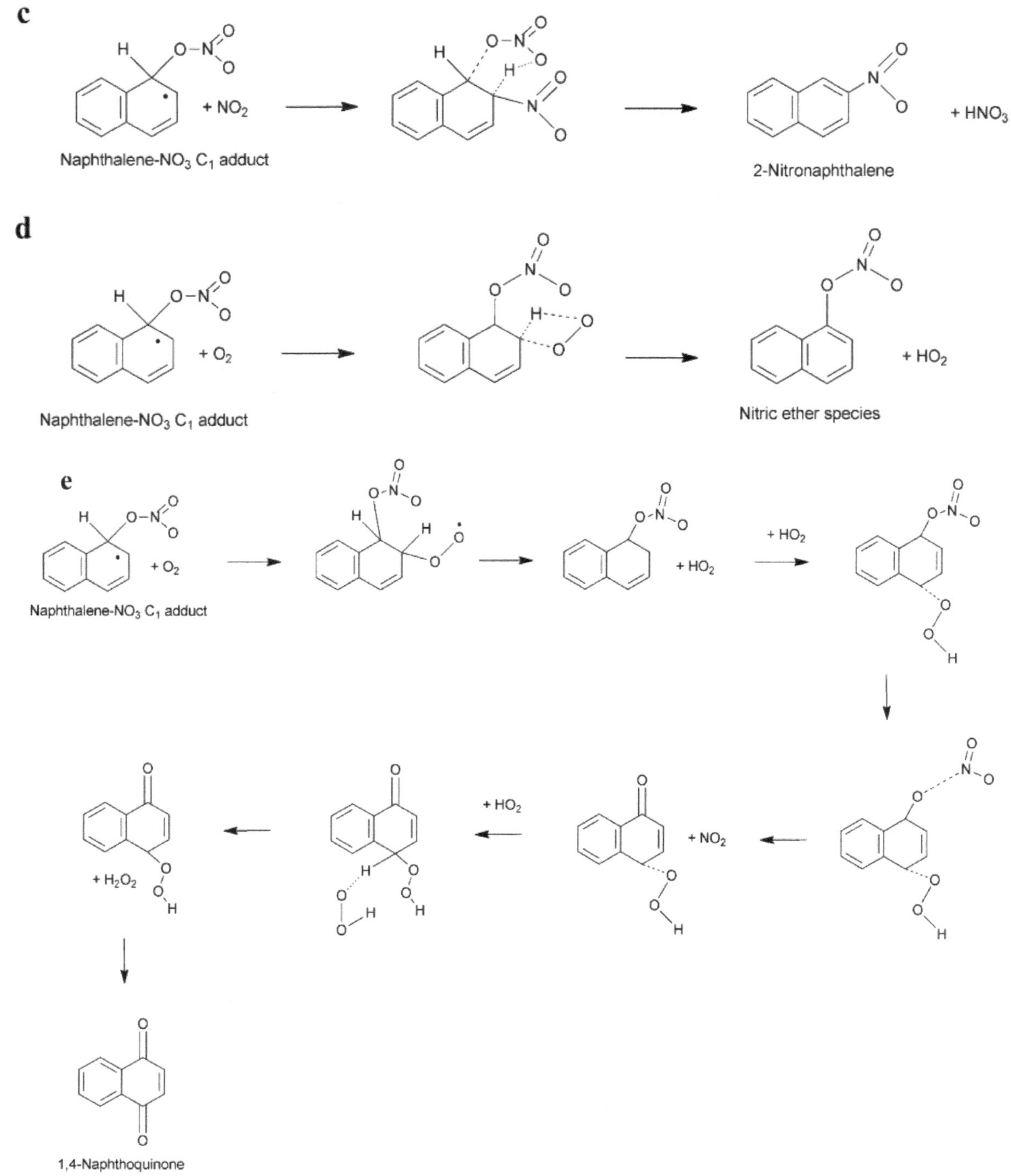

Scheme 8: Proposed reaction pathways for the further reaction of the PAH–NO_3 adducts: (a) unimolecular decomposition; (b) isomerisation followed by unimolecular decomposition; (c) reaction with NO_2; (d) reaction with O_2; (e) reaction with O_2–HO_2.[216]

It was also indicated that the Nap–NO_3 adduct will also undergo secondary reactions with NO_2, O_2 or O_2/HO_2 and these reaction pathways will be strongly competitive with the unimolecular decomposition pathway. Mechanisms for the proposed reactions are shown in Scheme 8c–e respectively. The energetically favorable reaction with NO_2 is expected to produce 2-nitronaphthalene and 1-nitronaphthalene isomers and HNO_3 for the reactions of C_1- and C_2-adducts respectively, while the reaction with O_2 is expected to produce a nitric ether species. Both of these pathways are expected to proceed via the formation and decomposition of an energy-rich intermediate.[216] The reaction of the Nap–NO_3 adduct with O_2/HO_2 is expected to produce 1,4-NQu. (Scheme 8e).

- Rate coefficients: Relatively fewer rate coefficient studies have been carried out for gas-phase PAH reactions with NO_3 than for the corresponding reactions with OH radicals. Measured or derived $k^{(2)}$ values for the gas phase reactions of PAHs with NO_3 radicals, along with the reaction conditions are presented in Table 5.

Table 5: Second-order rate coefficients $k^{(2)}$ for gas-phase reactions of PAH with the NO_3 radical

	k_{NO3} (cm^{-3} molecules^{-1} s^{-1}) (× [NO_2])	**k_{NO3} (s^{-1}) [NO_2] = 6.91 × 10^{11}molecules cm^{-3} ^{a}**	**Ref.**	**T(K)**	**Notes**
Nap	8.5×10^{-28}	1.1×10^{-16}	Pitts et al.[212]	298 ± 2	RR relative to $K_5(NO_3 + NO_2 - N_2O_5) = 3.41 \times 10^{-11}$ cm^{-3} molecules^{-1} s^{-1}
	4.8×10^{-28}	6.2×10^{-17}	Atkinson et al.[162]	298 ± 2	RR relative to $K_5(NO_3 + NO_2 - N_2O_5) = 3.41 \times 10^{-11}$ cm^{-3} molecules^{-1} s^{-1}
	3.3×10^{-28}	4.3×10^{-17}	Atkinson and Aschmann[174]	296 ± 2	RR relative to k(propene) = 9.45×10^{-15} cm^{-3} molecules^{-1} s^{-1}
	3.7×10^{-28}	4.7×10^{-17}	Atkinson et al.[163]	~297	RR relative to k(thioprene) = 9.93×10^{-14} cm^{-3} molecules^{-1} s^{-1}, measured over temp. range 272–297 K
	4.2×10^{-28}	5.5×10^{-17}	Atkinson et al.[163]	~297	RR relative to $K_5(NO_3 + NO_2 - N_2O_5) = 1.26 \times 10^{-27} e^{11275/T}$ cm^{-3} molecules^{-1} s^{-1}, measured over temp. range 272–297 K
	3.6×10^{-28}	4.6×10^{-17}	Atkinson[210]	298	Recommended value
1M-Nap	8.4×10^{-28}	1.1×10^{-16}	Atkinson and Aschmann[173]	298 ± 2	RR relative to k(naphthalene) = 3.6×10^{-28} cm^{-3} molecules^{-1} s^{-1}
	7.0×10^{-28}	9.0×10^{-17}	Atkinson and Aschmann[174]	296 ± 2	RR relative to k(trans-2-butene) = 3.89×10^{-13} cm^{-3} molecules^{-1} s^{-1}
	7.7×10^{-28}	9.9×10^{-17}	Atkinson[210]	298	Recommended value
	7.2×10^{-28}	9.2×10^{-17}	Phousongphouang and Arey[214]	298 ± 2	RR relative to k(naphthalene) = 3.65×10^{-28} cm^{-3} molecules^{-1} s^{-1}, derived from the same work
2M-Nap	1.1×10^{-27}	1.4×10^{-16}	Atkinson and Aschmann[173]	298 ± 2	RR relative to k(naphthalene) = 3.6×10^{-28} cm^{-3} molecules^{-1} s^{-1}
	1.1×10^{-27}	1.4×10^{-16}	Atkinson and Aschmann[174]	296 ± 2	RR relative to k(propene) = 9.45×10^{-15} cm^{-3} molecules^{-1} s^{-1}
	1.1×10^{-27}	1.4×10^{-16}	Atkinson[210]	298	Recommended value
	1.0×10^{-27}	1.3×10^{-16}	Phousongphouang and Arey[214]	298 ± 2	RR relative to k(naphthalene) = 3.65×10^{-28} cm^{-3} molecules^{-1} s^{-1}, derived from the same work
1E-Nap	9.8×10^{-28}	1.3×10^{-16}	Phousongphouang and Arey[214]	298 ± 2	RR relative to k(naphthalene) = 3.65×10^{-28} cm^{-3} molecules^{-1} s^{-1}, derived from the same work
2E-Nap	8.0×10^{-28}	1.0×10^{-16}	Phousongphouang and Arey[214]	298 ± 2	RR relative to k(naphthalene) = 3.65×10^{-28} cm^{-3} molecules^{-1} s^{-1}, derived from the same work
1,2DM-Nap	6.4×10^{-27}	8.3×10^{-16}	Phousongphouang and Arey[214]	298 ± 2	RR relative to k(2,7-DMN) = 21×10^{-28} cm^{-3} molecules^{-1} s^{-1}, derived from the same work
1,3DM-Nap	2.1×10^{-27}	2.7×10^{-16}	Phousongphouang and Arey[214]	298 ± 2	RR relative to k(naphthalene) = 3.65×10^{-28} cm^{-3} molecules^{-1} s^{-1}, derived from the same work
1,4DM-Nap	1.3×10^{-27}	1.7×10^{-16}	Phousongphouang and Arey[214]	298 ± 2	RR relative to k(naphthalene) = 3.65×10^{-28} cm^{-3} molecules^{-1} s^{-1}, derived from the same work

1,5DM-Nap	1.4×10^{-27}	1.8×10^{-16}	Phousongphouang and Arey[214]	298 ± 2	RR relative to k(naphthalene) = 3.65×10^{-28} cm^{-3} molecules^{-1} s^{-1}, derived from the same work
1,6DM-Nap	1.7×10^{-27}	2.1×10^{-16}	Phousongphouang and Arey[214]	298 ± 2	RR relative to k(naphthalene) = 3.65×10^{-28} cm^{-3} molecules^{-1} s^{-1}, derived from the same work
1,7DM-Nap	1.4×10^{-27}	1.7×10^{-16}	Phousongphouang and Arey[214]	298 ± 2	RR relative to k(naphthalene) = 3.65×10^{-28} cm^{-3} molecules^{-1} s^{-1}, derived from the same work
1,8DM-Nap	2.1×10^{-26}	2.7×10^{-15}	Phousongphouang and Arey[214]	298 ± 2	RR relative to k(2,7-DMN) = 21×10^{-28} cm^{-3} molecules^{-1} s^{-1}, derived from the same work
2,3DM-Nap	1.5×10^{-28}	1.9×10^{-17}	Atkinson and Aschmann[173]	298 ± 2	RR relative to k(naphthalene) = 3.6×10^{-28} cm^{-3} molecules^{-1} s^{-1}
	1.6×10^{-27}	2.1×10^{-16}	Atkinson and Aschmann[174]	296 ± 2	RR relative to k(propene) = 9.45×10^{-15} cm^{-3} molecules^{-1} s^{-1}
	1.6×10^{-27}	2.0×10^{-16}	Atkinson[210]	298	Recommended value
	1.5×10^{-27}	2.0×10^{-16}	Phousongphouang and Arey[214]	298 ± 2	RR relative to k(naphthalene) = 3.65×10^{-28} cm^{-3} molecules^{-1} s^{-1}, derived from the same work
2,6DM-Nap	2.1×10^{-27}	2.7×10^{-16}	Phousongphouang and Arey[214]	298 ± 2	RR relative to k(naphthalene) = 3.65×10^{-28} cm^{-3} molecules^{-1} s^{-1}, derived from the same work
2,7DM-Nap	2.1×10^{-27}	2.7×10^{-16}	Phousongphouang and Arey[214]	298 ± 2	RR relative to k(naphthalene) = 3.65×10^{-28} cm^{-3} molecules^{-1} s^{-1}
Ace		4.6×10^{-13} [b]	Atkinson and Aschmann[174]	296 ± 2	RR relative to k(trans-2-butene) = 3.89×10^{-13} cm^{-3} molecules^{-1} s^{-1}
	1.7×10^{-27}	2.1×10^{-16}	Atkinson and Aschmann[174]	296 ± 2	RR relative to k(trans-2-butene) = 3.89×10^{-13} cm^{-3} molecules^{-1} s^{-1}
Acy		5.5×10^{-12} [b]	Atkinson and Aschmann[174]	296 ± 2	RR relative to k(trans-2-butene) = 3.89×10^{-13} cm^{-3} molecules^{-1} s^{-1}
Fln		3.5×10^{-14} [c]	Kwok et al.[177]	297 ± 2	RR relative to k(1-butene) = 1.19×10^{-14} cm^{-3} molecules^{-1} s^{-1}
Phe		1.2×10^{-13} [d]	Kwok et al.[176]	296 ± 2	RR relative to k(1-butene) = 1.19×10^{-14} cm^{-3} molecules^{-1} s^{-1}
Flt	5.1×10^{-28}	6.6×10^{-17}	Atkinson et al.[163]	296 ± 2	RR relative to k(naphthalene) = 3.6×10^{-28} cm^{-3} molecules^{-1} s^{-1}
Pyr	1.6×10^{-27}	2.1×10^{-16}	Atkinson et al.[163]	296 ± 2	RR relative to k(naphthalene) = 3.6×10^{-28} cm^{-3} molecules^{-1} s^{-1}
1N-Nap	3.0×10^{-28}	3.9×10^{-17}	Atkinson[210]	298	Recommended value
2N-Nap	2.7×10^{-28}	3.5×10^{-17}	Atkinson[210]	298	Recommended value

[a] $[NO_2] = 6.91 \times 10^{11}$ molecules cm^{-3}; annual average, Harwell, UK (2011).

[b] $[NO_2] = <1.2 \times 10^{15}$ molecules cm^{-3}.

[c] $[NO_2] = (7.2–24) \times 10^{13}$ molecules cm^{-3}.

[d] $[NO_2] = (4.8–24) \times 10^{13}$ molecules cm^{-3}.

Atkinson[210] derived a $k^{(2)}$ value for the reaction of naphthalene with NO_3 based on a unit-weighted least squares analysis of three $k^{(2)}$ values from previous studies.[163,173,174] It was also shown that the $k^{(2)}$ value for the NO_3 reaction with naphthalene is lower than that of naphthalene-d_8 ($k_D/k_H \sim 1.2$),[203] indicating that there is very little deuterium isotope effect, suggesting that the H-atom abstraction mechanism is of minor importance.

Atkinson[210] recommended $k^{(2)}$ values for 1M-Nap, 2M-Nap and 2,3DM-Nap using unit-weighted averages of the previous values provided by Atkinson and Aschmann.[173,174] $k^{(2)}$ values measured by Phousongphouang and Arey[214] for 1M-Nap, 2M-Nap, and 2,3DM-Naps are in very good agreement with values recommended by Atkinson.[210] It is indicated that the $k^{(2)}$ for the reaction of naphthalene and alkyl-naphthalenes with NO_3 radicals is linearly dependent on the NO_2 mixing ratio over the range ~0.4–50 ppmV.[174,210,213] This is therefore consistent with the initial formation of the NO_3–alkylnaphthalene adduct.

It is indicated that, as for OH radical reactions, alkyl substitution enhances PAH reactivity towards NO_3 following the general order of reactivity DM-Naps > E-Naps > M-Naps > Nap.[214] The 'enhancement factor' reported for alkylnaphthalene compounds towards NO_3 generally parallels those for the reactions of OH, with relative enhancement ratios (NO_3/OH) of between 1.1 and 2.1 for most compounds. The exceptions are 1,2DM-Nap and 1,8DM-Nap, where the relative enhancement is 7.0 and 22 respectively. While potential explanations, including the steric influence of the methyl group positioning and the possible decrease of ring aromaticity, have been suggested, the precise reason for this enhanced reactivity is not currently known.

Atkinson[210] noted that the rate coefficient for Ace is similar to that of 2,3DM-Nap, suggesting that the cyclopenta-fused ring in Ace acts like two methyl substituent groups. As mentioned above, Ace is expected to react via H-atom abstraction from the $–CH_2–$ group(s) of the cyclopenta-fused ring in addition to the NO_3 addition mechanism.

To our knowledge, the only $k^{(2)}$ value derived for the gas phase reaction of NO_3 with fluorene was that reported by Kwok et al.[177] In this study, no evidence of an NO_2-dependent $k^{(2)}$ was observed, suggesting that the NO_3-addition mechanism does not dominate.[210] Indeed, it was suggested that the reaction proceeds via H-atom abstraction from the $–CH_2–$ group(s). $k^{(2)}$ values for the gas phase reactions of Flt and Pyr with NO_3 radicals have only been studied in one investigation.[163] For both compounds it was suggested that the reaction proceeds via initial addition of NO_3 to the aromatic ring, as detailed above.

NO_3 reactivity with Phe was studied by Kwok et al.[215] A linear relationship between $k^{(2)}$ and NO_2 concentration indicates that the reaction proceeds by the initial addition of NO_3 to the aromatic ring, forming the Phe–NO_3 adduct. Kinetic data[215] also suggested that an additional mechanism occurs, independent of NO_2 concentration. This may involve NO_3 addition to the C_9–C_{10} bond[215] or unimolecular decomposition of the NO_3–phenanthrene adduct.[216]

The $k^{(2)}$ values in Table 5 indicate that the gas phase reactions of LMW PAHs with NO_3 radicals proceed significantly slower (1–3 orders of magnitude) than the corresponding reactions with OH radicals. This would suggest that OH radical reactions represent a much more important loss process for PAHs than reactions with NO_3. The relative scarcity of rate coefficient data makes observations regarding the differences in NO_3 reactivity between individual PAHs more difficult. As with the OH radical reaction, it appears that Acy reacts faster than Ace and Nap, which can be attributed to the additional H-atom abstraction from the $>C=C<$ bond of the cyclopenta-fused ring. To our knowledge, no rate coefficient data are available for the reaction of NO_3 with Ant so no comparison can be made between this and the other 3 ring compounds.

- Reaction products: Reaction of PAHs through the more dominant NO_3 addition mechanism followed by preferential addition of NO_2 is expected to yield nitro-PAH isomers, while reactions involving NO_3 interaction with substituent groups through H-atom abstraction, for example Ace and Acy, and Fln, are not. As with the reactions involving OH radicals, the NO_3 radical product studies have focused primarily on the reactions of naphthalene with a smaller number of studies investigating other 3–4 ring compounds.

As with OH reactions, product studies show that the NO_3 initiated reaction of Nap forms the ring-retaining products naphthols, nitronaphthalenes and naphthoquinones. 1- and 2-nitronaphthalenes have been identified as reaction products in yields of 15–24% and 7–11% respectively.[163,173,182,212] The reaction mechanism is consistent with the expected NO_3 addition at the C_1 or C_2 of Nap to form the Nap–OH adduct, followed by addition of NO_2 (Scheme 8c). The formation yield ratio for 1N-Nap/2N-Nap observed in these studies is in good agreement, and is shown to be ~2.2.

Other lower yield products identified[182] include 1,4-NQu (1.9%), 1-hydroxy-2-nitronaphthalene (1.5%) and 2-hydroxy-1-nitronaphthalene (1.3%). The observed products are broadly consistent with the reaction mechanism detailed in Scheme 8. The formation of 1,4-NQu could be explained by pathways involving either (i) reaction of the Nap–NO_3 adduct with O_2/HO_2(Scheme 8e) or (ii) reaction of naphthols. It should be noted that the epoxynaphthalene and nitric ether species predicted by Qu et al.[216] have not been identified in product studies.

Atkinson et al.[213] identified naphthols and nitronaphthols. It was shown that the ratio of nitronaphthols/nitronaphthalenes was a factor of 1.6 higher when there was no initial NO_2addition compared to a reaction where NO_2 was initially present. This could indicate that the unimolecular decomposition of the PAH–NO_3 adduct (e) could be competitive with the assumed dominant reaction with NO_2.

Wang et al.[217] found that (with the exception of 1,4DM-Nap) the yields of EN-Naps and DM-Naps over the range of NO_2 mixing ratios of 0.2–2 ppmV are not linearly dependent upon NO_2 concentration and that the EN-Nap/DMN-Nap profile did not change significantly with varying NO_2 concentration over this range. It was shown that, as the NO_2 concentration increased by a factor of 8, the EN-Nap and DMN-Nap formation yield only increased by a factor of 3. This suggests that the OH-addition mechanism may not dominate for these compounds.

However, Zielinska et al.[218] reported that nitro-PAH products were formed from the gas phase reaction of 1M-Nap and 2M-Nap with NO_3 in yields of ~30%, approximately two orders of magnitude higher than the corresponding reaction initiated by OH. The MNN isomer distribution from the reaction of 1M-Nap was shown to be 1M3N-Nap > 1M5N-Nap > 1M4N-Nap > 1M8N-Nap ~ 1M6N-Nap > 1M7N-Nap > 1M2N-Nap. For the reaction with 2M-Nap the isomer distribution was 2M4N-Nap > 2M1N-Nap ~ 2M5N-Nap > 2M8N-Nap ~ 2M3N-Nap ~ 2M7N-Nap ~ 2M6N-Nap.

Wang et al.[219] found that the isomers with the highest mass formation rates are 2,7DM4NN, 1,2DM4NN, 2,6DM4N-Nap,

2,6DM1N-Nap, 1,6DM8N-Nap, 2,7DM1N-Nap, 1,6DM5N-Nap, 2E4N-Nap. The formation of these products was attributed to the addition of NO_3 to an alpha-carbon next to the $-CH_3$ group (in the e of the formation of 2,7DM4N-Nap, 2,6DM4N-Nap, 1,6DM8N-Nap, 2E4N-Nap), to an alpha-carbon on the methyl-substituted ring (in the e of 2,7DM1N-Nap, 2,6DM1N-Nap and 1,6DM5N-Nap) or ipso addition at C_1 (in the e of 1,2DM4N-Nap), followed by the addition of NO_2 at the para-position and removal of HNO_3. This proposed mechanism is therefore consistent with the NO_3 addition to the aromatic ring. It was also indicated that these reactions also yield ring-opened aromatic carbonyl (MW 170) products and alkylnaphthoquinones.[217]

For the gas phase reaction of Acy with NO_3, Atkinson and Aschmann[174] observed negligible levels of N-Acy isomers with the major products observed being oxygenated. Therefore a reaction pathway, possibly involving NO_3 addition to the C =C bond of the cyclopenta-fused ring, is proposed.[173,174] Arey et al.[180] observed the main reaction product of the reaction of Ace and NO_3 to be 4-NAce with a yield of ~40% with smaller amounts of the 3- and 5-isomers (~2%), in agreement with the NO_3 addition mechanism of to the aromatic ring and subsequent reaction with NO_2.

The gas phase reaction of Fln with NO_3 was shown to yield 9-fluorenone,[177] which is consistent with the observation that this reaction proceeds via initial H-atom abstraction from a $-CH_2-$ group of Fln, rather than through initial NO_3 addition to the aromatic ring.

For Ant, Phe and Pyr, nitro-PAH products are noted in low formation yields (<2%).[163,174,180] The formation of 2-nitroFlt from the reaction of Flt with N_2O_5 in CCl_4 and in the gas phase has been observed with a relatively high yield (24%).[163,220] The observation of these nitro-PAH products is in agreement with the expected NO_3 addition mechanism with further reaction with NO_2 (Scheme 8c).

Wang et al.[221] observed the presence of 9-Flr, dibenzopyranone, 9,10-PQu and 2,2′-diformylbiphenyl from the gas phase reaction of Phe with NO_3. The reaction was suggested to proceed analogously to that of the OH-initiated reaction, with NO_3 addition at the C_9–C_{10}-position expected. The calculated formation yield of 9,10-PQu was ~33%. Combining the measured formation yield with the room temperature rate coefficient,[176] the authors calculated a PQu formation rate of 800 pg m^{-3} h^{-1}, suggesting that nighttime NO_3-initiated reaction can contribute significantly to the atmospheric levels of PQu.[221]

The above studies indicate that for a number of PAHs (e.g. naphthalene, alkylnaphthalenes, acenaphthene, fluoranthene) the yield of nitro-PAH products from the NO_3-initiated reaction is notably higher than that of the corresponding reaction initiated by OH radicals, suggesting the Nap–NO_3 adduct reacts more preferentially with NO_2 than the Nap–OH adduct. This would suggest that, while NO_3 reactions may represent a less important atmospheric loss process for gas phase PAHs, these reactions are a more important source of nitro-PAH products. This indicates therefore that while during daylight hours the formation of PAH-derivatives will result from OH-initiated reactions, nighttime formation of these compounds from NO_3-initiated reactions will also be significant.

Gas Phase Reactions of PAHs with O_3

Ozone (O_3) forms part of a complex reaction cycle with volatile organic compounds, for example alkenes, alkanes, aromatic hydrocarbons, aldehydes and ketones, and other atmospheric gases such as O_2 and NO_x, which will lead to the formation of radical species, which influence the composition and oxidizing potential of the atmosphere.[148,222,223] In addition, direct reaction with ozone can be an important atmospheric loss process for volatile organic chemicals.[148,223]

The available $k^{(2)}$ values for the reactions of ozone with gas phase PAHs are shown in Table 6. Experimental studies have reported no loss, or very small loss of PAH during exposure studies with relatively high (~4 × 10^{13} molecules cm^{-3}) O_3 concentrations.[144,168,173,177] These studies therefore report estimated upper limits for $k^{(2)}$ values considerably lower (5–8 orders of magnitude) than the corresponding reactions with OH or NO_3.

Table 6: Rate coefficients reported for gas-phase reactions of PAH with ozone

	k_{O3} (cm^{-3} molecules^{-1} s^{-1})	Ref.	T (K)	Notes
Nap	<2.0 × 10^{-19}	Atkinson et al.[168]	294 ± 1	Upper limit
	<3.0 × 10^{-19}	Atkinson and Aschmann[144]	295 ± 1	Upper limit
1M-Nap	<1.3 × 10^{-19}	Atkinson and Aschmann[173]	298 ± 2	Upper limit
2M-Nap	<3.0 × 10^{-19}	Atkinson and Aschmann[144]	295 ± 1	Upper limit
	<4.0 × 10^{-19}	Atkinson and Aschmann[173]	295 ± 2	Upper limit
Ace	<5.0 × 10^{-19}	Atkinson and Aschmann[174]	296 ± 2	Upper limit
Acy	5.5 × 10^{-16}	Atkinson and Aschmann[174]	296 ± 2	
	1.6 × 10^{-16}	Reisen and Arey[155]	296 ± 2	RR, relative to k(2-methyl-2-butadiene) = 3.96 × 10^{-16} cm^{-3} molecules^{-3} s^{-1}
2,3DM-Nap	<4.0 × 10^{-19}	Atkinson and Aschmann[144]	295 ± 1	Upper limit
	<2.0 × 10^{-19}	Kwok et al.[177]	297 ± 2	Upper limit

Phe	4.0×10^{-19}	Kwok et al.[176]	296 ± 2	
1N-Nap	$<6.0 \times 10^{-19}$	Atkinson[207]	298 ± 2	Upper limit
2N-Nap	$<6.0 \times 10^{-19}$	Atkinson[207]	298 ± 2	Upper limit

Kwok et al.[176] did observe the reaction of phenanthrene with $(8.7–15.6) \times 10^{13}$ molecules cm^{-3} of O_3 and derived a $k^{(2)}$ value for this reaction, which was shown to be of the same order of magnitude as the upper limits calculated for other PAHs in previous studies. These observations therefore indicate that reactions of gas phase PAHs with O_3 will be negligible under atmospheric conditions and for most PAHs the reaction with ozone is too slow to be a significant degradation process or route to the formation of derivative products.

The relatively slow reaction of PAHs indicates that O_3 does not react to any observable extent with the aromatic ring or with saturated side-chain groups. However, it should be noted that the reaction of acenaphthylene with O_3 proceeds significantly (~3 orders of magnitude) faster than other PAHs. This is attributed to ozone reacting with the $>C=C<$ bond of the unsaturated cyclopenta-fused ring.[155,174]

Reisen and Arey[155] conducted a product study for the reaction of acenaphthylene with ozone. API-MS analysis identified a single secondary ozonide product. GC-MS analysis identified six other products: 1-naphthaldehyde, oxaacenaphthylen-2-one, 2-hydroxy-1-naphthaldehyde, 1,8-naphthalic anhydride (NtA) and two dialdehyde isomers (MW 184). These products were attributed to breakdown products of the secondary ozonide. A more comprehensive and quantitative product study is required to fully understand the reaction mechanism of this process.

Wang et al.[221] observed the presence of 9-fluorenone (9-Fln), dibenzopyranone, 9,10-phenanthrequinone (PQu) and 2,2′-diformylbiphenyl from the gas phase reaction of phenanthrene with O_3. The authors suggested that the mechanism would involve addition of O at the $C_9–C_{10}$ bond and rearrangement to a secondary ozonide. The calculated formation yield of 9,10-PQu was ~2%. Combining the measured formation yield with the room temperature rate coefficient,[176] the authors calculated a PQu formation rate of 0.2 pg m^{-3} h^{-1}, suggesting that reaction with O_3 is not a significant source of PQu in the atmosphere, compared with OH and NO_3 radical induced reactions.

Summary and Concluding Remarks

KEY CONCLUSIONS

- The main sink for gas phase PAHs appears to be reaction with the OH radical, with rate coefficients for these reactions shown to be considerably (up to 5 orders of magnitude) higher than the corresponding reactions with NO_3 for most 3–4 ring PAHs.
- For most PAHs, the dominant gas phase reaction mechanism involves addition of the OH or NO_3 radical to the aromatic ring to form an energy-rich adduct. Further reaction of this adduct with NO_2 will yield nitro-PAH products while adduct reaction with O_2 will form a peroxy-radical intermediate which can further react or decompose via a number of different pathways to yield ring-retaining products such as quinones, or a wide range of ring-opened products.
- Product studies suggest, for a number of PAHs, that the reaction is expected to proceed, at least in part due to the H-atom abstraction mechanism from C–H groups from the aromatic ring (fluorene), substituted alkyl groups (methyl- and ethyl-naphthalenes) or radical interaction with the $>C=C<$ bond of the cyclopenta-fused ring (acenaphthylene).
- While NO_3 reactions appear to be less significant than OH reactions as a PAH degradation process, considerably higher nitro-PAH yields suggest that night-time reactions of PAHs with NO_3 may be a significant contributor of these compounds in the atmosphere, in addition to day-time OH reactions.
- Reactions of PAHs with O_3 are considered to be of negligible importance in the atmosphere due to a slow reaction with the aromatic ring. The reaction of acenaphthylene is shown to be faster due to interaction of ozone with the $>C=C<$ bond of the cyclopenta-fused ring.
- While reasonable agreement is shown between rate coefficients for the reactions of OH, NO_3 and O_3 derived in different studies (absolute, relative, theoretical) for individual PAHs, discrepancies can often be explained by differences in reaction conditions and quantification methods used.
- There is shown to be considerable variation in reactivity with OH between individual PAH molecules. Reactivity is dependent upon the precise reaction mechanism and can be influenced by steric hindrance of reactive sites and the number and type of substituted groups. Alkyl groups are shown to have an 'activating effect' compared to the reactivity of 2- and 3-ring PAHs.
- The mechanisms of gas phase PAH reactions are still not fully understood. For most PAHs, the majority of reaction products remain unidentified and the precise reaction routes of these processes are still largely unclear. The applicability of observations from experimental studies in the ambient atmosphere is also uncertain (discussed in Section 5).
- Recommendations for future work: The following are areas we suggest for continued research in the area of gas phase PAH reactivity:
- The rates and products of the reactions of Ant with NO_3 and O_3 and of most (predominantly gaseous) 4-ring PAHs with OH, NO_3 and O_3 should be studied.
- A more comprehensive elucidation of the reaction products of the reactions of many PAHs with OH and NO_3 in order to gain a clearer understanding of the reaction mechanisms.
- Verifications of reaction rates based on atmospheric observations.

Heterogeneous Reactions of PAHs.

- Overview: Volatile and semi-volatile organic compounds associated with particulate matter can be influenced by heterogeneous photochemical reactions in the atmosphere.[224,225] Semi-volatile PAHs are shown to have a high affinity for soot and particulate matter.[96,104] Since ambient air sampling studies have revealed that PAHs and their derivative compounds are found to be associated with atmospheric particles[137,226–229] and given the wide range of particulate matter (soot, sands, mineral dusts, metal oxides, diesel exhausts, pollens) present in the atmosphere, it is important to understand the extent and nature of heterogeneous photochemical reactions that can lead to removal of PAHs from the atmosphere and/or their conversion to derivative compounds.

In the troposphere, direct photolysis of PAHs associated with solid surfaces is a potentially important process.[1,230,231] The chemistry of direct photolytic PAH decomposition has been discussed elsewhere for PAHs associated with organic aerosols,[232–238] inorganic substrates[239–248] and the air–ice interface.[249,250] Therefore, photolysis of surface-bound PAH will not be covered in this review. Here, we will discuss the present understanding of particle-associated PAHs with OH radicals, NO_3 radicals, NO_2 and O_3.

Heterogeneous PAH reactivity towards atmospheric oxidants has previously been reviewed by Van Cauwenberghe and Van Vaeck[230] and Van Cauwenberghe,[251] who provided an overview and extensive literature review of atmospheric reactions of PAH on various surfaces including primary combustion particles and ambient particulate matter with O_3 and NO_2. Finlayson-Pitts and Pitts[1] also provide an in depth discussion of PAH reactions with O_3 and NO_2 on particle surfaces. Here we provide an updated overview of heterogeneous PAH reactivity, incorporating more recent kinetic studies.

Heterogeneous PAH Reactions with OH Radicals.

- Overview: Heterogeneous reactions of OH radicals with PAHs adsorbed on different solid surfaces have been reported in a number of experimental studies.[252–257] $k^{(2)}$ values derived for OH reactions with PAHs associated with different substrate types are presented in Table 7.

Table 7: Second-order rate coefficients $k^{(2)}$ for heterogeneous reactions of PAH with OH, NO_2, O_3 and NO_3

PAH	OH reactions			NO_2 reactions			O_3 reactions			NO_3 reactions		
	k_{OH} ($cm^3 molec^{-1} s^{-1}$)	Ref.	Notes	k_{NO2} ($cm^3 molec^{-1} s^{-1}$)	Ref.	Notes	k_{O3} ($cm^3 molec^{-1} s^{-1}$)	Ref.	Notes	k_{NO3} ($cm^3 molec^{-1} s^{-1}$)	Ref.	Notes
Nap							$0.9 \times 10^{-18}/(1/c_{O3} + 10^{-15})$	Kahan et al.[296]	Octanol			
Phe	5.0×10^{-12}	Estève et al.[255]	Graphite particles	2.8×10^{-19}	Perraudin et al.[276]	Silica particles	2.4×10^{-17}	Perraudin et al.[298]	Graphite particles			
	3.2×10^{-13}	Estève et al.[256]	Diesel exhaust particles SRM 1650a	3.4×10^{-18}	Nguyen et al.[275]	Kerosene flame soot	2.3×10^{-17}	Perraudin et al.[298]	Silica particles			
				3.5×10^{-17}	Estève et al.[255]	Graphite particles	$2.3 \times 10^{-19}/(1/c_{O3} + 4.6 \times 10^{-16})$	Kahan et al.[296]	Octanol			
				1.1×10^{-17}	Estève et al.[256]	Diesel exhaust particles SRM 1650a	$1.0 \times 10^{-18}/(1/c_{O3} + 1.4 \times 10^{-16})$	Kahan et al.[296]	Ice			
				2.3×10^{-21}	Butler and Crossley[270]	Ethylene flame soot						
Ant	4.4×10^{-12}	Estève et al.[256]	Graphite particles	1.0×10^{-16}	Perraudin et al.[276]	Silica particles	9.8×10^{-17}	Perraudin et al.[298]	Graphite particles			
				3.4×10^{-18}	Nguyen et al.[275]	Kerosene flame soot	1.4×10^{-16}	Perraudin et al.[298]	Silica particles			
				6.9×10^{-17}	Estève et al.[255]	Graphite particles	$5.1 \times 10^{-18}/(1/c_{O3} + 1.96 \times 10^{-15})$	Kwamena et al.[292]	Octanol			
				1.0×10^{-16}	Ma et al.[278]	Silica particles	$1.3 \times 10^{-16}/(1/c_{O3} + 2.2 \times 10^{-15})$	Kwamena et al.[301]	Azelaic acid (wet)			
				5.3×10^{-17}	Ma et al.[278]	MgO particles	$1.0 \times 10^{-15}/(1/c_{O3} + 10^{-13})$	Kwamena et al.[301]	Phenylsiloxane oil			
							$1.2 \times 10^{-18}/(1/c_{O3} + 1.2 \times 10^{-18})$	Mmereki et al.[294]	Water			
Flt	1.4×10^{-14}	Bedjanianet al.[252]	Kerosene soot	3.2×10^{-21}	Perraudin et al.[276]	Silica particles	1.9×10^{-17}	Perraudin et al.[298]	Graphite particles			

	3.2×10^{-12}	Estève et al.[255]	Graphite particles	1.0×10^{-19}	Nguyen et al.[275]	Kerosene flame soot	1.5×10^{-17}	Perraudin et al.[298]	Silica particles			
	3.8×10^{-13}	Estève et al.[256]	Diesel exhaust particles SRM 1650a	2.9×10^{-17}	Estève et al.[255]	Graphite particles						
				1.0×10^{-17}	Estève et al.[256]	Diesel exhaust particles SRM 1650a						
				2.5×10^{-21}	Butler and Crossley[270]	Ethylene flame soot						
Pyr	1.6×10^{-14}	Bedjanianet al.[252]	Kerosene flame soot	2.0×10^{-17}	Perraudin et al.[276]	Silica particles	2.5×10^{-17}	Perraudin et al.[298]	Graphite particles	6.4×10^{-12}	Liuet al.[260]	Azelaic acid particles
	3.2×10^{-12}	Estève et al.[255]	Graphite particles	1.0×10^{-19}	Nguyen et al.[275]	Kerosene flame soot	5.9×10^{-17}	Perraudin et al.[298]	Silica particles			
	4.1×10^{-13}	Estève et al.[256]	Diesel exhaust particles SRM 1650a	5.1×10^{-17}	Estève et al.[255]	Graphite particles	9.3×10^{-17}	Miet et al.[297]	Silica particles			
	2.4×10^{-13}	Miet et al.[257]	Silica particles	1.5×10^{-17}	Estève et al.[256]	Diesel exhaust particles SRM 1650a	$2.2 \times 10^{-19}/(1/c_{O3} + 3.1 \times 10^{-16})$	Kahan et al.[296]	Octanol			
				4.8×10^{-21}	Butler and Crossley[270]	Ethylene flame soot						
				3.2×10^{-16}	Miet et al.[277]	Silica particles						
Chr	9.2×10^{-15}	Bedjanianet al.[252]	Kerosene flame soot	6.0×10^{-19}	Perraudin et al.[276]	Silica particles	1.5×10^{-17}	Perraudin et al.[298]	Graphite particles	4.0×10^{-12}	Liuet al.[260]	Azelaic acid particles
	5.0×10^{-12}	Estève et al.[255]	Graphite particles	1.0×10^{-19}	Nguyen et al.[275]	Kerosene flame soot	3.1×10^{-17}	Perraudin et al.[298]	Silica particles			
	4.4×10^{-13}	Estève et al.[256]	Diesel exhaust particles SRM 1650a	3.9×10^{-17}	Estève et al.[255]	Graphite particles						
				1.0×10^{-17}	Estève et al.[256]	Diesel exhaust particles SRM 1650a						
				2.6×10^{-21}	Butler and Crossley[270]	Ethylene flame soot						
BaA	9.2×10^{-15}	Bedjanianet al.[252]	Kerosene flame soot	6.7×10^{-18}	Perraudin et al.[276]	Silica particles	2.8×10^{-17}	Perraudin et al.[298]	Graphite particles	4.3×10^{-12}	Liuet al.[260]	Azelaic acid particles
	5.6×10^{-13}	Estève et al.[255]	Graphite particles	1.0×10^{-19}	Nguyen et al.[275]	Kerosene flame soot	8.7×10^{-17}	Perraudin et al.[298]	Silica particles			
	3.2×10^{-13}	Estève et al.[256]	Diesel exhaust particles SRM 1650a	3.3×10^{-17}	Estève et al.[255]	Graphite particles						
				1.3×10^{-17}	Estève et al.[256]	Diesel exhaust particles SRM 1650a						
				6.2×10^{-21}	Butler and Crossley[270]	Ethylene flame soot						

BkF	1.0×10^{-14}	Bedja-nianet al.[252]	Kerosene flame soot	2.2×10^{-18}	Perraudin et al.[276]	Silica particles	1.9×10^{-17}	Perraudin et al.[298]	Graphite particles			
	3.5×10^{-12}	Estève et al.[255]	Graphite particles	1.0×10^{-19}	Nguyen et al.[275]	Kerosene flame soot	3.6×10^{-17}	Perraudin et al.[298]	Silica particles			
				2.5×10^{-17}	Estève et al.[255]	Graphite particles						
BaP	1.1×10^{-14}	Bedja-nianet al.[252]	Kerosene flame soot	9.3×10^{-16}	Perraudin et al.[276]	Silica particles	5.3×10^{-17}	Perraudin et al.[298]	Graphite particles			
	4.1×10^{-12}	Estève et al.[255]	Graphite particles	1.0×10^{-19}	Nguyen et al.[275]	Kerosene flame soot	1.4×10^{-16}	Perraudin et al.[298]	Silica particles			
	2.9×10^{-13}	Estève et al.[256]	Diesel exhaust particles SRM 1650a	7.8×10^{-17}	Estève et al.[255]	Graphite particles	$2.0 \times 10^{-18}/(1/c_{O3} + 3.6 \times 10^{-16})$	Kahan et al.[296]	Octanol			
				1.5×10^{-17}	Estève et al.[256]	Diesel exhaust particles SRM 1650a	$<3.8 \times 10^{-18}/(1/c_{O3} + 10^{-16})$	Kwamena et al.[291,292]	NaCl			
				1.0×10^{-20}	Butler and Crossley[270]	Ethylene flame soot	$4.2 \times 10^{-15}/(1/c_{O3} + 2.8 \times 10^{-13})$	Kwamena et al.;[292]Pöschl et al.[290]	Soot			
							$1.7 \times 10^{-16}/(1/c_{O3} + 1.2 \times 10^{-15})$	Kwamena et al.[291]	Azelaic acid (wet)			
							$3.0 \times 10^{-16}/(1/c_{O3} + 9.50 \times 10^{-15})$	Kwamena et al.[292]	Silica particles			
BeP	1.1×10^{-14}	Bedja-nianet al.[252]	Kerosene flame soot	2.9×10^{-18}	Perraudin et al.[276]	Silica particles	1.6×10^{-17}	Perraudin et al.[298]	Graphite particles			
	4.7×10^{-12}	Estève et al.[255]	Graphite particles	1.0×10^{-19}	Nguyen et al.[275]	Kerosene flame soot	2.9×10^{-17}	Perraudin et al.[298]	Silica particles			
	4.7×10^{-13}	Estève et al.[256]	Diesel exhaust particles SRM 1650a	3.5×10^{-17}	Estève et al.[255]	Graphite particles						
				7.5×10^{-18}	Estève et al.[256]	Diesel exhaust particles SRM 1650a						
				2.8×10^{-21}	Butler and Crossley[270]	Ethylene flame soot						
Per	5.0×10^{-12}	Estève et al.[255]	Graphite particles	1.1×10^{-16}	Estève et al.[255]	Graphite particles						
IPy	3.5×10^{-13}	Estève et al.[256]	Diesel exhaust particles SRM 1650a	6.2×10^{-18}	Perraudin et al.[276]	Silica particles	1.9×10^{-17}	Perraudin et al.[298]	Graphite particles			
				1.0×10^{-19}	Nguyen et al.[275]	Kerosene flame soot	3.8×10^{-17}	Perraudin et al.[298]	Silica particles			
				7.5×10^{-18}	Estève et al.[256]	Diesel exhaust particles SRM 1650a						
BgP	5.9×10^{-12}	Estève et al.[255]	Graphite particles	4.7×10^{-17}	Perraudin et al.[276]	Silica particles						
				1.0×10^{-19}	Nguyen et al.[275]	Kerosene flame soot						
				3.8×10^{-17}	Estève et al.[255]	Graphite particles						
				7.9×10^{-21}	Butler and Crossley[270]	Ethylene flame soot						

BgF	8.4×10^{-15}	Bedjanianet al.[252]	Kerosene flame soot	1.0×10^{-19}	Nguyen et al.[275]	Kerosene flame soot						
AcP	1.0×10^{-14}	Bedjanianet al.[252]	Kerosene flame soot	1.0×10^{-19}	Nguyen et al.[275]	Kerosene flame soot						
DBahA	1.6×10^{-14}	Bedjanianet al.[252]	Kerosene flame soot	1.0×10^{-19}	Nguyen et al.[275]	Kerosene flame soot						
DBaeP	1.0×10^{-14}	Bedjanianet al.[252]	Kerosene flame soot	1.0×10^{-19}	Nguyen et al.[275]	Kerosene flame soot						
BbF	1.2×10^{-14}	Bedjanianet al.[252]	Kerosene flame soot	1.0×10^{-19}	Nguyen et al.[275]	Kerosene flame soot						
Cor	1.1×10^{-14}	Bedjanianet al.[252]	Kerosene flame soot	1.3×10^{-18}	Perraudin et al.[276]	Silica particles						
				1.0×10^{-19}	Nguyen et al.[275]	Kerosene flame soot						
				2.3×10^{-21}	Butler and Crossley[270]	Ethylene flame soot						
DBalP				1.8×10^{-16}	Perraudin et al.[276]	Silica particles	1.3×10^{-16}	Perraudin et al.[298]	Graphite particles			
							1.3×10^{-16}	Perraudin et al.[298]	Silica particles			
1-NP	1.0×10^{-13}	Miet et al.[257]	Silica particles	6.2×10^{-18}	Miet et al.[277]	Silica particles	2.2×10^{-17}	Miet et al.[297]	Silica particles	1.3×10^{-12}	Liuet al.[260]	Azelaic acid particles

Bertram et al.[253] studied the heterogeneous loss rate of OH radicals on various organic surfaces, coated on Pyrex tubes, using a flow-tube reactor coupled to a chemical ionization mass spectrometer. The average reaction probabilities (γOH_{ave}) for these organic surfaces were generally large, (>0.1) for relatively low OH concentrations (1×10^8 molecules cm^{-3}) and short exposure time (<5 minutes). For example for soot and solid pyrene surfaces, the γOH_{ave} values were 0.88 and 0.32, respectively. It was noted that the γOH_{ave} value was lower for an alumina surface than for organic surfaces. These results indicate that uptake of OH radicals to particle surfaces may be rapid and thus have the potential to contribute significantly to oxidative reactions of PAHs on organic surfaces in the atmosphere.

Estève et al.[254] used a discharge flow technique to study the heterogeneous reaction of OH with Phe adsorbed on a Pyrex surface, in the presence of NO_2 and in the absence of light. They investigated the reaction kinetics by measuring the total loss of OH, followed by laser-induced fluorescence (LIF). The efficiency of the heterogeneous reaction and the change in the surface properties with exposure time were demonstrated. In addition, oxidation products of the heterogeneous reaction of OH with Phe in the presence of NO_2, identified by GC-MS, were noted as mono-hydroxylated isomers of phenanthrene.

In their subsequent investigations, Estève et al.[255,256] studied the heterogeneous reactions of both OH radicals and NO_2 with PAHs associated with carbonaceous (graphite) particles and diesel particulate exhaust. This was achieved using an adapted fast flow reactor and by monitoring the decrease in particle-bound PAH concentration by GC-MS. This therefore allowed the calculation of pseudo-first order rate coefficients of these reactions from the experimental decays of PAH concentrations in relation to OH exposure time.

Rate coefficients derived for PAH reactions with OH on diesel exhaust particles were approximately an order of magnitude lower for most compounds than for reactions of PAHs adsorbed on graphite.[256] The matrix effects are detailed below (Section 3.2.2.6). These $k^{(2)}$ values fall approximately within the same order of magnitude for all PAHs investigated in the two studies.

Bedjanian et al.[252] investigated the reactions of 15 PAHs adsorbed on laboratory generated kerosene soot surfaces using a low-pressure discharge-flow reactor combined with an electron-impact mass spectrometer. An uptake coefficient of ≈0.2 was reported under the reaction conditions used. Pseudo-$k^{(1)}$ coefficients calculated in this study generally fell between those derived by Estève et al.[255] and Estève et al.,[256] suggesting that the substrate type influences PAH reactivity. However, the OH concentration reported in the study by Bedjanian et al.[252] was 1–2 orders of magnitude higher than that used in the studies by Estève and co workers[255,256] leading to much lower $k^{(2)}$ values being derived in this study.

In the above three investigations there is relatively little variation in reactivity between each individual PAH associated with the varying types of organic particulate matter. This was also observed in a laboratory investigation of PAH decay profiles caused by heterogeneous reactions with OH radicals occurring on natural atmospheric particles.[258] In each of the three studies detailed in Table 6 the $k^{(2)}$ values for all PAHs studied fall approximately within one order of magnitude. The above studies have demonstrated that rapid reaction with OH can potentially be an important potential pathway for degradation of PAHs associated with carbonaceous particles in the atmosphere.

To our knowledge, the only study to investigate the reaction of PAHs on silica particles was carried out by Miet et al.[257] The $k^{(2)}$ value derived for this reaction is lower than that of the corresponding reaction on diesel particles. More work is clearly needed to assess these reactions for a wider range of compounds on mineral substrates to compare with other surface reactions.

- Reaction mechanism and products: Ringuet et al.[258] investigated the reaction products formed from PAHs reactions with OH (+NO_2) occurring on natural atmospheric particles. Formation of NPAH derivative compounds was observed. For example, 1N-Pyr was observed, attributed to the heterogeneous reaction of Pyr as reported in NO_3 radical reaction studies.[259,260] This apparent discrepancy in radical-induced reaction products between the gas and particle phases has also been reported by Miet et al.[257] for OH reactions on silica. Furthermore, 2N-Pyr and 4N-Pyr were also reported as reaction products in this study, which were not reported elsewhere and had previously been thought to be solely gas phase reaction products.

The precise mechanism of surface-bound PAHs with OH radicals has not been elucidated and the lack of product evaluation makes it difficult to assess the possible reaction chemistry occurring. The non-linear dependence of the reaction rate on OH concentration[252] would indicate that this process is not a simple bi-molecular reaction. While the surface reaction of PAH towards OH is shown to be relatively fast, the presence of the 'plateau' in PAH degradation with exposure time would suggest that gas phase transport of OH to the soot surface is not the rate limiting step in these processes. Clearly more work is required to address these issues and investigate the mechanisms of these reactions in more detail.

Heterogeneous PAH Reactions with NO_3 Radicals.

- Overview: Heterogeneous reactions of PAHs with NO_3 radicals have received comparatively little attention in the literature compared with the corresponding reaction in the gas phase, and the reaction of surface-bound PAHs with OH, NO_2 and O_3. Initial investigations focused on the reactivity of pyrene and perylene[212,261] adsorbed on glass fibre filters towards N_2O, NO_3 radicals, NO_2 and N_2O_5. It was concluded that NO_3 did not react to a detectable extent with these compounds and that N_2O_5 was the main oxidising species. Kamens et al.[262]reported that the heterogeneous reaction rate coefficient for the reaction of PAHs with N_2O_5 ($\approx 2.5 \times 10^{14}$ molecules cm^{-3}) on wood and diesel soot is of the order of 10^{-18} cm^{-3} $molecules^{-1}s^{-1}$.

Recently, however, studies have been conducted that contradict these original assumptions and suggest that NO_3 does play an important role in heterogeneous PAH reactivity. Mak et al.[263] studied the reaction of NO_3 with methane and hexane soot and solid pyrene surfaces using a flow tube reactor and reported that the uptake of NO_3 on fresh soot and pyrene surfaces was fast (uptake coefficient >0.1). The authors suggested that NO_3 has the potential to rapidly oxidize soot surfaces and their reaction with PAHs potentially provides an important loss process for these compounds under certain atmospheric conditions.

Similarly, Karagulian and Rossi[264] investigated the reaction of soot and NO_3 and also reported rapid uptake (uptake coefficient of 5×10^{-2}). Furthermore, Gross and Bertram[265] studied the reactive uptake of NO_3 as well as other oxidants (including NO_2 and O_3) with three solid PAH surfaces. Their results suggested that NO_3 has the potential to be a more important sink for PAHs than reactions with NO_2 and O_3. In an investigation by Kwamena and Abbatt,[266] exposing anthracene and pyrene to a mixed flow of NO_3, NO_2 and N_2O_5, it was concluded that the nitro-PAH products formed were as the result of a reaction initiated by NO_3 or N_2O_5 and not through direct reaction with NO_2 or HNO_3.

- Reaction Mechanism and Products: The first reported quantitative heterogeneous reactivity data for PAH reaction with NO_3 were provided by Liu et al.[260] Rate coefficients were measured for the reactions of BaA, Chr and Pyr, as well as 1-nitropyrene, adsorbed on azelaic acid particles. The $k^{(2)}$ values measured for these reactions are shown in Table 7. These$k^{(2)}$ values for the NO_3 reactions are of the same order of magnitude or 1–3 orders of magnitude higher than for reactions with OH radicals and 5–7 orders of magnitude higher than the heterogeneous reactions with NO_2 and O_3, depending on the particle surface (discussed in the following sections). This suggests that heterogeneous reactions of PAHs with NO_3 may be very significant. However, more research in this area is required to assess the precise rates and mechanisms for a wider range of surface-bound PAHs.

Product studies for the heterogeneous reaction of PAHs with NO_3 radicals have been conducted for a number of PAHs on azelaic acid particles using Vacuum Ultraviolet Photoionization Aerosol Time of Flight Mass Spectrometry (VUV-ATOFMS).[259,260] Products identified included 9N-Ant, AQu, 1N-Pyr, 2-, 4- and 9N-Phe isomers, BaA-7,12-dione and 7N-BaA for the reactions of Ant, Pyr, Phe and BaA. The reaction of 1N-Pyr was also shown to produce 1,3-, 1,6-, and 1,8-DN-Pyr isomers.[260] While some of the products identified, e.g. AQu, are thought to be analogous to those formed from the gas phase reaction, the formation of 1N-Pyr is in contrast to gas phase reactions, in which 2N-Pyr is the major product.

Heterogeneous PAH Reactions with NO_2.

- Overview: The occurrence of heterogeneous reactions of PAHs with NO_2 was initially demonstrated through experimental studies exposing PAHs on various substrates to NO_2 in the presence of nitric acid, at ambient or elevated levels, resulting in the formation of nitro-PAH products. This has been observed on a number of different surfaces including filters,[267,268] wood smoke particles,[269] soot particles,[270] propane flame-generated soot,[271] coal fly ash,[272] silica plates,[273] metal oxides TiO_2, ZnO, Al_2O_3, Fe_2O_3, MgO, CaO, SiO_2 and graphite.[274] Details of these studies are summarized by Finlayson-Pitts and Pitts.[1]

More recently, full kinetic investigations of this process have been carried out with rate coefficient values for the reaction of NO_2 adsorbed on various particle surfaces calculated.[255,256,275,276] The $k^{(2)}$ values derived from these studies are presented in Table 7. In general, heterogeneous reactions of PAHs with NO_2 on carbonaceous particles are considerably slower than the corresponding reactions with OH radicals and O_3. Indeed, the reactions of PAHs with OH are 1–6 orders of magnitude faster than the reactions with NO_2, depending on the surface type and specific experimental conditions.

Estève and co-workers investigated the heterogeneous reactions of NO_2 with PAHs associated with graphite[255] and diesel exhaust particles.[256] It was shown that Pyr and BaP are the most reactive towards NO_2 on diesel exhaust and Ant, Per and BaP are the most reactive on graphite. Nguyen et al.[275] investigated the reactions of 17 PAHs associated with fresh kerosene flame soot over a temperature range (255–330 K) and the decay of PAHs due to the reaction of NO_2 was considered to be negligible, suggesting that PAH degradation by NO_2 alone is of minor importance in the atmosphere.

Perraudin et al.,[276] Miet et al.[277] and Ma et al.[278] studied the reaction of NO_2 with PAHs adsorbed on silica particles. Ma et al.[278] also investigated the reaction of gas phase NO_2 with anthracene MgO particles. $k^{(2)}$ values for this reaction were much more variable between individual PAH species than was observed for reactions with OH and O_3 or with NO_2 on carbonaceous substrates and are lower for most compounds by 1–5 orders of magnitude. The matrix effects are detailed below (Section 3.2.2.6).

- Reaction mechanism and products: The formation of nitro-PAH isomers from heterogeneous reactions of PAHs with NO_2[1] would indicate a direct molecular reaction between surface PAH and gaseous NO_2. Consistent with this, 1-nitropyrene has been identified as the only the product from the reaction of pyrene with NO_2 on silica,[277,279,280] alumina[279]and soot[84] particles. This is in contrast to gas-phase reactions with OH/NO_3 radicals followed by NO_2 where 2-nitropyrene is observed as the main reaction product.[188]

Miet et al.[277] suggested the mechanism for the nitration of pyrene due to the heterogeneous reaction with NO_2 on silica particles as being a simple elementary reaction between the two species, indicated by the linear relationship between pyrene reactivity and NO_2 concentration. The formation of 1-NPyr is attributed to electrophilic addition of NO_2 at C_1.[277] However, Inazu et al.[274] observed that the main products from the reaction of fluoranthene with NO_2 on various particle types were 1-, 2- and 7N-Flt isomers, indicating that an electrophilic nitration (which would primarily yield 3- or 8-isomers) does not occur.

The heterogeneous reaction of PAHs with NO_2 appears to be significantly enhanced by the presence of gaseous nitric acid.[267,268,272,281–283] For example, it was shown that the 24 h nitration yield for pyrene adsorbed on filters exposed to 10 ppmV NO_2 rose from 0.02% to 2.85% when traces of HNO_3 were present.[268] Grosjean et al.[281] studied the transformation of perylene and BaP deposited on glass and Teflon filters loaded with fly ash and diesel exhaust particles, exposed to 100 ppbV of NO_2. They reported no evidence for chemical transformations in these experiments. This could be attributed to the absence of HNO_3.[1]

In the study by Ma et al.[278] both 9N-Ant and 9,10-AQu were observed as the products from the nitration of anthracene with NO_2 on SiO_2 particles, whereas only 9,10-AQu was observed from the equivalent reaction on MgO particles. The difference between these two processes was attributed by the authors to the formation of HNO_3 on SiO_2 that does not occur on MgO particles.

For the reactions of NO_2 with PAHs on silica particles, it is proposed that HNO_3 is formed on the silica surface, which acts as a catalyst for the process.[280] Wang and co-workers proposed that the HNO_3 dissociates producing H^+ leading to the protonation of NO_2 and N_2O_4 to form HNO_2^+ and $HN_2O_4^+$. This initiates an electrophilic nitration process with reaction occurring at the C_1 position (the carbon with highest electron density), producing intermediate species that further react with NO_2 and/or decompose to form 1N-Pyr (Scheme 9).

a

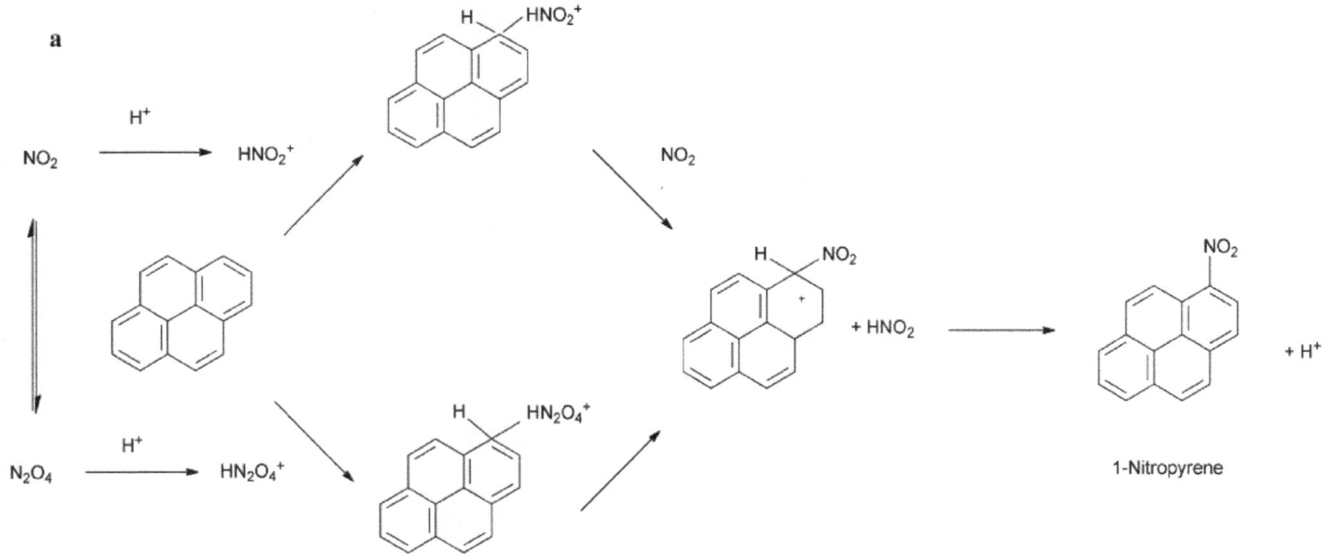

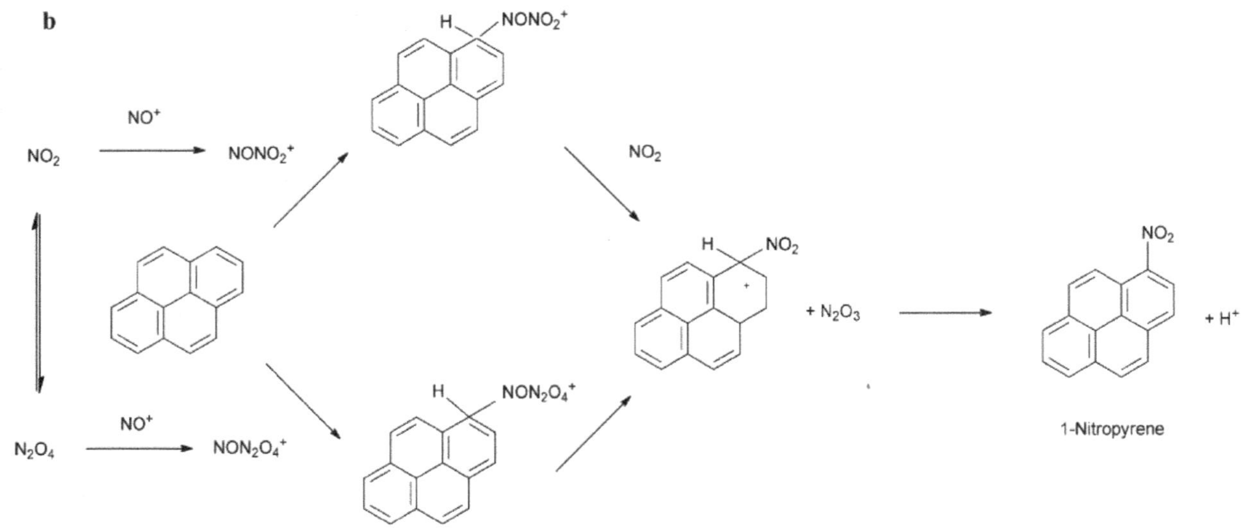

Scheme 9: Suggested mechanisms for the heterogeneous reaction of pyrene with NO_2.[280,284]

It was suggested that HNO_2 gas, which is also formed from NO_2 on the silica surface, also acts catalytically in the heterogeneous reaction of pyrene.[284] Furthermore, considering the combined catalytic effect of HNO_3 and HNO_2, a different electrophilic nitration mechanism was proposed (Scheme 10b) in which NO^+ is formed on the silica surface from the dissociation of HNO_2.[285] These species then reacts with NO_2 or N_2O_4 to form $NONO^{2+}$ and $NON_2O_4^+$, respectively, which react with surface bound Pyr, analogous to the reaction in Scheme 10a.

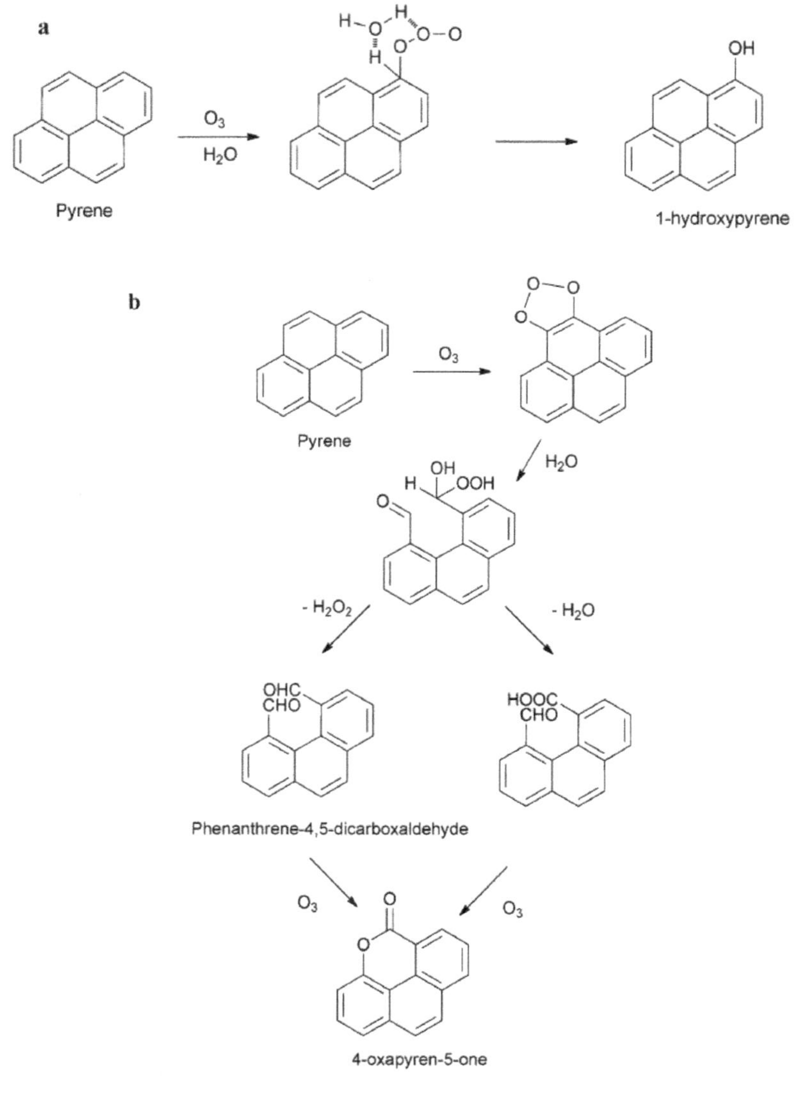

Scheme 10: Suggested mechanism for the heterogeneous reaction of pyrene with gaseous ozone, forming 1-hydroxypyrene (a), phenanthrene-4,5,dicarboxaldehyde and 4-oxapyren-5-one (b).[297]

Heterogeneous PAH Reactions with O_3.

- Overview: The occurrence of heterogeneous ozonolysis reactions of PAHs was first demonstrated in studies exposing PAHs on solid surfaces to O_3 and observing decay in the PAH profile and/or conversion to oxidised derivative compounds.[83,231,242,267,282,286–288] These studies indicated that under certain atmospheric conditions, PAHs adsorbed onto surfaces may undergo rapid heterogeneous reaction with gas phase O_3. Detailed review of these studies is provided by Finlayson-Pitts and Pitts.[1]

More recently, studies have focussed on full kinetic investigations of the ozonation of PAHs associated with different types of surfaces such as soot particles,[289,290] organic aerosols and sea salts,[291,292] air–water interface,[293–295] organic films,[296] mineral surfaces[297,298] and graphite.[298]

Perraudin et al.[298] studied the reaction of 13 PAHs with O_3 adsorbed on graphite and silica particles. Pseudo-$k^{(1)}$ values were calculated and shown to be proportional to gaseous O_3concentration. Therefore $k^{(2)}$ values were derived considering the concentration of O_3 (Table 7). The reactivity of PAHs followed the order Chr < BeP < Cor, BgP, Phe, Flt, BkF, Ipy < Pyr < BaA < BaP < Ant < DBalP and Flth < Cor < Phe < BeP < Chr < BkF < Ipy < Pyr < BgP < BaA < DBP < Ant, BaP for graphite particles and silica particles, respectively.

Wu et al.[299] had previously measured $k^{(2)}$ values for the reaction of O_3 with BaP adsorbed on silica particles of 1.78×10^{-17} cm^3 molecules^{-1} s^{-1} at room temperature, about 7 times higher than that measured by Perraudin et al.[298] This may be explained by the fact that the earlier study was not performed in the dark. This is consistent with a previous study by Katz et al.[286] where the reactivity of PAHs adsorbed on cellulose was shown to be more rapid under light irradiation conditions.

Miet et al.[297] investigated the reaction of O_3 with pyrene adsorbed on silica particles. The $k^{(2)}$ value derived from this study was approximately 5 times higher than that derived by Perraudin et al.[298] despite the reaction conditions (temperature, pressure, particulate concentration) being essentially identical in the two studies. Previously, Kamens et al.[300] investigated the reactions on wood smoke at low PAH loading with 0.52 ppmV O_3 and reported $k^{(2)}$ values (k, cm^3 molec^{-1} s^{-1}) of 4.1×10^{-18} (BPy), 7.2×10^{-18} (Chr), 1.0×10^{-18} (BaP), 1.3×10^{-18} (BaA), all roughly one order of magnitude lower than reported by Perraudin et al.[298] These observations clearly highlight the influence of the specific conditions and nature of the PAH and/or particle surface on the reaction kinetics of these processes.

- Reaction mechanism and products: The precise reaction mechanism(s) and influencing factors governing the heterogeneous reaction of surface-bound PAHs with O_3 have been investigated. A direct (Eley–Rideal) mechanism is possible, whereby a gas phase molecule collides directly with an adsorbed PAH molecule and facilitates a reaction. This would lead to a linear dependence between observed first order rate coefficient and the gaseous ozone concentration.

However, experimental studies suggest that the oxidation of surface-bound PAHs by gas phase ozone will follow Langmuir–Hinshelwood type kinetics.[266,290–294,301] These authors investigated the kinetics and products of these reactions as a function of ozone concentration and relative humidity (Table 8).

Table 8: Comparison of kinetics data for the heterogeneous reactions of BaP and anthracene on different aerosol substrates. Adapted from Kwamena et al. (ref. 301) and Mmereki et al. (ref. 294)

PAH	Langmuir adsorption equilibrium constant, K_{O3} (cm^3 × 10^{-15})	Pseudo-first order rate coefficient, k (s^{-1} × 10^3)	Surface	Ref.
BaP	280 ± 20	15 ± 1	Soot aerosols	Pöschl et al.[290]
	28	32[a]	Fused silica	Wu et al.[299]
	9.5	32[a]	Non activated silica	Alebic-Juretic et al.[302]
	2.80 ± 1.4	60 ± 18	Azelaic acid aerosol (72% RH)	Kwamena et al.[291]
	1.20 ± 0.4	48 ± 8	Azelaic acid aerosol (<1% RH)	
	<0.12[b]	32[a]	NaCl	
Ant	100 ± 40	10 ± 3	Phenylsiloxane oil aerosols	Kwamena et al.[301]
	6.4 ± 1.8	2.8 ± 0.9	Pyrex glass	Kwamena et al.[292]
	2.2 ± 0.9	57 ± 9	Azelaic acid aerosols	Kwamena et al.[301]
	2.143 ± 0.441	2.55 ± 0.17	Water	Mmereki et al.[294]
	1.922 ± 0.478	2.26 ± 0.20	Stearic acid film on water	
	0.681 ± 0.291	1.11 ± 0.12	Octanoic acid[d]	
	0.56	2	Octanol thin film	Kahan et al.[296]
	0.508 ± 0.088	2.59 ± 0.14	1-Octanol[c]	Mmereki and Donaldson[293]
	0.118 ± 0.036	0.48 ± 0.07	Hexanoic acid[e]	Mmereki et al.[294]

[a] k value used is an average from the studies of Pöschl et al.[290] and Kwamena et al.[291]

[b] Calculated assuming k_1 = **3 × 10**$^{-3}$ s^{-1}, which is the typical error of k_1 obtained for azlaic acid aersol kinetics under dry conditions. This was taken to be the fastest reaction occurring at the highest O_3 concentrations used (31 ppm).[291]

[c] 2.5×10^{-3} M. d 3.79×10^{-3} M. e 8×10^{-3} M.

The Langmuir–Hinshelwood mechanism involves one species (in this e the PAH molecule) being strongly adsorbed to the particle surface and the gas species (O_3) being in equilibrium between the gas phase and the particle surface.[291] The reaction therefore proceeds in two steps: (i) adsorption of the gas molecule, (ii) surface reaction of PAH and the gas molecule. The reaction rate is therefore dependent on the concentration of both PAH and O_3 and will exhibit a non-linear dependence on gas species concentration.

Earlier studies had indicated a linear relationship between ozone concentration and pseudo-first order reaction rate for ozone mixing ratios of 0.5–0.40 ppmV[302] and 0–1.5 ppmV[299] on silica. It has been suggested that for these studies, the Langmuir–Hinshelwood mechanism may have been reported if the experiments had been conducted over larger ozone concentration ranges.[291] A linear dependence of pseudo-first order reactivity on ozone concentration over the concentration range 0.4×10^{14}–3.3×10^{14} cm^3 $molecules^{-1}$ was observed by Perraudin et al.[298] for a range of PAHs, which may also indicate a simple direct mechanism. However, the suitability of a Langmuir–Hinshelwood-type mechanism was not tested in this study, so no definitive conclusion can be drawn.

The observation of Langmuir–Hinshelwood kinetics emphasizes the potential importance of the degree of ozone partitioning to the particle surface, as this will have a direct influence on the reaction rate. Therefore, in addition to the pseudo-first order rate coefficient (pseudo-$k^{(1)}$), which describes the surface reaction rate, the Langmuir–Hinshelwood mechanism is also characterized by the ozone partitioning coefficient (K_{O3}), which describes the degree of partitioning of ozone between the gas phase and the surface (Table 8).

There is notably more variation in K_{O3} values between these different reaction systems than for the pseudo-$k^{(1)}$ values. This may suggest that differences in heterogeneous reactivity of PAHs with O_3 observed on different substrates is due, to a large extent, to differences in the partitioning rate of O_3 to the surface rather than differences in the surface reaction mechanism(s). However, no direct correlation between K_{O3} and pseudo-$k^{(1)}$ is found in these data and the lack of $k^{(2)}$ values derived from these studies mean that no solid conclusions can be drawn regarding these reactions.

Differences in K_{O3} values have been attributed to differences in binding interactions of the initial ozone-substrate complex.[292] Therefore, K_{O3} will be influenced to a large degree by the specific type of particle and/or the coating of additional species (e.g. water, organic molecules etc.) that will affect the affinity of ozone to the surface and therefore influence PAH reactivity.[291] It has been suggested, based on adsorption enthalpy values, that ozone will interact more strongly with non-polar surfaces than with polar surfaces.[290] This would explain the higher rate of O_3 partitioning to non-polar soot and phenylsiloxane oil aerosols, for example, compared with polar NaCl and azelaic acid aerosols.

A number of studies have identified both ring-opened (e.g. dialdehydes, dicarboxylic acids, keto carboxylic acids), and ring-retaining (e.g. quinones) products from the heterogeneous reactions of a number of PAHs with O_3 on particle surfaces.[1] This suggests the significance of both bond-attack (ring-opening) and atom-attack (substitution) mechanisms at the most electrophilic positions.[297]

Van Cauwenberghe[251] proposed 2 mechanisms for the heterogeneous reaction of PAHs with O_3. (1) A one-step electrophilic–nucleophilic attack on olefin bonds with the highest electron density to initially form a primary ozonide. This will decompose to yield ring-opened products such as aldehydes and carboxylic acid compounds. (2) A two-step electrophilic attack that will ultimately lead to the formation of quinones.[1]

Miet et al.[297] tentatively identified several oxidation products formed from the heterogeneous reaction of pyrene with O_3 on silica including 1-hydroxypyrene, phenanthrene-4,5-dicarboxaldehyde and 4-oxapyren-5-one. The proposed mechanisms, suggested by the authors to explain the formation of these products are depicted in Scheme 10. It was suggested that hydroxypyrene would be formed from the reaction of O_3 with the most reactive carbon atom while phenanthrene-4,5-dicarboxaldehyde would be formed by ozonolysis of a C–C bond.[297] It was also suggested that 4-oxapyren-5-one would be formed by further reaction of phenanthrene-4,5-dicarboxaldehyde.

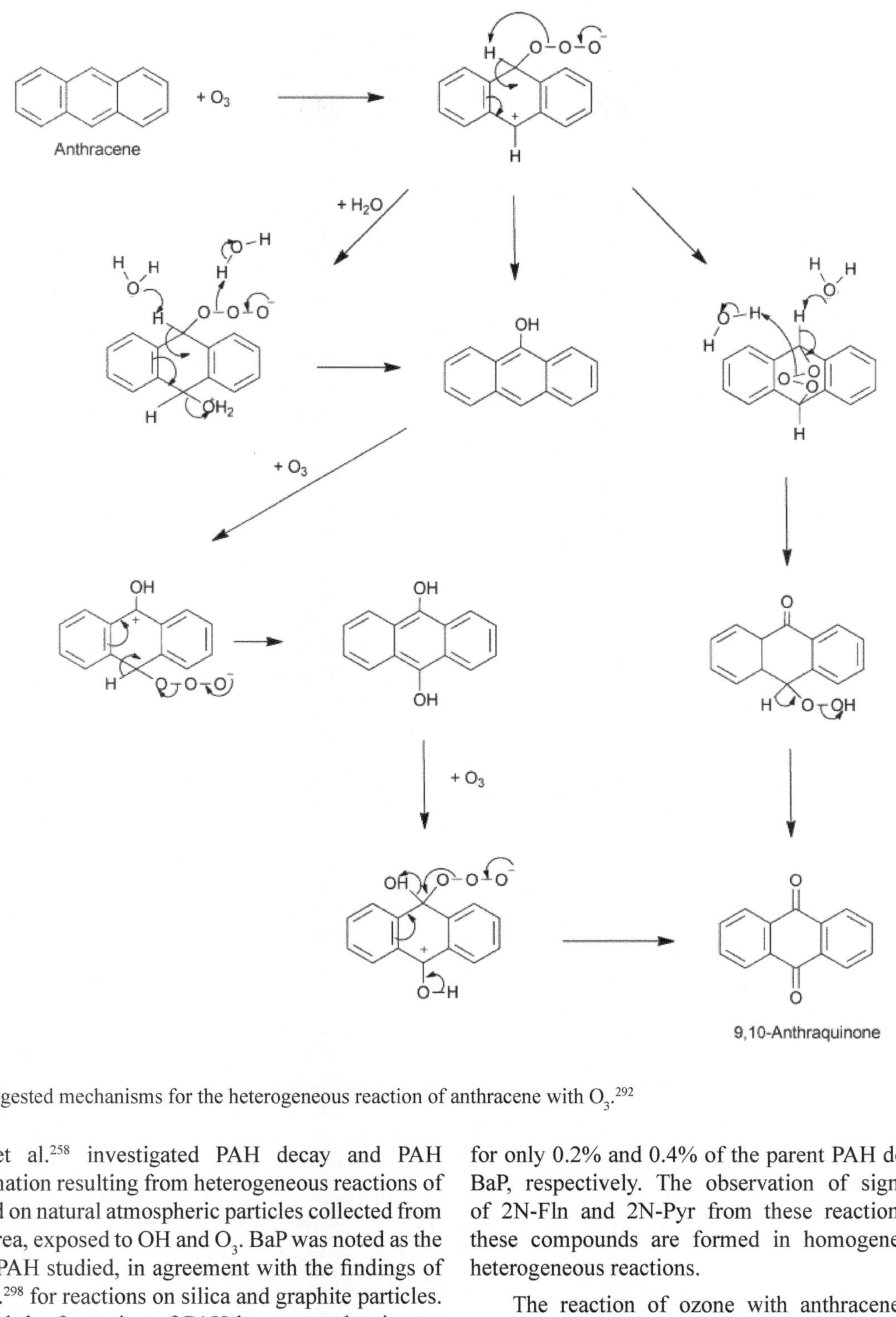

Scheme 11: Suggested mechanisms for the heterogeneous reaction of anthracene with O_3.[292]

Ringuet et al.[258] investigated PAH decay and PAH derivative formation resulting from heterogeneous reactions of PAHs adsorbed on natural atmospheric particles collected from a high traffic area, exposed to OH and O_3. BaP was noted as the most reactive PAH studied, in agreement with the findings of Perraudin et al.[298] for reactions on silica and graphite particles. PAH decay and the formation of PAH ketones and quinones e.g. 9-Flr and AQu were observed. However, it was noted that the rate of OPAH formation did not match with the rate of parent PAH degradation in these es, suggesting that other compounds/species are involved in these reactions.

The formation of NPAH compounds was also noted in these reaction studies, which is surprising as NO_2 was absent in this experiment. Organic nitrogen present was suggested as the precursor, which may react with O_3 to form a reactive species (e.g. NO_x, NO_2^-, NO_3^-) which then reacts with the PAH upon O_3 attack.

PAH degradation and NPAH formation was also observed for reaction of PAH with a mixture of NO_2 and O_3. However, it was noted that the formation of these products accounted for only 0.2% and 0.4% of the parent PAH decay for Fln and BaP, respectively. The observation of significant amounts of 2N-Fln and 2N-Pyr from these reactions indicates that these compounds are formed in homogeneous as well as heterogeneous reactions.

The reaction of ozone with anthracene adsorbed onto aqueous surfaces or Pyrex leads to the formation of 9,10-AQu only,[292,294] while both anthraquinone and anthrone are formed on silica particles.[303]

A possible mechanism for the reaction has been proposed, consistent with that described by Bailey.[304] This was suggested to involve two potential pathways (Scheme 11): the first of these involves three consecutive electrophilic attacks by ozone at the C_9 and C_{10} positions of anthracene with the first two ozone attacks being followed by the loss of oxygen and a proton transfer and the third attack followed by loss of oxygen and water.[294] The second of these pathways involves the formation of an ozonide due to O_3 addition across the C_9–C_{10} bond of Ant, which re-arranges to hydroperoxide, which finally dehydrates to form anthraquinone.[303] There is some disagreement between these studies as to which of these reaction pathways dominates. More work is clearly required to establish the mechanism responsible for the formation of quinones from heterogeneous reactions of PAHs with ozone.

The products of the heterogeneous reactions of BaP with ozone were investigated by Letzel et al.[305] They identified BaP-1,6-dione, BaP-3,6-dione and BaP-6,12-dione as the main reaction products, as previously observed by Pitts et al.,[267] as well as BaP-4,5-dione and 4-oxa-benzo[d,e,f]chrysene-5-one (B(def)C-lactone). The formation of these products would seem to be broadly consistent with the mechanism for the reaction of anthracene (Scheme 11).

The reaction yield of the production of anthraquinone from the reaction of anthracene with O_3 on a pyrex surface was calculated by Kwamena et al.[292] to be approximately 30% at O_3concentrations of up to 9.1×10^{15} molecules cm^{-3}. However at atmospherically relevant ozone mixing ratios (100 ppbV) the AQu formation yield was expected to be very small (<1%), suggesting that this reaction route may be of minor importance in the ambient atmosphere.

Factors Influencing Heterogeneous PAH Reactivity.

- Substrate type and conditions: It is clear from the discussion above that the chemical composition and specific conditions of the substrate to which PAHs are associated will have a critical role in the heterogeneous reactivity of these compounds towards oxidizing species. This 'substrate effect' has the potential to influence the degree of partitioning of the oxidant to the particle surface (as demonstrated for ozone reactions), the relative accessibility of the PAH within the particle matrix, as well as the precise reaction mechanism.
- Reaction with NO_2: Ramdahl et al.[279] showed that PAH reactivity on different surfaces follows the order silica > alumina > charcoal, for six PAHs exposed to 0.5 ppmV of NO_2in air containing water vapour and traces of HNO_3. It was estimated that PAH reactivity on alumina was 14–24% of that on silica. Similarly, Jäger and Hanus[306] observed the order of reactivity for the reaction of several PAHs with 1.3 ppmV of NO_2 to be silica gel > fly ash > aluminium oxide > carbon.

The enhanced reactivity of NO_2 with PAH adsorbed on silica relative to that associated with alumina was attributed to the catalytic effect of HNO_3, as discussed above.[280,284,285,307] The formation of nitric acid on the acidic silica surface is likely to be greater than that on neutral alumina surfaces which may explain the higher reactivity.[279] Similarly, Ma et al.[278] noted that the reaction rate on SiO_2 was almost two times faster than on MgO. The authors attributed this to the formation of HNO_3, which occurs on the surface of SiO_2 but not on MgO.

Inazu et al.[274] investigated the reaction of Flt with NO_2 on different particle types and the factors influencing the degradation of Flt and the yield of N-Flt*s. The amount of Flt degradation ranged from 20.9% for graphite to 79.5% for TiO_2 and followed the order TiO_2 > ZnO > Al_2O_3 > MgO > CaO > SiO_2 > Fe_2O_3 > graphite. The authors suggested three possible types of matrix effect to explain the different degradation rates and N-Flt yields on different substrate types.

One possible impact is suggested to be the photocatalytic generation of active species such as O_2^- and O^- on the particle surface.[274] This would apply only to TiO_2 and ZnO and Fe_2O_3, which have a band gap energy of less than 4.3 eV, explaining the higher degradation rates of Flt on TiO_2 and ZnO. The reason for Fe_2O_3 degradation being lower than expected is not currently known. While this phenomenon may explain the higher degradation of Flt on these compounds, the overall yield and selectivity of N-Flt products are lower than for other matrices such as MnO. This was attributed to the photoactivation of the support leading to the formation of oxidising agents as opposed to nitrating agents.

Another effect proposed to influence the Flt reactivity was potential fluoranthene–surface interaction due to the acidity/basicity of the matrix.[274] The order of matrix acidity was Al_2O_3 > SiO_2 > TiO_2 > Fe_2O_3 > ZnO > MgO > CaO. This would explain the relatively high degradation rate of Al_2O_3 resulting from cation radical formation. The selectivity of nitration on different supports was observed to be almost opposite to the acidity order. It is therefore suggested that basicity of the support facilitates N-Flt formation and hence explains the higher reaction yield and selectivity on MgO.

The photochemical activation of fluoranthene sorbed to a surface was not considered to be an important factor in these studies.[274] It was noted that irradiation and the presence of O_2 are important factors in heterogeneous reactions of PAHs but the specific surface area and particle size have a relatively small influence compared to the surface chemical properties.

- Reactions with O_3 and OH: The reactivity of PAHs associated with graphite, diesel exhaust and urban particles towards ozone, OH and NO_2 have been shown to reach a 'plateau', indicating that no further reaction occurs after a certain exposure time.[255,256,258] These heterogeneous processes proceed via a rapid surface reaction but a significant fraction of PAH is unavailable for the reaction due to poor diffusion of the oxidant through and/or inaccessibility of the PAH within the bulk particle. This was studied for the reaction of BaP sorbed to ammonium sulfate particles with O_3.[308] Solid EC coating and various liquid organic coatings (when exceeding 10 nm of thickness) prevented the reaction. These results provide a strong indication for matrix effects to enhance the LRTP of PAHs.

Interesting differences were noted between the reactions of O_3 with PAHs for graphite and silica particles: the plateau observed in the e of graphite was relatively similar for the different PAHs studied (40–60%), while the fraction of unreacted PAH was highly variable in the e of silica particles (10–90%) and appeared to be dependent on the reactivity of individual PAH molecules.[298]

For reactions with both NO_2 and OH degradation on diesel particles is slower or stops earlier than on graphite particles.[256] For example the amount of PAH degradation was 7–25% on diesel particles, compared to 10–75% on graphite particles (the same NO_2 concentration). These differences may be related to differences in particle surface area (108 m^2 g^{-1} for diesel vs.13.2 m^2 g^{-1} for graphite) and the corresponding presence of pores that could house adsorbed pollutant molecules, leading to differences in the relative accessibility of oxidants.[256] This difference may be caused by differences in particle generation: in contrast to physisorption of PAHs to (pure) graphite particles, PAHs present in diesel are generated simultaneously and incorporated in the carbonaceous core of the particles.[256]

It has been noted that the ozonation reactions of PAHs on solid substrates (e.g. silica and graphite) are considerably faster than the reactions on organic films. For example, the rate coefficients for the reaction of ozone with BaP and Anth were shown to be up to a factor of ten higher in the studies conducted by Kwamena and co-workers[292,301] than those calculated by Kahan et al.,[296] seemingly contradicting the above suggestion of enhanced affinity of ozone for organic surfaces. However, this observation has been attributed to the potential formation of PAH dimers on the solid surface which may react differently with the gaseous oxidant and thus alter the reaction kinetics.[296]

Ozone will have a higher affinity for non-polar surfaces, therefore will be more attracted to hydrophobic, organic particles like soot or azelaic acid and will have very low affinity for polar, ionic particles like NaCl.[291] For example, Kwamena et al.[291] noted that the partitioning efficiency for O_3 followed the order elemental carbon > fixed silica > non-activated silica > solid organic carbon > inorganic salts.

Therefore, the heterogeneous reaction of PAH with ozone could be enhanced by the presence of water. For example, Kwamena et al.[291] observed a partitioning coefficient for ozone to azelaic acid at high RH that was 230% higher than the value that was observed under dry conditions, leading to a pseudo-first order rate coefficient increase for the reaction of anthracene of 25%.[291] It was suggested that the partitioning of gas phase water to the particle provided a substrate to which O_3 had a greater affinity and hence led to enhanced O_3 partitioning. While theK_{O3} values for these two RH conditions differ by more than a factor 2, the actual pseudo-first order rate coefficient does not vary to such a degree (Table 8).

In contrast, however, in the presence of water both gas phase ozone loss and BaP decay rates were reduced,[290] attributed to the competitive adsorption of O_3 and H_2O. Additionally, the reaction kinetics of anthracene reacting with ozone on Pyrex tubes was unaffected by RH conditions.[292] Pyrex may be too hydrophobic for water adsorption and, hence, water may not have partitioned to the Pyrex surface during the high RH experiments.[292]

The enhanced reaction rate of anthracene with ozone on the octanol-coated aqueous surface compared to the pure water surface[293,294] can be explained by the fact that octanol is less polar than water and, hence, ozone will have a higher affinity for this surface. A similar explanation was given for the higher reaction rate for naphthalene on water droplets in the presence of fulvic acid.[295]

- Influence of PAH molecular structure: Where heterogeneous reactions of PAHs with NO_2 have been reported, the relative heterogeneous reactivity of NO_2 with different PAHs has been observed to follow the general order BaP > Pyr > Chr > Flt, Phe.[255,256,270,279] It is indicated that for the reactions of PAHs with O_3, Anth, BaA, BaP and DBalP are the most reactive compounds and Flt, Chr, BkF and Cor are the least reactive.[286,288,298]

It has been suggested that the nature of the PAH structure has the potential to influence heterogeneous reactivity of PAHs associated with silica particles, both for reactions with NO_2 and O_3.[276,298] The values of the derived second-order rate coefficients are considerably more variable between molecules for the reactions of PAHs with NO_2 (variation of ≈5 orders of magnitude) than the reactions with O_3 (variation of ≈1 order of magnitude).

The variability of reaction kinetics for different PAH structures observed for silica particles is in contrast to observations of PAHs adsorbed on carbonaceous particles (graphite, diesel exhaust, wood soot) where oxidation rates do not appear to be strongly dependent on the nature of the PAH for OH, NO_2 and O_3 reactions.[252,255,256,276,298] The derived $k^{(2)}$ values in these studies fall within the same order of magnitude for all PAH molecules assessed in these studies.

Similar rate coefficients of PAHs associated with carbonaceous particles, for example graphite, may be explained by chemical stabilization of the PAH due to delocalized π-electrons, which is effective regardless of PAH structure.[276] The interactions between PAHs and silica are not as strong as those for graphite, which may explain the enhanced differences in heterogeneous reactivity between the individual PAHs.[276]

Kwamena et al.[301] suggested that the values of the adsorption coefficient (K_{O3}) for the reactions of surface-bound PAH and ozone reflect the partitioning behavior of ozone to the surface irrespective of the PAH adsorbed onto it. This indicates that, once ozone is adsorbed onto the aerosol surface, there is an inherent barrier to the reaction regardless of the specific PAH that would result in similar rate coefficient values, although there is significant variation between calculated pseudo-first order rate coefficients in the reported studies. This barrier could be related to the level of mobility of the two reactants or to the transformation of ozone into more reactive forms on the substrate surface.

The lack of an observable trend between PAH structure and reactivity for heterogeneous reactions with OH is in contrast to the observations made for the corresponding reactions in the gas phase. In general, it is shown that the reactivity of PAHs with OH in the gas phase depends strongly on the structure of the specific molecule.[135,140,210] This clearly highlights the importance of the 'stabilizing effect' of organic particles on the reactivity of PAHs towards OH radicals. Similar comparisons for NO_2 and O_3 are not possible due to lack of data.

- Influence of PAH surface concentration: It is suggested that particle PAH reactivity with oxidants is influenced by the amount of PAH coated on the particle surface. Perraudin et al.[298] and Miet et al.[297] reported that the rate coefficient for the reaction of O_3 with pyrene adsorbed on silica particles can be up to 2 orders of magnitude higher when the particle surface concentration of pyrene was ≈50 μg g^{-1} compared to that of a surface concentration of ≈500 μg g^{-1}. The effect of this 'bulk shielding' due to multilayer coverage of BaP on soot particles was also shown to reduce the heterogeneous reactivity with ozone.[290] This demonstrates that heterogeneous reactivity of PAH with atmospheric oxidants is dependent on both the gaseous concentration of reactant and the particulate loading of PAH.

Perraudin et al.[298] commented that the enhanced heterogeneous reactivity of PAHs on silica particles at lower PAH particle loadings could partly be explained by the nature of the surface adsorption sites. They suggested that PAHs may be adsorbed on at least two different types of silica surface site, initially being coated on the more reactive sites, leading to higher reaction rates at lower surface concentrations. Miet et al.[297] noted that, while reactivity of PAH adsorbed on silica particles towards O_3 is linearly dependent on gaseous O_3 concentration for a constant PAH surface coating, the relationship for a constant O_3 concentration and different PAH particle loading is rather more complex. The authors attributed this to the possibility of PAH being adsorbed on two types of silica surface sites.

Similarly, Alebic-Juretic et al.[302] indicated that the rate coefficient for the reaction of ozone with PAHs adsorbed on silica gel surfaces is 1.3–2.3 times higher for less than a monolayer coverage of PAH than for a more than one monolayer. Persistance of BaP on soot and ammonium sulfate particles exposed to O_3 was attributed to this 'bulk shielding' effect[290,308] (Section 3.2.2.6.1.2). Therefore, instances of negligible reactivity reported for the reaction of PAHs exposed to ambient O_3 levels[83,281,309] may be explained by the lack of availability of BaP for reaction at the surface.

Butler and Crossley[270] and Nguyen et al.[275] reported a very low reactivity for PAHs on soot particles towards NO_2, in contrast to the studies of Estève et al.[255,256] for graphite and SRM 1650a diesel exhaust particles. It was noted by Nguyen et al.[275] that the typical PAH concentrations on SRM 1650a diesel exhaust are of the order of a few tens of $\mu g\ g^{-1}$ while those used by Butler and Crossley[270] and Nguyen et al.[275] are nearly two orders of magnitude higher.

A further explanation for the lower reactivity at higher PAH loading could be the presence of higher molecular weight oxidation products formed due to heterogeneous reactions.[273,290]These may have a lower volatility than the parent PAH and thus prevent the attacking gaseous oxidant from reaching the PAH that is 'hidden' underneath the exposed surface when there is a greater than monolayer coverage.

- Gas phase reaction vs. heterogeneous reaction: For semivolatile PAHs (mostly 3–4 ring compounds) comparisons for reactions with atmospheric oxidant gases can be carried out between reactions occurring in the gas phase and reactions occurring on various substrate types. This is possible only for the reactions of OH, NO_3 and O_3, as direct reactions of PAHs with gaseous NO_2 are not expected to occur in the gas phase.

It is suggested that reactivity of PAHs in the gas phase would be significantly greater than when associated with carbonaceous particles,[255,256] suggesting that the particle surface may have an inhibiting effect on the heterogeneous reaction with OH. It can be seen that $k^{(2)}$ values for PAH reactions with OH on carbonaceous particle surfaces are 1–3 orders of magnitude lower than those derived for gas phase reactions.

The presence of a 'plateau' in the experimental decays of PAHs in the reactions of OH indicates that a significant fraction of PAH is unavailable for reaction.[255,256] The higher level of this plateau observed in the e of diesel particles relative to that for graphite particles confirms the enhanced inhibiting factor of diesel particles compared to graphite. This 'stabilizing' effect of particles towards PAH reactivity is also reflected in the relatively similar rate coefficients observed for different PAHs associated with organic particles compared with the apparent PAH-dependent reactivity in the gas phase.

However, it is indicated that the reaction of PAHs with ozone is enhanced on particles in comparison with reactions in the gas phase. For example, Perraudin et al.[298] studied the reaction of PAHs with ozone adsorbed on graphite and silica particles and reported $k^{(2)}$ values approximately 2 orders of magnitude higher than those calculated for corresponding gas phase reactions (see Tables 2 and 6). Similar comparisons with studies investigating the reaction of ozone with individual PAHs on soot particles,[290] Pyrex,[292] organic aerosols[301] and the air–aqueous interface[293,294] are more difficult to make as second-order rate coefficients were not reported in these studies.

- Comparison of oxidizing species: absolute reaction rates. To assess the relative importance of the second order rate coefficients for the reactions of PAHs in the gas phase (Tables 3, 5 and 6) and associated with particle surfaces (Table 7), absolute reaction rates have been derived using the second order reaction rates derived from the literature and concentrations of atmospheric species typically found in the background atmosphere (Tables 9 and 10).

Table 9: Pseudo-first order reaction rate coefficients $k^{(1)}$ for the gas-phase reactions of PAHs with OH, NO_3 and O_3

	OH		NO_3		O_3	
	$k^{(1)}$ (s^{-1})	Ref.	$k^{(1)}$ (s^{-1})	Ref.	$k^{(1)}$ (s^{-1})	Ref.
Nap	4.8×10^{-5}	Phousongphouang and Arey[171]	5.5×10^{-8}	Pitts et al.[212]	1.4×10^{-7}	Atkinson et al.[168]
	4.3×10^{-5}	Atkinson[147]	3.1×10^{-8}	Atkinson et al.[162]	2.1×10^{-7}	Atkinson and Aschmann[144]
	4.6×10^{-5}	Brubaker and Hites[140]	2.1×10^{-8}	Atkinson and Aschmann[174]		
	5.3×10^{-5}	Klamt[172]	2.4×10^{-8}	Atkinson et al.[163]		
	4.7×10^{-5}	Biermann et al.[169]	2.7×10^{-8}	Atkinson et al.[163]		
	3.7×10^{-5}	Lorenz and Zellner[156]	2.3×10^{-8}	Atkinson[210]		
	4.3×10^{-5}	Klöpffer et al.[170]				
	4.8×10^{-5}	Atkinson et al.[168]				
	5.2×10^{-5}	Atkinson and Aschmann[144]				

1M-Nap	8.2×10^{-5}	Phousongphouang and Arey[171]	5.4×10^{-8}	Atkinson and Aschmann[173]	9.0×10^{-8}	Atkinson and Aschmann[173]
	1.1×10^{-4}	Atkinson and Aschmann[173]	4.5×10^{-8}	Atkinson and Aschmann[174]		
	1.2×10^{-4}	Klamt[172]	5.0×10^{-8}	Atkinson[210]		
			4.6×10^{-8}	Phousongphouang and Arey[214]		
2M-Nap	9.7×10^{-5}	Phousongphouang and Arey[171]	6.9×10^{-8}	Atkinson and Aschmann[173]	2.1×10^{-7}	Atkinson and Aschmann[144]
	1.1×10^{-4}	Atkinson and Aschmann[144]	7.0×10^{-8}	Atkinson and Aschmann[174]	2.8×10^{-7}	Atkinson and Aschmann[173]
	1.2×10^{-4}	Klamt[172]	7.0×10^{-8}	Atkinson[210]		
			6.6×10^{-8}	Phousongphouang and Arey[214]		
1E-Nap	7.3×10^{-5}	Phousongphouang and Arey[171]	6.3×10^{-8}	Phousongphouang and Arey[214]		
2E-Nap	8.0×10^{-5}	Phousongphouang and Arey[171]	5.2×10^{-8}	Phousongphouang and Arey[214]		
1,2DM-Nap	1.2×10^{-4}	Phousongphouang and Arey[171]	4.1×10^{-7}	Phousongphouang and Arey[214]		
1,3DM-Nap	4.4×10^{-5}	Banceu et al.[175]	1.4×10^{-7}	Phousongphouang and Arey[214]		
	1.5×10^{-4}	Phousongphouang and Arey[171]				
1,4DM-Nap	1.2×10^{-5}	Klamt[172]	8.4×10^{-8}	Phousongphouang and Arey[214]		
	1.2×10^{-4}	Phousongphouang and Arey[171]				
1,5DM-Nap	1.2×10^{-4}	Phousongphouang and Arey[171]	9.1×10^{-8}	Phousongphouang and Arey[214]		
1,6DM-Nap	1.3×10^{-4}	Phousongphouang and Arey[171]	1.1×10^{-7}	Phousongphouang and Arey[214]		
1,7DM-Nap	1.4×10^{-4}	Phousongphouang and Arey[171]	8.7×10^{-8}	Phousongphouang and Arey[214]		
1,8DM-Nap	1.3×10^{-4}	Phousongphouang and Arey[171]	1.4×10^{-6}	Phousongphouang and Arey[214]		
2,3DM-Nap	1.2×10^{-4}	Phousongphouang and Arey[171]	9.5×10^{-9}	Atkinson and Aschmann[173]		
	2.0×10^{-4}	Banceu et al.[175]	1.0×10^{-7}	Atkinson and Aschmann[174]	2.8×10^{-7}	Atkinson and Aschmann[144]
	1.5×10^{-4}	Atkinson and Aschmann[144]	1.0×10^{-7}	Atkinson[210]		
			9.8×10^{-8}	Phousongphouang and Arey[214]		
2,6DM-Nap	1.3×10^{-4}	Phousongphouang and Arey[171]	1.4×10^{-7}	Phousongphouang and Arey[214]		
2,7DM-Nap	1.4×10^{-4}	Phousongphouang and Arey[171]	1.4×10^{-7}	Phousongphouang and Arey[214]		
Ace	1.6×10^{-4}	Reisen and Arey[155]	2.3×10^{-4}	Atkinson and Aschmann[174]	3.5×10^{-7}	Atkinson and Aschmann[174]
	1.2×10^{-4}	Brubaker and Hites[140]	1.1×10^{-7}	Atkinson and Aschmann[174]		
	2.1×10^{-4}	Atkinson and Aschmann[174]				
	1.2×10^{-4}	Klöpffer et al.[170]				
	1.3×10^{-4}	Banceu et al.[175]				
	1.6×10^{-4}	Klamt[172]				
Acy	2.5×10^{-4}	Reisen and Arey[155]	2.7×10^{-3}	Atkinson and Aschmann[174]	3.8×10^{-4}	Atkinson and Aschmann[174]
	2.6×10^{-4}	Banceu et al.[175]			1.1×10^{-4}	Reisen and Arey[155]
	2.2×10^{-4}	Atkinson and Aschmann[174]				

Fln	3.2×10^{-5}	Kwok et al.[177]	1.8×10^{-5}	Kwok et al.[177]	1.4×10^{-7}	Kwok et al.[177]
	2.6×10^{-5}	Brubaker and Hites[140]				
	2.0×10^{-5}	Klamt[172]				
	2.6×10^{-5}	Klöpffer et al.[170]				
Phe	6.8×10^{-5}	Biermann et al.[169]	6.0×10^{-5}	Kwok et al.[176]	2.8×10^{-7}	Kwok et al.[176]
	6.2×10^{-5}	Atkinson[147]				
	5.1×10^{-5}	Klamt[172]				
	3.1×10^{-5}	Lorenz and Zellner[156]				
	2.5×10^{-5}	Kwok et al.[176]				
	5.4×10^{-5}	Brubaker and Hites[140]				
	6.4×10^{-5}	Lee et al.[159]				
1M-Phe	5.8×10^{-5}	Lee et al.[159]				
2M-Phe	1.3×10^{-4}	Lee et al.[159]				
3M-Phe	1.3×10^{-4}	Lee et al.[159]				
9M-Phe	1.5×10^{-4}	Lee et al.[159]				
Ant	2.2×10^{-4}	Biermann et al.[169]				
	3.8×10^{-4}	Brubaker and Hites[140]				
	2.6×10^{-5}	Kwok et al.[176]				
	4.0×10^{-4}	Klamt[172]				
	2.6×10^{-4}	Atkinson;[147] Biermann et al.[169]				
Flt	2.2×10^{-5}	Brubaker and Hites[140]	3.3×10^{-8}	Atkinson et al.[163]		
Pyr	1.0×10^{-4}	Atkinson et al.[163]	1.0×10^{-7}	Atkinson et al.[163]		
1N-Nap	1.1×10^{-4}	Atkinson[147]	1.9×10^{-8}	Atkinson[210]	4.1×10^{-7}	Atkinson[207]
2N-Nap	1.1×10^{-4}	Atkinson[147]	1.7×10^{-8}	Atkinson[210]	4.1×10^{-7}	Atkinson[207]

[a] $[OH] = 2 \times 10^{6}$ molecules cm^{-3}; global 12 h average (Atkinson and Arey[151]).

[b] $[NO_3] = 5 \times 10^{8}$ molecules cm^{-3}; global 12 h average (Atkinson and Arey[151]).

[c] $[O_3] = 6.9 \times 10^{11}$ molecules cm^{-3}; 2011 annual average, Harwell, UK.

Table 10: Pseudo-first order reaction rate coefficients $k^{(1)}$ for the heterogeneous reactions of PAHs with OH, NO_2, O_3 and NO_3

PAH	**OH reactions**			**NO_2 reactions**			**O_3 reactions**			**NO_3 reactions**		
	$k^{(1)}_{OH}$ (s^{-1})	**Ref.**	**Notes**	**$k^{(1)}_{NO_2}$ (s^{-1})**	**Ref.**	**Notes**	**$k^{(1)}_{O_3}$ (s^{-1})**	**Ref.**	**Notes**	**$k^{(1)}_{NO_3}$ (s^{-1})**	**Ref.**	**Notes**
Phe	1.0×10^{-5}	Estève et al.[255]	Graphite particles	1.9×10^{-7}	Perraudin et al.[276]	Silica particles	3.1×10^{-6}	Perraudin et al.[298]	Graphite particles			
	6.5×10^{-7}	Estève et al.[256]	Diesel exhaust particles	2.3×10^{-6}	Nguyen et al.[275]	Kerosene flame soot	3.0×10^{-6}	Perraudin et al.[298]	Silica particles			

				2.3×10^{-6}	Estève et al.[255]	Graphite particles						
				7.8×10^{-6}	Estève et al.[256]	Diesel exhaust particles						
				1.6×10^{-9}	Butler and Crossley[270]	Ethylene flame soot						
Ant	8.8×10^{-6}	Estève et al.[255]	Graphite particles	6.9×10^{-5}	Perraudin et al.[276]	Silica particles	1.3×10^{-5}	Perraudin et al.[298]	Graphite particles			
				2.3×10^{-6}	Nguyen et al.[275]	Kerosene flame soot	1.8×10^{-5}	Perraudin et al.[298]	Silica particles			
				4.8×10^{-5}	Estève et al.[255]	Graphite particles						
				6.9×10^{-5}	Ma et al.[278]	Silica particles						
				3.7×10^{-5}	Ma et al.[278]	MgO particles						
Flo	2.9×10^{-8}	Bedjanian et al.[252]		2.2×10^{-9}	Perraudin et al.[276]	Silica particles	2.5×10^{-6}	Perraudin et al.[298]	Graphite particles			
	6.5×10^{-6}	Estève et al.[255]	Graphite particles	6.9×10^{-8}	Nguyen et al.[275]	Kerosene flame soot	1.9×10^{-6}	Perraudin et al.[298]	Silica particles			
	7.6×10^{-7}	Estève et al.[256]	Diesel exhaust particles	2.0×10^{-5}	Estève et al.[255]	Graphite particles						
				6.9×10^{-6}	Estève et al.[256]	Diesel exhaust particles						
				1.7×10^{-9}	Butler and Crossley[270]	Ethylene flame soot						
Pyr	3.1×10^{-8}	Bedjanian et al.[252]	Kerosene flame soot	1.4×10^{-5}	Perraudin et al.[276]	Silica particles	3.2×10^{-6}	Perraudin et al.[298]	Graphite particles	3.2×10^{-3}	Liu et al.[260]	Azelaic acid particles
	6.5×10^{-6}	Estève et al.[255]	Graphite particles	6.9×10^{-8}	Nguyen et al.[275]	Kerosene flame soot	7.6×10^{-6}	Perraudin et al.[298]	Silica particles			
	8.2×10^{-7}	Estève et al.[256]	Diesel exhaust particles	3.5×10^{-5}	Estève et al.[255]	Graphite particles	4.2×10^{-5}	Miet et al.[297]	Silica particles			
	4.8×10^{-7}	Miet et al.[257]	Silica particles	1.0×10^{-5}	Estève et al.[256]	Diesel exhaust particles						
				3.3×10^{-9}	Butler and Crossley[270]	Ethylene flame soot						
				6.4×10^{-5}	Miet et al.[277]	Silica particles						
Chr	1.8×10^{-8}	Bedjanian et al.[252]	Kerosene flame soot	4.1×10^{-7}	Perraudin et al.[276]	Silica particles	1.9×10^{-6}	Perraudin et al.[298]	Graphite particles	2.0×10^{-3}	Liu et al.[260]	Azelaic acid particles
	1.0×10^{-5}	Estève et al.[255]	Graphite particles	6.9×10^{-8}	Nguyen et al.[275]	Kerosene flame soot	4.0×10^{-6}	Perraudin et al.[298]	Silica particles			
	8.8×10^{-7}	Estève et al.[256]	Diesel exhaust particles	2.6×10^{-5}	Estève et al.[255]	Graphite particles						

				6.9×10^{-6}	Estève et al.[256]	Diesel exhaust particles						
				1.8×10^{-9}	Butler and Crossley[270]	Ethylene flame soot						
BaA	1.8×10^{-8}	Bedjanian et al.[252]	Kerosene flame soot	4.6×10^{-6}	Perraudin et al.[276]	Silica particles	3.6×10^{-6}	Perraudin et al.[298]	Graphite particles	2.2×10^{-3}	Liu et al.[260]	Azelaic acid particles
	1.1×10^{-6}	Estève et al.[255]	Graphite particles	6.9×10^{-8}	Nguyen et al.[275]	Kerosene flame soot	1.1×10^{-5}	Perraudin et al.[298]	Silica particles			
	6.5×10^{-7}	Estève et al.[256]	Diesel exhaust particles	2.2×10^{-5}	Estève et al.[255]	Graphite particles						
				8.6×10^{-6}	Estève et al.[256]	Diesel exhaust particles						
				4.3×10^{-9}	Butler and Crossley[270]	Ethylene flame soot						
BkF	2.1×10^{-8}	Bedjanian et al.[252]	Kerosene flame soot	1.5×10^{-6}	Perraudin et al.[276]	Silica particles	2.5×10^{-6}	Perraudin et al.[298]	Graphite particles			
	7.1×10^{-6}	Estève et al.[255]	Graphite particles	6.9×10^{-8}	Nguyen et al.[275]	Kerosene flame soot	4.6×10^{-6}	Perraudin et al.[298]	Silica particles			
				1.7×10^{-5}	Estève et al.[255]	Graphite particles						
BaP	2.2×10^{-8}	Bedjanian et al.[252]	Kerosene flame soot	6.4×10^{-4}	Perraudin et al.[276]	Silica particles	6.8×10^{-6}	Perraudin et al.[298]	Graphite particles			
	8.2×10^{-6}	Estève et al.[255]	Graphite particles	6.9×10^{-8}	Nguyen et al.[275]	Kerosene flame soot	1.8×10^{-5}	Perraudin et al.[298]	Silica particles			
	5.9×10^{-7}	Estève et al.[256]	Diesel exhaust particles	5.4×10^{-5}	Estève et al.[255]	Graphite particles						
				1.0×10^{-5}	Estève et al.[256]	Diesel exhaust particles						
				6.9×10^{-9}	Butler and Crossley[270]	Ethylene flame soot						
BeP	2.2×10^{-8}	Bedjanian et al.[252]	Kerosene flame soot	2.0×10^{-6}	Perraudin et al.[276]	Silica particles	2.1×10^{-6}	Perraudin et al.[298]	Graphite particles			
	9.4×10^{-6}	Estève et al.[255]	Graphite particles	6.9×10^{-8}	Nguyen et al.[275]	Kerosene flame soot	3.7×10^{-6}	Perraudin et al.[298]	Silica particles			
	9.4×10^{-7}	Estève et al.[256]	Diesel exhaust particles	2.4×10^{-5}	Estève et al.[255]	Graphite particles						
				5.2×10^{-6}	Estève et al.[256]	Diesel exhaust particles						
				2.0×10^{-9}	Butler and Crossley[270]	Ethylene flame soot						
Per	1.0×10^{-5}	Estève et al.[255]	Graphite particles	7.3×10^{-5}	Estève et al.[255]	Graphite particles						
IPy	7.1×10^{-7}	Estève et al.[256]	Diesel exhaust particles	4.3×10^{-6}	Perraudin et al.[276]	Silica particles	2.5×10^{-6}	Perraudin et al.[298]	Graphite particles			

				6.9×10^{-8}	Nguyen et al.[275]	Kerosene flame soot	4.9×10^{-6}	Perraudin et al.[298]	Silica particles			
				5.2×10^{-6}	Estève et al.[256]	Diesel exhaust particles						
BgP	1.2×10^{-5}	Estève et al.[255]	Graphite particles	3.2×10^{-5}	Perraudin et al.[276]	Silica particles	2.5×10^{-6}	Perraudin et al.[298]	Graphite particles			
				6.9×10^{-8}	Nguyen et al.[275]	Kerosene flame soot	9.2×10^{-6}	Perraudin et al.[298]	Silica particles			
				2.6×10^{-5}	Estève et al.[255]	Graphite particles						
				5.5×10^{-9}	Butler and Crossley[270]	Ethylene flame soot						
BgF	1.7×10^{-8}	Bedjanian et al.[252]	Kerosene flame soot	6.9×10^{-8}	Nguyen et al.[275]	Kerosene flame soot						
AcP	2.1×10^{-8}	Bedjanian et al.[252]	Kerosene flame soot	6.9×10^{-8}	Nguyen et al.[275]	Kerosene flame soot						
DBA	3.2×10^{-8}	Bedjanian et al.[252]	Kerosene flame soot	6.9×10^{-8}	Nguyen et al.[275]	Kerosene flame soot						
DBaeP	2.1×10^{-8}	Bedjanian et al.[252]	Kerosene flame soot	6.9×10^{-8}	Nguyen et al.[275]	Kerosene flame soot						
BbF	2.4×10^{-8}	Bedjanian et al.[252]	Kerosene flame soot	6.9×10^{-8}	Nguyen et al.[275]	Kerosene flame soot						
Cor	2.2×10^{-8}	Bedjanian et al.[252]	Kerosene flame soot	9.2×10^{-7}	Perraudin et al.[276]	Silica particles	2.5×10^{-6}	Perraudin et al.[298]	Graphite particles			
				6.9×10^{-8}	Nguyen et al.[275]	Kerosene flame soot						
				1.6×10^{-9}	Butler and Crossley[270]	Ethylene flame soot						
DBalP				1.2×10^{-4}	Perraudin et al.[276]	Silica particles	1.7×10^{-5}	Perraudin et al.[298]	Graphite particles			
							1.7×10^{-5}	Perraudin et al.[298]	Silica particles			
1N-Pyr	2.0×10^{-7}	Miet et al.[257]	Silica particles	4.3×10^{-6}	Miet et al.[277]	Silica particles	2.9×10^{-6}	Miet et al.[297]	Silica particles	6.5×10^{-4}	Liu et al.[260]	Azelaic acid particles

[a] $[OH] = 2 \times 10^6$ molecules cm^{-3}; global 12 h average.[151]

[b] $[NO_2] = 1.3 \times 10^{11}$; 2011 annual average, Harwell, UK.

[c] $[O_3] = 6.9 \times 10^{11}$ molecules cm^{-3}; 2011 annual average, Harwell, UK.

[d] $[NO_3] = 5 \times 10^8$ molecules cm^{-3}; global 12 h average.[151]

These tables indicate that the gas phase reaction with OH remains the dominant loss process for PAHs and these reactions are approximately 3 orders of magnitude faster than the corresponding gas phase reactions with NO_3.

It appears that for heterogeneous processes, PAH reactions with NO_2 and O_3 may be more important than second order rate coefficients would suggest. Indeed the absolute rates for PAH reactions with NO_2 and O_3 are shown to be comparable to those of OH reactions, depending on the specific surface type and conditions.

Summary and Concluding Remarks

- Key conclusions.

- Heterogeneous reactivity of PAHs with atmospheric oxidants involves complex processes that have been shown to be influenced by a number of factors including the nature (chemical composition, surface area, porosity) of the matrix, presence of other species (e.g. nitric acid, water or other organic species), oxidant concentration, PAH molecule involved, and PAH surface concentration.
- The wide diversity in particle properties, including their chemical composition (organic, mineral, biogenic), sources (combustion, erosion, gas phase condensation), origin (natural, anthropogenic), the method of particle formation (temperature, pressure), physical properties (size, porosity, specific surface area), surface coatings (water, nitric acid, organic molecules), means that gaining a full understanding of this reactivity will be extremely difficult as the relative importance of these different factors is highly variable across reaction systems.
- Reaction with OH radicals is the dominant pathway for heterogeneous degradation of PAHs compared with reactions with NO_2 and O_3 with second order rate coefficients 1–7 orders of magnitude higher on carbonaceous particles.
- Particles are shown to exhibit a potential 'inhibiting factor' on the reactivity of PAHs due to slow diffusion of the oxidant or inaccessibility of PAHs in the bulk particle. This may turn PAHs to persistent compounds in air.
- Carbonaceous particles have the potential to stabilize the reactivity of PAH towards OH with slower rates than observed in the gas phase.
- Heterogeneous reactions of PAHs with O_3 are observed to proceed more quickly than the corresponding reactions in the gas phase, with second order rate coefficients ≈2 orders of magnitude greater for certain compounds and model matrices studied so far.
- A number of ring-opened and ring-retaining products have been identified from these heterogeneous reactions including nitro-PAHs and quinones. In some es, a difference has been noted between heterogeneous and gas phase reaction products. For example the observation of 1-nitropyrene from the reaction of particle-bound PAH is in contrast to the production of 2-nitropyrene from gas phase reactions.
- The heterogeneous reaction of PAH with ozone on a range of surfaces has been shown to follow Langmuir–Hinshelwood-type kinetics, where the pseudo-first order reaction rate is characterized by a dependence on gas phase concentration and available reaction sites and reactivity is strongly dependent on the degree of partitioning of ozone to the surface. This is shown to be strongly influenced by the specific nature of the particle surface.
- When atmospheric concentrations of the different oxidizing species are considered, the absolute heterogeneous reaction rates of PAHs with OH, NO_2 and O_3 are comparable, depending on the specific particle surface and reaction conditions.
- Heterogeneous reaction chemistry of PAHs is far from being fully understood with knowledge of the reaction mechanisms, products and the factors that influence the reaction kinetics still lacking. Reaction rates for ambient conditions are very difficult to extrapolate, also because many aerosol matrices have not been included in laboratory studies so far.
- Recommendations for future work: The following are areas we suggest for continued research in the area of heterogeneous PAH reactivity:
- Investigation of the mechanisms of the heterogeneous reactions of PAHs with NO_2, OH and NO_3 for various aerosol matrices and conditions.
- More comprehensive elucidation of reaction products of the heterogeneous reactions of PAHs, particularly for OH and NO_3 reactions.
- Investigation of the reactions with OH radicals on silica particles for comparison with other reaction systems.
- Verification of the reactivity and products based on field studies, with particular emphasis on matrix effects.

Atmospheric Implications

Photochemical lifetimes in the background atmosphere of gaseous PAHs, τ_{hom}, can be estimated based on the O_3 and OH kinetics (Sections 3.2.1.1 and 3.2.1.3), while the reactivity with oxidized N species will be largely limited to the high NO_x source areas. Present knowledge suggests τ_{hom} to range between ≈1 hour and ≈2 days for oxidant concentrations typical for the continental background in mid latitudes. As an implication of the diel variation of radical abundances, longer lifetimes result for night-time emission or transport in high latitudes.

Clearly, heterogeneous reactions (Section 3.2.2) can significantly contribute to the removal of semivolatile and non-volatile PAHs from the atmosphere and to the formation of mutagenic derivative compounds. A quantitative assessment of this contribution has to be considered preliminary and with great caution, because the representativity of model aerosols and laboratory conditions for ambient conditions is insufficient. For example, the matrix of secondary inorganic and organic aerosols has hardly been addressed. Furthermore, the processes ruling gas–particle partitioning are not fully understood, with the consequence that particulate mass fractions, θ, of the semivolatile PAHs in a given aerosol are uncertain, in particular in the cold (free troposphere, Arctic). Assuming representativity of studied model aerosols for ambient aerosols (namely silica for unspecific particulate mass, diesel and flame soot for BC) and a range of θ observed (although in the near ground atmosphere only) the PAHs' lifetimes towards heterogeneous photochemical degradation, τ_{het}, are estimated to range between 2 hours and >10 days for oxidant and BC concentrations typical for the continental background in mid latitudes. (For values >10 days atmospheric lifetime would be limited by physical sinks, i.e. deposition.) Total photochemical residence times, τ_{total}, resulting as $\tau_{total} = [(1 - \theta)/\tau_{hom} + \theta/\tau_{het}]^{-1}$, are listed in Table 9. Obviously, most of the uncertainty of τ_{total} for semivolatile and non-volatile PAHs results from the consideration of heterogeneous reactivity (related to the aerosol characteristics' variability). It enhances overall reactivity significantly for the semivolatile (partly partitioning to particles) PAHs, e.g. Flt and Chr, but less for Pyr (Table 9).

Zimmerman et al.[310] recently observed the formation of nitro-PAH isomers on PM samples exposed to a mixture of N_2O_5–NO_2–NO_3 in a chamber reaction study. The authors attributed this to a heterogeneous reaction with surface-adsorbed N_2O_5 as opposed to reaction with radical NO_3. It was proposed that such heterogeneous reactivity could lead to the formation of nitro-PAH isomers such as 1-NPyr during atmospheric transport of PM. However, the concentrations of NO_2 and N_2O_5 used in this study were factors of >100 higher than observed under ambient conditions so the applicability of this observation in the atmosphere is uncertain.

LONG-RANGE TRANSPORT

Slow photochemical degradation (Section 3.2) and limited deposition along transport (Section 3.1.2) permit long-range atmospheric transport of PAHs.

Observations at Remote Sites

Evidence of long-range transport of PAH comes from numerous observations in air and precipitation/deposition at remote terrestrial and marine sites (Table 11). Only a few monitoring stations at remote sites are operating and these are concentrated in Europe and North America. Levels at remote sites are at least one order of magnitude lower than within source areas[311,312] and some of the (often addressed) 16 USEPA priority PAHs were below the limits of quantification. Elevated PAH concentrations far from combustion sources were observed in long-range transported plumes of open fires, such as from biomass burning in Africa (over the Atlantic[313]) and boreal forest fires (in high latitudes[62,314]). Because of photodegradation being more efficient in summer and emissions being higher in winter, winter levels usually exceed summer levels at receptor sites. An exception, i.e. minima in winter, was observed on mountain sites in Oregon, USA[315] which can be explained by air mass trajectory statistics. Observations at remote sites are usually considered to prove the LRT potential. In some es, however, the influences of local emissions are difficult to exclude. For example, the Zeppelin observatory, near Ny Alesund, Svalbard, is considered to reflect the remoteness of the Arctic island's environment with usually very low levels observed (e.g. 46–75 pg m^{-3} in July and 260–1500 pg m^{-3} in January of recent years for the sum of 15 PAH; Table 11). However, samples collected at the station may have been influenced by smoldering combustion in underground coal mines in the years 2005–2006 and 2009–2010 and in a mine dump 2006–2008 in the region (located 150–200 km north of the station; Kallenborn, Norwegian Institute for Air Research, personal communication).

Table 11: Arithmetic mean of PAH measurements in air at remote sites, arithmetic mean concentration (sum of gaseous and particulate phase concentrations, pg m^{-3}). Sum of 15 PAHs, namely acenaphthylene, acenaphthene, fluorene, phenanthrene, anthracene, fluoranthene, pyrene, benzo[a]anthracene, chrysene, benzo[b]fluoranthene, benzo[k]fluoranthene, benzo[a]pyrene, indeno[123c,d]pyrene, dibenz[a,h]anthracene and benzo[ghi]perylene (i.e. 16 USEPA priority PAHs without naphthalene), unless otherwise specified

		n	c_{tot}	Ref.
Continental remote and mountain sites				
Tagish, Yukon, Canada 60°N/134°W	Oct 1993–Apr 1994/May–Sep 1993	7/5	300/61	Halsall et al.[318]
Redó, Pyrenees (2250 m a.s.l.)	Jun 1997/Feb 1997	7/3	≈1650/2750[a]	Fernández et al.[317]
Gossenkölle, Alps (2413 m a.s.l.)	Jul 1997/Mar 1997	6/4	≈2750/2000[a]	
Øvre Neådalsvatn, Central Norway (728 m a.s.l.)	1996–1998	8	1900	van Drooge et al.[325]
Izana, Tenerife (2367 m a.s.l.)	2000	12	240	
Skalnaté Pleso, High Tatras (1778 m a.s.l.)	2001–2002	9	5100	
Mt. Bachelor, Oregon, USA (44°N/122°W, 2763 m a.s.l.)	Spring and summer/fall and winter 2004–2006	26/10	≈1400/≈450	Primbs et al.[315]
Lake Nam Co, Tibetan Plateau (31°/91°E, 4730 m a.s.l.)	Oct 2006–Feb 2008	15	75–1470[b]	Xiao et al.[326]
Jungfraujoch, Alps (3573 m a.s.l.)	Jul–Oct 2006	1	1440[c,d]	Halse et al.[319] [e]
Moussala, Balkans (2925 m a.s.l.)		1	500[c,d]	
Chopok, High Tatras (2008 m a.s.l.)		1	1300[c,d]	
Zugspitze, Alps (2650 m a.s.l.)	Jun 2007/Feb 2008	8/14	71/1027	Lammel et al.[322]

Molopo Reserve, Barberspan (steppe), South Africa (1366 m a.s.l.)	Jan–Jun 2008	12	36–125[e]	Klánová et al.[312]
Tombouctou, (desert) Mali, West Africa (200 m a.s.l.)		6	55–160[e]	
Mt. Kenya (3650 m a.s.l.)		6	23–36[e]	
Marine sites				
Finokalia, Crete, Mediterranean 35°N/26°E	Feb 2000–Feb 2002	23	13 500	Tsapakis and Stephanou[25]
North Pacific (cruise) 52–62°N	Jul–Sep 2003	10	11 260	Ding et al.[26]
Biscay Bay, Atlantic (cruise) 46–49°N	Oct–Nov 2005	3	1370[f]	Nizzetto et al.[313]
Eastern North Atlantic (cruise) 25–31°N		4	230[f]	
Eastern Equatorial Atlantic (cruise) 0–25°S		13	170[f]	
Malta, Giordan Lighthouse, Med. 36°N/14°E	Aug–Nov 2006	1	2510[c,d]	Halse et al.[319] [e]
Mace Head, Ireland 53°N/10°W	Jul–Oct 2006	1	5690[c,d]	
Arctic and Antarctic sites				
Barrow, Alaska, US, 71°N	Mar 1979	3	≈1190[b]	Daisey et al.[316]
Aug 1979	5	157[b]		
Dye 3, Greenland 65°N/44°W	Winter-spring 1989 and spring-summer 1991	9	18–200[f]	Jaffrézo et al.[321]
Alert, Canadian Archipelago 82°N/62°W	Feb–Apr 1988	10	≈290–1100[b]	Patton et al.[324]
	Oct 1993–Apr 1994/May–Sep 1994	7/5	695/48	Halsall et al.[318]
	Jan–Dec 1992/1993/1994	52/52/52	900/570/370	Hung et al.[320]
	Jan–Dec 1998/1999/2000	52/52/52	430/180/150	
Dunai Island, Eastern Siberia 74°N/124°E	Nov 1993–Mar 1994/ May–Aug 1993	4/4	2450/122	Halsall et al.[318]
Zeppelinfjell, Spitsbergen 79°N12°E	Jul 1994–1996/Jan 1994–1996[g]	13/13	180–330/2400–6600[h]	EMEP[16]
	Jul 2008–2010/Jan 2008–2010[g]	13/13	46–75/260–1500	
Arctic Ocean (cruise) 65-80°N/150–180°W	Jul–Sep 2003	29	3700	Ding et al.[26]
Pallas, Finland 68°N/24°E	Jul–Oct 2006	1	600[c,d]	Halse et al.[319] [e]

[a] Sum of 24, including naphthalene.

[b] Sum of 11.

[c] Sum of 8.

[d] Passive sampling, equivalent air volume inferred from performance reference compound or following Melymuk et al.[323]

[e] Passive sampling, equivalent air volume inferred from side-by-side sampling (Klánová, personal communication).

[f] Sum of 9.

[g] Total sampling time of 8 d per month.

[h] Sum of 12.

Snow (firn) and ice core studies allow the assessment of temporal deposition trends. PAH deposition at high mountain sites in Europe and North America was traced back until 1700 and 1875, respectively.[327,328] Levels in Europe increased with fossil fuel use (coal), significantly by the year 1900.[327,329] PAH accumulates in high altitude soils and biota.[330,331] Contamination in the Arctic is comparable with high mountain sites in Europe: PAH levels in the range of 36–660 ng L^{-1} were reported from an ice core in the Canadian Archipelago attributed to the years 1963–1993[332] and 0.6–237 ng kg^{-1} in an ice core from Greenland attributed to 1987–1990[333] (Table 11). In the Arctic, meridional transports carrying PAHs are common in particular in the so-called haze season, i.e. spring-time.[318,334]

No continuous monitoring of PAHs in Antarctic air has been conducted so far. Deposited amounts are lower than in the Arctic, it seems. Snow contamination by PAH on the Ekström shelf ice, 2002–2005, was completely explained by emissions within the region (ships, research stations), whereas no influence of intercontinental transport was suggested.[335] Mostly due to the atmospheric dynamics of the region (polar vortex in the Southern Ocean), intercontinental atmospheric transport to the Antarctic continent is much less likely than in the Arctic. Levels in snow are 18–99 ng L^{-1} in surface and deep snow for 1987–1991 on the Antarctic Peninsula (Terra Nova Bay, 74–75°S, 163–164°E[336]) and 26–197 ng L^{-1} on the Ekström Shelf Ice in the Weddell Sea.[335]

PAH observations in the remote marine environment are very rare. Ship-based measurements are often unreliable because of blank problems related to fuel emissions from the ship.[337]Measurements in seawater demonstrate the deposition of PAH from combustion sources to the open ocean.[337]

Model Results

The transport of fluorene, phenanthrene, fluoranthene and benzo[a]pyrene from the UK to northern Russia has been studied using a Lagrangian box model.[338] Regional (North America) and trans-Pacific transport pathways of PAHs were documented based on back trajectories.[314,315,339–341] The atmospheric fate of PAHs was studied for the Los Angeles Basin using another multimedia model.[342] The potential influence of seasonally dependent environmental fate processes on the observed seasonality of concentrations of six PAHs in air was investigated for the UK using a multimedia box model (climatological atmospheric transports[343]). The regional and hemispheric transport and fate of benzo[a]pyrene was studied by atmospheric and multicompartment models with a focus on Europe[35,344–348] and East Asia.[349,350] Global distribution and fate of benzo[a]pyrene and 1–2 semivolatile PAHs were studied using an atmosphere general circulation model based multicompartment model[36,351] and using an assimilated meteorology driven atmospheric model.[352] Due to lack of consolidated kinetic data the heterogeneous reactivity of PAHs (i.e. basically total reactivity in the e of BaP) has been neglected so far in most model studies (assumption of zero degradability in the particulate phase).

Conventional atmospheric chemistry models neglect re-volatilization from surfaces. As benzo[a]pyrene volatility is low (vapour pressure, $p = 7.3 \times 10^{-7}$ Pa at 298 K) its tendency to re-volatilise upon deposition to surfaces is limited, but still secondary emissions are estimated to contribute 9% to total emissions on the European continent (2007 emissions[345]). However, parameterisations of air–soil exchange which are in use in multicompartment models are not well validated. Multicompartment models so far have been neglecting PAH cycling in the cryosphere (sea ice, land ice, temporary snow cover).

The LRT potential of PAHs has been quantified.[36,345,351,352] For benzo[a]pyrene, benzo[b]fluoranthene, benzo[k]fluoranthene, and indeno[1,2,3-cd]pyrene it was characterized to be regional and close to the one of polybrominated diphenylethers; i.e., areas of high emission are still reflected in the deposition pattern (e.g. Gusev et al.[35]). However, a large part of PAH transport occurs in the free troposphere, above the planetary boundary layer.[350,351] Atmospheric half-lives of 3–5 ring PAHs are of the order of hours or days and do vary among model studies considerably. This reflects uncertain or unknown rate coefficients and the uncertainty with regard to air–soil exchange and processes which determine partitioning, as well as different methods of representation of these processes. Accordingly, model estimates for the total environmental residence time, which weighs all the compartmental residence times including in soil and water, where degradability is slow, do vary largely across models, namely 2–66 months for benzo[a]pyrene.[353] Assuming similar degradability in the particulate and gas phase would lead to unrealistically short lifetimes.[36,352] Hence, modelling provides evidence that degradation kinetics in the particulate phase must be significantly slower than in the gas phase.

Gas–particle partitioning strongly influences the atmospheric cycling, the total environmental fate including compartmental distributions and the LRT potential. Using a global model including a dynamic aerosol sub-model with the components OM and soot, the comparison of various gas–particle partitioning models suggests that for BaP, Ant and Flt both adsorption (preferentially to soot) and absorption (into OM) processes are contributing, and that total environmental burdens would be significantly lower assuming adsorption alone.[36] Most of the total environmental burden of BaP, Ant and Flt, namely 80–90%, is found in soil (including vegetation) after a few years of continuous emission into the model world atmosphere, while the atmospheric burdens of these substances account for 2–4% and oceanic burdens for 5–12%.

The effective meridional spreading (MS) and the spatial range (SR), measures for the tendency of distributions to extend across latitudes, of benzo[a]pyrene were quantified to 1000 km in air by a zonal multimedia fate model but and 10 000 km by the general circulation model based multicompartment model.[353] Multimedia fate models underestimate the transport in the free troposphere, because temporal and spatial variability of wind is not represented and, hence, should underestimate the LRT potential. Multimedia models have been used in the context of substance screening for POPs criteria and accounting for the multicompartmental cycling of PAH on various spatial scales (e.g. Hauck et al.[354]). The LRT potential of semivolatile PAHs is higher than that of benzo[a]pyrene and other 5-ring PAHs. It is enhanced by re-volatilization. Re-volatilization was estimated to account for >10% of the Ant and Flt global total emissions throughout the year.[36] The numerical values of the various LRT metrics in use strongly depend on the evaluation methodology and on model design.

As a number of PAHs, e.g. Pyr (Section 3), are converted in photochemistry into long-lived derivatives which undergo LRT[1,355] and, eventually, are ubiquitously distributed, the assessment of PAHs' LRT potential and environmental risk should also include these derivatives. This aspect has been addressed in a box model study only.[342] As PAHs can enter the food web, integrated modelling is needed to assess the exposure of the ecosystems[356,357] and, hence, uptake along food chains.

Recommendations for Future Work

The assessment of LRTP relies on remote observations and related model results, i.e. residence times and LRTP.

- Monitoring at remote sites of continents other than Europe and North America
- Validation of air–soil exchange parameterizations used in models in order to better account for the LRT potential being enhanced by secondary emission (re-volatilization)
- Extend assessment of pollutants' LRT potential to derivatives of PAHs, in particular nitro-PAHs
- Include PAHs in food web modelling

OBSERVATIONAL EVIDENCE FOR ATMOSPHERIC REACTIONS OF PAHS IN THE ATMOSPHERE AND FORMATION OF NITRO- AND OXY-DERIVATIVES

Levels of oxy- and nitro-PAH compounds in the ambient atmosphere, as well as the extent and nature of their direct emission sources have been widely investigated in the last 30 years. Tables S1 and S2 of the ESI† provide an overview of specific oxy- and nitro-PAH compounds respectively, identified in atmospheric samples, in studies of different combustion emission sources, and in product studies of laboratory reaction experiments. A review of environmental occurrence of oxy-PAH in the atmosphere was recently provided by Walgraeve et al.,[358] which discussed the levels of these compounds in the atmosphere as well as their seasonality, partitioning behavior and environmental fate. To our knowledge, no such review exists for nitro-PAH compounds. It should be noted that Tables S1 and S2 (ESI†) do not provide details of measured concentrations. For details of this, the reader is directed to the specific references provided or the review by Walgraeve et al.[358]

As discussed in detail in Section 3, laboratory experiments strongly indicate that both gas phase and heterogeneous reactions of PAHs can occur under environmentally relevant conditions in the atmosphere and can contribute to the observed levels of these compounds. The potentially mutagenic nature of these reaction products is a concern from a public health perspective. However, the key questions remain: (i) to what extent do these reactions occur in the ambient atmosphere? (ii) What is the relative contribution of these reactions to observed levels of PAH-derivative compounds in the atmosphere compared with primary combustion emissions?

A number of ambient sampling approaches have been used to address these questions, in addition to or in conjunction with, laboratory reaction kinetics data. These sampling studies have been discussed previously.[135,151] Here we provide an updated review of the current evidence of PAH reactivity and the formation of nitro- and oxy-PAHs in the atmosphere. This includes a discussion of differences in isomer profiles, temporal concentration variations, air-mass trajectory studies, theoretical calculations and the impact of source reduction measures.

Isomer Profiles

As discussed in Section 3, the position on the aromatic ring where oxidation or nitration of a PAH compound occurs will determine the specific isomer of the oxy- or nitro-PAH derivative formed. Differences in the reaction mechanism during combustion processes and atmospheric photodecomposition processes can therefore produce distinct isomer product distributions. A key observation that can lead to an assessment of the relative contribution of direct emissions compared to atmospheric formation is therefore comparison of the isomer profiles of specific oxy- and nitro-PAH compounds observed in ambient sampling studies.

It has been noted that the key formation mechanism during combustion processes which produce nitro-PAH isomers is electrophilic nitration which creates products distinct from the isomers formed from radical-initiated reactions.[359] For example, electrophilic nitration products for the reactions of pyrene and fluoranthene are noted as 1N-Pyr and 3N-Flt, while gas phase reactions under simulated atmospheric conditions are shown to form mainly 2N-Pyr and 2N-Flt.[135]

As shown in Table, 2N-Flt and 2N-Pyr have been observed in ambient air samples in many different locations around the world but are generally not observed in direct emissions. 2N-Flt has been observed in a direct industrial emission from carbon electrode manufacture[360] but this is not considered to be a major contributor on a large scale.[361] Zhu et al.[362]also identified 2N-Flt in diesel vehicle emissions but with an emission rate significantly lower (~0.2%) than that of 1N-Pyr. The isomers 2N-Flt and 2N-Pyr are therefore commonly used as markers for gas phase reaction products. Conversely, 1N-Pyr has been identified in a number of incomplete emission sources including diesel- and gasoline-fuelled vehicles, coal-fired power plants and aluminium smelting but is not observed as a gas phase reaction product, making it suitable as a marker for direct emissions.

2N-Flt and 2N-Pyr have been reported to be present in the atmosphere at higher levels than 1N-Pyr.[360,361,363–372] The observation of 2N-Flt and 2N-Pyr associated with PM in ambient studies at these levels is therefore indicative of the occurrence of OH and/or NO_3 radical-induced reactions in the ambient atmosphere. Furthermore, measured concentrations of 2N-Flt and 2N-Pyr are generally consistent with those predicted based on measured rate coefficients and formation yields.[188,204,220,373]

Due to the distinct origins of the 2N-Flt and 1N-Pyr isomers, the 2N-Flt/1N-Pyr ratio can be used to assess the relative contribution from atmospheric reactions (OH and NO_3) compared to that of direct emissions.[355,364,365,372,374,375] It is suggested that, assuming comparable levels of Flt and Pyr are present and the same dispersion and photolytic loss rates of the two isomers, a 2N-Flt/1N-Pyr ratio of >5 indicates the dominance of atmospheric reactions while a ratio of <5 indicates the dominance of direct combustion emissions.[355,376]

It is indicated from laboratory studies that 2N-Flt can be formed via both OH and NO_3 initiated reactions,[163,188] while 2N-Pyr will be formed only by OH-initiated reactions.[163,220] The ratio between these two isomers can therefore be used as an indicator for the relative importance of OH (daytime) and NO_3 (night time) reaction pathways.[364,374,377] A value of between 5 and 10 would indicate the dominance of the OH reaction pathway, while a value of above 100 would suggest the enhanced importance of NO_3-initiated reactions.[376]

Values of the 2N-Flt/1N-Pyr and 2N-Flt/2N-Pyr ratios measured in a number of different locations and conditions are presented in Tables 12 and 13 respectively. It can be seen that the values of both isomer ratios are highly variable between sites and under different specific conditions. A 2N-Flt/1N-Pyr value of >5 is more commonly observed,[355,364,372,378] indicative of atmospheric formation dominating in these studies. Locations with lower 2N-Flt/1N-Pyr ratios are generally areas with a dominant specific source, such as heavily trafficked areas[367,374,379] and large urban centres.[364,371,380] It has been noted that the 2N-Flt/1N-Pyr ratio is generally higher at suburban sites relative to their proximate urban site.[364,374,375] This has been attributed to a longer exposure of the air mass to photochemical oxidants.[355]

Table 12: Summary of 2-NF/1-NP concentration ratios from ambient measurements

Location	Details	2NF/1NP	Ref.
Marseilles area, France	Urban and suburban	<5	Albinet et al.[134]
Marseilles area, France	Rural	>10	Albinet et al.[134]
Alpine Valley locations, South France	Mean summer value (one location)	>20	Albinet et al.[376]
Alpine Valley locations, South France	Mean winter value (all locations)	<10	Albinet et al.[376]
Baltimore, USA	Urban, winter	1–3	Bamford and Baker[364]
Baltimore, USA	Urban, summer	6–24	Bamford and Baker[364]
Baltimore, USA	Suburban	1–10	Bamford and Baker[364]
Baltimore, USA	Urban	8–30	Bamford and Baker[364]
Barcelona, Spain	Residential area	4	Bayona et al.[381]
Milan, Italy	Residential area	6.1	Cecinato et al.[382]
Rome, Italy	Residential area	1.4	Cecinato et al.[382]
Columbus, USA	Residential area	2.5	Chuang et al.[383]
Rome, Italy	Urban	6.7	Ciccioli et al.[355]
Milan, Italy	Urban	5.2	Ciccioli et al.[355]
Naples, Italy	Residential area	1	Ciccioli et al.[355]
Montelibretti, Italy	Suburban	9	Ciccioli et al.[355]
Madrid, Spain	Suburban	7	Ciccioli et al.[355]
C.Porziano, Italy	Suburban	12	Ciccioli et al.[355]
Birmingham, UK	Roadway tunnel	2.5	Dimashki et al.[379]
Ho Chi Minh City, Vietnam	Urban	21	Hien et al.[367]
Ho Chi Minh City, Vietnam	Traffic site	2.7	Hien et al.[367]
Copenhagen, Denmark	Traffic site	0.72	Feilberg et al.[374]
Tokyo, Japan	Urban (summer)	8.9	Kojima et al.[387]
Tokyo, Japan	Urban (winter)	5.4	Kojima et al.[387]
Kanazawa, Japan	Urban	1.8	Murahashi and Hayakawa[380]
Athens, Greece	Urban	2.1	Marino et al.[375]
Riverside, USA	Urban background	8.75	Pitts et al.[369]
Los Angeles, USA	Urban	3.9	Reisen and Arey[371]
Claremont, USA	Urban background	7.8	Ramdahl et al.[370]
St Louis, USA	SRM (1648)	3.5	Ramdahl et al.[370]
Washington DC, USA	SRM (1649)	3	Ramdahl et al.[370]
Aurskog, Norway	Rural residential	3.7	Ramdahl et al.[370]
Beijing, China	2008 Olympic Games	25–46	Wang et al.[397]
Houston, USA	Suburban	4.2	Wilson et al.[384]
Claremont, USA	Urban	21	Zielinska et al.[218]

Table 13: Summary of 2-NF/2-NP concentration ratios from ambient measurements

Location	Details	2NF/2NP	Ref.
Marseilles area, France	Rural	3.7	Albinet et al.[134]
Alpine Valley locations, France	Mean summer value (one location)	<60	Albinet et al.[376]
Alpine Valley locations, France	Mean winter value (all locations)	<10	Albinet et al.[376]
Baltimore, USA	Urban	5–57	Bamford and Baker[364]
Baltimore, USA	Suburban	7–60	Bamford and Baker[364]
Barcelona, Spain	Residential area	6	Bayona et al.[381]
Rome, Italy	Residential area	2.2	Cecinato et al.[382]
Milan, Italy	Residential area	4.6	Cecinato et al.[382]
Naples, Italy	Residential area	1.7	Ciccioli et al.[355]
Montelibretti, Italy	Suburban	4.5	Ciccioli et al.[355]
Madrid, Spain	Suburban	3.5	Ciccioli et al.[355]
C.Porziano, Italy	Suburban	6	Ciccioli et al.[355]
Copenhagen, Denmark	Urban and suburban	<10	Feilberg et al.[374]
Copenhagen, Denmark	Urban and suburban	14.2	Feilberg et al.[374]
Athens, Greece	Urban	1.9	Marino et al.[375]
Riverside, USA	Ambient POM	23.3	Pitts et al.[369]
Claremont, USA	Urban background	35	Ramdahl et al.[370]
St Louis, USA	SRM (1648)	9.3	Ramdahl et al.[370]
Washington DC, USA	SRM (1649)	12	Ramdahl et al.[370]
Aurskog, Norway	Rural residential	3.3	Ramdahl et al.[370]
Los Angeles and Riverside, USA	Winter	16 ± 7	Reisen and Arey[371]
Los Angeles and Riverside, USA	Summer	>35	Reisen and Arey[371]
Finokalia, Crete	Mean value from a diurnal study, marine background location	3.5	Tsapakis and Stephanou[377]
Beijing, China	2008 Olympic Games	3.4–4.8	Wang et al.[397]

The relatively low (<10) 2N-Flt/2N-Pyr ratios in most studies, are indicative of daytime OH-initiated reactions dominating over NO_3-initiated reactions, as indicated in laboratory reaction studies. Higher 2N-Flt 2N-Pyr ratios have been noted in rural areas compared to urban areas,[134,376] suggesting increased importance of NO_3 reactions, which may be attributed to lack of fresh inputs of NO.[364,376] It may be expected that, since NO_3 reactions occur almost entirely at night and OH reaction will occur during the day, there may be a distinct change in 2N-Flt/2N-Pyr over a diurnal cycle. However, Feilberg et al.[374] noted that there was only a small increase in the 2N-Flt/2N-Pyr value during night time sampling (see Table 13). The authors therefore suggest that NO_3-initiated reactions play a minor role in the formation of 2N-Flt in these night time samples and attributed the presence of 2N-Flt to transport and dilution of OH-initiated reaction products.

It should be noted that the 2N-Flt/1N-Pyr and 2N-Flt/2-NP ratios reflect simply the relative levels of these isomers in the atmosphere and while they can be used as a reasonable marker for OH and/or NO_3 initiated reactions in the atmosphere, it should be considered that these ratios can be influenced by a number of other factors which may alter the ratio. For example, the relative input and removal rates of particulate matter may be of importance. Similarly, meteorological factors such as changes in mixing height and levels of solar irradiation can affect the levels of these isomers by influencing the degree of their dispersion and photolytic loss respectively.[366] The use of these ratios should therefore be used with caution when assessing the relative importance of OH and/or NO_3-induced PAH reactivity, particularly over a relatively short sampling period. (Tables 12 and 13)[377,379–384]

The occurrence of radical-induced reactions has also been assessed by comparing the isomer distribution of naphthalene derivatives from ambient air samples with that of experimental reaction products. For example, it has been noted in several studies that the isomer profiles of N-Naps, MN-Naps and EN-Naps + DMN-Naps observed in ambient air are reasonably consistent with their formation from radical induced atmospheric reactions.[180,218,361,371,378,385]

As discussed by Atkinson and Arey (2007), MN-Naps profiles from a laboratory study of the reaction of 1- and 2M-Naps with OH radicals[180,371] strongly resemble that of an ambient air sample taken during the morning hours in Mexico City, suggesting that atmospheric reactions have a strong influence of the levels and isomer profile of these compounds in urban air. Similarly, night-time air samples provide evidence for the occurrence of NO_3-induced reactions in the atmosphere. The profile of MN-Naps from a chamber reaction of NO_3 radicals produces a different profile to that of the OH-induced reaction and closely resembles the profile of a night time air sample from a receptor site downwind of urban Los Angeles (see Atkinson and Arey[151] and Reisen and Arey[371] for a more detailed discussion of these observations).

Based on similar observations of alkylnitronaphthalene isomer profiles, Wang et al.[219] suggested specific isomers or isomer ratios as indicators for the occurrence of OH and/or NO_3reactions. For example, it was suggested that the isomers 2E1N-Nap, 1,7DM8N-Nap, 2,7DM1N-Nap, 1,6DM5N-Nap, 1,3DM4N-Nap and 1,2DM4N-Nap are indicative of NO_3-induced formation, while low levels of these compounds in comparison with 1,7DM5N-Nap suggest the dominance of OH-induced reactions.[219] Furthermore, relatively high ratios of 2,7DM4N-Nap/1,7DM5N-Nap and of 2M4N-Nap/1M5N-Nap are suggested to be indicative of the predominance of NO_3 reaction chemistry relative to OH reactions.[219]

Reisen and Arey[371] measured concentrations of PAHs and nitro-PAHs in California and noted that the formation ratios (nitro-PAH/PAH) for nitronaphthalenes and alkyl-substituted nitronaphthalenes followed the order N-Naps < MNNs < DMN-Naps + EN-Naps. These formation ratios were suggested to be due to atmospheric reactions of PAHs with OH radicals and were shown to be a factor ~10 higher in summer compared to winter, which was attributed to enhanced photochemical activity during summertime. The observed trend in nitro-PAH/PAH ratios for these compounds is consistent with the observed 'activating effect' of alkyl groups on OH and NO_3 reactivity as described by Phousongphouang and Arey[171,214], providing further evidence for the occurrence of atmospheric reactions, similar to those observed in laboratory reactions.

As discussed by Kameda,[386] the gas phase reaction of triphenylene with OH and NO_3 has been shown to form the nitro-PAH derivatives 1-nitrotriphenylene (1N-TPh) and 2-nitrophenylene (2N-TPh). These reactions preferentially form 2N-TPh. Kameda[386] noted that the ratio of 2N-TPh/1N-TPh observed in samples of airborne particles was >1.55, similar to that observed from gas phase OH- and NO_3-initiated reactions (2N-TPh/1N-TPh = 1.22 to >1.5). The 2N-TPh/1N-TPh ratio observed in diesel exhaust particulate samples was shown to be much lower (2N-TPh/1N-TPh = 0.37), suggesting that radical-induced reactions contribute to the observed levels of these compounds, particularly 2N-TPh and that direct combustion emissions predominantly contribute to the levels of 1N-TPh.

Studies have also used the ratio of oxy- or nitro-PAH to their 'parent' PAH to assess the importance of atmospheric reactions on the levels of these compounds. For example, in the review by Walgraeve et al.,[358] the ratio of oxy-PAH/'parent' PAH was assessed for ambient sampling studies in the literature. It was shown that during winter, 50% of these ratios was between 0.006 and 0.16 (19 studies[358]). In summer this ratio was reported to be about 20 times higher than the winter samples, with 50% of ratios between 0.54 and 3.6 (48 studies[358]). This suggests the increased importance of photochemical activity during summer leading to higher levels of oxy-PAHs, thus providing evidence for the importance of atmospheric reactions.

Temporal Concentration Patterns

Temporal (daily, seasonal and diurnal) variations in the concentrations of atmospheric pollutants will be influenced by a number of factors including the variation in specific combustion sources, mixing height level, photolytic loss rates and gas–particle partitioning. Temporal changes in the levels of oxy- and nitro-PAH compounds have been investigated to assess the impacts of these variations and can be viewed with reference to the relative contribution of photochemical reactions to the observed levels of these compounds.

Kojima et al.[387] investigated daily and seasonal variations in concentrations of nitro- and oxy-PAHs associated with airborne PM in Tokyo, Japan. They observed that summer/winter concentration ratios were higher for 2N-Flt (0.36) than for 1N-Pyr (0.19–0.27), consistent with 2N-Flt being a secondary reaction product and 1N-Pyr being a primary emission. In contrast, the summer/winter concentration ratios for oxy-PAHs were the same as or lower than that of PAHs, contrary to the expected lower concentrations resulting from the contribution of secondary reactions. The authors suggested that this discrepancy may result from the increased partitioning of semi-volatile oxy-PAHs from the particulate phase to the gas phase in summer resulting in relatively low concentrations associated with the PM samples, as previously observed by Albinet et al.[388] Kojima et al.[387] observed strong correlations between oxy-PAH concentrations and primary emissions such as PAHs and CO in winter, while a much weaker correlation was noted in summer. This may indicate the relative dominance of direct emissions of oxy-PAHs in winter and a greater importance of atmospheric reactivity in summer.

Reisen and Arey[371] noted that levels of PAHs in the Los Angeles area were elevated in winter relative to summer which can be attributed to lower mixing heights resulting in lower dispersion rates[364,371] or less efficient photolytic loss.[246,389] In contrast, nitro-PAHs at a site ~60 km downwind of Los Angeles had concentrations generally higher in summer, indicating the greater importance of photochemical reactions in summer, when levels of atmospheric oxidants (OH, NO_3, O_3) are higher,[178,389] therefore indicating enhanced secondary input of these compounds during summer. A similar observation was made by Eiguren-Fernandez et al.[390] for PAH–quinone compounds.

2N-Flt/1N-Pyr have been shown to be lower during winter, especially in urban areas,[376] indicating that atmospheric reactions are less important in winter relative to direct emissions. Similarly, in winter, the 2N-Flt/2N-Pyr ratio has been shown to be higher than in summer,[364,371,374] suggesting that NO_3-initiated reaction may play a more important role during winter. These two observations may be explained by lower intensity of photochemical activity in winter, leading to lower levels of OH radicals in the atmosphere during winter. Winter conditions may however be more favorable to the formation of the NO_3 radical, and to lesser photolytic losses of this species.

Variations in pollutant concentration over a 24 h period can be indicative of the nature and extent of their input and output mechanisms to the atmosphere. This is particularly relevant to the e of PAH-derivative compounds. While the diurnal variation of parent PAHs is likely to reflect direct source emission signals as modified by meteorology, the levels of nitro- and oxy-PAHs, which may result in part from atmospheric reactions, may display different patterns, possibly reflecting the relative balance of primary and secondary inputs. Furthermore, the nature of the atmospheric reactions of PAHs in the atmosphere, with OH reactions occurring during daylight and NO_3 reactions occurring at night, will lead to potentially distinct diurnal patterns in the levels of nitro- and oxy-PAHs.

Since OH radicals are only present during daylight hours night time/daytime concentration ratios provide an assessment of unreacted versus reacted air masses, and the relative values of these ratios should correspond to the relative reactivity of the individual compounds.

Arey et al.[391] measured 12 h 'daytime' and 'night time' concentrations of LMW PAHs during a photochemical air pollution episode in the Los Angeles Basin. The night time/daytime concentration ratios derived were shown to correlate with the OH radical reaction rate coefficients, for example with the largest ratios being observed for PAHs that are most reactive towards OH radicals. This indicates the occurrence of OH-induced reactions occurring in the atmosphere at similar rates to those predicted in laboratory studies. Similarly, Phousongphouang and Arey[171] compared night time/daytime concentration ratios of alkyl-nitronaphthalenes measured in ambient samples to their calculated rate coefficients for the gas phase reaction of OH radicals with alkyl-naphthalenes and observed a reasonable linear correlation with OH rate coefficients, supporting the expectation that OH reactions are dominating the chemistry of these compounds in the atmosphere.

Reisen and Arey[371] investigated diurnal variations of PAH and nitro-PAH levels to assess atmospheric reactivity by sampling over four time periods (morning, day, evening and night) in California, USA. It was shown that, in summer, PAH concentrations clearly decrease during the day and evening periods. This could be indicative of PAH loss due to atmospheric reactions with OH and the more reactive compounds were shown to decrease the most.[371] This 'reactive loss' was shown to be more significant in summer compared to winter.

Tsapakis and Stephanou[377] studied the diurnal pattern of nitro-PAH as well as PAH, OH radicals and O_3 at a background site in the eastern Mediterranean. The authors suggested that the diurnal pattern of PAHs was dominated by the input of local sources (volatilization from the sea in this e). 2N-Pyr and 2N-Flt were the most abundant nitro-PAHs identified and displayed a well-defined diurnal pattern, with a concentration maximum occurring at midday followed by a rapid decrease. This pattern closely matched the diurnal variation of OH radicals, suggesting that OH radical-initiated reactions are the dominant processes controlling the levels of these nitro-PAHs in the atmosphere. The subsequent rapid decrease in nitro-PAH concentration after the midday maximum was attributed to photolytic loss of nitro-PAH compounds.

Hien et al.[367] measured concentrations of 1N-Pyr and 2N-Flt in Ho Chi Minh City, Vietnam during both daytime and night time hours in residential and high traffic locations. They observed lower levels of both compounds during daylight hours in the residential area. The higher levels of 1N-Pyr during the night were attributed to photodecomposition of 1N-Pyr during the day, while higher levels of 2N-Flt during the night were suggested to be due to atmospheric formation from NO_3-induced reactions. In contrast, levels of 2N-Flt at the high traffic location were higher during the daytime, suggesting the importance of OH-radical induced reactions at these locations.

Quantitative Predictions and Atmospheric Trajectory Studies

A number of studies have attempted to make quantitative assessments of PAH reactivity in the atmosphere. These investigations can simply involve relating observed concentrations from sampling studies to laboratory reaction kinetics data and either comparing observed levels to those predicted theoretically or using measured concentration values to calculate a formation yield that can be related to observed laboratory studies. A number of studies have attempted to quantify the relative level of PAH-derivative compounds in the atmosphere originating from secondary input from photochemical reactions. This has been attempted by carrying out a sampling study along a specific air-mass trajectory with increasing distance from the source region towards a receptor site.

Arey et al.[392] initially predicted concentrations of 1/2N-Nap, 2N-Flt and 2N-Pyr, based on calculations using laboratory-derived OH reaction rate coefficients and nitro-PAH formation yields, measured parent PAH concentrations and incorporating photolytic loss of nitro-PAHs.[180] It was shown that there was a remarkably good agreement between the calculated values and those measured in the ambient atmosphere, suggesting the dominant input of these compounds from atmospheric reactions of parent PAHs compared to direct emissions.

More recently, as discussed by Atkinson and Arey[151] (and references therein), based on observed levels of 1- and 2-NNs in ambient sampling studies, the formation yields of the OH + naphthalene reaction were estimated using the [nitro-nap]/[nap] ratios, OH reaction rate coefficient and photolysis rate coefficient values from the literature and estimated average OH radical concentrations. It was shown that the calculated values for the formation yields of 1N-Nap and 2N-Nap were reasonably consistent with experimental data,[182,363] providing further evidence that PAH reactions with OH radicals and further reaction with NO_2 occur significantly in the ambient atmosphere.

The predicted PQu formation rates calculated by Wang et al.[221] are significant, compared to observed levels in ambient sampling studies, suggesting that atmospheric formation may be a significant contributor to the observed levels in the atmosphere. Eiguren-Fernandez et al.[390] investigated the changes in PQ concentration with decreasing distance from the highly polluted Los Angeles basin, following a prevailing wind trajectory towards a more remote downwind receptor site. A significant increase in PQu concentration was observed as the air parcel moved along this trajectory, suggesting the occurrence of photochemical reactions contributing to the observed levels of PQu.

Eiguren-Fernandez et al.[390] investigated the correlations of PQu concentrations to those of phenanthrene in the particulate phase, Phe(p) and the vapour phase, Phe(v). It was shown that in the source region, there was a good correlation between PQu and Phe(p), indicative of direct emissions dominating, while at the downwind locations, there was a good correlation between

PQu and Phe(v), suggesting the dominance of atmospheric reactions as the source of PQu in these locations. Eiguren-Fernandez et al.[390] also estimated the percentage contribution of atmospheric transformations to the observed levels of PQu at the downwind site, based upon ratios of benzo[ghi]perylene (a marker for vehicular emissions) to PQu at the source region and downwind sites (see Eiguren-Fernandez et al.[390] for details). It was estimated that ~90% of PQu measured in the Los Angeles basin was of secondary origin from photochemical reactions during atmospheric transport.

Kojima et al.[393] measured the concentrations of PAHs, oxy-PAHs and nitro-PAHs at a road-side location (A) and at sites 90 m (B) and 170 m (C) away from the roadside. Concentrations of PAHs and 1N-Pyr followed the order A > B > C (A : C ratio = 0.20–0.53 and 0.20–0.26 respectively) indicating that direct input from the traffic source and dilution with increasing distance from the road controlled their concentration. Oxy-PAH concentrations also decreased with increasing distance from the roadside, although not as significantly as PAHs and 1N-Pyr (A : C ratio = 0.42–0.69). This would suggest that their atmospheric concentrations are influenced by traffic emissions but also by an additional background source affecting all sites similarly.

In contrast, the concentration of 2N-Flt was approximately the same at all three sites (A : C ratio = 0.88–1.03), indicating the occurrence of input from atmospheric reactions. The authors attempted to quantify the contribution of atmospheric formation to the levels of oxy-PAHs, modifying the calculations used by Eiguren-Fernandez et al.[390] and based on the oxy-PAH and PAH concentrations at sites A and C. Their results indicated that percentage contribution of atmospheric formation at the downwind site ranged from 9% to 72% depending on the compound and season, with a higher secondary input contribution generally noted in summer. It was also noted that the percentage of secondary input followed the order 1,8-NtA > 9-Flr and 9,10-AQu > BA with values of >50% for 1,8-NtA, 9-Flr and 9,10-AQu in both summer and winter.

These findings therefore indicate that a considerable fraction of oxy-PAHs detected in atmospheric particulate matter in urban areas will originate from atmospheric reactions of primary PAH. However, the authors note that these calculations are sometimes inadequate and may not be applicable under all environmental conditions. The amount of oxy-PAH in the atmosphere and the relative importance of atmospheric reactions versus primary emissions will also be dependent on a number of other factors including the emission rates of PAH and oxy-PAH, meteorological factors such as mixing layer height, wind speed and direction, which influence dilution levels, and the degree of gas–particle partitioning of both PAH and oxy-PAH.[393]

Emission Reduction Related Studies

Urban air quality has been and will increasingly be deteriorating on most continents[394] and related limit or recommended values are often exceeded, even today. For example, in Europe, this has involved setting country-specific emission mitigation targets such as those set by the National Emission Ceiling Directive (2001/81/EC).[395]

The implementation of strict measures in certain areas to reduce emissions, either on a long-term or short-term period, provides an opportunity to investigate the impacts of these measures on the levels of oxy- and nitro-PAH compounds and to investigate the relative contribution of primary and secondary sources. To date, however, few investigations have addressed this area of research.

Kojima et al.[387] investigated the impact of government regulation regarding emissions from diesel vehicles on the levels of PAHs and nitro-PAHs, in Tokyo. The emission reduction measures in this e involved a 2003 regulation, enforcing the use of lower-emission vehicles or installation of PM reduction devices. Comparison of ambient PM samples taken in 1997–1998 and 2006–2007 revealed a significant decrease in the average annual concentrations of PAHs and 1-NP over this period, attributed to the reduction in primary vehicular emissions. In contrast, no decrease in concentration was noted for 2N-Flt and 2N-Pyr suggesting the importance of other inputs such as atmospheric reactions to influence the levels of these compounds.

During the 2008 Olympic Games, efforts were made to improve the air quality in Beijing by implementing restrictions on direct emissions from sources such as vehicular traffic, construction sites and coal-fired power plants.[396,397] This resulted in significant decrease in the levels of primary traffic-related pollutants such as CO and NO_2 during this source reduction period.[396] Reductions in concentrations were also noted for a number of nitro-PAH and oxy-PAH compounds ranging from 15.1% to 56.6% and 24.8% to 46.6% respectively.[397]

The mean 2N-Flt/1N-Pyr value of 25–46 measured over the source control and non-source control periods indicates the dominance of photochemical reactions over this time. However, this ratio was shown to decrease during the source control period, which was attributed to the prevailing meteorological conditions and the decreased traffic emissions during this time.[397]The relatively low 2N-Flt/2N-Pyr ratio of 3.4–4.8 calculated over this period suggested the dominance of daytime OH-initiated reactions.

It is evident that air monitoring studies in urban areas carried out before and after the implementation of pollution control measures have the potential to generate interesting observations regarding the differences in primary and secondary inputs of PAH-derivative compounds. More extensive work in this area in different urban locations and for a wider range of compounds is clearly needed.

Conclusions and Recommendations for Future Work

Field data suggest that gas phase reactions of PAHs with OH (and NO_3) radicals can contribute significantly to the observed levels of nitro- and oxy-PAHs – although for many atmospherically relevant parent PAHs this chemistry is still unknown. The mechanisms should be explored in order to identify precursors and sources, needed for emission mitigation.

- Compounds with multiple origins. Assessment of the relative primary and secondary inputs of PAH-derivatives in a large number of the studies discussed in this section has focused on marker compounds for either direct emissions (e.g. 1N-Pyr) or atmospheric reactions (e.g. 2N-Flt, 2N-Pyr). While this provides a tentative analysis for the relative contributions of OH-

and/or NO_3-initiated reactions, this approach will lead to definitive conclusion regarding source apportionment for only these specific compounds.

A large number of oxy- and nitro-PAH compounds identified in ambient air can result from both direct emissions and atmospheric reactions. This makes the task of assessing and potentially of quantifying the relative importance of their sources considerably more difficult. Among the wide range of PAH-derivative compounds in the atmosphere, sources in that sense have only been investigated for PQu and NM-Naps.

- The role of heterogeneous PAH reactions. Heterogeneous PAH reactions play an important role in determining the levels of PAH derivative compounds in the atmosphere, but so far the kinetics data and reaction products have been elucidated with the focus on gas phase processes. Heterogeneous PAH reactions are influenced by a wide range of factors in addition to those associated with gas phase systems. Furthermore, partitioning between gas and particulate phases of both parent PAHs and derivatives is subject to ambient conditions (temperature, PM composition) and its determination is subject to artefacts (due to heterogeneous PAH chemistry on sampling media; Section 3.1.1).

To assess exposure towards these pollutants, there is also a need to study the chemical sinks, i.e. kinetics and reaction products of the PAH derivatives in the atmosphere.

ABBREVIATIONS

Ace	Acenaphthene
Acy	Acenaphthylene
AcP	Acepyrene
Ant	Anthracene
AQu	Anthraquinone
BaA	Benzo[a]anthracene
BaP	Benzo[a]pyrene
BbF	Benzo[b]fluoranthene
BeP	Benzo[e]pyrene
BgF	Benzo[ghi]fluoranthene
BgP	Benzo[ghi]perylene
BkF	Benzo[k]fluoranthene
Chr	Chrysene
Cor	Coronene
DBA	Dibenzo[a,h]anthracene
DBaeP	Dibenzo[a,e]pyrene
DBahP	Dibenzo[a,h]pyrene
DBalP	Dibenzo[a,l]pyrene
DM-Nap	Dimethylnaphthalene
DM-Phe	Dimethylphenanthrene
E-Nap	Ethylnaphthalene
Flt	Fluoranthene
Fln	Fluorene
Flr	Fluorenone
FCA	Formycinnamaldehyde
IPy	Indeno[1,2,3-cd]pyrene
M-Nap	Methylnaphthalene
M-Phe	Methylphenanthrene
NQu	Naphthoquinone
NtA	Naphthalic anhydride
N-Nap	Nitronaphthalene
Phe	Phenanthrene
PQu	Phenanthrequinone
Per	Perylene
Pyr	Pyrene
TPh	Triphenylene

REFERENCES

1. B. J. Finlayson-Pitts and J. N. Pitts, Chemistry of the upper and lower atmosphere, Academic Press, San Diego, USA, 2000 Search .
2. J. M. Delgado-Saborit, C. Stark and R. M. Harrison, Environ. Int., 2011, **37**, 383–392 .
3. M. J. Wenborn, P. J. Coleman, N. R. Passant, E. Lymberidi, J. Sully and R. A. Weir, AEAT-3512/REMC/20459131/Issue 1, 1999.
4. WHO, Air quality guidelines, Global update 2005, World Health Organisation, Bonn, 2006.
5. European Union, Directive 2004/107/EC of the European Parliament and of the Council of 15 December 2004 23 relating to arsenic, cadmium, mercury, nickel and polycyclic aromatic hydrocarbons in ambient air, Official Journal 24 of the European Union, L23, 3–16, 2005.
6. EPAQS, A recommendation for a united kingdom air quality standard for polycyclic aromatic hydrocarbons, Department of the Environment, 1999.
7. R. M. Harrison, D. J. T. Smith and A. J. Kibble, Occup. Environ. Med., 2004, **61**, 799–805 .
8. A. J. Cohen, Environ. Health Perspect., 2000, **108**, 743–750 .
9. K. B. Okona-Mensah, J. Battershill, A. Boobis and R. Fielder, Food Chem. Toxicol., 2005, **43**, 1103–1116 .
10. G. Lammel, J. Novák, L. Landlová, A. Dvorská, J. Klánová, P. Čupr, J. Kohoutek, E. Reimer and L. Škrdlíková, in Urban Airborne Particulate Matter: Origins, Chemistry, Fate and Health Impacts, ed. F. Zereini and C. L. S. Wiseman, Springer, Berlin, 2010, pp. 39–62 Search .
11. T. Enya, H. Suzuki, T. Watanabe, T. Hirayama and Y. Hisamatsu, Environ. Sci. Technol., 1997, **31**, 2772–2776 .
12. T. Nielsen, A. Feilberg and M. L. Binderup, Environ. Sci. Pollut. Res., 1999, **6**, 133–137 .
13. A. Feilberg, T. Nielsen, M. L. Binderup, H. Skov and M. W. B. Poulsen, Atmos. Environ., 2002, **36**, 4617–4625 .
14. UNECE, United Nations Economic Commission for Europe, Convention on Long-range Transboundary Air Pollution, Protocol on Persistent Organic Pollutants,

Arhus, Denmark, 1998.

15. OSPAR, Oslo Paris Commission, Quality status report 2000, Chemistry, London, ch. 4, 2000, p. 108.
16. EMEP, Measurement data online, European Monitoring and Evaluation Programme, Geneva, 43, http://www.nilu.no/projects/ccc/emepdata.html, 2011.
17. IADN, Integrated Atmospheric Deposition Network, http://www.ec.gc.ca/rs-23 mn/default.asp?lang=En&n=BFE9D3A3-1, accessed 2.2.2012, 2011.
18. USEPA, 2007 US Environmental Protection Agency. EPI Suite v4.0, Exposure assessment tools and models, http://www.epa.gov/opt/exposure/pubs/episuite, 2007.
19. Y.-G. Ma, Y. D. Lei, H. Xiao, F. Wania and W.-H. Wang, J. Chem. Eng. Data, 2010, **55**, 819–825 .
20. T. E. M. ten Hulscher, L. E. van der Velde and W. A. Bruggeman, Environ. Toxicol. Chem., 1992, **11**, 1595–1603 .
21. Y. D. Lei, R. Chankalal, A. Chan and F. Wania, J. Chem. Eng. Data, 2002, **47**, 801–806 .
22. M. Odabasi, E. Cetin and A. Sofuoglu, Atmos. Environ., 2006, **40**, 6615–6625 .
23. M. F. Simcik, T. P. Franz, H. X. Zhang and S. J. Eisenreich, Environ. Sci. Technol., 1998, **32**, 251–257 .
24. M. Mandalakis, M. Tsapakis, A. Tsoga and E. G. Stephanou, Atmos. Environ., 2002, **36**, 4023–4035 .
25. M. Tsapakis and E. G. Stephanou, Environ. Pollut., 2005, **133**, 147–156 .
26. X. Ding, X.-M. Wang, Z.-Q. Xie, C.-H. Xiang, B.-X. Mai, L.-G. Sun, M. Zheng, G.-Y. Sheng, J.-M. Fu and U. Poeschl, Atmos. Environ., 2007, **41**, 2061–2072 .
27. J. He and R. Balasubramanian, Atmos. Environ., 2009, **43**, 4375–4383 .
28. E. Demircioglu, A. Sofuoglu and M. Odabasi, J. Hazard. Mater., 2011, **186**, 328–335 .
29. G. Spindler, T. Gnauk, A. Gruener, Y. Iinuma, K. Mueller, S. Scheinhardt and H. Herrmann, J. Atmos. Chem., 2012, **69**, 127–157 .
30. J. Paasivirta, S. Sinkkonen, P. Mikkelson, T. Rantio and F. Wania, Chemosphere, 1999, **39**, 811–832 .
31. Y. Zhang and S. Tao, Atmos. Environ., 2009, **43**, 812–819 .
32. T. Gocht, J. A. C. Barth, M. Epp, M. Jochmann, M. Blessing, T. C. Schmidt and P. Grathwohl, Appl. Geochem., 2007, **22**, 2652–2663 .
33. EEA, EMEP/EEA air pollutant emission inventory guidebook, European Environment Agency, Technical 28 Report No. 9/2009, Copenhagen, 2009.
34. S. S. Xu, W. X. Liu and S. Tao, Environ. Sci. Technol., 2006, **40**, 702–708 .
35. A. Gusev, S. Dutchak, O. Rozovskaya, V. Shatalov, V. Sokovyh, N. Vulykh, W. Aas and K. Breivik, Persistent Organic Pollutants in the Environment. EMEP Status Report 3/2011, Meteorological Synthesizing 38 Centre - East, Moscow, Russian Federation, 2011, p. 108.
36. G. Lammel, A. M. Sehili, T. C. Bond, J. Feichter and H. Grassl, Chemosphere, 2009, **76**, 98–106 .
37. J. Münch, F. Axenfeld, Historic emission data base of selected organic pollutants (POPs) in Europe (1970-42 95)-PCDD/F, B(a)P, PCB, g-HCH, HCB, DDE/DDT. Report to the European Commission (DG XII), 43 Environment and Climate Project No. ENV-CT96-0214, 1999.
38. I. Bewersdorff, A. Aulinger, V. Matthias and M. Quante, Meteorol. Z., 2009, **18**, 41–53 .
39. USEPA, U.S. Environmental Protection Agency. Locating and estimating air emissions from sources of 2 polycyclic organic matter. EPA-454/R-98-014. Office of Air Quality Planning and Standards, Research Triangle 3 Park, NC, USA, 1998.
40. J. M. Pacyna, K. Breivik, J. Munch and J. Fudala, Atmos. Environ., 2003, **37**, S119–S131 .
41. H. Denier van der Gon, M. van het Bolscher, A. Visschedijk and P. Zandveld, Atmos. Environ., 2007, **41**, 9245–9261 .
42. A. L. Robinson, R. Subramanian, N. M. Donahue, A. Bernardo-Bricker and W. F. Rogge, Environ. Sci. Technol., 2006, **40**, 7803–7810 .
43. E. Galarneau, Atmos. Environ., 2008, **42**, 8139–8149 .
44. M. Blumer, Sci. Am., 1976, **234**, 35–45 .
45. B. R. T. Simoneit, Int. J. Environ. Anal. Chem., 1985, **22**, 203–233 .
46. M. B. Yunker, R. W. Macdonald, R. Vingarzan, R. H. Mitchell, D. Goyette and S. Sylvestre, Org. Geochem., 2002, **33**, 489–515 .
47. A. Dvorská, G. Lammel and J. Klánová, Atmos. Environ., 2011, **45**, 420–427 .
48. M. Tobiszewski and J. Namiesnik, Environ. Pollut., 2012, **162**, 110–119 .
49. A. Dvorská, K. Komprdová, G. Lammel, J. Klánová and H. Plachá, Atmos. Environ., 2012, **46**, 147–154 .
50. B. R. T. Simoneit, J. J. Schauer, C. G. Nolte, D. R. Oros, V. O. Elias, M. P. Fraser, W. F. Rogge and G. R. s, Atmos. Environ., 1999, **33**, 173–182 .
51. J. J. Schauer, W. F. Rogge, L. M. Hildemann, M. A. Mazurek, G. R. s and B. R. T. Simoneit, Atmos. Environ., 1996, **30**, 3837–3855 .
52. D. R. Oros and B. R. T. Simoneit, Fuel, 2000, **79**, 515–536 .
53. T. Nielsen, Atmos. Environ., 1988, **22**, 2249–2254 .
54. F. Böhme, K. Welsch-Pausch and M. S. McLachlan, Environ. Sci. Technol., 1999, **33**, 1805–1813 .
55. M. S. McLachlan, Environ. Sci. Technol., 1999, **33**, 1799–1804 .
56. S. D. Choi, R. M. Staebler, H. Li, Y. Su, B. Gevao, T. Harner and F. Wania, Atmos. Chem. Phys., 2008, **8**, 4105–4113 .
57. A. D. St-Amand, P. M. Mayer and J. M. Blais, Atmos. Environ., 2009, **43**, 4283–4288 .
58. I. T. Cousins and K. C. Jones, Environ. Pollut., 1998, **102**, 105–118 .
59. W. Wilcke, J. Plant Nutr. Soil Sci., 2000, **163**, 229–248 .

60. A. J. Sweetman, M. Dalla Valle, K. Prevedouros and K. C. Jones, Chemosphere, 2005, **60**, 959–972 .

61. G. L. Daly, Y. D. Lei, L. E. tillo, D. C. G. Muir and F. Wania, Atmos. Environ., 2007, **41**, 7339–7350 .

62. R. Wang, S. Tao, B. Wang, Y. Yang, C. Lang, Y. Zhang, J. Hu, J. Ma and H. Hung, Environ. Sci. Technol., 2010, **44**, 1017–1022 .

63. R. M. Dickhut and K. E. Gustafson, Environ. Sci. Technol., 1995, **29**, 1518–1525 .

64. M. Odabasi, A. Sofuoglu, N. Vardar, Y. Tasdemir and T. M. Holsen, Environ. Sci. Technol., 1999, **33**, 426–434 .

65. R. Balasubramanian and J. He, Fate and transfer of persistent organic pollutants in a multi-media environment, in Urban Airborne Particulate Matter: Origins, Chemistry, Fate and Health Impacts, ed. F. Zereini and C. L. S. Wiseman, Springer, Heidelberg, 2010, pp. 277–307 Search .

66. J. Komprda, K. Kubosová, A. Dvorská, M. Scheringer, J. Klánová and I. Holoubek, J. Environ. Monit., 2009, **11**, 269–276 **RSC**.

67. D. M. Farmer, C. L. McNeil and B. D. Johnson, Nature, 1993, **361**, 620–623 .

68. J. DeWulf, H. Van Langenhove and B. Heireman, Water Res., 1998, **32**, 2106–2112 .

69. L. Lim, O. Wurl, S. Karuppiah and J. P. Obbard, Mar. Pollut. Bull., 2007, **54**, 1212–1219 .

70. C. Guitart, N. Garcia-Flor, J. C. Miquel, S. W. Fowler and J. Albaiges, J. Marine Syst., 2010, **79**, 210–217 .

71. A. Eschenbach, M. Kastner, R. Bierl, G. Schaefer and B. Mahro, Chemosphere, 1994, **28**, 683–692 .

72. P. B. Hatzinger and M. Alexander, Environ. Sci. Technol., 1995, **29**, 537–545 .

73. J. J. H. Haftka, H. A. J. Govers and J. R. Parsons, Environ. Sci. Pollut. Res., 2010, **17**, 1070–1079 .

74. P. Lehnik-Habrink, S. Hein, T. Win, W. Bremser and I. Nehls, J. Soils Sediments, 2010, **10**, 1487–1498 .

75. Y. Yang, N. Zhang, M. Xue and S. Tao, Environ. Pollut., 2010, **158**, 2170–2174 .

76. R. M. Allen-King, P. Grathwohl and W. P. Ball, Adv. Water Resour., 2002, **25**, 985–1016 .

77. S. Ribes, B. van Drooge, J. Dachs, O. Gustafsson and J. O. Grimalt, Environ. Sci. Technol., 2003, **37**, 2675–2680 .

78. M. Obst, P. Grathwohl, A. Kappler, O. Eibl, N. Peranio and T. Gocht, Environ. Sci. Technol., 2011, **45**, 7314–7322 .

79. S. Endo, P. Grathwohl, S. B. Haderlein and T. C. Schmidt, Environ. Sci. Technol., 2009, **43**, 3094–3100 .

80. B. Pan, S. Tao, D. Wu, D. Zhang, H. Peng and B. Xing, Chemosphere, 2011, **84**, 1578–1583 .

81. G. E. Schaumann, J. Plant Nutr. Soil Sci., 2006, **169**, 145–156 .

82. Y. Yang, N. Zhang, M. Xue, S. T. Lu and S. Tao, Environ. Pollut., 2011, **159**, 591–595 .

83. E. Brorstrom, P. Grennfelt and A. Lindskog, Atmos. Environ., 1983, **17**, 601–605 .

84. M. Carrara, J. C. Wolf and R. Niessner, Atmos. Environ., 2010, **44**, 3878–3885 .

85. M. Goriaux, B. Jourdain, B. Temime, J. L. Besombes, N. Marchand, A. Albinet, E. Leoz-Garziandia and H. Wortham, Environ. Sci. Technol., 2006, **40**, 6398–6404 .

86. I. G. Kavouras, J. Lawrence, P. Koutrakis, E. G. Stephanou and P. Oyola, Atmos. Environ., 1999, **33**, 4977–4986 .

87. M. Possanzini, A. Febo and A. Liberti, Atmos. Environ., 1983, **17**, 2605–2610 .

88. C. Schauer, R. Niessner and U. Poschl, Environ. Sci. Technol., 2003, **37**, 2861–2868 .

89. M. Tsapakis and E. G. Stephanou, Atmos. Environ., 2003, **37**, 4935–4944 .

90. C. E. Junge, Basic considerations about trace constituents in the atmosphere as related to the fate of global 16 pollutants, in Fate of Pollutants in the Air and Water Environments, Part I, ed. I. H. Suffet, Wiley, New York, 1977, vol. 17, pp. 7–26 Search .

91. J. F. Pankow, Atmos. Environ., 1987, **21**, 2275–2283 .

92. J. F. Pankow and T. F. Bidleman, Atmos. Environ., Part A, 1992, **26**, 1071–1080 .

93. K. U. Goss and R. P. Schwarzenbach, Environ. Sci. Technol., 1998, **32**, 2025–2032 .

94. H. Yamasaki, K. Kuwata and H. Miyamoto, Environ. Sci. Technol., 1982, **16**, 189–194 .

95. J. F. Pankow, L. M. Isabelle, D. A. Buchholz, W. T. Luo and B. D. Reeves, Environ. Sci. Technol., 1994, **28**, 363–363 .

96. E. Galarneau, T. F. Bidleman and P. Blanchard, Atmos. Environ., 2006, **40**, 182–197 .

97. K. E. Gustafson and R. M. Dickhut, Environ. Sci. Technol., 1997, **31**, 140–147 .

98. V. Subramanyam, K. T. Valsaraj, L. J. Thibodeaux and D. D. Reible, Atmos. Environ., 1994, **28**, 3083–3091.

99. S. Ozcan and M. E. Aydin, Atmos. Res., 2009, **93**, 715–722 .

100. K. Ravindra, L. Bencs, E. Wauters, J. de Hoog, F. Deutsch, E. Roekens, N. Bleux, P. Berghmans and R. Van Grieken, Atmos. Environ., 2006, **40**, 771–785 .

101. A. Finizio, D. Mackay, T. Bidleman and T. Harner, Atmos. Environ., 1997, **31**, 2289–2296 .

102. J. Dachs and S. J. Eisenreich, Environ. Sci. Technol., 2000, **34**, 3690–3697 .

103. R. Lohmann and G. Lammel, Environ. Sci. Technol., 2004, **38**, 3793–3803 .

104. M. T. O. Jonker and A. A. Koelmans, Environ. Sci. Technol., 2002, **36**, 3725–3734 .

105. B. T. Mader and J. F. Pankow, Environ. Sci. Technol., 2002, **36**, 5218–5228 .

106. S. Pongpiachan, K. Thamanu, K. F. Ho, S. C. Lee and P. Sompongchaiyakul, Southeast Asian J. Trop. Med. Public Health, 2009, **40**, 1377–1394 .

107. G. Lammel, J. Klánová, P. Ilić, J. Kohoutek, B. Gasić, I. Kovacić and L. Škrdlíková, Atmos. Environ., 2010, **44**, 5022–5027 .

108. K. U. Goss and R. P. Schwarzenbach, Environ. Sci. Technol., 2001, **35**, 1–9 .

109. M. H. Abraham, J. Chromatogr., 1993, **644**, 95–139 .

110. S. N. Atapattu and C. F. Poole, J. Chromatogr., A, 2008, **1195**, 136–145 .

111. C. M. Roth, K. U. Goss and R. P. Schwarzenbach, Environ. Sci. Technol., 2005, **39**, 6638–6643 .

112. K. U. Goss, Crit. Rev. Environ. Sci. Technol., 2004, **34**, 339–389 .

113. J. A. L. Cancio, A. V. tellano, S. S. Martin and J. F. S. Rodriguez, Water, Air, Soil Pollut., 2004, **154**, 127–138 .

114. L. P. Chrysikou, P. G. Gemenetzis and C. A. Samara, Atmos. Environ., 2009, **43**, 290–300 .

115. G. Kiss, Z. Varga-Puchony and Z. Hlavaj, Distribution of polycyclic aromatic hydrocarbons on atmospheric aerosol particles of different size, in Nucleation and Atmospheric Aerosols, ed. M. Kulmala and P. O. Wagner, Elsevier, Kidlington, 1996, pp. 501–503 Search .

116. J. Schnelle, T. Jänsch, K. Wolf, I. Gebefugi and A. Kettrup, Chemosphere, 1995, **31**, 3119–3127 .

117. C. Venkataraman and S. K. Friedlander, Environ. Sci. Technol., 1994, **28**, 563–572 .

118. Y. Kawanaka, Y. Tsuchiya, S.-J. Yun and K. Sakamoto, Environ. Sci. Technol., 2009, **43**, 6851–6856 .

119. Y. Kameda, J. Shirai, T. Komai, J. Nakanishi and S. Masunaga, Sci. Total Environ., 2005, **340**, 71–80 .

120. A. H. Miguel, T. W. Kirchstetter, R. A. Harley and S. V. Hering, Environ. Sci. Technol., 1998, **32**, 450–455.

121. L. Skrdlikova, L. Landlova, J. Klanova and G. Lammel, Atmos. Environ., 2011, **45**, 4305–4312 .

122. T. F. Bidleman, Environ. Sci. Technol., 1988, **22**, 361–367 .

123. T. Gocht, O. Klemm and P. Grathwohl, Atmos. Environ., 2007, **41**, 1315–1327 .

124. T. P. Franz and S. J. Eisenreich, Environ. Sci. Technol., 1998, **32**, 1771–1778 .

125. Y. Tasdemir and F. Esen, Atmos. Environ., 2007, **41**, 1288–1301 .

126. M. P. Ligocki, C. Leuenberger and J. F. Pankow, Atmos. Environ., 1985, **19**, 1609–1617 .

127. S. K. Sahu, G. G. Pandit and S. Sadasivan, Sci. Total Environ., 2004, **318**, 245–249 .

128. M. F. Simcik, Atmos. Environ., 2004, **38**, 491–501 .

129. W. G. N. Slinn, L. Hasse, B. B. Hicks, A. W. Hogan, D. Lal, P. S. Liss, K. O. Munnich, G. A. Sehmel and O. Vittori, Atmos. Environ., 1978, **12**, 2055–2087 .

130. M. P. Ligocki, C. Leuenberger and J. F. Pankow, Atmos. Environ., 1985, **19**, 1619–1626 .

131. J. H. Offenberg and J. E. Baker, Environ. Sci. Technol., 2002, **36**, 3763–3771 .

132. J. T. Hoff, F. Wania, D. Mackay and R. Gillham, Environ. Sci. Technol., 1995, **29**, 1982–1989 .

133. F. Wania, D. Mackay and J. T. Hoff, Environ. Sci. Technol., 1999, **33**, 195–197 .

134. A. Albinet, E. Leoz-Garziandia, H. Budzinski and E. Villenave, Sci. Total Environ., 2007, **384**, 280–292 .

135. R. Atkinson and J. Arey, Environ. Health Perspect., 1994, **102**, 117–126 .

136. F. Wania and D. Mackay, Environ. Sci. Technol., 1996, **30**, A390–A396 .

137. S. C. Zhang, W. Zhang, K. Y. Wang, Y. T. Shen, L. W. Hu and X. J. Wang, Environ. Monit. Assess., 2009, **151**, 197–207 .

138. A. Birgul, Y. Tasdemir and S. S. Cindoruk, Atmos. Res., 2011, **101**, 341–353 .

139. E. Terzi and C. Samara, Atmos. Environ., 2005, **39**, 6261–6270 .

140. W. W. Brubaker and R. A. Hites, J. Phys. Chem. A, 1998, **102**, 915–921 .

141. J. N. Pitts, Atmos. Environ., 1987, **21**, 2531–2547 .

142. J. Sasaki, J. Arey and W. P. Harger, Environ. Sci. Technol., 1995, **29**, 1324–1335 .

143. L. Wang, J. Arey and R. Atkinson, Environ. Sci. Technol., 2005, **39**, 5302–5310 .

144. R. Atkinson and S. M. Aschmann, Int. J. Chem. Kinet., 1986, **18**, 569–573 .

145. F. L. Eisele, G. H. Mount, D. Tanner, A. Jefferson, R. Shetter, J. W. Harder and E. J. Williams, J. Geophys. Res.: Atmos., 1997, **102**, 6457–6465 .

146. R. G. Prinn, J. Huang, R. F. Weiss, D. M. Cunnold, P. J. Fraser, P. G. Simmonds, A. McCulloch, P. A. L. Bernard, C. Harth, P. Salameh, S. O'Doherty, R. H. J. Wang, L. Porter and B. R. Miller, Science, 2001, **292**, 1882–1888 .

147. R. Atkinson, J. Phys. Chem. Ref. Data, Monogr., 1989, 1–246 Search .

148. R. Atkinson and J. Arey, Chem. Rev., 2003, **103**, 4605–4638 .

149. M. Krol, P. J. van Leeuwen and J. Lelieveld, J. Geophys. Res.: Atmos., 1998, **103**, 10697–10711 .

150. J. H. Seinfeld and S. N. Pandis, Atmospheric Chemistry and Physics, John Wiley & Sons, Inc., 1998, ISBN 0-471-17815-2 Search .

151. R. Atkinson and J. Arey, Polycyclic Aromat. Compd., 2007, **27**, 15–40 .

152. M. Krol and J. Lelieveld, J. Geophys. Res.: Atmos., 2003, **108**, 4125 .

153. R. G. Prinn, J. Huang, R. F. Weiss, D. M. Cunnold, P. J. Fraser, P. G. Simmonds, A. McCulloch, C. Harth, S. Reimann, P. Salameh, S. O'Doherty, R. H. J. Wang, L. W. Porter, B. R. Miller and P. B. Krummel, Geophys. Res. Lett., 2005, **32**, L07809 .

154. X. Qu, Q. Zhang and W. Wang, Chem. Phys. Lett., 2006, **429**, 77–85 .

155. F. Reisen and J. Arey, Environ. Sci. Technol., 2002, **36**, 4302–4311 .

156. K. Lorenz and R. Zellner, Ber. Bunsenges., 1983, **87**, 629–636 .

157. R. Ananthula, T. Yamada and P. H. Taylor, J. Phys. Chem. A, 2006, **110**, 3559–3566 .

158. F. Goulay, C. Rebrion-Rowe, J. L. Le Garrec, S. D. Le

Picard, A. Canosa and B. R. Rowe, J. Chem. Phys., 2005, **122**, 104308 .

159. W. Lee, P. S. Stevens and R. A. Hites, J. Phys. Chem. A, 2003, **107**, 6603–6608 .
160. R. Ananthula, T. Yamada and P. H. Taylor, Int. J. Chem. Kinet., 2007, **39**, 629–637 .
161. R. A. Perry, R. Atkinson and J. N. Pitts, J. Phys. Chem., 1977, **81**, 296–304 .
162. R. Atkinson, J. Arey, B. Zielinska and S. M. Aschmann, Environ. Sci. Technol., 1987, **21**, 1014–1022 .
163. R. Atkinson, J. Arey, B. Zielinska and S. M. Aschmann, Int. J. Chem. Kinet., 1990, **22**, 999–1014 .
164. R. Atkinson, S. M. Aschmann, J. Arey, B. Zielinska and D. Schuetzle, Atmos. Environ., 1989, **23**, 2679–2690.
165. R. Koch, R. Knispel, M. Elend, M. Siese and C. Zetzsch, Atmos. Chem. Phys., 2007, **7**, 2057–2071 .
166. N. Nishino, R. Atkinson and J. Arey, Environ. Sci. Technol., 2008, **42**, 9203–9209 .
167. A. Ricca and C. W. Bauschlicher, Chem. Phys. Lett., 2000, **328**, 396–402 .
168. R. Atkinson, S. M. Aschmann and J. N. Pitts, Environ. Sci. Technol., 1984, **18**, 110–113 .
169. H. W. Biermann, H. Mac Leod, R. Atkinson, A. M. Winer and J. N. Pitts, Environ. Sci. Technol., 1985, **19**, 244–248 .
170. W. Klöpffer, R. Frank, E. G. Kohl and F. Haag, Chem.-Ztg., 1986, **110**, 57–61 Search .
171. P. T. Phousongphouang and J. Arey, Environ. Sci. Technol., 2002, **36**, 1947–1952 .
172. A. Klamt, Chemosphere, 1993, **26**, 1273–1289 .
173. R. Atkinson and S. M. Aschmann, Atmos. Environ., 1987, **21**, 2323–2326 .
174. R. Atkinson and S. M. Aschmann, Int. J. Chem. Kinet., 1988, **20**, 513–539 .
175. C. E. Banceu, C. Mihele, D. A. Lane and N. J. Bunce, Polycyclic Aromat. Compd., 2001, **18**, 415–425 .
176. E. S. C. Kwok, W. P. Harger, J. Arey and R. Atkinson, Environ. Sci. Technol., 1994, **28**, 521–527 .
177. E. S. C. Kwok, R. Atkinson and J. Arey, Int. J. Chem. Kinet., 1997, **29**, 299–309 .
178. D. Helmig, J. Arey, R. Atkinson, W. P. Harger and P. A. McElroy, Atmos. Environ., Part A, 1992, **26**, 1735–1745 .
179. D. Helmig and W. P. Harger, Sci. Total Environ., 1994, **148**, 11–21 .
180. J. Arey, B. Zielinska, R. Atkinson and S. M. Aschmann, Int. J. Chem. Kinet., 1989, **21**, 775–799 .
181. J. Y. Lee and D. A. Lane, Atmos. Environ., 2009, **43**, 4886–4893 .
182. J. Sasaki, S. M. Aschmann, E. S. C. Kwok, R. Atkinson and J. Arey, Environ. Sci. Technol., 1997, **31**, 3173–3179 .
183. N. J. Bunce, L. Liu, J. Zhu and D. A. Lane, Environ. Sci. Technol., 1997, **31**, 2252–2259 .
184. K. E. Kautzman, J. D. Surratt, M. N. Chan, A. W. H. Chan, S. P. Hersey, P. S. Chhabra, N. F. Dalleska, P. O. Wennberg, R. C. Flagan and J. H. Seinfeld, J. Phys. Chem. A, 2010, **114**, 913–934 .
185. L. Wang, R. Atkinson and J. Arey, Environ. Sci. Technol., 2007, **41**, 2803–2810 .
186. N. Nishino, J. Arey and R. Atkinson, Environ. Sci. Technol., 2009, **43**, 1349–1353 .
187. J. Lee and D. A. Lane, Atmos. Environ., 2010, **44**, 2469–2477 .
188. J. Arey, B. Zielinska, R. Atkinson, A. M. Winer, T. Ramdahl and J. N. Pitts, Atmos. Environ., 1986, **20**, 2339–2345 .
189. D. Asaf, D. Pedersen, V. Matveev, M. Peleg, C. Kern, J. Zingler, U. Platt and M. Luria, Environ. Sci. Technol., 2009, **43**, 9117–9123 .
190. S. S. Brown, W. P. Dube, J. Peischl, T. B. Ryerson, E. Atlas, C. Warneke, J. A. de Gouw, S. T. Hekkert, C. A. Brock, F. Flocke, M. Trainer, D. D. Parrish, F. C. Feshenfeld and A. R. Ravishankara, J. Geophys. Res.: Atmos., 2011, **116**, D24305 .
191. A. Geyer, B. Alicke, S. Konrad, T. Schmitz, J. Stutz and U. Platt, J. Geophys. Res.: Atmos., 2001, **106**, 8013–8025 .
192. U. Platt, G. Lebras, G. Poulet, J. P. Burrows and G. Moortgat, Nature, 1990, **348**, 147–149 .
193. J. M. Roberts, C. J. Hahn, F. C. Fehsenfeld, J. M. Warnock, D. L. Albritton and R. E. Sievers, Environ. Sci. Technol., 1985, **19**, 364–369 .
194. A. M. Winer, R. Atkinson and J. N. Pitts, Science, 1984, **224**, 156–159 .
195. C. A. Cantrell, J. A. Davidson, K. L. Busarow and J. G. Calvert, J. Geophys. Res.: Atmos., 1986, **91**, 5347–5353 .
196. C. A. Cantrell, W. R. Stockwell, L. G. Anderson, K. L. Busarow, D. Perner, A. Schmeltekopf, J. G. Calvert and H. S. Johnston, J. Phys. Chem., 1985, **89**, 139–146 .
197. W. R. Stockwell and J. G. Calvert, J. Geophys. Res., C: Oceans Atmos., 1983, **88**, 6673–6682 .
198. W. P. L. Carter, A. M. Winer and J. N. Pitts, Environ. Sci. Technol., 1981, **15**, 829–831 .
199. N. Carslaw, L. J. Carpenter, J. M. C. Plane, B. J. Allan, R. A. Burgess, K. C. Clemitshaw, H. Coe and S. A. Penkett, J. Geophys. Res.: Atmos., 1997, **102**, 18917–18933 .
200. D. Mihelcic, D. Klemp, P. Musgen, H. W. Patz and A. Volzthomas, J. Atmos. Chem., 1993, **16**, 313–335 .
201. A. Geyer, R. Ackermann, R. Dubois, B. Lohrmann, T. Muller and U. Platt, Atmos. Environ., 2001, **35**, 3619–3631 .
202. R. Atkinson and A. C. Lloyd, J. Phys. Chem. Ref. Data, 1984, **13**, 315–444 .
203. R. Atkinson, E. C. Tuazon and J. Arey, Int. J. Chem. Kinet., 1990, **22**, 1071–1082 .
204. R. Atkinson, A. M. Winer and J. N. J. Pitts, Atmos. Environ., 1986, **20**, 331–340 .
205. R. A. Graham and H. S. Johnston, J. Phys. Chem., 1978, **82**, 254–268 .

206. F. Magnotta and H. S. Johnston, Geophys. Res. Lett., 1980, **7**, 769–772 .

207. R. Atkinson, J. Phys. Chem. Ref. Data, Monogr., 1994, **2**, 216 Search .

208. M. W. Malko and J. Troe, Int. J. Chem. Kinet., 1982, **14**, 399–416 .

209. T. J. Wallington, R. Atkinson, A. M. Winer and J. N. Pitts, Int. J. Chem. Kinet., 1987, **19**, 243–249 .

210. R. Atkinson, J. Phys. Chem. Ref. Data, 1991, **20**, 459–507 .

211. J. F. Noxon, J. Geophys. Res., C: Oceans Atmos., 1983, **88**, 1017–1021 Search .

212. J. N. Pitts, B. Zielinska, J. A. Sweetman, R. Atkinson and A. M. Winer, Atmos. Environ., 1985, **19**, 911–915.

213. R. Atkinson, E. C. Tuazon, I. Bridier and J. Arey, Int. J. Chem. Kinet., 1994, **26**, 605–614 .

214. P. T. Phousongphouang and J. Arey, Environ. Sci. Technol., 2003, **37**, 308–313 .

215. E. S. C. Kwok, R. Atkinson and J. Arey, Int. J. Chem. Kinet., 1994, **26**, 511–525 .

216. X. G. Qu, Q. Z. Zhang and W. X. Wang, Chem. Phys. Lett., 2006, **432**, 40–49 .

217. L. Wang, S. M. Aschmann, R. Atkinson and J. Arey, Environ. Sci. Technol., 2009, **43**, 2766–2772 .

218. B. Zielinska, J. Arey, R. Atkinson and P. A. McElroy, Environ. Sci. Technol., 1989, **23**, 723–729 .

219. L. Wang, R. Atkinson and J. Arey, Environ. Sci. Technol., 2010, **44**, 2981–2987 .

220. B. Zielinska, J. Arey, R. Atkinson, T. Ramdahl, A. M. Winer and J. N. Pitts, J. Am. Chem. Soc., 1986, **108**, 4126–4132 .

221. L. Wang, R. Atkinson and J. Arey, Atmos. Environ., 2007, **41**, 2025–2035 .

222. M. S. Alam, M. Camredon, A. R. Rickard, T. Carr, K. P. Wyche, K. E. Hornsby, P. S. Monks and W. J. Bloss, Phys. Chem. Chem. Phys., 2011, **13**, 11002–11015 **RSC**.

223. R. Atkinson, Atmos. Environ., 2000, **34**, 2063–2101 .

224. U. Schurath and K. H. Naumann, Pure Appl. Chem., 1998, **70**, 1353–1361 .

225. D. Vione, V. Maurino, C. Minero, E. Pelizzetti, M. A. J. Harrison, R. I. Olariu and C. Arsene, Chem. Soc. Rev., 2006, **35**, 441–453 .

226. G. Andreou and S. Rapsomanikis, J. Hazard. Mater., 2009, **172**, 363–373 .

227. P. Di Filippo, C. Riccardi, D. Pomata and F. Buiarelli, Atmos. Environ., 2010, **44**, 2742–2749 .

228. Z. P. Gu, J. L. Feng, W. L. Han, L. Li, M. H. Wu, J. M. Fu and G. Y. Sheng, J. Environ. Sci., 2010, **22**, 389–396 .

229. S. Harrad and L. Laurie, J. Environ. Monit., 2005, **7**, 722–727 **RSC**.

230. K. van Cauwenberghe and L. van Vaeck, Mutat. Res., 1983, **116**, 1–20 .

231. L. van Vaeck and K. van Cauwenberghe, Atmos. Environ., 1984, **18**, 323–328 .

232. T. D. Behymer and R. A. Hites, Environ. Sci. Technol., 1988, **22**, 1311–1319 .

233. M. Jang and S. R. McDow, Environ. Sci. Technol., 1995, **29**, 2654–2660 .

234. M. Jang and S. R. McDow, Environ. Sci. Technol., 1997, **31**, 1046–1053 .

235. R. M. Kamens, Z. Guo, J. N. Fulcher and D. A. Bell, Environ. Sci. Technol., 1988, **22**, 103–108 .

236. S. R. McDow, M. S. Jang, Y. Hong and R. M. Kamens, J. Geophys. Res.: Atmos., 1996, **101**, 19593–19600.

237. S. R. McDow, Q. R. Sun, M. Vartiainen, Y. S. Hong, Y. L. Yao, T. Fister, R. Q. Yao and R. M. Kamens, Environ. Sci. Technol., 1994, **28**, 2147–2153 .

238. S. R. McDow, M. Vartiainen, Q. R. Sun, Y. S. Hong, Y. L. Yao and R. M. Kamens, Atmos. Environ., 1995, **29**, 791–797 .

239. T. D. Behymer and R. A. Hites, Environ. Sci. Technol., 1985, **19**, 1004–1006 .

240. R. Dabestani, K. J. Ellis and M. E. Sigman, J. Photochem. Photobiol., A, 1995, **86**, 231–239 .

241. R. T. Dabestani, J. T. Barbas, K. J. Ellis and M. E. Sigman, Abs. Am. Chem. Soc., 1995, **210**, 148-ENVR Search .

242. D. A. Lane and M. Katz, in Fate of Pollutants in the Air and Water Environments, Part 2, ed. A. Suffet, Wiley-Interscience, New York, 1977, pp. 133–154 Search .

243. M. E. Sigman, R. Arce, C. Rayes, J. T. Barbas and R. Dabestani, Abs. Am. Chem. Soc., 1997, **213**, 220-ENVR Search .

244. J. T. Barbas, R. Dabestani and M. E. Sigman, J. Photochem. Photobiol., A, 1994, **80**, 103–111 .

245. J. T. Barbas, M. E. Sigman, A. C. Buchanan and E. A. Chevis, Photochem. Photobiol., 1993, **58**, 155–158 .

246. J. T. Barbas, M. E. Sigman and R. Dabestani, Environ. Sci. Technol., 1996, **30**, 1776–1780 .

247. R. A. Yokley, A. A. Garrison, E. L. Wehry and G. Mamantov, Environ. Sci. Technol., 1986, **20**, 86–90 .

248. W. A. Korfmacher, D. F. S. Natusch, D. R. Taylor, G. Mamantov and E. L. Wehry, Science, 1980, **207**, 763–765 .

249. T. F. Kahan and D. J. Donaldson, J. Phys. Chem. A, 2007, **111**, 1277–1285 .

250. T. F. Kahan, R. Zhao, K. B. Jumaa and D. J. Donaldson, Environ. Sci. Technol., 2010, **44**, 1302–1306 .

251. K. A. van Cauwenberghe, in Handbook of Polycyclic Aromatic Hydrocarbons, ed. A. Bjørseth and T. Ramdahl, Dekker, New York, 1985, pp. 351–384 Search .

252. Y. Bedjanian, M. L. Nguyen and G. Le Bras, Atmos. Environ., 2010, **44**, 1754–1760 .

253. A. K. Bertram, A. V. Ivanov, M. Hunter, L. T. Molina and M. J. Molina, J. Phys. Chem. A, 2001, **105**, 9415–9421 .

254. W. Estève, H. Budzinski and E. Villenave, Polycyclic Aromat. Compd., 2003, **23**, 441–456 .

255. W. Estève, H. Budzinski and E. Villenave, Atmos. Environ., 2004, **38**, 6063–6072 .

256. W. Estève, H. Budzinski and E. Villenave, Atmos. Environ., 2006, **40**, 201–211 .

257. K. Miet, H. Budzinski and E. Villenave, Polycyclic Aromat. Compd., 2009, **29**, 267–281 .

258. J. Ringuet, A. Albinet, E. Leoz-Garziandia, H. Budzinski and E. Villenave, Atmos. Environ., 2012, **61**, 15–22 .

259. Y. Zhang, B. Yang, J. Gan, C. Liu, X. Shu and J. Shu, Atmos. Environ., 2011, **45**, 2515–2521 .

260. C. Liu, P. Zhang, B. Yang, Y. Wang and J. Shu, Environ. Sci. Technol., 2012, **46**, 7575–7580 .

261. J. N. Pitts, J. A. Sweetman, B. Zielinska, R. Atkinson, A. M. Winer and W. P. Harger, Environ. Sci. Technol., 1985, **19**, 1115–1121 .

262. R. M. Kamens, J. Guo, Z. Guo and S. R. McDow, Atmos. Environ., Part A, 1990, **24**, 1161–1173 .

263. J. Mak, S. Gross and A. K. Bertram, Geophys. Res. Lett., 2007, **34**, L10804 .

264. F. Karagulian and M. J. Rossi, J. Phys. Chem. A, 2007, **111**, 1914–1926 .

265. S. Gross and A. K. Bertram, J. Phys. Chem. A, 2008, **112**, 3104–3113 .

266. N. O. A. Kwamena and J. P. D. Abbatt, Atmos. Environ., 2008, **42**, 8309–8314 .

267. J. N. Pitts, K. A. van Cauwenberghe, D. Grosjean, J. P. Schmid, D. R. Fitz, W. L. Belser, G. B. Knudson and P. M. Hynds, Science, 1978, **202**, 515–519 Search .

268. H. Tokiwa, R. Nakagawa, K. Morita and Y. Ohnishi, Mutat. Res., 1981, **85**, 195–205 .

269. Z. Guo and R. M. Kamens, J. Atmos. Chem., 1991, **12**, 137–151 .

270. J. D. Butler and P. Crossley, Atmos. Environ., 1981, **15**, 91–94 .

271. E. Brorstrom-Lunden and A. Lindskog, Environ. Sci. Technol., 1985, **19**, 313–316 .

272. M. M. Hughes, D. F. S. Natusch, D. R. Taylor and M. V. Zeller, in Polynuclear Aromatic Hydrocarbons: Chemistry and Biological Effects, ed. A. Bjørseth and A. J. Dennis, Battelle Press, Columbus, 1980 Search .

273. C. H. Wu and H. Niki, Environ. Sci. Technol., 1985, **19**, 1089–1094 .

274. K. Inazu, T. Kobayashi and Y. Hisamatsu, Chemosphere, 1997, **35**, 607–622 .

275. M. L. Nguyen, Y. Bedjanian and A. Guilloteau, J. Atmos. Chem., 2009, **62**, 139–150 .

276. E. Perraudin, H. Budzinski and E. Villenave, Atmos. Environ., 2005, **39**, 6557–6567 .

277. K. Miet, K. Le Menach, P. M. Flaud, H. Budzinski and E. Villenave, Atmos. Environ., 2009, **43**, 837–843 .

278. J. Ma, Y. Liu and H. He, Atmos. Environ., 2011, **45**, 917–924 .

279. T. Ramdahl, A. Bjorseth, D. M. Lokensgard and J. N. Pitts, Chemosphere, 1984, **13**, 527–534 .

280. H. M. Wang, K. Hasegawa and S. Kagaya, Chemosphere, 1999, **39**, 1923–1936 .

281. D. Grosjean, Atmos. Environ., 1983, **17**, 2112–2114 .

282. A. Lindskog, E. Brorström-Lundén and A. Sjödin, Environ. Int., 1985, **11**, 125–130 .

283. R. A. Yokley, A. A. Garrison, G. Mamantov and E. L. Wehry, Chemosphere, 1985, **14**, 1771–1778 .

284. H. M. Wang, K. Hasegawa and S. Kagaya, Chemosphere, 2000, **41**, 1479–1484 .

285. H. M. Wang, K. Hasegawa and S. Kagaya, J. Environ. Sci. Health, Part A: Toxic/Hazard. Subst. Environ. Eng., 2000, **35**, 765–773 .

286. M. Katz, C. Chan, H. Tosine and T. Sakuma, in Polynuclear Aromatic Hydrocarbons, ed. P. W. Jones and P. Leber, Ann Arbor Science Publishers, Ann Arbor, MI, 1979, pp. 171–189 Search .

287. J. N. Pitts, D. M. Lokensgard, P. S. Ripley, K. A. van Cauwenberghe, L. van Vaeck, S. D. Shaffer, A. J. Thill and W. L. Belser, Science, 1980, **210**, 1347–1349 .

288. J. N. Pitts, H. R. Paur, B. Zielinska, J. Arey, A. M. Winer, T. Ramdahl and V. Mejia, Chemosphere, 1986, **15**, 675–685 .

289. S. Lelievre, Y. Bedjanian, N. Pouvesle, J. L. Delfau, C. Vovelle and G. Le Bras, Phys. Chem. Chem. Phys., 2004, **6**, 1181–1191 **RSC**.

290. U. Pöschl, T. Letzel, C. Schauer and R. Niessner, J. Phys. Chem. A, 2001, **105**, 4029–4041 .

291. N. O. A. Kwamena, J. A. Thornton and J. P. D. Abbatt, J. Phys. Chem. A, 2004, **108**, 11626–11634 .

292. N. O. A. Kwamena, M. E. Earp, C. J. Young and J. P. D. Abbatt, J. Phys. Chem. A, 2006, **110**, 3638–3646 .

293. B. T. Mmereki and D. J. Donaldson, J. Phys. Chem. A, 2003, **107**, 11038–11042 .

294. B. T. Mmereki, D. J. Donaldson, J. B. Gilman, T. L. Eliason and V. Vaida, Atmos. Environ., 2004, **38**, 6091–6103 .

295. S. Raja and K. T. Valsaraj, J. Air Waste Manage. Assoc., 2005, **55**, 1345–1355 .

296. T. F. Kahan, N. O. A. Kwamena and D. J. Donaldson, Atmos. Environ., 2006, **40**, 3448–3459 .

297. K. Miet, K. Le Menach, P. M. Flaud, H. Budzinski and E. Villenave, Atmos. Environ., 2009, **43**, 3699–3707 .

298. E. Perraudin, H. Budzinski and E. Villenave, J. Atmos. Chem., 2007, **56**, 57–82 .

299. C. H. Wu, I. Salmeen and H. Niki, Environ. Sci. Technol., 1984, **18**, 603–607 .

300. R. M. Kamens, J. M. Perry, D. A. Saucy, D. A. Bell, D. L. Newton and B. Brand, Environ. Int., 1985, **11**, 131–136 .

301. N. O. A. Kwamena, M. G. Staikova, D. J. Donaldson, I. J. George and J. P. D. Abbatt, J. Phys. Chem. A, 2007, **111**, 11050–11058 .

302. A. Alebic-Juretic, T. Cvitas and L. Klasinc, Environ. Sci. Technol., 1990, **24**, 62–66 .

303. E. Perraudin, H. Budzinski and E. Villenave, Atmos. Environ., 2007, **41**, 6005–6017 .

304. P. S. Bailey, Ozonation in Organic Chemistry: Vol. II, Nonolefinic Compounds, Academic Press, New York, 1982, pp. 82–84 Search .

305. T. Letzel, E. Rosenberg, R. Wissiack, M. Grasserbauer and R. Niessner, J. Chromatogr., A, 1999, **855**, 501–514 .

306. J. Jäger and V. Hanus, J. Hyg., Epidemiol., Microbiol., Immunol., 1980, **24**, 1–15 Search .

307. H. Y. Chen, Y. G. Teng and J. S. Wang, Sci. Total Environ., 2012, **414**, 293–300 .

308. S. Zhou, A. K. Y. Lee, R. D. McWhinney and J. P. D. Abbatt, J. Phys. Chem. A, 2012, **116**, 7050–7056 .

309. R. W. Coutant, L. Brown, J. C. Chuang, R. M. Riggin and R. G. Lewis, Atmos. Environ., 1988, **22**, 403–409 .

310. K. Zimmermann, N. Jariyasopit, S. L. Massey Simonich, S. Tao, R. Atkinson and J. Arey, Environ. Sci. Technol., 2013, **47**, 8434–8442 .

311. F. M. Jaward, N. J. Farrar, T. Harner, A. J. Sweetman and K. C. Jones, Environ. Toxicol. Chem., 2004, **23**, 1355–1364 .

312. J. Klánová, P. Čupr, I. Holoubek, J. Borůvková, P. Přibylová, R. Kareš, T. Tomsej and T. Ocelka, J. Environ. Monit., 2009, **11**, 1952–1963 **RSC**.

313. L. Nizzetto, R. Lohmann, R. Gioia, A. Jahnke, C. Temme, J. Dachs, P. Herckes, A. Di Guardo and K. C. Jones, Environ. Sci. Technol., 2008, **42**, 1580–1585 .

314. S. A. Genualdi, R. K. Killin, J. Woods, G. Wilson, D. Schmedding and S. L. M. Simonich, Environ. Sci. Technol., 2009, **43**, 1061–1066 .

315. T. Primbs, A. Piekarz, G. Wilson, D. Schmedding, C. Higginbotham, J. Field and S. M. Simonich, Environ. Sci. Technol., 2008, **42**, 6385–6391 .

316. J. M. Daisey, R. J. McCaffrey and R. A. Gallagher, Atmos. Environ., 1981, **15**, 1353–1363 .

317. P. Fernández, J. O. Grimalt and R. M. Vilanova, Environ. Sci. Technol., 2002, **36**, 1162–1168 .

318. C. J. Halsall, L. A. Barrie, P. Fellin, D. C. G. Muir, B. N. Billeck, A. R. White, L. Lockhart, F. Y. Rovinsky, E. Y. Kononov and B. Pastukhov, Environ. Sci. Technol., 1997, **31**, 3593–3599 .

319. A. K. Halse, M. Schlabach, S. Eckhardt, A. Sweetman, K. C. Jones and K. Breivik, Atmos. Chem. Phys., 2011, **11**, 1549–1564 .

320. H. Hung, P. Blanchard, C. J. Halsall, T. F. Bidleman, G. A. Stern, P. Fellin, D. C. G. Muir, L. A. Barrie, L. M. Jantunen, P. A. Helm, J. Ma and A. Konoplev, Sci. Total Environ., 2005,**342**, 119–144 .

321. J. L. Jaffrézo, P. Masclet, M. P. Clain, H. Wortham, S. Beyne and H. Cachier, Atmos. Environ., Part A, 1993, **27**, 2781–2785 .

322. G. Lammel, J. Klánová, J. Kohoutek, R. Prokeš, L. Ries and A. Stohl, Environ. Pollut., 2009, **157**, 3264–3271 .

323. L. Melymuk, M. Robson, P. A. Helm and M. L. Diamond, Atmos. Environ., 2011, **45**, 1867–1875 .

324. G. W. Patton, M. D. Walla, T. F. Bidleman and L. A. Barrie, J. Geophys. Res.: Atmos., 1991, **96**, 10867–10877 .

325. B. L. van Drooge, P. Fernandez, J. O. Grimalt, E. Stuchlik, C. J. Torres Garcia and E. Cuevas, Environ. Sci. Pollut. Res., 2010, **17**, 1207–1216 .

326. H. Xiao, S. Kang, Q. Zhang, W. Han, M. Loewen, F. Wong, H. Hung, Y. D. Lei and F. Wania, J. Geophys. Res.: Atmos., 2010, **115**, D16310 .

327. J. Gabrieli, P. Vallelonga, G. Cozzi, P. Gabrielli, A. Gambaro, M. Sigl, F. Decet, M. Schwikowski, H. Gäggeler, C. Boutron, P. Cescon and C. Barbante, Environ. Sci. Technol., 2010,**44**, 3260–3266 .

328. S. Usenko, D. H. Landers, P. G. Appleby and S. L. Simonich, Environ. Sci. Technol., 2007, **41**, 7235–7241 .

329. T. Gocht, K. M. Moldenhauer and W. Puttmann, Appl. Geochem., 2001, **16**, 1707–1721 .

330. L. K. Ackerman, A. R. Schwindt, S. L. M. Simonich, D. C. Koch, T. F. Blett, C. B. Schreck, M. L. Kent and D. H. Landers, Environ. Sci. Technol., 2008, **42**, 2334–2341 .

331. X. Wang, T. Yao, Z. Cong, X. Yan, S. Kang and Y. Zhang, Chin. Sci. Bull., 2007, **52**, 1405–1413 .

332. A. J. Peters, D. J. Gregor, C. F. Teixeira, N. P. Jones and C. Spencer, Sci. Total Environ., 1995, **160–161**, 167–179 .

333. J. L. Jaffrézo, M. P. Clain and P. Masclet, Atmos. Environ., 1994, **28**, 1139–1145 .

334. V. Hoyau, J. L. Jaffrézo, P. H. Garrigues, M. P. Clain and P. Masclet, Polycyclic Aromat. Compd., 1996, **8**, 35–44 .

335. P. Kukučka, G. Lammel, A. Dvorská, J. Klánová, A. Möller and E. Fries, Environ. Chem., 2010, **7**, 504–513.

336. P. G. Desideri, L. Lepri, L. Checchini and D. Santianni, Int. J. Environ. Anal. Chem., 1994, **55**, 33–46 .

337. R. Lohmann, R. Gioia, K. C. Jones, L. Nizzetto, C. Temme, Z. Xie, D. Schulz-Bull, I. Hand, D. M. Griffin, E. Morgan and L. Jantunen, Environ. Sci. Technol., 2009, **43**, 5633–5639 .

338. C. J. Halsall, A. J. Sweetman, L. A. Barrie and K. C. Jones, Atmos. Environ., 2001, **35**, 255–267 .

339. W. D. Hafner and R. A. Hites, Environ. Sci. Technol., 2003, **37**, 3764–3773 .

340. C. Lang, S. Tao, W. Liu, Y. Zhang and S. Simonich, Environ. Sci. Technol., 2008, **42**, 5196–5201 .

341. U. M. Sofowote, H. Hung, A. K. Rastogi, J. N. Westgate, P. F. Deluca, Y. Su and B. E. McCarry, Atmos. Environ., 2011, **45**, 967–976 .

342. D. Yaffe, Y. Cohen, J. Arey and A. J. Grosovsky, Risk Anal., 2001, **21**, 275–294 .

343. K. Prevedouros, E. Brorström-Lundén, C. J. Halsall, K. C. Jones, R. G. M. Lee and A. J. Sweetman, Environ. Pollut., 2004, **128**, 17–27 .

344. A. Aulinger, V. Matthias and M. Quante, J. Appl. Meteo. Climatol., 2007, **46**, 1718–1730 .

345. A. Gusev, O. Rozovskaya, V. Shatalov, V. Sokovyh, W. Aas and K. Breivik, Persistent Organic Pollutants in the Environment. EMEP Status Report 3/2009, Meteorological Synthesizing Centre – East, Moscow, Russian Federation, 2009, p. 84.

346. V. Matthias, A. Aulinger and M. Quante, Atmos. Environ., 2009, **43**, 4078–4086 .

347. V. Shatalov, A. Gusev, S. Dutchak, O. Rozovskaya, V. Sokovykh, N. Vulykh, W. Aas and K. Breivik, Persistent organic pollutants in the environment. EMEP Status Report 3/2010, Geneva, 2010, p. 100.

348. C. Silibello, G. Calori, M. P. Costa, M. G. Dirodi, M.

Mircea, P. Radice, L. Vitali and G. Zanini, Atmos. Pollut. Res., 2012, **3**, 399–407 .

349. Y. Zhang, H. Shen, S. Tao and J. Ma, Atmos. Environ., 2011, **45**, 2820–2827 .

350. Y. Zhang, S. Tao, J. Ma and S. Simonich, Atmos. Chem. Phys., 2011, **11**, 11993–12006 .

351. A. M. Sehili and G. Lammel, Atmos. Environ., 2007, **41**, 8301–8315 .

352. C. L. Friedman and N. E. Selin, Environ. Sci. Technol., 2012, **46**, 9501–9510 .

353. G. Lammel, W. Klöpffer, V. S. Semeena, E. Schmidt and A. Leip, Environ. Sci. Pollut. Res., 2007, **14**, 153–165 .

354. M. Hauck, M. A. J. Huijbregts, J. M. Armitage, I. T. Cousins, A. M. J. Ragas and D. van de Meent, Chemosphere, 2008, **72**, 959–967 .

355. P. Ciccioli, A. Cecinato, E. Brancaleoni, M. Frattoni, P. Zacchei, A. H. Miguel and P. D. Vasconcellos, J. Geophys. Res.: Atmos., 1996, **101**, 19567–19581 .

356. E. Alonso, N. Tapie, H. Budzinski, K. Lemenach, L. Peluhet and J. V. Tarazona, Environ. Sci. Pollut. Res., 2008, **15**, 31–40 .

357. D. Marinov, S. Dueri, I. Puillat, R. Carafa, E. Jurado, N. Berrojalbiz, J. Dachs and J. M. Zaldivar, Mar. Pollut. Bull., 2009, **58**, 1554–1561 .

358. C. Walgraeve, K. Demeestere, J. Dewulf, R. Zimmermann and H. Van Langenhove, Atmos. Environ., 2010, **44**, 1831–1846 .

359. T. Nielsen, Environ. Sci. Technol., 1984, **18**, 157–163 .

360. A. Liberti and P. Ciccioli, HRC & CC, J. High Resolut. Chromatogr. Chromatogr. Commun., 1986, **9**, 492–501 .

361. J. Arey, B. Zielinska, R. Atkinson and A. M. Winer, Atmos. Environ., 1987, **21**, 1437–1444 .

362. J. P. Zhu, X. L. Cao, R. Pigeon and K. Mitchell, J. Air Waste Manage. Assoc., 2003, **53**, 67–76 .

363. R. Atkinson, J. Arey, B. Zielinska, J. N. Pitts and A. M. Winer, Atmos. Environ., 1987, **21**, 2261–2262 .

364. H. A. Bamford and J. E. Baker, Atmos. Environ., 2003, **37**, 2077–2091 .

365. P. Ciccioli, A. Cecinato, E. Brancaleoni, R. Draisci and A. Liberti, Aerosol Sci. Technol., 1989, **10**, 296–310 .

366. P. Ciccioli, A. Cecinato, R. Cabella, E. Brancaleoni and P. Buttini, Atmos. Environ., Part A, 1993, **27**, 1261–1270 .

367. T. T. Hien, L. T. Thanh, T. Kameda, N. Takenaka and H. Bandow, Atmos. Environ., 2007, **41**, 7715–7725 .

368. T. Nielsen, B. Seitz and T. Ramdahl, Atmos. Environ., 1984, **18**, 2159–2165 .

369. J. N. Pitts, J. A. Sweetman, B. Zielinska, A. M. Winer and R. Atkinson, Atmos. Environ., 1985, **19**, 1601–1608 .

370. T. Ramdahl, B. Zielinska, J. Arey, R. Atkinson, A. M. Winer and J. N. Pitts, Nature, 1986, **321**, 425–427 .

371. F. Reisen and J. Arey, Environ. Sci. Technol., 2005, **39**, 64–73 .

372. W. Wang, N. Jariyasopit, J. Schrlau, Y. Jia, S. Tao, T.-W. Yu, R. H. Dashwood, W. Zhang, X. Wang and S. L. M. Simonich, Environ. Sci. Technol., 2011, **45**, 6887–6895 .

373. J. A. Sweetman, B. Zielinska, R. Atkinson, T. Ramdahl, A. M. Winer and J. N. Pitts, Atmos. Environ., 1986, **20**, 235–238 .

374. A. Feilberg, M. W. B. Poulsen, T. Nielsen and H. Skov, Atmos. Environ., 2001, **35**, 353–366 .

375. F. Marino, A. Cecinato and P. A. Siskos, Chemosphere, 2000, **40**, 533–537 .

376. A. Albinet, E. Leoz-Garziandia, H. Budzinski, E. Villenave and J. L. Jaffrezo, Atmos. Environ., 2008, **42**, 43–54 .

377. M. Tsapakis and E. G. Stephanou, Environ. Sci. Technol., 2007, **41**, 8011–8017 .

378. F. Reisen, S. Wheeler and J. Arey, Atmos. Environ., 2003, **37**, 3653–3657 .

379. M. Dimashki, S. Harrad and R. M. Harrison, Atmos. Environ., 2000, **34**, 2459–2469 .

380. T. Murahashi and K. Hayakawa, Anal. Chim. Acta, 1997, **343**, 251–257 .

381. J. M. Bayona, M. ellas, P. Fernandez, A. M. Solanas and J. Albaiges, Chemosphere, 1994, **29**, 441–450 .

382. A. Cecinato, J. Sep. Sci., 2003, **26**, 402–408 .

383. J. C. Chuang, G. A. Mack, M. R. Kuhlman and N. K. Wilson, Atmos. Environ., Part B, 1991, **25**, 369–380 .

384. N. K. Wilson and T. R. McCurdy, Atmos. Environ., 1995, **29**, 2575–2584 .

385. P. Gupta, W. P. Harger and J. Arey, Atmos. Environ., 1996, **30**, 3157–3166 .

386. T. Kameda, J. Health Sci., 2011, **57**, 504–511 .

387. Y. Kojima, K. Inazu, Y. Hisamatsu, H. Okochi, T. Baba and T. Nagoya, Atmos. Environ., 2010, **44**, 2873–2880 .

388. A. Albinet, E. Leoz-Garziandia, H. Budzinski, E. Villenave and J. L. Jaffrezo, Atmos. Environ., 2008, **42**, 55–64 .

389. R. M. Kamens, H. Karam, J. H. Guo, J. M. Perry and L. Stockburger, Environ. Sci. Technol., 1989, **23**, 801–806 .

390. A. Eiguren-Fernandez, A. H. Miguel, R. Lu, K. Purvis, B. Grant, P. Mayo, E. Di Stefano, A. K. Cho and J. Froines, Atmos. Environ., 2008, **42**, 2312–2319 .

391. J. Arey, R. Atkinson, B. Zielinska and P. A. McElroy, Environ. Sci. Technol., 1989, **23**, 321–327 .

392. J. Arey, R. Atkinson, S. M. Aschmann and D. Schuetzle, Polycyclic Aromat. Compd., 1990, **1**, 33–50 .

393. Y. Kojima, K. Inazu, Y. Hisamatsu, H. Okochi, T. Baba and T. Nagoya, Asian J. Atmos. Environ., 2010, **4**, 1–8 .

394. A. Pozzer, P. Zimmermann, U. M. Doering, J. van Aardenne, H. Tost, F. Dentener, G. Janssens-Maenhout and J. Lelieveld, Atmos. Chem. Phys., 2012, **12**, 6915–6937 .

395. M. Giannouli, E.-A. Kalognomou, G. Mellios, N. Moussiopoulos, Z. Samaras and J. Fiala, Atmos. Environ., 2011, **45**, 4753–4762 .

396. H. Cai and S. Xie, Sci. Total Environ., 2011, **409**, 1935–1948 .

397. W. Wang, S. Simonich, B. Giri, Y. Chang, Y. Zhang, Y. Jia, S. Tao, R. Wang, B. Wang, W. Li, J. Cao and X. Lu, Sci. Total Environ., 2011, **409**, 2942–2950 .

398. Ref. 398–472 are cited in the ESI†; J. O. Allen, N. M. Dookeran, K. Taghizadeh, A. L. Lafleur, K. A. Smith and A. F. Sarofim, Environ. Sci. Technol., 1997, **31**, 2064–2070 .

399. C. A. Alves and C. A. Pio, J. Braz. Chem. Soc., 2005, **16**, 1017–1029 .

400. P. tells, F. J. Santos and M. T. Gaiceran, J. Chromatogr., A, 2003, **1010**, 141–151 .

401. A. K. Cho, E. Di Stefano, Y. You, C. E. Rodriguez, D. A. Schmitz, Y. Kumagai, A. H. Miguel, A. Eiguren-Fernandez, T. Kobayashi, E. Avol and J. R. Froines, Aerosol Sci. Technol., 2004, **38**, 68–81 .

402. M. Y. Chung, R. A. Lazaro, D. Lim, J. Jackson, J. Lyon, D. Rendulic and A. S. Hasson, Environ. Sci. Technol., 2006, **40**, 4880–4886 .

403. O. Delhomme, M. Millet and P. Herckes, Talanta, 2008, **74**, 703–710 .

404. A. Eiguren-Fernandez, A. H. Miguel, E. Di Stefano, D. A. Schmitz, A. K. Cho, S. Thurairatnam, E. L. Avol and J. R. Froines, Aerosol Sci. Technol., 2008, **42**, 854–861 .

405. N. Kishikawa, M. Nakao, Y. Ohba, K. Nakashima and N. Kuroda, Chemosphere, 2006, **64**, 834–838 .

406. R. Koeber, J. M. Bayona and R. Niessner, Environ. Sci. Technol., 1999, **33**, 1552–1558 .

407. J. Konig, E. Balfanz, W. Funcke and T. Romanowski, Anal. Chem., 1983, **55**, 599–603 .

408. J. Lintelmann, K. Fischer, E. Karg and A. Schroppel, Anal. Bioanal. Chem., 2005, **381**, 508–519 .

409. J. Lintelmann, K. Fischer and G. Matuschek, J. Chromatogr., A, 2006, **1133**, 241–247 .

410. Y. Liu, M. Sklorz, J. Schnelle-Kreis, J. Orasche, T. Ferge, A. Kettrup and R. Zimmermann, Chemosphere, 2006, **62**, 1889–1898 .

411. J. D. McDonald, B. Zielinska, J. C. Sagebiel and M. R. McDaniel, Aerosol Sci. Technol., 2002, **36**, 1033–1044 .

412. C. Neususs, M. Pelzing, A. Plewka and H. Herrmann, J. Geophys. Res.: Atmos., 2000, **105**, 4513–4527 .

413. S. Nicol, J. Dugay and M. C. Hennion, Chromatographia, 2001, **53**, S464–S469 .

414. E.-J. Park, D.-S. Kim and K. Park, Environ. Monit. Assess., 2008, **137**, 441–449 .

415. N. Re-Poppi and M. Santiago-Silva, Atmos. Environ., 2005, **39**, 2839–2850 .

416. E. Schnelle-Kreis, M. Sklorz, A. Peters, J. Cyrys and R. Zimmermann, Atmos. Environ., 2005, **39**, 7702–7714 Search .

417. J. Schnelle-Kreis, I. Gebefugi, G. Welzl, T. Jaensch and A. Kettrup, Atmos. Environ., 2001, **35**, S71–S81 .

418. M. D. Sienra, Atmos. Environ., 2006, **40**, 2374–2384 .

419. B. R. T. Simoneit, X. Bi, D. R. Oros, P. M. Medeiros, G. Sheng and J. Fu, Environ. Sci. Technol., 2007, **41**, 7294–7302 .

420. M. Sklorz, J.-J. Briede, J. Schnelle-Kreis, Y. Liu, J. Cyrys, T. M. de Kok and R. Zimmermann, J. Toxicol. Environ. Health, Part A, 2007, **70**, 1866–1869 .

421. M. Strandell, S. Zakrisson, T. Alsberg, R. Westerholm, L. Winquist and U. Rannug, Environ. Health Perspect., 1994, **102**, 85–92 .

422. M. Tsapakis, E. Lagoudaki, E. G. Stephanou, I. G. Kavouras, P. Koutrakis, P. Oyola and D. von Baer, Atmos. Environ., 2002, **36**, 3851–3863 .

423. A. Valavanidis, K. Fiotakis, T. Vlahogianni, V. Papadimitriou and V. Pantikaki, Environ. Chem., 2006, **3**, 118–123 .

424. N. Yassaa, B. Y. Meklati, A. Cecinato and F. Marino, Environ. Sci. Technol., 2001, **35**, 306–311 .

425. T. Kameda, T. Goto, A. Toriba, N. Tang and K. Hayakawa, J. Chromatogr., A, 2009, **1216**, 6758–6761 .

426. M. P. Ligocki and J. F. Pankow, Environ. Sci. Technol., 1989, **23**, 75–83 .

427. P. M. Fine, G. R. s and B. R. T. Simoneit, Environ. Sci. Technol., 2002, **36**, 1442–1451 .

428. E. M. Fitzpatrick, A. B. Ross, J. Bates, G. Andrews, J. M. Jones, J. Aspden, H. Phylaktou, M. Pourkashanian and A. Williams, Process Saf. Environ. Prot., 2007, **85**, 430–440 .

429. M. P. Fraser, G. R. s and B. R. T. Simoneit, Environ. Sci. Technol., 1998, **32**, 2051–2060 .

430. M. D. Hays, P. M. Fine, C. D. Geron, M. J. Kleeman and B. K. Gullett, Atmos. Environ., 2005, **39**, 6747–6764 .

431. Y. Iinuma, E. Brüggemann, T. Gnauk, K. Müller, M. O. Andreae, G. Helas, R. Parmar and H. Herrmann, J. Geophys. Res.: Atmos., 2007, **112**, D08209 .

432. C. A. Jakober, S. G. Riddle, M. A. Robert, H. Destaillats, M. J. Charles, P. G. Green and M. J. Kleeman, Environ. Sci. Technol., 2007, **41**, 4548–4554 .

433. C. A. Jakober, M. A. Robert, S. G. Riddle, H. Destaillats, M. J. Charles, P. G. Green and M. J. Kleeman, Environ. Sci. Technol., 2008, **42**, 4697–4703 .

434. J. Oda, I. Maeda, T. Mori, A. Yasuhara and Y. Saito, Environ. Technol., 1998, **19**, 961–976 .

435. J. Oda, S. Nomura, A. Yasuhara and T. Shibamoto, Atmos. Environ., 2001, **35**, 4819–4827 .

436. X. Rao, R. Kobayashi, R. White-Morris, R. Spaulding, P. Frazey and M. J. Charles, J. AOAC Int., 2001, **84**, 699–705 .

437. N. Re-Poppi and M. R. Santiago-Silva, Chromatographia, 2002, **55**, 475–481 .

438. W. F. Rogge, L. M. Hildemann, M. A. Mazurek, G. R. s and B. R. T. Simoneit, Environ. Sci. Technol., 1993, **27**, 2736–2744 .

439. W. F. Rogge, L. M. Hildemann, M. A. Mazurek, G. R. s and B. R. T. Simoneit, Environ. Sci. Technol., 1993, **27**, 1892–1904 .

440. W. F. Rogge, L. M. Hildemann, M. A. Mazurek, G. R. s and B. R. T. Simoneit, Environ. Sci. Technol., 1993, **27**, 636–651 .

441. W. F. Rogge, L. M. Hildemann, M. A. Mazurek, G. R. s and B. R. T. Simoneit, Environ. Sci. Technol., 1997, **31**, 2731–2737 .

442. W. F. Rogge, L. M. Hildemann, M. A. Mazurek, G. R. s and B. R. T. Simoneit, Environ. Sci. Technol., 1998, **32**, 13–22 .

443. G. Shen, S. Wei, Y. Zhang, R. Wang, B. Wang, W. Li, H. Shen, Y. Huang, Y. Chen, H. Chen, W. Wei and S. Tao, Atmos. Environ., 2012, **60**, 234–237 .

444. S. Sidhu, B. Gullett, R. Striebich, J. Klosterman, J. Contreras and M. de Vito, Atmos. Environ., 2005, **39**, 801–811 .

445. M. Stefanova, S. P. Marinov, A. M. Mastral, M. S. Callen and T. Garcia, Fuel Process. Technol., 2002, **77**, 89–94 .

446. B. Zielinska, J. Sagebiel, J. D. McDonald, K. Whitney and D. R. Lawson, J. Air Waste Manage. Assoc., 2004, **54**, 1138–1150 .

447. H. A. Bamford, D. Z. Bezabeh, M. M. Schantz, S. A. Wise and J. E. Baker, Chemosphere, 2003, **50**, 575–587 .

448. K. Hayakawa, T. Murahashi, M. Butoh and M. Miyazaki, Environ. Sci. Technol., 1995, **29**, 928–932 .

449. H. Kakimoto, M. Kitamura, Y. Matsumoto, S. Sakai, F. Kanoh, T. Murahashi, K. Akutsu, R. Kizu and K. Hayakawa, J. Health Sci., 2000, **46**, 5–15 .

450. H. Kakimoto, H. Yokoe, Y. Matsumoto, S. Sakai, F. Kanoh, T. Murahashi, K. Akutsu, A. Toriba, R. Kizu and K. Hayakawa, J. Health Sci., 2001, **47**, 385–393 .

451. L. C. Marr, K. Dzepina, J. L. Jimenez, F. Reisen, H. L. Bethel, J. Arey, J. S. Gaffney, N. A. Marley, L. T. Molina and M. J. Molina, Atmos. Chem. Phys., 2006, **6**, 1733–1745 .

452. H. F. Nassar, N. Tang, T. Kameda, A. Toriba, M. I. Khoder and K. Hayakawa, Atmos. Environ., 2011, **45**, 7352–7359 .

453. M. D. Sienra and N. G. Rosazza, Atmos. Res., 2006, **81**, 265–276 .

454. N. Tang, T. Hattori, R. Taga, K. Igarashi, X. Y. Yang, K. Tamura, H. Kakimoto, V. F. Mishukov, A. Toriba, R. Kizu and K. Hayakawa, Atmos. Environ., 2005, **39**, 5817–5826 .

455. N. Tang, M. Tabata, V. F. Mishukov, V. Sergineko, A. Toriba, R. Kizu and K. Hayakawa, J. Health Sci., 2002, **48**, 30–36 .

456. B. L. Valle-Hernandez, V. Mugica-Alvarez, E. Salinas-Talavera, O. Amador-Munoz, M. A. Murillo-Tovar, R. Villalobos-Pietrini and A. De Vizcaya-Ruiz, Sci. Total Environ., 2010,**408**, 5429–5438 .

457. B. Zielinska, J. Arey, R. Atkinson and A. M. Winer, Atmos. Environ., 1989, **23**, 223–229 .

458. J. C. Ball and W. C. Young, Environ. Sci. Technol., 1992, **26**, 2181–2186 .

459. R. M. Campbell and M. L. Lee, Anal. Chem., 1984, **56**, 1026–1030 .

460. W. M. Draper, Chemosphere, 1986, **15**, 437–447 .

461. T. L. Gibson, Atmos. Environ., 1982, **16**, 2037–2040 .

462. K. Hayakawa, M. Butoh, Y. Hirabayashi and M. Miyazaki, Jpn. J. Toxicol. Environ. Health, 1994, **40**, 20–25 .

463. K. Hayakawa, M. Butoh and M. Miyazaki, Anal. Chim. Acta, 1992, **266**, 251–256 .

464. IARC, International Agency for Research on Cancer, Monographs on the Evaluation of Carcinogenic Risks to Humans, Volume 46: Diesel and Gasoline Engine Exhausts and Some Nitroarenes, http://monographs.iarc.fr/ENG/Monographs/vol46/volume46.pdf, 1989.

465. G. Karavalakis, S. Stournas and E. Bakeas, Sci. Total Environ., 2009, **407**, 3338–3346 .

466. T. Murahashi, R. Kizu, H. Kakimoto, A. Toriba and K. Hayakawa, J. Health Sci., 1999, **45**, 244–250 .

467. M. C. Paputa-Peck, R. S. Marano, D. Schuetzle, T. L. Riley, C. V. Hampton, T. J. Prater, L. M. Skewes, T. E. Jensen, P. H. Ruehle, L. C. Bosch and W. P. Duncan, Anal. Chem., 1983,**55**, 1946–1954 .

468. S. M. Rappaport, Z. L. Jin and X. B. Xu, J. Chromatogr., 1982, **240**, 145–154 .

469. M. A. Ratcliff, A. J. Dane, A. Williams, J. Ireland, J. Luecke, R. L. McCormick and K. J. Voorhees, Environ. Sci. Technol., 2010, **44**, 8343–8349 .

470. D. Schuetzle, Environ. Health Perspect., 1983, **47**, 65–80 .

471. D. Schuetzle, T. L. Riley, T. J. Prater, T. M. Harvey and D. F. Hunt, Anal. Chem., 1982, **54**, 265–271 .

472. R. Westerholm and K. E. Egebäck, Environ. Health Perspect., 1994, **102**, 13–23 .

CHAPTER 6

MECHANISTIC AND KINETIC ASPECTS OF PENTOSE DEHYDRATION TOWARDS FURFURAL IN AQUEOUS MEDIA EMPLOYING HOMOGENEOUS CATALYSIS

Bart Danon, Gianluca Marcotullio and Wiebren de Jong

INTRODUCTION

Nowadays, biomass as a renewable and potentially sustainable resource is seriously considered to contribute to the production of green chemicals, products and biofuels *via* the oil refinery analogue of a biorefinery.[1–6] Second and higher generation biomasses, based on (co-)processing of non-edible plant parts to generate such products, are attractive in this respect as they do not cause unacceptable interferences with the food and feed production chain. In a biorefinery, fractionation of the biomass' main organic constituents, the polymeric carbohydrates (hemicellulose and cellulose) and lignin, can be performed in different ways.[2,7] Many studies have been performed concerning the conversion of non-first generation biomass carbohydrates to ethanol using hydrolytic pre-treatment and subsequent fermentation.[8] Lignin is usually converted to produce steam *via* combustion, but studies towards valorization of this major biomass constituent are also accumulating,[9] though up to now only ligno-sulphonate production in the context of the pulp and paper industry has been commercialized.[10] However, the valorization of the hemicellulose derived pentose sugars to higher value added chemicals than ethanol has been subjected to substantially fewer studies. Furfural is one such chemical. It consists of a furan-ring with an aldehyde side group, see also Figure 1.

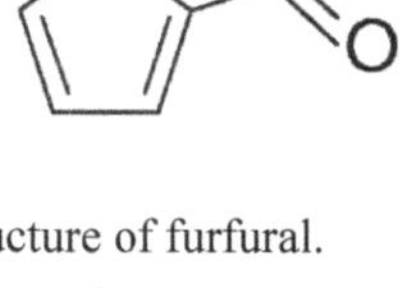

Figure 1: Chemical structure of furfural.

Citation: Bart Danon Gianluca Marcotullio and Wiebren de Jong, "Mechanistic and kinetic aspects of pentose dehydration towards furfural in aqueous media employing homogeneous catalysis," DOI: 10.1039/C3GC41351A.

Its formation was discovered by Döbereiner back in 1821, but it only reached the stage of the first industrial production processes after about one century when in 1922 Quaker Oats started industrial production in the USA (Cedar Rapids, Iowa) based on the agricultural food residue oat husk.[1,11,12] Furfural is considered to be an attractive platform chemical,[13,14] and therefore it is suitable in a biorefinery context for the production of a myriad of product types, *e.g.* solvents (tetrahydrofuran, THF), plastics (in particular, polyamides), resins *via* furfuryl alcohol (already in current practice) and fuel additives (*e.g.* methyl-tetrahydrofuran used in novel fuel formulations,[15] or rather methylfuran that can be blended into petrol as has been demonstrated by road testing[16]) with huge market potential. Current production levels are of the order of 300 ktons per year globally with a major production capacity in China[17] and furfural production is mainly targeted at furfuryl alcohol production which is used to produce foundry resins.[18] Furfural itself finds application in the oil refining sector for the selective extraction of butadiene and lube oils[19] and in novel applications such as pesticides and nematicides in agricultural practice.

The production of furfural based on biomass, mainly nonfood residues of food crops and wood, is already an old practice. Furfural is one of the few chemicals that are not manufactured by a petro-chemical route as biomass chemical processing is cheaper. A substantial number of processes that were and are still operated have been described well by Zeitsch.[12] Recently, an updated overview of current furfural production processes was presented.[20] Acid catalysed hydrolysis and dehydration of the pentosan derived sugars forms the basis of such industry. Most processes use sulphuric acid as the mineral acid, but hydrochloric acid has also been used in different low to high concentrations. The product recovery is conventionally accomplished *via*steam stripping with a relatively high steam and thus energy consumption, in the range of 25–35 tons per ton of furfural produced.[21] Industrial furfural production today thus still largely relies on the batch dehydration of pentosan-rich biomass using sulphuric acid, with yields typically hovering around 50%.[18] Such reactor configurations cause the furfural once formed to stay too long in the acid aqueous medium so that follow-up degradation reactions take place. A crucial key to furfural production optimization lies in understanding the mechanism and related kinetics of its formation from the hemicellulose derived sugars, like xylose and arabinose to name the most prevalent, and in particular in the minimization of side and loss reactions. This paper will therefore review studies that have been performed about C5 sugar conversion to furfural and furfural loss reactions under homogeneously acid catalysed aqueous reaction conditions. Heterogeneous catalysis is not addressed in this work; a recent overview was given by Dashtban *et al.*[22] First, the mechanisms of the formation of furfural from pentoses that have been described in the literature will be critically reviewed. Then, studies concerning the kinetics of furfural formation and degradation reactions to solid humins will be summarized and discussed, including an outlook on the application to the formation from biomasses.

MECHANISTIC ASPECTS

Mechanisms of Furfural Formation

The discussion around the mechanistic aspects of furfural formation might seem of only academic interest, although a deeper understanding of such a mechanism is also crucial for planning an appropriate catalytic strategy in the furfural industry.

More than one reaction mechanism has been proposed in different studies based on different techniques, and under disparate reaction conditions. Firstly, the reaction was believed to start from the acyclic form of the pentoses, either *via* a 1,2-enediol intermediate **2** and subsequent dehydration (see Figure 2)[23,24] or directly *via* a 2,3-(α,β-)unsaturated aldehyde **4**, see Figure 3.[25,26]More recently, other authors believed the reaction to take place starting from the pyranose form of the pentoses, by the action of H^+ on the O-2 of the pyranose ring, leading to the 2,5-anhydroxylose furanose intermediate **7** which is subsequently dehydrated to furfural, see Figure 4.[27–30]

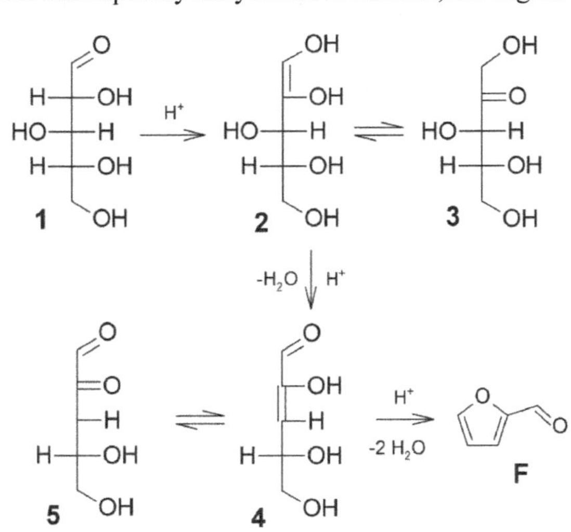

Figure 2: Xylose dehydration mechanism *via* enolization.[23,24]

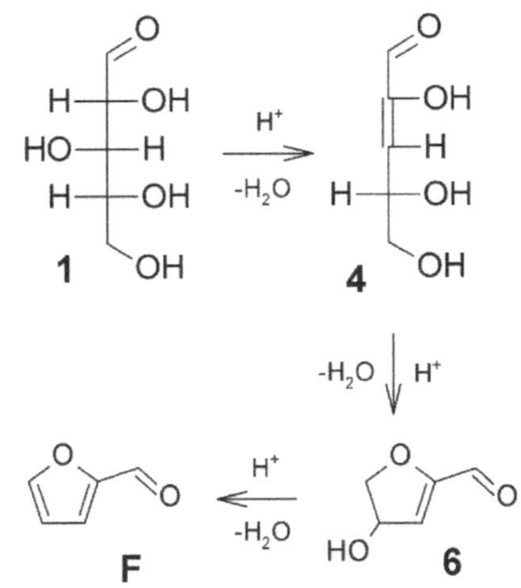

Figure 3: Xylose dehydration mechanism *via* β-elimination.[25,26]

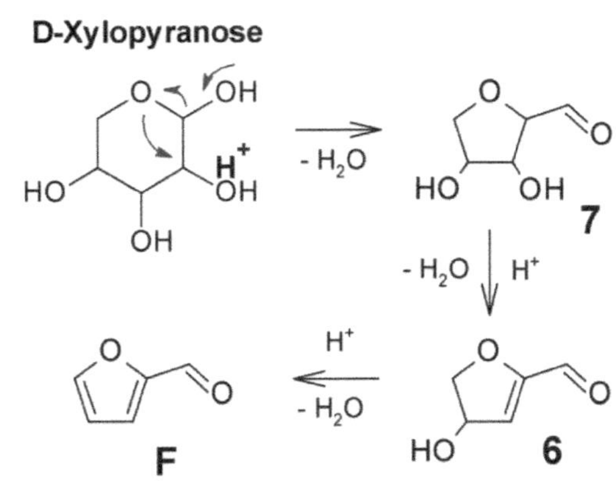

Figure 4: Xylose dehydration mechanism *via* cyclic intermediates.[27–30]

Published experimental evidence exists, mainly based on the reaction of labeled model compounds and on isotope exchange techniques, which is decisive in elucidating some aspects of this discussion. In the first place it has been shown that when reacting xylose-1-^{14}C in 12% aqueous HCl, the carbon at the C-1 position is found nearly completely at the 'aldehydic' carbon of the final product, 2-furaldehyde-^{14}C.[26] In fact all the mechanisms proposed so far adhere to this evidence. Using similar isotope-exchange techniques, in more than one work it has been proven that, under strongly acid conditions, and in the presence of deuterated/tritiated water, nearly no carbon-bound tritium/deuterium is found in the furfural generated.[23,31,32] This suggests that no solvent hydrogen is exchanged during the course of the reaction, contrary to the well-known mechanism of aldose–ketose isomerization under alkaline conditions.[33,34] In this perspective, Harris and Feather[31] provided evidence for the intramolecular hydrogen transfer reaction, *i.e.* the direct conversion of D-glucose-2-^{3}H into D-fructose-1-^{3}H in 1 M aqueous H_2SO_4, although such transfer was only partial at lower acidity (pH 3). More recently, Binder *et al.* reported similar results,[32] showing 48% deuterium incorporation at the furfural C-1 position when xylose-2-^{2}H was converted in the presence of water, concluding that 1,2-enediol **2** is not involved in intramolecular hydrogen-transfer, and aldose dehydration might always take place *via* the intermediate isomerization to the ketose **3**. This is supported by the fact that ketopentoses react much faster in water,[23,35–37] in analogy with the lower reactivity of glucose compared to fructose.[38]

Ahmad *et al.* reported similar evidence,[23] showing a decreasing deuterium content at the furfural aldehydic carbon when reacting pentoses in deuterated water at 96 °C at decreasing pD from 4.5 to 1.5, although explaining the only partial deuterium exchange at C-1 at higher pD using the fact that the 1,2-enediol **2** is faster tautomerizing to the ketose **3** than dehydrating. The authors confirmed this interpretation with the gradual exchange of H-la and H-lb in the ketose. Importantly, in the same work it was remarked that no hydrogen was exchanged in the aldoses; hence these were converted irreversibly to 1,2-enediol. Such an indication is in agreement with the general aldose reluctance to isomerization under acidic conditions,[31] and leads to the conclusion that, since enolization reactions are normally reversible, enolization or isomerization can be considered the rate limiting step in the formation of furfural.[39]

Remarkably, the original mechanism proposed by Hurd *et al.* in 1932 is also consistent with the foregoing experimental evidence.[25] In this e the β-elimination forming a 2,3-unsaturated aldehyde intermediate **4** conforms to the absence of solvent hydrogen in furfural, and resembles a well-known reaction in organic chemistry whereby β-hydroxyaldehydes readily lose water in acidic solution. On the other hand this theory fails to explain intramolecular hydrogen-transfer.

As far as the furanose 2,5-anhydroxylose intermediate is concerned (Figure 4), this mechanism presents more than one weak point compared to the others. Notably, it cannot explain the observed intramolecular hydrogen-transfer, as well as the partial deuterium exchange at C-1 and C-3 reported at lower acidity.[23] Moreover, it is not straightforward to extend this mechanism to the pentuloses, at the same time explaining their significantly higher reactivity compared to the corresponding pentoses.

In conclusion, on the one hand it seems reasonable that pentoses react with furfural *via* their acyclic form. In fact ketopentoses, which present a significantly higher proportion of acyclic form in water solution when compared to aldopentoses,[40,41] react much faster in water yielding furfural in proportion to the acidity of the solution.[23,35–37] This is in agreement with the explanation normally given for the higher reactivity of fructose compared to glucose in acidic solution.[38] Furthermore, the same authors putting forward the theory of the 2,5-anhydroxylose intermediate to furfural[29] showed a very low energy of activation related to protonation at O-5 and subsequent ring opening.

Reasonably, the presence of an electron-rich carbonyl group in the acyclic sugars is crucial for dehydration, as enols are very good electron donors favoring the elimination of water. In fact, the dehydration of sorbitol and xylitol requires much more severe conditions to take place compared to glucose and xylose, and they do not yield the corresponding furfuryl-alcohols as one could wrongly expect. In analogy furan cannot be formed from erythritol.[25] The reaction of sugar alcohols in acid solution leads to 1,4-cyclization, yielding rather stable compounds such as 1,4-anhydroxylitol,[42–46] see Figure 5. The incomplete dehydration of sugar alcohols, and their relative reluctance to undergo such a reaction, is indicative of the importance of the enol functions in the intermediate dehydrations to furfural. Lastly, recent kinetic studies from our group[47] have shown a positive entropy of activation during acid-catalyzed dehydration of xylose, indicating increasing degrees of freedom across the activated state, imputed to protonation and ring opening.

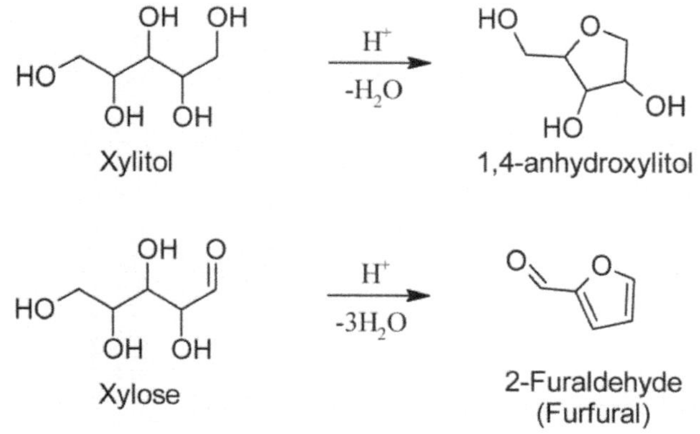

Figure 5: Acid catalyzed dehydration of xylitol compared to xylose.

On the other hand, although glucose–fructose[31,38,48] and xylose–xylulose[32,35–37] isomerizations have been often reported to take place in the course of aldoses conversion into furfurals, there is no common agreement on the key intermediate of the reaction, this being the ketose sugar **3**, the 1,2-enediol **2**, or the 2,3-unsaturated aldehyde **4**. From the foregoing evidence, when sugars are reacted under acidic conditions it seems reasonable to accept the mechanism proposed by Feather and Ahmad *et al.*,[23,24] although neither direct aldose–ketose isomerization nor β-elimination can be ruled out. It is difficult to select one prevailing mechanism; the mechanisms rather seem to coexist, gaining importance one over the other according to the different reaction conditions and the nature of the solvent/catalyst system.

In Figure 6 an attempt is made to group the mechanisms proposed. At lower acidities, enolization and subsequent isomerization is favored, although the formation of furfurals is retarded and there are more options for side reactions (*via***3e**). Under these conditions ketoses may be detected,[31] and enol intermediates such as **2** and **4** generally tautomerize respectively to **3** and **5**faster than they are dehydrated.[23] Further enolization to **3e** has been reported to be irreversible,[23] seemingly leading to side reactions. Also β-elimination cannot be excluded under these conditions, offering a direct path to **4**, hence reducing the possibilities for side reactions *via***3e**.

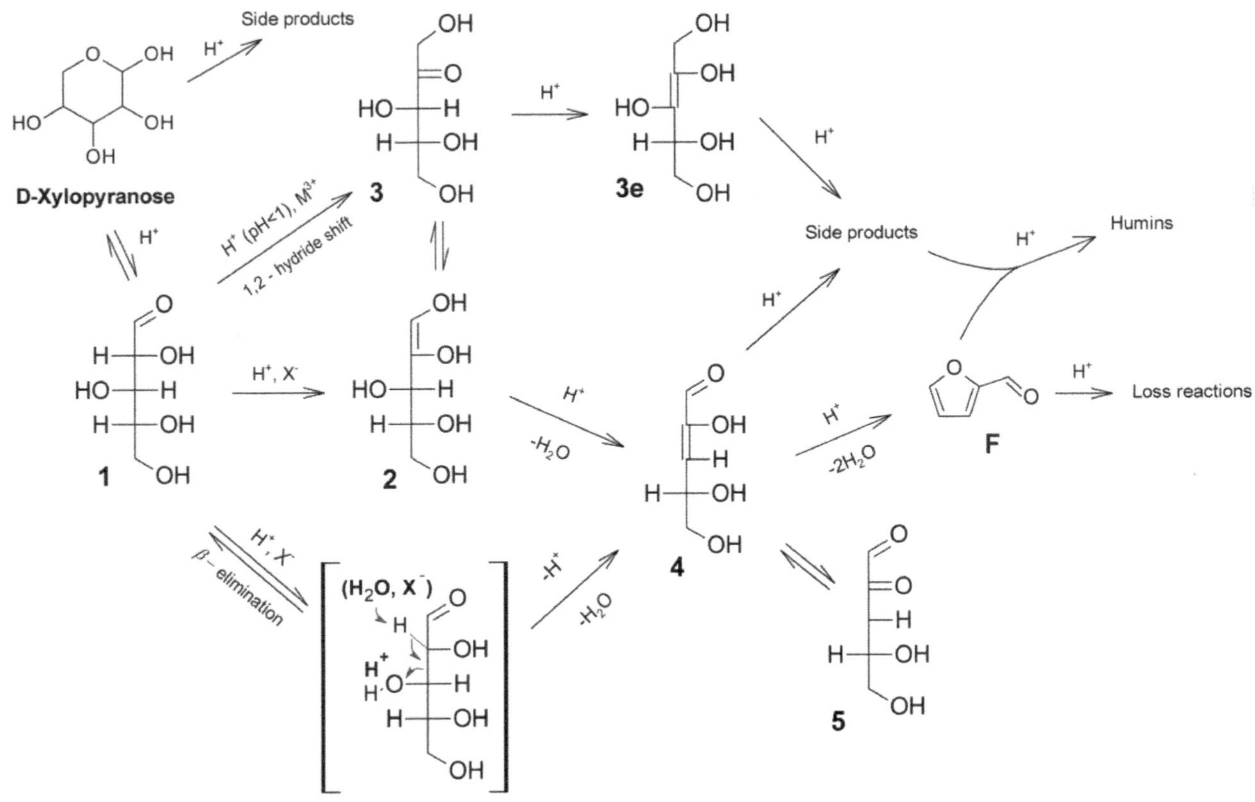

Figure 6: Plausible mechanism of xylose reaction to furfural in acid media. X^- indicates halides ions, and M^{3+} indicates metal cations.

At lower pH (especially pH < 1) direct isomerization *via* intramolecular hydrogen-transfer takes place, and the ketose is readily dehydrated, likely *via* an enol intermediate **2**.

Side and Loss Reactions

There are many options for side and loss reactions during furfural formation which are primarily responsible for limiting the furfural final yield.[49] Firstly, pentoses may be degraded to low molecular weight products, likely generated from the fragmentation of reaction intermediates such as **4** and **3e**. These compounds are primarily organic acids,[50] and aldehydes such as acetaldehyde, formaldehyde and crotonaldehyde.[27,51] Furfural is also degraded to an appreciable extent under the same conditions it is normally produced,[52–54]yielding both insoluble resinous products and other products largely unidentified, except for reductic acid.[55] Moreover, when the initial sugar concentration is relatively high, second-order condensation reactions also take place to a significant extent.[56] These reactions yield soluble and insoluble substances, involving small aldehydes and furfural reacting with each other. The products of such reactions have been studied by Rice *et al.*,[51,57,58] showing that the humic substance formed, both the soluble and the insoluble fraction, presents common infrared spectra, with the only exception of the resinous product formed when reacting furfural alone. The infrared spectra of this humic substance showed the presence of hydroxyl, carbonyl, and possibly carbon–carbon double bonds, and its solubility appears to be linked primarily to the molecular weight rather than to its chemical structure.[57]

Catalytic Strategies Emerging from the Observation of the Reaction Mechanism

Combined Acid–Base Catalysis

Considering aldose enolization or isomerization as the rate limiting step to furfurals, similar conclusions may be drawn. Both enolization and isomerization are normally favored by basic conditions.[31,48] Evidently, enolization/isomerization followed by three dehydrations is a combination of reactions normally favored respectively by basic and acid conditions; thus furfural formation from aldoses cannot be considered a simple dehydration reaction, obeying, as such, the rules of the specific acid catalysis. In fact, it was observed in the past that only poor furfural yields were obtained when xylose was reacted using typical dehydrating agents, such as pure $ZnCl_2$ or phosphorous pentoxide, whereas when an aqueous solution of $ZnCl_2$ was employed, the furfural yield approached that obtained using strong acid solutions.[25] A combination of homogeneous Brønsted acid and base in a one-pot reaction is clearly not an option, although alternative solutions are being proposed.

Based on similar reasoning, Takagaki *et al.*[59,60] attempted to promote the isomerization of aldoses and subsequent dehydration to furfurals by using a combination of acid and basic heterogeneous catalysts in one-pot reactions. When a combination of hydrotalcite (base) and Amberlyst-15 (acid) were used, results showed significant improvements both in terms of furfural selectivity and xylose conversion compared to only solid acid or homogeneous acid catalysis. In particular, good results are achieved especially at lower temperatures and in polar aprotic solvents such as dimethylformamide. A similar approach has been adopted for pentoses by Choudhary *et al.*[35,36]

although making use of Sn-beta zeolites or a Lewis acid as isomerization catalysts and conducting the reaction in aqueous solution. Interestingly, a significant drop in activation energy was observed[36] in the reaction sequence involving xylose–xylulose isomerization (96.7 kJ mol^{-1} in aqueous $CrCl_3$), and xylulose dehydration (64.9 kJ mol^{-1} in aqueous HCl), compared to the known acid-catalyzed xylose dehydration (133 kJ mol^{-1}).[36,39,47] Overall a significantly improved yield was reported (76%) at 413 K when a biphasic system water–toluene was employed.

Effect of Halide Salts

Quantitative conversion of xylose to furfural is possible, and it has been commonly experienced for decades in the old standard methods for the estimation of pentosan content of plant material.[11,12,61–63] Nearly 100% furfural yield may be attained by distilling atmospherically xylose or raw biomass in a 12 wt% HCl solution saturated with NaCl. Salts such as NaCl have a double positive effect in the production of furfural, firstly by salting-out the reaction product in biphasic systems,[64–66] and secondly by directly enhancing the furfural selectivity and rate of formation.[39,67,68] Starting from this observation the latter effect has been recently studied in monophasic aqueous reactive systems,[39] attributing to the uncommon weak-base behavior of Cl^- ions in aqueous acidic solutions (pH 1.3) the capacity of promoting the 1,2-enediol intermediate, and thus furfural selectivity and yield. In a similar fashion, such Cl^- assistance to enolization by α-hydrogen abstraction may promote elimination to **4**. In a following study[69] halides in general have been shown to positively affect furfural formation proportionally to their nucleophilic character in aqueous dilute acid solutions, whereby I^- followed by Br^- and Cl^- have been shown to promote a more selective dehydration to furfural. Synergistic effects have also been observed, especially when using I^- and Cl^- containing salts.[69]

Effect of Bi- and Trivalent Metal Ions

$FeCl_3$ in water has been often shown to be significantly more effective than a strong acid solution of the same pH during biomass hydrolysis and furfural formation.[65,67,68,70,71] Gravitis[72] reported the metal cations to catalyze the reaction of biomass derived carbohydrates with furfural in proportion to their ionization potential, mentioning an increasing effectiveness for K^+, Na^+, Ca^{2+}, Mg^{2+}, and Fe^{3+}. Furthermore, some metal chlorides in ionic liquids have also been shown to affect the dehydration of sugars,[73] particularly with regard to Cr^{3+} salts, both in polar aprotic solvents[32,74] and in aqueous solution.[36,37]

It is known that aqueous solutions of $FeCl_3$ present a Brønsted acid character due to the hydrolysis of Fe^{3+} resulting in the formation of different kinds of complexes:[75]

$$[Fe(H_2O)_6]^{3+} \rightleftarrows H^+ + [Fe(OH)(H_2O)_5]^{2+}$$

$$2[Fe(OH)(H_2O)_5]^{2+} \rightleftarrows [(H_2O)_5FeOFe(H_2O)_5]^{4+} + H_2O$$

The H^+ molar concentration of a 100 mM $FeCl_3$ aqueous solution under ambient conditions is about 17 mM, *i.e.* 17% of an equimolar HCl solution. On the other hand, it has been noticed[76] that the pH of such solutions drops significantly after heating, forming a dark precipitate consisting of iron oxides such as $–Fe_2O_3$ and –FeOOH.[77] Consequently the H^+concentration of the $FeCl_3$ liquid solution increases according to the simplified reaction:

$$3Fe^{3+} + 5H_2O \rightleftarrows 9H^+ + Fe_2O_3\downarrow + FeOOH\downarrow$$

For these reasons aqueous $FeCl_3$, already at 100 °C, can be regarded as a source of HCl and iron oxides, lowering accordingly the pH of the solution. With similar reasoning the pH drop of $CrCl_3$ solutions after heating has been justified,[36] and a similar behavior may be expected from similar trivalent metal chlorides. In this way the peculiar catalytic effect of metal chlorides may be partly justified by HCl formation, although other aspects are also involved.

When xylose reacted in *N,N*-dimethylacetamide in the presence of $CrCl_3$, isomerization into xylulose (**3**) *via* 1,2-hydrogen transfer has been observed, with further dehydration into furfural, see Figure 7.[32] Evidence of xylose isomerization to xylulose and lyxose in the presence of $CrCl_3$ and $AlCl_3$ in aqueous solutions has also been shown.[36,37] Hence bi- and trivalent metal ions such as Cr^{2+}, Cr^{3+}, Fe^{3+}, and Al^{3+} have been proven to be appealing isomerization catalysts, already at very low concentration (*e.g.* 6 mM). When suitably combined with Brønsted acids they strongly favor furfural formation, although side reactions are not always easy to control limiting the overall furfural yields.

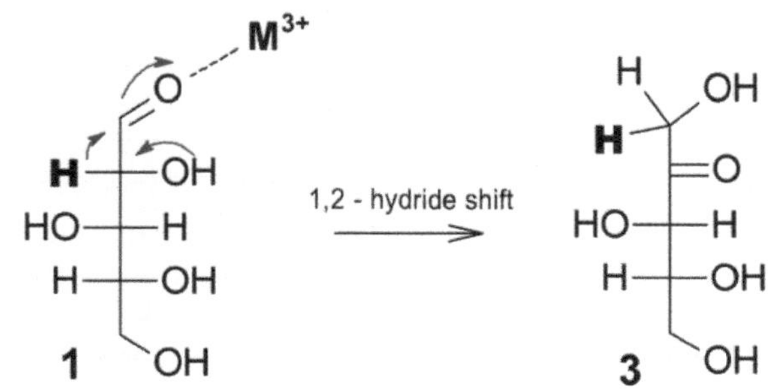

Figure 7: Proposed isomerization of xylose *via* 1,2-hydride shift.[32]

KINETIC ASPECTS

In this section kinetic studies on furfural loss reactions and furfural formation from pentoses will be subsequently discussed. The focus is on studies in water and with homogeneous catalysts. At the end of the section, a short overview of studies on furfural production from biomass is also presented. These papers are not discussed in detail, but merely mentioned for further reference. Heterogeneous catalysis and reaction in ionic liquids are not included in this review.

Furfural Loss Reactions

Firstly, the kinetic studies on furfural degradation in aqueous dilute acidic solutions are reviewed. Most studies follow a reaction mechanism including a single degradation reaction, which is presented in Figure 8. Unless indicated otherwise, this reaction is assumed to follow first-order kinetics. In Table 1 the reported Arrhenius parameters are summarized for this reaction, following

$$k = A \cdot e^{\left[\frac{-E}{RT}\right]} \quad (1)$$

where k is the reaction rate constant (s^{-1}), A is the pre-exponential factor (s^{-1}), E is the activation energy (kJ mol^{-1}), R

is the universal gas constant (kJ mol^{-1} K^{-1}) and T is the temperature (K).

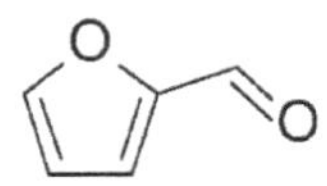

Figure 8: Most employed reaction mechanism for the degradation of furfural.

Table 1: Kinetic studies on furfural loss reactions. $[F]_0$ = initial furfural concentration (mM); A = pre-exponential factor (s^{-1}); E = activation energy (kJ mol^{-1}); $[H^+]$ = hydrogen ion concentration (M); $a_H{}^+$ = hydrogen ion activity (–)

Ref.	Solvent	$[F]_0$ (mM)	Catalyst (mM)	Temperature (°C)	ln(A) (s^{-1})	E (kJ mol^{-1})
53	H_2O	10–20	H_2SO_4 (50)	150–210	12.36	83.6
78	H_2O	47.8	H_2SO_4 (3.1–400)	160–240	$[H^+]\cdot 17.15$	92.3
54	H_2O	10	HCl (100)	150–169	3.07	48.1
52	H_2O	60.4–72.5	H_2SO_4 (36.4–145.5)	150–200	$a_H{}^+\cdot 26.64$	125.1
79	H_2O	50	HCOOH (50–80)	160–200	(Uncatalyzed) 13.82	75.5
					(Acid-catalyzed) $[H^+]\cdot 31.44$	135.0
80	H_2O	160	HCl (100)	140–160	8.43	67.6
81	H_2O	34	None	180–220	7.59	58.8
56	H_2O	50	HCl (50) + NaCl (500)	160–200	16.84	102.1

Williams and Dunlop quantitatively investigated the loss reactions of furfural in dilute aqueous solutions.[53] Both sulphuric acid and hydrochloric acid have been used as the catalyst, but only for the experiments with H_2SO_4 the Arrhenius parameters have been determined. The rate of degradation showed a proportional correlation with the hydrogen ion concentration. Moreover, the loss reaction showed dependency not only on the hydrogen ion concentration and the temperature, but also on the furfural concentration (which indicates a higher reaction order). However, the experimental results still followed an (apparent) first-order behavior and were therefore perceived as pseudo-unimolecular. Two different degradation products were identified: formic acid and a resinous tar. The formic acid is assumed to be formed *via* the hydrolytic fission of the aldehyde group of furfural. For the formation of the resins, condensation polymerization reactions were postulated. Root *et al.* performed a more extensive kinetic study on the furfural loss reactions.[78] Experiments starting with aqueous furfural solutions with various concentrations of sulphuric acid and at various temperatures were performed in order to calculate the kinetic parameters. The measured furfural concentrations showed an apparent first-order disappearance rate. The calculated pre-exponential factor was defined as a function of the hydrogen ion concentration (mol L^{-1}) and was $[H^+]\cdot\exp(17.15)$ s^{-1}. The corresponding activation energy was calculated to be 92.36 kJ mol^{-1}. However, it was also observed that the initial furfural concentration had a significant influence on the disappearance reaction rate; at higher initial furfural concentrations the degradation reactions were faster. A mechanism was proposed that could explain these observations. In this mechanism (following the work of Schoenemann *et al.*, see the next section[82]) it was postulated that the furfural was degraded both *via* (self-)degradation (with first-order kinetics) and *via* a reaction of furfural with some kind of degradation intermediate (with second-order kinetics). It was shown that for the proposed mechanism, incorporating these two routes for furfural disappearance, at low furfural concentrations the reactions showed apparent first-order behavior. Moreover, Zeitsch compared the furfural resinification reaction rates with the disappearance rate of xylose obtained by Root *et al.*, and showed that at higher temperatures (above 200 °C) resinification plays a minor role.[12,78]

At the turn of the millennium, Rose *et al.* have determined the Arrhenius parameters for furfural decomposition in dilute hydrochloric acid solutions.[54] The kinetic results were compared with the results of Williams and Dunlop with sulphuric acid; the calculated activation energy with HCl was lower, which was attributed to the fact that H_2SO_4 was not fully dissociated under these conditions.

Subsequently, Marcotullio *et al.* elaborated the assumption of partially dissociated H_2SO_4 by substituting the hydrogen ion concentration with the hydrogen ion activity ($a_H{}^+$) in the equation for the reaction rate constant.[52] For this purpose an extensive study on the kinetics of furfural degradation in dilute sulphuric acid solutions (36–145 mM) at temperatures between 150 and 200 °C has been performed. From these experiments the activation energy (125.1 kJ mol^{-1}) and pre-exponential factor (exp(26.64) s^{-1}) were derived. Moreover, the authors did not observe any formic acid formation during the destruction of the furfural.

Lamminpää *et al.* determined the kinetics of furfural degradation in aqueous solutions as a part of a study on furfural production from xylose with formic acid as the catalyst.[79] First order reaction kinetics were assumed, based on a relatively few measurements. Both the temperature and pH were varied, and because it was observed that the pH had a larger influence on the furfural destruction rate at higher temperature, this rate was divided into an uncatalyzed and an acid-catalyzed part. For both parts, an activation energy and a pre-exponential factor have been determined. The activation energy of the uncatalyzed reaction is in good agreement with the value found by Jing *et al.* (see below).[81] In the e of the acid-catalyzed reaction the activation energy is higher than that in most of the other studies. This could be due to the fact that in the other studies only one (acid-catalyzed) reaction was assumed. Therefore, this single reaction also includes the uncatalyzed reaction, resulting in a lower activation energy.

Also, Weingarten *et al.* determined the kinetics of furfural degradation as part of a study on furfural production from xylose, but then in a biphasic reactor.[80] It was assumed that the furfural only degraded in the aqueous phase and not in the organic phase. Furfural degradation experiments were performed with 160 mM furfural in a 100 mM HCl aqueous solution at temperatures between 140 and 160 °C. The experimental data were assumed to follow first-order kinetics and the activation energy and pre-exponential factor were estimated to be 67.58 kJ mol^{-1} and exp(8.43) s^{-1}, respectively.

Jing *et al.* have investigated the degradation of furfural in high temperature liquid water (without the addition of any catalyst).[81] The kinetics were determined based on experiments in the temperature range of 180 to 220 °C and at a constant pressure of 10 MPa and were assumed to be first-order. The estimated value for the activation energy was 58.8 kJ mol^{-1} and that for the pre-exponential factor was exp(7.59) s^{-1}.

Danon *et al.* have performed a study on the influence of glucose on the degradation of furfural.[56] Furfural (50 mM) degradation experiments have been performed in a 50 mM HCl and a 500 mM NaCl solution at temperatures between 160 and 200 °C. Experiments with and without glucose present in the reaction mixture have been performed. Both a first- and a second-order kinetic model could reasonably well predict the experimental results for pure furfural degradation. For comparison purposes, only the first-order model is included in Table 1. When glucose was present in the reaction mixture, clear second-order reaction kinetics were observed for the degradation of furfural. A Diels–Alder reaction was propounded for this second-order behavior.

Finally, the paper published by Curtis and Hatt in 1948 needs to be included in this discussion.[83] The main added value of this paper is the liquid–vapor equilibrium data presented in the first two parts. However, in the last part, the reaction rates of loss reactions of furfural in pure water and in sulphuric acid solutions are presented. The reaction rates were only determined at 180 °C and the analysis methodology might seem outdated by twenty-first century standards. However, their conclusions, such as the observation of first-order reaction kinetics and the proportionality of the disappearance rate to the acid concentration, clearly anticipated the results obtained in later studies.

It can be concluded that the degradation of furfural has not been studied extensively; however, the presented results do allow for some important observations. First, many authors note that the reaction order of the degradation reaction(s) might be different from unity; however, the majority conclude that the experimental data are well represented by a first-order kinetic model. Then, a comparison of the activation energies (excluding the saline e) indicates that the highest values are obtained in studies where H_2SO_4 was used as the catalyst. The activation energies predicted for the uncatalyzed reaction and with HCl as the catalyst are in the same range, and they are lower than those with H_2SO_4.

Furfural Formation from Pentoses

In this section, the kinetic studies on the dehydration of monomeric pentoses to furfural are presented. Since most studies assume a reaction mechanism with two first-order reactions (R1 and R2) for the disappearance of pentose (see Figure 9), the values of the reaction rate constants for these types of reactions are presented in Table 2, following

$$R = \frac{d[X]}{dt} = -k[X] \tag{2}$$

where R is the reaction rate (mol L^{-1} s^{-1}), [X] is the xylose concentration (mol L^{-1}) and k is the reaction rate constant (s^{-1}).

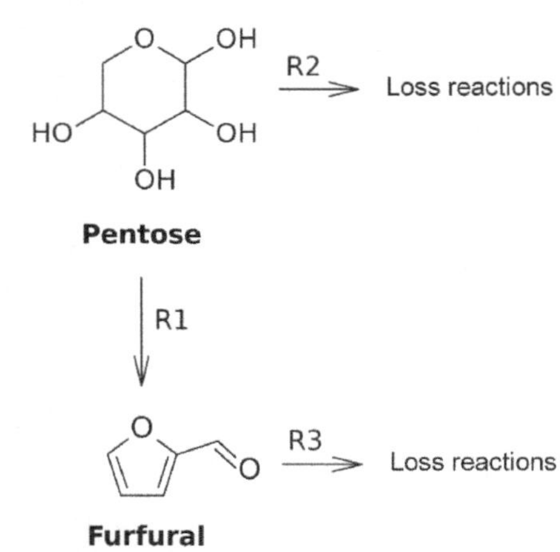

Figure 9: Most employed reaction mechanism for the dehydration of a pentose towards furfural.

Table 2: Kinetic studies on furfural formation from pentoses. n.d. = not determined; ~ indicates that the value is interpolated from a figure; FA = fumaric acid; MA = maleic acid; MIBK = methyl isobutyl ketone; [X] = xylose concentration (M); [F] = furfural concentration (M); b.p. = boiling point; CPME = cyclopentyl methyl ester

Ref.	Pentose (mM)	Solvent	Catalyst (mM)	Temp. (°C)	k_1 (s^{-1})	k_2 (s^{-1})	Max. yield (%)
81	Xylose (72)	H_2O	None	200	1.98×10^{-4}	2.05×10^{-4}	~50
84	Xylose (200)	H_2O	None	200	4.30×10^{-6}	$[X]\cdot 7.73 \times 10^{-9}$ (k_{2a})	~52
						6.39×10^{-6} (k_{2b})	

85	Xylose (50)	H_2O	None	350	2.8×10^{-1}	9.7×10^{-1}	6
85	Xylose (50)	H_2O	None	400	9.7×10^{-1}	6.88	8
86	Arabinose (14–350)	H_2O	None	200	1.56×10^{-4}	1.74×10^{-4}	8
87	Arabinose (33)	H_2O	None	170	5.81×10^{-3} $(k_1 + k_2)$		n.d.
87	Arabinose (33)	H_2O	FA (50)	170	4.52×10^{-3} $(k_1 + k_2)$		n.d.
87	Arabinose (33)	H_2O	MA (50)	170	5.81×10^{-3} $(k_1 + k_2)$		n.d.
87	Arabinose (33)	H_2O	H_2SO_4 (50)	170	1.56×10^{-2} $(k_1 + k_2)$		n.d.
25	Xylose (3)	H_2O	HCl (5200)	107	3.2×10^{-2} $(k_1 + k_2)$		93
25	Xylose (3)	H_2O	H_2SO_4 (3750)	109	1.1×10^{-2} $(k_1 + k_2)$		69
82	Xylose (84–350)	H_2O	HCl (277)	150	3.8×10^{-4}	[X] $0.16 \cdot [F]^{0.5}$ 3.0×10^{-5}	~46
78	Xylose (667)	H_2O	H_2SO_4 (50)	200	4.17×10^{-3}	n.d.	~48
30	Ribose (0.1–1.0)	H_2O	HCl (1000)	80	1.67×10^{-6}	5.45×10^{-7}	48
30	Xylose (0.1–1.0)	H_2O	HCl (1000)	80	6.99×10^{-7}	1.12×10^{-7}	38
30	Lyxose (0.1–1.0)	H_2O	HCl (1000)	80	6.77×10^{-7}	2.92×10^{-7}	35
30	Arabinose (0.1–1.0)	H_2O	HCl (1000)	80	3.12×10^{-7}	0.28×10^{-7}	26
27	Xylose (50)	H_2O	H_2SO_4 (20)	250	n.d.	n.d.	64
50	Xylose (67)	H_2O	None	190	2.75×10^{-4} $(k_1 + k_2)$		~13
50	Xylose (67)	H_2O	H_2SO_4 (50)	190	4.05×10^{-4} $(k_1 + k_2)$		~52
50	Xylose (67)	H_2O	NaOH (100)	220	1.03 $(k_1 + k_2)$		~49
39	Xylose (35)	H_2O	HCl (50) + KCl (340)	200	5.49×10^{-3}	1.24×10^{-3}	74
39	Xylose (35)	H_2O	HCl (50) + NaCl (340)	200	5.28×10^{-3}	1.17×10^{-3}	73
39	Xylose (35)	H_2O	HCl (50) + $CaCl_2$ (170)	200	5.10×10^{-3}	1.20×10^{-3}	72
39	Xylose (35)	H_2O	HCl (50) + $FeCl_3$ (113)	200	$>6.00 \times 10^{-2}$	$>4.00 \times 10^{-2}$	62
39	Xylose (35)	H_2O	HCl (50) + NaCl (856)	200	1.19×10^{-2}	1.07×10^{-2}	81
69	Xylose (35)	H_2O	H_2SO_4 (50) + KCl (500)	200	6.28×10^{-3}	1.11×10^{-3}	77
69	Xylose (35)	H_2O	H_2SO_4 (50) + KBr (500)	200	5.54×10^{-3}	7.7×10^{-4}	80
69	Xylose (35)	H_2O	H_2SO_4 (50) + KI (500)	200	5.36×10^{-3}	6.5×10^{-4}	81
69	Xylose (35)	H_2O	H_2SO_4 (50) + KCl/KI (500/250)	200	7.46×10^{-3}	3.7×10^{-4}	87
79	Xylose (20–80)	H_2O	HCOOH (50)	200	6.85×10^{-4}	2.28×10^{-4}	~65
68	Xylose (13)	H_2O	$FeCl_3$ (49)	180	2.05×10^{-3} $(k_1 + k_2)$		~23
68	Xylotriose (11)	H_2O	$FeCl_3$ (49)	180	3.72×10^{-2} $(k_1 + k_2)$		~24

80	Xylose (740)	H_2O	HCl (100)	170	6.11 × 10^{-4}	[F]·2.04 × 10^{-3}	30
80	Xylose (740)	H_2O–MIBK (1 : 1)	HCl (100)	170	6.11 × 10^{-4}	[F]·2.04 × 10^{-3}	85
37	Xylose (266)	H_2O	HCOOH (326)	180	n.d.	n.d.	74
67	Xylose (667)	H_2O–toluene (1 : 15)	H_2SO_4 (1019) + NaCl (2.4 g)	b.p.	n.d.	n.d.	83
88	Xylose (740)	H_2O–CPME (1 : 2.33)	H_2SO_4 (105)	180	n.d.	n.d.	~59
89	Xylose (33)	H_2O	HCl (100)	180	n.d.	n.d.	60

For those studies that only present the Arrhenius constants, the values of the kinetic rate parameters at specific temperatures have been calculated for comparison purposes.

The order in which the studies are discussed is roughly chronological. However, first, those studies where no additional catalysts were employed are presented. Then, the studies with homogeneous catalysts are discussed. The overview will conclude with several studies in biphasic media (with an aqueous and an organic phase) and some papers that do not include a proper kinetic study, but still seem relevant for the present review.

Jing *et al.* have concisely investigated furfural formation from xylose in an aqueous system.[81] Over a temperature range between 180 and 200 °C and at 10 MPa, experiments were performed with 72 mM xylose as the initial concentration. Reasonable yields were obtained with values between 30 and 50%. Additionally, Arrhenius parameters were estimated for the reactions using Arrhenius plots. The kinetic parameters estimated for the degradation of furfural (as discussed in the previous section) were used for the estimation of the kinetic parameters of the dehydration reactions.

Kim *et al.* performed a similar study on xylose dehydration in high temperature water.[84] At temperatures between 140 and 240 °C feed streams with 200 mM xylose in water have been investigated in a continuous tubular reactor. The assumed reaction mechanism is comparable to that presented in Figure 9, although reaction R2 is assumed to consist of two reactions, one following first-order and the other following second-order kinetics, k_{2b} and k_{2a}, respectively. The highest furfural yields were obtained between 200 and 220 °C. In additional experiments various feed flows (which result in a variation of the residence time) and various initial xylose concentrations were employed. It is shown that lower feed flow (0.016 mL min^{-1} which coincides with a residence time of 3.7 h) and lower initial xylose concentrations result in higher furfural yields.

Also Aida *et al.* have investigated the reaction kinetics of xylose in sub- and supercritical water without any catalysts.[85] Xylose degradation was investigated at 350 and 400 °C and at pressures between 40 and 100 MPa. Based on the identified products in the experiments a complex reaction mechanism was proposed including the Lobry de Bruyn–Alberta van Ekenstein transformations (LBET) and retro-aldol reactions. However, no furfural degradation reactions were included in the reaction mechanism. The method of chemical analysis was extensive, including a total organic carbon detector, two HPLC systems and a GC-MS analysis. The LBET reactions mainly led to the formation of xylulose and the retro-aldol reactions cover the otherwise termed xylose loss reactions. In all experiments the majority of the retrieved products were from retro-aldol reactions. The highest furfural yields were obtained at the highest pressures (100 MPa), although these yields were still relatively low: 6 and 8% at 350 and 400 °C, respectively. For comparison with the other studies reported in this overview the kinetic rate parameters of the retro-aldol reactions are summed to form k_2. Finally, the authors have also included a sensitivity analysis of the estimated kinetic parameters.

Gairola *et al.* have kinetically modeled the hydrothermal dehydration of arabinose (as the second most abundant pentose in hemicellulose) towards furfural.[86] Temperatures between 180 and 260 °C were employed at initial arabinose concentrations of 14, 67 and 350 mM. The authors recognize the fact that for optimal modeling the kinetic parameters of furfural degradation should be determined separately from the determination of the other kinetic parameters. For this purpose they use the declining furfural concentration in those experiments where after a certain reaction time the sugars had fully disappeared. The authors state that this method is to be preferred over separate kinetic experiments with pure furfural, since in the more complex reaction mixture the furfural degrades faster. However, this statement is not substantiated with any experimental data. Because the estimated Arrhenius parameters for arabinose are of the same order as those reported by Jing *et al.* for xylose,[81] it is propounded that it is justified to use solely xylose as the model compound for biomass hydrolysis. This seems to be a coarse assumption, since, as can be readily deduced from Table 2, xylose reacts (at 200 °C) around 30% faster than arabinose (compare *e.g.* with the results of Jing[81]).

Although Kootstra *et al.* had a diametrically opposite objective of the other studies mentioned here (they tried to minimize the furfural production during acidic hydrolysis of pentosans because furfural is a known inhibitor for yeasts in subsequent sugar-to-ethanol fermentation processes), their study still contains some interesting kinetic results for the degradation of arabinose in acidic solutions.[87] The kinetics of arabinose (33 mM) degradation in acidic solutions, with 50 mM fumaric, maleic and sulphuric acid, and in pure water were investigated at 150 and 170 °C. The rate of degradation with sulphuric acid was higher, but, surprisingly, with fumaric or maleic acid the rate was comparable to or even lower than that in pure water. Some tentative explanations are propounded for this difference; however, in these considerations a possible alkaline catalysis in pure water has been disregarded. Their main conclusion from the kinetic study, that arabinose is more stable under these conditions compared to xylose or glucose, does agree with the trends observed in other studies. No furfural yields were reported.

Already in 1932 Hurd and Isenhour performed an extensive study on the kinetics of the formation of furfural from pentoses in acidic solutions.[25] A setup with continuous furfural removal by steam distillation was used. The kinetic rate parameters of xylose degradation have been determined for solutions with hydrochloric and sulphuric acid at various (relatively high) concentrations. In Table 2 the two experiments with the highest furfural yields for the two different acids are included. Especially with HCl the yield was exceptionally high (93%), probably due to the continuous extraction of the formed furfural. But it has to be also taken into account that especially the identification methods have been improved significantly during the interlapsed time (which is almost a century).

In 1948 Dunlop published a landmark paper on the formation of furfural from xylose.[90] In this paper the disappearance rate of xylose was determined at two different HCl concentrations (50 and 100 mM) at 160 °C. Since in the applied reaction mechanism no direct loss reactions from the pentoses are assumed, only the (total) disappearance rate of xylose was determined. The results show that this rate follows first-order kinetics and that it is proportional to the hydrogen ion concentration.

Schoenemann *et al.* have presented the kinetic values of xylose dehydration towards furfural in a larger context of designing chemical reactors.[82] The proposed reaction mechanism differs from that presented in Figure 9. It is assumed that xylose dehydrates to furfural *via* an intermediate and that furfural degrades both individually ('*Verharzung*') and in reaction with this intermediate ('*Kondensation*'), see also Figure 10. However, in the rate expression of this combined intermediate and furfural ('*Kondensation*') reaction the concentration of xylose was included, instead of the intermediate (probably because the latter could not be measured). From calculations based on their experiments this reaction had reaction orders of 1.16 and 0.5 for xylose and furfural, respectively.

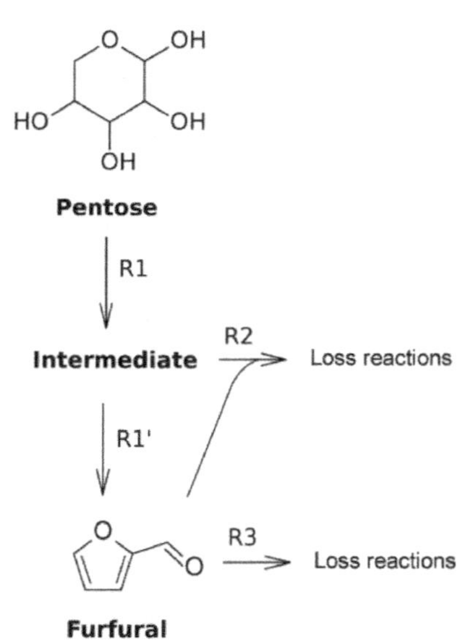

Figure 10: Reaction mechanism for the dehydration of a pentose towards furfural including a combined intermediate and furfural ('*Kondensation*') reaction.[82]

The study by Root *et al.* includes an extensive kinetic evaluation of the dehydration of xylose to furfural.[78] Xylose dehydration has been investigated over a wide range of temperatures (160–280 °C) and initial xylose (20.8–1332 mM) and sulphuric acid (50–400 mM) concentrations. They follow a mechanism where furfural both degrades separately and in reaction with an intermediate from xylose, analogous to Schoenemann.[82] However, only values for the rate constant k_1 have been presented. The assumed mechanism is further verified by experiments starting with both xylose and furfural in the initial reaction solution. Although these results are not presented, it is stated that the value of k_1 was not influenced by the addition of any furfural to the initial reaction solution. In the Arrhenius equation for the rate constant k_1 additional parameters were added for the temperature dependency of the specific gravity of water, the actual acid–water ratio and the activity coefficient of the hydrogen ions. Also, several correlation terms were added to the differential concentration expressions in order to achieve better predictions over the entire investigated range of conditions.

It is noted here that Zeitsch has presented the kinetic results of the three previously discussed studies in his extensive monograph on furfural production and utilization.[12] The kinetic values of Root *et al.* are used to obtain some indications for the design of a reactor.[78]

Garrett and Dvorchik have investigated the dehydration of the different aldopentoses towards furfural.[30] HCl was used as the catalyst in concentrations ranging from 125 to 1000 mM at reaction temperatures between 60 and 80 °C. Also the initial pentose concentration was varied from 0.1 to 1 mM. It is noted that under these conditions (especially at these relatively low temperatures) the reactions are very slow, and thus maximum furfural yields were obtained after 200 to 800 hours. The most important result of this kinetic study was the comparison of the reactivities of the different aldopentoses. Based on the values of k_1 and k_2 that were fitted to the experimental data, the order of reactivity of the aldopentoses was determined to be ribose > xylose ~ lyxose > arabinose. In Table 2 the modeled values of both k_1 and k_2 are presented, since only for k_1 the experimental values have been published. However, the experimental and modeled values for k_1 did agree very reasonably. Antal *et al.* have minutely evaluated the reaction mechanism of xylose dehydration.[27] Based on their mechanistic considerations a complex kinetic model was proposed. In this model different reaction pathways are incorporated for the three different isomers of xylose (xylofuranose, xylopyranose and acyclic xylose). The ratio between these isomers was included among the to be estimated parameters of the model. The time-dependent H^+ concentration was calculated using the algebraic charge conservation relationship. In the experimental series the initial xylose and sulphuric acid concentrations as well as the reaction time were varied. The highest furfural yield of 64% was obtained with 20 mM of sulphuric acid at 250 °C after 50 seconds. Furthermore, it is noted that under many different conditions also small amounts of pyruvaldehyde were measured. Finally, it was found that pressures between 6.3 and 34.5 MPa did not significantly influence the reaction chemistry.

Oefner *et al.* have compared xylose degradation in pure water, acidic and alkaline solutions.[50] Over a temperature range between 180 and 220 °C this reaction was investigated in pure water, in water with various concentrations of H_2SO_4 (0.5, 5 or 50 mM) or with 100 mM NaOH. Besides furfural, also other degradation products (pyruvic, formic, glycolic, lactic and acetic acid) have been identified with isotachophoretic analyses. Only the first-order xylose degradation reaction (the sum of R1 and R2 in Figure 9) has been kinetically modeled. For this reaction the kinetic parameters have been determined using an Arrhenius plot. The value of the activation energy for xylose degradation in an alkaline solution turned out to be significantly different from those of xylose degradation in either pure water or acidic solutions (63.7 *versus* 119.4 and 120.7 kJ mol^{-1}, respectively). Moreover, it was shown that in the alkaline experiments the yields of the organic acids were significantly higher than those in the other experiments.

Marcotullio *et al.* have performed an extensive experimental study on the influence of different salts as a secondary catalyst on the dehydration of xylose to furfural. The experimental results are presented as the selectivity and maximum yields based on modeled kinetic rate constants. In 2010 a study was presented using HCl as the primary catalyst and several different chloride-containing salts as the secondary catalyst.[39] From these experiments, in which the temperature, acid and chloride concentration were kept constant (see also Table 2), it can be concluded that the cations have a relatively small influence on the maximum furfural yield (for K^+, Na^+, Ca^{2+} or Fe^{3+}, the maximum yields are 74.1, 72.8, 72.3 and 62.0, respectively). Only in the e of $FeCl_3$ the yield was significantly lower; however, in this experiment the reaction was much faster. The highest maximum furfural yield (81.3%) is obtained using high concentrations of NaCl (856 mM) as the secondary catalyst. Based on these results, it was postulated that the chloride ion catalyzes the rate-limiting enolization step. In a second study published in 2011, xylose was dehydrated to form furfural using H_2SO_4 as the primary catalyst.[69] In these experiments different potassium-containing salts are added as the secondary catalysts. Both the calculated and measured maximum furfural yields are presented, which agree very reasonably (the latter are presented in Table 2). Again, the differences between the maximum furfural yields of the experiments with different halides (KCl, KBr and KI) are small. See section 2.2.2 for a more detailed discussion on the effects of halide containing salts on the formation of furfural.

Besides the pure furfural degradation experiments, Lamminpää *et al.* also studied the formation of furfural from xylose with formic acid as the catalyst.[79] The experimental results were used to compare three different reaction mechanisms, *i.e.*, a mechanism as presented in Figure 9, a second mechanism as presented in Figure 10 and a third mechanism including the loss reactions R2 from both Figure 9 and 10. It has to be noted that in this study the intermediate concentration has been (numerically) included in the kinetic modeling. In the kinetic expressions the H^+ concentration was incorporated using equilibrium equations for the dissociated formic acid. Moreover, the kinetic rate expressions have been divided into an acid-catalyzed and a solvent (water) catalyzed part, resulting in two separate rate constants. It was noted that the solvent-catalyzed rate parameters were not determined in separate experiments. The general conclusion of the comparison of the three different mechanisms with the experimental data is that the loss reaction of xylose alone is prevailing over the loss reaction of an intermediate with furfural. The mechanism presented in Figure 9 yielded optimal predictions. At 200 °C the H^+ concentration was around 50 mM (derived from their Figure 2).[79] It was noted here that the contribution of the acid-catalyzed reaction was about 5 times that of the solvent-catalyzed reaction. The maximum furfural yields have only been presented graphically and were thus extrapolated for the experiment at 200 °C from their Figure 9.[79]

Liu *et al.* have investigated the degradation of xylose and xylotriose in aqueous solutions with various inorganic salts and no acid catalyst present.[68] Of the investigated salts $FeCl_3$ resulted in the largest increase in the sugar degradation rate compared to reference experiments in pure water and in a dilute sulphuric acid solution. The order of influence (from small to large) was NaCl, KCl, $CaCl_2$, $MgCl_2$ and $FeCl_3$. It has to be noted that all the salts were added at equal weight percentages (0.8 wt%), thus resulting in different concentrations of the ions (especially the chloride ions). All experiments were performed at 180 °C. The kinetic study focused only on the degradation rate of the sugars, which effectively means that values were calculated for ($k_1 + k_2$). The order of this degradation reaction was assumed to be first-order. Not only the rate constants for the degradation of xylose and xylotriose increased significantly with the addition of $FeCl_3$, but also the fraction of unaccounted losses (compared to the amount of furfural formed) increased. The maximum furfural yields (with $FeCl_3$ present) were 23 and 24% for xylose and xylotriose (after 20 and 15 minutes), respectively.

The kinetics of xylose dehydration in a biphasic system was investigated by Weingarten *et al.*[80] In this study a mixture of water and methyl isobutyl ketone (MIBK, 1 : 1) was heated using a microwave. For comparison purposes, some monophasic (with only an aqueous phase) experiments with both conventional and microwave assisted heating were also performed. It turned out that the microwave heating, under these conditions, did not result in significant differences in the kinetics of xylose degradation and furfural formation compared to conventional heating. In the kinetic model it is assumed that the dehydration and degradation reactions take place only in the aqueous phase; therefore, the same kinetic values are reported for both the monophasic and biphasic experiments. Also, it is assumed (based on rather small differences between the xylose conversion rates in the monophasic and biphasic systems) that xylose is degrading in a reaction with furfural. In the kinetic model for the biphasic system a submodel was included to predict the mass transfer of furfural to the organic phase. At 170 °C, maximum furfural yields of 30 and 80% were measured for monophasic and biphasic systems, respectively. The higher yield in the biphasic system was attributed to the simultaneous furfural production in and extraction from the aqueous phase. However, the kinetic model for the biphasic system was unfortunately not able to predict reliable concentrations of xylose and furfural for the experiments at higher temperatures. This might be due to the fact that no separate degradation reaction of xylose was incorporated into the reaction mechanism.

Finally, several papers concerning the dehydration of monosaccharides towards furfural that do not include a kinetic study need to be mentioned here. Firstly, Yang *et al.* have investigated the dehydration of xylose to furfural with formic acid as the catalyst.[37] According to the authors the optimal conditions were 266 mM xylose, 217 mM HCOOH and 180 °C, under which a furfural yield of 74% was achieved. Rong *et al.* performed a study on the dehydration of xylose to furfural in a biphasic mixture (water and toluene). Both sulphuric acid and either NaCl or $FeCl_3$ were added as the catalysts. A maximum yield of 83% was obtained in a water–toluene mixture of 1 : 15 (on volume basis), 667 mM xylose, 1019 mM H_2SO_4 and 2.4 g NaCl. Campos Molina *et al.* have investigated xylose dehydration in sulphuric acid solutions with NaCl as an additional catalyst and cyclopentyl methyl ester (CPME) as an extracting co-solvent.[88] A maximum furfural yield of 59% was obtained for 740 mM xylose and 105 mM H_2SO_4 at 180 °C. Yemiş *et al.* studied the production of furfural from xylose in a microwave-assisted process with hydrochloric acid as the catalyst serving as a reference experiment for their investigation of xylan and biomass hydrolysis, see also the next section.[89]

It can be concluded that many different kinetic studies have been performed on the dehydration of monomeric pentoses to furfural. However, there are hardly any round robin studies. This hampers a meticulous comparison of the individual results of Table 2 and merely allows for some trend conclusions. Firstly, most authors accept first-order kinetics for all the dehydration reactions of the pentoses, with only some discussions on the order and reactants of the (direct) loss reaction of the pentose. Then, the addition of an acidic catalyst seems to both accelerate the reactions and improve the furfural yields. Also the addition of a secondary (saline) catalyst further improves the rates and yields, although with $FeCl_3$ a kind of optimum seems to have been surpassed, since its addition results in very fast reactions but lower furfural yields, compared to the addition of other salts. Finally, in those studies that employ some kind of furfural extraction from the reacting media, the final yields are significantly higher.

Furfural Formation from Biomass

In this section an overview of studies that are concerned with the hydrolysis of biomass resources is included. In order to position the biomass in the reaction network, a simplified reaction mechanism is presented in Figure 11. In the following, those studies that properly focus on furfural production will be discussed concisely. The majority of these studies do not include a kinetic study, and when they do, the assumed mechanisms and reactions are highly diverse. Therefore, in Table 3 it is merely mentioned whether kinetics are included in these papers. Another useful source of information on this subject can be found in the mini review by Dutta *et al.*[102]

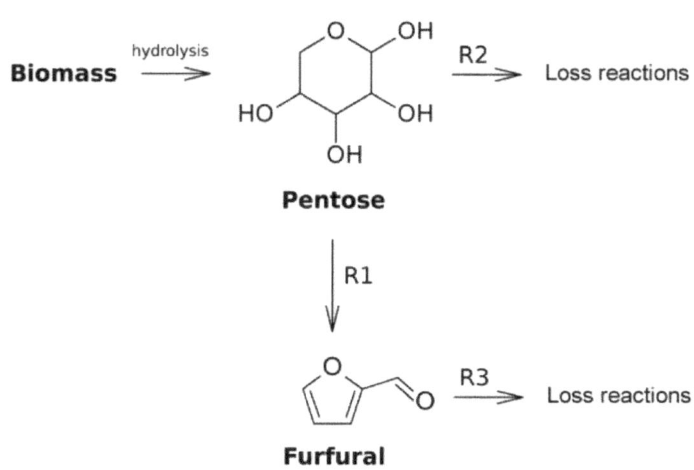

Figure 11: Simplified reaction mechanism for the formation of furfural from biomass.

Table 3: Kinetic studies on furfural formation from biomass. SLR = solid to liquid ratio; LSR = liquid to solid ratio; AcOH = acetic acid; CPME = cyclopentyl methyl ester; THF = tetrahydrofuran; NEH = Northeastern Hardwood

Ref.	Biomass resource	Solvent	Catalyst	Temperature (°C)	Max. yield	Kinetics
91	*Eucalyptus globulus* (8.5 g g^{-1})	H_2O–AcOH (5 : 95)	HCl (0.4 wt%)	130	4.48 g L^{-1}	Yes
92	Sorghum straw (0.1 g g^{-1} H_2O)	H_2O	H_3PO_4 (6 wt%)	134	13.7 g L^{-1}	Yes
93	Date-palm mid-ribs (50–100 mL g^{-1})	H_2O	H_2SO_4 (5–15 wt%)	100–140	11–53% of potential	Yes
86	Wheat straw (0.14 g mL^{-1})	H_2O	None	230	29 mol%	No
86	Brewery waste (0.28 g mL^{-1})	H_2O	None	230	13 mol%	No
89	Xylan (1 : 200 SLR)	H_2O	HCl (100 mM)	180	37 wt% of pentose	No

89	Flax shives (1 : 100 SLR)	H_2O	HCl (100 mM)	180	72 wt% of pentose	No
94	Wheat straw (150 mL g^{-1})	H_2O	HCl (pH 0.6)	155	51 wt% of pentose	No
88	*Cynara cardunculus* (4 wt%)	H_2O–CPME (1 : 2.33)	H_2SO_4 (1 wt%)	170	95 mol%	No
88	*Cynara cardunculus* (4 wt%)	H_2O–CPME (1 : 2.33)	H_2SO_4 (1 wt%) + NaCl (0.4 wt%)	170	99 mol%	No
95	Olive stones (8.5 wt%)	H_2O	H_2SO_4 (50–250 mM)	220–240	50–65% of potential	No
96	Rice hulls (25 mL g^{-1})	H_2O	H_2SO_4 (15%) + ZnO (0.1 g)	110	13.2 wt%	No
97	Rice husk (4–160 mL g^{-1})	H_2O	H_2SO_4 (1–7 wt%)	100–180	6–50% of potential	No
98	Rice straw (0.05 g mL^{-1})	H_2O	H_2SO_4 (0.5%)	150	59 g kg^{-1} straw	No
98	Rice straw (0.05 g mL^{-1})	H_2O–THF (1 : 1)	H_2SO_4 (0.5%) + NaCl (30 wt%)	150	118 g kg^{-1} straw	No
64	NEH (3.2% wt xylose)	H_2O–THF (1 : 2.3)	HCl (0.44 M) + NaCl (±20 g)	210	92.2% of potential	No
99	Corn cob (0.6 LSR)	H_2O	$FeCl_3$ (20 mM) + AcOH (3%)	180	68% of potential	No
100	Corn cob (0.6 LSR)	H_2O	$FeCl_3$ (60 mM) + AcOH (2%)	160–200	50–69% of potential	No
100	Corn cob (0.6 LSR)	Concentrated seawater	$FeCl_3$ (60 mM) + AcOH (2%)	190	73% of potential	No
101	Corn cob (1 : 8 SLR)	H_2O	H_2SO_4 (1.75 vol%)	200	13.19 g L^{-1}	No
101	Corn cob (1 : 8 SLR)	H_2O	HCl (2 vol%)	180	13.90 g L^{-1}	No

The following three studies include reaction kinetics. Abad *et al.* have investigated the production of furfural from *Eucalyptus globulus* wood samples in acetic acid–water–HCl solutions.[91] Sorghum straw was hydrolyzed using phosphoric acid as the catalyst by Vázquez *et al.*[92] The hydrolysis of midribs of the date-palm tree was studied in sulphuric acid by Bamufleh *et al.*[93]

Then, many other studies have been performed on the production of furfural from biomass without a kinetic study. Gairola *et al.* have presented, in addition to their kinetic study on arabinose dehydration as discussed in the previous section, experimental results of the production of furfural from wheat straw and a brewery waste with supercritical CO_2 extraction of the produced furfural.[86] Also Yemiş *et al.* have included, besides a yield study on furfural production from xylose, the results of the microwave assisted hydrolysis of xylan from birchwood, two types of straw and flax shives.[89] In another paper by these authors wheat straw was hydrolyzed to furans under similar conditions.[94] Campos Molina *et al.* have investigated *Cynara cardunculus* (Cardoon) in sulphuric acid solutions with NaCl as an additional catalyst and cyclopentyl methyl ester as a co-solvent.[88] Mansilla *et al.* have performed a study on the hydrolysis of rice hulls in both a one- and a two-stage process and with different metallic oxides as the catalyst, of which the dehydration experiment with the highest furfural yield is reported in Table 3, which is after a pre-hydrolysis step.[96] Montané *et al.* investigated the hydrolysis of olive stones to furfural in dilute sulphuric acid at high temperatures.[95] Rice husk has been hydrolysed in a process including furfural extraction by supercritical carbon dioxide by Sangarunlert *et al.*[97] Amiri investigated the hydrolysis of rice straw to both hydroxymethylfurfural and furfural in either single-phase dilute sulphuric acid solutions or in biphasic systems.[98] Xing *et al.* studied the formation of furfural from hydrolysates of Northeastern hardwood in a two-stage (including a prehydrolysis step) and biphasic system, employing both HCl and NaCl as the catalysts.[64] Cut corn cob was hydrolyzed by Mao *et al.* to furfural in an aqueous solution co-catalyzed by $FeCl_3$ and acetic acid, of which the latter is introduced in the reactor by steam injection of an acetic acid solution which simultaneously served as the furfural extractor.[99] In a follow-up publication, additional experiments under these conditions, but with (concentrated) seawater as the solvent, were presented.[100] Sánchez *et al.*have investigated liquors obtained from corn cob autohydrolysis with sulphuric and hydrochloric acid as the catalysts, where the autohydrolysis can be regarded as a pre-hydrolysis step.[101]

A quick review of the results presented in Table 3 yields similar conclusions as have been drawn based on the results of the dehydration of the monosaccharides. Firstly, the employed reaction conditions in the different studies hardly show any overlap between them, making a meticulous comparison very difficult. However, as far as a comparison is possible, it can be concluded that the addition of both a primary (acidic) catalyst and a secondary (saline) catalyst appears to improve the final furfural yield from biomass substrates. Moreover, also the addition of an extracting (organic) phase significantly improves the furfural yields.

Finally, an extensive amount of literature is dedicated to the kinetics of the hydrolysis of biomass with the objective to maximize the yield of monosaccharides. The kinetic models of these studies can be roughly divided into two parts. Firstly, the more complex kinetic models for the hydrolysis of the pentosan or hemicellulosic part of biomass include furfural degradation reactions.[103–107] The second (and largest) part of these models regard any further reactions of the monomeric pentoses as a single degradation reaction (and term the product of this reaction either as degradation products or as furfural).[108–115]

CONCLUSIONS

Furfural formation in aqueous acidic media has been reviewed from a mechanistic and a kinetic point of view. After carefully reviewing the vast, and sometimes contradictory, experimental evidence published during roughly a century of research, a comprehensive reaction mechanism has been proposed. This mechanism presents more than one route, all starting from acyclic xylose, and involving alternately 1,2-enolization, β-elimination or isomerization *via* 1,2-hydride shift as key steps. Among recent research trends, those studies that employ combined acid–base catalysts, soluble halide salts and trivalent cations in aqueous solutions appear to be most promising.

As far as the kinetics of furfural formation from pentoses and furfural disappearance in aqueous acidic media are concerned, published experimental data span over a relatively wide range of conditions. Analogously, and partly due to the ongoing discussion on the exact reaction mechanism, on the reactants and on the order of the reactions, many different kinetic models have been exploited in these studies, making it hard to compare the results directly or to group the results in one comprehensive model. Therefore, especially for the part on furfural formation, it is recommended to use the data presented here only under the specific conditions as used by each author. However, an attempt has been made to present the published kinetic data in such a manner as to allow a global comparison. Additionally, a very concise overview of research performed on direct furfural production from lignocellulosic materials has also been included. In conclusion, an intricate set of reactions accompanying furfural formation from pentoses, despite appearing well established in some aspects, is yet to be fully unraveled, especially as regards the complex set of side and loss reactions seemingly involving intermediates of reactions largely unknown. Such uncertainties are reflected in the contradictory kinetic models exploited and kinetic data presented in the literature, which still prevent a common and coherent interpretation.

REFERENCES

1. B. Kamm, P. Gruber and M. Kamm, *Biorefineries Industrial Processes and Products; Status Quo and Future Directions*, Wiley-VCH Verlag GmbH & Co, 2006 Search .
2. V. Menon and M. Rao, *Prog. Energy Combust. Sci.*, 2012, **38**, 522–550 .
3. M. Aresta, A. Dibenedetto and F. Dumeignil, *Biorefinery: from biomass to chemicals and fuels*, De Gruyter, 2012 Search .
4. B. Kamm and M. Kamm, *Appl. Microbiol. Biotechnol.*, 2004, **64**, 137–145 .
5. H. Ohara, *Appl. Microbiol. Biotechnol.*, 2003, **62**, 474–477 .
6. G. Huber, S. Iborra and A. Corma, *Chem. Rev.*, 2006, **106**, 4044–4098 .
7. F. Carvalheiro, L. Duarte and F. Girio, *J. Sci. Ind. Res.*, 2008, **67**, 849–864 .
8. M. Galbe and G. Zacchi, *Appl. Microbiol. Biotechnol.*, 2002, **59**, 618–628 .
9. S. Cheng, I. D'Cruz, Z. Yuan, M. Wang, M. Anderson, M. Leitch and C. Xu, *J. Appl. Polym. Sci.*, 2011, **121**, 2743–2751 .
10. J. Van Dam, B. De Klerk-Engels, P. Struik and R. Rabbinge, *Ind. Crops Prod.*, 2005, **21**, 129–144 .
11. A. P. Dunlop and F. N. Peters, *The Furans*, Reinhold Publishing Corporation, New York, 1953 Search .
12. K. J. Zeitsch, *The chemistry and technology of furfural and its many by-products*, Elsevier, 2000, vol. 13 Search .
13. J. J. Bozell and G. R. Petersen, *Green Chem.*, 2010, **12**, 539–554 **RSC**.
14. G. Dautzenberg, M. Gerhardt and B. Kamm, *Holzforschung*, 2010, **65**, 439–451 Search .
15. Y. Kar and H. Deveci, *Energy Sources*, 2006, **28**, 909–921 .
16. J.-P. De Lange, E. Van der Heide, J. Van Buijtenen and R. Price, *ChemSusChem*, 2012, **5**, 150–166 .
17. A. Corma, S. Iborra and A. Velty, *Chem. Rev.*, 2007, **107**, 2411–2502 .
18. A. Mamman, J.-M. Lee, Y.-C. Kim, I. Hwang, N.-J. Park, Y. Hwang, J. S. Chang and J.-S. Hwang, *Biofuels Bioprod. Bioref.*, 2008, **2**, 438–454 .
19. A. De Lu, L. Rodríguez, P. Sánchez and A. Carnicer, *Sep. Sci. Technol.*, 1993, **28**, 2465–2477 .
20. W. de Jong and G. Marcotullio, *Int. J. Chem. React. Eng.*, 2010, **8**, A69 Search .
21. D. Win, *Aust. J. Technol.*, 2005, **8**, 185–190 Search .
22. M. Dashtban, A. Gilbert and P. Fateh, *J. Sci. Technol. Forest Prod. Process.*, 2012, **2**, 44–53 Search .
23. T. Ahmad, L. Kenne, K. Olsson and O. Theander, *Carbohydr. Res.*, 1995, **276**, 309–320 .
24. M. S. Feather, D. W. Harris and S. B. Nichols, *J. Org. Chem.*, 1972, **37**, 1606–1608 .
25. C. D. Hurd and L. L. Isenhour, *J. Am. Chem. Soc.*, 1932, **54**, 317–330 .
26. W. A. Bonner and M. R. Roth, *J. Am. Chem. Soc.*, 1959, **81**, 54545456 Search .
27. M. J. Antal Jr., T. Leesomboon, W. S. Mok and G. N. Richards, *Carbohydr. Res.*, 1991, **217**, 71–85 .

28. X. Qian, M. Nimlos, M. Davis, D. Johnson and M. Himmel, *Carbohydr. Res.*, 2005, **340**, 2319–2327 .

29. M. R. Nimlos, X. Qian, M. Davis, M. E. Himmel and D. K. Johnson, *J. Phys. Chem. A*, 2006, **110**, 11824–11838 .

30. E. Garrett and B. Dvorchik, *J. Pharm. Sci.*, 1969, **58**, 813–820 .

31. D. W. Harris and M. S. Feather, *Carbohydr. Res.*, 1973, **30**, 359–365 .

32. J. B. Binder, J. Blank, A. Cefali and R. T. Raines, *ChemSusChem*, 2010, **3**, 1268–1272 .

33. H. S. Isbell, H. L. Frush, W. R. Wade and C. Hunter, *Carbohydr. Res.*, 1969, **9**, 163–175 .

34. S. J. Eitelman and D. Horton, *Carbohydr. Res.*, 2006, **341**, 2658–2668 .

35. V. Choudhary, A. B. Pintar, S. I. Sandler, D. G. Vlachos and R. F. Lobo, *ACS Catal.*, 2011, **1**, 1724–1728 .

36. V. Choudhary, S. I. Sandler and D. G. Vlachos, *ACS Catal.*, 2012, **2**, 2022–2028 .

37. W. Yang, P. Li, D. Bo and H. Chang, *Carbohydr. Res.*, 2012, **357**, 53–61 .

38. A. Rosatella, S. Simeonov, R. Frade and C. Afonso, *Green Chem.*, 2011, **13**, 754–793 **RSC**.

39. G. Marcotullio and W. de Jong, *Green Chem.*, 2010, **12**, 1739–1746 **RSC**.

40. J. Wu, A. S. Serriani and T. Vuorinen, *Carbohydr. Res.*, 1990, **206**, 1–12 .

41. K. N. Drew, J. Zajicek, G. Bondo, B. Bose and A. S. Serriani, *Carbohydr. Res.*, 1998, **307**, 199–209 .

42. A. Wisniewski, J. Szafranek and J. Sokolowski, *Carbohydr. Res.*, 1981, **97**, 229–234 .

43. A. Wisniewski, J. Sokolowski and J. Szafranek, *J. Carbohydr. Chem.*, 1983, **2**, 293–304 .

44. M. Kurszewska, E. Skorupowa, J. Madaj and A. Wisniewski, *J. Carbohydr. Chem.*, 2004, **23**, 169–177 .

45. F. Fenouillot, A. Rousseau, G. Colomines, R. Saint-Loup and J.-P. Pascault, *Prog. Polym. Sci.*, 2010, **35**, 578–622 .

46. G. Flèche and M. Huchette, *Starch/Staerke*, 1986, **1**, 26–30 .

47. G. Marcotullio, PhD thesis, Delft University of Technology, 2011.

48. M. Watanabe, Y. Aizawa, T. Iida, T. M. Aida, C. Levy, K. Sue and H. Inomata, *Carbohydr. Res.*, 2005, **340**, 1925–1930 .

49. H. E. Hoydonckx, W. M. Van Rhijn, W. Van Rhijn, D. E. De Vos and P. A. Jacobs, *Ullmann's Encyclopedia of Industrial Chemistry*, Wiley VCH Verlag GmbH & Co. KGaA, 2000 Search .

50. P. Oefner, A. Lanziner, G. Bonn and O. Bobleter, *Monatsh. Chem.*, 1992, **123**, 547–556 .

51. F. Rice and L. Fishbein, *J. Am. Chem. Soc.*, 1956, **78**, 1005–1009 .

52. G. Marcotullio, M. A. Tavares Cardoso, W. de Jong and A. H. M. Verkooijen, *Int. J Chem. React. Eng.*, 2009, **7**, A67 Search .

53. D. Williams and A. Dunlop, *Ind. Eng. Chem.*, 1948, **40**, 239–241 .

54. I. Rose, N. Epstein and A. Watkinson, *Ind. Eng. Chem. Res.*, 2000, **39**, 843–845 .

55. M. S. Feather, *J. Org. Chem.*, 1969, **34**, 1998–1999 .

56. B. Danon, L. van der Aa and W. de Jong, *Carbohydr. Res.*, 2013, **375**, 145–152 .

57. F. Rice, *J. Org. Chem.*, 1957, **23**, 465–468 .

58. F. Rice and A. Johnson, *J. Org. Chem.*, 1958, **23**, 1966–1969 .

59. A. Takagaki, M. Ohara, S. Nishimura and K. Ebitani, *Chem. Lett.*, 2010, 838–840 .

60. A. Takagaki, M. Ohara, S. Nishimura and K. Ebitani, *Chem. Commun.*, 2009, 6276–6278 **RSC**.

61. E. E. Hughes and S. F. Acree, *J. Res. Natl. Bur. Stand.*, 1938, **21**, 327–336 .

62. E. I. Fulmer, L. M. Christensen, R. M. Hixon and R. L. Foster, *J. Phys. Chem.*, 1936, **40**, 133141 Search .

63. V. Voorhees, *Ind. Eng. Chem. Anal. Ed.*, 1943, **15**, 378378 Search .

64. R. Xing, W. Qi and G. Huber, *Energy Environ. Sci.*, 2011, **4**, 2193–2205 .

65. T. vom Stein, P. M. Grande, W. Leitner and P. Domínguez de María, *ChemSusChem*, 2011, **4**, 1592–1594 .

66. E. I. Gürbüz, S. G. Wettstein and J. A. Dumesic, *ChemSusChem*, 2012, **5**, 383–387 .

67. C. Rong, X. Ding, Y. Zhu, Y. Li, L. Wang, Y. Qu, X. Ma and Z. Wang, *Carbohydr. Res.*, 2012, **350**, 77–80 .

68. C. Liu and C. Wyman, *Carbohydr. Res.*, 2006, **341**, 2550–2556 .

69. G. Marcotullio and W. de Jong, *Carbohydr. Res.*, 2011, **346**, 1291–1293 .

70. L. Liu, J. Sun, C. Cai, S. Wang, H. Pei and J. Zhang, *Bioresour. Technol.*, 2009, **100**, 5865–5871 .

71. L. Liu, J. Sun, M. Li, S. Wang, H. Pei and J. Zhang, *Bioresour. Technol.*, 2009, **100**, 5853–5858 .

72. J. Gravitis, N. Vedernikov, J. Zandersons and A. Kokorevics, *ACS Symp. Ser.*, 2001, **784**, 110–122 .

73. H. Zhao, J. E. Holladay, H. Brown and Z. C. Zhang, *Science*, 2007, **316**, 1597–1600 .

74. J. B. Binder and R. T. Raines, *J. Am. Chem. Soc.*, 2009, **131**, 1979–1985 .

75. F. Cotton and G. Wilkinson, *Advanced Inorganic Chemistry*, Wiley Interscience, 5th edn, 1988 Search .

76. G. Marcotullio, E. Krisanti, J. Giuntoli and W. de Jong, *Bioresour. Technol.*, 2011, **102**, 5917–5923 .

77. B. Voigt and A. Göbler, *Cryst. Res. Technol.*, 1986, **21**, 1177–1183 .

78. D. Root, J. Seaman, J. Harris and W. Neill, *Forest Prod. J.*, 1959, **9**, 158–165 .

79. K. Lamminpää, J. Ahola and J. Tanskanen, *Ind. Eng. Chem. Res.*, 2012, **51**, 6297–6303 .

80. R. Weingarten, J. Cho, W. Conner and G. Huber, *Green Chem.*, 2010, **12**, 1423–1429 **RSC**.

81. Q. Jing and X. Lu, *Chin. J. Chem. Eng.*, 2007, **15**, 666–669 .

82. K. Schoenemann and H. Hofmann, *Chem. Ing. Technol.*, 1957, **29**, 665–674 .

83. R. Curtis and H. Hatt, *Aust. J. Sci. Res.*, 1948, **1**, 213–235 .

84. S. Kim, M. Lee, E. Park, S. Lee, H. Lee, K. Park and M.-J. Park, *React. Kinet. Mech. Cat.*, 2011, **103**, 267–277 .

85. T. Aida, N. Shiraishi, M. Kubo, M. Watanabe and R. Smith Jr., *J. Supercrit. Fluids*, 2010, **55**, 208–216 .

86. K. Gairola and I. Smirnova, *Bioresour. Technol.*, 2012, **123**, 592–598 .

87. A. Kootstra, N. Mosier, E. Scott, H. Beeftink and J. Sanders, *Biochem. Eng. J.*, 2009, **43**, 92–97 .

88. M. Campos Molina, R. Mariscal, M. Ojeda and M. López Granados, *Bioresour. Technol.*, 2012, **126**, 321–327 .

89. O. Yemiş and G. Mazza, *Bioresour. Technol.*, 2011, **102**, 7371–7378 .

90. A. P. Dunlop, *Ind. Eng. Chem.*, 1948, **40**, 204–209 .

91. S. Abad, J. Alonso, V. Santos and J. Parajó, *Bioresour. Technol.*, 1997, **62**, 115–122 .

92. M. Vázquez, M. Oliva, S. Téllez-Luis and J. Ramírez, *Bioresour. Technol.*, 2007, **98**, 3053–3060 .

93. H. Bamufleh, Y. Alhamed and M. Daous, *Ind. Crops Prod.*, 2013, **42**, 421–428 .

94. O. Yemiş and G. Mazza, *Bioresour. Technol.*, 2012, **109**, 215–223 .

95. D. Montané, J. Salvadó, C. Torras and X. Farriol, *Biomass Bioenergy*, 2002, **22**, 295–304 .

96. H. Mansilla, J. Baeza, S. Urzuá, G. Maturana, J. Villaseñor and N. Durán, *Bioresour. Technol.*, 1998, **66**, 189–193 .

97. W. Sangarunlert, P. Piumsomboon and S. Ngamprasertsith, *Korean J. Chem. Eng.*, 2007, **24**, 936–941 .

98. H. Amiri, K. Karimi and S. Roodpeyma, *Carbohydr. Res.*, 2010, **345**, 2133–2138 .

99. L. Mao, L. Zhang, N. Gao and A. Li, *Bioresour. Technol.*, 2012, **123**, 324–331 .

100. L. Mao, L. Zhang, N. Gao and A. Li, *Green Chem.*, 2013, **15**, 727–737 **RSC**.

101. C. Sánchez, L. Serrano, M. Andres and J. Labidi, *Ind. Crops Prod.*, 2013, **42**, 513–519 .

102. S. Dutta, S. De, B. Saha and M. I. Alam, *Catal. Sci. Technol.*, 2012, **2**, 2025–2036 .

103. T. Carrasco and C. Roy, *Wood Sci. Technol.*, 1992, **26**, 189–208 Search .

104. G. Garrote, H. Domínguez and J. Parajó, *Process Biochem.*, 2001, **36**, 571–578 .

105. D. Nabarlatz, X. Farriol and D. Maontané, *Ind. Eng. Chem. Res.*, 2004, **43**, 4124–4131 .

106. J. Morinelly, J. Jensen, M. Browne, T. Co and D. Shonnard, *Ind. Eng. Chem. Res.*, 2009, **48**, 9877–9884 .

107. E. Guerra-Rodríguez, O. Portilla-Rivera, L. Jarquín-Enríquez, J. Ramírez and M. Vázquez, *Biomass Bioenergy*, 2012, **36**, 346–355 .

108. N. Eken-Saraçoğlu, S. F. Mutlu, G. Dilmaç and H. Çavuşoğlu, *Bioresour. Technol.*, 1998, **65**, 29–33 .

109. B. Lavarack, G. Griffin and D. Rodman, *Catal. Today*, 2000, **63**, 257–265 .

110. B. Lavarack, G. Griffin and D. Rodman, *Biomass Bioenergy*, 2002, **23**, 367–380 .

111. R. Aguilar, J. Ramárez, G. Garrote and M. Vázquez, *J. Food Eng.*, 2002, **55**, 309–318 .

112. A. Herrera, S. Téllez-Luis, J. Ramírez and M. Vázquez, *J. Cereal Sci.*, 2003, **37**, 267–274 .

113. E. Canettieri, G. Rocha, J. Carvalho Jr. and J. Silva, *Ind. Eng. Chem. Res.*, 2007, **46**, 1938–1944 .

114. Y. Lu and N. Mosier, *Biotechnol. Prog.*, 2007, **23**, 116–123 .

115. X. Liu, M. Lu, N. Ai, F. Yu and J. Ji, *Ind. Crops Prod.*, 2012, **38**, 81–86 .

CHAPTER 7

MECHANISMS OF SN2 REACTIONS

Carsten Hennig and Stefan Schmatz

INTRODUCTION

State-resolved differential cross sections for chemical reactions, containing information on angular distributions of the scattered products and thus describing a reactive process at a very detailed level, provide the most stringent test for a quantitatively accurate theoretical model for elementary chemical reactions.[1–6] Only a few reactive systems could be studied with quantum-state resolution in scattering experiments,[7–14] where the flux of the reaction products is measured that go into different scattering angles and final ro-vibrational states for different initial levels. Differential cross sections often exhibit complicated interference patterns[1–5,15] which are difficult to analyze. Ion–molecule reactions, studied experimentally employing crossed-beam imaging,[15] an approach widely used for reactions between neutral species,[3,16] are also employed for the investigation of reactions between neutrals and anions, based on the technique of velocity map imaging.[17] A velocity map imaging spectrometer in conjunction with a low-energy ion source has been used successfully to study gas-phase S_N2 reactions by Wester and co-workers.[18]

Several reaction mechanisms are possible in the prototypical S_N2 reaction $X^- + CH_3Y \rightarrow XCH_3 + Y^-$ with halogen atoms X and Y.[19–25] In the direct mechanism, the nucleophile X^- directly attacks the cone built by the carbon and the three hydrogen atoms and forms the new X–C bond in a direct collision while simultaneously the other bond, C–Y, is broken. In the indirect mechanism, on the other hand, a relatively long-lived resonance state is produced, corresponding to an intermolecular complex that is bound by long-range ion–dipole forces. This complex carries out rotations before dissociating back into the reactants or finally forming the products. By investigating the importance of different reaction channels, new insight can be gained into this fundamental chemical reaction mechanism.

Citation: Carsten Hennig and Stefan Schmatz, "Mechanisms of S_N2 reactions," DOI: 10.1039/c5cp04312c.

In our previous quantum-mechanical studies on nucleophilic bimolecular substitutions,[26–32] reaction probabilities and total reaction cross sections have been calculated and discussed in detail. In ref. 32, differential cross sections have been computed based on rotationally resolved time-independent quantum-mechanical scattering calculations for $Cl^- + CH_3Br \rightarrow ClCH_3 + Br^-$. The results show almost isotropic cross sections for reactant molecules with high rotational quantum numbers. Backward scattering is disfavored for reaction out of states with small rotational excitation, in particular the rovibrational ground state. This is due to a quantum-mechanical effect, the interference of partial waves, that in general can be rationalized by simple classical arguments. In particular for higher vibrational excitation, an umbrella effect that favors the backward direction can be explained by the strong enhancement of the reactivity by opening a direct reaction mechanism. In non-central collisions, or in other words collisions with initial rotational motion, a torque due to the ion–dipole interaction is exerted on the molecule which after a rotation by about 90° completes the reaction.

Connor and co-workers for the first time applied an approximate version of Fuller's nearside/farside theory of angular scattering – originally developed to understand nuclear heavy-ion differential elastic cross sections – to molecular collisions.[33] While the nearside/farside theory provides valuable insight into structured angular distributions, a disadvantage of a nearside/farside decomposition is that it is not unique.

In the present paper, we report on a nearside/farside analysis in the prototypical substitution reaction $Cl^- + CH_3Br \rightarrow ClCH_3 + Br^-$. From a theoretical point of view, it is necessary to extract converged complex S-matrix elements from a quantum scattering calculation, not only the squared moduli in which the phase information is lost. While the latter are sufficient for the calculation of integral scattering cross sections (and can be obtained by averaging without taking into account the phases), differential cross sections are computed by first summing up the S-matrix elements weighted by

Legendre polynomials and squaring the moduli of that sums afterwards. The complex S-matrix elements are obtained from the scattering wavefunction after projecting the asymptotic wavefunction onto channel Jacobi coordinates.

The paper is organized as follows: in Section II we report on the theory behind the nearside/farside analysis as applied to a molecular collision, while numerical details are contained in Section III. The results and their discussion are presented in Section IV. Finally, Section V gives our conclusions. Throughout, all energies are quoted in cm^{-1}.

THEORY

The computation of S-matrix elements for the reaction $Cl^- + CH_3Br$ (v,j) → $ClCH_3$ (v′,j′) + Br^- is described in detail in ref. 31 and 32. In ref. 32, state-resolved differential cross sections at total energy E for vanishing helicity have been computed from S-matrices $S^J_{nm}(E)$ employing the usual formula

$$\frac{d\sigma_{nm}}{d\Omega}(E,\theta) = \frac{1}{4k_n^2}|f_{nm}(E,\theta)|^2, \quad (1)$$

with the scattering amplitude f(E,θ) given by a partial wave summation

$$f_{nm}(E,\theta) = \sum_{J=0}^{\infty} a_{nm,J}(E) P_J(\cos(\theta)) \quad (2)$$

and coefficients $a_{nm,J}(E)$ according to

$$a_{nm,J}(E) = (2J+1)S^J_{nm}(E). \quad (3)$$

Here, k_n denotes the wavenumber of the initial state and P_J is the usual Legendre polynomial of the first kind of order J. θ = 0 denotes forward scattering, and θ = π corresponds to backward scattering.

The differential cross sections themselves can be directly related to experimental results. To shed more light on the underlying reaction mechanisms, we employed a nearside/farside analysis introduced by Fuller[34] which has successfully been applied to reactive scattering.[33,35] In a nearside/farside analysis, the scattering amplitude is decomposed into a sum of two subamplitudes moving clockwise ($f^-(\theta)$) and counter-clockwise ($f^+(\theta)$) around the interaction zone:

$$f(\theta) = f^+(\theta) + f^-(\theta) \quad (4)$$

with $f^\pm(\theta)$ defined by

$$f^\pm(\theta) = \sum_{J=0}^{\infty} a_J P_J^\mp(\cos(\theta)). \quad (5)$$

In a semiclassical picture, $f^-(\theta)$ can be associated with trajectories scattered from the nearside of the target (with respect to the detector), while $f^+(\theta)$ corresponds to trajectories on the farside of the target. The two semiclassical trajectories contribute to the same scattering angle. Applying this picture to the Cl/Br exchange reaction, a direct reaction mechanism is expected to contribute mainly to the nearside scattering amplitude, while an indirect mechanism (with the formation of a long-lived complex) contributes to both the nearside and the farside amplitude. This is schematically shown in Figure 1. Here, it should be noted that semiclassical trajectories can be easily displayed for simple non-reactive atom–atom collisions, where the change of the scattering vector is monitored. In this e, nearside and farside trajectories leading to the same scattering angle show different angles with respect of the rotation of the scattering vector (cf. Figure 1 in ref. 36). However, when more than two particles are involved, the task of identifying nearside and farside contributions is not trivial. For non-reactive atom–diatom collisions, a similar procedure can be applied employing the projection of the Jacobi vector onto the fixed plane perpendicular to the total angular momentum vector; its precession angle during the course of the collision is then monitored.[36] This formalism can be extended relatively simply to reactive atom–diatom collisions.[36] Thus, the trajectories shown in Figure 1 are not easily comparable to curves directly related to geometrical data obtained from a conventional semiclassical calculation.

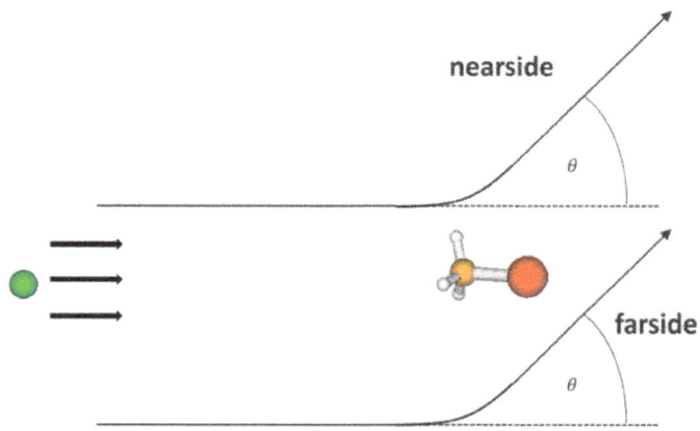

Figure 1: Nearside and farside scattering for the S_N2 reaction $Cl^- + CH_3Br \rightarrow ClCH_3 + Br^-$. Two semiclassical trajectories with the same scattering angle, i.e. equally contributing to the differential cross section are shown. The nearside trajectory runs nearer to the detector which is thought to be positioned at the angle θ. See the text for comments on this simplified picture since for collisions beyond the simple elastic atom–atom e the semiclassical nearside/farside theory is not trivial.

However, the decomposition of the scattering amplitude into subamplitudes is not unique. The final differential cross section is composed of contributions from nearside scattering, farside scattering and interferences of both. In the present analysis, we use the canonical decomposition

$$P_J^\pm(\cos(\theta)) = \frac{1}{2}\left(P_J(\cos(\theta)) \mp \frac{2i}{\pi} Q_J(\cos(\theta))\right) \quad (6)$$

with $Q_J(\cos(\theta))$ being Legendre polynomials of the second kind. Substituting eqn (4) into eqn (1), we obtain a decomposition of the differential cross section into a farside (supeript +), a nearside (−) and an interference (i) component:

$$\frac{d\sigma_{nm}}{d\Omega}(E,\theta) = \frac{d\sigma^+_{nm}}{d\Omega}(E,\theta) + \frac{d\sigma^-_{nm}}{d\Omega}(E,\theta) + \frac{d\sigma^i_{nm}}{d\Omega}(E,\theta), \quad (7)$$

where the nearside/farside components are given by

$$\frac{d\sigma^\pm_{nm}}{d\Omega}(E,\theta) = \frac{1}{4k_n^2}\left|f^\pm_{nm}(E,\theta)\right|^2 \quad (8)$$

while the interference component results from

$$\frac{d\sigma^i_{nm}}{d\Omega}(E,\theta) = \frac{1}{4k_n^2}\left(f^+_{nm}(E,\theta)\overline{f^-_{nm}(E,\theta)} + \overline{f^+_{nm}(E,\theta)}f^-_{nm}(E,\theta)\right). \quad (9)$$

with the bars denoting complex conjugation.

NUMERICAL DETAILS

The S_N2 reaction is treated as a pseudo-triatomic system with the methyl group regarded as a united atom. The internal geometry of this group is optimised for each set of Jacobi coordinatesR, r and γ. Starting with the exact Hamiltonian operator in Jacobi coordinates and restricting ourselves to vanishing helicity, we drop the Coriolis coupling term (centrifugal sudden approximation, CS[37,38]) and obtain the Hamiltonian for total angular momentum $J > 0$

$$\hat{H}_{CS} = \hat{T}^{J=0} + \hat{T}^{J}_{CS} + \hat{V}, \tag{10}$$

with the influence of total angular momentum approximated in the kinetic energy contribution $\hat{T}^{J}_{CS}$ according to

$$\hat{T}^{J}_{CS} = \delta_{j'j}\frac{\hbar^2}{2\mu_1 R^2}J(J+1) \tag{11}$$

where j denotes the angular momentum of the pseudo-diatom.

In the framework of time-independent quantum scattering theory the Schrödinger equation for fixed total energy E and fixed J is solved employing an expansion of the wavefunction in products of translational functions and eigenfunctions of the surface Hamiltonian. The numerical solution of the scattering problem makes use of the technique of R-matrix propagation.[39]We employ the wave functions and overlap matrices computed for $J = 0$ and add the above term $\hat{T}^{J}_{CS}$ as an energetic correction to the Hamiltonian in the propagation of the wavefunction along the hyperradius. Asymptotically we impose appropriate boundary conditions and finally obtain the complex scattering matrix.

We made use of the same parameters as employed in ref. 31 and 32. To be able to compare the results, we carry out the calculations at the same energies for nearside/farside state-resolved differential cross sections as in ref. 32: $E_{dif} = 350$, 850, 1400 and 1900 cm^{-1}; to smoothen many fluctuations that are characteristic of this resonance-driven reaction, for each energy E_{dif}, differential cross sections were actually computed at 21 equidistant values $E_{dif} - 1.0$ cm^{-1}, $E_{dif} - 0.9$ cm^{-1},..., $E_{dif} +$ 1.0 cm^{-1} and then averaged. In addition, for each angle θ_{dif} the differential cross sections have been averaged over intervals of $[\theta_{dif} - \theta_{av}, \theta_{dif} + \theta_{av}]$, with $\theta_{av} = 5°$.

RESULTS AND DISCUSSION

In Figure 2, 3, 5 and 6, the initial-state selected differential reaction cross sections and the corresponding nearside/farside decompositions are graphically displayed. They are summed over all product states for total energies E_{tot} of 350, 850, 1400 and 1900 cm^{-1}. $\sin(\theta)(d\sigma/d\Omega)(E_{tot})$ is shown, which is the integrand employed in the calculation of the total cross sections $\sigma_{tot}(E_{tot})$, making the contributions to σ_{tot} visible. As a consequence, the data for $\theta = 0$ and $\theta = 180°$ are exactly zero, as long as the differential cross sections remain finite. This is not seen in the plots, first because this occurs only at precisely these two angles and second because of data averaging. The data are smoothly averaged over intervals of 10° with the exception of 0 and 180°. Forward scattering with the newly formed Br^- ion moving to the left corresponds to $\theta = 0$, while backward scattering with that ion moving to the right is associated with $\theta = 180°$.

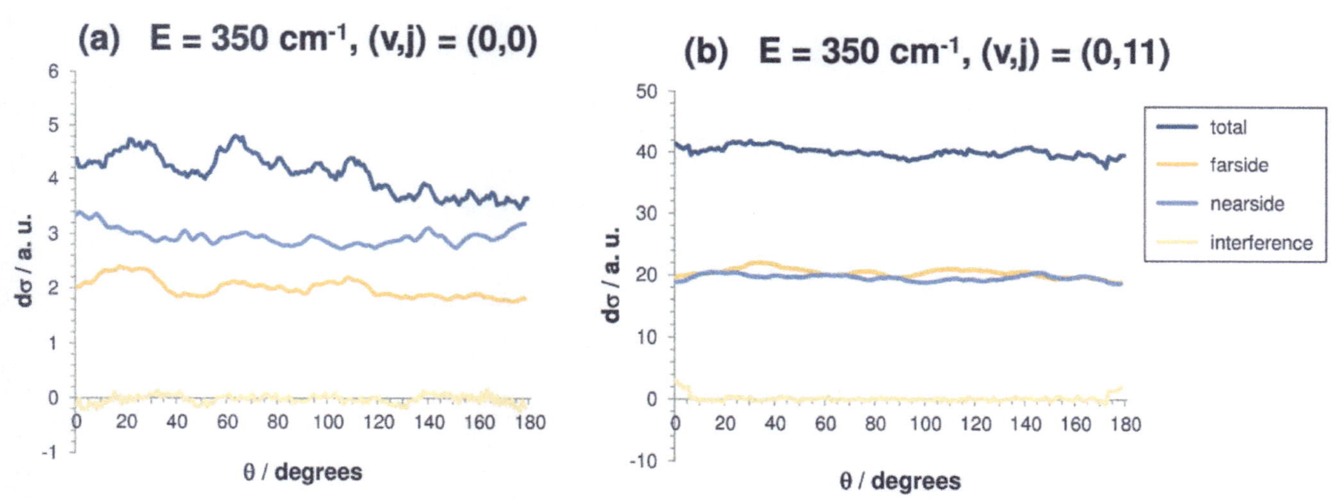

Figure 2: Differential reaction cross sections for $Cl^- + CH_3Br(v,j) \rightarrow ClCH_3 + Br^-$ at E = 350 cm^{-1} and decomposition in nearside and farside contributions for selected initial rovibrational states (v,j). Also shown is the interference contribution. (a) (v,j) = (0,0); (b) (v,j) = (0,11).

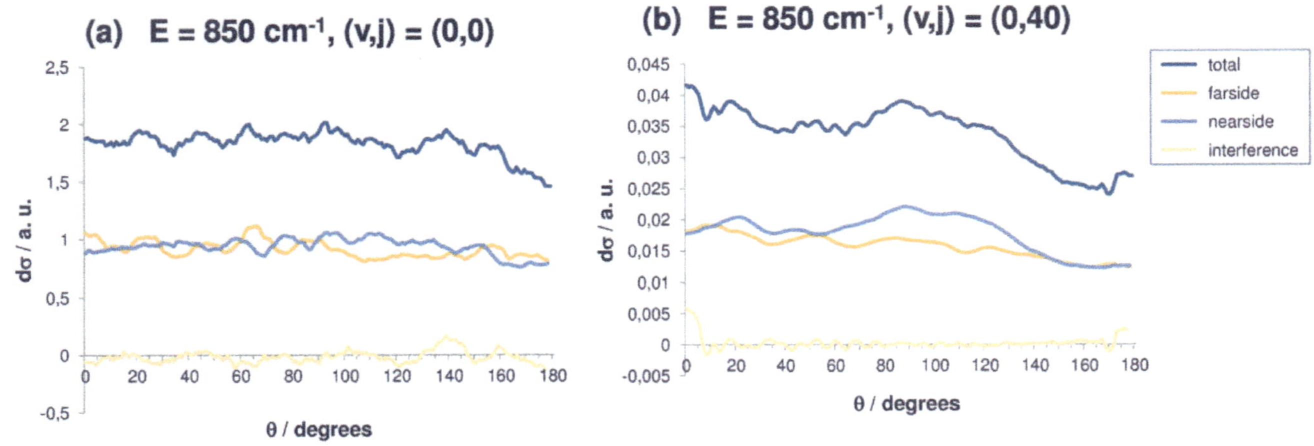

Figure 3: Differential reaction cross sections for $Cl^- + CH_3Br(v,j) \rightarrow ClCH_3 + Br^-$ at E = 850 cm^{-1} and decomposition in nearside and farside contributions for selected initial rovibrational states (v,j). Also shown is the interference contribution. (a) (v,j) = (0,0); (b) (v,j) = (0,40).

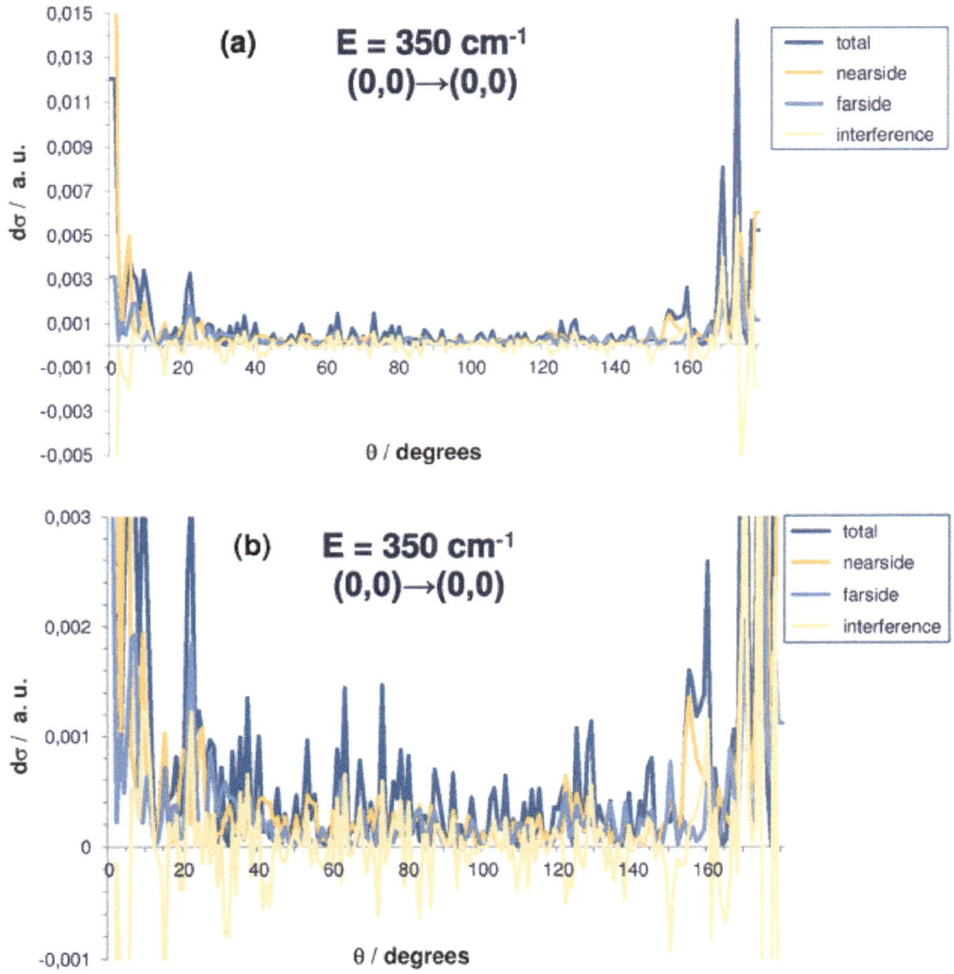

Figure 4: (a) State-to-state specific differential reaction cross section for $Cl^- + CH_3Br\ (0,0) \rightarrow ClCH_3\ (0,0) + Br^-$ at E = 350 cm^{-1} and decomposition in nearside, farside and interference contributions. (b) Same as (a), but detail with enlarged scale along the vertical coordinate axis. Note the negative and positive parts in the interference contribution.

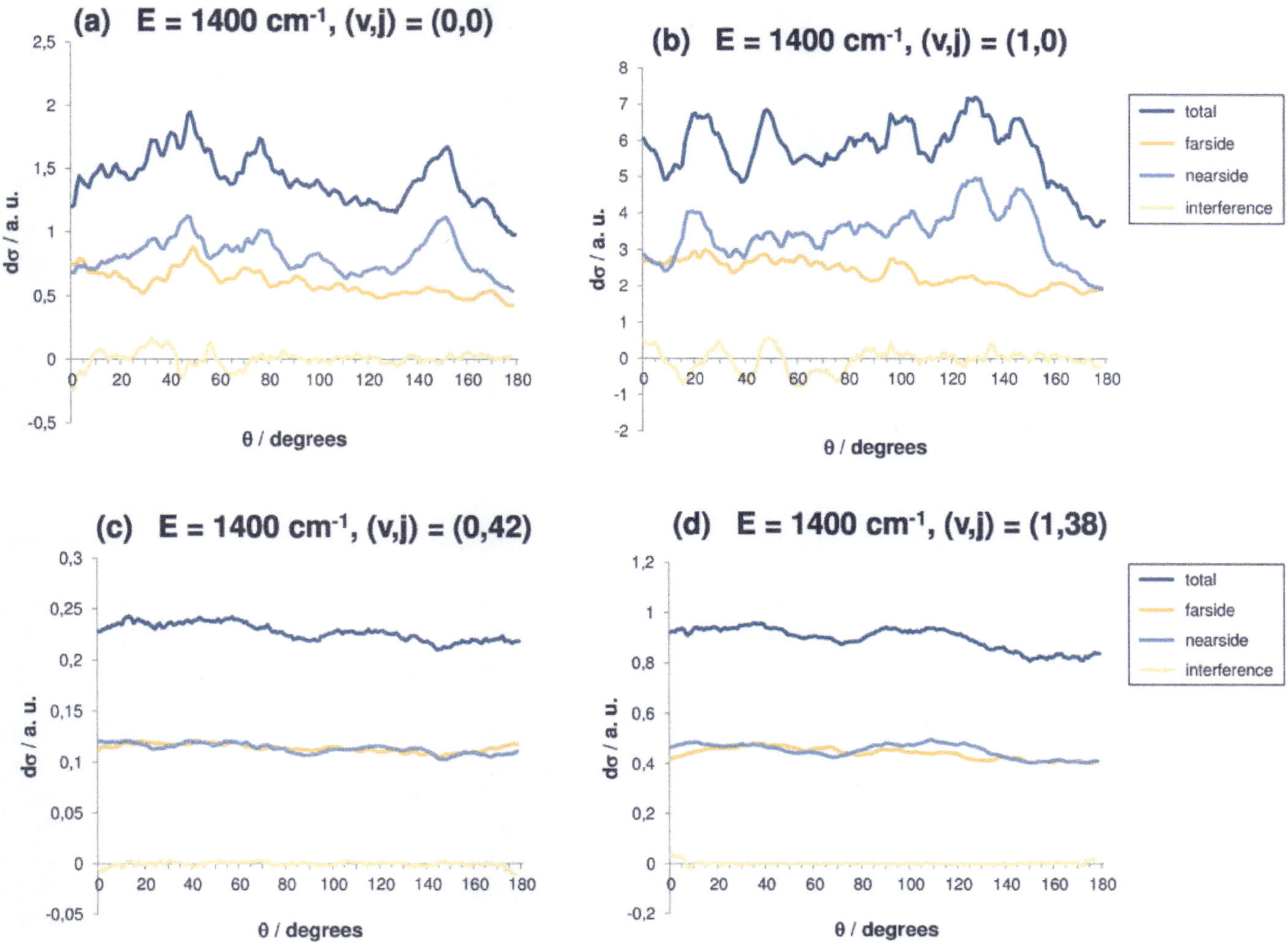

Figure 5: Differential reaction cross sections for $Cl^- + CH_3Br(v,j) \rightarrow ClCH_3 + Br^-$ at $E = 1400\ cm^{-1}$ and decomposition in nearside and farside contributions for selected initial rovibrational states (v,j). Also shown is the interference contribution. (a) (v,j) = (0,0); (b) (v,j) = (1,0); (c) (v,j) = (0,42); (d) (v,j) = (1,38).

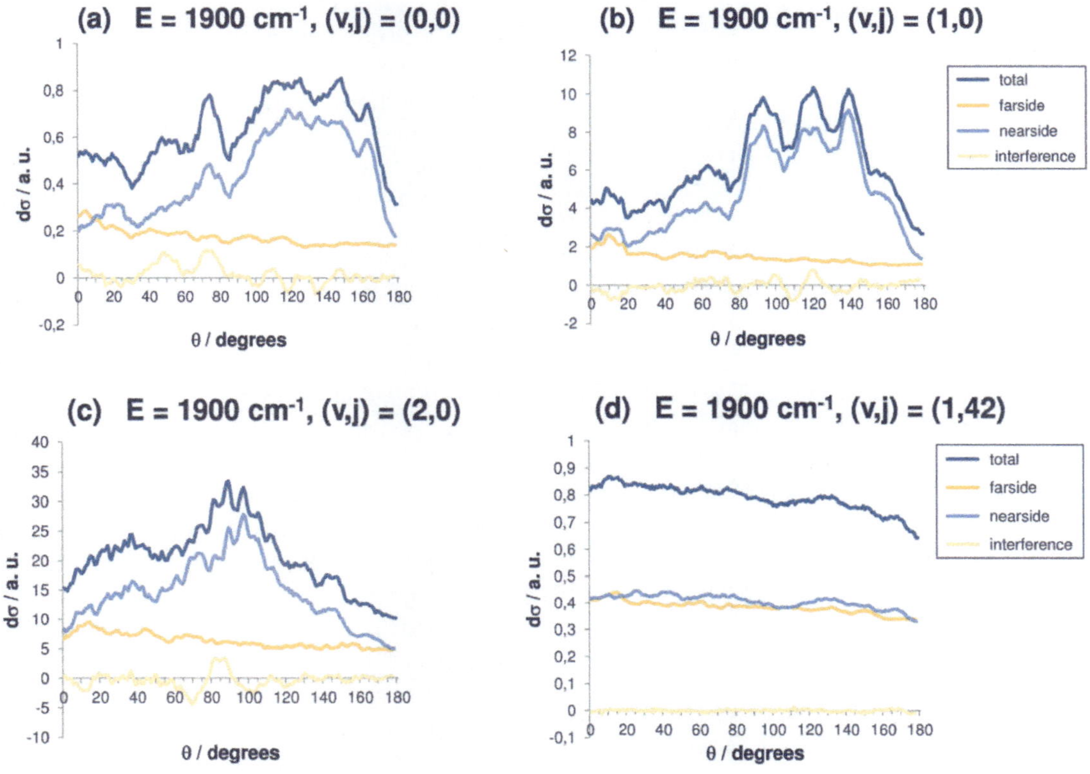

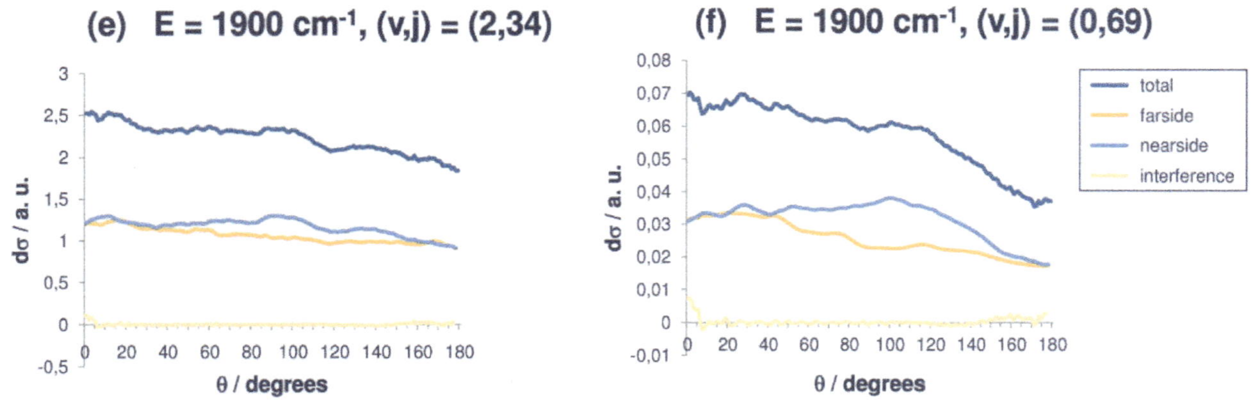

Figure 6: Differential reaction cross sections for $Cl^- + CH_3Br(v,j) \rightarrow ClCH_3 + Br^-$ at E = 1900 cm^{-1} and decomposition in nearside and farside contributions for selected initial rovibrational states (v,j). Also shown is the interference contribution. (a) (v,j) = (0,0); (b) (v,j) = (1,0); (c) (v,j) = (2,0); (d) (v,j) = (1,42); (e) (v,j) = (2,34); (f) (v,j) = (0,69).

Figure 2 shows the differential reaction cross sections for a total energy of 350 cm^{-1} and different initial rotational excitations (v = 0; j = 0, 11). At this relatively small energy, it can be seen clearly that $(d\sigma/d\Omega)\sin(\theta)$ increases with j (note the scaling) and becomes increasingly more isotropic. For small j, many oscillations typical of resonance features can be observed. Here, the fact that the cross sections become larger with j can be explained mainly by an energetic effect. Rotationally excited initial states are higher in energy, and thus the systems have – at a given total energy – less translational energy available which, however, enters the equation for the calculation of the cross section in the denominator. For the initial state (0,0), back scattering is slightly less effective. For low total energies, the nearside/farside analysis reveals some structure in the differential cross sections with a slight preference of small scattering angles. For initially vibrationless and rotationless states the nearside scattering dominates the farside contribution by 20%, partly by more than 50%, pointing towards a preference of the direct reaction mechanism over complex-formation.

However, if rotation is excited, the relative difference between nearside and farside contribution becomes negligible. The interference contribution between the nearside and the farside amplitude is very small. For higher rotational excitation, this effect becomes even more pronounced, indicating that if the methyl bromide molecule rotates, the reaction occurs mainly via complex-formation and not via the direct reaction pathway. So, the differential cross section becomes isotropic for high initial rotational excitation.

In agreement with our previous results,[30] the differential cross sections are largest for small energies. At higher total energies (850 cm^{-1}, Figure 3), the (0,0) differential cross section shows a similar behavior, i.e. back scattering being slightly less probable, but the average differential cross section has become smaller by roughly a factor of two. Also, the oscillations in the differential cross sections are less pronounced. For initial rotational excitation, the differential cross sections decrease and become more isotropic, so that finally almost no anisotropy can be observed (the cross section for angular momentum quantum number j = 40 is shown in Figure 3(b)). Isotropic cross sections point to the existence of long-living resonance states which shows that the correlation with the initial angle disappears. For reaction out of the ground state and out of the state with rotational excitation j = 40, the reaction cross sections are anisotropic, the coarse-grained fluctuation is more pronounced for j = 40 while the fine structure is less distinct. The scale of the cross sections in Figure 3 should be noted. Again, nearside and farside scattering contributions are equally strong and overall the interference contribution obviously is negligible.

However, it should be kept in mind that the curves shown are averaged due to strong oscillating patterns in the cross sections for this resonance-driven reaction. Also in the integral cross sections calculated within a model with the number of degrees of freedom reduced to only two, many extremely narrow resonances are present; an even more pronounced resonance structure is found in the differential cross section. Furthermore, the differential cross sections shown in the present paper are summed over product states, and this summation process smoothens the curves presented in this paper.

The raw data, i.e. state-to-state cross sections before averaging over energies and angles, show prominent interference patterns. This can be seen from Figure 4 for the state-to-state differential cross section at a fixed energy of 350 cm^{-1} as a function of the scattering angle (the (0,0) → (0,0) transition is shown). In particular, while the total, nearside and farside contributions are strictly positive due to their definition, the interference term is real, but can be negative as can be clearly seen from Figure 4. After averaging, the interference contribution becomes negligible. This to a large extent is due to the annihilation of negative and positive components. That justifies the interpretation of this term as an interference type component of the differential cross section and helps to identify the reaction mechanisms in a semiclassical picture and to establish the link to experimental results.

For even higher total energy (Figure 5(a)), 1400 cm^{-1}, similar features can be observed so that the trend becomes even clearer. However, the coarse oscillations for (0,0) are again more pronounced. Figure 5(b) shows the same for additional initial vibrational excitation, which is possible at this given total energy. The differential cross sections increase as expected from our previous collinear studies,[30] while the behavior with respect to angular excitation remains as before. The dip for backward scattering becomes more pronounced and the oscillations are coarser. For reaction out of the ground state and the state with one quantum in the C–Br stretching vibration, the differential cross sections show structure, but not a clear preference of a particular angular regime. Nearside scattering dominates, in particular for (v,j) = (1,0) and for wide angles, whereas for small angles the contributions of nearside and farside scattering are very similar. A relatively strong interference contribution is observed. Clearly, the direct mechanism is enhanced when the broken bond is excited (v = 1, Figure 5(a)). High rotational energy means low translational energy. The cross sections are almost perfectly isotropic and the nearside and farside contributions are very similar.

Again, for energies of 1900 cm^{-1} (Figure 6(a–f)), the differential cross sections tend to be more isotropic for larger values of j. The cross section for the initial state (0,0) becomes more and more interesting (Figure 6(a)), because the peak for backward scattering ($\theta = 180°$) is only one third as high as the maximum value located in the range between 100° and 160°. The structural features become even coarser and the typical resonance structure is lost. Without initial excitation (Figure 6(a)), the nearside part is strongly dominating. The scattering occurs in the angular range between 100° and 170°. The farside contribution remains almost constant, which points towards a direct reaction mechanism. Additional vibrational excitation (Figure 6(b)) shifts this range to 90°–140° with backward scattering less than one third of the maximum cross section. An even larger effect is seen in Figure 6(b) with a pronounced interference pattern. Two quanta in the vibrational mode cause a further shift down to 80°–115°. For two quanta in the C–Br stretching vibration, (v,j) = (2,0) (Figure 6(c)), an angular maximum at 90° is observed. For even higher initial energies (1900 cm^{-1}) the differential cross sections are shown in Figure 6(d). For high rotational excitation (v = 42), the same effect as already discussed occurs. The same can be said for Figure 6(e). From Figure 6(f), we see that the forward scattering is reduced and the nearside contribution again dominates.

To rationalize the calculated behavior of the differential reaction cross sections, in particular the isotropy for large j-values, it is useful to go back to simple classical-mechanical arguments. For a fastly rotating target molecule, all directions for dissociation of the intermediate complex are possible and equally probable. A complex that lives long enough (i.e. the resonance of which is narrow enough with respect to energy) can perform several rotations before the actual reaction takes place. So, no direction is favored, yielding isotropic differential cross sections.

It should be discussed why cross sections for backward scattering are smaller. Remember that in our previous collinear studies[28–30] only backward scattering was possible, with the chloride anion attacking the methyl group, forming a bond with the carbon atom, and after the umbrella flip of the methyl group bouncing back to the direction where the chloride ion came from. In the full three-dimensional picture, this direction is strongly disfavored because the chlorine anion approaches the CH_3Br molecule off from the collinear axis. In the classical-mechanical picture, due to the strong ion–dipole interaction a torque is exerted on the molecule, and it begins to carry out a rotation or, when the ion is very close to the molecule, a libration. However, due to the shape of the potential, no full rotational cycle can be performed if the available energy is not high enough, and an angle of about 90° seems most probable. This explains why exact backward scattering is such a disfavored event. The strong electrostatic forces have a substantial influence on the differential cross sections in the es without initial rotational excitation. The molecule orients towards the ion in the e of any deviation from the central collision.

CONCLUSIONS

We have carried out a nearside/farside analysis on the nucleophilic bimolecular substitution reaction $Cl^- + CH_3Br(v,j) \rightarrow ClCH_3(v',j') + Br^-$ as it was first introduced in molecular scattering by Connor and co-workers based on an idea of Fuller. It is revealed that in gas-phase S_N2 reactions the structure in the differential cross sections at higher energies is related to the nearside scattering contribution, pointing towards a direct reaction mechanism where the attacking ion directly approaches the open cone formed by the carbon and the three connected hydrogen atoms. The indirect mechanism, on the other hand, is described by complex-formation and subsequent dissociation after carrying out rotational motion. The differential cross section becomes isotropic because no angle is favored. This mechanism clearly is connected with the farside contribution which introduces no structure in the differential reaction cross section.

ACKNOWLEDGEMENTS

This work was financially supported by the Deutsche Fohungsgemeinschaft (DFG), grant SCHM 1651/1-3. Additional support from the Fonds der chemischen Industrie is gratefully acknowledged.

REFERENCES

1. D. C. Clary, Proc. Natl. Acad. Sci. U. S. A., 2008, **105**, 12649 .
2. D. C. Clary, Science, 2008, **321**, 789 .
3. K. Liu, Annu. Rev. Phys. Chem., 2001, **52**, 139 .
4. X. Yang, Annu. Rev. Phys. Chem., 2007, **58**, 433 .
5. W. Hu and G. C. Schatz, J. Chem. Phys., 2006, **125**, 132301 .
6. J. Mikosch, M. Wei and G. Nyman, J. Phys. Chem. A, 2010, **114**, 9617 Search .
7. P. D. D. Monks, J. N. L. Connor and S. C. Althorpe, J. Phys. Chem. A, 2006, **110**, 741 .
8. D. Dai, C. C. Wang, S. A. Harich, X. Wang, X. Yang, S. D. Chao and R. T. Skodje, Science, 2003, **300**, 1730.
9. F. J. Aoiz, L. Banares, B. Martinez-Haya, J. tillo, D. E. Manolopoulos, K. Stark and H.-J. Werner, J. Phys. Chem. A, 1997, **101**, 6403 .
10. M. Qiu, Z. Ren, L. Che, D. Dai, S. A. Harich, X. Wang, X. Yang, C. Xu, D. Xie, M. Gustafsson, R. T. Skodje, Z. Sun and D. H. Zhang, Science, 2006, **311**, 1440 .
11. J. N. L. Connor, Phys. Chem. Chem. Phys., 2004, **6**, 377 .
12. N. Balucani, D. Skouteris, L. Cartechini, G. Cappozza, E. Segeloni, P. avecchia, M. H. Alexander, G. Capecchi and H.-J. Werner, Phys. Rev. Lett., 2003, **91**, 13201 .
13. D. H. Zhang and J. Z. H. Zhang, J. Chem. Phys., 1994, **101**, 1146 .
14. E. J. Rackham, T. Gonzales-Lezana and D. E. Manolopoulos, J. Chem. Phys., 2003, **119**, 12895 .

15. J. Mikosch, M. Weidemüller and R. Wester, Int. Rev. Phys. Chem., 2010, **29**, 589 .

16. P. avecchia, Rep. Prog. Phys., 2000, **63**, 355 .

17. A. T. J. B. Eppink and D. H. Parker, Rev. Sci. Instrum., 1997, **68**, 3477 .

18. J. Mikosch, U. Frühling, S. Trippel, D. Schwalm, M. Weidemüller and R. Wester, Phys. Chem. Chem. Phys., 2006, **8**, 2990 .

19. W. L. Hase, Science, 1994, **266**, 998 .

20. M. L. Chabinyc, S. L. Craig, C. K. Regan and J. I. Brauman, Science, 1998, **279**, 1882 .

21. J. K. Laerdahl and E. Uggerud, Int. J. Mass Spectrom. Ion Processes, 2002, **214**, 277 .

22. S. Schmatz, ChemPhysChem, 2004, **5**, 600 .

23. J.-L. Le Garrec, B. R. Rowe, J. L. Queffelec, J. B. A. Mitchell and D. C. Clary, J. Phys. Chem., 1997, **107**, 1021 .

24. J. Mikosch, S. Trippel, C. Eichhorn, R. Otto, U. Lourderaj, J. X. Zhang, W. L. Hase, M. Weidemüller and R. Wester, Science, 2008, **319**, 183 .

25. J. Zhang and W. L. Hase, J. Phys. Chem. A, 2010, **114**, 9635 .

26. S. Schmatz and D. C. Clary, J. Chem. Phys., 1998, **109**, 8200 .

27. S. Schmatz and D. C. Clary, J. Chem. Phys., 1999, **110**, 9483 .

28. C. Hennig and S. Schmatz, J. Chem. Phys., 2004, **121**, 220 .

29. C. Hennig and S. Schmatz, J. Chem. Phys., 2005, **122**, 234307 .

30. C. Hennig and S. Schmatz, Chem. Phys. Lett., 2007, **446**, 250 .

31. C. Hennig and S. Schmatz, J. Chem. Phys., 2009, **121**, 224303 .

32. C. Hennig and S. Schmatz, Phys. Chem. Chem. Phys., 2012, **14**(37), 12982 .

33. J. N. L. Connor, P. McCabe, D. Sokolovski and G. C. Schatz, Chem. Phys. Lett., 1993, **206**, 119 .

34. R. C. Fuller, Phys. Rev. C: Nucl. Phys., 1975, **12**, 1561 .

35. A. J. Dobbyn, P. McCabe, J. N. L. Connor and J. F. tillo, Phys. Chem. Chem. Phys., 1999, **1**, 1115 .

36. D. Sokolovski and J. N. L. Connor, Chem. Phys. Lett., 1999, **305**, 238 .

37. R. T. Pack, J. Chem. Phys., 1974, **60**, 633 .

38. P. McGuire and D. J. Kouri, J. Chem. Phys., 1974, **60**, 2488 .

39. E. B. Stechel, R. B. Walker and J. C. Light, J. Chem. Phys., 1978, **69**, 3518 .

CHAPTER 8

STUDYING CHEMICAL REACTIVITY IN A VIRTUAL ENVIRONMENT

Moritz P. Haag and Markus Reiher

INTRODUCTION

A detailed understanding of the atomic rearrangements occurring during a chemical reaction is at the heart of chemistry. Computational investigations based on first principles of quantum mechanics can provide such a microscopic picture. They also complement the findings of experiments by making information on reaction steps accessible for which experimental data are not available.

Usually, the elucidation of reaction mechanisms starts with the set up of structures which are likely to be close to either the reactants, the products or the transition state structures. Subsequent optimization of the electronic structure and of the atomic arrangement is computationally expensive, and it is by no means guaranteed that these optimizations produce the desired mechanism. A trial-and-error protocol has to be repeated until all potentially relevant reaction steps have been identified. In most es the build-up of the starting structures is carried out in conventional molecule editors using a computer mouse and a three dimensional ball-and-stick representation of the structure.

Despite the remarkable progress in the development of fast algorithms for electronic structure calculations and the continuously increasing computer power, it is still a challenge to predict reaction mechanisms of molecular systems consisting of a few hundred atoms. The reason for this is the inherently high complexity induced in the potential energy surface by the large number of degrees of freedom of such systems. Already for an elementary reaction step, a transition-state search requires starting structures from which local optimizations can be initiated and such optimizations may easily fail. Then, only little additional information is provided on how to improve on such a failed search.

Citation: Moritz P. Haag and Markus Reiher, "Studying chemical reactivity in a virtual environment," DOI: 10.1039/c4fd00021h.

In addition, technical hurdles need to be overcome. Firstly, the build-up of three dimensional chemical structures with two dimensional input devices is a slow and cumbersome procedure. Secondly, manipulating the system and receiving the resulting response from the calculations are sequential and delayed processes that often require different computer programs. This separation is not only due to the electronic structure optimization being computationally expensive, but also due to the fact that there exists currently no general implementation to have the manipulation of the system and the instantaneous presentation of the quantum chemical reactivity response in the same program.

Both issues can efficiently be resolved by having all the processes united in a single virtual environment. Such a virtual environment allows for a deeper immersion of the scientist into the problem. In addition to the visual presentation the complex output of quantum chemical calculations can be perceived more intuitively by addressing further human senses such as the haptic sense. The main advantage then is the seamless combination of manipulation and output presentation. It allows the chemist to exploit prior knowledge about the system and his/her 'chemical intuition' for the exploration of reaction mechanisms.

Studying complex phenomena in a more intuitive way is not entirely new to chemists and biologists. For molecular dynamics,[1,2] protein docking[3] and structural biology[4] interactive approaches and even haptic interactions have been applied. The idea of an interactive build-up of molecular assemblies with continuous energy minimization in the background for a quick assembly of reasonable structures has recently been implemented for parameterized analytical potentials and classical force fields in the molecule editors SAMSON[5] and AVOGADRO.[6]

In 2007, we initiated a research program to explore chemical reaction mechanisms interactively with quantum chemical approaches. We started with the haptic exploration of interpolated Born–Oppenheimer potential energy surfaces within the framework of what we call Haptic Quantum

Chemistry.[7] During such a haptic exploration scientists are able to feel the quantum mechanical forces exerted on the atoms, when they are moved with an input device capable of providing force-feedback. We further extended this approach to larger systems and presented a protocol to refine potential energy surfaces guided by the haptic exploration.[8] To be able to treat structural relaxation in complex systems we also investigated the possibility to calculate the necessary forces directly.[9] Recently, we have investigated real-time reactivity exploration[10] in the framework of the SAMSON environment, for which a fast semi-empirical method for structure optimizations of hydrocarbons is already available.[11,12]

In this work, we explore the physical meaning of externally controlled configurational changes in molecular assemblies. We work out the requirements for treating molecular assemblies in virtual environments and how much of the underlying physical theory can be recovered in an interactive exploration of chemical reactivity. The goal is to provide an intuitive and at the same time chemically meaningful experience of a reactive molecular system.

The interactive study of reaction mechanisms in a virtual environment as described in this work is envisaged as being part of an even larger framework comprising additional automatic or semi-automatic tools supporting the investigation of complete reaction networks. We, therefore, start in section 2 with a description of this bigger picture to clarify the context of the methodology described in this work. The main ingredient for an interactive study of chemical reactivity is a real-time availability of first-principles forces, which will be detailed in section 3. We continue then by dissecting the main components of a reactivity exploration in a virtual environment in section 5. In section 6 the manipulation and the presentation of molecular systems are discussed in detail. This is then followed by an analysis of the physical laws governing a molecular system in a virtual environment in section 7. We then conclude the work and provide an outlook at the end of this paper.

VIRTUAL ENVIRONMENT FOR THE INVESTIGATION OF REACTION MECHANISMS

The interplay between the various components of the virtual framework introduced in this work for the investigation of reaction networks is summarized in Figure 1. The figure highlights the essential components of the framework and serves as a blue print for a quantum chemical tool for the exploration of reaction networks.

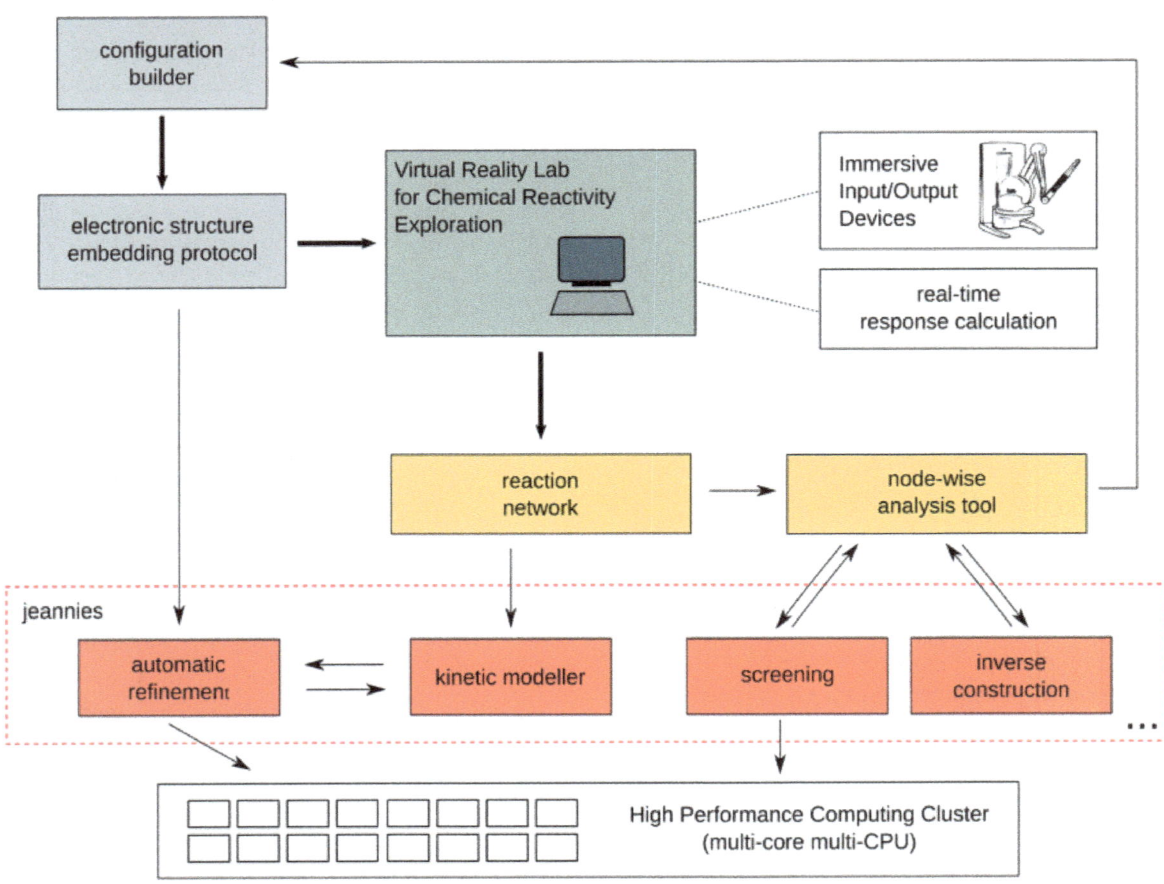

Figure 1: Design of a virtual environment for the investigation of reaction mechanisms and networks.

The components of this virtual reactivity laboratory are as follows: The terminal in green color is a symbol for the visual interface at which the relevant data can be displayed to the operator. The display may be a simple screen that could be placed right next to a bench in a wet lab or it can be as advanced as a virtual reality cave. The 'relevant data' will be the reaction network consisting of nodes and links, which is undergoing a rolling improvement that does not depend on whether the operator is in front of the terminal or not. Instead, service programs take care of as many tedious tasks and decisions as possible. We call these service programs 'jeannies'. They are supposed to run constantly in the background and independently take as many decisions as possible. The decisions taken are displayed to the user in order to give him or her the opportunity to steer these tasks when deemed necessary.

Information on the nodes and links of the reaction network are graphically displayed upon clicking on these entities. They will then open in a new frame or window on the terminal. The most important frame is the molecule editor which allows the investigation and manipulation of molecular structures to be taken from or eventually integrated into the reaction network. For the manipulation of molecular configurations, new jeannies are set up (denoted in grey and red in Figure 1), which may fulfil highly system-dependent tasks — from supplementing heavy-atom only PDB structures with hydrogen atoms to the automated construction of reactive complexes, that might lead to new nodes in the reaction network, to inverse and rational design tools[13,14] for the construction of totally new reactants and structures.

Many other advanced jeannies can be envisaged. For instance, a kinetic modeler that is able to solve large systems of coupled kinetic differential equations — analogously to algorithms developed in systems biology and bio-informatics — is decisive for an automatic evaluation of the reaction network. Crucial regions in the network can be identified in terms of their kinetic relevance and can then be refined with accurate quantum chemical methods. Also these refinement calculations can be automatically launched on a high-performance compute-cluster (though the operator may actively suppress them if he or she considers them not useful for the exploration).

REAL-TIME FORCE CALCULATION FROM FIRST PRINCIPLES

A key for an interactive exploration of chemical reactivity is the calculation of the quantum mechanical response of the molecular system upon a modification of its structure in real time. In the following we will refer to such modifications as structure manipulations. They include all manipulations that change the relative position of atoms in the system. The primary quantum mechanical response to such changes is a changed electronic structure of the molecular system leading to a different potential energy (electronic energy in the Born–Oppenheimer approximation). From the electronic structure energies and forces can be obtained. The secondary response is the change in the arrangement of the atoms not manipulated, if structural relaxation is allowed. It is important to emphasize that only electronic-structure methods based on the first principles of quantum mechanics are suitable for reactivity studies because bonds are broken and formed, which cannot be reliably predicted by the pre-defined potentials of a (classical) force-field.

To obtain the quantum mechanical response, the electronic structure has to be calculated first. The forces can then be obtained analytically from the results of this optimization procedure. The response of the atomic structure is calculated by minimizing the forces on the unmanipulated atoms. To provide the result in real-time there is an upper time limit for all steps. However, in most es the electronic structure optimization is the limiting factor. Depending on external factors — such as the details of the optional structure optimization procedure, the user-defined speed of structure manipulation or the required adiabaticity of the manipulation — the limit can vary considerably. However, an electronic structure (and force) calculation requiring on the order of one hundred milliseconds will be sufficiently fast.

In order to make the electronic structure optimization sufficiently fast, usually additional approximations have to be introduced. A general rule is: The faster the optimization is, the more severe are the approximations. Thus, in addition to the time limit it is necessary to specify a lower limit for the quality. To do so we recall the goals of an instantaneous exploration of chemical reactivity in a virtual environment, namely to explore the potential energy surface and identify the structures corresponding to the reactants, products, and transition states of elementary reactions, keeping in mind that they are only a starting point for further refinement using the standard methodology developed in quantum chemistry. Hence, the potential energy surface, *i.e.*, the electronic structure method employed to obtain it, should reproduce all main features of the exact surface. Thus, local minima and transition states should be detectable, while their absolute position and depths or heights are not required with high accuracy.[8] Clearly, not in every situation are all of the features needed. A first exploration to familiarize with the molecular system can be of low quality and only later, parts of the surface are examined more accurately. More information on this iterative character of a haptic exploration of potential energy surfaces to continuously refine them can be found in ref. 8.

Pre-Calculated *ab initio* Surfaces and Interpolation

The interpolation of pre-calculated potential energies is a fast option to obtain the forces. We found[7] that the interpolating moving least squares (IMLS) method[15–18] is very well suited for haptic rendering. Other possible schemes are based on splines,[19] the modified Shepard method[20] or neural networks.[21] IMLS is a local method which performs the fitting at each required position giving configurations in the proximity higher weights. This feature enables the introduction of a cut-off procedure to reduce the number of data points needed for the interpolation, which is beneficial if the data set for the whole system is very large. Another advantage is that, because of the analytical expressions obtained at each point, the calculation of the forces is trivial as it involves only derivatives of a polynomial. However, an interpolation solely based on gradients is also possible. Furthermore, IMLS provides a good starting point for an automated refinement procedure of the surfaces.[22] To refine the surfaces of contiguous haptic explorations an efficient procedure is available that is guided by the findings of previous explorations.[8]

No matter what technique is chosen, interpolation allows one to achieve a very high accuracy of the potential energy surface and the forces because the raw data can be obtained with high accuracy *ab initio* methods. However, the application of interpolation is also limited. If the dimensionality of the hyper-surface increases, too many points have to be calculated and the interpolation becomes slow. In addition, if relaxation is allowed, the configuration space spanned by the manipulation is no longer unique but depends on the history of the exploration. Usually, addition or abstraction reactions involving only atoms or diatomic molecules in systems where relaxation does not play a major role are good candidates for the haptic exploration based on interpolation of pre-calculated surfaces.[8]

Classical Reactive Force Fields

For es which require a direct calculation of the forces due to large relaxation effects (introducing significant exploration history effects) during the reaction, reactive classical force fields are the fastest methods available. The bond-order potentials like the Tersoff potential,[23] the Brenner potential[24] or the Finnis–Sinclair potentials[25] are examples for such methods. Other general reactive force fields are the empirical valence bond potential by Warshel and co-workers[26] and the ReaxFF[27,28] force field. For a recent overview see ref. 29. All of these methods have in common that their simple formalism permits a very fast force calculation for a given atom arrangement. However, the reactive force fields, like all classical force fields, are highly parameterized and of limited transferability.

Standard Semi-Empirical Methods

The next step towards a first-principles treatment of the electronic structure are semi-empirical methods. Since very many different variants have been developed in the past few decades, we focus on the ones suitable for the interactive study of reactive systems.

The simplest method of this family is the atom superposition and electron delocalization molecular orbital (ASED-MO) method,[30] which is based on the extended Hückel method[31] to treat the valence electrons of atoms. It applies atomic-repulsion corrections[32] in order to obtain correct structures, which is the main deficiency of the extended Hückel method. This method has already been shown to be feasible for the interactive study of structures containing only carbon and hydrogen.[11,12] It has the advantage of being non-iterative as the Fock operator does not depend on the molecular orbital coefficients owing to the severe approximation involved. Moreover, only a few simple integrals have to be evaluated.

The next more complex group of semi-empirical methods invokes the neglect of differential diatomic overlap (NDDO)[33,34] and the modified neglect of diatomic overlap (MNDO)[35,36]approximations. The PM6[37] and OMx[38] family of methods are recent members of this group. They are based on the Hartree–Fock theory but apply approximations such as neglecting electron–electron interaction integrals and replacing many of the remaining ones by empirically adjusted parameters.

Also density functional tight-binding (DFTB) methods like the most recent DFTB3[39–41] approach are suitable semi-empirical methods for a fast and quite reliable force calculation. Although DFTB methods involve fit parameters they are not truly semi-empirical in a strict sense since all these parameters are determined from full Kohn–Sham density functional theory (DFT) calculations. As the name suggests DFTB methods can be understood as a DFT analogue to NDDO/MNDO-based methods.[42]

All these semi-empirical methods are electronic structure methods (in contrast to reactive force field approaches). From the electronic structure (eventually, from the orbitals) quantum chemical properties can be calculated. Furthermore, all apply a finite basis set expansion of the orbitals. Hence, the optimized molecular orbital coefficients provide a reasonable estimate for more elaborate electronic structure methods to refine the results of the exploration. The connection to more accurate methods is therefore straightforward.

Minimal Hartree–Fock and Density Functional Methods

The main idea behind this group of methods is to reduce the computational cost of standard quantum chemical calculations based on Hartree–Fock or density functional theory to a justifiable minimum. Since this group of methods does not introduce system-dependent parameters, they are generally applicable and, with increasing computational effort, gradually improvable towards the reference method (*i.e.*, towards the Hartree–Fock limit and the Kohn–Sham DFT result at the basis-set limit free of numerical inaccuracies). It is possible to apply algorithms developed in the field of linear-scaling electronic structure methods. However, to achieve very fast calculations mainly the size of the basis set is decreased, since this provides an easy to apply and very efficient way[9] to reduce the computational cost with a built-in natural strategy for systematic improvement. Moreover, effective core potentials (ECPs) reduce the number of explicitly treated electrons and thus also the number of basis functions.

We have explored this methodology taking DFT with a minimal basis set and ECPs and applying the usual integral screening and density-fitting techniques to accelerate the calculations. We have applied this minimal set-up first to calculate data points for surface interpolation[8] and then to show that reasonable energies and forces can be calculated for small systems even in real time.[9] Interestingly, such minimal approaches are experiencing a renaissance in quantum chemistry. The corrected small basis set Hartree–Fock method proposed by Sure and Grimme[43]for the fast calculation of mass spectroscopy data takes Hartree–Fock theory as a starting point with a minimal basis set called MINIX. To minimize the errors electronic core potentials and three different atom pair-wise correction terms with pre-fitted parameters have been applied.

The recent development in the field of quantum chemistry to utilize the computer power offered by graphical processing units or other specially tailored hardware, will make first-principles force calculations possible with increasing accuracy. Elaborated screening techniques to avoid unnecessary integral evaluations and sophisticated algorithms to evaluate the remaining integrals will also contribute to this development.[9]

ANALYSIS OF CONTEMPORARY SEMI-EMPIRICAL METHODS

In the following we discuss timings for state-of-the-art implementations and accuracy studies of semi-empirical and density functional tight-binding methods. These results complement our previous work in which we demonstrated the general applicability of interpolation approaches[7,8] and minimal one-determinant Hartree–Fock and DFT methods.[9] With semi-empirical and density functional tight-binding methods the size of molecules that can be considered in real-time calculations is clearly much larger than the size of the reactive systems studied in ref. 9(Br_2 approaching ethene and a chlorine anion approaching H_3CF).

Runtime Study

In order to study contemporary semi-empirical methods and density-functional tight-binding methods for their potential to provide real-time energies and forces, we performed single-point energy calculations for different test systems for a variety of NDDO-based semi-empirical Hamiltonians (AM1,[44] RM1,[45] PM3,[46] PM6,[37] PM6-D3,[37,47] PM7[48]) and the third-order self-consistent-charge density-functional tight-binding method DFTB3.[39–41] Since the possibilities to obtain timings from the available implementations are limited and are not comparable among different programs, the timings were taken by measuring the overall execution time. To average out the influence of varying background tasks performed by the operating system 100 consecutive runs were carried out and averaged.

Our set of test molecules of different sizes and different elemental composition is shown in Figure 2. The two alanine oligopeptides represent typical biological molecules. The system denoted as FeGP (iron-guanylylpyridinol) is a model for the active site of [Fe] hydrogenase, where the methylenetetrahydromethanopterin (HMPT) is a cofactor which we chose to model by structure e in Figure 2. Both structures have been taken from ref. 52. The so-called compound A (*cf.* structure A in Figure 3, of ref. 53) is a large organic molecule. The Schrock complex with an approaching N_2 molecule[54–60] is an example of a large transition metal complex employed in homogeneous nitrogen-fixation catalysis.

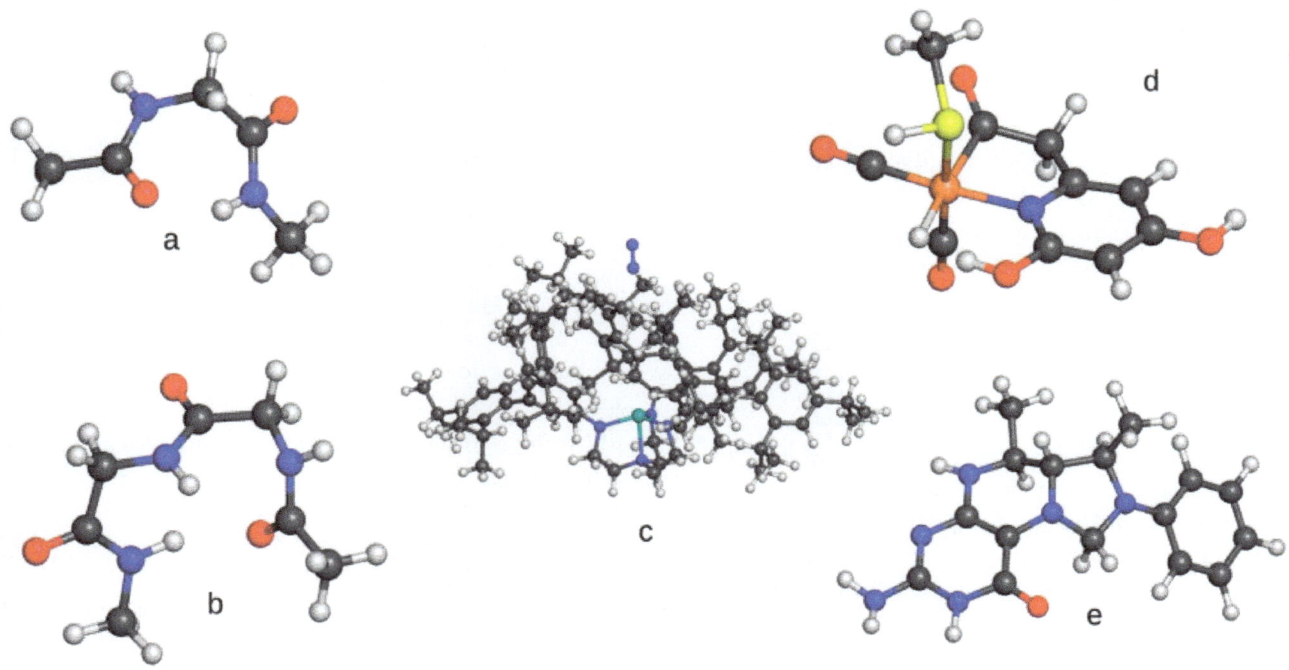

Figure 2: Molecular systems used for the timings. (a) Alanine dimer, (b) Alanine trimer, (c) Schrock trisamidoamine Mo complex with N_2, (d) FeGP, (e) model for HMPT. Structures a, b, e have been obtained from a DFTBA[49,50] optimization in Gaussian 09[51] Rev. d.01. Structure d has been taken from ref. 52, structure c from ref. 8.

All calculations started from the default molecular orbital (MO) coefficients guess provided by the respective program. The timings obtained are compiled in Table 1. Those molecules containing only C, N, O and H atoms required a similar number of iterations to reach self-consistency. However, a comparison of the required iterations for Ala_3 and the FeGP model system shows that the iteration number has only a small influence on the execution time. Although the times measured here contain the overhead of reading and writing to disk, printing information to output files and setting up the SCF iterations, they are all within the time range of what we would require for real-time calculations. Only the very large Schrock complex requires several seconds to optimise an electronic structure, which is mainly due to its size.

Table 1: Overall execution times in milliseconds averaged over 100 consecutive single point energy calculations on a Linux workstation with an Intel® Xeon® CPU @ 3.40 GHz. The internal standard guess MOs have been adopted. The number of cycles to reach self-consistency are given in parentheses. MOPAC2012[61] for AM1, RM1, PM3, PM6, PM6-D3 and PM7. DFTB+ [62] v.1.2.2 for DFTB3 with the 3OB[63] parameter set. Missing timings (denoted by '–') are due to a lack of appropriate parameters for certain elements of the periodic table

	AM1	RM1	PM3	PM6	PM6-D3	PM7	DFTB3
Ala_2	248 ± 5	249 ± 9	249 ± 7	248 ± 9	251 ± 6	249 ± 10	344 ± 37
19 atoms	(14)	(14)	(14)	(14)	(14)	(15)	(11)
C, H, N, O							
Ala_3	249 ± 5	249 ± 6	250 ± 6	252 ± 5	252 ± 6	252 ± 5	342 ± 32
26 atoms	(14)	(14)	(14)	(14)	(14)	(14)	(13)

C, H, N, O							
FeGP	–	–	–	258 ± 6	259 ± 6	259 ± 5	–
29 atoms	(–)	(–)	(–)	(236)	(236)	(61)	(–)
Fe, S, C, O, N, H							
HMPT	255 ± 6	255 ± 4	255 ± 7	256 ± 4	257 ± 4	257 ± 5	354 ± 40
43 atoms	(14)	(14)	(17)	(20)	(20)	(15)	(12)
C, H, N, O							
Compound A	275 ± 21	276 ± 29	273 ± 6	289 ± 54	295 ± 63	279 ± 34	562 ± 32
84 atoms	(13)	(13)	(17)	(25)	(26)	(14)	(10)
C, H, N, O							
Schrock-N_2	–	–	–	9197 ± 498	9414 ± 574	5719 ± 103	–
280 atoms	(–)	(–)	(–)	(142)	(142)	(95)	(–)
C, H, N, Mo							

The timings given here are only upper limits. In an implementation tailored to the needs of real-time quantum chemistry execution time can be reduced by restructuring the algorithms considering the following points.

- Separate all system specific initialization and perform it at the beginning.
- Take MO coefficients from previous optimizations as a starting point for the next.
- Eliminate every output that is written to disc or to the console.
- Avoid any reading from or writing to disc.
- Perform as many steps as possible in-core.
- Choose convergence thresholds according to the resolution of the output device(s).

Since all methods investigated compare similarly in terms of providing real-time energies, they may be selected according to their accuracy, their ability of being systematically improvable, and to the possibility of including them in a subsystem approach. Because of its close connection to DFT the DFTB approach is well suited for subsystem approaches using DFT.

Quality Assessment

The ability of an electronic-structure method to yield qualitatively correct potential energy surfaces is of paramount importance for the study of chemical reactivity. As already stated, absolute barrier heights, depths, and accurate positions of local minima and first-order saddle points are not ultimately important as long as these features are present and can be detected.

To investigate the semi-empirical and density-functional tight-binding methods for their ability to yield such qualitatively correct surfaces, we chose a rotor-like molecule with three anthracene blades that undergoes a rearrangement with a challenging plateau-like barrier and that has recently been studied in our group with more accurate first-principles methods.[53] The energy profile from ref. 53 serves as a reference. The reference potential energy surface features two local minima (A and B) which are connected by two transition states (TS1 and TS2) with a shallow minimum in between (MinAB). For the corresponding structures see Figure 3.

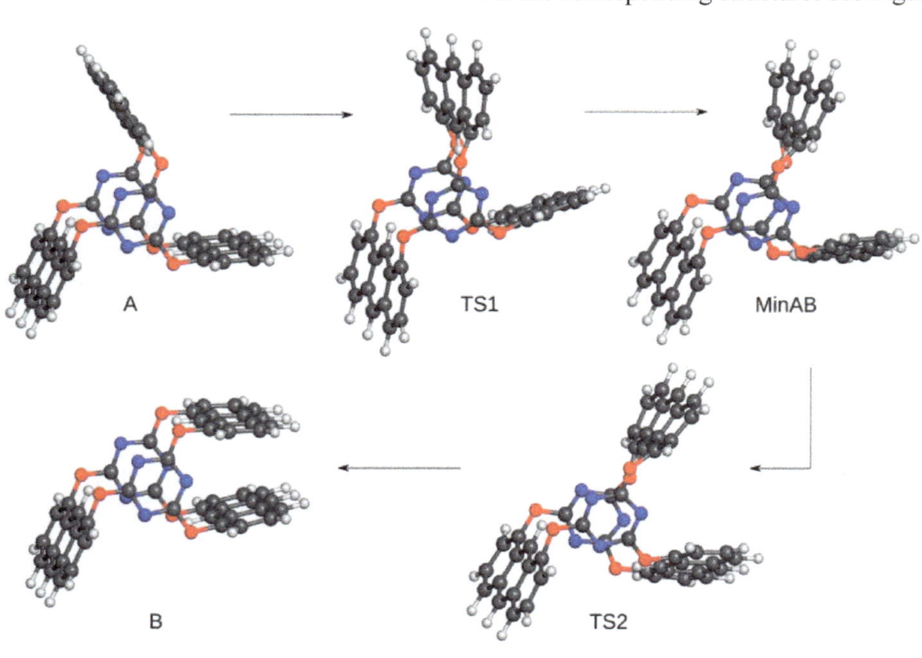

Figure 3: Reference (starting) structures taken from ref. 53 for the reaction profiles in Figure 4 and Table 2.

Since every semi-empirical Hamiltonian produces a different (approximate) potential energy surface the energies have been calculated after a structure optimization employing the respective electronic structure method (*cf.*Table 2). The NDDO-based calculations were performed with MOPAC2012.[61,64] The DFTB3 structures were optimised in Gaussian09 d.01[51] with the DFTBA[49] method and the DFTB3[39–41] values were obtained from a single-point energy calculation on these structures with the DFTB+[62] (v.1.2.2) program. The results are collected inTable 2 and selected energy profiles are shown in Figure 4.

Table 2: Energy profile of the reaction A → TS1 → MinAB → TS2 → B relative to the energy of A in kcal mol^{-1}. Below each energy the root mean square deviation (RMSD) with respect to the reference structure is given in Å. The BP86, D3, def2-TZVP reference profile has been taken from ref. 53

		A	TS1	MinAB	TS2	B
BP86-D3/def2-TZVP[53]	**E**	**0.0**	**20.1**	**19.6**	**20.5**	**−3.7**
AM1	E	0.0	17.3	0.4	22.0	0.4
	RMSD	0.118	0.354	1.334	0.265	0.506
RM1	E	0.0	23.4	15.1	32.0	0.6
	RMSD	0.099	0.405	0.276	0.268	0.359
PM3	E	0.0	16.2	13.7	18.3	0.4
	RMSD	0.135	0.314	0.392	0.194	0.612
PM6	E	0.0	13.2	12.6	13.8	0.5
	RMSD	0.123	0.196	0.128	0.123	0.488
PM6-D3	E	0.0	13.0	12.3	13.3	−1.4
	RMSD	0.115	0.196	0.123	0.125	0.361
PM7	E	0.0	18.3	12.1	30.9	−8.3
	RMSD	0.085	0.405	0.237	0.273	0.090
DFTBA	E	0.0	21.8	0.0	18.6	0.3
	RMSD	0.122	0.731	1.199	0.194	0.353
DFTB3	E	0.0	19.2	0.0	17.9	0.2

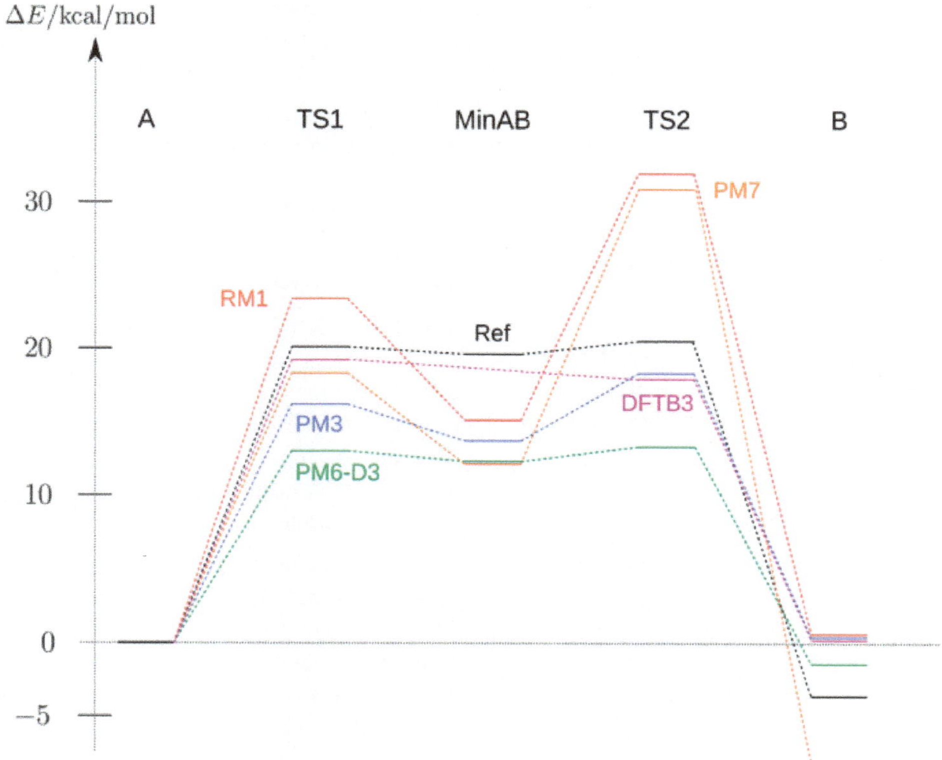

Figure 4: Energy profile of the reaction A → TS1 → MinAB → TS2 → B relative to the energy of A of the corresponding method. Only a selection of the methods from Table 2 is shown. The reference BP86, DFT-D3, def2-TZVP profile has been taken from ref. 53.

Except for AM1 all semi-empirical methods show the local minima and saddle points of the reference surface (*cf.* Table 2). As expected, the heights or depths vary from method to method but the overall structure of the energy profile is conserved throughout. The transition state structures deviate more from the reference structures than the local minima, as is expected for semi-empirical methods.[48] The correct sign of the energy difference between product and reactant is only reproduced by PM6-D3[37,47] and PM7.[48] This is caused by a strong dispersion interaction effect, which is explicitly treated in a post-SCF manner in PM6-D3 and is directly incorporated in the PM7 Hamiltonian.[37,47,48] All NDDO-based methods have the first transition state energetically below the second one, as in the reference calculation.

PM6-D3 shows the best performance of all methods analysed. Although the heights of the TS1, MinAB and TS2 are significantly lower compared to the DFT reference the relative energies of the transition states and the middle minimum as well as the relative energy between product and reactant are reproduced. The best method for obtaining the correct structures of reactant and product (A and B) is PM7 with RMSD values below 0.1Å.

Interestingly, it was not possible to locate the stable minimum MinAB with the DFTB methods, but they reproduced the two transition states and the minima A and B quite well.

The overall performance of all methods studied here is astonishingly good, *i.e.* they perfectly meet the quality requirements stated in the beginning. The fact that the semi-empirical approaches without dispersion correction overestimate the stability of the MinAB structure is no drawback as the detection of a potential intermediate is feasible. Applied in an interactive chemical reactivity study these methods would show the desired features of the potential energy surface and would guide the user to the correct structures. The relatively small RMSD values with respect to the DFT reference (*cf.* Table 2) indicate that, if the user is able to correctly locate the minima, they will be perfectly suited as starting structures for a refinement employing a more accurate method.

REACTIVITY EXPLORATION IN VIRTUAL ENVIRONMENTS

In the following we describe in detail how a virtual reactivity exploration experience should proceed. We work out what is needed from the chemist's perspective in order to successfully study chemical reactivity.

Preparatory Steps

The procedure starts with the construction of the molecular structure of interest. This construction requires the definition of the type and position of atoms and the determination of the overall charge and spin multiplicity. To arrive at a stable configuration the energy can be continuously minimised in the background while the molecular structure is constructed.[5,6] However, applying a first-principles method can be problematic for placing additional atoms one by one. It may produce unwanted structural distortions upon continuous structure optimization or may already show a non-converging electronic structure optimization and artificially large gradients are the result. Classical force fields can be more suitable at this stage.

The Exploration Procedure

After the build-up of the structure the operator exploits *ab initio* force calculations to start the exploration. A possible first step is to grab one or more atoms and distort the structure by pulling at them. If the device with which the manipulation is performed features force–feedback functionality, the force acting on the manipulated atoms and hence the reactivity can be directly felt (haptic quantum chemistry[7]). Without haptic interaction the user has to rely on the visual information to infer the reactivity. In that e the user needs to be informed about how likely the applied distortion is by visual elements like force arrows on the atoms or a panel that indicates high energy configurations, which is neither a direct sensation nor is it practical for large molecular systems (the user would drown in a sea of force vectors).

In the e of intermolecular reactions, the starting point is a stable configuration with no intermolecular interactions, *e.g.*, at the dissociation limit. This means that many orientational changes are possible without a change in potential energy and therefore without forces. The reactive fragments are oriented such that a suitable initial position is obtained to start the reactive exploration. From there the scientist may record all configurations visited during the exploration. The recording can be stopped and restarted several times which results in a multitude of different paths. Usually both, the initial and the end configuration, are local minima that are stored separately from the paths. During the exploration the chemist is guided by the forces and by the structural changes occurring as a result of the structure manipulations. The exploration is therefore driven by a combination of the operator's prior knowledge, the information provided through the forces, and the structural changes upon structural manipulations.

Such an exploration can be performed to study the shape of the potential energy surface in a localized region or to directly find possible reaction paths between already known stable configurations (local minima).

Output for Reaction Networks

The structures recorded during the exploration together with the associated energies and gradients form the primary (raw) output data. They are what we recently defined as 'core quantities' in our definition of Real-time Quantum Chemistry.[9] Additional data such as dipole moments, the electron density and other quantum chemical properties can also be part of the output if calculated on the fly.

The structures, energies and gradients can be analyzed and the distinct points with zero gradients can be extracted. They can be considered as nodes of a reaction network that emerges successively during an exploratory study. The paths between the structures with zero gradients (stationary points) connect the nodes as links. If they correspond to a minimum energy path, the associated transition state structure node can be marked as such. They are usually the main objective of the exploration. If, however, the stationary points are far apart in configuration space, no unique transition structure will be found for any connecting path, which can then be taken as a hint to search for additional local minimum structures in between (new nodes). This then produces a more fine-grained reaction network for which new connecting paths are defined. They are the starting point for further transition state searches—either by haptic exploration or by automatized protocols launched in the background.

Output for Reaction Paths

Obtaining all relevant nodes will be mandatory if a complete reaction network is needed. For the study of reaction mechanisms, however, the links connecting the nodes are the prime objective, since they contain kinetic information about how the atoms rearrange during a reaction. The analysis of these paths can be seen as an interactive analogue to the Reaction Path Hamiltonian methods[65–67] or the Unified Reaction Valley[67,68] methods. A very close connection exists to the Reaction Force approach[69] since the operator can follow these paths directly if a haptic device is employed. The output of the exploration can be used as a starting point for a more detailed analysis of the reaction dynamics or kinetics with standard approaches[70–72]eventually applying more accurate electronic structure methods.

Experimental Realization of Manual Force Exertion

The virtual exploration of chemical reactivity is in fact connected to experimental realizations for the manual manipulations of molecular matter. To study chemical reactivity the resisting forces rendered to the user are an indirect descriptor of the topology of the potential energy surface. Such forces can be directly measured in special experiments. In these experimental setups mechanical forces can be applied to single molecules to change their structure and even to alter their reactivity. There exists a variety of experimental techniques for the realization of this mechanochemistry. Examples are optical/magnetic tweezers, the bio-membrane force probe technique, hydrodynamic techniques and atomic force microscopy. The manipulations applied by the user in the virtual environment can thus be directly linked to exerting real forces on parts of the molecular assembly. For a review of these experimental techniques see ref. 73. In fact, such experiments allow the characterization of chemical bonds in a direct way[74,75] and the virtual environment can become a convenient means to steer such experiments.

The mechanical manipulation of molecular matter is called mechanochemistry. Theoretical work on mechanochemistry has been performed on ring opening reactions.[76,77] In this example it has been shown how additional mechanical forces alter the potential energy profiles of reactions that involve the rupture of chemical bonds. A review on the theoretical concepts and the computational tools has recently been provided by Ribas-Arino and Marx.[78]

INTERACTION WITH MOLECULAR SYSTEMS IN VIRTUAL ENVIRONMENTS

The core objective of an interactive and real-time treatment of molecular systems is the instantaneous calculation of energy and gradients, *i.e.* the response of a given atom arrangement.[9]Since a plethora of structures, energies and gradients are produced very quickly and in a cumulative manner, they can no longer be mapped out in huge data files as these can hardly be condensed to intuitive insights. Immersion into the virtual environment of the molecular system allows the scientist to perceive the vast amount of data more easily and intuitively.[79]Immersion refers to the process of creating a perception in a virtual reality.

Immersive Perception

The visual presentation of the structure plays an important role for immersion. Hence, it should be done in the way elaborated in chemistry more than hundred years ago, namely in terms of balls as atoms and sticks as bonds. This is the default way to present the structures as applied in molecular visualization tools like VMD,[80] PyMOL[81] or Chimera[82] to name only a few. Also conventional molecular structure builders like GaussView,[83] the graphical user interface of ADF,[84] Avogadro[6] and SAMSON[5] can serve as examples.

Suitable visual output devices for an efficient immersion are devices that are able to give a comprehensive and intuitive picture of the virtual world. The larger the visual devices are, the more they exclude the perception of reality around the user. Panoramic screens or head-mounted displays[85–87] are best suited for this purpose. Rendering the atoms, bonds and all volumetric data with ray tracing is very helpful in this regard. Also current three dimensional presentation techniques like polarized or active shutter systems are beneficial, since they allow for a much easier perception of the spatial configuration of the molecular system.

To intensify the immersion by addressing the tactile sense requires the rendering of the occurring forces by haptic devices. Haptic pointer devices, for instance, can render forces in three to six dimensions. They can be employed to probe the forces acting on different sites of the molecular assembly. The scientist picks an atom and feels the resulting force response of the system. Also two or more haptic pointer devices can be utilized simultaneously to understand the interplay of forces. Such devices can be purchased at modest prices, which makes them attractive for virtual environments employed in chemical laboratories.

A more elaborate way to render forces and to address the human tactile sense are external skeletons. Since they render the force directly for each finger, they provide an even more natural experience for the tactile sense. However, these devices are still very expensive and currently not useful for a large scale deployment in chemical laboratories.

Additional data like force vectors on atoms not probed by a haptic device, the total energy, partial charges, isosurfaces of electron densities or molecular orbitals need to be seamlessly integrated as well. The most intuitive form to do so is to visualize them in three dimensions together with the molecular structure. Since these quantities as well as the structures change very fast during the manual exploration efficient implementations using graphical processing units and multi-core central processing units are needed.[88,89]

Global scalar quantities like the electronic energy could be rendered by addressing even more senses. For example, the auditory sense could be utilized for this purpose. High-frequency sounds could signal a configuration with very high energy and thus regions in configuration space that are not accessible in a given thermal setting. However, a change of background color achieves the same and is certainly less annoying. Accordingly, the tactile sense remains the most important one besides human vision.

Immersive Interaction

Human–computer interaction is a key ingredient in immersive technologies as it allows the user to participate in the virtual environment. For the study of chemical reactivity, interaction is mandatory because it drives the reactivity exploration.

Input devices facilitate the human–computer interaction and therefore play an important role. In an immersive virtual environment they allow for a manipulation of the virtual objects. Therefore, the real-world position of the device is transferred into virtual-world coordinates of a virtual representation of the device. With that the virtual objects, in our e, atoms or atom groups, can be moved. A general requirement for devices to be suitable for immersion is that a physical movement of the operator is transferred directly into a movement in the virtual world. Transformations applied in between, *e.g.* to avoid range limitations of the physical device or to stabilize motions, need to be transparent for the operator, since the positions of virtual objects are controlled with them.

Since the forces resulting from such manipulations can be very sensitive with respect to position, a precise position control is most important. A lack of precision renders the control over the manipulation nontransparent which means that physical movements of the device are not consistent with the virtual movement. This also implies that the devices should not loose accuracy if the user quickly moves the devices. Fluctuations in the position due to instabilities caused by the user is also a reason for a less precise position. Especially freely movable input devices require attention to avoid unintentional small amplitude motions. The stabilization procedure can take place already in the hardware, but also on the software side. Stabilization can be efficiently achieved by scaling down the input movement.

The interaction with large molecular assemblies in a virtual environment is challenging because they are complex and dense (many objects in a small space). Moving in and interacting with such an environment should be done with devices that allow a manipulation directly in three dimensions. Input devices with less dimensions, like an ordinary computer mouse, allow only a two dimensional position control and are therefore less intuitive to handle. Such input devices acting in two dimensions require additional visual cues and sometimes also additional input with another device to be able to manipulate three dimensional objects.

Three dimensional input devices can be divided into two categories. To the first category belong devices in which a physical object like a ball or a pen is moved. The haptic pointer device, which we have been using in the framework of haptic quantum chemistry, features a pen connected *via* joints to a machine which calculates position and orientation and sends them to the computer. Such devices allow a relatively precise control over position in three dimensions (below 0.01mm according to manufacturer specifications[90,91]).

The second category contains devices which do not require to move a physical object. They directly capture the motion of the user's hand. Such motion sensing devices provide an even more natural form of three-dimensional input. Modern motion sensing devices can have an accuracy down to the sub-millimeter range.[92] A problem with such devices is the detection of two fingers very close to each other. Hence, grabbing atoms with such devices is difficult, since it involves two fingers coming close to each other. Pointing at the atoms and moving them by just one finger is easier to implement, but is also less intuitive. The accuracy of the fingertip positions can be improved by wearing passive or active markers that make the detection easier.

Currently haptic pointer devices are more suitable since input and output can be combined in one such device.

Rendering Forces in Reactive Potentials

The rendering of the forces is a very important component of the virtual environment proposed in this work, since they convey significant information. One has to distinguish between the force rendered by the device, which is felt by the user, and the force calculated by the chosen electronic structure method. The former we may call 'haptic force' and the latter 'molecular force'.

In a straightforward implementation both forces would be exactly the same. However, the molecular forces are in the range of 1 nN, whereas the forces which can be felt by the user are on the order of 1 N. It is immediately clear that both forces cannot be the same but have to be transformed. This transformation needs to fulfil two requirements in order to be suitable for studying chemical reactivity. As a guide through the configuration space avoiding unphysical configurations the rendered forces need to be intuitive and smooth. Yet, the transformation between the molecular and the haptic force needs to be as transparent as possible, since from the force direction and magnitude the user infers the shape of the potential energy surface. The most transparent transformation is to multiply the molecular force by a constant factor to obtain the haptic force. This is how we transform the nanoscale molecular forces to a haptic force that is experienced by the user.

The shape of potential energy surfaces occurring in reactive chemical systems requires additional transformations. Shallow and deep minima, steep walls and flat hills may lie in close proximity on a reactive potential energy surface. This results in a wide range of forces that need to be rendered. Since the range of forces which can be rendered by the devices is limited and also the tactile sense has a limited resolution, the forces need to be scaled. For instance, in strongly repulsive regions huge forces quickly exceed the limits of the devices. Hence, the forces need to be scaled down. In almost flat regions, however, where shallow minima need to be detectable the force requires to be scaled up. A constant force scaling[93] as mentioned before is therefore not suited. A variable gain scheme[94] can account for both situations without the need to change the scaling manually.

The third reason why one needs to transform forces is more technical. To avoid instabilities and to compensate for slow simulation loops often frequency filters are applied. However, a too dominant filter can lead to an in-transparent force rendering. To obtain a stable control scheme usually the position is not directly controlled but the user applies an additional force that is incorporated into the simulation.[95,96] This, however, hides the true forces acting on the manipulated atom from the user, which makes it unsuitable for studying chemical reactivity, where the forces transport a substantial part of the desired information.

Moving Objects *vs.* Probing Reactivity

For the manipulation of molecular assemblies one may distinguish between two different intentions. The first intention is to move parts of the assembly around to either see them from a different perspective or to arrange them in a way better suited to start the reactivity exploration. During such a rearrangement the internal structure of the molecules stays intact. The second intention is to probe the chemical reactivity, which involves the breaking and/or formation of chemical bonds. The virtual environment has to accommodate both scenarios.

A seamless transition from moving a molecule to probing reactivity is needed when atoms come closer and the forming or breaking of bonds becomes possible. The problem can be illustrated by considering the abstraction of an atom from a molecule *versus* just pulling at an atom to move the molecule when no further constraints are applied. Whether pulling at an atom results in an abstraction or just in a movement of the whole molecule depends on how fast the system responds in terms of the relaxation described at the beginning of this section. A manipulation within an infinitely fast relaxing system would correspond to a reversible adiabatic process. If the virtual environment is designed to mimic this scenario there would be no net change in energy and therefore also no reactivity. A non-adiabatic process is emulated when the relaxation rate is slower than the manipulation by the user. In this scenario moving a molecule by just pulling at an atom would not be possible without breaking a chemical bond.

This dilemma can be resolved in different ways depending on the limits imposed by the underlying force calculations: either by adjusting the relaxation rate or by adding additional constraints. The first is only possible if the relaxation (structure optimization) is always faster than the manipulation. Then one can adjust the rate such that the users can switch between adiabatic and non-adiabatic manipulations by adjusting their manipulation speed. Pulling fast at an atom induces the abstraction, while pulling slowly just moves the molecule. Complementary to that would be to change the relaxation rate instead.

If an adjustment of the relaxation rate is not possible or unwanted, additional constraints can be introduced. Possible constraints are to fix atoms in space, preserve internal structures or pin the orientation and center of mass of a molecule. The user can realize them employing additional input devices (*e.g.*, one hand holds a metal complex in position and the other probes the reactivity of a possible ligand) or by keeping the corresponding degrees of freedom fixed during the relaxation procedure.

In both solutions employing additional input devices is the preferred way since then the constraints can easily be altered or removed if necessary. By contrast, changing the parameters of the relaxation procedure, *e.g.*, changing the rate or the constraints, requires interruption of the exploration. Accordingly, this should only be done if the constraints are expected to remain constant during the exploration, *e.g.*, when certain parts of the molecular assembly are known to be non-reactive.

Restricting the Manipulation to Avoid High Energy Configurations

If in a molecular structure every atom is freely movable, then the manipulation can easily lead to configurations with very high energy. For instance, if bond lengths are decreased below the equilibrium distance of two fragments, very high electronic energies result. Since high energy configurations may not be accessible by the system under real conditions at finite temperature such manipulations have to be avoided or at least made hard to perform.

A natural way of restriction is provided by force–feedback devices. The rendered force guides the user and allows only reasonable manipulations, since a rapid increase in energy is associated with a steep gradient pushing the user in the opposite direction. By this an artificial blocking of certain manipulations can be avoided.

Another possibility is to perform a very fast relaxation of the remaining degrees of freedom. The energy is thereby removed from the system so that the system always maintains a configuration with a reasonable energy. For the physical implications of such an energy removal see next section.

PHYSICAL LAWS IN VIRTUAL ENVIRONMENTS

Clearly, any useful virtual environment designed to immerse the chemist into the reactive molecular assembly has to obey certain physical laws. They are a trade-off between an intuitive presentation and the 'true' physics. For the former most of the requirements have already been outlined in the sections above. Here, we will show which consequences for the design of the virtual environment follow from the physical theory.

Potential Energy and Force Field of the Molecular Systems

Although the atoms are visually represented as hard spheres they are described as classical point-like objects in the calculations. Also the interaction of the operator with the molecular system necessitates a classical treatment of the atoms, since at all times they have a completely determined position in space, *i.e.*, their movement corresponds to classical trajectories. Accordingly, the virtual environment implies a 'macroscopic' picture of the molecular entities which is classical by definition. Hence, quantum mechanical effects of the atoms, such as tunneling, are not directly representable. One may account for them by discontinuous atom position dislocations when induced by a tunneling probability measure in separate studies if the potential energy surface is sufficiently well known. However, this will only be necessary for potential proton transfer steps.

The dynamics of the electrons are not explicitly represented in the virtual environment, but they largely define the interactions between the atoms. This separated treatment of the electron and atom dynamics is based on the Born–Oppenheimer approximation.[97] In fact, the electrons give rise to the potential in which the atoms move. The electronic energy is then defined by the electronic Schrödinger equation

$$\hat{H}_{el}\Psi_{el} = E_{el}\Psi_{el} \quad (1)$$

and the (approximate) nuclear Schrödinger equation reads

$$[\hat{T}_{nuc} + E_{el}]\Psi_{nuc} = E_{nuc}\Psi_{nuc}. \quad (2)$$

The electronic energy E_{el} ($\{\boldsymbol{R}_I\}$) is a parametric function of the nuclear coordinates $\{\boldsymbol{R}_I\}$. A first-principles treatment of the electrons requires that E_{el} ($\{\boldsymbol{R}_I\}$) has to be calculated by solving equation (1). This has to be done at discrete configurations of the atoms.

Despite its calculation as a discrete function, E_{el} can be considered as a continuous function of all atom coordinates. We may define a continuous variable $\boldsymbol{x}$ in $3M$ dimensional space as

$$\boldsymbol{x} = (\boldsymbol{R}_1, \boldsymbol{R}_2, \ldots, \boldsymbol{R}_M)^T \text{ with } \boldsymbol{R}_I = (x_I, y_I, z_I)^T. \quad (3)$$

Accordingly, $E_{el}(\boldsymbol{x})$ is a hyper-surface in a $3M$-dimensional Cartesian configuration space spanned by the coordinates of all atoms. There is a conservative force field connected to this surface. It describes the forces acting on each atom A and is given by the negative gradient $\boldsymbol{g}_A$ of the potential energy $E_{el}(\boldsymbol{x})$

$$\boldsymbol{f}_A(\boldsymbol{x}) = -\boldsymbol{g}_A \text{ with } \boldsymbol{g}_A = \nabla_A E_{el}(\boldsymbol{x}). \quad (4)$$

The gradients (or forces) can be calculated directly from the electronic structure. Only this analytic differentiation is sufficiently fast for immersive techniques.

From this setting it follows naturally that the molecular system has to be described classically with respect to the atom movements and quantum mechanically with respect to the electronic structure and derived properties—a standard and well-investigated approximation for mechanistic studies in complex molecular systems. Interestingly, this partition into a classical and a quantum mechanical description reveals another way to determine which information has to be presented visually and which information needs to be rendered by addressing other senses. The part described by classical mechanics is presented visually and the highly non-intuitive quantum behavior of the electrons is expressed and presented as forces rendered to the tactile sense.

A fully quantum mechanical treatment of the atom movement would also be possible, but the visualization of such a situation cannot be easily realized in a virtual environment where molecules are represented by ball-and-stick arrangements. Instead, nuclear probability distributions need to be considered, which are related to the molecular structures in the Born–Oppenheimer approximation.[98,99] However, the tremendous computational demands for such a full quantum treatment are prohibitive, apart from the fact that the force on an atom is then no longer straightforwardly defined.

Coordinates of the Configuration Space

The operator manipulates the molecular structures by changing the Cartesian coordinates of one or a few atoms at a time. Hence, the Cartesian coordinate system employed above is the most natural way to define the configuration space. Alternatively, coordinates based on the internal degrees of freedom of the molecular system can be chosen. But for the discussions to follow, the coordinate system does not need to be specified.

We assume that the collection of coordinates can be separated into different sets (subsystems). The first separation follows naturally from the way the user interacts with the system. For every manipulation we can define a set of coordinates $\boldsymbol{s}$ which are strictly defined by the manipulation (system coordinates) and the set of all remaining (environment) coordinates $\boldsymbol{e}$,

$$(\boldsymbol{R}_1, \boldsymbol{R}_2, \ldots, \boldsymbol{R}_M) \rightarrow (\boldsymbol{s}, \boldsymbol{e}). \quad (5)$$

Since the set of environment coordinates is not constrained by the manipulation, different ways to treat them are possible. They may be completely frozen, they may be subdivided into subsystems with reduced degrees of freedom or they may be optimized. Completely frozen inactive coordinates are only reasonable if they are not very much affected by the manipulation (detected by forces defined as gradient components with respect to these coordinates). Although the environment is frozen the active (system) part movement is not just ballistic, since interactions can still occur. Partly frozen coordinates (frozen subsystems) are useful, when some groups of atoms are expected to be internally rigid. The most general treatment is to optimize the inactive coordinates according to a certain protocol, which is discussed in the following section.

Energy Deposition, Redistribution and Dissipation

Most of the manipulations performed to study reactivity involve a change in electronic energy. If a low temperature is to be maintained any excess energy has to be removed from the system. The energy deposited by the structure manipulation of the operator may be restricted according to a single measure proportional to room temperature. Any excess energy can result in a relaxation of all unconstrained coordinates. That is to minimize the energy by varying the unconstrained coordinates,

$$E_{min}(\boldsymbol{s}) = \min_{\boldsymbol{e}} E_{el}(\boldsymbol{s}, \boldsymbol{e}) . \quad (6)$$

Removing the excess energy from the free coordinates is an artificial dissipation of energy. If always the direction with minimum resistance is chosen, the path is equal to the so called minimum energy path between reactants and products.

The redistributed energy gives rise to a motion of the atoms along the unconstrained degrees of freedom. Apart from structure relaxation, this motion can also be treated by the usual methods of molecular dynamics (MD) and corresponds then to a steered BO-MD[100,101] scheme similar to steered classical MD.[102] Under certain conditions, the work performed by the operator is equal to the free energy change of an equilibrium process according to the Jarzynski equality.[103]

Reaction Path and Reaction Force

The structures $\boldsymbol{x}$ recorded during a virtual exploration correspond to points in the configuration space for a pseudo-one-dimensional reaction coordinate $\xi(\boldsymbol{x})$, the arc length of the reaction path. With each point in configuration space along the path ξ a set of forces is associated

$$\boldsymbol{x}(\xi) \rightarrow \boldsymbol{f}(\xi). \quad (7)$$

If ξ describes a minimum energy path (MEP), there are only forces along the constrained coordinates and the forces in the (orthogonal) environment coordinates are reduced or zero by structure minimization. Therefore, the work W along the MEP can be obtained from

$$W = \int_{\mathrm{MEP}} f^{(s)}(\xi)\, d\xi\ , \tag{8}$$

where $f^{(s)}$ denotes the forces along the constrained (system) coordinates.

Virtual Inertia and Friction

The correct physical dynamics of an object *in vacuo* requires that a force exerted on it accelerates it and the object moves then with constant velocity when the force is switched off. In the e of molecules, any structure manipulation is performed by exerting forces which would lead to a continuous linear movement after the force vanished at the end of the manipulation. The operator would be constantly occupied with stopping unintentionally accelerated molecules. The energy minimization discussed before can remove excess energy from the inactive coordinates and therefore also from the overall translation. In any e, a molecule which had been pulled at should not move after it has been released by the operator (provided that no other forces are present).

The technical implementation of the energy dissipation process, however, gives rise to a sensation of viscosity when a molecule is grabbed and pulled around in the virtual environment. The removal of energy by relaxing all unconstrained degrees of freedom requires a certain amount of time. This delays the relaxation and, hence, pulling at a molecule shows a resisting force. This resisting force leads to a sensation of friction as if the molecule would be dragged through a viscous liquid. It appears that this artifact is in fact useful, since the user naturally expects inertia when accelerating atoms of fragments, although the haptic exploration does not consider any mass-dependent reaction to the exerted force (action).

Adiabatic and Non-Adiabatic Manipulations

Depending on the rate of the relaxation process relative to the speed of manipulation, the manipulations are either adiabatic or non-adiabatic. In the adiabatic e the relaxation is faster than the manipulation and hence the path described by the manipulation including relaxation is a minimum energy path. If the relaxation is slower than the manipulation, a non-adiabatic manipulation is performed and no minimum energy path results.

Implications of Modified Forces

As has already been discussed in section 6 an intuitive interaction with molecular systems in virtual environments often requires the manipulation of the calculated forces before they are rendered to the user. In a more formal way the forces felt by the operator f_h are a modified version of the molecular force f_m as

$$f_h = C\,(|f_m|)f_m, \tag{9}$$

where C is a force-dependent scaling factor and is in general a function of the absolute value of the molecular force. Since the forces are derived from the potential, also the potential is implicitly modified. Hence, the scientist explores not the true potential energy surface.

As we have already discussed in our work on the haptic exploration of potential energy surfaces[8] the requirement for the scaled surface is that it preserves the main features like the positions of minima and first-order saddle points. The steepness of walls or valleys is not that important if their relative depth or height is approximately preserved. Especially, schemes with a constant scaling factor[93] for the forces (C = const) need to meet this requirement. Clearly, the reaction network explored in such a way can be refined by accurate quantum chemical methods in the background.

In variable gaining schemes[94] the scaling factor is a function of the absolute value of the haptic force. To preserve the existence of all features allowing only a modification of their relative heights or depths, the function C is subject to some conditions. (1) $C = 0$ only at $|\mathbf{f}_m| = 0$ and nowhere else and (2) C must be strictly positive. Only then are no artificial saddle points created. With such a function, for instance, a shallow minimum is detectable but its depth may be exaggerated.

It is important to give the operator the freedom to change the scaling during an exploration. Since during an exploration all structures, gradients and energies can be recorded the true potential energy surface can be reconstructed afterwards.

Influence of Interactive Manipulations

Every input device implicitly introduces a constraint by keeping certain atoms at fixed positions. Additional constraints may be imposed by the parameters of the optimization procedures. Hence, in the course of an exploration many constraints can be simultaneously active. In fact every constraint that is introduced restricts the part of configuration space that can be explored.

However, since every constraint introduces additional forces, the force on a certain atom, can be significantly different with and without a constraint. If, for instance, some atom positions are fixed either by an additional input device or by just excluding them from the relaxation procedure, new constraints are introduced. These constraints in turn alter the behavior of the system and can be regarded as additional forces. It is clear that the design of such constraints is a crucial ingredient for the manual exploration of reaction mechanisms.

Local Reactivity *vs.* Conformational Entropy and Large Amplitude Motions

We envisage a virtual exploration of chemical reactivity that involves the making and breaking of chemical bonds featuring a bond energy that is at least one order of magnitude larger than the thermal energy. Only under these conditions can a manual exploration of individual molecular structures be meaningful. A typical example is the formation of coordinative bonds to a transition metal ion. The thermodynamics and kinetics of such a process in solution is usually dominated by the changes in electronic energy, while temperature and entropic effects may be neglected in a first step. This is the reason why computational studies on transition-metal catalyzed reactions can be successfully carried out by stationary quantum chemical methods. We may call such situations dominated by 'local reactivity'. Temperature and entropic effects are then considered within simplified models for the degrees of freedom (*e.g.*, by harmonic potentials). Usually, they lead to small modifications of the electronic energy profile. Significant departures of the free energy surface from the Born–Oppenheimer potential energy surface can be observed if the number of reactants changes in an elementary reaction step in a gas-phase process because of non-negligible contributions from the translational entropy change. This situation changes when many weak contacts among conformationally flexible reactants and surrounding molecules become important such that entropy changes upon chemical transformations can pile up. Then, resorting to statistical methods as provided by molecular dynamics or stochastic simulation schemes becomes mandatory. Other chemical processes that may require such techniques are the occurrence of structural rearrangements on long time scales (such as large amplitude motions that may induce massive overall structural changes in a macromolecular assembly). However, sampling approaches that may account for these processes can be called upon within the virtual reactivity environment described here. In fact, the hardware demands are currently just exorbitantly larger. As a consequence, the first implementation of a virtual reactivity lab may only deal with typical local-reactivity problems.

GENERAL DISCUSSION AND CONCLUSIONS

The remarkable theoretical and algorithmic achievements in quantum chemistry in the past decades made it possible to assign an electronic energy and molecular properties to a molecular structure in a reasonable time. Traditional research along these lines seeks to improve on accuracy as well as on increasing the size of molecules for which such calculations are still feasible. We have argued that a different focus on algorithmic developments will lead to a paradigm shift in this field. This produces a virtual environment for the intuitive and direct study of chemical reaction mechanisms. Its ingredients are (i) the possibility to calculate quantum-mechanical information about a reactive system of at most a few hundred atoms in real time[9] and (ii) the possibility to explore large amounts of data by physically experiencing their contents through the tactile human sense.[7] The latter can be realized directly[9] or through an intermediate interpolation layer.[7,8] It is important to understand that such a reactivity exploration tool requires new concepts for the step-by-step exploration of a reaction path.[10] For reactivity exploration, algorithmic tools need to be developed[10] that are significantly advanced compared to interactive structure optimizations, which aim to aid the set-up of configurations in molecule editors. For the latter, the final result, *i.e.*, a minimum structure on the Born–Oppenheimer surface counts, while considering physical principles for the way in which such a minimum structure is found is not decisive.

In this work, we undertook the next step and elaborated on a new route of development in quantum chemistry towards ubiquitous computing that deeply immerses a chemist into his/her molecular target. This virtual environment poses new challenges for the organization and presentation of tons of quantum chemical raw data. Generating alpha-numerical input files for complicated electronic structure calculations and processing the alpha-numerical output is one of the main obstacles prohibiting ubiquitous interactive computing in chemistry. This traditional, but pedestrian way of reactivity exploration is likely to die out, when ubiquitous computing, big-data handling, and immersive technology becomes available within a single implementation. Before such a virtual reactivity lab can be made available, a thorough analysis of its components is mandatory.

A central part is certainly how the quantum mechanical response is calculated in such an environment. The three main requirements for a suitable electronic structure method are: (1) it needs to be fast enough for a real-time execution, (2) it has to be based on the first principles of quantum mechanics for bond breaking and making and (3) the quality has to be such that the main features of the potential energy surface are detectable. As a first step one has to resort to approximate electronic structure methods. Hence, we discussed and evaluated contemporary semi-empirical methods suitable to study chemical reactivity in real time. With respect to both, time and accuracy, the semi-empirical methods investigated here permit an interactive study of chemical reactivity within a virtual environment. Combining the results of this work with the results we obtained in our previous studies[7–9] we now have a range of methods at hand, which allow the calculation of real-time quantum mechanical responses. These methods range from interpolation techniques, semi-empirical methods up to minimal DFT and HF calculations.

Furthermore, we describe in detail here the design of a virtual environment tailored for the exploration of chemical reaction mechanisms and the subsequent study of reaction networks. We show how the quantities presented in the proposed virtual environment are connected to those manifest in a quantum mechanical description. The introduced approximations, which are necessary to allow for an interactive study of chemical reactivity have been analyzed and the inherent limits have been worked out. We also discuss routes on how these limits can be partly alleviated for the incorporation of, *e.g.*, tunneling or entropic effects. The scaling of the forces (and therefore also of the potential energy surface) in a virtual environment has been analyzed, too. Carefully chosen, they do not impair the physics of the system but can even enhance the experience. In fact, an interactive exploration of the Born–Oppenheimer potential energy surface given by the electronic energy can be carried out with a minimal quantum chemical model that recovers the rough position of minima and first-order saddle points and thus the specific structure of the surface (*i.e.*, without introducing artifacts like spurious or missing stable intermediates).

The operator does not need to analyze any intermediate results. Instead, good candidates for stable structures are stored as nodes of a reaction network, which the operator sees emerging on the screen. Connecting lines of these nodes can be supplemented with a barrier height from an automated transition state search, which can be launched in the background by one of many service programs, which we have called 'jeannies' here. If the jeannie realizes that there is no unique transition state, the connecting line will most likely not represent an elementary reaction step and this fact can be visualized. The jeannie may hand this information to another service program responsible for the optimization of local minimum structures (within the structural interval defined by the two nodes in the reaction network). Since real-time quantum chemical data is a prerequisite for the whole virtual environment, the local-minimum and the transition-state searches do not need to be very efficient, but they need to be very stable. This poses a whole new set of constraints on such algorithms.

While the operator observes the expanding reaction network on the screen and can investigate structures on nodes as well as the connecting transition-state structure by a simple mouse click, he/she may choose to steer the construction of the emerging network. Certain branches of the network may thus be shut down and excluded from further exploration at will when deemed necessary. Accepted nodes may be refined by more accurate quantum chemical calculations, automatically launched in the background by another jeannie. While the accurate optimization of a minimum structure at a node is certainly valuable for obtaining reliable structural information, the associated electronic energy is only of secondary importance. Instead, the energetical position of the node within the reaction network is of primary importance. In this systems-chemical view of complex chemical processes, a rolling kinetic modelling of all elementary reaction steps emerging in the network will determine, which nodes and links are most important. These are the ones for which more accurate quantum chemical calculations are then launched by the jeannie.

While we have already realized some of the components necessary for the chemical virtual reality lab, many ideas described here await their implementation and we will continue to finally provide such a tool to the chemistry and materials science communities. Currently, the chemical reactivity of hydrocarbons can already be explored in real time and reaction networks for these compounds can be mapped out.[10] If the fully-fledged tool is completed, it will represent a new type of predictive and creative tool for chemical research *in silico*. Imagine a catalytic reaction mediated by some catalyst. Modelling a catalytic cycle is currently a very laborious and time-consuming undertaking. Because of the huge effort required, side reactions, of which there are orders of magnitude more than there are reaction steps in the catalytic cycle, are usually omitted. The virtual reality lab would be able to deal with side reactions, too. The data can be naturally absorbed into the components of the reaction network introducing new configurations to be constructed and finally added to the existing nodes. Natural reaction partners of the intermediates in the catalytic cycle are all molecules occurring in the cycle (including the intermediates themselves) as well as standard molecules from the environment like solvent molecules (especially water) and dioxygen. The plethora of structures needed for such a screening of potentially degradative reactions can again be set up and screened automatically by jeannies. Clearly, another useful ingredient for this endeavour is a jeannie that aids designing structures for reactive systems with specific functionality.[13,14]

Hence, the virtual environment envisaged here will be much more than a simple tool to control and manage quantum chemical calculation (as by standard graphical user interfaces to quantum chemistry program packages). Instead, it will be a new tool for truly creative work in chemistry that can challenge the work of experimental chemists.

ACKNOWLEDGEMENTS

This work has been supported by ETH Zurich (grant number: ETH-08 11-2).

REFERENCES

1. J. F. Prins, J. Hermans, G. Mann, L. S. Nyland and M. Simons, *Future Gener. Comput. Syst.*, 1999, **15**, 485–495 .
2. A. Bolopion, B. Cagneau, S. Redon and S. Régnier, *Intelligent Robots and Systems*, 2009. IROS 2009. IEEE/RSJ International Conference on, 2009, pp. 237–242 .
3. N. Férey, J. Nelson, C. Martin, L. Picinali, G. Bouyer, A. Tek, P. Bourdot, J. Burkhardt, B. Katz, M. Ammi, C. Etchebest and L. Autin, *Virtual Reality*, 2009, **13**, 273–293 .
4. J. Heyd and S. Birmanns, *Virtual Reality*, 2009, **13**, 245–255 .
5. M. Bosson, S. Grudinin, X. Bouju and S. Redon, *J. Comput. Phys.*, 2012, **231**, 2581–2598 .
6. M. D. Hanwell, D. E. Curtis, D. C. Lonie, T. Vandermeersch, E. Zurek and G. R. Hutchison, *J. Cheminf.*, 2012, **4**, 1–17 .
7. K. H. Marti and M. Reiher, *J. Comput. Chem.*, 2009, **30**, 2010–2020 .
8. M. P. Haag, K. H. Marti and M. Reiher, *ChemPhysChem*, 2011, **12**, 3204–3213 .
9. M. P. Haag and M. Reiher, *Int. J. Quantum Chem.*, 2013, **113**, 8–20 .
10. M. P. Haag, A. C. Vaucher, M. Bosson, S. Redon and M. Reiher, *ChemPhysChem* , submitted [arXiv: 1405.4036].
11. M. Bosson, C. Richard, A. Plet, S. Grudinin and S. Redon, *J. Comput. Chem.*, 2012, **33**, 779–790 .
12. M. Bosson, S. Grudinin and S. Redon, *J. Comput. Chem.*, 2013, **34**, 492–504 .
13. T. Weymuth and M. Reiher, *Int. J. Quantum Chem.*, 2014, **114**, 838–850 .
14. T. Weymuth and M. Reiher, *Int. J. Quantum Chem.*, 2014, **114**, 823–837 .
15. R. Dawes, D. L. Thompson, Y. Guo, A. F. Wagner and M. Minkoff, *J. Chem. Phys.*, 2007, **126**, 184108 .
16. Y. Guo, L. B. Harding, A. F. Wagner, M. Minkoff and D. L. Thompson, *J. Chem. Phys.*, 2007, **126**, 104105 .
17. G. G. Maisuradze and D. L. Thompson, *J. Phys. Chem. A*, 2003, **107**, 7118–7124 .
18. G. G. Maisuradze, D. L. Thompson, A. F. Wagner and M. Minkoff, *J. Chem. Phys.*, 2003, **119**, 10002–10014 .
19. J. M. Bowman, J. S. Bittman and L. B. Harding, *J. Chem. Phys.*, 1986, **85**, 911–921 .
20. M. A. Collins, *Theor. Chem. Acc.*, 2002, **108**, 313–324 .
21. L. M. Raff, M. Malshe, M. Hagan, D. I. Doughan, M. G. Rockley and R. Komanduri, *J. Chem. Phys.*, 2005, **122**, 084104 .
22. R. Dawes, D. L. Thompson, A. F. Wagner and M. Minkoff, *J. Chem. Phys.*, 2008, **128**, 084107 .
23. J. Tersoff, *Phys. Rev. B*, 1988, **37**, 6991–7000 .
24. D. W. Brenner, *Phys. Rev. B: Condens. Matter*, 1990, **42**, 9458–9471 .
25. M. W. Finnis and J. E. Sinclair, *Philos. Mag. A*, 1984, **50**,

45–55 .

26. A. Warshel and R. M. Weiss, *J. Am. Chem. Soc.*, 1980, **102**, 6218–6226 .
27. A. C. T. van Duin, S. Dasgupta, F. Lorant and W. A. Goddard, *J. Phys. Chem. A*, 2001, **105**, 9396–9409 .
28. K. D. Nielson, A. C. T. van Duin, J. Oxgaard, W.-Q. Deng and W. A. Goddard, *J. Phys. Chem. A*, 2005, **109**, 493–499 .
29. S. Mishra and M. Meuwly, *Kinetics and Dynamics*, SpringerNetherlands, 2010, vol. 12, pp. 137–155 .
30. A. B. Anderson, *Int. J. Quantum Chem.*, 1994, **49**, 581–589 .
31. R. Hoffmann, *J. Chem. Phys.*, 1963, **39**, 1397–1412 .
32. A. B. Anderson, *J. Chem. Phys.*, 1975, **62**, 1187–1188 .
33. J. A. Pople, D. P. Santry and G. A. Segal, *J. Chem. Phys.*, 1965, **43**, S129–S135 .
34. J. A. Pople, D. L. Beveridge and P. A. Dobosh, *J. Chem. Phys.*, 1967, **47**, 2026–2033 .
35. M. J. S. Dewar and W. Thiel, *J. Am. Chem. Soc.*, 1977, **99**, 4907–4917 .
36. M. J. S. Dewar and W. Thiel, *J. Am. Chem. Soc.*, 1977, **99**, 4899–4907 .
37. J. J. P. Stewart, *J. Mol. Model.*, 2007, **13**, 1173–1213 .
38. W. Weber and W. Thiel, *Theor. Chem. Acc.*, 2000, **103**, 495–506 .
39. M. Elstner, D. Porezag, G. Jungnickel, J. Elsner, M. Haugk, T. Frauenheim, S. Suhai and G. Seifert, *Phys. Rev. B: Condens. Matter*, 1998, **58**, 7260–7268 .
40. Yang, H. Yu, D. York, Q. Cui and M. Elstner, *J. Phys. Chem. A*, 2007, **111**, 10861–10873 .
41. M. Gaus, Q. Cui and M. Elstner, *J. Chem. Theory Comput.*, 2011, **7**, 931–948 .
42. N. Otte, M. Scholten and W. Thiel, *J. Phys. Chem. A*, 2007, **111**, 5751–5755 .
43. R. Sure and S. Grimme, *J. Comput. Chem.*, 2013, **34**, 1672–1685 .
44. M. J. Dewar, E. G. Zoebisch, E. F. Healy and J. J. Stewart, *J. Am. Chem. Soc.*, 1985, **107**, 3902–3909 .
45. G. B. Rocha, R. O. Freire, A. M. Simas and J. J. P. Stewart, *J. Comput. Chem.*, 2006, **27**, 1101–1111 .
46. J. J. P. Stewart, *J. Comput. Chem.*, 1989, **10**, 209–220 .
47. S. Grimme, J. Antony, S. Ehrlich and H. Krieg, *J. Chem. Phys.*, 2010, **132**, 154104 .
48. J. J. P. Stewart, *J. Mol. Model.*, 2013, **19**, 1–32 .
49. G. Zheng, H. A. Witek, P. Bobadova-Parvanova, S. Irle, D. G. Musaev, R. Prabhakar, K. Morokuma, M. Lundberg, M. Elstner, C. Köhler and T. Frauenheim, *J. Chem. Theory Comput.*, 2007, **3**, 1349–1367 .
50. G. Zheng, M. Lundberg, J. Jakowski, T. Vreven, M. J. Frisch and K. Morokuma, *Int. J. Quantum Chem.*, 2009, **109**, 1841–1854 .
51. M. J. Frisch, G. W. Trucks, H. B. Schlegel, G. E. Scuseria, M. A. Robb, J. R. Cheeseman, G. Scalmani, V. Barone, B. Mennucci, G. A. Petersson, H. Nakatsuji, M. Caricato, X. Li, H. P. Hratchian, A. F. Izmaylov, J. Bloino, G. Zheng, J. L. Sonnenberg, M. Hada, M. Ehara, K. Toyota, R. Fukuda, J. Hasegawa, M. Ishida, T. Nakajima, Y. Honda, O. Kitao, H. Nakai, T. Vreven, J. A. Montgomery, Jr., J. E. Peralta, F. Ogliaro, M. Bearpark, J. J. Heyd, E. Brothers, K. N. Kudin, V. N. Staroverov, R. Kobayashi, J. Normand, K. Raghavachari, A. Rendell, J. C. Burant, S. S. Iyengar, J. Tomasi, M. Cossi, N. Rega, J. M. Millam, M. Klene, J. E. Knox, J. B. Cross, V. Bakken, C. Adamo, J. Jaramillo, R. Gomperts, R. E. Stratmann, O. Yazyev, A. J. Austin, R. Cammi, C. Pomelli, J. W. Ochterski, R. L. Martin, K. Morokuma, V. G. Zakrzewski, G. A. Voth, P. Salvador, J. J. Dannenberg, S. Dapprich, A. D. Daniels, O. Farkas, J. B. Foresman, J. V. Ortiz, J. Cioslowski and D. J. Foxls, *Gaussian 09*, Gaussian Inc. Wallingford CT 2009 .
52. A. R. Finkelmann, M. T. Stiebritz and M. Reiher, *J. Phys. Chem. B*, 2013, **117**, 4806–4817 .
53. M. J. Kory, M. Bergeler, M. Reiher and A. D. Schlüter, *Chem.–Eur. J.*, 2014 DOI:10.1002/chem.201400364.
54. D. V. Yandulov and R. R. Schrock, *Science*, 2003, **301**, 76–78 .
55. R. Schrock, *Angew. Chem., Int. Ed.*, 2008, **47**, 5512–5522 .
56. M. Reiher, B. Le Guennic and B. Kirchner, *Inorg. Chem.*, 2005, **44**, 9640–9642 .
57. B. Le Guennic, B. Kirchner and M. Reiher, *Chem.–Eur. J.*, 2005, **11**, 7448–7460 .
58. S. Schenk, B. Le Guennic, B. Kirchner and M. Reiher, *Inorg. Chem.*, 2008, **47**, 3634–3650 .
59. S. Schenk and M. Reiher, *Inorg. Chem.*, 2009, **48**, 1638–1648 .
60. S. Luber, J. Neugebauer and M. Reiher, *J. Chem. Phys.*, 2009, **130**, 064105 .
61. J. J. P. Stewart, *MOPAC2012*, http://openmopac.net, Stewart Computational Chemistry.
62. B. Aradi, B. Hourahine and T. Frauenheim, *J. Phys. Chem. A*, 2007, **111**, 5678–5684 .
63. M. Gaus, A. Goez and M. Elstner, *J. Chem. Theory Comput.*, 2013, **9**, 338–354 .
64. J. D. C. Maia, G. A. Urquiza Carvalho, C. P. Mangueira, S. R. Santana, L. A. F. Cabral and G. B. Rocha, *J. Chem. Theory Comput.*, 2012, **8**, 3072–3081 .
65. W. H. Miller, N. C. Handy and J. E. Adams, *J. Chem. Phys.*, 1980, **72**, 99–112 .
66. B. A. Ruf and W. H. Miller, *J. Chem. Soc., Faraday Trans. 2*, 1988, **84**, 1523–1534 **RSC**.
67. E. Kraka, *Wiley Interdiscip. Rev.: Comput. Mol. Sci.*, 2011, **1**, 531–556 .
68. Z. Konkoli, E. Kraka and D. Cremer, *J. Phys. Chem. A*, 1997, **101**, 1742–1757 .
69. A. Toro-Labbé, S. Gutiérrez-Oliva, J. S. Murray and P. Politzer, *Mol. Phys.*, 2007, **105**, 2619–2625 .
70. A. Laio and M. Parrinello, *Proc. Natl. Acad. Sci. U. S. A.*, 2002, **99**, 12562–12566 .
71. C. Dellago, P. G. Bolhuis, F. S. Csajka and D. Chandler, *J. Chem. Phys.*, 1998, **108**, 1964–1977 .

72. R. Alves, F. Antunes and A. Salvador, *Nat. Biotechnol.*, 2006, **24**, 667–672 .

73. M. K. Beyer and H. Clausen-Schaumann, *Chem. Rev.*, 2005, **105**, 2921–2948 .

74. M. Grandbois, M. Beyer, M. Rief, H. Clausen-Schaumann and H. E. Gaub, *Science*, 1999, **283**, 1727–1730 .

75. S. W. Schmidt, A. Kersch, M. K. Beyer and H. Clausen-Schaumann, *Phys. Chem. Chem. Phys.*, 2011, **13**, 5994–5999 **RSC**.

76. M. T. Ong, J. Leiding, H. Tao, A. M. Virshup and T. J. Martinez, *J. Am. Chem. Soc.*, 2009, **131**, 6377–6379 .

77. J. Ribas-Arino, M. Shiga and D. Marx, *Angew. Chem., Int. Ed.*, 2009, **48**, 4190–4193 .

78. J. Ribas-Arino and D. Marx, *Chem. Rev.*, 2012, **112**, 5412–5487 .

79. M. Schirski, C. Bischof and T. Kuhlen, *Proceedings of the 13th Eurographics conference on Virtual Environments*, Aire-la-Ville, Switzerland, Switzerland, 2007, pp. 69–76 .

80. W. Humphrey, A. Dalke and K. Schulten, *J. Mol. Graphics*, 1996, **14**, 33–38 .

81. Schrödinger, LLC, The PyMOL Molecular Graphics System, http://www.pymol.org.

82. E. F. Pettersen, T. D. Goddard, C. C. Huang, G. S. Couch, D. M. Greenblatt, E. C. Meng and T. E. Ferrin, *J. Comput. Chem.*, 2004, **25**, 1605–1612 .

83. R. Dennington, T. Keith and J. Millam, *GaussView Version 5*, Semichem Inc., Shawnee Mission, KS, 2009 .

84. G. te Velde, F. M. Bickelhaupt, E. J. Baerends, C. Fonseca Guerra, S. J. A. van Gisbergen, J. G. Snijders and T. Ziegler, *J. Comput. Chem.*, 2001, **22**, 931–967 .

85. J. C. Chung, M. R. Harris, F. P. Brooks, H. Fuchs, M. T. Kelley, J. Hughes, M. Ouh-young, C. Cheung, R. L. Holloway and M. Pique, *Proc. SPIE–Int. Soc. Opt. Eng.*, 1989, **1083**, 42–52 .

86. E. Patrick, D. Cosgrove, A. Slavkovic, J. A. Rode, T. Verratti and G. Chiselko, *Proceedings of the SIGCHI conference on Human Factors in Computing Systems*, New York, NY, USA, 2000, pp. 478–485 .

87. T. Shibata, *Displays*, 2002, **23**, 57–64 .

88. J. E. Stone, J. Saam, D. J. Hardy, K. L. Vandivort, W.-m. W. Hwu and K. Schulten, *Proceedings of 2nd Workshop on General Purpose Processing on Graphics Processing Units*, New York, NY, USA, 2009, pp. 9–18 .

89. J. E. Stone, D. J. Hardy, I. S. Ufimtsev and K. Schulten, *J. Mol. Graphics Modell.*, 2010, **29**, 116–125 .

90. SensAble Technologies, Inc., http://www.sensable.com.

91. Force Dimension, http://www.forcedimension.com.

92. F. Weichert, D. Bachmann, B. Rudak and D. Fisseler, *Sensors*, 2013, **13**, 6380–6393 .

93. Y.-G. Lee and K. W. Lyons, *Computer-Aided Design*, 2004, **36**, 75–90 .

94. A. Bolopion, B. Cagneau, S. Redon and S. Régnier, *World Haptics Conference (WHC)*, 2011 IEEE, 2011, pp. 469–474 .

95. A. Bolopion, B. Cagneau, S. Redon and S. Régnier, *J. Mol. Graphics Modell.*, 2010, **29**, 280–289 .

96. A. Bolopion, B. Cagneau, S. Redon and S. Régnier, *Advanced Intelligent Mechatronics (AIM)*, 2010 IEEE/ASME International Conference on, 2010, pp. 329–334 .

97. M. Born and R. Oppenheimer, *Ann. Phys.*, 1927, **389**, 457–484 .

98. E. Mátyus, J. Hutter, U. Müller-Herold and M. Reiher, *Phys. Rev. A: At., Mol., Opt. Phys.*, 2011, **83**, 052512 .

99. E. Mátyus, J. Hutter, U. Müller-Herold and M. Reiher, *J. Chem. Phys.*, 2011, **135**, 204302 .

100. D. Marx and J. Hutter, *Ab Initio Molecular Dynamics: Basic Theory and Advanced Methods*, Cambridge University Press, 2009 .

101. B. Kirchner, P. J. Dio and J. Hutter, *Multiscale Molecular Methods in Applied Chemistry*, Springer Berlin Heidelberg, 2012, vol. 307, pp. 109–153 .

102. J. E. Stone, J. Gullingsrud and K. Schulten, *Proceedings of the 2001 Symposium on Interactive 3D Graphics*, New York, NY, USA, 2001, pp. 191–194 .

103. C. Jarzynski, *Phys. Rev. Lett.*, 1997, **78**, 2690–2693 .

CHAPTER 9

ENERGETICS OF ELEMENTARY REACTION STEPS RELEVANT FOR CO OXIDATION: CO AND O_2 ADSORPTION ON MODEL PD NANOPARTICLES AND PD(1 1 1)

Matthias Peter , Sergey Adamovsky , Jose Manuel Flores Camacho and Swetlana Schauermann

INTRODUCTION

Confinement of matter at a nanoscopic scale is well known to result in new specific properties of a material. These new properties inherent to nanoparticles are particularly important for a number of practical applications of heterogeneous catalysis, such as *e.g.* CO oxidation over transition metals.[1,2] This reaction has attracted much attention in recent decades both in surface science and ambient pressure studies and many phenomena have already been well understood.[3–7] Despite the comprehensive understanding and general agreement on the microscopic mechanisms of CO oxidation over many transition metals, less information is available on the thermodynamics of the underlying surface processes. Particularly for metallic nanoparticles, the quantitative information on the adsorption and reaction enthalpies of elementary reaction steps, such as dissociative oxygen adsorption, CO adsorption on clean and oxygen-containing surfaces, CO and recombination with an adsorbed oxygen atom, is missing. It is a current scientific challenge to obtain detailed quantitative information on the binding strength of the gaseous adsorbates and on the reaction enthalpies as a function of the particle size and the particle surface composition, since these parameters essentially determine the surface coverages under the operation conditions and with this strongly affect the absolute reaction rates. The information would allow a deeper fundamental understanding of how the surface binds reactants and guides them through various elementary steps to the products, and provide a basis for a rational design of new catalytic and functional materials.

In previous studies related to CO oxidation, the energetics of surface processes were probed mainly by desorption-based methods such as temperature programmed desorption (TPD) and equilibrium adsorption isotherm measurements.[7–13] However, these methods can be applied correctly only for fully reversible adsorption–desorption processes such as *e.g.* CO adsorptionand desorption on clean metals. In contrast, this condition does not apply for CO recombination with an oxygen atom or oxygen dissociative adsorption, which can be additionally accompanied by some side processes such as *e.g.* oxygen diffusion into the subsurface,[12–15] refaceting,[11,14] particle reconstruction,[7,16,17] as well as surface[17–19] and bulk oxide formation.[18,20]These obstacles and also the need for assumptions on the pre-exponential factors, which have to be made for the kinetic modeling of the desorption process, strongly limit the quantitative determination of binding and reaction energies related to CO oxidation by traditional desorption-based methods. A strategy to overcome these shortcomings is a direct calorimetric measurement of adsorption and reaction enthalpies under isothermal conditions.[21–24]

In this work, we report on the first direct calorimetric measurement of binding energies of gas phase molecules relevant to CO oxidation on model Pd nanoparticles and compare them to Pd(111) single crystal. The energetics of surface processes were probed on well-defined Pd nanoparticles supported on a thin Fe_3O_4 crystalline oxide film prepared *in situ* under ultra-high vacuum (UHV) conditions. We apply a recently developed UHV single crystal adsorption calorimeter (SCAC) based on molecular beam techniques[24] to investigate the energetics of dissociative oxygen adsorption and CO adsorption on clean and O-precovered Pd nanoparticles and on Pd(111), as a reference. With this approach we address the question: how do the reduced dimensions of a metallic substrate (extended laterally stiff single crystal *vs.* small atomically flexible nanoparticles) and its surface composition affect the adsorption strength of the reactive species?

EXPERIMENTAL PROCEDURE

The adsorption experiments were performed at the Fritz-Haber-Institute (Berlin) in an ultra-high vacuum (UHV) single crystal adsorption calorimetry (SCAC) apparatus based on molecular beam techniques described in detail elsewhere.[24] An effusive doubly differentially pumped multi-channel array source was employed to produce an oxygen (Linde, purity

Citation: Matthias Peter , Sergey Adamovsky , Jose Manuel Flores Camacho and Swetlana Schauermann, "Energetics of elementary reaction steps relevant for CO oxidation: CO and O_2 adsorption on model Pd nanoparticles and Pd(111)," DOI: 10.1039/c3fd00001j.

99.999%) beam at the intensity 1×10^{14} molecules cm^{-2} s^{-1} (5.3×10^{-7} mbar on the sample surface) or CO beam (Linde, purity 99.995%) at the intensity 1×10^{14} molecules cm^{-2} s^{-1} (5.3×10^{-7} mbar on the sample surface). The beam was cut into pulses of 266 ms length by a remote-controlled chopper, after that the gas impinged on a sample prepared on an ultrathin (1 μm) single crystal (Pd(111) or model supported catalysts grown on Pt(111)). The small change in the temperature of the sample caused by adsorption of oxygen was measured by a heat detector, consisting of a 9 μm-thick β-polyvinylidene fluoride (β-PVDF) pyroelectric ribbon coated with Au on both sides. The energy calibration was performed by applying pulses of laser light (HeNe, Linos, wavelength 632.8 nm, 5 mW), which passes through the same path as the molecular beam and is chopped in a way identical to the molecular flux. Simultaneously, the fraction of the molecules adsorbed in a single beam pulse (*i.e.*, the sticking coefficient) was defined by the modified King–Wells method[25] with a quadrupole mass spectrometer (QMS Hiden, HAL 301/3F) in a non-line-of-sight geometry.

The Pd/Fe_3O_4/Pt(111) model catalyst was prepared by growing a thin (~50 Å) Fe_3O_4(111) film on a Pt(111) single crystal by repeated cycles of Fe (>99.99%, Goodfellow) physical vapor deposition at a substrate temperature of ~120 K and subsequent oxidation at 875 K (see refs. 26 and 27 for details). The cleanliness and the quality of the oxide film were checked byAES and LEED. Pd particles were grown *in situ* by physical vapor deposition from a Pd rod (Goodfellow, >99.9%) using a commercial evaporator (Focus, EFM 3) with a deposition rate of 0.3 Å min^{-1} (2.04×10^{14} atoms cm^{-2} min^{-1}). The nominal deposition thickness value amounts to 4.0 Å, or to 2.8×10^{15} atoms cm^{-2}. Directly after Pd deposition, the sample was annealed to 600 K and cooled to 300 K, and a microcalorimetric measurement of oxygen or CO adsorption heats was performed. The evaporation rates of Fe and Pd were calibrated by a quartz-crystal microbalance (QCM, Sigma instruments).

RESULTS AND DISCUSSION

Combined Energy and Sticking Coefficient Measurement: CO Adsorption on supported Pd Nanoparticles and Pd(111)

The model surface employed in this study consisted of Pd nanoparticles supported on a well-ordered thin Fe_3O_4 film grown on a Pt(111) single crystal (Figure 1; for details of the preparation procedure and structural characterization by scanning tunneling microscopy (STM) see refs. 26 and 27). On average, a Pd particle on this surface is about 4 nm in diameter and contains ~720 Pd atoms; the particle density on the support amounts to 3.8×10^{12} particles cm^{-2}. According to STM data, the particles are crystalline and exhibit a nearly hexagonal shape with a flat top facet, which indicates that the particles grow in (111) orientation and their sides are terminated with (111) and (100) facets.

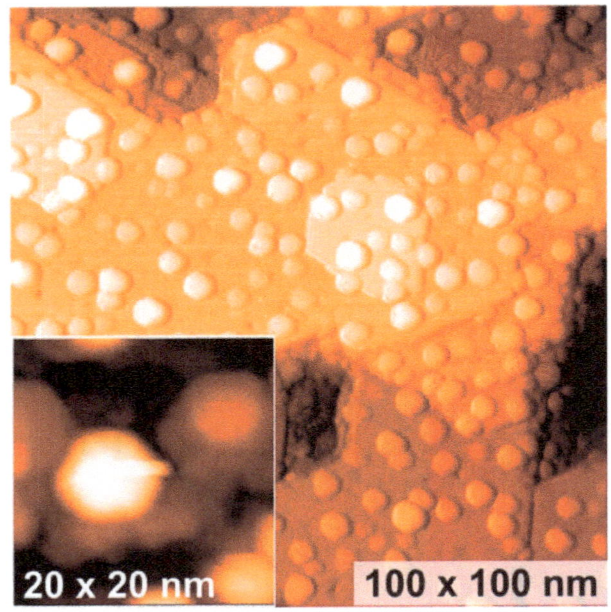

Figure 1: STM image of the Pd/Fe_3O_4/Pt(111) model catalyst used in this study, nominal Pd deposition thickness 4 Å, from ref. 27.

In the calorimetric experiment, two parameters were measured simultaneously and independently: the sticking coefficient and the corresponding adsorption energy (differentialadsorption heat). The details of the calorimetric measurement were comprehensively discussed elsewhere.[24] Briefly, we prepare a model catalyst on an ultrathin (1 μm) Pt(111) single crystal and transfer it *in situ* to the calorimeter. In a calorimetric experiment a pulse of gas from a chopped molecular beam impinges on an ultrathin single crystal surface. A fraction of the molecules striking the surface adsorb and this heat input causes a transient temperature rise (~10 mK), which is measured by a pyroelectric heat detector. The absolute energy calibration is performed by applying pulses of laser light of known energy, which passes through the molecular beam and is chopped in a way identical to the molecular flux. Two parameters, critically influencing the calibration – the absolute power of the laser light and the reflectivity of the sample – are measured *in situ*. Simultaneously, the absolute fraction of the molecules adsorbed in a single beam pulse is measured in a sticking coefficient measurement by the King–Wells method[25] in order to determine the absolute value of deposited energy per number of adsorbed molecules.

Figure 2 shows the complete calorimetric data set obtained at 300 K: the sticking coefficients (Figure 2a and 2b) and the differential adsorption heats (Figure 2c and 2d) plotted as a function of CO surface coverage for Pd nanoparticles and Pd(111) single crystal. Note that at this temperature CO does not permanently chemisorb on the Fe_3O_4 support[28] but on the Pd particles only and the experimental time scale is long enough in order to let the molecules diffuse over the surface and find the strongest adsorption site.[29]

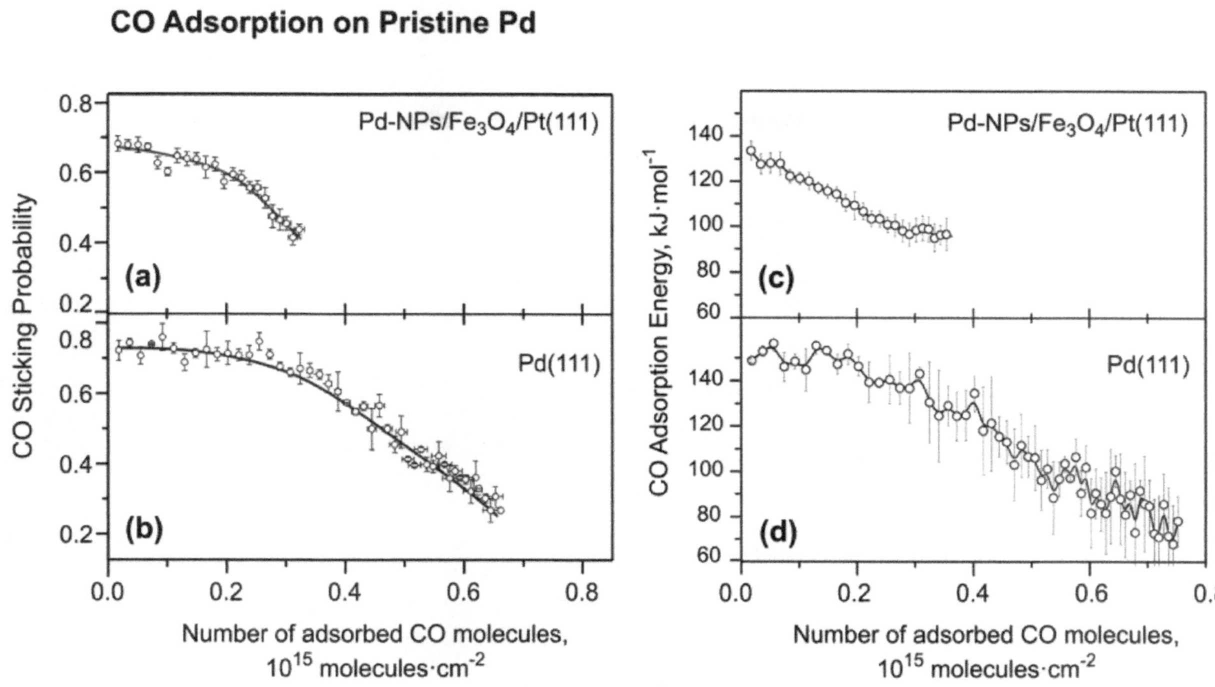

Figure: 2 Sticking probability plotted as a function of number of adsorbed CO molecules for Pdnanoparticles (a) and for Pd(111) (b). Differential heats of adsorption plotted as a function of number of adsorbed CO molecules for Pd nanoparticles (c) and Pd(111) (d). All data were obtained at 300 K. The data are shown as an average of four to six independent measurements on freshly prepared surfaces. The error bars correspond to the error of the mean.

The initial CO sticking coefficient on Pd(111) (Figure 2b) reaches the value 0.72 ± 0.03 and remains nearly constant until the CO surface coverage of ~2.7 × 10^{14} CO molecules cm^{-2} (or 0.18 CO molecules per surface Pd atom) is achieved. After that it decreases to a quasi steady state value[21] of about 0.3 for a CO coverage of 0.76 × 10^{15} molecules cm^{-2} (or 0.5 CO molecules per surface Pd atom) in saturation. The apparent non-zero sticking coefficient close to saturation arises from the partial CO desorption between the beam pulses and additional adsorption of CO when the next pulse hits the surface (see full discussion in refs. 21,23,24). The CO saturation coverage on Pd(111) is in excellent agreement with the previously reported value 0.5 with reference to the number of surface Pd atoms, which was obtained by a number of independent surface-sensitive techniques (see *e.g.* refs. 30–33). The value of the initial sticking coefficient measured in this study is somewhat lower than the previously reported values of 0.8–1.0[34] and 0.95 ± 0.05.[35] Qualitatively similar dependence of the sticking coefficient on surface CO coverage was observed on Pd nanoparticles, with the initial sticking coefficient amounting to 0.68 ± 0.02 and the saturation coverage of 0.4 × 10^{15} molecules cm^{-2} (Figure 2a). The final CO surface coverage expressed as a number of adsorbed CO molecules per surface Pd atom is ~0.48, which is in good agreement with the value 0.5 measured on Pd(111)[30–33] and Pd(100)[36] single crystal surfaces at room temperature. This result shows that the absolute adsorption capacity of investigated Pd nanoparticles is close to that of the single crystal surfaces and indicates that no dramatic changes in the adsorption mechanism occur upon reduction of the particle sizes down to 4 nm. It is apparent that the general adsorption mechanism follows a typical precursor type behavior, which was previously observed on the Pd single crystal surfaces (compare *e.g.*refs .34 and 35).

The sticking coefficient data can be rationalized on a basis of the so-called "capture zone" effect.[37–41] This phenomenon is based on the existence of two adsorption channels: the adsorbate molecules can either directly impinge on the metal particles and chemisorb or they can be trapped in a weakly bound state on the support and reach the particles *via* surface diffusion. The capture zone is defined as an area from which impinging adsorbates can be collected by the particles. This effect was predicted and later experimentally confirmed to considerably enhance the adsorption rate on the particles.[5,7] In the present experiments, the capture zone effect was found to play an important role in the overall adsorption rate. The magnitude of this effect can be evaluated from the comparison of the measured initial sticking probability data on the supported particles and the estimate of the highest possible initial sticking probability on the particles of the same size in absence of the contribution from the support. For the Pd particles investigated in this study (average particle size of ~4 nm and the island density 3.8 × 10^{12} particles cm^{-2}), about 48% of the Fe_3O_4 surface area is covered by Pd particles. This means that only ~48% of the impinging molecules arrive directly on the metal surface and about 52% interact first with the support. If one assumes the highest possible sticking coefficient on metallic Pd to be 1.0 and fully neglects the contribution from the capture zone effect, an initial sticking coefficient close to 0.48 can be expected (corresponding to the scenario when 48% of the molecules completely stick to the metal particles). However, the experimentally measured initial sticking coefficient 0.68 is a factor of 1.4 higher than the estimated upper limit for the direct sticking on the metal particles. This observation can be rationalized only in the scope of the "capture zone" model, where the additional 20% of the impinging molecules become first weakly trapped on the support and reach the Pd particles, where they strongly chemisorb, within their residence time.

Thus, the capture zone effect leads to an enhancement of the adsorption rate by a factor of at least 1.4. If one assumes the experimentally measured value of the sticking coefficient 0.74 on Pd(111), the enhancement factor increases to ~1.9. Note that CO does not chemisorb on Fe_3O_4 at the experimental temperature of 300 K,[28] so that the enhanced adsorption cannot be explained simply by CO chemisorption on the support. Similar enhancement of the initial sticking probability due to the "capture zone effect" has been reported for a number of model surfaces consisting of Pd nanoparticles supported on various oxides.[26,40,41]

The differential adsorption heats obtained on Pd nanoparticles and Pd(111) are plotted as a function of CO surface coverage in Figure 2c and 2d. Each curve is an average of 4 to 6 independent measurements on freshly prepared model systems. On Pd(111), the heat of CO adsorption initially amounts to 149 ± 4 kJ mol^{-1}. After adsorption of about 0.76 × 10^{15} CO molecules cm^{-2} the surface of the particles reaches saturation (with the corresponding CO surface coverage of 0.5 ML with respect to the number of surface Pd atoms) and the adsorption heat decreases to the final non-zero value 75 ± 9 kJ mol^{-1}, which is related to the adsorption–desorption equilibrium established in quasi steady state regime.[21] Note that the final heat value is subject to large statistical errors since a very small absolute number of CO molecules is added in a single CO pulse, which is difficult to measure. Two reasons can account for the decreasing adsorption energy with growing CO coverage: intermolecular repulsion of neighboring CO molecules and increasing competition for the d-electrons of Pd nanoclusters participating in the CO–Pd bonding. For the Pd(111) single crystal surface, both the initial heat of adsorption as well as the saturation coverage at 300 K were found to be in a very good quantitative agreement with the data reported earlier in the literature.[42–45]

A qualitatively similar dependence of the adsorption heat on CO coverage was observed for Pd nanoparticles. On this surface, the initially high adsorption heat 134 ± 4 kJ mol^{-1}continuously decreases with growing CO coverage and reaches a lower steady state value ~90 kJ mol^{-1} after adsorption of ~0.4 × 10^{15} molecules cm^{-2}. Remarkably, the initial adsorption energy is 15 kJ mol^{-1} lower than the corresponding value on the Pd(111) surface despite the fact that Pd nanoparticles contain a large fraction of low-coordinated surface sites, such as edges and corners, which are often discussed to be strong binding sites. Previously, we comprehensively investigated the dependence of the CO initial adsorption energy on Pd as a function of the cluster size in the broader size range from 2 to 8 nm (or from 120 to 4900 Pd atoms per particle).[46] The initial adsorption heats were compared in the lowest coverage limit, where the adsorption heat is determined by the interaction of individual CO molecules with metal, and is not perturbed by the CO intermolecular repulsion or by the competition for the d-electrons. In the whole investigated size region the adsorption heat was observed to strongly decrease with decreasing particle size: from 126 ± 3 kJ mol^{-1} on 8 nm-sized Pd particles to 106 ± 1 kJ mol^{-1}on the smallest ~2 nm clusters. Additionally, all investigated particles showed smaller initial adsorption heat as compared to the single crystal surface (149 ± 3 kJ mol^{-1}).

It has to be pointed out that the surface of the Pd particles investigated in this work exposes mainly (111) terraces along with a smaller fraction of (100) facets and low-coordinated defect sites such as edges and corners. According to theoretical calculations, such irregular sites exhibit higher CO adsorption energies than Pd(111).[47] Experimentally, there is a general agreement that the CO adsorption energy on the (100) plane is about 10–15 kJ mol^{-1} higher than on the (111) plane (see *e.g.* ref. 42). However, the dramatic effect of low-coordination defect sites (like step edges) on the CO adsorption energy, such as observed for Pt(111),[48] is *not* seen on Pd(111).[42,44] Ramsier, Lee and Yates[44] even concluded from comparing CO TPD and adsorption energies for different stepped Pd facets that "structure of the Pd surface is only of minor importance in CO adsorption/desorption kinetics." These experimental results suggest that the degree of coordination of the Pd surface atoms on the stepped surfaces is not as important in determining the CO adsorption energy as on surfaces of other metals like Pt. Therefore, the increasing fraction of low-coordination Pd atoms with decreasing particle size may not be the dominant effect in determining how the CO adsorption energy varies with Pd particle size. Instead, other size-dependent properties of a Pd nanocluster have to be responsible for the obtained decrease of the initial adsorption enthalpy of carbon monoxide with decreasing the particle size.

First, the decrease of the chemisorption energy on Pd nanoparticles can result from the contraction of the lattice parameter of a small metal nanoparticle. Previously, it has been demonstrated experimentally[49,50] that the interatomic bond length in small metal particles decreases with decreasing particle size. It is now well established also by computational studies of various metals including Pd[51–53] that relaxed structures of metal clusters exhibit shorter interatomic distances than bulk crystals. This phenomenon has been rationalized as a result of decreasing average coordination number of the atoms with the higher surface-to-volume ratio of the cluster.[51–53] Particularly for Pd, it was shown that the average calculated Pd–Pd nearest-neighbor distance decreases linearly with decreasing linear dimension of a nanoparticle in the range of 55–260 atoms per cluster.[54] This lattice contraction in the small metal clusters was theoretically shown to result in a reduction of the adsorbate binding energy. Particularly for CO adsorption on the Pd clusters consisting of 55–260 atoms, it was demonstrated that the clusters with optimized structures show systematically lower adsorption energies by about 10–15 kJ mol^{-1} than the clusters with a bulk terminated geometry, *i.e.* with all interatomic distances d(Pd–Pd) = 275 pm.[55] This finding agrees with the principle of bond order conservation:[56,57] in the contracted clusters, one expects weaker adsorption bonds and stronger binding within the adsorbate as a result of better saturated valences of the substrate atoms. The decrease of the adsorbate binding energy with the lattice constant can be also rationalized in terms of the strain effect, which is associated with a downward shift of the valance d-band with decreasing lattice parameter.[55,58] Such effects were found not to be restricted to 3-fold hollow sites on (111) facets, but reflect a general trend that also holds for adsorption at other sites of the particles such as bridge sites and edges.[55]

A second reason for the decrease of adsorption heat of a gas-phase molecule on the small metal clusters is a feasible weakening of the dispersion force (van der Waals interaction) that is induced by dynamic response of bulk electrons of the metal to charge density fluctuations in an adsorbed molecule. Previously, it has been shown that the electron population at the Fermi edge, which is mainly relevant for this interaction, drastically changes with the cluster size in the range of a few nanometers.[59] The gradual decrease of the initial adsorption heat with the decreasing cluster size can be reasonably understood, since smaller clusters contain less electrons available for dynamic response. Similar behavior has been previously observed in TPDstudies for methane on Pd[60] and NO on Ag nanoclusters[61] and was also attributed to the reduced polarizability of the small metal nanoparticles.

Combined Energy and Sticking Coefficient Measurement: O_2 Dissociative Adsorption on Supported Pd Nanoparticles and Pd(111)

Dissociative oxygen adsorption was investigated on the same two Pd surfaces at 300 K, at which oxygen adsorbs dissociatively and does not stick to the oxide support. Figure 3 shows the obtained sticking coefficients (Figure 3a and 3b) and the adsorption energies (Figure 3c and 3d) plotted as a function of O surface coverage.

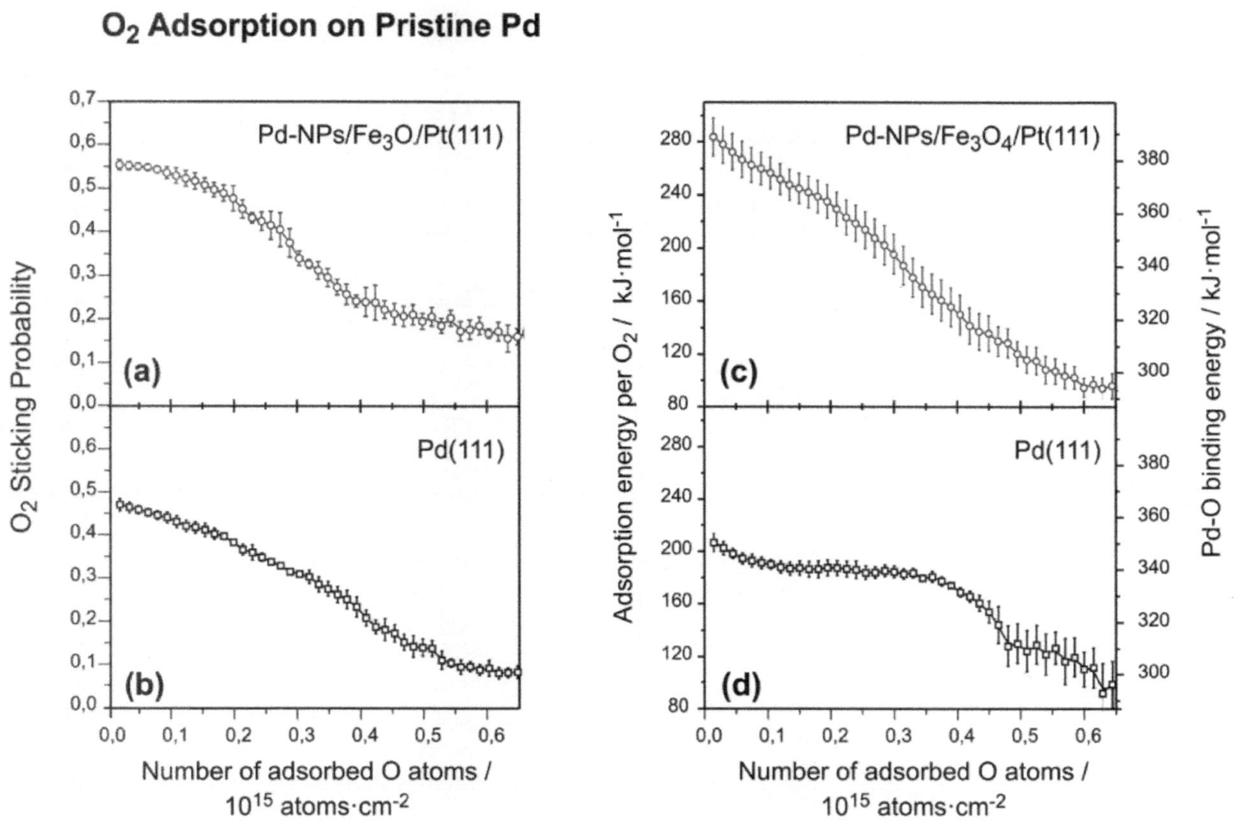

Figure 3: Sticking probability of oxygen plotted as a function of number of adsorbed O atoms for Pd nanoparticles (a) and for Pd(111) (b). Differential adsorption heat per mole of O_2 (left axis) and Pd–O binding energy (right axis) plotted as a function of the number of adsorbed O atoms. The sticking coefficients and adsorption heats were obtained at 300 K and are shown as an average of four to six independent measurements on freshly prepared surfaces. The error bars correspond to the error of the mean.

On Pd(111), the initial O_2 sticking probability reaches 0.47 ± 0.01 and steadily decreases until the O surface coverage approaches the value of $\sim 0.5 \times 10^{15}$ O atoms cm^{-2} (or ~0.3 O atoms per surface Pd atom, Figure 3b). The measured saturation coverage is in agreement with the literature data ranging from 0.25 to 0.4 observed for different O_2 exposures on this surface.[9,10,12,13,15] Previously, formation of O coverages larger than 0.25 was observed for high O_2 exposures and a number of reasons were suggested to be responsible for this phenomenon, such as *e.g.* slow formation of compressed O overlayers or O diffusion into the subsurface region.[12–15] Apparently, similar processes might take place also under our reaction conditions and they appear to be more pronounced on Pd nanoparticles. The detailed analysis of oxygen adsorption in the high coverage regime both on Pd(111) and Pd nanoparticles will be the subject of a forthcoming publication.[62] Here, we will mainly focus on oxygen coverages up to 0.25, at which these side processes do not play any significant role.

On Pd nanoparticles, a qualitatively similar dependence of the sticking coefficient on surface O coverage was observed, with the initial sticking coefficient amounting to 0.55 ± 0.01 and the saturation coverage lying between 0.4×10^{15} and 0.5×10^{15} O atoms cm^{-2} (Figure 3a). Note that the initial sticking coefficient of O_2 on the Pd clusters is even higher than on Pd(111) despite the fact that only ~48% of the surface is covered by Pd. To rationalize this high dissociative sticking coefficient, we estimated the capture zone effects on this surface. If one assumes the same value of the initial sticking coefficient as on Pd(111) (0.47), the fraction of the molecules that impinge directly on the Pd nanoparticles and remain adsorbed amounts to 0.23. The rest of the adsorbed molecules (relative fraction ~0.32) must first be trapped on the Fe_3O_4 support and then diffuse to the metal particles within their residence time and dissociate on it (note that O_2 neither dissociatively nor molecularly adsorbs on the oxide support under our reaction conditions). This scenario does not seem to be very likely though, since it implies that a molecule impinging on the surface has a higher probability to be trapped on the Fe_3O_4 support and then undergo the dissociative adsorption on Pd particles than directly stick and dissociate on metallic nanoparticles (0.62 *vs.* 0.47, correspondingly). More realistically, O_2 molecules have higher initial probability to dissociate on Pd nanoparticles than on Pd(111) as the first type of surface contains a large number of potentially highly reactive low-coordinated adsorption sites. To evaluate the feasibility of this hypothesis, we estimated the magnitude of the capture zone effect for the boundary e when the initial sticking coefficient of O_2 on Pd nanoparticles is equal to unity. In this e, the fraction of the molecules impinging directly on the particles and dissociating there should amount to 0.48 and only the remaining fraction of O_2 amounting to 0.07 has to arrive from the support and dissociate on the metal, which might be a realistic scenario. Based on the available data set, the exact quantification of the initial sticking coefficient on Pd nanoparticles is not possible, since the experimentally measured value is a convolution of the intrinsic initial sticking coefficient with the capture zone effects. The strategy to tackle this problem would be *e.g.* a systematic variation of the particles sizes and density on the support, thus allowing an estimation of the relative contribution of the capture zone effect. The corresponding experiments are currently under way. Based on the available data, we can only conclude that the intrinsic initial sticking coefficient on Pd nanoparticles lies in the range between 0.47 and 1.0.

Figure 3c and 3d show the dependence of the differential O_2 adsorption heat (left axis) and the calculated Pd–O bonding strength (right axis) plotted as a function of O atoms surface coverage for Pd nanoparticles and Pd(111).

Both model surfaces show a very strong coverage dependence of oxygen binding energies. For Pd(111), the initial adsorption energy on the clean surface amounts to ~205 ± 7 kJ mol^{-1}, which drops to about 100 kJ mol^{-1} close to saturation. As discussed above, two reasons might potentially account for the decreasing adsorption energy with increasing O coverage: interatomic repulsion of neighboring O atoms, and increasing competition for the d-electrons of Pd atoms participating in the O–Pd bonding. A qualitatively similar dependence of the adsorption energy on the O coverage was observed for Pd nanoparticles. However, there is a very pronounced difference in the initial adsorption energy between the Pd clusters and Pd(111). While O binds with ~205 kJ mol^{-1} on the pristine Pd(111) surface, the initial binding energy rises to ~280 ± 13 kJ mol^{-1} on the pristine Pd nanoparticles (both values are differential adsorption heats for O_2 molecule). Note that on average only about 7 O atoms per Pd nanoparticle (comprising ~320 surface Pd atoms) contribute to the measured initial adsorption energy, which allows us to consider this energy value as a limiting e for the interaction of an individual O atom with a pristine Pd nanoparticle. Also note that neither subsurface O diffusion nor surface oxide formation can occur at the experimental low temperature conditions (300 K) and low oxygen coverages, which could otherwise affect the measured values.

The energy value obtained by SCAC for Pd(111) in the lowest coverage limit (205 kJ mol^{-1} at the surface coverage 0.01×10^{15} O atoms cm^{-2}) is in good agreement with literature values of O adsorption energies at low coverages situated in the range of 210 to 230 kJ mol^{-1}.[10,13,15] In contrast, the very high initial energy value obtained for Pd nanoparticles (275 kJ mol^{-1} at 0.01×10^{15} O atoms cm^{-2}) cannot be compared to any value reported in TPD studies either for low Miller index Pd surfaces, or for stepped Pd crystals, which typically span the range between 190 and 230 kJ mol^{-1}.[9,13,15,63] Thus, this high initial adsorption energy must be related to some structural feature inherent to Pd nanoparticles, which cannot be reproduced even by the steps of the high Miller index Pd surfaces. Such adsorption sites on Pd nanoparticles might be the low coordinated surface sites such as particle edges and corners. In previous STMstudies on supported Pd nanoparticles, oxygen was shown to modify such sites first, indicating thus that they are the strongest adsorption sites on a Pd nanoparticle.[64] To check the feasibility of this explanation, we carried out CO titration experiments by infrared reflection absorption spectroscopy (IRAS) to identify the preferential adsorption sites of oxygen on Pd nanoparticles. In previous IRAS studies, the CO molecule was shown to be a suitable probe molecule for probing different adsorption sites on Pd nanoparticles, allowing independent identification of the regular adsorption sites on the (111) terraces and the low-coordinated adsorption sites such as edges and corners.[65] With this spectroscopic approach, it was experimentally confirmed that oxygen adsorption first modifies the low coordinated surface sites such as edges and corners. For more details, the reader is referred to ref. 66. From these observations we can unambiguously attribute the unexpectedly high initial adsorption energy (280 kJ mol^{-1}), measured in a calorimetric experiment on supported Pd nanoparticles, to oxygen adsorption at edge and corner sites. It remains, however, an open question as to why the edge sites of Pd nanoclusters interact so strongly with adsorbed oxygen. Further theoretical work is needed in order to rationalize the microscopic nature of this phenomenon, which can presumably originate from the reconstruction or refaceting of the particles edges that was previously reported in the literature.[64,67] It is also feasible that the edge sites exhibit a higher activity in dissociating of O_2 molecules, which can account for the experimentally observed high values of the initial sticking coefficient on Pd nanoparticles.

CO Adsorption on supported Pd Nanoparticles and Pd(111) Precovered by O

Under the reaction conditions, both CO and oxygen interact with the surface, which is partially or nearly fully covered by co-reactants. The presence of co-adsorbates can significantly change the binding energies and consequently strongly affect the surface coverages under the operation conditions. For the reactions involving a dissociative step, such as CO oxidation, even small changes in the surface coverage might result in switching to a "poisoning" regime, where high concentration of one reactant (CO) can nearly completely suppress dissociation of the other one (O_2).[5,7] Therefore it is a very important issue and a current scientific challenge to accurately determine the binding energies in the systems of two co-adsorbates and understand how this binding strength depends on the structure of the underlying metal substrate. To approach this goal, we measured the binding energies of CO on both Pd surfaces pre-saturated with oxygen. Oxygen deposition was carried out at 300 K to enable dissociation of molecular species, and immediately afterwards CO adsorption energies were measured at 100 K, at which CO does not react with O to produce CO_2.

The results of these experiments are displayed in Figure 4a and 4b for Pd nanoparticles and Pd(111), respectively. They show that both surfaces precovered with oxygen exhibit similaradsorption capacity for CO to the clean Pd surfaces. The measured saturation values are ~0.75×10^{15} molecules cm^{-2} on Pd(111) and ~0.50×10^{15} molecules cm^{-2} on O/Pd nanoparticles(compare to the values 0.76×10^{15} molecules cm^{-2} and 0.4×10^{15} molecules cm^{-2} on clean Pd surfaces, correspondingly). However, it should be noted that CO adsorption on the clean surfaces was investigated at 300 K, while the experiments on the O-precovered surfaces were carried out at 100 K to avoid surface reaction between adsorbed CO and O. According to previous experimental and theoretical studies on Pd(111), CO occupies two hollow sites (hcp and fcc) at 300 K forming a $c(2 \times 4)_{2CO}$ overlayer with the saturation coverage of 0.5 CO molecules per surface Pd atom.[68,69] At 100 K, the CO saturation coverage increases on the clean Pd(111) surface to 0.75 due to occupation of an additional on-top adsorption site per unit cell.[70,71] On Pd(111) precovered with oxygen, CO forms a new mixed $p(2 \times 2)_{CO+O}$ phase in the temperature range 60–110 K.[72] STM results in combination with DFT calculations suggest that this overlayer structure is consistent with the occupation of a (fcc) hollow site by one oxygen atom per unit cell and by either one or two CO molecules per unit cell.[72] In the latter e, for which the unit cell contains one oxygen atom and two CO molecules, the CO saturation coverage should amount to 0.5 per surface Pd atom, which is in good agreements with the experimentally measured value ~0.51.

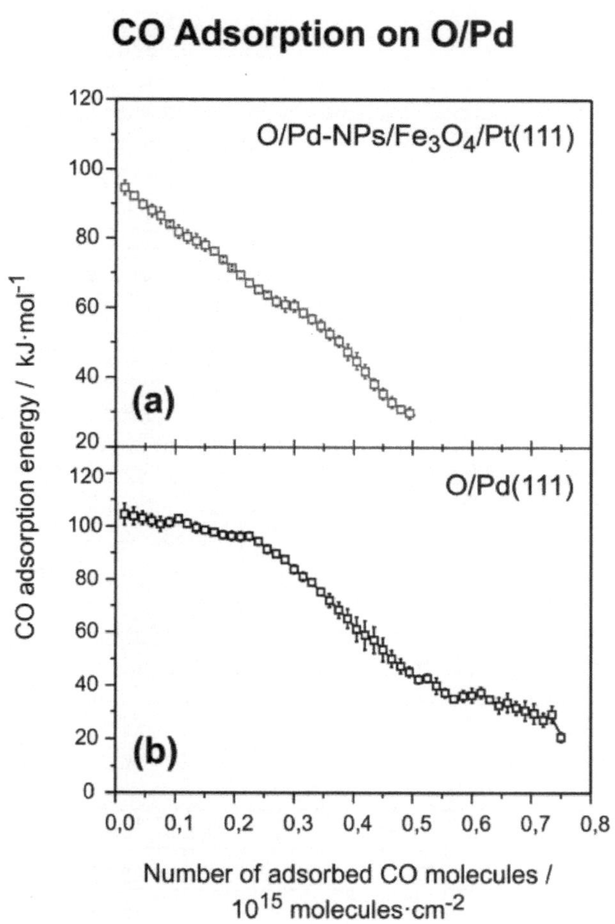

\

Figure 4 Differential heats of CO adsorption plotted as a function of number of adsorbed CO molecules for Pd nanoparticles (a) and Pd(111) (b) pre-covered with a monolayer of O atoms. Oxygen was deposited at 300 K, CO adsorption experiments were carried out at 100 K.

On O/Pd(111), the initial heat of CO adsorption amounts to 104 ± 4 kJ mol^{-1}, which is ~45 kJ mol^{-1} lower than the CO adsorption energy on the pristine Pd(111) surface. The initial CO adsorption heat on Pd nanoparticles precovered with oxygen reaches 95 ± 2 kJ mol^{-1}. This value is ~39 kJ mol^{-1} lower than the initial heat of CO adsorption on the clean Pd nanoparticles. Interestingly, both O-presaturated surfaces show a similar decrease of the CO adsorption energy by ~40–45 kJ mol^{-1}. Moreover, the same energy difference (~40 kJ mol^{-1}) in CO adsorption heat was observed over the whole range of CO surface coverages on both surfaces (data not show).

Two major observations can be made from the comparison of clean *vs.* O-precovered Pd surfaces with respect to the CO adsorption energies. First, the presence of a monolayer of co-adsorbed oxygen atoms lowers the CO binding energy by about 40–45 kJ mol^{-1} both on the single crystal surface and on Pd nanoparticles. This value can be considered as a measure for the repulsive interaction between oxygen and CO on Pd. Second, the difference between CO binding energies on Pd(111) and Pd nanoparticles of about 10–15 kJ mol^{-1} remains preserved even on the O-covered surfaces. Previously, the decreasing CO adsorption energy with decreasing particle size on clean Pd particles was attributed primarily to the contraction of the Pd lattice in small clusters. The presented data demonstrate that the same trend is valid also for Pd surfaces covered with oxygen with approximately the same magnitude.

CONCLUSIONS

In conclusion, we investigated the interaction of CO and O_2 molecules with well-defined Pd nanoclusters and with the extended Pd(111) surface by a direct calorimetric method. The main focus of this study lies in a detailed understanding of how a nanometer scale confinement of matter affects the binding strength of gaseous adsorbates relevant to CO oxidation.

CO and oxygen adsorption were investigated both on clean and O-precovered Pd nanoparticles and Pd(111) and corresponding sticking coefficients and adsorption energies were determined as a function of adsorbate coverages.

The CO adsorption kinetics on Pd nanoparticles investigated by the King–Well method show a strong enhancement of the adsorbate flux onto Pd clusters due to a "capture zone" effect, which involves weak trapping of the adsorbates on the support and diffusion to the metal particles. The CO adsorption rate on nanoparticles was estimated to be enhanced by factor of 1.4–1.9 depending on the assumptions on the initial sticking coefficient on metallic Pd. For oxygen adsorption, the "capture zone effect" was found to also play a role; however, its exact quantification was difficult because of possibly higher activity of Pd nanoparticles towards dissociative O_2 adsorption as compared to Pd(111).

The initial binding energy of carbon monoxide was found to be lower on Pd nanoparticles than on Pd(111) (134 *vs.* 149 kJ mol^{-1}, correspondingly) despite the high fraction of low-coordinated surface sites present on nanoclusters. This observation can be rationalized as a consequence of Pd lattice contraction in small particles leading to weakening of chemisorptive interaction. Additionally, a reduced polarizability of small Pd nanoparticles can result in a reduction of van der Waals attraction. On the O-precovered Pd surfaces, the CO binding energy was found to be reduced by about 40–45 kJ mol^{-1} as compared to the corresponding pristine Pd surfaces in the full range of CO coverages. Also in the presence of a monolayer of oxygen, CO binds weaker on Pd nanoclusters than on Pd(111) by about 10–15 kJ mol^{-1}. This result suggests that the strain effect observed for pristine Pd nanoclusters remains preserved also on the particles pre-covered with adsorbates, particularly oxygen.

The measured initial oxygen adsorption heat on Pd(111) amounts to 205 ± 7 kJ mol^{-1}. On Pd nanoparticles, this values increases by ~75 kJ mol^{-1}; this energy can be attributed to strong oxygen binding at the low coordinated adsorption sites of Pd nanoparticles such as edges as corners, which was also confirmed by spectroscopic data. The unexpectedly strong binding energy of oxygen at the particles edges exceeds all literature values of oxygen adsorption energies at stepped single crystal surfaces. Apparently, the local adsorption environment of Pd–O bond plays a very important role in determining the binding strength in this adsorption system. This result is in a sharp contrast to the CO–Pd e, where the presence of the low-coordinated surface sites on a Pd nanoparticle was observed to not result in stronger CO–Pd binding energy. Instead, weakening of CO interaction energy with Pd nanoparticles was detected, which can be most likely attributed to the contraction of the Pd lattice. Both Pd surfaces exhibit very strong coverage dependence of O binding energy covering the range up to 180 kJ mol^{-1}.

In summary, the reduced dimensions of the Pd substrate can affect the binding strength of the adsorbates involved in CO oxidation in two principle ways: (i) *via* the change of the local adsorption environment that can result *e.g.* in stronger adsorbate bonding *e.g.* at the particle edges and (ii) *via* contraction of the Pd lattice in small clusters and a concomitant weakening of chemisorptive interaction. The latter effect is expected to play a role not only for CO as shown in this study but also for other adsorbates (provided that the local adsorption environment remains preserved). The presented experimental data (summarized in Table 1) provide important benchmarks for upcoming theoretical calculation.

Table 1: Summary of the adsorption energies of the CO and oxygen adsorbates on the clean O-precovered Pd surfaces obtained in this study

	Adsorption enthalpy, kJ mol^{-1} (low coverage → quasi saturation[a])	
	Pd(111)	**Pd nanoparticles (~4 nm)/Fe_3O_4/Pt(111)**
CO on pristine Pd, 300 K	149 ± 4 → 75	134 ± 4 → 90
O_2 on pristine Pd (per O_2 molecule), 300 K	205 ± 7 → 100	280 ± 13 → 100
CO on O (1 ML)/Pd, 100 K	104 ± 4 → 30	94 ± 2 → 30

[a] Note that the absolute accuracy of the enthalpy values in the quasi saturation regime is lower than for the lowest adsorbate coverage because of the experimental difficulties in determining very low sticking coefficients.

ACKNOWLEDGEMENTS

Support by the Fonds der Chemischen Industrie is gratefully acknowledged. The authors thank Hans-Joachim Freund, Charles T. Campbell (UW) and Beatriz Roldan (UCF) for valuable discussions.

REFERENCES

1. G. Ertl, H. Knoezinger and J. Weitkamp, in *Handbook of Heterogeneous Catalysis*, ed. G. Ertl, H. Knoezinger and J. Weitkamp, VCH, Weinheim, 1997 .
2. J. M. Thomas and W. J. Thomas, *Principle and Practice of Heterogeneous Catalysis*, VCH, Weinheim, 1997 .
3. T. Engel and G. Ertl, *Adv. Catal.*, 1979, **28**, 41 .
4. S. M. McClure and D. W. Goodman, *Chem. Phys. Lett.*, 2009, **469**, 1 .
5. J. Libuda and H.-J. Freund, *Surf. Sci. Rep.*, 2005, **57**, 157 .
6. J. Wintterlin, S. Völkening, T. V. W. Janssens, T. Zambelli and G. Ertl, *Science*, 1997, **278**, 1931 .

7. C. R. Henry, *Surf. Sci. Rep.*, 1998, **31**, 231 .

8. I. Stará and V. Matolín, *Surf. Sci.*, 1994, **313**, 99 .

9. G. Zheng and E. I. Altman, *Surf. Sci.*, 2002, **504**, 253 .

10. H. Conrad, G. Ertl, J. Küppers and E. E. Latta, *Surf. Sci.*, 1977, **65**, 245 .

11. R. Westerström, J. Gustafson, A. Resta, A. Mikkelsen, J. N. Andersen, E. Lundgren, N. Seriani, F. Mittendorfer, M. Schmidt, J. Klikovits, P. Varga, M. D. Ackermann, J. W. M. Frenken, N. Kasper and A. Stierle, *Phys. Rev. B*, 2007, **76**, 155410 .

12. F. P. Leisenberger, G. Koller, M. Sock, S. Surnev, M. G. Ramsey, F. P. Netzer, B. Klötzer and K. Hayek, *Surf. Sci.*, 2000, **445**, 380 .

13. G. Zheng and E. Altman, *Surf. Sci.*, 2000, **462**, 151 .

14. A. Vlad, A. Stierle, R. Westerström, S. Blomberg, A. Mikkelsen and E. Lundgren, *Phys. Rev. B*, 2012, **86**, 35407 .

15. B. Klötzer, K. Hayek, C. Konvicka, E. Lundgren and P. Varga, *Surf. Sci.*, 2001, **482–485**, 237 .

16. P. Nolte, A. Stierle, N. Kasper, N. Y. Jin-Phillipp, N. Jeutter and H. Dosch, *Nano Lett.*, 2011, **11**, 4697 .

17. R. Westerström, M. E. Messing, S. Blomberg, A. Hellman, H. Grönbeck, J. Gustafson, N. M. Martin, O. Balmes, R. van Rijn, J. N. Andersen, K. Deppert, H. Bluhm, Z. Liu, M. E. Grass, M. Hävecker and E. Lundgren, *Phys. Rev. B*, 2011, **83**, 115440 .

18. J. Han, G. Zhu, D. Y. Zemlyanov and F. H. Ribeiro, *J. Catal.*, 2004, **225**, 7 .

19. E. Lundgren, G. Kresse, C. Klein, M. Borg, J. N. Andersen, M. de Santis, Y. Gauthier, C. Konvicka, M. Schmidt and P. Varga, *Phys. Rev. Lett.*, 2002, **88**, 246103 .

20. T. Schalow, M. Laurin, B. Brandt, S. Schauermann, S. Guimond, H. Kuhlenbeck, D. E. Starr, S. K. Shaikhutdinov, J. Libuda and H.-J. Freund, *Angew. Chem., Int. Ed.*, 2005, **44**, 7601 .

21. W. A. Brown, R. Kose and D. A. King, *Chem. Rev.*, 1998, **98**, 797 .

22. J. Stuckless, N. Frei and C. T. Campbell, *Rev. Sci. Instrum.*, 1998, **69**, 2427 .

23. C. T. Campbell and O. Lytken, *Surf. Sci.*, 2009, **603**, 1365 .

24. J.-H. Fischer, J. Hartmann, J. A. Farmer, J. M. Flores-Camacho, C. T. Campbell, S. Schauermann and H.-J. Freund, *Rev. Sci. Instrum.*, 2011, **82**, 24102 .

25. D. King and M. Wells, *Surf. Sci.*, 1972, **29**, 454 .

26. T. Schalow, B. Brandt, D. E. Starr, M. Laurin, S. K. Shaikhutdinov, S. Schauermann, J. Libuda and H.-J. Freund, *Phys. Chem. Chem. Phys.*, 2007, **9**, 1347 **RSC**.

27. T. Schalow, B. Brandt, D. E. Starr, M. Laurin, S. Schauermann, S. K. Shaikhutdinov, J. Libuda and H.-J. Freund, *Catal. Lett.*, 2006, **107**, 189 .

28. C. Lemire, R. Meyer, V. Henrich, S. K. Shaikhutdinov and H.-J. Freund, *Surf. Sci.*, 2004, **572**, 103 .

29. T. Mitsui, M. K. Rose, E. Fomin, D. F. Ogletree and M. Salmeron, *Phys. Rev. Lett.*, 2005, **94**, 036101 .

30. A. M. Bradshaw and F. M. Hoffmann, *Surf. Sci.*, 1978, **72**, 513 .

31. J. Szanyj, W. K. Kuhn and D. W. Goodman, *J. Vac. Sci. Technol., A*, 1993, **11**, 1969 .

32. S. Surnev, M. Sock, M. G. Ramsey, F. P. Netzer, M. Wiklund, M. Borg and J. N. Andersen, *Surf. Sci.*, 2000, **470**, 171 .

33. T. Gießel, O. Schaff, C. J. Hirschmugl, V. Fernandez, K.-M. Schindler, A. Theobald, S. Bao, R. Lindsay, W. Berndt, A. M. Bradshaw, C. Baddeley, A. F. Lee, R. M. Lambert and D. P. Woodruff, *Surf. Sci.*, 1998, **406**, 90 .

34. H. Conrad, G. Ertl, J. Kocha and E. E. Latta, *Surf. Sci.*, 1974, **43**, 462 .

35. T. Engel, *J. Chem. Phys.*, 1978, **69**, 373 .

36. A. Ortega, F. M. Hoffmann and A. M. Bradshaw, *Surf. Sci.*, 1982, **119**, 79 .

37. K. Tsu and M. Boudart, *2nd Actes du Congr. Int. Cat. (Technip, Paris)*, 1961, **1**, 593 .

38. Y. Kim and M. Boudart, *Langmuir*, 1991, **7**, 2999 .

39. E. Gillet, S. Channakhone, V. Matolin and M. Gillet, *Surf. Sci.*, 1985, **152–153**, 603 .

40. V. Matolin and E. Gillet, *Surf. Sci.*, 1986, **166**, L115 .

41. M. Bowker, P. Stone, R. Bennett and N. Perkins, *Surf. Sci.*, 2002, **497**, 155 .

42. H. Conrad, G. Ertl, J. Koch and E. E. Latta, *Surf. Sci.*, 1974, **43**, 462 .

43. G. Ertl and J. Koch, *Z. Naturforsch.*, 1970, **25a**, 1906 .

44. R. D. Ramsier, K.-W. Lee and J. T. Yates, Jr., *Surf. Sci.*, 1995, **322**, 243 .

45. T. Engel, *J. Chem. Phys.*, 1978, **69**, 373 .

46. J.-H. Fischer-Wolfarth, J. A. Farmer, J. M. Flores-Camacho, A. Genest, I. V. Yudanov, N. Rösch, C. T. Campbell, S. Schauermann and H.-J. Freund, *Phys. Rev. B*, 2010, **81**, 241416 .

47. I. V. Yudanov, R. Sahnoun, K. M. Neyman, N. Rösch, J. Hoffmann, S. Schauermann, V. Johánek, H. Unterhalt, G. Rupprechter, J. Libuda and H.-J. Freund, *J. Phys. Chem. B*, 2003,**107**, 255 .

48. C. T. Campbell, G. Ertl, H. Kuipers and J. Segner, *Surf. Sci.*, 1981, **107**, 207 .

49. S. A. Nepijko, M. Klimenkov, M. Adelt, H. Kuhlenbeck, R. Schlögl and H.-J. Freund, *Langmuir*, 1999, **15**, 5309 .

50. R. Lamber, S. Wetjen and N. I. Jaeger, *Phys. Rev. B*, 1995, **51**, 10968 .

51. S. Krüger, S. Vent, F. Nötermann, M. Staufer and N. Rösch, *J. Chem. Phys.*, 2001, **115**, 2082 .

52. P. Nava, M. Sierka and R. Ahlrichs, *Phys. Chem. Chem. Phys.*, 2003, **5**, 3372 **RSC**.

53. O. Häberlen, S.-C. Chung, S. Krüger and N. Rösch, *J. Chem. Phys.*, 1997, **106**, 5189 .

54. I. V. Yudanov, R. Sahnoun, K. M. Neyman and N. Rösch, *J. Chem. Phys.*, 2002, **117**, 9887 .

55. I. V. Yudanov, M. Metzner, A. Genest and N. Rösch, *J. Phys. Chem. C*, 2008, **112**, 20269 .

56. E. Shustorovich and H. Sellers, *Surf. Sci. Rep.*, 1998, **31**, 1 .

57. R. A. van Santen and G. J. Kramer, *Chem. Rev.*, 1995, **95**, 637 .

58. B. Hammer, Y. Morikawa and J. K. Nørskov, *Phys. Rev. Lett.*, 1996, **76**, 2141 .

59. S. Sandel, J. Libuda, P. Brühwiler, S. Andersson, A. J. Maxwell, M. Bäumer, N. Mårtensson and H.-J. Freund, *J. Vac. Sci. Technol., A*, 1996, **14**, 1546 .

60. K. Watanabe, Y. Matsumoto, M. Kampling, K. Al-Shamery and H.-J. Freund, *Angew. Chem., Int. Ed.*, 1999, **38**, 2192 .

61. D. Mulugeta, K. H. Kim, K. Watanabe, D. Menzel and H.-J. Freund, *Phys. Rev. Lett.*, 2008, **101**, 146103 .

62. P. W. Davies and R. M. Lambert, *Surf. Sci.*, 1981, **110**, 227 .

63. K. Hujrup Hansen, Ž. Šljivančanin, E. Lęgsgaard, F. Besenbacher and I. Steinsgaard, *Surf. Sci.*, 2002, **505**, 25 .

64. K. Wolter, O. Seiferth, H. Kuhlenbeck, M. Bäumer and H.-J. Freund, *Surf. Sci.*, 1998, **399**, 190 .

65. M. Peter, J. M. Flores Camacho, S. Adamovsky, L. K. Ono, K.-H. Dostert, C. P. O'Brien, B. Roldan Cuenya, S. Schauermann and H.-J. Freund, *Angew. Chem., Int. Ed.*, 2013 DOI:10.1002/anie.201209476.

66. P. Nolte, A. Stierle, N. Kasper, N. Y. Jin-Phillipp, N. Jeutter and H. Dosch, *Nano Lett.*, 2011, **11**, 4697 .

67. T. Gießel, O. Schaff, C. J. Hirschmugl, V. Fernandez, K.-M. Schindler, A. Theobald, S. Bao, R. Lindsay, W. Berndt, A. M. Bradshaw, C. Baddeley, A. F. Lee, R. M. Lambert and D. P. Woodruff, *Surf. Sci.*, 1998, **406**, 90 .

68. D. Loffreda, D. Simon and P. Sautet, *Surf. Sci.*, 1999, **425**, 68 .

69. L. Surnev, M. Sock, M. G. Ramsey, F. P. Netzer, M. Wiklund, A. Borg and J. N. Andersen, *Surf. Sci.*, 2000, **470**, 171 .

70. M. K. Rose, T. Mitsui, J. C. Dunphy, A. Borg, D. F. Ogletree, M. Salmeron and P. Sautet, *Surf. Sci.*, 2002, **512**, 48 .

71. J. Méndez, S. H. Kim, J. Cerdá, J. Wintterlin and G. Ertl, *Phys. Rev. B*, 2005, **71**, 085409 .

CHAPTER 10

CONTROL OF CHEMICAL REACTIVITY BY TRANSITION-STATE AND BEYOND

Hua Guo and Kopin Liu

INTRODUCTION

The textbook description of a chemical reaction typically invokes the concept of a transition state, through which the transformation of reactants to products is transpired. The transition state was first proposed by Eyring, Wigner, Evans, and Polanyi as an activated complex in a reaction, which, once reached, dissociates irreversibly to the products.[1–3] As a result, it serves as a bottleneck for the reaction flux, thus controlling reaction kinetics. This conceptual construct is not only very useful for understanding chemical reactivity, but also forms the basis of the powerful transition-state theory (TST).[1–3] TST is based on the premise that the rate coefficient of a bimolecular reaction is proportional to the reaction flux passing through a dividing surface that separates the reactants from products.[4] A convenient location of the dividing surface is at the top of the free-energy barrier, in which the recrossing is minimal. It should however be noted that the classical TST is not quantitatively accurate. Quantum mechanical effects, such as energy quantization and tunneling can be taken into account using semi-classical approximations.[5] Nowadays, the validity and applicability of TST are firmly established.

Unfortunately, the precise location of the transition state cannot be readily defined. Even for the same reaction, the position of the free-energy barrier might change with temperature. Indeed, the most sophisticated version of TST is variational, which minimizes the rate coefficients to account for recrossing.[5] Experimental detection of the transition state is also difficult because of the ultrashort lifetime of the activated complex.[6] To better understand the transition state and its influence on reaction kinetics and dynamics, it is thus best approaching the problem from a theoretical perspective, from which the reaction is simply the result of nuclear motion on a potential energy surface (PES).[4,7] The PES is the sum of the nuclear repulsion and electronic energy within the Born–Oppenheimer or adiabatic approximation, which separates the nuclear motion from the electronic one based on the large mass disparity between the two. For most reactions occurring at low temperature, only the ground electronic state is involved. Once the PES is known, the reaction dynamics (and equivalently ro-vibrational spectroscopy) can be characterized, at least in principle, by solving the nuclear Schrödinger equation. In this context, spectroscopy and scattering can be regarded as experimental ways to probe the PES.

For a typical activated bimolecular reaction, there is always a saddle point on the PES that is the highest point along the reaction coordinate. This stationary point defines the topographic features nearby in the transition-state region for the reaction. Specifically, a first-order saddle point features one coordinate associated with a negative second derivative while all other coordinates possess positive second derivatives. The vector for the normal mode with an imaginary frequency represents the reaction coordinate that traverses the barrier.[4] In TST, the dividing surface is often defined at the saddle point perpendicular to the reaction coordinate.[5]

It is now well established that the transition-state controls the overall reactivity, which undeores the success of TST in predicting the rate coefficients. There is also an increasing body of evidence that suggests that the transition state also plays an important role in regulating reaction dynamics.[8–12] A most common manifestation of the transition-state control of reaction dynamics is mode specificity, namely the differing ability of various reactant modes in promoting the reaction.[13,14] As described below, technological advances have enabled one to prepare the reactant in a single quantum state, and the measurement of its reactivity shed valuable light on how effective it helps to overcome the reaction barrier.[15] An equally informative approach is to probe the product energy disposal with quantum resolution, which can now be achieved with exquisite detail with various laser and product imaging techniques,[16–19] also discussed below. In the same time, new developments in quantum reactive scattering methods and PES construction have also advanced to a level that detailed

Citation: Hua Guo and Kopin Liu, "Control of chemical reactivity by transition-state and beyond," DOI: 10.1039/c6sc01066k.

comparison with state-to-state experimental data becomes possible.[11,20–24] The accumulation of detailed quantum-state resolved data allows a much more detailed understanding of the transition-state control of the reaction dynamics.

Of course, reaction dynamics can also be affected by other factors. One important quantum mechanical factor is resonances, which are metastable states formed on the PES, many near the transition state. These states can impact reaction dynamics in a significant way, as demonstrated by ample experimental[16,25,26] and theoretical evidence.[11,27] However, discussion of the impact of resonances is beyond the scope of this review.

In this perspective, we discuss recent advances in our understanding of transition-state control of reaction dynamics, from both experimental and theoretical perspectives. While such control is indeed dominant, an important point that is becoming widely recognized is that this ideal textbook description of reactions requires some refinements, because of important and prevalent features on the PES. For example, shallow wells in the entrance channel could exert sufficient forces on the approaching reactant molecules, leading to modifications of the transition-state control of the reactivity and dynamics. By the same token, potential wells in the exit channel may also exercise influence on the product energy disposal. On the other hand, the anisotropy of the PES suggests that the reactivity depends on the approach of the reactants relative to each other, as some preferred angles of approach may avoid unproductive collisions. A better understanding of these factors will not only deepen our understanding of reaction dynamics, but also potentially allow control of reaction yield and product branching.

TRANSITION STATE AND ITS CONTROL OF REACTION DYNAMICS

As alluded to above, an important manifestation of transition-state control of reaction dynamics is mode specificity.[12,13] The most celebrated model to rationalize mode specificity is the Polanyi rules, which were distilled from studies of atom–diatom reactions with various reaction energies and barrier locations.[8] For reactions with a reactant-like, or early, barrier, relative translation energy between the two collision partners is the most effective in overcoming the barrier. On the other hand, for reactions with a product-like, or late, barrier, vibrational excitation has a higher efficacy in enhancing the reactivity than the translational energy. Polanyi's model, which emphasizes the importance of the location of the transition state in enhancing reactivity for activated reactions, has been widely used for predicting the relative efficacy of vibrational and translational excitations in promoting the reaction.

While intuitive and insightful, the Polanyi rules only offer qualitative predictions because the earliness/lateness of the barrier location is difficult to quantify. More importantly, they do not, strictly speaking, apply to reactions involving polyatomic reactants. As a result, these empirical rules provide little guidance on the relative efficacies of the internal, namely vibrational and rotational, modes of the reactants in promoting the reaction. Recently, the Polanyi rules have been extended to polyatomic reactions by examining the coupling of reactant modes with the reaction coordinate at the transition state.[28,29] In this Sudden Vector Projection (SVP) model, the time scale of the collision is assumed to be much shorter than that required for intramolecular vibrational energy redistribution (IVR) in the reactants, a condition that is generally satisfied for the reactions discussed in this perspective. As a result, the reaction can be considered in the sudden limit, in which the energy deposited into a particular reactant mode is preserved until the transition state is reached. In this sudden limit, the coupling between a reactant mode and the reaction coordinate at the transition state is approximated by the projection of the corresponding normal mode vectors. If the projection is large, the reactant mode is strongly coupled with the reaction coordinate. In this e, energy deposited into this mode flows readily into the reaction coordinate, thus enhancing the reactivity. On the other hand, if the projection is small, the coupling with the reaction coordinate is weak, and the energy flow is limited. This leads to a low ability to enhance the reactivity.

The SVP model is thus formally quantitative and treats all reactant modes on an equal footing. Like the Polanyi rules, the SVP model emphasizes the importance of the transition state, represented by the saddle point of the underlying PES, in mode specificity. However, the latter uses the projection of reactant mode on the reaction coordinate at the transition state, rather than the location of the barrier, as the descriptor of the coupling strength. It has been shown that the coupling strength is closely correlated with the position of the barrier and the predictions of the SVP model are completely consistent with the Polanyi rules for atom–diatomic reactions.[28]

By invoking microscopic reversibility, the mode specificity of the reverse reaction dictates how energy is disposed in the product modes of the forward reaction. Thus, the SVP model can also be used to predict product energy disposal.[28,29] In the same spirit, a product mode that has a large projection on to the reaction coordinate at the transition state is expected to be excited, and vice versa.

It should be noted that the idea of attributing mode specificity to coupling with the reaction coordinate at the transition sate has been percolating in the reaction dynamics community for some time. For example, Schatz and Ross have proposed a Franck–Condon model to approximate reactive scattering using inelastic scattering wavefunctions.[30] Franck–Condon models have also been widely used to predict product state distributions in reactions.[31–33] More recently, Skodje and coworkers have proposed a Franck–Condon model to approximately compute the reactive S-matrix elements,[34] and Manthe and coworkers devised an exact scheme for state-to-state reactive scattering based on transition state wave packets.[35] The SVP model can be considered as a simplified sudden model, in which the Franck–Condon factors are approximated by projections of normal mode vectors onto the reaction coordinate vector at the transition state.[28] Its strength is its simplicity, as it requires neither PESs nor dynamical calculations. Only properties of the transition state are needed. Similar ideas of projecting reactant vectors to reaction coordinate at the transition state have been put forth by Wang and Bowman[36] and by Siebrand et al.[37] for treating mode-specific tunneling.

The SVP model has been applied to several prototypical reactions and in most es its predictions have been borne out.[9,38] The general success of this transition-state based model attests the importance of the transition state in controlling reaction dynamics. In particular, the SVP model correctly predicts the mode specificity in some reactions for which a naïve extension of Polanyi rules failed, some of which are discussed in more details below. However, it is expected that the SVP model will fail if the reaction is not sudden, in which the IVR may be prevalent. This limit is more appropriately described with an adiabatic model, such as the one based on reaction path Hamiltonian.[39]

EXPERIMENTAL AND THEORETICAL ADVANCES

With the advent of tunable infrared (IR) lasers, two molecular beam approaches have been developed to investigate the impact of the vibrational excitation of reactants on chemical reactions under single-collision conditions. The PHOTOLOC method pioneered by Zare's group[40] employs a co-expansion of a gas mixture. A pulsed nanosecond laser initiates the reaction by photolyzing one of the gaseous molecules (the precursor) to generate the reactive radicals (typically the Cl or H atoms). Another IR laser is employed to prepare the vibrational excited reactants. A third probe laser is fired tens of nanosecond later to interrogate the internal states of reaction products via the resonance-enhanced multiphoton ionization (REMPI) scheme. A velocity-sensitive time-of-flight (TOF) spectrometer then registers the temporal profile of the desired ions, which can be converted to the speed distribution of the probed product state, assuming a negligible change of the recoil velocity of the REMPI-ions from the neutral products. The measured speed distribution represents a one-dimensional (1D) projection of the desired three-dimensional (3D) state-tagged angular distribution, which can be recovered by the law of cosines.[40]

The crossed molecular beam approach, in particular that equipped with a time-sliced velocity map imaging detector,[41] offers a powerful alternative. The velocity-map imaging technique takes advantage of multiplex detection,[42,43] and the time-sliced version[41,44,45] not only yields higher resolution to reveal the correlated state distributions of the coproducts from the raw image,[18,46,47] but also alleviates much ambiguity in analyzing an (intrinsically) cylindrically asymmetric image by simplifying the density-to-flux correction[48] from a 3D problem (for a conventional crushed image) to a 2D one.[41] The effect of vibrational excitation on reactivity has also been investigated using a crossed-beam machine with the traditional universal mass spectroscopic detection method.[49]

Both PHOTOLOC and crossed-beam approaches could be operated in the TOF mass-spectrometric mode by scanning the probe laser wavelength to obtain the integral cross section (ICS) information on product channel branching ratios and state distributions. To obtain the vibrational enhancement factor, i.e., the relative reactivity $\sigma^{\#}/\sigma_0$ of vibrationally excited (denoted by the supeript "#") versus ground-state (the subscript "0") reactants, one needs to determine the IR pumping efficiency $(n^{\#}/n_0)$.[50–52] For illustration, the product signal from the ground-state reaction (i.e., IR off) can be expressed as

$$S^{off} = cn_0\sigma_0. \quad (1)$$

As the IR laser is on, the signals become

$$S^{on} = c[(n_0 - n^{\#})\sigma_0 + n^{\#}\sigma^{\#}], \quad (2)$$

where c is a proportional constant, n_0 and $n^{\#}$ denote the concentrations of the ground state (IR-off) and laser excited reactants, respectively, and σ_0 and $\sigma^{\#}$ are the corresponding reaction cross sections. Eqn (1) and (2) then lead to

$$S^{on}/S^{off} = 1 - (n^{\#}/n_0)(1 - \sigma^{\#}/\sigma_0). \quad (3)$$

Clearly, knowledge of $n^{\#}/n_0$ is required in order to derive the desired quantity of $\sigma^{\#}/\sigma_0$ at a given collisional energy. It is instructive to note that the measured ratio of S^{on}/S^{off} gives an immediate impression about relative values of $\sigma^{\#}$ to σ_0: (i) if $S^{on}/S^{off} > 1$ (i.e., signal enhancement upon IR irradiation), then $\sigma^{\#} > \sigma_0$, (ii) $S^{on} \sim S^{off}$ (i.e., signal strengths unchanged) implies $\sigma^{\#} \sim \sigma_0$, and (iii) $S^{on}/S^{off} < 1$ (i.e., signal depletion) yields $\sigma^{\#} < \sigma_0$.

Theoretically, significant progress has also been made in three areas. The first has to do with the emergence of highly accurate electronic structure theories such as multi-reference configuration interaction (MRCI) and coupled cluster (CC) methods. These methods, particularly in the explicit correlated (F12) version, can now provide chemically accurate (<1 kcal mol^{-1}) energies with reasonable computational costs.[53] The second advance is the efficient and accurate analytical representation of PESs from these electronic structure calculations. It is now possible to develop globally accurate high-dimensional PESs for reactive systems.[11,23,24] Finally, quantum reactive scattering on these accurate PESs can now be performed for tetra-atomic systems, some with state-to-state resolution. These advances allow a direct comparison with experimental measurements, as in the recent study of the HD + OH → H_2O + D reaction. In this exemplary e, the calculated and measured product angular distributions, namely the differential cross sections (DCSs), of this four-atom reaction are compared.[54] As shown in Figure 1, the agreement is excellent. Thanks to these developments, quantum reactive scattering has moved beyond the atom–diatom reactions. For a detailed discussion of the recent advances in quantum reactive scattering, the reader is referred to the recent review on this topic.[11]

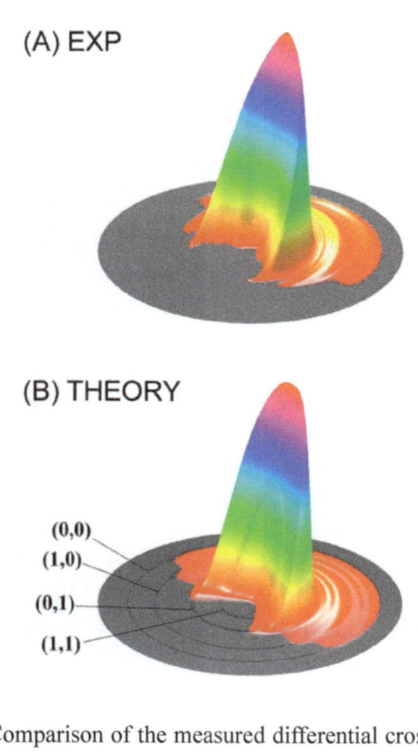

Figure 1: Comparison of the measured differential cross section (A) with that obtained from full-dimensional quantum dynamical calculations (B) for the HD + OH → H_2O + D reaction at the collision energy of 6.9 kcal mol^{-1}. Adapted from ref. 54 with permission.

E STUDIES

X + $H_2O \leftrightarrow$ HX + OH Reactions

The hydrogen abstraction of water by H/Cl has served as a prototypical model to understand the transition-state control of reactivity. Earlier theoretical studies have predicted that the stretching vibrations of the H_2O reactant greatly enhance the reactivity,[55] which was confirmed by later experiments.[56–59] These two endothermic reactions have late barriers, consistent with Hammond's postulate,[60] which states that in a series of reactions, the more endothermic (exothermic) the reaction, the more the transition state will resemble the products (reactants). As a result, the strong vibrational enhancement of reactivity can be rationalized by the Polanyi rules.[8] In Figure 2, projections of various reactant modes onto the reaction coordinate at the transition state are shown for the two reactions, which illustrate that this strong mode specificity is also predicted by the SVP model.[29] Interestingly, both the symmetric and antisymmetric stretching modes of H_2O have roughly the same efficacies, as predicted by the SVP model (Figure 2). As discussed latter, this is due to the local-mode nature of the stretching vibrational modes of H_2O.[61] The observed mode specificity has later been reproduced by exact quantum dynamics (QD) studies.[62–64] An interesting corollary of mode specificity is the so-called bond selectivity. For example, HOD in which the OH or OD vibration was excited was found to react by preferential cleavage of the excited bond.[65,66] Obviously, the mass difference between H and D renders the OH and OD local modes uncoupled, and the bond cleavage proceeds via the corresponding transition state on the PES. The bond selectivity is reproduced by QD calculations,[67–70] and well understood in terms of the SVP model.[29]

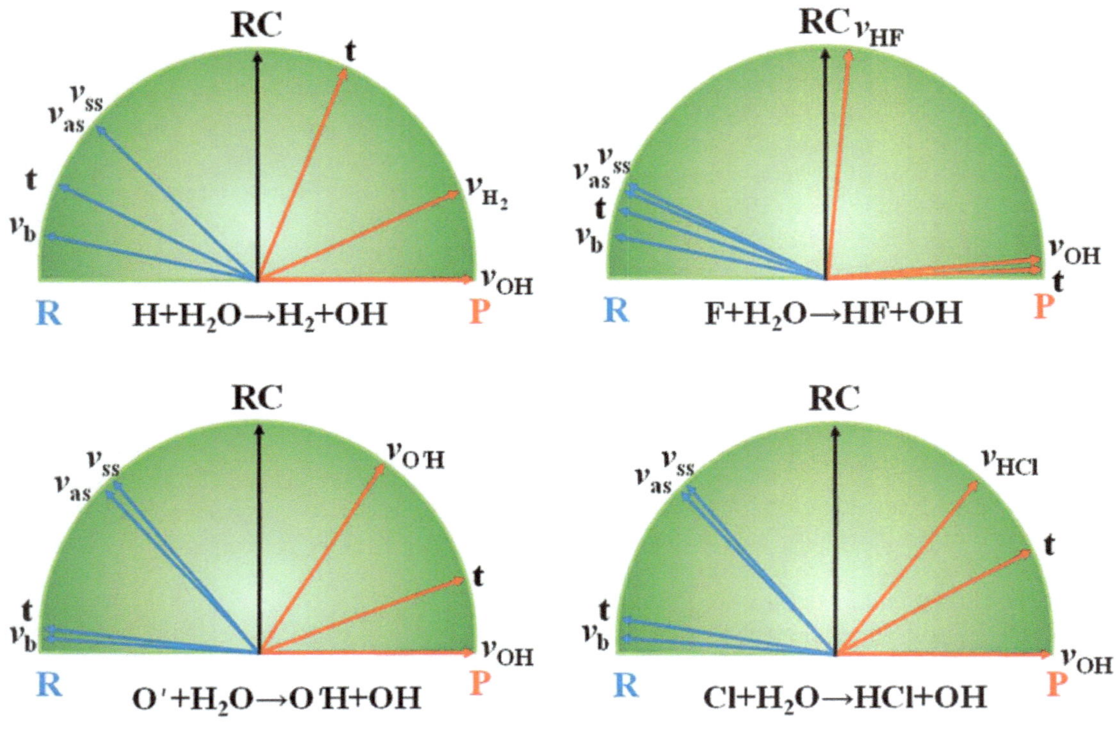

Figure 2: SVP projections of reactant (R) vectors and product (P) vectors onto the reaction coordinate (RC) at the transition state for the three reactions. Adapted from ref. 29 with permission.

While the mode specificity and bond selectivity of the H/Cl + H_2O/HOD reactions are consistent with Polanyi's predictions, the situation is quite different for the F + H_2O reaction. In contrast to the former two reactions that have late barriers, the latter has a low and early barrier. The Polanyi rules thus predict that the reaction would be enhanced more effectively by the translational energy. However, our QD calculations based on an ab initio based PES indicated that except for low collision energies all three vibrational modes of H_2O enhance the reaction more effectively than translation.[71] This apparent inconsistency was resolved by the SVP model, as shown in Figure 2, which found that while the couplings between the vibrational modes and the reaction coordinate at the transition state are weak, they are stronger than that for the translational mode.[29] The inability of the Polanyi rules to predict the mode specificity in this and other polyatomic reactions[72] suggests that the location of the barrier may not necessarily be the best descriptor of mode specificity. Rather, the coupling between the reactant mode and reaction coordinate at the transition state, as suggested by the SVP model, is perhaps a more reliable way to predict mode specificity. The SVP model also correctly predicted the strong vibrational excitation in the HF product and the spectator nature of the OH moiety in this reaction, in good agreement with both experiment[73] and theory.[74,75] The SVP values for the HF and OH vibrational modes are close to the extremes (~1.0 and ~0.0), underlying their respective coupling strengths with the reaction coordinate at the transition state.[29]

More detailed evidence for transition-state control of reaction dynamics can be found at the state-to-state level. One of the interesting observations is that the OD product in the H + D_2O reaction is mostly found in its ground vibrational state, even when the stretching vibrations of D_2O is excited with one stretching quantum.[58] This is termed the "memory loss" effect, as the product distributions appear to be independent of the reactant excitation.[76] On one hand, it suggests that the non-reactive OD bond is a spectator in the reaction. Perhaps more interestingly, this "memory loss" effect should be considered in the local-mode picture in which the D_2O normal-mode vibrations is expressed as a linear combination of symmetrized local-mode basis.[61]In this local-mode picture, the cleavage of a particular OD bond almost exclusively proceeds via one of the two equivalent transition states on the PES where this OD bond is elongated.[77]As a result, the other non-reactive OD bond retains its vibrational ground state in the product. This is exactly the same reason that the two stretching modes of H_2O promote the H + H_2O reaction with roughly the same efficacy, as alluded above.

Very recently, the transition-state control of the H_2 + OH reaction, the reverse reaction of H + H_2O, was investigated at the state-to-state level.[78] This reaction produces stretching excited H_2O, mostly with one stretching quantum at low energies, which can be understood as a result of the fact that the H_2O stretching modes are strongly coupled with the reaction coordinate at the transition state. Indeed, the symmetric and antisymmetric stretching modes of the H_2O products are found to have roughly the same population, consistent with the local-mode picture discussed above. More interestingly, the H_2O product is dominated by the (002) state when the OH reactant is in its first excited vibrational state, as shown in Figure 3. A local-mode analysis revealed that the (002) normal mode state corresponds to the $|11\rangle$ local mode state, in which the H_2O product contains one quantum of excitation in each OH bond. One OH quantum is apparently from the excited spectator OH moiety, while the other derives from the strong coupling of the H_2O stretching vibrations with the reaction coordinate at the transition state. An intriguing observation here is that the product state distribution of this reaction depends sensitively on the vibrational excitation of the OH reactant. This state-to-state mode specificity appears to be quite different from the "memory loss" effect in the reverse reaction (H + H_2O), as discussed above. The key to reconcile these two extremes is to recognize that the non-reactive OH moiety is a spectator in the reverse reaction, which has near-zero coupling with the reaction coordinate at the transition state. As a result, energy deposited into this mode is unlikely to leak out to the reaction coordinate, leading to no enhancement of reactivity. In other words, its energy is sequestered during the reaction. On the other hand, the water stretching vibrations are active modes in the forward reaction that have significant coupling with the reaction coordinate at the transition state, resulting in facile energy flow from these modes to the reaction coordinate and enhanced reactivity. In the meantime, the same energy flow depletes the energy in these modes, resulting in the loss of memory. In other words, the "memory loss" effect suggests that the product energy disposal is almost completely determined by the transition state, as advocated by the SVP model, and illustrated in Figure 4.

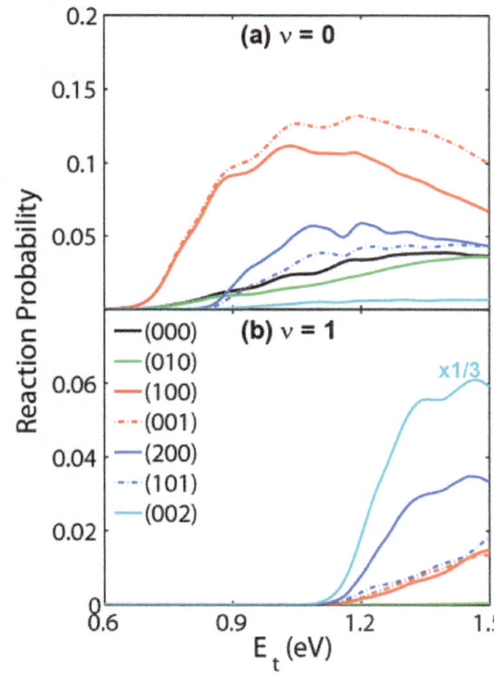

Figure 3: Vibrational state distribution (J = 0) of H_2O in various normal mode states produced by reaction H_2 + OH → H + H_2O. When OH is in the ground vibrational state (a), the H_2O product is dominated by the (100) and (001) state. However, the H_2O product from the H_2 + OH (v = 1) reaction (b) is dominated by the (002) state. Adapted from ref. 29 with permission.

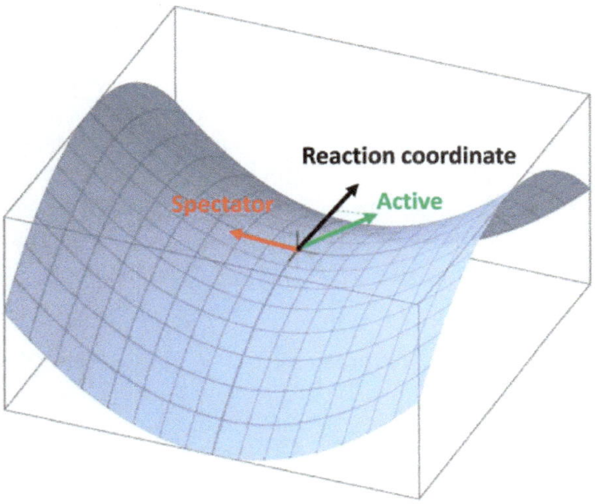

Figure 4: Illustration of the projection of an active and spectator modes onto the reaction coordinate at the transition state.

Interestingly, the manner in which the transition state regulates the reaction can be visualized by the transition-state wave packet method.[35,79] It has been shown that quantum states in the transition-state region of the PES, represented in that method by the eigenstates of the thermal flux operator, controls the reaction flux. In the e of X + H_2O reaction, recent quantum state-to-state calculations have shown that such control is exercised in a coherent manner, in which several such thermal flux eigenstates act concertedly to regulate the flow of the reaction flux.[80,81] Later work further indicates that OH excited thermal flux eigenstates are responsible for the energy sequestration of vibrational energy in the spectator OH mode in such reactions.[78]

The H/F + H_2O → H_2/HF + OH reactions have also been probed via transition-state spectroscopy. To this end, the stable species H_3O^- or FH_2O^- are photodetached to produce a wave packet near the transition state of the corresponding reaction. Very different wave packet dynamics were observed for these two reactions. In the H_3O^- e, the dynamics are very fast and the photoelectron spectrum has no structure.[82,83] On the other hand, the photoelectron spectrum of FH_2O^- is highly structured, featuring both reactive and non-reactive Feshbach resonances.[84,85]

It is important to realize that the transition-state control of the reactivity is not always absolute. Modifications of such control can be induced by either potential or kinetic sources. The former is facilitated by features on PESs. While the H + H_2O ↔ H_2 + OH reaction has very shallow pre- and post-reaction barriers, the other X + H_2O reactions have significant wells along the reaction pathway. The pre-reaction well might exert strong forces on the reactants, whereas the post-reaction well could impact the final state distribution of the products. An interesting e is the pre-reaction well of the F + H_2O reaction, which is relatively deep due to its hemi-bond (two-center, three-electron bond) character, as shown in Figure 5.[86] For this low barrier reaction, the stereodynamic forces at low collision energies are strong enough to cause a significant enhancement of the reactivity.[87] This modification has also a quantum mechanical character, manifesting in the form of tunneling.[88] Similar stereodynamic forces have been seen in many other systems, including Cl + HD[89] and F + CHD_3 reactions.[90,91]

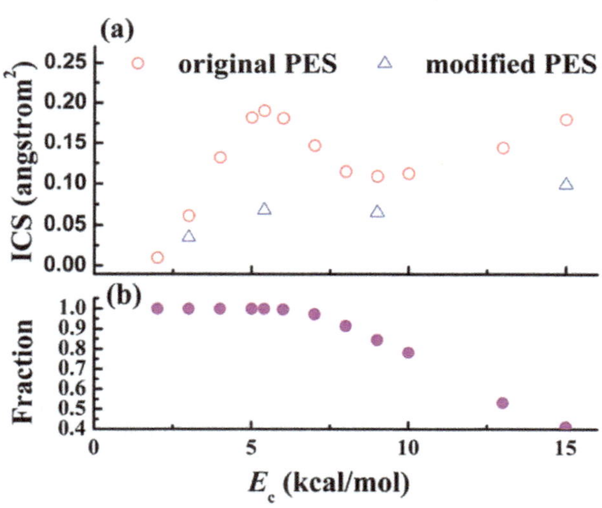

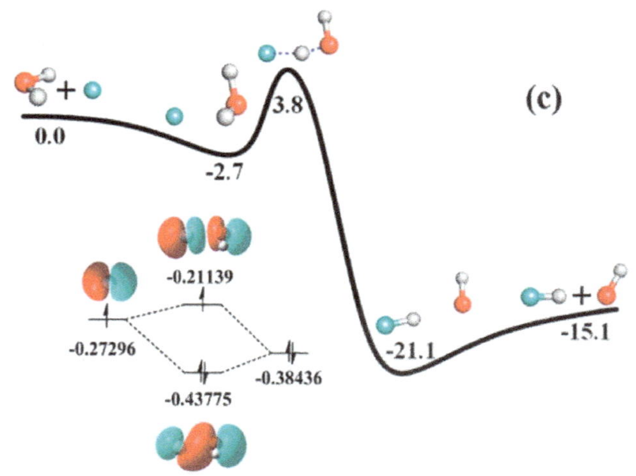

Figure 5: (a) Calculated reaction cross-sections of the F + H_2O → HF + OH reaction based on an ab initio PES (original) and a modified PES (modified) in which the F–H_2O well is artificially removed. The peak near 6 kcal mol^{-1} is thus attributed to the F–H_2O well, through which the majority of the reactive trajectories are steered towards the reaction barrier, as shown in (b). The strong stereodynamic forces in this reaction is due to the F–H_2O pre-reaction well, shown in (c), which has a hemi-bond character. Reproduced with permission from ref. 86 and 87.

X + CH_4 ↔ HX + CH_3 Reactions

This series of six-atom reactions, with X = H, O(^{3}P), F, and Cl, has become a benchmark to advance our understanding of polyatomic chemical reactivity not only because of their fundamental interests in offering a wide range of diverse PES topographies, but also for their important roles in combustion, atmospheric, and astrophysical chemistry.[12,23,38] The reaction of H + CH_4 exhibits a high barrier of 14.7 kcal mol^{-1} with a small endothermicity of 2.8 kcal mol^{-1}. At the transition state the H–H distance is slightly elongated from the equilibrium value in H_2, while the C–H distance is significantly stretched from the bond length in methane.[92–95] By Hammond's postulate[60] the transition-state structure is product-like, and by the Polanyi rules[8]the reaction is characterized by a late barrier. Upon one quantum excitation of the asymmetric stretching mode ($v_3 = 1$) of CH_4, a rate enhancement of ~3 was found experimentally in H + CH_4 at the same collision energy,[96] consistent with the Polanyi rule for a late barrier reaction. Several theoretical calculations have since confirmed this vibrational effect on reactivity.[97–101]The SVP model also indicates a moderate coupling of the v_3 mode with the reaction coordinate at the transition state, and thus a vibrational enhancement.[38] This vibrational effect undeores the transition-state control of reactivity. The SVP model further predicts a slightly stronger coupling for the symmetric stretching mode v_1;[38] but experimental confirmation has not yet been reported for this mode. Bond selectivity has also been observed[102] and confirmed theoretically[103,104] in the H + CHD_3 reaction, in which the local C–H stretching mode (v_1 = 1) excitation in CHD_3 promotes the H-atom abstraction channel. This bond selectivity can readily be rationalized by the SVP model as the v_1 mode is essentially the reaction coordinate.[38]

The reverse reaction H_2 + CH_3 has been studied theoretically.[105,106] The results from reduced-dimensional QD calculations indicated that the H_2 stretching and CH_3 umbrella modes, as well as the translational energy, strongly promote the reaction, while the CH_3 symmetric stretching mode behaves as a spectator exerting little effect. The observed mode specificity is confirmed by full-dimensional quasi-classical trajectory (QCT) calculations.[105] Again, the SVP model provides rationalization of the mode specificity in terms of the coupling strengths of reactant modes with the reaction coordinate at the transition states.[105]

In contrast to the above H + CH_4 e, the reaction of F + CH_4 is highly exothermic by 31.8 kcal mol^{-1} with a low barrier of about 0.7 kcal mol^{-1} (with zero-point-energy corrected) in the entrance valley,[107]i.e., an early barrier. According to the Polanyi rules,[8] the initial translation energy should be more efficacious than the vibrational excitation of methane in promoting the reaction rate. In addition, with a reactant-like transition-state structure, the vibrational motion of reactant would have little coupling to the reaction coordinate, as also indicated by the SVP model.[38] Hence, the vibrational modes should behave as a spectator in F + methane, as such one might have expected little reactivity change upon the vibrational excitation of methane. Counter-intuitively, the excitation of the C–H stretching mode in CHD_3 was found to inhibit the C–H bond rupture leading to the HF + CD_3 product in F + CHD_3 (v_1 = 1). As shown in Figure 6,[90] the IR-on REMPI spectrum clearly displays a negative impact to the formation of the dominant CHD_2 (v = 0) products, which demonstrates, from eqn (3), that the stretch-excited reactivity should be smaller than the ground state reactivity. In fact, significant depletions of the product REMPI signals were observed for all, including the CH_3 + DF channel,[90,108,109] but the CHD_2 (v_1 = 1) state. The latter can be regarded as the spectator when the F-atom abstracts one of the unexcited D-atoms, leaving the initially deposited vibration energy in the excited C–H bond of the CHD_2 products. Similar inhibitory results was reported in the F + CH_4 (v_3 = 1) reaction.[110]

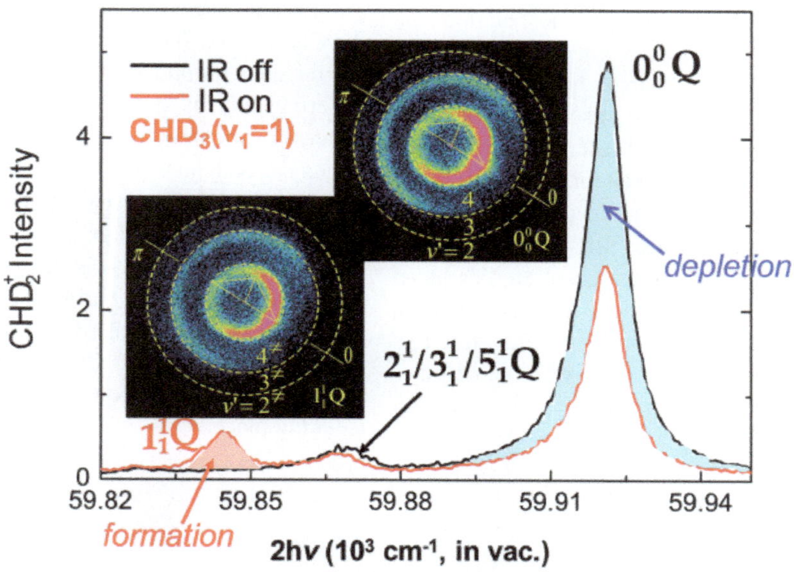

Figure 6: Two normalized REMPI spectra of the probed CHD_2 products, with IR-on (red line) and IR-off (black), at E_c = 3.6 kcal mol^{-1}. A significant reduction of the signals for the origin band 0^0_0 (blue-shaded area) and the formation of a new band 1^1_1 (red-shaded area) are observed. Two product images, both with IR-on, are shown for probing the 1^1_1 Q (left-lower) and 0^0_0 Q (right-upper) bands, respectively. Superimposed on the images are the scattering directions; the 0° angle refers to the initial CHD_3 beam direction in the center-of-mass frame (modified with permission from ref. 90).

This surprising effect was speculated as the result of the stereodynamic forces exerting on the approaching reactants by the pre-reaction van der Waals (vdW) wells. The experimental conjecture was later borne out in a QCT calculation on an accurate PES,[91] which highlights the possibility that anisotropic forces modifies the transition-state control of reaction dynamics for low, early-barrier reactions. However, more recent theoretical investigations, also by QCT but on different PESs in the entrance valley, suggested no inhibitory effect by CHD_3 (v_1 = 1).[111,112] The source of the theory-experiment discrepancy is unclear at present. On the experimental side, it could arise from some other excited methyl products, which are inaccessible to the current REMPI detection method and yet with much higher abundances than those states being probed thus far. On the theory side, as other experimental evidence strongly suggests a resonance-mediated pathway in F + methane,[113,114] the QCT calculations may not capture the quantum nature of this reaction. Alternatively, theory and experiment may refer to subtly different quantities. Whereas the experiment results correspond to the relative reactivity of F + CHD_3 (v_1 = 1, $|J, K\rangle$ = $|2, 0\rangle$ and $|2, \pm1\rangle$) to that of F + CHD_3 (v = 0, T_{rot} ~ 9 K), the theory refers to the single rotational $|J, K\rangle$ = $|0, 0\rangle$ state for both CHD_3 (v_1 = 1) and CHD_3 (v = 0). As will be discussed below in the Cl + CHD_3 (v_1 = 1, $|J, K\rangle$) reaction, the initially selected rotational $|J, K\rangle$ -states can result in profound effects on reactivity. More studies, both experiment and theory, are needed to reconcile the discrepancy.

It is interesting to note that in the above F + H_2O e, which is also very exothermic with a low barrier and a relatively deep pre-reaction well, the stretching excitation of water is predicted to significantly enhance the reactivity,[71] in opposite to the F + CH_4 reaction. In both es, the transition-state control of reactivity is strongly modulated by the long-range anisotropic interactions in the entrance valley, yet resulting in vastly different outcomes. The effects of bending excitation of CD_4 and CHD_3 in reaction with F-atom have also been studied,[115–117] in which both experiment and theory (QCT) show that bending excitation activates the reactivity at low collision energies and become inactivated at higher collision energies. Experimental results on both the integral and differential cross sections further suggested signature for quantum resonances in bend-excited reactions, which await further theoretical investigations.[116,117]

Unlike the late barrier in the H + CH_4 reaction and early barrier in the F + CH_4 reaction, the reaction between O(^{3}P) and CH_4 presents an interesting e in which the transition state has a central location with a barrier height of ~14 kcal mol^{-1}.[118] The Polanyi rules thus provide little guidance on the mode specificity in such reactions. Reduced-dimensional QD and full-dimensional QCT calculations revealed significant mode specificity,[119,120] which can be rationalized by the SVP model.[38] The theoretical results are found to be consistent with the experimental data.[121,122]

The hydrogen abstraction reaction between Cl and methane is perhaps one of the most studied polyatomic reactions in both experiment and theory. It is slightly endothermic by 1.2 kcal mol^{-1} and with a late (adiabatic) barrier height of 3.4 kcal mol^{-1}.[123] In accord with the Polanyi rules as well as the SVP model, both experiments and theoretical calculations indicate reactivity promotion by stretching excitations of methane reactants. A number of reviews have been devoted to the mode specificity and bond selectivity of this reaction,[9–14] to which the interested readers are referred. More recent studies have turned the attention to its stereochemical aspects.[10,40] Two complementary approaches have been taken to elucidate the effects of the collisional geometry on chemical reactivity: by rotational state selectivity of the vibrationally excited methane and by prealigning the excited methane in space for a given rovibrationally selected state. These studies allow unprecedented control of collision events, eliminating averaging in the rotational and orientational levels. They shed valuable light on the "chemical shape"[124] of the reactant molecules as they react with their collisional partners.

Figure 7 illustrates an experimental setup for measuring the reactivity effect of aligned CHD_3 (v_1 = 1, |J, K⟩ = |1, 0⟩).[125] As depicted, four beams (two molecular beams, IR-excitation and UV-probe lasers) lie in the same x–z plane, and the initial relative velocity can be arranged being perpendicular (parallel) to the IR (UV) laser. Because the $v_1 = 1 \leftarrow 0$ transition is a parallel band with the transition dipole moment lying along the C–H bond, varying the linear polarization of the IR laser effectively changes the direction of the excited C–H bond with respect to the approaching Cl-atom in the collision frame. Figure 8 presents the raw data. The IR-off image represents the ground state reaction, yielding a pair of CD_3 (v = 0) + HCl (v = 0) products, i.e.the $(0_0, 0)_g$ state pair (the subscript "g" denotes the ground-state reaction). Two IR-on images correspond to the CD_3 (v = 0) signals from the stretch-excited reaction (plus the residual ground state signals from the unexcited CHD_3), differing only by the IR-polarization directions. The distinct patterns of the new features in the two IR-on images signify a strong stereodynamic effect in the Cl + CHD_3 (v_1 = 1, |J, K⟩ = |1, 0⟩) reaction. The right-lower panel of Figure 8 shows the IR-polarization dependencies of the two product pairs, $(0_0, 0)_s$ and $(0_0, 1)_s$, from the stretch-excited reactions. Interestingly, the two product state pairs display a striking out-of-phase oscillation. A more detailed analysis offers direct experimental evidence for a collinear Cl–H–C transition-state structure and allows a 3D visualization of how the chemical transformation is taking place.[126–128]

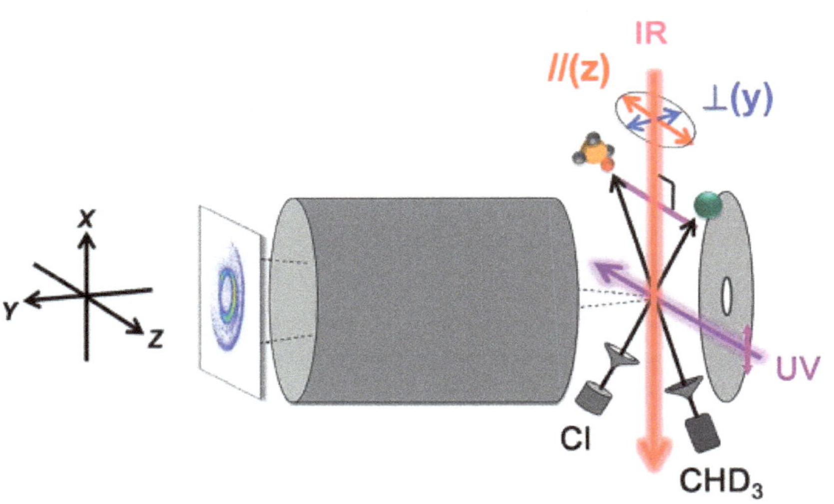

Figure 7: Experimental setup for state-selected, aligned molecular beam scattering experiments. All four beams (two molecular beams, IR-pump, and UV-probe lasers) lie in the xz plane. The REMPI-tagged ions are velocity-mapped and guided along the y-direction. A linearly polarized IR laser directed perpendicularly to the relative velocity vector of the two molecular beams prepares the vibrationally excited CHD_3 at the scattering center. Reagent alignment is controlled by IR laser polarization direction, "//" refers to an end-on attack and "⊥" to a side-on approach. Time-sliced velocity-map image, shown at the left, reveals the alignment effects on product pair-correlated distribution (modified with permission from ref. 125).

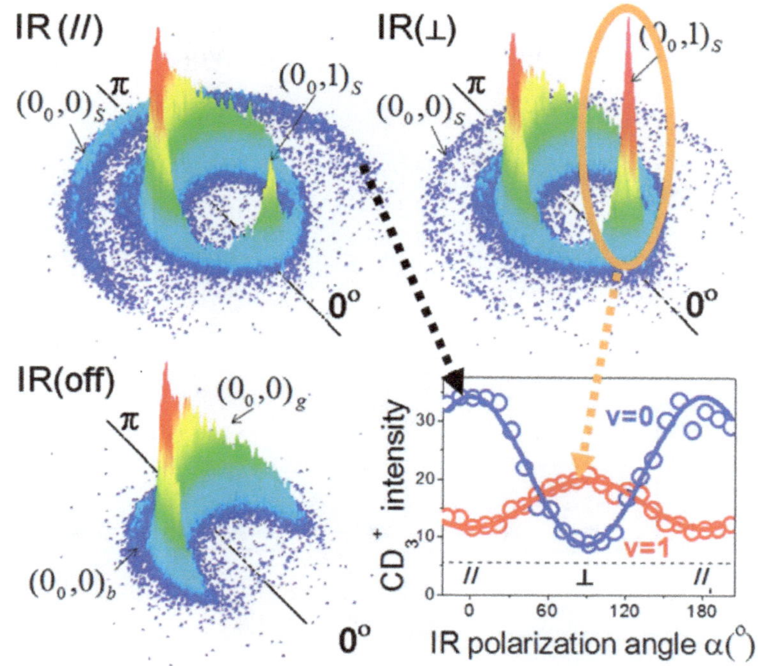

Figure 8: Three CD_3 (v = 0) product images with a superimposed axis indicating the scattering direction; the 0° angle refers to the initial CHD_3 beam direction in the center-of-mass frame of Cl + CHD_3 reaction. The distinct ring-like features are dictated by the conservation of energy and momentum. The pair-correlated labeling is defined as follows: the numbers in the parentheses denote the (v_{CD3}, v_{HCl}) products; the outer subscript indicates the CHD_3 reactant state ("g" for ground state, "b" for bend-excited state, and "s" for the stretch-excited state). The lower-right panel shows the polarization-angle dependence of the probed signals from the stretch-excited reaction. The v = 0 data correspond to the outer-ring feature labeled $(0_0, 0)_s$; for the v = 1 data, only the forward signals for $(0_0, 1)_s$, as encircled, are counted (modified with permission from ref. 125).

Moreover, the experiment also showed a maximal alignment effects allowed by quantum mechanics.[127] The implication is that the asymptotically aligned CHD_3 (v_1 = 1) reactant must retain its directional properties en route to the reaction barrier; otherwise, a weaker polarization dependency should have been observed. This is quite intriguing in that the interaction between Cl and CHD_3 also exhibits vdW wells in the entrance valley with the well depths deeper than the above F + CHD_3 system. Yet, a significant reorientation of the approaching F-atom was surmised in the latter reaction as discussed above. In that regard, we note that the vibrationally enhanced reactivity has also been observed in O(^{3}P) + CHD_3 (v_1 = 1) by enlarging the cone of the reactive acceptance due to the focusing effects of the long-range anisotropic interactions en route to the barrier.[121] The three seemly analogous reactions all have similar vdW topographies in the entrance valley, yet display vastly contrasting stereodynamic behaviors. Will it be subtly related to the diverse barrier locations or other features in PESs? The challenge is to gain deeper insights into these perplexing findings and to comprehend the underlying origin, which remains a challenge for theory.

An alternative approach to elucidate the geometric dependency of reactivity is to examine the rotational mode specificity. Exemplified in Figure 9 is the relative reactivity for the ro-vibrationally selected CHD_3 (v_1 = 1, |J, K⟩) reactants with Cl-atoms.[129,130] A clear J- and K-dependency is revealed: a rotating CHD_3 (v_1 = 1) reactant promotes the reactivity, and for a given J the tumbling rotation (K = 0) yields higher reactivity than the spinning rotation (|K| = J). Concurrent QCT and reduced dimensionality QD calculations confirmed the observed trends and offered an intuitively appealing explanation: the range of the attack angles near the barrier opens up (i.e., the bending potential at the transition state becomes softer) for the rotating reactants.[110] It is worth noting that the initial rotational energy difference is very small for those low J/K states; for example, $E_{J=2}$ is merely 0.055 kcal mol^{-1} above the rotationless state. Hence, the observed rotational enhancement factor of ~2 is quite significant. In particular, it is about the same magnitude, on the equivalent amount of total energy, as the reported vibrational enhancement factor in Cl + CHD_3 (v_1 = 1).[131,132] As the experimental vibrational factors reported in literatures often are rotationally ill-defined or in some es for the J = 2 state, it becomes somewhat ambiguous, in view of the rotational mode specificity illustrated here, as to how to untangle the "pure" vibrational enhancement from a collective effect of ro-vibrational excitation. This serves as a cautionary note for theory-experiment comparison and for future studies.

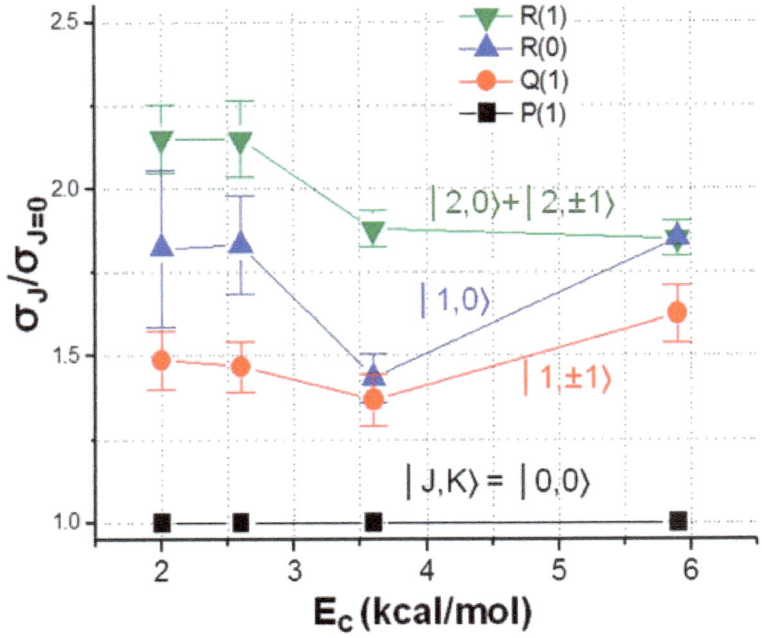

Figure 9: Dependency of relative reactivity on the rotational states of vibrationally excited CHD_3 ($v_1 = 1$) reactants in reaction with Cl atom. At each E_c, the reactivity of the rotational ground state is set at unity (modified with permission from ref. 129).

IMPLICATIONS IN LARGER REACTIONS

The principles discussed above for relatively small reactive systems are applicable to more complex reactions. One such example is the 1,3 dipolar cycloaddition reaction between a 1,3 dipole and a dipolarophile to form a five-membered ring, via a concerted transition state.[133] This classical organic reaction is a key step in click chemistry.[134] Recently, Houk and coworkers have carried out extensive direct dynamics calculations, starting the trajectories from the transition state towards the reactant channel.[135,136] They found that the bending vibration of the 1,3 dipoles is always excited in these trajectories. Based on microscopic reversibility, it was argued that the bending excitation in the 1,3 dipoles enhances reactivity of these reactions, which is consistent with the fact that these reactions require heating. Interestingly, the involvement of the 1,3 dipole bending vibration in such reactions can actually be predicted by the SVP model, without the expensive direct dynamics calculations. In a recent study,[137] we have shown that reactant bending modes have large overlaps with the reaction coordinate at the transition state, in support of the conclusion of Houk and coworkers. In addition, the SVP model also predicts enhancement effects for several other reactant modes, which are to be confirmed.

This example suggests that the transition state exerts strong control of dynamics even in organic reactions involving large molecules.[138] One distinct advantage of the SVP model is that its predictions only require information on the transition state as well as on the reactants. The reaction coordinate can be readily determined using any ab initio method, without the complete PES, which is increasingly more difficult to obtain as the system becomes larger. As a result, this model is particularly useful for ever more complex reactions.

CONCLUSIONS

Recent advances in quantum state resolved reaction dynamics have revealed an unprecedented level of detail for bimolecular reactions. Experimentally, it has now become possible to selectively excite the reactants into a single quantum state and to align them, in addition to measuring the product state distributions with the most exquisite detail. Theoretically, accurate PESs can now be constructed based on a large number of ab initio points, upon which QD and QCT calculations can be carried out. The accumulation of data in both theory and experiment has led to a much more in-depth understanding of how the transition-state controls the reactivity. The examples discussed in this perspective have demonstrated convincingly that such a control lies at the heart of the observed mode specificity and bond selectivity. Our knowledge on these phenomena is best illustrated by a simple transition-state based construct, such as the SVP model, in which the ability of a particular reactant mode in promoting the reaction is attributed to the projection of the corresponding normal mode vector onto the reaction coordinate at the transition state. The SVP model, which can be considered as a generalization of the venerable Polanyi rules, particularly for reactions involving polyatomic molecules, has been shown to be generally successful in its predictions, as long as the sudden approximation, namely the assumption that the collision time is much shorter than the IVR rate, is valid. While the transition-state control of reactivity is in general quite dominant, evidence has also been presented for the modifications of the transition state control of reactivity. These modifications stem from features on the PES that become non-negligible under specific conditions. The most important point is perhaps that the lessons learned in reactions involving small molecules can be extended to larger reactive systems, of which the transition-state control of their reactivity has seldom been explored.

ACKNOWLEDGEMENTS

HG thanks the US Department of Energy (Grant no. DE-FG02-05ER15694) for continuing support. KL acknowledges the financial supports by Academia Sinica and the Ministry of Science and Technology of Taiwan.

REFERENCES

1. H. Eyring, J. Chem. Phys., 1935, **3**, 107–115 .
2. E. Wigner, J. Chem. Phys., 1937, **5**, 720–725 .
3. M. G. Evans and M. Polanyi, Trans. Faraday Soc., 1939, **35**, 178–185 .
4. J. I. Steinfeld, J. S. Francisco and W. L. Hase, Chemical Kinetics and Dynamics, Prentice Hall, Englewood Cliffs, NJ, 1989 Search .
5. D. G. Truhlar, B. C. Garrett and S. J. Klippenstein, J. Phys. Chem., 1996, **100**, 12771–12800 .
6. J. C. Polanyi and A. H. Zewail, Acc. Chem. Res., 1995, **28**, 119–132 .
7. R. D. Levine, Molecular Reaction Dynamics, Cambridge University Press, Cambridge, 2005 Search .
8. J. C. Polanyi, Science, 1987, **236**, 680–690 .
9. H. Guo and B. Jiang, Acc. Chem. Res., 2014, **47**, 3679–3685 .
10. K. Liu, J. Chem. Phys., 2015, **142**, 080901 .
11. D. H. Zhang and H. Guo, Annu. Rev. Phys. Chem., 2016 DOI:10.1146/annurev-physchem-040215–112016.
12. K. Liu, Annu. Rev. Phys. Chem., 2016 DOI:10.1146/annurev-physchem-040215–112522.
13. F. F. Crim, Acc. Chem. Res., 1999, **32**, 877–884 .
14. F. F. Crim, Proc. Natl. Acad. Sci. U. S. A., 2008, **105**, 12654–12661 .
15. R. N. Zare, Science, 1998, **279**, 1875–1879 .
16. K. Liu, Annu. Rev. Phys. Chem., 2001, **52**, 139–164 .
17. N. Balucani, G. Capozza, F. Leonori, E. Segoloni and P. avecchia, Int. Rev. Phys. Chem., 2006, **25**, 109–163.
18. K. Liu, Phys. Chem. Chem. Phys., 2007, **9**, 17–30 .
19. X. Yang, Annu. Rev. Phys. Chem., 2007, **58**, 433–459 .
20. S. C. Althorpe and D. C. Clary, Annu. Rev. Phys. Chem., 2003, **54**, 493–529 .
21. F. J. Aoiz, L. Bañares and V. J. Herrero, Int. Rev. Phys. Chem., 2005, **24**, 119–190 .
22. H. Guo, Int. Rev. Phys. Chem., 2012, **31**, 1–68 .
23. G. Czakó and J. M. Bowman, J. Phys. Chem. A, 2014, **118**, 2839–2864 .
24. J. Li, B. Jiang, H. Song, J. Ma, B. Zhao, R. Dawes and H. Guo, J. Phys. Chem. A, 2015, **119**, 4667–4687 .
25. F. Fernandez-Alonso and R. N. Zare, Annu. Rev. Phys. Chem., 2002, **53**, 67–99 .
26. K. Liu, Adv. Chem. Phys., 2012, **149**, 1–46 .
27. Z. Sun, B. Zhao, S. Liu and D. H. Zhang, in Molecular Quantum Dynamics, From Theory to Application, ed. F. Gatti, Springer, Heidelberg, 2014 Search .
28. B. Jiang and H. Guo, J. Chem. Phys., 2013, **138**, 234104 .
29. B. Jiang and H. Guo, J. Am. Chem. Soc., 2013, **135**, 15251–15256 .
30. G. C. Schatz and J. Ross, J. Chem. Phys., 1977, **66**, 1021–1036 .
31. D. Wang and J. M. Bowman, Chem. Phys. Lett., 1993, **207**, 227–235 .
32. U. Havalee and M. Shapiro, J. Chem. Phys., 1976, **64**, 2826–2839 .
33. R. Schinke, Photodissociation Dynamics, Cambridge University Press, Cambridge, 1993 Search .
34. M. Gustafsson and R. T. Skodje, J. Chem. Phys., 2006, **124**, 144311 .
35. R. Welsch, F. Huarte-Larrañaga and U. Manthe, J. Chem. Phys., 2012, **136**, 064117 .
36. Y. Wang and J. M. Bowman, J. Chem. Phys., 2013, **139**, 154303 .
37. W. Siebrand, Z. Smedarchina and A. Fernández-Ramos, J. Chem. Phys., 2013, **139**, 021101 .
38. B. Jiang and H. Guo, J. Chin. Chem. Soc., 2014, **61**, 847–859 .
39. W. H. Miller, N. C. Handy and J. E. Adams, J. Chem. Phys., 1980, **72**, 99–112 .
40. W. R. Simpson, A. J. Orr-Ewing, T. P. Rakitzis, S. A. Kandel and R. N. Zare, J. Chem. Phys., 1995, **103**, 7299–7312 .
41. J. J. Lin, J. Zhou, W. Shiu and K. Liu, Rev. Sci. Instrum., 2003, **74**, 2495–2500 .
42. D. W. Chandler and P. L. Houston, J. Chem. Phys., 1987, **87**, 1445–1446 .
43. A. T. J. B. Eppink and D. H. Parker, Rev. Sci. Instrum., 1997, **68**, 3477–3484 .
44. C. R. Gebhardt, T. P. Rakitzis, P. C. Samartzis, V. Ladopoulos and T. N. Kitsopoulos, Rev. Sci. Instrum., 2001, **72**, 3848–3853 .
45. D. Townsend, M. P. Minitti and A. G. Suits, Rev. Sci. Instrum., 2003, **74**, 2530–2539 .
46. J. J. Lin, J. Zhou, W. Shiu and K. Liu, Science, 2003, **300**, 966–969 .
47. J. G. Zhou, J. J. Lin, W. C. Shiu and K. Liu, Phys. Chem. Chem. Phys., 2006, **8**, 3000–3006 .
48. D. M. Sonnenfroh and K. Liu, Chem. Phys. Lett., 1991, **176**, 183–190 .
49. D. L. Proctor and H. F. Davis, Proc. Natl. Acad. Sci. U. S. A., 2008, **105**, 12673–12677 .
50. S. Yan, Y.-T. Wu and K. Liu, Phys. Chem. Chem. Phys., 2007, **9**, 250–254 .
51. J. Riedel, S. Yan, H. Kawamata and K. Liu, Rev. Sci. Instrum., 2008, **79**, 033105 .
52. Y. Cheng, H. Pan, F. Wang and K. Liu, Phys. Chem. Chem. Phys., 2014, **16**, 444–452 .
53. T. Shiozaki and H.-J. Werner, Mol. Phys., 2013, **111**, 607–630 .
54. C. Xiao, X. Xu, S. Liu, T. Wang, W. Dong, T. Yang, Z. Sun, D. Dai, X. Xu, D. H. Zhang and X. Yang, Science, 2011, **333**, 440–442 .
55. G. C. Schatz, J. Chem. Phys., 1981, **74**, 1133–1139 .
56. A. Sinha, M. C. Hsiao and F. F. Crim, J. Chem. Phys., 1991, **94**, 4928–4935 .
57. A. Sinha, J. D. Thoemke and F. F. Crim, J. Chem. Phys., 1992, **96**, 372–376 .
58. M. J. Bronikowski, W. R. Simpson and R. N. Zare, J.

Phys. Chem., 1993, **97**, 2204–2208 .

59. J. D. Thoemke, J. M. Pfeiffer, R. B. Metz and F. F. Crim, J. Phys. Chem., 1995, **99**, 13748–13754 .
60. G. S. Hammond, J. Am. Chem. Soc., 1955, **77**, 334–338 .
61. M. S. Child and L. Halonen, Adv. Chem. Phys., 1984, **57**, 1–58 .
62. D. H. Zhang, M. A. Collins and S.-Y. Lee, Science, 2000, **290**, 961–963 .
63. B. Fu and D. H. Zhang, J. Chem. Phys., 2013, **138**, 184308 .
64. H. Song and H. Guo, J. Phys. Chem. A, 2015, **119**, 6188–6194 .
65. A. Sinha, M. C. Hsiao and F. F. Crim, J. Chem. Phys., 1990, **92**, 6333–6334 .
66. M. J. Bronikowski, W. R. Simpson and R. N. Zare, J. Phys. Chem., 1993, **97**, 2194–2203 .
67. D. H. Zhang and J. C. Light, J. Chem. Soc., Faraday Trans., 1997, **93**, 691–697 .
68. J. Li, H. Song and H. Guo, Phys. Chem. Chem. Phys., 2015, **17**, 4259–4267 .
69. B. Fu and D. H. Zhang, J. Chem. Phys., 2015, **142**, 064314 .
70. H. Song, S.-Y. Lee, Y. Lu and H. Guo, J. Phys. Chem. A, 2015, **119**, 12224–12230 .
71. J. Li, B. Jiang and H. Guo, J. Am. Chem. Soc., 2013, **135**, 982–985 .
72. H. Song and H. Guo, J. Phys. Chem. A, 2015, **119**, 826–831 .
73. A. M. Zolot and D. J. Nesbitt, J. Chem. Phys., 2008, **129**, 184305 .
74. J. Li, R. Dawes and H. Guo, J. Chem. Phys., 2012, **137**, 094304 .
75. J. Li, B. Jiang and H. Guo, J. Chem. Phys., 2013, **138**, 074309 .
76. R. Welsch and U. Manthe, J. Phys. Chem. Lett., 2015, **6**, 338–342 .
77. S. Liu and D. H. Zhang, Chem. Sci., 2016, **7**, 261–265 .
78. B. Zhao, Z. Sun and H. Guo, J. Am. Chem. Soc., 2015, **137**, 15964–15970 .
79. B. Zhao, Z. Sun and H. Guo, J. Chem. Phys., 2014, **140**, 234110 .
80. B. Zhao, Z. Sun and H. Guo, J. Chem. Phys., 2014, **141**, 154112 .
81. B. Zhao, Z. Sun and H. Guo, J. Chem. Phys., 2015, **142**, 241101 .
82. E. de Beer, E. H. Kim, D. M. Neumark, R. F. Gunion and W. C. Lineberger, J. Phys. Chem., 1995, **99**, 13627–13636 .
83. D. H. Zhang, M. Yang, M. A. Collins and S.-Y. Lee, Proc. Natl. Acad. Sci. U. S. A., 2002, **99**, 11579–11582.
84. R. Otto, J. Ma, A. W. Ray, J. S. Daluz, J. Li, H. Guo and R. E. Continetti, Science, 2014, **343**, 396–399 .
85. J. Ma and H. Guo, J. Phys. Chem. Lett., 2015, **6**, 4822–4826 .
86. J. Li, Y. Li and H. Guo, J. Chem. Phys., 2013, **138**, 141102 .
87. J. Li, B. Jiang and H. Guo, Chem. Sci., 2013, **4**, 629–632 .
88. B. Zhao and H. Guo, J. Phys. Chem. Lett., 2015, **6**, 676–680 .
89. D. Skouteris, J. F. tillo and D. E. Manolopoulos, Comput. Phys. Commun., 2000, **133**, 128 .
90. W. Zhang, H. Kawamata and K. Liu, Science, 2009, **325**, 303–306 .
91. G. Czakó and J. M. Bowman, J. Am. Chem. Soc., 2009, **131**, 17534–17535 .
92. X. Zhang, B. J. Braams and J. M. Bowman, J. Chem. Phys., 2006, **124**, 021104 .
93. T. Wu, H.-J. Werner and U. Manthe, J. Chem. Phys., 2006, **124**, 164307 .
94. J. C. Corchado, J. L. Bravo and J. Espinosa-García, J. Chem. Phys., 2009, **130**, 184314 .
95. Y. Zhou, B. Fu, C. Wang, M. A. Collins and D. H. Zhang, J. Chem. Phys., 2011, **134**, 064323 .
96. J. P. Camden, H. A. Bechtel, D. J. A. Brown and R. N. Zare, J. Chem. Phys., 2005, **123**, 134301 .
97. M. L. Wang and J. Z. H. Zhang, J. Chem. Phys., 2002, **116**, 6497–6504 .
98. Z. Xie, J. M. Bowman and X. Zhang, J. Chem. Phys., 2006, **125**, 133120 .
99. C. Rangel, J. C. Corchado and J. Espinosa-García, J. Phys. Chem. A, 2006, **110**, 10375–10383 .
100. G. Schiffel and U. Manthe, J. Chem. Phys., 2010, **133**, 174124 .
101. R. Liu, H. Xiong and M. Yang, J. Chem. Phys., 2012, **137**, 174113 .
102. J. P. Camden, H. A. Bechtel, D. J. A. Brown and R. N. Zare, J. Chem. Phys., 2006, **124**, 034311 .
103. Z. Xie and J. M. Bowman, Chem. Phys. Lett., 2006, **429**, 355–359 .
104. Y. Zhou, C. Wang and D. H. Zhang, J. Chem. Phys., 2011, **135**, 024313 .
105. Y. Wang, J. Li, L. Chen, Y. Lu, M. Yang and H. Guo, J. Chem. Phys., 2015, **143**, 154307 .
106. Z. Zhang, J. Chen, M. Yang and D. H. Zhang, J. Phys. Chem. A, 2015, **119**, 12480–12484 .
107. G. Czakó, B. C. Shepler, B. J. Braams and J. M. Bowman, J. Chem. Phys., 2009, **130**, 084301 .
108. J. Yang, D. Zhang, B. Jiang, D. Dai, G. Wu, D. Zhang and X. Yang, J. Phys. Chem. Lett., 2014, **5**, 1790–1794 .
109. J. Yang, D. Zhang, Z. Chen, F. Blauert, B. Jiang, D. Dai, G. Wu, D. Zhang and X. Yang, J. Chem. Phys., 2015, **143**, 044316 .
110. H. Kawamata, W. Q. Zhang and K. Liu, Faraday Discuss., 2012, **157**, 89–100 .
111. J. Palma and U. Manthe, J. Phys. Chem. A, 2015, **119**, 12209–12217 .
112. J. Espinosa-Garcia, J. Phys. Chem. A, 2016, **120**, 5–13 .
113. J. Zhou, J. J. Lin and K. Liu, J. Chem. Phys., 2004, **121**, 813–818 .

114. J. Zhou, J. J. Lin and K. Liu, Mol. Phys., 2010, **108**, 957–968 .

115. G. Czakó, Q. A. Shuai, K. Liu and J. M. Bowman, J. Chem. Phys., 2010, **133**, 131101 .

116. F. Wang and K. Liu, J. Phys. Chem. Lett., 2011, **2**, 1421–1425 .

117. F. Wang and K. Liu, J. Phys. Chem. A, 2013, **117**, 8536–8544 .

118. G. Czakó and J. M. Bowman, Proc. Natl. Acad. Sci. U. S. A., 2012, **109**, 7997–8001 .

119. R. Liu, M. Yang, G. Czakó, J. M. Bowman, J. Li and H. Guo, J. Phys. Chem. Lett., 2012, **3**, 3776–3780 .

120. G. Czakó, R. Liu, M. Yang, J. M. Bowman and H. Guo, J. Phys. Chem. A, 2013, **117**, 6409–6420 .

121. F. Wang and K. Liu, Chem. Sci., 2010, **1**, 126–133 .

122. J. Zhang and K. Liu, Chem.–Asian J., 2011, **6**, 3132–3136 .

123. G. Czakó and J. M. Bowman, J. Chem. Phys., 2012, **136**, 044307 .

124. R. D. Levine, J. Phys. Chem., 1990, **94**, 8872–8880 .

125. F. Wang, J.-S. Lin and K. Liu, Science, 2011, **331**, 900–903 .

126. F. Wang and K. Liu, Chin. J. Chem. Phys., 2013, **26**, 705–709 .

127. F. Wang, K. Liu and T. P. Rakitzis, Nat. Chem., 2012, **4**, 636–641 .

128. F. Wang, J.-S. Lin and K. Liu, J. Chem. Phys., 2014, **140**, 084202 .

129. R. Liu, F. Wang, B. Jiang, G. Czakó, M. Yang, K. Liu and H. Guo, J. Chem. Phys., 2014, **141**, 074310 .

130. F. Wang, H. Pan and K. Liu, J. Phys. Chem. A, 2015, **119**, 11983–11988 .

131. S. Yan, Y. T. Wu, B. Zhang, X.-F. Yue and K. Liu, Science, 2007, **316**, 1723–1726 .

132. F. Wang, J.-S. Lin, Y. Cheng and K. Liu, J. Phys. Chem. Lett., 2013, **4**, 323–327 .

133. K. V. Gothelf and K. A. Jørgensen, Chem. Rev., 1998, **98**, 863–910 .

134. H. C. Kolb, M. G. Finn and K. B. Sharpless, Angew. Chem., Int. Ed., 2001, **40**, 2004–2021 .

135. L. Xu, C. E. Doubleday and K. N. Houk, Angew. Chem., Int. Ed., 2009, **48**, 2746–2748 .

136. L. Xu, C. E. Doubleday and K. N. Houk, J. Am. Chem. Soc., 2010, **132**, 3029–3037 .

137. A. Li and H. Guo, Chem. Phys. Lett., 2015, **624**, 102–106 .

138. G. L. Barnes and W. L. Hase, Nat. Chem., 2009, **1**, 103–104 .

CHAPTER 11

COMPETITION-DRIVEN SELECTION IN COVALENT DYNAMIC NETWORKS AND IMPLEMENTATION IN ORGANIC REACTIONAL SELECTIVITY

P. Kovaříček, A. C. Meister, K. Flídrová , R. Cabot , K. Kovaříčková and J.-M. Lehn

INTRODUCTION

Dynamic covalent libraries (DCLs)[1–11] are formed by reversible combinatorial linkage of molecular components generating an equilibrium mixture of constituents under thermodynamic control. The reversibility of the linkage allows for constitutional exchange giving access to all possible combinations of components. By application of a stimulus, the DCL is able to undergo constitutional adaptation whereby some species are amplified at the expense of others that are depleted. These features make dynamic covalent chemistry (DCC) attractive for application in material science,[12–20] surface modification,[21] synthesis of nanoarchitectures,[22–25] design of receptors[7] and sensors[26] as well as in the search for biologically active compounds.[2,27–32] In fact, only "virtual presence"[1] of, *i.e.* access to, all combinations is required as the amplified species can be formed upon the application of the stimulus – in the terms of the "lock and key" principle, one may say that the lock assembles its key.

Imines, resulting from the reversible condensation of a carbonyl component with an amine, are of particular interest for setting up DCLs, due to the ease of formation and exchange, usually under mild conditions. The heteroatomic imine bond is also satisfactorily orthogonal[33–36] to many other reversible bonds and its dynamic nature can be "frozen" by reduction or other transformations.[37] Simplification of the complex mixture of species in the library is achieved through adaptive sorting,[38–41] for example in course of crystallisation,[42–44] oxidation,[45]distillation[46,47] or coordination.[48–52]

Herewith we report the operation of a *reactional self-sorting* process in DCLs of imines of increasing complexity, driven by competition among the reagents in mixtures of aldehydes and oligoamines components and leading to improved selectivity in product formation and/or in site of reaction. The aldehydes in the studied DCLs are moderately selective in reactions with different amines to form various products such as imines, aminals or amino-lactones in the e of *o*-carboxybenzaldehyde.[53,54] Competition between an increasing number of aldehydes for a given mixture of amines leads to increased selectivity of each of them. It leads to a simplification of the final composition resulting from the complexity of the DCL and its collapse into a system of reduced multiplicity (and lower entropy), amounting to a state that may be termed "simplexity" in a process of general type from chemical to biological systems.[55,56] In fact, any recognition process, be it reactional or interactional (see Figure 3 in ref. 57) results in the reduction of the complexity of a mixture, of its simplification, through the operation of competitive selection. On the supramolecular/interactional level it is expressed in the "instructed mixture paradigm", whereby the behaviour of mixtures is driven by the instructions (molecular information) present in its members resulting in self-recognition (or self-sorting), a self-process,[58] belonging to the general phenomenon of self-organization.[9,59] The term "simplexity" does not carry the meaning that the system in itself is less complex, as there is in fact an increased complexity (increased number of reacting species) underlying and resulting in the simplification (increased selectivity) observed. We also propose a formal approach for the representation and comparison of DCLs (see below and ESI†) which has been used for evaluating the efficiency of selection in the self-sorting processes investigated here within the realm of dynamic covalent chemistry (DCC).[1–5,28–30,57]

Aldehyde–Amine Dynamic Covalent Libraries

In view of the ubiquity of amines and imines in organic chemistry and in biochemistry, as well as the ability to control each partner in the reaction, condensations between aldehydes and amines were taken as model reactions in order to probe reaction selectivity in competitive dynamic reversible systems.

Citation: P. Kovaříček, A. C. Meister, K. Flídrová , R. Cabot , K. Kovaříčková and J.-M. Lehn, "Competition-driven selection in covalent dynamic networks and implementation in organic reactional selectivity," DOI: 10.1039/c5sc04924e.

At first, different aldehydes were reacted with different amines in order to test their differences in reactivity and the nature of the product formed. Salicylaldehyde (SALAL), pyridine-2-carboxaldehyde (PYRAL) and 2-carboxybenzaldehyde (CAXAL) were selected as prototypical aldehydes. On the other hand, isopentylamine (IPA), *N,N'*-dimethyl-1,3-diaminopropane (Me_2PDA) and piperidine (PIP) were selected as their amine counterparts. As solvent a mixture of d_6-DMSO with 1% of D_2O was used to guarantee solubility of all partners, in which the water addition maintains constant water content during the aldehyde–amine condensation. On the basis of previous studies, differences in reaction outcome was expected for the various aldehyde–amine pairs,[53] anticipating that: (a) SALAL would have a preference for primary amines to form imines,[53] (b) PYRAL would react with diamines to provide a cyclic aminal[53] and (c) CAXAL would react with secondary amines to afford amino-lactones through trapping of the intermediary iminium by the neighbouring carboxylate group.[54] These anticipated selectivities towards formation of different condensation products were verified in separate experiments: SALAL reacted with IPA to provide the expected imine, PYRAL gave the six-membered cyclic aminal with Me_2PDA, and CAXAL indeed reacted with PIP to form the amino-lactone. These reactions proceeded with quantitative conversion and the resulting preferentially formed products may be considered as "matching pairs", effecting reactional recognition, like supramolecular systems effect interactional recognition (Figure 1).[4,57]

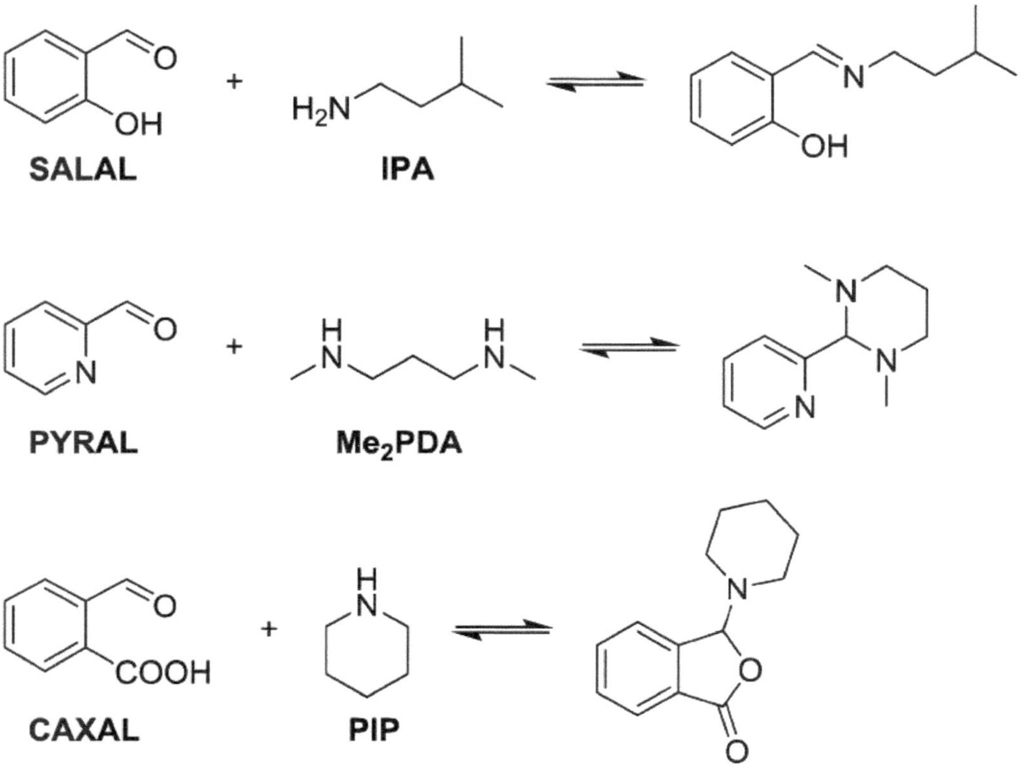

Figure 1: Schemes of anticipated reaction outcome of different aldehydes and amines with their acronyms used in the study. SALAL forms preferentially imines and PYRAL cyclic aminals. CAXAL is capable of reacting at a secondary amine site with formation of a lactone ring.

[2 × 2] Aldehyde–Amine Mixtures

Selection by Preferential Imine/Lactone Formation

In order to explore selection in a [2 × 2] competition, the e of a mixture of the two aldehydes SALAL–CAXAL and the two amines IPA–PIP was examined in detail.

Checking first the combinations opposite to the "matching pairs" (see above) showed that both aldehydes reacted with both amines: when CAXAL was mixed with 1 eq. of IPA, equilibrium was reached almost instantaneously giving a dynamic mixture of the imine and lactone formed with IPA (details in ESI, Section 3.2†); on the other hand, SALAL reacted withPIP in approximately 30% conversion to the aminal formed from two amine molecules and one aldehyde. In subsequent experiments, the propensity towards the formation of the "matching pair" product was tested in presence of a competing amine, *i.e.* each aldehyde was reacted with an equimolar mixture of the two amines. SALAL showed a very high preference for imine formation reacting solely with IPA and leaving PIP unreacted in the solution. However, the conversion of the aldehyde in this e was not complete, approximately 10% remaining unreacted. In contrast, CAXAL under the same reaction conditions provided two products: the lactone on the PIP ring was formed in about 39% yield and the dynamic imine–lactone product from IPA was present in about 61% yield (details in the ESI, Section 3.2†). Finally, the full [2 × 2] library was considered: SALAL and CAXAL were mixed with both IPA and PIP in equimolar amounts and the resulting product mixture was examined using NMR spectroscopy. Equilibrium was reached after 5 hours and the mixture contained only the two "matching pair" species, the imine SALAL–IPA and the lactone CAXAL–PIP, both in quantitative conversion.

This remarkable simplicity achieved in the comparatively complex mixture, contrasts with the ability of both aldehydes to react with both amines, and *vice versa* each amine with the two aldehydes. It demonstrates the occurrence of a selectivity enhancement of a poor selector by competition with a strong selector. Specifically, a species reacting non-selectively with several compounds is brought to become selective towards a given product, when put in presence of a competing species of high selectivity which monopolises one of the components. The process amounts to a *competitive selectivity enhancement*. The formation of the matching pair products can also be seen as resulting from agonist amplification[9,57] by competition *via* component exchange through an underlying network of equilibrating reactions (Figure 2).

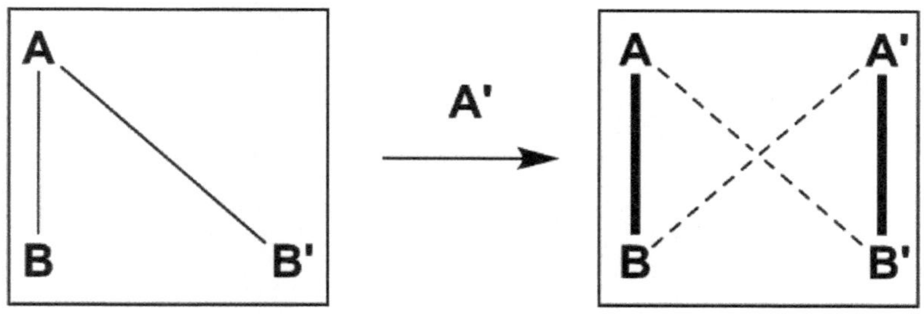

Figure 2: Schematic representation of the process of competitive amplification (enhanced selection) induced by increasing the complexity of the mixture: while **A** is capable of reacting with both **B** and **B′**, addition of **A′** leads to enforced formation of the **A′B′** product which traps the **B′** component and therefore depletes the **AB′** constituent. Due to the underlying network of equilibrating reactions this also leads to amplified formation of the**AB** constituent in a type of agonist amplification process of **AB** by the formation of **A′B′**. The amplified agonistic species are represented by thick lines on the vertices of the square while the depleted species are denoted by the striped diagonals.

Stoichiometry is of course a crucial parameter providing the enhanced selection of the inherent self-sorting process. In the present e, the components react in 1 : 1 ratio to form the preferred product. Therefore, the initial mixture must contain the components in this ratio so that the condensation of the matching pair in quantitative conversion depletes all of the starting materials, which in turn do not interfere with the formation of the second matching pair in the complex system. This is clearly demonstrated by the fact that whereas the aforementioned mixture SALAL–CAXAL–IPA–PIP can give, in principle, 14 products (accounting for 2 imines, 4 homo- and 2 heteroaminals, 4 hemiaminals and 2 lactones), only two products are observed in the mixture containing all four species in equimolar amounts.

Selection by Aminal Formation

PYRAL offers the opportunity of considering selection by the formation of aminal species, different from imine and lactone. Indeed, PYRALefficiently forms both five- and six-membered-ring aminals with *N,N'*-dimethyl-1,2-diaminoethane (Me_2EDA) or *N,N'*-dimethyl-1,3-diaminopropane (Me_2PDA), reaching quantitative conversion when reacted separately with either of the two diamines. On the other hand, when it was reacted with an equimolar mixture of Me_2EDA and Me_2PDA, the five-membered ring was formed preferentially (77% conversion to the aminal of Me_2EDA). Both SALAL and CAXAL reacted as well with these diamines giving the corresponding aminals in high conversion (about 90% for SALAL and quantitative for CAXAL). Libraries similar to the previous one were examined using PYRAL and its matching diamine (Me_2EDA and Me_2PDA) combined either with the SALAL–IPA or the CAXAL–PIP pair. The experiments (described in detail in the ESI, Section 3.2 and 3.3†) revealed that although all aldehyde–amine combinations were forming efficiently, high selectivity for the matching pairs was observed when all constituents of the library were mixed in equimolar ratio. In particularly, for the SALAL–PYRAL–IPA–Me_2PDA mixture the SALAL–IPA imine and the PYRAL–Me_2PDA aminal were formed in 80% conversion (and the non-matching pairs SALAL–Me_2PDA and PYRAL–IPA in 20%), while in the e of the PYRAL–CAXAL–Me_2EDA–PIP mixture the PYRAL–Me_2EDA aminal and the CAXAL–PIP lactone were present in 97% conversion (and the non-matching pairs were formed only in 3%).

Selection in Extended Aldehyde–Amine Dynamic Combinatorial Libraries

The three [2 × 2] libraries described above form, when combined, a [3 × 3] dynamic library composed of three aldehydes and three amine partners. In this vein, a mixture of SALAL, PYRAL and CAXAL was combined with a mixture of IPA, Me_2PDA and PIP (all equimolar) and the NMR spectra were recorded (Figure 3). Equilibration of the mixtures was accelerated by heating at 60 °C overnight. The equilibrated samples contained the imine of SALAL and IPA in 88% conversion and the aminal of SALAL and Me_2PDA in 12%, whereas the imine to aminal ratio of PYRAL was exactly the opposite, and CAXAL provided quantitatively the lactone withPIP. The system thus displayed selectivity in product formation in each e as well as a remarkable selectivity inversion when comparing the two es, which display imine/aminal ratios of 88/12 = 7.

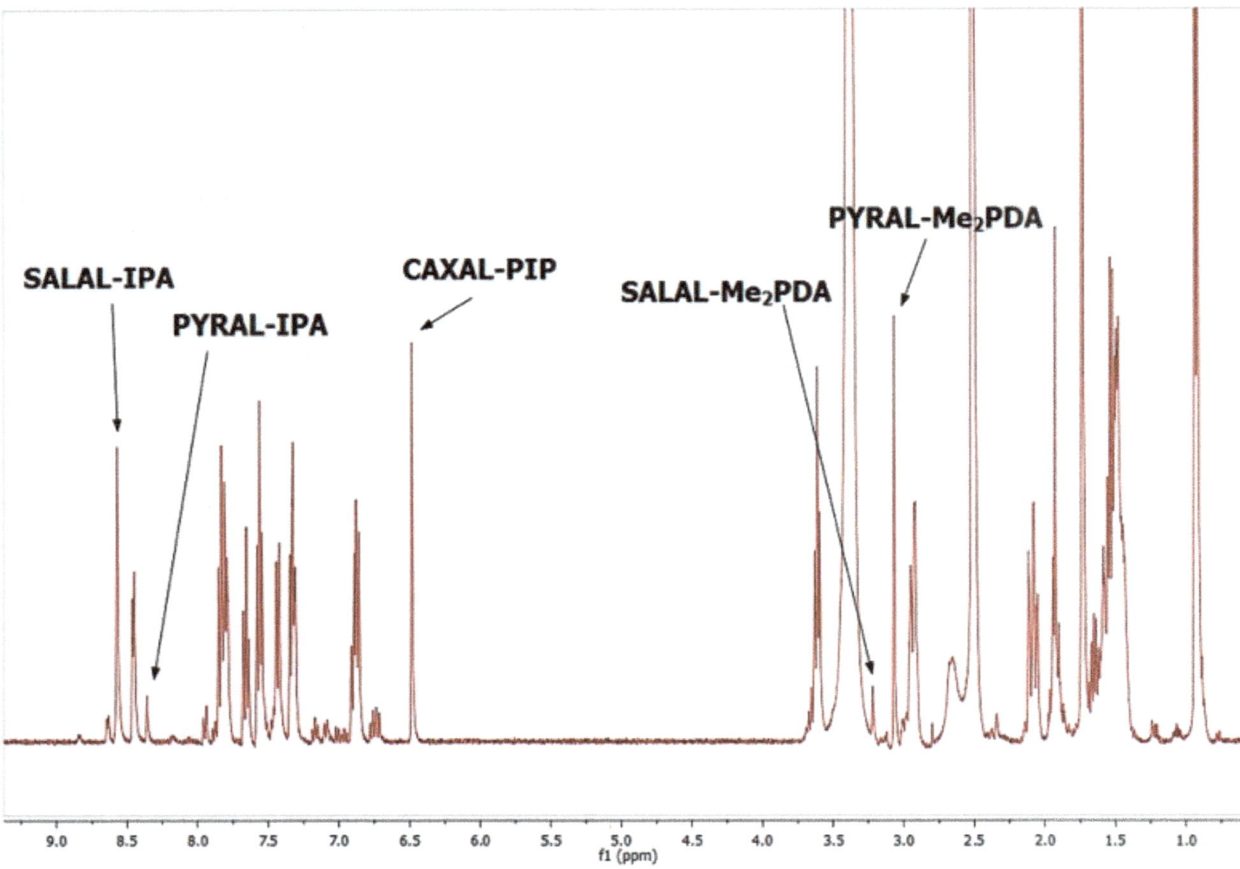

Figure 3: ^{1}H-NMR spectrum of the [3 × 3] aldehyde–amine library with assignment of the peaks. Three major species are formed: the imine of SALAL and IPA, the aminal of PYRALand Me_2PDA, and the lactone of CAXAL and PIP, the latter being the only product of this aldehyde.

As the final concentration of species can play a role in the selection process,[60] we repeated the library experiments at several concentrations: 4, 10, 20, 60 and 100 mM of the components. The equilibrium composition of the library in this range was constant showing no concentration dependency.

The library was also assessed for the effect of pH. To this end, the pre-equilibrated library was examined on addition up to 4 equivalents of acid ($MeSO_3H$) or up to 2 eq. of base (*t*-BuOK). Addition of 1 eq. of acid leads to almost complete hydrolysis of the aminals of both SALAL and PYRAL, while on addition of the second equivalent of $MeSO_3H$ the imine hydrolysis reached 60%. Continued titration by acid led to complete hydrolysis of the imines and extensive hydrolysis of the CAXAL derived lactone, finally giving only the hydrolysed products, after addition of 4 equivalents of acid. In contrast, upon titration by base, disappearance of the lactone signals was observed. Interestingly, this lactone depletion was not accompanied by the emergence of the aldehyde peak of CAXAL, but rather of a new imine of CAXAL with IPA and liberation of SALAL was observed. This example clearly displays the operation of the underlying network of interconnected equilibrating reactions which leads to the complex adaptation of the dynamic library to a stimulus.

The selectivity enhancement displayed here for [$n \times n$] dynamic libraries represents a process of competitive selection, whereby simplification results from competition within the set of equilibrating constituents undergoing component exchange. This behaviour relates also to the process of co-evolution in a dynamic library, leading to the synergistic expression of given constituents.[9,61]

We have developed a formal method for quantification of selection in dynamic combinatorial libraries and especially for (self)-sorting processes which involve coupled reactions. It is based on a matrix representation of the combinatorial library including the particular e of the selection induced by increased complexity of the mixture.

Kinetics in Dynamic Reaction Selectivity

The key aspect of the field of dynamic covalent chemistry (DCC) is that it is governed by thermodynamics and therefore relates to the equilibrium composition. However, the equilibration of the mixture proceeds *via* component exchanges, each with a given rate constant, and the equilibrium is thus established when the sum of all reaction rates is zero. Kinetics are therefore intimately involved in the status of the composition of a dynamic set at a given time.[62] It is in particular the e when considering the vastly different rates of C=N double bond formation between different carbonyl groups and different types of amino compounds.[63] Such kinetic features can be briefly illustrated on the library consisting of SALAL, CAXAL, IPA and PIP.

When SALAL was reacted with IPA (20 mM, buffered d_6-DMSO, see ESI, Section 3,† for details) the formation of the imine followed second order kinetics with the rate constant of 0.01 M^{-1} s^{-1} and 90% overall conversion to the imine (note that in the reaction without a buffer the conversion was quantitative). When the experiment was repeated in presence of 1 eq. ofCAXAL, significantly slower rate ($k = 0.001$ M^{-1} s^{-1}) and only 77% conversion of SALAL was observed (Chart 1), showing that competition of the two aldehydes for one amine affects both the equilibrium (conversion) and the kinetics (rate of formation). Finally, when the experiment was performed in presence of 1 eq. of both CAXAL and PIP, forming the corresponding lactone, the original SALAL conversion of 90% was restored and the rate of formation was even accelerated ($k = 0.03\ M^{-1}\ s^{-1}$) due probably to the catalytic effect of the secondary amine.[64] Thus, as expected, on the way towards the simplexity state, the evolution of the fractions of the different library constituents depends on the kinetics until equilibrium has been reached.

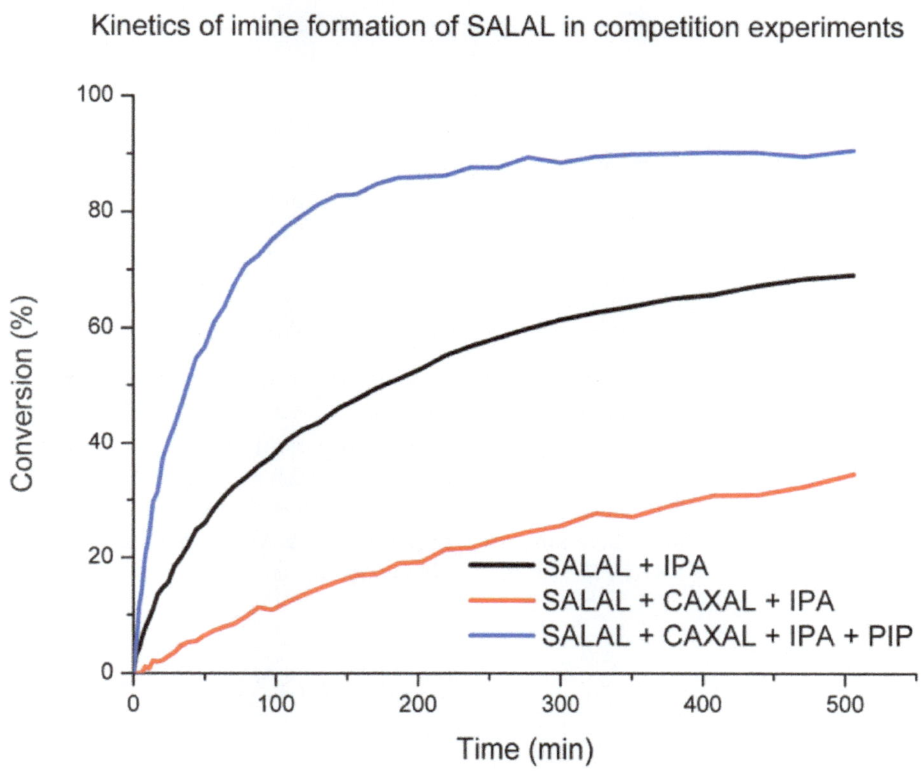

Chart 1: Relative percentage of the imine formed by the reaction of **SALAL** + **IPA** plotted as a function of time. The rate is significantly lower when **CAXAL** is added to the mixture as a result of competition of two aldehydes for one amine. Interestingly, when **PIP** is also added, the rate is higher than the previous two rates, indicating catalysis due to **PIP**. Kinetic experiments were performed in solution 20 mM for each compound in d_6-DMSO + 1% D_2O with 0.8 M triethanolamine buffer.

Reactional Organization along Oligoamine Amine Chains

Selection in aldehyde–amine DCLs described above arises from the preference of a given aldehyde for its matching amine partner through equilibration involving reaction both with different amines and with different sites in an oligo(poly) amine. Combining different amine structural motives within one molecule opens the way to intramolecularly organise aldehyde residues along an oligo(poly)amine chain in a given sequence. Thus, using the described "matching pairs" a multivalent polyamine molecule containing both primary amine end groups and secondary amines along the chain could serve as the organisational scaffold for the positioning of aldehyde residues under functional recognition (Figure 4).[4,57,65]

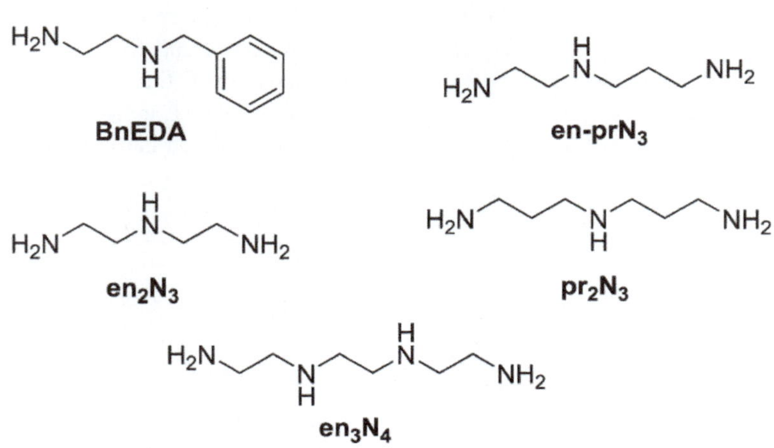

Figure 4: Structures and acronyms of polyamines used in the multivalency- based sorting experiments.

In the simpleste of *N*-benzylethylenediamine (BnEDA), the chain has one primary and one secondary nitrogen, expected to represent the reactivity of both IPA and PIP respectively. When an equimolar mixture of SALAL and CAXAL was reacted with 1 eq. BnEDA the dominant species in the solution (70%) was the expected imine–lactone: the imine of SALALformed on the primary nitrogen and the lactone of CAXAL closed on the secondary nitrogen (Figure 5). This predominant formation of a single product is remarkable given the fact that the mixture can in principle generate a large number of other species.

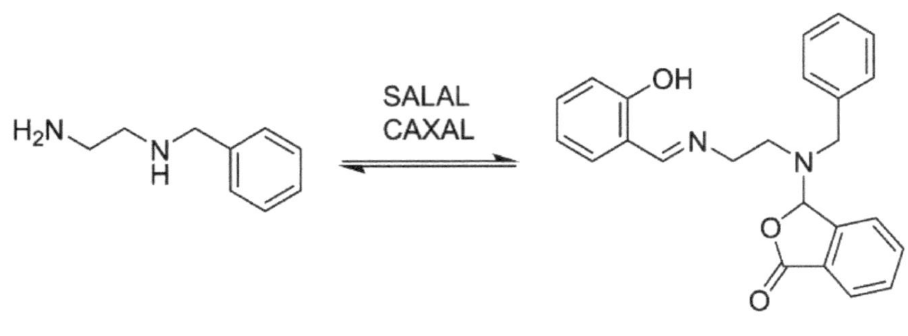

Figure 5: Simultaneous reaction of BnEDA with the imine-forming SALAL and the lactone-forming CAXAL with formation of the expected imine–lactone product in about 70% conversion. The imine is formed on the primary amine group and the lactone on the secondary one.

With extended chains with three nitrogen sites, such as diethylenetriamine (en_2N_3) or bis(3-aminopropyl)amine (pr_2N_3), selective formation of aminals with PYRAL can be assessed. Closure of the aminal ring bridges two nitrogen sites, terminal and central, while the third one, the other primary amine, is available for imine formation. Thus, when either of the triamines was reacted with 2 eq. of PYRAL, the imine–aminal was the major product formed. In contrast, when the triamines were reacted with 2 eq. of SALAL, the terminal bis-imine was the only product. Remarkably, reaction of the two aldehydes mixed in 1 : 1 ratio with the two triamines, resulted in selective formation of the expected products, the SALAL-imine and PYRAL-aminal, with respectively 77% and 84% conversion for en_2N_3 and pr_2N_3 (Figure 6).

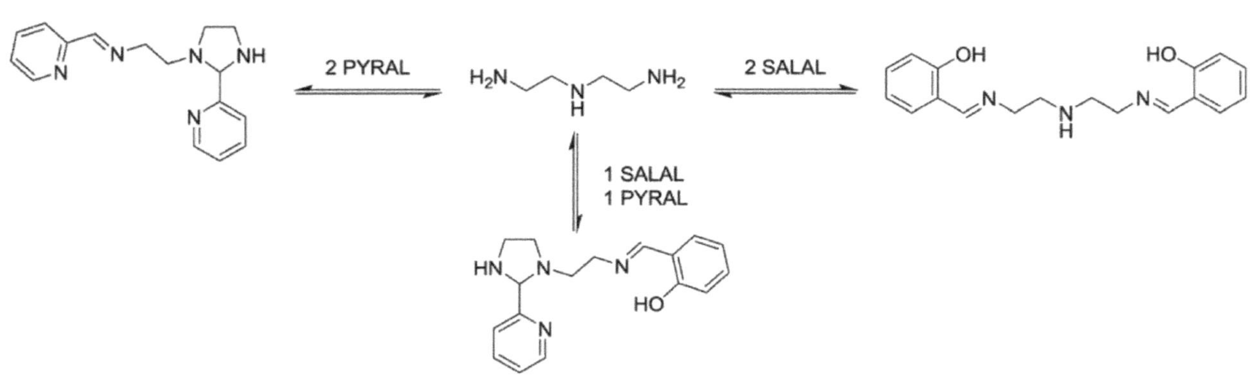

Figure 6: Selective intramolecular organization of two aldehyde residues *via* imine formation of**SALAL** and aminal formation of **PYRAL** affording the imine–aminal (bottom). Similar results are obtained with ethylene or propylene spacers between nitrogen atoms.

The non-symmetric triamines **en-prN$_3$** and spermidine were also examined in the reaction with the two aldehydes. When **en-prN$_3$** was reacted with 1 eq. of PYRAL and 1 eq. ofSALAL the desired imine–aminal species with five membered aminal ring formed in 85% conversion, with trace amounts of six-membered isomer. When spermidine was reacted under the same conditions, only one size of the aminal ring can be formed as seven-membered rings are much more difficult to form compared to the six-membered ones, and indeed the NMR spectrum revealed that the six-membered PYRAL aminal bearing the SALAL-imine on the C_4-arm was formed in overall conversion of 71% (Figure 7a).

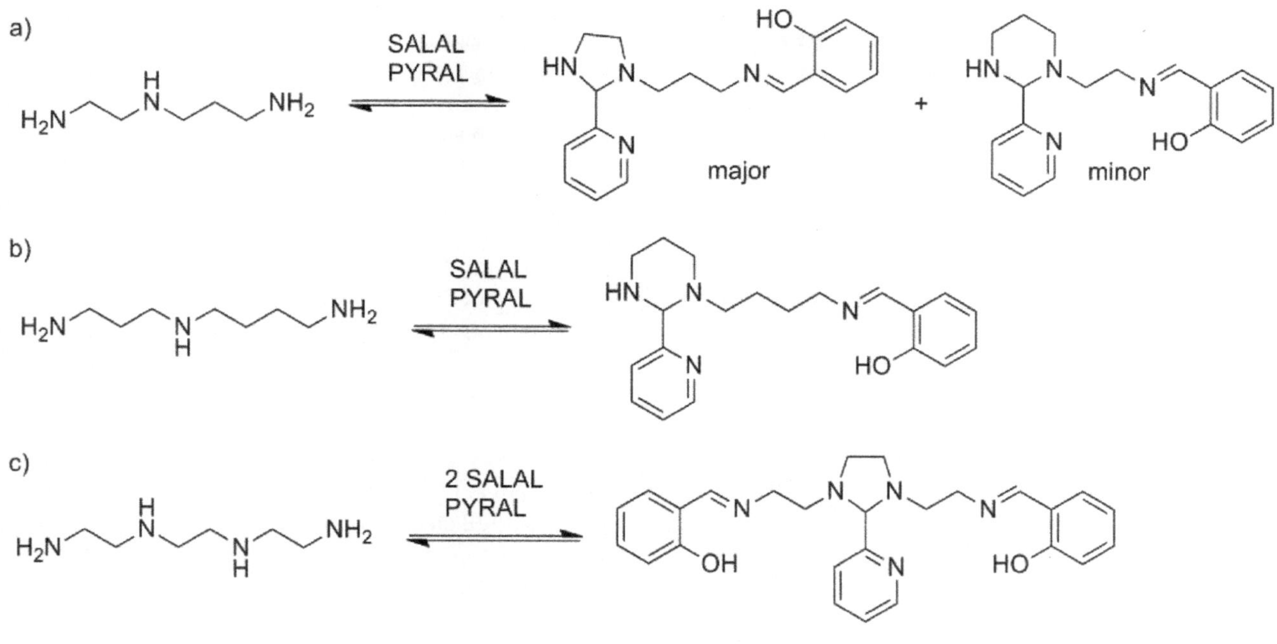

Figure 7: Reactional selectivity with multivalent polyamines: (a) reaction of en-prN_3 with a mixture of SALAL and PYRAL (1 eq. each) gave a mixture of two imine–aminal isomers; (b) spermidine afforded the expected imine–aminal in 71% conversion; (c) en_3N_4 gave with about 60% conversion, the product bearing imines at the termini of the chain, accompanied by the aminal of PYRAL bridging the two central secondary amine sites.

In a further extension to four amino sites, the reaction of triethylenetetramine (**en_3N_4**) with 2 equivalents of SALAL gave a complex mixture of products: four different imine signals (three sharp and one broad) and three aminal peaks were observed in the NMR spectrum, corresponding to all possible combinations of imine–aminal structures formed in an essentially statistical fashion. However, when 1 equivalent of PYRAL was added to the mixture, preferential formation of its aminal on the middle secondary amino groups enhanced the formation of the imines of SALAL at the extremities leading to an equilibrium conversion of about 60% (Figure 7c).

These experiments clearly demonstrate that despite the ambiguous reactivity of both the amine sites and the aldehydes (imine–aminal-lactone equilibria), the system of higher complexity involving all components led to a pronounced competition-enhanced selectivity towards a preferred species, as compared to less complex mixtures. These results provide a remarkable illustration of selectivity amplification, *i.e.* simplification induced by an increase in complexity.

SELECTIVE DYNAMIC PROTECTION OF AMINO GROUPS BY REACTION WITH ALDEHYDES

Specific reversible reaction of different carbonyl reagents with different amino groups may offer a strategy that can be exploited for the selective dynamic protection of amines. The synthesis of complicated multifunctional molecules often requires to perform a selective reaction with a given functional group in presence of other similar groups. Such specific addressing of a given group in complex molecules is enabled by the use of protecting groups (PG).[66,67] However, introduction and removal of a PG necessitates two more synthetic steps accompanied by purification procedures. It is therefore of much interest to develop "protecting-group-free" synthetic methodologies.[68,69] Selective dynamic protecting groups (DPGs) could in principle offer an attractive intermediate approach whereby the desired functional group would be protected directly in the reaction mixture under thermodynamic control without isolation and purification, yet the dynamic nature provides reversibility for comparatively easy dynamic deprotection, for instance *via* transimination. Selective DPGs can be fully complementary to traditional PGs in organic synthesis[66] as there are several types of dynamic linkages which have been shown to be orthogonal to each other,[33–36,70] thus providing a pool of reagents for different functional groups.[71,72] Together with dynamic kinetic resolution,[45] the present work demonstrates the contribution that the implementation of Dynamic Covalent Chemistry (DCC) can make to the field of organic synthesis.

Amine protection with carbonyl compounds enables C-alkylations,[73–76] O-alkylations[77,78] or C=C double bond dihydroxylation.[79] As protecting aldehydes, salicylaldehyde[80–83] and formaldehyde[84–86] have been used for selective imine or aminal formation respectively followed by the typical acidic hydrolysis work up.[73–76] Deprotection can be effected by transimination on application of hydrazides[87] or hydroxylamine derivatives.[82,88]

Selective Derivatization of Amines

Selective dynamic protection of primary amino groups. Common amine protection groups such as Boc, Cbz, Fmoc, *etc.* are introduced by reactions presenting often low selectivity among different amine sites using the corresponding chloroformate or anhydride. A selective version of this protocol employs trifluoroacetate protection of primary amine,[89] but removal of the CF_3CO group may be problematic and therefore lowers overall yields in these reactions. To explore the application of a DPG strategy, we first examine the e of *N*-methyl-1,3-diaminopropane (MeDAP). It contains both a primary and a secondary amino group and can, in principle, form aminal with an aldehyde thus comprising possible challenges in selective derivatisation of polyamines. Previous results have shown that SALAL forms selectively imines as the thermodynamic product with primary amines, as was confirmed also for MeDAP(quantitative conversion in less than 2 hours). The imine formation strongly differentiates the two nitrogen sites in their reactivities: while the secondary amino group is not altered, the primary one is engaged in the imine, which presents in addition a hydrogen bond with the neighbouring OH group. As a result, only the secondary amino group is free to react effectively with electrophiles such as acyl chlorides or isocyanates giving the corresponding amides or ureas while the imine function is not affected.

The dynamic nature of the reversible imine bond forming reaction gives the opportunity for facile protecting group removal in conditions which do not affect acylated amines or similar derivatives. Imine hydrolysis under acidic conditions can be replaced by much milder thermodynamically driven imine exchange. To this end, addition of a hydrazine, a hydrazide or an alkoxyamine leads to cleavage of the imine to give a hydrazone, an acylhydrazone or an oxime (respectively) of the carbonyl partner in quantitative conversion with liberation of the free primary amino group. The double implementation of the features of DCC gives the advantage to perform the three-step reaction sequence, protection–derivatisation–deprotection, in a "one pot" fashion without the need for isolation and purification of the intermediates (Figure 8). A number of such processes have been explored (Figure 8; Table 1). To be successful, this selective DPG approach requires of course high imine formation, high transimination on deprotection, as well as stability of the imine in the conditions used for the derivatization. Comprehensive method optimisation, synthetic details and full characterisation of all products are provided in the ESI, Section 4.† In the following text, the reported yields represent isolated amount of pure products.

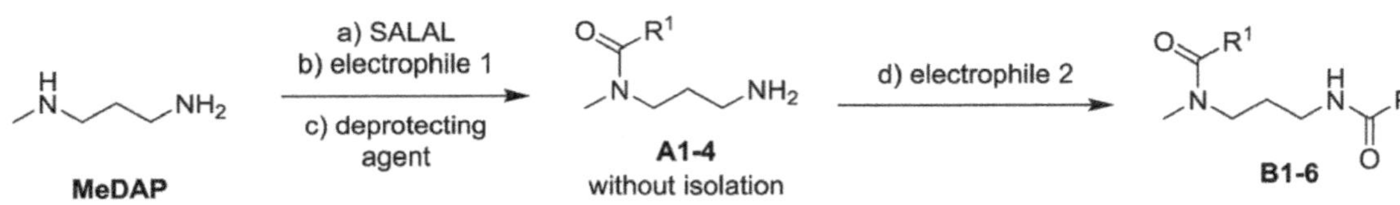

Figure 8: Illustration of the use of the DPG strategy for selective derivatization. Selective acylation of MeDAP is enabled by the selective imine formation using SALAL to protect the primary amine in presence of a secondary one. The reaction is under thermodynamic control which eliminates the need for purification of the protected intermediate. The dynamic nature of the imine bond also allows for easy protecting group removal and the resulting mono-derivatised A1–4 can be reacted "one pot" with a second electrophile to give doubly derivatized products B1–6. The imine formation of SALAL directs the selectivity of the first acylation and thus the reaction sequence. Complete conversion of the imine exchange reaction (used in the deprotection step) provides acylhydrazone and oxime derivatives of**SALAL**, which are under given conditions benign byproducts. The sequence of operations allows in principle for multi-step reaction performed "one pot" and requiring purification only after the last step.

Table 1: Products and isolated yields of the selective derivatisation of **MeDAP** using **SALAL** as protecting group for the primary amine group. Deprotection was effected by imine exchange and the reactions were conducted in "one pot" fashion followed by a single isolation and purification step. Cbz-GlyONp = *N*-benzyloxycarbonyl-glycine-*p*-nitrophenyl ester; Boc-LeuOSu =*N*-*t*-butyloxycarbonyl-leucine-*N'*-hydroxysuccinimide ester; Boc-AlaOSu = *N*-*t*-butyloxycarbonyl-alanine-*N'*-hydroxysuccinimide ester; CbzCl = benzyl chloroformate

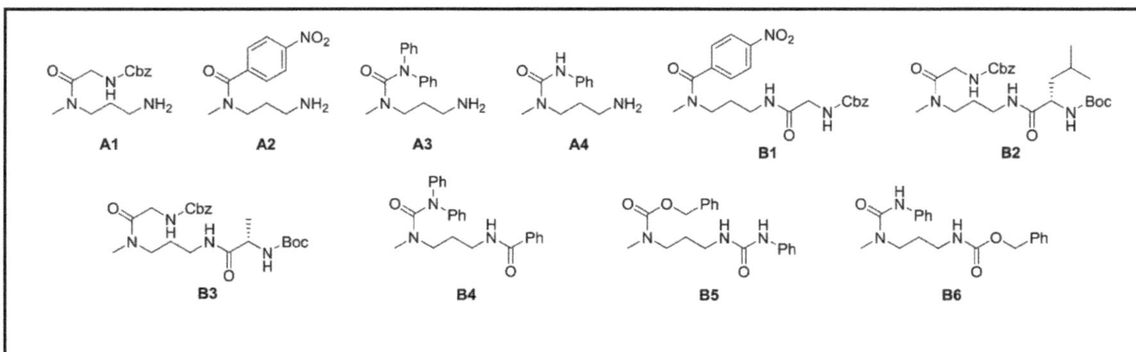

Ref.	Electrophile 1	Electrophile 2	Solvent	Deprotecting agent	Yield
A1	Cbz-GlyONp	—	CH_3CN	$BnONH_3Cl$	82%
A2	*p*NO_2PhCOCl	—	CH_3CN	$BnONH_3Cl$	69%
A3	Ph_2NCOCl	—	CH_3CN	$PhCONHNH_2$	70%
A4	PhNCO	—	CH_3CN	$BnONH_3Cl$	86%
B1	*p*NO_2PhCOCl	Cbz-GlyONp	CH_3CN	$PhCONHNH_2$	78%
B2	Cbz-GlyONp	Boc-LeuOSu	CH_3CN	$PhCONHNH_2$	83%

B3	Cbz-GlyONp	Boc-AlaOSu	CH_3CN	$PhCONHNH_2$	82%
B4	Ph_2NCOCl	PhCOCl	EtOH	$BnONH_3Cl$	75%
B5	CbzCl	PhNCO	CH_3CN	$PhCONHNH_2$	71%
B6	PhNCO	CbzCl	EtOH	$BnONH_3Cl$	82%

In the reaction of MeDAP with the *p*-nitrophenyl activated ester of protected glycine (Cbz-GlyONp) using dynamic protection by 1 eq. SALAL, the *N*-Cbz-glycyl substituent was introduced on the secondary amine, giving isolated yield of 82% for the full three-step one-pot reaction sequence. Importantly, when the reaction was performed without the addition ofSALAL, it afforded 85% isolated yield of a product consisting of a mixture of regioisomers due to unsufficient difference in reactivity of the two nitrogen atoms. The regioisomers were present in 2 : 1 ratio in favour of the product of acylation on the secondary amine. The product of double acylation of MeDAP was isolated as well in about 6% yield. We explored the versatility of the protocol by varying the reagents, the solvent and also the nature of the deprotecting agent. The results are summarized in Table 1 (entries A1–A4). Good isolated yields in the range of 69–86% of the desired products were obtained after the complete three-step reaction sequence.

The operation of DCC under thermodynamic control has been exploited here for the protection and deprotection steps. The establishment of equilibrium offers the possibility to perform sequences of reactions in systems of increasing complexity. Thus, the equilibrium mixtures in the previous experiments before isolation of products consist of the oxime or acylhydrazone ofSALAL together with a MeDAP derivative displaying a free primary amino group (complete conversion in deprotection). We have therefore investigated the possibility to perform a controlled sequential derivatisation in a "one pot" fashion (Figure 8), in which the imine formation by SALAL directs the sequence of derivatization to give products bearing the residue introduced first on the secondary amine group and the second one on the primary amine site. The reaction was repeated for various combinations of electrophiles, in two solvents and with two deprotecting agents giving yields of the desired products in the range of 71–83% (92–95% per step, Table 1 entries B1–B6). In contrast, when the reaction was performed in absence of the SALAL-imine protecting group using *N,N*-diphenylcarbamoyl chloride and benzoyl chloride as electrophiles, a mixture of all four possible products was obtained in yields from 16 to 34%.

Inverted Sequence of Derivatisation

In contrast to **SALAL**, which yields selectively imines, PYRAL preferentially gives aminals.[53] Aminal formation between a primary and a secondary amine site transforms the latter into a tertiary amine and the former into a secondary amine site, which thus is available for reaction with an electrophile. As a consequence, acylation leads to a regioselectivity opposite to that obtained when SALAL is used, resulting in an opposite sequence of acyl groups in the e of sequential double acylation. Such a process has been performed with MeDAP as shown in Figure 9.

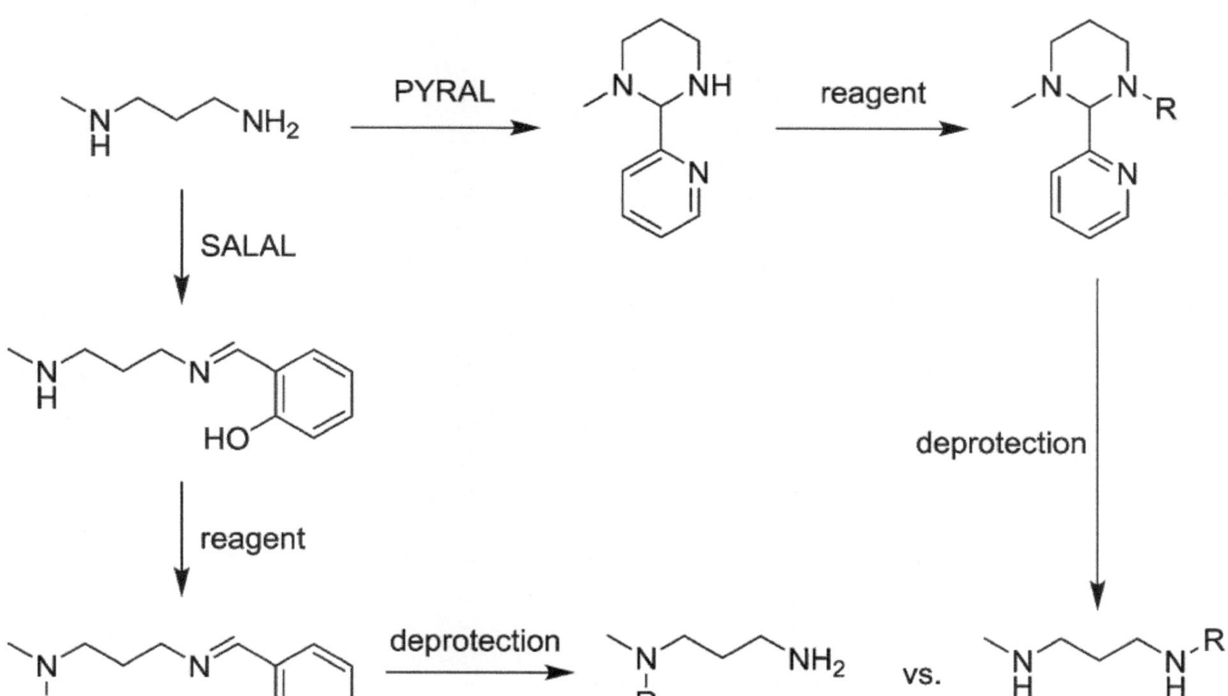

Figure 9: Representation of the generation of opposite sequences of substitution on consecutive double acylation of MeDAP when either SALAL or PYRAL is used as dynamic protecting group. Opposite regioisomers are obtained as a result of the preferential formation of imine and aminal condensation products by SALAL and PYRAL respectively. Both imines and aminals are formed reversibly which allows for mild deprotection with hydroxylamine or hydrazide derivatives.

When PYRAL was mixed with MeDAP, efficient formation of the aminal within 2 hours was confirmed by NMR. The equilibrated solution was then treated with phenyl isocyanate forming the urea moiety and the protecting group was removed by the exchange reaction with *O*-benzylhydroxylamine which converts the aminal of PYRAL completely to the corresponding oxime. The resulting product **C1** was isolated in 74% yield, accompanied with 2% of the opposite regioisomer (Table 2). It is thus possible to perform a double derivatisation of the diamine in a "one pot" fashion by addition of a second electrophile after deprotection. To this end, we have used phenyl isocyanate as the first electrophile and the activated ester of*N*-protected phenylalanine or benzyl chloroformate in the second acylation (isolated yields 76 and 74%, respectively), demonstrating the potential of the approach implementing dynamic protecting groups and mild deprotection procedures compared to conventional protecting groups used in peptide synthesis.

Table 2: Inverted selectivity of derivatisation of **MeDAP**. Formation of the six-membered aminal ring by the condensation with **PYRAL** leaves only the terminal amino group available for the reaction with electrophiles. Isolated yields for one pot multi-step procedures. Boc-PheOSu = *N-t*-butyloxycarbonyl-phenylalanine-*N'*-hydroxysuccinimide ester; CbzCl = benzyl chloroformate

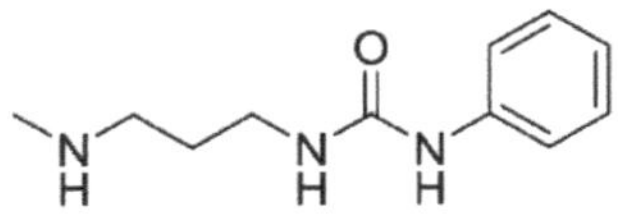

C1

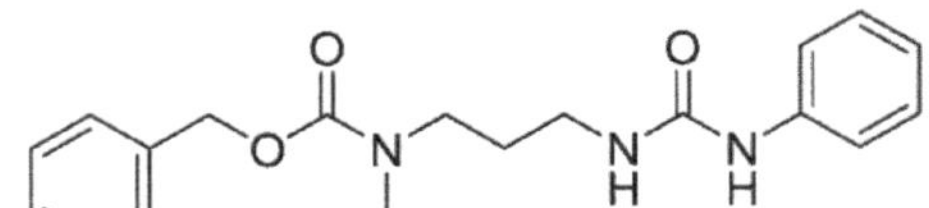

C2

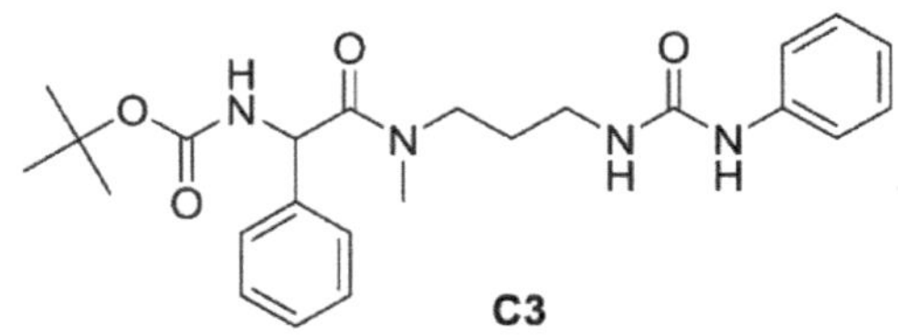

C3

Ref.	Reagent 1	Reagent 2	Solvent	Deprotecting agent	Yield
C1	PhNCO	—	CH_3CN	$BnONH_3Cl$	74%
C2	PhNCO	CbzCl	CH_3CN	$BnONH_3Cl$	74%
C3	PhNCO	Boc-PheOSu	CH_3CN	$BnONH_3Cl$	76%

The reaction employing PYRAL protection *via* aminal formation was studied with several different electrophiles used in the first acylation, but the isolated product always consisted of the diamine derivatised on the secondary amine (due to aminal–imine equilibrium, see above). Even after extensive optimisation of reaction conditions (details in the ESI, Section 4.1.3†) and replacement of PYRAL with formaldehyde, reported in the literature as aminal forming reagent,[84,85] the terminal regioisomer was only obtained with phenyl isocyanate. Literature reports describe the use of aminal forming aldehyde to drive the selectivity of the Michael addition of acrylonitrile to polyamines.[85,90] In this vein, MeDAP was reacted first with 1 eq. ofPYRAL (or SALAL) and then 1 eq. of acrylonitrile was added. Formation of the product of Michael addition (Figure 10a) was followed by NMR (at r.t. in d_4-methanol) revealing that the reaction time in presence of PYRAL was much longer (35% conversion after 24 hours) than in the e of SALAL protected version (quantitative in 24 hours). Moreover, while SALALdrove the reaction with full selectivity for the addition on the secondary nitrogen, the reaction mixture with PYRAL consisted of both regioisomers in ratio 1 : 2.5 in favour of the same product as with SALAL (due to the aminal–imine dynamic interconversion). When the acrylonitrile Michael addition was repeated without any protecting aldehyde, again a mixture of products was obtained, but in this e favouring the addition on the primary amine. This finding indicates kinetic resolution between primary and secondary amines in the acrylonitrile Michael addition, which was supported by the reaction of 1,3-diaminopropane with 2 eq. of acrylonitrile providing a quantitative yield of the *N,N'*-bis(2-cyanoethyl)-1,3-diaminopropane (Figure 10b, details in the ESI, Section 4.4†).[91] In conclusion, the PYRAL protection is a suitable approach in the e of highly reactive isocyanate species, presumably because it reacts faster than is the rate of intramolecular aminal–imine interconversion. In the e of acyl chlorides this equilibrium, although strongly shifted towards the aminal, leads to non-selective derivatisation of both amino groups. Acrylonitrile, reacts preferentially with primary amines and SALAL protection is needed to achieve the selective reaction on the secondary nitrogen.

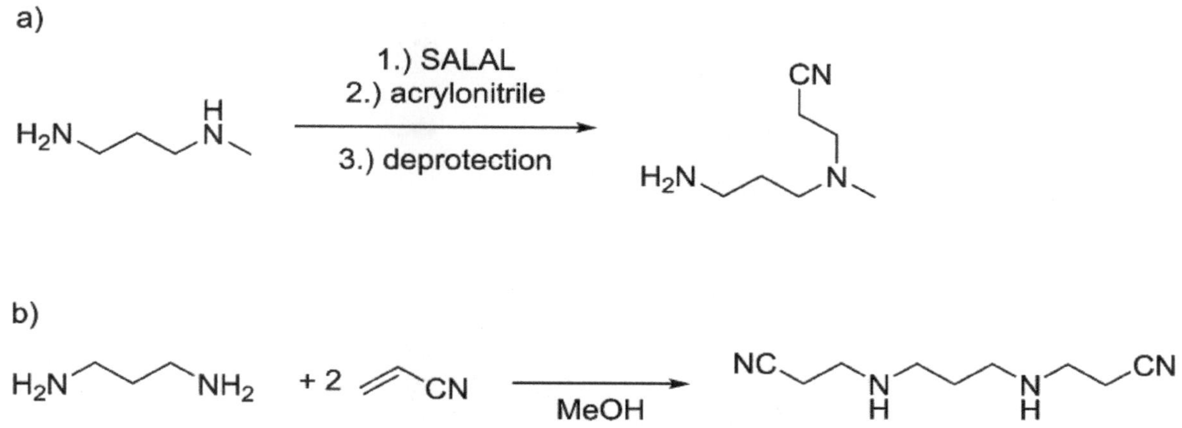

Figure 10: (a) Michael addition of acrylonitrile proceeds faster on the primary amino group. Selective reaction of the secondary amino group can be achieved by protection with **SALAL**. (b) Double Michael reaction of acrylonitrile (2 eq.) with 1,3-diaminopropane proceeds with complete selectivity on primary amino groups in full conversion without any protecting group.

Selective Derivatisation of Oligoamines

Selective dynamic protection of amino groups was further explored with challenging polyamine substrates. To demonstrate the principle and to draw a comparison with known alternative protocols, we have examined the monoderivatisation of diethylenetriamine **en$_2$N$_3$** at the central secondary amine group. The traditional protection/reaction/deprotection approach[89] employs trifluoroacetyl protection of the terminal NH_2 functions, followed by reaction with Boc_2O and removal of the trifluoroacetate groups by reflux in ammonia solution with overall 63% yield including at least two chromatographic purification steps. In the present e, SALAL is used as the protecting agent (2 eq.) and the reaction is performed at room temperature, in "one pot" fashion, with a single isolation and purification step giving the desired product in an overall yield of 80% (see ESI for the synthetic protocol†). Selective DPGs offer large versatility which is not easily available with conventional protecting groups. If pr$_2$N$_3$ is protected by 1 eq. of PYRAL forming the aminal and then reacted with an electrophile, selective monoderivatisation of one of the termini is achieved in good yields (70%, Figure 11, see ESI, Section 4.5,† for synthetic details). In this e, the PYRAL protection is crucial since reproducing the reaction without the protecting aldehyde led to a complex mixture of all possible products in essentially statistic ratio.

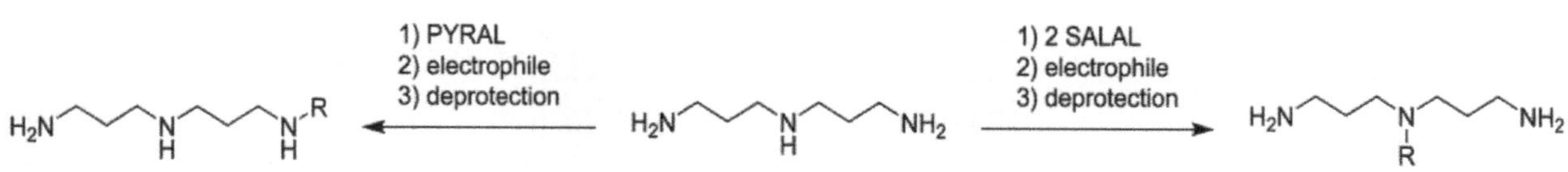

Figure 11: Selective derivatisation of polyamines in the e of **pr$_2$N$_3$**. **en$_2$N$_3$** gives similar results. (Right) End-capping of the polyamine chain with **SALAL** protection drives the reaction to the central secondary nitrogen, whereas (left) linking two nitrogens in the aminal form of **PYRAL** leads to selective derivatisation of only one of the terminal primary amine.

Finally, the sequential derivatisation shown above for diamines was also investigated with the biologically relevant triamine spermidine. In this e, the reaction sequence started with protection by 1 eq. of **PYRAL** forming the aminal and thus leaving only one primary amino group available for the subsequent reaction with the activated ester of an amino acid (Cbz-GlyONp). The protecting group was removed by exchange reaction with benzyloxyamine and without isolation, the subsequent protection by 1 eq. of **SALAL** was introduced. The selective imine formation at the other terminal NH_2 group of the starting polyamine left the central secondary amine free, thus allowing for selective derivatisation in the middle of the chain by benzyl chloroformate. Deprotection was again performed by imine exchange with benzyloxyamine and the last remaining nitrogen atom was thereafter derivatized using *p*-nitrobenzoyl chloride (Figure 12). This 7-step, one-pot reaction sequence performed at r.t. under ambient atmosphere afforded after a single purification step the desired triply derivatized product in 31% yield (85% calc. per step), demonstrating the potential of dynamic selective protecting groups in organic synthesis.

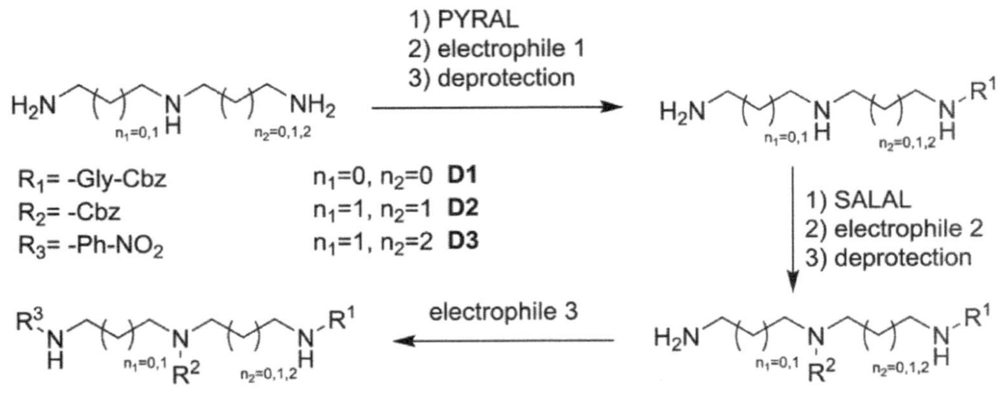

Figure 12: Selective derivatisation of oligo(poly)amines is enabled by preferential product formation of a given aldehyde with the amine under thermodynamic control. Mild deprotection by imine exchange reaction, again thermodynamically driven, allows performing a sequence of three derivatisation steps without isolation and purification of intermediates.

The presented strategy of selective DPGs showed remarkable versatility when **SALAL** was used as the protecting agent. On the other hand, inverted sequence of derivatisation by employing **PYRAL** was limited to es in which the acylating reagent reacts faster than the rate of equilibration of the protecting group between several species. These two approaches are thus complementary and when combined, provide a powerful tool to perform selective and/or sequential functionalization of polyamines.

CONCLUSIONS

Dynamic covalent chemistry (DCC) operates at thermodynamic equilibrium achieved by component exchange through reversible covalent reactions. The condensation of carbonyl compounds with amines is of special interest as it can result in the formation of N–C–N aminals in addition to the usual C=N imines, depending on the nature of the carbonyl component. It offers thus a richer palette of constituents that may be exploited towards the design of selective reaction pathways of interest for organic synthetic strategies. It allows for substrate selectivity in mixtures of amines as well as for the programming of reaction sequences towards the control of regioselective positioning of carbonyl residues along an oligo(poly)amine chain. These features lead to the concept of *selective dynamic protecting groups*, whereby primary and secondary amine functional groups may be reversibly derivatized in processes presenting selectivity between different amine compounds as well as between different amino sites within a polyamine *via* a network of underlying equilibrating reactions.

The selective DPG concept has been exploited towards the regioselective derivatization of various amines with different reagents ranging from isocyanates to activated esters of amino acids. The establishment of equilibrium in the reaction of the substrate with the protecting group as well as in the deprotection step eliminates the need for isolation and purification of intermediates and allows to run reactions in "one-pot" fashion. Furthermore, the thermodynamic control over the reversible condensations opens the possibility to perform sequences or networks of reactions in systems of increasing complexity, as it was shown in the e of selective sequential derivatisation of polyamines. Exploitation of other reversibly formed species, such as acetals or boroxines, can provide useful alternatives in total synthesis of complex products. Altogether the present work represents an extension of the thermodynamically driven DCC, implementing *dynamic organic reactivity*, into the traditionally kinetically governed realm of organic synthesis and provides ground for further exploration of its potential.

Finally, in a broad perspective, the data presented provide a remarkable illustration of simplification within a given constitutional dynamic library, induced by an increase in complexity, which leads to enhanced competition within the set of equilibrating constituents undergoing component exchange *via* a network of interconnected reactions engaged in agonistic and antagonistic relationships with feedback loops. In general terms, higher complexity results in simplification through competition.

ACKNOWLEDGEMENTS

We thank the ERC Advanced grant "SUPRADAPT" for financial support. A. C. M. thanks the French Embassy in Berlin for a post-doctoral fellowship. P. K. thanks the Université de Strasbourg for a doctoral fellowship. K. K. thanks the Erasmus exchange programme for an international internship.

NOTES AND REFERENCES

1. J.-M. Lehn, *Chem.–Eur. J.*, 1999, **5**, 2455–2463 .
2. O. Ramstrom and J.-M. Lehn, *Nat. Rev. Drug Discovery*, 2002, **1**, 26–36 .
3. P. T. Corbett, J. Leclaire, L. Vial, K. R. West, J.-L. Wietor, J. K. M. Sanders and S. Otto, *Chem. Rev.*, 2006, **106**, 3652–3711 .
4. J.-M. Lehn, *Chem. Soc. Rev.*, 2007, **36**, 151–160 .
5. S. Ladame, *Org. Biomol. Chem.*, 2008, **6**, 219–226 .
6. R. A. R. Hunt and S. Otto, *Chem. Commun.*, 2011, **47**, 847–858 .
7. S. R. Beeren and J. K. M. Sanders, *J. Am. Chem. Soc.*, 2011, **133**, 3804–3807 .
8. S. Hamieh, V. Saggiomo, P. Nowak, E. Mattia, R. F. Ludlow and S. Otto, *Angew. Chem., Int. Ed.*, 2013, **52**, 12368–12372 .
9. J.-M. Lehn, *Angew. Chem., Int. Ed.*, 2013, **52**, 2836–2850 .

10. S. P. Black, J. K. M. Sanders and A. R. Stefankiewicz, *Chem. Soc. Rev.*, 2014, **43**, 1861–1872 .
11. J.-M. Lehn, *Angew. Chem., Int. Ed.*, 2015, **54**, 3276–3289 .
12. C. Wang, G. Wang, Z. Wang and X. Zhang, *Chem.–Eur. J.*, 2011, **17**, 3322–3325 .
13. D. N. Bunck and W. R. Dichtel, *J. Am. Chem. Soc.*, 2013, **135**, 14952–14955 .
14. J.-M. Lehn, *Prog. Polym. Sci.*, 2005, **30**, 814–831 .
15. E. Moulin, G. Cormos and N. Giuseppone, *Chem. Soc. Rev.*, 2012, **41**, 1031–1049 .
16. T. Aida, E. W. Meijer and S. I. Stupp, *Science*, 2012, **335**, 813–817 .
17. J. Li, J. M. A. Carnall, M. C. A. Stuart and S. Otto, *Angew. Chem., Int. Ed.*, 2011, **50**, 8384–8386 .
18. T. Ono, S. Fujii, T. Nobori and J.-M. Lehn, *Chem. Commun.*, 2007, 4360–4362 .
19. T. Ono, S. Fujii, T. Nobori and J.-M. Lehn, *Chem. Commun.*, 2007, 46–48 .
20. C. B. Minkenberg, F. Li, P. van Rijn, L. Florusse, J. Boekhoven, M. C. A. Stuart, G. J. M. Koper, R. Eelkema and J. H. van Esch, *Angew. Chem., Int. Ed.*, 2011, **50**, 3421–3424 .
21. A. Ciesielski, M. El Garah, S. Haar, P. Kovaříček, J.-M. Lehn and P. Samorì, *Nat. Chem.*, 2014, **6**, 1017–1023 .
22. A. Granzhan, T. Riis-Johannessen, R. Scopelliti and K. Severin, *Angew. Chem., Int. Ed.*, 2010, **49**, 5515–5518 .
23. K. Acharyya, S. Mukherjee and P. S. Mukherjee, *J. Am. Chem. Soc.*, 2013, **135**, 554–557 .
24. F. B. L. Cougnon, N. Ponnuswamy, N. A. Jenkins, G. D. Pantoş and J. K. M. Sanders, *J. Am. Chem. Soc.*, 2012, **134**, 19129–19135 .
25. S. J. Rowan and J. F. Stoddart, *Org. Lett.*, 1999, **1**, 1913–1916 .
26. G. J. Mohr, C. Demuth and U. E. Spichiger-Keller, *Anal. Chem.*, 1998, **70**, 3868–3873 .
27. T. Hotchkiss, H. B. Kramer, K. J. Doores, D. P. Gamblin, N. J. Oldham and B. G. Davis, *Chem. Commun.*, 2005, 4264–4266 .
28. *Dynamic combinatorial chemistry in drug discovery, bioorganic chemistry, and materials science*, ed. B. L. Miller, Wiley, Hoboken, N.J, 2010.
29. M. Hochgürtel and J.-M. Lehn, in *Fragment-based Approaches in Drug Discovery*, ed. W. Jahnke and D. A. Erlanson, Wiley-VCH Verlag GmbH & Co. KGaA, 2006, pp. 341–364.
30. A. Herrmann, *Chem. Soc. Rev.*, 2014, **43**, 1899–1933 .
31. S. Duan, W. Yuan, F. Wu and T. Jin, *Angew. Chem., Int. Ed.*, 2012, **51**, 7938–7941 .
32. J. D. Cheeseman, A. D. Corbett, J. L. Gleason and R. J. Kazlauskas, *Chem.–Eur. J.*, 2005, **11**, 1708–1716 .
33. V. Goral, M. I. Nelen, A. V. Eliseev and J.-M. Lehn, *Proc. Natl. Acad. Sci. U. S. A.*, 2001, **98**, 1347–1352 .
34. M. Schmittel and K. Mahata, *Angew. Chem., Int. Ed.*, 2008, **47**, 5284–5286 .
35. A. Wilson, G. Gasparini and S. Matile, *Chem. Soc. Rev.*, 2014, **43**, 1948–1962 .
36. M. L. Saha, S. De, S. Pramanik and M. Schmittel, *Chem. Soc. Rev.*, 2013, **42**, 6860–6909 .
37. L. A. Wessjohann, D. G. Rivera and F. León, *Org. Lett.*, 2007, **9**, 4733–4736 .
38. J.-B. Lin, X.-N. Xu, X.-K. Jiang and Z.-T. Li, *J. Org. Chem.*, 2008, **73**, 9403–9410 .
39. Q. Ji, R. C. Lirag and O. Š. Miljanić, *Chem. Soc. Rev.*, 2014, **43**, 1873–1884 .
40. Y.-M. Legrand, A. van der Lee and M. Barboiu, *Inorg. Chem.*, 2007, **46**, 9540–9547 .
41. M. L. Saha and M. Schmittel, *Org. Biomol. Chem.*, 2012, **10**, 4651–4684.
42. C.-F. Chow, S. Fujii and J.-M. Lehn, *Chem. Commun.*, 2007, 4363–4365 .
43. P. N. W. Baxter, J.-M. Lehn and K. Rissanen, *Chem. Commun.*, 1997, 1323–1324 .
44. M. Hutin, C. J. Cramer, L. Gagliardi, A. R. M. Shahi, G. Bernardinelli, R. Cerny and J. R. Nitschke, *J. Am. Chem. Soc.*, 2007, **129**, 8774–8780 .
45. K. Osowska and O. S. Miljanić, *J. Am. Chem. Soc.*, 2011, **133**, 724–727 .
46. K. Osowska and O. Š. Miljanić, *Angew. Chem., Int. Ed.*, 2011, **50**, 8345–8349 .
47. Q. Ji and O. Š. Miljanić, *J. Org. Chem.*, 2013, **78**, 12710–12716 .
48. S. De, K. Mahata and M. Schmittel, *Chem. Soc. Rev.*, 2010, **39**, 1555–1575 .
49. K. Mahata and M. Schmittel, *J. Am. Chem. Soc.*, 2009, **131**, 16544–16554 .
50. M. Schmittel and K. Mahata, *Chem. Commun.*, 2008, 2550–2552 .
51. I. Kocsis, D. Dumitrescu, Y.-M. Legrand, A. van der Lee, I. Grosu and M. Barboiu, *Chem. Commun.*, 2014, **50**, 2621–2623 .
52. J. R. Nitschke and J.-M. Lehn, *Proc. Natl. Acad. Sci. U. S. A.*, 2003, **100**, 11970–11974 .
53. P. Kovaříček and J.-M. Lehn, *J. Am. Chem. Soc.*, 2012, **134**, 9446–9455 .
54. P. Kovaříček and J.-M. Lehn, *Chem.–Eur. J.*, 2015, **21**, 9380–9384 .
55. (*a*) For general considerations on the notion of "simplexity" in particular in a biological context, see A. Berthoz, La simplexité, Odile Jacob, 2009; (*b*) For an elaboration in the context of organic synthesis, see: Ph. Compain, L'actualité Chimique, 2003, april-may issue, 129–134.
56. (*a*) For an early e, see the competitive selection in the generation of double and triple helicates in an interconverting set of ligands and metal ions: see R. Krämer, J.-M. Lehn and A. Marquis-Rigault, *Proc. Natl. Acad. Sci. U. S. A.*, 1993, **90**, 5394–5398 ; (*b*) For the effect of increased complexity in self-sorting, see: K. Mahata and M. Schmittel, *Beilstein J. Org. Chem.*, 2011, **7**, 1555–1561.

57. J.-M. Lehn, in *Constitutional Dynamic Chemistry*, ed. M. Barboiu, Springer, Berlin Heidelberg, 2012, pp. 1–32.
58. J.-M. Lehn, *Supramolecular chemistry: concepts and perspectives*, VCH, Weinheim, 1995.
59. J.-M. Lehn, *Proc. Natl. Acad. Sci. U. S. A.*, 2002, **99**, 4763–4768 .
60. I. Saur and K. Severin, *Chem. Commun.*, 2005, 1471–1473 .
61. S. Ulrich and J.-M. Lehn, *J. Am. Chem. Soc.*, 2009, **131**, 5546–5559 .
62. C. Bohne, *Chem. Soc. Rev.*, 2014, **43**, 4037–4050 .
63. M. Chaur, S. Kulchat, J.-M. Lehn, and work in progress.
64. M. Ciaccia, R. Cacciapaglia, P. Mencarelli, L. Mandolini and S. Di Stefano, *Chem. Sci.*, 2013, **4**, 2253–2261 .
65. C. Fasting, C. A. Schalley, M. Weber, O. Seitz, S. Hecht, B. Koksch, J. Dernedde, C. Graf, E.-W. Knapp and R. Haag, *Angew. Chem., Int. Ed.*, 2012, **51**, 10472–10498 .
66. P. G. M. Wuts and T. W. Greene, *Greene's protective groups in organic synthesis*, Wiley-Inteience, Hoboken, N.J, 4th edn, 2007.
67. K. C. Nicolaou, *Classics in total synthesis: targets, strategies, methods*, VCH, Weinheim, New York, 1996.
68. P. S. Baran, T. J. Maimone and J. M. Richter, *Nature*, 2007, **446**, 404–408 .
69. K. C. Nicolaou, *J. Org. Chem.*, 2009, **74**, 951–972 .
70. C.-H. Wong and S. C. Zimmerman, *Chem. Commun.*, 2013, **49**, 1679–1695 .
71. Y. Li and X. Liu, *Chem. Commun.*, 2014, **50**, 3155–3158 .
72. S. Mundinger, U. Jakob and W. Bannwarth, *Chem.–Eur. J.*, 2014, **20**, 1258–1262 .
73. P. Bey and J. P. Vevert, *Tetrahedron Lett.*, 1977, **18**, 1455–1458 .
74. B. W. Metcalf and P. ara, *Tetrahedron Lett.*, 1975, **16**, 3337–3340 .
75. R. Polt and M. A. Peterson, *Tetrahedron Lett.*, 1990, **31**, 4985–4986 .
76. J. M. Hornback and B. Murugaverl, *Tetrahedron Lett.*, 1989, **30**, 5853–5856 .
77. M. Bergmann and L. Zervas, *Ber. Dtsch. Chem. Ges. B*, 1931, **64**, 975–980 .
78. M. A. Peterson and R. Polt, *J. Org. Chem.*, 1993, **58**, 4309–4314 .
79. E. J. Corey, A. Guzman-Perez and M. C. Noe, *J. Am. Chem. Soc.*, 1995, **117**, 10805–10816 .
80. J. N. Williams Jr and R. M. Jacobs, *Biochem. Biophys. Res. Commun.*, 1966, **22**, 695–699 .
81. J. C. Sheehan and V. J. Grenda, *J. Am. Chem. Soc.*, 1962, **84**, 2417–2420 .
82. A. R. Khomutov, A. S. Shvetsov, J. J. Vepsäläinen and A. M. Kritzyn, *Tetrahedron Lett.*, 2001, **42**, 2887–2889 .
83. J. D. Prugh, L. A. Birchenough and M. S. Egbertson, *Synth. Commun.*, 1992, **22**, 2357–2360 .
84. J. S. McManis and B. Ganem, *J. Org. Chem.*, 1980, **45**, 2041–2042 .
85. B. Ganem, *Acc. Chem. Res.*, 1982, **15**, 290–298 .
86. J. C. Bradley, J. P. Vigneron and J. M. Lehn, *Synth. Commun.*, 1997, **27**, 2833–2845 .
87. D. R. Mootoo and B. Fraser-Reid, *Tetrahedron Lett.*, 1989, **30**, 2363–2366 .
88. K.-J. Fasth, G. Antoni and B. Langström, *J. Chem. Soc., Perkin Trans. 1*, 1988, 3081–3084 .
89. S. Srinivasachari, Y. Liu, G. Zhang, L. Prevette and T. M. Reineke, *J. Am. Chem. Soc.*, 2006, **128**, 8176–8184 .
90. B. Frydman, S. Bhattacharya, A. Sarkar, K. Drandarov, S. Chesnov, A. Guggisberg, K. Popaj, S. Sergeyev, A. Yurdakul, M. Hesse, H. S. Basu and L. J. Marton, *J. Med. Chem.*, 2004,**47**, 1051–1059 .
91. S. K. Sharma, Y. Wu, N. Steinbergs, M. L. Crowley, A. S. Hanson, R. A. ero and P. M. Woster, *J. Med. Chem.*, 2010, **53**, 5197–5212 .

CHAPTER 12

DESIGN AND SYNTHESIS OF ANALOGUES OF NATURAL PRODUCTS

Martin E. Maier

INTRODUCTION

Natural product research represents an important part of our scientific culture. From natural products (NPs) a lot of biology could be learned. Their diverse structures have enriched the collection of known organic molecules. Studies related to their biosynthesis have provided valuable insights into genetics and enzymology. Moreover, many NPs turned out to be useful molecules or lead compounds in the fragrance, crop science or pharmaceutical industries. For the latter this is well documented. Thus, an analysis of the sources of new drugs from 1981–2010 indicated that only 36% of the new chemical entities (NCEs) were discovered without inspiration from a natural product.[1] Despite the obvious importance, interest in natural product isolation by the pharmaceutical industry has declined in recent years. For this, several reasons might be responsible. Certainly, natural product isolation is time consuming and elaborate. A big challenge for the discovery of new NPs is dereplication, that is, to recognize and weed out already known substances. In parallel the last decade has seen a downward trend in the NCEs per year from 2001 to 2010. According to a recent hypothesis this might be due to the heavy reliance of the industry on combinatorial chemistry.[1] Also, it seems that computational techniques like virtual screening are not yet able to produce good lead compounds. Basically, we are far from the vision that a drug candidate can be produced by rational design on a computer or by virtual screening of the chemical space. We still do not know enough about molecular recognition and what it takes for a ligand to trigger a signal on a receptor. Certainly, drug discovery is much more difficult than to make a plaster of a shoe print. It is rather comparable to horse racing or football betting, where knowledge and experience are advantageous.

Citation: Martin E. Maier, "Design and synthesis of analogues of natural products," DOI: 10.1039/c5ob00169b.

Therefore it might be worthwhile to reconsider natural products as the inspiration and the starting point of novel drugs.[2] Several papers have analyzed differences between NPs and compounds from medicinal chemistry. Thus, classical medicinal chemistry is characterized by a lot of C–N bond formation. In contrast, NP syntheses strongly rely on C–C bond forming reactions. These are often accompanied by creation of OH groups, for example through allylation or aldol reactions. As a consequence, the number of asymmetric centers is higher in NPs. In addition, most of the rings in NPs are partly or completely saturated. With regard to physicochemical or drug-like properties of NPs, an analysis by A. Ganesan of 24 NPs that became approved drugs between 1970 and 2006 revealed that these fall into two categories.[3] NPs of the first category adhere to the Lipinski Rule of Five. NPs of the second category just follow the log $P < 5$ rule and the rule regarding the number of H-bond donors <5. The obvious advantage of NPs over non-natural molecules is that during their biosynthesis they have already interacted with different enzymes and proteins.[4] Accordingly, their core scaffolds can be considered privileged structures. In other words, out of all possible small organic molecules NPs only occupy certain areas of chemical space that seems predestined for interaction with proteins.

The analysis and visualization of chemical space is a topic of bioinformatics. Thus, software programmes were devised that are able to generate all theoretically possible molecules following basic rules for covalent bonds and functional groups. For example, the chemical universe database GDB-17 includes all molecules up to a size of 17 atoms considering C, N, O, S and halogens. It contains 166.4 billion molecules.[5] In principle this database should contain many NP-like structures. To explore such large collections of molecules, descriptors are calculated for each molecule followed by projecting multidimensional property space in lower dimensions for visualization purposes.[6,7] In another instance, a representative universal library spanning the small molecule chemical space was constructed from the known chemical universe by stochastic search, including mutation and crossover for generating new molecules from known ones. This library contains 8.9 million structures.[8] All these libraries appear useful for virtual screening and bioinformatics analysis. However, hits would still have to be synthesized in order to find out whether they show biological activity or not.

Because of the reasons mentioned above, NPs appear good starting points for analogue synthesis. Chemistry related to NPs also provides the opportunity to test new reactions and synthetic strategies. In the following we discuss strategies that have been used in the synthesis of NP analogues. Of course only representative examples could be considered, since countless papers deal with the synthesis of NP analogues.

DIVERTED TOTAL SYNTHESIS (DTS)

The synthesis of NP analogues is probably as old as the field of total synthesis. In the simplest approach a NP can be converted to an analogue by modifications of its functional groups. In a total synthesis programme, modifications are possible on advanced intermediates. Moreover, total synthesis allows deep seated structural modifications by using modified building blocks in fragment-based synthesis. The term diverted total synthesis (DTS) was suggested by Danishefsky to describe this strategy.[9] This includes the preparation of modified NPs as probes for elucidating the mode of action.[10] To streamline and simplify the synthesis of NP analogues, attempts have also been made to implement solid-phase strategies.[11] Famous es from the last one or two decades for diverted total synthesis include the epothilone story.[12] Two prominent epothilones are sagopilone (**2**) even though it is not marketed, and the amide analogue **3**(ixempra), which is in clinical use. The development of carfilzomib (**5**) from epoxomicin (**4**)[13] or the synthesis of the powerful antibiotic eravacycline (**7**) taking tetracycline (**6**) as the lead structure (Figure 1) can also be mentioned.[14,15] Related to DTS is the concept of diversity-oriented synthesis.[16] While the compound collections obtained by this strategy are typically different from natural products, some examples are known where NP-like structures were obtained.[17]

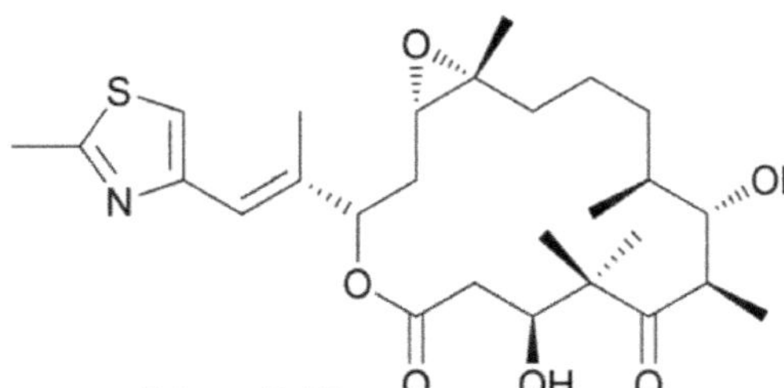

epothilone B (**1**)

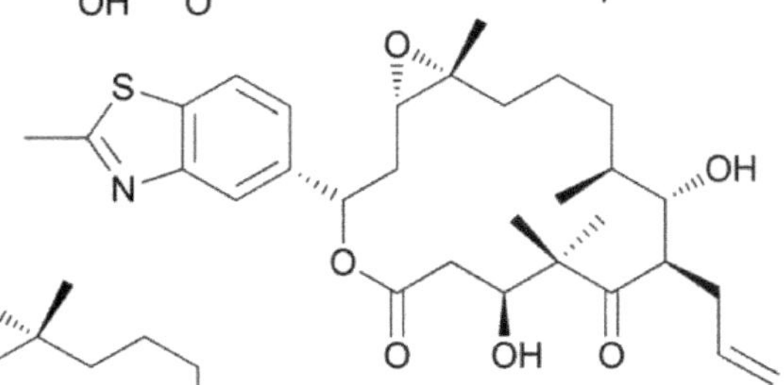

sagopilone (**2**)

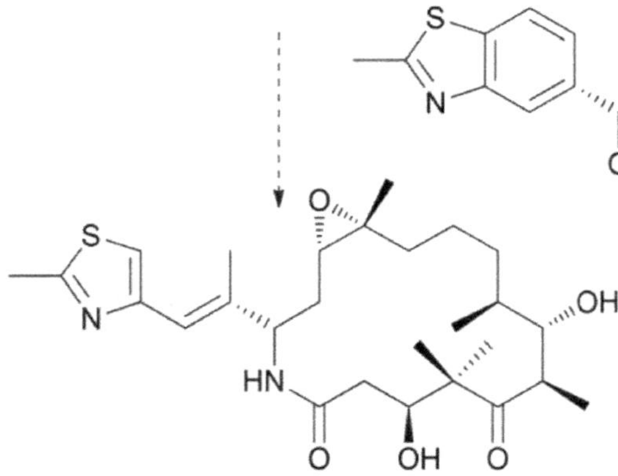

ixempra (ixabepilone) (**3**)

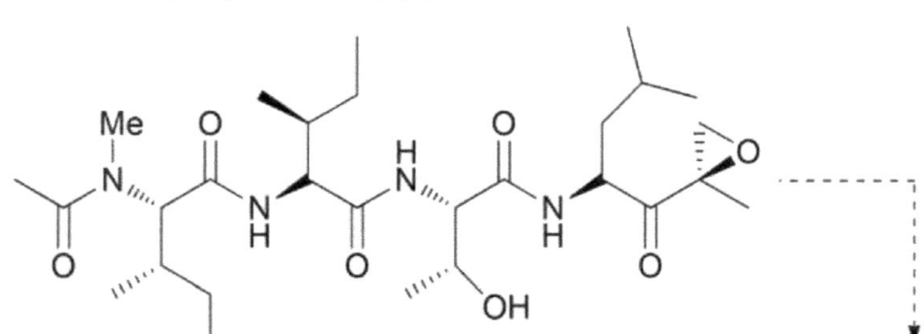

epoxomicin (**4**)

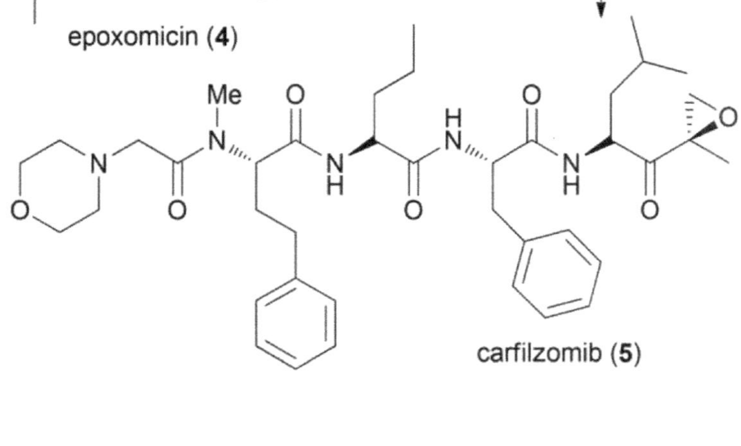

carfilzomib (**5**)

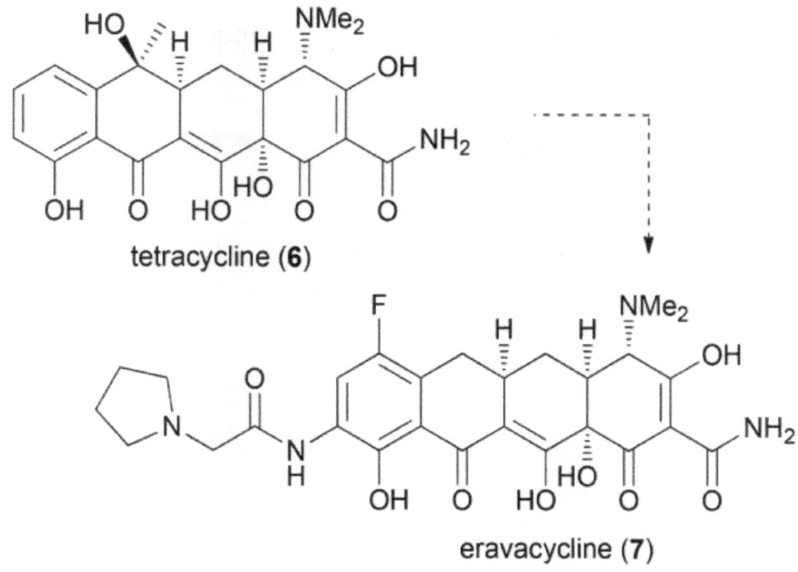

Figure 1: Some recent examples of the synthesis of promising natural product analogues.

Latrunculin Analogues

An illustrative e for DTS was put forward by the Fürstner group in connection with the total synthesis of the two marine macrolides latrunculin A (**8**) and B (**9**) (Figure 2).[18] The latrunculins were isolated from a sponge collected in the red sea.[19] They display a range of biological activities with the dominant one being the disruption of actin polymerization. For this reason the latrunculins are important tools for chemical biology. Key structural features include a 16- or 14-membered macrolactone ring that contains several double bonds. The side chain features an acylthiazolidinone substituent. The keto function of **8** forms a hemiacetal with the 13-OH of the macrolactone ring. Studies on semisynthetic latrunculin analogues revealed that*N*-alkylated derivatives are inactive.[20] Carbonate derivatives of 17-OH (hemiacetal OH) display reduced affinity to actin.[21]

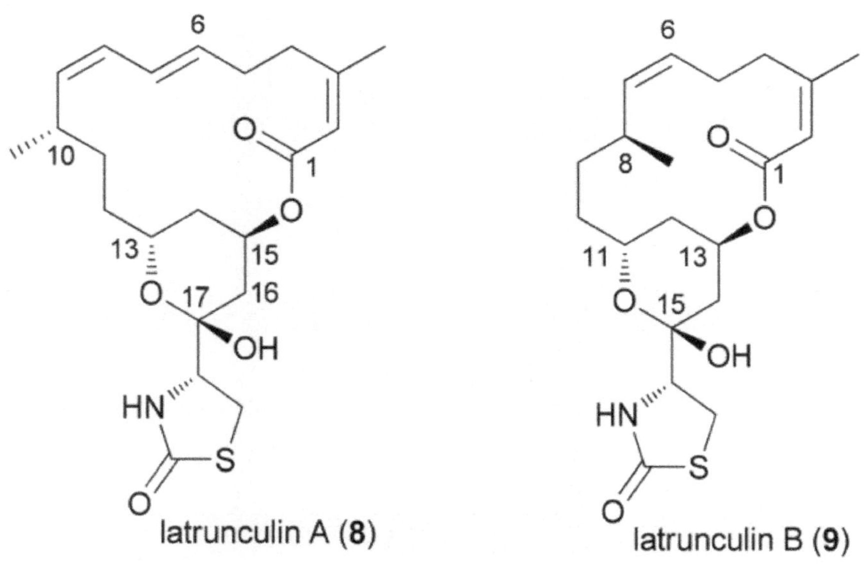

Figure 2: Structures of the two latrunculins.

Due to the convergent synthetic strategy used by the Fürstner group twelve latrunculin analogues could be prepared that addressed the role of the methyl groups in the macrocyclic ring and the configurations in the thiazolidinone ring.[18] The general synthetic strategy, exemplified by the synthesis of latrunculin B (**9**), is shown in Scheme 1.[22] An aldol reaction between alkynal **10** and acetylthiazolidinone **11** followed by cleavage of the silyl protecting group and acetalization furnished pyran **12**. This key fragment was then connected with several unsaturated carboxylate salts **13**, yielding esters like **14**. Formation of the macrolactone could be realized by a ring-closing alkyne metathesis (rcam). Routine steps then led to the target molecules.

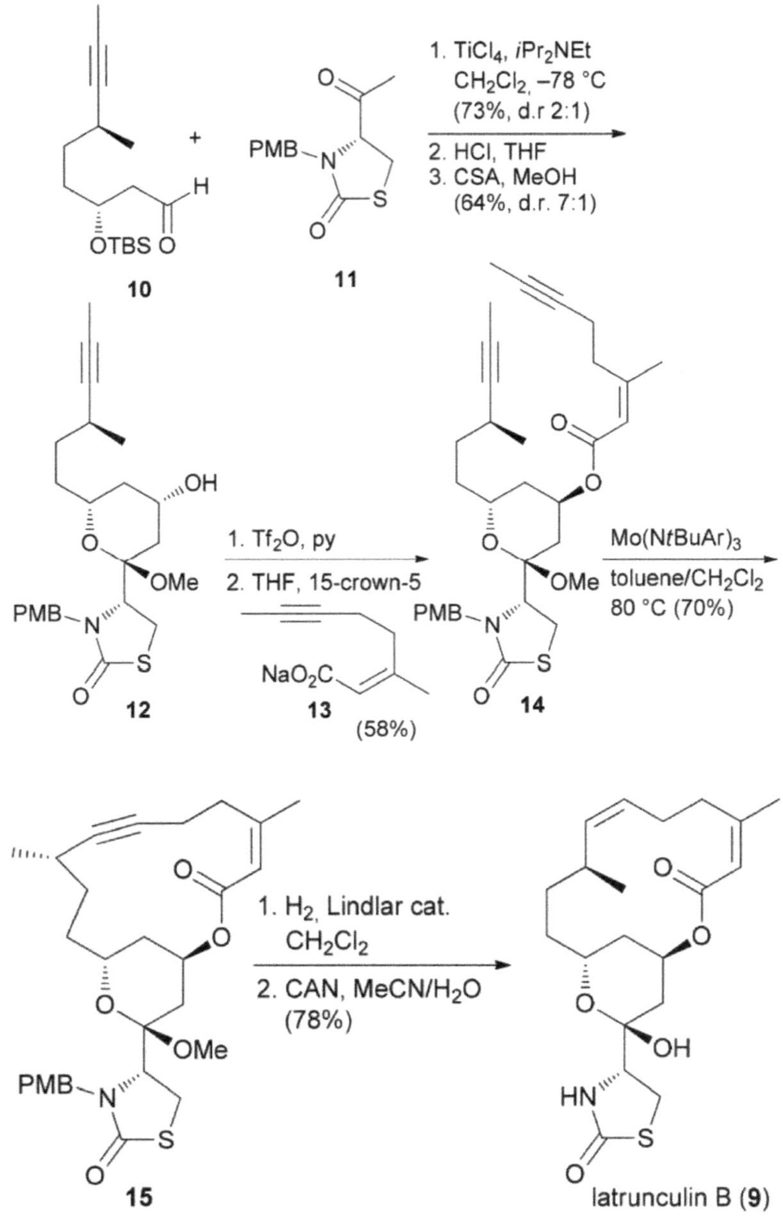

Scheme 1: Synthetic strategy for the synthesis of natural and non-natural latrunculins. Ar= 3,5-dimethylphenyl

The five most active compounds, listed with decreasing microfilament activity, are shown in Figure 3. While the activity of latrunculin A (**8**) could not be surpassed, the bis-nor-methyl-latrunculin B analogue **16** turned out to be more active than **9**.

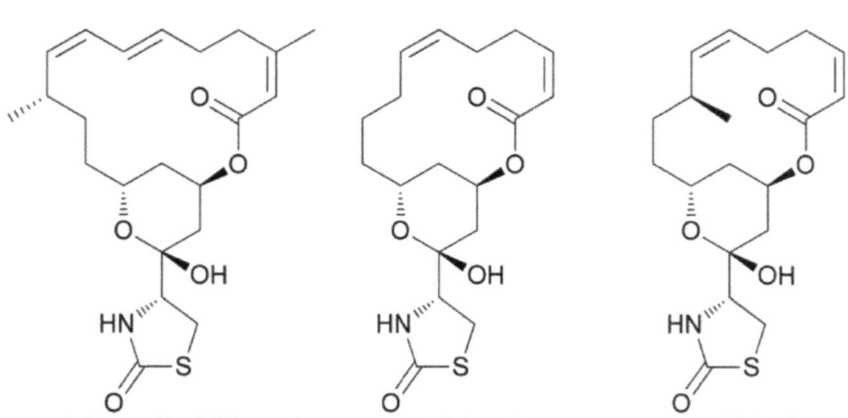

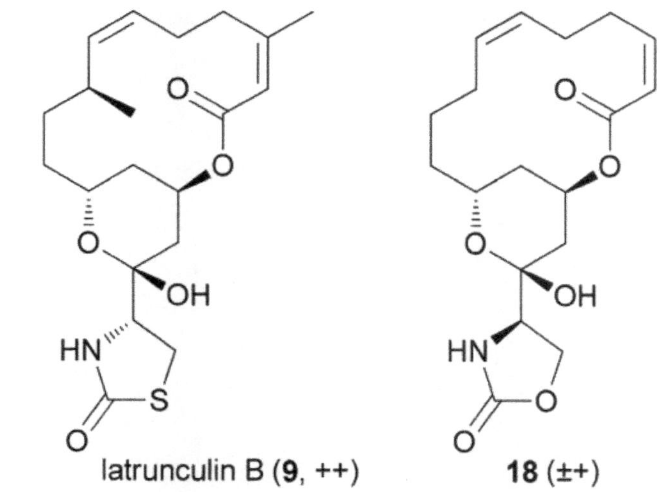

Figure 3: The most active analogues prepared by the Fürstner group together with their microfilament disrupting activity (at 10 μM effective concentration). Abbreviations: ± weak effect, + significant effect, ++ strong effect, +++ very strong effect (less than 20% viable cells).

Spongistatin Analogues

With highly complex NPs, the production of a larger collection of analogues is a rather elaborate effort. But even with smaller numbers of analogues, useful information regarding SAR can be obtained. Thus, among two analogues prepared, compound **20**, which is the E-ring dehydrated version of spongistatin 1 (**19**), was even more potent than the NP itself (IC_{50} values are given for HCT116 colon cell lines) (Figure 4).[23] Compound **20** was formed as a by-product in the deprotection of a 35-OTBS ether with aqueous HF-MeCN. This example highlights the role of serendipitous discoveries in chemistry. One should note that already the NP is active in the low picomolar range. So to beat this activity is truly remarkable. In contrast the analogue with a truncated side chain (ending with C46) was much less active (407 nM).

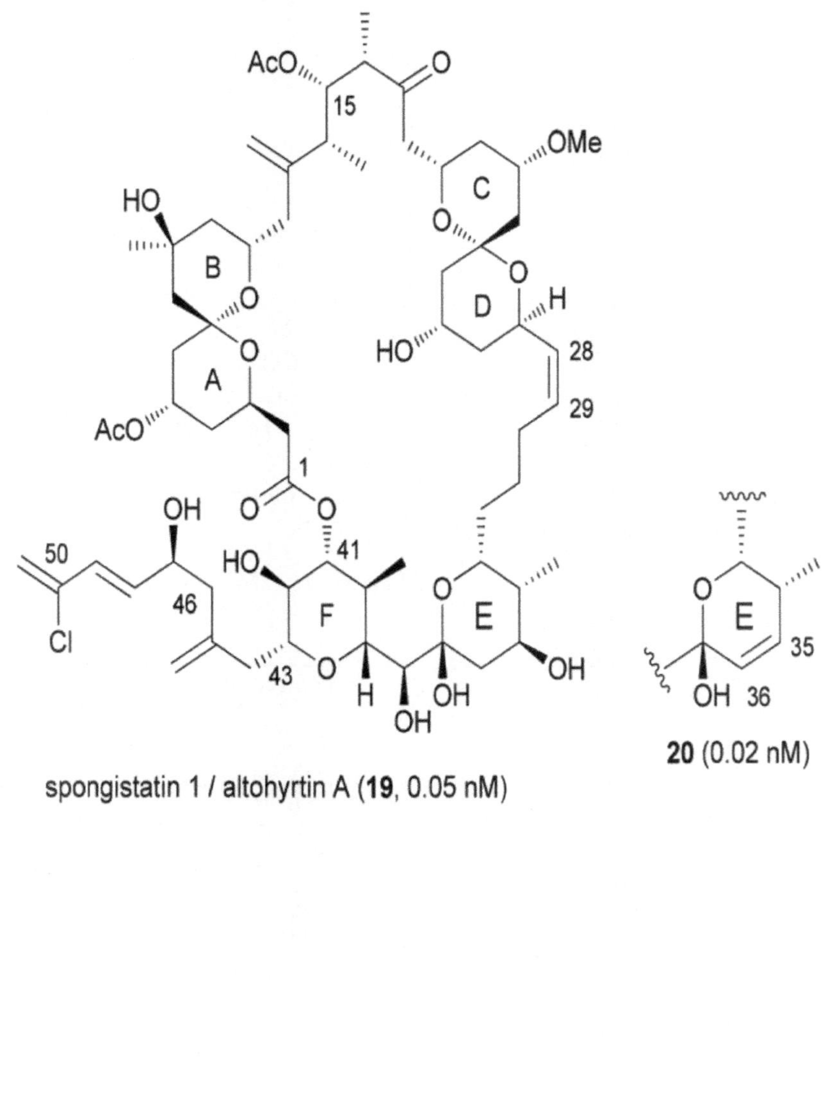

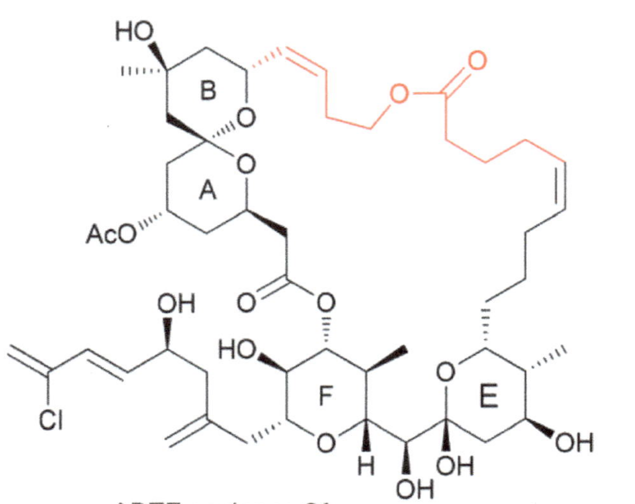

ABEF analogue **21**

Figure 4: Structure of (+)-spongistatin 1 (**19**) and the analogues **20** and **21**.

The spongistatins are bis(spiroacetal) macrolides, which were isolated from marine sponges. Their potent antitumor activity is due to inhibition of mitosis by binding in the vinca binding region of tubulin.[24]

Another option for DTS of complex macrolides is to replace the sections of the macrocycle that are not involved in binding to the target proteins with simpler fragments. Based on the hypothesis that the small-molecule recognition domains most likely include the more rigid parts of a macrocycle, Smith, III, *et al.* designed spongistatin analogues with a simple tether between the ABEF rings (Figure 4).[25] While not as active as the NP lead, the ABEF analogue **21** still displayed nanomolar activity, for example an IC_{50} of 60.5 nM *versus* 0.06 nM (for **19**) against U937 lymphoma cells.

Pironetin Analogues

A common strategy in the design of NP analogues is to truncate or simplify the structure down to the essential elements. In a recent study, Marco *et al.* prepared four analogues of the NP pironetin (**22**) (Scheme 2).[26] Unlike many molecules that bind to microtubules, pironetin exerts its antitumor activity by binding to α-tubulin and not to β-tubulin. Well known NPs that interact with β-tubulin include colchicine, paclitaxel and the epothilones. The microtubule dynamics can be disturbed either by breaking up microtubules or by stabilization them. Pironetin inhibits the assembly of tubulin to microtubules. It seems that the Lys352 residue of α-tubulin adds to the enoate double bond. In addition, hydrogen bonding of Asn258 to the pyrone carbonyl and the 9-OMe group contributes to the binding. SAR studies on pironetin point to the importance of a 9-OH or 9-OMe group and a 7-OH group. The Marco group removed the alkyl groups at C4, C8 and C10, thus simplifying the syntheses significantly. The terminus of the side chain was varied in the analogues. The synthetic strategy is outlined in Scheme 2. Thus, repetitive Brown allylation reactions came to use for the creation of the three stereocenters. A final acrylate formation, followed by ring-closing metathesis (rcm) and deprotection, delivered the desired analogues.

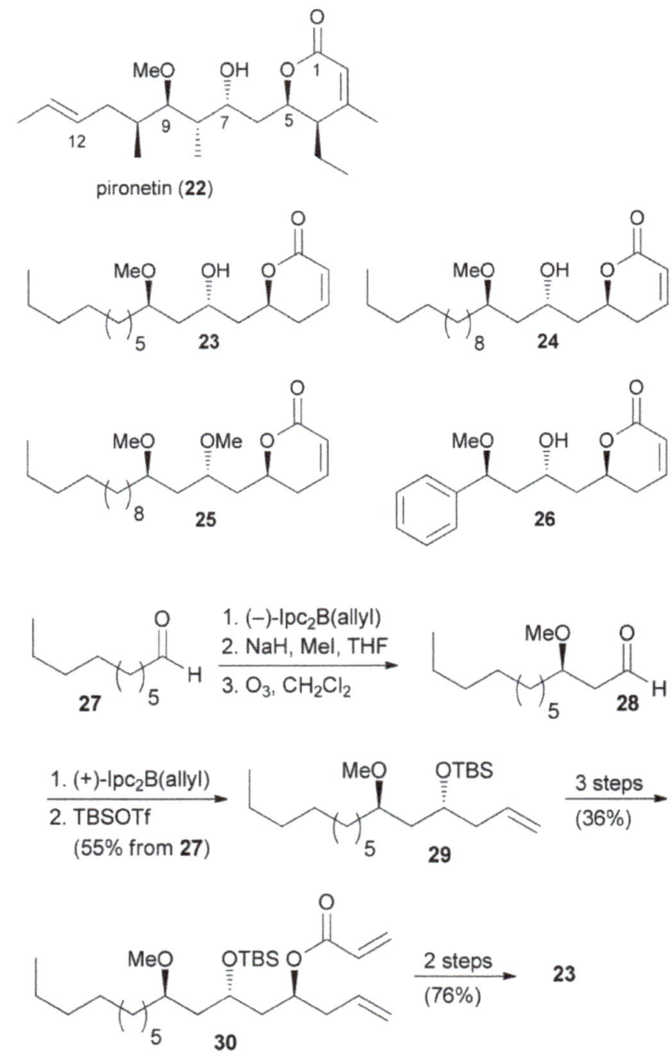

Scheme 2: Synthesis of pironetin analogues.

The growth inhibitory activity of the analogues was determined on ovarian cancer cell lines (Table 1). While pironetin was highly potent, the analogues were at least three orders of magnitude less active, with IC_{50} values in the micromolar range. Cell cycle analysis indicated that analogue **26** shares a common mode of action with pironetin (**22**), whereas analogues **23–25** possibly work by another mechanism. This study shows that the omission of alkyl groups may cause a reduction of the hydrophobic contribution to binding. Moreover, a loss of conformational control may be detrimental.

Table 1: Growth inhibitory activity of pironetin (**22**) and analogues determined on the A2780 ovarian cancer cell line[a]

Compound	**IC_{50} [μM]**
Pironetin (**22**)	0.0062
23	39.7
24	14.1
25	9.2
26	54.7

[a] In this and other tables the biological activity data for the compounds are given without standard deviation.

Benzolactone Enamide Analogues

Our own work regarding the synthesis of NP analogues mainly falls in the category DTS. Thus, in es where we found an efficient synthetic strategy for a NP, we looked for opportunities in the sequence to branch out towards analogues.

An early example from our group addressed apicularen A analogues.[27,28] Apicularen A (**32**) is a prominent member of the benzolactone enamide class of NPs.[29] Besides apicularen A, the family comprises the salicylihalamides **31**, the oximidines, like **33**, and lobatamide C (**34**) (Figure 5). Common structural features include a macrocyclic salicylate ring with an enamide side chain on the aliphatic part of the macrolactone.

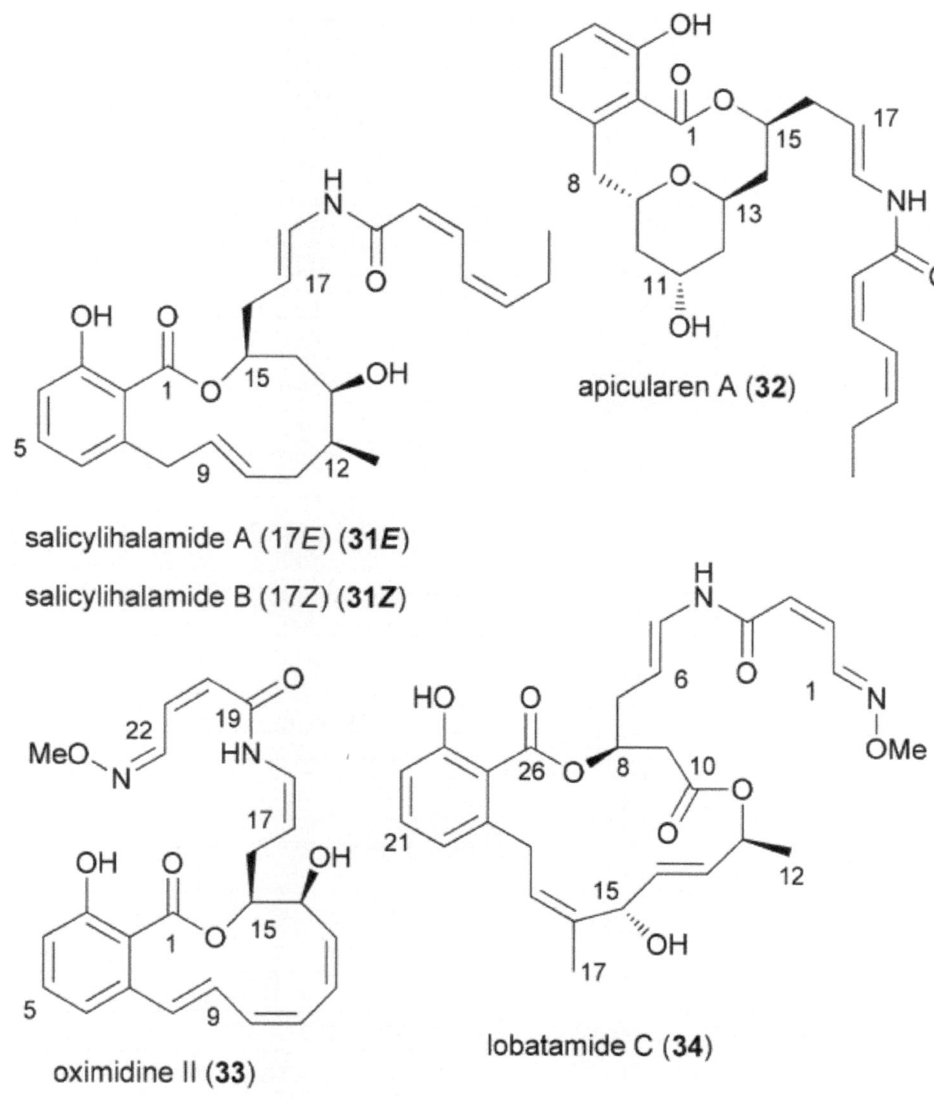

Figure 5: Structures of some prominent benzolactone enamides.

The benzolactone enamides are potent inhibitors of vacuolar (H^+)-ATPases. These are multicomponent proton pumps that are responsible for acidification of single membrane intracellular compartments like endosomes and lysosomes.[30] Biochemical studies showed that salicylihalamide binds to the V_0 domain of the proton pump.[31] While most of the total syntheses of apicularen A pass through a *trans* pyran prior to macrolactonization,[32] the strategy we developed features a macrolactonization to a 12-membered macrolactone ring followed by a transannular etherification (Scheme 3).[33,34] The precursor **35** does indeed show a striking similarity to the salicylihalamide core structure.

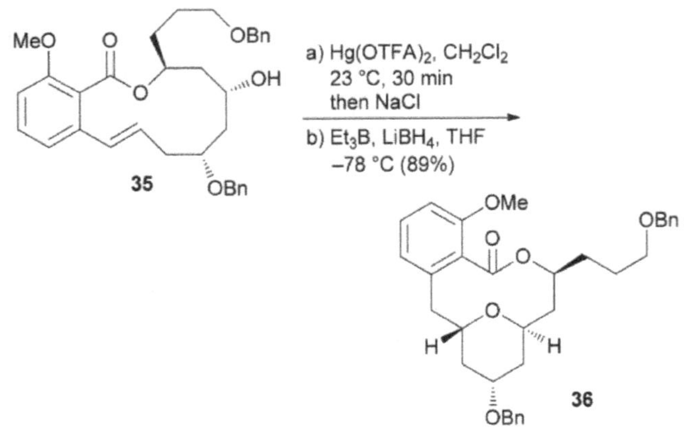

Scheme 3: Key transannular cyclization towards the core structure **36** of apicularen A.

From model studies on the transannular cyclization and from the total synthesis campaign several aldehydes could be made available. These were used for attachment of the enamide side chain *via* formation of the hemiaminal followed by dehydration as shown for the conversion of **37** to **40**. In all, five analogues **40**, **41**, **42**, **43** and **44** were prepared (Scheme 4). Out of these,**40** and **41** contain the intact enamide side chain but have changes in the macrolactone core.

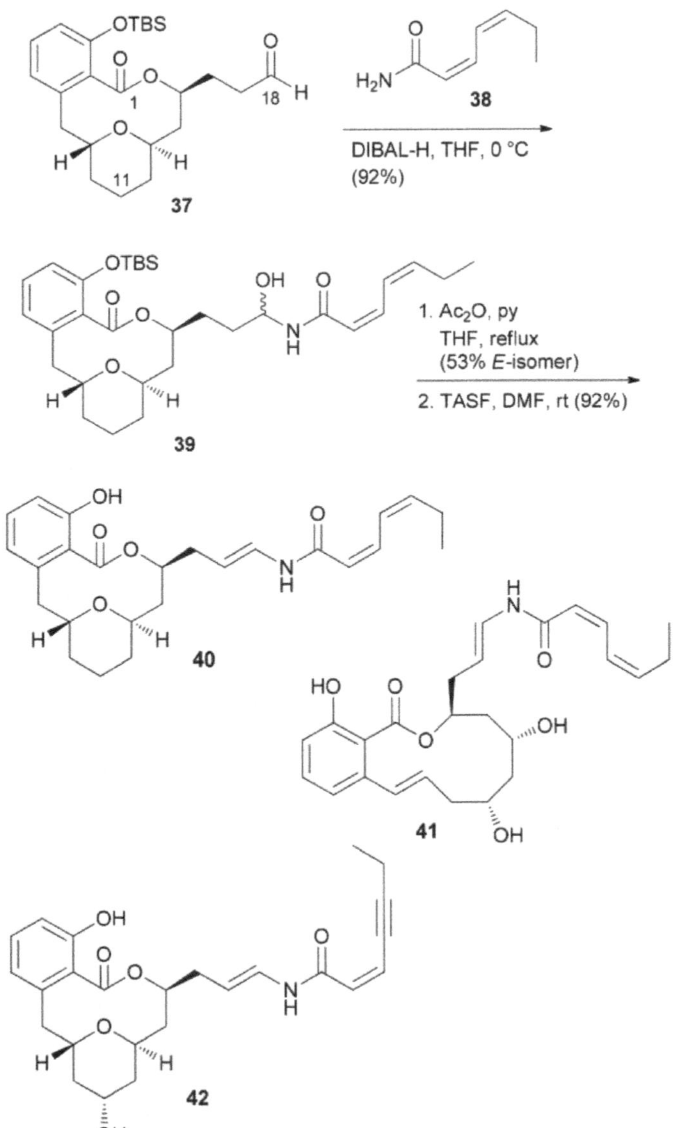

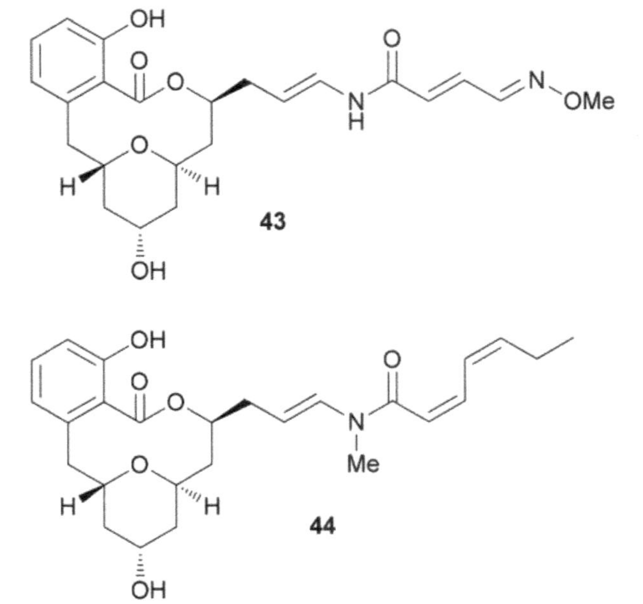

Scheme 4: Structures of some apicularen A analogues. TASF = tris(dimethylamino)sulfonium fluoride.

These compounds turned out to be quite active (Table 2). On the other hand, compounds **42**, **43** and **44** feature the natural core structure but a different enamide part. These analogues were roughly three orders of magnitude less cytotoxic (Table 2). Other studies concur with these findings.[35] Thus, it seems that variations in the acyl part of the enamide are tolerated provided that the chain is neither too short nor too long. On the other hand, the macrolactone part seems to be less sensitive to structural modifications.

Table 2: Cytotoxicity values (IC_{50}, nM) of some apicularen A analogues against various cell lines

Compound	Cell line		
	L-929	KB-3-1	KB-V1
32	6.8	2.3	15.9
40	19	7.1	2.4
41	200	180	1360
42	4550	1820	9100
43	1800	1130	9000
44	770	660	1980
Sali A (31*E*)	57	6.8	9.1

For the other prominent benzolactone enamide, salicylihalamide A (**31*E***), our group contributed with several approaches to the macrolactone core. These strategies are summarized in Scheme 5. The first approach relied on an intramolecular Suzuki coupling of ester **46** to macrolactone **45**. Here a diastereoselective hydroboration generated the stereocenter at C12 prior to the coupling step.[36] This strategy was also successful in the reverse order. Thus, hydroboration of fragment **48** followed by coupling with vinyl iodide **47** gave rise to seco compound **49**. Deprotection to the corresponding hydroxy acid set the stage for an intramolecular Mitsunobu esterification.[37] Possibly a more efficient strategy is to close the macrolactone ring by ring-closing metathesis. In this variant the stereocenters at C12 and C13 could be created in an efficient manner by an Evans aldol reaction.[38] Using this strategy for the synthesis of the aliphatic sector **51**, we prepared macrolactone **45** in very good yield.

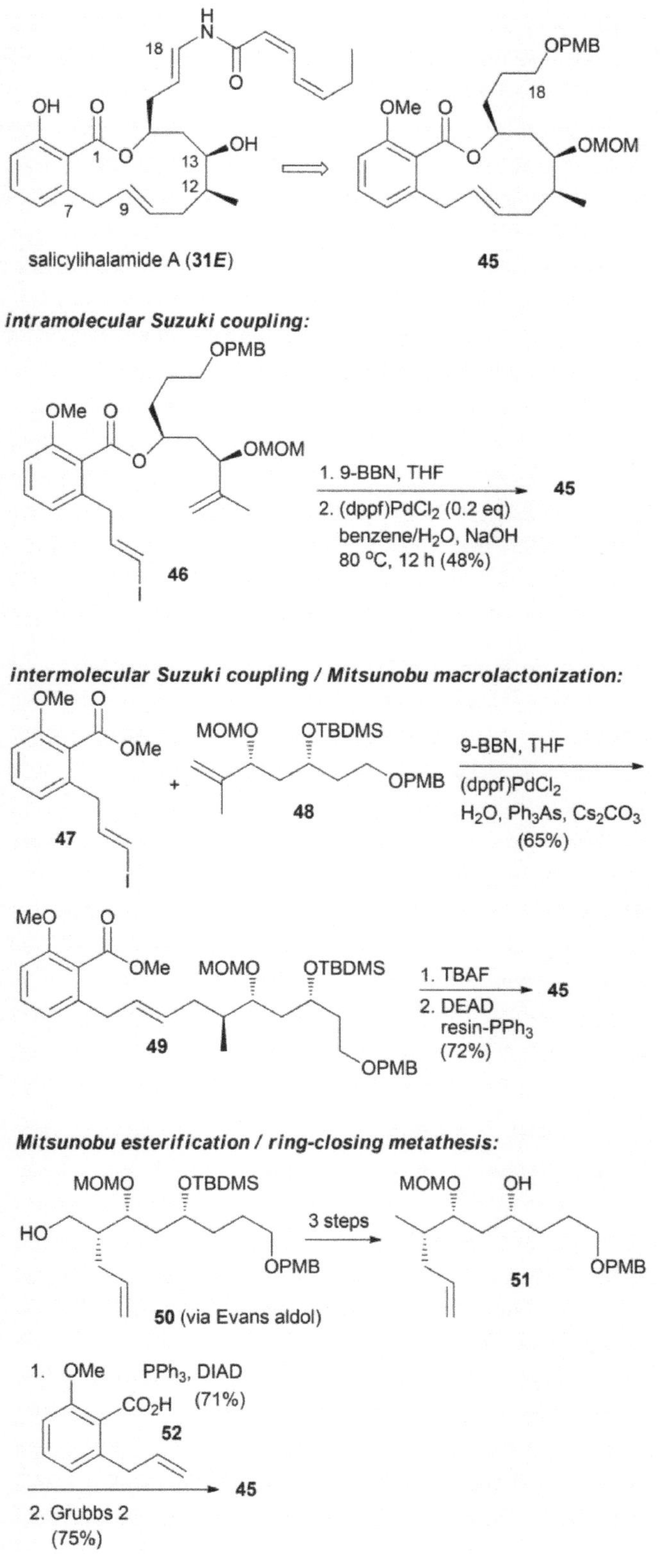

Scheme 5: Synthetic strategies developed in our group towards the synthesis of salicylihalamide congeners.

In the e of a terminal alkyne its protection as TIPS derivative **53** was required in order to achieve good yields in the rcm reaction. But the alkynyl group allowed for a formal total synthesis of salicylihalamide A by converting it to a vinyl iodide (Scheme 6).[39] In addition, we prepared the pyridine substituted derivative **55** by Sonogashira cross-coupling. In a simpler substrate lacking the C12 and C13 substituents, we used the side chain alkyne for click reactions with a range of azides. This way the triazoles **56** and **57** were obtained (Scheme 6).[40] These analogues showed cytotoxicity in the micromolar range, pointing to the crucial role of the enamide side chain in **31*E***.

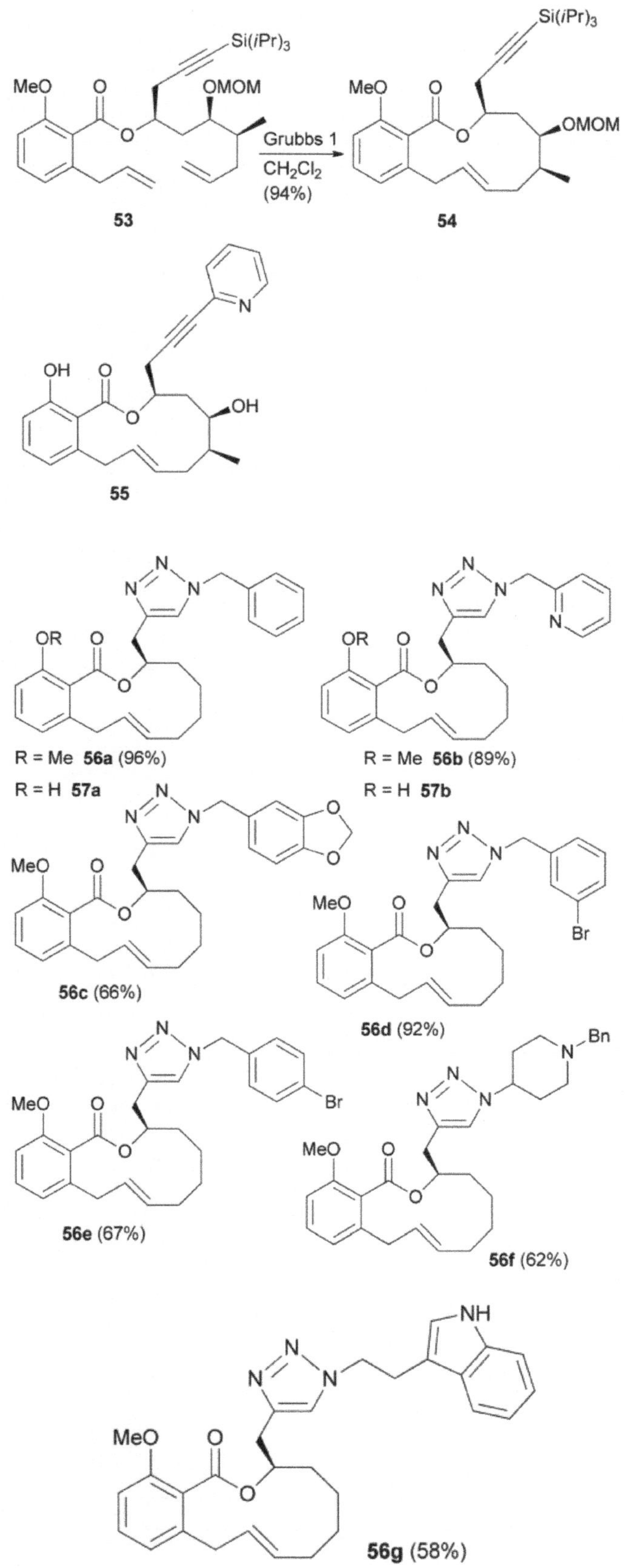

Scheme 6: Benzotriazole analogues of salicylihalamide A prepared from an alkyne precursor by a click reaction.

The ring-closing metathesis strategy for the synthesis of salicylihalamide A (**31*E***) has been pioneered by De Brabander *et al.*[41,42] They prepared a range of analogues where the enamides were created by nucleophilic addition to vinyl isocyanates. One of the more promising analogues, called saliphenylhalamide (**66**, SaliPhe), has been explored as a potential lead for anticancer therapy.[43] The methyl ketone **58** was added to aldehyde **59** in a boron-aldol reaction in a non-selective reaction (Scheme 7). The mixture was carried on through the next steps. Silylation and *syn*-selective reduction was followed by Mitsunobu esterification and ring-closing metathesis. At the stage of the phenol **63** the unwanted diastereomer could be removed by chromatography. The final steps established the enamide *via* addition of lithium phenylacetylide to the intermediate vinyl isocyanate.

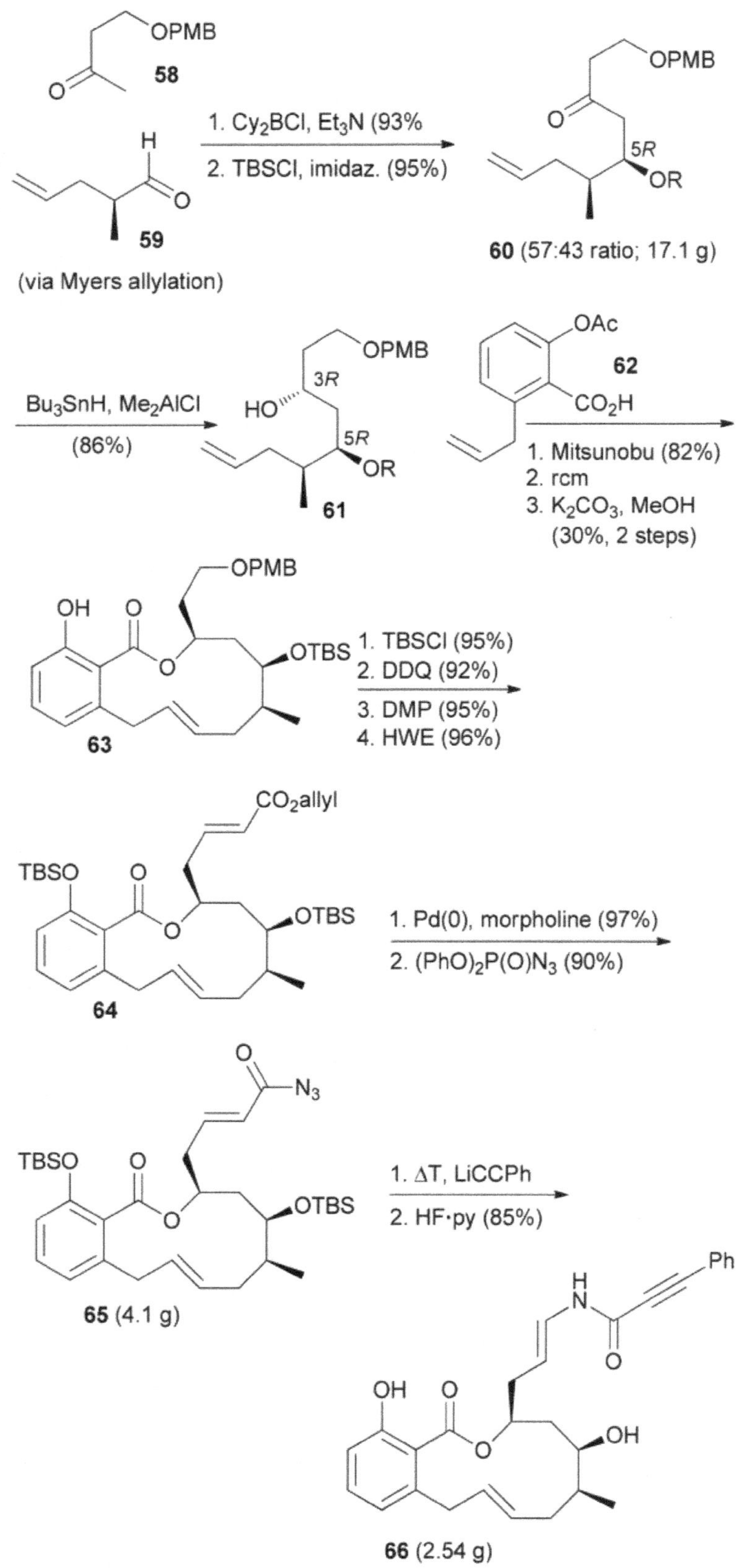

Scheme 7: Synthetic route towards saliphenylhalamide **66**.

In a recent paper SaliPhe and other V-ATPase inhibitors were tested as antiviral compounds.[44] It could be shown that V-ATPase is required for human papillomavirus (HPV) infection. Addition of inhibitors affects uncoating/disassembly but not the process of endocytosis. If the toxicity towards the host could be controlled, the inhibition of V-ATPases might be an interesting antiviral strategy.

Cruentaren A Analogues

Unlike the benzolactone enamides, which target V-ATPase, the macrolactone cruentaren A (**67**) inhibits mitochondrial F-ATPase (Scheme 8). This protein is quite important since it uses a proton gradient to promote the synthesis of ATP. A recent study described that the inhibition of F_1F_0-ATPase by **67** disrupts the synthase-Hsp90 interaction.[45] As Hsp90 is an important chaperone,[46] cruentaren A can indirectly cause client protein degradation. Structural features of this NP, which was isolated from the myxobacterium *Byssovorax cruenta*,[47] include a 12-membered lactone with a *Z*-double bond and an unsaturated side chain amide. With its biological activity and our previous experience in the synthesis of benzolactones, cruentaren A (**67**) appeared as a promising target. Up to now, four total syntheses for **67** have been published.[48–51] Three of them, including our own, rely on a ring-closing alkyne metathesis[52] (rcam) to establish the macrocyclic ring. Major differences in these routes are in the preparation of the building blocks, a benzoic acid like **68**, and a homopropargylic alcohol such as **69**. For example, we prepared fragment **68***via* an Evans aldol reaction forming the C9–C10 bond (Scheme 9). Here the 10-Me group came from a reduction of a carboxylic function. The other fragment **69**, featuring four stereocenters, was constructed by an aldol reaction (C15–C16) and a Marshall–Tamaru reaction. In the latter transformation a chiral allenyl metal was added to an aldehyde generating a 2-substituted butynyl structure. The triple bond was then converted *via* a hydroboration/reduction step to the hydroxyethyl moiety. The two fragments **68** and **69** were combined to give ester **70** after converting the acid **68** to the corresponding carbonyl imidazolide. Using the Schrock catalyst **71** an excellent yield for the rcam reaction could be realized. The alkyne in the macrolactone ring was kept for the elaboration of the side chain. This prevented an undesired translactonization to the six-membered lactone. Conversion of the advanced intermediate**73** to the propargylic amine **76** was followed by condensation with acid **77**. After deprotection steps, a final Lindlar hydrogenation provided cruentaren A (**67**) (Scheme 8).

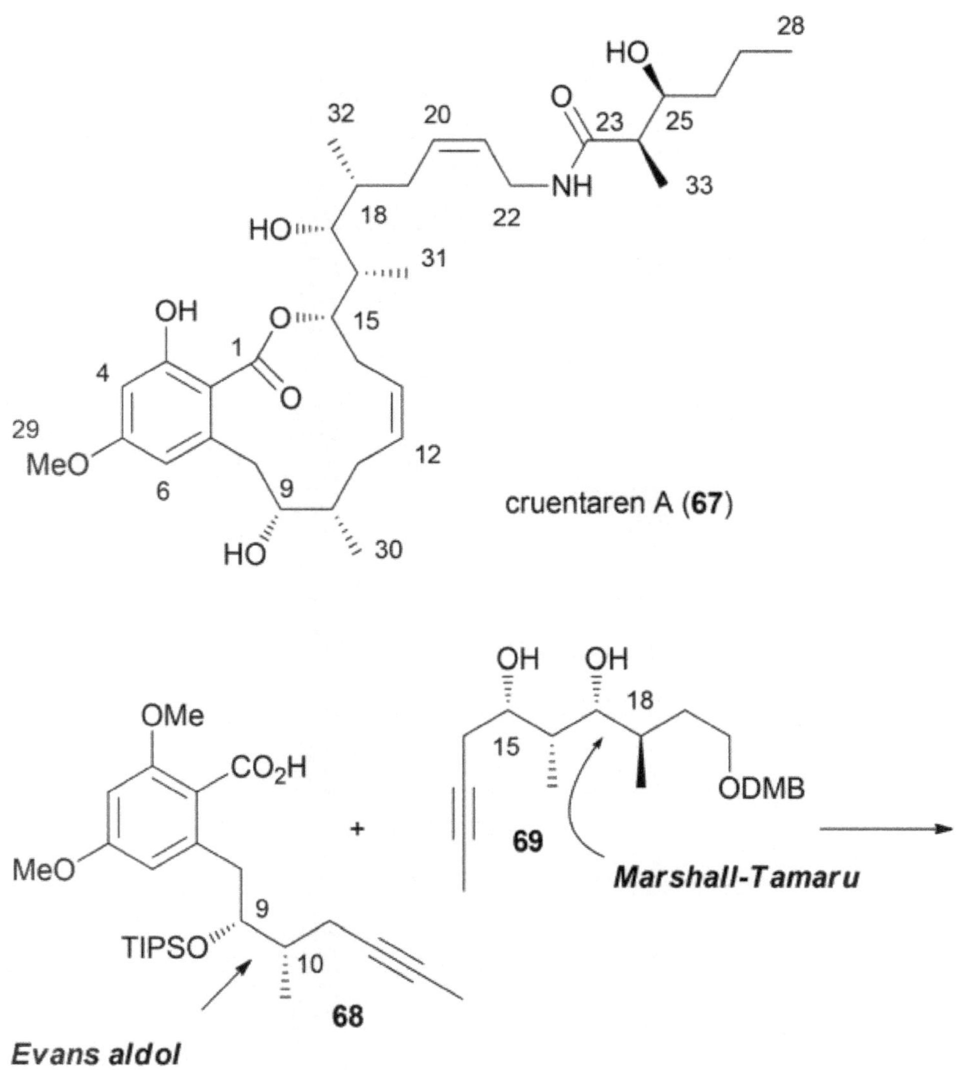

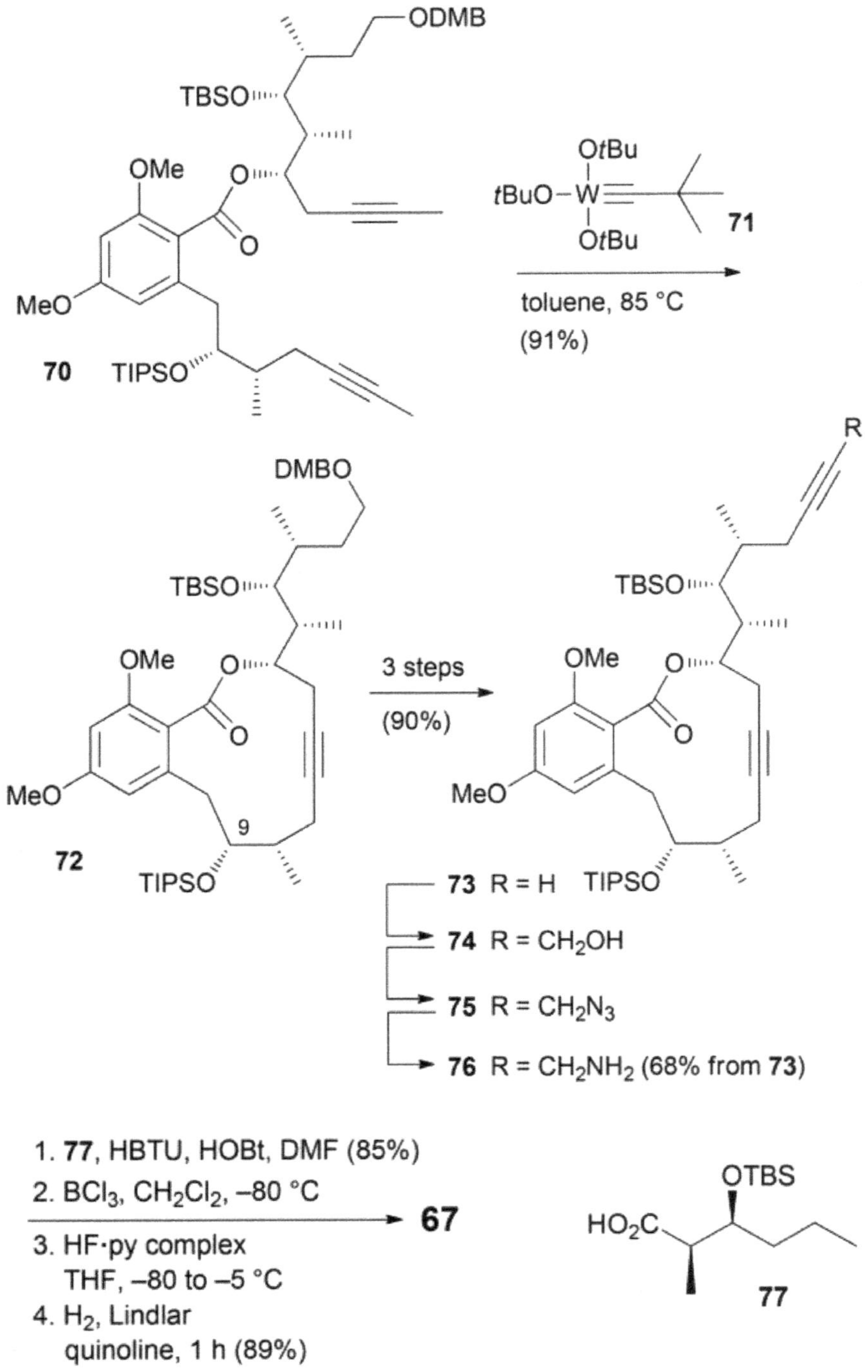

Scheme 8: Overview of the synthetic route to cruentaren A (**67**).

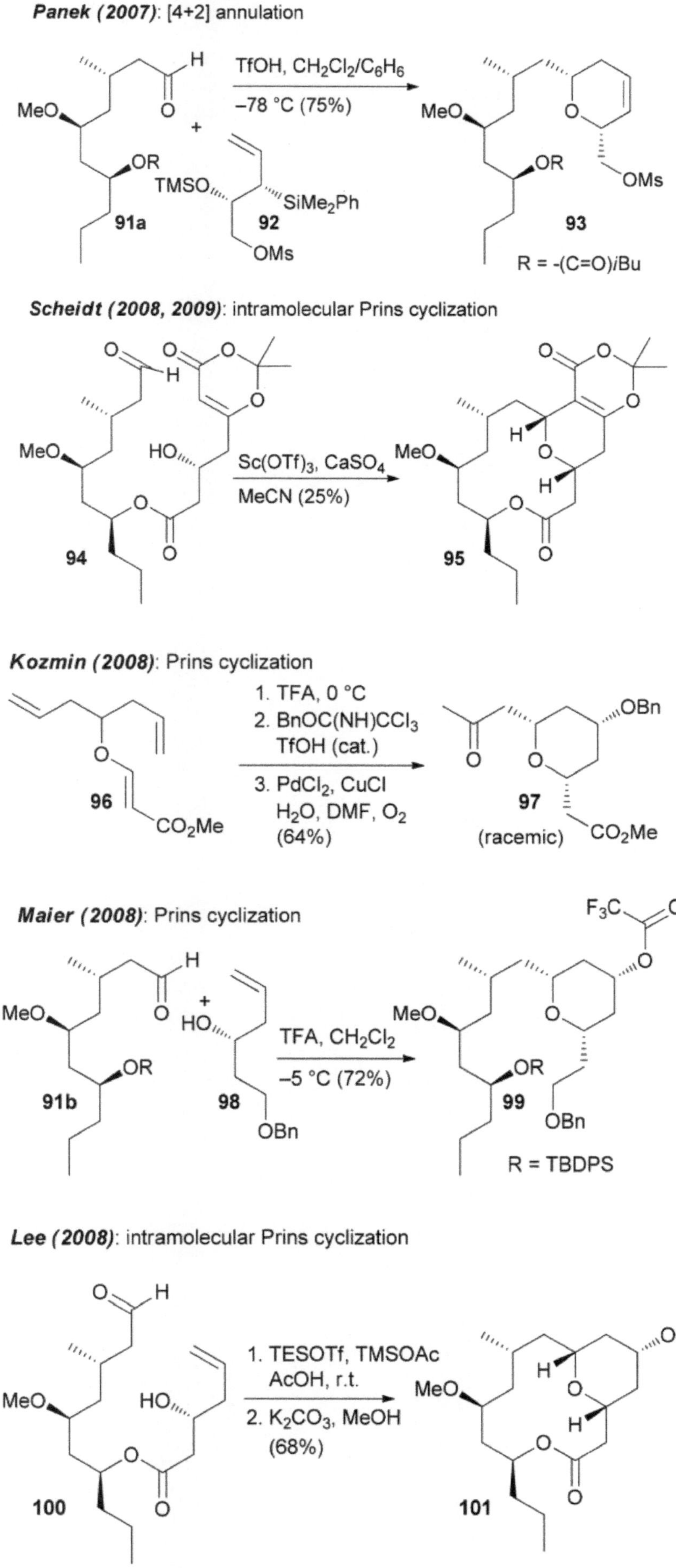

Scheme 9: Summary of the strategies towards the pyran system of neopeltolide, part 1.

The efficient route to amine **76** allowed access to sufficient material and the preparation of several amide analogues (Figure 6).[53] All of them were somewhat less active with **78c** being the most potent one. This points to the importance of the hydroxyl group of the amide. Compound **83** with two triple bonds was not very active (IC_{50} = 11.1 μM). The 3-OH group also significantly contributes to the biological properties as can be seen in the reduced activity of 3-OMe-cruentaren A (IC_{50} = 28 nM *vs.* 0.7 nM for **67**).

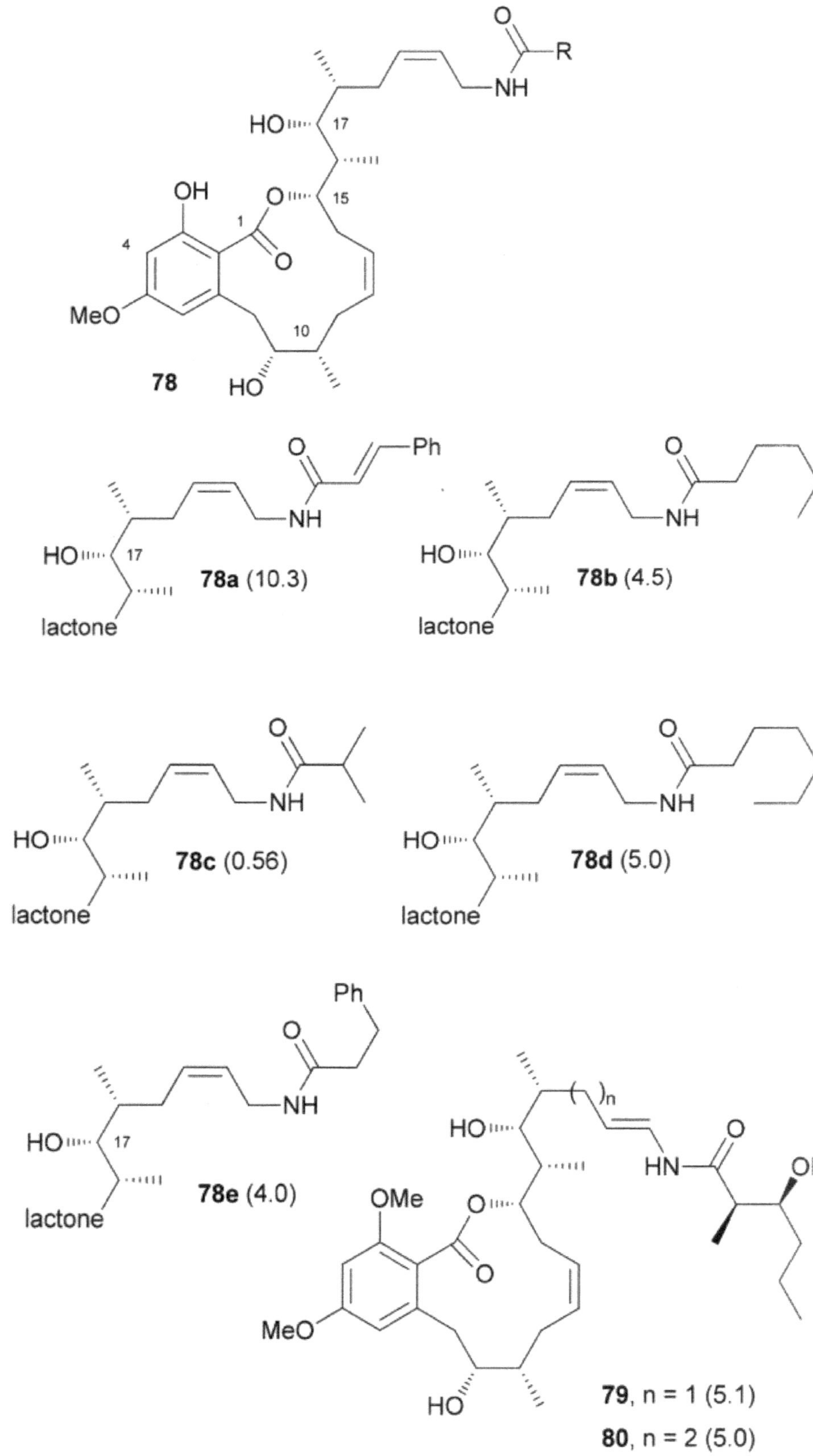

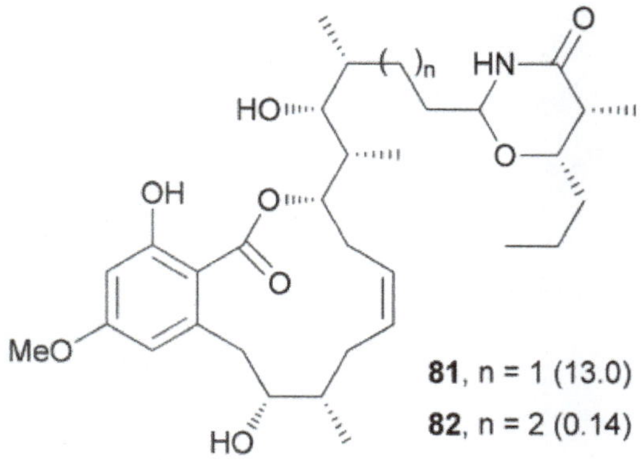

81, n = 1 (13.0)

82, n = 2 (0.14)

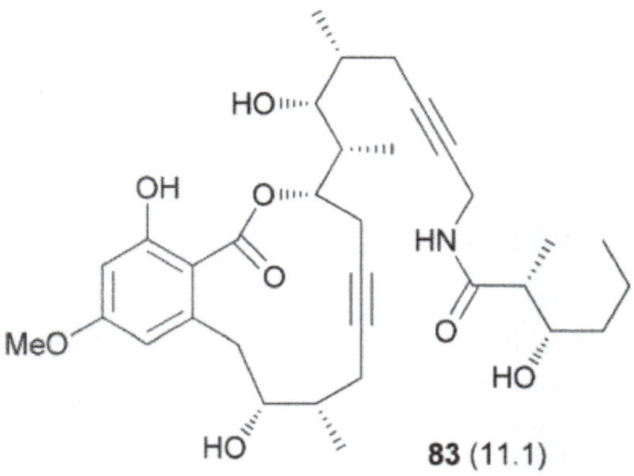

83 (11.1)

Figure 6: Structures of some cruentaren A analogues. The IC_{50} values (L929 cell line, μM) are given in brackets after the compound numbers.

In addition to the allylamide derivatives we also wanted to prepare enamide derivatives. Although speculative it might be that the allylamide isomerizes to an enamide that then could form a highly electrophilic acyliminium ion, similar to the benzolactone enamides. It was possible to prepare vinyl iodides by hydrozirconation and iodination. A cross-coupling reaction with the side chain amide was used to create the enamide functionality. Cleavage of the silicon protecting groups and Lindlar reduction of the triple bond furnished analog **79**. Unfortunately, the enamide did not survive the demethylation step with BCl_3. Rather under these conditions the 1,3-oxazinan-4-ones **81** and **82** were formed, respectively. Out of these, **82** showed good growth inhibitory activity (IC_{50} = 140 nM). Enamides **79** and **80** themselves were only moderately active (Figure 6).

The Fürstner group also prepared a range of analogues.[54] Among them were stereoisomers of the side chain hydroxy acid and the 9-epimer **87** (Figure 7). In addition, the 9-deoxy analogue**89** was prepared. Among the prepared analogues **85**, **86** and **87** were the most active ones. It seems that a hydroxyl group at C9 is quite important for the activity. Moreover, either a *syn* configuration in the amide side chain or at least the correct configuration of the secondary hydroxyl group is crucial for activity.

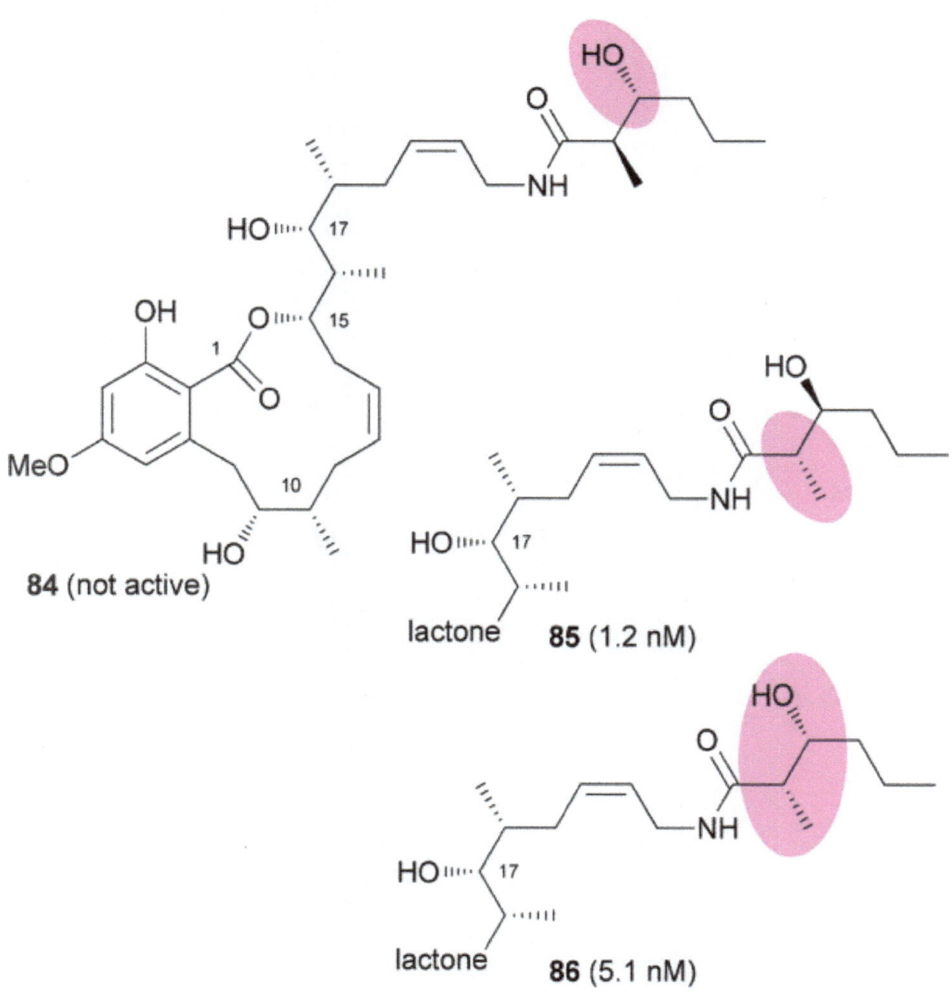

84 (not active)

85 (1.2 nM)

86 (5.1 nM)

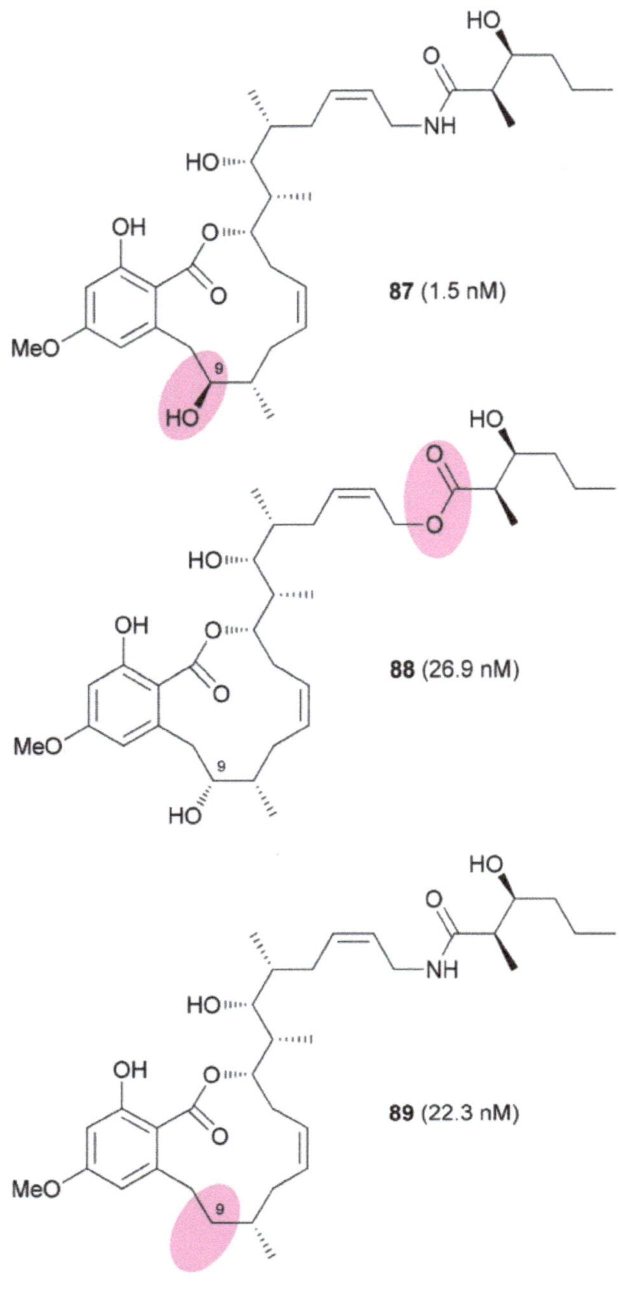

Figure 7: Structures of cruentaren A analogues prepared by the Fürstner group. IC_{50} values (against L 929 cells) are given in brackets.

Neopeltolide Analogues

Neopeltolide (**90**) (Figure 8) is another macrolide for which we developed an efficient synthetic route that allowed us to prepare a range of analogues.[55] This macrolide was isolated from a sponge *Daedalopelta* of the Neopeltodae family.[56] It exhibits potent cytotoxic activity with IC_{50} values in the nanomolar range against a number of cancer cell lines.

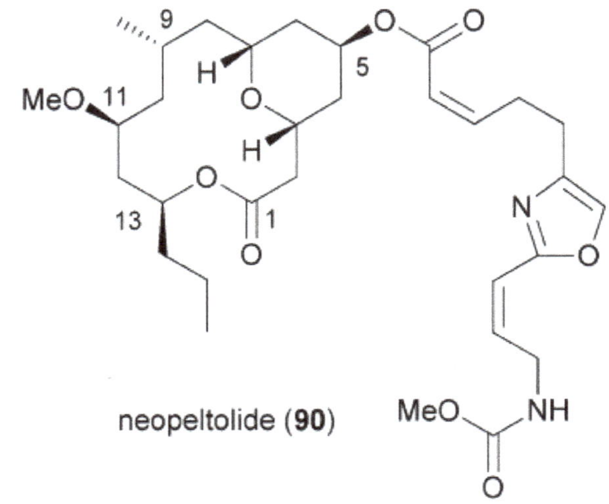

Figure 8: Structure of (+)-neopeltolide (**90**).

It is certainly not among the most complicated structures but it does pose certain challenges in constructing the macrolactone and the oxazole-containing side chain. So far, ten total syntheses of neopeltolide have been published.[57] The first two, by Panek and Scheidt, also corrected the initial published structure that had been misassigned at positions 11 and 13. In addition to the total syntheses at least ten formal total syntheses have been reported in the literature.[58] All these papers nicely illustrate the state of the art in the synthesis of 2,6-*cis*-tetrahydropyrans and the stereoselective generation of stereocenters in 1,3-distance. Schemes 9 and 10 summarize the key steps that have been used in the synthesis of the pyran ring. In several of the syntheses a Prins cyclization came to use. Other strategies feature a hetero-Diels–Alder reaction.

Paterson (2008): hetero Diels-Alder reaction

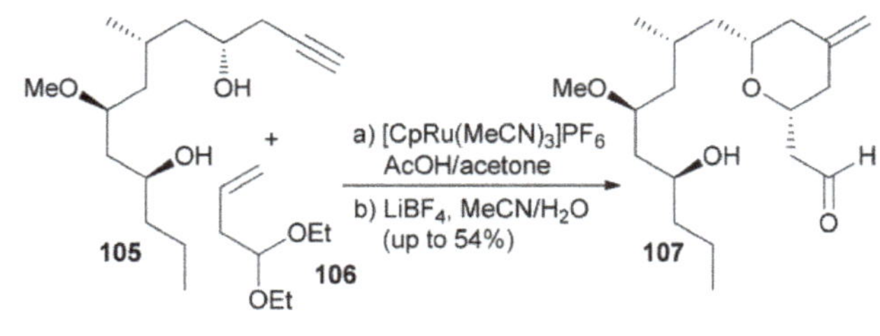

Roulland (2009): Ru(II)-catalyzed ene-yne coupling

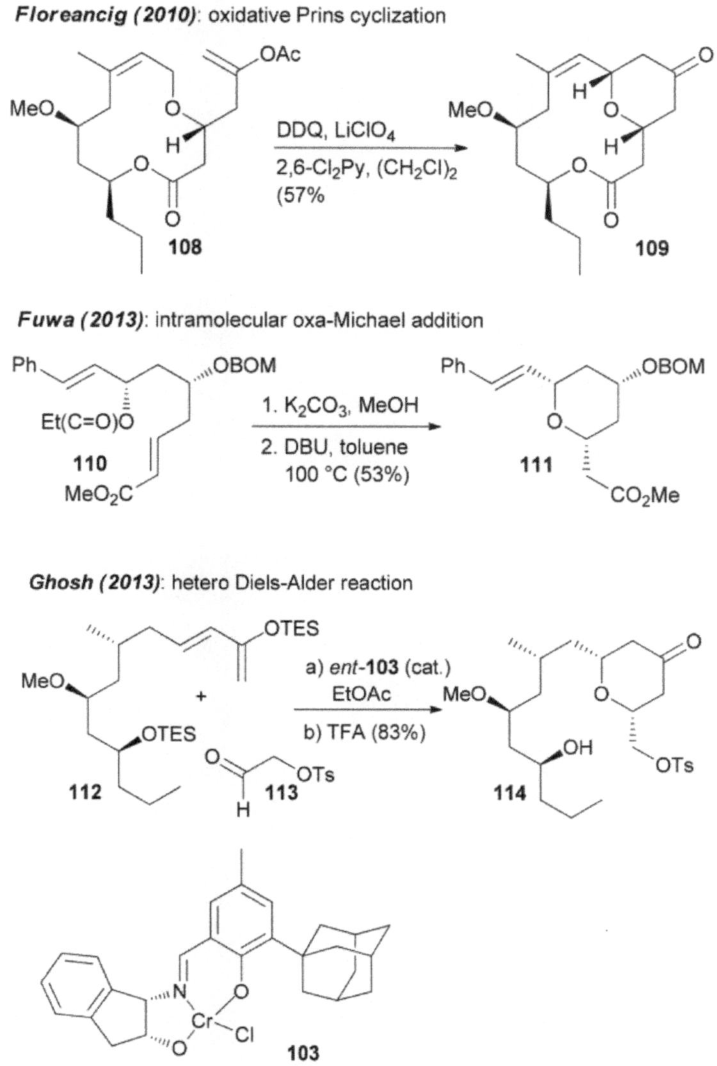

Scheme 10: Summary of the strategies towards the pyran system of neopeltolide, part 2.

Regarding the mode of action of neopeltolide a major contribution came from the Kozmin group.[57e] Using a yeast haploinsufficiency screen the cytochrome *bc*1 complex was identified as the molecular target of neopeltolide (**90**) and the related NP leucandrolide A (**115**) (Figure 9). Thus, these compounds interfere with the crucial energy supply chain in a cell that explains their strong toxicity. The fact that leucandrolide A (**115**), which has a macrolactone core different from **90**, addresses the same cellular target shows that some variations in the macrolactone should be possible without a significant reduction in potency. So far several groups have contributed to SAR studies of **90**.[59] The key findings are highlighted in Figure 9. Thus, the pyran as well as the intact oxazole side chain are essential for activity. They basically represent the minimal structural units. In addition, the axial orientation of the ester at C5 significantly contributes to the activity. The region bridging C7 and C3 seems to tolerate changes to some degree, for example, deletion of substituents, dehydrogenation or epimerization at C9 or C11.

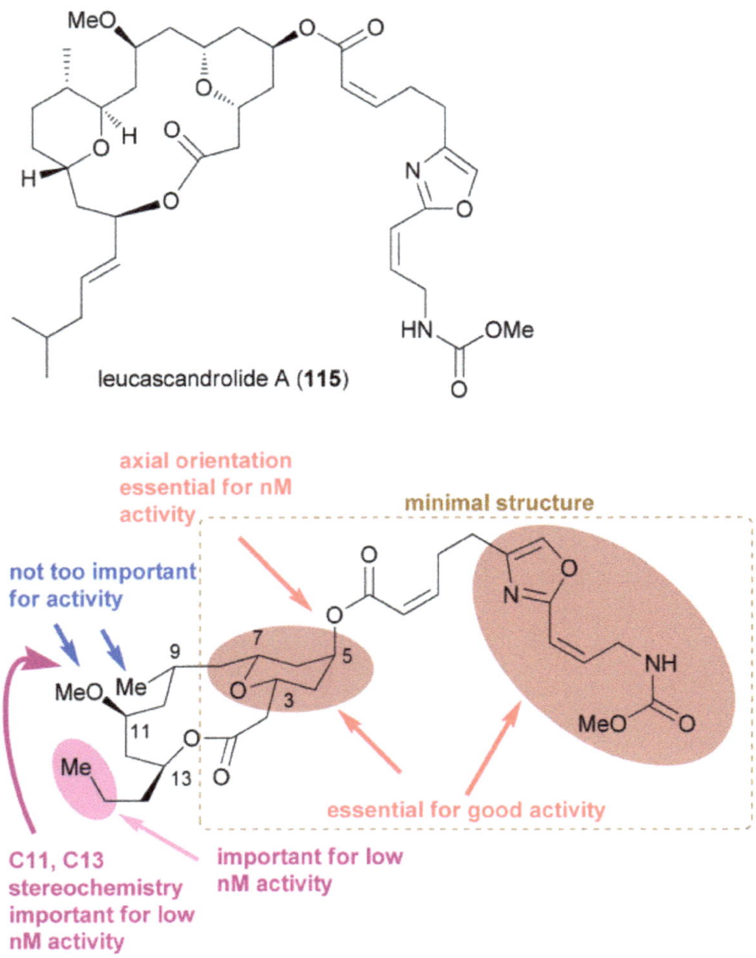

Figure 9: Structure of the related NP leucandrolide (**115**) and a summary of the SAR for neopeltolide. The lower part of the figure was adapted from ref. 59*d*.

Some specific analogues are shown in Figure 10. The exact binding side in the target cytochrome however is not known. Due to the high toxicity, most likely neopeltolide will remain as a tool for biology. Its potential as drug is probably low owing to the toxicity.

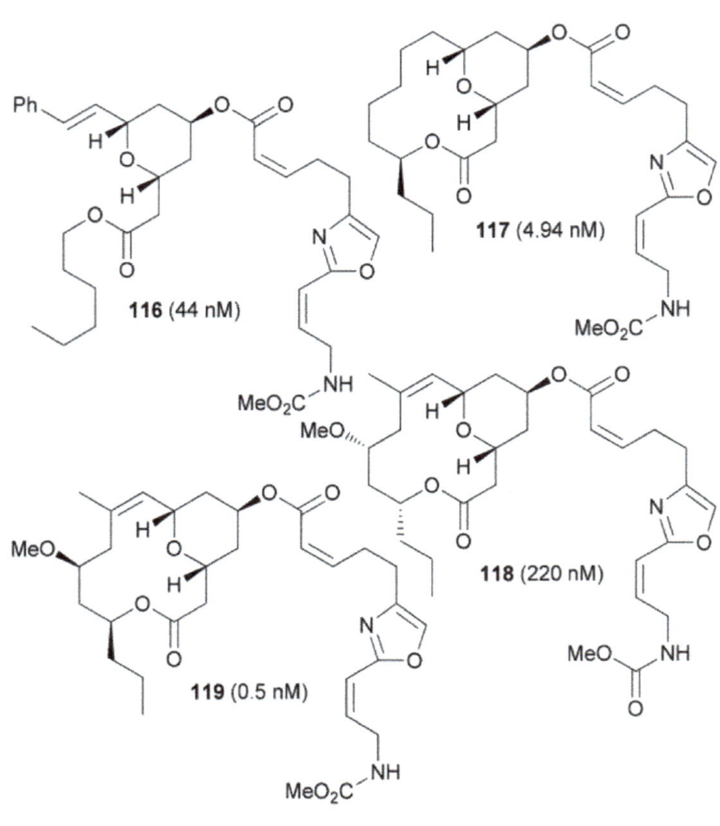

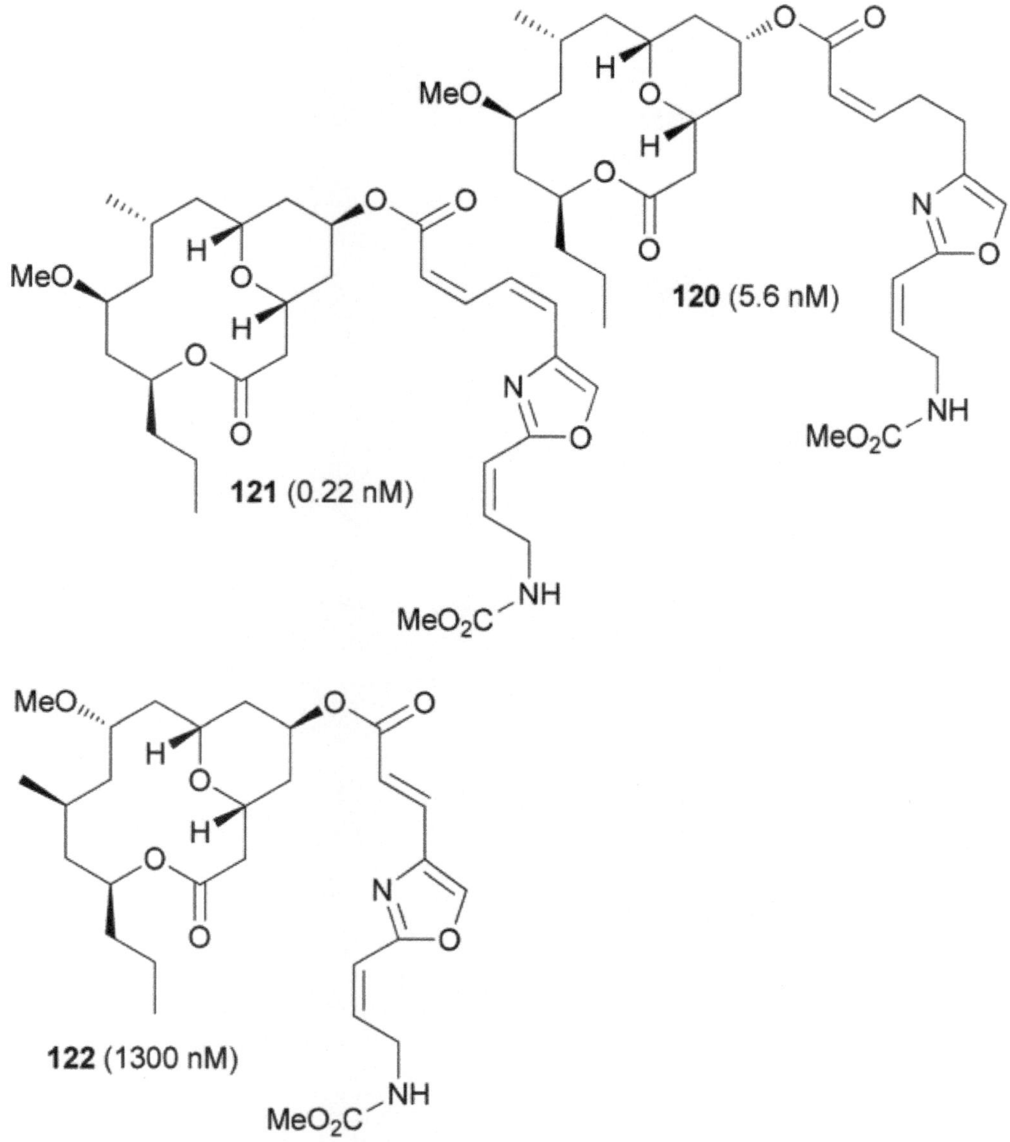

Figure 10: Representative neopeltolide analogues. The IC_{50} values were determined on the A549 cell line. Compounds **116–119** (ref. 59*d*), and compounds **120–122** (ref. 55).

Chondramide A Analogues

Among natural products, those which consist of obvious subunits are ideal for the synthesis of analogues. In this regard peptides, cyclic peptides or cyclic depsipeptides can be mentioned. The latter represent a unique class of secondary metabolites that in a way bridge the peptide with the polyketide world.[60] They may contain unusual amino acids like D-configured or *N*-methylated ones. In addition, they incorporate a hydroxyl acid from the polyketide pathway. This building block connects the peptide part giving rise to an ester bond. Some prominent examples include the jasplakinolides[61] (**123**) and the structurally related chondramides (Figure 11).[62] While the jasplakinolides have been isolated from a marine sponge,[63] the chondramides were found in myxobacteria. The tripeptide segments of the jasplakinolides and the chondramides are quite similar, containing an L-alanine, an *N*-methyl-D-tryptophan and an L-β-tyrosine. However, the ring sizes are different since jasplakinolide contains an 8-hydroxy acid, whereas the chondramides feature a 7-hydroxy acid. The lagunamides[64] and viequeamide A[65] (**130**) are some recent examples of cyclic depsipeptides. Both show cytotoxic properties.

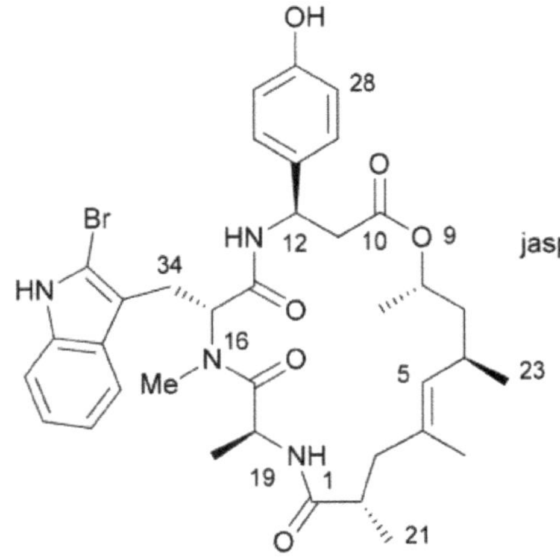

jasplakinolide (**123**)

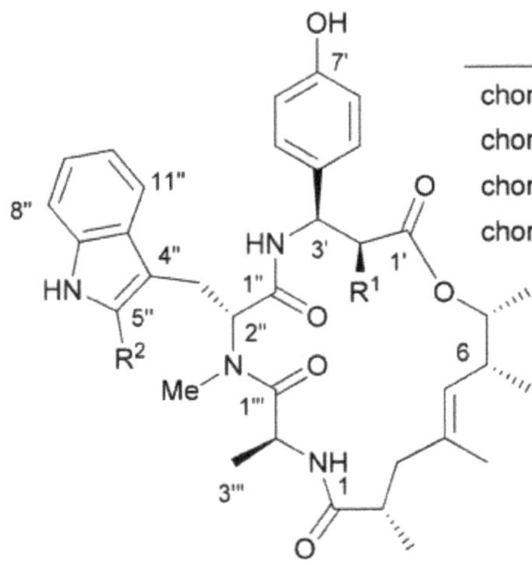

	R^1	R^2
chondramide A (**124**)	OMe	H
chondramide B (**125**)	OMe	Cl
chondramide C (**126**)	H	H
chondramide D (**127**)	H	Cl

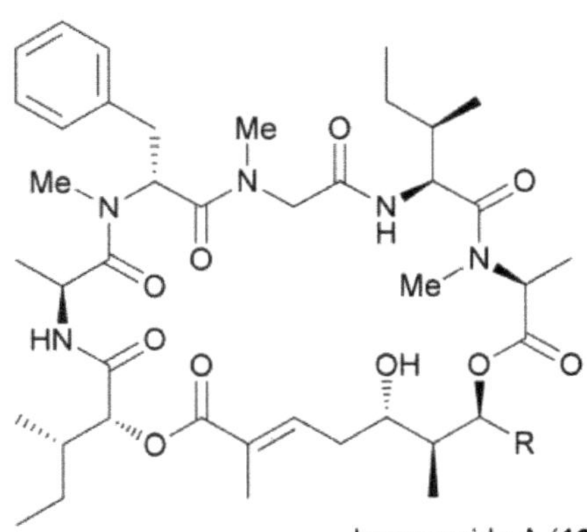

lagunamide A (**128**), R = 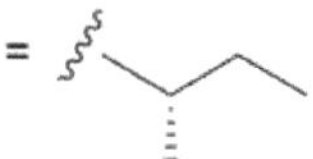

lagunamide B (**129**), R = 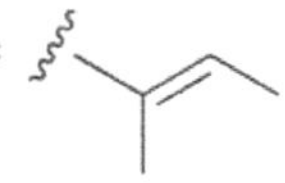

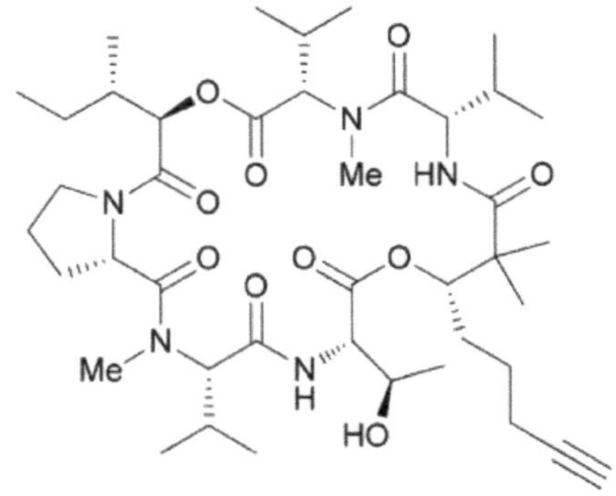

viequeamide A (**130**)

Figure 11: Structures of some cyclodepsipeptides.

For jasplakinolide several total syntheses have been described.[66] Most of them close the macrocyclic ring by formation of the ester or an amide bond. The Waldmann group utilized a ring-closing metathesis strategy.[67] The first total syntheses of chondramide C were achieved by the Waldmann and Kalesse groups.[68,69] They also secured the stereochemistry in the ω-hydroxy acid.

Since cyclodepsipeptides are still peptide-like, they often modulate protein–protein interactions. The potent antitumor activity of the jasplakinolides and chondramides is a consequence of their stabilizing effect on F-actin filaments.[70]

In our own work related to the chondramides we initially developed a concise synthesis of the ω-hydroxy acid based on a vinylogous aldol reaction and an asymmetric alkylation.[71]Thereafter, we achieved a total synthesis of chondramide A.[72] Unlike chondramide C which contains a β-tyrosine, chondramide A features a 2-methoxy-β-tyrosine. Its configuration was determined as (2*S*,3*S*). This amino acid derivative could be prepared by a Sharpless dihydroxylation followed by a selective Mitsunobu substitution of the benzylic hydroxyl group with hydrazoic acid (Scheme 11). With the tyrosine derivative **134** and the tryptophan **131** in hand, assembly of the tripeptide segment was started. Thereafter the tripeptide acid derived from **135**was combined with the ω-hydroxy ester **136** by Mitsunobu esterification. The synthesis could be completed with macrolactam formation in the presence of *N,N,N′,N′*-tetramethyl-*O*-(benzotriazol-1-yl) uronium tetrafluoroborate (TBTU) and HOBT in reasonable yield.

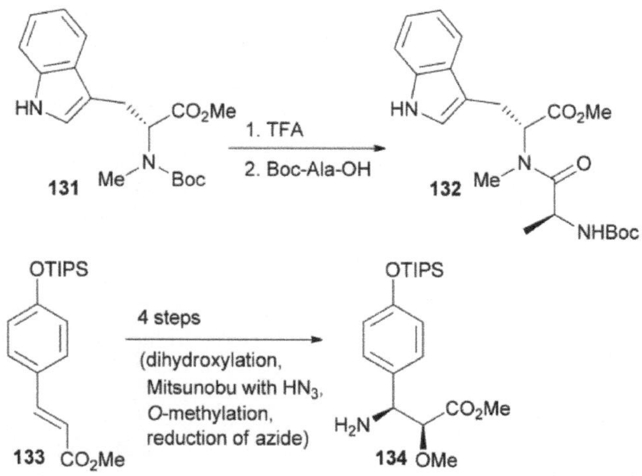

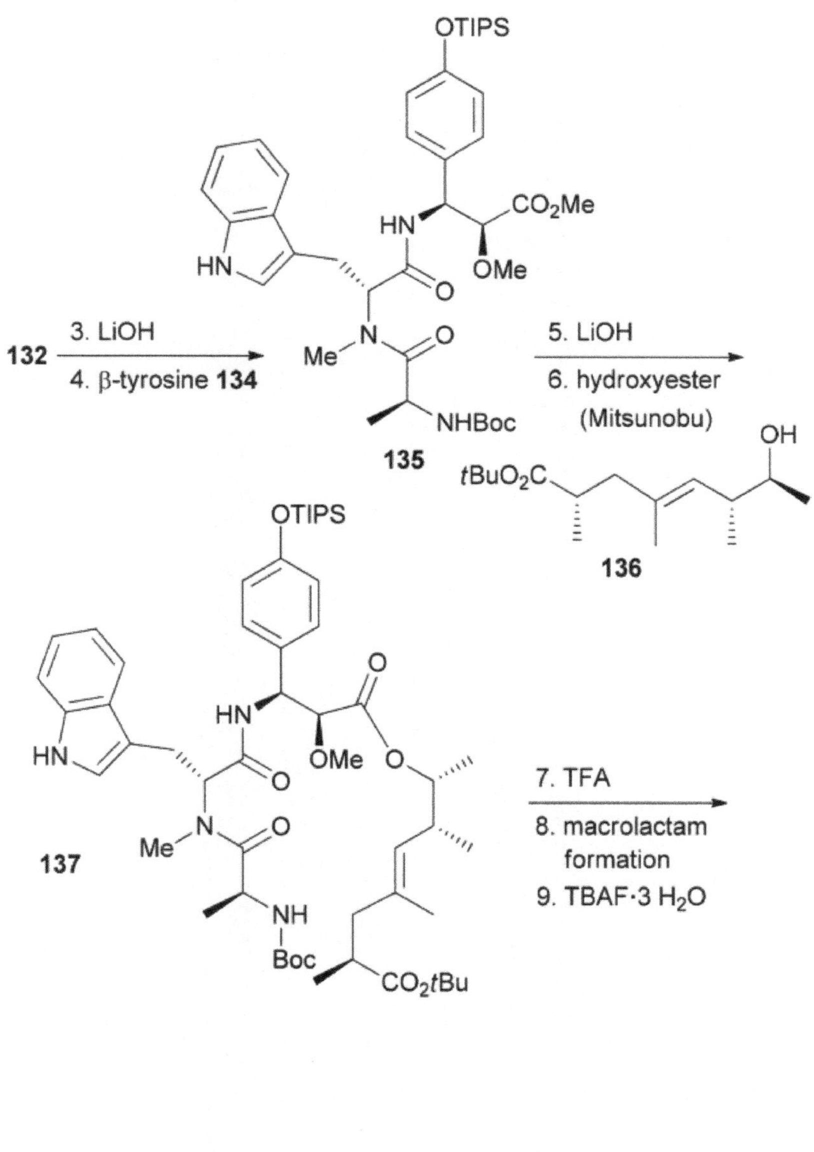

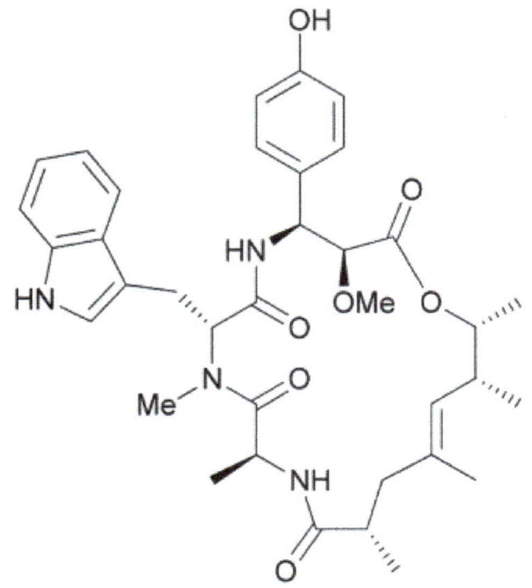

chondramide A (**124**)

Scheme 11: Summary of the synthetic strategy to chondramide A (**124**).

With efficient routes to the building blocks and a reliable strategy for assembly and ring closure, the synthesis of chondramide A analogues was undertaken.[73,74] One reason was to contribute to the SAR. Another reason was to exploit differences in the actin filaments between parasites and mammalian cells. For example, in the regions of contact, sequence alignment indicated residue differences at positions 200, 270 and 272. In *Toxoplasma gondii* they are G, K and A, whereas human actin has S, M and S at these positions. Altogether, we synthesized ten chondramide A analogues where the phenolic hydroxyl group was replaced with some other substituents. Most of the analogues retained high cytotoxicity in the low nanomolar region. Among the most potent ones were the –CN, –F and $–NO_2$ substituted β-tyrosine analogues. The *para*-position also tolerates hydroxymethyl and hydroxyethyl substitution. On the other hand replacement of the OH group with a phenyl or carboxamide caused a reduction of activity. Scheme 12 also summarizes the SAR from the contributions of the Waldmann, Kalesse and the Maier group illustrated with jasplakinolide.[75]

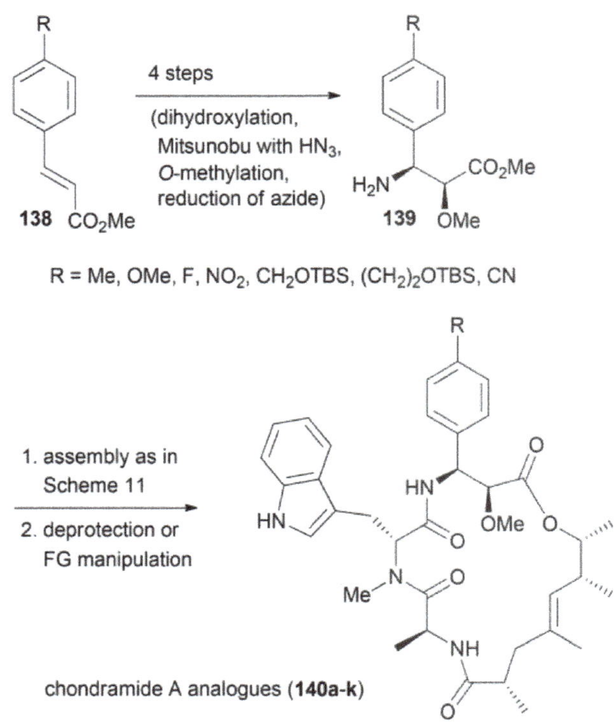

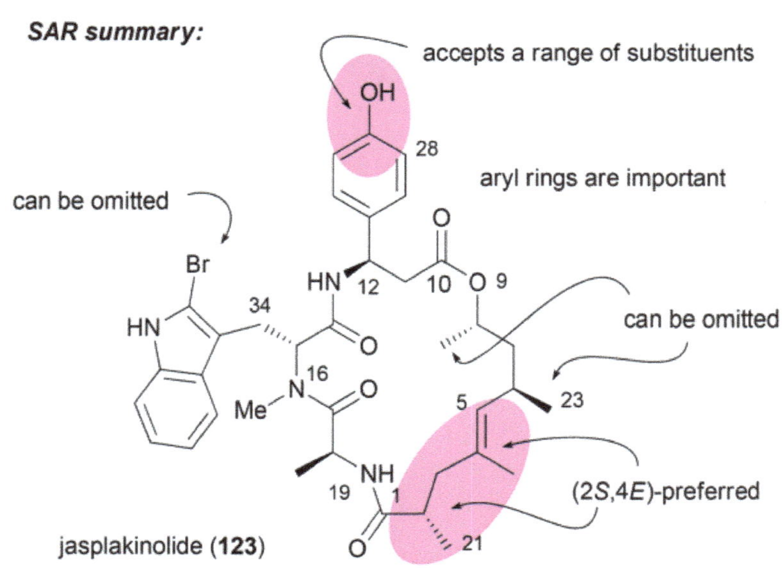

Scheme 12: Synthesis of various 2-methoxy-β-tyrosine analogues and their incorporation into the chondramide A system. Summary of the SAR for the chondramide/jasplakinolide class of cyclodepsipeptides illustrated on the structure of jasplakinolide (**123**).

These chondramide A analogues were indeed able to block parasite invasion into host cells (Table 3).[76] Although they lack specificity for parasite *vs.* host actin other analogues might be able to disturb actin dynamics in parasites without harming the host.

Table 3: Biological activities of chondramide A and the analogues on human primary foreskin fibroblast (HFF) cells together with growth inhibitory data on *T. gondii*

Compound	R	EC_{50} (HFF) [μM]	EC_{50} (*T. gondii*) [μM]
124	OH	0.109	0.51
140b	Me	0.027	0.53
140c	OMe	0.066	0.69
140d	F	0.029	0.32
140e	NO_2	0.040	0.61
140f	NH_2	0.057	0.66
140g	CH_2OH	0.052	0.74
140h	$(CH_2)_2OH$	0.059	0.60
140i	CN	0.027	0.38
140j	$CONH_2$	1.424	>45
140k	Ph	0.213	1.34

Pladienolide B

The relevance of NPs stems also from the fact that ocionally compounds are isolated that lead to the discovery of biological processes which offer new opportunities in the treatment of diseases. Recent examples include NPs that bind to components of the spliceosome. In retrospect this is not too surprising since the spliceosome is responsible for the correct removal of introns from mRNA precursors in eukaryotic cells.[77] The spliceosome is assembled from five *small nuclear ribonucleoproteins* (snRNP; U1, U2, U4, U5, U6) and associated protein factors. The intron removal requires sequence recognition. This is guided by certain motifs in the introns like branch sequences and polypyrimidine regions. It seems that the U2 snRNP has a key role in that it pairs with the branch sequence. This snRNP itself is composed of the U2 snRNA, U2 snRNP specific proteins (sm proteins) and two splicing factors SF3a and SF3b. The latter is a multiprotein complex made up of splicing associated proteins (SAP) 155, 145 and 130. These are also known as SF3b1, which seems to be the target of pladienolide B,[78] SF3b2 and SF3b3. The exons can be joined in various ways as a result of alternative splicing. Thus, one gene can code for several proteins. Erroneous splicing is a key feature of cancer cells. Accordingly, inhibition of splicing appears as a promising anticancer strategy, at least for some types of cancer.

So far several natural products are known to inhibit the splicing process (Figure 12). They include polyketides of the pladienolide family, like the natural products pladienolide B (**141**), pladienolide D (**142**)[79] and FD-895 (**144**).[80] Molecules like GEX1A (herboxidiene) (**145**)[81] have a side chain similar to the one in pladienolide B. Another group of spliceosome inhibitors represented by FR901464 (**146**)[82] features two pyran rings. A related compound is thailanstatin (**148**).[83] The Eisai company, which had discovered the pladienolides, early on described the structure of the carbamate derivative E7107 (**143**). This compound is also active indicating that modifications at C7 are tolerated. It even advanced to human trials.[84,85] A dose-limiting study revealed common side effects of antitumor compounds, for example, nausea, vomiting and diarrhoea. However, since some patients also suffered from bilateral optic neuritis (visual disturbance due to inflammation of the optic nerve) the human trials were discontinued.

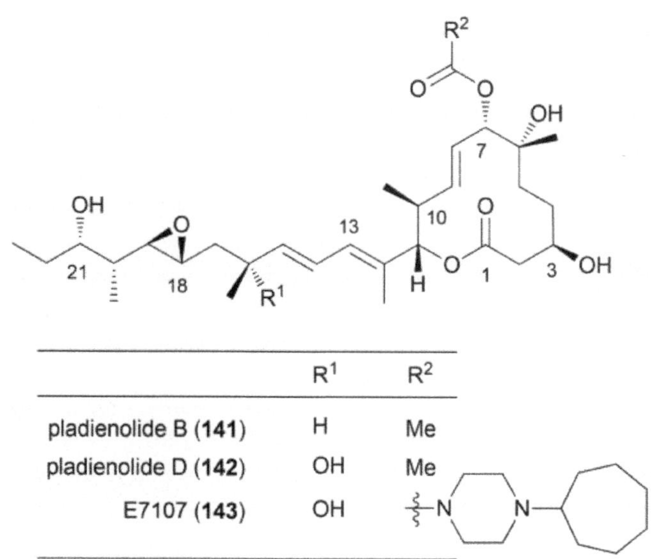

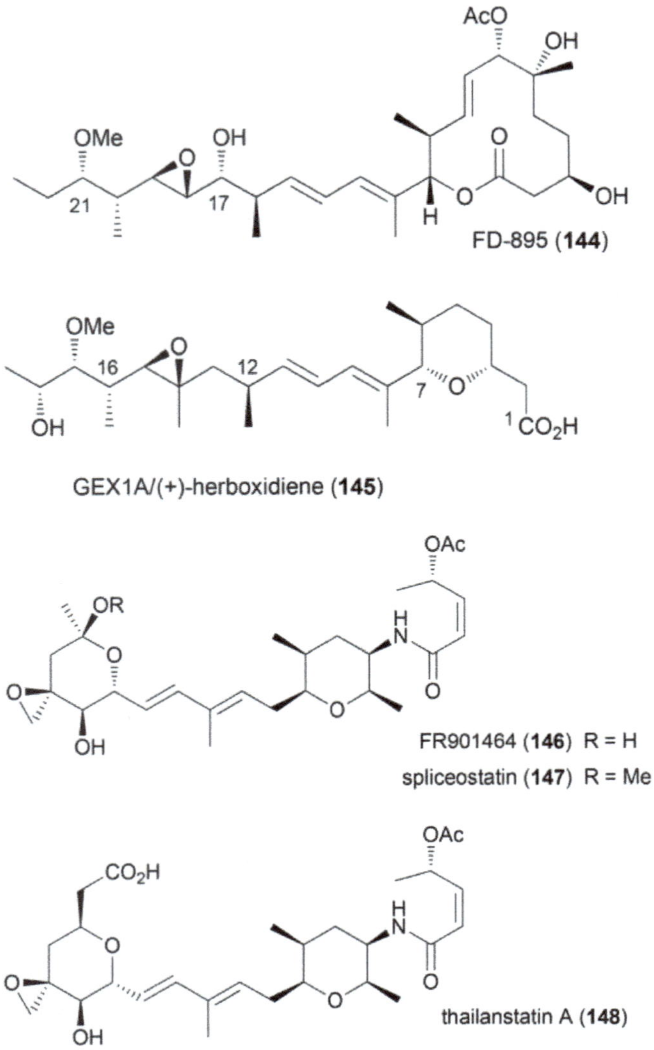

Figure 12: Structures of some important spliceosome inhibitors.

In order to discover improved and simpler SF3b inhibitors, SAR studies are clearly important. For FR901464 a range of analogues have been prepared by the groups of Koide and Webb. For example on meayamycin (**149**), an analogue that has two methyl groups at C1, SAR studies revealed the importance of the (*Z*)-pentenamide side chain with the allylic acetate, the methyl groups on the pyran ring, the diene region and the spiro epoxide (Figure 13).[86] Changes or omissions on these parts produced much less active compounds. Related analogues, called sudemycins, were developed by Webb *et al.*[87] These analogues also contained the pharmacophore common to FR901464 and the pladienolides, namely the spiro epoxide at a certain distance to the allylic acetate.[88] In some cell lines these analogues achieved IC_{50} values in the low micromolar range.

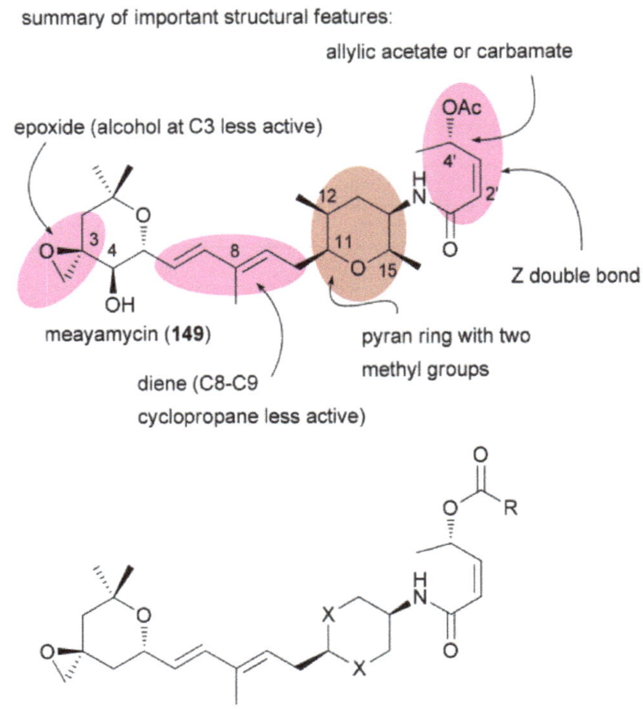

	X	R
sudemycin C1 (**150**):	CH_2	$CH(Me)_2$
sudemycin E (**151**):	O	$CH(Me)_2$
sudemycin F1 (**152**):	O	$N(Me)_2$

Figure 13: Key structural feature important for the activity of FR901464 and analogues inspired by FR901464.

Since pladienolide appears to be more challenging from a synthetic point of view, fewer analogues are known. A simplified pladienolide analogue **154** was made by Webb *et al.*[89] It lacks the methyl groups at C10, C15 and C22 as well as the 3-OH (Figure 14). While this compound turned out to be essentially inactive, the more hydrophobic compound **153** (21-*O-tert*-butyldimethylsilyl) was at least moderately active (IC_{50} 10.7 µM, SK-MEL-2, melanoma cell line). This result clearly undeores the role of methyl groups as contributors to hydrophobic binding and as conformational control elements. Research by Chandrasekhar showed that aryl analogues of pladienolide B, like **155**, with a truncated side chain are also active.[90] However, this seems not to be in line with the pharmacophore model of Webb. Compound **156** with a truncated side chain lacking the epoxide and with the wrong configuration at C7 was also completely inactive.[91] Pladienolide D analogue **157** lacking the 6-OH turned out to be quite active.[92]

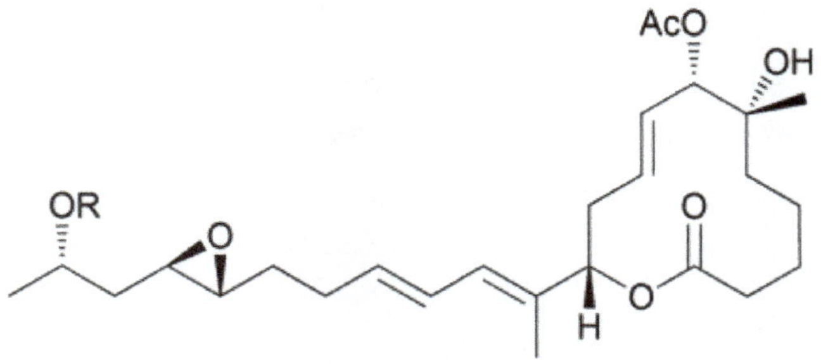

153 R = TBS [moderately active; 10.7 μM (SK-MEL-2 melanoma)]
154 R = H (inactive)

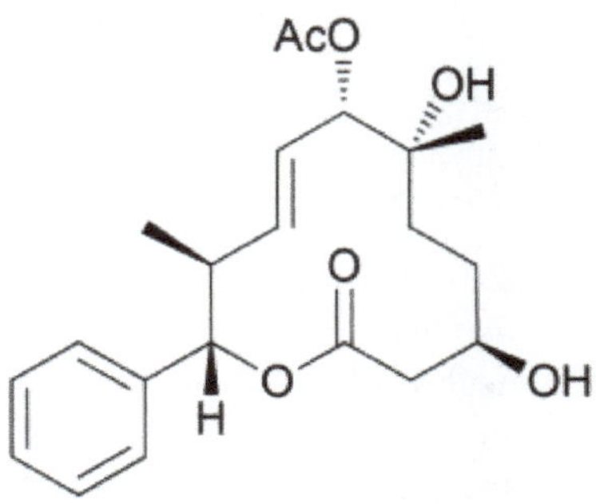

155 (1.5 μM, A549)

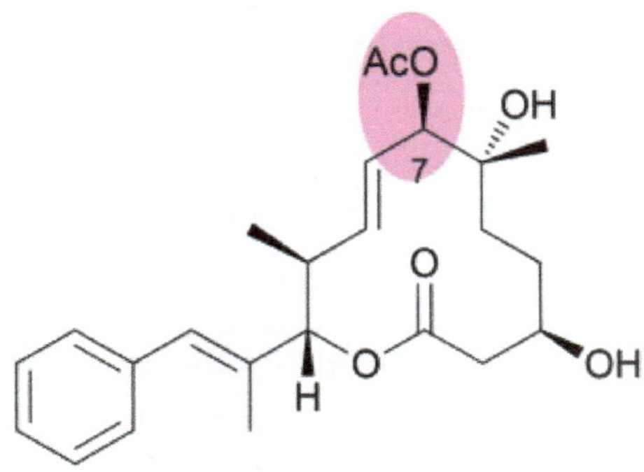

156 (inactive)

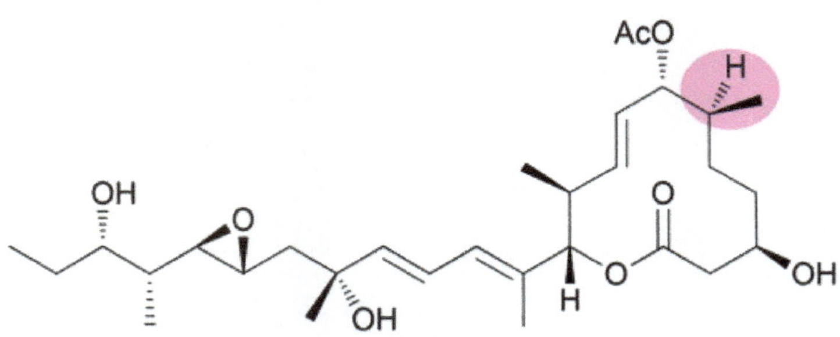

6-deoxypladienolide D (**157**) (active)

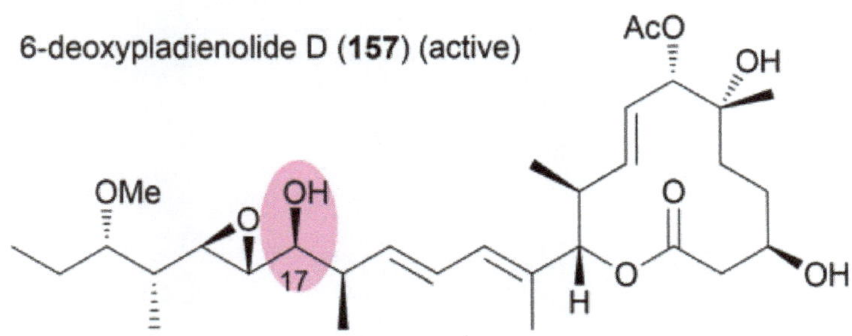

17-epi-FD-895 (**158**) (0.8 nM, HCT-116)
FD-895 (**144**) (23.5 nM, HCT-116)

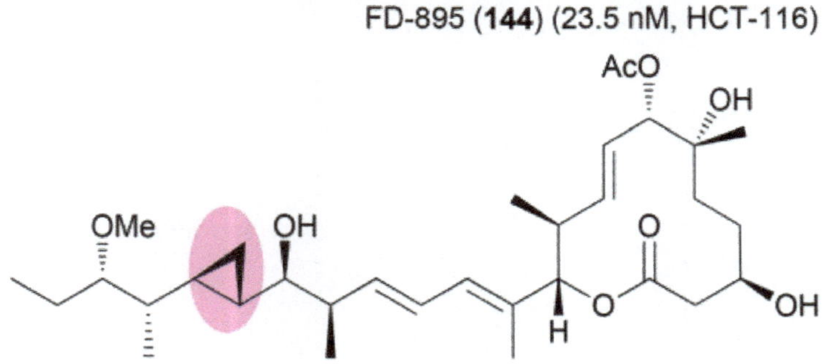

FD-895 cyclopropyl analogue (**159**) (42.9 nM, HCT-116)

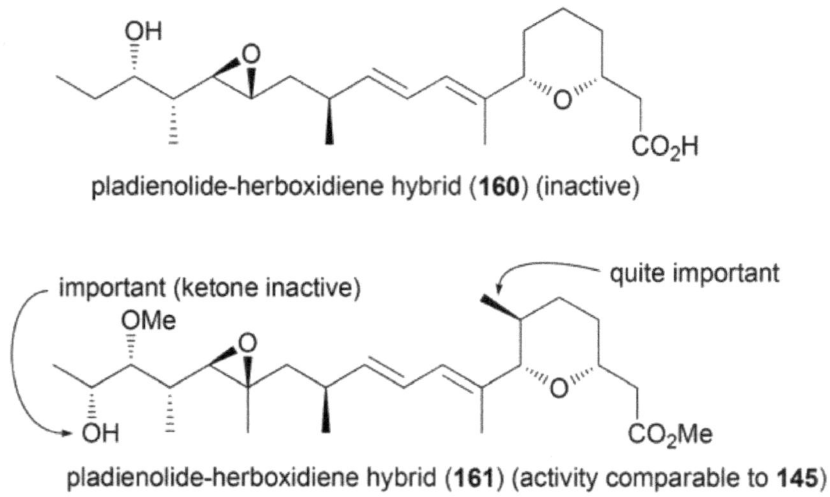

Figure 14: Pladienolide B analogues and SAR information gathered from them. Compounds **153**and **154** (ref. 89), compound **155** (ref. 90), compound **156** (ref. 91), compound **157** (ref. 92), compounds **158** and **159** (ref. 93), and compounds **160** and **161** (ref. 94).

In the course of the elucidation of the side chain stereochemistry of FD-895 several diastereomers were prepared.[93] With 17-*epi* derivative **158** an analogue was found that was almost 20 times more active than FD-895 itself. As can be seen with the less active cyclopropyl analogue **159**, the epoxide does contribute to the biological activity. Studies on pladienolide–herboxidiene hybrid compounds highlighted the importance of the methyl substituent in the pyran ring and the role of the hydroxyl group in the side chain.[94] However, one should say that the findings partly contradict themselves. Thus whereas **160** with a free hydroxyl group is inactive, the free hydroxyl group of **161** seems to be crucial.

Focusing on the synthesis of pladienolides, the most challenging seems to be creation of building blocks that contain the C6–C7 diol and the formation of the macrolactone. One strategy utilized by Kotake is the asymmetric dihydroxylation of (*Z*)-alkene **162**.[95] An esterification of the derived alkenoic acid **164** with alkenol **165** provided substrate **166** for a ring-closing metathesis (Scheme 13). The Burkart group first prepared monoprotected diol **169** by a Brown allylation of aldehyde **168**. Thereafter the tertiary alcohol at C6 was created *via* addition of MeMgBr to the alkoxyketone derived from **169**. As before, the ring was closed *via* rcm (Scheme 13).

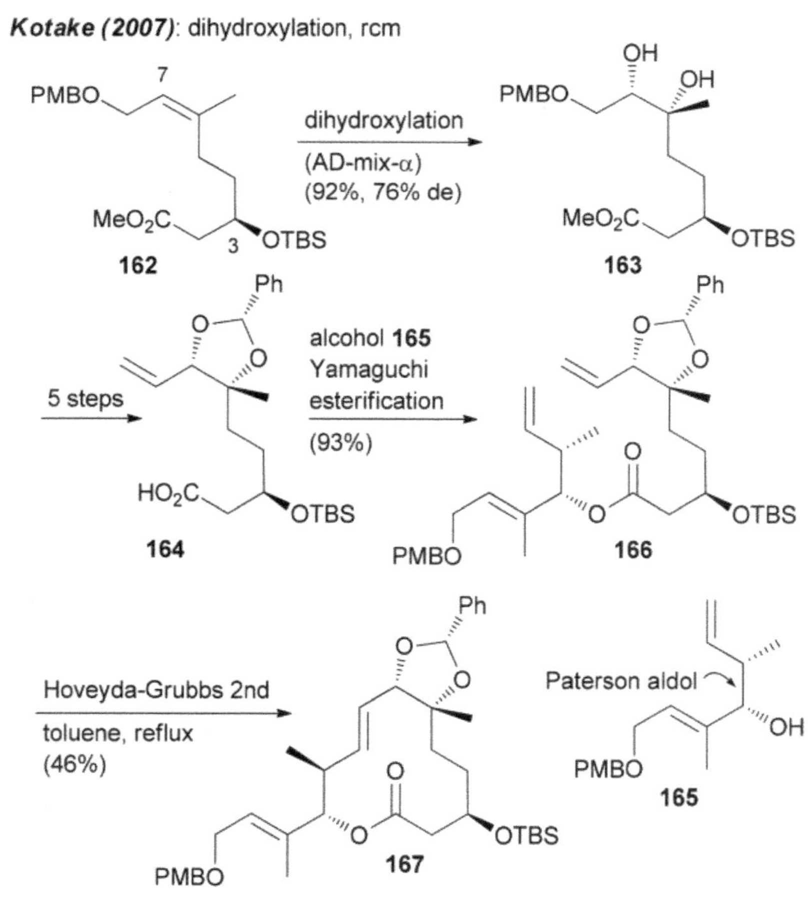

Burkart (2012): Brown allylation, addit. of MeMgBr to alkoxyketone

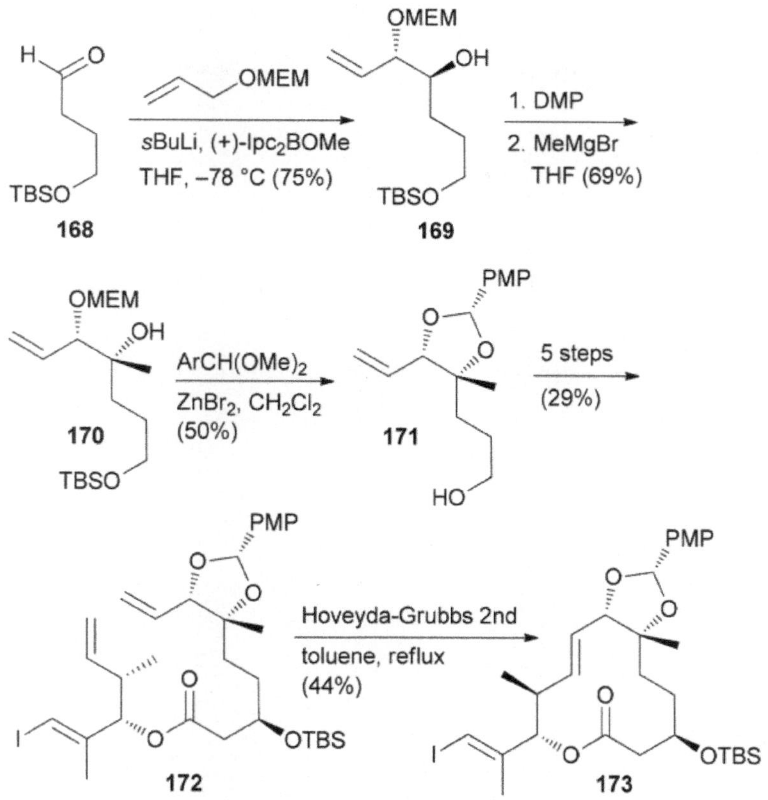

Scheme 13: Synthetic strategies to create building blocks containing the C6,C7- diol.

The chelation controlled Grignard addition to an α-alkoxyketone was also used by Ghosh *et al.*[96] However, here the synthesis of the required ketone started with divinylmethanol **174**. Opening of the derived epoxide **175** with the dienolate of **176** and asymmetric reduction of the keto ester **177** gave diol derivative **178** (Scheme 14). Oxidation and reaction with the Grignard reagent led to enoate **179**. This was further elaborated to macrolactone **181**.

Ghosh (2012): alkylation of β-keto ester with epoxide, asymm. reduction

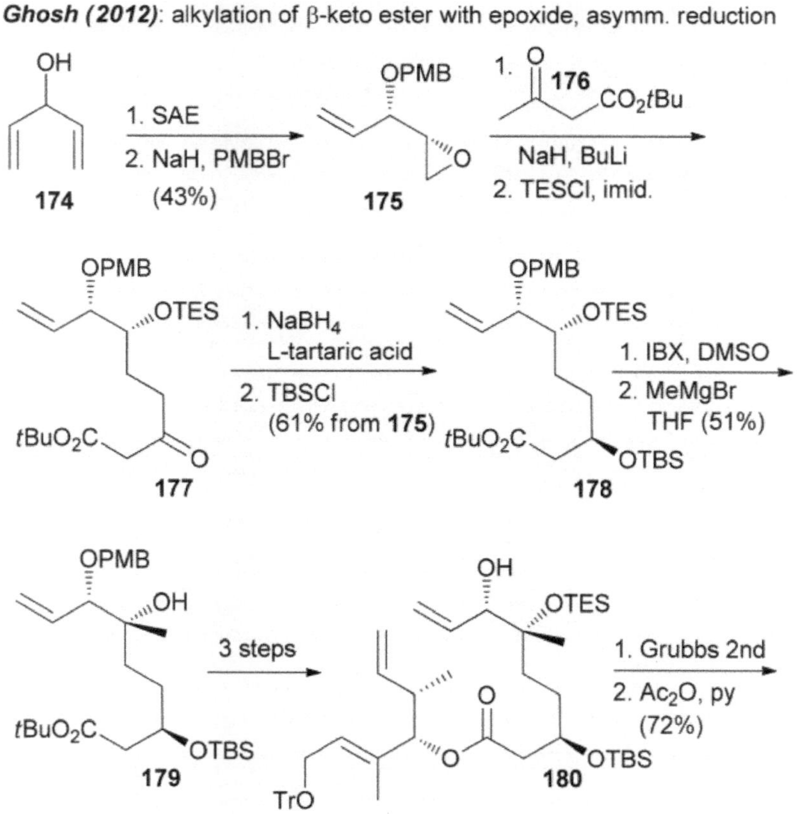

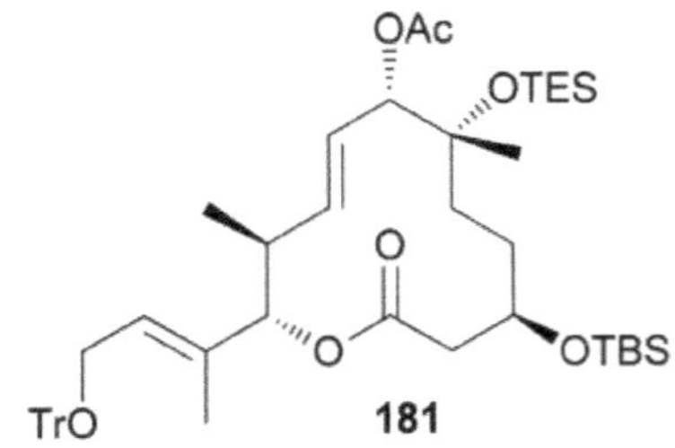

Chandrasekhar (2013): SAE, epoxide opening

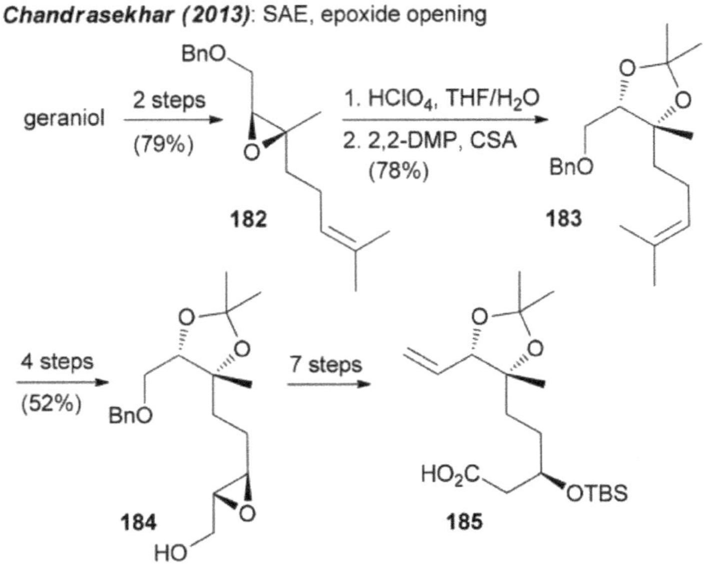

Scheme 14: Other strategies to create building blocks containing the C6,C7-diol.

A very simple and clever way to set up the C6–C7 region was put forward by Chandrasekhar.[90] Thus, a Sharpless asymmetric epoxidation (SAE) led to epoxide **182** (Scheme 15). Its hydrolytic opening followed by protection of the diol gave compound **183**. As can be seen, further elaboration of **183** to acid **185** did require a number of steps. In this work also an rcm strategy came to use for the creation of the macrolactone.

Maier (2011, 2014): intermolecular HWE, Shiina macrolactonization

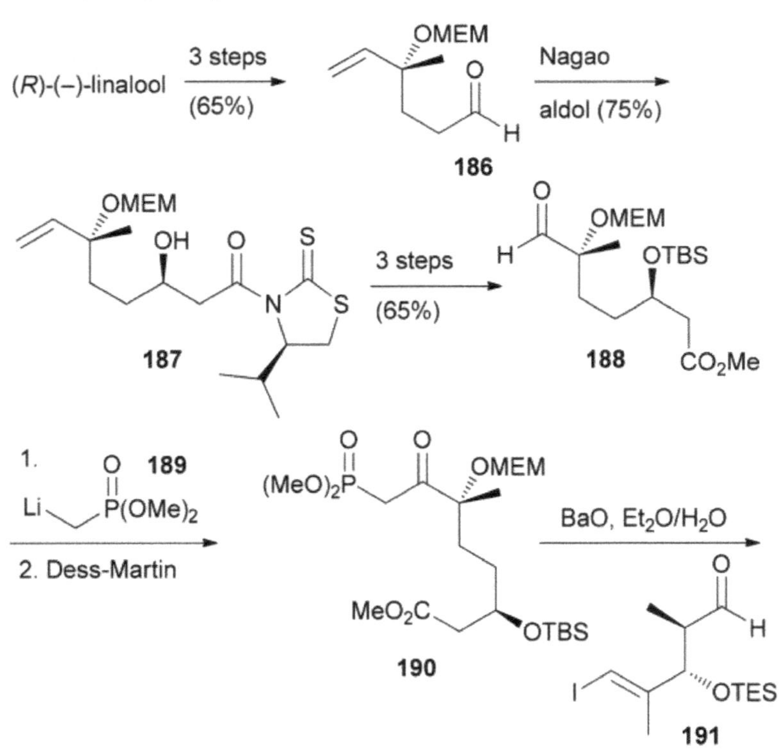

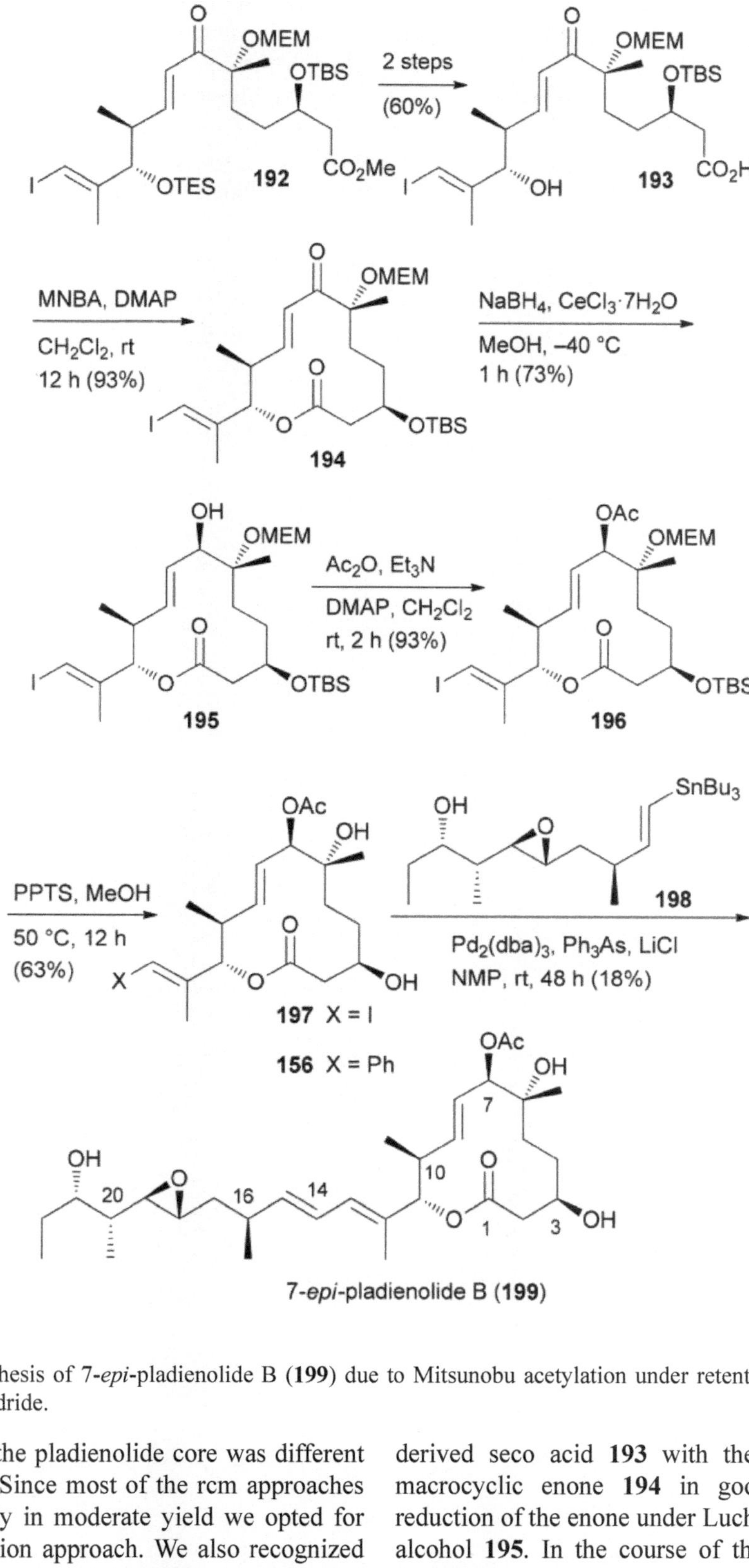

Scheme 15: Unexpected synthesis of 7-*epi*-pladienolide B (**199**) due to Mitsunobu acetylation under retention of configuration. MNBA = 2-methyl-6-nitrobenzoic anhydride.

Our own approach to the pladienolide core was different but not without problems. Since most of the rcm approaches gave the macrolactone only in moderate yield we opted for a classical macrolactonization approach. We also recognized the match of the stereocenter in (*R*)-(−)-linalool with that of C6 in the pladienolides. To combine the subunits for the seco acid, a HWE reaction was planned followed by reduction of the enone. Accordingly, (*R*)-(−)-linalool was converted to aldehyde **186** which upon an acetate aldol reaction furnished hydroxy acid derivative **187** (Scheme 15). Manipulation of the carboxylic function and the double bond delivered aldehyde **188**. Addition of lithiated phosphonate **189** and oxidation gave rise to keto phosphonate **190**. Under optimized conditions phosphonate **190** underwent a HWE reaction with aldehyde **191** providing enone **192**. Cyclization of the derived seco acid **193** with the Shiina reagent[97] afforded macrocyclic enone **194** in good yield. Unfortunately, a reduction of the enone under Luche conditions gave the 7-*epi*-alcohol **195**. In the course of the synthesis of pladienolide B analogue **156** (Scheme 15), after a Stille cross-coupling with phenyltributylstannane, the C7 allylic alcohol was acetylated under Mitsunobu conditions.[91] However, as we later found out, the Mitsunobu reaction had occurred with retention of configuration. Thus, acetylation of **195** either under Mitsunobu or classical acetylation conditions gave the same allylic acetate **196**. Cleavage of the MEM and TBS protecting groups delivered core structure **197**. A final Stille cross-coupling with vinylstannane **198** furnished pladienolide B analogue**199**.[98] Its ^{1}H NMR spectrum showed distinct differences to that from pladienolide B (**141**).

Since it was not possible to reduce the enone in the macrolactone **194** in the desired way, attempts were made to reduce the enone function on an acyclic precursor. After some trials we found that the chelation controlled reduction of enone **192** delivered the desired allylic alcohol **200** as the major diastereomer (Scheme 16). In a similar manner to that for the 7-*epi*analogue, ester **201** was converted to macrolactone **204** and then to pladienolide B (**141**). The biological evaluation of both compounds revealed a dramatic effect of the configuration at C7. Thus, whereas pladienolide B was highly active (IC_{50} = 7.5 nM, L929 mouse fibroblast cells), the 7-*epi*-epimer **199** was devoid of any activity.[98] This analogue showed that an acetate or carbamate at C7 with the correct configuration is an essential part of the pharmacophore.

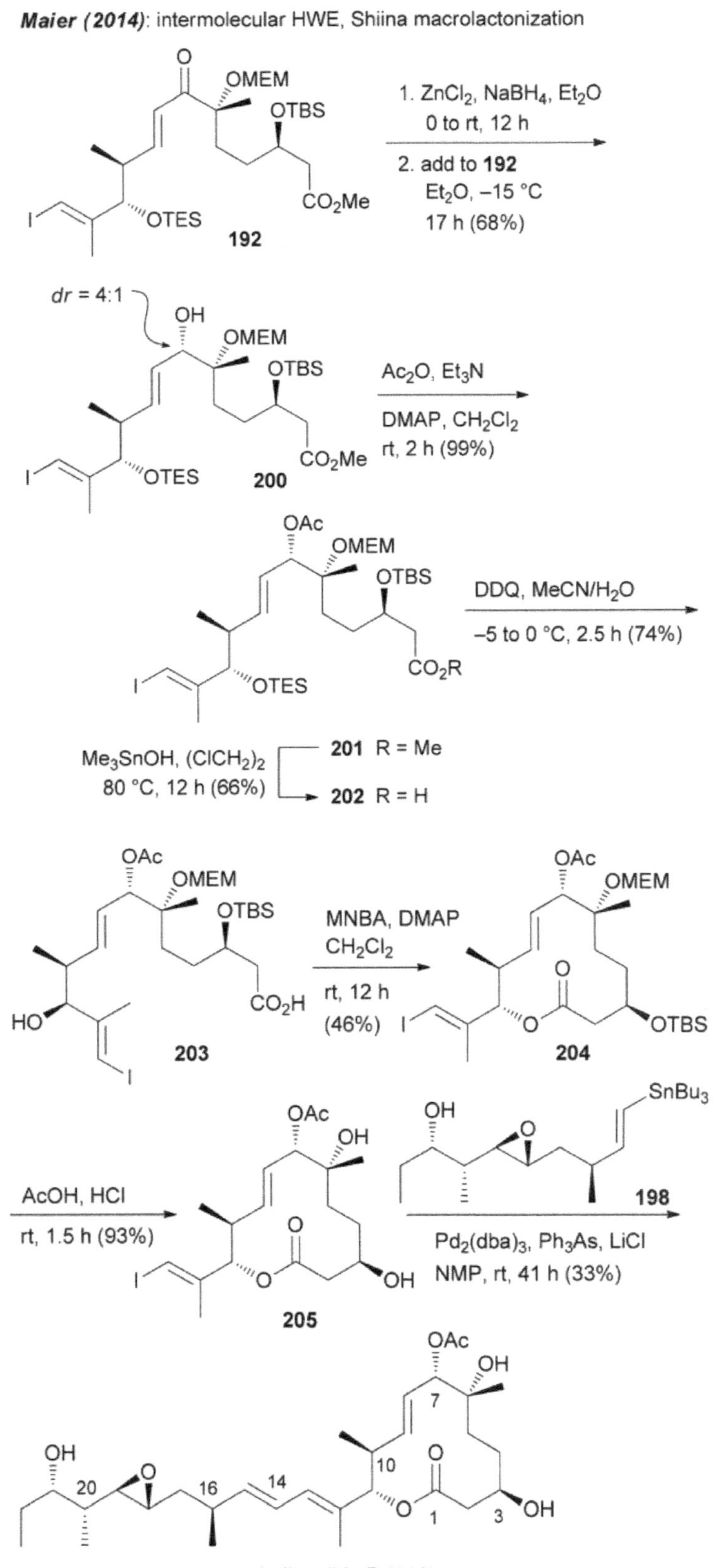

Scheme 16: Synthesis of pladienolide B (**141**) *via* chelation controlled reduction of enone**192.**

FUNCTION-ORIENTED SYNTHESIS (FOS)

The idea of function-oriented synthesis is to circumvent some inherent disadvantages of biologically active NPs. Thus, often they are difficult to obtain from natural sources, in particular if they originate from the marine environment. If their structures are highly complex, they are generally not amenable to total synthesis in larger amounts. Also, in many es they are not suitable for direct use in humans before some chemical modifications. The ultimate goal of FOS is the design of simpler analogues of NPs where the activity is optimized towards the intended application. At the same time the analogues should allow for straightforward syntheses.[99] The concept of FOS is kind of obvious and trivial. It is similar to DTS but the emphasis here is on function. The analogue of a FOS programme might be much simpler and structurally much different from the NP. The Wender group demonstrated this strategy with the synthesis of several bryostatin analogues. The bryostatins comprise 20 marine macrolides, which were isolated from the bryozoan *Bugula neritina*.[100] They are polyketides featuring three pyran rings and several stereocenters. They essentially have the same macrocyclic core structure but most of them differ in the substituents at C7 and C20.

The bryostatins are characterized by a range of biological activities, but prevalent is their ability to modulate the activity of protein kinase C. This leads to inhibition of cell growth and angiogenesis. So far several syntheses of bryostatins were reported.[101] The shortest one, the synthesis of bryostatin 7 (**207**), requires 36 steps in total (TS) with 20 steps for the longest linear sequence (LLS) (Scheme 17).[102] Based on the notion that the relevant pharmacophoric regions include the upper lipophilic A and B rings plus the polar carbonyl and hydroxyl groups at C1, C19 and C26, respectively, the Wender group synthesized many highly potent but simpler bryostatin analogues. Among the most potent derivatives was macrolactone **210** (Scheme 17). Esterification of acid **209** with alcohol **208** followed by macrocyclization *via* intramolecular acetal formation gave analogue **210** (32 TS), (17 LLS).[103] The design of these analogues was probably inspired by the good availability of the lower fragment **208**.[104] This allowed for connection with various simplified upper fragments.[105]

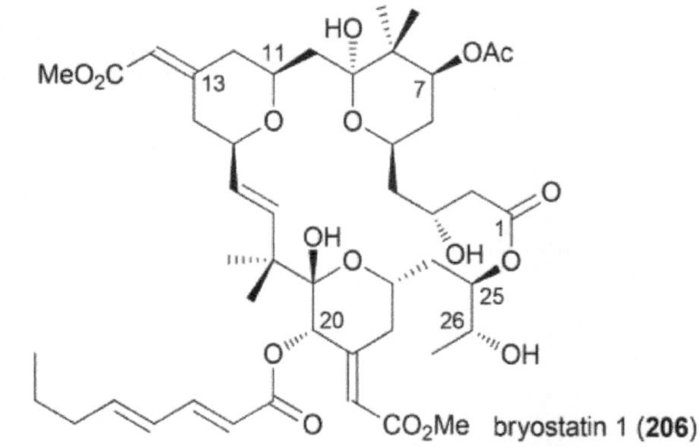

bryostatin 1 (**206**)

K_i = 1.35 nM

bryostatin 1: 31 steps (LLS), 58 (TS)

bryostatin 7 (**207**, 20-OAc): 20 steps (LLS), 36 (TS)

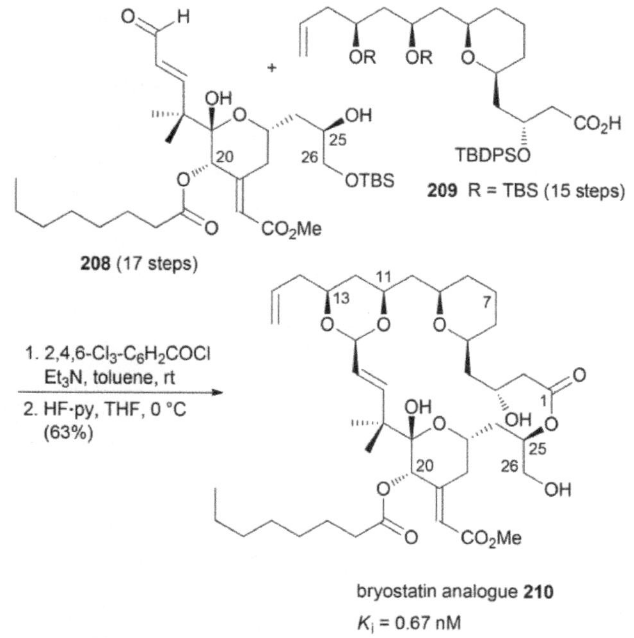

Scheme 17: Synthesis of bryostatin analogue **210** from the lower fragment **208** and the diol derivative **209**.

Although not extensively covered in this account, one should mention the discovery of eribulin mesylate whose design was inspired by the marine-derived natural product halichondrin B. This rather complex polyketide is a potent antitumor agent which works by inhibition of microtubule dynamics. In the course of synthetic studies using a convergent approach it was discovered that the biological activity resides in the right part of the molecule.

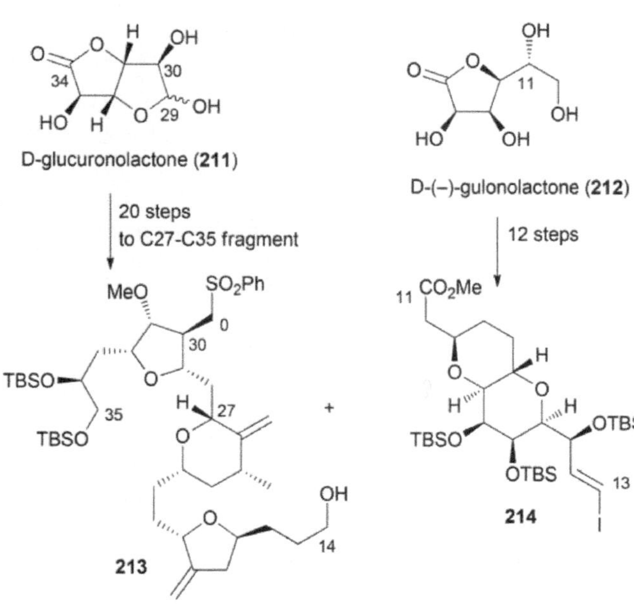

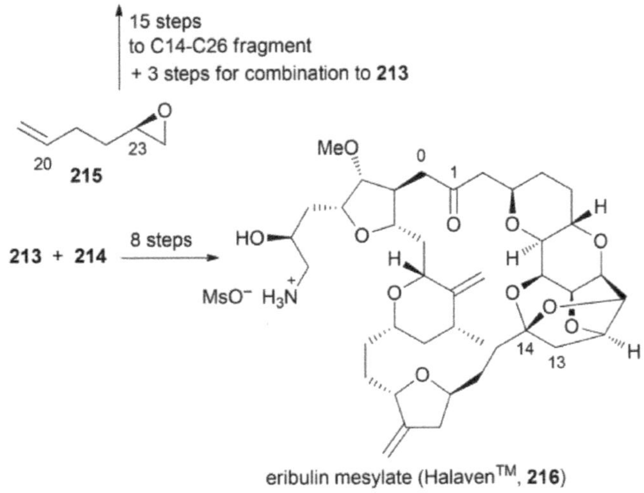

Scheme 18: Synthesis of eribulin mesylate (**216**) on the production scale.

Among simplified macrolactone, macrocyclic ether, macrolactam and macrocyclic ketone analogues, the latter was found to be the best compromise.[106,107] Eribulin mesylate (**216**) has been approved for clinical use in 2010 in the USA (Scheme 18). Key fragments were prepared using a chiron approach starting with D-glucuronolactone (**211**) and D-(−)-gulonolactone (**212**). In addition, epoxide **215** served as a starting material. First the C0–C1 bond was formed by addition of the anion of **213** to the aldehyde derived from **214**. The macrocyclic ring was closed between C13–C14 using a Nozaki–Hiyama–Kishi (NHK) reaction. Altogether, three key bond formations relied on the NHK coupling (C19–C20, C26–C27, C13–C14). This e nicely undeores the high level of organic synthesis in some pharmaceutical companies.

Taking the benzolactone hypothemycin (**218**) as a lead, the Winssinger group designed simpler mimetics that react irreversibly with certain kinases that have an exposed cysteine in the so-called DFG binding motif. Thus, analogues like **218** feature the resorcyclic acid part as a hinge binder and an electrophilic trap to address the cysteine (Figure 15).[108]

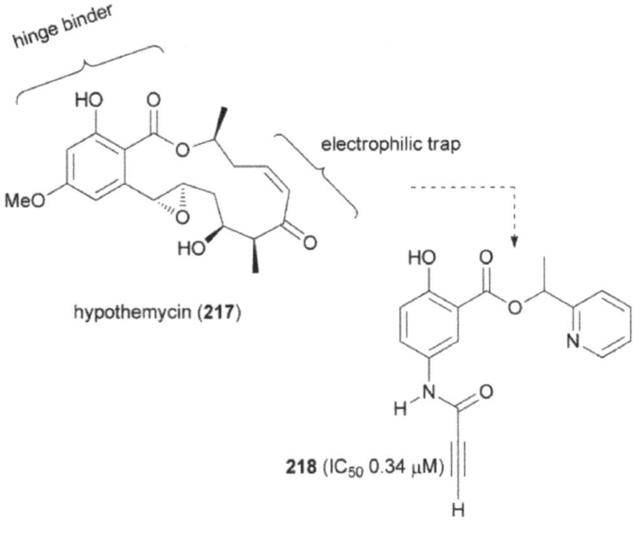

Figure 15: Design of affinity-based probes for kinases with cysteine in the binding region. The cellular inhibition of EGFR autophosphorylation in A431 cells after stimulation with EGF is given for the most active compound **218**.

BIOLOGY-ORIENTED SYNTHESIS (BIOS)

The principle of BIOS is to use scaffolds contained in NPs as starting points for the synthesis of compound libraries.[109] This concept is based on the notion that NP scaffolds and possibly substructures of NP core structures define privileged structures. In other words, evolutionarily formed NP scaffolds match highly conserved subfolds of ligand sites.

For the selection of suitable NP-like scaffolds the Waldmann group conceived two complementary approaches. The first is a hierarchical structural organization of scaffolds contained in NPs. After removal of side chains that do not contain rings, in each truncation step one peripheral ring is removed according to certain guidelines. This way, parent–child relationships are generated with the parent being the simpler structure. This leads to tree-like branches of NP scaffolds. One such branch is shown in Figure 16. Representative NPs for each scaffold include digitoxigenin (**219**), the aglycon of the cardiac glycoside digitoxin, dehydroepiandrosterone (DHEA, **220**), the tetradecahydrophenanthrene-type diterpene klysimplexin T[110] (**221**), nakamurol A[111] (**222**), and gabosine A[112] (**223**).

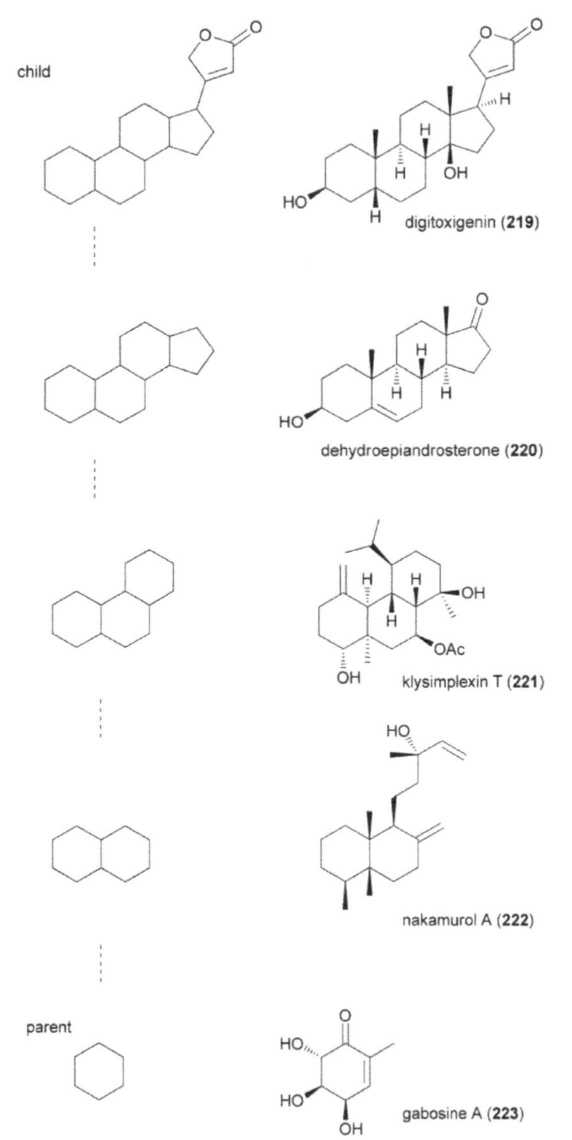

Figure 16: Example of a branch from the NP scaffold tree (left column). The parent is always the simpler structure in a chain. Representative NPs for each scaffold are shown in the right column.

If one argues that a parent (simpler) scaffold to a child scaffold which contains an active representative might also show some activity if properly functionalized, then a parent scaffold might form the basis of a compound library. Of particular interest are branches that contain gaps, meaning core structures in a sequence that do not have a corresponding NP. Software to visualize the hierarchical organization of NPs is freely available.[113] Of particular use is this kind of clustering if biological activity data are available for molecules, representing scaffolds in the branches. In this e, so-called virtual scaffolds located between a complex and a simpler one for both of which compounds are known with related activity should be ideally suited for library development.

A second approach put forward by the Waldmann group starts with proteins. In this protein structure similarity clustering (PSSC), protein folds are clustered according to their similarity. The hypothesis is that a ligand for a member of a protein cluster might also bind to other members of a cluster.[114] Therefore, it should be possible to use such NP scaffolds, their parents or grand parents for library construction to address members of a protein cluster that might be relevant for medicinal application.

Recent examples from the Waldmann group are still based on the BIOS approach but at the same time are substantially chemistry driven.[115,116] This means that substructures of NPs are used as guidelines which can be constructed or functionalized by novel and simple chemistry. For example, taking the tetrahydroisoquinoline NP noscapine (**224**) as a lead, a compound collection of tetrahydroisoquinolines **226** was prepared by reacting iminium ions **225** with acetylene nucleophiles (Scheme 19). Not surprisingly, several derivatives were found that, like**227**, interfere with microtubule polymerization. In the assays performed, alkyne **227** showed the highest activity.

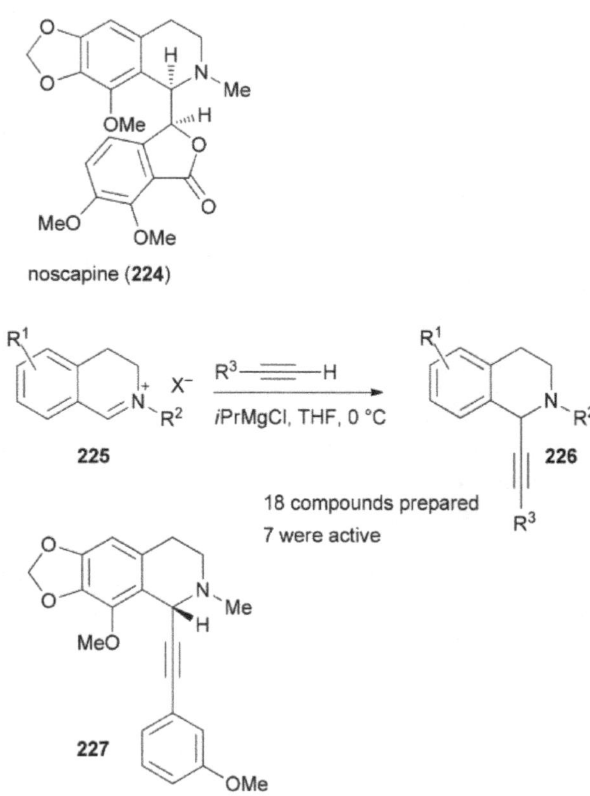

Scheme 19: Synthesis of a tetrahydroisoquinoline collection.

If a class of NPs comprises a large number of members they also should be promising leads for analogue syntheses. For example, iridoids are widespread in nature and display a range of biological activities.[117] The Waldmann group utilized an asymmetric [3 + 2]-cycloaddition between azomethine ylides, generated from glycine imines **229**, and pyranones **230** (Scheme 20). The latter are available from furfuryl alcohols *via* oxidative rearrangement. In the presence of a Cu(I) salt, the ligand **231** and DBU, the cycloaddition proceeded with good dia- and enantioselectivities.[118] Out of a collection of 115 derivatives, nine were able to inhibit Wnt signalling with an IC_{50} below 10 μM. The Wnt pathway plays a role in cell differentiation and stem-cell renewal. Other representatives from this collection turned out to be inhibitors of the hedgehog signalling pathway.

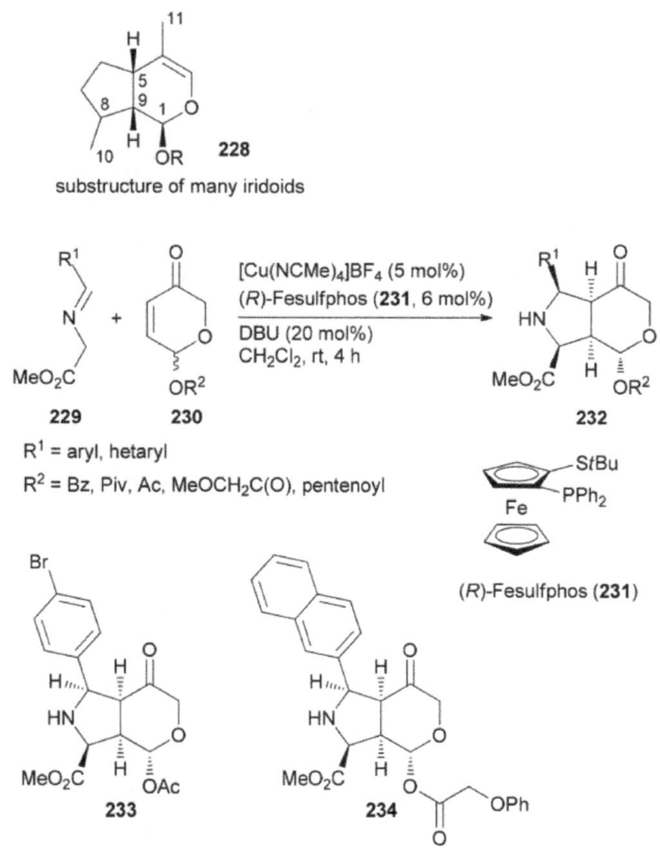

Scheme 20: Synthesis of a library of fused heterocycles **232** inspired by the structure of iridoids.

COMPLEXITY TO DIVERSITY (CTD)

Due to the fact that NPs have unique shapes, ring systems and diverse functional groups they are in principle interesting starting points for the preparation of NP analogues. The problem is to selectively address the various functional groups of a NP in order to avoid the formation of complex mixtures. According to Hergenrother, who coined the term complexity to diversity (CtD) one can distinguish two categories.[119] One strategy entails reactions on a single functional group to generate a new structure. A classical singular example is the preparation of ansa-steroids by a domino Diels–Alder/retro-Diels–Alder reaction between ergosterol derivatives and propynal.[120,121] For example, steroid **235** could be opened to macrocycle **237** utilizing this concept (Scheme 21).[122] A second strategy addresses several functional groups on a NP to create new derivatives.

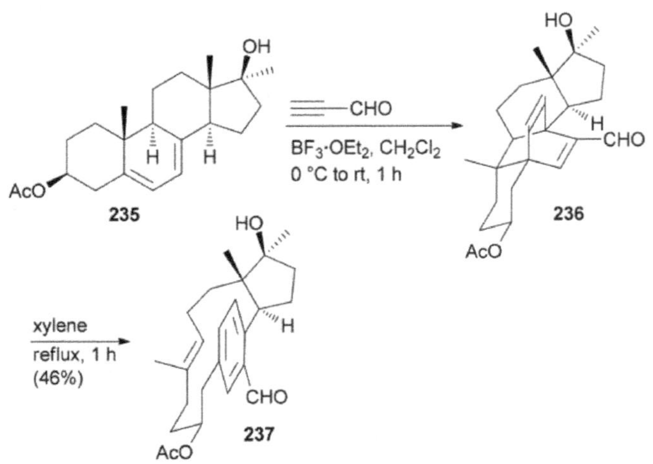

Scheme 21: Synthesis of macrocyclic paracyclophane **237** from diene **235***via* a Diels–Alder/retro-Diels–Alder sequence.

Selective modifications of the diterpene abietic acid (**238**) enabled the construction of 84 different compounds.[123] The initial challenge is to differentiate the two double bonds and the carboxylic group. This is possible by dihydroxylation. From diol **240** a new ring system can be obtained by oxidative cleavage followed by intramolecular aldol condensation (Scheme 22).

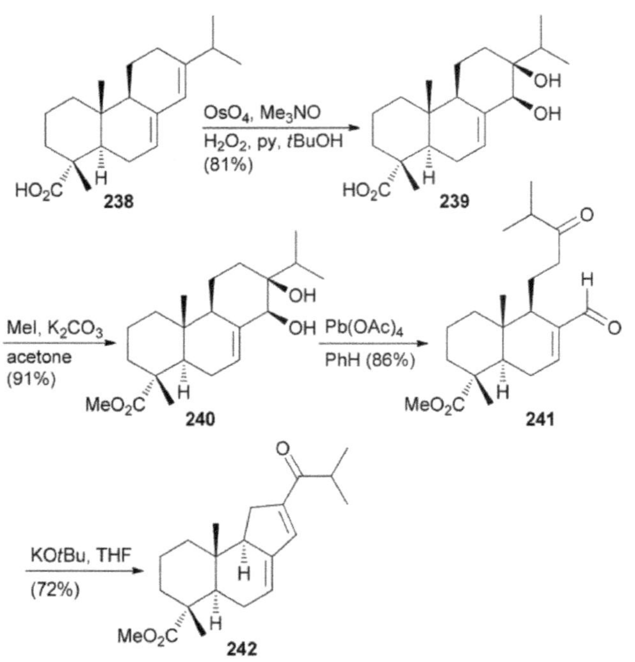

Scheme 22: Synthesis of tricyclic compound **242** by the ring distortion strategy.

Manipulation of the B-ring is possible after protection of the diol in **240** as a cyclic carbonate. Ring expansion was illustrated with cyclopropanation followed by electrocyclic ring opening induced by $AgBF_4$ in MeOH (Scheme 23). Ring system **245** resulted from dihydroxylation of the double bond, oxidative cleavage and intramolecular aldol reaction.

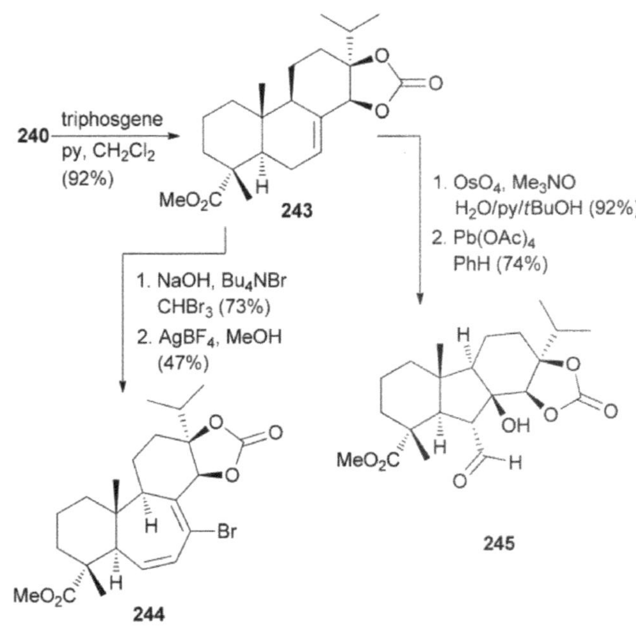

Scheme 23: Conversion of alkene **240** to compounds **244** and **245**.

The CtD strategy can be applied to all kinds of natural products. For example, the alkaloid quinine (**246**) was also subjected to controlled structural changes (Scheme 24).[124] Thus, treatment of **246** with thionochloroformate resulted in ring cleavage between N1 and C2. The putative intermediate thiocarbonate rearranged to the polycyclic *S*-alkyl thiocarbonate **247**whose structure could be confirmed by X-ray crystallography. On the other hand, acid induced Hoffmann elimination led to piperidine **248**. After protection of the amine, Petasis methylenation of the keto function followed by ring-closing metathesis resulted in octahydroisoquinoline derivative **250** (Scheme 24). Further unique structures were accessible by Grignard additions to the quinoline ring of **246**.

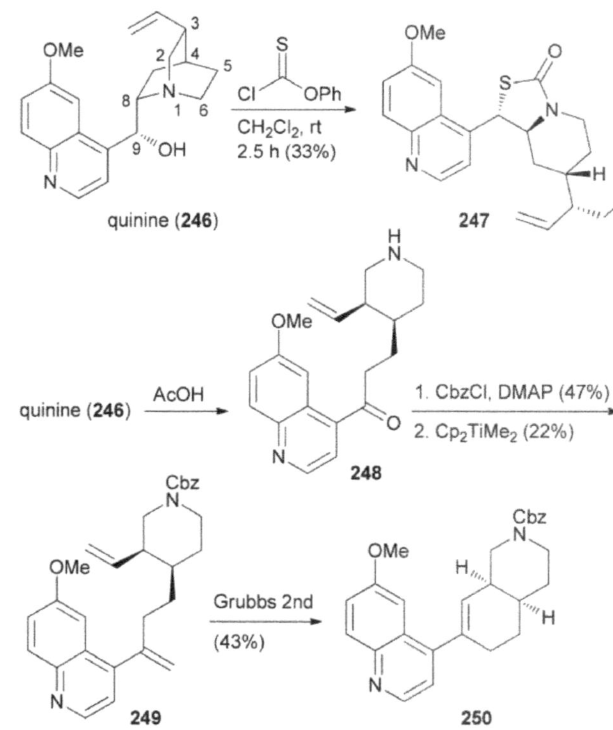

Scheme 24: CtD strategies applied to the alkaloid quinine (**246**).

Several related functional groups in a molecule like an enone or epoxide can be engaged in a domino sequence with a bis-nucleophile to furnish derivatives with increased structural complexity. An illustrative e is the remodeling of fumagillol (**251**) (Scheme 25). In the presence of the Lewis acid La(OTf)$_3$ and an arylamine a 5-*exo*-bis-epoxide opening takes place to give annulated pyrrolidine derivatives **252**.[125] A subsequent reaction of the bicyclic compounds **252** with DDQ (1 equiv.) in the presence of a nucleophile led to a stereoselective addition resulting in complex NP analogues **253**. In this example, a terpene is transformed to alkaloid-like compounds.

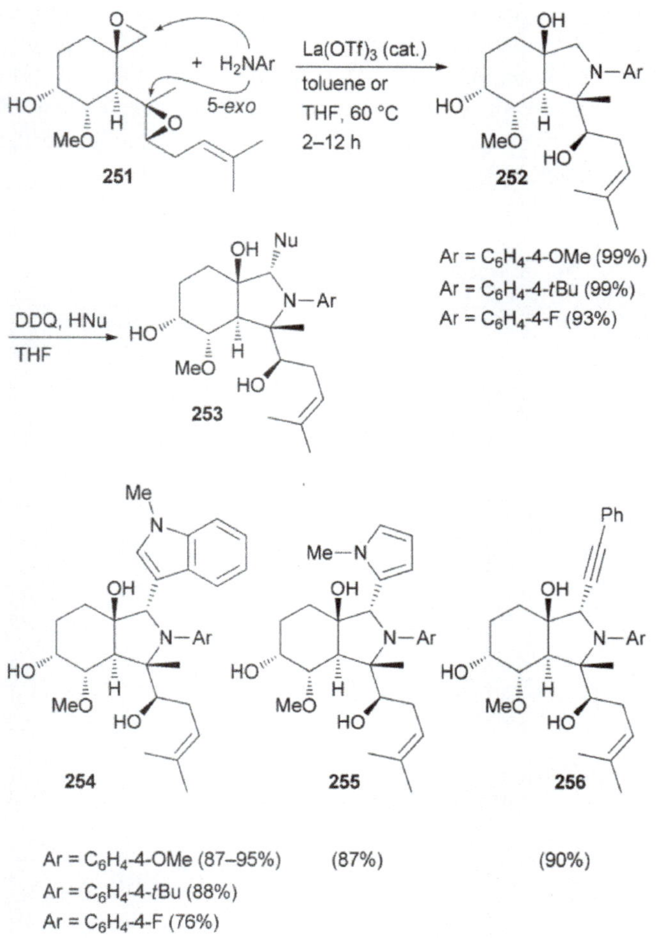

Scheme 25: CtD-type manipulations on fumagillol (**251**).

The CtD strategy appears to be promising and worthwhile for easily available natural products. However, so far it seems that no useful biological activity has been reported for any of the CtD derived NP analogues.

HYBRID MOLECULES

The principle of this concept is to combine partial or whole structures of molecules to create new, possibly more active, molecular entities.[126] It is a quite simple and, in certain ways, naive approach. But in biology this concept is widespread. This is very often the e when the subunits of a molecule undertake different functions, for example redox processes or target binding, like DNA recognition. Thus, in the anthracycline antibiotic daunorubicin (**257**), three functional units can be identified (Figure 17): a DNA intercalating domain, a domain interacting with the enzyme topoisomerase II, and finally, the sugar part that binds to the minor groove. In fact, the aglycon has 70–100 times lower antitumor activity than daunorubicin. A more complicated structure which can be considered a hybrid molecule is calicheamicin γ^{I}_{1}.[127]

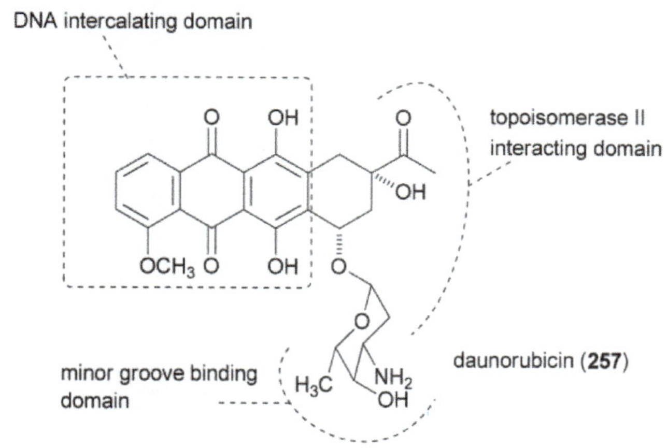

Figure 17: Structure of daunorubicin with its three functional subunits.

The many natural products that are made up of identical parts may also be classified as hybrid molecules. Some prominent examples, including rhizopodin (**258**), blepharocalyxin D (**259**), incarvilleatone (**260**), blennolide G (**261**), cephalostatin 1 (**262**) and chaetocin A (**263**), are shown in Figure 18.

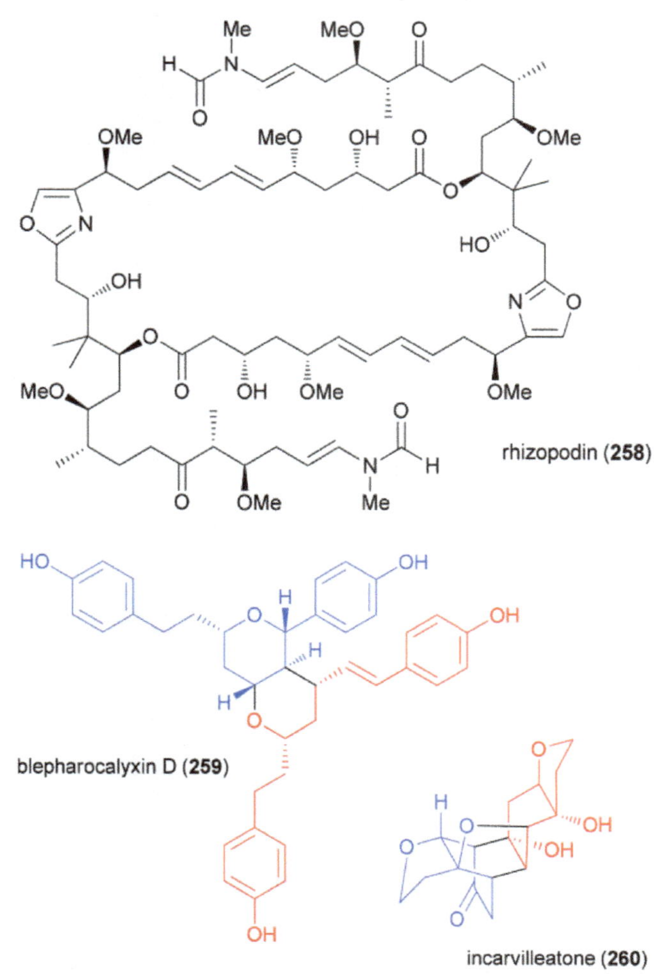

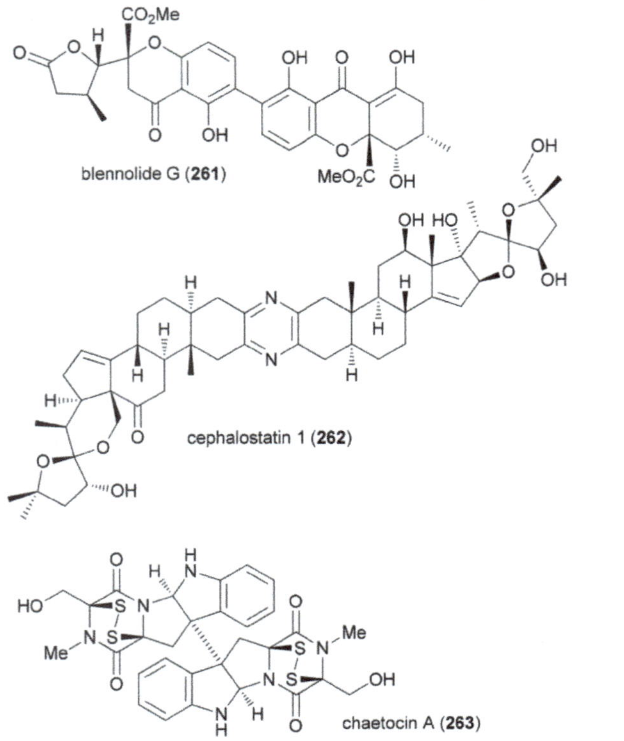

Figure 18: Some examples of dimeric natural products. In es where the subunits are less obvious they are colored.

The dopamine derivative aspongamide A (**265**) is an example of a trimeric natural product. It is made up of three *N*-acetyl-dopamine (**264**) molecules (Figure 19). This compound was isolated from the insect *Aspongopus chinensis*, which is found in southern China.[128] Dried bodies of this insect are used as traditional medicine for the treatment of pain, stomach problems and kidney diseases. This insect was investigated with the aim to find natural products that are active against chronic kidney disease (CKD). This disease is a consequence of diabetes. The trimer **265** turned out to inhibit smad3 phosphorylation, which plays a role in pathogenesis of CKD.

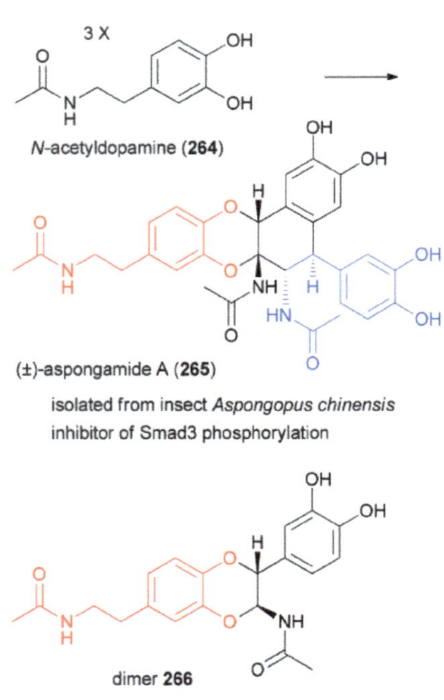

Figure 19: Aspongamide A (**259**) as an example of a trimeric natural product.

The use of hybrid molecules is one of the pillars of bioconjugation. For example, target identification of a bioactive molecule often relies on constructs where the small molecule is linked *via* a T-piece adapter to a photoreactive group and biotin.[129] In these es it is known from the outset that the biological activity of such constructs is usually reduced, but the purpose of course is different. One such hybrid molecule is depicted in Figure 20. It contains the microbial product GEX1A (herboxidiene, **145**). This NP was originally isolated as a herbicide.[130] Later it turned out to also have antitumor activity *in vivo*. Its side chain is quite similar to that of the polyketide pladienolide B (**141**), a compound that also displays potent antitumor activity. As pladienolide B targets the SF3b1 subunit of the spliceosome, it is not too surprising that construct **267** led to the discovery that GEX1A also targets a subunit of SF3b.[131] In particular, the development of click reactions and chemical ligation methods facilitated the construction of hybrid molecules.

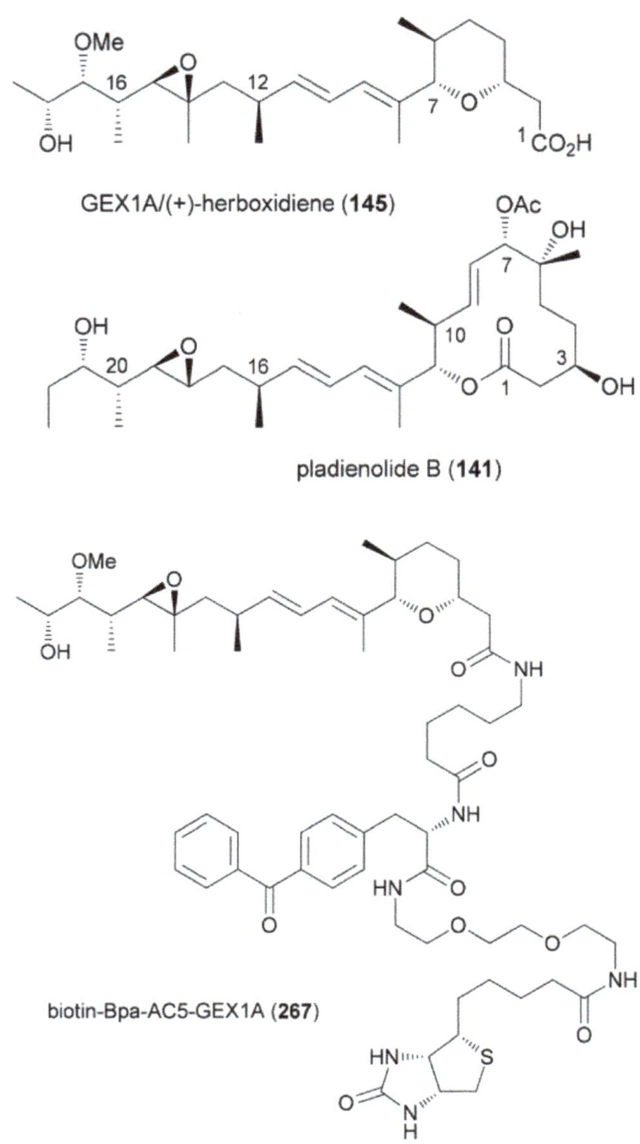

Figure 20: Hybrid **261** that led to the discovery that GEXA1A targets the spliceosome.

According to Meunier three categories of hybrid molecules can be distinguished (Figure 21).[132] It might be that both parts of the hybrids act on one target. This does not need to be the same binding pocket.

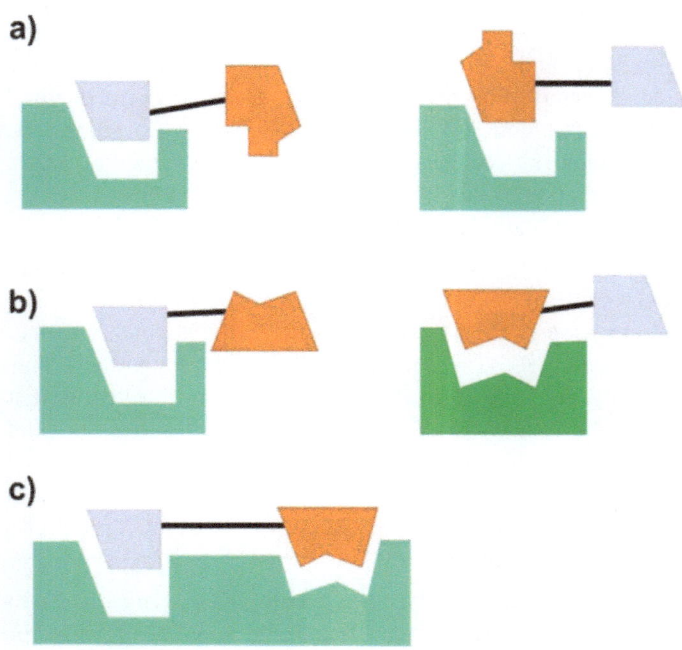

Figure 21: Categories of hybrid molecules according to their mode of action. (a) Both parts act on the same target (not necessarily on the same site); (b) each part acts on a different target; (c) both parts bind to one target at closely related binding sites.

A e in point is the so-called trioxaquines that contain an aminoquinoline and a 1,2,4-trioxane. The latter is inspired by the natural product artemisinin (**268**) which is a powerful antimalarial drug. The conjugate PA1103/SAR116242 (**270**) and related molecules are considered "double-edged sword" drugs (Figure 22). Thus, the aminoquinoline sector can stack with heme, whereas the trioxane part can alkylate heme after reductive activation. Indeed these hybrids are active on chloroquine-resistant strains of *Plasmodium falciparum*. Compound **270** was eventually selected as an antimalarial drug candidate for further development.[133]

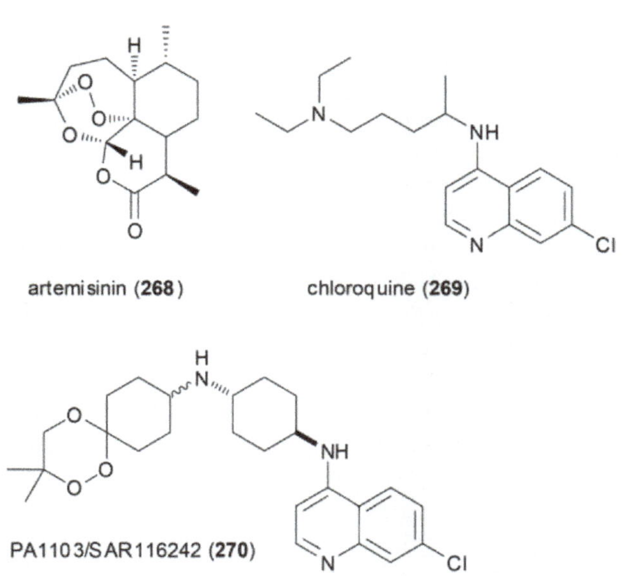

Figure 22: Trioxaquines like **270** are patterned after artemisinin (**268**) and chloroquine (**269**). Both parts act on heme of the parasite.

Hybrid molecules of the second category act with each subunit on different and non-related targets. If the linker is cleaved in the biological system, the hybrids are like dual prodrugs. The parts of the hybrid molecule may also act at the same time on two connected targets, as illustrated with category (c). Examples include fragment based drug discovery[134] or DNA ligands, like the above-mentioned daunorubicin. If the exact mode of action of a hybrid is not known, it is of course difficult to categorize it according to Figure 21.

Macrocycles frequently have been turned into hybrid molecules. An option is to borrow a side chain from an active molecule and to attach it to a macrocyclic core. For example, some of the 16-membered macrosphelides are able to induce apoptosis in human lymphoma U937 cells. In an attempt to enhance their potency, hybrid macrosphelides were designed that contain the side chain of the epothilones (Scheme 26).[135] The synthesis of hybrid **272** started from homoallylic alcohol **273**, which was extended by three esterification reactions to triester **274**. A ring-closing metathesis reaction and simple functional group manipulations led to NP analogue **272**. This compound showed significantly higher apoptosis-inducing activity than the parent macrosphelides.

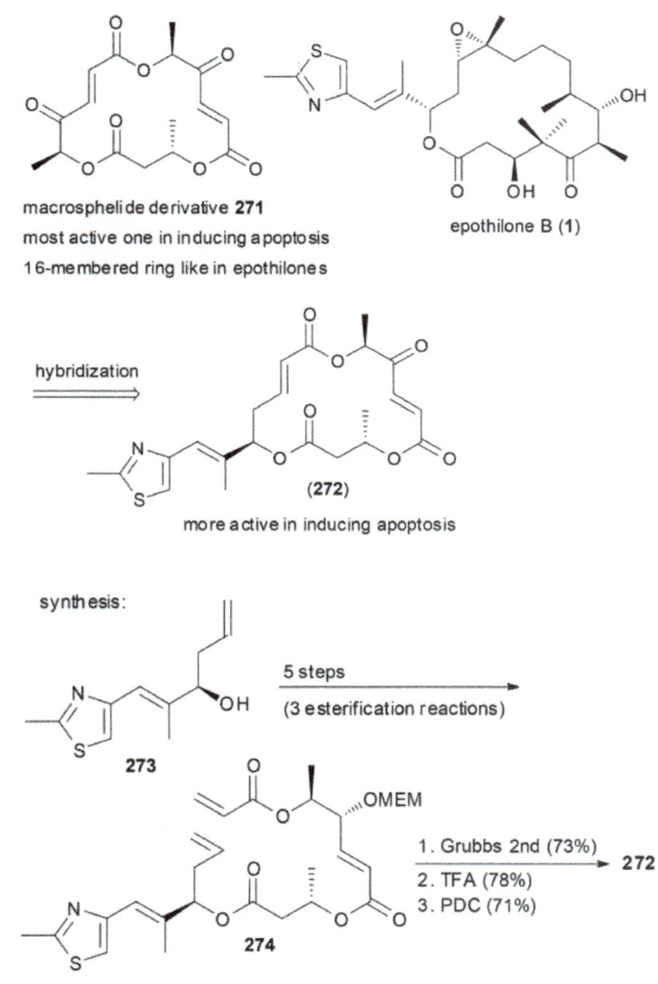

Scheme 26: Design and synthesis of macrosphelide–epothilone hybrids.

A more dramatic modification is the replacement of a region in a macrocycle with a section from another NP. This is best done by total synthesis. The Kihning group designed hybrids that contain the aromatic core of geldanamycin (**275**) and the maytansine aliphatic part (Scheme 27).[136] Geldanamycin (**275**) and maytansine (**279**) are representatives of the ansamycin antibiotics. The subfamily of the maytansinoids consists of several ansamitocins as well as maytansine. The geldanamycin[137] subfamily includes macbecin I (**276**), herbimycin (**277**), and reblastin (**278**). While the maytansinoids share the same ring size for the macrolactam with the geldanamycins, they differ in their mode of action. Whereas the maytansinoids bind to and block the polymerization of tubulin, the cytotoxicity of the geldanamycins is due to their interaction with the heat shock protein Hsp90α. Therefore, the synthesis of hybrids like **281** seemed to be of interest and to make sense.

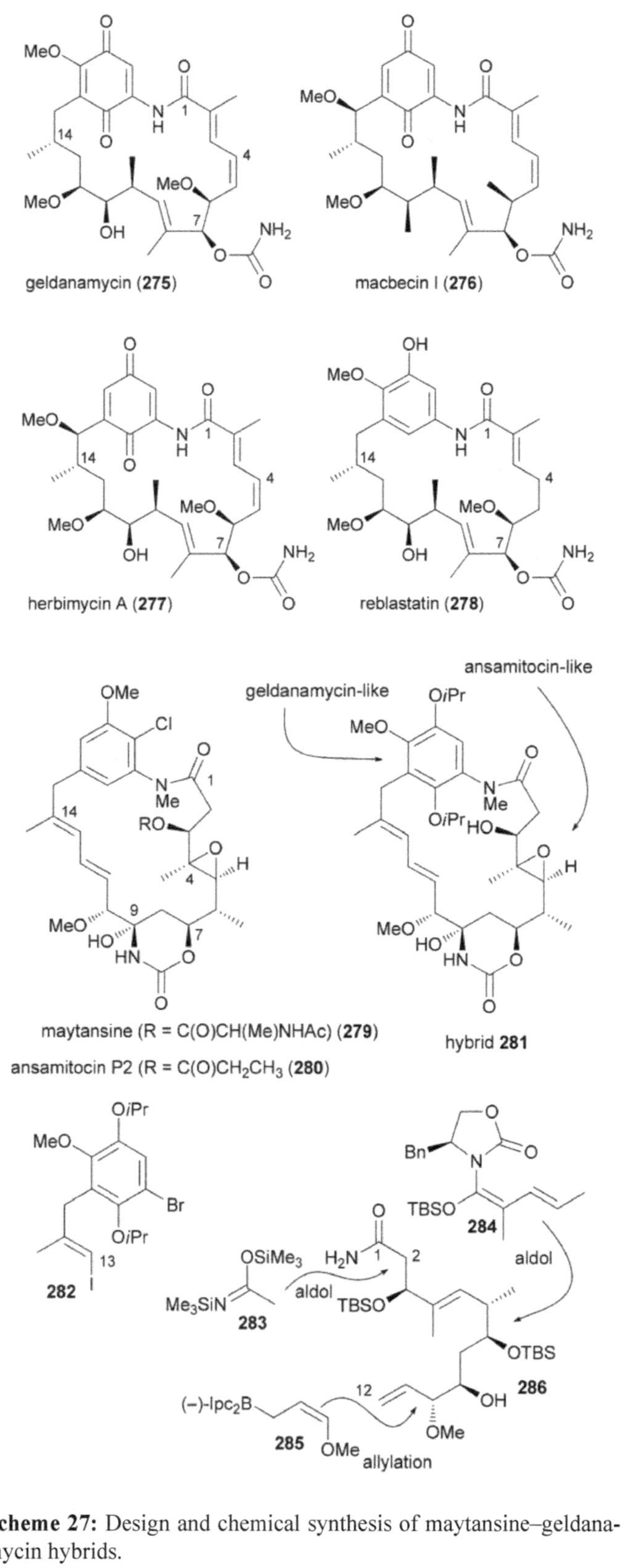

Scheme 27: Design and chemical synthesis of maytansine–geldanamycin hybrids.

The aliphatic C1–C12 part **286** could be obtained from acrolein in a sequence of 12 steps utilizing asymmetric allylation and aldol reactions to create C–C bonds and the stereocenters (Scheme 27). A Heck coupling between vinyl iodide **282** and building block **286** was followed by an intramolecular copper-catalyzed amidation reaction. After the generation of the hydroxyketone functionality at C7–C9 a biotransformation attached the carbamoyl group at 7-OH.

Due to the isoproxy groups that could not be removed, hybrid **281** exists as a mixture of separable atropisomers. The minor isomer displayed significant activity against cancer cell lines (KB-3-1: 2.2 μg mL^{-1}; SKOV-3: 4.5 μg mL^{-1}). This activity might be connected to a mechanism that is different from the ansamycins. On the one hand **281** does not bind to Hsp90α. On the other hand its conformation is different from that of the maytansinoids. This example shows that it is difficult to predict reliably at the outset of a design the activity of a hybrid molecule.

The replacement of a larger section of a macrolactone with a fragment obtained from an amino acid is another strategy for obtaining hybrids of a NP. For example, latrunculin derived hybrid macrocyclic carbamates **287** were prepared using this rationale (Figure 23).[138] So far, however, biological data for these and other compounds are lacking.

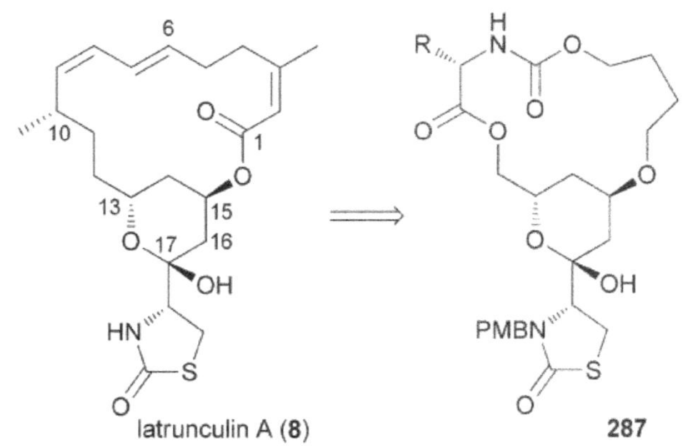

Figure 23: Hybrid structures designed based on latrunculin A.

Molecules that bind to microtubules represent an important class of anticancer agents. Such molecules can stop cell cycle progression either by breaking up or stabilizing microtubules. Based on combretastatin A-4 (**288**), which causes disruption of microtubules through binding to β-tubulin, and pironetin (**22**), which inhibits tubulin assembly by targeting α-tubulin, the Marco group designed hybrids consisting of combretastatin A-4 (CoA4, **288**) and simplified pironetin analogues linked together *via* a tether of variable length (Figure 24).[139] The simplified pironetins **291** and **292** contain at least the essential α-pyrone moiety. Compound **292** additionally features the 7-OH and the 9-OMe groups. The idea was to find improved compounds that can either bind α- or β-tubulin. Even though the exact distance between the two binding sites is not known, compounds might be found that address both sites simultaneously.

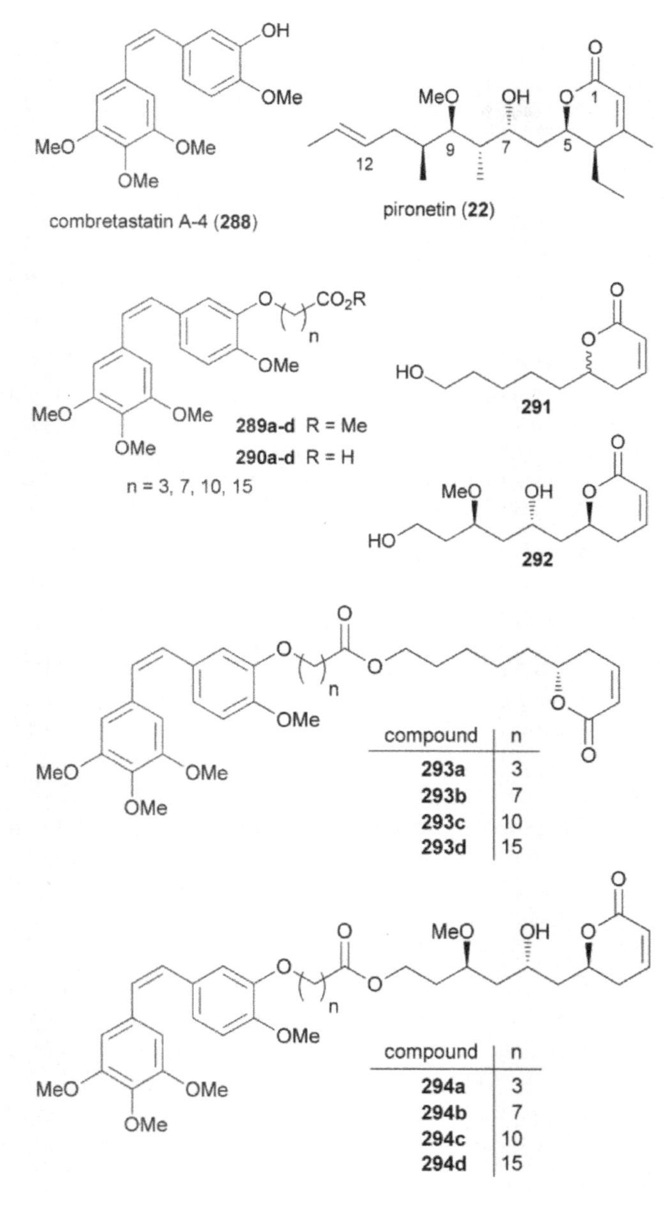

Figure 24: Structures of pironetin analogue/combretastatin hybrids and the corresponding precursors.

As described above (Scheme 2) the secondary alcohol functions were generated by Brown allylation reactions on the corresponding aldehydes. CoA4 derivatives equipped with a linker were prepared from CoA4 (**288**) by etherification with ω-bromo esters to give esters **289a–d** followed by saponification to the acids **290a–d**. A Yamaguchi esterification served to prepare the hybrids **293a–d** and **294a–d** (Figure 24).

The screening for cytotoxicity was performed with two tumor cell lines (human colon adenocarcinoma HT-29 and breast adenocarcinoma MCF-7 cell lines) and, for comparison, with the normal human embryonic kidney cell line HEK-293. As it turned out all compounds were active in the low to medium micromolar range including the parent NPs (Table 4). Compounds**289b**, **289c**, **289d** and **290d** were active in both cell lines. Worth mentioning are compounds *ent*-**293a**, **294c**, **289c**, **289d** and **290d** that show pronounced selectivity for the tumor cell lines. Among the hybrids **293a–d** and **294a–d** that differ from each other in the length of the connecting spacer, the ones with a ten-carbon chain show maximum activity (**293c** and **294c**). For all hybrids it seems that there is no simultaneous binding to both α- and β-tubulin. Comparing the hybrids **293a–c** and their enantiomers to the pironetin analogue points to the fact that the cytotoxicity mainly originates from the CoA4 region. This example shows that a rational design of hybrids remains rather difficult.

Table 4: Cytotoxicity of the hybrids from Figure 24 against various cell lines given as IC_{50} values [μM][a]

Compound	HT-29	MCF-7	HEK-293
CoA4 (**288**)	4.2	1	25
Pironetin (**22**)	8	18	46
293a	12	52	39
293b	32	>300	80
293c	10	>300	>300
293d	107	>300	>300
ent-**293a**	*3.5*	*17*	*114*
ent-**293b**	19	>300	>300
ent-**293c**	38	69	59
ent-**293d**	64	>300	72
294a	15.4	31	22
294b	44	>300	>300
294c	*2.9*	*5.6*	*22*
294d	50.8	21	81
289a	60.1	10.4	121
290a	50	2.9	67
289b	3.5	9.6	9.6
290b	16	17	>300
289c	*8*	*3*	*>300*
290c	103	5	>300
289d	*5*	*9.6*	*>300*
290d	*1.9*	*4.4*	*39*
291	109	>300	35
ent-**291**	81	109	119
292	25.4	2.5	>300

[a] Compounds with high toxicity towards both tumoral cell lines are marked in italics.

BIOSYNTHESIS INSPIRED SYNTHESIS OF NP ANALOGUES

In many es the biosynthesis of a NP is known. At least hypotheses regarding key transformations do exist in general. This is often provided whenever an alkaloid or terpene with a novel structure is published.[140,141] For NPs from microorganisms a lot is known about the biosynthetic machinery and the genes that code for the structure of a NP. This is the e with polyketides from bacteria and partly from fungi.[142] These genes are arranged in a sequence of modules. A loading module catches the starting carboxylic acid derivative, which is bound to the biosynthesis machinery as a thioester. Every module extends the chain by two carbon atoms *via* a Claisen condensation. The modules may contain additional domains that perform typical steps common to fatty acid biosynthesis, namely, reduction of the 3-ketoester to a 3-hydroxy ester (KR = keto reductase), elimination of water to give an enoate (DH = dehydratase) and conjugate reduction of the enoate (ER = enoyl reductase). Typical extender units ar malonyl-CoA and methylmalonyl-CoA. Some other ones ar known such as methoxymalonyl-CoA or allylmalonyl-CoA Thus, by simply looking at the structures of the seco acid c a polyketide macrolactone, the domains, besides the ACP an KS that perform the Claisen condensations, can be predictec This is shown in Figure 25 for the immunosuppressive FK50 (**295**).[143] Module 4 uses allylmalonyl-CoA, and modules 7 an 8 use methoxymalonyl-CoA as extender units.

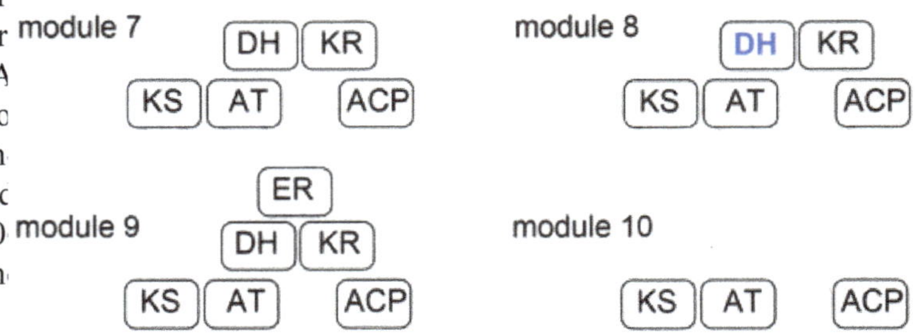

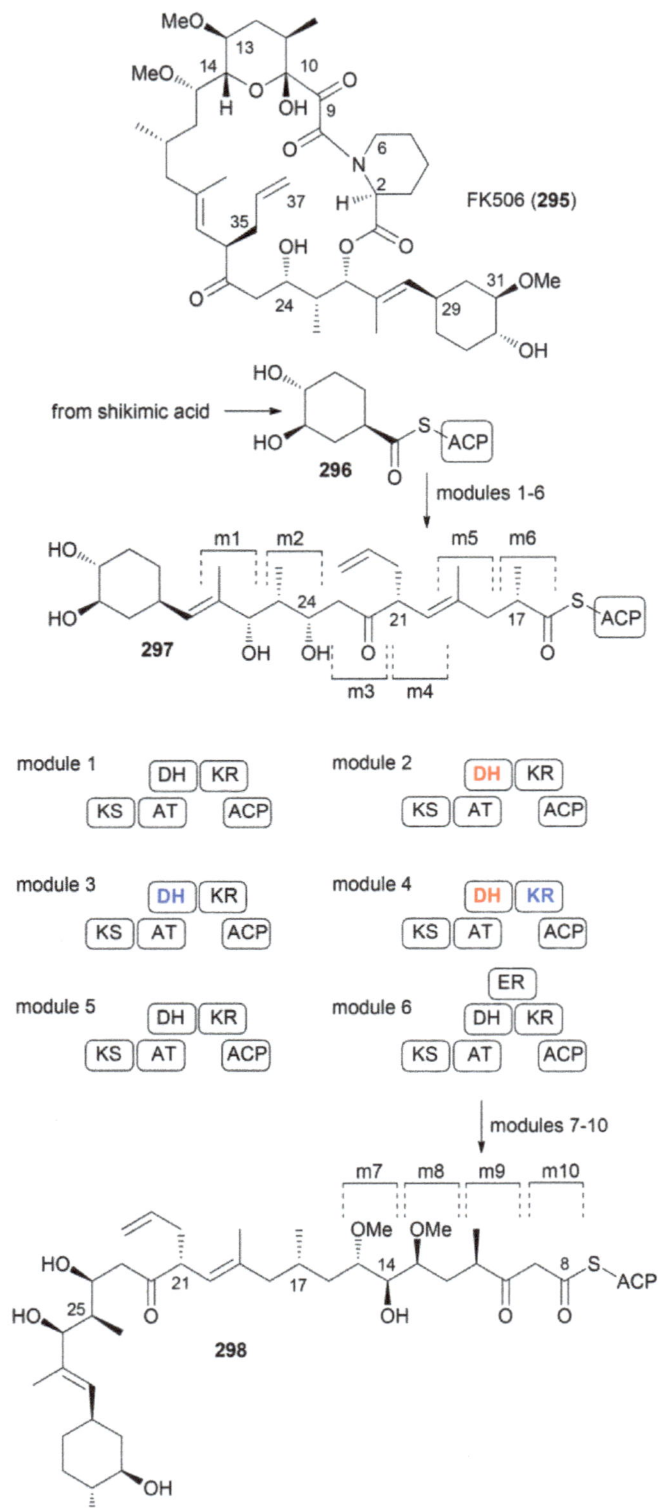

Figure 25: Biosynthesis of the polyketide FK506 (tacrolimus). Domains in modules might be inactive (red), non-functional (blue) or completely absent. KS, ketoacyl synthase (performs Claisen condensation; prior to the Claisen condensation the growing chain is attached to KS, the extender unit to ACP); AT, acyl transferase; DH, dehydratase; ER, enoyl reductase; KR, keto reductase; ACP, acyl carrier protein.

Mutasynthesis

The detailed knowledge about the biosynthesis of polyketides can be exploited for the synthesis of NP analogues. In the e of mutasynthesis a mutant organism is generated where the synthesis of a natural building block has been blocked by deletion of the corresponding gene. Frequently, this refers to the starting unit of a polyketide which typically is very different from the extender units.[144] Using mutasynthesis, the Kihning group produced a range of geldanamycin derivatives that by total synthesis would not be available so efficiently.[145] Thus, by feeding various 3-aminobenzoic acids to the mutant *S. hygroscopicus* K390-61-1, in which the synthesis of the natural starting unit was blocked, a range of novel geldanamycin analogues could be obtained (Scheme 28). The acids shown were accepted well and the process could be scaled-up. One limitation of mutasynthesis lies in the fact that only certain building blocks are accepted by the acyltransferase (AT) of the loading module.

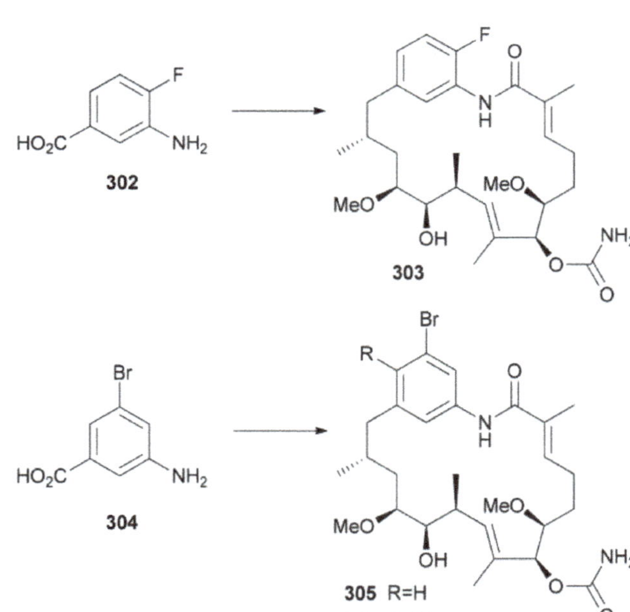

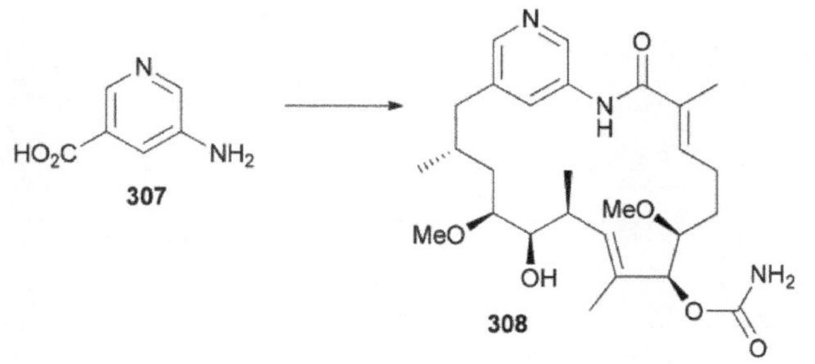

Scheme 28: Mutasynthesis of nonbenzoquinone geldanamycin analogues by mutasynthesis. The analogues were obtained by feeding the acids to the blocked mutant of (K390-61-1) of *S. hygroscopicus.*

Mutasynthesis can be combined with chemical synthesis (semisynthesis) if building blocks are incorporated that allow for subsequent selective transformations. This has been illustrated with the synthesis of ansamitocin analogues.[146] For example, feeding of 3-amino-4-bromobenzoic acid (**309**) to the AHBA (−) blocked mutant of *Actinosynnema pretiosum* produced the ansamitocin P3 derivative **310**. This compound turned out to be a suitable substrate for Pd-catalyzed cross-coupling reactions (Scheme 29).[147]

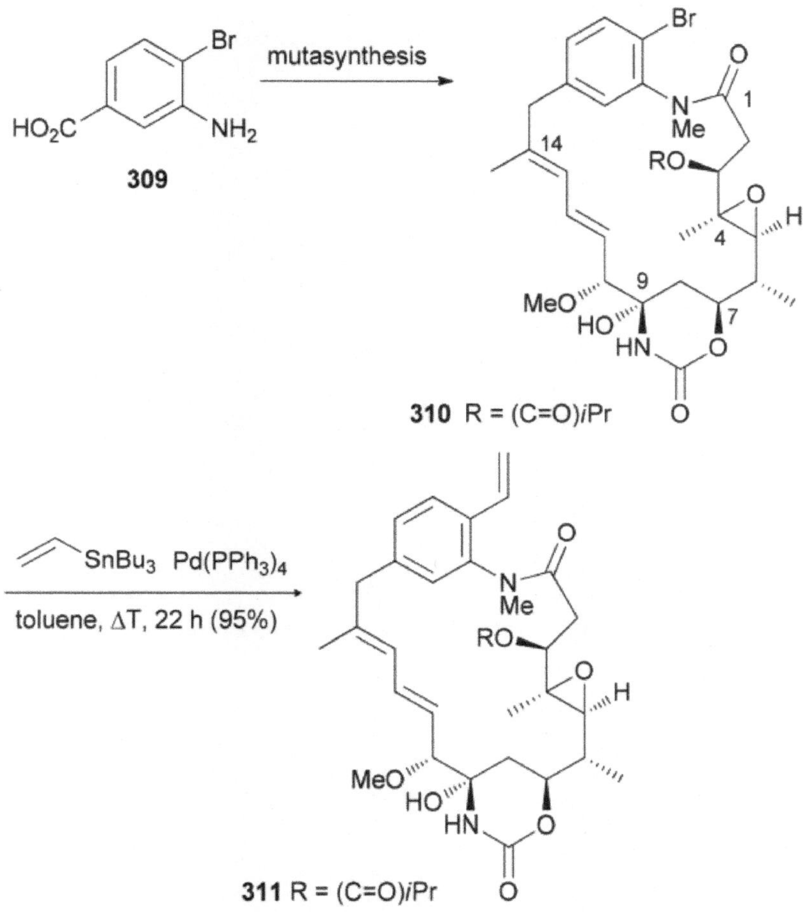

Scheme 29: Mutasynthesis followed by chemical modification. Both compounds were quite active [IC_{50} (pmol mL^{-1}), KB-3-1 cell line, **310**: 0.46, **311**: 0.67].

In e an extender unit in a polyketide significantly differs from the standard acetate and propionate building blocks, mutasynthesis of analogues is possible by knocking out the synthesis of the unusual extender unit and adding substituted malonyl-CoA extender units. For example, the allyl group at C21 of the immunosuppressive FK506 is incorporated into the polyketide chain *via* the allyl malonyl-CoA extender unit **317** (Scheme 30). Investigations by Moore *et al.* showed that it is synthesized from 3-ethylacrylate **314**. Reductive carboxylation is followed by introduction of the terminal double bond. For the final dehydrogenation the gene tcsD is responsible. By knocking out the tcsB gene that is involved in the synthesis of this building block (Claisen condensation to **313**), a deletion mutant of *Streptomyces* sp. KCTC 11604BP was generated and used for the mutasynthesis of some FK506 analogues.[148] By feeding *trans*-2-hexenoic acid (**320**), 4-methylpentanoic acid (**322**) and 4-fluorocrotonic acid (**324**) the known analogue **321** and the new analogues **323** and **325** were obtained. Biological testing showed 36-methyl-FK506 (**323**) to exhibit improved neurite outgrowth activity. However, the *in vivo* production levels turned out to be rather modest (0.06 mg mL^{-1} for **323**).

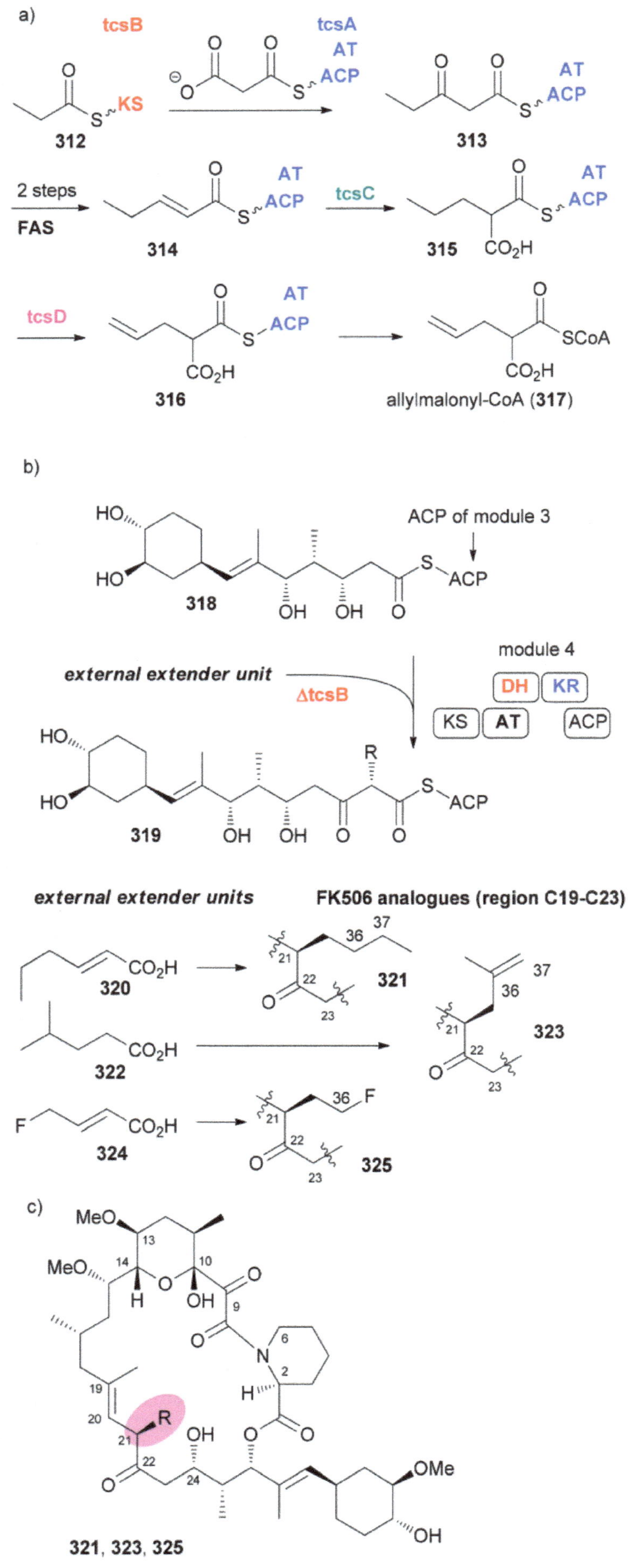

Scheme 30: Mutasynthesis of FK506 analogues **321**, **323**, and **325** by feeding the acids **320**,**322**, and **324**, respectively, to the ΔtcsB strain of *Streptomyces* sp. KCTC 11604BP.

In order to replace parts of a NP that originate from common extender units, pathway engineering is necessary. For example, a range of geldanamycin analogues could be produced by replacing the original acyltransferase (AT) domains in various modules of the geldanamycin polyketide synthase (GdmPKS) with a malonyl AT domain from the rapamycin PKS. Thus, changing AT1 which incorporates normally methylmalonyl-CoA led to analogue **326** (Figure 26).[149] Other obtained analogues are also shown. In all es production yields were sufficient for the characterization of the analogues and activity assays. Analogue **329** displayed a four-fold enhancement of the binding affinity for Hsp90.

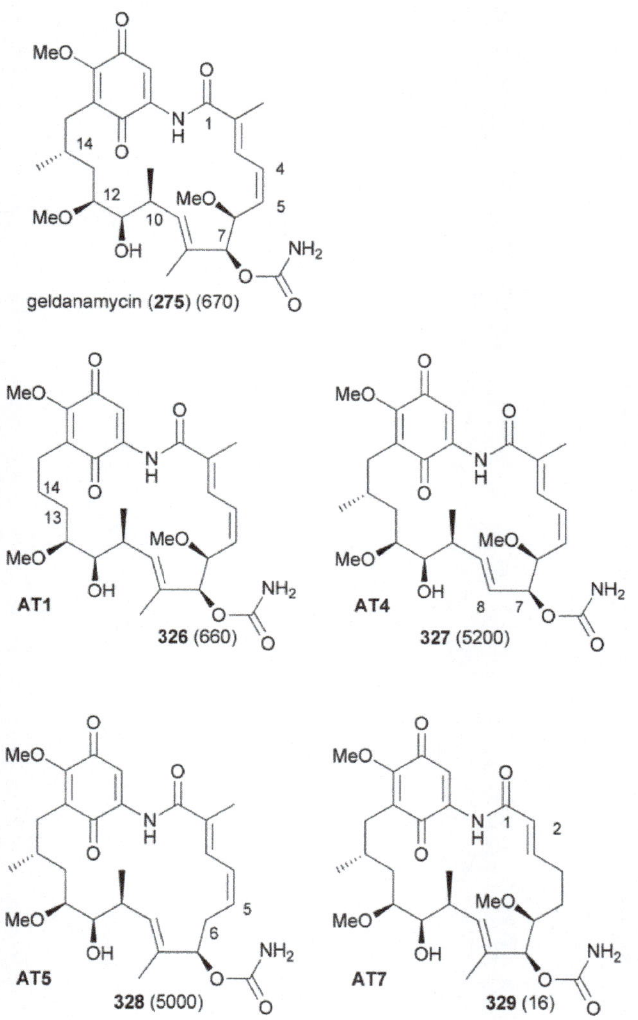

Figure 26: Geldanamycin analogues obtained by AT substitutions in the GdmPKS. AT1 incorporates chain atoms 13 and 14, AT4 atoms 7 and 8, AT5 atoms 5 and 6, and AT7 atoms 1 and 2. Values for Hsp90 K_d (nM) are given in brackets.

If all the genes required for the synthesis of an unusual extender unit are engineered into a producer where the particular biosynthesis of the extender unit to be replaced is blocked, then novel NP analogues can be obtained as well. The group of Moore cloned the three genes anlF, anlG and anlE that produce the isobutyrylmalonyl-CoA building block in an ansalactam producer strain into the integrative vector pSET152. Gene expression was placed under the control of the constitutive ermE* promoter in vector pLD6. These plasmids were conjugated into the mutant *Streptomyces* sp. KCTC 11604BP ΔtcsB so that background production of FK506 was eliminated.[150] With the gene tcsD, which is responsible for dehydrogenation, still intact, good production levels of 36-methyl-FK506 (**323**) could be realized (Scheme 31).

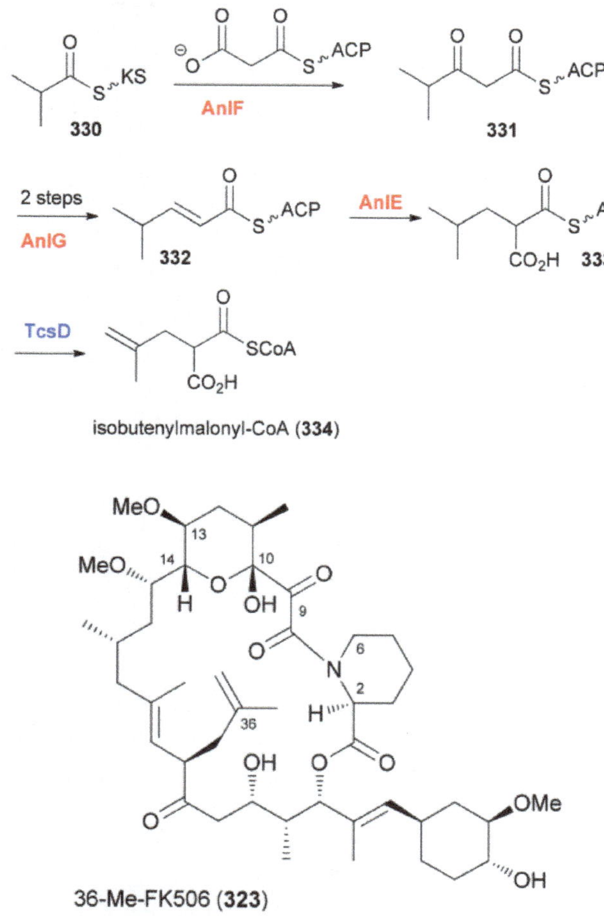

Scheme 31: Production of 36-methyl-FK506 *via* heterologous expression of the anlE-anlF-anlG gene sette in a *Streptomyces* sp. KCTC 11640BP ΔtcsB mutant strain. The red enzymes are from an ansalactam A producer.

Silent Genes

The sequencing of bacterial genomes and the associated genetic analysis did provide some stunning facts. Thus, it is possible to predict the stereochemistry of secondary hydroxyl functions from the gene sequence of the corresponding keto reductases and the stereochemistry of methyl-bearing centers.[151] Another finding was that in many polyketide synthases some genes are silent. That is, some enzymes are not operative. The exact reason for this is unknown. In order to shed some light on this phenomenon the total synthesis of NPs encoded by the complete polyketide synthases is a worthwhile task. This has recently been done for the macrolide soraphen A.[152] For some time this NP was a promising antitumor and antifungal compound due to its ability to inhibit eukaryotic acetyl-coenzyme A carboxylase. According to the genetic analysis, modules 5 and 8 contain inactive domains (Figure 27).[153,154] Module 5 should produce a saturated section. Since the ER is not active an enoate double bond should remain (C8–C9). However soraphen A (**335**) has a double bond at C9–C10. This could either be the result of a deconjugation on the enoate or of a postketide transformation. Module 8 is lacking an ER and features inactive KH and DH domains.

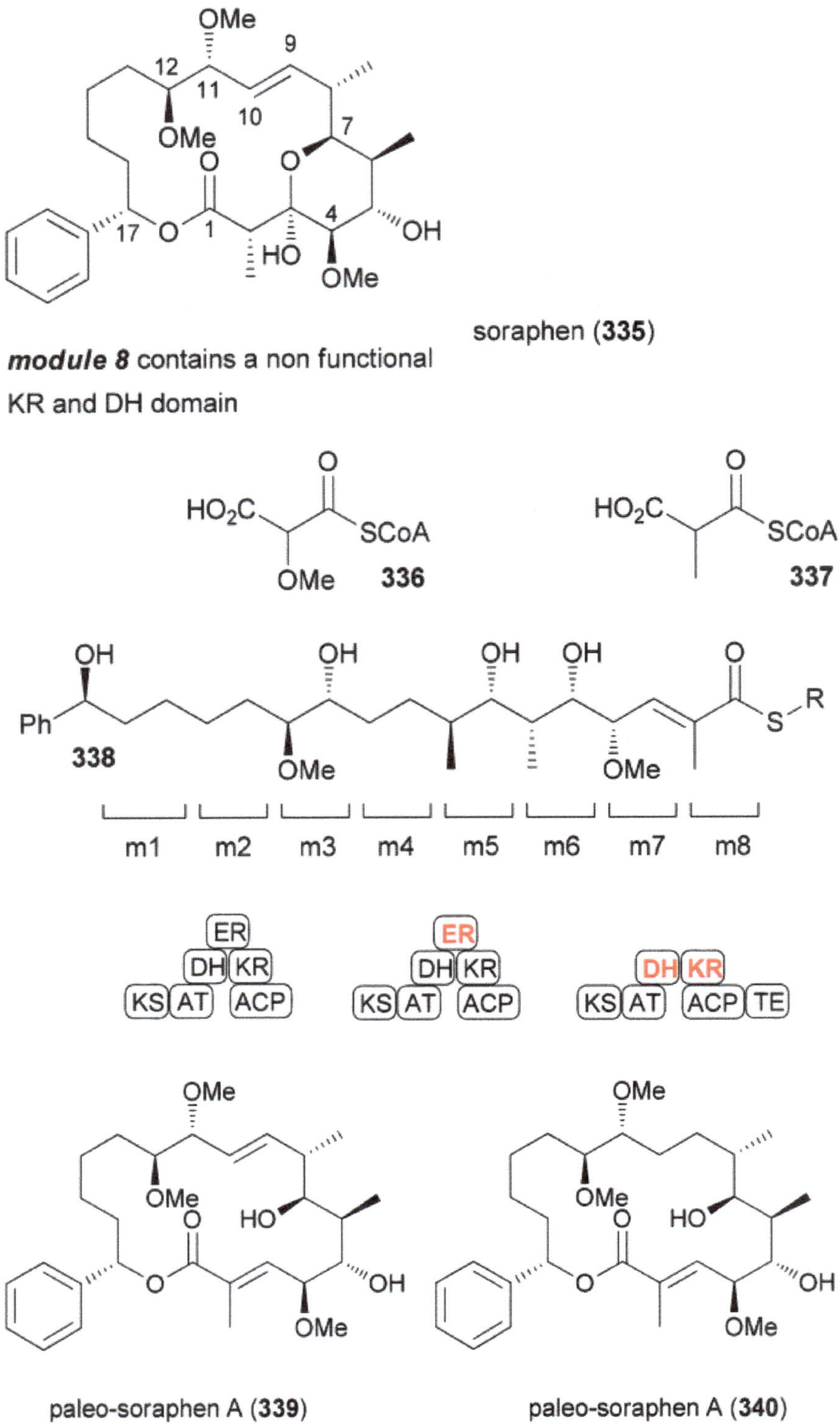

Figure 27: Structure of soraphen and arrangement of modules 1–8. The domains for modules 2, 5 and 8 are indicated together with the structure of the seco acid **338** corresponding to the full functional domains. KS = keto synthase, AT = acyl transferase, DH = dehydratase, KR = keto reductase, ER = enoyl reductase, TE = thioesterase.

Based on the structure of soraphen A and the gene sequence, Kalesse *et al.* prepared the so-called paleo-soraphens A (**339**) and B (**340**) (Scheme 32). First, pentenal **341** was convertedvia an organocatalytic α-hydroxylation to building block **342**. A cross-metathesis between **342** and **343** followed by hydrogenation delivered the C11–C17 part **345**. The C3–C10 region **349**containing an array of five vicinal stereocenters could be constructed from Roche aldehyde **347** which was elongated on both sides using an Evans aldol reaction and a Marshall reaction, respectively. The vinyl iodide function resulted from hydrozirconation/iodination of the corresponding terminal alkyne. For the coupling of aldehyde **345** with vinyl iodide **349** a Nozaki–Hiyama–Kishi reaction came to use. After oxidation of the resulting alcohols (2 : 1 mixture), enone **350** was obtained. A chelation-controlled reduction [$Zn(BH_4)_2$, Et_2O, −55 °C, 90%, d.r. > 25 : 1) delivered the *anti*-stereochemistry at C11/C12. Methylation and reductive ester cleavage furnished the advanced building block **351**.

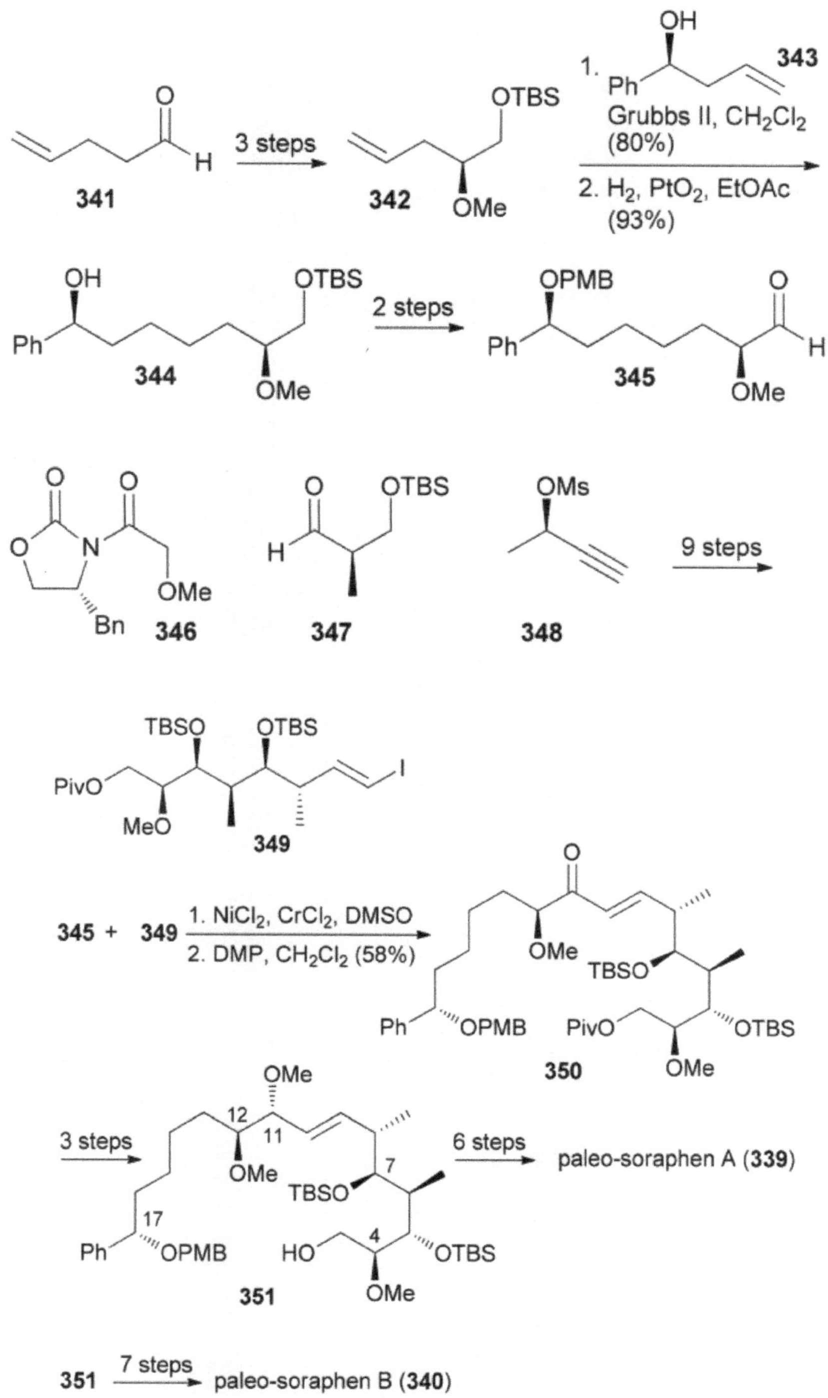

Scheme 32: Key transformations in the synthesis of the paleo-soraphens A and B.

From there, six routine steps including chain extension by Wittig reaction and Shiina macrolactonization led to paleo-soraphen A (**339**). One additional step, namely the hydrogenation of the double bond of **351** in the presence of the Crabtree catalyst, was required to reach paleo-soraphen B (**340**).

Biological tests showed both compounds to have much lower antifungal activity (Table 5). The cytotoxic activity was also almost gone.

Table 5: Antifungal activity of the paleo-soraphens in comparison with soraphen A (**325**) given as IC_{50} values (μg mL^{-1})

Compound	***Pythium debaryanum***	***Botrytis cinerea***	***Mucor hiemalis***
Paleo-soraphen A (**339**)	0.4	>40	21
Paleo-soraphen B (**340**)	>40	>40	16
Soraphen A (**335**)	0.045	0.0035	0.0006

The mode of action of the two paleo-soraphens was subsequently probed by impedance profiling. The principle of this method is that molecules with a similar mode of action will show similar time-dependent impedance curves. A hierarchical cluster analysis indicated paleo-soraphen A to be in close proximity to soraphen A. In contrast, paleo-soraphen B turned up in a cluster with camptothecin, a well-known topoisomerase inhibitor.

While in the e at hand this concept did not lead to highly potent compounds it appears promising for the selection of targets for synthesis programmes. If this observation should be general, it might also reveal regions of NPs that are important for the biological activity. Thus, the parts which originate from modules with silent domains would then be important for activity.

Propionate-Scanning

It is obvious that the molecular scaffold and substituents on a NP determine its affinity and selectivity to a cellular target. The scaffold together with operating conformational effect secures directionality and distance of functional groups that are involved in direct interactions with the target protein. Many NPs are characterized by a molecular built. This means that they are assembled from small building blocks as in the e of peptides, polyketides or terpenes. In a way they can be considered natural oligomers. However, in general it is not known what determines the selection of a building block for a certain position. To what extent this is random or driven by evolution remains largely unclear. In peptides an alanin scan is often used to address the importance of amino acids at certain positions.[155,156]

In polyketides two main building blocks, acetate and propionates, are common. A propionate results in a methyl substituent on the backbone. There are polyketides that consist of only one type of building block. For example, zearalenone (**353**) and curvularin (**355**) are solely acetate based NPs (Figure 28). On the other hand erythronolide B (**356**) only contains propionate building blocks.

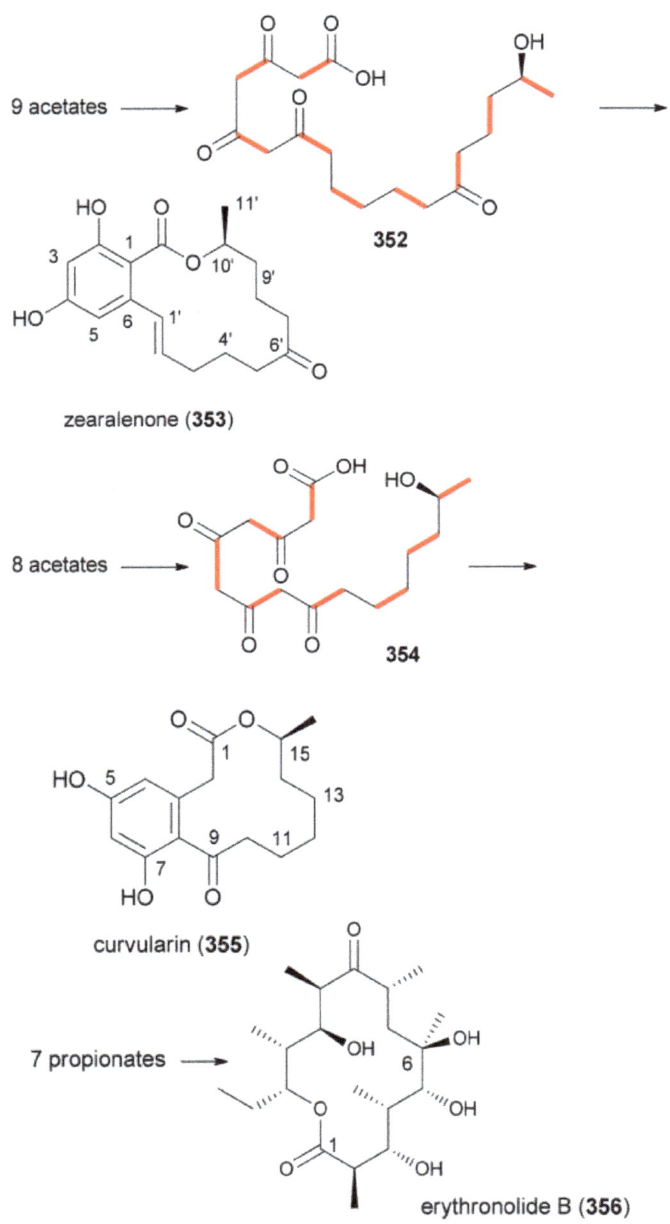

Figure 28: Structures of natural products that are only made up of one type of building block. Key polyketide intermediates for zearalenone and curvularin are also shown. Red bold bonds indicate the acetate carbons. The 6-OH in **356** originates from an oxidation.

Typically polyketide NPs are made up of both types of building blocks. Thus, the antitumor compound dictyostatin (**357**) is constructed from seven acetates and six propionates (Figure 29). It is also worth mentioning that plants and microorganisms themselves produce NP analogues in large numbers. Thus, with all natural product classes rather similar compounds have been isolated over the years. In fact, it is nowadays not that easy to discover really novel natural products. With regard to polyketides, epothilone A (**358**) and B (**1**) represent a pair of macrolactones that only differ in one methyl group. The myxobacterium strain *Sporangium cellulosum* produces 22 mg mL^{-1} epothilone A (**358**) and 11 mg mL^{-1} epothilone B (**1**).[157] The additional methyl group (propionate instead of acetate) in epothilone B results in about ten times stronger antitumor activity. In the e of the statins also several similar compounds are known that only differ in a few substituents. While compactin (**359**), isolated from the fungus *Penicillium citrinum*, has too many side effects, lovastatin (mevacor) (**360**) from *Aspergillus terreus* and pravastatin (mevalotin) (**361**) from the bacterium *Nocardia autotrophica* are in clinical use. Due to their inhibitory effect on HMG reductase, the statins are used as cholesterin lowering drugs. However, one should mention that designed NP mimetics like atorvastatin have a higher market share. These examples undeore the importance of substituents for biological activity.

Our goal in this project was to investigate the role of an additional methyl group in the simple lactones zearalenone (**353**) and curvularin (**355**). The plan was not to use genetic engineering but rather organic synthesis. We wanted to consider biosynthetic aspects, meaning that the methyl group would be positioned on the scaffold where it would make sense from a biosynthesis point of view. Thus, replacing the second acetate with propionate led to propionate analogues like **362** and **363** (Figure 30).[158] In tests for cytotoxicity several analogues showed activity in the low micromolar range, comparable with or better than zearalenone itself. Thermal shift assays performed on Hsp90 revealed a rough correlation of the T_m-shifts[159] with the IC_{50} values. A higher T_m-shift means stronger binding to Hsp90.

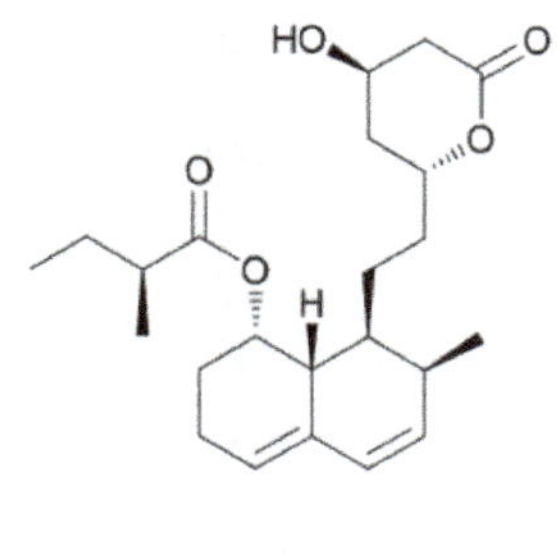

compactin (mevastatin) (**359**)

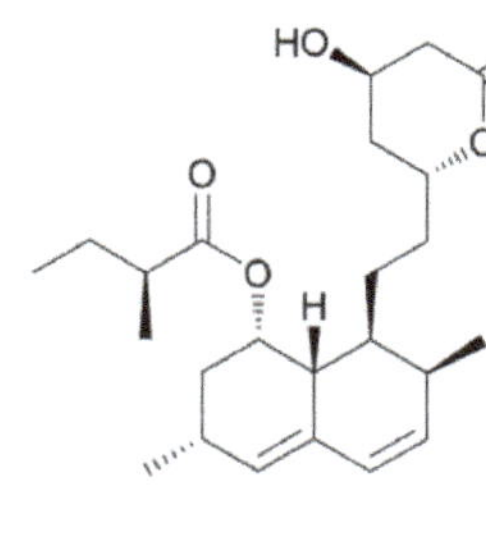

lovastatin (**360**)

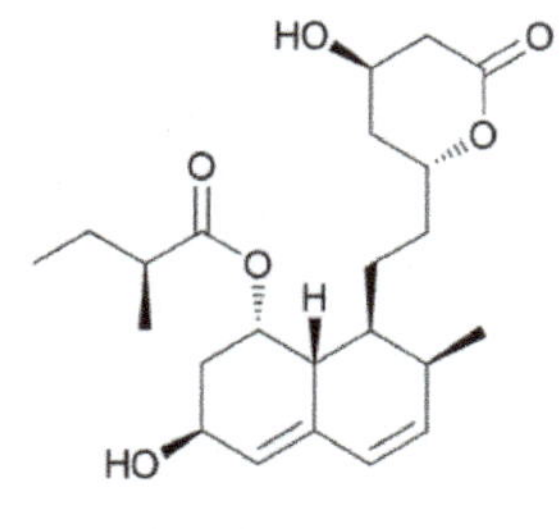

pravastatin (**361**)

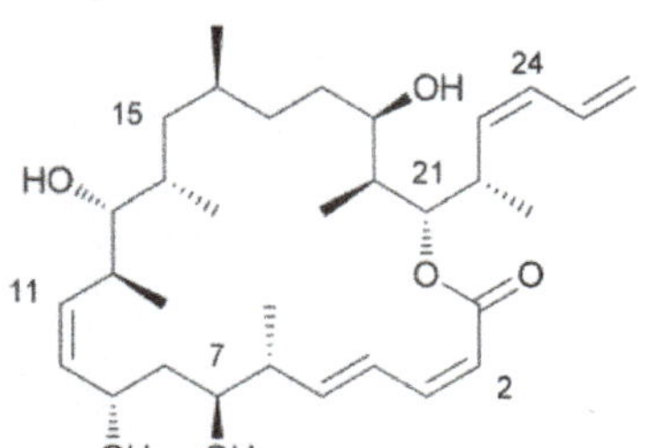

dictyostatin (**357**)

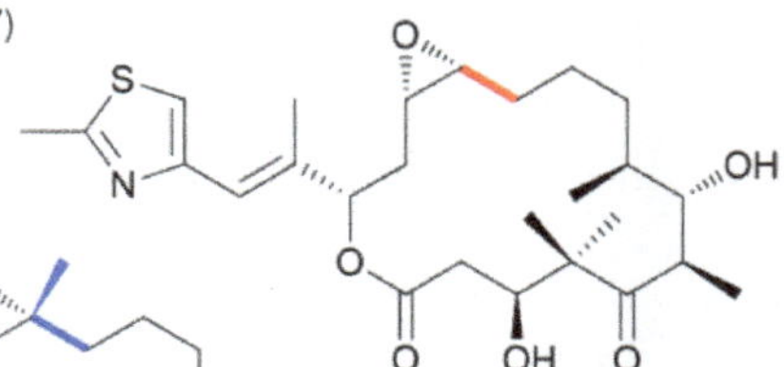

epothilone A (**358**)

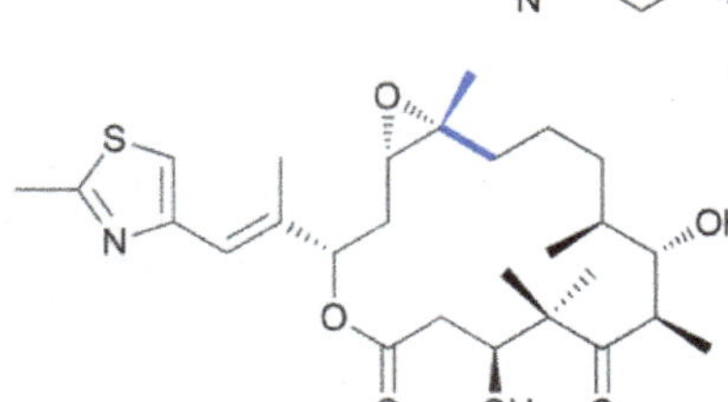

epothilone B (**1**)

Figure 29: Structure of dictyostatin (**357**), a mixed polyketide, and examples of NP analogues produced by nature.

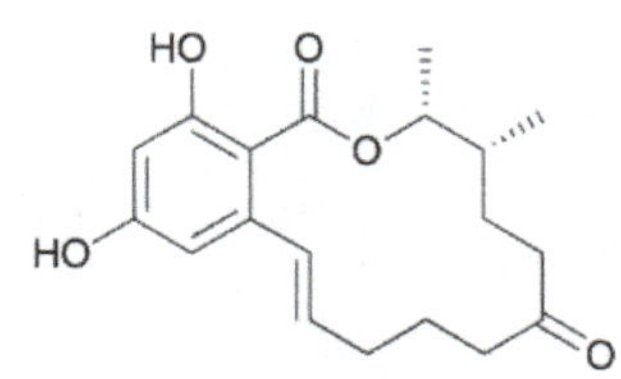

362 (10 μM, 3.9)

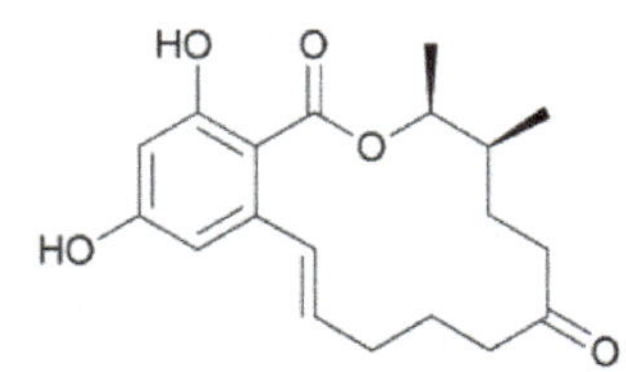

ent-**362** (6.9 μM, 3.6)

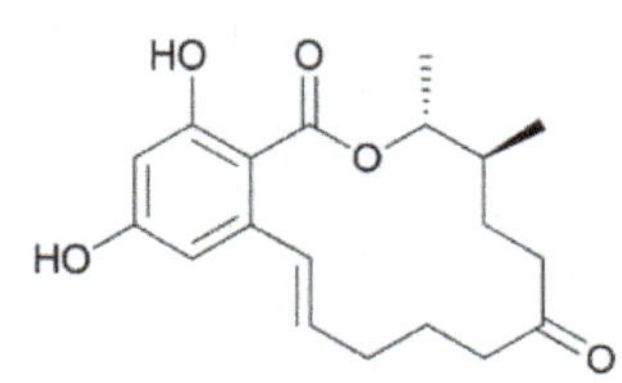

363 (15 μM, 3.2)

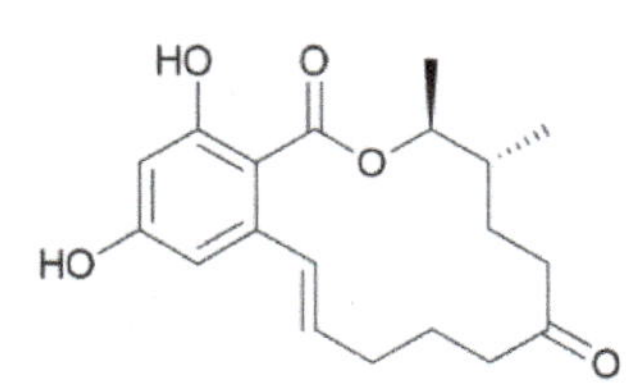

ent-**363** (14 μM, 2.5)

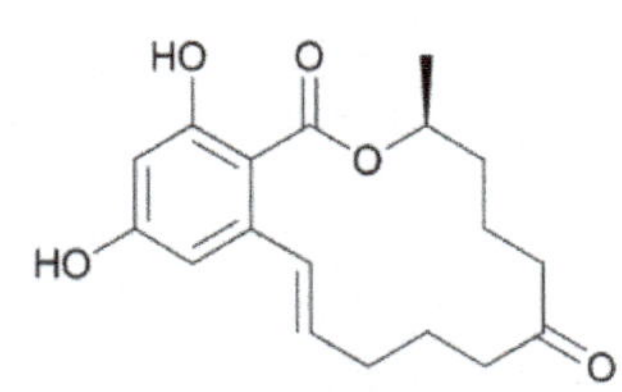

zearalenone (**353**, 9.4 μM. 2.9)

Figure 30: Zearalenone analogues with a propionate as a second building block. The IC_{50} values against L929 mouse fibroblasts and the T_m-shift values against Hsp90 are given in brackets.

Shifting the methyl group to the third acetate produced rather weak analogues (Figure 31). Thus, it seems that a methyl group at this position interferes with binding to target proteins relevant for the cytotoxic effects. The corresponding aryl methyl ethers showed even weaker activity.

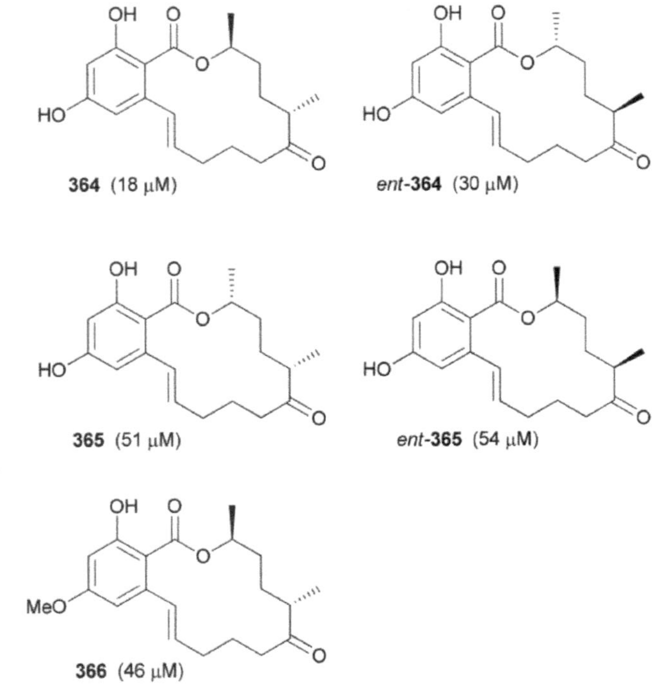

Figure 31: Zearalenone analogues with a propionate as the third building block. The IC_{50} values against L929 mouse fibroblasts are given in brackets. Here T_m-shift values were not determined.

A propionate at position four (5′-Me) led to a rather interesting analogue **367** (Figure 32). As evaluated by a thermal shift assay it binds to fewer kinase targets than zearalenone.[160] Binding to Hsp90 was completely blocked by the additional methyl group. Instead, analogue **367** was found to bind to human carbonyl reductase 1 (CBR 1) in the presence of the oxidized cofactor $NADP^+$ (IC_{50} = 210 nM).

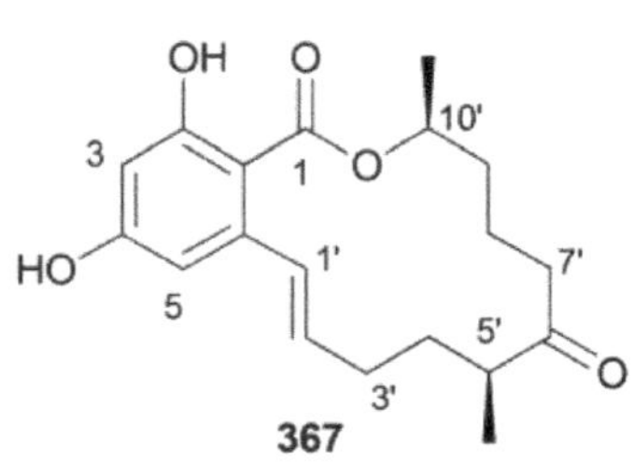

T_m-shifts against some kinases:

	PIM	SLK	LOK	Hsp90
zearalenone (**353**)	3.0	4.4	4.6	3.3
analogue **367**	2.2	3.5	-	-

Figure 32: Zearalenone analogue **367** does not bind to Hsp90 and is less strong than other kinases according to DSF. DSF = differential scanning fluorometry.

Shifting the methyl group to the next acetate (3′-Me) also produced an analogue 368 that was more active in the cell proliferation assay than zearalenone (353) (Figure 33).[161] The dissociation constant of 368 for Hsp90 was determined with 130 nM.

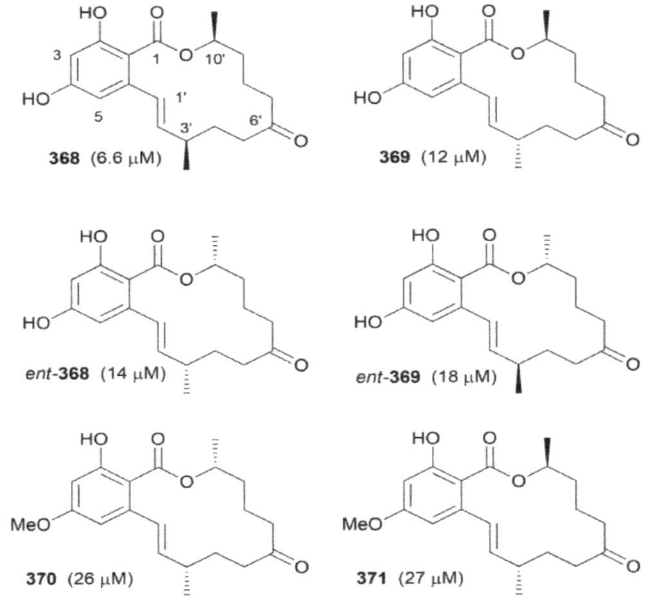

Figure 33: Zearalenone analogues with a propionate as the fifth building block. The IC_{50} values against L929 mouse fibroblasts are given in brackets.

While we could make these analogues, each substitution pattern required a separate synthetic strategy. Key steps for each group of analogues are summarized in the following schemes. Thus, the synthesis of compounds of type **362** started with the Frater alkylation product **372** which *via* alkynoate **374** was converted to Weinreb amide **375** (Scheme 33). Grignard addition eventually led to alkenol **376**. Its esterification with acid **377** under Mitsunobu conditions furnished ester **378**. Ring-closing metathesis [Grubbs 2nd, toluene, 80 °C, 5 h (85%)] and deprotection steps led to analogue **362**.

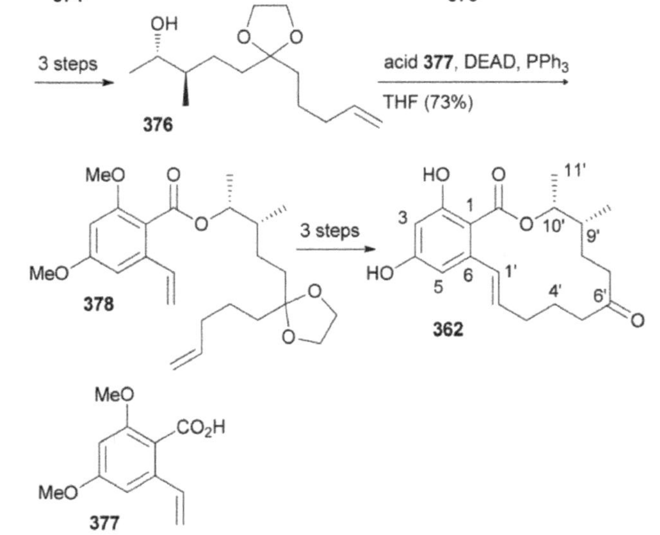

Scheme 33: Synthesis strategy towards zearalenone analogue **362**.

For benzolactones of type **364** an rcm-strategy could be used as well (Scheme 34). In the assembly of the aliphatic building block **381** an Evans aldol reaction between 5-hexenal and carboxylic acid derivative **379** set the stereochemistry at C7′. In some of these analogues, the C7′ center was prone to epimerization.

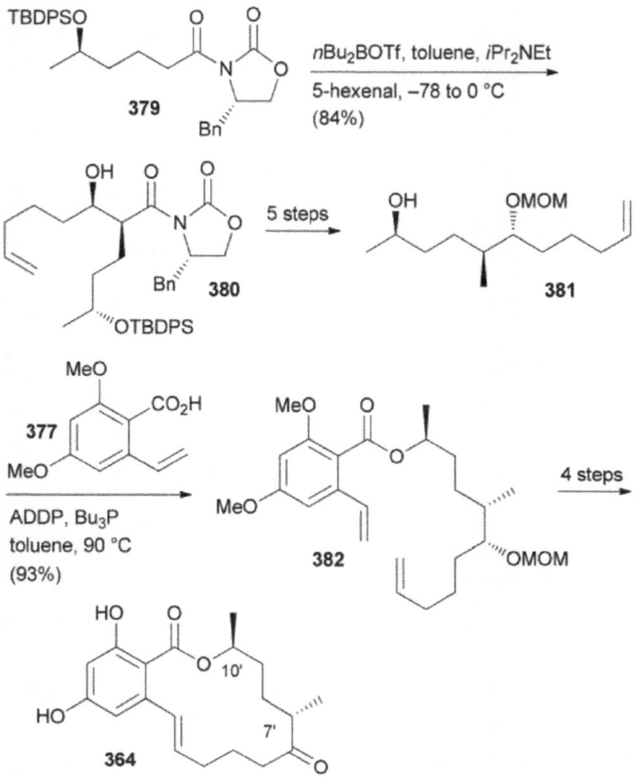

Scheme 34: Synthesis strategy towards zearalenone analogue **358**. ADDP = azodicarboxylic acid dipiperidide.

A different strategy was employed for the synthesis of analogue **367**. Thus, alkynol **383** resulting from a Marshall–Tamaru reaction served as the precursor for alkenol **384** (Scheme 35).[160] Its hydroboration and coupling with vinyl iodide **385** led to hydroxy acid **387**. Macrolactone formation under Mitsunobu conditions followed by routine steps delivered zearalenone analogue **367**.

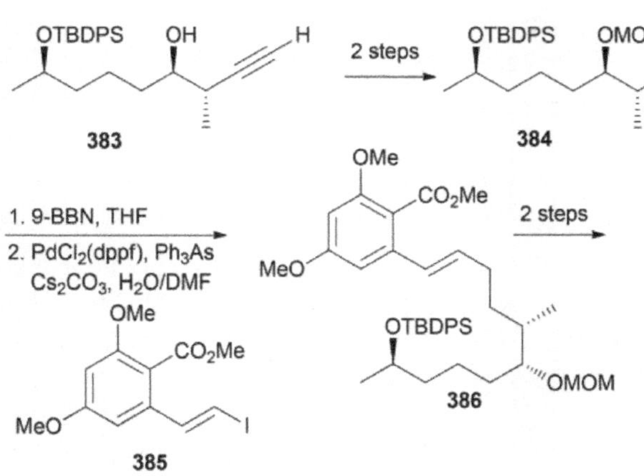

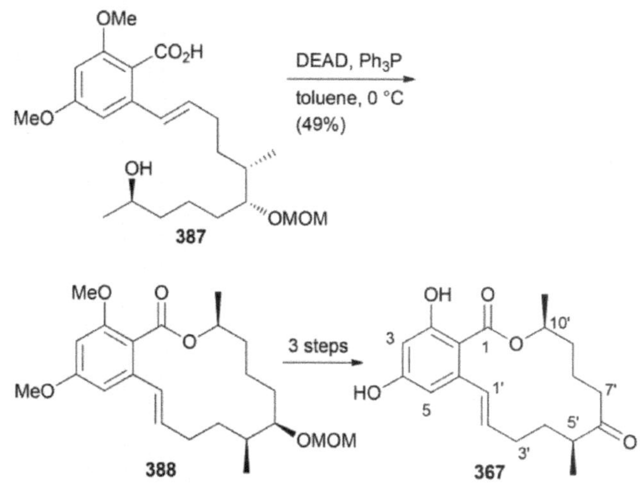

Scheme 35: Synthesis strategy towards zearalenone analogue **367**.

Yet another approach was chosen for analogue **368** (Scheme 36). Here aldehyde **390** was converted to propargylic alcohol **391**. Protection and hydrogenation gave the building block**392**. The derived sulfone **393** was combined with aldehyde **394***via* Julia–Kocienski olefination. Carboxylation followed by protecting group manipulation and macrolactonization furnished benzolactone **368**. However, even for simple compounds like the polyketide zearalenone the synthesis effort is substantial. Even though we just studied zearalenone analogues it can be concluded that methyl groups modulate biological activity. This is also known from classical drug discovery programmes.[162] But with analogue **367** we found one e where the additional methyl group redirects the molecule to another target protein. Such es seem quite interesting. A similar scanning could be performed with hydroxyl functions or combinations of methyl and hydroxyl groups.

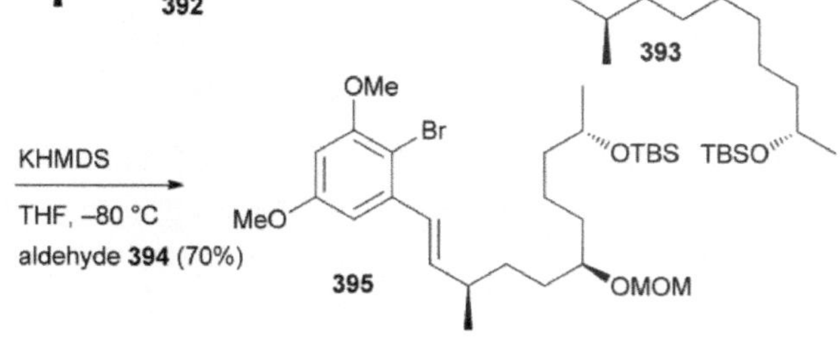

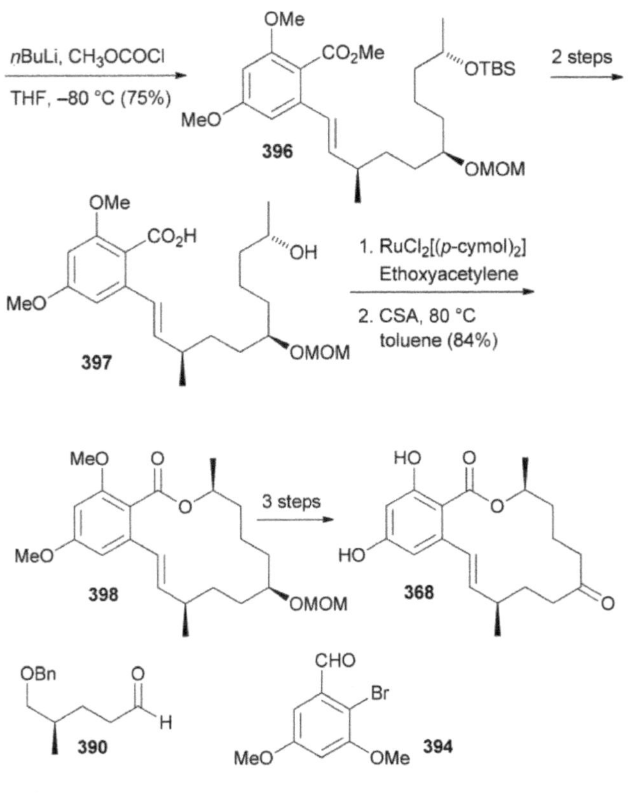

Scheme 36: Synthesis strategy towards zearalenone analogue **368**.

CONCLUSIONS

This review shows that natural product analogues can be prepared by organic synthesis. However, even for simple compounds like, for example, the polyketide zearalenone, the synthesis effort required is substantial. But it is less a problem how we can prepare a certain molecule, but rather a problem concerning which molecules chemists should prepare. After the era of combinatorial chemistry the goal should be to prepare as few as possible molecules to achieve a molecule with the desired properties. Thus, some kind of preselection seems highly desirable in order to avoid the bottleneck organic synthesis. However, even in e the target is known, reliable and accurate docking of flexible NPs or similar compounds has so far appeared impossible. Thus, for many years to come, organic chemists will work like hunters and gatherers. While prediction of the biological activity of a designed molecule will remain difficult, strategies based on genomics appear promising.[163] This would require chemists to acquire the relevant skills. Another option would be to use a brute force approach and to put some emphasis on the development of docking software. Thus, with a suitable structure generator that would generate virtual NP analogues based on biosynthesis rules, subsequent docking might be able to suggest promising compounds for synthesis. Until then the design and synthesis of NP analogues will rely largely on trial and error, supplemented with hunches and experience of scientists. In addition, one should not forget the role of serendipity for discovery and the advancement of science.

ACKNOWLEDGEMENTS

Financial support by the Federate State of Baden-Württemberg, the Fonds der Chemischen Industrie, and in particular the DFG is gratefully acknowledged. Part of this work was carried out within the framework of EU COST Action CM0804 (Chemical Biology with Natural Products). I also thank Dr Florenz Sasse (HZI Braunschweig) and his team for performing cellular assays with our compounds for many years.

NOTES AND REFERENCES

1. (*a*) D. J. Newman and G. M. Cragg, *J. Nat. Prod.*, 2012, **75**, 311 G. M. Cragg and D. J. Newman, *Biochim. Biophys. Acta, Gen. Subj.*, 2013, **1830**, 3670 ; (*b*) B. B. Mishra and V. K. Tiwari, *Eur. J. Med. Chem.*, 2011, **46**, 4769 .
2. (*a*) A. L. Harvey, *Drug Discovery Today*, 2008, **13**, 894 ; (*b*) C. Cordier, D. Morton, S. Murrison, A. Nelson and C. O'Leary-Steele, *Nat. Prod. Rep.*, 2008, **25**, 719 ; (*c*) K. Grabowski, K.-H. Baringhaus and G. Schneider, *Nat. Prod. Rep.*, 2008, **25**, 892 ; (*d*) N. I. Vasilevich, R. V. Kombarov, D. V. Genis and M. A. Kirpichenok, *J. Med. Chem.*, 2012, **55**, 7003 .
3. A. Ganesan, *Curr. Opin. Chem. Biol.*, 2008, **12**, 306 .
4. E. Kellenberger, A. Hofmann and R. J. Quinn, *Nat. Prod. Rep.*, 2011, **28**, 1483 .
5. L. Ruddigkeit, L. C. Blum and J.-L. Reymond, *J. Chem. Inf. Model.*, 2012, **53**, 56 .
6. M. Awale, R. van Deursen and J.-L. Reymond, *J. Chem. Inf. Model.*, 2013, **53**, 509 .
7. M. Gütlein, A. Karwath and S. Kramer, *J. Cheminf.*, 2012, **4**, 7, DOI:10.1186/1758-2946-4-7.
8. A. M. Virshup, J. Contreras-Garcia, P. Wipf, W. Yang and D. N. Beratan, *J. Am. Chem. Soc.*, 2013, **135**, 7296 .
9. R. M. Wilson and S. J. Danishefsky, *J. Org. Chem.*, 2006, **71**, 8329 .
10. For a review, see: A. M. Szpilman and E. M. Carreira, *Angew. Chem., Int. Ed.*, 2010, **49**, 9592 .
11. J. P. Nandy, M. Prakesch, S. Khadem, P. T. Reddy, U. Sharma and P. Arya, *Chem. Rev.*, 2009, **109**, 1999 .
12. For recent reviews, see: (*a*) K.-H. Altmann, F. Z. Gaugaz and R. Schiess, *Mol. Diversity*, 2011, **15**, 383 ; (*b*) M. J. Edelman and M. Shvartsbeyn, *Clin. Lung Cancer*, 2012, **13**, 171 ; (*c*) S. Kaiser, J. J. Muller, P. E. Froehlich, S. Cristina Baggio Gnoatto and A. M. Bergold, *Anti-Cancer Agents Med. Chem.*, 2013, **13**, 1057 .
13. K. B. Kim and C. M. Crews, *Nat. Prod. Rep.*, 2013, **30**, 600 .
14. P. M. Wright, I. B. Seiple and A. G. Myers, *Angew. Chem., Int. Ed.*, 2014, **53**, 8840–8869 .
15. One can find countless es for DTS in the literature, for example on dictyostatin: (*a*) W. Zhu, M. Jiménez, W.-H. Jung, D. P. Camarco, R. Balachandran, A. Vogt, B. W. Day and D. P. Curran, *J. Am. Chem. Soc.*, 2010, **132**, 9175 ; (*b*) I. Paterson, G. J. Naylor, N. M. Gardner, E. Guzmán and A. E. Wright, *Chem. – Asian J.*, 2011, **6**, 459 ; (*c*) L. L. Vollmer, M. Jiménez, D. P. Camarco, W. Zhu, H. N. Daghestani, R. Balachandran, C. E. Reese, J. S. Lazo, N. A. Hukriede, D. P. Curran, B. W. Day and A. Vogt, *Mol. Cancer Ther.*, 2011, **10**, 994 ; (*d*) J. Gallon, J. Esteban, S. Bouzbouz, M. Campbell, S. Reymond and J. Cossy, *Chem. – Eur. J.*, 2012, **18**, 11788 .

16. For some reviews, see: (*a*) M. D. Burke and S. L. Schreiber, *Angew. Chem., Int. Ed.*, 2004, **43**, 46 ; (*b*) T. E. Nielsen and S. L. Schreiber, *Angew. Chem., Int. Ed.*, 2008,**47**, 48 ; (*c*) R. J. Spandl, A. Bender and D. R. Spring, *Org. Biomol. Chem.*, 2008, **6**, 1149 ; (*d*) C. Cordier, D. Morton, S. Murrison, A. Nelson and C. O'Leary-Steele, *Nat. Prod. Rep.*, 2008, **25**, 719 ; (*e*) W. R. J. D. Galloway, A. Isidro-Llobet and D. R. Spring, *Nat. Commun.*, 2010, **1**, 1 ; (*f*) C. Serba and N. Winssinger, *Eur. J. Org. Chem.*, 2013, 4195 .

17. For example: (*a*) C. Chen, X. Li, C. S. Neumann, M. M. C. Lo and S. L. Schreiber, *Angew. Chem., Int. Ed.*, 2005, **44**, 2249 ; (*b*) S. Su, D. E. Acquilano, J. Arumugasamy, A. B. Beeler, E. L. Eastwood, J. R. Giguere, P. Lan, X. Lei, G. K. Min, A. R. Yeager, Y. Zhou, J. S. Panek, J. K. Snyder, S. E. Schaus and J. A. Porco Jr., *Org. Lett.*, 2005, **7**, 2751 ; (*c*) V. Duplan, C. Serba, J. Garcia, G. Valot, S. Barluenga, M. Hoerlé, M. Cuendet and N. Winssinger, *Org. Biomol. Chem.*, 2014, **12**, 370 .

18. A. Fürstner, D. Kirk, M. D. B. Fenster, C. Aiessa, D. De Souza, C. Nevado, T. Tuttle, W. Thiel and O. Müller, *Chem. – Eur. J.*, 2007, **13**, 135 .

19. Y. Kashman, A. Groweiss and U. Shmueli, *Tetrahedron Lett.*, 1980, **21**, 3629 .

20. S. Kudrimoti, S. A. Ahmed, P. R. Daga, A. E. Wahba, S. I. Khalifa, R. J. Doerksen and M. T. Hamann, *Bioorg. Med. Chem.*, 2009, **17**, 7517 .

21. K. A. El Sayed, M. A. Khanfar, H. M. Shallal, A. Muralidharan, B. Awate, D. T. A. Youssef, Y. Liu, Y.-D. Zhou, D. G. Nagle and G. Shah, *J. Nat. Prod.*, 2008, **71**, 396 .

22. A. Fürstner, D. De Souza, L. Turet, M. D. B. Fenster, L. Parra-Rapado, C. Wirtz, R. Myott and C. W. Lehmann, *Chem. – Eur. J.*, 2007, **13**, 115 .

23. I. Paterson, D. Y. K. Chen, M. J. Coster, J. L. Aceña, J. Bach and D. J. Wallace, *Org. Biomol. Chem.*, 2005, **3**, 2431 .

24. For a review, see: J. Pietruszka, *Angew. Chem., Int. Ed.*, 1998, **37**, 2629 .

25. A. B. Smith III, C. A. Risatti, O. Atasoylu, C. S. Bennett, J. Liu, H. Cheng, K. TenDyke and Q. Xu, *J. Am. Chem. Soc.*, 2011, **133**, 14042 .

26. M. Carda, J. Murga, S. Díaz-Oltra, J. García-Pla, J. Paños, E. Falomir, C. Trigili, J. F. Díaz, I. Barasoain and J. A. Marco, *Eur. J. Org. Chem.*, 2013, 1116 .

27. A. F. Petri, F. Sasse and M. E. Maier, *Eur. J. Org. Chem.*, 2005, 1865 .

28. Among the benzolactones, radicicol served as a prominent lead structure for analogue synthesis because of its binding to Hsp90: (*a*) E. Moulin, V. Zoete, S. Barluenga, M. Karplus and N. Winssinger, *J. Am. Chem. Soc.*, 2005, **127**, 6999 ; (*b*) X. Lei and S. J. Danishefsky, *Adv. Synth. Catal.*, 2008, **350**, 1677 ; (*c*) R. R. A. Kitson and C. J. Moody, *J. Org. Chem.*, 2013, **78**, 5117 .

29. The benzolactone enamides might be considered a subclass of the resorcylic acid lactones. These are covered in the following review: N. Winssinger and S. Barluenga, *Chem. Commun.*, 2007, 22 .

30. For some reviews regarding the biology of V-ATPases, see: (*a*) E. J. Bowman and B. J. Bowman, *J. Bioenerg. Biomembr.*, 2005, **37**, 431 ; (*b*) K. W. Beyenbach and H. Wieczorek, *J. Exp. Biol.*, 2006, **209**, 577 ; (*c*) M. Huss and H. Wieczorek, *J. Exp. Biol.*, 2009, **212**, 341 ; (*d*) M. Perez-Sayans, J. M. Somoza-Martin, F. Barros-Angueira, J. M. G. Rey and A. Garcia-Garcia, *Cancer Treat. Rev.*, 2009, **35**, 707 .

31. X.-S. Xie, D. Padron, X. Liao, J. Wang, M. G. Roth and J. K. De Brabander, *J. Biol. Chem.*, 2004, **279**, 19755 .

32. For total syntheses, see: (*a*) A. Bhattacharjee, O. R. Seguil and J. K. De Brabander, *Tetrahedron Lett.*, 2001, **42**, 1217 ; (*b*) K. C. Nicolaou, D. W. Kim and R. Baati,*Angew. Chem., Int. Ed.*, 2002, **41**, 3701 ; (*c*) Q. Su and J. S. Panek, *J. Am. Chem. Soc.*, 2004, **126**, 2425 ; (*d*) S. S. Palimkar and J. Uenishi, *Org. Lett.*, 2010, **12**, 4160 .

33. A. F. Petri, A. Bayer and M. E. Maier, *Angew. Chem., Int. Ed.*, 2004, **43**, 5821 .

34. Other groups also used transannular pyran formation: (*a*) F. Hilli, J. M. White and M. A. Rizzaa, *Tetrahedron*, 2011, **67**, 5054 ; (*b*) Y.-H. Jung, Y.-J. Kim, J. Lee and J. Tae, *Chem. – Asian J.*, 2007, **2**, 656 .

35. (*a*) K. C. Nicolaou, D. W. Kim, R. Baati, A. O'Brate and P. Giannakakou, *Chem. – Eur. J.*, 2003, **9**, 6177 ; (*b*) S. S. Palimkar, J. Uenishi and H. Ii, *J. Org. Chem.*, 2011, **77**, 388 .

36. M. Bauer and M. E. Maier, *Org. Lett.*, 2002, **4**, 2205 .

37. C. Herb and M. E. Maier, *J. Org. Chem.*, 2003, **68**, 8129 .

38. C. Herb, A. Bayer and M. E. Maier, *Chem. – Eur. J.*, 2004, **10**, 5649 .

39. C. Herb, F. Dettner and M. E. Maier, *Eur. J. Org. Chem.*, 2005, 728 .

40. J. Ritschel, F. Sasse and M. E. Maier, *Eur. J. Org. Chem.*, 2007, 78 .

41. S. Lebreton, X.-S. Xie, D. Ferguson and J. K. De Brabander, *Tetrahedron*, 2004, **60**, 9635 .

42. For the synthesis of some other salicylihalamide analogues, see: (*a*) Y. Sugimoto, K. Konoki, M. Murata, M. Matsushita, H. Kanazawa and T. Oishi, *J. Med. Chem.*, 2009, **52**, 798 ; (*b*) S. Tang and K. L. Erickson, *J. Nat. Prod.*, 2008, **71**, 898 .

43. S. Lebreton, J. Jaunbergs, M. G. Roth, D. A. Ferguson and J. K. De Brabander, *Bioorg. Med. Chem. Lett.*, 2008, **18**, 5879 .

44. K. H. Müller, G. A. Spoden, K. D. Scheffer, R. Brunnhöfer, J. K. De Brabander, M. E. Maier, L. Florin and C. P. Muller, *Antimicrob. Agents Chemother.*, 2014, **58**, 2905 .

45. J. A. Hall, B. R. Kusuma, G. E. L. Brandt and B. S. J. Blagg, *ACS Chem. Biol.*, 2014, **9**, 976 .

46. For a recent review, see: J. Franke, S. Eichner, C. Zeilinger and A. Kihning, *Nat. Prod. Rep.*, 2013, **30**, 1299 .

47. L. Jundt, H. Steinmetz, P. Luger, M. Weber, B. Kunze,

H. Reichenbach and G. Höfle, *Eur. J. Org. Chem.*, 2006, 5036 .

48. V. V. Vintonyak and M. E. Maier, *Angew. Chem., Int. Ed.*, 2007, **46**, 5209 .

49. A. Fürstner, M. Bindl and L. Jean, *Angew. Chem., Int. Ed.*, 2007, **46**, 9275–9278 .

50. M. Fouché, L. Rooney and A. G. M. Barrett, *J. Org. Chem.*, 2012, **77**, 3060 .

51. B. R. Kusuma, G. E. L. Brandt and B. S. J. Blagg, *Org. Lett.*, 2012, **14**, 6242 .

52. For a recent review, see: A. Fürstner, *Angew. Chem., Int. Ed.*, 2013, **52**, 2794 .

53. (*a*) V. V. Vintonyak and M. E. Maier, *Org. Lett.*, 2008, **10**, 1239 ; (*b*) V. V. Vintonyak, M. Calà, F. Lay, B. Kunze, F. Sasse and M. E. Maier, *Chem. – Eur. J.*, 2008, **14**, 3709 .

54. M. Bindl, L. Jean, J. Herrmann, R. Müller and A. Fürstner, *Chem. – Eur. J.*, 2009, **15**, 12310 .

55. V. V. Vintonyak, B. Kunze, F. Sasse and M. E. Maier, *Chem. – Eur. J.*, 2008, **14**, 11132 .

56. A. E. Wright, J. C. Botelho, E. Guzmán, D. Harmody, P. Linley, P. J. McCarthy, T. P. Pitts, S. A. Pomponi and J. K. Reed, *J. Nat. Prod.*, 2007, **70**, 412 .

57. (*a*) W. Youngsaye, J. T. Lowe, F. Pohlki, P. Ralifo and J. S. Panek, *Angew. Chem., Int. Ed.*, 2007, **46**, 9211 ; (*b*) S. K. Woo, M. S. Kwon and E. Lee, *Angew. Chem., Int. Ed.*, 2008, **47**, 3242–3244 ; (*c*) D. W. Custar, T. P. Zabawa and K. A. Scheidt, *J. Am. Chem. Soc.*, 2008, **130**, 804 ; (*d*) H. Fuwa, S. Naito, T. Goto and M. Sasaki, *Angew. Chem., Int. Ed.*, 2008, **47**, 4737–4739 ; (*e*) O. A. Ulanovskaya, J. Janjic, M. Suzuki, S. S. Sabharwal, P. T. Schumacker, S. J. Kron and S. A. Kozmin, *Nat. Chem. Biol.*, 2008, **4**, 418 ; (*f*) I. Paterson and N. A. Miller, *Chem. Commun.*, 2008, 4708 ; (*g*) X. Guinchard and E. Roulland, *Org. Lett.*, 2009, **11**, 4700 ; (*h*) Y. Cui, W. Tu and P. E. Floreancig, *Tetrahedron*, 2010, **66**, 4867 ; (*i*) H. Fuwa, A. Saito and M. Sasaki, *Angew. Chem., Int. Ed.*, 2010, **49**, 3041 ; (*j*) A. K. Ghosh, K. A. Shurrush and Z. L. Dawson, *Org. Biomol. Chem.*, 2013, **11**, 7768 .

58. For the most recent one, see: G. V. M. Sharma, S. V. Reddy and K. V. S. Ramakrishna, *Org. Biomol. Chem.*, 2012, **10**, 3689 .

59. (*a*) H. Fuwa, A. Saito, S. Naito, K. Konoki, M. Yotsu-Yamashita and M. Sasaki, *Chem. – Eur. J.*, 2009, **15**, 12807 ; (*b*) D. W. Custar, T. P. Zabawa, J. Hines, C. M. Crews and K. A. Scheidt, *J. Am. Chem. Soc.*, 2009, **131**, 12406 ; (*c*) Y. Cui, R. Balachandran, B. W. Day and P. E. Floreancig, *J. Org. Chem.*, 2012, **77**, 2225 ; (*d*) H. Fuwa, M. Kawakami, K. Noto, T. Muto, Y. Suga, K. Konoki, M. Yotsu-Yamashita and M. Sasaki, *Chem. – Eur. J.*, 2013, **19**, 8100 ; (*e*) H. Fuwa, T. Noguchi, M. Kawakami and M. Sasaki, *Bioorg. Med. Chem. Lett.*, 2014, **24**, 2415 .

60. For a review, see: G. S. Bagavananthem Andavan and R. Lemmens-Gruber, *Mar. Drugs*, 2010, **8**, 810 .

61. Y.-Y. Xu, C. Liu and Z.-P. Liu, *Curr. Org. Synth.*, 2013, **10**, 67 .

62. (*a*) B. Kunze, R. Jansen, F. Sasse, G. Höfle and H. Reichenbach, *J. Antibiot.*, 1995, **48**, 1262 ; (*b*) R. Jansen, B. Kunze, H. Reichenbach and G. Höfle, *Liebigs Ann.*, 1996, 285 .

63. (*a*) T. M. Zabriskie, J. A. Klocke, C. M. Ireland, A. H. Marcus, T. F. Molinski, D. J. Faulkner, C. Xu and J. C. Clardy, *J. Am. Chem. Soc.*, 1986, **108**, 3123 ; (*b*) P. Crews, L. V. Manes and M. Boehler, *Tetrahedron Lett.*, 1986, **27**, 2797 .

64. A. Tripathi, J. Puddick, M. l. R. Prinsep, M. Rottmann and L. T. Tan, *J. Nat. Prod.*, 2010, **73**, 1810 .

65. (*a*) Isolation: P. D. Boudreau, T. Byrum, W.-T. Liu, P. C. Dorrestein and W. H. Gerwick, *J. Nat. Prod.*, 2012, **75**, 1560 ; (*b*) Total synthesis: D. Wang, S. Song, Y. Tian, Y. Xu, Z. Miao and A. Zhang, *J. Nat. Prod.*, 2013, **76**, 974 .

66. See, for example: A. K. Ghosh and D. K. Moon, *Org. Lett.*, 2007, **9**, 2425 and references cited therein.

67. T.-S. Hu, R. Tannert, H.-D. Arndt and H. Waldmann, *Chem. Commun.*, 2007, 3942 .

68. H. Waldmann, T.-S. Hu, S. Renner, S. Menninger, R. Tannert, T. Oda and H.-D. Arndt, *Angew. Chem., Int. Ed.*, 2008, **47**, 6473–6477 .

69. U. Eggert, R. Diestel, F. Sasse, R. Jansen, B. Kunze and M. Kalesse, *Angew. Chem., Int. Ed.*, 2008, **47**, 6478–6482 .

70. (*a*) F. Sasse, B. Kunze, T. M. A. Gronewold and H. Reichenbach, *J. Natl. Cancer Inst.*, 1998, **90**, 1559 ; (*b*) L.-G. Milroy, S. Rizzo, A. Calderon, B. Ellinger, S. Erdmann, J. Mondry, P. Verveer, P. Bastiaens, H. Waldmann, L. Dehmelt and H.-D. Arndt, *J. Am. Chem. Soc.*, 2012, **134**, 8480 .

71. A. Schmauder, S. Müller and M. E. Maier, *Tetrahedron*, 2008, **64**, 6263 .

72. A. Schmauder, L. D. Sibley and M. E. Maier, *Chem. – Eur. J.*, 2010, **16**, 4328 .

73. A. Zhdanko, A. Schmauder, C. I. Ma, L. D. Sibley, D. Sept, F. Sasse and M. E. Maier, *Chem. – Eur. J.*, 2011, **17**, 13349 .

74. For the synthesis of jasplakinolide analogues, see: (*a*) A. K. Ghosh, Z. L. Dawson, D. K. Moon, R. Bai and E. Hamel, *Bioorg. Med. Chem. Lett.*, 2010, **20**, 5104 ; (*b*) L.-G. Milroy, S. Rizzo, A. Calderon, B. Ellinger, S. Erdmann, J. Mondry, P. Verveer, P. Bastiaens, H. Waldmann, L. Dehmelt and H.-D. Arndt, *J. Am. Chem. Soc.*, 2012, **134**, 8480 .

75. R. Tannert, L.-G. Milroy, B. Ellinger, T.-S. Hu, H.-D. Arndt and H. Waldmann, *J. Am. Chem. Soc.*, 2010, **132**, 3063 .

76. C. I. Ma, K. Diraviyam, M. E. Maier, D. Sept and L. D. Sibley, *J. Nat. Prod.*, 2013, **76**, 1565 .

77. For some reviews, see: (*a*) S. Bonnal, L. Vigevani and J. Valcárcel, *Nat. Rev. Drug Discovery*, 2012, **11**, 847 ; (*b*) S. M. Dehm, *Cancer Res.*, 2013, **73**, 5309 ; (*c*) J. Zhang and J. L. Manley, *Cancer Discovery*, 2013, **3**, 1228 ; (*d*) A. G. Matera and Z. Wang, *Nat. Rev. Mol. Cell Biol.*, 2014, **15**, 108 .

78. A. Yokoi, Y. Kotake, K. Takahashi, T. Kadowaki, Y.

Matsumoto, Y. Minoshima, N. H. Sugi, K. Sagane, M. Hamaguchi, M. Iwata and Y. Mizui, *FEBS J.*, 2011, **278**, 4870 .

79. (*a*) T. Sakai, T. Sameshima, M. Matsufuji, N. Kawamura, K. Dobashi and Y. Mizui, *J. Antibiot.*, 2004, **57**, 173 ; (*b*) T. Sakai, N. Asai, A. Okuda, N. Kawamura and Y. Mizui, *J. Antibiot.*, 2004, **57**, 180 .
80. M. Seki-Asano, T. Okazaki, M. Yamagishi, N. Sakai, Y. Takayama, K. Hanada, S. Morimoto, A. Takatsuki and K. Mizoue, *J. Antibiot.*, 1994, **47**, 1395 .
81. B. G. Isaac, S. W. Ayer, R. C. Elliott and R. J. Stonard, *J. Org. Chem.*, 1992, **57**, 7220 .
82. (*a*) H. Nakajima, B. Sato, T. Fujita, S. Takase, H. Terano and M. Okuhara, *J. Antibiot.*, 1996, **49**, 1196 ; (*b*) H. Nakajima, S. Takase, H. Terano and H. Tanaka, *J. Antibiot.*, 1997, **50**, 96 .
83. (*a*) X. Liu, S. Biswas, M. G. Berg, C. M. Antapli, F. Xie, Q. Wang, M.-C. Tang, G.-L. Tang, L. Zhang, G. Dreyfuss and Y.-Q. Cheng, *J. Nat. Prod.*, 2013, **76**, 685 ; (*b*) H. He, A. S. Ratnayake, J. E. Janso, M. He, H. Y. Yang, F. Loganzo, B. Shor, C. J. O'Donnell and F. E. Koehn, *J. Nat. Prod.*, 2014, **77**, 1864 .
84. M. Sato, N. Muguruma, T. Nakagawa, K. Okamoto, T. Kimura, S. Kitamura, H. Yano, K. Sannomiya, T. Goji, H. Miyamoto, T. Okahisa, H. Mikasa, S. Wada, M. Iwata and T. Takayama, *Cancer Sci.*, 2014, **105**, 110 .
85. S. M. Dehm, *Clin. Cancer Res.*, 2013, **19**, 6064 .
86. S. Osman, B. J. Albert, Y. Wang, M. Li, N. L. Czaicki and K. Koide, *Chem. – Eur. J.*, 2011, **17**, 895 .
87. L. Fan, C. Lagisetti, C. C. Edwards, T. R. Webb and P. M. Potter, *ACS Chem. Biol.*, 2011, **6**, 582 .
88. C. Lagisetti, A. Pourpak, T. Goronga, Q. Jiang, X. Cui, J. Hyle, J. M. Lahti, S. W. Morris and T. R. Webb, *J. Med. Chem.*, 2009, **52**, 6979 .
89. M. K. Gundluru, A. Pourpak, X. Cui, S. W. Morris and T. R. Webb, *Med. Chem. Commun.*, 2011, **2**, 904 .
90. V. P. Kumar and S. Chandrasekhar, *Org. Lett.*, 2013, **15**, 3610 .
91. S. Müller, T. Mayer, F. Sasse and M. E. Maier, *Org. Lett.*, 2011, **13**, 3940 .
92. K. Arai, S. Buonamici, B. Chan, L. Corson, A. Endo, B. Gerard, M.-H. Hao, C. Karr, K. Kira, L. Lee, X. Liu, J. T. Lowe, T. Luo, L. A. Marcaurelle, Y. Mizui, M. Nevalainen, M. W. O'Shea, E. S. Park, S. A. Perino, S. Prajapati, M. Shan, P. G. Smith, P. Tivitmahaisoon, J. Y. Wang, M. Warmuth, K.-M. Wu, L. Yu, H. Zhang, G.-Z. Zheng and G. F. Keaney, *Org. Lett.*, 2014, **16**, 5560 .
93. R. Villa, M. K. Kashyap, D. Kumar, T. J. Kipps, J. E. tro, J. J. La Clair and M. D. Burkart, *J. Med. Chem.*, 2013, **56**, 6576 .
94. C. Lagisetti, M. V. Yermolina, L. K. Sharma, G. Palacios, B. J. Prigaro and T. R. Webb, *ACS Chem. Biol.*, 2013, **9**, 643 .
95. R. M. Kanada, D. Itoh, M. Nagai, J. Niijima, N. Asai, Y. Mizui, S. Abe and Y. Kotake, *Angew. Chem., Int. Ed.*, 2007, **46**, 4350 .
96. A. K. Ghosh and D. D. Anderson, *Org. Lett.*, 2012, **14**, 4730 .
97. I. Shiina, H. Fukui and A. Sasaki, *Nat. Protocols*, 2007, **2**, 2312 .
98. S. Müller, F. Sasse and M. E. Maier, *Eur. J. Org. Chem.*, 2014, 1025 .
99. For the concept and some reviews, see: (*a*) P. A. Wender, V. A. Verma, T. J. Paxton and T. H. Pillow, *Acc. Chem. Res.*, 2008, **41**, 40 ; (*b*) P. A. Wender, *Tetrahedron*, 2013, **69**, 7529 ; (*c*) P. A. Wender, *Nat. Prod. Rep.*, 2014, **31**, 433 .
100. G. R. Pettit, C. L. Herald, D. L. Doubek, D. L. Herald, E. Arnold and J. Clardy, *J. Am. Chem. Soc.*, 1982, **104**, 6846 .
101. For a review about the synthesis of bryostatins, see: K. J. Hale and S. Manaviazar, *Chem. – Asian J.*, 2010, **5**, 704 .
102. Y. Lu, S. K. Woo and M. J. Krische, *J. Am. Chem. Soc.*, 2011, **133**, 13876 .
103. P. A. Wender, J. C. Horan and V. A. Verma, *Org. Lett.*, 2006, **8**, 5299 .
104. P. A. Wender, J. Baryza, C. E. Bennett, C. Bi, S. E. Brenner, M. O. Clarke, J. C. Horan, C. Kan, E. Lacôte, B. Lippa, P. G. Nell and T. M. Turner, *J. Am. Chem. Soc.*, 2002, **124**, 13648 .
105. P. A. Wender, J. L. Baryza, S. E. Brenner, B. A. DeChristopher, B. A. Loy, A. J. Schrier and V. A. Verma, *Proc. Natl. Acad. Sci. U. S. A.*, 2011, **108**, 6721 .
106. M. J. Yu, W. Zheng, B. M. Seletsky, B. A. Littlefield and Y. Kishi, in *Annual Reports in Medicinal Chemistry*, ed. J. E. Macor, Academic Press, New York, 2011, ch. 14, vol. 46, pp. 227–241.
107. M. J. Yu, W. Zheng and B. M. Seletsky, *Nat. Prod. Rep.*, 2013, **30**, 1158 .
108. C. Zambaldo, K. K. Sadhu, G. Karthikeyan, S. Barluenga, J.-P. Daguer and N. Winssinger, *Chem. Sci.*, 2013, **4**, 2088 .
109. For some reviews, see: (*a*) M. Kaiser, S. Wetzel, K. Kumar and H. Waldmann, *Cell. Mol. Life Sci.*, 2008, **65**, 1186 ; (*b*) S. Wetzel, R. S. Bon, K. Kumar and H. Waldmann, *Angew. Chem., Int. Ed.*, 2011, **50**, 10800 ; (*c*) S. Rizzo and H. Waldmann, *Chem. Rev.*, 2014, **114**, 4621 ; (*d*) H. van Hattum and H. Waldmann, *J. Am. Chem. Soc.*, 2014, **136**, 11853 .
110. B.-W. Chen, C.-H. Chao, J.-H. Su, C.-W. Tsai, W.-H. Wang, Z.-H. Wen, C.-Y. Huang, P.-J. Sung, Y.-C. Wu and J.-H. Sheu, *Org. Biomol. Chem.*, 2011, **9**, 834 .
111. S. Díaz, J. Cuesta, A. González and J. Bonjoch, *J. Org. Chem.*, 2003, **68**, 7400 .
112. For a review, see: D. H. Mac, S. Chandrasekhar and R. Grée, *Eur. J. Org. Chem.*, 2012, 5881 .
113. (*a*) S. Wetzel, K. Klein, S. Renner, D. Rauh, T. I. Oprea, P. Mutzel and H. Waldmann, *Nat. Chem. Biol.*, 2009, **5**, 581 ; (*b*) K. Klein, O. Koch, N. Kriege, P. Mutzel and T. Schäfer, *Mol. Inf.*, 2013, **32**, 964 .
114. (*a*) B. M. McArdle, M. R. Campitelli and R. J. Quinn, *J. Nat. Prod.*, 2006, **69**, 14 ; (*b*) D. Camp, R. A. Davis,

M. Campitelli, J. Ebdon and R. J. Quinn, *J. Nat. Prod.*, 2011, **75**, 72 .

115. S. Rizzo and H. Waldmann, *Chem. Rev.*, 2014, **114**, 4621 .

116. See also: I. D. Jenkins, F. Lacrampe, J. Ripper, L. Alcaraz, P. Van Le, G. Nikolakopoulos, P. de Almeida Leone, R. H. White and R. J. Quinn, *J. Org. Chem.*, 2008, **74**, 1304 .

117. B. Dinda, S. Debnath and R. Banik, *Chem. Pharm. Bull.*, 2011, **59**, 803 .

118. H. Takayama, Z.-J. Jia, L. Kremer, J. O. Bauer, C. Strohmann, S. Ziegler, A. P. Antonchick and H. Waldmann, *Angew. Chem., Int. Ed.*, 2013, **52**, 12404–12408 .

119. K. C. Morrison and P. J. Hergenrother, *Nat. Prod. Rep.*, 2014, **31**, 6 .

120. P. K. Chowdhury, A. Prelle, D. Schomburg, M. Thielmann and E. Winterfeldt, *Liebigs Ann. Chem.*, 1987, 1095–1099 .

121. For a recent application, see: J.-P. Krieger, G. Ricci, D. Lesuisse, C. Meyer and J. Cossy, *Angew. Chem., Int. Ed.*, 2014, **53**, 8705–8708 .

122. S. Bäurle, T. Blume, A. Mengel, C. Parchmann, W. Skuballa, S. Bäsler, M. Schäfer, D. Sülzle and H.-P. Wrona-Metzinger, *Angew. Chem., Int. Ed.*, 2003, **42**, 3961–3964 .

123. R. J. Rafferty, R. W. Hicklin, K. A. Maloof and P. J. Hergenrother, *Angew. Chem., Int. Ed.*, 2014, **53**, 220–224 .

124. R. W. Huigens III, K. C. Morrison, R. W. Hicklin, T. A. Flood Jr., M. F. Richter and P. J. Hergenrother, *Nat. Chem.*, 2013, **5**, 195 .

125. A. J. Grenning, J. K. Snyder and J. A. Porco, *Org. Lett.*, 2014, **16**, 792 .

126. For reviews, see: (*a*) G. Mehta and V. Singh, *Chem. Soc. Rev.*, 2002, **31**, 324 ; (*b*) L. F. Tietze and H. P. Bell, *Angew. Chem., Int. Ed.*, 2003, **42**, 3996 ; (*c*) K. Suzuki, *Chem. Rec.*, 2010, **10**, 291 .

127. For a review, see: K. C. Nicolaou, *J. Org. Chem.*, 2009, **74**, 951 .

128. Y.-M. Yan, J. Ai, Y.-N. Shi, Z.-L. Zuo, B. Hou, J. Luo and Y.-X. Cheng, *Org. Lett.*, 2014, **16**, 532 .

129. For a review, see: S. Ziegler, V. Pries, C. Hedberg and H. Waldmann, *Angew. Chem., Int. Ed.*, 2013, **52**, 2744 .

130. B. G. Isaac, S. W. Ayer, R. C. Elliott and R. J. Stonard, *J. Org. Chem.*, 1992, **57**, 7220 .

131. M. Hasegawa, T. Miura, K. Kuzuya, A. Inoue, S. Won Ki, S. Horinouchi, T. Yoshida, T. Kunoh, K. Koseki, K. Mino, R. Sasaki, M. Yoshida and T. Mizukami, *ACS Chem. Biol.*, 2011,**6**, 229 .

132. B. Meunier, *Acc. Chem. Res.*, 2007, **41**, 69 .

133. F. Coslédan, L. Fraisse, A. Pellet, F. o. Guillou, B. Mordmüller, P. G. Kremsner, A. Moreno, D. Mazier, J.-P. Maffrand and B. Meunier, *Proc. Natl. Acad. Sci. U. S. A.*, 2008, **105**, 17579 .

134. For some reviews, see: (*a*) P. J. Hajduk and J. Greer, *Nat. Rev. Drug Discovery*, 2007, **6**, 211 ; (*b*) C. W. Murray and D. C. Rees, *Nat. Chem.*, 2009, **1**, 187 ; (*c*) D. E. Scott, A. G. Coyne, S. A. Hudson and C. Abell, *Biochemistry*, 2012, **51**, 4990 ; (*d*) M. Baker, *Nat. Rev. Drug Discovery*, 2013, **12**, 5 .

135. Y. Matsuya, T. Kawaguchi, K. Ishihara, K. Ahmed, Q.-L. Zhao, T. Kondo and H. Nemoto, *Org. Lett.*, 2006, **8**, 4609 .

136. G. Jürjens and A. Kihning, *Org. Lett.*, 2014, **16**, 3000 .

137. For a review, see: J. Franke, S. Eichner, C. Zeilinger and A. Kihning, *Nat. Prod. Rep.*, 2013, **30**, 1299 .

138. M. Aeluri, B. Dasari and P. Arya, *Org. Lett.*, 2015, **17**, 472 ; See also: S. K. R. Guduru, R. Jimmidi, G. S. Deora and P. Arya, *Org. Lett.*, 2015, **17**, 480 .

139. S. Torijano-Gutiérrez, C. Vilanova, S. Díaz-Oltra, J. Murga, E. Falomir, M. Carda and J. A. Marco, *Eur. J. Org. Chem.*, 2014, 2284 .

140. For some recent examples, see: (*a*) H. Lou, S. Zheng, T. Li, J. Zhang, Y. Fei, X. Hao, G. Liang and W. Pan, *Org. Lett.*, 2014, **16**, 2696 ; (*b*) L. Song, H. Yao and R. Tong, *Org. Lett.*, 2014, **16**, 3740 .

141. (*a*) Y. J. Hong and D. J. Tantillo, *J. Am. Chem. Soc.*, 2014, **136**, 2450 ; (*b*) M. Isegawa, S. Maeda, D. J. Tantillo and K. Morokuma, *Chem. Sci.*, 2014, **5**, 1555 .

142. (*a*) M. A. Fischbach, C. T. Walsh and J. Clardy, *Proc. Natl. Acad. Sci. U. S. A.*, 2008, **105**, 4601 ; (*b*) J. Young and R. E. Taylor, *Nat. Prod. Rep.*, 2008, **25**, 651 ; (*c*) C. Hertweck, *Angew. Chem., Int. Ed.*, 2009, **48**, 4688 .

143. H. Motamedi and A. Shafiee, *Eur. J. Biochem.*, 1998, **256**, 528 .

144. For some recent reviews, see: (*a*) A. Kihning, F. Taft and T. Knobloch, *Org. Biomol. Chem.*, 2007, **5**, 3245 ; (*b*) J. Kennedy, *Nat. Prod. Rep.*, 2008, **25**, 25 ; (*c*) A. Kihning and F. Hahn, *Angew. Chem., Int. Ed.*, 2012, **51**, 4012 .

145. S. Eichner, H. G. Floss, F. Sasse and A. Kihning, *ChemBioChem*, 2009, **10**, 1801 .

146. For a review, see: F. Taft, S. Eichner, T. Knobloch, K. Harmrolfs, J. Hermane and A. Kihning, *Synlett*, 2012, 1416 .

147. F. Taft, M. Brünjes, H. G. Floss, N. Czempinski, S. Grond, F. Sasse and A. Kihning, *ChemBioChem*, 2008, **9**, 1057 .

148. S. Mo, D. H. Kim, J. H. Lee, J. W. Park, D. B. Basnet, Y. H. Ban, Y. J. Yoo, S.-w. Chen, S. R. Park, E. A. Choi, E. Kim, Y.-Y. Jin, S.-K. Lee, J. Y. Park, Y. Liu, M. O. Lee, K. S. Lee, S. J. Kim, D. Kim, B. C. Park, S.-g. Lee, H. J. Kwon, J.-W. Suh, B. S. Moore, S.-K. Lim and Y. J. Yoon, *J. Am. Chem. Soc.*, 2011, **133**, 976 .

149. K. Patel, M. Piagentini, A. Rascher, Z.-Q. Tian, G. O. Buchanan, R. Regentin, Z. Hu, C. R. Hutchinson and R. McDaniel, *Chem. Biol.*, 2004, **11**, 1625 .

150. A. Lechner, M. C. Wilson, Y. H. Ban, J.-y. Hwang, Y. J. Yoon and B. S. Moore, *ACS Synth. Biol.*, 2012, **2**, 379 .

151. (*a*) A. Kitsche and M. Kalesse, *ChemBioChem*, 2013, **14**, 851 ; (*b*) O. Hartmann and M. Kalesse, *Angew. Chem., Int. Ed.*, 2014, **53**, 7335 .

152. H.-H. Lu, A. Raja, R. Franke, D. Landsberg, F. Sasse and M. Kalesse, *Angew. Chem., Int. Ed.*, 2013, **52**, 13549 .

153. J. Ligon, S. Hill, J. Beck, R. Zirkle, I. Molnár, J. Zawodny, S. Money and T. Schupp, *Gene*, 2002, **285**, 257 .

154. S. C. Wenzel, R. M. Williamson, C. Grünanger, J. Xu, K. Gerth, R. A. Martinez, S. J. Moss, B. J. Carroll, S. Grond, C. J. Unkefer, R. Müller and H. G. Floss, *J. Am. Chem. Soc.*, 2006,**128**, 14325 .

155. Representative examples: (*a*) Y. Chen, M. Bilban, C. A. Foster and D. L. Boger, *J. Am. Chem. Soc.*, 2002, **124**, 5431 ; (*b*) J. Nam, D. Shin, Y. Rew and D. L. Boger, *J. Am. Chem. Soc.*, 2007, **129**, 8747 .

156. For a review, see: V. J. Hruby, *Nat. Rev. Drug Discovery*, 2002, **1**, 847 .

157. G. Höfle, N. Bedorf, H. Steinmetz, D. Schomburg, K. Gerth and H. Reichenbach, *Angew. Chem., Int. Ed. Engl.*, 1996, **35**, 1567 .

158. M. Ugele, F. Sasse, S. Knapp, O. Fedorov, A. Zubriene, D. Matulis and M. E. Maier, *ChemBioChem*, 2009, **10**, 2203 .

159. F. H. Niesen, H. Berglund and M. Vedadi, *Nat. Protocols*, 2007, **2**, 2212 .

160. T. J. Zimmermann, F. H. Niesen, E. S. Pilka, S. Knapp, U. Oppermann and M. E. Maier, *Bioorg. Med. Chem.*, 2009, **17**, 530 .

161. C. Rink, F. Sasse, D. Matulis, A. Zubriene and M. E. Maier, *Chem. – Eur. J.*, 2010, **16**, 14469 .

162. For reviews, see: (*a*) C. S. Leung, S. S. F. Leung, J. Tirado-Rives and W. L. Jorgensen, *J. Med. Chem.*, 2012, **55**, 4489 ; (*b*) H. Schönherr and T. Cernak, *Angew. Chem., Int. Ed.*, 2013, **52**, 12256–12267 .

163. J. R. Doroghazi, J. C. Albright, A. W. Goering, K.-S. Ju, R. R. Haines, K. A. Tchalukov, D. P. Labeda, N. L. Kelleher and W. W. Metcalf, *Nat. Chem. Biol.*, 2014, **10**, 963 .

CHAPTER 13 THERMAL DECOMPOSITION OF CARBOXYLATE IONIC LIQUIDS

Matthew T. Clough, Karolin Geyer, Patricia A. Hunt, Jürgen Mertes and Tom Welton

INTRODUCTION

Ionic liquids are versatile materials, with applications as electrolytes,[1] engineering fluids,[2] pre-treatment agents for wood grinding/biomass extraction[3] and, fundamentally, as solvents and catalysts for synthetic chemical reactions.[4] They are liquid at, or near, room temperature.[5] Fully composed of cationic and anionic species, the unsymmetrical and charge-diffuse nature of the ions contributes to their low melting points. More recently, it has been suggested that there may be an entropic factor that also contributes to the low melting points of ionic liquids.[6]Dialkylimidazolium ionic liquids are generally preferred for their low melting points and ease of handling and preparation.[7]

The ionic liquid 1-ethyl-3-methylimidazolium acetate, $[C_2C_1im][OAc]$, has been popular in biomass pre-treatment studies.[8] These processes are carried out at elevated temperatures (120–170 °C) for extended periods of time (up to 24 h).[9] Degradation of the ionic liquid will reduce the efficacy of the process, and the build up of volatile or corrosive decomposition products poses a hazard, particularly in large-scale industry.

In addition, such processes are only economic if the ionic liquid can be recycled and used multiple times. Hence, a thorough understanding of the long-term thermal decomposition behavior of $[C_2C_1im][OAc]$, and related compounds, including knowledge of the upper operational temperature limit and the precise mechanisms that are occurring, is crucial.[10]

Thermogravimetric Analysis (TGA) has been used extensively to study the thermal stability of ionic liquids.[11] TGA experiments can be broadly divided into two categories: (i) 'temperature-ramped' or 'scanning' TGA experiments measure weight % as a function of sample temperature, with a constant heating rate (typically 10 °C min^{-1} or 20 °C min^{-1}); (ii) 'isothermal' TGA experiments record weight % as a function of time, whilst the sample is maintained at a constant temperature.

Temperature-ramped TGA experiments allow rapid comparison of thermal stability between compounds in a series. The T_{onset} value is determined from extrapolation of the steepest point of the thermograph, and has been used to quantify the thermal stabilities of ionic liquids.[11d,12] However, the T_{onset} value is known to vary with experimental conditions, including heating rate, sample quantity and pan material.[11c,13] Nevertheless, temperature-ramped TGA is a suitable technique for assessing the relative thermal stabilities of a series of compounds, when care is taken to keep experimental conditions uniform. It is well established that the T_{onset} value provides a large overestimate of ionic liquid thermal stability.[14] The '$T_{x/z}$' parameter was proposed, where x represents a given decomposition extent (e.g. 1%) in z length of time (e.g. 10 hours). Hence, the $T_{0.01/10}$ parameter represents the temperature at which exactly 1% weight loss occurs in 10 hours. MacFarlane and co-workers determined the $T_{0.01/10}$ values for several imidazolium, pyrrolidinium and phosphonium ionic liquids, and found them to be lower by typically 110 °C compared to the T_{onset} temperatures.[15]

To date, the thermal decomposition of $[C_2C_1im][OAc]$ has been partially characterised.[16] In this study, we investigate the thermal decomposition of $[C_2C_1im][OAc]$ and related compounds (Figure 1) using a variety of experimental and computational techniques. The emphasis is on determining the long-term thermal stability of the ionic liquids, and understanding the thermodynamics, kinetics, and chemical mechanisms of degradation.

Citation: Matthew T. Clough, Karolin Geyer, Patricia A. Hunt, Jürgen Mertes and Tom Welton, "Thermal decomposition of carboxylate ionic liquids," DOI: 10.1039/c3cp53648c.

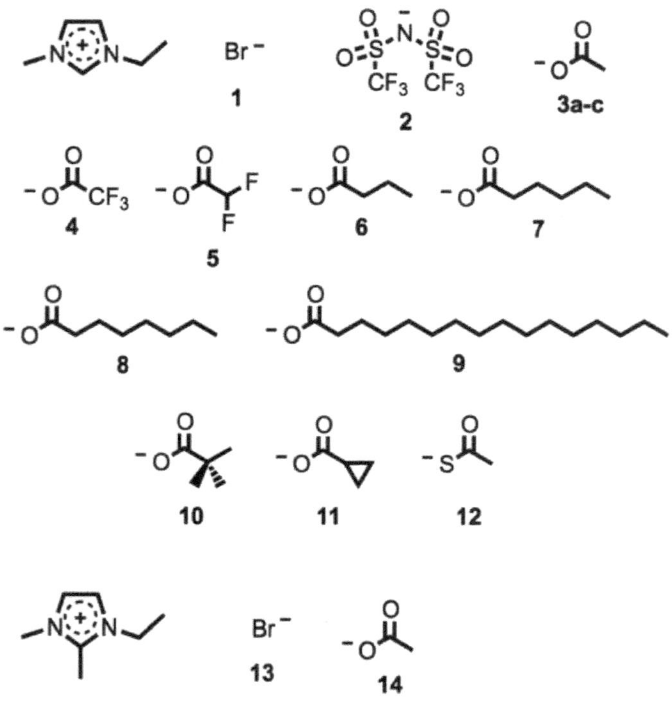

Figure 1: Ionic liquids investigated in this study, incorporating the 1-ethyl-3-methylimidazolium $[C_2C_1im]^+$ cation (**1–12**) and 1-ethyl-2,3-dimethylimidazolium $[C_2C_1C_1im]^+$ cation (**13** and**14**).

EXPERIMENTAL

The ionic liquids investigated in this study are shown in Figure 1. With the exception of **3c**, which was purchased, all compounds were prepared in our laboratory from commercial starting materials, taking care to ensure purity at each step. Compounds incorporating the bromide, **1**, and bis(trifluoromethanesulfonyl) imide, **2**, anions have been previously studied, and are included in the study for comparative purposes.[14b,17] Each of the carboxylate ionic liquids **4–12** can be regarded as an analogue of $[C_2C_1im][OAc]$, **3**, involving systematic functionalization of the acetate anion by fluorination (**4** and **5**), alkyl chain extension (**6–9**), including branching (**10**), a strained cyclopropane ring (**11**), and by incorporating a sulfur atom into the carboxylate group to give the thioacetate anion (**12**). Three separate samples of 1-ethyl-3-methylimidazolium acetate, $[C_2C_1im][OAc]$, denoted **3a–c**, represent the same compound prepared by three different synthetic procedures. A metathesis reaction involving $[C_2C_1im]Br$, **1**, and silver(I) acetate in aqueous solution yielded **3a**. Compound **3b** was prepared by the neutralization of an aqueous solution of $[C_2C_1im][OH]$ with acetic acid. The aqueous $[C_2C_1im][OH]$ was generated from a metathesis of $[C_2C_1im][HSO_4]$ with barium hydroxide in aqueous solution, which precipitates insoluble barium sulfate.[18] Ionic liquid **3c** was a purchased commercial sample of $[C_2C_1im][OAc]$.[19]

Thermogravimetric Analysis (TGA) was carried out on a PerkinElmer 'Pyris 1' TGA apparatus. Temperature-ramped TGA experiments were performed on 4–8 mg of the ionic liquid, using a heating rate of 10 °C min^{-1}, and platinum pans. Unless otherwise stated, nitrogen was used as the inert carrier gas. Isothermal TGA was performed on 7–8 mg of the ionic liquid. Thermogravimetric Analysis–Mass Spectrometry (TGA-MS) experiments were performed on 20–60 mg of the ionic liquid, using ceramic pans. A heating rate of 10 °C min^{-1} and a CP Grade helium flow of 20 ml min^{-1} were employed. The 'Pyris 1' TGA was connected to a 'Hiden Analytical HPR 20' Mass Spectrometer with a ceramic heated capillary. Electron Ionization (EI) mass spectrometry was employed, with an ionization energy of 70 eV.

Differential Scanning Calorimetry (DSC) experiments were performed on 1–9 mg of the ionic liquid, using aluminium pans. In order to allow volatile degradants to escape, and to avoid the risk of the pans bursting under pressure, a small incision was made in the top of both the sample and the reference pan. The pan was heated initially to 100 °C for 30 minutes, in order to remove water. The sample pan was then re-weighed and the DSC program was initiated. DSC traces were collected in the range 30–410 °C, with a heating rate of 2.5 °C min^{-1}. After the first heating step, the sample was cooled to 30 °C at a rate of 20 °C min^{-1}. The sample was then maintained at 30 °C for 15 minutes, followed by a second heating step using an identical program to the first heating step. The second heating step would indicate the phase behavior of any residual decomposition products present in the sample pan following the first heating stage.

RESULTS AND DISCUSSION

The thermal stabilities of ionic liquids **1–14** were compared using temperature-ramped TGA (Figure 2). The temperature-ramped TGA experiments were performed in triplicate for each ionic liquid, in order to assess the reproducibility of the T_{onset} measurement. For all ionic liquids in this investigation, the T_{onset} values were reproducible to within ±2 °C. The middle value for each ionic liquid is tabulated in Table 1. The three samples of $[C_2C_1im][OAc]$, **3a–c**, give identical T_{onset} values within the experimental error (Table 1). This implies that the commercial sample, **3c**, generated on large scale, is as stable at high temperatures as the samples prepared in the laboratory either via silver(I) or barium metathesis reactions, **3a** and **3b** respectively.

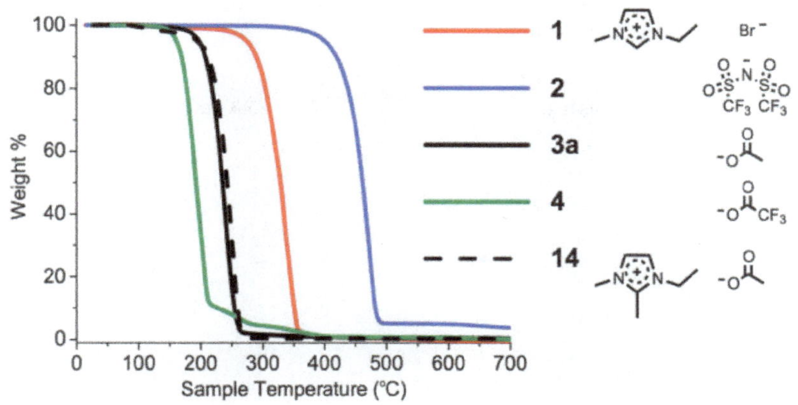

Figure 2: Temperature-ramped TGA thermographs for selected $[C_2C_1im]^+$ and $[C_2C_1C_1im]^+$ionic liquids, **1**, **2**, **3a**, **4** and **14**, with heating rate 10 °C min^{-1}.

Table 1 T_{onset} values for ionic liquids **1–14**, measured under a nitrogen atmosphere and with a heating rate of 10 °C min^{-1}

Ionic liquid		T_{onset} ± 2 (°C)
1	$[C_2C_1im]Br$	301
2	$[C_2C_1im][NTf_2]$	436
3a	$[C_2C_1im][OAc]$ Ag route	214
3b	$[C_2C_1im][OAc]$ Ba route	216
3c	$[C_2C_1im][OAc]$ Commercial	216
4	$[C_2C_1im][CF_3CO_2]$	172
5	$[C_2C_1im][CHF_2CO_2]$	263
6	$[C_2C_1im][CH_3(CH_2)_2CO_2]$	213
7	$[C_2C_1im][CH_3(CH_2)_4CO_2]$	212
8	$[C_2C_1im][CH_3(CH_2)_6CO_2]$	213
9	$[C_2C_1im][CH_3(CH_2)_{14}CO_2]$	216
10	$[C_2C_1im][C(CH_3)_3CO_2]$	212
11	$[C_2C_1im][c\text{-}(C_3H_5)CO_2]$	219
12	$[C_2C_1im][SAc]$	188
13	$[C_2C_1C_1im]Br$	299
14	$[C_2C_1C_1im][OAc]$	221

Trifluorination of the acetate anion in compound **4** leads to lower thermal stability than the non-fluorinated compound **3a–c** (Figure 2). Pibiri and co-workers examined the thermal stabilities of a series of protic perfluoroalkyl-carboxylate ionic liquids using TGA, and found these to be low.[20]

By contrast, ionic liquid **5**, which incorporates the difluoroacetate anion, is found to be significantly more stable than both the non-fluorinated and trifluorinated compounds, **3** and **4**, with a T_{onset} of 263 °C. Therefore, $[C_2C_1im][CHF_2CO_2]$, **5**, was found to be an analogue of the acetate ionic liquid, **3**, with improved thermal stability.

Extension of the carboxylate alkyl chain, in compounds **6–9**, has a minimal effect on the T_{onset} thermal decomposition temperature of the ionic liquid (Figure 2). Surprisingly, even the 16-carbon palmitate compound, $[C_2C_1im][CH_3(CH_2)_{14}CO_2]$, **9**, gave a thermograph very similar to the acetate ionic liquid, **3a**. Branching of the carboxylate alkyl chain, illustrated by compounds **10** and **11**, also has a very small effect on the T_{onset} value. There is a small increase in the basicity of the carboxylate anion with the addition of electron-donating alkyl substituents, indicated by the pK_a values of the conjugate carboxylic acids (acetic acid 4.76; butyric acid 4.83; cyclopropanecarboxylic acid 4.83; hexanoic acid 4.85; octanoic acid 4.89; trimethylacetic/pivalic acid 5.03).[21] Nevertheless, this brings about no substantial change in the T_{onset} value of the ionic liquid. Therefore, temperature-ramped TGA experiments indicate that there is no substantial difference in the thermal stability of $[C_2C_1im]^+$ carboxylate ionic liquids with differing length/branched alkyl chains, and that the decomposition mechanism involves the reactive carboxylate functional group.

The sulfur analogue, $[C_2C_1im][SAc]$, **12**, is substantially less thermally stable than the acetate equivalent, **3**. Thermal decomposition of **12** occurs through a nucleophilic substitution mechanism (Figure 7, below), because of the high nucleophilicity and low basicity of the thioacetate anion (pK_a thioacetic acid = 3.33).[21] The negative charge on the thioacetate anion is known to be centerd almost entirely on the sulfur atom, due to the large difference in the energies of the thiol and thione forms.[22] It was observed that over a period of six months, the yellow thioacetate ionic liquid, **12**, gradually darkened when stored at 4 °C under nitrogen (Figure 3).

(a)

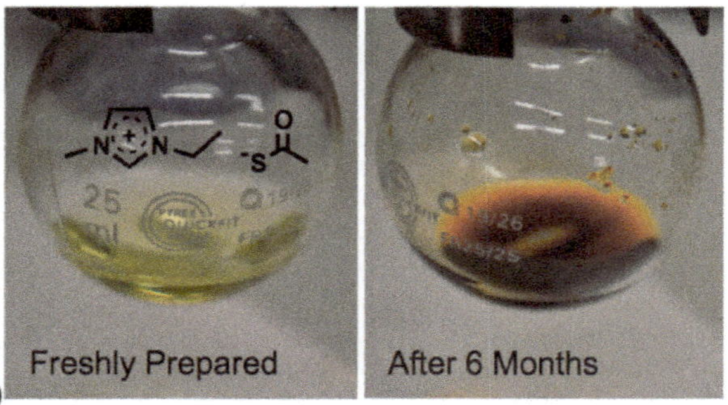

(b)

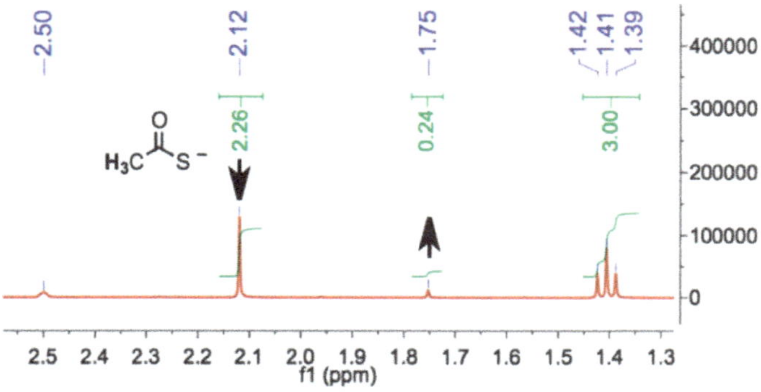

Figure 3: (a) Observed decomposition of [C_2C_1im][SAc], **12**, after six months, when stored under nitrogen at 4 °C. (b) Aliphatic region of the ^{1}H NMR spectrum of [C_2C_1im][SAc], **12**, after six months, demonstrating the disappearance of the thioacetate singlet at 2.12 ppm, and the appearance of a peak at 1.75 ppm.

The color change suggests poor long-term stability. Moreover, the ^{1}H NMR spectrum of the darkened liquid showed a decrease in the integration of the thioacetate methyl peak, accompanied by the appearance of a new small singlet at 1.75 ppm. ^{1}H NMR and LSIMS mass spectrometry experiments revealed that partial hydrolysis of thioacetate to acetate had occurred. The convoluted synthesis and offensive odor of **12** indicate that it will not be a suitable candidate for many ionic liquid applications.

The identity of the flow gas could potentially affect the thermal stabilities of ionic liquids **1–14**. Therefore, subsequent temperature-ramped TGA experiments were performed for the archetypal ionic liquid [C_2C_1im][OAc], **3c**, under compressed air and CO_2 flow gases, in addition to the usual inert nitrogen.. Prior to each scan, the ionic liquid was heated to 80 °C for 10 hours in the TGA apparatus under a gentle flow of each gas, to ensure equilibration of the dissolved gases in the ionic liquid. The presence of oxygen gas in compressed air introduces the possibility of oxidation or combustion occurring. Moreover, employing CO_2 as the carrier gas may indicate the presence of an N-heterocyclic carbene (NHC) species forming in the ionic liquid, since CO_2 could act as a carbene trap.[23] Interestingly, no change in the thermal stability of **3c** was observed under the compressed air environment and a T_{onset} value of 216 °C was recorded, suggesting that neither oxidation or combustion are thermal decomposition mechanisms. However, for the experiment employing a CO_2 carrier gas, an initial gain of mass was observed for the first two hours of the 10 hour 80 °C equilibration period, followed by gradual loss of mass for the remainder of the equilibration step. The sample then gave a T_{onset} value of 225 °C in the subsequent temperature-ramped step, slightly higher than for **3c** in the nitrogen experiment. This result suggests the formation of the [C_2C_1im$^+$-CO_2^-] complex previously observed by Rogers and co-workers.[23] Therefore, it is crucial to consider the nature of the gaseous environment when employing ionic liquid **3c**.

The T_{onset} temperature, as determined by a step-tangent method, is well known to overestimate the thermal stability of ionic liquid compounds.[14,15] Therefore, the $T_{0.01/10}$ value was determined for [C_2C_1im][OAc], **3a**, and [$C_2C_1C_1$im][OAc], **14**, from isothermal TGA experiments using the method of MacFarlane and co-workers.[14b] Isothermal TGA graphs for **3a** are shown below (Figure 4).

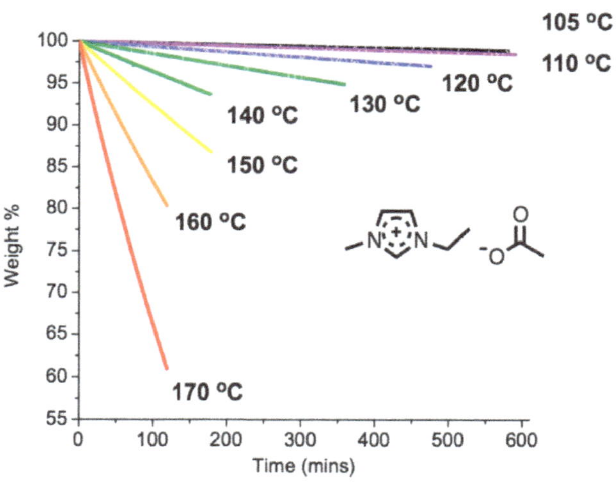

(a)

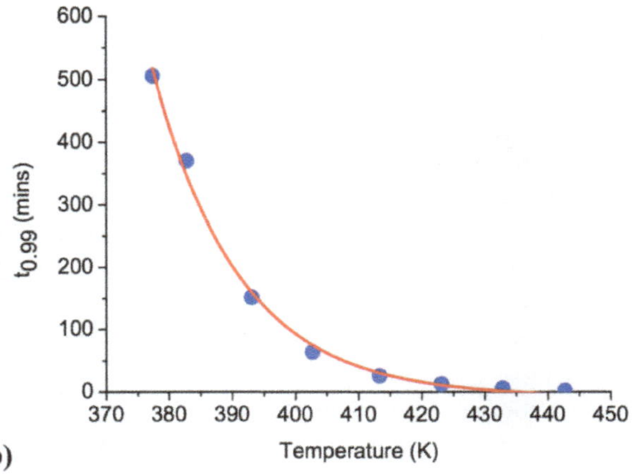

(b)

Figure 4: (a) Isothermal TGA analysis of $[C_2C_1im][OAc]$, **3a**. (b) The $t_{0.99}$ (time taken for 1% decomposition to occur) values were plotted against temperature (T) in Kelvin, giving an exponential fit. The fitted curve was extrapolated backwards to find $T_{0.01/10}$ corresponding to $t_{0.99}$ = 600 min/10 hours. $T_{0.01/10}$ was determined as 102 °C.

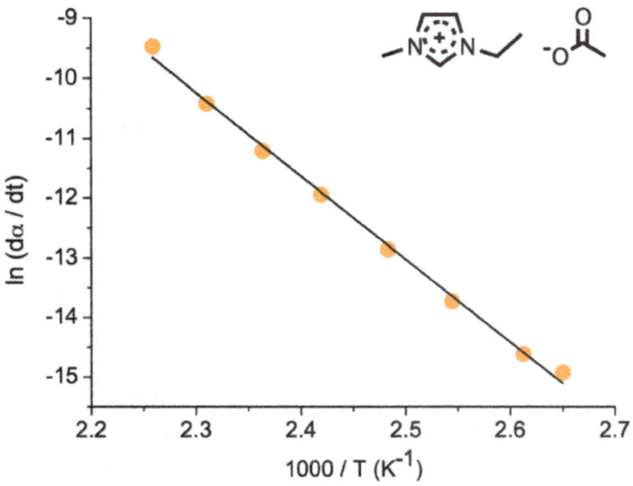

Figure 5: Thermal decomposition kinetics for $[C_2C_1im][OAc]$, **3a**, determined from isothermal TGA experiments. E_a = 115.8 kJ mol^{-1}, A = 2.98 × 10^9.

For **3a** and **14**, the $T_{0.01/10}$ parameter was measured as 102 °C and 99 °C, respectively. For **3a**, this value is significantly lower than the operating temperature of many biomass processes employing this ionic liquid (typically upwards of 120 °C).[9] In addition, biomass procedures are often carried out for periods substantially longer than 10 hours. Therefore, these procedures are likely to lead to substantial degradation of this acetate ionic liquid.

The $T_{0.01/10}$ value is a reasonable approximation of long-term thermal stability, although the ability to accurately predict the rate of decomposition at any given temperature is necessary. The Arrhenius equation, (1), representing the ionic liquid decomposition, can be determined from the isothermal TGA data.[24] Hence, the activation energy, E_a, and the pre-exponential factor, A, can be calculated. This allows rapid and accurate prediction of the decomposition rate at any temperature. The Arrhenius equations representing decomposition of $[C_2C_1im][OAc]$,**3a**, and $[C_2C_1C_1im][OAc]$, **14**, have been determined using this isothermal method.

Where ionic liquids thermally decompose via the reaction of one cation with one anion, thereby generating volatile degradants, the concentration of cations and anions in the bulk ionic liquid does not change. Therefore in these circumstances, decomposition is pseudo zeroth order. The rate of a decomposition of this type can be expressed in terms of the molar proportion of the ionic liquid that has decomposed as a function of time, dα/dt (1). Plotting ln(dα/dt) against 1/T gives a linear fit with a gradient equal to $-E_a/R$ and a y-axis intercept equal to ln(A).

$$d\alpha/dt = A\exp(-E_a/RT) \quad (1)$$

The plot of ln(dα/dt) against 1000/T for **3a** is shown in Figure 5..

The activation energy, E_a, was found to be 115.8 kJ mol^{-1} for **3a** and 99.6 kJ mol^{-1} for **14**. Although the $T_{0.01/10}$ values for ionic liquids **3a** and **14** are very similar (102 and 99 °C, respectively), the activation energies differ by approximately 16 kJ mol^{-1}. However, the pre-exponential factors differ by two orders of magnitude (2.98 × 10^9 for **3a**, 2.19 × 10^7 for **14**). The pre-exponential factor, A, often referred to as the 'frequency factor', is intrinsically linked to the frequency of molecular collisions occurring, whether or not the collision leads to a reaction. The differing values of A suggest significantly fewer collisions are occurring for **14** than for **3a**, although the lower E_a value for **14** indicates that a greater proportion of collisions will result in a decomposition reaction. The differing values of A may be a result of the reduced mobility of the 1-ethyl-2,3-dimethylimidazolium cation, caused by the additional steric bulk and van der Waals interactions of the C^2-methyl substituent.

Ionic liquid **14** has a measured melting point of 90–92 °C, whereas **3a** is a liquid at room temperature. Therefore, at temperatures high enough for **14** to be in the liquid state, decomposition will already be occurring at a rate of approximately 1% in 24 hours. This will restrict the suitability of ionic liquid **14** for many applications.

When describing thermal decomposition in an open system, such as in the TGA apparatus, the contribution to mass loss from evaporation of the ionic liquid must also be considered. Despite once being heralded as completely non-volatile, ionic liquids are now known to exert a vapor pressure.[25] Successful distillation of high thermal stability $[NTf_2]$-containing ionic liquids has been demonstrated, by employing ultra-high vacuum (UHV) conditions.[25b,d] Heym and colleagues observed that for $[C_4C_1im][NTf_2]$, evaporation dominated thermal decomposition at low heating rates in thermogravimetric experiments.[25e] The enthalpy of vaporization at 298 K, $\Delta_{vap}H_{298}$, for this ionic liquid has been measured as 133–140 kJ mol^{-1}, depending on the chosen experimental method.[25a]

For the dialkylimidazolium and trialkylimidazolium carboxylate ionic liquids investigated here, it is unclear whether thermal decomposition is occurring exclusively, or whether vaporization of the intact ionic liquid is also involved. However, Shimizu and co-workers have calculated the $\Delta_{vap}H$ value for $[C_4C_1im][OAc]$ (using the relationship $\Delta_{vap}H = U_{cohesive} + RT$), and found this value to be 100 kJ mol^{-1} higher than for $[C_4C_1im][NTf_2]$.[25j] Furthermore, Chandran et al. obtained a $\Delta_{vap}H$ value of 187.6 kJ mol^{-1} for $[C_4C_1im][OAc]$ from MD simulations.[25g] The T_{onset} temperatures for all alkyl and fluoroalkyl carboxylate ionic liquids here are at least 170 °C lower than for $[C_2C_1im][NTf_2]$, **2** (Table 1).

Therefore, there is likely to be a contribution from ionic liquid vaporization for ionic liquid **2**.[25e] However, on the basis of low thermal decomposition temperatures and high vaporization enthalpies, the contribution to mass loss from vaporization is likely to be negligible for all alkyl and fluoroalkyl carboxylate ionic liquids, **3–14**, in this study. Instead, all mass loss is attributed to the evaporation of volatile thermal decomposition products.

An understanding of the safe operating temperatures of an ionic liquid is crucial. In addition, knowledge of the thermal decomposition products is important; the production of volatile or corrosive degradants represents a safety hazard, and may lead to a reduction in the efficacy of a process. Certain ionic liquids are known to thermally decompose by nucleophilic or basic attack of the anion on the cation,[11a,15,17b,26] or by initial degradation of the anion into reactive species, which subsequently react with the cation.[17a,c,27] The formation of NHC species from dialkylimidazolium ionic liquids is recognized as another potential decomposition route.[28] Ionic liquids incorporating the tetrafluoroborate, $[BF_4]^-$, and hexafluorophosphate, $[PF_6]^-$, anions are susceptible to hydrolysis, liberating highly corrosive hydrogen fluoride, HF.[29] Carboxylate anions are both moderately nucleophilic and basic, and several chemically-plausible decomposition mechanisms can therefore be proposed for the 1-ethyl-3-methylimidazolium carboxylate ionic liquids (A–E, Figure 6).

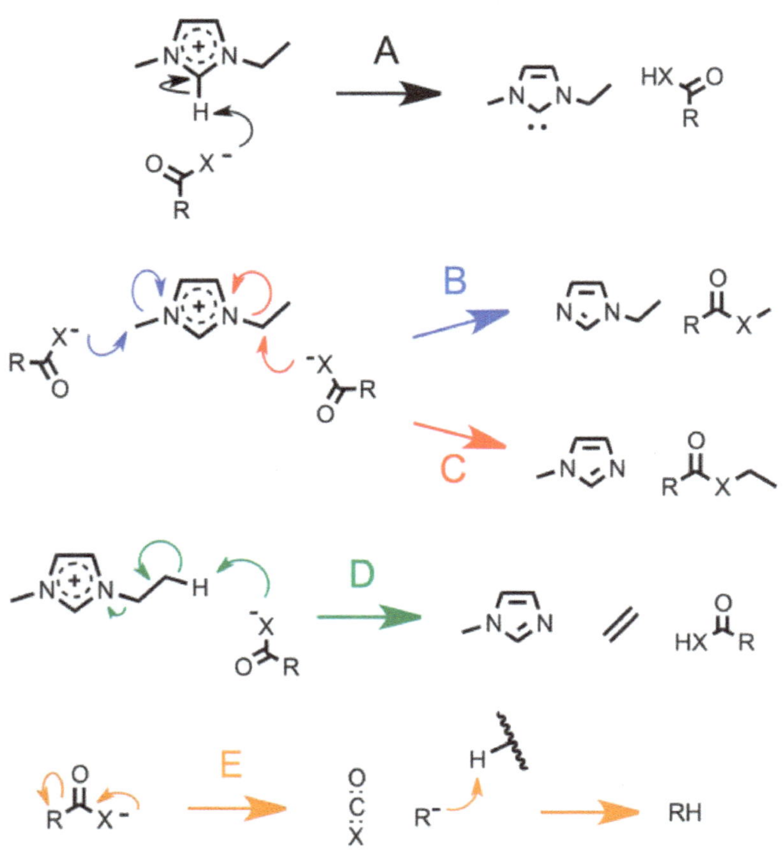

R = CH_3, X = O	**3a**
R = CF_3, X = O	**4**
R = $(CH_2)_6CH_3$, X = O	**8**
R = CH_3, X = S	**12**

Figure 6: Postulated decomposition mechanisms for $[C_2C_1im]^+$ ionic liquids **3a**, **4**, **8** and **12**: (A) N-heterocyclic carbene (NHC) formation; (B) + (C) S_N2 nucleophilic substitution; (D) E2 Hofmann elimination; (E) decarboxylation of the anion.

Thermogravimetric Analysis (TGA) coupled with Mass Spectrometry (MS) allows the stream of inert gas to be analyzed, and chemical products from the ionic liquid thermal decomposition can be detected. The detection of peaks at m/z values corresponding to parent ions or fragment ions of certain decomposition products is indicative of a particular mechanism or mechanisms occurring.

Dialkylimidazolium and trialkylimidazolium ionic liquids incorporating bis(trifluoromethanesulfonyl)imide (**2**), acetate (**3a**, **14**), trifluoroacetate (**4**), octanoate (**8**) and thioacetate (**12**) anions have been investigated in this TGA-MS study. In order for the mass spectral peaks to be of sufficient intensity, a larger quantity (20–60 mg) of ionic liquid was used than in the regular temperature-ramped TGA experiments. High purity helium was used as the carrier gas. Electron Ionisation (EI) mass spectra were recorded at intervals throughout the experiment, in the range m/z 10–300. A 70 eV ionisation energy was employed. The main peaks of interest were of m/z 40 and above.

Fragmentation of small-molecule decomposition products is possible under the harsh conditions of electron ionisation mass spectrometry (EIMS). Therefore, the parent ion corresponding to a given thermal decomposition product may not always be observed, even if the decomposition product is present in the inert flow gas. Data from the 'National Institute of Standards and Technology' (NIST) Chemistry WebBook was used to identify expected fragmentation of small-molecule decomposition products.[30]

For each ionic liquid, a single mass spectrum was selected from the TGA-MS data, recorded at a time point during the main thermal decomposition period of the ionic liquid. Major peaks were then assigned as parent peaks, or fragments, of decomposition products. Example MS histograms for $[C_2C_1im][OAc]$, **3a**, $[C_2C_1im][CF_3CO_2]$, **4**, and $[C_2C_1im][SAc]$, **12**, are shown in Figure 7. Subsequently, the intensity of particular m/z peaks could be plotted as a function of temperature, and overlaid with the TGA thermograph. An example TGA-MS graph of this type for **3a** is shown in Figure 8.

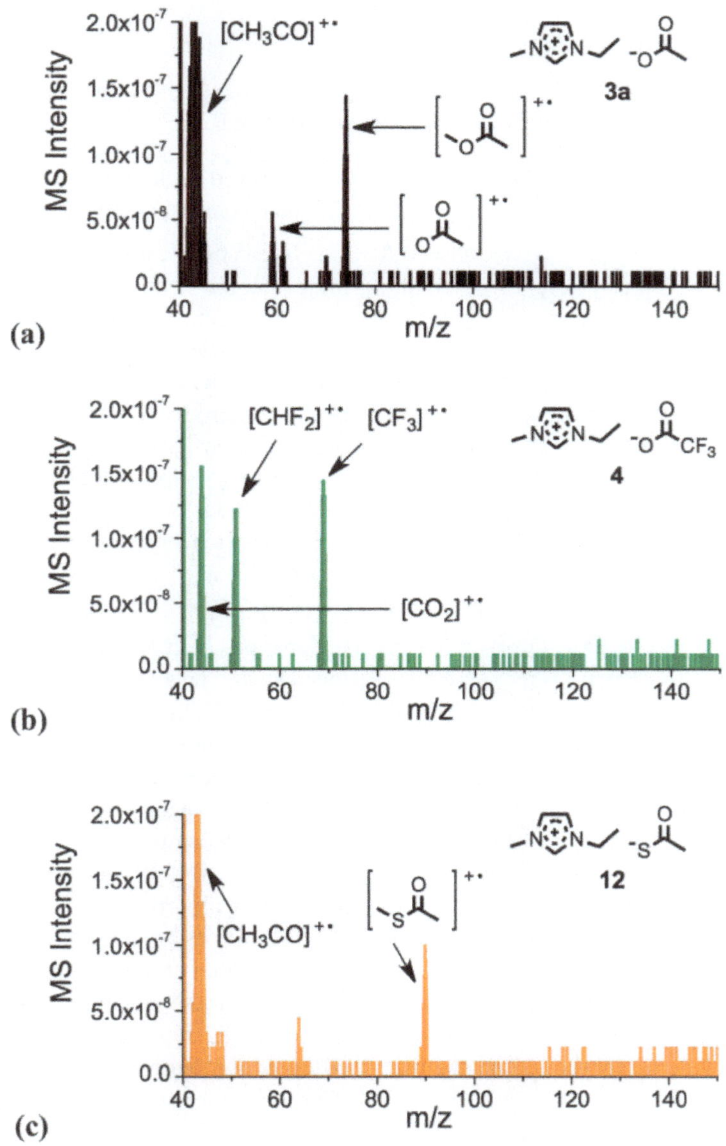

Figure 7: TGA-MS data for ionic liquids $[C_2C_1im][OAc]$, **3a**, (a), $[C_2C_1im][CF_3CO_2]$, **4**, (b), and $[C_2C_1im][SAc]$, **12**, (c). The displayed MS histograms were recorded during the main thermal decomposition period of the ionic liquid. Peaks were assigned to parents or fragments of expected decomposition products. Spectra are shown in the region 40–150 m/z, for clarity.

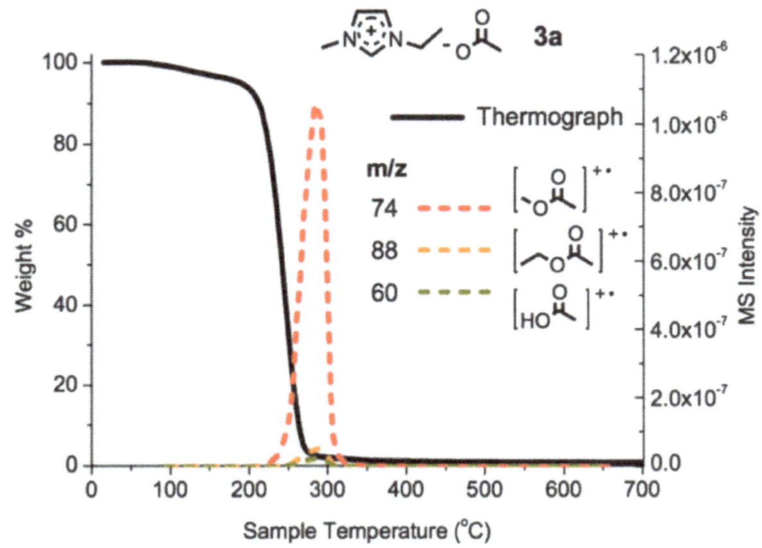

Figure 8: TGA-MS investigation of $[C_2C_1im][OAc]$, **3a**. The solid black line represents the TGA thermograph; colored dashed lines represent the MS intensities for selected m/z values; the assigned decomposition products are shown.

The thermal decomposition of $[C_2C_1im][NTf_2]$, **2**, has been previously studied in detail, and a decomposition mechanism was suggested.[17c] Chen and colleagues observed a peak at m/z64, which was assigned as $[SO_2]^{+\cdot}$. In addition, their experiments showed smaller peaks corresponding to a number of other degradants, including 1-methylimidazole, 1-ethylimidazole, ethene, CH_3NH_2 and CHF_3. They therefore proposed a mechanism for the decomposition of **2**, involving an anion-centered decomposition to generate reactive radical intermediates ($CF_3^{\cdot}$, $NH_2^{\cdot}$, $F^{\cdot}$), which then react with the $[C_2C_1im]^+$ cation. The results of our TGA-MS studies on compound **2** are consistent with their findings; peaks at m/z 64 and m/z 69 were seen during the main decomposition period, which were assigned as $[SO_2]^{+\cdot}$ and $[CF_3]^+$, respectively. Therefore, the presence of these peaks suggests degradation of the $[NTf_2]^-$ anion during the decomposition of ionic liquid **2**.

In the TGA-MS experiments for the archetypal carboxylate ionic liquid $[C_2C_1im][OAc]$, **3a**, the onset of thermal decomposition is accompanied by the detection of a large peak at m/z 74 in the mass spectrometer, which is assigned to the parent ion of methyl acetate, $[CH_3CO_2CH_3]^{+\cdot}$ (Figure 7a and 8). This strongly suggests an S_N2 mechanism occurring at the methyl substituent of the $[C_2C_1im]^+$ cation, because it is unlikely that any other mechanism could produce substantial amounts of this degradant (Figure 6). The peak at m/z 59 was assigned to $[CH_3CO_2]^+$, a known fragment of methyl acetate, but not of acetic acid or ethyl acetate in EI mass spectrometry.[30] This is further evidence for the S_N2 mechanism occurring at the methyl substituent.

A small peak corresponding to ethyl acetate, $[CH_3CO_2CH_2CH_3]^{+\cdot}$ (m/z 88), was also observed, indicative of nucleophilic attack at the more sterically crowded ethyl substituent (Figure 8). These mechanisms are consistent with the nucleophilic character of the acetate anion. An acetic acid parent peak (m/z 60) is detected as a minor decomposition product (Figure 8). This suggests that either an E2 Hofmann elimination mechanism or NHC formation is occurring to a small extent. No fragment at m/z 60 is expected in the EI mass spectrum of either methyl acetate or ethyl acetate,[30] therefore our observed peak at m/z 60 cannot be regarded as a fragment of either ester decomposition product.

Interestingly, no significant peaks were observed for the molecular ion of 1-methylimidazole (m/z 82), 1-ethylimidazole (m/z 96), or for the N-heterocyclic carbene 1-ethyl-3-methylimidazol-2-ylidene (m/z 110) that would form in the event of NHC decomposition (mechanism A, Figure 6). The neutral heterocycle 1-ethylimidazole should be present as a product of the S_N2 thermal decomposition mechanism occurring at the methyl substituent of the 1-ethyl-3-methylimidazolium cation of **3a**. Attempts to confirm 1-ethylimidazole as a decomposition product of **3a**, by high-temperature ^1H NMR experiments, are described below.

The low T_{onset} value for $[C_2C_1im][CF_3CO_2]$, **4**, can be understood in terms of an anion-centered decarboxylation mechanism, enabled by the stability of the CF_3^- anion or $CF_3^{\cdot}$ radical that would be formed. The TGA-MS data for this compound supports this hypothesis, as indicated by the presence of peaks at m/z 44, $[CO_2]^{+\cdot}$, m/z 69, $[CF_3]^+$, and m/z 51, $[CHF_2]^+$. The peak atm/z 69 could be attributed to multiple plausible decomposition products of $[C_2C_1im][CF_3CO_2]$, **4**, including CHF_3, trifluoroacetic acid, methyl trifluoroacetate and ethyl trifluoroacetate.[30]However, the fragment at m/z 51, $[CHF_2]^+$, strongly suggests the presence of CHF_3 as a major decomposition product of ionic liquid **4**, since it is not a known fragment of the potential acid or ester products generated by S_N2 or E2 mechanisms.[30] Although the peak at m/z 44 is observed throughout the duration of the TGA-MS experiment (CO_2 is a minor contaminant of the helium carrier gas), there is a visible spike in intensity corresponding to the onset of mass loss. The TGA-MS data strongly suggests an anion-centered decarboxylation mechanism for the thermal decomposition of ionic liquid **4**.

The TGA-MS results for the octanoate ionic liquid, **8**, were compared with those for the acetate ionic liquid, **3a**, in order to understand the effect of alkyl chain extension on the prevailing mechanisms of thermal decomposition. For **8**, a peak was observed at m/z 158, which has been assigned as the parent peak of methyl octanoate, $[CH_3(CH_2)_6CO_2CH_3]^{+\cdot}$.

This suggests an S_N2 nucleophilic attack of the octanoate anion at the methyl substituent of the $[C_2C_1im]^+$ cation (mechanism B, Figure 6). Lower intensity peaks were observed that can be assigned to fragments of octanoic acid and ethyl octanoate. Unlike the TGA-MS experiment for **3a**, the parent peaks of 1-methylimidazole (m/z 82) and 1-ethylimidazole (m/z 96) were observed in the spectrum of $[C_2C_1im][CH_3(CH_2)_6CO_2]$, **8**. The presence of these two neutral imidazole products is further evidence for S_N2-type decomposition mechanisms.

For the ionic liquid $[C_2C_1im][SAc]$, **12**, a large peak at m/z 90 was observed, corresponding to the parent ion of methyl thioacetate, $[CH_3COSCH_3]^{+\cdot}$. This is strong evidence of an S_N2 mechanism occurring at the methyl substituent of the $[C_2C_1im]^+$ cation, analogous to the decomposition of **3a**. This is unsurprising, given the known high nucleophilicity of the thioacetate anion.

For the C^2-methyl substituted ionic liquid, $[C_2C_1C_1im]$[OAc], **14**, the NHC decomposition route (mechanism A, Figure 6) is not possible. Instead, a proton may be abstracted from the C^2methyl group, generating acetic acid and a neutral species, 1-ethyl-3-methyl-2-methylene-2,3-dihydroimidazole (m/z 124). However, no detectable peak was observed at m/z 124 during the decomposition of ionic liquid **14**, suggesting that this mechanism does not occur to any measurable extent. Instead, the major observed decomposition product in the TGA-MS experiment for ionic liquid **14** was methyl acetate, $[CH_3CO_2CH_3]^{+\cdot}$, at m/z 74. Therefore, it is likely that the decomposition of both acetate ionic liquids, **3a** and **14**, occurs predominantly via an S_N2 nucleophilic attack at the imidazolium methyl substituent.

The TGA-MS results therefore indicate the occurrence of S_N2 nucleophilic substitution decomposition mechanisms (**3a**, **8**, **12** and **14**), or anion-centered decomposition, generating reactive intermediates which subsequently attack the $[C_2C_1im]^+$ cation (**2** and **4**). For each of the ionic liquids investigated in this TGA-MS study, there was no substantive evidence of the anion behaving as a base.

The decomposition of $[C_2C_1im]$[OAc], **3a**, was explained in terms of an S_N2 nucleophilic substitution mechanism at the methyl substituent of the $[C_2C_1im]^+$ cation, on the basis of the large observed peak for methyl acetate in the TGA-MS experiment. However, the other major expected decomposition product, 1-ethylimidazole, was not observed.

In an attempt to monitor the formation of 1-ethylimidazole and other decomposition products, a neat sample of **3a** was heated to 120 °C for 12 hours in a sealed, thick-walled NMR tube. A ^1H NMR spectrum was recorded every 30 minutes. The high-temperature ^1H NMR experiment was performed unlocked, in the absence of any NMR solvent. Therefore, to determine the chemical shifts of the ionic liquid and decomposition product peaks, two separate control NMR spectra were obtained, whereby pure $[C_2C_1im]$[OAc] had been deliberately spiked with a small quantity of an expected decomposition product, 1-methylimidazole and 1-ethylimidazole respectively. A DMSO-d_6 capillary was used as a reference in each reference sample.

During the 12-hour experiment, no additional peaks were observed. It is possible that the broadened ionic liquid peaks engulfed the decomposition product peaks, or that the decomposition products were partially in the gas phase. However, upon cooling **3a** to room temperature, new peaks not present in the ionic liquid prior to the high-temperature experiment were observed in the ^1H NMR spectrum (Figure 9). Several peaks in the region 7.4–6.1 ppm were observed at approximately 1–2% of the integration of the ionic liquid aromatic peaks. The chemical shifts of these aromatic peaks matched the neutral heterocyclic molecules 1-methylimidazole and 1-ethylimidazole (Figure 9a). In addition, a triplet peak was observed at 0.69 ppm, which has been assigned to the terminal $-CH_3$ unit of the ethyl chain of 1-ethylimidazole (Figure 9b). A considerably smaller peak was observed for the aliphatic methyl group of 1-methylimidazole. This lends evidence towards an S_N2 nucleophilic substitution reaction occurring at both alkyl substituents of the 1-ethyl-3-methylimidazolium cation, but predominantly at the less hindered methyl substituent.

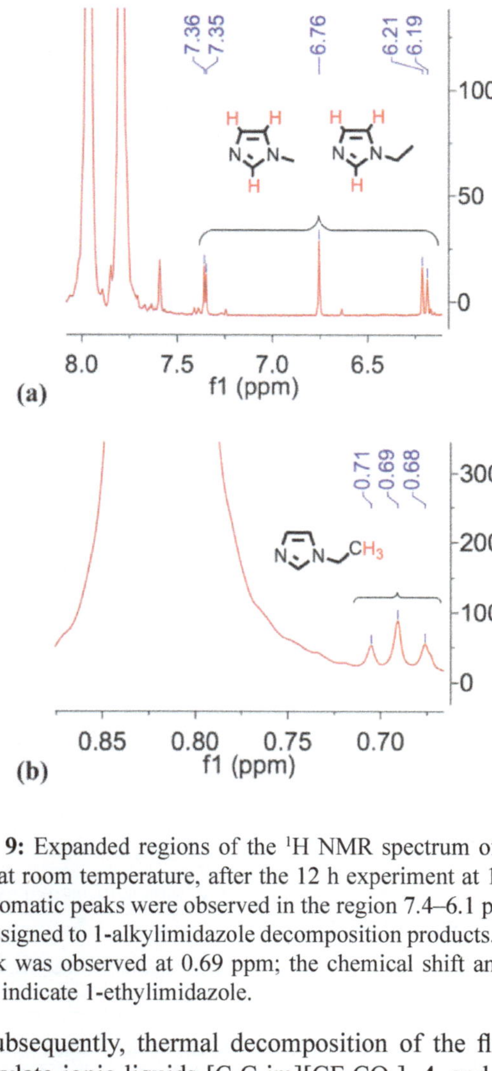

Figure 9: Expanded regions of the ^1H NMR spectrum of $[C_2C_1im]$[OAc] at room temperature, after the 12 h experiment at 120 °C. (a) New aromatic peaks were observed in the region 7.4–6.1 ppm, which were assigned to 1-alkylimidazole decomposition products. (b) A triplet peak was observed at 0.69 ppm; the chemical shift and splitting pattern indicate 1-ethylimidazole.

Subsequently, thermal decomposition of the fluoroalkyl carboxylate ionic liquids $[C_2C_1im][CF_3CO_2]$, **4**, and $[C_2C_1im][CHF_2CO_2]$, **5**, was investigated using ^1H and ^{19}F NMR. Small samples (ca. 20 mg) of **4** and **5** were placed in 25 ml sealed glass pressure vessels, and maintained at 100 °C for 18 hours, before being allowed to cool to room temperature. The residues were dissolved in DMSO-d_6, and ^1H and ^{19}F NMR spectra were recorded.

The ^{19}F NMR experiments did not reveal any fluorine-containing thermal decomposition products for either ionic liquid **4** or **5**. Two expected decomposition products from the anion decarboxylation of **4** are CHF_3 and CO_2, which are both gaseous products (Figure 6). Therefore, if formed, these thermal decomposition products would have escaped when the pressure tubes were opened. The high T_{onset} value for $[C_2C_1im][CHF_2CO_2]$, **5**, may explain why no fluorine-containing products were observed for this ionic liquid after the 18 hour 100 °C heating period.

By contrast, small peaks were observed in the ^1H NMR spectrum for **4** that were not present in the pure ionic liquid before commencing the experiment. These peaks, at 7.18, 6.87 and 3.64 ppm, were subsequently assigned to the C^4, C^5 and $-CH_3$ protons on the neutral heterocycle 1-methylimidazole, by comparing the spectrum of **4** after the heating period against a reference ^1H NMR spectrum of 1-methylimidazole in DMSO-d_6. These ^1H and ^{19}F NMR results are compatible with the previous observation that $[C_2C_1im][CF_3CO_2]$, **4**, decomposes viaan anion-centered decarboxylation (Figure 7b). Therefore, it is likely that the CF_3^- anion, generated by the decarboxylation mechanism, then removes a β-proton from the ethyl substituent on the $[C_2C_1im]^+$ cation in an E2-type mechanism, generating the observed 1-methylimidazole.

Smaller peaks were observed at the same chemical shifts in the ^1H NMR of **5**, following the 18 hour heating period at 100 °C, indicating that a very small quantity of 1-methylimidazole had also been generated from the thermal decomposition of this ionic liquid. 1-Methylimidazole was not employed at any stage of the synthesis of compounds **4** and **5**, therefore it cannot be regarded as an impurity.

The thermal decomposition of ionic liquids **3b**, **4**, **11** and **14** was subsequently investigated using Differential Scanning Calorimetry (DSC), and $\Delta H_{decomposition}$ values were measured. In order to correlate the DSC peaks to specific decomposition processes observed in the TGA, a temperature-ramped TGA experiment was performed for each of the four ionic liquids at the same slow heating rate of 2.5 °C min^{-1}. These TGA thermographs were subsequently superimposed on the DSC traces. An example graph for $[C_2C_1im][OAc]$, **3b**, is displayed in Figure 10. $\Delta H_{decomposition}$ data for the four ionic liquids is shown in Table 2.

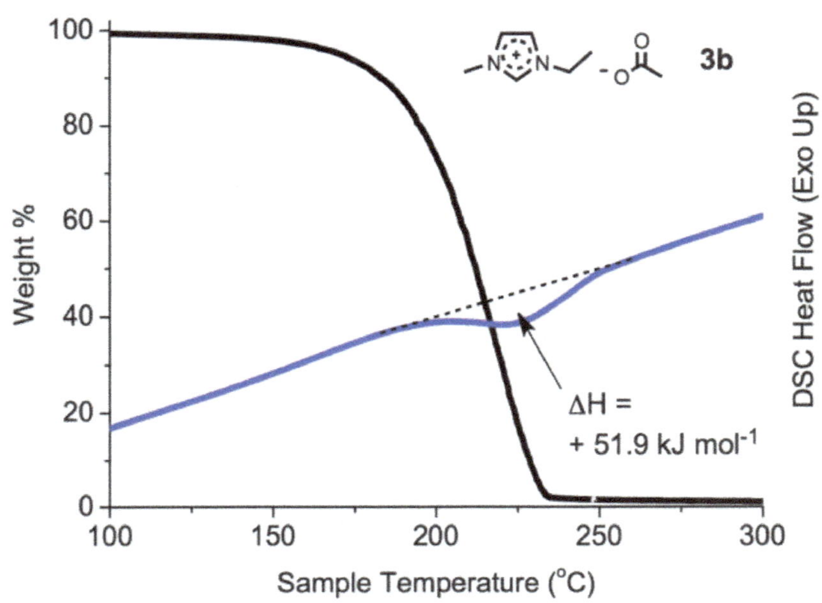

Figure 10" DSC trace for the decomposition of $[C_2C_1im][OAc]$, **3b**, shown in the range 100–300 °C (blue). A temperature-ramped TGA thermograph is superimposed (black).

Table 2: $\Delta H_{decomposition}$ values for four ionic liquids, measured using Differential Scanning Calorimetry (DSC). Each ionic liquid exhibited broad endothermic decomposition

Ionic liquid		$\Delta H_{decomposition}$ (kJ mol^{-1})
3b	$[C_2C_1im][OAc]$	51.9
4	$[C_2C_1im][CF_3CO_2]$	15.7, 10.5
11	$[C_2C_1im][c\text{-}(C_3H_5)CO_2]$	64.7
14	$[C_2C_1C_1im][OAc]$	18.0

The DSC profile for the first heating step of **3b** displayed one broad endothermic peak, between approximately 200–250 °C. This endothermic peak occurs in the same temperature range as the TGA weight loss, and has been assigned as decomposition of the ionic liquid, with $\Delta H_{decomposition}$ = +51.9 kJ mol^{-1} (Figure 10).

The DSC profile representing the decomposition of $[C_2C_1im][CF_3CO_2]$, **4**, demonstrated a two-step process, corresponding to the two visible decomposition stages in the temperature-ramped TGA thermograph. Both steps were endothermic in nature and of similar magnitude, +15.7 and +10.5 kJ mol^{-1}, respectively.

Ionic liquid **11**, containing the strained cyclopropane ring, exhibits endothermic decomposition in a similar fashion to the acetate ionic liquid **3b**. The measured $\Delta H_{decomposition}$ value of +64.7 kJ mol^{-1} for **11** is comparable to the value for **3b**.

No additional peaks were observed in the DSC decomposition profile for **11**, suggesting that opening of the cyclopropane ring does not occur. A similar shallow, broad endothermic decomposition DSC trace was obtained for the C^2-methyl ionic liquid, $[C_2C_1C_1im][OAc]$, **14**, and $\Delta H_{decomposition}$ was measured as $+18.0\ kJ\ mol^{-1}$.

The four ionic liquids investigated by DSC (**3a**, **4**, **11** and **14**) appeared to thermally decompose entirely to volatile products; the aluminium DSC pans were opened after the experiments, and there was no observable decomposition residue. Accordingly, the second heating scan of these DSC experiments displayed no noticeable peaks. However, one very small endothermic peak was observed in the cooling stage for $[C_2C_1im][CF_3CO_2]$, **4**, at 247 °C. This suggests that a small quantity of decomposition residue was present after the first heating stage for this ionic liquid.

In summary, the DSC experiments indicated that for the four investigated imidazolium ionic liquids incorporating alkyl (**3a**, **11**, **14**) and fluoroalkyl (**4**) carboxylate anions, thermal decomposition is an endothermic event with positive values of $\Delta H_{decomposition}$. In addition, there was an absence of any decomposition residue for the three non-fluorinated ionic liquids investigated, although a small amount of residue may have been formed from the decomposition of $[C_2C_1im][CF_3CO_2]$, **4**.

COMPUTATIONAL RESULTS

The liquid structure of 1-ethyl-3-methylimidazolium acetate, $[C_2C_1im][OAc]$, and the interaction of this ionic liquid with water and CO_2, have been previously explored using experimental and computational techniques.[31]

In this investigation, DFT was used in order to understand the ion pair structuring and, subsequently, the decomposition mechanisms of the two acetate ionic liquids, $[C_2C_1im][OAc]$ and $[C_2C_1C_1im][OAc]$. All calculations were performed at the B3LYP-D/6-311++G(d,p) level of theory.

Figure 11 and 12 represent the low energy ion pair conformers for $[C_2C_1im][OAc]$ and $[C_2C_1C_1im][OAc]$ obtained in this study, and the naming convention of the ion pairs used in the following discussion. Tables 3 and 4 summarise the ΔG and ΔE (BSSE and ZPE corrected) energies of these ion pair conformers. In addition, the NBO charges for the lowest energy ion pair of $[C_2C_1im][OAc]$ and $[C_2C_1C_1im][OAc]$ are listed.

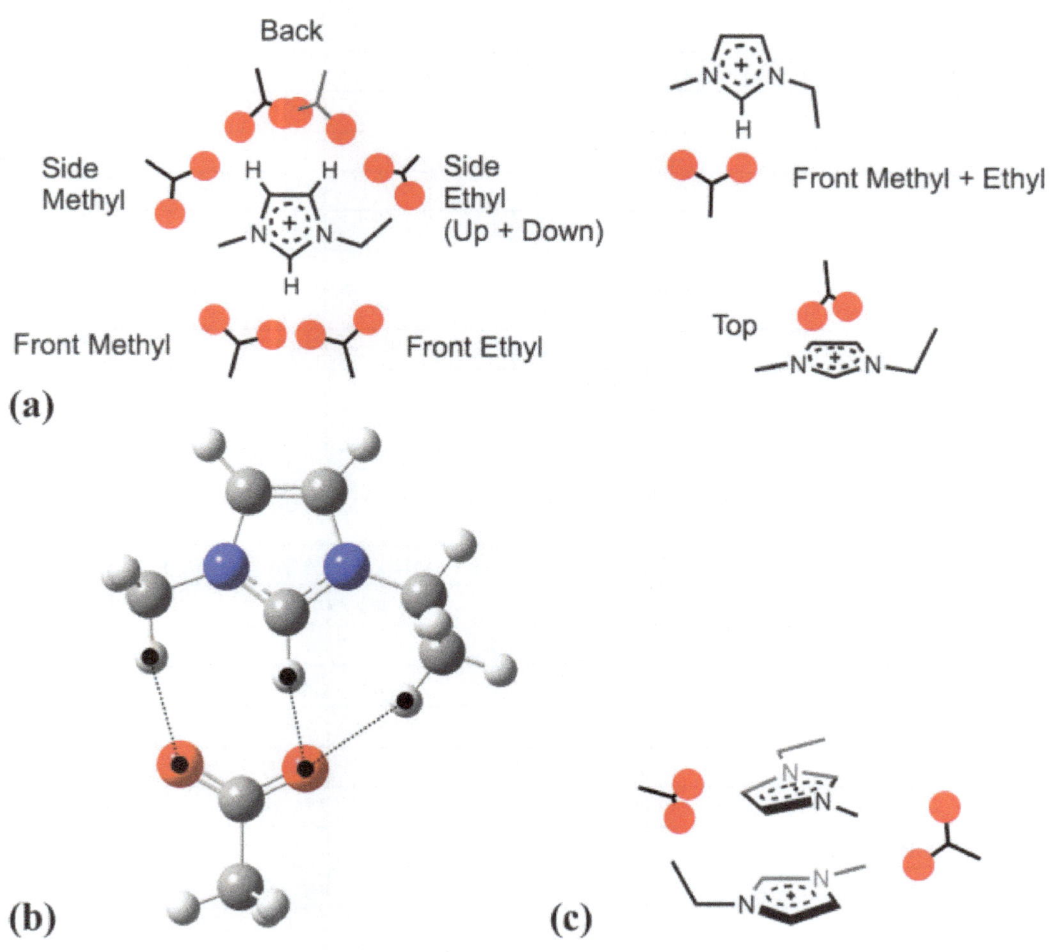

Figure 11: (a) Diagram representing the location of primary cation–anion interaction for $[C_2C_1im][OAc]$. (b) Structure of the lowest energy 'Front Methyl + Ethyl' ion pair, showing primary H-bond interactions. (c) Diagram representing a dimer conformer of $[C_2C_1im][OAc]$.

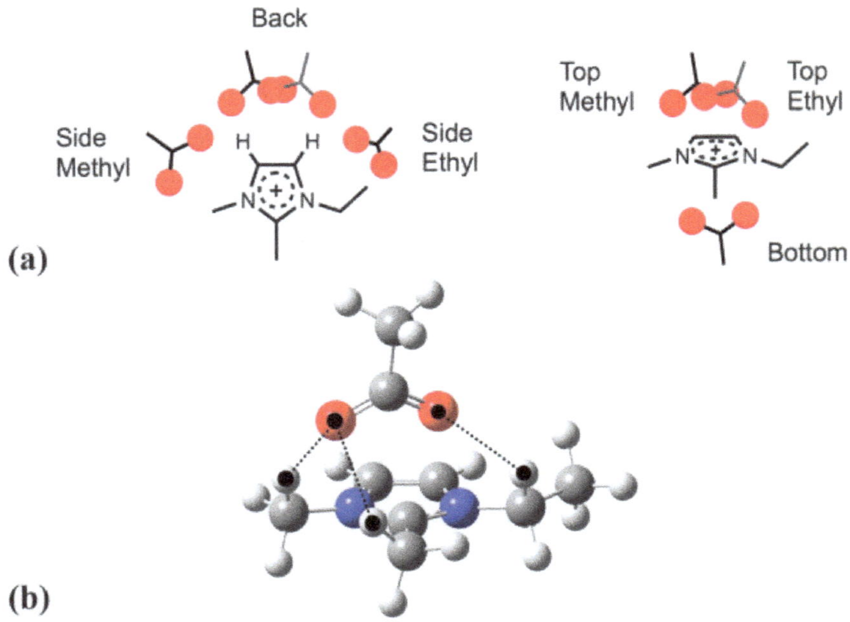

Figure 12: (a) Diagram representing the location of primary cation–anion interaction for $[C_2C_1C_1im][OAc]$. (b) Structure of the lowest energy 'Top Methyl 1' ion pair, showing primary H-bond interactions.

Table 3: Ion pair energies for $[C_2C_1im][OAc]$. ΔE values are zero-point and BSSE corrected

Ion pair $[C_2C_1im][OAc]$	ΔE (kJ mol^{-1})	ΔG (kJ mol^{-1})
From methyl + ethyl	0.0	0.0
Front ethyl	4.9	3.2
Front methyl	4.9	3.5
Top	9.4	11.0
Side ethyl (down)	41.4	42.0
Side methyl	46.0	43.8
Side ethyl (up)	44.9	46.9
Back methyl	59.6	57.2
Back ethyl	55.7	57.3

Table 4: Ion pair energies for $[C_2C_1C_1im][OAc]$. ΔE values are zero-point and BSSE corrected

Ion pair $[C_2C_1C_1im][OAc]$	ΔE (kJ mol^{-1})	ΔG (kJ mol^{-1})
Top methyl 1	0.0	0.0
Top methyl 2	−1.8	0.3
Bottom	−0.6	1.0
Top ethyl	−1.0	5.8
Side ethyl	29.9	24.9
Side methyl	33.2	34.1
Back ethyl	46.4	41.9
Back methyl	48.7	43.1

The results suggest that the acetate anion is preferentially located in the plane of the imidazolium ring, in agreement with previous studies on this ionic liquid.[31a,d] Anion sites for our ion pair conformers of $[C_2C_1im][OAc]$ resemble those previously determined for the prototypical ionic liquid 1-butyl-3-methylimidazolium chloride, $[C_4C_1im]Cl$.[32] Whereas the chloride anion is monoatomic, acetate can bind strongly through the two oxygen atoms of the carboxylate group. As a result, the low energy 'Front' conformers are more centralised, with the acetate binding both to the C^2 proton and α-/β-protons on the methyl and ethyl side chains. In the front and back ion pair conformers, both oxygen atoms of the acetate anion remain essentially in the plane of the imidazolium ring, whereas in the top and side conformers the acetate lies out of the plane of the ring. The two distinct 'Side Ethyl' conformers are denoted 'up' (acetate and ethyl chain both lie above the plane of the imidazolium ring) and 'down' (acetate and ethyl chain on opposite sides of the plane of the ring).

The lowest energy conformer (Figure 11) located in this investigation is the 'Front Methyl + Ethyl' structure, similar to the structure reported by Hollóczki and co-workers.[28d] In this conformer, the ethyl chain rotates forward in order to bind to the acetate anion through a β-proton. Two other 'Front' conformers, 'Front Methyl' and 'Front Ethyl' (Table 3), are approximately 3.5 kJ mol^{-1} higher in energy, and lack this interaction with the ethyl β-proton. A 'Top' ion pair structure was obtained, whereby the acetate anion lies above the plane of the imidazolium ring, with an energy approximately 10 kJ mol^{-1} greater than the 'Front Methyl + Ethyl' structure. Attempts to optimise the top conformer at the B3LYP/6-311++G(d,p) level of theory were unsuccessful, indicating that the inclusion of dispersion is necessary for locating this ion pair.

Bowron and co-workers observed a high probability of the acetate anion lying in the plane of the imidazolium ring for [C_2C_1im][OAc], which resulted in a decrease in anion density above and below the ring.[31d] As a result, cation–cation π-stacking was possible. Interestingly, in our investigation no front ion pair conformers were located where the acetate lies perpendicular to the plane of the imidazolium ring.

Conformational analysis of the ethyl side chain was performed for the low energy 'Front Methyl + Ethyl' and 'Top' ion pair conformers of [C_2C_1im][OAc]. In each e, the C^2–N–C_α–C_β dihedral angle was rotated in increments of 30°, and all other coordinates were optimised. The change in energy and structure of the ion pair was recorded as a function of the dihedral angle. Graphs representing the conformational analysis around the 'Front Methyl + Ethyl' and 'Top' conformers are displayed. The structure of the front ion pair was largely unchanged upon rotation of the ethyl chain, with the acetate anion remaining bound to the C^2 proton and α-proton of the methyl side chain. Small rotational energy barriers of <10 kJ mol^{-1} were observed for ethyl rotation of the front conformer. For the top ion pair conformer, rotation of the ethyl chain carried the acetate anion towards the front of the imidazolium ring, giving structures analogous to the 'Front Methyl' and 'Front Methyl + Ethyl' minima.

In a recent molecular dynamics study, Brehm and co-workers investigated the preferred location of the acetate anion, relative to the [C_2C_1im]$^+$ cation, by examining the angle between the ring normal vector and the connecting line from the center of the ring to the position of the acetate carboxylate carbon atom.[31i] Their results indicated that in the neat ionic liquid, the acetate is largely associated with the plane of the ring (angle approximately 90°). However, their results also showed some probability of the acetate being located out of the ring plane by up to approximately 30°. To this end, we have investigated a sample of dimer structures for [C_2C_1im][OAc] (stable geometries involving two cations and two anions). Structures were optimized as for the ion pairs, at the B3LYP-D/6-311++G(d,p) level of theory. Although our calculations have not yet yielded an exhaustive list of dimer structures, the results do indicate the possibility of the acetate anions being located out of the plane of the imidazolium rings. An example optimized dimer structure is displayed in Figure 11c.

For the ionic liquid 1-ethyl-2,3-dimethylimidazolium acetate, [$C_2C_1C_1$im][OAc], the C^2 proton has been replaced by a methyl substituent, and thus there are no equivalent stable 'Front' ion pair conformers. Instead, the low energy conformers are 'Top' and 'Bottom' structures, whereby the acetate anion sits above or below the plane of the imidazolium ring. Four distinct 'Top' and 'Bottom' structures were located, with relative energies within a 6 kJ mol^{-1} range. The lowest energy ion pair structure obtained, shown in Figure 12b, involves the H-bond interaction of the acetate anion with protons on the C^2 methyl substituent and α-protons on both the methyl and ethyl side chains. The ethyl side chain is rotated almost into the plane of the ring to facilitate this interaction.

Plausible thermal decomposition mechanisms of [C_2C_1im][OAc] and [$C_2C_1C_1$im][OAc] have been explored using DFT. The decomposition routes under investigation are displayed in Figure 13. Mechanisms A–D are analogous to those shown for the TGA-MS results (Figure 6). For each of the two ionic liquids, possible decomposition mechanisms include the S_N2 attack of the acetate anion at either the N-methyl (mechanisms B1 and B2) or N-ethyl (mechanisms C1 and C2) substituents on the imidazolium cation, or the E2 Hofmann elimination mechanism (mechanisms D1 and D2) occurring at the ethyl substituent. The decomposition mechanisms at the C^2 position (mechanisms A1 and A2) are different for the two acetate ionic liquids; for [C_2C_1im][OAc], the formation of an NHC is possible, whereby the C^2 proton is removed by the basic acetate anion (mechanism A1). For [$C_2C_1C_1$im][OAc], the carbene-forming mechanism is not possible. Instead the basic acetate anion may remove a proton from the C^2 methyl substituent, generating acetic acid and the neutral species, 1-ethyl-3-methyl-2-methylene-2,3-dihydroimidazole (mechanism A2).

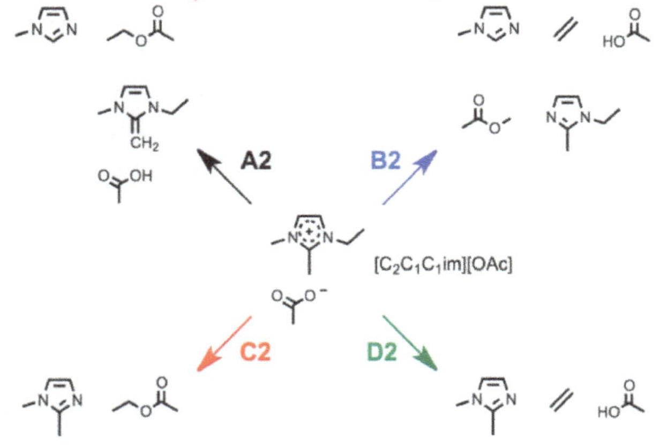

Figure 13: Decomposition routes investigated for the two acetate ionic liquids in this computational study: [C_2C_1im][OAc], A1–D1, and [$C_2C_1C_1$im][OAc], A2–D2.

Decomposition reaction profiles have been studied starting from the lowest energy ion pair of $[C_2C_1im][OAc]$ and $[C_2C_1C_1im][OAc]$, ('Top Methyl + Ethyl' and 'Top Methyl 1', respectively). The respective S_N2 and E2 decomposition mechanisms (B1–D1, B2–D2) are analogous for the two acetate ionic liquids, and are therefore discussed first. The transition state structures for mechanisms B1, C1 and D1 are shown in Figure 14, and the relevant ΔE, ΔG, ΔH and TΔS energies are displayed in Table 5. The decomposition mechanisms at the C^2 position (A1 and A2) are less trivial, and are discussed separately shortly.

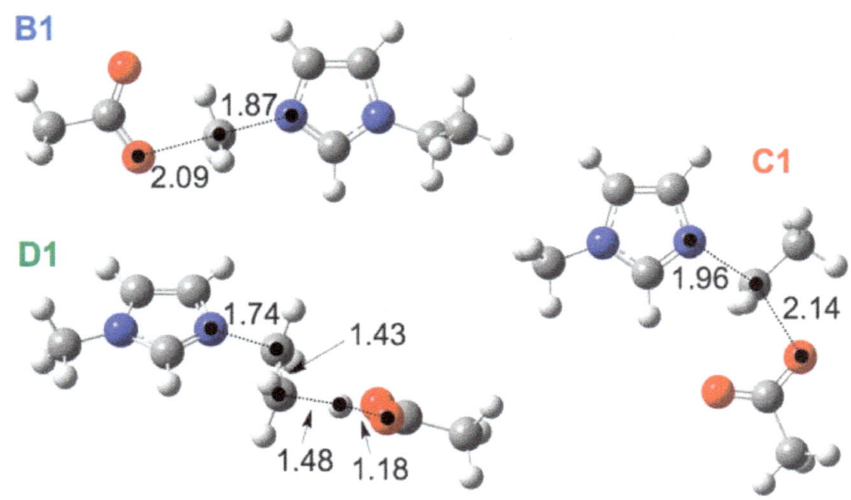

Figure 14: Transition states for $[C_2C_1im][OAc]$, representing the S_N2 nucleophilic substitution (B1 and C1) and E2 Hofmann elimination (D1) mechanisms, atomic distances are in angstroms.

Table 5: Transition state and product energies for the S_N2 and E2 mechanisms of $[C_2C_1im][OAc]$ and $[C_2C_1C_1im][OAc]$, relative to the lowest energy ion pair conformer of each ionic liquid. ΔE values are zero-point corrected

Mechanism	Transition state (kJ mol^{-1})				Product (kJ mol^{-1})			
	ΔE	*ΔG*	*ΔH*	*TΔS*	*ΔE*	*ΔG*	*ΔH*	*TΔS*
Bl	139.1	134.6	139.5	4.9	4.0	−35.3	0.6	35.9
C1	135.6	135.4	136.2	0.8	0.2	−39.6	−3.0	36.7
D1	156.4	156.1	156.8	0.7	61.3	−25.8	63.3	89.1
B2	119.8	118.5	119.1	0.5	2.3	−41.1	−1.0	40.1
C2	116.7	119.3	114.8	−4.6	−0.1	−41.7	−5.2	36.5
D2	138.6	137.7	138.6	0.9	61.0	−27.8	61.1	88.9

For both acetate ionic liquids, the $\Delta H_{activation}$ energies for the two S_N2 nucleophilic substitution mechanisms are approximately 20 kJ mol^{-1} lower than for the E2 Hofmann elimination mechanism (Table 5). In addition, the calculated $\Delta H_{activation}$ energies for the S_N2 mechanisms are 15–24 kJ mol^{-1} higher than the experimentally determined E_a values from the isothermal TGA experiments (E_a = 115.8 kJ mol^{-1} for $[C_2C_1im][OAc]$, 99.6 kJ mol^{-1} for $[C_2C_1C_1im][OAc]$). The TGA-MS data (Figure 7 and 8) suggests that S_N2 attack occurs predominantly at the methyl substituent of the cation, however the calculations find little difference in the activation barriers for attack at the N-methyl or N-ethyl substituents. These results indicate that there is no inherent large energetic difference between these reactions and that steric effects are most likely responsible for the more facile reactivity at the N-methyl substituent.

The DSC investigation of $[C_2C_1im][OAc]$ and $[C_2C_1C_1im][OAc]$ indicated that thermal decomposition was an endothermic event ($\Delta H_{decomposition}$ = +51.7 and +18.0 kJ mol^{-1}, respectively, Figure 10). By contrast, our gas phase calculations suggested that the $\Delta H_{product\text{-}reactant}$ values for the S_N2 mechanisms of these two ionic liquids are close to zero. However, the DSC experiments involve both the thermal decomposition of the ionic liquid and an associated phase change of the decomposition products from the ionic liquid to gaseous molecules. Therefore the difference in $\Delta H_{product\text{-}reactant}$ values from the gas phase calculations and DSC experiments give an indication as to the magnitude of $\Delta_{vap}H$ associated with the vaporization of neutral ester and imidazole products from the S_N2 thermal decomposition mechanisms. It is unclear whether thermal decomposition occurs from within the liquid phase, followed by subsequent vaporization of the neutral species, or whether decomposition is occurring primarily at the surface of the ionic liquid.

The potential for N-heterocyclic carbene (NHC) formation in dialkylimidazolium ionic liquids is now widely recognised.[28] This phenomenon has been exploited in catalytic applications.[33] In the e of [C_2C_1im][OAc], removal of the C^2 proton by the basic acetate anion would generate acetic acid and the NHC species, 1-ethyl-3-methylimidazol-2-ylidene (mechanism A1, Figure 13). A stable minimum exists involving the interaction of acetic acid and the NHC, which is referred to here as the 'Carbene-Acid Pair' (CAP). Hollóczki and colleagues performed DFT calculations on the ion pair and CAP structure of [C_2C_1im][OAc] using a range of functionals and basis sets, and found the energy difference to be less than 20 kJ mol^{-1} in each e.[28d] Their calculations indicated that the C^2 proton is very labile in the gas phase. Moreover, mass spectrometry experiments indicated the presence of acetic acid (m/z60) and the isolated neutral carbene species (m/z 110) as decomposition products of [C_2C_1im][OAc] in the gas phase.

In our investigation, structures relating to the NHC decomposition route were optimized, and are shown in Figure 15. The C^2–O(acetate) distances, displayed in blue, are not exactly equal to the sum of the C^2–H and O–H distances because of the slightly non-linear bond angle. Bond distances are in angstroms.

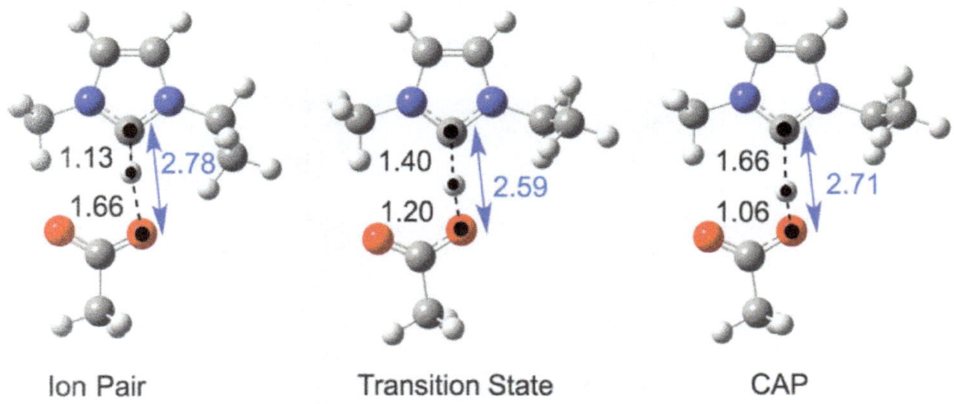

Figure 15: Gas phase 'Front Methyl + Ethyl' ion pair, transition state and carbene-acid pair (CAP) structures for mechanism A1.

The SCF energies showed the TS to be lower in energy than the CAP product; 0.0, 12.6 and 13.7 kJ mol^{-1} for the ion pair, TS and CAP respectively. When ZPE corrections were taken into account this effect was enhanced; 0.0, 1.7 and 11.5 kJ mol^{-1} for the ion pair, TS and CAP respectively. These energies suggest that the transition state is lower in energy than the CAP, which is implausible.

However, the optimized TS structure has a single strong negative frequency of −728 cm^{-1}. Moreover, the intrinsic reaction coordinate (IRC) protocol was employed, and demonstrated that this TS structure does link the ion pair and CAP.

In order to understand why the TS falls at a lower energy than the CAP product, the structures in Figure 15 were investigated further. The gas phase calculations (Figure 15) indicate a contraction of the C^2–O(acetate) distance in the TS (2.59 Å) relative to the ion pair (2.78 Å) and CAP (2.71 Å). Thus, the C^2-carbon–oxygen distance was fixed at a range of values between 2.4 and 2.8 angstroms (Å), and the hydrogen atom moved in increments of 0.05 Å, computing the potential energy surface (PES) shown in Figure 16.

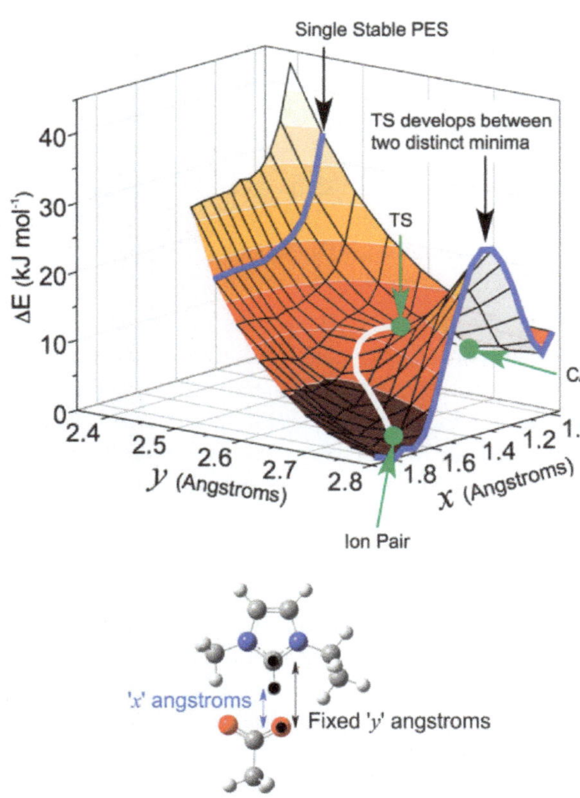

Figure 16: 3D energy surface for the proton transfer between the imidazolium C^2 carbon atom and the acetate oxygen atom of [C_2C_1im][OAc] (mechanism A1). The white line represents the lowest energy pathway from the ion pair through the transition state to the CAP (critical points shown as green circles).

This PES (Figure 16) reveals that at short C^2–O distances, (y) in the range 2.4–2.6 Å, a very broad flat single energy well exists spanning all of the ion-pair, TS and CAP structures (rear blue line in Figure 16). As the C²–O distance is increased towards 2.8 Å, the ion pair is stabilised and a barrier develops between the ion-pair and CAP. The barrier increases as C²–O increases in the range 2.6–2.8 Å.

A scan performed with a fixed C²–O distance, 'y', of 2.8 Å shows the presence of a true TS (26.5 kJ mol^{-1}), front blue line in Figure 16. The CAP is approximately 9 kJ mol^{-1} higher in energy than the ion pair. Thus, a more reasonable reaction profile has been obtained.

The white line on Figure 16 shows the likely reaction path, starting from the ion pair minimum through the lowest energy TS (which occurs in the region of the PES where the barrier is just developing) and finishing in the CAP product. Thus, the TS essentially 'vanishes' before a CAP minimum has fully developed.

The energy, E_v, of the transition state vibrational mode (−728 cm^{-1} or 8.7 kJ mol^{-1}) is comparable to the energy differences between the ion pair, transition state and CAP (<14 kJ mol^{-1}). To compare the energies of these structures on a more equal footing the vibrational energy associated with the reaction coordinate has been removed from the zero point corrected energies for the reactant and product. (Because the TS vibration is imaginary it has already been excluded from the zero point corrected energy of the TS.) This corresponds to the C–H vibration in the ion pair (2507 cm^{-1} or 30.0 kJ mol^{-1}) and the O–H vibration in the CAP (2139 cm^{-1} or 25.6 kJ mol^{-1}). The resulting effective energies (ΔE_{eff}) are 0.0 kJ mol^{-1} for the ion-pair, 31.7 kJ mol^{-1} for the TS and 15.9 kJ mol^{-1} for the CAP. The relative magnitudes of these effective energies provide a more reasonable reaction profile for mechanism A1, with a transition state of higher energy than the ion pair and CAP.

Altogether, these results show that a TS exists at the C²–O distances of the ion-pair and CAP (~2.76 Å), but that the two ions must contract significantly (reduce C²–O distance, 'y') to follow the lowest energy path for the reaction.

Despite the known labile nature of the C² proton of $[C_2C_1im][OAc]$ in the gas phase, no strong evidence for an NHC decomposition pathway under inert conditions was observed in our experimental investigation. Acetic acid and compounds relating to the NHC species were observed as only very minor peaks in the TGA-MS experiments; instead, the major observed decomposition products were indicative of S_N2 nucleophilic substitution mechanisms.

To further explore NHC formation in the liquid phase, mechanism A1 was investigated using DFT and a Conductor-type Polarisable Continuum Model (CPCM).[34] The CPCM calculations approximate the liquid phase environment and provide a rudimentary approximation of the bulk polar liquid and hence stabilization due to a polar environment. Solvents benzene (C_6H_6, $\varepsilon_r = 2.28$), dichloromethane (CH_2Cl_2, $\varepsilon_r = 8.93$) and methanol (CH_3OH, $\varepsilon_r = 33.0$) were used to demonstrate the effects of a medium increasingly capable of stabilizing an ionic solute.[21,34] The values of the dielectric constant for the molecular solvents CH_3OH and CH_2Cl_2 cover the range anticipated to be typical for ionic liquids.[35]

No stable structures were obtained using the CPCM environment for the TS or CAP structures. This is consistent with recent ab initio MD simulations of the ionic liquid $[C_2C_1im][OAc]$.[28h] Subsequently, scans were performed with a fixed C²–O distance, 'y', of 2.76 Å in the gas phase and using CPCM conditions, Figure 17. The fixed coordinate prevents contraction of the 'y' distance, and therefore these scans do not perfectly represent the proton transfer mechanism. However, frequency analysis of the maximum structure in the gas phase scan (Figure 17) revealed the expected (single imaginary) vibrational mode corresponding to mechanism A1.

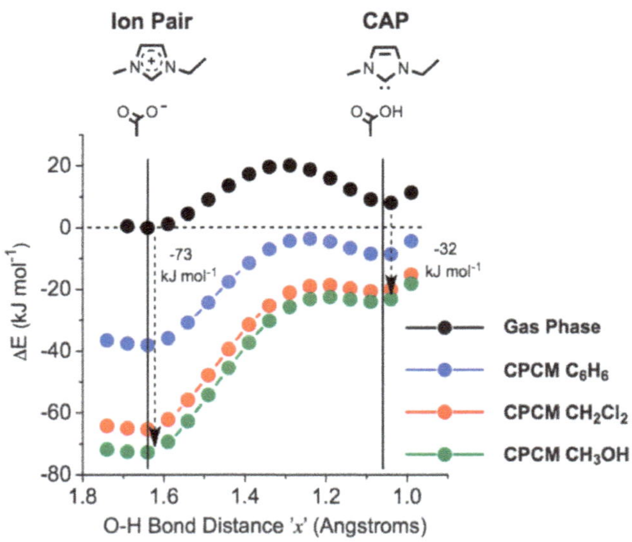

Figure 17 Change in ΔE during the proton shift between the C² carbon atom and the acetate oxygen atom (mechanism A1), in the gas phase, and with CPCM solvent environments of C_6H_6 ($\varepsilon_r = 2.28$), CH_2Cl_2 ($\varepsilon_r = 8.93$) and CH_3OH ($\varepsilon_r = 33.0$). The C²–O distance, 'y', is fixed at 2.76 Å.

The CPCM scans of Figure 17 suggest that the energy of the ion pair is stabilized to a far greater extent than the neutral CAP and TS in a polarised medium. These results are supported by the 'charge-screening' effects observed in ab initio MD simulations of the ionic liquid $[C_2C_1im][OAc]$.[28h] In the gas phase scan, the CAP is ~8 kJ mol^{-1} higher in energy than the ion pair. Introducing a weakly polar solvent environment ($\varepsilon_r = 2.28$) lowers the ion pair energy by 38 kJ mol^{-1} relative to the gas phase calculation, compared to approximately 17 kJ mol^{-1} for the CAP. For the more polar solvent environment ($\varepsilon_r = 33.0$), the energy of the ion pair is lowered by ~73 kJ mol^{-1} relative to the gas phase calculation, and in contrast the energy of the CAP is lowered by only 32 kJ mol^{-1}.

The potential energy surface within the polar medium is extremely shallow in the region of the CAP, with a very small energy barrier for the reverse reaction to re-form the ion pair (for example, $\Delta E_{activation}$ = <2 kJ mol^{-1} for the CH_3OH solvent, Figure 17). Thus, the PES has a 'shelf-like' feature that develops into a small barrier in less polar environments; such a low barrier will be difficult to observe experimentally, and will result in a highly reversible mechanism. This feature may explain why the CAP can be trapped with CO_2,[23] but is not observed as a primary decomposition route at high temperatures under inert (nitrogen, helium or compressed air) conditions.

In addition, these results indicate that it may be possible to encourage NHC formation by diluting the ionic liquid with a neutral non-polar solvent, reducing the 'charge-screening effect'. Alternatively, forcing the C^2–O distance to contract, perhaps under high pressure, may facilitate carbene formation.

For $[C_2C_1C_1im][OAc]$, the abstraction of a proton from the C^2-methyl group by the acetate anion generates acetic acid and 1-ethyl-3-methyl-2-methylene-2,3-dihydroimidazole (mechanism A2, Figure 13). This mechanism was also initially characterized by gas phase DFT calculations, and subsequently using CPCM..

The calculations suggest that even in the gas phase, the ion pair is of lower energy than the neutral products. Again, the CPCM calculations using CH_2Cl_2 and CH_3OH solvents bring about a large stabilization of the ion pair, and a significantly smaller stabilization of the transition state and product energies. Similar to $[C_2C_1im][OAc]$, the CPCM calculations for $[C_2C_1C_1im][OAc]$ indicate that the ion pair is thermodynamically more stable than the neutral products when in a polarized medium. Moreover, the very small energy barrier for the reverse reaction ($\Delta G_{activation} \leq 3$ kJ mol^{-1}, CPCM calculation using a CH_3OH solvent) indicates that the reaction is likely to be highly reversible.

Therefore, for the two decomposition mechanisms occurring at the front of the imidazolium ring (A1 and A2, Figure 13), the CPCM calculations indicate stabilization of the ion pair relative to the neutral products in an ionic environment. In addition, for each mechanism there is a minimal energy barrier (<3 kJ mol^{-1}) for the reverse reaction. These results suggest that mechanisms A1 and A2, if occurring in the bulk liquid, are likely to be highly reversible. Our TGA-MS experimental data corroborates with this observation, because the expected neutral decomposition products generated by these two mechanisms were not observed.

Moreover, the measured activation energies, E_a, for thermal decomposition of both acetate ionic liquids were in reasonable agreement with the calculated $\Delta H_{activation}$ values for the S_N2 thermal decomposition mechanisms. Ester products corresponding to the S_N2 mechanisms were detected as major decomposition products in the TGA-MS experiments, and neutral imidazole products were clearly observed in the high-temperature ^{1}H NMR experiment.

Decomposition mechanism A1 of $[C_2C_1im][OAc]$ is likely to occur in the gas phase, as observed by the mass spectrometry experiment of Hollóczki and colleagues.[28d] In addition, studies indicate that the NHC is highly accessible in the liquid phase.[23] However, our combined experimental and computational investigation suggests that thermal decomposition of $[C_2C_1im][OAc]$ from the liquid phase does not primarily occur via the NHC route, when under inert conditions (nitrogen, helium or compressed air gases).

CONCLUSIONS

A broad series of $[C_2C_1im]^+$ and $[C_2C_1C_1im]^+$ carboxylate ionic liquids has been synthesized, and the thermal stabilities were compared using temperature-ramped TGA. The difluoroacetate ionic liquid, $[C_2C_1im][CHF_2CO_2]$, was found to be the most stable imidazolium carboxylate ionic liquid of the series, with a T_{onset} value approximately 50 °C higher than the non-fluorinated ionic liquid, $[C_2C_1im][OAc]$. The high thermal stability of $[C_2C_1im][CHF_2CO_2]$ suggests that it may be a suitable replacement for the trifluoroacetate analogue in applications operating at high temperatures.[36] Extension and branching of the anion alkyl chain, from two to 16 carbon atoms, had a minimal effect on the thermal stability of the ionic liquid, as determined from temperature-ramped TGA.

The Arrhenius parameters for decomposition of $[C_2C_1im][OAc]$ and $[C_2C_1C_1im][OAc]$ have been experimentally determined, and allow rapid prediction of the rate of decomposition of the ionic liquid at any temperature, in an open system. The safe operating temperature limit for $[C_2C_1im][OAc]$ was found to be substantially lower than previously thought, and significantly lower than the operating temperature of many literature biomass procedures. In light of this result, certain biomass experiments that employ $[C_2C_1im][OAc]$ should be modified. If the basic acetate anion is essential for the efficacy of the process, then the procedure must be carried out at a reduced temperature. Alternatively, low-melting inorganic mixtures of lithium, potassium and caesium acetates have been recently reported, which have improved thermal stability.[37] Where the acetate anion is not essential, a more stable ionic liquid incorporating a different anion may be employed.

Continuing to operate biomass processing procedures using $[C_2C_1im][OAc]$ at temperatures of 120 °C and above for extended periods of time will result in the build-up of volatile neutral products, posing a hazard and preventing effective reuse/recycle of the ionic liquid. Our results demonstrating the lower than previously believed maximum operating temperatures for this ionic liquid have important implications for these experiments and procedures will need to be modified in light of these.

It has been demonstrated that ionic liquids incorporating the acetate, octanoate and thioacetate anions decompose primarily via S_N2-type nucleophilic substitution mechanisms, largely at the imidazolium N-methyl substituent. A high-temperature ^{1}H NMR experiment showed the formation of the expected neutral 1-alkylimidazole products. However, the ionic liquids incorporating the fluorinated anions were observed to decompose by initial decarboxylation of the anion.

Ion pair structures and plausible decomposition mechanisms of $[C_2C_1im][OAc]$ and $[C_2C_1C_1im][OAc]$ were explored using Density Functional Theory methods. The activation barriers for the S_N2 nucleophilic decomposition mechanisms were found to be 20 kJ mol^{-1} lower than the E2 Hofmann elimination reactions, and the calculated $\Delta H_{activation}$ energies were in reasonable agreement with the experimentally determined E_a values from the isothermal TGA experiments. The CPCM calculations for decomposition mechanisms at the front of the imidazolium rings suggest that each ionic liquid is thermodynamically more stable than the neutral products that form, and that these mechanisms are likely to be highly reversible. Whilst the NHC species may be highly accessible in $[C_2C_1im][OAc]$ and other dialkylimidazolium carboxylate ionic liquids, our experimental and computational results suggest that NHC formation is not a significant decomposition mechanism at elevated temperatures and under inert conditions.

ACKNOWLEDGEMENTS

The authors gratefully acknowledge the help of Dr Richard L. Thompson and W. Douglas Carswell from the University of

Durham with the TGA-MS experiments, and Prof. John M. Seddon and Dr Arwen I. I. Tyler for assistance with the DSC experiments. We are thankful to Claire Ashworth, Dr Jason P. Hallett, Dr Kevin R. J. Lovelock and Dr Klemens Massonne for valuable conversations. The authors are thankful to BASF SE for the sponsorship of Matthew T. Clough.

REFERENCES

1. M. Armand, F. Endres, D. R. MacFarlane, H. Ohno and B. Scrosati, Nat. Mater., 2009, **8**, 621 .
2. H. Zhao, Chem. Eng. Commun., 2006, **193**, 1660 .
3. A. Brandt, M. J. Ray, T. Q. To, D. J. Leak, R. J. Murphy and T. Welton, Green Chem., 2011, **13**, 2489 .
4. J. P. Hallett and T. Welton, Chem. Rev., 2011, **111**, 3508 .
5. T. Welton, Coord. Chem. Rev., 2004, **248**, 2459 .
6. (a) I. Krossing, J. M. Slattery, C. Daguenet, P. J. Dyson, A. Oleinikova and H. Weingärtner, J. Am. Chem. Soc., 2006, **128**, 13427 ; (b) P. A. Hunt, J. Phys. Chem. B, 2007, **111**, 4844 .
7. J. S. Wilkes and M. J. Zaworotko, J. Chem. Soc., Chem. Commun., 1992, 965 .
8. (a) A. Brandt, J. P. Hallett, D. J. Leak, R. J. Murphy and T. Welton, Green Chem., 2010, **12**, 672 ; (b) A. Brandt, J. Gräsvik, J. P. Hallett and T. Welton, Green Chem., 2013, **15**, 550 .
9. (a) W. Li, N. Sun, B. Stoner, X. Jiang, X. Lu and R. D. Rogers, Green Chem., 2011, **13**, 2038 ; (b) D. B. Fu and G. Mazza, Bioresour. Technol., 2011, **102**, 8003 .
10. (a) C. Maton, N. De Vos and C. V. Stevens, Chem. Soc. Rev., 2013, **42**, 5963 ; (b) S. Sowmiah, V. Srinivasadesikan, M.-C. Tseng and Y.-H. Chu, Molecules, 2009, **14**, 3780 .
11. (a) B. K. M. Chan, N.-H. Chang and M. R. Grimmett, Aust. J. Chem., 1977, **30**, 2005 ; (b) M. R. R. Prasad, K. Krishnan, K. N. Ninan and V. N. Krishnamurthy, Thermochim. Acta, 1997, **297**, 207 ; (c) H. L. Ngo, K. LeCompte, L. Hargens and A. B. McEwen, Thermochim. Acta, 2000, **357–358**, 97 ; (d) C. P. Fredlake, J. M. Crosthwaite, D. G. Hert, S. N. V. K. Aki and J. F. Brennecke, J. Chem. Eng. Data, 2004, **49**, 954 ; (e) D. M. Fox, J. W. Gilman, H. C. De Long and P. C. Trulove, J. Chem. Thermodyn., 2005, **37**, 900 ; (f) A. Seeberger, A.-K. Andresen and A. Jess, Phys. Chem. Chem. Phys., 2009, **11**, 9375 ; (g) L. A. Daily and K. M. Miller, J. Org. Chem., 2013, **78**, 4196 .
12. (a) J. M. Crosthwaite, M. J. Muldoon, J. K. Dixon, J. L. Anderson and J. F. Brennecke, J. Chem. Thermodyn., 2005, **37**, 559 ; (b) I. H. J. Arellano, J. G. Guarino, F. U. Paredes and S. D. Arco, J. Therm. Anal. Calorim., 2011, **103**, 725 .
13. M. Kosmulski, J. Gustafsson and J. B. Rosenholm, Thermochim. Acta, 2004, **412**, 47 .
14. (a) D. M. Fox, W. H. Awad, J. W. Gilman, P. H. Maupin, H. C. De Long and P. C. Trulove, Green Chem., 2003, **5**, 724 ; (b) K. J. Baranyai, G. B. Deacon, D. R. MacFarlane, J. M. Pringle and J. L. Scott, Aust. J. Chem., 2004, **57**, 145 .
15. T. J. Wooster, K. M. Johanson, K. J. Fraser, D. R. MacFarlane and J. L. Scott, Green Chem., 2006, **8**, 691 .
16. (a) F. Wendler, L.-N. Todi and F. Meister, Thermochim. Acta, 2012, **528**, 76 ; (b) F. Wendler, L.-N. Todi, F. Meister, Thermal investigations of polysaccharide blend solutions in imidazolium ionic liquids, 241st ACS National Meeting, March 27–31, Anaheim, USA, 2011; (c) A. W. T. King, A. Parviainen, P. Karhunen, J. Matikainen, L. K. J. Hauru, H. Sixta and I. Kilpeläinen, Adv., 2012, **2**, 8020 .
17. (a) M. C. Kroon, W. Buijs, C. J. Peters and G.-J. Witkamp, Thermochim. Acta, 2007, **465**, 40 ; (b) S. D. Chambreau, J. A. Boatz, G. L. Vaghjiani, C. Koh, O. Kostko, A. Golan and S. R. Leone, J. Phys. Chem. A, 2012, **116**, 5867 ; (c) Y. Chen, Y. Cao, Y. Shi, Z. Xue and T. Mu, Ind. Eng. Chem. Res., 2012, **51**, 7418 .
18. J. L. Ferguson, J. D. Holbrey, S. Ng, N. V. Plechkova, K. R. Seddon, A. A. Tomaszowska and D. F. Wassell, Pure Appl. Chem., 2012, **84**, 723 .
19. 1-Ethyl-3-methylimidazolium acetate, BASF Basionics™ 'BC 01', -No. 143314-17-4.
20. I. Pibiri, A. Pace, S. Buscemi, V. Causin, F. Rastrelli and G. Saielli, Phys. Chem. Chem. Phys., 2012, **14**, 14306 .
21. CRC Handbook of Chemistry and Physics, Internet Version, ed. D. R. Lide, Taylor and Francis, Boca Raton, FL, 87th edn, 2007, http:/www.hbcpnetbase.com.
22. R. Fausto, L. A. E. Batista De Carvalho and J. J. C. Teixeira-Dias, THEOCHEM, 1990, **207**, 67 .
23. G. Gurau, H. Rodríguez, S. P. Kelley, P. Janiczek, R. S. Kalb and R. D. Rogers, Angew. Chem., Int. Ed., 2011, **50**, 12024 .
24. J. H. Flynn and L. A. Wall, Polym. Lett., 1966, **4**, 323 .
25. (a) J. M. S. S. Esperança, J. N. Canongia Lopes, M. Tariq, L. M. N. B. F. Santos, J. W. Magee and L. P. N. Rebelo, J. Chem. Eng. Data, 2010, **55**, 3 ; (b) M. J. Earle, J. M. S. S. Esperança, M. A. Gilea, J. N. Canongia Lopes, L. P. N. Rebelo, J. W. Magee, K. R. Seddon and J. A. Widegren, Nature, 2006, **439**, 831 ; (c) A. Deyko, K. R. J. Lovelock, J.-A. Corfield, A. W. Taylor, P. N. Gooden, I. J. Villar-Garcia, P. Licence, R. G. Jones, V. G. Krasovskiy, E. A. Chernikova and L. M. Kustov, Phys. Chem. Chem. Phys., 2009, **11**, 8544 ; (d) A. W. Taylor, K. R. J. Lovelock, A. Deyko, P. Licence and R. G. Jones, Phys. Chem. Chem. Phys., 2010, **12**, 1772 ; (e) F. Heym, B. J. M. Etzold, C. Kern and A. Jess, Phys. Chem. Chem. Phys., 2010, **12**, 12089 ; (f) K. R. J. Lovelock, A. Deyko, P. Licence and R. G. Jones, Phys. Chem. Chem. Phys., 2010, **12**, 8893 ; (g) A. Chandran, K. Prakash and S. Senapati, Chem. Phys., 2010, **374**, 46 ; (h) A. Deyko, K. R. J. Lovelock, P. Licence and R. G. Jones, Phys. Chem. Chem. Phys., 2011, **13**, 16841 ; (i) S. P. Verevkin, D. H. Zaitsau, V. N. Emel̓yanenko and A. Heintz, J. Phys. Chem. B, 2011, **115**, 12889 ; (j) K. Shimizu, M. Tariq, M. F. Costa Gomes, L. P. N. Rebelo and J. N. Canongia Lopes, J. Phys. Chem. B, 2010, **114**, 5831 ; (k) S. P. Verevkin, D. H. Zaitsau, V. N. Emel'yanenko, A. V. Yermalayeu, C. Schick, H. Liu, E. J. Maginn, S. Bulut, I. Krossing and R. Kalb, J. Phys. Chem. B, 2013, **117**, 6473 .

26. (a) A. G. Glenn and P. B. Jones, Tetrahedron Lett., 2004, **45**, 6967 ; (b) M. C. Kroon, W. Buijs, C. J. Peters and G.-J. Witkamp, Thermochim. Acta, 2007, **465**, 40 ; (c) H. Ohtani, S. Ishimura and M. Kumai, Anal. Sci., 2008, **24**, 1335 ; (d) N. Meine, F. Benedito and R. Rinaldi, Green Chem., 2010, **12**, 1711 .

27. J. Huang and A. F. Hollenkamp, J. Phys. Chem. C, 2010, **114**, 21840 .

28. (a) J. Dupont and J. Spencer, Angew. Chem., Int. Ed., 2004, **43**, 5296 ; (b) T. L. Amyes, S. T. Diver, J. P. Richard, F. M. Rivas and K. Toth, J. Am. Chem. Soc., 2004, **126**, 4366 ; (c) S. T. Handy and M. Okello, J. Org. Chem., 2005, **70**, 1915 ; (d) O. Hollóczki, D. Gerhard, K. Massonne, L. Szarvas, B. Németh, T. Veszprémi and L. Nyulászi, New J. Chem., 2010, **34**, 3004 ; (e) O. Hollóczki and L. Nyulászi, Org. Biomol. Chem., 2011, **9**, 2634 ; (f) Z. Kelemen, O. Hollóczki, J. Nagy and L. Nyulászi, Org. Biomol. Chem., 2011, **9**, 5362 ; (g) H. Rodríguez, G. Gurau, J. D. Holbrey and R. D. Rogers, Chem. Commun., 2011, **47**, 3222 ; (h) O. Hollóczki, D. S. Firaha, J. Friedrich, M. Brehm, R. Cybik, M. Wild, A. Stark and B. Kirchner, J. Phys. Chem. B, 2013, **117**, 5898 .

29. (a) R. P. Swatloski, J. D. Holbrey and R. D. Rogers, Green Chem., 2003, **5**, 361 ; (b) M. G. Freire, C. M. S. S. Neves, I. M. Marrucho, J. A. P. Coutinho and A. M. Fernandes, J. Phys. Chem. A, 2010, **114**, 3744 ; (c) S. Steudte, J. Neumann, U. Bottin-Weber, M. Diedenhofen, J. Arning, P. Stepnowski and S. Stolte, Green Chem., 2012, **14**, 2474 .

30. National Institute of Standards and Technology (NIST) Chemistry WebBook, http://webbook.nist.gov/chemistry/.

31. (a) N. R. Dhumal, H. J. Kim and J. Kiefer, J. Phys. Chem. A, 2009, **113**, 10397 ; (b) J. Kiefer, K. Obert, J. Fries, A. Bosmann, P. Wasseheid and A. Leipertz,Appl. Spectrosc., 2009, **63**, 1041 ; (c) M. B. Shiflett and A. Yokozeki, J. Chem. Eng. Data, 2009, **54**, 108 ; (d) D. T. Bowron, C. D'Agostino, L. F. Gladden, C. Hardacre, J. D. Holbrey, M. C. Lagunas, J. McGregor, M. D. Mantle, C. L. Mullan and T. G. A. Youngs, J. Phys. Chem. B, 2010, **114**, 7760 ; (e) W. Shi, K. Damodaran, H. B. Nulwala and D. R. Luebke, Phys. Chem. Chem. Phys., 2012, **14**, 15897 ; (f) S. Stevanovic, A. Podgoršek, A. A. H. Pádua and M. F. Costa Gomes, J. Phys. Chem. B, 2012, **116**, 14416 ; (g) M. Brehm, H. Weber, A. S. Pensado, A. Stark and B. Kirchner, Phys. Chem. Chem. Phys., 2012, **14**, 5030 ; (h) S. Zhang, Y. Zhang, X. Ma, L. Lu, Y. He and Y. Deng, J. Phys. Chem. B, 2013, **117**, 2764 ; (i) M. Brehm, H. Weber, A. S. Pensado, A. Stark and B. Kirchner, Z. Phys. Chem., 2013, **227**, 177 .

32. (a) P. A. Hunt, B. Kirchner and T. Welton, Chem.–Eur. J., 2006, **12**, 6762 ; (b) P. A. Hunt and I. R. Gould, J. Phys. Chem. A, 2006, **110**, 2269 ; (c) P. A. Hunt, I. R. Gould and B. Kirchner, Aust. J. Chem., 2007, **60**, 9 .

33. M. E. M. Berger, D. Assenbaum, N. Taccardi, E. Spiecker and P. Wasseheid, Green Chem., 2011, **13**, 1411 .

34. J. Tomasi, B. Mennucci and R. Cammi, Chem. Rev., 2005, **105**, 2999 .

35. (a) C. Wakai, A. Oleinikova, M. Ott and H. Weingärtner, J. Phys. Chem. B, 2005, **109**, 17028 ; (b) H. Weingärtner, Angew. Chem., Int. Ed., 2008, **47**, 654 .

36. C. Shi, Q. Ge, F. Zhou, M. Cai, X. Wang, X. Fang and X. Pan, Sol. Energy, 2009, **83**, 108 .

37. S. Bajus, A. Deyko, A. Bösmann, F. Maier, H.-P. Steinrück and P. Wasseheid, Dalton Trans., 2012, **41**, 14433 .

CHAPTER 14

FREE RADICALS, NATURAL ANTIOXIDANTS, AND THEIR REACTION MECHANISMS

Satish Balasaheb Nimse and Dilipkumar Pal

INTRODUCTION AND BACKGROUND

Antioxidants are molecules that inhibit or quench free radical reactions and delay or inhibit cellular damage.[1] Though the antioxidant defenses are different from species to species, the presence of the antioxidant defense is universal. Antioxidants exists both in enzymatic and non-enzymatic forms in the intracellular and extracellular environment.

Normal biochemical reactions, increased exposure to the environment, and higher levels of dietary xenobiotics result in the generation of reactive oxygen species (ROS) and reactive nitrogen species (RNS).[2] ROS and RNS are responsible for the oxidative stress in different pathophysiological conditions.[3] Cellular constituents of our body are altered in oxidative stress conditions, resulting in various disease states. The oxidative stress can be effectively neutralized by enhancing cellular defenses in the form of antioxidants.[4,5] Certain compounds act as *in vivo* antioxidants by raising the levels of endogenous antioxidant defenses. Expression of genes encoding the enzymes such as superoxide dismutase (SOD), catalase (CAT), glutathione peroxidase (GSHPx) increases the level of endogenous antioxidants.[6]

Antioxidants can be categorized in multiple ways. Based on their activity, they can be categorized as enzymatic and non-enzymatic antioxidants. Enzymatic antioxidants work by breaking down and removing free radicals. The antioxidant enzymes convert dangerous oxidative products to hydrogen peroxide (H_2O_2) and then to water, in a multi-step process in presence of cofactors such as copper, zinc, manganese, and iron. Non-enzymatic antioxidants work by interrupting free radical chain reactions. Few examples of the non-enzymatic antioxidants are vitamin C, vitamin E, plant polyphenol, carotenoids, and glutathione.[7]

The other way of categorizing the antioxidants is based on their solubility in the water or lipids. The antioxidants can be categorized as water-soluble and lipid-soluble antioxidants. The water-soluble antioxidants (*e.g.* vitamin C) are present in the cellular fluids such as cytosol, or cytoplasmic matrix. The lipid-soluble antioxidants (*e.g.* vitamin E, carotenoids, and lipoic acid) are predominantly located in cell membranes.

The antioxidants can also be categorized according to their size, the small-molecule antioxidants and large-molecule antioxidants. The small-molecule antioxidants neutralize the ROS in a process called radical scavenging and carry them away. The main antioxidants in this category are vitamin C, vitamin E, carotenoids, and glutathione (GSH). The large-molecule antioxidants are enzymes (SOD, CAT, and GSHPx) and sacrificial proteins (albumin) that absorb ROS and prevent them from attacking other essential proteins.

To understand the mechanism of action of antioxidants, it is necessary to understand the generation of free radicals and their damaging reactions. This review elaborates the generation and damages that free radicals create, mechanism of action of the natural antioxidant compounds and assays for the evaluation of their antioxidant properties. The reaction mechanisms of the antioxidant assays are discussed. The scope of this article is limited to the natural antioxidants and the *in vitro* assays for evaluation of their antioxidant properties.

GENERATION OF FREE RADICALS

The generation of ROS (Table 1) begins with rapid uptake of oxygen, activation of NADPH oxidase, and the production of the superoxide anion radical ($O_2^{\cdot-}$, eqn (1)),

$$2O_2 + NADPH \xrightarrow{\text{(oxidase)}} 2O_2^{\cdot-} + NADP^+ + H^+ \quad (1)$$

Citation: Satish Balasaheb Nimse and Dilipkumar Pal, "Free radicals, natural antioxidants, and their reaction mechanisms," DOI: 10.1039/c4ra13315c.

Table: 1 List of the ROS[165]

Symbol	Name
$^1O_2$	Singlet oxygen
$O_2^{\cdot-}$	Superoxide anion radical
$^{\cdot}OH$	Hydroxyl radical
$RO^{\cdot}$	Alkoxyl radical
$ROO^{\cdot}$	Peroxyl radical
H_2O_2	Hydrogen peroxide
LOOH	Lipid hydroperoxide

The $O_2^{\cdot-}$ is then rapidly converted to H_2O_2 (eqn (2)) by SOD

$$2O_2^{\cdot-} + 2H^+ \xrightarrow{(SOD)} H_2O_2 + O_2 \quad (2)$$

These ROS can act by either of the two oxygen dependent mechanisms resulting in the destruction of the microorganism or other foreign matter. The reactive species can also be generated by the myeloperoxidase–halide–H_2O_2 system. The enzyme myeloperoxidase (MPO) is present in the neutrophil cytoplasmic granules. In presence of the chloride ion, which is ubiquitous, H_2O_2 is converted to hypochlorous (HOCl, eqn (3)), a potent oxidant and antimicrobial agent.[8]

$$Cl^- + H_2O_2 + H^+ \xrightarrow{(MPO)} HOCl + H_2O \quad (3)$$

ROS are also generated from $O_2^{\cdot-}$and H_2O_2 *via* 'respiratory burst' by Fenton (eqn (4)) and/or Haber–Weiss (eqn (5)) reactions.[9]

$$H_2O_2 + Fe^{2+} \rightarrow {}^{\cdot}OH + OH^- + Fe^{3+} \quad (4)$$

$$O_2^{\cdot-} + H_2O_2 \rightarrow {}^{\cdot}OH + OH^- + O_2 \quad (5)$$

The enzyme nitric oxide synthase produce reactive nitrogen species (RNS), such as nitric oxide ($NO^{\cdot}$) from arginine (eqn (6)).

$$\text{L-Arg} + O_2 + \text{NADPH} \rightarrow NO^{\cdot} + \text{citrulline} \quad (6)$$

An inducible nitric oxide synthase (iNOS) is capable of continuously producing large amount of $NO^{\cdot}$, which act as a $O_2^{\cdot-}$quencher. The $NO^{\cdot}$ and $O_2^{\cdot-}$ react together to produce peroxynitrite ($ONOO^-$, eqn (7)), a very strong oxidant, hence, each can modulate the effects of other. Although neither $NO^{\cdot}$ nor $O_2^{\cdot-}$ is a strong oxidant, peroxynitrite is a potent and versatile oxidant that can attack a wide range of biological targets.[10]

$$NO^{\cdot} + O_2^{\cdot-} \rightarrow ONOO^- \quad (7)$$

Peroxynitrite reacts with the aromatic amino acid residues in the enzyme resulting in the nitration of the aromatic amino acids. Such a change in the aminoacid residue can result in the enzyme inactivation. However, nitric oxide is an important cytotoxic effector molecule in the defense against tumor cells, various protozoa, fungi, helminthes, and mycobacteria.[11,12] The other sources of free radical reactions are cyclooxygenation, lipooxygenation, lipid peroxidation, metabolism of xenobiotics, and ultraviolet radiations.[13]

DAMAGING REACTIONS OF FREE RADICALS

ROS (Table 1) induced oxidative stress is associated with the chronic diseases such as cancer, coronary heart disease (CHD), and osteoporosis.[14] Free radicals attack all major classes of biomolecules, mainly the polyunsaturated fatty acids (PUFA) of cell membranes. The oxidative damage of PUFA, known as lipid peroxidation is particularly destructive, because it proceeds as a self-perpetuating chain reaction.[15,16]

The general process of lipid peroxidation can be envisaged as depicted bellow (eqn (8)–(11)), where LH is the target PUFA and $R^{\cdot}$ is the initializing, oxidizing radical. Oxidation of the PUFA generates a fatty acid radical ($L^{\cdot}$) (eqn (8)), which rapidly adds oxygen to form a fatty acid peroxyl radical ($LOO^{\cdot}$, eqn (9)). The peroxyl radicals are the carriers of the chain reactions. The peroxyl radicals can further oxidize PUFA molecules and initiate new chain reactions, producing lipid hydroperoxides (LOOH) (eqn (10) and (11)) that can break down to yet more radical species.[17]

$$LH + R^{\cdot} \rightarrow L^{\cdot} + RH \quad (8)$$

$$L^{\cdot} + O_2 \rightarrow LOO^{\cdot} \quad (9)$$

$$LOO^{\cdot} + LH \rightarrow LOOH + L^{\cdot} \quad (10)$$

$$LOOH \rightarrow LO^{\cdot} + LOO^{\cdot} + \text{aldehydes} \quad (11)$$

Lipid hydroperoxides always break down to aldehydes. Many of these aldehydes are biologically active compounds, which can diffuse from the original site of attack and spread the attack to the other parts of the cell.[18,19] Lipid peroxidation has been widely associated with the tissue injuries and diseases.[20]

Oxygen metabolism generates $^{\cdot}OH$, $O_2^{\cdot-}$, and the non-radical H_2O_2. The $^{\cdot}OH$ is highly reactive and reacts with biological molecules such as DNAs, proteins, and lipids, which results in the chemical modifications of these molecules. There are several research reports on the oxidative damage of DNA due to the $^{\cdot}OH$.[21–23]

The $^{\cdot}OH$ reacts with the basepairs of DNA, resulting in the oxidative damage of the heterocyclic moiety and the sugar moiety in the oligonucleotides by a variety of mechanisms. This type of oxidative damage to DNA is highly correlated to the physiological conditions such as mutagenesis, carcinogenesis, and aging.[24,25] The addition reactions yield OH-adduct radicals of DNA bases (Scheme 1), whereas the allyl radical of thymine and carbon-centered sugar radicals (Scheme 2) are formed from the abstraction reactions.

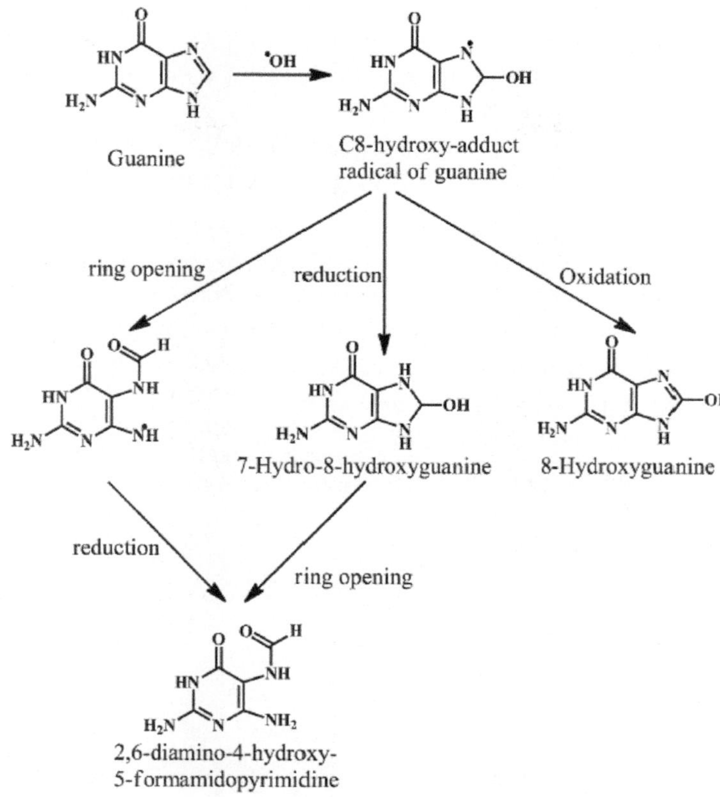

Scheme 1: Reaction of hydroxyl radical with guanine.

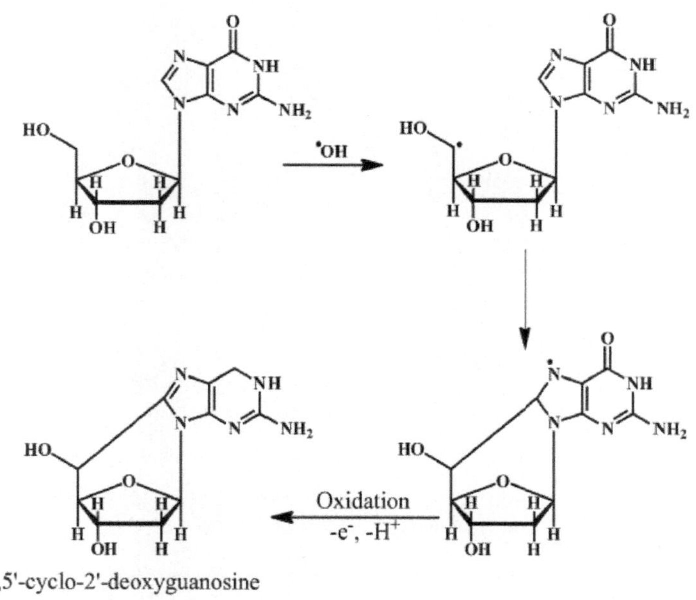

Scheme 2: Reaction of hydroxyl radical with the sugar moiety of DNA.

As shown in the Scheme 1, the ˙OH reacts with the guanine of the DNA to produce the C-8-hydroxy-adduct radical of guanine, which is converted to the 2,6-diamino-4-hydroxy-5-formamidopyrimidine upon reduction and ring opening reactions. However, the C-8-hydroxy-adduct radical of guanine is converted to the 8-hydroxyguanine upon oxidation reaction. The ˙OH radical reacts with the heterocyclic moiety of the thymine and cytosine at C5- and C6-positions, resulting in the C5–OH and C6–OH adduct radicals, respectively. The oxidation reaction of these adduct radicals with water (followed by deprotonation) results in the formation of the cytosine glycol and thymine glycol, respectively.[26] Overall, the reactions of the ˙OH with the DNA bases result in the impaired dsDNA.

As shown in the Scheme 2, the ˙OH reacts with the sugar moiety of DNA by abstracting an H-atom from rom C5 carbon atom. One unique reaction of the C5′-centered radical of the sugar moiety in DNA is the addition to the C8-position of the purine ring in the same nucleoside (*e.g.* guanine). This intramolecular cyclization results in the formation of the 8,5′-cyclopurine-2′-deoxynucleosides. The reactions of carbon-centered sugar radicals result in the DNA strand breaks and base-free sites by a variety of mechanisms.

Proteins are oxidatively damaged by the combined action of activated oxygen species and the trace metal ions such as Fe^{2+} and Cu^{2+}. The amino acid's lysine, proline, histidine, and arginine have been found to be the most sensitive to oxidative damage. Recent studies indicate that, a wide range of residue modifications can occur including formation of peroxides,[27,28]and carbonyls.[29] Generation of the carbonyl residue is a useful measure of oxidative damage to proteins. Thus, the oxidative damage to tissue results in the increased amount of oxidized protein. A detailed review by Cooke *et al.* provides important informations on the oxidative DNA damage, mechanisms, mutations, and related diseases.[30]

Low levels of antioxidants have been associated with the heart disease and cancer.[31,32] Antioxidants provide protection against a number of disease processes such as aging, allergies, algesia, arthritis, asthma, atherosclerosis, autoimmune diseases, bronchopulmonary dyspepsia, cancer. The other disorders to which antioxidants provide protection are cataract, cerebral ischemia, diabetes mellitus, eczema, gastrointestinal inflammatory diseases, genetic disorders.[33] Following section elaborates the mechanism of action of the radical scavenging activities of various natural antioxidant molecules.

MODULATION OF FREE RADICALS BY NATURAL ANTIOXIDANTS

Two types of antioxidants namely the enzymatic antioxidants and nonenzymatic antioxidants modulate the free radical reactions. Body protects itself from ROS by using enzymatic antioxidant mechanisms.[34] The antioxidant enzymes reduce the levels of lipid hydroperoxide and H_2O_2, thus they are important in the prevention of lipid peroxidation and maintaining the structure and function of cell membranes. Examples of the enzymatic antioxidants (Figure 1, Table 2) are CAT, GSHPx, SOD, and peroxiredoxin I–IV (I–IV).

$$2H_2O_2 \xrightarrow{(catalase)} 2H_2O + O_2 \qquad (12)$$

$$ROOH + 2GSH \xrightarrow{(GSHPx)} ROH + GSSG \qquad (13)$$

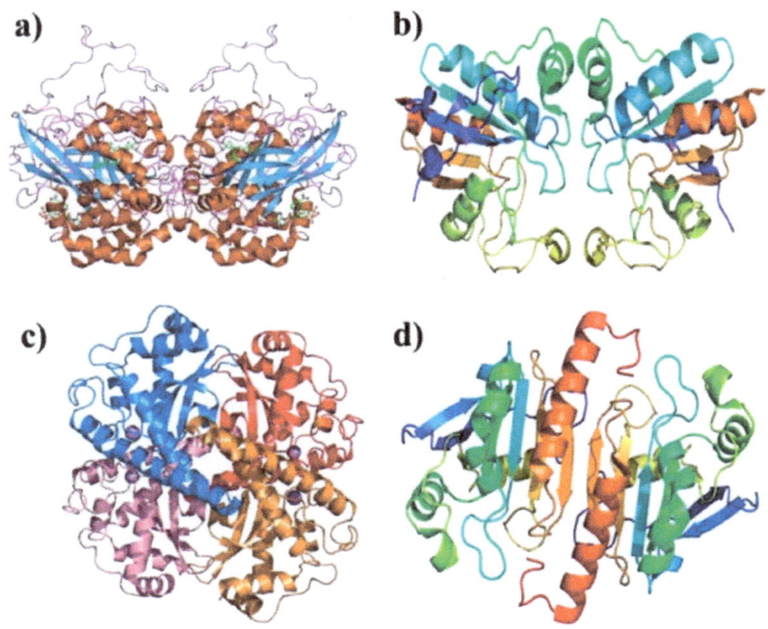

Figure 1: (a) CAT, (b) GSHPx, (c) SOD, and Prx-I.

Table 2: Enzymatic antioxidants, their cellular locations and the reactions they carry out

Enzymatic antioxidant	Cellular location	Substrate	Reaction
Mn/Cu/Zn SOD	Mitochondrial matrix (Mn SOD) cytosol (Cu/Zn SOD)	$O_2^{\cdot-}$	$O_2^{\cdot-} \rightarrow H_2O_2$
CAT	Peroxisomes cytosol	H_2O_2	$2H_2O_2 \rightarrow O_2 + H_2O$
GSHPx	Cytosol	H_2O_2	$H_2O_2 + GSH \rightarrow GSSG + H_2O$
Prx–I	Cytosol	H_2O_2	$H_2O_2 + TrxS_2 \rightarrow Trx(SH)_2 + H_2O$

SOD's located in the cytosol and mitochondria, catalytically convert the $O_2^{\cdot-}$ into oxygen and H_2O_2 in presence of the metal ion cofactors such as copper (Cu), zinc (Zn), or manganese (Mn).[35] The enzyme CAT present in the peroxisome, converts H_2O_2 to water and oxygen (eqn (12)).[36,37] GSHPx are found both in the cytoplasm and extracellularly in almost every human tissue. GSHPx convert the H_2O_2 into the water (Table 2). The enzyme GSHPx has strong activity towards both H_2O_2 and fatty acid hydroperoxides (eqn (13)).[38,39] The enzyme peroxyredoxin catalyze the reduction of H_2O_2, organic hydroperoxides and the peroxynitrite ($ONOO^-$). The different expression profiles, subcellular locations, and substrates of the antioxidant enzymes reveal the complex nature of the ROS biology. Clearly, the antioxidant enzymes play a major role in the prevention of oxidative damage. As demonstrated in the Scheme 3, CAT, GSHPx, and SOD show synergistic effect in the scavenging of $O_2^{\cdot -}$.

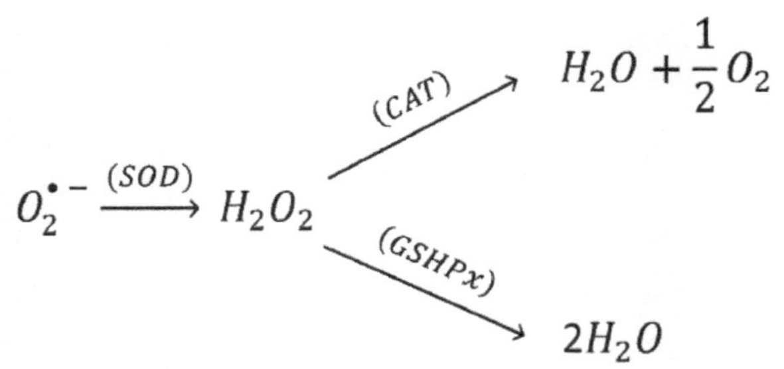

Scheme 3: Radical scavenging activity of SOD, CAT, and GSHPx.

The enzymatic antioxidants and their mechanism of antioxidant activity has been explained in details in several review articles.[40–42] Therefore, this article focuses mainly on the nonenzymatic antioxidants of natural origin.

The nonenzymatic antioxidants are of two types, the natural antioxidants and the synthetic antioxidants. However, the scope of this article is limited to the natural antioxidants; hence the synthetic antioxidants will not be considered for the discussion.

Vitamins

Vitamin E **1**,[43] vitamin C **2**,[44] vitamin A **3**.

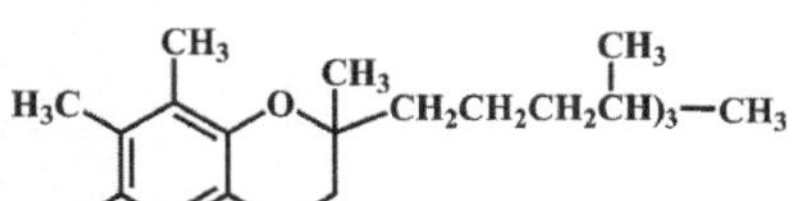

1

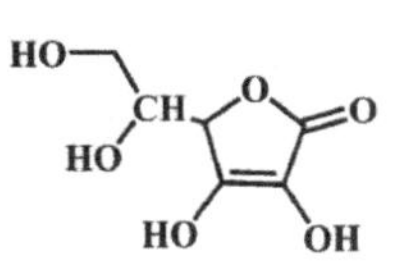

2

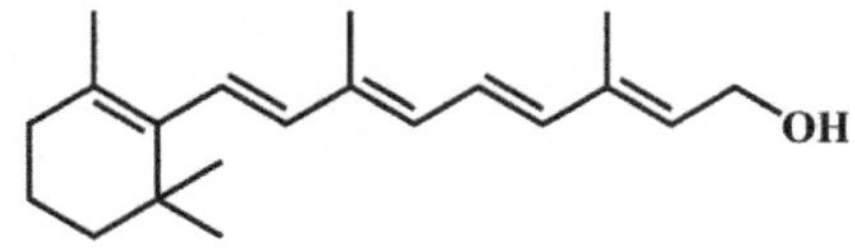

3

Vitamin E (α-tocopherol) **1**, is an efficient lipid soluble antioxidant that functions as a 'chain breaker' during lipid peroxidation in cell membranes and various lipid particles including low-density lipoprotein (LDL). It functions to intercept lipid peroxyl radicals (LOO˙) and to terminate the lipid peroxidation chain reactions (eqn (14)).

$$LOO^{\cdot} + \alpha\text{-tocopherol–OH} \rightarrow LOOH + \alpha\text{-tocopherol–O}^{\cdot} \quad (14)$$

The resultant tocopheroxyl radical is relatively stable and in normal circumstances, insufficiently reactive to initiate lipid peroxidation itself, which is an essential criterion of a good antioxidant.[45–47] It should be noted that, vitamin E exerts antioxidant effects by scavenging lipid peroxyl radicals *in vivo* as well as *in vitro* systems. However, vitamin E is not an efficient scavenger of ˙OH and alkoxyl radicals (˙OR) *in vivo*.[48]

Vitamin C or ascorbic acid **2**, is a water-soluble free radical scavenger. Moreover, it regenerates vitamin E in cell membranes in combination with GSH or compounds capable of donating reducing equivalents.[49–51] Vitamin C, changes to the ascorbate radical (Scheme 4) by donating an electron to the lipid radical in order to terminate the lipid peroxidation chain reaction. The pairs of asçorbate radicals react rapidly to produce one molecule of ascorbate and one molecule of dehydroascorbate. The dehydroascorbate does not have any antioxidant capacity. Hence, dehydroascorbate is converted back into the ascorbate by the addition of two electrons. The last stage of the addition of two electrons to the dehydroascorbate has been proposed to be carried out by oxidoreductase.

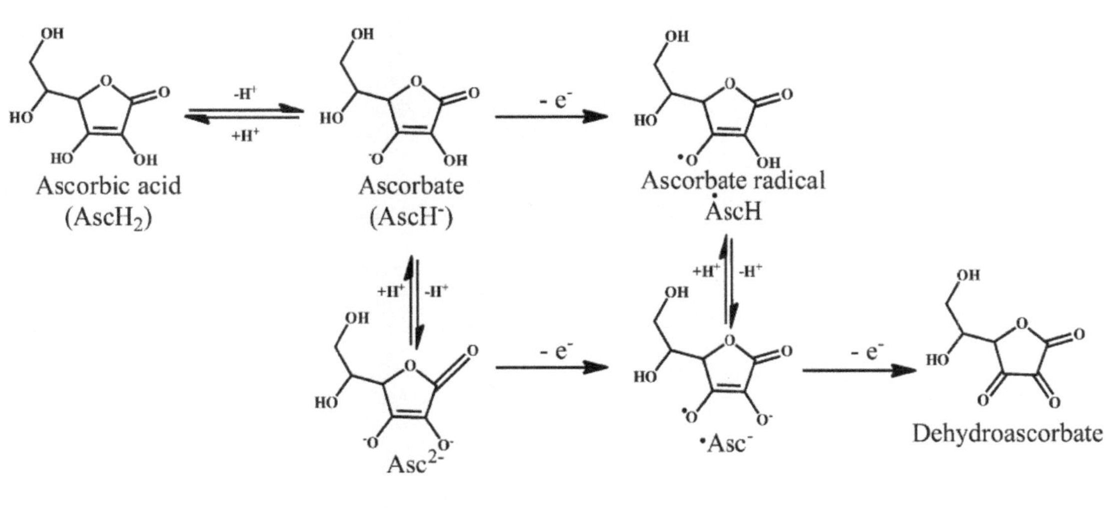

Scheme 4: Mechanism of radical scavenging activity of ascorbic acid **2**.

Antioxidant potential of vitamin A **3** was first described by Monaghan and Schmitt,[52] who reported that vitamin A can protect lipids against rancidity. Several reviews have appeared to outline the basic structural and metabolic characteristics of vitamin A and information about its potential as antioxidants in relation to the heart diseases.[53,54] Vitamin A has a vital antioxidant contribution in protecting human LDL against copper-stimulated oxidation (Scheme 5).[55,56]

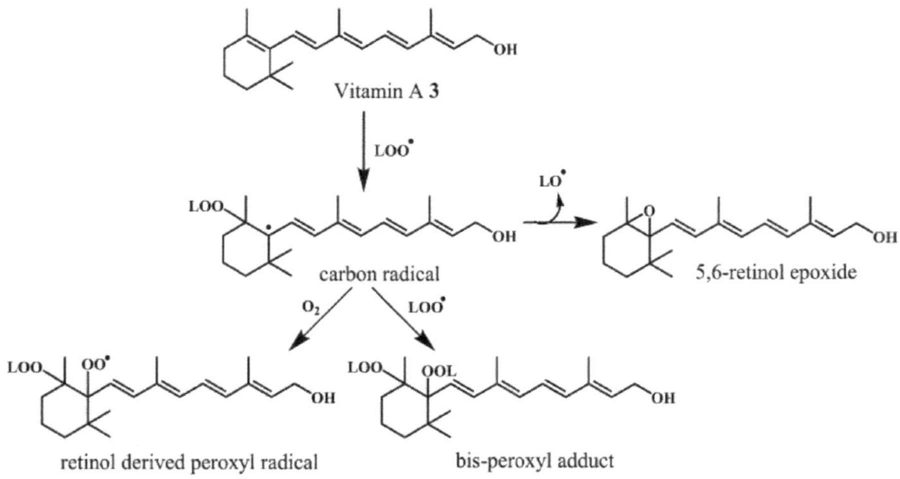

Scheme 5: Mechanism of radical scavenging activity of vitamin A **3**.

Bioflavonoids

Flavonol **4** (*e.g.* quercetin **5**, myricetin **6**), flavone **7** (*e.g.* apigenin **8**, luteoline **9**), flavonolols **10** (*e.g.* taxifolin **11**), flavan-3-ols **12** (*e.g.* catechin **13**, epigallocatechin **14**), flavonone **15** (*e.g.*hesperetin **16**, naringenin **17**), anthocyanidin **18** (*e.g.* cynidin **19**, delphidin **20**), isoflavone **21** (*e.g.* genistein **22**, daidzein **23**).[57,58]

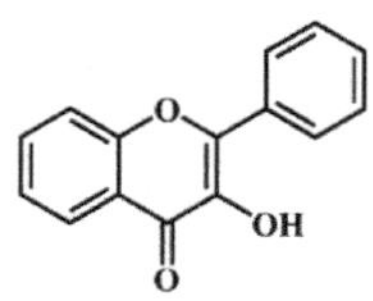

4

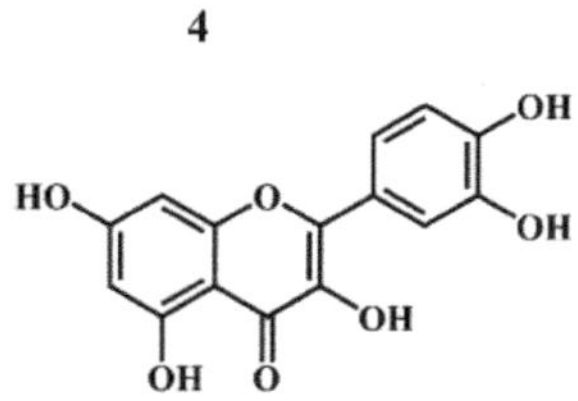

5

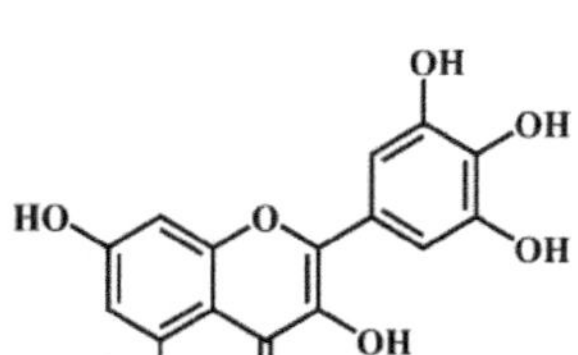

6

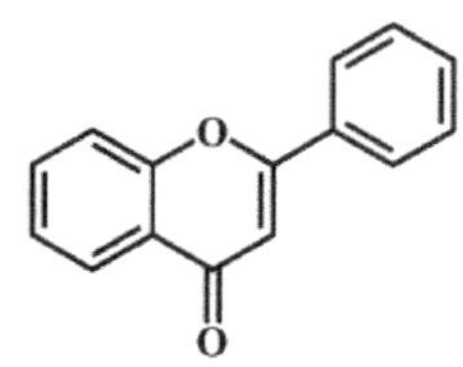

7

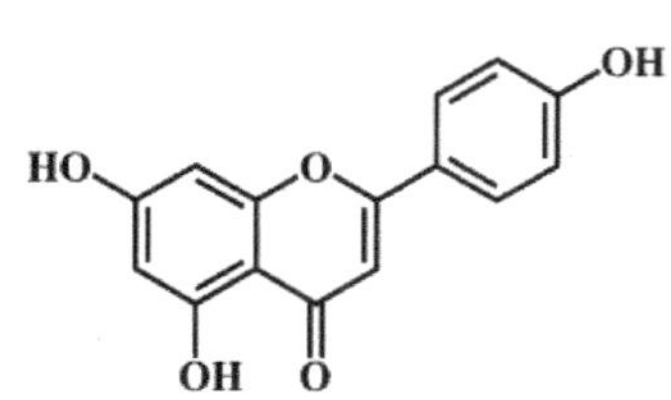

8

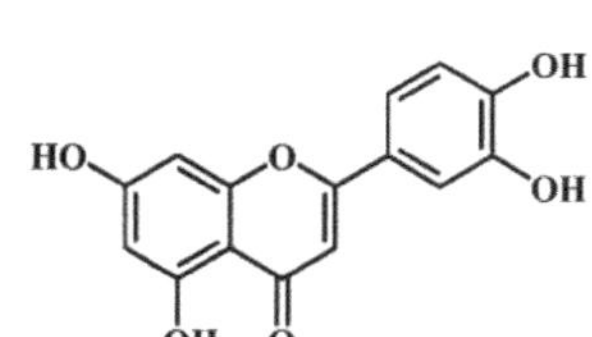

9

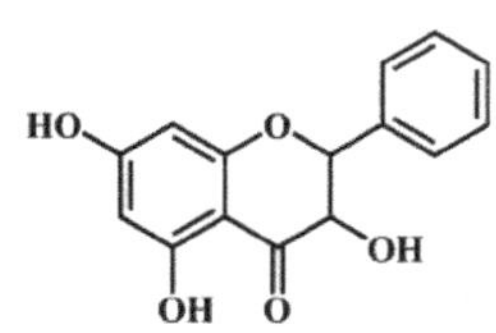

10

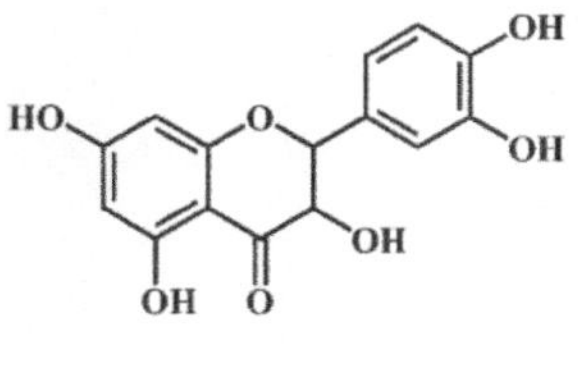

11

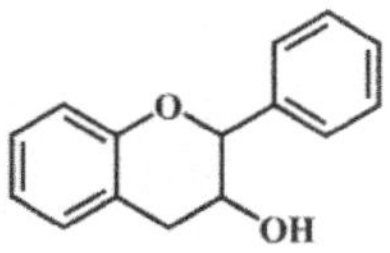

12

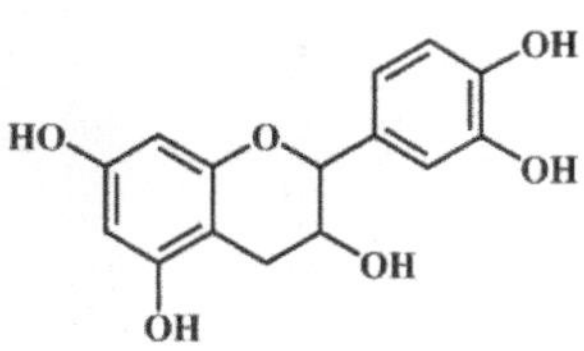

13

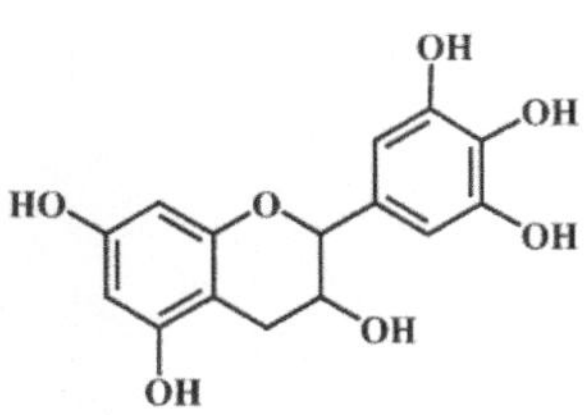

14

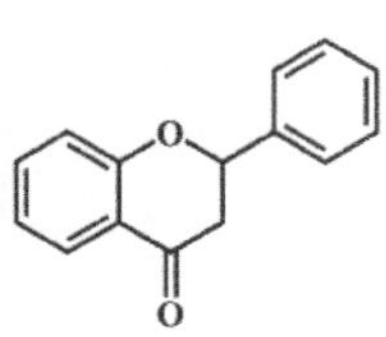

15

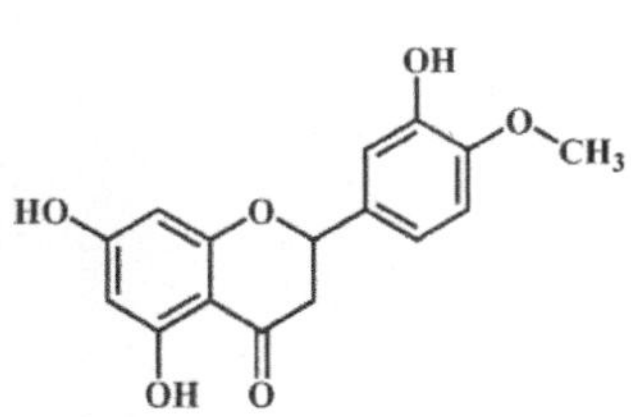

16

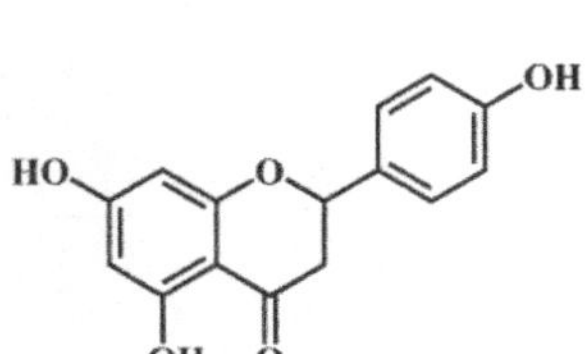

17

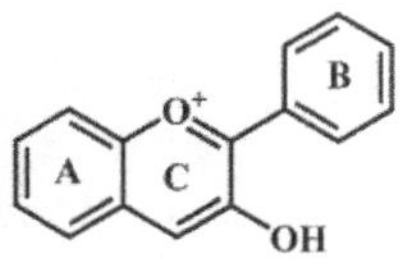

18

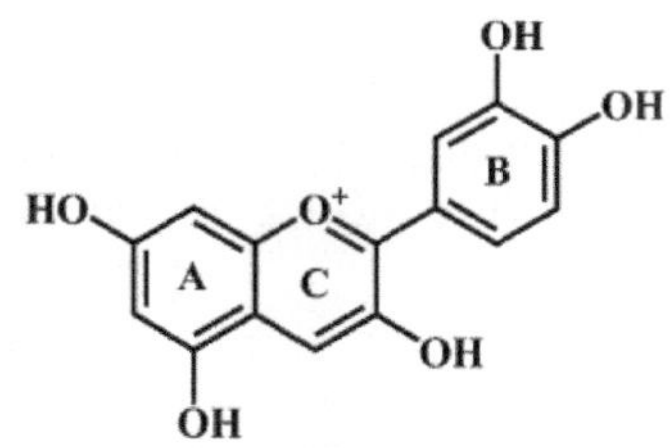

19

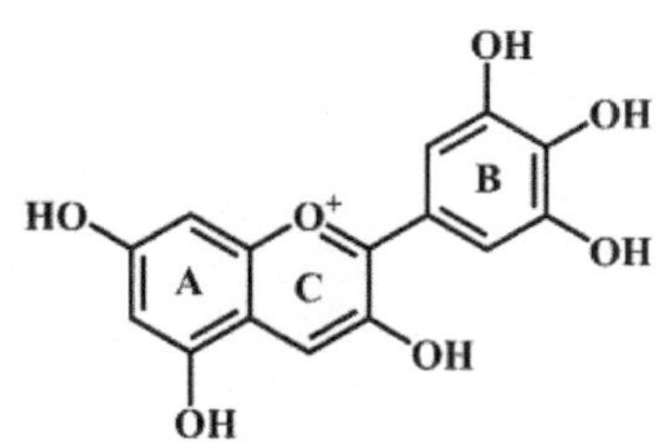

20

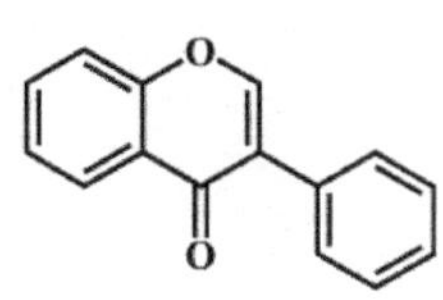

21

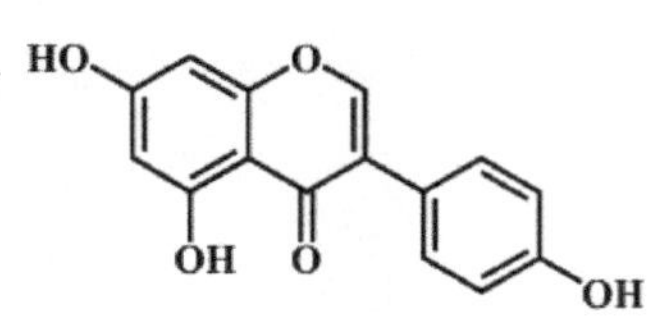

22

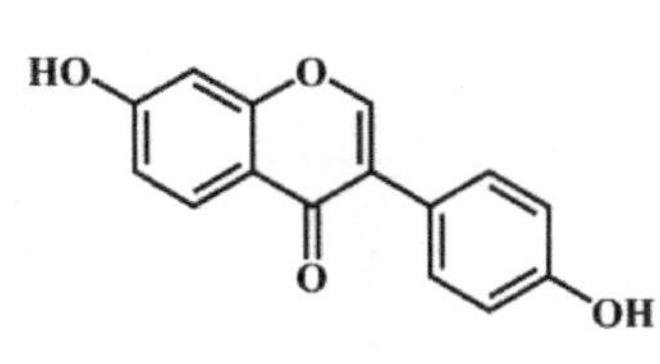

23

Bioflavonoids are a group of natural benzo-γ-pyran derivatives (**4–23**) and are found to possess strong antioxidant activities.[59,60] Bioflavonoids widely distributed in fruits and vegetables, are reported to exert multiple biological effects including free radical-scavenging activity. It has been reported that the bioflavonoids have a protective effect on the DNA damage induced by the hydroxyl radicals.[61] One of the mechanisms that explains the protective effect of the flavonoids on the DNA is the involvement of the chelating metal ions, such as copper or iron. The flavonoids complexed with the copper or iron prevent the generation of the ROS.[62–64]

Quercetin **5** is a flavonol, known to protect DNA from oxidative damage resulting from the attack of $^{\cdot}OH$, H_2O_2, and $O_2^{\cdot-}$ on the DNA oligonucleotides (Scheme 6).[65] On the contrary, quercetin is also reported to be carcinogenic agent.[66,67] According to the reports, quercetin has opposite effects on DNA damage induced by cupric ion depending on the concentration of cupric ion (Scheme 7).[68,69] At the low concentration of cupric ions (≤25 μM), quercetin exhibit a protective role. While, at higher concentration of cupric ion (≥25 μM), quercetin enhances the damage to DNA by ROS. Therefore, it is very important to consider the concentration of the chelating metal ions, such as copper or iron while evaluating the protective or degenerative effects of quercetin and other bioflavonoids.

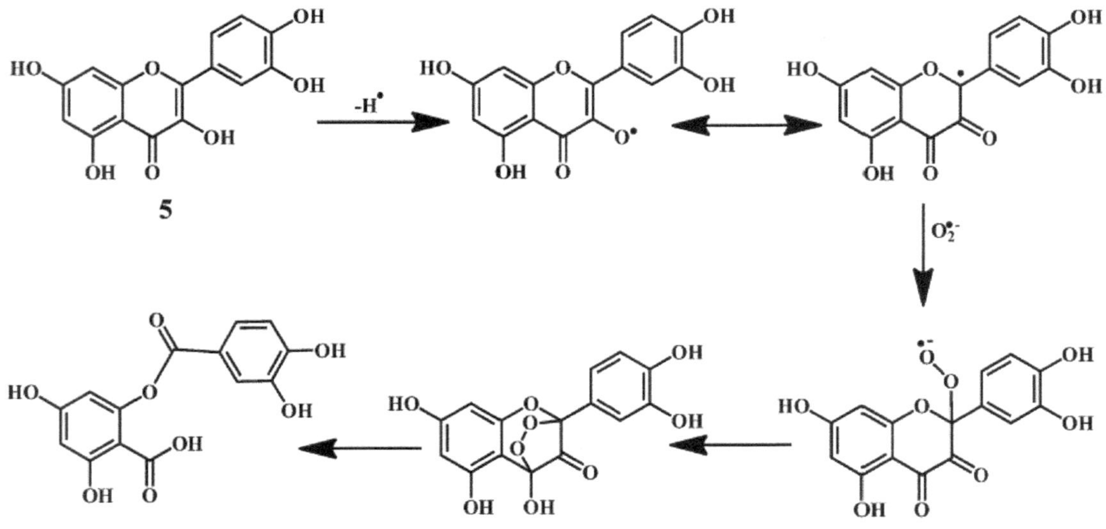

Scheme 6: Mechanism of superoxide anion radical scavenging activity of quercetin 5.

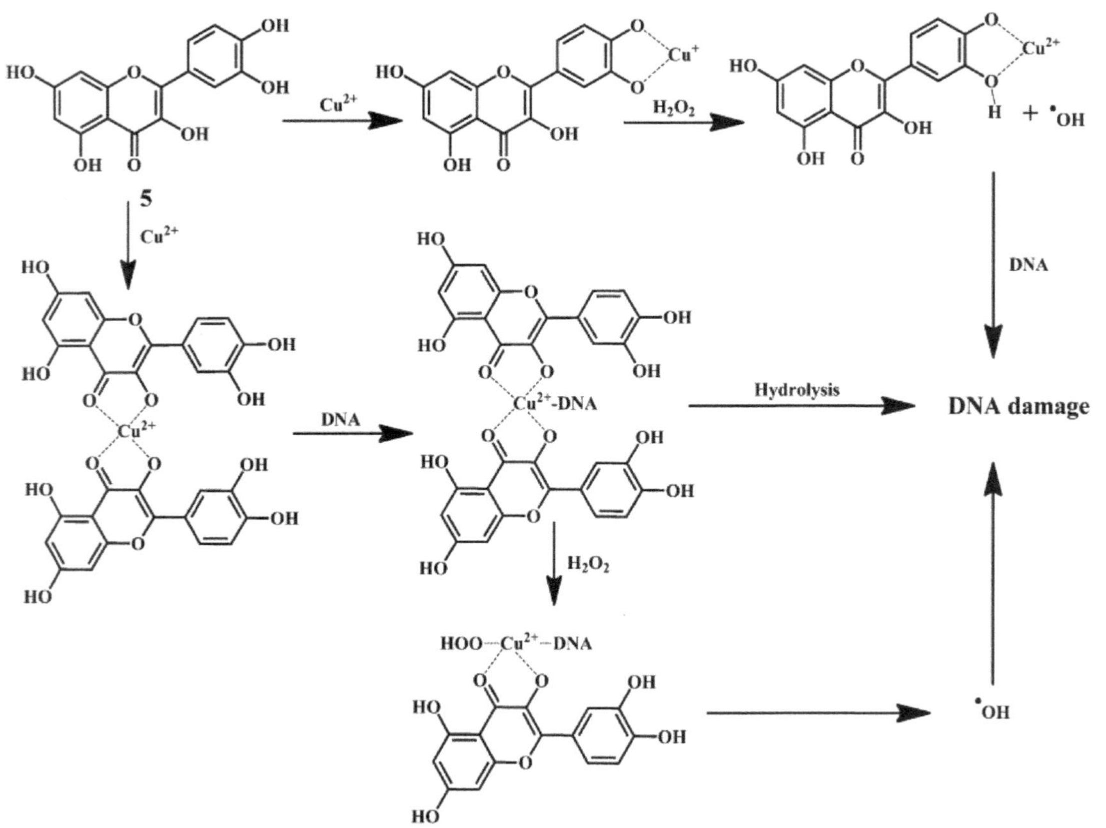

Scheme 7: Mechanism of DNA damage induced by quercetin copper complex.

Anthocynidine, a class of flavonoids are potential antioxidants and their effectiveness in the inhibition of the lipid oxidation is related to their metal ion-chelating activity (Scheme 8) and free-radical scavenging activity (Scheme 9). Three structural groups are important determinants of the radical-scavenging activity of anthocynidines **18–20**.[70] First, the *ortho*-dihydroxy structure in the B-ring. Second, the 2,3 double bond in conjugation. Third, the 4-oxofunction in the C-ring. Flavonoids form complexes with the metal ions by using the 3- or 5-hydroxyl and 4-ketosubstituents or hydroxyl groups in *ortho* position in the B-ring.[71]

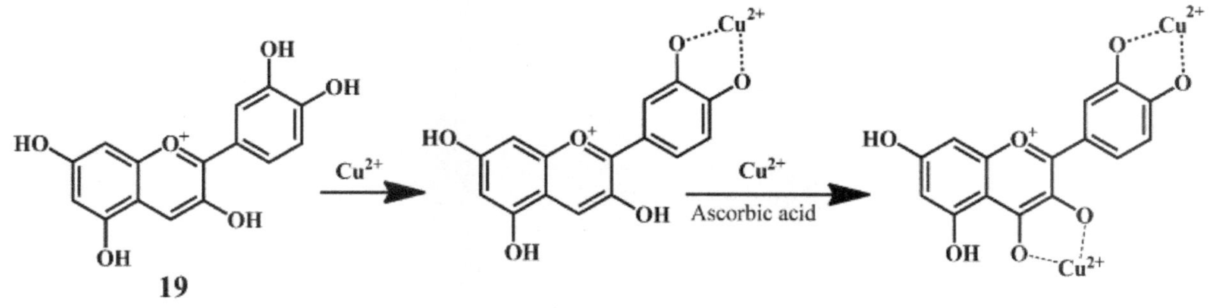

Scheme 8: Metal ion (Cu^{2+}) chelating activity of anthocynidine (cynidin **19**).

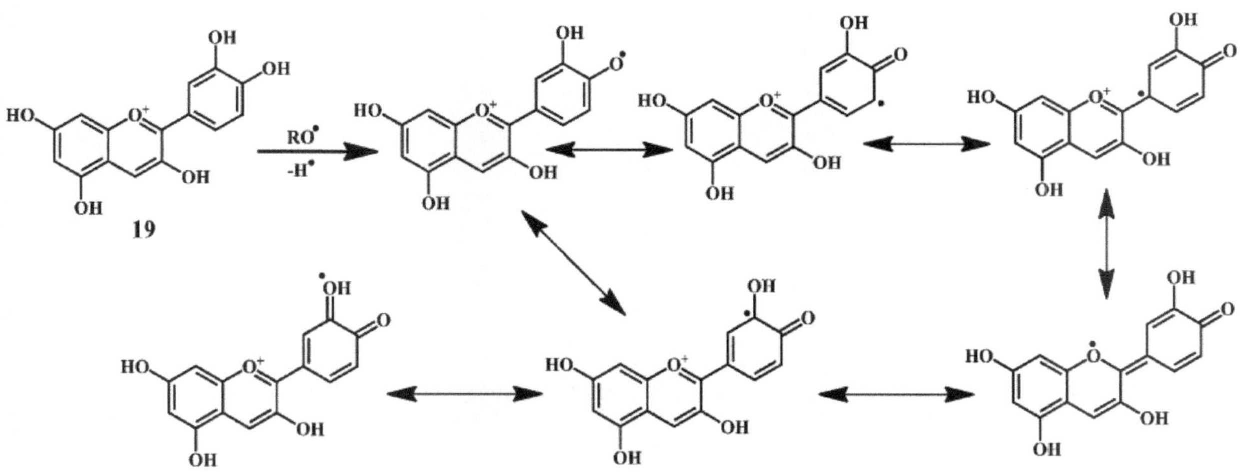

Scheme 9: Mechanism of radical scavenging activity of cynidin **19**.

As shown in the Scheme 9, the anthocynidins (cynidin **19**) can donate an electron (accompanied by a hydrogen nucleus) to a free radical from –OH groups attached to the phenolic rings.[72–74] This electron stabilizes and inactivates the free radical. In this process, the polyphenolic reducing agent changes to an aroxyl radical, which is comparatively more stable due to resonance than the free radical that it has reduced. The overall result is the termination of damaging oxidative chain reactions.

Carotenoids[75,76]

- Carotines: lycopene 26, β-carotene 27,
- Xanthophyll: zeaxanthine 28, lutein 29.

Carotenoids are among the most common lipid soluble phytonutrients. Lycopene **24** and β-carotene **25** are the prominent carotenoids among other 600 different compounds.[77] The biosynthetic pathway as shown in Scheme 10 demonstrates the synthesis of carotenoids **26–29** from phytoene **25**, which is synthesized from two molecules of geranylgeranyl pyrophosphate**24**. Carotenoids are well known to scavenge the peroxyl radicals more efficiently as compared to any other ROS. The peroxyl radicals generated in the process of lipid peroxidation can damage the lipids in the cell wall. Scavenging of peroxyl radicals can disrupt the reaction sequence and prevent the damage to cellular lipids. The long unsaturated alkyl chains in carotenoids make them highly lipophilic. Carotenoids are known to play an important role in the protection of cellular membranes and lipoproteins against the ROS due to their peroxyl radical scavenging activity.[78,79] Carotenoids deactivate the peroxyl radicals by reacting with them to form resonance stabilized carbon-centered radical adducts.

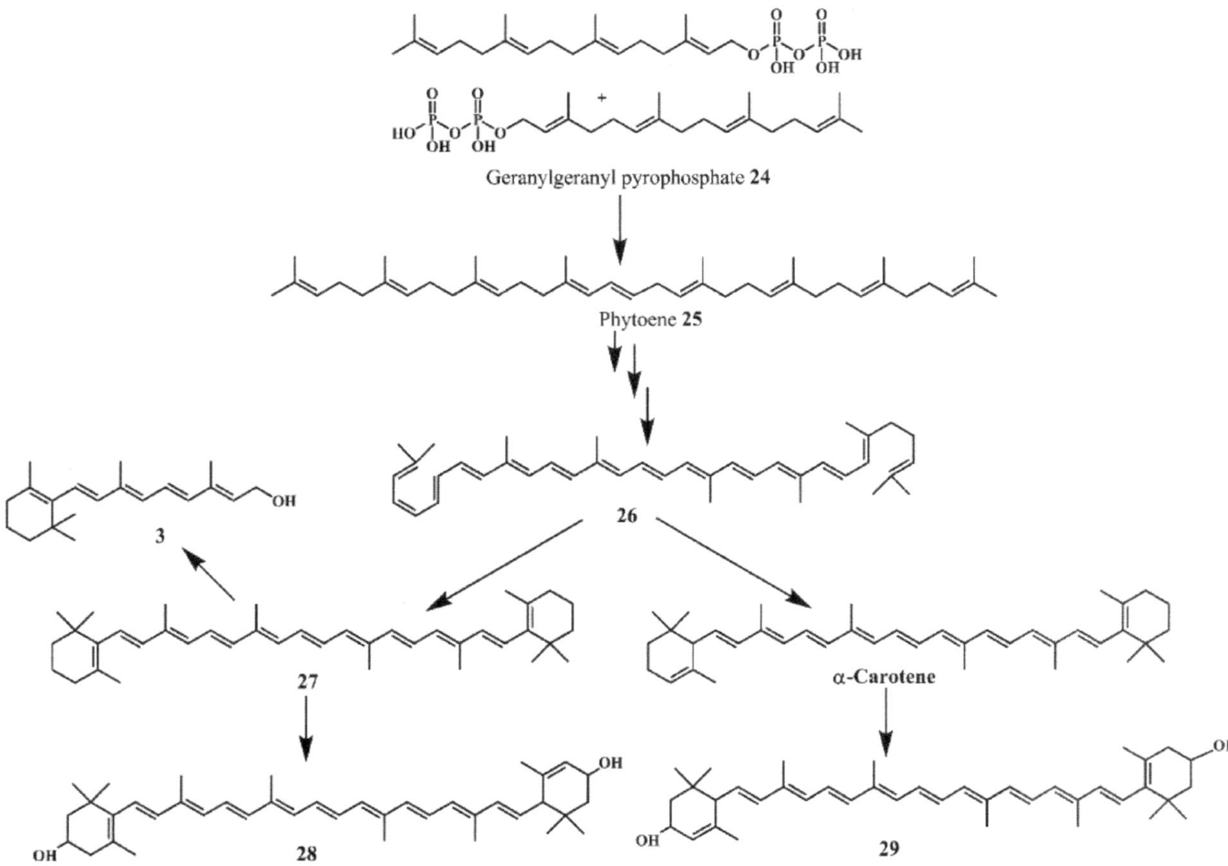

Scheme 10: Biosynthetic pathway for the synthesis of carotenoids **26–29**.

Lycopene **24**, is the most potent antioxidant naturally present in many fruits and vegetables. The high number of conjugated double bonds in lycopen endows it the singlet oxygen quenching ability. Lycopene demonstrate the strong singlet oxygen quenching ability as compared to the α-tocopherol **1** or β-carotene **25**.[80] β-Carotene **12** is a naturally occurring orange-colored carotenoid, abundantly found in the yellow-orange fruits and in dark-green leafy vegetables.[81,82] β-Carotene demonstrates potential antioxidant property due to its chemical structure and the interaction with biological membranes.[83] It is well-known that, the β-carotene quenches singlet oxygen with higher efficiency as compared to the α-tocopherol.[84] In addition, it is also known that the (*Z*)-isomers of the β-carotene possess antioxidant activity *in vitro*.[85,86] Furthermore, the β-carotene can be converted into the two molecules of vitamin A by the β-carotene-15,15′-dioxygenase catalyzed cleavage.

Hydroxycinnamates

The examples are ferulic acid **30**, caffeic acid **31**, sinapic acid **32**, *p*-coumaric acid **33**.

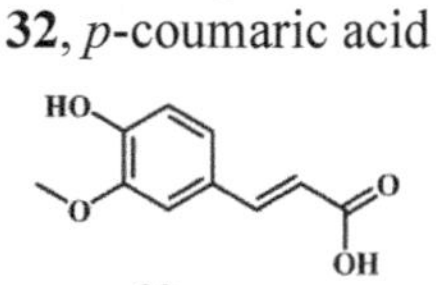

30

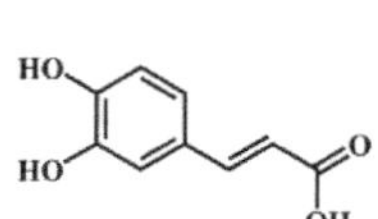

31

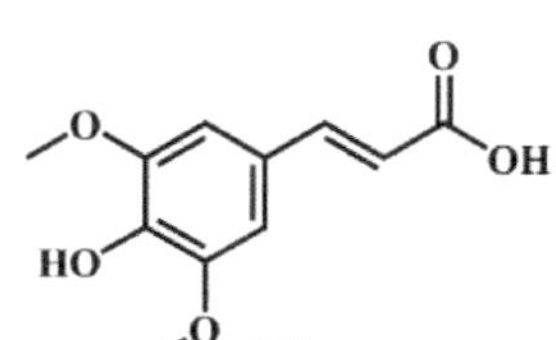

32

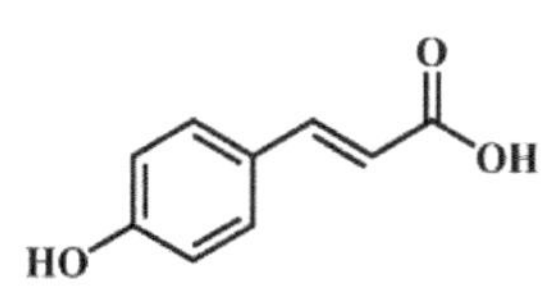

33

It is widely accepted that, the dietary antioxidants that protect LDL from oxidation can prevent the atherosclerosis and coronary heart disease. Hydroxycinnamic acids **30–33** and their conjugates prevent oxidative damage to the LDL.[87] The *in vitro* studies involving human LDL as the oxidizing substrate showed that the hydroxycinnamic acids have higher antioxidant activity as compared to the corresponding hydroxybenzoic acids.[88] The antioxidant activity of the derivatives of the hydroxycinnamates is clearly correlated with the hydroxylation and methylation patterns of the aromatic ring. The antioxidant efficiency of the free hydroxycinnamates on the human LDL oxidation *in vitro*, decreases in the order of caffeic acid **31** > sinapic acid **32** > ferulic acid **30** > *p*-coumaric acid **33**.

The presence of the *o*-dihydroxy group in the phenolic ring (as in caffeic acid) enhances the antioxidant activity of hydroxycinnamic acids toward human LDL oxidation *in vitro*.[89] The radical scavenging antioxidant mechanism of the hydroxycinnamic acids are similar to that of the flavanoids because of their ability to donate an hydroxyl hydrogen and resonance stabilization of the resulting antioxidant radicals. The *o*-dihydroxy substituents also allow the metal ion chelation similar to that of flavanoids.

Other Natural Antioxidants

Theaflavin **34**, theaflavin-3-gallate **35**, allicin **36**, piperine **37**, curcumin **38**.[90,91]

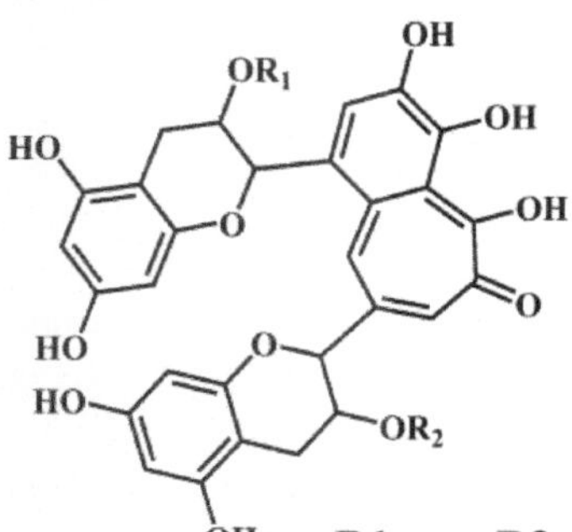

	R1	R2
34	H	H
35	Gallate	H

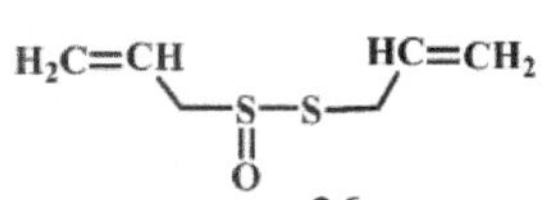

36

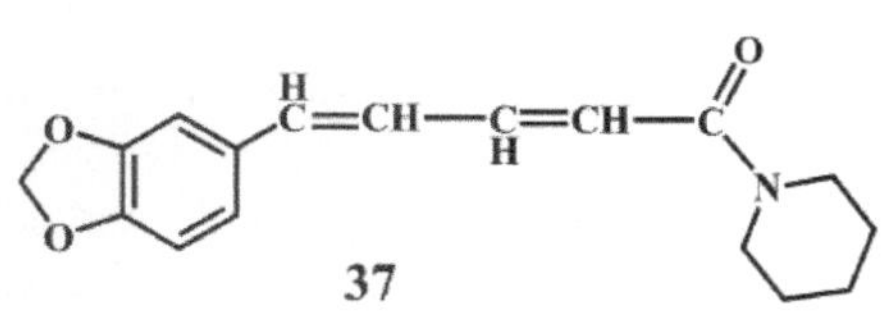

37

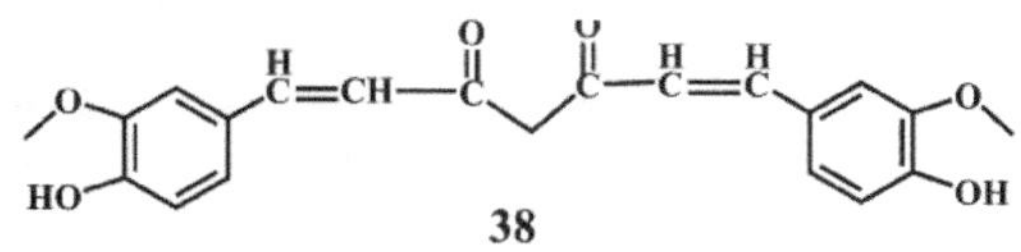

38

Theaflavin **34** and theaflavin-3-gallate **35** possesses *in vitro* antioxidative properties against lipid peroxidation in the erythrocyte membranes and microsomes. They also suppress the mutagenic effects induced by H_2O_2.[92] Theaflavins inhibit the H_2O_2 induced cleavage and mutagenicity of the DNA single-strand.[93,94] In general, theaflavins scavenge the free radicals to produce antioxidative and antimutagenic effects. Apart from the aromatic hydroxyl groups of theaflavins, the gallic acid moiety is essential for their antioxidant activity. The theaflavin-3-gallate **35** is a stronger antioxidant than that of theaflavin **34**. Moreover, the digallate derivatives of theaflavin demonstrate the increased antioxidant activity.

Allicin (diallyl thiosulfinate) **36** is the biologically active compound mainly found in the garlic extracts. Allicin is known to possess various biological activities including the antibacterial, antifungal, and inhibition of cancer promotion.[95] Moreover, allicin is known to reduce serum cholesterol and triglyceride levels as well as atherosclerotic plaque formation and platelet aggregation.[96] Until now, a variety of biological effects of allicin were attributed to antioxidant activity.[97] However, recently it has been found that the active ingredients responsible for the antioxidant property of garlic is 2-propenesulfenic acid and not the allicin.[98] Thiosulfinates undergoes Cope elimination to form sulfenic acids, thioaldehydes or thioketones. The S–S bond in the thiosulfinate is much weaker than the S–C bond in a sulfoxide. Hence, this process can occur at room temperature. Cope elimination is even more susceptible for the allyl (and benzyl) thiosulfinates, such as allicin **36**, because of the weak β C–H bond of the allyl moiety. Allicin is known to undergo Cope elimination at room temperature to give 2-propenesulfenic acid and thioacrolein as shown in the Scheme 11.[99,100]

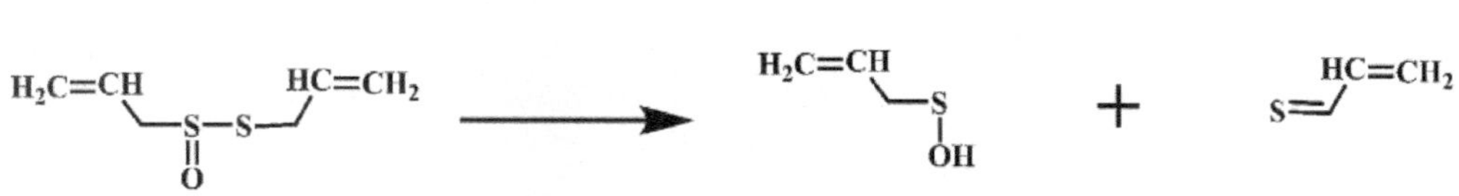

36 **2-propene sulfonic acid** **thioacrolein**

Scheme 11: Cope elimination products of allicilin **36**.

The Scheme 12a demonstrate the mechanism of the radical-scavenging activity of the allicin. The radical-scavenging activity of allicin involves H-atom transfer to a peroxyl radical from the methylene of the allyl group on the divalent sulfur. Scheme 12b demonstrate an alternative mechanism, where the radical-scavenging activity of allicin can be accounted for 2-propenesulfenic acid, which is produced from allicin by Cope elimination.[101] 2-Propenesulfenic acid is reported to be over 1000 times more reactive toward ˙OOH radicals than allicin (2.60×10^7 *vs.* 7.38×10^3 L mol^{-1} s^{-1}, at 298 K).[102]

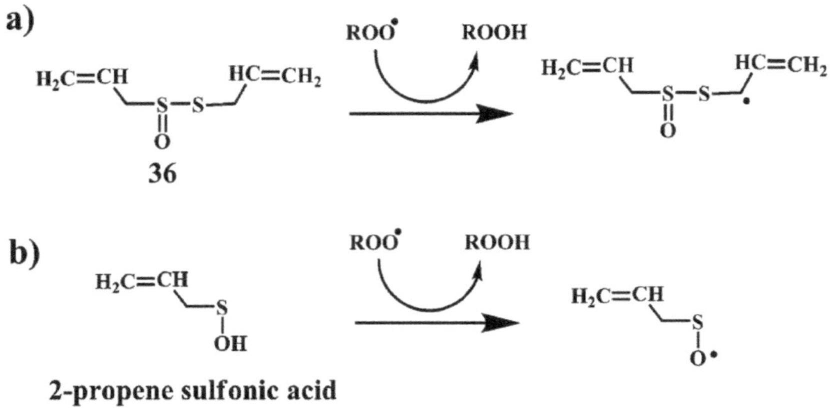

Scheme 12: Mechanism for the radical-trapping activity of (a) allicin and (b) 2-propenesulfenic acid.

Piperine (1-piperoylpiperidine) 37, is an alkaloid present in fruits of black pepper (*Piper nigrum*), long pepper (*Piper longum*), and other piper species (family: Piperaceae). Piperine possesses many pharmacological activities, including anti-inflammatory and analgesic effect,[103] anti-ulcer activities,[104] antidepressant effect,[105] cognitive enhancing effect,[106] cytoprotective effect, and antioxidant activity.[107] It is interesting to notice that, the higher concentration of piperine results in the increased production of the ˙OH. Whereas, in low concentrations piperine acts as an antioxidant.[108] Piperine demonstrates synergistic antioxidant activity by doubling the absorption of dietary curcumin **38**.[109]

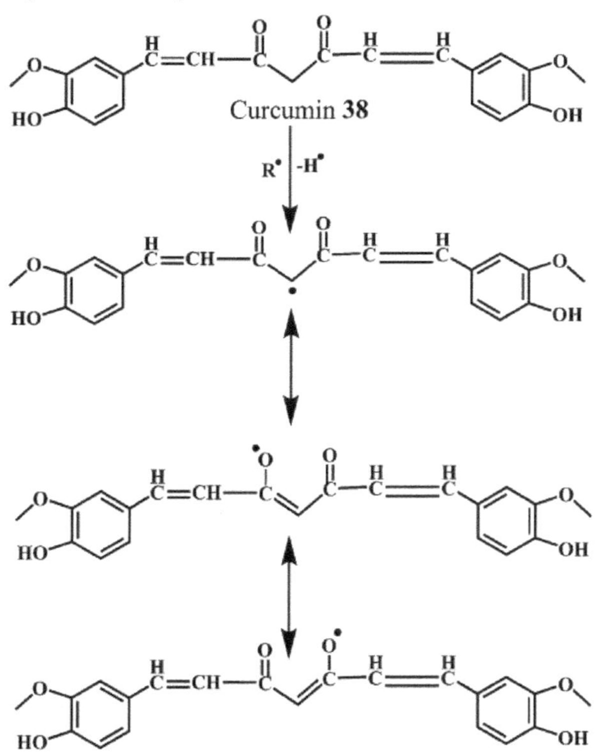

Scheme 13: Mechanism of radical scavenging activity of curcumin **38** initiated by methylenic moiety.

Curcumin **38**, a lipid soluble active principle of turmeric is a bis-α, β-unsaturated β-diketone that exhibit›s keto–enol tautomerism.[110] Curcumin **38**, shows remarkable antioxidant activity, and it has been found to be an excellent free radical scavenger.[111] Curcumin has a chain breaking antioxidant ability comparable to that of the vitamin E.

As shown in the Schemes 13 and 14, the free radical scavenging activity of curcumin is correlated to the phenolic OH group and the CH_2 group of the β-diketone moiety. The free radical can undergo electron transfer or abstract H-atom from either of these two sites. However, pulse radiolysis and other biochemical methods credited the antioxidant activity of curcumin to its phenolic OH group.[112]

The Scheme 14 depicts the mechanism for the autoxidation of curcumin initiated by hydrogen abstraction from one of the phenolic hydroxyl groups.[113] The phenoxyl radical moves into the carbon chain leaving a quinone methide that is eventually quenched by the water molecules. The methide radical performs a 5-exo-cyclization with the double bond to give the cyclopentadione ring and generating the carbon-centered radical.

The reaction of curcumin with the molecular oxygen (O_2) results in the peroxyl radical. The peroxyl radical is then reduced to the hydroperoxide by abstracting a hydrogen atom from another curcumin molecule, propagating the autoxidation chain reaction. Subsequently, the hydroperoxide loses water and rearranges into the spiro-epoxide. The hydrolysis of the epoxide by the (water-derived) hydroxyl group results in the formation of the final bicyclopentadione product.[114] It has been found that the copper complex of curcumin (curcumin–Cu(II)) show promising SOD activity, with improved antioxidant efficacy.[115,116]

The mechanism of the $O_2^{\cdot-}$ scavenging activity of the curcumin–Cu(II) complex is depicted in the Scheme 15. When $O_2^{\cdot-}$ are allowed react with the curcumin–Cu(II) complex, a major fraction of $O_2^{\cdot-}$ reacts with Cu^{2+} moiety, while only a small fraction reacts with curcumin. The reaction causes reduction of Cu^{2+} to Cu^{+}. The Cu^{+} undergoes subsequent oxidation by another molecule of $O_2^{\cdot-}$, thereby regenerating the parent complex.

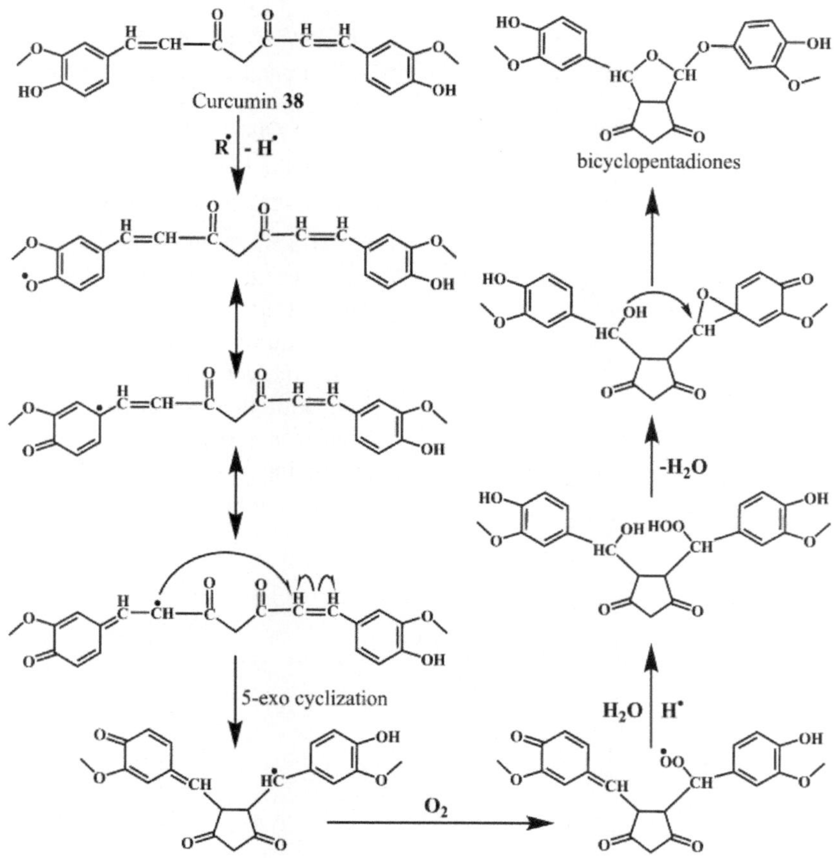

Scheme 14: Mechanism of radical scavenging activity of curcumin **38** initiated by phenolic moiety.

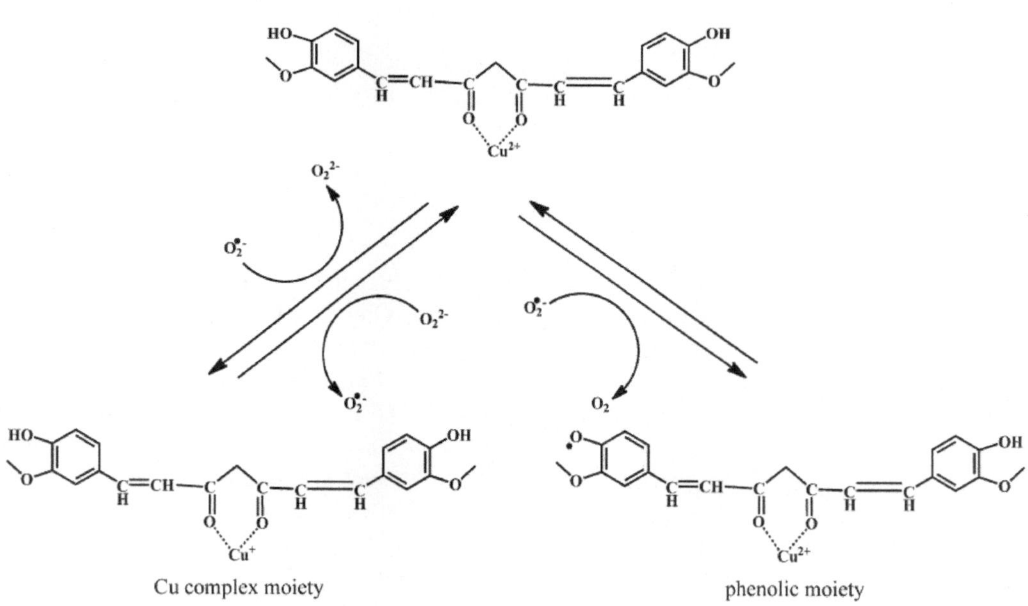

Scheme 15: Mechanism of radical scavenging activity of curcumin–Cu(II) complex.

Therefore, the catalytic activity comes mainly from the reversible redox reactions within the Cu^{2+}/Cu^{+} couple in the complex. However, in presence of the excess $O_2^{\cdot -}$, the phenolic moiety undergoes oxidation resulting in the production of the phenoxyl radicals. Then these phenoxyl radicals can generate new products or react with reduced copper ions of the complex resulting in the regeneration of the complex.

Physisological Antioxidants

Uric acid **39** in plasma, and GSH **40**.

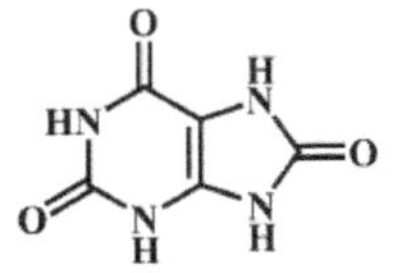

39

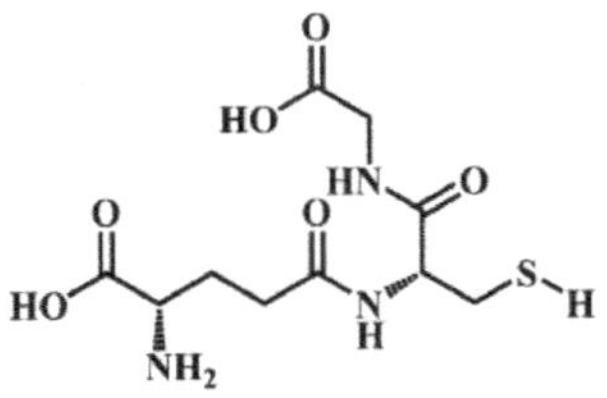

40

Uric acid **39** in plasma possesses strong radical scavenging activity.[117,118] Uric acid is the most abundant aqueous antioxidant found in humans. It contributes for as much as two-thirds of all free radical scavenging activities in the plasma.[119] Uric acid is a powerful scavenger of carbon-centered radicals and peroxyl radicals in the hydrophilic environment. However, it loses it's radical scavenging activity within lipid membranes.[120]

Uric acid is an exceptional scavenger of peroxynitrite ($ONOO^-$) in the extracellular fluid.[121] However, it is important to note that the uric acid cannot scavenge the $O_2^{\bullet -}$. Moreover, uric acid requires the presence of ascorbic acid (Scheme 16) and thiols for the complete scavenging of peroxynitrites. Neither of these antioxidants (ascorbic acid, thiols) alone can prevent reaction of peroxynitrite with tetrahydrobiopterin, which leads to uncoupling of nitric oxide ($NO^\bullet$) synthase.[122] This indicates that the uric acid plays a crucial role in the scavenging of the peroxynitrite.

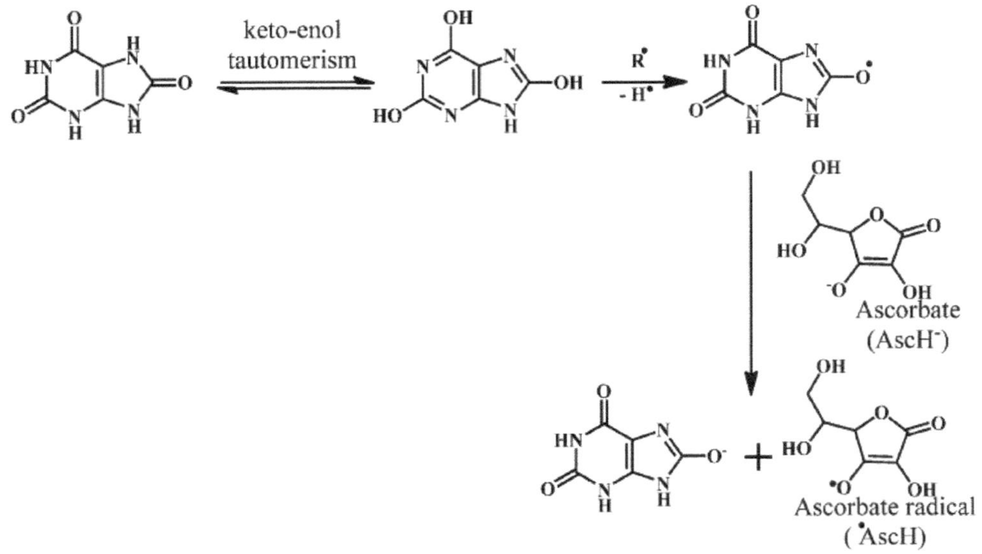

Scheme 16: Mechanism of radical scavenging activity of uric acid.

GSH **40** in cell cytosol, together with its related enzymes, comprises a system that maintains the intracellular reducing environment, which acts as primary defense against excessive generation of harmful ROS.[123,124] The oxygen radical scavenging activity of GSH directly expedites the ROS neutralization and the repair of ROS-induced damage.[125]

As shown in the Scheme 17, three groups of enzymes can be identified in the GSH catalytic cycle: glutathione oxidase, glutathione reductase, and GSHPx. Glutathione oxidase and GSHPx catalyze the oxidation of GSH to GSH disulfide (GSSG). Whereas, glutathione reductase is responsible for the regeneration of GSH from GSSG in an NADPH-dependent process.[126] Cells can produce GSSG or convert it to GSH by using NADPH in the presence of the glutathione reductase. However, the *de nova* synthesis of glutathione from its amino acid constituents is required for the elevation of glutathione as an adaptive response to oxidative stress. The presence of the sulfhydryl group in glutathione allows it to serve as an antioxidant.

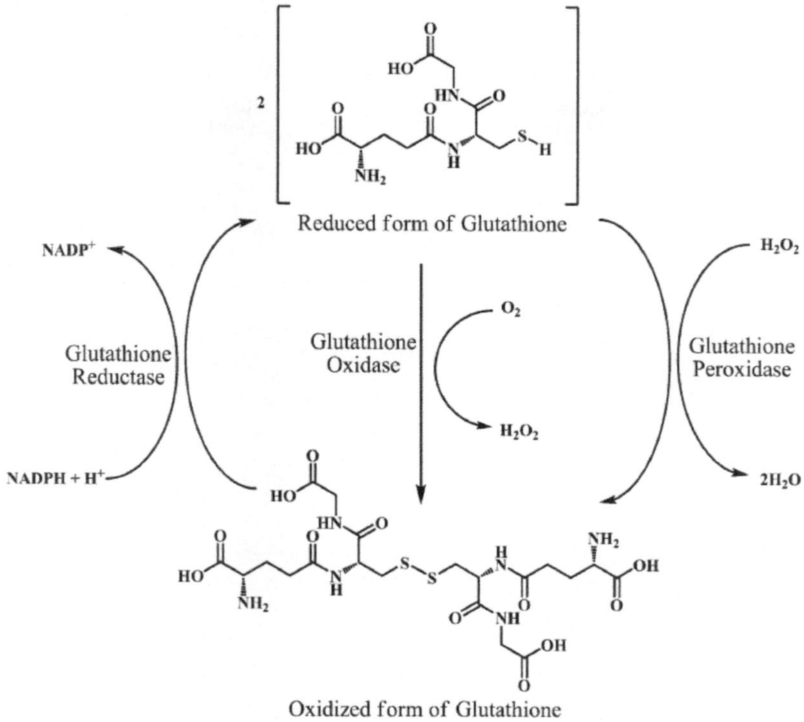

Scheme 17: Interconversion of glutathione in its reduced form (GSH) and oxidized form (GSSG) by the action of glutathione oxidase, glutathione reductase, and glutathione peroxidase enzymes.

Fungal Antioxidants

The microorganisms such as *Ganoderma lucidum*,[127] *Ganoderma applanatum*, *Meripilus giganteus*, *Flammulina velutipes*, and *Endophytic Fungi*[128,129] possess a very efficient antioxidative system consisting of enzymatic (peroxidases, lace, catalase, and superoxide dismutase) and nonenzymatic elements (phenolic derivatives or polysaccharides).

The synthetic antioxidants are the second type of nonenzymatic antioxidants. Cinnamic acid derivatives[130,131] **41**, **42**, melatonin **43**, selegiline **44**, are the few examples of the synthetic antioxidants.[132,133]

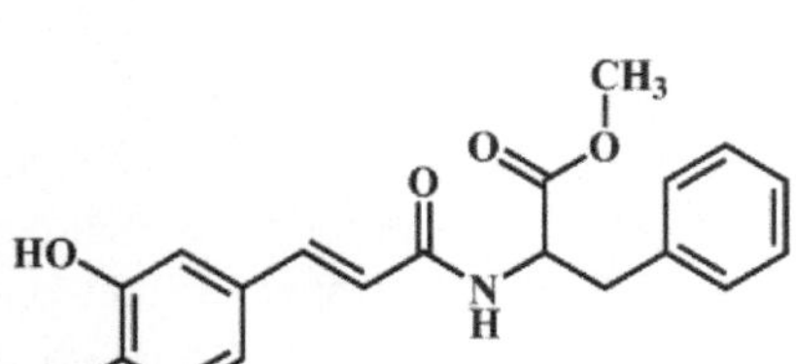

41

42

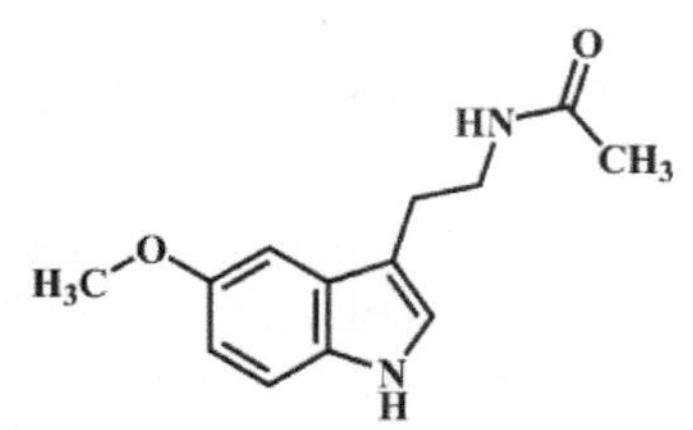

43

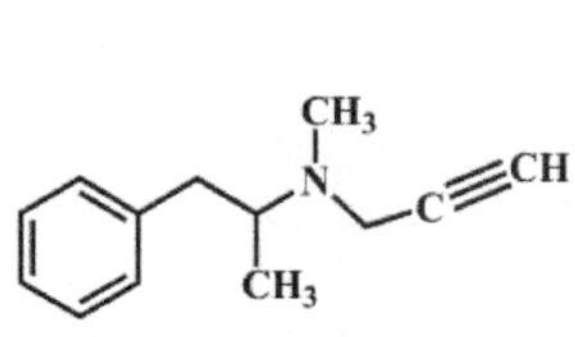

44

IN VITRO METHODS FOR EVALUATION OF ANTIOXIDANT ACTIVITY

Various *in vitro* methods are available for the evaluation of antioxidant activity of different compounds.[134–137]

Assay of Superoxide Anion Radical Scavenging Activity

SOD is an antioxidant enzyme involved in scavenging the ROS.[138] SOD converts the $O_2^{\cdot-}$ to H_2O_2. The H_2O_2 is then converted to the O_2 and H_2O in the reaction catalyzed by GSHPx and CAT.[139] There are several classes of SOD, which include intracellular copper, zinc SOD (Cu, Zn SOD/SOD_1), mitochondrial manganese SOD (Mn SOD/SOD_2), and extracellular Cu, Zn SOD (EC SOD/SOD_3).

The method for the evaluation of the $O_2^{\cdot -}$ scavenging activity of antioxidants is explained here by using PMS–NADH–NBT system, which is composed of *N*-methylphenazine methosulphate (PMS), nitroblue tetrazolium chloride (NBT), and NADH (a reduced form of nicotineamide-adenine-dinucleotide).

As shown in the Scheme 18, the $O_2^{\cdot -}$ produced in the coupling reaction of PMS–NADH in presence of dissolved oxygen reduces NBT. The decrease of absorbance at 560 nm with antioxidant indicates the consumption of $O_2^{\cdot -}$ in the reaction mixture. The $O_2^{\cdot -}$ scavenging activity can be measured as described by Robak and Gryglewski.[140] Gallic acid, BHA, ascorbic acid, α-tocopherol, and curcumin can be used as positive controls in this assay.

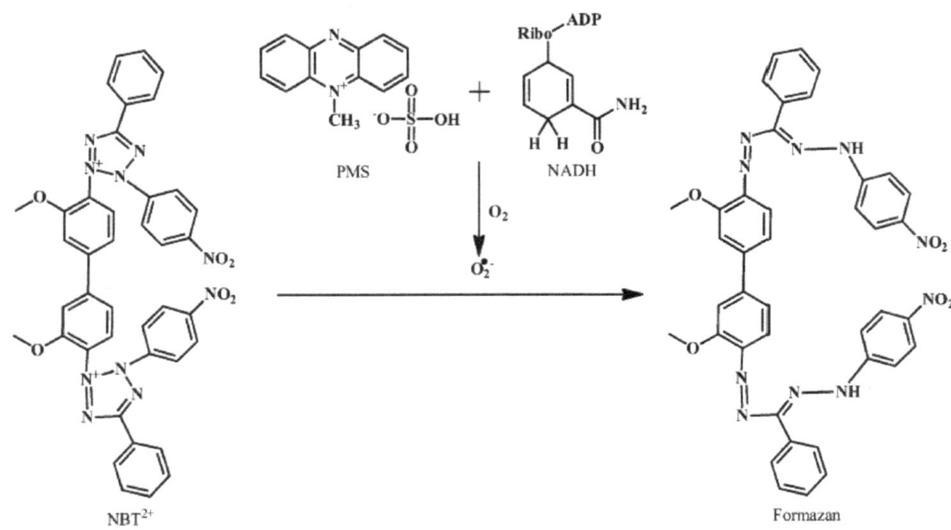

Scheme 18: Reduction of NBT by superoxide anion radical produced in PMS– NADH reaction.

Assay of 2,2-diphenyl-1-picrylhydrazyl (DPPH) Free Radical Scavenging Activity

Evaluation of the antioxidant activity of any compound can be carried out either by *in vitro* or *in vivo* models.[141,142] DPPH is a stable free radical that can accept an electron or hydrogen radical to become a stable diamagnetic molecule.

Due to its odd electron, the methanolic solution of DPPH shows a strong absorption band at 517 nm. As shown in the Scheme 19, the DPPH radical reacts with suitable reducing agent producing new bond, thus changing the color of solution. The solution loses color with the increase in the concentration of antioxidant as the electrons taken up by DPPH radical from the antioxidant.[143,144] Such reactivity has been used to test the ability of compounds/plant extracts to act as free radical scavengers.[145] Reduction of the DPPH radicals can be monitored spectrophotometrically by the decrease in absorbance at 517 nm.

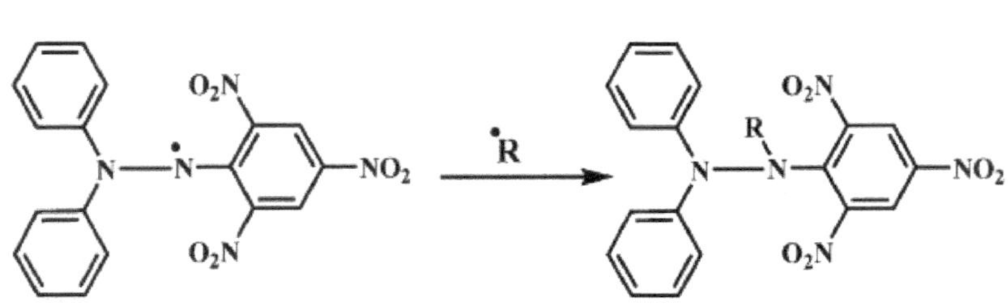

DPPH

Scheme 19: Reaction of DPPH radical with other radicals ($^{\cdot}R = {^{\cdot}H}$, alkyl radical *etc.*).

Assay for Total Reactive Oxygen Potential (TRAP) and Total Antioxidant Reactivity (TAR)

Luminol enhanced chemiluminescence is used to measure TRAP and TAR.[146,147] When the luminol is allowed to react with the free radical source, a steady chemiluminescence is observed that can be directly correlated to the rate of luminol oxidation.[148] The addition of free radical scavengers reduces the chemiluminescence intensity.[149] The effect of antioxidants on the induced chemiluminescence intensity of luminol by radicals derived from the thermolysis of 2,2′-azobis-2-amidinopropanedihydrochloride (AAPH) can be employed to monitor the TRAP and TAR levels.[150]

As shown in the Scheme 20, the AAPH undergoes thermal decomposition in solution to produce two carbon-centered amidino propane (AP) radicals, which can add O_2 to form peroxyl radicals. However, the carbon-centered radicals usually predominate.[151] The amidino propane (AP) radical takes up a proton from luminol to produce a luminol radical. The luminol radical reacts with de-protonated H_2O_2 to yield a short-lived hydroperoxide intermediate (LO_2H^-), which rapidly decomposes into the excited state 3-aminophthalic acid (AP*). The AP* loses energy in the form of chemiluminescence to give ground state 3-aminophthalic acid.[152]

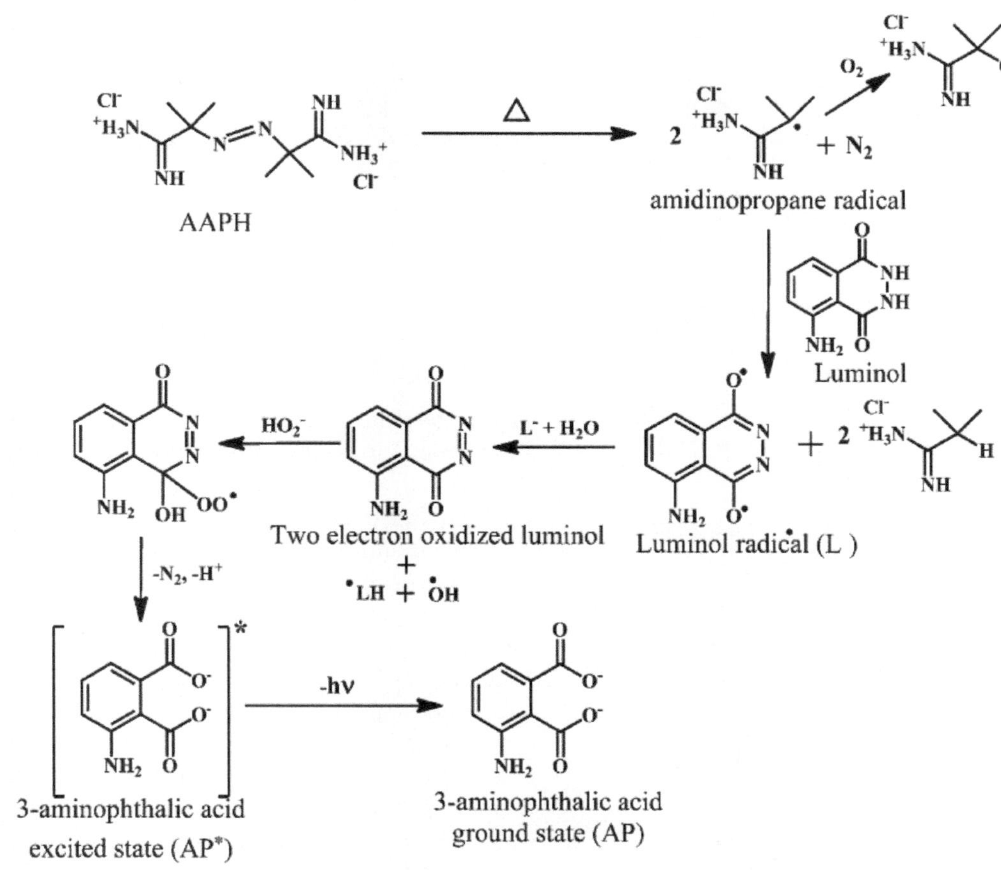

Scheme 20: Mechanism of AAPH induced chemiluminescence of luminol.

In vitro Antioxidant Evaluation by Phospholipids Peroxidation

Lipid peroxidation is an oxidative degradation of lipids.[153] In this process, the free radical takes up the electrons from the lipids in cell membranes, which results in the cell damage.[154,155]The tentative mechanism for this free radical chain reaction involved in the phospholipid peroxidation is depicted in the Scheme 21.

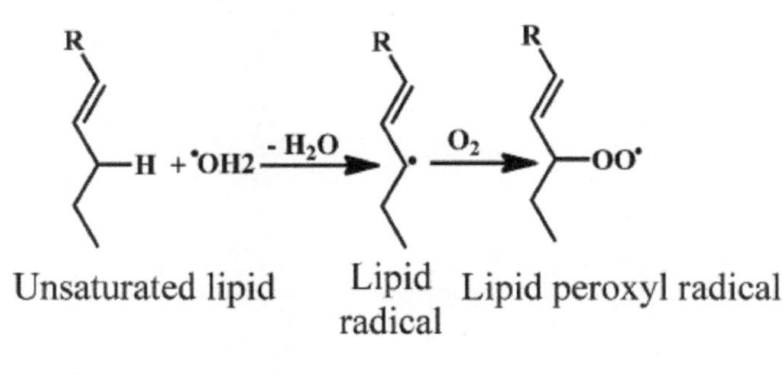

Scheme 21 Phospholipid peroxidation of unsaturated lipids.

The activity of test compound to inhibit peroxidation of membrane lipids at pH 7.4 is tested using phospholipids. The interference of the test drug with color development is determined by adding a previously determined concentration of the test compound to the TBA reagents and used to determine the extent of peroxidation of animal phospholipids.[156,157] In this assay, the antioxidant activity is a measure of concentration-dependent inhibition of a phospholipid peroxidation.

In vitro Antioxidant Evaluation by Deoxyribose Assay

The ˙OH in presence of ascorbic acid attack the sugar deoxyribose to generate the product that on heating with thiobarbituric acid (TBA) or thiobarbiturate reactive substances (TBARS), at low pH, yield a chromogen. Therefore, the deoxyribose assay can be used to detect ˙OH scavenging activity of test compounds.

The reaction of deoxyribose and ˙OH has been discussed extensively in the literature.[158,159] The ˙OH attack deoxyribose to form products that react with TBA upon heating at low pH and yield a pink chromogen. Scheme 22 depicts the proposed mechanism of chromogen formation from reaction of deoxyribose and ˙OH followed by reaction with TBARS.

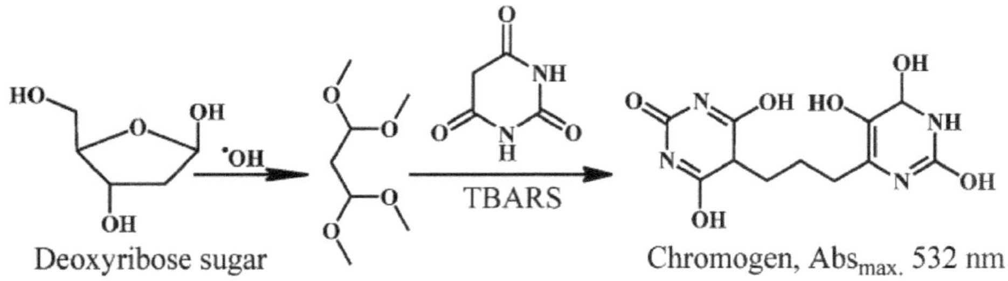

Scheme 22: Reaction of deoxyribose sugar with hydroxyl radical in presence of TBARS.

In general, the *in vivo* assays for testing potential antioxidants are more expensive because they require complex cellular testing systems or full clinical trials. However, it is very important to proceed to cellular assays after screening antioxidant activity with an *in vitro* method in order to obtain information on some aspects like uptake, bioavailability, and metabolism.[160] The new definition of an antioxidant, a redox-active compound or mixture able to modulate the redox status of the cell, makes it critical to use *in vivo* assays in order to evaluate the antioxidant activity of a compound.[161] There are several reports on the *in vivo* assays for the evaluation of the antioxidant activity.[162,163] However, we have limited the scope of this review to the *in vitro* assays for the evaluation of the antioxidant activities of natural antioxidants. There are several other reports, which elaborate the advantages and disadvantages of various methods for the evaluation of antioxidant activity.[164,165]

CURRENT TRENDS AND FUTURE DIRECTIONS

In recent years, there has been upsurge in the novel approaches for the study of free radicals and antioxidants in relation with the improvement of human health. Multiple studies have showed that the neuronal and behavioral changes occur with ageing, even in the absence of degenerative disease. Recent studies have found the association between the lower status of dietary antioxidants and decline in the cognitive function. The evidences from the experimental, clinical, and epidemiological studies indicate that the consumption of foods containing high levels of dietary antioxidants may prevent or reduce the risk of cognitive deterioration. Tempol, an example of a new class of SOD mimetic drugs, alleviates acute and chronic pain. These drugs substantially reduce the tissue damage incurred by inflammation. The speculations of the relations between radical damage and disease conditions need to the support of by more secure data. The knowledge on the mechanisms of various physiological radical reactions and the mechanisms of the antioxidants in scavenging those free radicals will open up the path for more potent drug molecules.

Many investigators found that, increasing the level of defense mechanisms against oxidative stress could extend an organism's health span. Therefore, few setbacks in the antioxidant research with the molecules showing strong antioxidant activity *in vitro* and non-antioxidant effects in cells and tissues should not discourage the important research in this field. Finally, the collective effort is must be undertaken for the understanding of the mechanisms in the free radical scavenging activities of known antioxidants to derive the potent antioxidants.

CONCLUSION

ROS, the radical derivatives of oxygen are the most important free radical in biological systems. The ROS are the harmful byproducts generated during the normal cellular functions. Increasing intake of natural antioxidants may help to maintain a tolerable antioxidant status, perhaps the normal physiological functioning. The reported chemical evidence suggests that the dietary antioxidants help in the disease prevention. The antioxidant compounds react in one-electron reactions with free radicals *in vitro* and prevent the oxidative damage. Therefore, it is very important to understand the reaction mechanism of antioxidant with the free radicals. The reaction mechanisms can be used to evaluate the antioxidant activity of various naturally occurring antioxidant compounds. This review elaborates the mechanism of action of the natural antioxidant compounds and assays for the evaluation of their antioxidant activities. The reaction mechanisms of the antioxidant assays are briefly discussed (165 references). The scope of this article is limited to the natural antioxidants and the *in vitro* assays for evaluation of their antioxidant properties.

ABBREVIATIONS

ROS	Reactive oxygen species
RNS	Reactive nitrogen species
SOD	Superoxide dismutase
CAT	Catalase
GSHPx	Glutathione peroxidase
MPO	Myeloperoxidase
$O_2^{\bullet -}$	Superoxide anion radical
H_2O_2	Hydrogen peroxide
$^{\bullet}OH$	Hydroxyl radical
EDRF	Endothelium derived relaxation factor
CHD	Coronary heart disease
PUFA	Polyunsaturated fatty acids
DNA	Deoxyribonucleic acid
Prx	Peroxiredoxin
LDL	Low density lipoprotein
GSH	Glutathione
GSSG	Glutathione disulfide
PMS	*N*-Methylphenazine methosulphate
NBT	Nitroblue tetrazolium chloride
DPPH	2,2-Diphenyl-1-picrylhydrazyl
TRAP	Total reactive oxygen potential

TAR	Total antioxidant reactivity
AAPH	2,2′-Azobis-2-amidinopropanedihydro-chloride
TBA	Thiobarbituric acid
TBARS	Thiobarbiturate reactive substances

ACKNOWLEDGEMENTS

This research was supported by the Hallym University Research Fund (HRF-201411-014).

NOTES AND REFERENCES

1. I. S. Young and J. V. Woodside, *J. Clin. Pathol.*, 2001, **54**, 176–186 .
2. K. Bagchi and S. Puri, *East. Meditrr. Health. J.*, 1998, **4**, 350–360 .
3. Y. W. Kim and T. V. Byzova, *Blood*, 2014, **123**, 625–631 .
4. H. Sies, *Exp. Physiol.*, 1997, **82**, 291–295 .
5. D. K. Pal and S. B. Nimse, *Asian J. Chem.*, 2006, **13**, 3004–3008 .
6. C. E. Thomas and B. Kalyanaraman, in *Oxygen Radical and the Disease Process*, Harvard Academic Publishers, Netherlands, 1997 .
7. F. Shahidi and Y. Zhong, *Eur. J. Lipid Sci. Technol.*, 2010, **112**, 930–940 .
8. B. M. Babior, *Blood*, 1999, **93**, 1464–1476 .
9. J. A. Knight, in *Free radicals, antioxidants, aging and disease*, AACC Press, Washington, 1999 .
10. L. Zhu, C. Gunn and J. S. Beckman, *Arch. Biochem. Biophys.*, 1992, **298**, 452–457 .
11. S. Moncanda, R. M. J. Palmer and E. A. Higgs, *Pharmacol. Rev.*, 1991, **43**, 109–142 .
12. C. F. Nathan and J. B. Hibbs Jr, *Curr. Opin. Immunol.*, 1991, **3**, 65–70 .
13. F. Shahidi and Y. Zhong, *Chem. Soc. Rev.*, 2010, **39**, 4067–4079 .
14. A. V. Rao and S. Agarwal, *Nutr. Res.*, 1999, **19**, 305–326 .
15. A. M. Jenkinson, A. R. Collins, S. J. Duthie, K. W. J. Wahle and G. G. Duthie, *FASEB J.*, 1999, **13**, 2138–2142 .
16. Y. Park, S. Nam, H. J. Yi, H. J. Hong and M. Lee, *Nutr. Res.*, 2009, **29**, 812–818 .
17. H. Esterbauer, M. Dieber-Rotheneder, G. Waeg, G. Striegl and G. Juergens, *Chem. Res. Toxicol.*, 1990, **3**, 77–92 .
18. W. A. Pryor and N. A. Porter, *Free Radical Biol. Med.*, 1990, **8**, 541–543 .
19. T. P. A. Devasgayam, K. K. Boloor and T. Ramasarma, *Indian J Biochem. Biophys.*, 2003, **40**, 300–308 .
20. H. Esterbauer, R. J. Schanr and H. Zollen, *Free Radical Biol. Med.*, 1991, **11**, 81–121 .
21. M. Dizdaroglu, P. Jaruga, M. Birincioglu and H. Rodriguez, *Free Radical Biol. Med.*, 2002, **32**, 1102–1115.
22. B. Halliwell and J. M. C. Gutteridge, in *Free radicals in biology and medicine*, Oxford University Press, Oxford, 4th edn, 2007 .
23. C. von Sonntag, in *Free-radical-induced DNA damage and its repair*, Springer, Hiedelberg, 2006 .
24. M. Dizdaroglu, *Mutat. Res.*, 1992, **275**, 331–342 .
25. A. P. Breen and J. A. Murphy, *Free Radical Biol. Med.*, 1995, **18**, 1033–1077 .
26. M. Dizdaroglu and P. Jaruga, *Free Radical Res.*, 2012, **46**, 382–419 .
27. J. A. Simpson, S. Narita, S. Gieseg, S. Gebicki, J. M. Gebicki and R. T. Dean, *Biochem. J.*, 1992, **282**, 621–624 .
28. S. Gieseg, S. Duggan and J. M. Gebicki, *Biochem. J.*, 2000, **350**, 215–218 .
29. E. R. Stadtman, *Free Radical Biol. Med.*, 1990, **9**, 315–325 .
30. M. S. Cooke, M. D. Evans, M. Dizdaroglu and J. Lunec, *FASEB J.*, 2003, **17**, 1195–1214 .
31. P. M. Kris-Etherton, K. D. Hecker, A. Bonanome, S. M. Coval, A. E. Binkoski, K. F. Hilpert, A. E. Griel and T. D. Etherton, *Am. J. Med.*, 2002, **113**, 71–88 .
32. D. Pal, S. Banerjee and A. Ghosh, *J. Adv. Pharm. Technol. Res.*, 2012, **3**, 16–24 .
33. E. A. Lissi, M. Salim-Hanna, C. Pascual and M. Deltillo, *Free Radical Biol. Med.*, 1995, **18**, 153–158 .
34. M. Koruk, S. Taysi, M. C. Savas, O. Yilmaz, F. Akcay and M. Karakok, *Ann. Clin. Lab. Sci.*, 2004, **34**, 57–62 .
35. D. R. Gough and T. G. Cotter, *Cell Death Dis.*, 2011, **2**, e213 .
36. C. D. Zhan, R. K. Sindhu, J. Pang, A. Ehdaie and N. D. Vaziri, *J. Hypertens.*, 2004, **22**, 2025–2033 .
37. J. R. Stone and S. Yang, *Antioxid. Redox Signaling*, 2006, **8**, 243–270 .
38. E. Cabiscol, J. Tamarit and J. Ros, *Int. Microbiol.*, 2000, **3**, 3–8 .
39. J. R. Arthur, *Cell Mol. Life Sci.*, 2000, **57**, 1825–1835 .
40. J. M. Matés, *Toxicology*, 2000, **153**, 83–104 .
41. M. Valko, C. J. Rhodes, J. Moncol, M. Izakovic and M. Mazur, *Chem.-Biol. Interact.*, 2006, **160**, 1–40 .
42. J. M. Matés, C. Perez-Gomez and I. N. De tro, *Clin. Biochem.*, 1999, **32**, 595–603 .
43. S. Tafazoli, J. S. Wright and P. J. O'Brien, *Chem. Res. Toxicol.*, 2005, **18**, 1567–1574 .
44. A. Vojdani, M. Bazargan, E. Vojdani and J. Wright, *Cancer. Detect. Prev.*, 2000, **24**, 508–523 .
45. P. K. Witting, J. M. Upston and R. Stocker, *Biochemistry*, 1997, **36**, 1251–1258 .
46. P. Morlière, L. K. Patterson, C. M. M. Santos, A. M. S. Silva, J. Mazière, P. Filipe, A. Gomes, E. Fernandes, M. B. Q. Garciah and R. Santusc, *Org. Biomol. Chem.*, 2012, **10**, 2068–2076 .
47. R. Stocker, V. W. Boury and B. Frei, *Proc. Natl. Acad. Sci. U. S. A.*, 1991, **88**, 1646–1650 .

48. E. Niki, *Free Radical Biol. Med.*, 2014, **66**, 3–12 .

49. E. Niki, *Am. J. Clin. Nutr.*, 1991, **54**, 1119S–1124S .

50. K. L. Retsky, M. W. Freeman and B. Frei, *J. Biol. Chem.*, 1993, **268**, 1304–1309 .

51. C. Oh, M. Li, E. Kim, J. S. Park, J. Lee and S. W. Ham, *Bull. Korean Chem. Soc.*, 2010, **31**, 3513–3514 .

52. B. R. Monaghan and F. O. Schmitt, *J. Biol. Chem.*, 1932, **96**, 387–395 .

53. R. S. Parker, *FASEB J.*, 1996, **10**, 542–551 .

54. A. V. Vieira, W. J. Schneider and P. M. Vieira, *J. Endocrinol.*, 1995, **146**, 201–207 .

55. M. A. Livrea, L. Tesoriere, A. Bongiorno, A. M. Pintaudi, M. Ciaccio and A. Riccio, *Free Radical Biol. Med.*, 1995, **18**, 401–409 .

56. L. Tesoriere, D. D'arpa, R. Re and M. A. Livrea, *Arch. Biochem. Biophys.*, 1997, **343**, 13–18 .

57. P. G. Pietta, *J. Nat. Prod.*, 2000, **63**, 1035–1042 .

58. C. A. Rice-Evans, N. J. Miller and G. Paganga, *Free Radical Biol. Med.*, 1996, **20**, 933–956 .

59. P. Pietta, *J. Nat. Prod.*, 2000, **63**, 1035–1042 .

60. R. J. Nijveldt, E. van Nood, D. E. C. van Hoorn, P. G. Boelens, K. van Norren and P. A. van Leeuwen, *Am. J. Clin. Nutr.*, 2001, **74**, 418–425 .

61. A. Russo, R. Acquaviva, A. Campisi, V. Sorrenti, C. D. Giacomo, G. Virgata, M. L. Barcellona and A. Vanella, *Cell Biol. Toxicol.*, 2000, **16**, 91–98 .

62. F. V. Rubens de Souza and F. W. De Giovani, *Redox Rep.*, 2004, **9**, 97–104 .

63. T. Armida, T. Maurizio, T. Andrea and B. Sergio, *J. Mol. Struct.*, 2005, **44**, 759–766 .

64. J. Zhou, L. F. Wang, J. Y. Wang and N. Tang, *Transition Met. Chem.*, 2001, **26**, 57–93 .

65. V. Krishnamachari, L. H. Levine and P. W. Pareá, *J. Agric. Food Chem.*, 2002, **50**, 4357–4363 .

66. A. Das, J. H. Wang and E. J. Lien, *Prog. Drug Res.*, 1994, **42**, 133–166 .

67. S. J. Duthie, J. W. ohnson and V. L. Dobson, *Mutat. Res.*, 1997, **390**, 141–151 .

68. T. Jun, Z. Liancai and W. Bochu, *Int. J. Pharmacol.*, 2007, **3**, 19–26 .

69. D. Galaris and A. Evangelou, *Crit. Rev. Oncol. Hematol.*, 2002, **42**, 93–103 .

70. S. S. Pekkarinen, I. M. Heinonen and A. I. Hopia, *J. Sci. Food Agric.*, 1999, **79**, 499–506 .

71. M. G. Miguel, *J. Appl. Pharm. Sci.*, 2011, **01**, 07–15 .

72. A. tañeda-Ovando, M. L. Pacheco-Hernández, M. E. Páez-Hernández, J. A. Rodríguez and G.-C. Vidal, *Food Chem.*, 2009, **113**, 859–871 .

73. R. J. Nijveldt, E. van Nood, D. E. C. van Hoorn, P. G. Boelens, K. van Norren and P. A. M. van Leeuwen, *Am. J. Clin. Nutr.*, 2001, **74**, 418–425 .

74. S. A. van Acker, D. J. van den Berg, M. N. Tromp, D. H. Griffioen, W. P. van Bennekom, W. J. van der Vijgh and A. Bast, *Free Radical Biol. Med.*, 1996, **20**, 331–342 .

75. W. Stahl and H. Sies, *Mol. Aspects Med.*, 2003, **24**, 345–351 .

76. L. Mueller and V. Boehm, *Molecules*, 2011, **16**, 1055–1069 .

77. J. A. Olson and N. I. Krinsky, *FASEB J.*, 1995, **9**, 1547–1550 .

78. H. Sies and W. Stahl, *Am. J. Clin. Nutr.*, 1995, **62**, 1315S–1321S .

79. W. Stahl and H. Sies, *Mol. Aspects Med.*, 2003, **24**, 345–351 .

80. J. G. Erhardt, C. Meisner, J. C. Bode and C. Bode, *Am. J. Clin. Nutr.*, 2003, **78**, 1219–1224 .

81. A. V. Rao and L. G. Rao, *Pharmacol. Res.*, 2007, **55**, 207–216 .

82. N. I. Krinsky and E. J. Johnson, *Mol. Aspects Med.*, 2005, **26**, 459–516 .

83. G. Riccioni, *Curr. Atheroscler. Rep.*, 2009, **11**, 434–439 .

84. P. Di Mascio, S. Kaiser and H. Sies, *Arch. Biochem. Biophys.*, 1989, **274**, 532–538 .

85. G. Levin, M. Yeshurun and S. Mokady, *Nutr. Cancer*, 1997, **27**, 293–297 .

86. V. Böhm, N. L. Puspitasari-Nienaber, M. G. Ferruzzi and S. J. Schwartz, *J. Agric. Food Chem.*, 2002, **50**, 221–226 .

87. M. F. Andreasen, A. Landbo, L. P. Christensen, Å. Hansen and A. S. Meyer, *J. Agric. Food Chem.*, 2001, **49**, 4090–4096 .

88. F. Natella, M. Nardini, M. Di Filici and C. Scaccini, *J. Agric. Food Chem.*, 1999, **47**, 1453–1459 .

89. A. S. Meyer, J. L. Donovan, D. A. Pearson, A. L. Waterhouse and E. N. Frankel, *J. Agric. Food Chem.*, 1998, **46**, 1783–1787 .

90. V. P. Menon and A. R. Sudheer, *Adv. Exp. Med. Biol.*, 2007, **595**, 105–125 .

91. J. S. Wright, *J. Mol. Struct.: THEOCHEM*, 2002, **59**, 207–217 .

92. N. J. Miller, C. telluccio, L. Tijburg and C. Rice-Evans, *FEBS Lett.*, 1996, **392**, 40–44 .

93. M. Shiraki, Y. Hara, T. Osawa, H. Kumon, T. Nakayama and S. Kawakishi, *Mutat. Res.*, 1994, **323**, 29–34 .

94. L. K. Leung, Y. Su, R. Chen, Z. Zhang, Y. Huang and Z. Y. Chen, *J. Nutr.*, 2001, **131**, 2248–2251 .

95. S. Ankri, T. Miron, A. Rabinkov, M. Wilchek and D. Mirelman, *Antimicrob. Agents Chemother.*, 1997, **41**, 2286–2288 .

96. K. C. Agarwal, *Med. Res. Rev.*, 1996, **16**, 111–124 .

97. K. Prasad, V. A. Laxdal, M. Yu and B. L. Raney, *Mol. Cell. Biochem.*, 1995, **148**, 183–189 .

98. V. Vaidya, K. U. Ingold and D. A. Pratt, *Angew. Chem., Int. Ed.*, 2009, **48**, 157–159 .

99. E. Block, *Angew. Chem., Int. Ed.*, 1992, **31**, 1135–1178 .

100. P. T. Lynett, K. Butts, V. Vaidya, G. E. Garretta and D. A. Pratt, *Org. Biomol. Chem.*, 2011, **9**, 3320–3330 .

101. R. Amorati, P. T. Lynett, L. Valgimigli and D. A. Pratt, *Chem.–Eur. J.*, 2012, **18**, 6370–6379 .

102. A. Galano and M. Francisco-Marquez, *J. Phys. Chem. B*, 2009, **113**, 16077–16081 .

103. S. K. Gupta, P. Bansal, R. K. Bhardwaj and T. Velpandian, *Pharmacol. Res.*, 2000, **41**, 657–662 .

104. Y. F. Bai and H. Xu, *Acta Pharmacol. Sin.*, 2000, **21**, 357–359 .

105. S. A. Lee, S. S. Hong, X. H. Han, J. S. Hwang, G. J. Oh, K. S. Lee, M. K. Lee, B. Y. Hwang and J. S. Ro, *Chem. Pharm. Bull.*, 2005, **53**, 832–835 .

106. J. Wattanathorn, P. Chonpathompikunlert, S. Muchimapura, A. Priprem and O. Tankamnerdthai, *Food Chem. Toxicol.*, 2008, **46**, 3106–3110 .

107. K. Selvendiran, J. P. Singh, K. B. Krishnan and D. Sakthisekaran, *Fitoterapia*, 2003, **74**, 109–115 .

108. G. Shoba, D. Joy, T. Joseph, M. Majeed, R. Rajendran and P. S. Srinivas, *Planta Med.*, 1998, **64**, 353–356.

109. P. Anand, A. B. Kunnumakkara, R. A. Newman and B. B. Aggarwal, *Mol. Pharm.*, 2007, **4**, 807–818 .

110. P. Ananda, S. G. Thomas, A. B. Kunnumakkara, C. Sundaram, K. B. Harikumar, B. Sung, S. T. Tharakan, K. Misra, I. K. Priyadarsini, K. N. Rajasekharan and B. B. Aggarwal,*Biochem. Pharmacol.*, 2008, **76**, 1590–1611 .

111. K. I. Priyadarsini, D. K. Maity, G. H. Naik, M. Sudheer, M. K. Unnikrishnan, J. G. Satav and H. Mohan, *Free Radical Biol. Med.*, 2003, **35**, 475–484 .

112. M. Salem, S. Rohanib and E. R. Gillies, *Adv.*, 2014, **4**, 10815–10829 .

113. O. N. Gordon and C. Schneider, *Trends Mol. Med.*, 2012, **18**, 361–363 .

114. M. Griesser, V. Pistis, T. Suzuki, N. Tejera, D. A. Pratt and C. Schneider, *J. Biol. Chem.*, 2011, **286**, 1114–1124 .

115. N. Sreejayan, M. N. A. Rao, K. I. Priyadarsini and T. P. A. Devasagayam, *Int. J. Pharmacol.*, 1997, **151**, 127–130 .

116. A. Barik, B. Mishra, L. Shen, H. Mohan, R. M. Kadam, S. Dutta, H. Zhang and K. I. Priyadarsini, *Free Radical Biol. Med.*, 2005, **39**, 811–822 .

117. B. Stinefelt, S. S. Leonard, K. P. Blemings, X. Shi and H. Klandorf, *Ann. Clin. Lab. Sci.*, 2005, **35**, 37–45 .

118. A. Galano and J. R. Alvarez-Idaboy, *Adv.*, 2011, **1**, 1763–1771 .

119. W. S. Warning, *QJM: An International Journal of Medicine*, 2002, **95**, 691–693 .

120. S. Muraoka and T. Miura, *Pharmacol. Toxicol.*, 2003, **93**, 284–289 .

121. G. L. Squadrito, R. Cueto, A. E. Splenser, A. Valavanidis, H. Zhang, R. M. Uppu and W. A. Pryor, *Arch. Biochem. Biophys.*, 2000, **376**, 333–337 .

122. N. Kuzkaya, N. Weissmann, D. G. Harrison and S. Dikalov, *Biochem. Pharmacol.*, 2005, **70**, 343–354 .

123. P. Ahmadpoor, E. Eftekhar, J. Nourooz-Zadeh, H. Servat, K. Makhdoomi and A. Ghafari, *Iran J. Kidney Dis.*, 2009, **3**, 22–27 .

124. R. B. Pereira, C. Sousa, A. Costa, P. B. Andrade and P. Valentão, *Molecules*, 2013, **18**, 8858–8872 .

125. A. Pastore, G. Federici, E. Bertini and F. Piemonte, *Clin. Chim. Acta*, 2003, **333**, 19–39 .

126. P. C. Huber and W. P. Almeida, *Quim. Nova*, 2008, **31**, 1170–1179 .

127. M. A. Karaman, N. M. Mimica-Dukić and M. N. Matavulj, *Arch. Biol. Sci.*, 2005, **57**, 93–100 .

128. R. B. Jalgaonwala, B. V. Mohite and R. T. Mahajan, *J. Microbiol. Biotechnol. Res.*, 2011, **1**, 21–32 .

129. N. Sadrati, H. Daoud, A. Zerroug, S. Dahamna and S. Bouharati, *J. Plant Prot. Res.*, 2013, **53**, 128–136 .

130. F. Vélez-González, D. Ortegón-Reyna, A. A. Ramos-Organillo, A. L. Peraza-Campos, M. T. Sumaya-Martínez, R. Tapia-Benavides and F. J. Martínez-Martíneza, *ARKIVOC*, 2008, **V**, 55–64 .

131. S. Lee, J. Han, H. Kim, E. Kim, T. Jeong, W. S. Lee and K. Cho, *Bioorg. Med. Chem. Lett.*, 2004, **14**, 4677–4681 .

132. F. Natella, M. Nardini, M. D. Felice and C. Scaccini, *J. Agric. Food Chem.*, 1999, **47**, 1453–1459 .

133. D. Point, P. Coudert, F. Leal, C. Rubat, V. Sautou-Miranda, J. Chopineau and J. Couquelet, *Il Farmaco*, 1998, **53**, 85–88 .

134. D. Krishnaiah, R. Sarbatly and R. Nithyanandam, *Food Bioprod. Process.*, 2011, **89**, 217–233 .

135. D. Pal and S. Mitra, *J. Adv. Pharm. Technol. Res.*, 2010, **1**, 268–272 .

136. M. N. Alam, N. J. Bristi and M. Rafiquzzaman, *Saudi Pharm. J.*, 2013, **21**, 143–152 .

137. M. Carocho and I. C. F. R. Ferreira, *Food Chem. Toxicol.*, 2013, **51**, 15–25 .

138. J. A. Imlay, *Nat. Rev. Microbiol.*, 2013, **11**, 443–454 .

139. A. Meyer-Isaksen, *Trends Food Sci. Technol.*, 1995, **6**, 300–304 .

140. J. Robak and R. J. Gryglewski, Flavonoids are scavengers of superoxide anions, *Biochem. Pharmacol.*, 1988, **37**, 837–841 .

141. C. W. Choi, S. C. Kim, S. S. Hwang, B. K. Choi, H. J. Ahn, M. Y. Lee, S. H. Park and S. K. Kim, *Plant Sci.*, 2002, **163**, 1161–1168 .

142. C. Sánchez-Moreno, *Food Sci. Technol. Int.*, 2002, **8**, 121–137 .

143. C. A. Calliste, P. Trouillas, D. P. Allais, A. Simon and J. L. Duroux, *J. Agric. Food Chem.*, 2001, **49**, 3321–3327 .

144. C. A. Calliste, D. Kozlowski, J. L. Duroux, Y. Champavier, A. J. Chulia and P. Trouillas, *Food Chem.*, 2010, **118**, 489–496 .

145. H. Chang, Y. Ho, M. Sheu, Y. Lin, M. Tseng, S. Wu, G. Huang and Y. Chang, *Bot. Stud.*, 2007, **48**, 407–417.

146. P. Evelson, M. Travacio, M. Repetto, J. Escobar, S. Llesuy and E. A. Lissi, *Arch. Biochem. Biophys.*, 2001, **388**, 261–266 .

147. H. Leontowicz, S. Gorinstein, A. Lojek, M. Leontowicz, M. Ciz, R. Soliva-Fortuny, Y. Park and S. Jung, *J. Nutr. Biochem.*, 2002, **13**, 603–610 .

148. M. G. Repetto and S. F. Llesuy, *Braz. J. Med. Biol. Res.*,

2002, **35**, 523–534 .

149. P. Miciński, K. Pawlicki, E. Wielgus, M. Bochenek, P. Gogol and B. Ficek, *Reprod. Biol.*, 2011, **11**, 135–144 .
150. R. He, Y. Li, X. Li, R. Yi, X. Wang, B. Tsoi, K. K. H. Lee, K. Abe, X. Yang and H. Kurihara, *PLoS One*, 2013, **8**, e57732 .
151. F. S. Hosseinian, A. D. Muir, N. D. Westcott and E. S. Krol, *Org. Biomol. Chem.*, 2007, **5**, 644–654 .
152. S. Kohtani, K. Yoshida, T. Maekawa, A. Iwase, A. Kudo, H. Miyabe and R. Nakagaki, *Phys. Chem. Chem. Phys.*, 2008, **10**, 2986–2992 .
153. H. Esterbauer, J. H. Gebicki and G. Jürgens, *Free Radical Biol. Med.*, 1992, **13**, 341–390 .
154. E. Niki, Y. Yoshida, Y. Saito and N. Noguchi, *Biochem. Biophys. Res. Commun.*, 2005, **338**, 668–676 .
155. T. A. Dix and J. Aikens, *Chem. Res. Toxicol.*, 1993, **6**, 2–18 .
156. V. Mišík, K. Ondriaš and P. Balgavý, *Gen. Physiol. Biophys.*, 1992, **11**, 317–325 .
157. K. Marshall, R. J. Reiter, B. Poeggeler, O. I. Aruoma and B. Halliwell, *Free Radical Biol. Med.*, 1996, **21**, 307–315 .
158. O. I. Aruoma, in *Free radicals in tropical diseases*, Harwood Academic Publishers, 1993 .
159. O. I. Aruoma, *Free Radical Biol. Med.*, 1996, **20**, 675–705 .
160. C. L. Alarcón and A. Denicola, *Anal. Chim. Acta*, 2013, **763**, 1–10 .
161. K. L. Wolfe and R. H. Liu, *J. Agric. Food Chem.*, 2007, **55**, 8896–8907 .
162. D. Honzel, S. G. Carter, K. A. Redman, A. G. Schauss, J. R. Endres and G. S. Jensen, *J. Agric. Food Chem.*, 2008, **56**, 8319–8325 .
163. J. A. Royall and H. Ischiropoulos, *Arch. Biochem. Biophys.*, 1993, **302**, 348–355 .
164. M. Antolovich, P. D. Prenzler, E. Patsalides, S. McDonald and K. Robards, *Analyst*, 2002, **127**, 183–198 .
165. R. Apak, S. Gorinstein, V. Böhm, K. M. Schaich, M. Özyürek and K. Güçlü, *Pure Appl. Chem.*, 2013, **85**, 957–998 .

CHAPTER 15 ALDOL REACTIONS MEDIATED BY A TETRAHEDRAL BORONATE

Tobias Müller, Kristina Djanashvili , Isabel W. C. E. Arends , Joop A. Peters and Ulf Hanefeld

The aldol addition is an important C–C bond forming reaction.[1] A key step in the mechanism is nucleophilic attack of a deprotonated ketone (enolate) on an aldehyde to form a β-hydroxyketone. Various inorganic bases, including LiOH, NaOH, Na_2CO_3 and $Ca(OH)_2$, have been applied for the initial deprotonation of the ketone.[2–5] However, the base catalyst is often also active in the elimination reaction, the dehydration of the β-hydroxyketone. Suppression of the aldol elimination is of great interest, as β-hydroxyketones are versatile building blocks in, for instance, the synthesis of diols, amino alcohols, lactones and polyketides.[6–10]

Boronates are known to mediate aldol reactions by acting both as an activator and as a template for the reactants, the ketone and the aldehyde.[11–14] An interesting modification is the addition of a highly reactive trimethyl silyl enol ether to aldehydes in water with sodium dodecyl sulfate as a surfactant and diarylborinic acid **1** as the catalyst (Figure 1).[15] More recently, it was shown that the activation of the ketone as silyl enol ether is unnecessary when salt **2** as the catalyst is applied (Figure 1). In this system, which is limited to aqueous solutions, intramolecular interaction of the neighbouring nitrogen and boron atoms stabilizes the tetrahedral boronate. Side reactions, such as dehydration, remain a problem; particularly when the reaction is favoured by stabilization of the elimination product as a result of mesomerism.

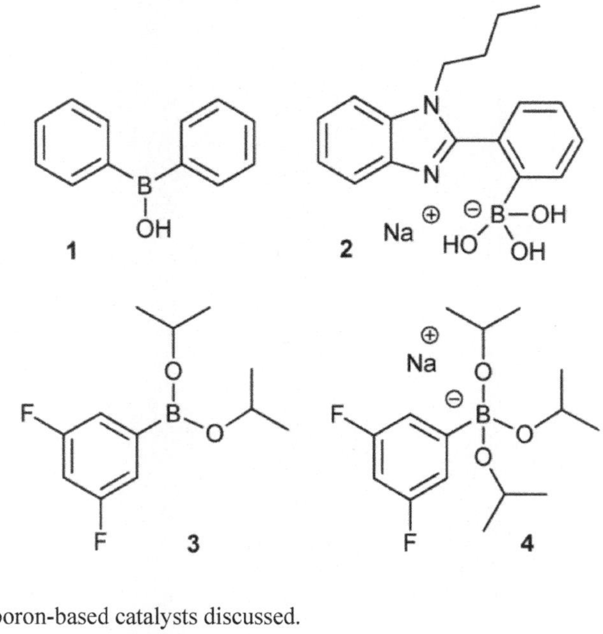

Figure 1: Molecular structures of the boron-based catalysts discussed.

Citation: Tobias Müller , Kristina Djanashvili , Isabel W. C. E. Arends , Joop A. Peters and Ulf Hanefeld, "Aldol reactions mediated by a tetrahedral boronate," DOI: 10.1039/c2cc37047f.

Here, we report the results of an investigation on the feasibility of the application of phenylboronate **4** as a catalyst in aldol reactions in organic media (Scheme 1). For comparison, the corresponding boronic acid **3** and the classic base sodium isopropoxide were included in this study. Compounds **3** and **4** were synthesized from 3,5-difluorophenylboronic acid by reaction with the appropriate amount of isopropanol and sodium isopropoxide and were isolated as stable solids with a purity greater than 99%. Both new compounds show good solubility and thermal stability in organic solvents; they did not show any dissociation as demonstrated by their characteristic ^{11}B NMR chemical shifts for trigonal and tetragonal B-atoms, respectively. The high stability of these B-compounds may be ascribed to the electron-withdrawing fluoro-substituents in the aromatic ring. These substituents also contribute to the high solubility of these complexes in organic solvents, which allows us to perform the aldol reaction with one of the reactants acting as solvent. In the present investigation, acetone was used as the solvent and the reactions were monitored by NMR and GC-MS. To rule out the possible deactivation of the catalyst(**4**) by water, the reaction was performed in the presence of equimolar amounts of it (relative to **4**). Neither product formation nor selectivity was influenced by water.

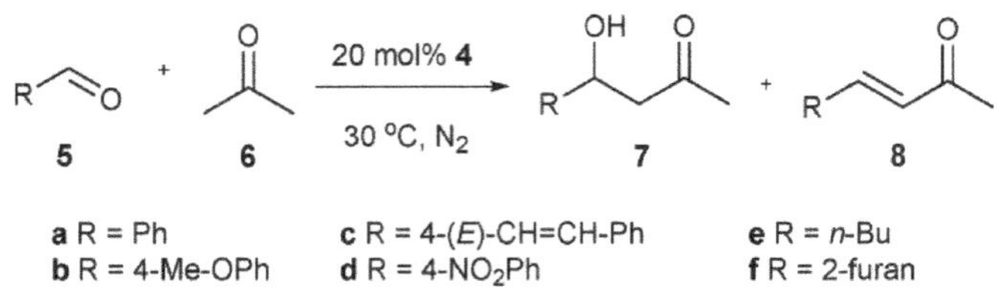

Scheme 1: The aldol reaction with a series of aldehydes and acetone.

In the presence of NaO*i*Pr as the catalyst and at 30 °C, benzaldehyde (**5a**) and an excess of acetone (**6**) reacted very fast to give aldol **7**, which was subsequently dehydrated to α,β-unsaturated ketone **8** with a lower reaction rate (Figure 2B). The rates of both the aldol reaction and the elimination reaction appeared to be strongly dependent on the concentration of thecatalyst (Table 1, entries 1–4). In addition to the elimination side product, higher molecular weight side product **9** resulting from dimerization was observed as well. Decreasing the temperature to 5 °C did not improve the selectivity towards the desired aldol product (Figure 2B). Decreasing temperature and catalyst concentration at the same time did not inhibit the elimination reaction either.

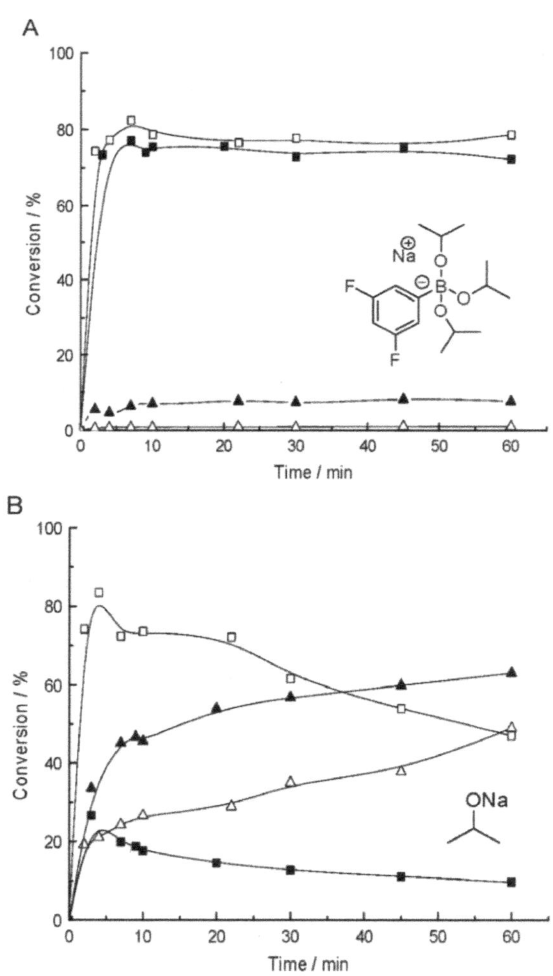

Figure 2: Reaction between benzaldehyde and acetone, catalyzed by 20 mol% of boronate salt **4**(A) and sodium 2-propanolate (B), monitored by ^{1}H NMR at 30 °C (solid data points) and 5 °C (empty data points). The aldol and the elimination product formation is presented by squares and triangles, respectively.

Table 1: The aldol reactions of 2 mmol of **5a** and 20 mmol of **6** (see Scheme 1) for 20 min at 30 °C. Conversions and yields are determined by ^{1}H NMR

Entry	4 (mol%)	NaO*i*Pr (mol%)	Conv.^{a} (%)	Yield 7 (%)	Yield 8 (%)
1	—	20	>99	18	57
2	—	10	>99	23	55
3	—	5	97	40	40
4	—	2	13	13	<1
5	20	—	96	76	8
6	10	—	90	80	10
7	5	—	39	35	4
8	10	10	>99	11	45

^{a} **Dimerization side-product,** PhC=C(C=O)C=CPh (9) was detected by GC-MS.

Trigonal B-compound **3** (up to 20 mol%) was completely inactive in the aldol reaction of **5a** and **6**, which suggests that a base is essential for the reaction to proceed. By contrast, application of tetragonal boronate **4** resulted in high conversion to the aldol within a few minutes (Table 1, entries 5–7); the reaction rate towards the aldol was about the same as that with NaO*i*Pr as the catalyst, but now the subsequent elimination reaction was much slower. With 20 mol% **4**, only 8% elimination product was obtained. By decreasing the reaction temperature to 5 °C, similar results were obtained (Figure 2A). Combining 10 mol% **4** and 10 mol% of NaO*i*Pr resulted in full conversion of the benzaldehyde. The aldol product was formed which was dehydrated by NaO*i*Pr to the condensation product and some higher molecular weight products resulting from dimerization (Table 1, entry 8).

A series of aldehydes was tested in the aldol reaction in the presence of catalyst **4** (Scheme 1 and Table 2). The reactions proceeded within a few minutes with high conversions. Aromatic aldehydes and furfural, a bio-based platform chemical, all gave rise to both aldol adduct **7** and the corresponding dehydration product **8** with a high selectivity towards **7**, compared to the results obtained with the reported boron catalysts **1** and **2**.[15,16] The highest selectivity was obtained with aliphatic aldehyde **5e**: 66% of the β-hydroxyketone **7e** and less than 1% of the dehydration product **8e** were formed (Table 2, entry 5). For the aromatic aldehydes conversion, selectivity to the β-hydroxyketone seems to be related to the nature of substituents and the bond delocalization. Similar lower yields and selectivities for **5b** and **c** were observed earlier.[15,16]

Table 2: The aldol reactions of 2 mmol of **5** and 20 mmol of **6** (Scheme 1), catalysed by 20 mol% of **4** at 30 °C. Conversions and yields are determined by ^{1}H NMR

Entry	R	*t* (min)	Conv.^{a} (%)	Yield 7 (%)	Yield 8 (%)
1	**a**	20	96	76	8
2	**b**	20	80	38	24
3	**c**	10	88	47	41
4	**d**	5	95	52	<5
5	**e**	30	95	66	<1
6	**f**	15	99	40	21

^{a} **Dimerization side-product,** PhC=C(C=O)C=CPh (9), was detected by GC-MS.

We suggest a reaction mechanism as depicted schematically in Scheme 2. Under the reaction conditions applied, the dissociation of compound **4** towards $^{-}$O*i*Pr and corresponding **3** is negligible in acetone. This is confirmed by its ^{11}B NMR spectrum, which shows exclusively a resonance for tetrahedral boron at around −15.49 ppm. No resonance related to compound **3** was observed. Since the reaction rate of the aldol reaction, in the presence of 2 mol% NaO*i*Pr, is much lower than with 20 mol% **4**, it is likely that undissociated **4** rather than isopropoxide is the actual catalyst in this system (Table 1, entries 4 and 5). After deprotonation of acetone by **4**, the resulting enolate exchanges with *i*PrOH. The resulting tetrahedral boron-enolate **10** reacts with the aldehyde to form the aldol product. In the presence of *i*PrOH the aldol product **7** is released and the tetrahedral boronate **4** returns into the catalytic cycle.

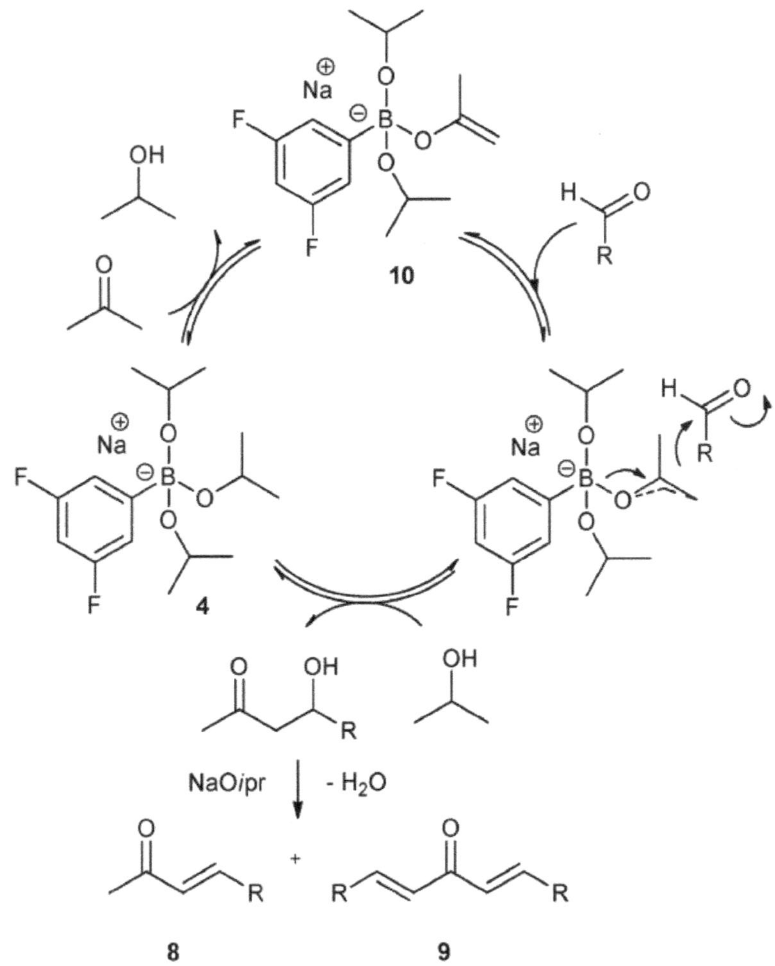

Scheme 2: Proposed mechanism for the aldol reaction catalyzed by tetrahedral boronate salt4, including the elimination reaction in the presence of a strong nucleophilic base, NaO*i*Pr.

The occurrence of an intermediate tetrahedral boronate with both aldehyde and acetone coordinated to the B-atom, as suggested by Evans *et al.* for other B-activated aldol reactions,[11] is unlikely in the present e. ^{11}B NMR spectra of a sample of **4** and benzaldehyde showed a single resonance at −16 ppm, thus no coordination of benzaldehyde to **4** (related to 0.1 M boric acid solution in D_2O at 0 ppm). This also rules out the mechanism, similar to Evans, suggested by Shibasaki *et al.*[11]

In contrast, acetone coordinates to tetrahedral boronate **4**, which could be confirmed by both ^{11}B NMR and Raman spectroscopy . The latter spectra demonstrate the appearance of an additional boron ester peak as a result of the interaction between **4** and acetone (solid line) compared to the spectra of **4** measured in its solid form (dashed line). Furthermore, the ^{11}B NMR spectra (inset) show an additional resonance at around −17.1 ppm, confirming the interaction of **4** with acetone.

In contrast ^{11}B NMR experiments with **3** show that no tetrahedral boronate was formed due to coordination of acetone (**6**) or benzaldehyde (**5a**) to phenyl boronic ester **3**. In both es, the only observable resonance was at 8 ppm (related to H_3BO_3 (0.1 M) at 0 ppm) corresponding to the trigonal boronic acid ester **3**. This is in agreement with its catalytic inertness in the aldol reaction. The advantage of the designed catalytic tetrahedral boronate **4** is its steric bulkiness, which results in a reduced rate of water elimination reaction in the aldol formation compared to that when isopropoxide is applied as the base catalyst.

In conclusion, we present tetrahedral sodium triisopropyl 3,5-difluorophenylboronate (**4**) as a base, which can catalyse the aldol reaction. It is soluble in organic media, allowing the use of a substrate, acetone, as solvent. The reaction proceeds rapidly, with high conversion and good selectivity towards β-hydroxyketone. The tetrahedral boronate acts as a base, suitable to form boron-enolate with acetone, but is not able to dehydrate the formed aldol.

This research has been performed within the framework of the CatchBio program. The authors gratefully acknowledge the support of the Smart Mix Program of the Dutch Ministry of Economic Affairs and the Dutch Ministry of Education, Culture and Science. The authors are thankful to L. Panella (DSM), P. Alsters (DSM), J. G. de Vries (DSM), B. Kaptein (DSM), G. Kemperman (MSD) for fruitful discussions, and S.A. Kulkarni (TU Delft, P&E/IRS) for the Raman measurements.

NOTES AND REFERENCES

1. J. Podlech, *Angew. Chem., Int. Ed.*, 2010, **49**, 6490–6495 .
2. P. S. Gradeff, *US Patent*, 1974, 3840601; AN 1975:43622 .
3. P. W. D. Mitchell, *US Patent*, 1989, 4874900; AN 1989:574454 .

4. K. Weissermel and H.-J. Arpe, *Industrial Organic Chemistry*, Wiley-VCH, Weinheim, 4th edn, 2003 .
5. Z. Zhang, Y. W. Dong and G. W. Wang, *Chem. Lett.*, 2003, **32**, 966–967 .
6. (*a*) S. K. Karmee and U. Hanefeld, *ChemSusChem*, 2011, **4**, 1118–1123 ; (*b*) A. M. Frey, S. K. Karmee, K. P. de Jong, J. H. Bitter and U. Hanefeld, *ChemCatChem* DOI:10.1002/cctc.201200282.
7. R. Mestres, *Green Chem.*, 2004, **6**, 583–603 .
8. J. Sukumaran and U. Hanefeld, *Chem. Soc. Rev.*, 2005, **34**, 530–542 .
9. P. Clapes, W.-D. Fessner, G. A. Sprenger and A. K. Samland, *Curr. Opin. Chem. Biol.*, 2010, **14**, 154–167 .
10. C. Hertweck, *Angew. Chem., Int. Ed.*, 2009, **48**, 4688–4716 .
11. (*a*) D. A. Evans, E. Vogel and J. V. Nelson, *J. Am. Chem. Soc.*, 1979, **101**, 6120–6123 ; (*b*) T. Nitabaru, N. Kumagai and M. Shibasaki, *Tetrahedron Lett.*, 2008, **49**, 272–276 ; (*c*) T. Nitabaru, A. Nojiri, M. Kobayashi, N. Kumagai and M. Shibasaki, *J. Am. Chem. Soc.*, 2009, **131**, 13860–13869 .
12. D. A. Evans, J. Bartroli and T. L. Shih, *J. Am. Chem. Soc.*, 1981, **103**, 2127–2129 .
13. D. A. Evans, J. V. Nelson, E. Vogel and T. R. Taber, *J. Am. Chem. Soc.*, 1981, **103**, 3099–3111 .
14. (*a*) G. Li, H.-X. Wei, B. S. Phelps, D. W. Purkiss and S. H. Kim, *Org. Lett.*, 2001, **3**, 823–826 ; (*b*) U. Hanefeld, A. M. Hooper and J. Staunton, *Synthesis*, 1999, 401–403 ; (*c*) A. Koskinen, *Asymmetric Synthesis of Natural Products*, John Wiley & Sons, Chichester, 1993, pp. 69–75 ; (*d*) R. C. Harris, A. L. Cutter, K. J. Weissman, U. Hanefeld, M. C. Timoney and J. Staunton, *J. Chem. Res.*, 1998, 283 ; (*e*) A. L. Wilkinson, U. Hanefeld, B. Wilkinson, P. F. Leadlay and J. Staunton, *Tetrahedron Lett.*, 1998, **39**, 9827–9830 ; (*f*) K. J. Weissman, M. Bycroft, A. L. Cutter, U. Hanefeld, E. J. Frost, M. C. Timoney, R. Harris, S. Handa, M. Roddis, J. Staunton and P. F. Leadlay, *Chem. Biol.*, 1998, **5**, 743–754 ; (*g*) C. Körner, P. Starkov and T. D. Sheppard, *J. Am. Chem. Soc.*, 2010, **132**, 5968–5969 .
15. Y. Mori, J. Kobayashi, K. Manabe and S. Kobayashi, *Tetrahedron*, 2002, **58**, 8263–8268 .
16. K. Aelvoet, A. S. Batsanov, A. J. Blatch, C. Grosjean, L. G. F. Patrick, C. A. Smethurst and A. Whiting, *Angew. Chem.*, 2008, **120**, 780–782 (*Angew. Chem., Int. Ed.* , 2008 , **47** , 768–770) .

CHAPTER 16

METAL CATALYZED DEFUNCTIONALIZATION REACTIONS

Atanu Modak and Debabrata Maiti

INTRODUCTION

Defunctionalization reactions have become an indispensable aspect in the advancement of chemical synthesis despite affording a less functionalized entity from a more functionalized organic material. The key challenge of the defunctionalization reaction lies in the high bond dissociation energy of the C–X (X = carbon or heteroatom) bond. Metal mediated methods emerged as a savior in addressing this issue as the metal destabilizes the C–X bond through single or two electron transfer processes. Moreover, the remarkable electrophilicity of the metal allows them to form the metal carbanion complex which is the key intermediate of defunctionalization as well as cross coupling reactions. In the e of the defunctionalization reaction the metal carbanion intermediate is intercepted by a hydrogen equivalent to result in the required defunctionalized product, whereas in the cross coupling reaction the other coupling partner comes into the picture. Therefore, defunctionalization has an eminent significance in the functionalization reaction in terms of a mechanistic point of view (Scheme 1).

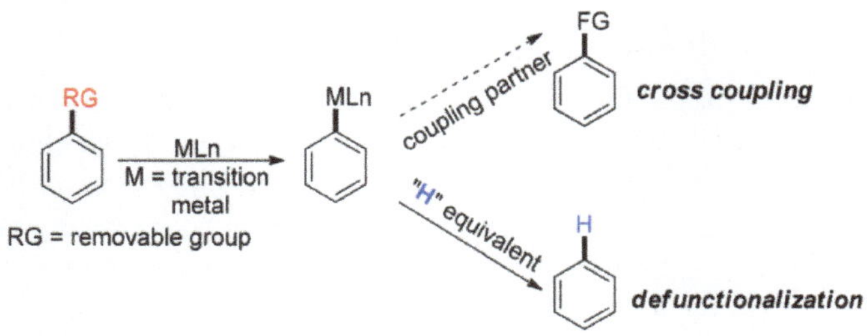

Scheme 1: Schematic diagram of transition metal mediated defunctionalization.

Besides, defunctionalization reactions also have a direct impact in synthetic organic chemistry, facilitating functional groups to be utilized temporarily in a particular synthetic transformation.[1,2] In some diverse aspects, defunctionalized substrates are considered more useful than their functionalized form *e.g.* preparation of useful materials through biomass degradation,[3–9] converting environmentally hazardous entities to relatively harmless forms through desulfurization[10,11] or dechlorination[12–16] *etc.*

In this mini review, we have highlighted the chronological development of metal assisted defunctionalization reactions from the stoichiometric to the catalytic stage with their application in synthetic organic chemistry (Scheme 2).

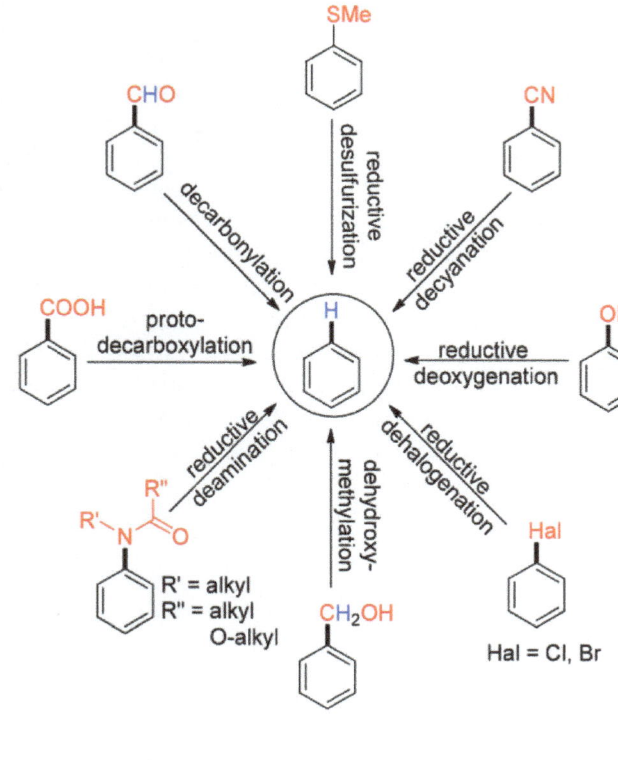

Scheme 2: Scope of defunctionalization.

Citation: Atanu Modak and Debabrata Maiti, "Metal catalyzed defunctionalization reactions," DOI: 10.1039/c5ob01949d.

DECARBOXYLATION

Decarboxylation of aromatic carboxylic acids has a great significance in the up-gradation of coal derived substrates, as aromatic carboxylic acids are the major air oxidized products of coal.[17] In synthetic chemistry, the carboxylic moiety serves as the directing group through metal-chelation assisted C–H activation or electronically assisted aromatic functionalization. Therefore, an efficient decarboxylation of aromatic carboxylic acids can make the carboxylic group a traceless directing group in aromatic C–H functionalization. A number of metal mediated/catalyzed methods are known for the effective protodecarboxylation of carboxylic acids which occurred through a metal assisted anionic or radical decarboxylation.

Mercury Mediated Decarboxylation

Mercury mediated decarboxylation of aromatic carboxylates has been known for a long time. In 1901, Pesci first reported the mercury mediated decarboxylation of an aromatic anhydride to produce an aromatic mono acid.[18] Since then a number of groups have worked on mercury as a decarboxylating agent.[19–23] Unfortunately, all of these methods involve high reaction temperatures, use of large amounts of toxic metals or a long reaction time. Recently in 2006 Moseley and coworkers reported the mercury mediated decarboxylation of naphthoic anhydride at a faster reaction rate under microwave conditions (Scheme 3).[24]

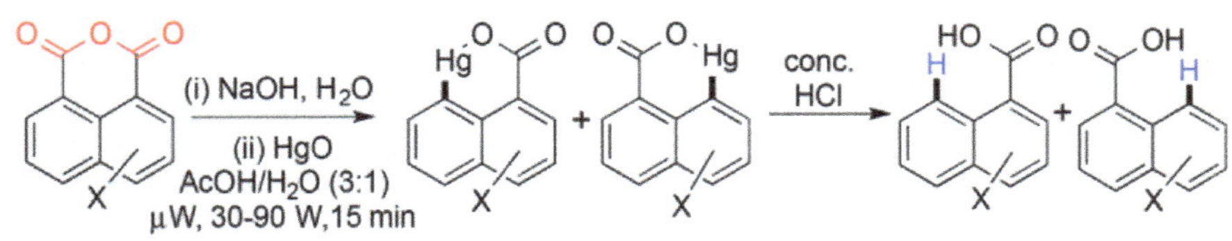

Scheme 3: Mercury mediated decarboxylation of naphthoic anhydride under microwave conditions.

Copper salt as a Decarboxylating Agent

After mercury, copper was the second metal which was exploited for effective protodecarboxylation. The first copper mediated decarboxylation was performed in 1930 by Shepard and coworkers using a stoichiometric copper–quinoline combination.[25] They successfully decarboxylated halogenated furan carboxylic acid at very high temperature. Later, the decarboxylation method was further developed by the Nilsson,[26] Cohen[27,28] and Sheppard[29] groups for activated aromatic carboxylic acids. According to their study an N-heterocyclic amine type solvent/ligand is essential for stabilization of the copper-decarboxylate intermediate. In 2007, Gooßen and coworkers for the first time reported the copper catalyzed decarboxylation of deactivated carboxylic acids with the help of a bis-*N*-chelating ligand at 170 °C.[30] Decarboxylation of deactivated acids was achieved by introducing a sterically hindered ligand like 4,7-diphenyl-1,10-phenanthroline. A quick version of the protodecarboxylation protocol was also developed by the Goossen group under microwave conditions with a catalytic amount of Cu/phenanthroline.[31] According to the mechanistic study and DFT calculation, it was proposed that copper directly inserts into the aryl carboxylate bond and then the hydrolytic cleavage of the copper–carbon bond results in the protodecarboxylated product (Scheme 4).[30]

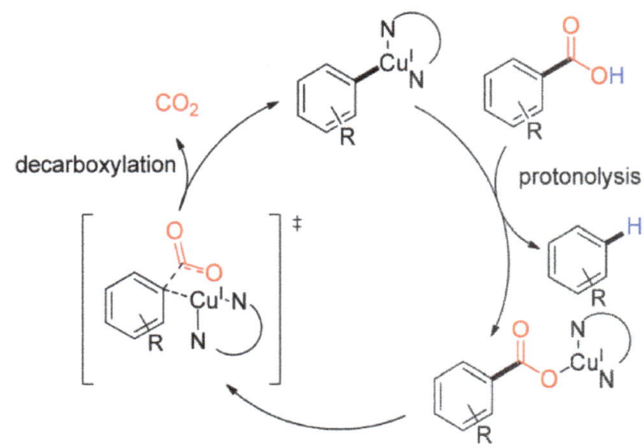

Scheme 4: Catalytic cycle of the decarboxylation in a copper–phenanthroline system.

Palladium Catalyzed Decarboxylation

Schleyer successfully achieved decarboxylation in the gas phase at a very high temperature with a nickel (180 °C) and palladium (330 °C) catalyst under a hydrogen gas atmosphere.[32] Later in 2004 Matsubara and coworkers reported a palladium/charcoal catalyzed hydrothermal condition for decarbonylation and decarboxylation.[33] A high deuterium containing entity ensued from a completely non-deuterated substrate due to H–D exchange under their hydrothermal reaction conditions using deuterium oxide as the solvent.[33] In 2007 Kozlowski reported a palladium catalyzed method for the decarboxylation of electron rich bis-*ortho*-substituted aromatic carboxylic acids using trifluoroacetic acid as the proton source.[34] In the reaction mechanism, a four-membered palladium species was proposed as an intermediate prior to the decarboxylation (Scheme 5).[35]

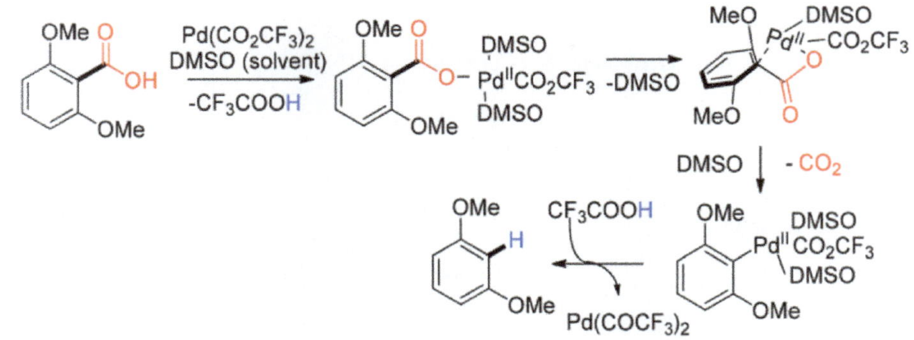

Scheme 5: Proposed mechanism for palladium catalyzed decarboxylation of bis-*ortho*-substituted aromatic carboxylic acid.

Ruthenium Catalyzed Directed Decarboxylation

Murai and coworkers in 2001 demonstrated the chelation assisted decarboxylation of esters using the $Ru_3(CO)_{12}$ catalyst through acyl C–O bond fission (Scheme 6).[36] According to their proposed reaction mechanism, a heteroatom containing directing group stimulates the metal center for site-selective cleavage of the acyl C–O bond through a tetrahedral intermediate followed by a fast decarbonylation yielding the resultant hydrocarbon (Scheme 6).

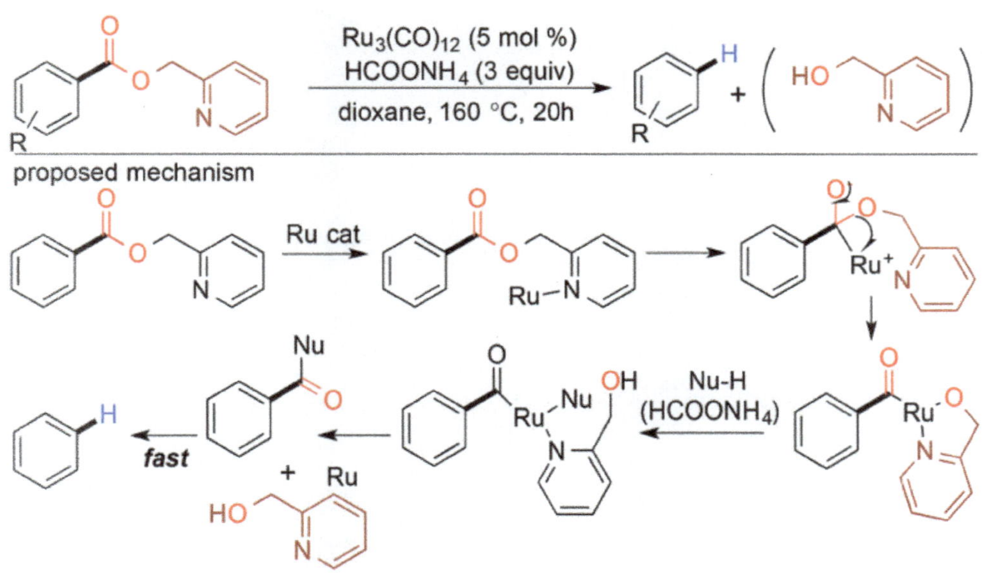

Scheme 6: Chelation assisted decarboxylation of esters.

Silver Catalyzed Decarboxylation

Silver catalyzed protodecarboxylation of heteroaromatic carboxylic acids was explored by Larossa and coworkers in 2009.[37] From their substrate scope study it was observed that the α heteroatom is a prerequisite for decarboxylation (Scheme 7). In 2012 they nicely showed that a deutero-substrate can be prepared through deuteron-decarboxylation of the corresponding carboxylic acid using D_2O as the D^+ source.[38] In the same year, Greaney and coworkers established a new approach for protodecarboxylation under oxidative radical conditions using catalytic Ag(I)/$K_2S_2O_8$ (Scheme 8).[39] A variety of benzoic acid derivatives with electron rich and electron donating functional groups were decarboxylated successfully. According to their hypotheses, first Ag(I) is *in situ* oxidized to Ag(II) in the presence of $K_2S_2O_8$, later the carboxylate anion is oxidized to the carboxylated radical by *in situ* generated Ag(II). After that, a successful decarboxylation yielded an aryl radical from the carboxylate radical. Finally, the aryl radical abstracted a hydrogen radical from acetonitrile to deliver the protodecarboxylated product.

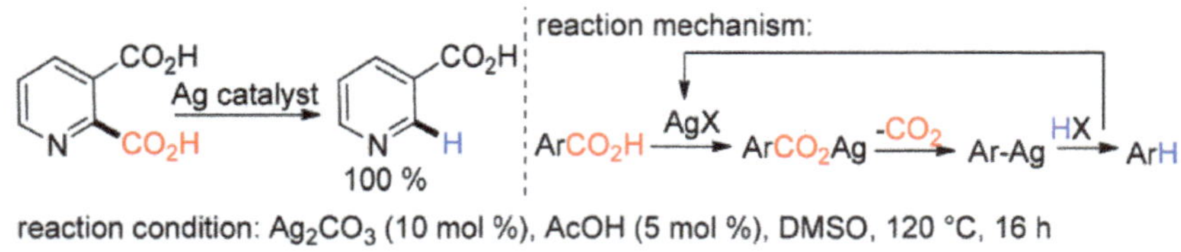

Scheme 7: Silver catalyzed decarboxylation of hetero-aromatic carboxylic acid.

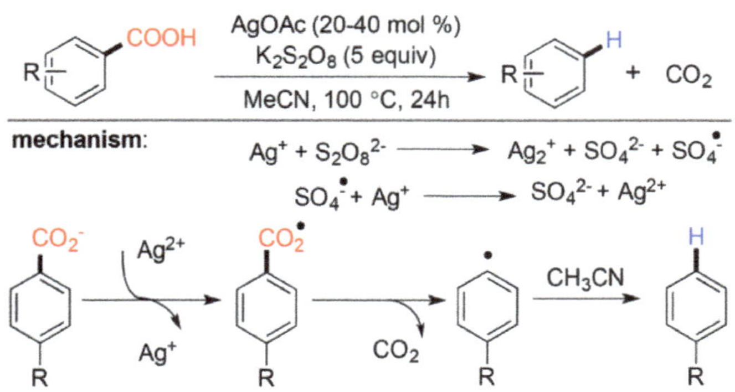

Scheme 8: Silver catalyzed protodecarboxylation under oxidative radical conditions.

Rhodium Catalyzed Decarboxylation

A rhodium catalyzed protodecarboxylation was demonstrated by Zhao and coworkers using (cod)Rh(μ-OH)$_2$ as a catalyst in 2010.[40] Unfortunately, their substrate scope was limited to*ortho*-substituted arenecarboxylic acids. Non-activated benzoic acid and *ortho*-methoxy benzoic acid remained unreactive under their optimized conditions.

Gold Mediated Decarboxylation

In 2011, Nolan and coworkers first explored the gold mediated decarboxylation of aromatic carboxylic acids with an N-heterocyclic carbene–gold(I) complex.[41] After their initial breakthrough, in 2013 they reported the first gold catalyzed protodecarboxylation of aromatic carboxylic acids.[42] Both electron-rich and electron-poor benzoic acid derivatives were successfully employed for decarboxylation. According to the reaction mechanism, protodecarboxylation takes place through a gold carboxylate and gold aryl intermediate (Scheme 9).

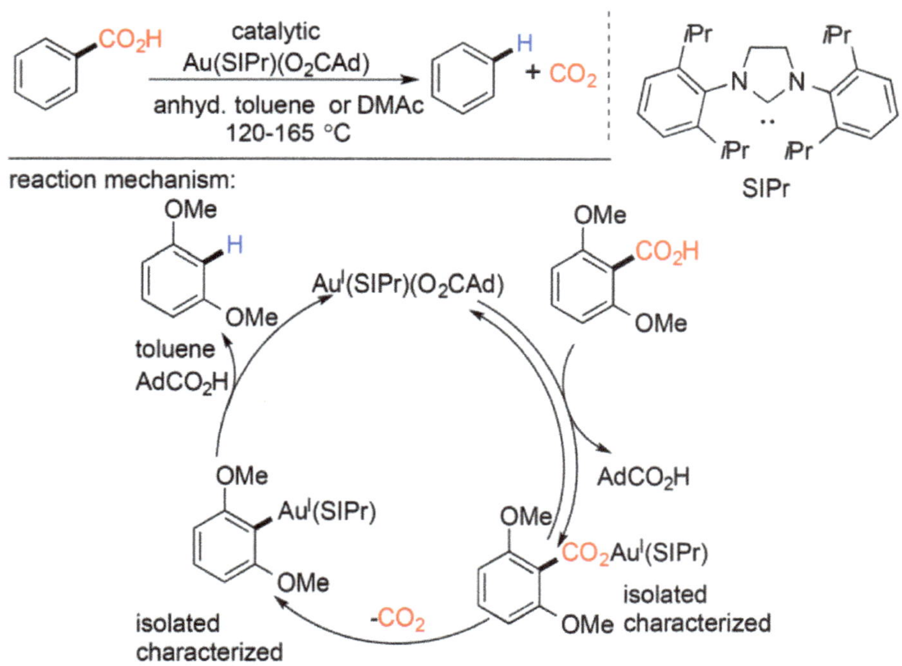

Scheme 9: Gold catalyzed protodecarboxylation.

Decarboxylation under Photochemical Conditions

Under photochemical conditions decarboxylation of aliphatic acids was achieved by Bakac and coworkers.[43] In the presence of the Fe^{3+} ion, an aqueous solution of propionic acid was converted to a mixture of hydrocarbons and CO_2 by near UV photolysis using anaerobic conditions (Scheme 10).

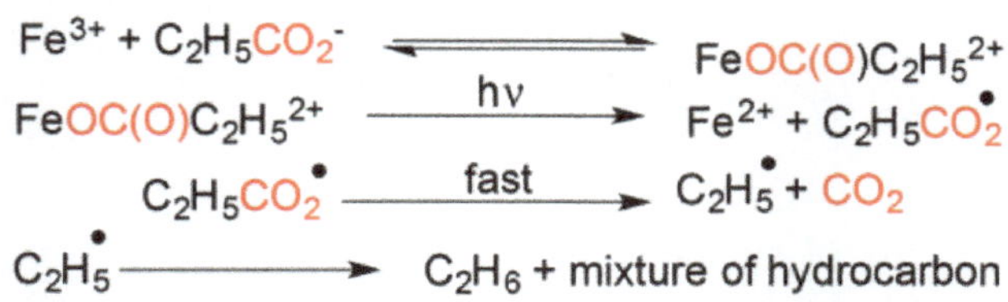

Scheme 10: Decarboxylation under photochemical conditions.

DECARBONYLATION

Transition metal catalyzed decarbonylation of aldehydes has a great importance in synthetic as well as biological chemistry. In a biological system, the cyanobacterial aldehyde decarbonylase enzyme causes the decarbonylation of a fatty aldehyde resulting in the corresponding alkane or alkene and formic acid.[44] The synthetic alternative methods of such a reaction can solve the energy crisis producing low boiling point fuel. In terms of the reaction pathway, the decarbonylation of aldehydes in synthetic chemistry differs from nature as it emits carbon monoxide instead of formic acid. On the other hand, the aldehyde group possesses worthwhile features like the *ortho* directing ability in aromatic systems, increasing the electrophilicity of dienophiles for facile addition reaction, *in situ* carbon monoxide generation *etc.* In this regard, the decarbonylation of aldehydes has a potential significance in organic synthesis for utilizing the aldehyde group temporarily in a particular transformation.

Palladium as a Decarbonylating Agent

The first decarbonylation of an aldehyde was done in 1960 by Wilt with heterogeneous Pd/charcoal at a high temperature (more than 180 °C).[45] A few years later, in 1965 Tsuji and coworkers reported the palladium catalyzed decarbonylation of an acylchloride and an aldehyde, but this protocol is limited due to the high temperature requirement.[46] Recently, Maiti and coworkers demonstrated a palladium acetate catalyzed efficient, mild and ligand free protocol for the decarbonylation of aldehydes.[47] The reaction exhibited tolerance against a wide range of functional groups such as cyano, ester, keto, nitro and acid. Heterocyclic aldehydes with a single heteroatom and two heteroatoms gave good yields of the corresponding decarbonylated products. It was observed that there was no trace of the β–H eliminated side product formed against simple alkyl aldehydes (Scheme 11). Under the microwave conditions, the decarbonylation of the aldehyde was also reported by the same group.[48] A biologically active molecule eulatachromene was synthesized by utilizing the deformylation strategy judiciously (Scheme 12).[48] The same strategy was also used by the Bräse group, where they used the rhodium catalyzed method for decarbonylation.[49]

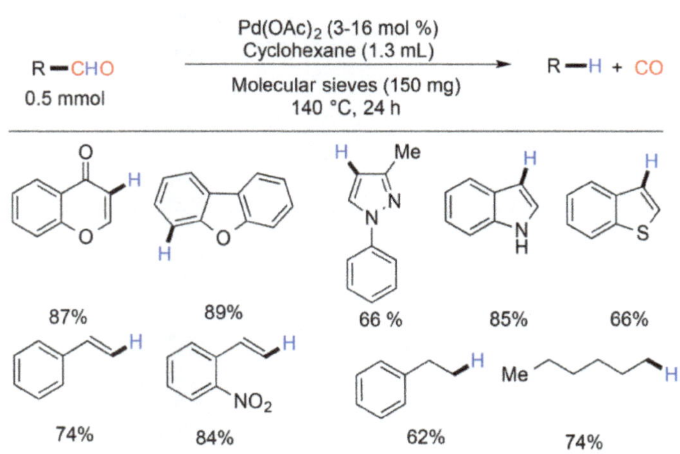

Scheme 11: Palladium acetate catalyzed decarbonylation.

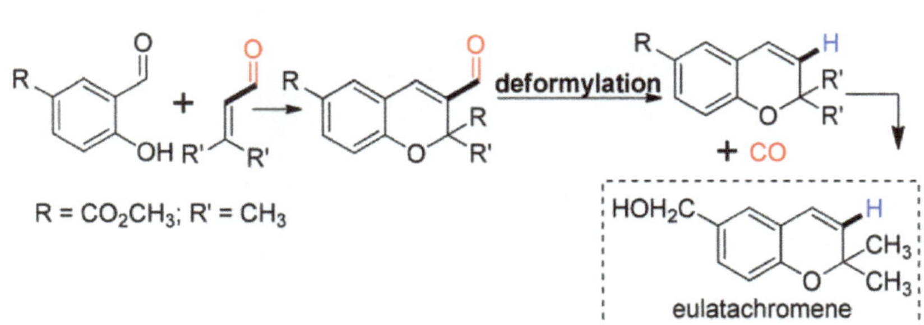

Scheme 12: Synthesis of eulatachromene.

In 2013, Fu and co-workers reported a heterogeneous decarbonylation of biomass derived feedstocks, 5-hydroxymethylfurfural (HMF) to furfuryl alcohol (FFA) with catalytic palladium nanoparticles under mild conditions (Scheme 13). Furfuryl alcohol has a great significance in the preparation of furan based light fuels.[4] Later, in 2015 Thomas Rauchfuss and co-workers demonstrated the Pd/C catalyzed decarbonylation of a furfural derivative with the intention to upgrade cellulosic biomass.[5]

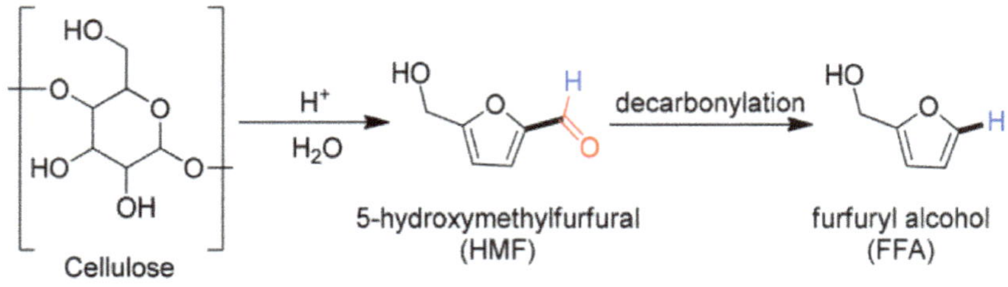

Scheme 13: Furfuryl alcohol synthesis from cellulose by decarbonylation of aldehyde.

Rhodium as a Decarbonylating Agent

Rhodium is most abundantly used in decarbonylation reactions compared to other metals. The first report on rhodium as a decarbonylating agent was communicated in 1965 by Tsuji and Ohno, using a stoichiometric amount of Willkinson's catalyst.[50] Later, the catalytic decarbonylation of an aldehyde was reported by Doughty and Pignolet in 1978 with Rh(dpp)Cl as the catalyst.[51] In 1992 O'Connor demonstrated a rhodium catalyzed efficient method for the decarbonylation of a primary aldehyde, where diphenylphosphoryl azide acted as a carbon monoxide trap.[52] Crabtree and coworkers in 1999 discovered yet another efficient way of decarbonylation of primary and aryl aldehydes under mild conditions using the catalytic [Rh(CO)(triphos)][SbF_6] complex.[53] In 2006, the Madsen group reported a method for the deformylation of aldehydes with the combination of $RhCl_3 \cdot 3H_2O$ and the phosphine ligand with a wide substrate scope.[54]

Following experimental and computational studies they proposed a catalytic cycle for the decarbonylation of aldehydes, where the aldehyde first oxidatively adds to the metal center.[55] In the next step, migratory extrusion occurs and then reductive elimination results in the defunctionalized product. Subsequently, the substrate binds with the metal center by the oxygen donor site, replacing the carbon monoxide and repeats the catalytic cycle. The migratory extrusion step is established as the turnover limiting step (Scheme 14). Madsen also applied the decarbonylation strategy to the unprotected aldose using a catalytic amount of $Rh(dppp)_2Cl$ and obtained alditol which is one carbon lesser than the starting material.[6] This one step degradation of aldose through the decarbonylation reaction has a great significance in glycochemistry. Recently, Kappe and coworkers engineered a continuous-flow protocol for the decarbonylation of aldehydes using a catalytic amount of $Rh(OAc)_2$/dppe.[56]

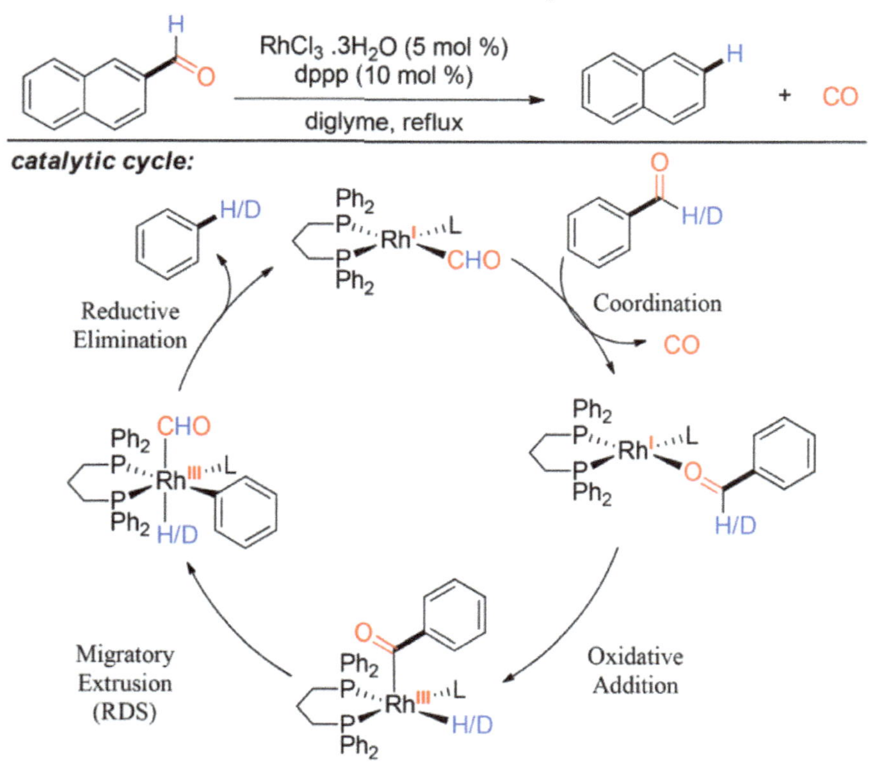

Scheme 14: Catalytic cycle for rhodium catalyzed decarbonylation of aldehyde.

Decarbonylation Using Other Metals

After palladium and rhodium, ruthenium also proved to be capable for decarbonylation reactions. In 1980, Dolphin and co-workers discovered a ruthenium–porphyrin based complex for stoichiometric decarbonylation.[57] An iridium catalyzed decarbonylation method was reported by Tsuji and coworkers in 2008, wherein a variety of functional groups were tolerated under their mild reaction conditions.[58]

DESULFURIZATION

Reductive desulfurization is one of the key steps in manufacturing nonpolluting fuel from natural resources. In view of synthetic organic chemistry, the reductive desulfurization has a great significance owing to temporary advantages coming from the sulfur moiety on its neighboring atoms.[59] There are two strategies for reductive desulphurization, one is the direct desulfurization of thio ether and the other being a two-step process where first thio ether gets oxidized to sulfone or sulfoxide followed by the reductive cleavage of the sulfur–carbon bond. The former strategy is more challenging owing to the higher C–S bond dissociation energy in thio ether compared to that in sulfone. The desulfurization of the sulfone moiety has been achieved through simple reduction in the presence of a hydride source. But, in the e of direct desulphurization of thio ether, special conditions are required for activating the C–S bond (Scheme 15). A number of transition metal mediated methodologies have been reported for direct desulfurization of thio ethers. In this review we mainly focused on the desulfurization of (hetero)aromatic thio ethers which occurs through the metal catalyzed reductive cleavage of the C(sp^2)–SR bond.

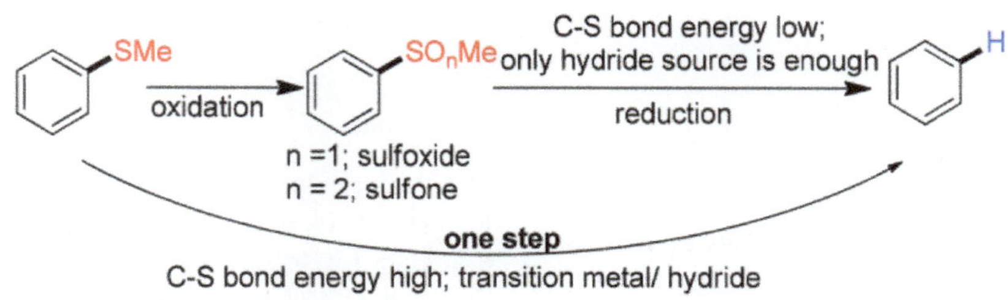

Scheme 15: Schematic diagram of reductive desulfurization of phenyl methyl thioether.

Nickel as a Desulfurizing Agent

In 1940 Mozingo and coworkers first reported the desulfurization of thio ether and S-containing amino acids using RANEY® nickel.[60] After that, several reports have been published on desulfurization using nickel containing reagents like RANEY® nickel, nickel/borohydride,[61,62] nickel-containing complex reducing agent (NiCRA) *etc.*[10] But, all these methods require a high metal/substrate ratio and also render poor chemoselectivity. In 2012, Martin and coworkers demonstrated a general and convenient protocol for cleaving the carbon–sulphur bond of aryl methyl thioethers with a catalytic amount of $Ni(COD)_2$.[63] They have also nicely depicted the synthetic applicability of the desulfurization reaction where an aryl sulfide was used as a temporary directing group (Scheme 16).

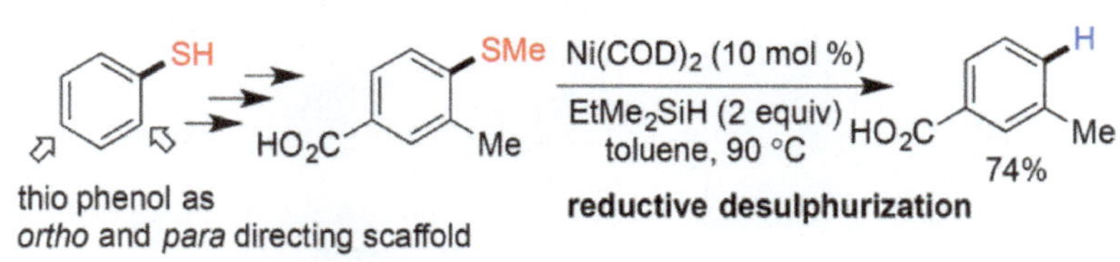

Scheme 16: Application of reductive desulfurization.

Palladium as a Desulfurizing Agent

The combination of catalytic palladium/charcoal and a hydride source is capable of reductive desulfurization under heterogeneous conditions. In 1994, hydrazine-Pd/C was used for catalytic desulfurization of methylthioether of pyridimine derivatives.[64] Later Graham[65] and Nakada[66] independently developed the palladium catalyzed reductive desulfurization of alkyl aryl/heterocyclic thio-ethers using triethyl silane as a reducing agent. After their initial report, Nakada and coworkers developed an efficient method for the synthesis of imidazolinium salts by the Pd-catalyzed reduction of their corresponding thioureas with triethylsilane and trialkylsilyl triflate (Scheme 17).[67]

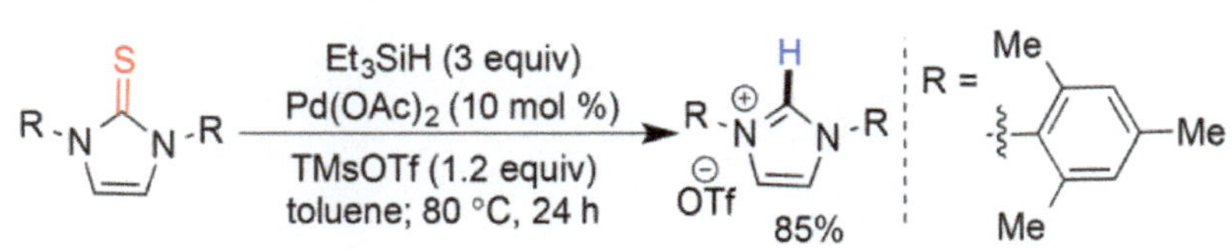

Scheme 17: Imidazolinium salt preparation through reductive desulfurization.

Rhodium as a Desulfurizing Agent

Recently in 2013, Willis and co-workers reported a rhodium catalyzed reductive desulfurization of *ortho*-keto aryl methyl sulfides under mild conditions. The excellent tolerance of the keto group enabled them to use the methyl sulfide moiety as a traceless directing scaffold for hydroacylation of alkanes and alkynes (Scheme 18).[68]

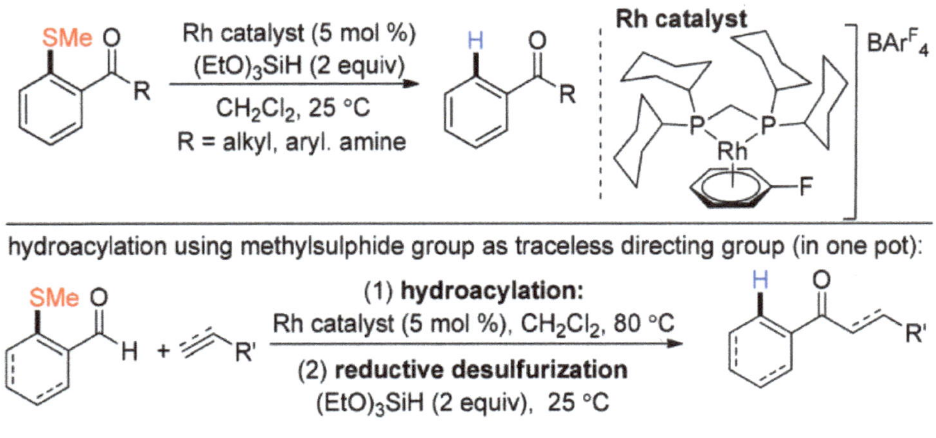

Scheme 18: Rhodium catalyzed reductive desulfurization of methyl phenyl thio ether.

Other Metal Mediated Desulfurization

Except for palladium and nickel mediated reductive desulfurization protocols, RANEY® copper,[69] zinc/HCl,[70] red-Al,[71,72] Al/$HgCl_2$,[73] and Zn/AcOH/Ac_2O [74] are known. But, all these protocols are inadequate for broader use due to poor chemoselectivity, limited substrate scope, harsh reaction conditions, excess metal waste *etc.*

DECYANATION

An efficient methodology for removal of the cyano group permits exploration of the expedient feature of the cyano moiety temporarily in the synthesis of complex molecules. The cyano group has been widely used in chelation assisted C–H activation through an end on and side on directing fashion.[75–79] Due to its high electron withdrawing ability, it can enhance the α-acidity for the desired organic transformation.[80,81] Despite the high C–CN bond dissociation energy,[82] decyanation has been achieved with the help of metal mediated reduction. Reductive decyanation with alkali metals is well known but these processes are limited due to the harsh reaction conditions used. In contrast, a transition metal complex/hydride source combination is accepted as a celebrated class for its milder and easily accessible reaction conditions. From a mechanistic point of view, there are two categories of the reaction pathway known for transition metal catalyzed C–CN bond cleavage.[83] One of them is the direct insertion of a low valent metal in the C–CN bond through oxidative addition and the other being silyl metal-assisted carbon cyano bond cleavage through an iminoacyl intermediate (Scheme 19).

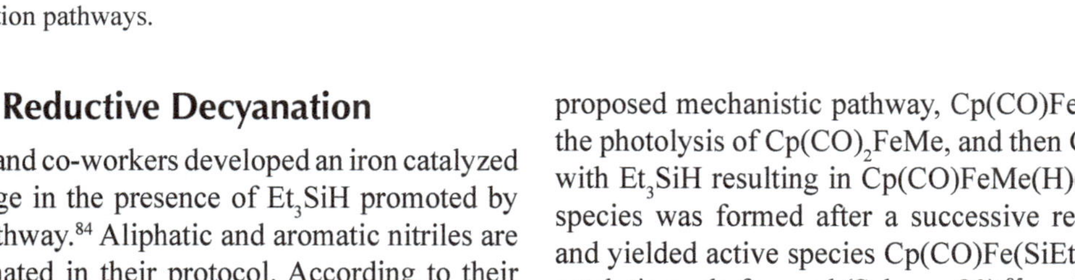

Scheme 19: Decyanation pathways.

Iron Catalyzed Reductive Decyanation

In 2005, Nakazawa and co-workers developed an iron catalyzed C–CN bond cleavage in the presence of Et_3SiH promoted by a photochemical pathway.[84] Aliphatic and aromatic nitriles are successfully decyanated in their protocol. According to their proposed mechanistic pathway, Cp(CO)FeMe is generated by the photolysis of $Cp(CO)_2FeMe$, and then Cp(CO)FeMe reacts with Et_3SiH resulting in $Cp(CO)FeMe(H)(SiEt_3)$. Decyanated species was formed after a successive reductive elimination and yielded active species $Cp(CO)Fe(SiEt_3)$ which carried the catalytic cycle forward (Scheme 20).[85]

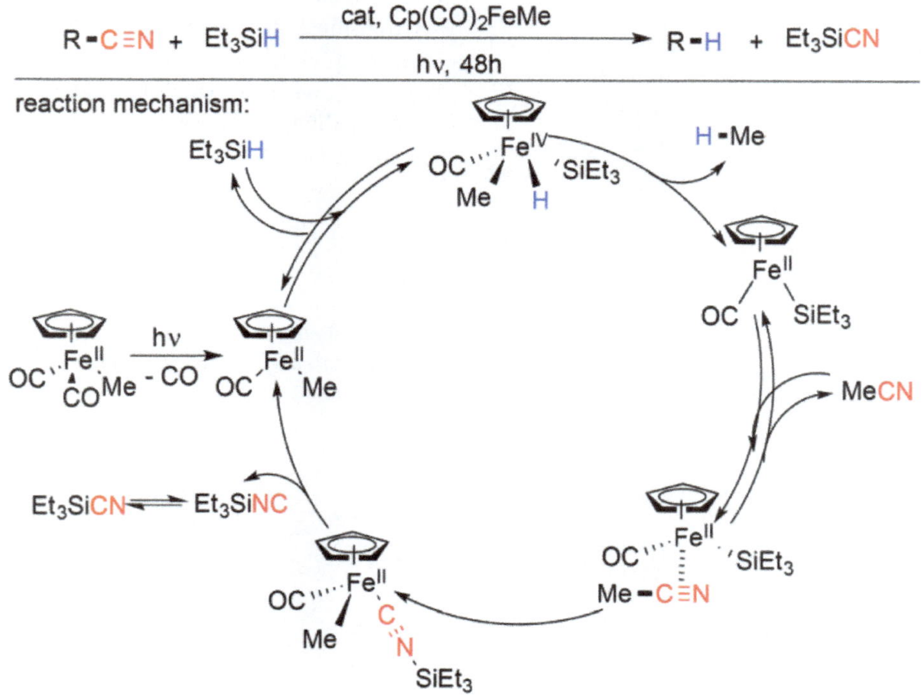

Scheme 20: Proposed mechanism for $Cp(CO)_2FeMe$ catalyzed reductive decyanation.

Rhodium Catalyzed Reductive Decyanation

In 2009, Chatani and coworkers reported Rh-catalyzed reductive decyanation of nitriles using hydrosilane as the reducing agent.[86] Utilization of hydrosilane as a mild reducing agent allowed a large functional group tolerance towards ethers, amines, amides and esters. In the e of primary alkyl cyanides, the complete suppression of β-hydrogen elimination was achieved using their catalytic system. They also demonstrated the cyano moiety as a transient steering group in regioselective functionalization (Scheme 21). An iminoacyl moiety was proposed as the key intermediate prior to C–CN bond cleavage.[87]

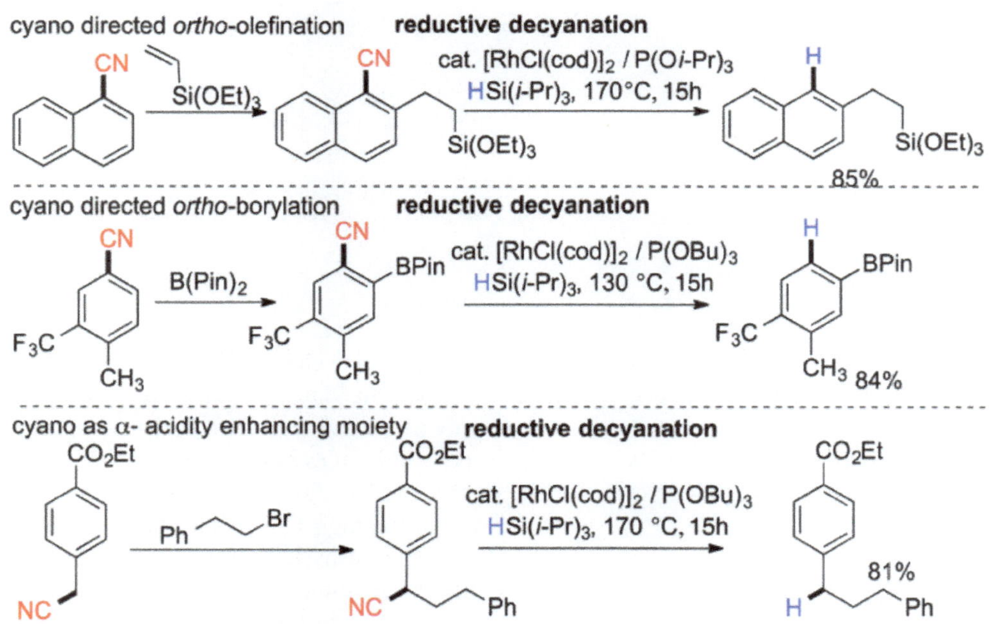

Scheme 21: Application of reductive decyanation as a traceless steering group.

Nickel Catalyzed Decyanation

Recently in 2013, Maiti and co-workers studied the Ni-catalyzed decyanation of inert carbon–cyano bonds with $(Me_2SiH)_2O$ as the hydride source (Scheme 22).[88] It was found that alkyl, aryl and heteroaryl nitriles had been decyanated successfully. The direct insertion of the metal between the C–CN bond through oxidative addition was proposed. The resultant species was transmetalated with $(Me_2SiH)_2O$ and finally reductive elimination yielded the decyanated compound. $AlMe_3$ was used as a Lewis acid which formed an adduct with the nitrile moiety facilitating the oxidative addition of the C–CN bond (Scheme 22).

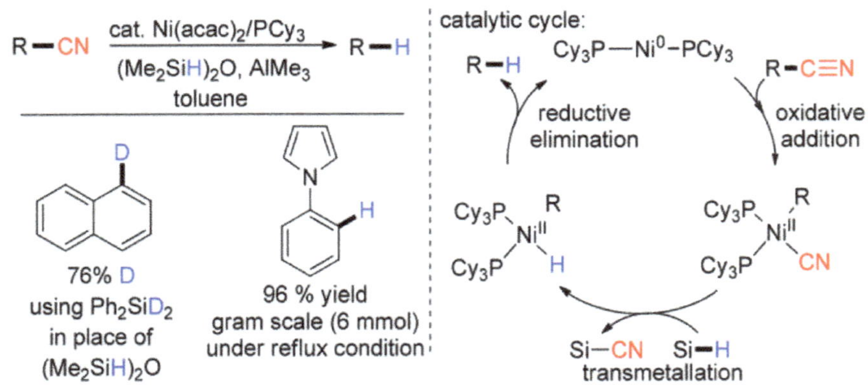

Scheme 22: $Ni(acac)_2$ catalyzed reductive decyanation.

In the same year, Maiti and coworkers reported the Ni-catalyzed hydrogenolysis of inactivated carbon–cyano bonds using catalytic $Ni(cod)_2/PCy_3$ and $AlMe_3$ (3 equiv.) under 1 bar pressure of H_2 gas.[89] The role of $AlMe_3$ was not only to activate the inert C–CN bond for facile oxidative addition of $Ni(cod)_2$ but also to assist in the removal of HCN produced under the reaction conditions (Scheme 23).

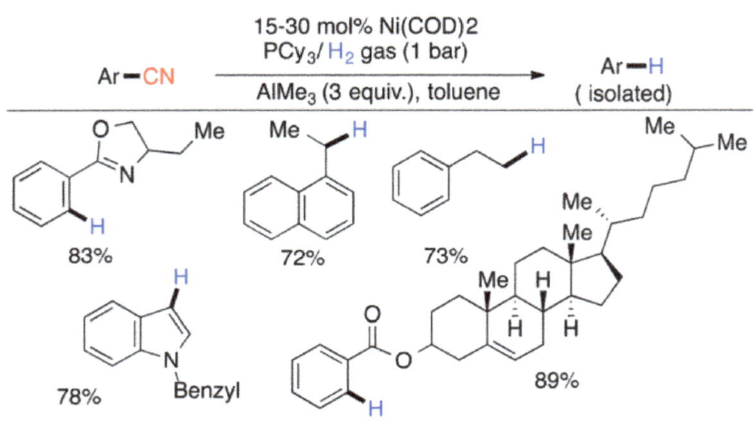

Scheme 23: Nickel catalyzed hydrogenolysis of an inactivated carbon–cyano bond.

Thereafter, Enthaler and coworkers developed a nickel–phosphine precatalyst, which is capable of catalytic hydrodecyanation in the presence of *tert*-butylmagnesium chloride as the hydride source.[90] Later, the same group improved their catalytic system, replacing the hydride source with $LiBH_4$.[91]

DEOXYGENATION OF PHENOL

Deoxygenation of phenol derivatives through the transition metal catalyzed hydrogenolysis of the aromatic carbon–oxygen bond has a great significance in synthetic chemistry. The electron donating property of the phenol or ether linkage favors the *ortho*/*para* functionalization in an aromatic moiety.[92] On the other hand, the ether moiety can also act as a directing scaffold through chelation assisted carbometallation.[93,94] Thus, deoxygenation can be an interesting strategy for utilizing the beneficial characteristics of the phenol or ether moiety. On the other hand, an efficient deoxygenation protocol can provide a number of deoxygenated analogues of biologically active molecules as the phenol moiety is ubiquitous in natural products. The direct deoxygenation of phenol is quiet difficult due to the high C–O bond strength of phenol. So, the well-known pathway to remove the phenol group is to transform the phenol moiety to electronically poor ester groups like triflates, sulphonates, carbamates *etc.*[95–97] Then, the metal mediated reductive cleavage of the C–O bond results in the deoxygenated product. In other words, the reductive cleavage of the C–O bond in the ether moiety is quite challenging. An efficient method for reductive dealkoxylation or aryloxylation has immense significance in biomass derived fuel production, as ether linkages are quite common in biomass, *e.g.* lignin.

Reductive Dealkoxylation and Dearyloxylation

Nickel has proven itself as the most appropriate metal for the reductive deoxygenation of ether for a long time. In 1935, Duzee and Adkins discovered the deoxygenation/hydrogenolysis of alkyl and aryl ethers in the presence of RANEY® nickel under an H_2 atmosphere (150–250 atm).[98] But, the disadvantages of this method lie in the uncontrolled hydrogenation of the aromatic ring before and after the hydrogenolysis of ethers. In 2010, Martin and coworkers reported the first metal catalyzed protocol for the reductive scission of inert C–O bonds of aryl methyl ethers.[99] This methodology showed tolerance to a broad range of functional groups like silyl, esters, amides, acetals, amines, nitrogen containing heterocycles, *etc*. They also showed that regioselective demethoxylation can be achieved for a particular methoxy group, in the presence of more than one methoxy group in the substrate (Scheme 24).

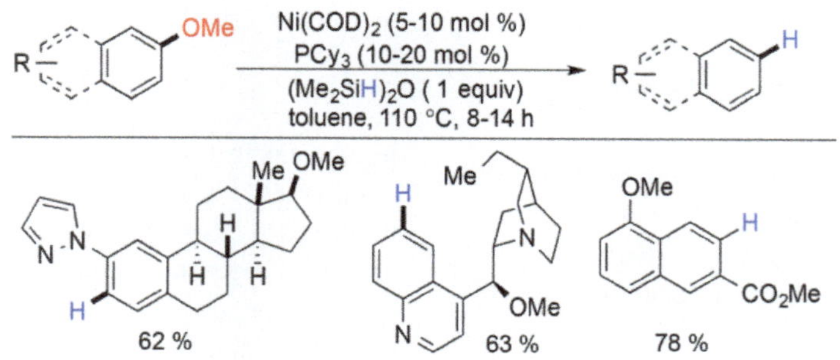

Scheme 24: $Ni(COD)_2$ catalyzed demethoxylation of arylmethyl ethers.

Based on the experimental observation and theoretical calculations, they proposed a reaction mechanism for nickel catalyzed reductive demethoxylation, where at first $Ni(COD)_2$comproportionates to the key catalytic species Ni(I)–$SiEt_3$ in the presence of PCy_3, and Et_3SiH. After that Ni(I)–$SiEt_3$ coordinates with 2-methoxy naphthalene in an η^2-fashion and following a migratory insertion results in the benzyl nickel species.[100] Subsequently, a successive migration of the Ni center releases $MeOSiR_3$. Finally a reversible σ-bond metathesis delivers the demethoxylated product and regenerates the catalytic cycle (Scheme 25).

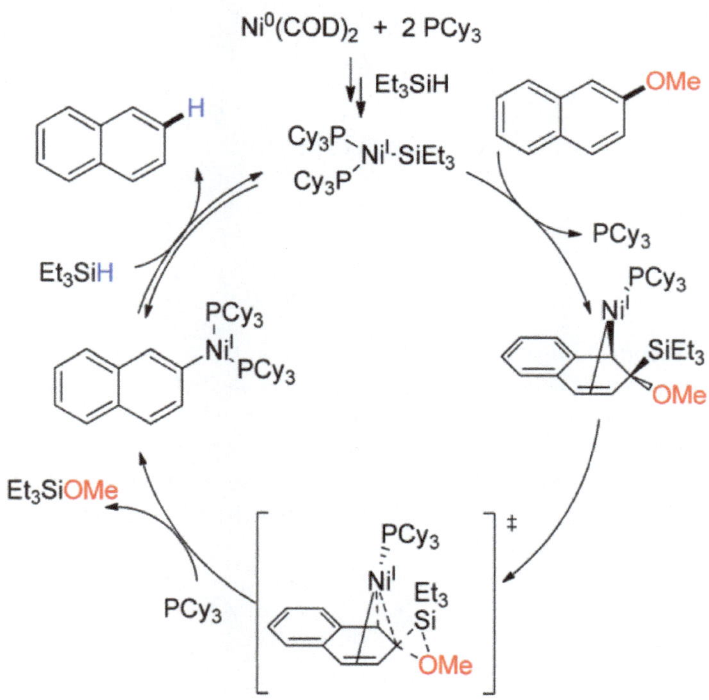

Scheme 25: Proposed mechanism for $Ni(COD)_2$ catalyzed demethoxylation of arylmethyl ethers.

Chatani›s group reported the reductive cleavage of aryl-oxygen bonds in alkoxy and pivaloxy arenes using a catalytic $Ni(COD)_2/PCy_3$ combination and $HSiMe(OMe)_2$ as a mild reducing agent.[101] Under the reaction conditions, the ethoxy and isopropoxy group had been cleaved, but the bulky alkoxy substituent showed lower reactivity. They also judiciously presented the importance of the reductive deoxygenation of aryl pivalates, by using the pivalate moiety as a transient directing group in regioselective aromatic functionalization (Scheme 26).

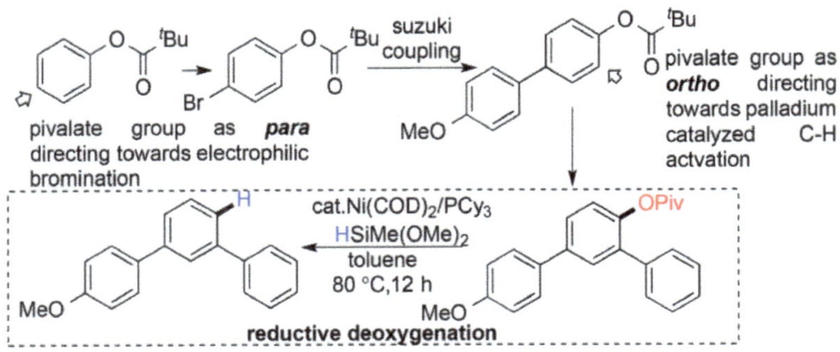

Scheme 26: Synthetic application of reductive deoxygenation.

Hartwig and Sergeev in 2011 discovered the first hydrogenolysis of aryl ethers with catalytic $Ni(COD)_2$ and SIPr·HCl as the ligand.[7] A strong base, NaO^{t}Bu, was used for this transformation, which was thought to help in the C–O bond cleavage or dihydrogen activation forming anionic nickel complexes. The preferential scission of the C–O bond adjacent to the more electron-deficient aryl ring in unsymmetrical diaryl ethers had been attained under their hydrogenolysis conditions. The hydrogenolysis of aryl *n*-hexyl ethers and alkyl benzyl ethers was achieved successfully (Scheme 27).

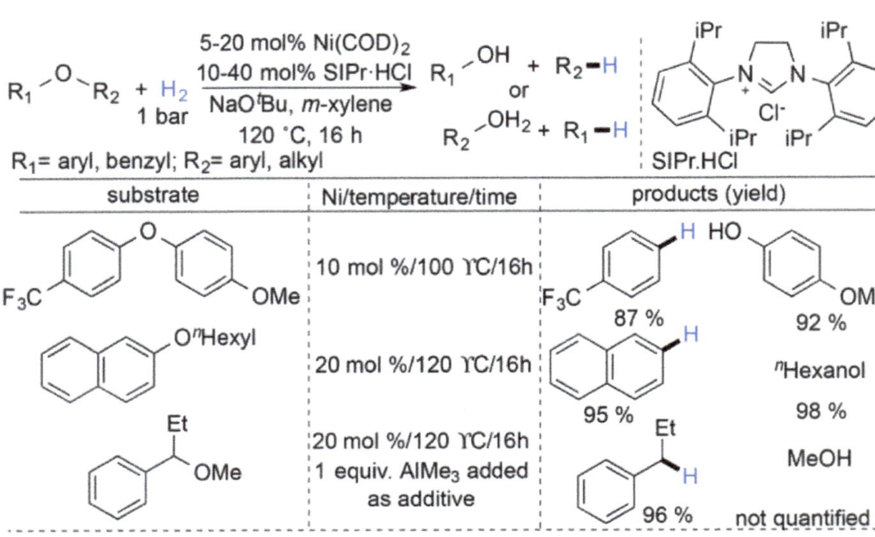

Scheme 27 Hydrogenolysis of aryl ethers with catalytic $Ni(COD)_2$ and SIPr·HCl as the ligand.

A year later, the same group reported a ligand free hydrogenolysis of the C–O bond using $Ni(CH_2TMS)_2(TMEDA)$ as the catalyst which was proven to be more active than their previously reported catalytic conditions.[102] Agapie and coworkers studied the mechanism of the hydrogenolysis reaction of alkyl aryl ethers to gain knowledge about the source of hydrogen.[103] The isotope labeling study showed that the α-hydrogen of the alkyl aryl ether was the hydrogen source for reduction in the $Ni(COD)_2$-NHC catalytic system (Scheme 28). Recently, Chatani and coworkers developed external reductant free nickel catalyzed reaction conditions for the reductive deoxygenation of alkyl aryl ethers.[104] Their proposed reaction mechanism is similar to the one reported by Agapie.[103]

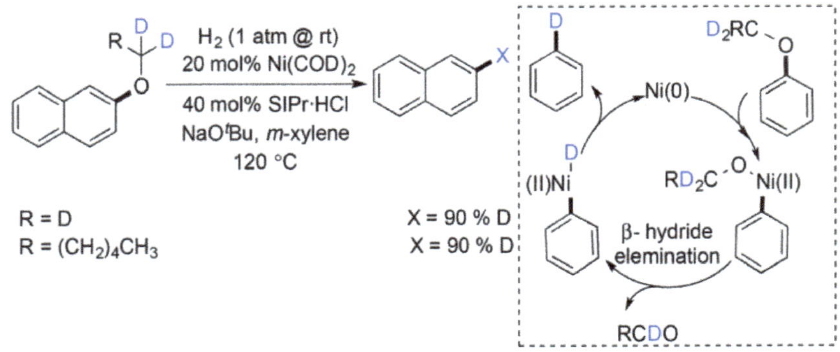

Scheme 28: Proposed mechanism for hydrogenation of aryl alkyl ether.

In 2013 Kawanami and coworkers reported a Rh/C catalyzed hydrogenolysis of diphenyl ether derivatives in the presence of H_2 gas (0.5 MPa) in super critical CO_2–water medium.[105]Wang achieved the selective reductive cleavage of inert aryl C–O bonds by using a catalytic amount of $Fe(acac)_3$ with $LiAlH_4$ as the reducing agent and NaO^{t}Bu as the base.[106] The experimental studies established that the catalytic reaction occurred through an active heterogeneous species. The same group showed that $Co(acac)_2$ is also capable for the reductive cleavage of inert aryl C–O bonds under similar reaction conditions to the iron catalyzed reaction.[107] Recently, Stephenson and coworkers reported the depolymerization of lignin under photolytic conditions using $[Ir(ppy)_2(dtbbpy)]PF_6$ as the photocatalyst.[108]

Reductive Deoxygenation of the Sulfonate Ester of Phenol

Palladium is the most acknowledged transition metal for the reductive deoxygenation of aryl sulfonate esters in the presence of a hydrogen source. Catalytic palladium on charcoal under a hydrogen atmosphere is one of the well-known reagent combinations for the hydrogenation of aryl nonaflates.[109] Afterwards, the reductive deoxygenation of aryl triflates was reported by the Ortar,[110] Wulff,[111] Chen[112] groups independently using trialkylammonium formate as the hydrogen donor in the presence of a catalytic amount of palladium salt/ phosphine ligand.[113]Later, in 1989, Saa and his group showed that naphthyl triflate can be deoxygenated in the presence of a catalytic amount of $(Ph_3P)_2PdCl_2$/dppp in DMF solvent, with Bu_3N as the hydride donor.[114] Using, triethyl silane as the hydride source, Kotsuki reported the reductive deoxygenation of aryl and enol triflates in the presence of a catalytic amount of a palladium(II) salt and bi-dentate phosphine ligand combination.[115] Later Neumeyer successfully employed Katsuki›s method in synthesizing pharmacologically important molecules from opioid derivatives.[116]Afterwards, Lipshutz found a simple, mild Pd(0) catalyzed deoxygenation protocol for aryl perfluorosulphonates with $Me_2NH{\cdot}BH_3$ as the hydrogenating source.[117]

Holmes and Pan in 2001 reported a novel method for deoxygenation of phenol with the help of a solid supported perfluoroalkylsulphonyl (PFS) fluoride linker.[118] The PFS linker was attached with the phenol, resulting in an aryl sulphonate, and a palladium catalyzed cleavage of the aryl C–O bond followed. Using the same strategy, Revell and Ganesan in 2004 developed a tetrafluoroarylsulphonate linker as a traceless functionality for the solid-phase deoxygenation of phenol (Scheme 29).[119]

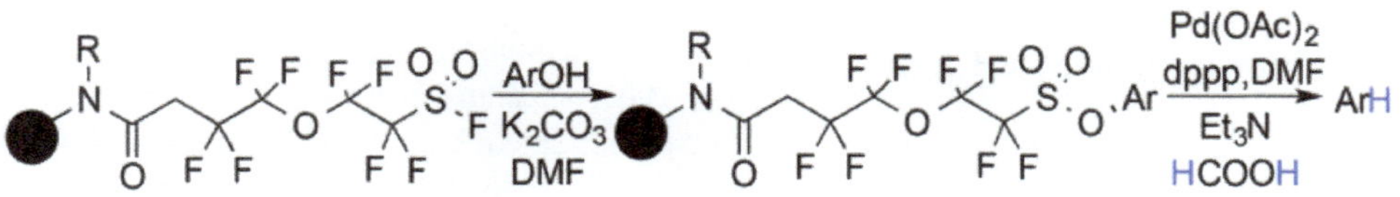

Scheme 29: Solid supported deoxygenation of phenol.

Sajiki and co-workers published a Pd/C catalyzed deoxygenation protocol for aryl triflates and mesylates with a Mg metal–MeOH combination as the hydrogen source.[120] A stoichiometric amount of NH_4OAc was used for better conversion. According to their mechanistic study it was hypothesized that two consecutive single electron transfer (SET) processes from Pd(0) to aryl sulfonates resulted in Pd(II) and an aryl anion.[121] After a protonation by methanol, the aryl anion formed the deoxygenated arene. Palladium(II) transformed to Pd(0) after a successive reduction with magnesium. NH_4OAc was speculated to exert a dual activating effect caused by acetic acid and ammonia. Acetic acid helped the magnesium species for better solubility in methanol and ammonia coordinated to Pd/C as the ligand (Scheme 30).

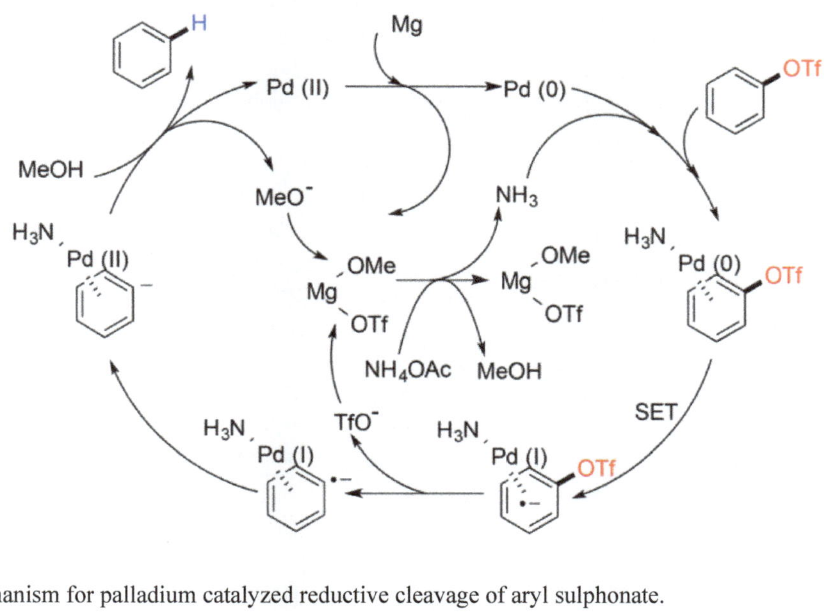

Scheme 30: Proposed mechanism for palladium catalyzed reductive cleavage of aryl sulphonate.

After their initial contribution, Sajiki reported the catalytic Pd/C-diethylamine mediated hydrodeoxygenation of aryl triflates (Scheme 31).[122] Unfortunately, their protocol was limited due to the poor tolerance towards functional groups like olefins, benzyl esters, and nitro/aromatic ketones.

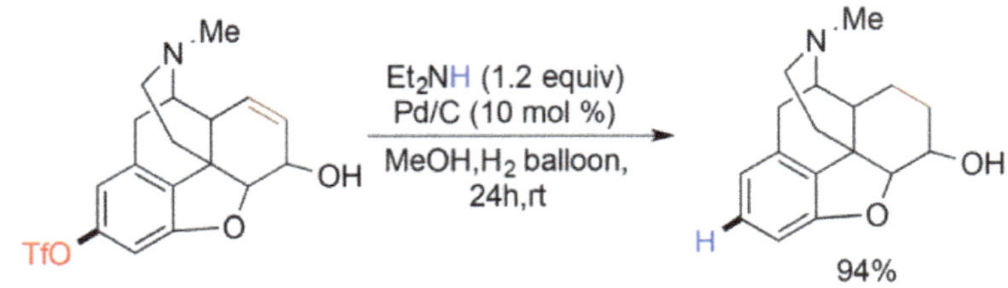

Scheme 31: Reductive deoxygenation with catalytic Pd/C–diethylamine combination..

Recently, Kwong and coworkers developed a palladium catalyzed facile deoxygenation of aryl tosylates using isopropanol as a mild reducing agent.[123]

In reductive deoxygenation reactions, nickel has also been treated as a celebrated class. In 1975 Kratzl reported a RANEY® nickel assisted deoxygenation of aryl sulphonates.[124] Later Sasaki developed the nickel catalyzed deoxygenation of aryl sulphonates.[125] A $NiBr_2(PPh_3)_2$/bidentate phosphine ligand catalyzed method for deoxygenation with a stoichiometric amount of Zn/KI in methanol solvent was reported.[126] Recently Lipshutz and coworkers explored nickel/graphite as a catalyst for the reductive deoxygenation of aryl sulphonates using $Me_2NH{\cdot}BH_3$as the hydride source.[127] They have also tuned the reaction under microwave conditions with a shorter reaction time.[128]

REDUCTIVE DEHALOGENATION

The reductive removal of the halo moiety has been treated as one of the tempting research areas for a long time. Halo-moieties are used as a blocking group in a particular organic transformation.[129–131] Hence, an easy selective removal of this halo moiety allows the synthetic chemist to utilize the halo group as a fertile blocker. On the other hand, reductive dehalogenation has a great significance for environmental cause. Some halogenated, mainly chlorinated arenes are toxic and considered as persistent organic pollutants, but the dehalogenated ones are not.[12–16] These halogenated arenes are resistant to biodegradation also. So an efficient artificial degradation through reductive dehalogenation is considered as a savior from these toxic halogenated materials. Transition metals are well known for the activation of the C–halo bond. Therefore, a number of reductive dehalogenation methods are known with a transitional metal and a hydride source combination.

First Row Transition Metal Complexes for Reductive Dehalogenation

During early stages, reductive dehalogenation was performed by Masamune using a lithium–copper hydride complex.[132,133] Afterwards, a copper-hydride species $(KCuH_2)_n$ was prepared by Negishi, which showed the reductive dehalogenation of aryl halides.[134] Later, Tashiro and coworkers reported reductive dehalogenation with the help of a Ni–Al alloy in alkaline aqueous solution.[131,135] In 1975, Tyrlik[136] and Tufariello[137] independently showed a Mg–Ti complex as a dehalogenating reagent, where solvent molecules (THF and H_2O, respectively) were speculated as the hydrogen source. An admixture of $LiAlH_4$ and a catalytic/stoichiometric amount of first transition metal chloride salt was used for reductive dehalogenation by Ashby.[138]Roth and coworkers reported a Ni(0) catalyzed mild and simple protocol for reductive debromination using $NaBH_4$ as the hydride source.[139] In another approach, Ni(0) catalyzed reductive dehalogenation in the presence of Zn and H_2O was developed by Colon.[140] Using MgH_2 as the reducing agent, Ni(II) catalyzed reductive dehalogenation was reported by Carfagna and coworkers.[141] In 2001, Lipshutz and co-workers examined the Ni/C catalyzed reductions of aryl chlorides in the presence of $Me_2NH{\cdot}BH_3$ and K_2CO_3.[142] Aryl chlorides containing electron rich and electron donating functional groups and a heterocyclic chloride were successfully dehalogenated under their reaction conditions. They have efficiently converted trichlorobiphenyl to biphenyl through this protocol.

Palladium as a Reductive Dehalogenating Agent

Palladium emerged as one of the efficient metals for reductive dehalogenation due to its renowned reactivity towards the C–halo bond through oxidative addition. Heck exploited Pd/C as a useful catalyst for reductive dehalogenation in the presence of diethyl ammonium formate.[143] Palladium catalyzed reductive dehalogenation was further developed by Helquist[144,145] and Blum[146] independently. They discovered $Pd(PPh_3)_4$ catalyzed reductive dehalogenation of organic halides using sodium methoxide[145] or sodium formate[144] or benzyl alcohol[146] as the hydride source. Reductive dehalogenation using the palladium carbene complex (IPr)Pd(allyl)Cl was explored by Nolan and coworkers under both conventional heating and microwave conditions using isopropanol as the hydrogen equivalent.[147] According to their proposed catalytic cycle, Pd(0) was formed in the presence of a base, then oxidative addition of the aryl halide occurred. The Ar–[Pd]–H complex was formed in the presence of isopropanol releasing acetone. Finally, reductive elimination produced the dehalogenated product (Scheme 32).

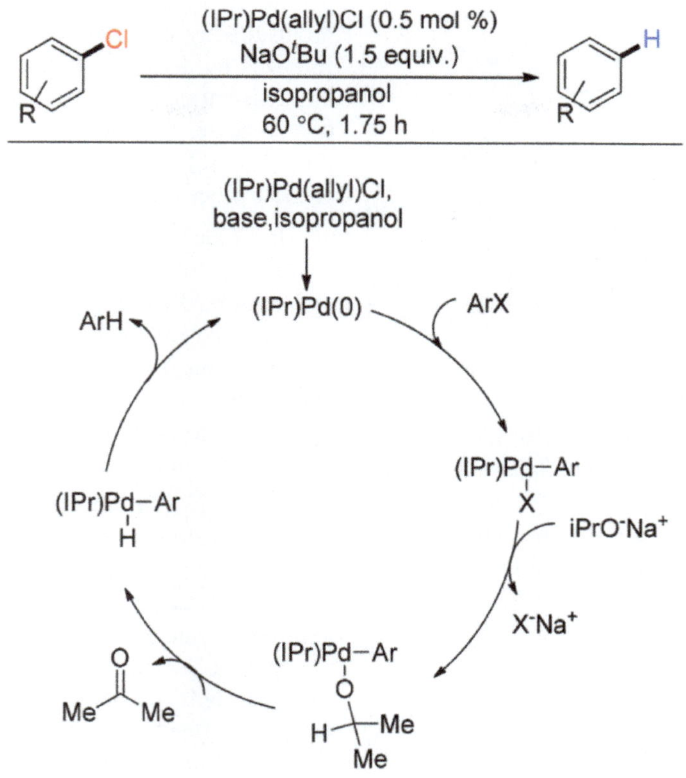

Scheme 32: Proposed mechanism for palladium catalyzed reductive dehalogenation..

In 2010, Jimenez and coworkers presented the Pd/C catalyzed hydrogenation of aryl chlorides and bromide.[130] A broad range of functional groups including phenol, ketone, carboxylic acid, amide, nitro, cyano moieties *etc.* were tolerated under their reaction conditions. Recently, Chelucci and coworkers reported palladium catalyzed mild reaction conditions for the dehalogenation of heterocyclic chlorides and bromide using $NaBH_4$–TMEDA as the hydride source.[148] Regioselective hydrodebromination was also achieved in the presence of more than one bromide group (Scheme 33).

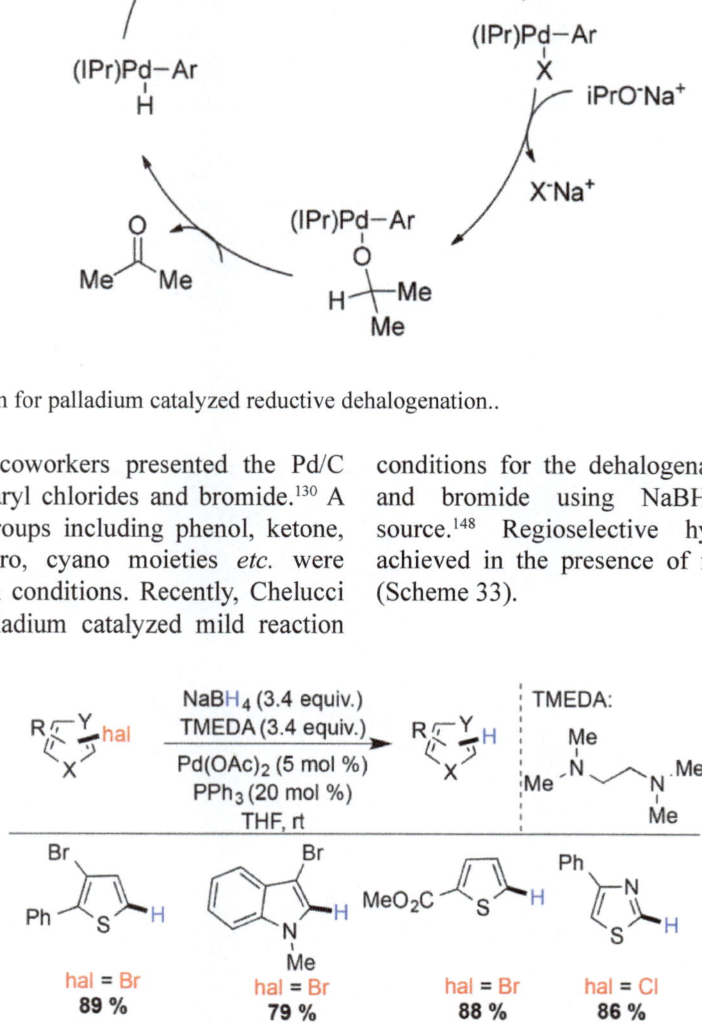

Scheme 33: Palladium catalyzed reductive dehalogenation of heterocylic chloride.

In 2012, Sajiki and coworkers published the Pd/C catalyzed dechlorination of polychlorinated biphenyls in the presence of Mg metal/MeOH at room temperature.[149] A broad range of aryl chlorides with a variety of functional groups was tolerated in their reaction protocol. The authors proposed a reaction pathway which was similar to their previously reported reductive deoxygenation of aryl triflates.[121]

Other metals in Reductive Dehalogenation Reaction

In 1999, Esteruelas *et al.* reported the rhodium catalyzed reductive dechlorination of aryl chlorides, using triethylsilane as the hydride source.[150] Though a great effort has been made on the reductive dechlorination of polychloroarenes, their substrate scope was rarely addressed to exhibit the functional group compatibility. Miura and coworkers demonstrated an $In(OAc)_3$-catalyzed reduction of organic bromides and iodide with $PhSiH_3$ in ethanol.[151] A radical mechanism was proposed for reductive dehalogenation (Scheme 34).

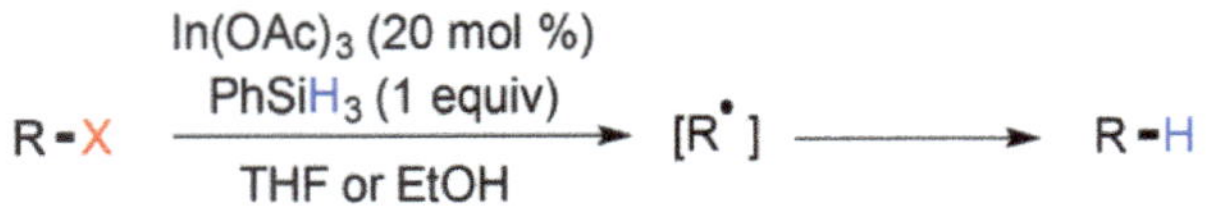

Scheme 34: Reductive dehalogenation under radical conditions.

Corey Stephenson achieved reductive dehalogenation under photochemical conditions in 2009 by using $Ru(bpy)_3Cl_2$ as the electron-transfer photoredox catalyst. Benzyl halide and α-halo carbonyl compounds were selectively dehalogenated without affecting any other functional group present in the molecule.[152] According to the proposed mechanism, the reaction proceeds through single-electron transfer from the ammonium formate complex to the excited Ru(II)* yielding Ru(I) and the radical cation of the complex. Subsequently, the electron rich Ru(I) complex reduces the alkyl halide to the alkyl radical. In the final step the alkyl radical abstracts the hydrogen radical from either the amine or the formate of the ammonium formate radical cation.

DEHYDROXYMETHYLATION

Dehydroxymethylation is the one pot transformation of a primary alcohol to a one carbon less defunctionalized moiety *i.e.* removal of the CH_2OH group. Dehydroxymethylation permits the synthetic chemist to remove the hydroxymethylene group directly, where the alcohol is oxidized to an aldehyde first followed by decarbonylation of the aldehyde. The successful implementation of one reaction system which can deliver these two processes sequentially is desirable for the development of this strategy. In biological systems also, dehydroxymethylation is a key process in DNA demethylase activity.[153]

In 2006 Madsen reported the dehydroxymethylation reaction employing two separate catalysts for oxidation and decarbonylation.[54] Later, they reported the Ir-catalyzed dehydrogenative decarbonylation of primary alcohols with the liberation of syngas.[154] This is achieved with a catalyst generated *in situ* from $[Ir(coe)_2Cl]_2$ and racemic BINAP in a mesitylene solution saturated with water. Lewis acids such as LiCl were found to enhance the rate of transformation.

In 2013, Maiti and co-workers generalized the dehydroxymethylation reaction using one catalyst for the successful transformation where catalytic $Pd(OAc)_2$ was used as both an oxidant and a decarbonylating agent.[155] The tolerance to a broad range of functional groups proved the utility of this method. Selective dehydroxymethylation was performed successfully with a natural product derived substrate, bearing the hydroxymethyl group. They have also demonstrated the importance of dehydroxymethylation, by taking advantage of the CH_2OH group as an impermanent directing group (Scheme 35).[155,156]

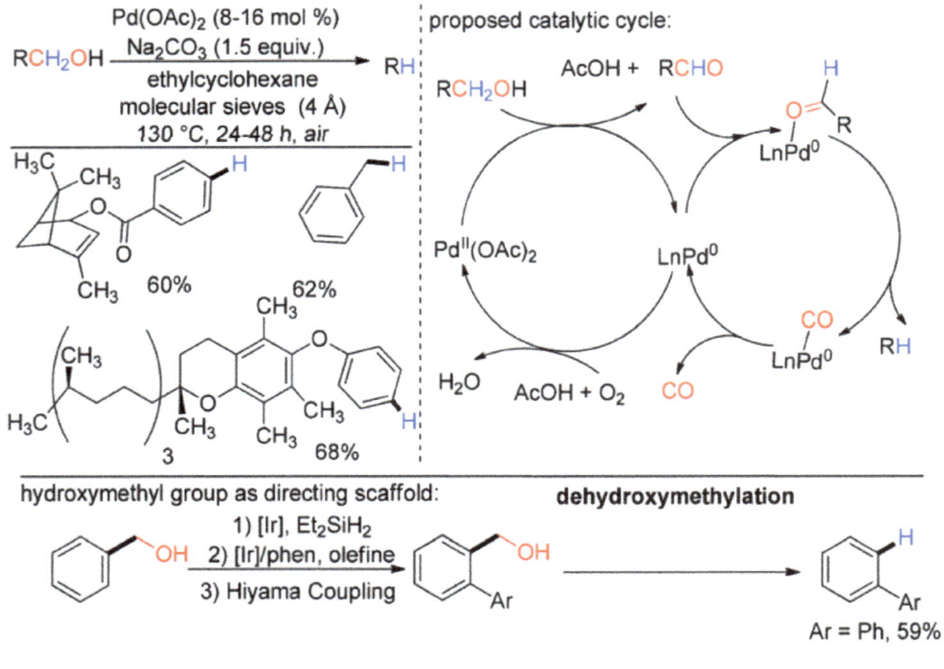

Scheme 35: Palladium catalyzed dehydroxymethylation.

DEAMINATION

The transition metal catalyzed reductive deamination was achieved through C–N bond cleavage. An electron withdrawing moiety with the nitrogen center was essential for successful implementation. In 2014, Chatani and coworkers have described the Ni-catalyzed reductive aromatic C–N bonds in *N*-aryl amides and carbamates. A catalytic amount of $Ni(cod)_2/PCy_3$ with HB(pin) was required (Scheme 36). Sterically demanding substrates bearing a C–N bond at the 1-position of the naphthalene, as well as those containing an *ortho* substituent underwent the reductive cleavage reaction to give the desired products. Several other fused aromatic systems also proceeded smoothly under the optimized conditions, whereas benzene derivatives were found to be much less reactive.[157]

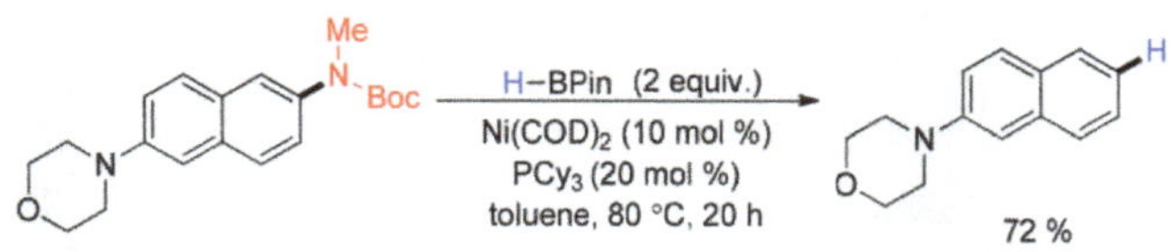

Scheme 36: Nickel catalyzed reductive deamination.

CONCLUSION

The development of novel strategies for carbon–carbon/heteroatom bond cleavage can lead to exploration of diverse catalytic systems. The development of inexpensive transition-metal catalysts, mild conditions, green and sustainable oxidants, and a wide substrate scope remained the major focus of these studies. Efficient defunctionalization strategies are emerging which will play a vital role in the production of energy *viz.* bio-fuels are being made from organic sources like biomass. Defunctionalization is also important in the synthesis of drugs and biomimetics, as they form the core in a wide range of pharmaceuticals.

ACKNOWLEDGEMENTS

We thank SERB-India (EMR/2014/000164) for funding. Financial support received from CSIR-India (fellowship to A. M.) is gratefully acknowledged.

NOTES AND REFERENCES

1. F. Zhang and D. R. Spring, *Chem. Soc. Rev.*, 2014, **43**, 6906–6919 .
2. G. Rousseau and B. Breit, *Angew. Chem., Int. Ed.*, 2011, **50**, 2450–2494 .
3. A.-L. Marshall and P. J. Alaimo, *Chem. – Eur. J.*, 2010, **16**, 4970–4980 .
4. Y.-B. Huang, Z. Yang, M.-Y. Chen, J.-J. Dai, Q.-X. Guo and Y. Fu, *ChemSusChem*, 2013, **6**, 1348–1351 .
5. J. Mitra, X. Zhou and T. Rauchfuss, *Green Chem.*, 2015, **17**, 307–313 .
6. R. N. Monrad and R. Madsen, *J. Org. Chem.*, 2007, **72**, 9782–9785 .
7. A. G. Sergeev and J. F. Hartwig, *Science*, 2011, **332**, 439–443 .
8. J. Zakzeski, P. C. A. Bruijnincx, A. L. Jongerius and B. M. Weckhuysen, *Chem. Rev.*, 2010, **110**, 3552–3599 .
9. J. C. Serrano-Ruiz and J. A. Dumesic, *Energy Environ. Sci.*, 2011, **4**, 83–99 .
10. S. Becker, Y. Fort, R. Vanderesse and P. Caubere, *J. Org. Chem.*, 1989, **54**, 4848–4853 .
11. J. J. Eisch, L. E. Hallenbeck and M. A. Lucarelli, *Fuel*, 1985, **64**, 440–442 .
12. S. Agarwal, S. R. Al-Abed and D. D. Dionysiou, *Environ. Sci. Technol.*, 2007, **41**, 3722–3727 .
13. E. Hadnagy, K. H. Gardner and L. M. Rauch, *J. Environ. Sci. Health, Part A*, 2007, **42**, 685–695 .
14. S. Agarwal, S. R. Al-Abed, D. D. Dionysiou and E. Graybill, *Environ. Sci. Technol.*, 2009, **43**, 915–921 .
15. R. DeVor, K. Carvalho-Knighton, B. Aitken, P. Maloney, E. Holland, L. Talalaj, S. Elsheimer, C. A. Clausen and C. L. Geiger, *Chemosphere*, 2009, **76**, 761–766 .
16. A. Ido, S. Ishihara, A. Kume, T. Nakanishi, Y. Monguchi, H. Sajiki and H. Nagase, *Chemosphere*, 2013, **90**, 57–64 .
17. R. Hayatsu, R. G. Scott and R. E. Winans, Oxidation in Organic Chemistry 5-D, in *Oxidation of Coal*, ed. W. Trahanovsky, Elsevier, 2012, ch. IV.
18. L. Pesci, *Atti Real. Accad. Lincei*, 1901, 10.
19. G. J. Leuck, R. P. Perkins and F. C. Whitmore, *J. Am. Chem. Soc.*, 1929, **51**, 1831–1836 .
20. D. Dodd and M. D. Johnson, *J. Chem. Soc. B*, 1970, 1337–1343 .
21. F. C. Whitmore and R. P. Perkins, *J. Am. Chem. Soc.*, 1929, **51**, 3352–3353 .
22. M. S. Newman and M. C. Vander Zwan, *J. Org. Chem.*, 1973, **38**, 319–321 .
23. G. B. Deacon, S. J. Faulks and G. N. Pain, in *Adv. Organomet. Chem*, ed. F. G. A. Stone and W. Robert, Academic Press, 1986, vol. 25, pp. 237–276 .
24. J. D. Moseley and J. P. Gilday, *Tetrahedron*, 2006, **62**, 4690–4697 .
25. A. F. Shepard, N. R. Winslow and J. R. Johnson, *J. Am. Chem. Soc.*, 1930, **52**, 2083–2090 .
26. M. Nilsson, *Acta Chem. Scand.*, 1966, **20**, 423–426 .
27. T. Cohen and R. A. Schambach, *J. Am. Chem. Soc.*, 1970, **92**, 3189–3190 .
28. T. Cohen, R. W. Berninger and J. T. Wood, *J. Org. Chem.*, 1978, **43**, 837–848 .
29. A. Cairncross, J. R. Roland, R. M. Henderson and W. A. Sheppard, *J. Am. Chem. Soc.*, 1970, **92**, 3187–3189 .
30. L. J. Gooßen, W. R. Thiel, N. Rodríguez, C. Linder and B. Melzer, *Adv. Synth. Catal.*, 2007, **349**, 2241–2246 .
31. L. J. Goossen, F. Manjolinho, B. A. Khan and N. Rodríguez, *J. Org. Chem.*, 2009, **74**, 2620–2623 .
32. W. F. Maier, W. Roth, I. Thies and P. V. R. Schleyer, *Chem. Ber.*, 1982, **115**, 808–812 .
33. S. Matsubara, Y. Yokota and K. Oshima, *Org. Lett.*, 2004, **6**, 2071–2073 .
34. J. S. Dickstein, C. A. Mulrooney, E. M. O'Brien, B. J. Morgan and M. C. Kozlowski, *Org. Lett.*, 2007, **9**, 2441–2444 .
35. J. S. Dickstein, J. M. Curto, O. Gutierrez, C. A. Mulrooney and M. C. Kozlowski, *J. Org. Chem.*, 2013, **78**, 4744–4761 .
36. N. Chatani, H. Tatamidani, Y. Ie, F. Kakiuchi and S. Murai, *J. Am. Chem. Soc.*, 2001, **123**, 4849–4850 .
37. P. Lu, C. Sanchez, J. Cornella and I. Larrosa, *Org. Lett.*, 2009, **11**, 5710–5713 .
38. R. Grainger, A. Nikmal, J. Cornella and I. Larrosa, *Org. Biomol. Chem.*, 2012, **10**, 3172–3174 .
39. S. Seo, J. B. Taylor and M. F. Greaney, *Chem. Commun.*, 2012, **48**, 8270–8272 .
40. Z.-M. Sun, J. Zhang and P. Zhao, *Org. Lett.*, 2010, **12**, 992–995 .

41. S. Dupuy, F. Lazreg, A. M. Z. Slawin, C. S. J. Cazin and S. P. Nolan, *Chem. Commun.*, 2011, **47**, 5455–5457 .
42. S. Dupuy and S. P. Nolan, *Chem. – Eur. J.*, 2013, **19**, 14034–14038 .
43. J. M. Carraher, O. Pestovsky and A. Bakac, *Dalton Trans.*, 2012, **41**, 5974–5980 .
44. T. Patra, S. Manna and D. Maiti, *Angew. Chem., Int. Ed.*, 2011, **50**, 12140–12142 .
45. J. Hawthorne and M. Wilt, *J. Org. Chem.*, 1960, **25**, 2215–2216 .
46. J. Tsuji, K. Ohno and T. Kajimoto, *Tetrahedron Lett.*, 1965, **6**, 4565–4568 .
47. A. Modak, A. Deb, T. Patra, S. Rana, S. Maity and D. Maiti, *Chem. Commun.*, 2012, **48**, 4253–4255 .
48. Akanksha and D. Maiti, *Green Chem.*, 2012, **14**, 2314–2320 .
49. M. C. Bröhmer, N. Volz and S. Bräse, *Synlett*, 2009, 1383–1386.
50. J. Tsuji and K. Ohno, *Tetrahedron Lett.*, 1965, **6**, 3969–3971 .
51. D. H. Doughty and L. H. Pignolet, *J. Am. Chem. Soc.*, 1978, **100**, 7083–7085 .
52. J. M. O'Connor and J. Ma, *J. Org. Chem.*, 1992, **57**, 5075–5077 .
53. C. M. Beck, S. E. Rathmill, Y. J. Park, J. Chen, R. H. Crabtree, L. M. Liable-Sands and A. L. Rheingold, *Organometallics*, 1999, **18**, 5311–5317 .
54. M. Kreis, A. Palmelund, L. Bunch and R. Madsen, *Adv. Synth. Catal.*, 2006, **348**, 2148–2154 .
55. P. Fristrup, M. Kreis, A. Palmelund, P.-O. Norrby and R. Madsen, *J. Am. Chem. Soc.*, 2008, **130**, 5206–5215 .
56. B. Gutmann, P. Elsner, T. Glasnov, D. M. Roberge and C. O. Kappe, *Angew. Chem., Int. Ed.*, 2014, **53**, 11557–11561 .
57. G. Domazetis, B. Tarpey, D. Dolphin and B. R. James, *J. Chem. Soc., Chem. Commun.*, 1980, 939–940 .
58. T. Iwai, T. Fujihara and Y. Tsuji, *Chem. Commun.*, 2008, 6215–6217 .
59. J. Rentner, M. Kljajic, L. Offner and R. Breinbauer, *Tetrahedron*, 2014, **70**, 8983–9027 .
60. R. Mozingo, D. E. Wolf, S. A. Harris and K. Folkers, *J. Am. Chem. Soc.*, 1943, **65**, 1013–1016 .
61. A. Zumbrunn, *Synthesis*, 1998, 1357–1361 .
62. I. Kawasaki, N. Sakaguchi, A. Khadeer, M. Yamashita and S. Ohta, *Tetrahedron*, 2006, **62**, 10182–10192 .
63. N. Barbero and R. Martin, *Org. Lett.*, 2012, **14**, 796–799 .
64. S. Niwas, P. Chand, V. P. Pathak and J. A. Montgomery, *J. Med. Chem.*, 1994, **37**, 2477–2480 .
65. T. H. Graham, W. Liu and D.-M. Shen, *Org. Lett.*, 2011, **13**, 6232–6235 .
66. T. Matsumura, T. Niwa and M. Nakada, *Tetrahedron Lett.*, 2012, **53**, 4313–4316 .
67. T. Matsumura and M. Nakada, *Tetrahedron Lett.*, 2014, **55**, 1412–1415 .
68. J. F. Hooper, R. D. Young, A. S. Weller and M. C. Willis, *Chem. – Eur. J.*, 2013, **19**, 3125–3130 .
69. W. Pfleiderer, *Tetrahedron*, 1988, **44**, 3373–3378 .
70. J. J. Baldwin, E. L. Engelhardt, R. Hihmann, G. S. Ponticello, J. G. Atkinson, B. K. Wasson, C. S. Sweet and A. Scriabine, *J. Med. Chem.*, 1980, **23**, 65–70 .
71. H. Yoshioka, Y. Matsuya, T. Choshi, E. Sugino and S. Hibino, *Chem. Pharm. Bull.*, 1996, **44**, 709–714 .
72. T. Choshi, A. Tonari, H. Yoshioka, K. Harada, E. Sugino and S. Hibino, *J. Org. Chem.*, 1993, **58**, 7952–7954 .
73. T. Sugimoto, N. Nishioka, S. Murata and S. Matsuura, *Heterocycles*, 1987, 26.
74. C. Yamazaki, H. Arima and S. Udagawa, *J. Heterocycl. Chem.*, 1996, **33**, 41–44 .
75. B. Du, X. Jiang and P. Sun, *J. Org. Chem.*, 2013, **78**, 2786–2791 .
76. D. Leow, G. Li, T.-S. Mei and J.-Q. Yu, *Nature*, 2012, **486**, 518–522 .
77. M. C. Reddy and M. Jeganmohan, *Chem. Commun.*, 2015, **51**, 10738–10741 .
78. M. Bera, A. Modak, T. Patra, A. Maji and D. Maiti, *Org. Lett.*, 2014, **16**, 5760–5763 .
79. M. Bera, A. Maji, S. K. Sahoo and D. Maiti, *Angew. Chem., Int. Ed.*, 2015, **54**, 8515–8519 .
80. T. D. Krizan and J. C. Martin, *J. Am. Chem. Soc.*, 1983, **105**, 6155–6157 .
81. J.-M. Mattalia, C. Marchi-Delapierre, H. Hazimeh and M. Chanon, *ARKIVOC*, 2006, 90–118 .
82. M. Tobisu and N. Chatani, *Chem. Soc. Rev.*, 2008, **37**, 300–307 .
83. H. Kinuta, H. Takahashi, M. Tobisu, S. Mori and N. Chatani, *Bull. Chem. Soc. Jpn.*, 2014, **87**, 655–669 .
84. H. Nakazawa, K. Kamata and M. Itazaki, *Chem. Commun.*, 2005, 4004–4006 .
85. H. Nakazawa, M. Itazaki, K. Kamata and K. Ueda, *Chem. – Asian J.*, 2007, **2**, 882–888 .
86. M. Tobisu, R. Nakamura, Y. Kita and N. Chatani, *J. Am. Chem. Soc.*, 2009, **131**, 3174–3175 .
87. M. Tobisu, R. Nakamura, Y. Kita and N. Chatani, *Bull. Korean Chem. Soc.*, 2010, **31**, 582–587 .
88. T. Patra, S. Agasti, Akanksha and D. Maiti, *Chem. Commun.*, 2013, **49**, 69–71 .
89. T. Patra, S. Agasti, A. Modak and D. Maiti, *Chem. Commun.*, 2013, **49**, 8362–8364 .
90. M. Weidauer, C. I. Someya, E. Irran and S. Enthaler, *Asian J. Org. Chem.*, 2013, **2**, 150–156 .
91. S. Enthaler, M. Weidauer, E. Irran, J. D. Epping, R. Kretschmer and C. I. Someya, *J. Organomet. Chem.*, 2013, **745–746**, 262–274 .
92. D.-G. Yu, F. de Azambuja and F. Glorius, *Angew. Chem., Int. Ed.*, 2014, **53**, 7710–7712 .
93. Á. Iglesias, R. Álvarez, Á. R. de Lera and K. Muñiz, *Angew. Chem., Int. Ed.*, 2012, **51**, 2225–2228 .
94. J. Oyamada, M. Nishiura and Z. Hou, *Angew. Chem., Int. Ed.*, 2011, **50**, 10720–10723 .

95. E. Vowinkel and C. Wolff, *Chem. Ber.*, 1974, **107**, 907–914 .

96. J. M. Herrmann and B. König, *Eur. J. Org. Chem.*, 2013, 7017–7027 .

97. P. Sebok, T. Timar, T. Eszenyi and T. Patonay, *J. Org. Chem.*, 1994, **59**, 6318–6321 .

98. E. M. v. Duzee and H. Adkins, *J. Am. Chem. Soc.*, 1935, **57**, 147–151 .

99. P. Álvarez-Bercedo and R. Martin, *J. Am. Chem. Soc.*, 2010, **132**, 17352–17353 .

100. J. Cornella, E. Gómez-Bengoa and R. Martin, *J. Am. Chem. Soc.*, 2013, **135**, 1997–2009 .

101. M. Tobisu, K. Yamakawa, T. Shimasaki and N. Chatani, *Chem. Commun.*, 2011, **47**, 2946–2948 .

102. A. G. Sergeev, J. D. Webb and J. F. Hartwig, *J. Am. Chem. Soc.*, 2012, **134**, 20226–20229 .

103. P. Kelley, S. Lin, G. Edouard, M. W. Day and T. Agapie, *J. Am. Chem. Soc.*, 2012, **134**, 5480–5483 .

104. M. Tobisu, T. Morioka, A. Ohtsuki and N. Chatani, *Chem. Sci.*, 2015, **6**, 3410–3414 .

105. M. Chatterjee, A. Chatterjee, T. Ishizaka and H. Kawanami, *Catal. Sci. Technol.*, 2015, **5**, 1532–1539 .

106. Y. Ren, M. Yan, J. Wang, Z. C. Zhang and K. Yao, *Angew. Chem., Int. Ed.*, 2013, **52**, 12674–12678 .

107. Y.-L. Ren, M. Tian, X.-Z. Tian, Q. Wang, H. Shang, J. Wang and Z. C. Zhang, *Chem. Lett.*, 2014, **52**, 36–39 .

108. J. D. Nguyen, B. S. Matsuura and C. R. J. Stephenson, *J. Am. Chem. Soc.*, 2014, **136**, 1218–1221 .

109. L. R. Subramanian, A. G. Martinez, A. H. Fernandez and R. M. Alvarez, *Synthesis*, 1984, 481–485 .

110. S. Cacchi, P. G. Ciattini, E. Morera and G. Ortar, *Tetrahedron Lett.*, 1986, **27**, 5541–5544 .

111. G. A. Peterson, F.-A. Kunng, J. S. McCallum and W. D. Wulff, *Tetrahedron Lett.*, 1987, **28**, 1381–1384 .

112. Q.-Y. Chen, Y.-B. He and Z.-Y. Yang, *J. Chem. Soc., Chem. Commun.*, 1986, 1452–1453 .

113. P. Ganji and H. Ibrahim, *J. Org. Chem.*, 2012, **77**, 511–518 .

114. J. M. Saa, M. Dopico, G. Martorell and A. Garcia-Raso, *J. Org. Chem.*, 1990, **55**, 991–995 .

115. H. Kotsuki, P. K. Datta, H. Hayakawa and H. Suenaga, *Synthesis*, 1995, 1348–1350 .

116. C. D. Hupp and J. L. Neumeyer, *Tetrahedron Lett.*, 2010, **51**, 2359–2361 .

117. B. H. Lipshutz, D. J. Buzard and R. W. Vivian, *Tetrahedron Lett.*, 1999, **40**, 6871–6874 .

118. Y. Pan and C. P. Holmes, *Org. Lett.*, 2001, **3**, 2769–2771 .

119. J. D. Revell and A. Ganesan, *Chem. Commun.*, 2004, 1916–1917 .

120. H. Sajiki, A. Mori, T. Mizusaki, T. Ikawa, T. Maegawa and K. Hirota, *Org. Lett.*, 2006, **8**, 987–990 .

121. A. Mori, T. Mizusaki, T. Ikawa, T. Maegawa, Y. Monguchi and H. Sajiki, *Chem. – Eur. J.*, 2007, **13**, 1432–1441 .

122. A. Mori, T. Mizusaki, T. Ikawa, T. Maegawa, Y. Monguchi and H. Sajiki, *Tetrahedron*, 2007, **63**, 1270–1280 .

123. W. K. Chow, C. M. So, C. P. Lau and F. Y. Kwong, *Org. Chem. Front.*, 2014, **1**, 464–467 .

124. W. Lonsky, H. Traitler and K. Kratzl, *J. Chem. Soc., Perkin Trans. 1*, 1975, 169–170 .

125. K. Sasaki, M. Sakai, Y. Sakakibara and K. Takagi, *Chem. Lett.*, 1991, **20**, 2017–2018 .

126. K. Sasaki, T. Kubo, M. Sakai and Y. Kuroda, *Chem. Lett.*, 1997, **26**, 617–618 .

127. B. H. Lipshutz, B. A. Frieman, T. Butler and V. Kogan, *Angew. Chem., Int. Ed.*, 2006, **45**, 800–803 .

128. T. A. Butler, E. C. Swift and B. H. Lipshutz, *Org. Biomol. Chem.*, 2008, **6**, 19–25 .

129. A. Dobbs, *J. Org. Chem.*, 2001, **66**, 638–641 .

130. A. Ramanathan and L. S. Jimenez, *Synthesis*, 2010, 217–220 .

131. M. Tashiro, G. Fukata and K. Oe, *Org. Prep. Proced. Int.*, 1975, **7**, 183–188 .

132. S. Masamune, P. A. Rossy and G. S. Bates, *J. Am. Chem. Soc.*, 1973, **95**, 6452–6454 .

133. S. Masamune, G. S. Bates and P. E. Georghiou, *J. Am. Chem. Soc.*, 1974, **96**, 3686–3688 .

134. T. Yoshida and E.-i. Negishi, *J. Chem. Soc., Chem. Commun.*, 1974, 762–763 .

135. G.-B. Liu, T. Tsukinoki, T. Kanda, Y. Mitoma and M. Tashiro, *Tetrahedron Lett.*, 1998, **39**, 5991–5994 .

136. S. Tyrlik and I. Wolochowicz, *J. Chem. Soc., Chem. Commun.*, 1975, 781–782 .

137. T. R. Nelsen and J. J. Tufariello, *J. Org. Chem.*, 1975, **40**, 3159–3160 .

138. E. C. Ashby and J. J. Lin, *J. Org. Chem.*, 1978, **43**, 1263–1265 .

139. S.-T. Lin and J. A. Roth, *J. Org. Chem.*, 1979, **44**, 309–310 .

140. I. Colon, *J. Org. Chem.*, 1982, **47**, 2622–2625 .

141. C. Carfagna, A. Musco, R. Pontellini and G. Terzoni, *J. Mol. Catal.*, 1989, **57**, 23–28 .

142. B. H. Lipshutz, T. Tomioka and K. Sato, *Synlett*, 2001, 0970–0973 .

143. N. A. Cortese and R. F. Heck, *J. Org. Chem.*, 1977, **42**, 3491–3494 .

144. P. Helquist, *Tetrahedron Lett.*, 1978, **19**, 1913–1914 .

145. A. Zask and P. Helquist, *J. Org. Chem.*, 1978, **43**, 1619–1620 .

146. A. Zoran, Y. Sasson and J. Blum, *J. Mol. Catal.*, 1984, **27**, 349–353 .

147. O. Navarro, H. Kaur, P. Mahjoor and S. P. Nolan, *J. Org. Chem.*, 2004, **69**, 3173–3180 .

148. G. Chelucci, S. Baldino and A. Ruiu, *J. Org. Chem.*, 2012, **77**, 9921–9925 .

149. S. Ishihara, A. Ido, Y. Monguchi, H. Nagase and H. Sajiki, *J. Hazard. Mater.*, 2012, **229–230**, 15–19 .

150. M. A. Esteruelas, J. Herrero, F. M. López, M. Martín and L. A. Oro, *Organometallics*, 1999, **18**, 1110–1112 .

151. K. Miura, M. Tomita, Y. Yamada and A. Hosomi, *J. Org. Chem.*, 2007, **72**, 787–792 .

152. J. M. R. Narayanam, J. W. Tucker and C. R. J. Stephenson, *J. Am. Chem. Soc.*, 2009, **131**, 8756–8757 .

153. K. I. Ladwein and M. Jung, *Angew. Chem., Int. Ed.*, 2011, **50**, 12143–12145 .

154. E. P. K. Olsen and R. Madsen, *Chem. – Eur. J.*, 2012, **18**, 16023–16029 .

155. A. Modak, T. Naveen and D. Maiti, *Chem. Commun.*, 2013, **49**, 252–254 .

156. E. M. Simmons and J. F. Hartwig, *J. Am. Chem. Soc.*, 2010, **132**, 17092–17095 .

157. M. Tobisu, K. Nakamura and N. Chatani, *J. Am. Chem. Soc.*, 2014, **136**, 5587–5590 .

CHAPTER 17 MILD METAL-CATALYZED C–H ACTIVATION: EXAMPLES AND CONCEPTS

T. Gensch, M. N. Hopkinson, F. Glorius and J. Wencel-Delord

INTRODUCTION

At the beginning of this century, the availability of a variety of coupling reactions which use non-activated C–H bonds as reaction sites seemed like an unrealistic dream. However, over the last 15 years, the field of C–H activation has become one of the most rapidly developing areas of homogeneous catalysis,[1] reshaping the landscape of both organometallic catalysis and synthetic chemistry.[2] The first years of the C–H activation boom were mainly devoted to the discovery, sometimes quite unpredictably, of new catalytic systems allowing the direct insertion of a metal catalyst into a C–H bond followed by the functionalization of the resulting organometallic intermediates. These early works proved unambiguously that a carefully designed catalytic system is able to engage C–H bonds in organometallic reactions. However, the cleavage of high energy C–H bonds (≈110 kcal mol^{-1} for C(aryl)–H bonds) typically requires harsh reaction conditions resulting in limited substrate scope and low functional group tolerance. As such, C–H activation has not yet found widespread applications in the late stage functionalization of complex molecules, which often contain many functionalities not tolerant of harsh reaction conditions. The driving force of current research is the development of the next generation of C–H transformations, which will help to make these processes truly useful synthetic tools. Alongside improving the efficiency and selectivity of C–H transformations, many groups have focused on developing reactions that proceed under milder conditions. "Mildness", however, is a relative concept that is multi-faceted and can be interpreted in many ways. In 2011, we defined a "truly mild C–H activation reaction" as one that proceeds "at or below ambient temperature, under neutral conditions and in the absence of strong oxidants and reductants."[1e] These reaction conditions can inherently lead to improved functional group tolerance, selectivity and overall applicability of the process. Previously, we reviewed examples of C–H activation reactions published up to 2011 which could be conducted under comparatively mild conditions.[1e] However, over the past five years, interest in this topic has grown rapidly and a large number of very challenging transformations have been disclosed in the literature which occur under surprisingly mild conditions.

Herein we highlight the current state of the art and detail the great advances that have been made in the field of mild C–H activation since 2011. In this regard, we also aim to outline the major concepts that can lead to improvements in the mildness of C–H transformations, and have sorted the processes into several categories accordingly. We hope that a description and discussion of these strategies will serve as a useful guide for the design of C–H transformations and inspire the next generation of mild reactions. In many respects, the progress achieved has resulted from a greater understanding of the mechanistic aspects of C–H activation. As shown in Scheme 1, a typical transformation between a C–H activation substrate (termed the "substrate") and an appropriate reactant (termed the "reaction partner") can be simplistically broken down into four general mechanistic steps: the C–H activation process itself (I), functionalization of the subsequent organometallic species (II), release of the product molecule (III) and finally, if required, regeneration of the active catalyst (IV). For a given reaction, any one of these steps may be "mildness-limiting" and different strategies can be applied to address them. Furthermore, some approaches may enhance certain features of mildness, such as lowering the reaction temperature or removing the requirement for acidic additives, at the expense of other aspects.

Citation: T. Gensch, M. N. Hopkinson, F. Glorius and J. Wencel-Delord, "Mild metal-catalyzed C-H activation: examples and concepts," DOI: 10.1039/c6cs00075d.

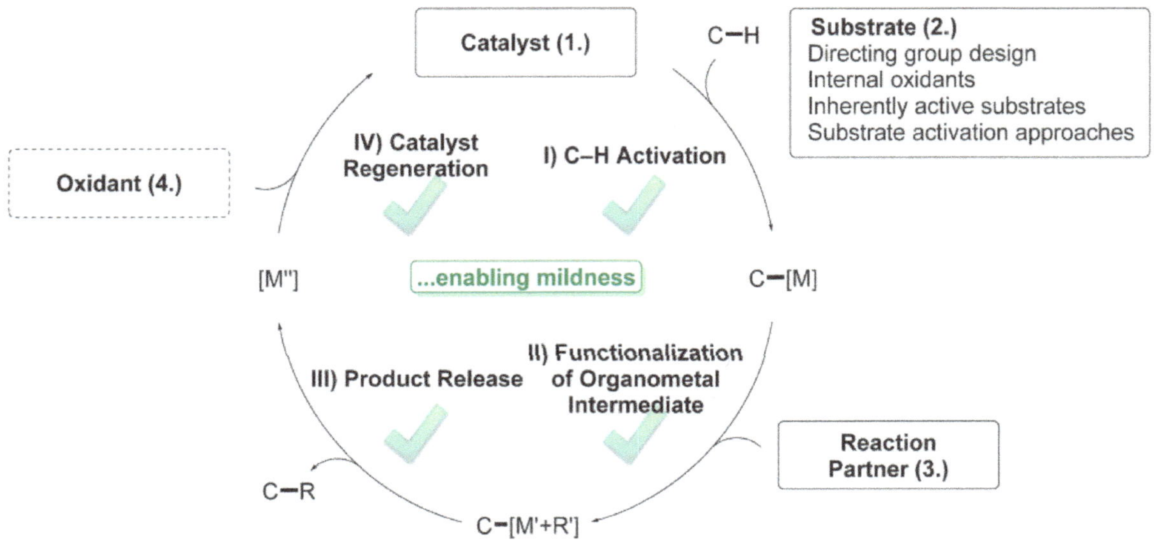

Scheme 1: Generalized mechanistic scheme for C–H activation catalysis and categorization of concepts enabling mildness.

In this review, the various concepts that have proved useful for the development of mild C–H transformations are presented under the four headings shown in Scheme 1. In the first section, C–H transformations where modifications to the transition metal catalyst have allowed for mild reaction conditions are discussed. The catalyst is involved throughout every stage of a C–H transformation and optimization of the active catalyst structure can duly enhance any of the four general mechanistic steps shown in Scheme 1. This can be achieved by fine tuning the intrinsic steric and electronic properties of the catalyst by adding ancillary ligands and external additives, or simply by judicious selection of the transition metal itself. Section 2 focuses on the substrate molecule undergoing C–H activation. Installing a strongly-coordinating directing group, for example, can aid the C–H activation step (I) by improving binding to the metal catalyst, while more weakly-coordinating directing groups may enhance the subsequent step in the cycle (II) by facilitating interaction between the generated organometallic species and the reaction partner. Directing groups also act as ligands on the transition metal catalyst and therefore affect each step of the cycle. Furthermore, internal oxidants incorporated into the directing group allow for oxidative transformations to be conducted without a strong external oxidant, fulfilling one of the criteria for mildness as defined above. Other strategies for improving the mildness of C–H transformations have included activating the substrate in ways not involving the directing group. For example, coordination of chromium carbonyl fragments to arenes has been demonstrated to facilitate aryl C–H bond cleavage. Intuitively, mild transformations have been reported utilizing inherently reactive substrates. The selection of activated reaction partners has also been effective for transformations where functionalization of the organometallic intermediate (step II) is "mildness-limiting". Examples of mild C–H transformations for which this strategy has proved useful are highlighted in Section 3 and include room temperature reactions employing Grignard reagents and nitrene precursors. Lastly, in Section 4, oxidative C–H coupling reactions which avoid the use of strong external oxidants are presented. It should be noted, however, that it is often difficult to definitively identify the underlying feature of the process responsible for rendering it mild and many of the examples make use of several of the concepts outlined above. As in our first review on this topic, the discussion is limited to transformations occurring *via* metal-catalyzed inner-sphere C–H bond cleavage directly leading to the formation of C–M intermediates. We do not target an exhaustive presentation of all publications that fulfill at least some of the criteria for mildness but instead highlight seminal examples which best demonstrate the current state of the art or illustrate the concepts under discussion. In this respect, we hope to demonstrate to the scientific community that C–H activation is fast becoming a mature field and can now be considered a truly efficient tool for the synthesis of complex molecules.

TUNING OF THE CATALYST›S PROPERTIES

The metal catalyst is involved throughout each stage of a C–H transformation and duly plays a fundamental role in determining its mildness. The identity and oxidation state of the metal largely dictates which C–H activation mechanism operates in any given reaction.[3] High-valent late transition metals such as Pd^{II}, Rh^{III}, Ir^{III} and Ru^{II} typically react *via* electrophilic pathways such as concerted metalation-deprotonation (CMD). Low-valent, electron-rich metals such as Rh^{I} and Ir^{I}, on the other hand, often undergo oxidative addition into C–H bonds. This latter process can greatly benefit from the addition of strongly binding ligands such as phosphines, N-heterocyclic carbenes or bidentate nitrogen-containing molecules which typically increase the electron-density at the metal center and provide a handle for tuning the reactivity. By contrast, C–H transformations catalyzed by high-valent metals generally proceed in the absence of additional donor ligands. This fact can be rationalized by noting that electrophilic C–H activation mechanisms usually require the catalysts to be rather electron-poor in nature. Strong σ-donor ligands can render the metal center overly electron-rich and inhibit C–H bond activation. In addition, directing groups (DGs) typically installed onto the substrate molecules to ensure reactivity and regioselectivity act as ligands on the metal center. The limited number of coordination sites around the metal atom also means that strongly binding ancillary species may prevent coordination of the reactants, hindering efficient turnover of the catalytic cycle. Consequently, opportunities to finely tune the intrinsic properties of these high-valent catalysts with the intention of improving their reactivity under mild reaction conditions are somewhat limited. Extensive efforts by the scientific community, however, have recently enabled a greater understanding of the mechanistic aspects of C–H activation and have opened the door toward advanced catalytic systems. In the following section, examples of C–H transformations where tuning of the coordination sphere of both low-valent and high-valent metals has allowed for mild reaction conditions are presented.

Pd-Based Catalytic Systems

Among the metals employed as catalysts in direct C–H activation reactions, precious second- and third-row transition metals such as Pd, Rh, Ir and Ru have attracted by far the most attention. Naturally, these metals were the first to be used in pioneering examples of mild, low temperature C–H transformations.[1] One of the first approaches toward this challenging goal involved modifying the electronic properties of the catalyst by rendering it significantly more electrophilic. The use of trifluoroacetic acid (for the sake of clarity, we have adopted the abbreviation TfaOH for trifluoroacetic acid) as an additive in Pd-catalyzed transformations has shown particular promise as a way to generate highly activated $Pd(OTfa)^+$ moieties *in situ*. Although this approach sacrifices one aspect of mildness by employing a strongly acidic additive, it has been shown to enable several impressive room temperature transformations. An overview of this strategy was provided in our previous review, however several recent examples that further illustrate the power of this concept are discussed here.

One of the most important challenges in the field of C–H activation concerns the development of "undirected" processes.[1h] These reactions, which involve the direct functionalization of substrates without the assistance of a directing group, are particularly difficult because the catalyst is not pre-coordinated by the substrate and there is consequently no increased local concentration in close proximity to the metal. Due to these reasons, the dehydrogenative coupling reactions between simple arenes and either anilides (involving two separate C–H activation events) or activated olefins (Fujiwara–Moritani reactions) recently reported by Song and You, which occur smoothly at room temperature, are particularly interesting (Scheme 2).[4]The success of these processes relies on the strongly electrophilic character of the active $Pd(OTfa)_2$ catalyst formed *in situ* from $Pd(OAc)_2$ and trifluoroacetic acid. Kinetic isotope effect (KIE) studies indicate that C–H scission of the simple arene in these transformations probably occurs *via* an electrophilic palladation (S_EAr) pathway. The highly electron-deficient trifluoroacetate-ligated Pd species is well-suited to this kind of mechanism and allows the dehydrogenative C–C couplings to take place at room temperature. It should be noted, however, that the classification of this S_EAr-type process as a C–H activation reaction is debatable according to the definition given in the introduction to this review. In practice, it is often difficult to determine the operating C–H scission mechanism for transformations mediated by highly Lewis-acidic metal species, and different substrate structures and reaction media may promote different pathways.[5] A diverse array of biaryls, arylcoumarins, arylquinolones and styrenes could be prepared with moderate to high regioselectivities according to the intrinsic electronic and steric properties of the benzene derivatives. Although these room temperature reactions utilize an inexpensive Pd source and persulfate oxidant, a large excess of the benzene derivative is required (it is typically used as solvent), limiting their practical utility.

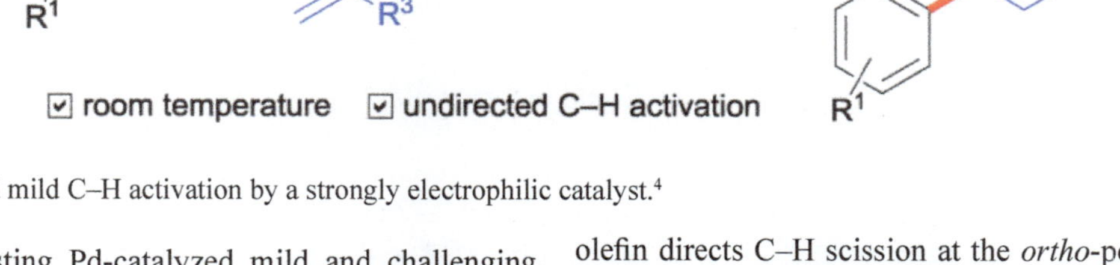

Scheme 2: Undirected mild C–H activation by a strongly electrophilic catalyst.[4]

Another interesting Pd-catalyzed mild and challenging C–H functionalization was disclosed in 2012 by Cheng.[6] In this example, an olefin moiety was employed as a π-coordinating group to direct insertion of a metal catalyst into the *ortho*-C–H bond of an aromatic substrate (Scheme 3). The presence of the TfaOH additive, leading to the *in situ* generation of the strongly electrophilic $Pd(OTfa)^+$ species, was crucial to facilitate π-coordination of the C=C bond. Binding of the olefin directs C–H scission at the *ortho*-position and leads to the formation of stabilized chelates. No product formation was observed in the absence of the TfaOH additive or when weaker congeners such as AcOH or PivOH were employed instead. As well as proceeding smoothly at room temperature, this oxidative coupling reaction fulfils another criterion of mildness by employing a relatively benign oxidant: O_2 itself is capable of regenerating the Pd^{II} catalyst after β-hydride elimination and no additional strong oxidant is required.

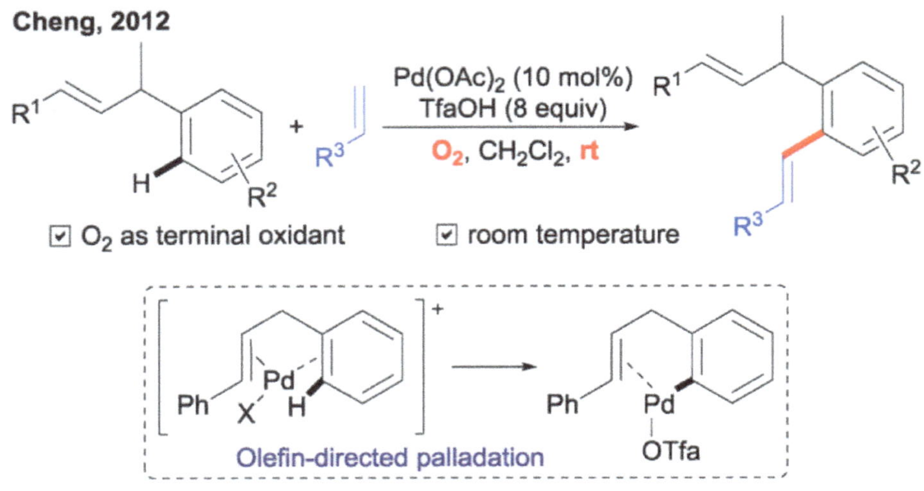

Scheme 3: Olefin-directed C–H activation.[6]

Following the same strategy involving the generation of highly electrophilic Pd species, Wang used a substoichiometric amount of methanesulfonic acid to promote *ortho*-alkoxylation of acetanilide derivatives.[7] In addition to its role in decreasing the electron-richness of the catalyst, the strong organic acid additive also acts as a labile ligand at the Pd center, facilitating coordination of the alcohol coupling partner. This approach allows for the efficient preparation of the C–O coupling products at ambient temperature.

The use of specifically-designed ligands to accelerate Pd-catalyzed C–H activation was initially explored by Yu, who in the middle of the last decade surmised that weakly-donating mono-protected amino acid (MPAA) ligands could drastically impact direct functionalization reactions.[1i,8] To date, MPAA ligands are arguably the most prominent ligands employed in Pd-catalyzed C–H activation reactions, facilitating both enantioselective and very mild transformations of a diverse array of substrates. Mechanistic investigations suggest that the particular potential of these additives may result from their bidentate coordination of the palladium catalyst and the role of the ligand N–H moiety as an intramolecular proton shuttle in the C–H bond cleavage step.[9] Such ligand-improved reactivity is clearly illustrated in the successful C–H transformations of generally unreactive electron-deficient arenes.[10] The authors discovered that the presence of a finely-tuned MPAA ligand not only enhances the turnover number (TON) of the C–H olefination of diversely-substituted phenylacetic acid derivatives, but also leads to the generation of a more active catalytic species (Scheme 4). This accordingly permits the functionalization of electron-deficient substrates which are totally unreactive in similar transformations conducted in the absence of such ligands. A detailed investigation of the reaction kinetics uncovered that the MPAA prevents build-up of mixed acetate–phenylacetate resting states of the palladium catalyst.[10b]

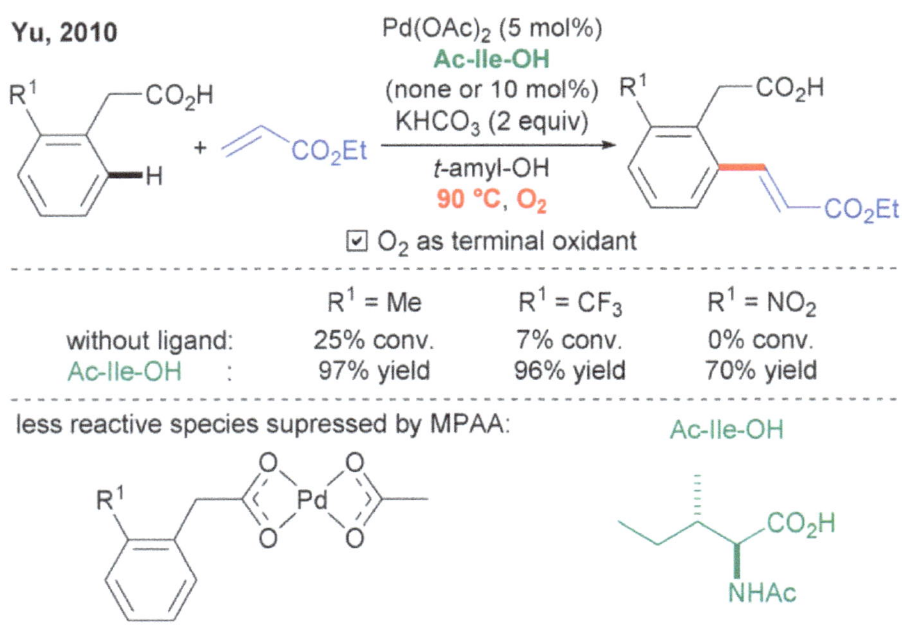

Scheme 4: Mono-protected amino acids in mild C–H activation.[10]

Maybe the most appealing feature of MPAA ligands is their chirality.[11] Recently, MPAAs were employed to access chiral amines *via* two conceptually distinct transformations: desymmetrization and kinetic resolution of racemic substrates.[12] In 2013, Yu and co-workers judiciously used the MPAA *N*-benzoyl-leucine (Bz-Leu-OH) to enhance the iodination reaction of a symmetric diarylmethyl amine.[12a] Given the stereogenic nature of the ligand, an enantioselective outcome was envisaged provided the chirally substituted catalyst species is able to differentiate between two prochiral aromatic moieties (Scheme 5). An extensive optimization study showed that both the *N*-protecting group and the hydrocarbon substituent on the ligand have a dramatic impact on the enantioselectivity and efficiency of this transformation. Under the optimized reaction conditions, the targeted chiral amines were delivered in moderate to good yields (51–85%) and with excellent chiral induction. Importantly, the efficiency of this transformation at room temperature maximizes the level of enantioselectivity. Notably, the background racemic reaction catalyzed by a non-ligated catalyst is avoided by the addition of DMSO or DMF, which bind to the uncoordinated Pd^{II} species and inhibit the reaction. In addition to its mildness, the tolerance of this transformation toward air, the use of I_2 as the iodine source, and its efficiency at gram-scale further highlight the overall synthetic utility of this transformation.

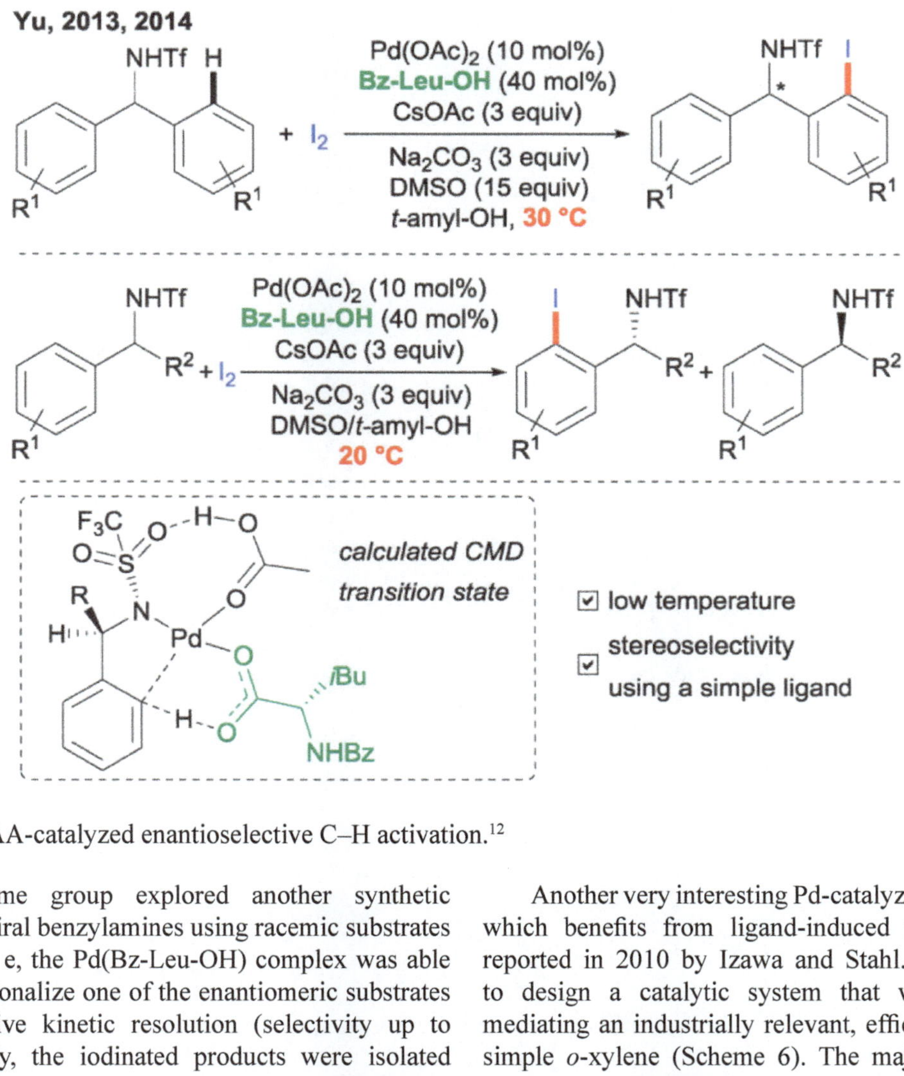

Scheme 5: Pd–MPAA-catalyzed enantioselective C–H activation.[12]

Later, the same group explored another synthetic approach toward chiral benzylamines using racemic substrates (Scheme 5). In this e, the Pd(Bz-Leu-OH) complex was able to selectively functionalize one of the enantiomeric substrates in a highly selective kinetic resolution (selectivity up to 244).[12b] Accordingly, the iodinated products were isolated in yields up to 49% and moderate to good ee's (65–93%), and the remaining starting materials were recovered with excellent optical purities (92–99% ee). Furthermore, the recovered optically pure substrates, when resubmitted to the standard reaction conditions but using the MPAA ligand with the opposite configuration, were smoothly converted into the other enantiomers of the iodinated benzylamines. As such, this process allows for the preparation of both enantiomers of the iodinated products in high optical purities at ambient temperature (20 °C). A computational study on the mechanism of this transformation identified a beneficial coordination of acetic acid or DMSO to both the triflate residue on nitrogen and palladium leading to a fixed geometry where interactions between the benzylic substituent and the triflate determine enantioselectivity.[12c]

Another very interesting Pd-catalyzed C–H transformation which benefits from ligand-induced reactivity tuning was reported in 2010 by Izawa and Stahl.[13] The authors sought to design a catalytic system that would be capable of mediating an industrially relevant, efficient homocoupling of simple *o*-xylene (Scheme 6). The major challenge inherent to this transformation involves controlling the regio- and chemoselectivity to allow the selective formation of the desired biaryl units. Consequently a dual Pd and Cu catalyst system was developed and, hypothesizing that *N*-coordinating ligands may be beneficial for reactivity and selectivity, a range of pyridine additives were tested. Derivatives bearing an electron-withdrawing substituent at the 2-pyridyl position led to improved reactivity and, after a detailed optimization study, 2-fluoropyridine was selected as the optimal ligand. When combined with a low loading of $Pd(OAc)_2$ as the pre-catalyst (0.1 mol%), $Cu(OTf)_2$ (0.1 mol%) and TfaOH as an additive, in acetic acid as the solvent, this ligand led to a radically improved set of conditions for the coupling reaction delivering the desired biaryl product in 94% selectivity compared to the previously reported 40%. Moreover, excellent regioselectivity

(88%) and efficiency (TON = 80) were achieved at a reaction temperature substantially lower (80 °C) than the previously reported 140 °C, while oxygen could be used as the terminal oxidant. Although there is still some way to go before this challenging process can be considered truly mild as defined in this review, the substantial improvements achieved in this study demonstrate how judicious use of ligands can greatly improve the mildness of C–H transformations. In this e, the electron-withdrawing character of the fluorine substituent at the 2-position of the pyridine is thought to maintain some degree of electrophilicity of the Pd catalyst upon coordination of the ligand. Pyridine derivatives which are even less basic, however, are unable to modulate the reactivity of the Pd catalyst and there is seemingly a balance that must be found. The coordination sphere of the active Pd catalyst also incorporates both acetate and trifluoroacetate anionic ligands, which are necessary to ensure smooth CMD metalation.[3]

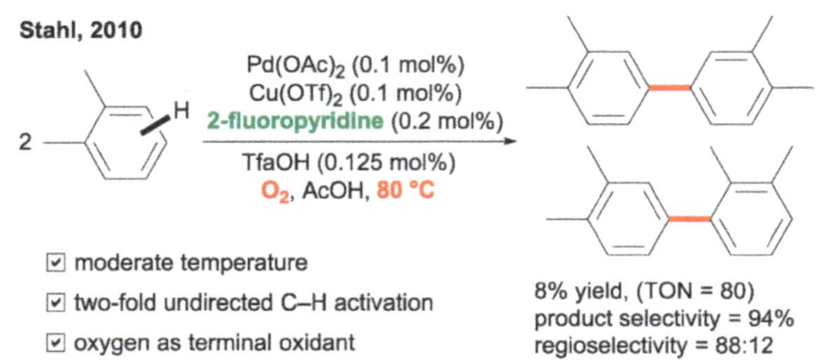

Scheme 6: Homocoupling of *o*-xylene under oxygen.[13]

A conceptually very different approach toward mild Pd-catalyzed C–H activation and subsequent C–B bond formation was disclosed by Kuninobu and Takai.[14] This original idea consisted of redefining the role of the directing group, which typically coordinates the metal catalyst, boosting reactivity and regioselectivity. In this particular e, the Lewis basic nitrogen of the pyridine DG was used to form a Lewis base/Lewis acid adduct with the boron atom of the 9-borabicyclo[3.3.1]nonane (9-BBN) reaction partner (Scheme 7). The Pd^0 catalyst subsequently undergoes an oxidative addition into the B–H bond of the borane leading to a Pd^{II} species whose electronic properties are modified by the boryl ligand. Subsequent intramolecular C–H metalation directed to the *ortho*-position leads, after loss of H_2 and C–B bond-forming reductive elimination, to the *ortho*-borylated products at room temperature. Notably, improved yields were obtained when performing this reaction at 25 °C, compared to the initially used 135 °C. In contrast to most other transformations employing high-valent Pd catalysts discussed in this review, which involve electrophilic metalation or CMD mechanisms, the C–H activation step in this process is thought to occur *via* σ-bond metathesis. As such, this transformation showes the potential of this somewhat underexplored pathway as an approach to achieve mild C–H transformations.

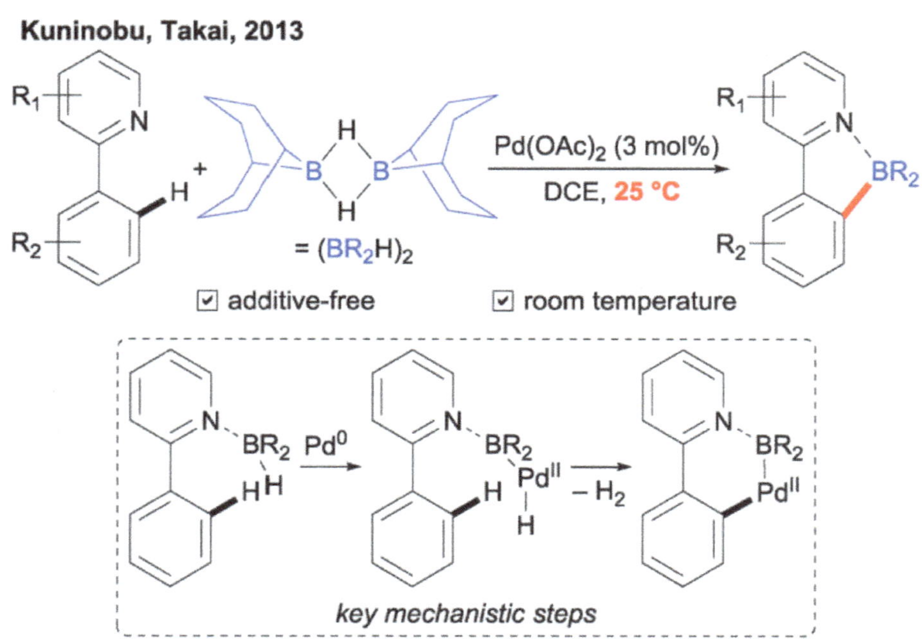

Scheme 7: Mild Pd-catalyzed C–H borylation.[14]

During the development of Pd-catalyzed C–H activation chemistry, it was realized that many transformations occurring in the presence of strong oxidants could in fact proceed *via* Pd^{IV} or Pd^{III} intermediates.[15] The recognition of this possibility sparked the development of methods that deliberately make use of the distinct reactivity of these higher oxidation states with regards to chemoselectivity and facilitated reductive elimination of C–heteroatom bonds. In the vast majority of the C–H activation transformations involving Pd^{IV}, the C–H activation itself most likely occurs at a Pd^{II} species, with oxidation to Pd^{IV} and subsequent product-releasing reductive elimination regenerating Pd^{II} (Scheme 8, right part). In some es, the distinction between this order of events and an alternative re-oxidation of Pd^{0} species after a reductive elimination from Pd^{II} cannot be made clearly when no mechanistic evidence is presented and is probably also dependent on the choice of oxidant. On the other hand, Pd^{IV} intermediates can potentially be involved in the C–H activation step itself (Scheme 8, left part).[16] This scenario has fascinating implications, as different selectivities and reaction conditions should be expected when changing the nature of the catalyst for any given mechanistic step as drastically as changing its oxidation state. To date, very few examples that clearly demonstrate C–H activation by a Pd^{IV} species have been reported. Judging by the current evidence, Pd^{IV} C–H activation catalysis has the potential to proceed at low temperatures with distinct regioselectivity, although the presence of oxidants capable of generating such a high oxidation state at palladium could be a limitation to overall mildness.

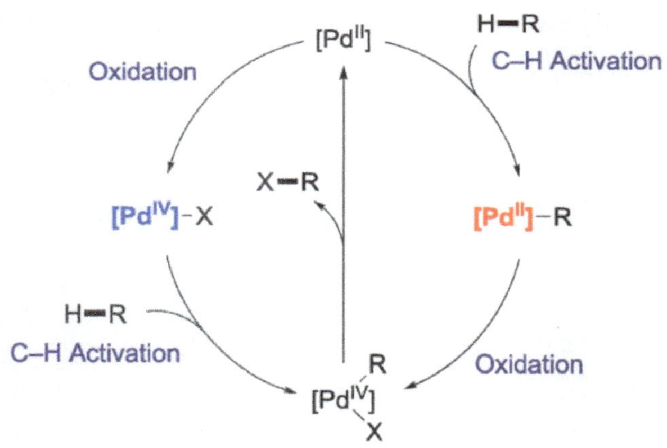

Scheme 8: Alternative pathways in Pd^{II}/Pd^{IV}-catalyzed C–H activation.[16]

In a seminal study, the group of Sanford proposed C–H activation at Pd^{IV} for the first time in 2006. An oxidative coupling process involving two-fold C–H activation of aryl pyridine derivatives proceeded at room temperature using Oxone ($KHSO_5$) as an oxidant (Scheme 9).[17] On the basis of selectivities obtained in experiments using stoichiometric amounts of preformed palladacycles, they proposed a sequence of Pd^{II} and Pd^{IV} mediated C–H activation events. Subsequent studies on the reactivity of isolated aryl Pd^{II} and Pd^{IV} complexes further substantiated the existence of Pd^{IV} mediated C–H activation.[18]

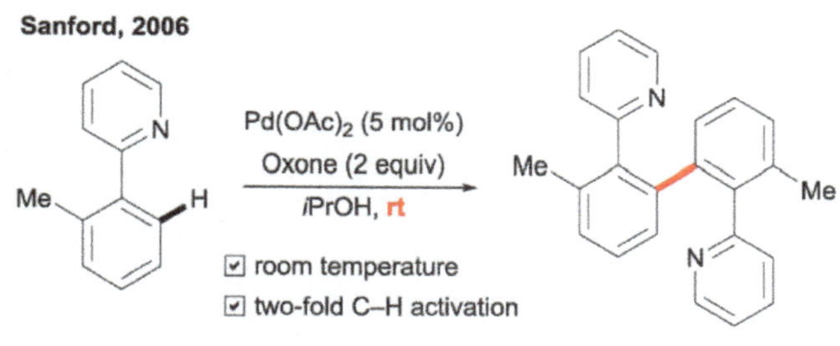

Scheme 9: Oxidative arene dimerization involving proposed C–H activation at Pd^{IV}.[17]

The group of Michael reported an unexpected result during a study on the diamination of olefins. When the reaction was carried out in toluene as solvent in the presence of *N*-fluorobenzenesulfonimide (NFSI), an aminoarylation product was obtained instead of the expected diamination at room temperature (Scheme 10).[19] This product results from an undirected C–H activation of toluene with a very high regioselectivity for the *para*-position. Compared with other undirected C–H activation methods, the low reaction temperature of this process is very uncommon. A number of other arenes could also be employed with high *para*-selectivity, delivering arylated pyrrolidine derivatives in good yields. Based on mechanistic evidence obtained from stoichiometric experiments and KIE studies, the authors suggested that undirected C–H activation of toluene occurs after pre-coordination at a Pd^{IV} intermediate. Yu's group later published a procedure for the *ortho*-arylation of benzamides *via* two-fold C–H activation with a number of simple arenes using $Pd(OAc)_2$ as catalyst and NFSI as oxidant (Scheme 10) at a somewhat higher temperature.[20] Again, very high *para*-selectivity for the non-directed C–H functionalization was observed. This stands in contrast to the relatively low levels of regioselectivity observed in undirected Pd^{II}-catalyzed transformations, and led the authors to propose a mechanistic scenario involving C–H activation at a Pd^{IV} center in analogy to Michael's work. Notably, when $K_2S_2O_8$ was used instead of NFSI, a lower level of *para*-selectivity was observed. It has been noted that NFSI can be considered a by-standing oxidant for promoting chemoselective reductive elimination because the ligated fluoride anions generated upon oxidation do not participate in competing reductive eliminations.[21]

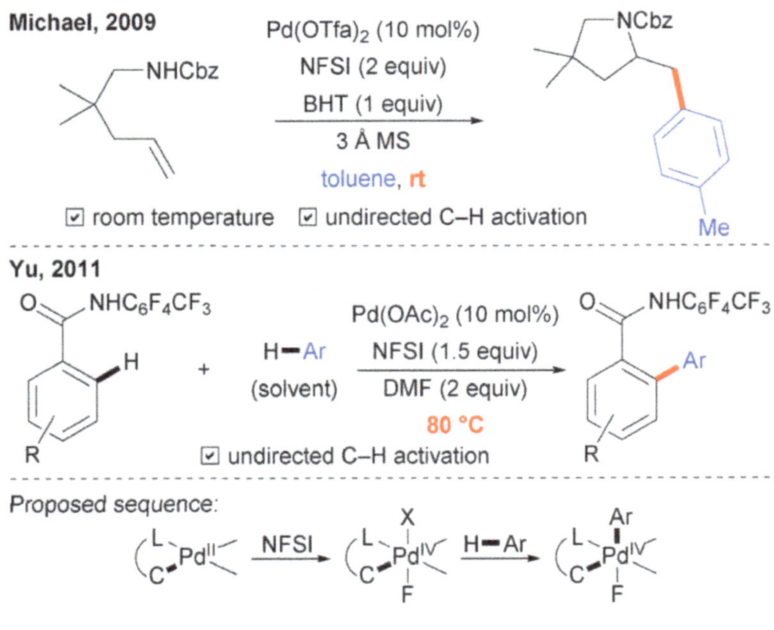

Scheme 10: Highly *para*-selective undirected C–H activation proposed to occur at Pd^{IV}intermediates.[19,20] BHT = 2,6-bis(1,1-dimethylethyl)-4-methylphenol.

The group of Szabó reported an allylic C–H acyloxylation reaction that proceeds at 0 °C using (bis(trifluoroacetoxy) iodo)benzene (PIFA) as an oxidant and acyloxylating reagent (Scheme 11).[22] The corresponding acetate-substituted analog of the reagent, which is known to be less competent in oxidizing Pd^{II},[18a] failed to give any product while an independently prepared sample of a potential Pd^{II}–allyl intermediate decomposed in the presence of PIFA, with only traces of product being formed. Based on these observations, a Pd^{IV} species was proposed to mediate the actual C–H activation event, which then leads to allylic trifluoroacetate products useful as active substrates in Pd^{0}-catalyzed allylic functionalization reactions. The exceptionally low reaction temperature may result from the favorable combination of a potent Pd^{IV} catalyst, an intrinsically activated substrate (*vide infra*) and a reactive coupling partner (*vide infra*).

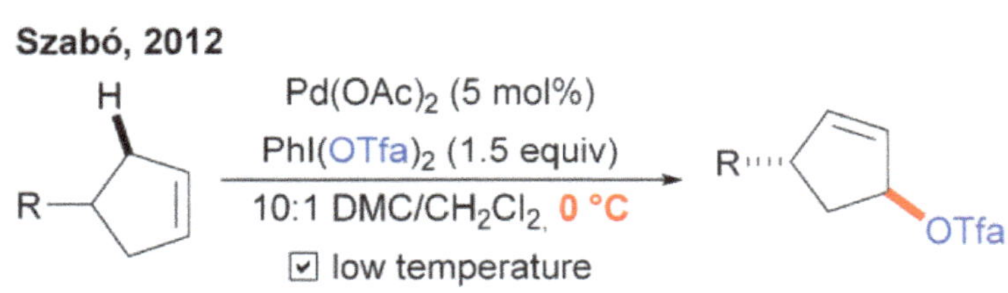

Scheme 11: Allylic C–H trifluoroacetoxylation *via* proposed allylic C–H activation at Pd^{IV}. DMC = dimethylcarbonate.[22]

Rh-Based Catalytic Systems

In addition to Pd, Rh is among the most powerful and general catalysts employed in direct C–H functionalization reactions. In particular, Cp^*Rh^{III} ($Cp^* = \eta^5$-pentamethylcyclopentadienyl) stands out as a highly potent catalyst for a range of diverse transformations.[1j,23] Although historically rather harsh reaction conditions were required to allow direct activation of nonreactive C–H bonds with this catalyst, several astute strategies such as the use of oxidizing DGs and/or very reactive coupling partners have recently been disclosed to achieve such transformations at room temperature. These discoveries are presented in following sections of this review. Moreover, a few research groups have focused on exploring the electronic and/or steric properties of the Cp* ligand with the goal of developing more reactive or selective Rh catalysts. As for comparable high-valent Pd-based catalytic systems, highly electrophilic Rh^{III} catalysts are desirable to enable a CMD-type C–H activation at ambient temperature. In this context, interesting results were achieved by Tanaka and co-workers, who designed an electron-deficient congener of the Cp* ligand: the diethoxy carbonyl trimethyl cyclopentadienyl (Cp^E).[24] The authors anticipated that the more electrophilic Cp^ERh^{III} species could outcompete the standard Cp^*Rh^{III} catalyst when reacted with electron-rich substrates such as anilides and benzyl alcohols (Scheme 12).[25] Indeed, alkyne annulation of such compounds occurred smoothly at room temperature. Notably, only a small amount of $Cu(OAc)_2 \cdot H_2O$ (20 mol%) used as a co-oxidant was necessary because air could be employed as an ideal terminal oxidant. The corresponding indoles and isochromenes were delivered in yields generally ranging from 70% to 95% while the same transformations with the standard Cp*Rh catalyst were significantly less efficient. More recently, the same electron-deficient Rh^{III} catalyst was also employed in an oxidative Heck olefination of anilides at room temperature originally reported by Glorius with Cp^*Rh^{III} at 120 °C.[26] As KIE studies revealed that the C–H bond cleavage step is probably turnover-limiting, the finely-tuned electronic properties of the Cp^ERh^{III} catalyst are believed to improve the metalation step for these electron-rich substrates.

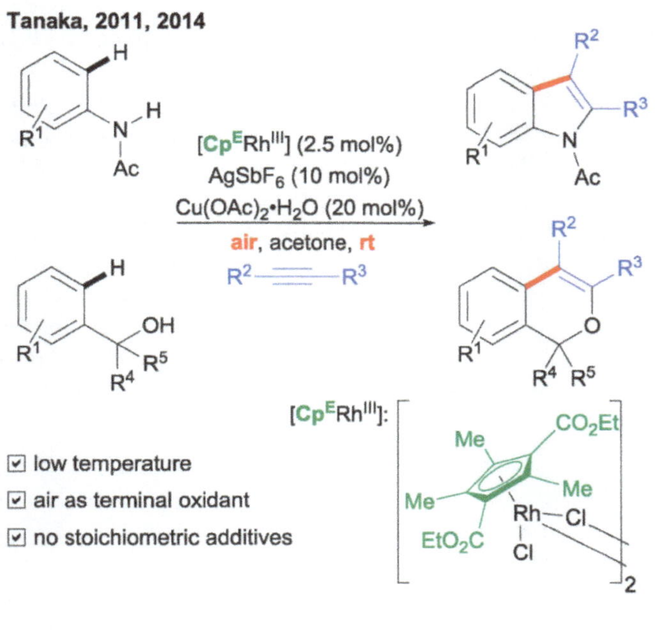

Scheme 12: Electron-deficient Rh^{III}-catalyst for the functionalization of electron-rich substrates.[24,25]

Notably, several alternatives to the standard $Cp^{*}Rh^{III}$-catalyst have been designed in particular by the groups of Rovis,[27a,b] Satoh and Miura[27c] and Cramer.[27d] In most es, however, these have been employed to improve selectivity rather than reactivity and aspects of these are presented in following sections of this review.

A different approach to catalyst design aimed at generating extremely reactive catalytic species was proposed by Sawamura.[28] The key feature of this work is its use of a Rh^{I} pre-catalyst. Low-valent metal catalysts such as Rh^{I} and Ir^{I} generally react with C–H bonds *via* an oxidative addition pathway and, consequently, increasing their electron richness through the addition of strongly coordinating ligands can be beneficial for reactivity. Following such a hypothesis, the authors discovered that Rh^{I} precatalysts such as $[Rh(cod)(OH)]_2$ (cod = 1,5-cyclooctadiene), may be very efficiently coordinated by a silica-supported bridgehead mono-phosphine, generating a very efficient heterogeneous catalyst (Scheme 13A). This supported species was capable of mediating the direct borylation of aromatic substrates bearing N-containing DGs. Impressively, the reaction occurred smoothly at room temperature with total consumption of a starting material observed within only one hour. Notably, both the immobilization and constrained geometry of the ligand were essential to ensure good efficiency of the overall transformation. The heterogeneous ligand architecture allows for binding of only one phosphine to the catalyst and hence provides a vacant coordination site necessary to bind the substrate.

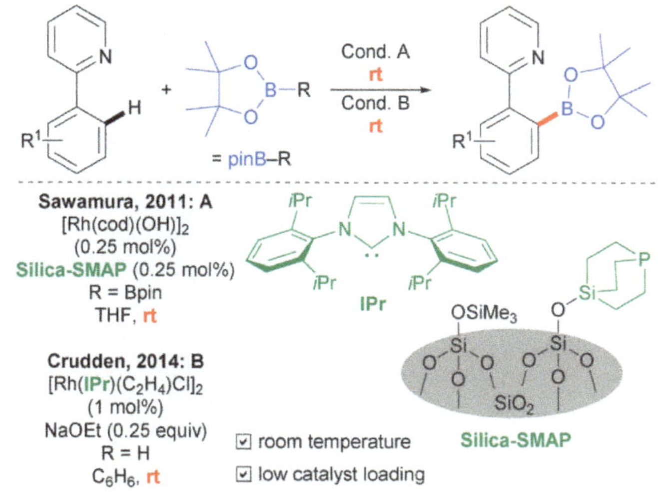

Scheme 13: Rh^{I}-catalyzed directed borylation.[28,29]

More recently, Crudden *et al.* discovered that a much simpler, homogeneous catalytic system based on a Rh^{I} precatalyst and a strongly-binding and sterically-encumbered N-heterocyclic carbene (NHC) ligand is an appealing alternative to Sawamura's system (Scheme 13B).[29]

An additional example of Rh^{I}-catalyzed direct functionalization occurring at relatively low temperature was disclosed by Cheng and Hartwig (Scheme 14).[30] The authors discovered that a catalytic system based on a Rh^{I} pre-catalyst coordinated by a 2,2′-bisphosphino biaryl ligand, combined with a hydrogen acceptor (olefin) and a silane allows for direct C–Si couplings with simple benzene derivatives. The regioselectivity of this dehydrogenative silylation is controlled by the steric properties of the arene substrate with functionalization occurring predominantly at the less sterically-encumbered position. The presence of the strongly-coordinating ligand is essential to achieve a reasonable level of reactivity.

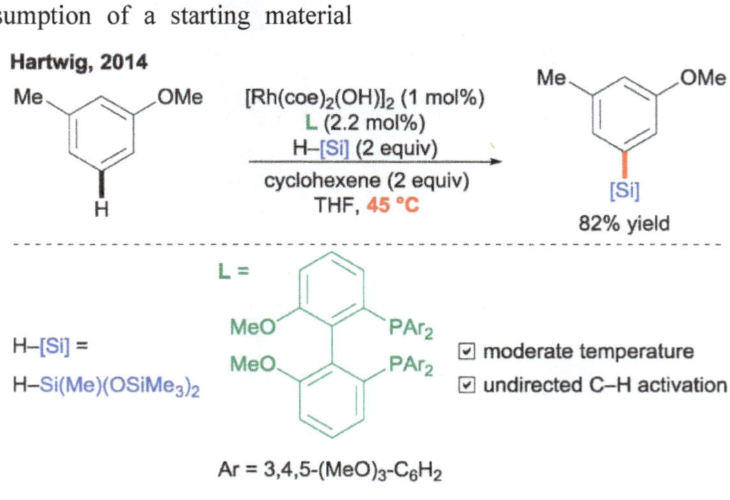

Scheme 14: Rh^{I}-catalyzed undirected C-silylation.[30] coe = cyclooctene.

These studies demonstrate that electron-rich Rh[I]-catalysts are able perform C–H activation at surprisingly moderate temperatures.

Ir-Based Catalytic Systems

Notwithstanding the importance of several Rh-based catalytic systems that have been employed in mild C–H borylation reactions, historically these processes have been most successfully achieved using Ir catalysts. Already at the beginning of this century, Ishiyama, Miyaura and Hartwig discovered that undirected C–H cleavage/C–B bond formation could occur at temperatures as low as room temperature in the presence of a $[Ir(cod)Cl]_2$/4,4′-di-*tert*-butyl-2,2′-bipyridine (dtbpy) catalyst system (Scheme 15). Exceptionally high TON up to 24 800 for borylation of C_6D_6 at 100 °C with 0.0030 mol% catalyst loading were achieved.[31] Importantly, it is believed that an Ir^{III} species bearing the *N,N*-bidentate ligand and three boryl motifs is the active catalyst performing the undirected C–H activation. It seems that the presence of a strongly-coordinating and highly donating ligand is crucial to enhance the electron-richness of the Ir atom bearing three boron moieties. Accordingly, $[Ir(dtbpy)(Bpin)_3]$ appears to be an extremely powerful catalyst, accelerating a rate-limiting C–H bond cleavage not only under mild reaction conditions but also without DG assistance.

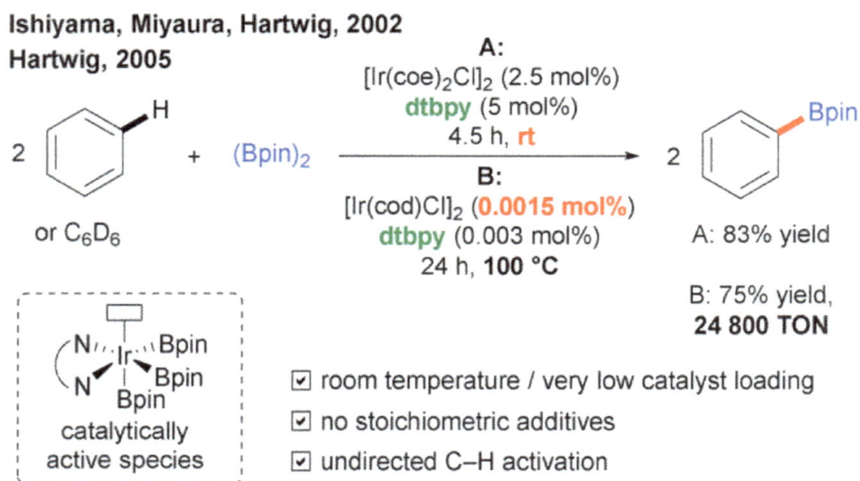

Scheme 15: Pioneering work on Ir-catalyzed borylation.[31]

Following these seminal works, numerous related catalytic systems aimed at the borylation of various scaffolds have been disclosed. Firstly, significant efforts have been devoted to identifying a catalytic system capable of performing the desired coupling at room temperature and using an equimolar amount of the economically more viable pinacolborane (H-Bpin) reaction partner.[32] In this context, $[Ir(cod)(OMe)]_2$ turned out to be the optimal pre-catalyst, and the critical role of electron-rich bipyridine ligands in favoring the formation of the active C–H activation catalyst with a relatively unhindered coordination sphere was acknowledged. Further design of an optimal ligand led to the selection of 3,4,7,8-tetramethyl-1,10-phenanthroline (tmphen) as an excellent auxiliary, frequently outcompeting its dtbpy congener. Detailed mechanistic investigations revealed that the improved reactivity of the tmphen ligand probably results from both its increased electron-richness compared to dtbpy and its conformational rigidity, which enforces η^2-coordination to the Ir center.[33] An additional key advantage of this second-generation ligand is that now both $(Bpin)_2$ and H-Bpin species may be used as borylating agents, allowing for improved atom economy and economic viability.

Importantly, such direct transformations are of great synthetic usefulness as they can be employed to prepare functionalized heterocyclic compounds.[34] Indeed, the general protocol for the Ir/tmphen-catalyzed borylation is compatible with various heteroarenes such as azoles, benzoxazoles, benzothiazoles, benzimidazoles, azaindoles and pyridines, among others (Scheme 16). Frequently, these transformations occur readily at room temperature, and in a regioselective manner. In-depth studies by the group of Hartwig rationalized the regioselectivity of these transformations with various substrates. The borylation of benzoxazoles and benzothiazoles was found to be under electronic control and duly occurs at the most acidic position. By contrast, the reactivity of pyridine, quinolone and other azines is strongly influenced by the presence of the basic nitrogen. Functionalization occurs predominantly at the beta or gamma positions with such heterocycles. Finally, if N–H-containing substrates are submitted to the standard reaction conditions, C–H activation occurs at a position distal to the N-atom because rapid N–H borylation creates an unfavorable steric environment adjacent to the nitrogen and directs C–H borylation to the less hindered position. An additional advantage of this strategy results from the very low catalyst loadings. Good reactivity is maintained using very small amounts of the Ir species, rendering this approach attractive for possible industrial and large-scale applications. Notably, some reactivity enhancement may also be observed when the typically-employed THF solvent is replaced by MTBE (methyl *tert*-butyl ether), as demonstrated in the direct functionalization of pyridines and other azines.[35]

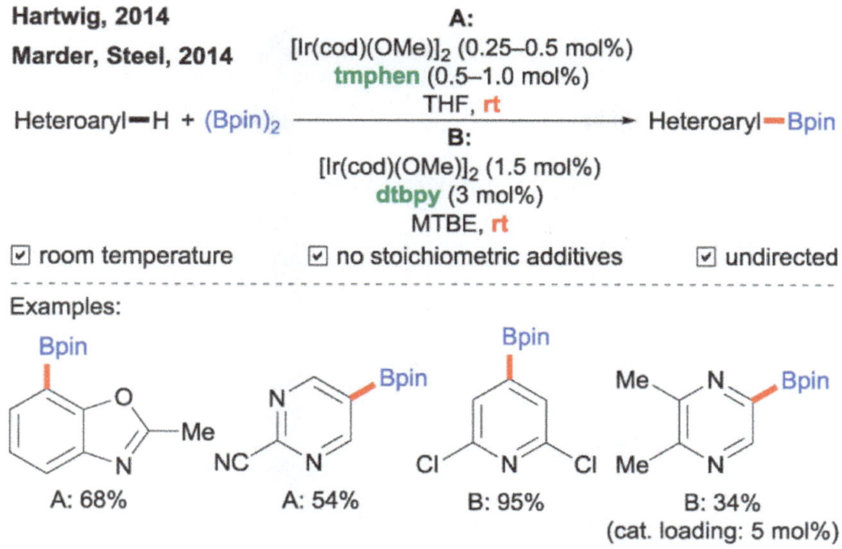

Scheme 16: Direct borylation of heteroarenes.[34,35]

In addition to the direct functionalization of (hetero)arenes, a closely related catalytic system has been shown to also enable borylation of secondary benzylic C–H bonds, cyclopropanes and other aliphatic derivatives.[36] However, as the metalation of C(sp^3)–H bonds is generally much more challenging than C(sp^2)–H bonds, higher reaction temperatures, typically above 80 °C, are required to ensure reasonable catalytic activity.

The low reactivity of aliphatic substrates could be overcome to some extent using Sawamura's heterogeneous catalytic system. This group discovered that a silica-supported phosphine ligand efficiently coordinates the $[Ir(cod)(OMe)]_2$ pre-catalyst to afford a very powerful heterogeneous borylation catalyst (Scheme 17). Several homogeneous phosphine ligands showed significantly lower reactivity even at 60 °C (no reaction at rt), probably due to the lack of a vacant coordination site on the metal atom. Accordingly, the immobilized catalytic system permits mild and selective *ortho*-functionalization of a range of various arenes bearing an ester DG.[37] Furthermore, the scope of this transformation could be extended to other starting materials; in particular to alkylpyridine derivatives.[38] In some es, the targeted C–H bond cleavage and C–B bond formation were feasible at 25 °C, but, in general, increasing the reaction temperature to 60 °C led to improved outcomes.

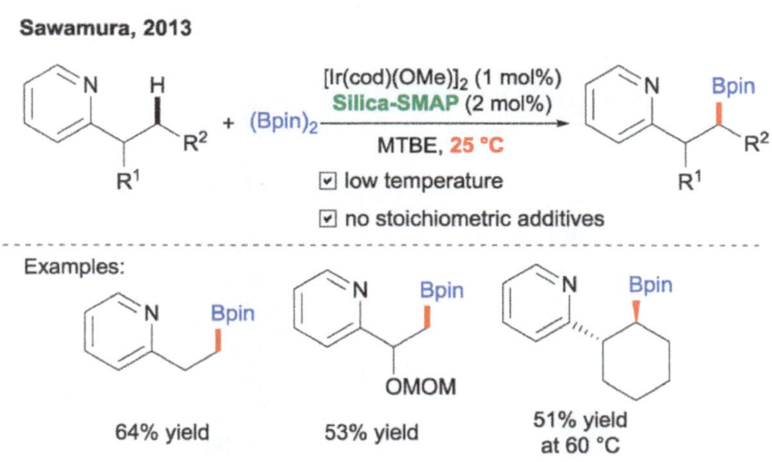

Scheme 17: Silica-supported Ir-based heterogeneous catalyst for the direct borylation of C(sp^3)–H bonds.[38]

Very recently, Kuninobu and Kanai very astutely illustrated that the extremely mild character of these borylation reactions and their compatibility with *N,N*-bidentate ligands can facilitate very challenging and original transformations (Scheme 18).[39] They hypothesized that embedding a pendant hydrogen-bond donor moiety on the bipyridine ligand should enable secondary interactions between this ligand and a C–H substrate bearing a hydrogen-bond acceptor DG. Such ligand–substrate coordination could result in otherwise inaccessible regioselectivity in the borylation process. Furthermore, the additive-free nature of the typical borylation reaction was also crucial to maintain supramolecular ligand–substrate interactions. Following such a hypothesis, several bipyridine ligands with embedded linkers and an additional urea moiety were designed, and an optimal ligand leading to a highly *meta*-selective C–B coupling was discovered. The high regioselectivity of this transformation is maintained with a variety of different DGs such as tertiary esters, amides and phosphorus-containing moieties, while both aromatic and heteroaromatic substrates could be employed. Notably, the coupling is efficient at temperatures as low as 25 °C.

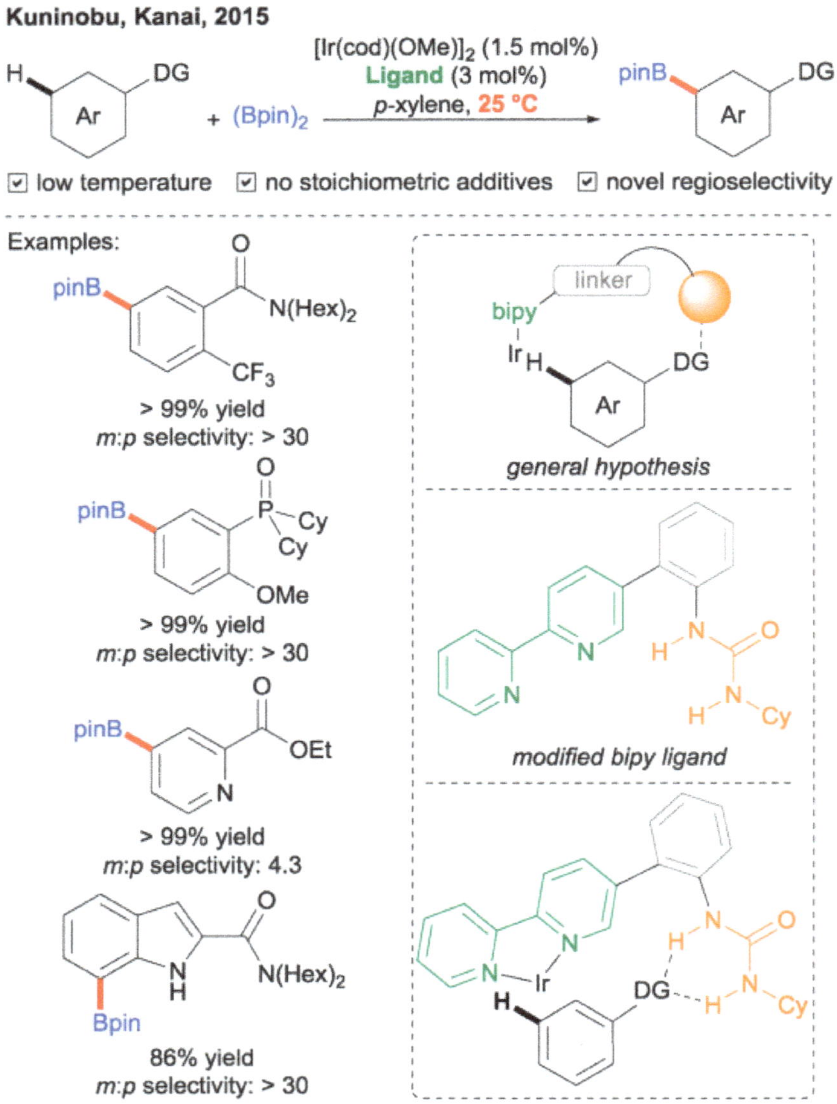

Scheme 18 *meta*-Selective Ir-catalyzed borylation.[39]

Ru-Based Catalytic Systems

Despite the growing interest of the scientific community in Ru-based C–H activation reactions,[1g,40] transformations catalyzed by this metal occurring under mild conditions are rather rare. One of the most interesting examples was, however, reported in 2010 by Kakiuchi and Murai (Scheme 19).[41] Mechanistic investigations of the catalytic reactivity of a common $RuH_2(CO)(PPh_3)_3$ pre-catalyst in the direct olefination of aromatic ketones revealed that a key step of this transformation involves the generation of the catalytically active species from the pre-catalyst. Indeed, heating $RuH_2(CO)(PPh_3)_3$ in the presence of an olefin coupling partner afforded a new metalacyclic species as a mixture of four geometric isomers. These Ru species are believed to result from C–H activation of a PPh_3 ligand, while the carbonyl ligand remains bound to the Ru center. Intriguingly, while generation of this active catalyst requires high temperature (90 °C), once formed, the metalacycle itself is highly efficient, mediating the C–H olefination transformation at room temperature. Apparently, conversion of the simple PPh_3ligand into a bidentate C^P ligand upon heating, although thermodynamically unfavored, drastically modifies the electronic properties of the Ru atom, rendering it more active in the C–H bond cleavage step. The efficiency of this transformation was illustrated by the mild and additive-free coupling of acetophenone derivatives with trimethylvinylsilane. It is important to note that, even at room temperature, the C–H activation step is not rate-determining, and that Ru-species are capable of breaking C–H bonds with remarkable ease.

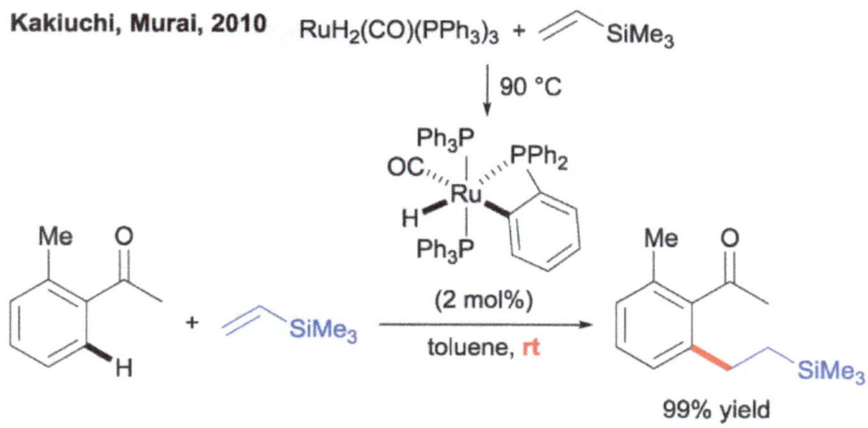

Scheme 19: Early example of Ru-catalyzed C–H cleavage occurring at ambient temperature.[41]

Another important example of a mild Ru-catalyzed transformation was reported a few years later by Zhao.[42] The improved catalytic reactivity of Ru was achieved not by using a bidentate C^P ligand coordinated to a Ru–hydride species, but rather by employing a strong σ-donating NHC ligand (Scheme 20). It is important to highlight here that the use of such a strong ligand is possible because low valent Ru^0 species are probably the active catalysts in the C–H scission step. As such, an oxidative addition pathway is most likely operating and C–H activation may be aided by strongly donating ligands in a similar fashion to the previously described Rh^I-catalyzed transformations. The authors observed that when benzophenone imine was reacted with an alkyne in the presence of a [Ru(cod)(η^3-methallyl)$_2$] pre-catalyst and IPr, a formal [3+2] carbocyclization occurred at room temperature delivering indenamine products. The remarkable reactivity enhancement observed with the NHC ligand (no reaction was observed without it) was rationalized by its significant electron-richness and the resulting ability to promote insertion of the π-system of the alkyne into the Ru–C bond. The alkyne insertion step is believed to be rate-determining in this system, showing the aptitude of the Ru/NHC complex as a catalyst for aromatic C–H bond activation. Furthermore, the redox-neutral nature of this transformation precludes the need for an external oxidant. In addition, the low reaction temperature and the lack of acidic or basic additives and waste generation mean that this process can be considered a truly mild C–H transformation.

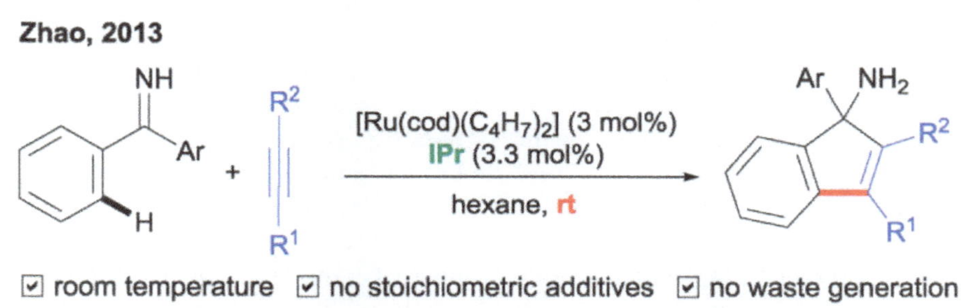

Scheme 20: Ru/NHC-catalyzed [3+2]annulation of ketimines and alkynes.[42]

First-Row Transition Metal Catalysts

Until the beginning of this century, the field of C–H activation was largely dominated by second-row transition metal catalysts. The use of cost-effective, alternative systems employing other, more abundant metals such as manganese, iron, cobalt and nickel was rather anecdotal and harsh reaction conditions were frequently required to achieve any reactivity. However, over the last few years, major advances toward mild transformations have been achieved.[43]

Co-Based Catalytic Systems

The field of mild C–H activation was strikingly reshaped when the potential of complex catalytic systems based on cobalt complexes was discovered in 2010 by Yoshikai.[44] The Yoshikai group first revealed the unexpected potential of a catalytic system combining a Co^{II} precursor, mono-phosphine ligand, pyridine additive and a Grignard reagent reductant to promote an*ortho*-directed olefination of phenyl pyridine derivatives with alkynes. Subsequently, it was observed that a closely related transformation of aryl imines occurred smoothly at the remarkably low reaction temperature of 20 °C (Scheme 21).[45] A detailed optimization study revealed that all components of this catalytic system are crucial for the success of the overall transformation. Firstly, the electronic properties of the cobalt catalyst need to be finely tuned by adding an ancillary ligand. Triarylphosphine ligands bearing electron-withdrawing substituents were the most promising auxiliaries, but pushing the electron-poor character of these ligands to the extreme, such as when using $P(3,5\text{-}(CF_3)_2\text{–}C_6H_3)_3$ or $P(C_6F_5)_3$, led to a total shut down in reactivity. Moreover, no product formation was observed when a trialkyl phosphine (PCy_3) or bidentate phosphines (dppe, dppp, Xantphos) were used. The presence of a sub-stoichiometric amount of a Grignard reagent was essential. Methyl, *tert*-butyl and aryl RMgX species promoted the desired transformation rather poorly and the best results were obtained with 50 mol% of tBuCH$_2$MgBr. As the reaction rate dropped significantly when the amount of Grignard reagent was decreased to 40 mol%, it can be reasonably surmised that the role of this additive is not simply limited to the *in situ* reduction of Co^{II} to generate catalytically active low-valent Co^0 species. In addition, the organometallic reagent is thought to react with the cobalt catalyst and form organocobalt(0)-ate species. Finally, the presence of the Lewis basic additive, pyridine, which plays an as yet undefined role, led to a further improvement in efficiency. Under such finely-tuned reaction conditions, the desired direct coupling occurred smoothly with a large range of aryl imines bearing a variety of substituents, including several sensitive functionalities such as amide, cyano and bromo groups. Both symmetric and non-symmetric, as well as aromatic and alkyl-substituted alkyne-coupling partners were tolerated. This pioneering work clearly showes the exceptional potency of low valent Co/PAr$_3$ species to promote C–H bond cleavage at ambient temperature.

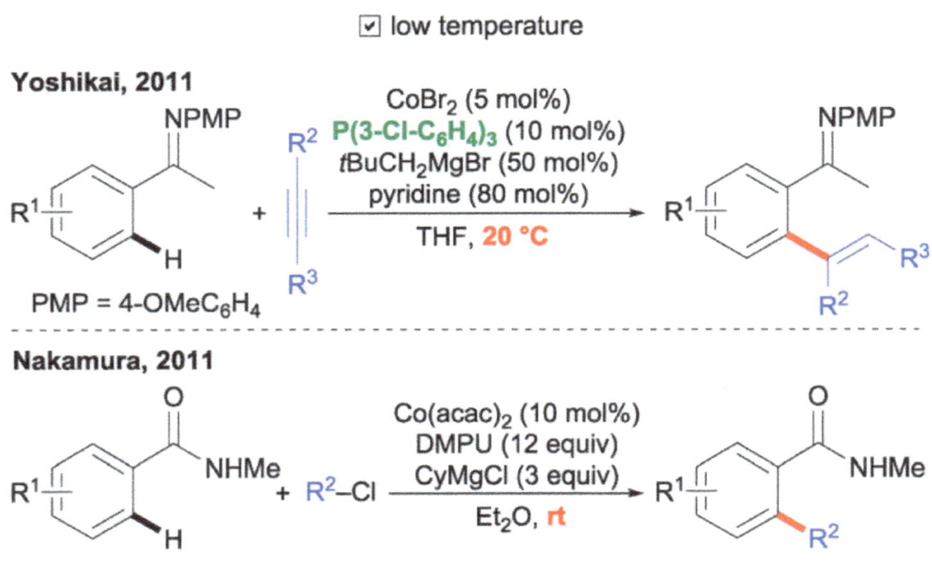

Scheme 21: Low-valent Co-catalyzed C–H activation.[45,46]

Almost simultaneously but independently, Nakamura and co-workers reported a related Co-based catalytic system for the mild C–H functionalization of benzamides (Scheme 21).[46] In this e, DMPU (1,3-dimethyl-3,4,5,6-tetrahydro-2(1*H*)-pyrimidinone) played the role of ancillary ligand facilitating a challenging C(sp^2)–C(sp^3) coupling using alkyl chlorides as coupling partners. As with Yoshikai's system, a Grignard reagent (CyMgCl) was required to ensure reactivity, and it was speculated that this reactant plays multiple roles during the catalytic cycle: deprotonating the benzamide nitrogen, removing the hydrogen atom in the C–H cleavage step, aiding reduction of the Co^{II} pre-catalyst and limiting side-reactions involving homocoupling or resulting from β-hydride elimination.

This pioneering work of Yoshikai and Nakamura became a springboard for the discovery of several closely related transformations[43j,47] such as the cobalt-catalyzed alkynylation and alkylation of heterocycles.[48] Moreover, unexpected results were obtained during the development of a hydroarylation of phenylpyridine derivatives with styrenes (Scheme 22).[49] When a Co^{II} precursor was used in combination with Me$_3$SiCH$_2$MgCl and PCy$_3$, branched 1,1-diarylethane derivatives were afforded. However, an almost complete reversal in the regioselectivity of the process toward linear products was observed using an NHC ligand. This interesting switch in selectivity probably results from the increased steric congestion imparted by the NHC ligand, which leads to an increased preference for insertion into the less hindered end of the terminal alkene. When phenylpyridine substrates were employed, a reaction temperature of 60 °C was required, while substrates bearing imine DGs reacted smoothly at lower temperatures (40 °C–rt).[50] In addition, coupling with vinylsilanes and aliphatic olefins at room temperature or 60 °C, respectively, could also be performed when monophosphine ligands were replaced by *N,N*-bidentate ancillaries such as phenanthroline or neocuproine, affording the linear products.[51] On the other hand, NHCs were found to be superior ligands to mono-phosphines for the *ortho*-arylation or benzamides with chloroarenes,[52] smoothly delivering the desired products at RT whereas a reaction temperature of 80 °C was required using PEt$_3$ as ligand. A similar system was also successfully applied in alkylation reactions with alkyl chloride and bromide coupling partners[53] and in a direct ring-opening process with aziridines.[54]

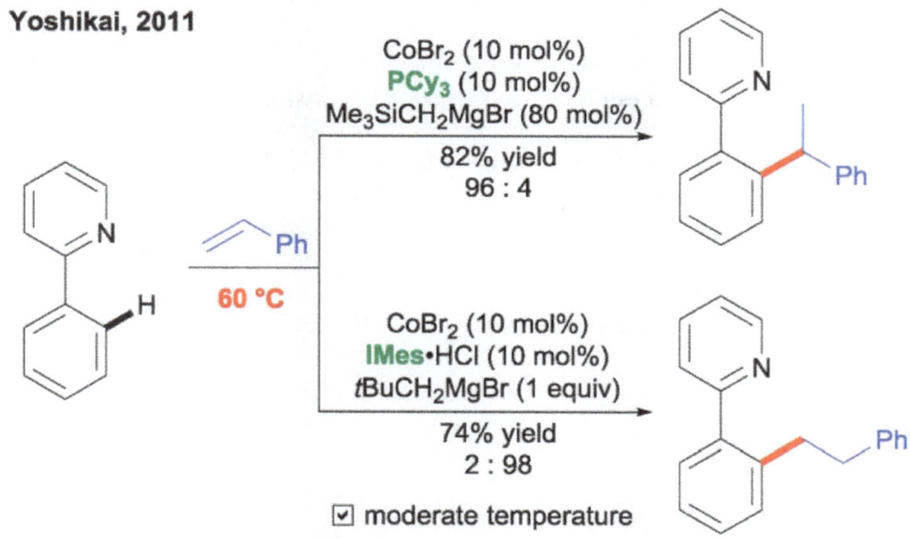

Scheme 22: Selectivity switch in Co-catalyzed styrene hydroarylation.[49]

Significant contributions to the field of mild Co-catalyzed C–H activation were also made by the group of Ackermann.[43k,l] In 2012, they discovered that a Co^{II} pre-catalyst in combination with a Grignard reagent and NHC ligand in DMPU smoothly converts phenylpyridine and indole-type substrates into arylated and benzylated derivatives using sulfamate, carbamate and phosphate coupling partners (Scheme 23).[55] Although these transformations generally required a reaction temperature of 60 °C, in some es reasonable catalytic efficiency was also observed at room temperature. Similarly, a closely related catalytic system was used to perform alkylation reactions using alkyl chlorides as coupling partners. This process was successful even with secondary alkyl derivatives, which are often challenging substrates in this kind of transformation.[56] Finally, the Ackermann group showed very recently that a Co/NHC/RMgCl catalytic system may also be used to perform direct room temperature C–H alkenylation reactions involving C–O bond cleavage of alkenyl acetate, phosphate, carbonate or carbamate coupling partners (Scheme 23).[57] The stereoconvergent nature of this transformation results in the exclusive formation of the *E* diastereomers.

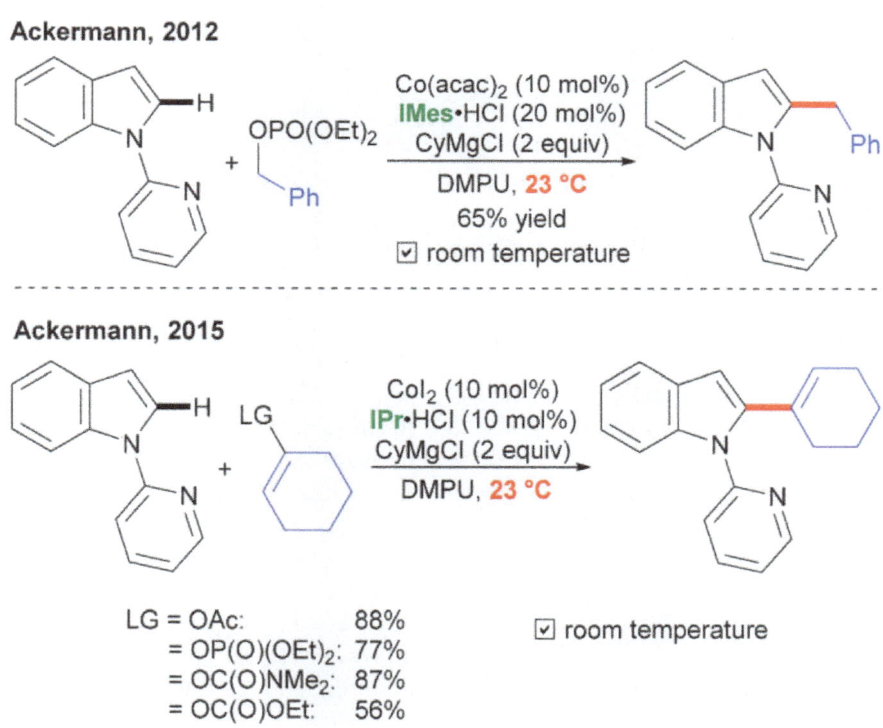

Scheme 23: Co-catalyzed benzylation and alkenylation.[55,57]

All the examples discussed above clearly demonstrate the versatility of low-valent Co species as catalysts for mild C–H activation. Careful modulation of the electronic properties and coordination environment of cobalt catalysts can be achieved through judicious selection of ligands, which can include mono-phosphines, NHCs or *N,N*-bidentate scaffolds. Notably, the fundamental C–H cleavage step appears to be remarkably easy with low valent Co catalysts, irrespective of the ligand. Another interesting feature of this catalytic system, in particular in the e of alkylation reactions, is the potential involvement of radical intermediates and single electron transfer (SET) events in the C–H functionalization mechanism. Indeed, it is believed that SET between the cyclometalated intermediate and an alkyl halide may occur, generating an alkyl radical species with the final product being formed *via* radical coupling (Scheme 24).[43j]In addition, C–H transformations of aromatic substrates bearing OMe, F, CN or Cl groups in the *meta*-position with this metal demonstrate unusual regioselectivity favoring functionalization at the more hindered *ortho*-position between these substituents and the DG. Finally, an additional advantage of this mild C–H activation system is its compatibility with cost-efficient yet typically unreactive chlorinated coupling partners.

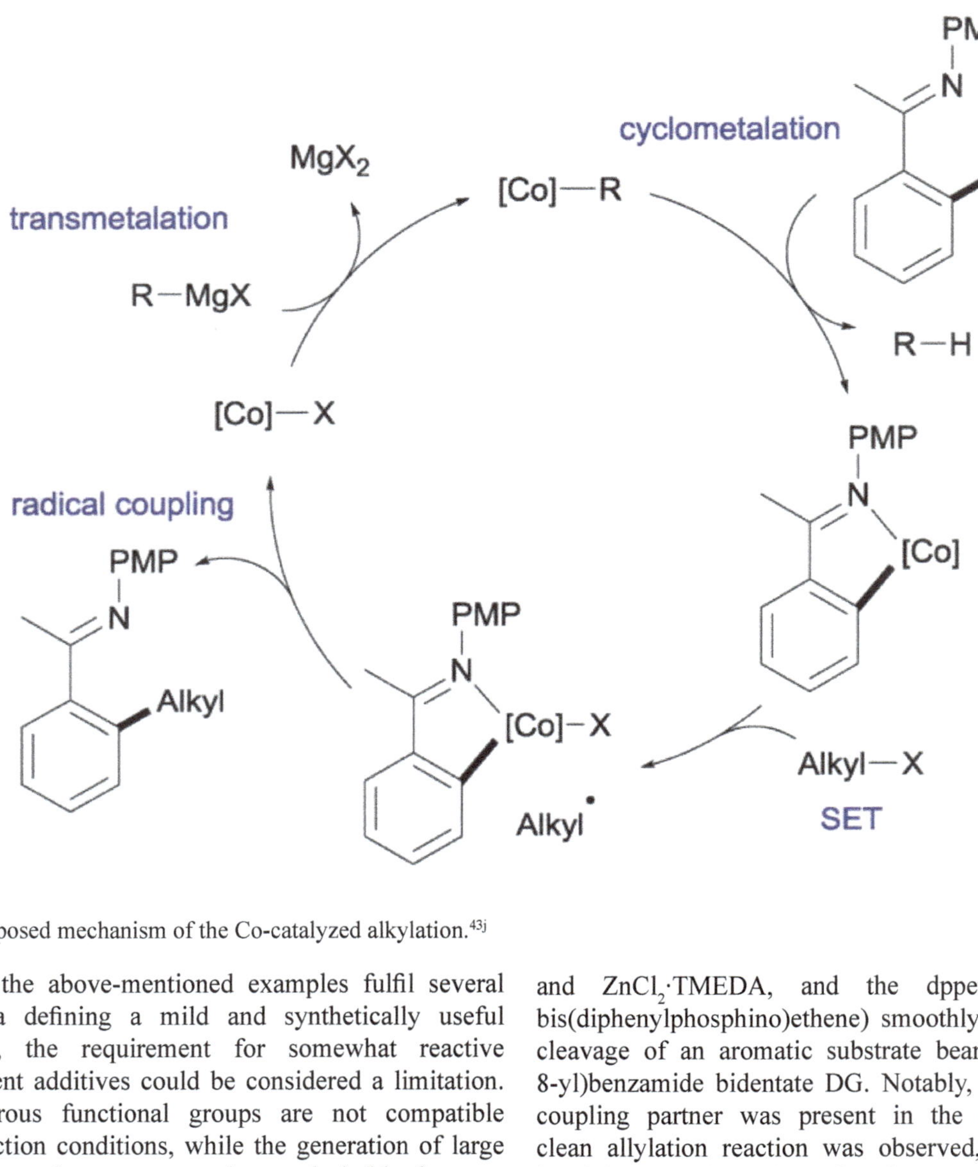

Scheme 24: Proposed mechanism of the Co-catalyzed alkylation.[43j]

Although the above-mentioned examples fulfil several of the criteria defining a mild and synthetically useful transformation, the requirement for somewhat reactive Grignard reagent additives could be considered a limitation. Indeed, numerous functional groups are not compatible with these reaction conditions, while the generation of large amounts of by-products as waste is not desirable from an ecological perspective.

Fe-Based Catalytic Systems

Considering the potential of Co catalysts in mild C–H activation reactions, the scientific community has duly become interested in investigating the reactivity of Fe-based catalysts in these kinds of transformations. In 2013, Ilies and Nakamura reported that, in a similar fashion to Co, low-valent Fe-species also enable rather mild direct functionalization of aryl C–H bonds (Scheme 25).[58] An Fe^{III}-pre-catalyst, in combination with $(t\mathrm{BuCH_2})_2\mathrm{Zn}$ generated *in situ* from $t\mathrm{BuCH_2MgBr}$ and $\mathrm{ZnCl_2{\cdot}TMEDA}$, and the dppen ligand (*cis*-1,2-bis(diphenylphosphino)ethene) smoothly enabled C–H bond cleavage of an aromatic substrate bearing an *N*-(quinolin-8-yl)benzamide bidentate DG. Notably, when an allyl ether coupling partner was present in the reaction mixture, a clean allylation reaction was observed, and side reactions involving the formation of *ortho*-neopentylated derivatives or C–C coupling products between the neopentyl and allylic moieties were efficiently suppressed. This reaction, which occurred rapidly at 70 °C, was found to be strongly dependent on the choice of organozinc reagent and ligand. When using $\mathrm{Me_2Zn}$ or $\mathrm{Ph_2Zn}$, the side-reactions were predominant, while, in the absence of a ligand or in the presence of bypyridine, mono-phosphine or diphosphine additives bearing a flexible backbone, the reactivity was significantly decreased. A further step forward toward mild Fe-catalyzed C–H activation was achieved a few months later when the same group disclosed that replacing the bidentate DG with a mono-coordinating pyrazole DG further increases the reactivity of the catalytic system (Scheme 25).[59] Accordingly, the *ortho*-allylation of 1-arylpyrazoles using the $Fe(acac)_3$ pre-catalyst, $(tBuCH_2)_2Zn$ as a base and reducing agent, and dtbpy as a ligand could be successfully performed at 0 °C. Such a low-temperature C–H bond cleavage is remarkable and clearly demonstrates the potential of low-valent first-row transition metal species as catalysts for C–H transformations. In contrast to the Co-based catalytic systems, which favor functionalization at the more hindered *ortho*-positions of *meta*-substituted arenes, this Fe-based catalyst is more sensitive to steric constraints and mediates allylation at the most accessible *ortho*-position.

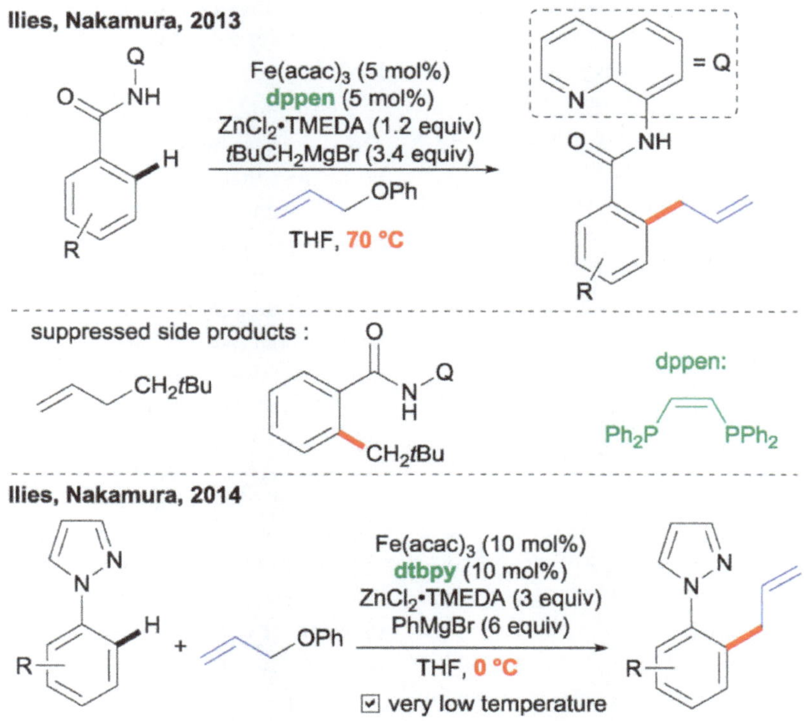

Scheme 25: Iron-catalyzed C–H allylation.[58,59]

In parallel to the reports mentioned above, Cook and co-workers investigated Fe-catalyzed C–H alkylation reactions of aryl substrates bearing widely-employed 8-aminoquinoline DGs (Scheme 26).[60] The catalytic system combining an $Fe(acac)_3$ precursor, dppe as ligand and PhMgBr promoted the desired coupling at 65 °C with remarkable efficiency, delivering the desired products after only a few minutes. Importantly, this catalytic system is also compatible with secondary alkyl precursors, allowing very challenging, hindered C(sp^2)–C(sp^3) couplings. A closely related catalytic system applied to C(sp^2)–C(sp^2) and C(sp^2)–C(sp^3) couplings was disclosed by Ackermann and co-workers in 2014 (Scheme 26). In these processes, a triazolyldimethyl amide (TAM) was designed as a highly efficient bidentate DG (termed a "substrate"-activator).[61]

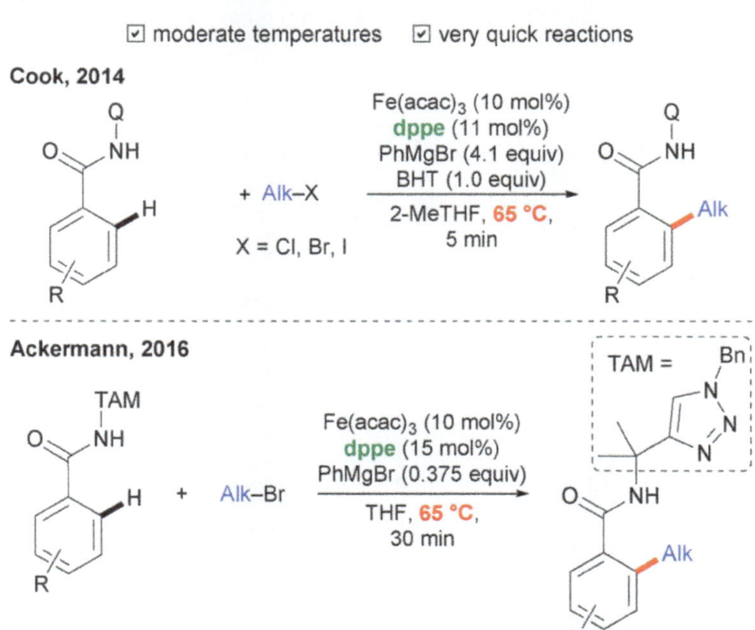

Scheme 26: Low-valent Fe-catalyzed C(sp^2)–C(sp^3) couplings.[60,61]

Notably, despite the obvious potential of these Fe-based catalytic systems, at present they suffer from the same limitations as the Co catalysts described in the section above, resulting from the use of reactive Grignard reagents.

SUBSTRATE

Directing Group Design

The most widely-applied synthetic strategy to overcome the intrinsic regioselectivity problems encountered in C–H activation involves embedding a coordinating moiety within a substrate.[62] These "directing groups" bind to the metal catalyst and lead to an effective increase in the local concentration of one particular C–H bond at the metal center. Upon coordination, the DG plays the role of a ligand and may significantly influence the electronic properties of the metal catalyst. In particular, the DG can greatly affect the electrophilicity of the metal species and may potentially allow for the stabilization of high oxidation state intermediates. Moreover, the choice of DG has a direct impact on the rate of the C–H cleavage step and greatly influences the stability of the resulting metalacyclic intermediate.

In early works on C–H activation, the majority of catalytic systems employed strongly σ-donating or π-accepting nitrogen-, sulphur- or phosphorus-containing DGs. It was recognized that such strongly coordinating groups favor rapid C–H bond scission and lead to the formation of thermodynamically stable metalacyclic intermediates. As a result, many metalacyclic species could be isolated and fully characterized.[63] However, the high stability of the catalytic intermediates can also be problematic for achieving catalytic turnover and may result in low compatibility with a range of electrophilic and nucleophilic coupling partners. In addition, the presence of such strongly coordinating DGs can be synthetically restrictive due to the limited modifiability of these groups. In contrast, the use of commonly-present functionalities such as carboxylates, ketones, esters or amides, which are weaker coordinating moieties, would greatly increase the synthetic value of C–H activation. Although such weak interactions between the substrate and the metal catalyst could render the C–H cleavage more challenging, the resulting weakly-bound metalacyclic intermediate may be significantly more reactive toward the coupling partners in the other steps of the catalytic cycle. Accordingly, for es where mechanistic steps subsequent to C–H activation are rate-limiting, milder conditions and a more diverse array of potential transformations could be envisaged. Following this hypothesis, many research groups have focused on designing catalytic systems involving weakly-coordinating DGs.[64]

The amide moiety is one of the most commonly used DGs and is compatible with a variety of transition metal catalysts. Surmising that weaker coordination between this DG and a catalyst may boost the reactivity of a catalytic system, Yu devised a class of secondary *N*-aryl amide DGs bearing electron-withdrawing substituents on the aromatic ring which increase the acidity of the N–H bond. Deprotonation of the amide under weakly basic conditions promotes coordination to the metal in a η^1 manner *via* N, optimizing the geometry of the key intermediates and facilitating the C–H bond cleavage. Accordingly, the electron-deficient amide moiety bearing a 2,3,5,6-tetrafluoro-4-(trifluoromethyl)phenyl substituent was established as the most potent DG and substrates bearing this group have been widely employed in various C–H activation reactions. An illustrative example of a remarkably mild transformation that demonstrates the superior reactivity of this weakly-coordinating auxiliary was reported in 2013.[65] The authors studied a Ru-catalyzed C–H amination reaction and discovered that the desired C–N coupling with *N*-benzoyloxyamines occurred smoothly at room temperature when an arene substrate bearing this electron-deficient amide auxiliary was employed (Scheme 27). Notably, only the weak base potassium carbonate was required for the deprotonation of the amide. Consequently, the scope of this transformation is very broad and the process is even compatible with heterocyclic substrates including pyrazole, thiophene, benzothiophene, furan, benzofuran and indole. Following the same hypothesis, Fabis and co-workers employed *N*-tosylamide as an appealing DG for Pd-catalyzed C–H methoxylation and halogenation reactions of arenes.[66]

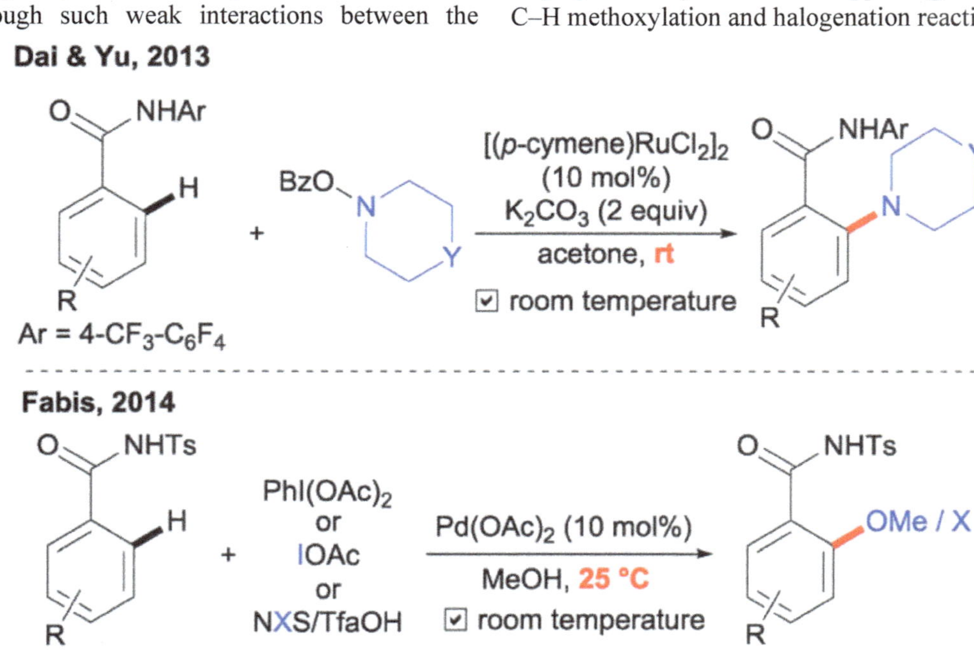

Scheme 27: Weakly coordinating amides as directing groups.[65,66]

Importantly, in both es, the formation of metalacyclic intermediates could be monitored at room temperature by ^{1}H NMR. Unfortunately, as no kinetic isotope effect (KIE) studies were conducted for either reaction, it is not possible to comment on the relative rate of the different catalytic steps. It could be reasonably speculated, however, that the weakly coordinating amide DG permits facile C–H bond cleavage and the resulting Ru- or Pd-based metalacyclic intermediates are subsequently oxidized to afford Ru^{IV} or Pd^{IV} species capable of undergoing fast reductive elimination.

Inspired by the potential of the amide DG bearing an electron deficient aromatic substituent, Han and collaborators hypothesized that the installation of such a coordinating moiety could enable the direct functionalization of arylphosphine derivatives.[67] As expected, the desired direct arylation could be achieved using arylboronic acid derivatives as coupling partners. Remarkably, among five different amide DGs tested, only the electron poor moiety $CONHC_6F_5$ led to conversion of the starting material at 40 °C. As this mild C–H functionalization reaction generated stereogenic phosphinamides, the same group subsequently disclosed a closely related protocol for an enantioselective transformation (Scheme 28).[68] Inspired by Yu's work, an MPAA was selected as a chiral ligand. After an extensive optimization of the reaction conditions, including investigations into the organoboron reaction partner, oxidant, base and solvent, an efficient catalytic system was developed. Notably, in this study, the aromatic substituent of the amide DG was modified, and the $-C_6F_5$ moiety initially used was replaced by another electron-deficient ring, 4-CN–C_6F_4. Interestingly, facile C–H cleavage occurred readily at 40 °C, but the resulting metalacycle was sufficiently stable to be isolated and fully characterized. This species can also be used as the active catalyst.[67]

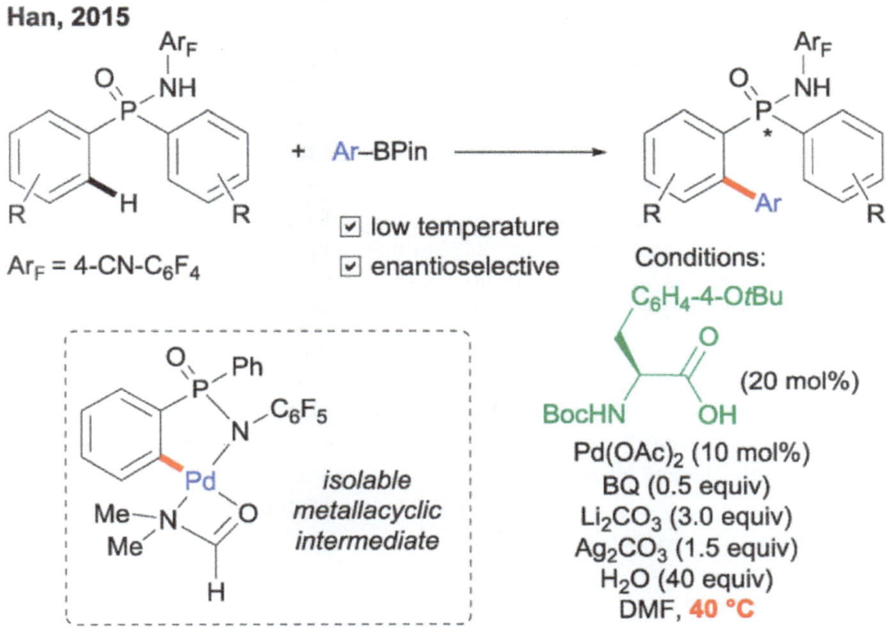

Scheme 28: Enantioselective arylation of phosphonamides.[68]

Also exploring the strategy of improving a catalytic system by tuning the properties of DGs, Gaunt and co-workers became interested in the use of imine moieties.[69] They hypothesized that the less electron withdrawing character of this auxiliary, compared to analogous carbonyl groups, would lower the electron deficiency of an aromatic ring and hence enable cyclometalation under mild reaction conditions. Indeed, the cyclopalladation of benzaldimines at room temperature is well described.[63a] However, employing imines in Pd-catalyzed C–H transformations presents an intrinsic difficulty due to the instability of these groups under mildly acidic conditions. The amine derivatives generated by hydrolysis bind strongly to the metal and lead to catalyst poisoning. The authors presumed that imine hydrolysis could be prevented by increasing the steric bulk of the amine component. Accordingly, arylation of a benzaldimine derived from 2,6-diisopropylaniline was achieved at room temperature using aryltrifluoroborates as the reaction partners (Scheme 29). The scope of this transformation is large with regard to both the Ar–BF_3K derivative and the benzaldimines, yielding the desired biaryl compounds in moderate to good yields.

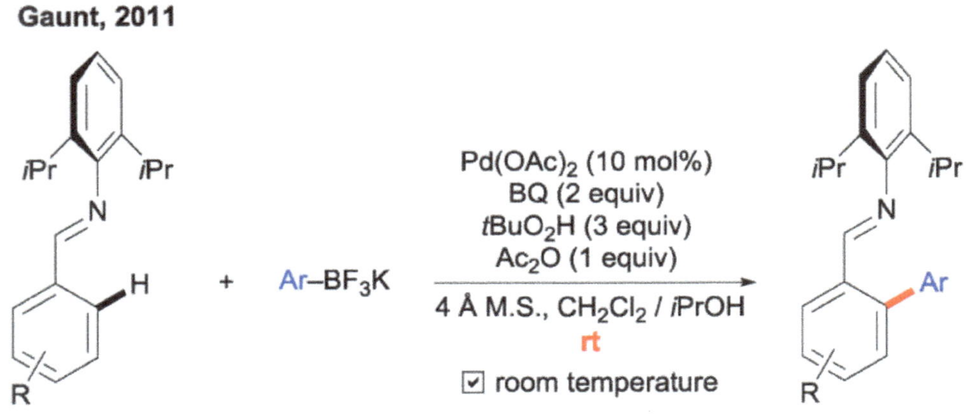

Scheme 29: Imine-directed arylation.[69]

The coordinating properties of the DG may also be influenced by the choice of reactants, additives and solvents used for a targeted C–H activation reaction.[70] Such "*in situ* adjustment" was observed in the e of sulfoxide-directed atropodiastereoselective C–H activation.[71] In this study, biaryl scaffolds bearing a sulfoxide moiety playing a role of both DG and chiral auxiliary were employed as substrates in an oxidative Heck (Fujiwara–Moritani) reaction. Rather surprisingly, a large solvent-effect on both the efficiency and the level of chiral induction of this transformation was observed. While the oxidative Heck reaction performed in 1,2-dichloroethane (DCE) required a rather high reaction temperature of 80 °C to attain a reasonable conversion of the starting material and was only rather moderately diastereoselective, using the highly polar, polyfluorinated alcohol HFIP (1,1,1,3,3,3-hexafluoroisopropanol) as solvent led to an exceptional increase in both reactivity and atroposelectivity (Scheme 30). Since mechanistic studies revealed the formation of a strong hydrogen bond between the sulfoxide DG and HFIP solvent, it can be presumed that such DG/solvent interactions amend the electronic properties of the DG, rendering it less coordinating (weaker DG), while simultaneously modifying the chiral environment around the catalyst. Thus, the sulfoxide-directed Fujiwara–Moritani reaction conducted in HFIP permitted the synthesis of versatile atropisomeric scaffolds in generally excellent yield and very high optical purity. Notably, a related direct acetoxylation and iodination, when performed in HFIP, delivered the corresponding diastereoselective C–O and C–I coupling products at room temperature.[72] KIE studies conducted for both C–C and C–O couplings suggested that, in the first e, the metalation step is rate-determining, whereas under acetoxylation conditions, C–H cleavage is reversible, and the reductive elimination is believed to be the kinetically limiting step. Such divergent outcomes clearly show that sulfoxide directed C–H activation is relatively fast at room temperature in HFIP and the rate of the overall transformation is dependent on the kinetics of each step in the cycle (reductive elimination to afford C–C bonds is arguably much easier than C–O bond formation).

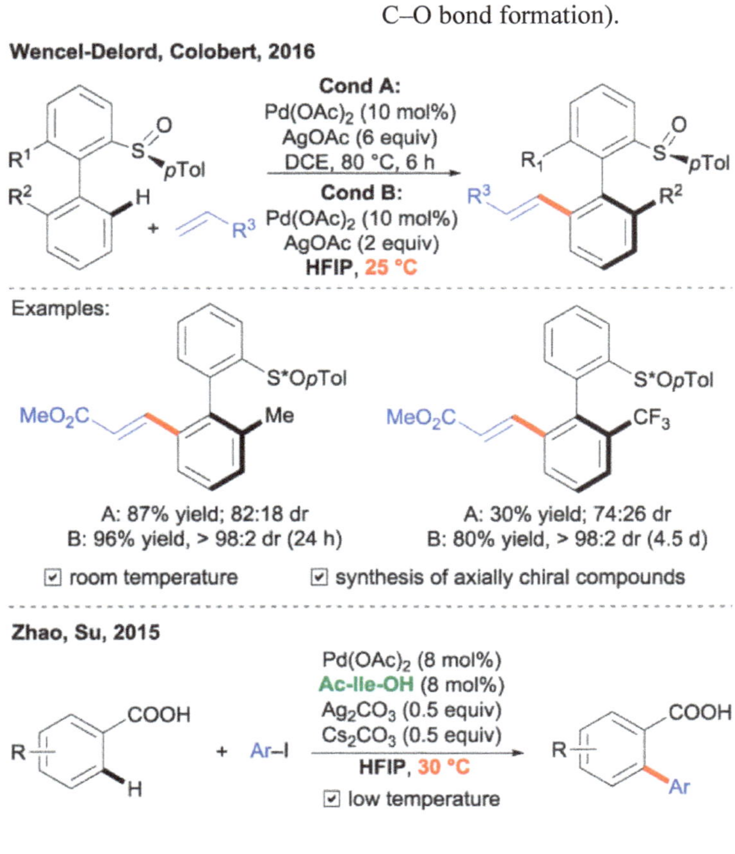

Scheme 30: HFIP-assisted C–H activation at low temperatures.[71,73]

A related mild arylation of benzoic acids with a ligand-supported Pd catalyst was reported by Zhao and Su.[73] In analogy to the sulfoxide-directed reaction, this direct functionalization of electron-deficient aromatics was efficiently conducted in HFIP while no reaction occurred in other solvents such as 2,2,2-trifluoroethanol (TFE), water, acetic acid, dimethylformamide or benzene (Scheme 30). Although the key role of HFIP in this reaction was not clearly elucidated, the same type of "H-bond substrate activation" by interaction of HFIP with the DG could be expected.[74]

Another example of how design of DGs can lead to the development of mild C–H transformations involves the use of N–N moieties in Rh^{III}-catalyzed reactions. A seminal *N*-nitroso-directed olefination of arenes was disclosed by Zhu (Scheme 31).[75] The particular potential of the nitroso-DG can be attributed to the variety of coordination modes, such as η^1-*O*-binding, η^1-*N*-binding and bridged μ-η^1:η^1-*N,O*-binding, by which this group may interact with metal centers. Accordingly, the reactivity of the cyclometalated intermediates can be regulated through adjustments in the coordination mode of the DG. This *N*-nitroso directing group imparts sufficient stabilization to the metalacyclic intermediate formed upon C–H activation to render it isolable. The nature of the substituent on the C–H substrate drastically affects the outcome of the C–H cleavage event as irreversible *ortho*-rhodation was observed for electron-poor substrates and reversible *ortho*-rhodation was observed for electron-rich substrates. This discrepancy could suggest that, although the rate-determining step is different in each e, the cyclometalation remains sufficiently efficient to be performed under moderate temperatures even with more challenging, electron-deficient substrates.

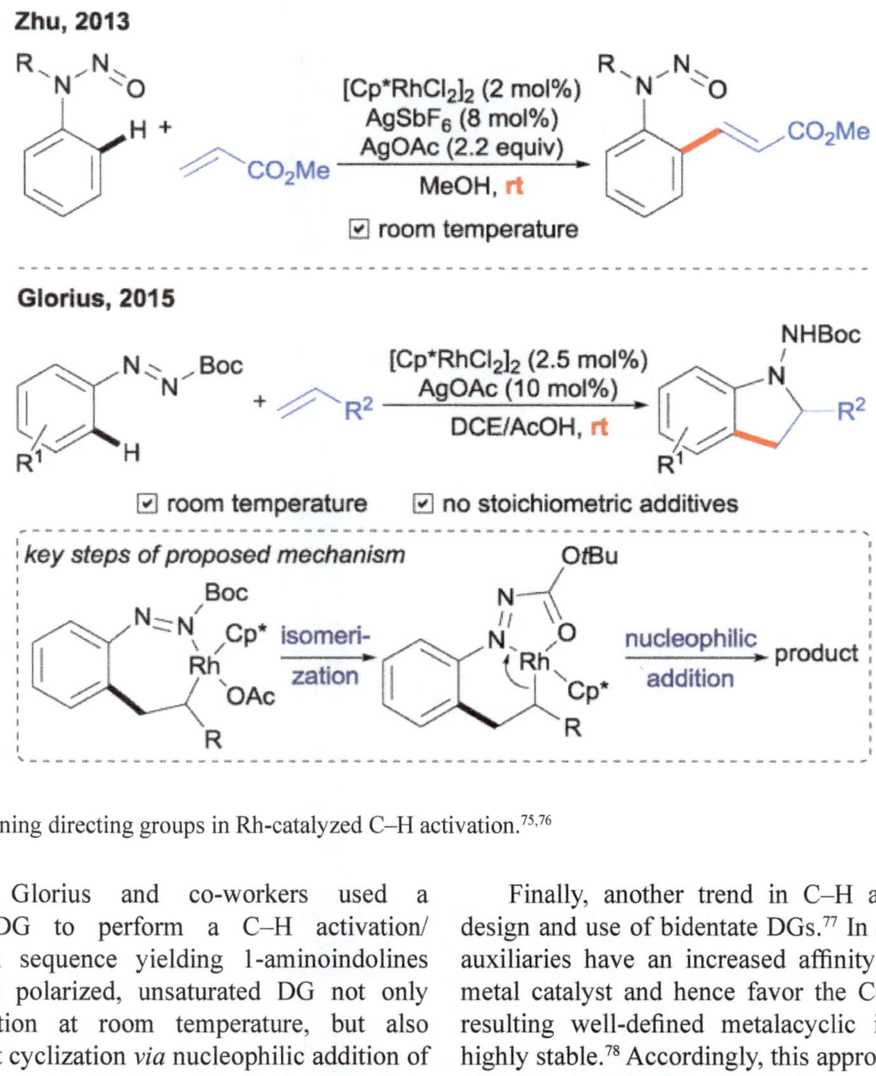

Scheme 31: N–N-containing directing groups in Rh-catalyzed C–H activation.[75,76]

Very recently, Glorius and co-workers used a diazenecarboxylate DG to perform a C–H activation/nucleophilic addition sequence yielding 1-aminoindolines (Scheme 31).[76a] This polarized, unsaturated DG not only enabled cyclometalation at room temperature, but also underwent subsequent cyclization *via* nucleophilic addition of a C(sp^2)–Rh species onto the N=N moiety. In this particular e, the C–H scission remains the slowest step of the catalytic cycle, while the rhodacycle is relatively stable (isolable in 80% yield). Notably, the presence of a coordinating substituent, such as a *tert*-butyloxycarbonyl (Boc) group, on the DG is crucial to prevent problematic β-hydride elimination and this moiety might stabilize the olefin insertion product by chelation. A computational study furthermore suggests that this chelation also facilitates the isomerization to a 6-membered rhodacycle.[76b]

Finally, another trend in C–H activation involves the design and use of bidentate DGs.[77] In general, such bidentate auxiliaries have an increased affinity for coordination of a metal catalyst and hence favor the C–H cleavage step. The resulting well-defined metalacyclic intermediates are also highly stable.[78] Accordingly, this approach is appealing for the mild functionalization of especially unreactive C–H bonds. In particular, several C–H transformations of aliphatic C(sp^3)–H bonds have been reported although generally rather high reaction temperatures are still necessary to ensure reasonable catalytic efficiency.[79]

A clear illustration of such a phenomenon was reported in 2012 by Sahoo.[80] This group discovered the particular potential of the *S*-methyl-*S*-2-pyridyl-sulfoximine (MPyS) bicoordinating DG[81] to enhance direct functionalization of unactivated primary β-C(sp^3)–H bonds (Scheme 32). The challenging cyclometalation of an aliphatic C–H bond was triggered by facile coordination of palladium by the two coordinating moieties of MPyS (the pyridyl and sulfoximine nitrogens). Notably, the choice of this DG resulted in an efficient targeted C(sp^3)-acetoxylation at ambient temperature whereas other *N,N* and *N,S*-bidentate DGs failed to afford the oxidized product. The exceptional ability of this bidentate DG to facilitate C–H scission is illustrated by the fact that C–H activation is reversible in this system. The facile cleavage and easy recovery of the MPyS further highlights the synthetic value of this protocol. Following this work, Sahoo developed a related catalytic system for MPyS-directed acetoxylation of C(sp^2)–H bonds which could be followed by an intramolecular cyclization to afford benzofuranones.[82]

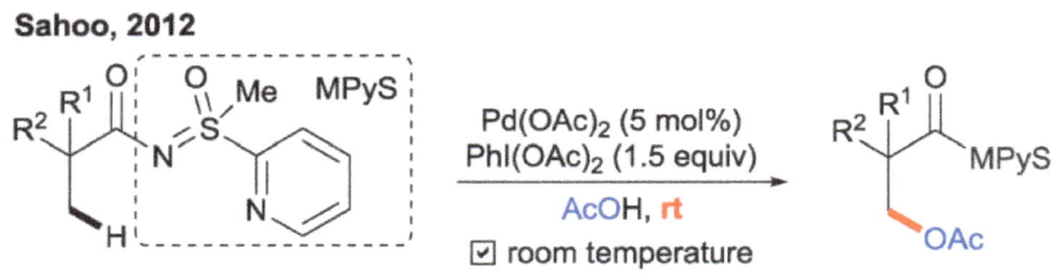

Scheme 32: A bidentate directing group as a handle for mild Pd-catalyzed aliphatic C– H activation.[80]

The bidentate 8-aminoquinoline moiety was also employed to facilitate direct Pd-catalyzed trifluoromethylthiolation of C(sp^3)–H bonds under relatively mild reaction conditions (Scheme 33).[83]

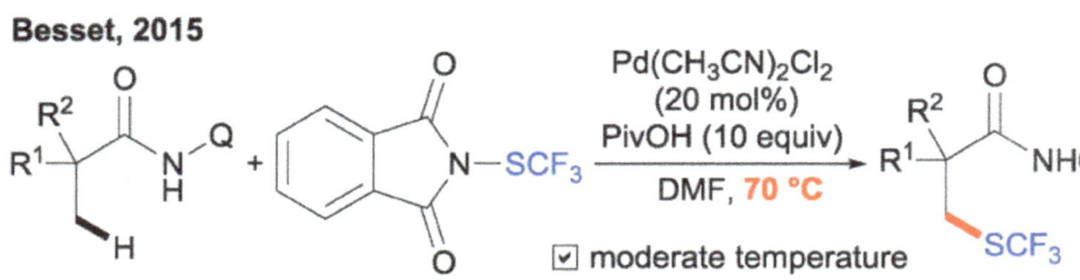

Scheme 33: C(sp^3)–H trifluoromethylthiolation with a bidentate DG.[83]

Recently, Daugulis discovered that a bidentate DG may also be a useful handle to promote cobalt-catalyzed C–H activation of aromatic substrates under very mild conditions.[84] When reacted in the presence of a Co^{II} precatalyst, $Mn(OAc)_3{\cdot}2H_2O$ as a co-oxidant and a base, the 8-aminoquinoline-substituted benzamide underwent mild *ortho*-metalation and subsequent coupling with alkenes or carbon monoxide to yield heterocyclic compounds (Scheme 34).[85] These transformations are believed to proceed *via* a Co^{III} intermediate formed by oxidation of $Co(OAc)_2$ in the presence of the aminoquinoline amide ligand. An advantage of this high-valent catalytic system is its excellent functional group tolerance with even iodo-, nitro-, heteroaryl- and cyano-groups being tolerated.

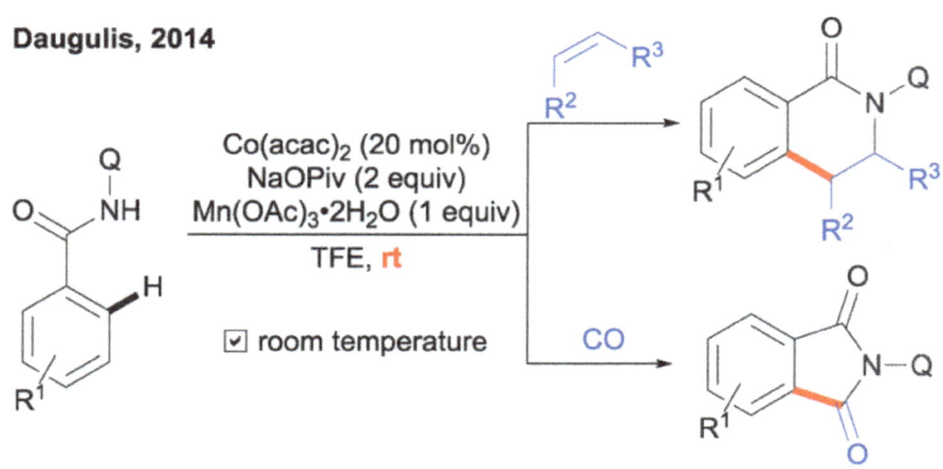

Scheme 34: Co-catalyzed functionalization of aromatic substrates bearing bidentate directing groups.[85]

Internal Oxidants

The development of directing groups that serve as internal oxidants has revolutionized oxidative C–C bond forming C–H activation, especially under Rh^{III} catalysis. In this context, an internal oxidant is considered a moiety in the C–H activation substrate that is both a DG and an oxidant for the catalyst species either after or before a reductive step in the catalytic cycle. In practice, these moieties contain cleavable N–N or N–O bonds. Generally, milder conditions can be employed compared to similar reactions using external oxidants and no extra metal waste is formed. The decreased reaction temperatures can be a consequence of the intramolecular nature of the re-oxidation step. The low-valent metal species is formed in the vicinity of the internal oxidant and re-oxidation can occur immediately in an inner-sphere sense. In some es, the order of events can also be reversed when the internal oxidant triggers an oxidatively induced reductive elimination. This specificity of the oxidation mechanism is beneficial for the tolerance of the reaction toward oxidatively-sensitive functional groups. Additionally, some oxidizing DGs aid the C–H activation step in their role as directing groups, as exemplified by the use of hydroxamates in redox-neutral transformations. As the nature of the directing group changes during the course of the oxidation, complete control for mono-functionalization is generally inherent to this concept. Overreaction can otherwise be a limitation with certain directing groups that bind strongly to the catalyst. In many methods involving this approach, the initial metalacycle functionalization is followed up by a cyclization event, forming interesting heterocyclic products which incorporate part of the former directing group. On the downside, extra steps may be required to install a suitable oxidizing directing group, potentially involving oxidation steps and thus merely shifting the use of an oxidant to another part of the synthesis. Additionally, in most es a stoichiometric by-product is formed by the cleavage of the N–N or N–O bond. The topic of internal oxidants as directing groups has been reviewed recently[86] and we will therefore limit the discussion of this enabling technology to a selection of examples that demonstrate exceptional mildness.

The concept of oxidizing directing groups was introduced to the field of C–H activation by the group of Neuville and Zhu in 2008[87a] and shortly thereafter by the groups of Cui and Wu[87b] and Hartwig[87c] in reactions demonstrating two fundamentally different uses of internal oxidants. In the first example, a Pd^0-catalyzed annulation of acyl oximes with arynes, the oxidation of Pd^0 by addition into the N–O bond initiates the reaction.[87a] Hartwig's Pd^0-catalyzed internal arene amination with oxime esters is related to this mode of action.[87c] On the other hand, the second example was a Pd^{II}-catalyzed alkenylation of quinoline-*N*-oxides with acrylates where the re-oxidation of Pd^0 after reductive elimination closes the catalytic cycle.[87b]While enabling oxidative transformations without stoichiometric external oxidants, the targeted transformations still required very high reaction temperatures. A real breakthrough toward milder C–H activation was disclosed by the group of Guimond and Fagnou who found that benzhydroxamates are particularly appealing substrates for Rh^{III} catalysis (Scheme 35).[88]Following their initial report on an isoquinolone synthesis from *O*-methyl benzhydroxamates and alkynes at 60 °C, it was shown that the use of *O*-pivaloyl benzhydroxamates as substrates enabled alkyne and alkene insertion and subsequent cyclization to the corresponding (dihydro)isoquinolones at temperatures as low as room temperature.[89] The alkene insertion, however does not yield dihydroisoquinolones from *O*-methyl benzhydroxamates, demonstrating how striking differences in reactivity can arise from apparently small changes in the directing group substitution. Later, computational studies were able to rationalize this reactivity (Scheme 36).[90] First of all, the ability of the pivalate to stabilize catalytic intermediates through chelation with the carbonyl group is thought to be responsible for the increased reactivity of the *O*-pivaloyl benzhydroxamates. Furthermore, this chelation prevents β-hydride elimination after the alkene insertion step. The lack of such stabilization leads to the formation of olefinated products with *O*-methyl benzhydroxamate substrates (Scheme 36, right side). Intriguingly, the authors identified an oxidative acyloxy migration to a nitrene intermediate formally featuring Rh^V as the favored pathway with *O*-pivaloyl benzhydroxamates (Scheme 36, left side). Subsequent reductive nitrene insertion into the Rh–C bond and protodemetalation then afford the cyclized product.[90] On the other hand, no nitrene intermediate is involved in the computed mechanism for the cyclization with alkynes. Instead, a sequence of C–N reductive elimination and N–O oxidative addition succeeds the alkyne insertion.[91] In contrast to the reactivity of *O*-methyl benzhydroxamates under Rh^{III} catalysis, cyclization reactions can be performed with alkynes and alkenes under Ru^{II} catalysis at low temperatures using these substrates, furnishing related (dihydro)isoquinolone products.[92]

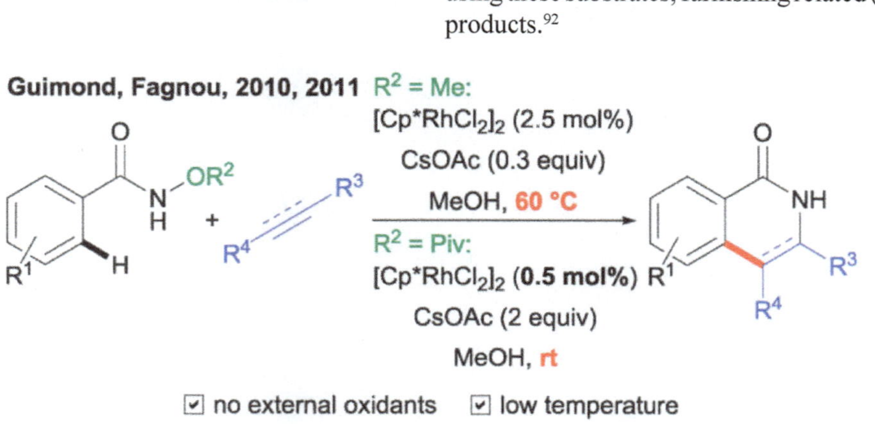

Scheme 35: Seminal studies employing hydroxamate directing groups in Rh^{III}-catalyzed C–H activation.[88,89b]

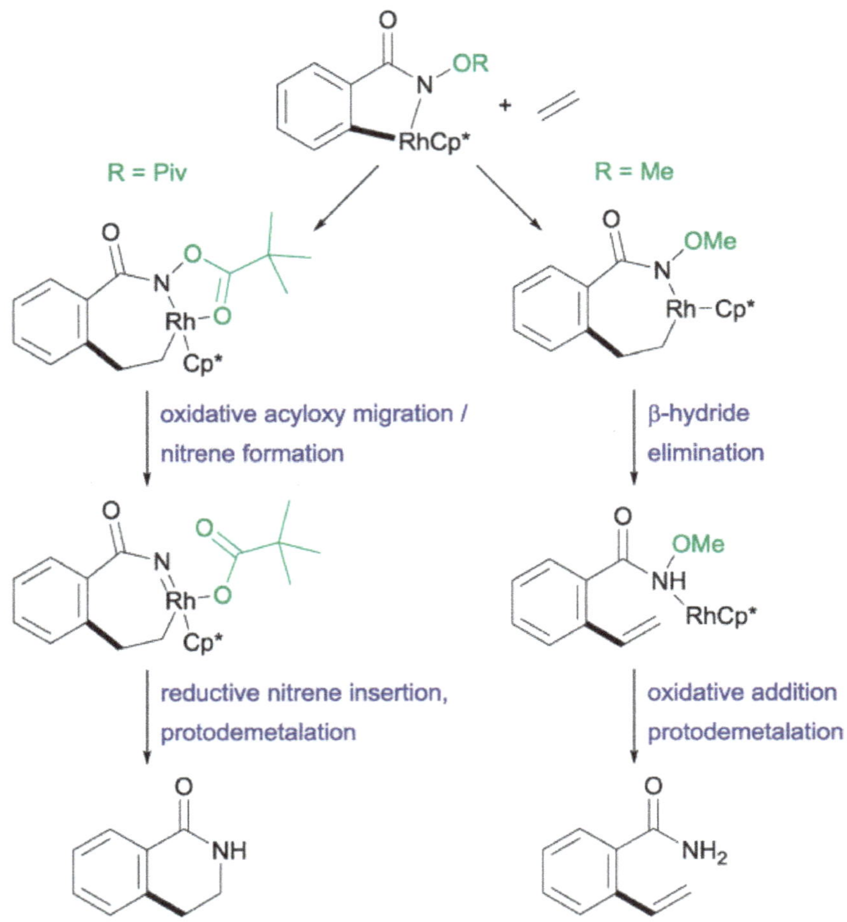

Scheme 36: Computed mechanism of the Cp*Rh^{III}-catalyzed reaction of benzhydroxamates with alkenes.[90]

Subsequent work has showed the versatility of the *O*-pivaloyl hydroxamic acid directing group in Rh^{III} catalysis for the mild synthesis of a number of heterocyclic motives (Scheme 37). This diversity is exemplified by the synthesis of isoquinolone derivatives with synthetically valuable moieties such as *exo*-double bonds[93] and borates[94] with predictable selectivities, as well as by the development of isoindolone and benzoazepinone syntheses using diazo compounds as carbene precursors.[95]

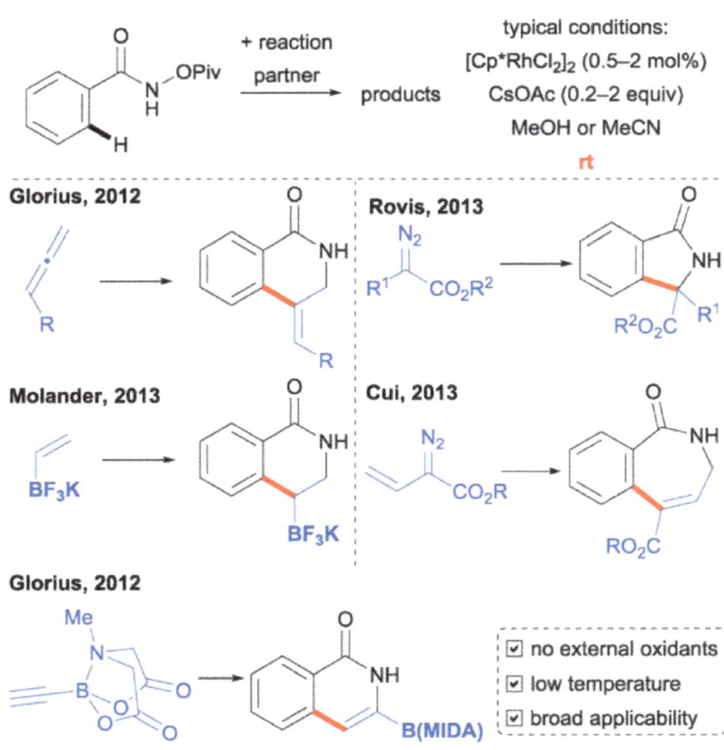

Scheme 37: Examples of Cp*Rh^{III}-catalyzed transformations of *O*-pivaloyl benzhydroxamates at room temperature.[93–95]

Notably, the mildness of the transformations involving oxidizing directing groups has encouraged the development of enantioselective Rh^{III}-catalyzed C–H transformations. Based on the dihydroisoquinolone synthesis from *O*-pivaloyl benzhydroxamic acids originally presented by Glorius[89a] and Guimond,[89b] the groups of Cramer, and Ward and Rovis independently presented two concepts for the synthesis of enantioenriched dihydroisoquinolones from benzhydroxamates and olefins (Scheme 38).[96] Ward and Rovis developed an artificial metalloenzyme containing the $CpRh^{III}$ fragment bound through a biotin tag on the cyclopentadienyl ligand to a streptavidin mutant.[96a] The compatibility with enzyme catalysis demonstrates the mildness of the transformation achieved using the oxidizing directing group. Cramer's group focused on the development of chiral cyclopentadienyl derivatives surmising that selectivity could be achieved using a shielding element and symmetrical positional locks on the ligand.[96b] Using this design, they achieved the enantioselective annulation of olefins and *O*-*tert*-butoxycarbonyloxy-benzhydroxamates in ee's up to 94%. Further development led to a second generation of ligands bearing an annulated binaphthyl motif on the cyclopentadienyl, which have been used in an enantioselective carbene insertion reaction delivering isoindolones at room temperature.[27d,96c]

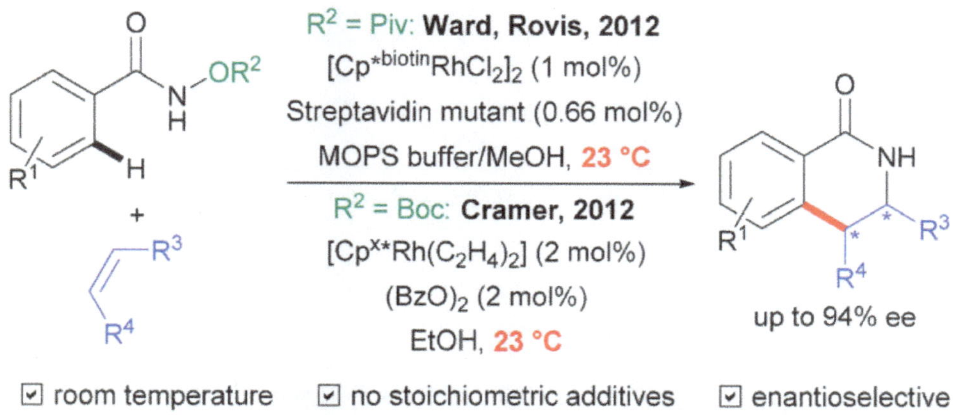

Scheme 38: Enantioselective Rh^{III}-catalyzed C–H functionalization.[96]

The diverse reactivity of phenoxy- and enoxyamides with different catalyst systems under mild conditions further illustrates the potential of oxidizing directing groups in combination with strategic planning and finely tuned reaction conditions. The group of Liu and Lu achieved an intramolecular functional group transfer under Rh catalysis. After alkyne insertion, the acetamide group is transferred from oxygen to carbon *via* O–N bond cleavage to afford enamides as products in what is effectively a carboamination of the alkyne. This process thus avoids the generation of a stoichiometric by-product during the course of the oxidation (Scheme 39).[97a] Under slightly different reaction conditions, the authors instead obtained cyclized benzofuran products, which are also produced in a ruthenium-catalyzed variant of this reaction.[97b] Notably, under iridium catalysis, terminal alkynes could be used as reaction partners, yielding alkynylated phenol products (Scheme 39).[97c]

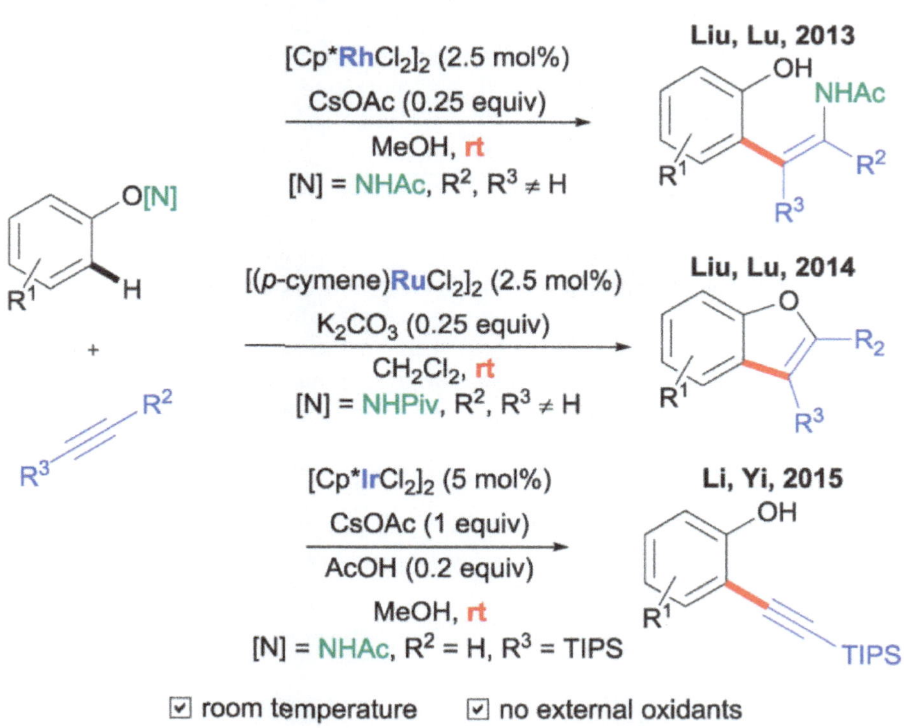

Scheme 39: Versatile reactivity of phenoxyamides.[97]

The Rovis group encountered an intriguing result in the reaction between *N*-enoxyphthalimides and alkenes under rhodium catalysis. Instead of generating β-olefinated ketones, they obtained cyclopropanes in a formal [2+1] annulation where the C–H activation substrate serves as one-carbon component (Scheme 40). Through variation of the cyclopentadienyl ligand on rhodium, they achieved the diastereoselective synthesis of *trans*-disubstituted cyclopropenes.[98a] Further tuning the cyclopentadienyl ligand substitution and reaction conditions, however, resulted in a high selectivity for *trans*-carboamination products from the same starting materials, again incorporating the former oxidizing directing group completely into the product.[98b]

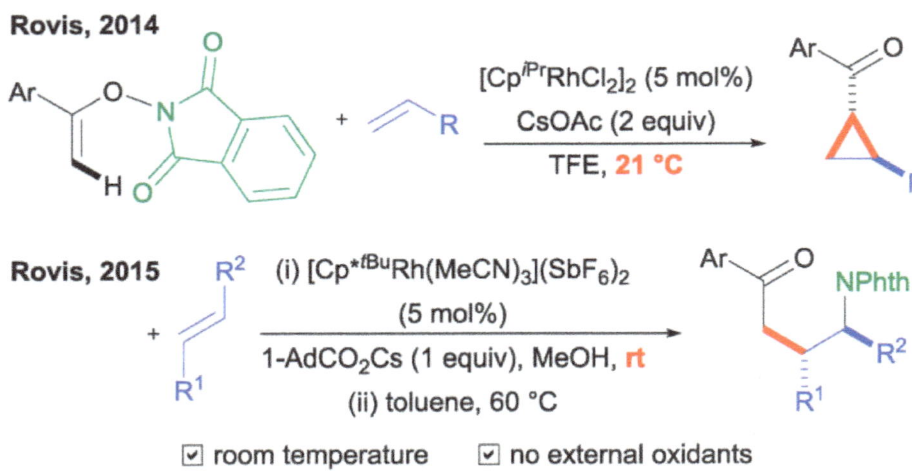

Scheme 40: Divergent reactivity of enoxyphthalimides with alkenes under Rh^{III} catalysis.[98]

Intrinsically Active Substrates

Mild reaction conditions can be a consequence of the intrinsically high reactivity of a substrate either toward C–H activation or in the subsequent functionalization of the organometallic intermediate generated upon C–H activation. In this regard, deprotonative C–H activation is facilitated in substrates containing acidic C–H bonds. On the other hand, metalation at electron-rich positions can lead to strong C–metal bonds, thus driving a C–H metalation equilibrium. Furthermore, electron-rich substrates can afford highly nucleophilic organometallic intermediates, which readily undergo functionalization with electrophilic reaction partners. In this section, selected examples of mild C–H transformations which employ active substrates are presented although it should be noted that many related reactions have been reported and the selection is not exhaustive.

Heterocycles are a highly diverse class of compounds that have received considerable interest as substrates for direct C–H functionalization due to the ubiquity of heterocycles in bioactive compounds. Among the successfully employed compounds, there are both examples of electron-poor heterocycles containing acidic C–H bonds such as (benz)oxazoles ($pK_a < 30$ at C2), and examples of electron-rich, nucleophilic compounds such as indoles. A multitude of methods involving a variety of different C–H activation pathways including deprotonation by an external base, radical mechanisms and electrophilic addition have been developed for a variety of metal catalysts employing intrinsically activated compounds.[99] Determining the operative mechanism of any given transformation can sometimes be difficult, and it may depend on the precise reaction conditions.

The palladium-catalyzed arylation of indoles with diaryliodonium salts can proceed under very mild conditions. In contrast to electron rich phosphine–Pd^0 catalysts that had previously been employed in Pd$^{0/II}$ cycles for the arylation of indoles, the use of more electrophilic Pd^{II} catalysts resulted in increased reactivity and very mild conditions. The proposed mechanism involves an initial electrophilic palladation at the most nucleophilic C3-position of indole, followed by a 1,2-migration of the palladium to the 2-position, which leads to a more stable C–Pd bond. After the original report of this method by the group of Sanford, more recent work by the groups of Ackermann and Fairlamb was focused on using highly-functionalized tryptophan derivatives and oligopeptides for the late-stage functionalization of biologically relevant compounds (Scheme 41).[100] They found that solvents such as water and ethyl acetate can replace the acetic acid reaction medium that was originally employed, thus achieving very mild reaction conditions. Overall, this process is additive-free, uses a benign solvent and proceeds smoothly at room temperature.

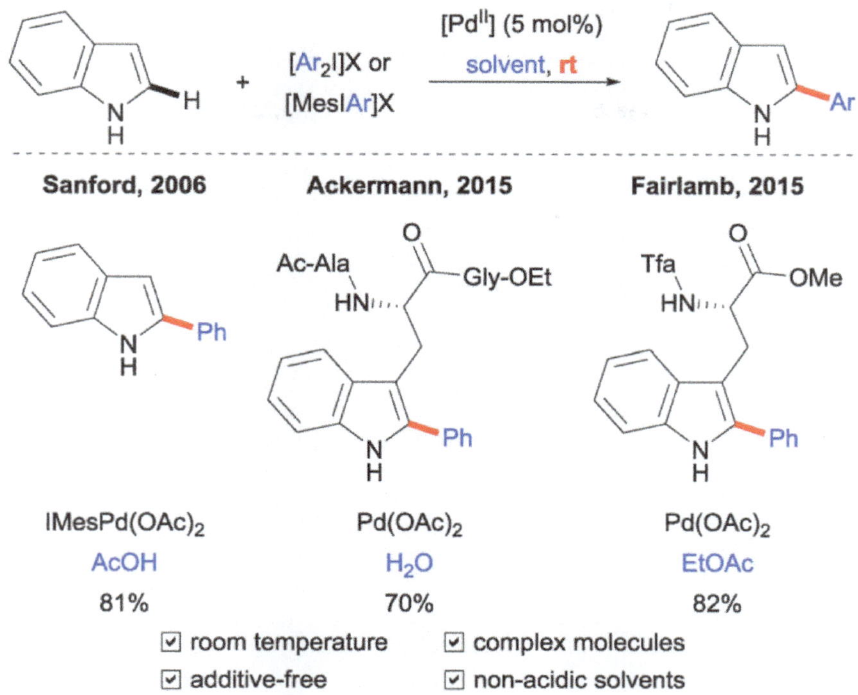

Scheme 41: Mild Pd-catalyzed indole C–H arylation.[100]

The group of Bergman and Ellman disclosed a Rh^{III}-catalyzed, acetamide-directed C–H activation process followed by isocyanate insertion (Scheme 42).[101] The reaction proceeds under additive-free conditions at room temperature using enamide substrates. By contrast, conducting the same reaction with aromatic acetanilide substrates required a higher reaction temperature of 75 °C, demonstrating the increased reactivity of the enamides. Under the reaction conditions, deuterium scrambling was observed at the arene C–H bond of acetanilide, but not for the enamide C–H bond of cyclohexenyl acetamide. Thus, while arene C–H activation is reversible under these conditions, the enamide C–H activation appears to be irreversible. This is most likely due to acceleration of the subsequent nucleophilic addition step to the isonitrile as a result of the increased nucleophilicity of the enamide relative to the arene.

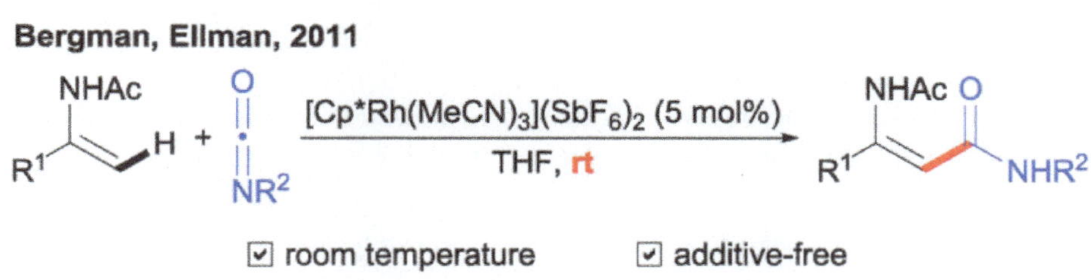

Scheme 42: Mild Rh^{III}-catalyzed enamide C–H activation.[101]

Allylic C–H functionalization has been developed into a mature and versatile field in its own right, often using a catalyst system consisting of Pd^{II} precursors, (bis-)sulfoxide ligands and benzoquinone as a redox-promoter and/or oxidant.[102] Generally, these reactions can be performed at 45 °C or less and proceed with high selectivity for the branched over the linear functionalization products. Remarkably, applications in the late stage syntheses of complex molecules have proven the applicability of this mild methodology.[103] Although a highly optimized catalyst system is employed with finely-tuned ligand properties, the overall "mildness-determining feature" of this class of transformations is the increased reactivity of allylic C–H bonds. The allyl anion is stabilized by conjugation, thus facilitating deprotonation of the substrate. Additionally, the stabilizing η^3-binding mode of Pd in the resulting organometallic intermediate is a further driving force for the C–H cleavage. In a computational study, the C–H activation step itself was found to proceed *via* intramolecular CMD with acetate, in analogy to Pd-catalyzed C–H activations of other substrate classes.[104] We wish to demonstrate the mildness and versatility of the allylic C–H functionalization with two recent examples. Firstly, the allylic alkylation of various alkenes with α-nitroketones was achieved at 45 °C under neutral conditions by taking advantage of a finely-tuned ligand system.[105] A bis(sulfoxide) ligand promotes the allylic C–H abstraction step, DMSO promotes the consequent functionalization and 2,6-dimethylbenzoquinone (DMBQ) promotes the re-oxidation of the palladium (Scheme 43). White›s group dubbed this mechanism involving the proposed switching between specialized weakly coordinating ligands at each intermediate "Serial Ligand Catalysis". Also, an intramolecular cyclization of *N*-Boc protected homoallylic amines was achieved by the addition of a phosphoric acid that assists in the allylic C–H cleavage and activates the resulting allyl–Pd intermediate for attack by the weakly nucleophilic carbamate oxygen. In this process, oxazolidinones were obtained with novel regioselectivities.[106]

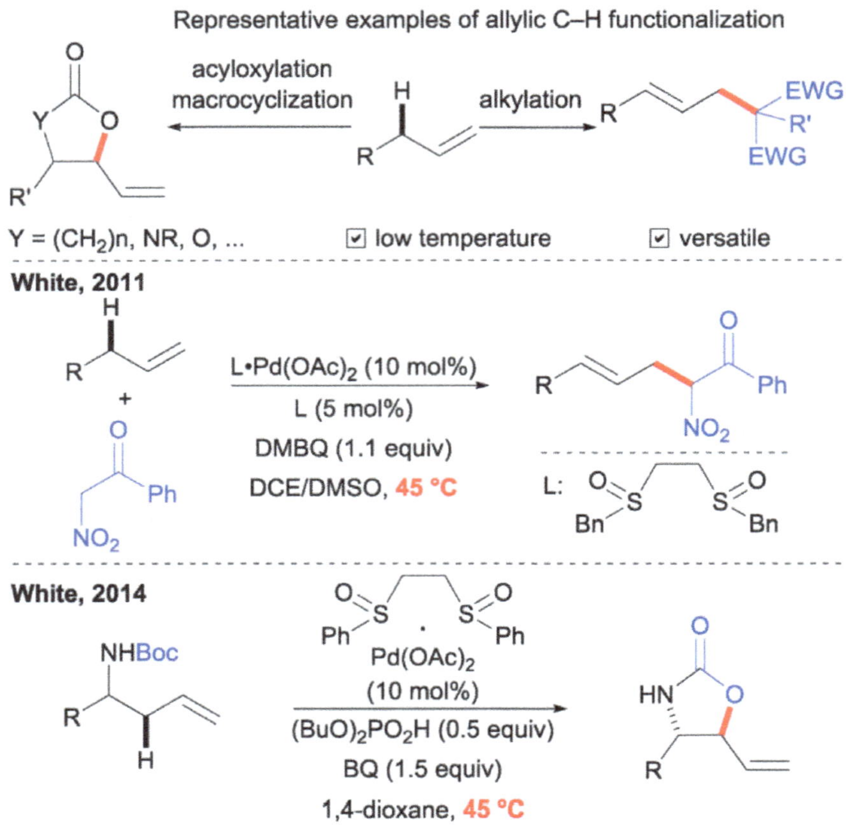

Scheme 43: Selective allylic C–H functionalization.[105,106]

The group of Mascareñas and Gulías observed an intriguing result during their studies into Rh^{III}-catalyzed annulation reactions of *ortho*-vinyl phenols with alkynes. When they used geminally disubstituted olefins as substrates, a dearomatizing (3+2) annulation occurred, furnishing spirocyclic products (Scheme 44).[107] In deuteration studies in the absence of the alkyne reaction partner, both terminal hydrogens of the olefin underwent H/D-scrambling. However, in the presence of the reaction partner, no deuteration was observed, indicating that C–H cleavage is irreversible under the reaction conditions. The authors proposed a mechanism involving formation of an *ortho*-quinone methide prior to C–H abstraction. Thus, in this special substrate class, the Rh-catalyst can activate C–H bonds by generating a cationic intermediate with highly acidic allylic protons, rather than through an agostic interaction as in CMD-type mechanisms. Due to steric repulsion, the 8-membered metalacycle obtained after alkyne insertion was proposed to isomerize to a more stable spirometalacycle.

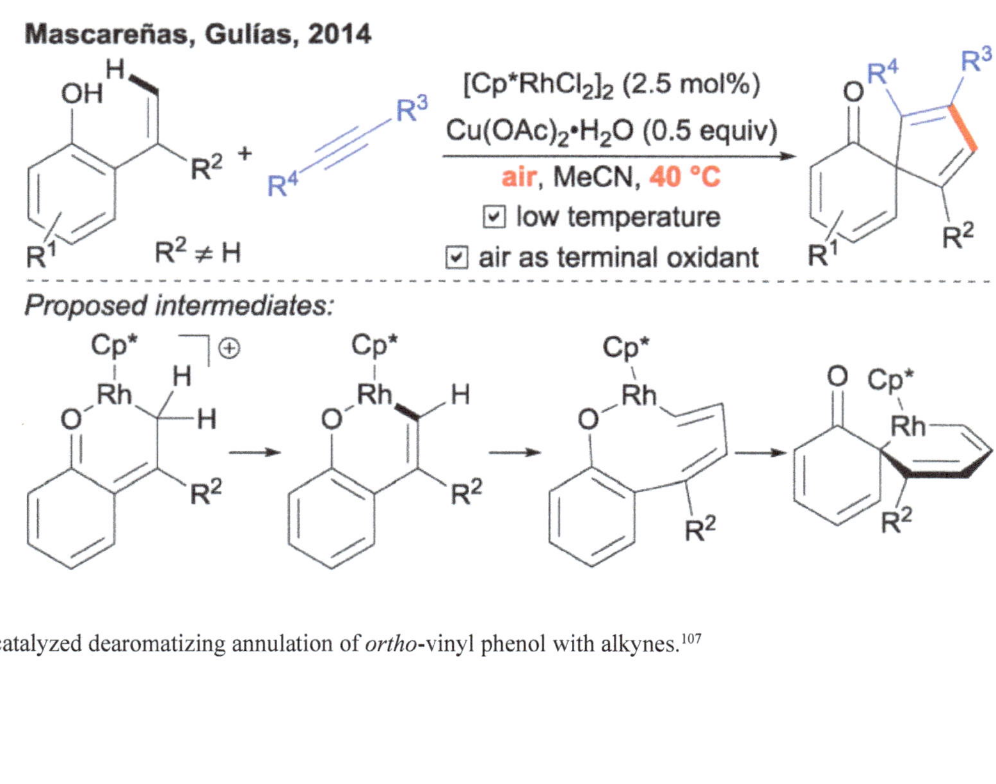

Scheme 44: Rh-catalyzed dearomatizing annulation of *ortho*-vinyl phenol with alkynes.[107]

Substrate Activation

A very appealing yet rarely applied strategy for achieving mildness in C–H activation is the reversible, temporary activation of one of the reaction partners toward the desired reaction. Despite the potential power of this concept, very few reports successfully employing an *in situ* substrate activation strategy have appeared to date. In 2013, two groups independently reported the C–H arylation of η^6-arene $Cr^0(CO)_3$ complexes with palladium. For several years, the coordination of arenes to Cr^0 fragments has been known to activate the arene toward nucleophilic aromatic substitution and to increase the acidity of the aryl C–H bonds.[108] Moreover, the stoichiometric C–H cyclometalation of arene Cr^0 complexes with a directing group such as pyridine on the arene had been demonstrated with a number of metals previously, although the conditions employed in these studies were not intrinsically milder than those using non-coordinated arenes. Fuchter and co-workers reasoned that the increased C–H bond acidity should correlate with an increased reactivity toward C–H activation with Pd^{II} and demonstrated a stoichiometric two-step protocol for a room temperature cyclometalation and subsequent arylation with aryl boronic acids (Scheme 45).[109] The group of Larrosa developed a related method for the catalytic direct arylation of arene Cr^0 complexes using a Pd^0 pre-catalyst and aryl iodides as coupling partners (Scheme 46).[110] They did not employ a chelating directing group on the coordinated arene substrates, and instead took advantage of the *ortho* directing effect of a fluorine substituent on palladation. In competition experiments with fluorobenzenes, it was shown that the chromium complexes display increased reactivity toward C–H activation by at least two orders of magnitude *versus* the non-complexed arenes. A distortion/interaction analysis based on DFT calculations suggests that the increased reactivity is due to more facile bending of the C–H bond to be activated in the chromium complexes compared to the free arenes. Contrary to the initial hypothesis, the electronic changes to the arene ring upon complexation to chromium are actually detrimental to the C–H activation step.

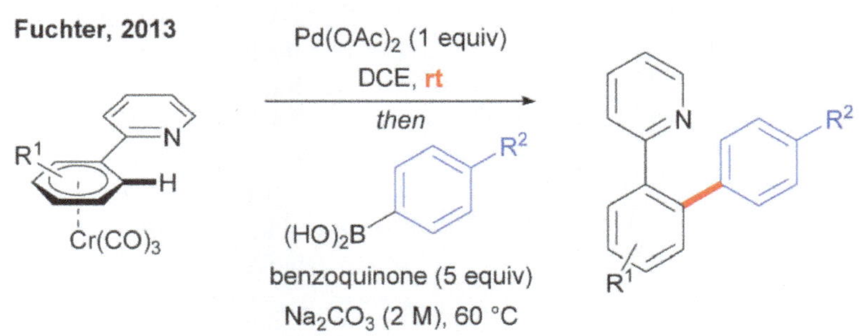

Scheme 45: Direct arylation of η^6-arene $Cr^0(CO)_3$ complexes *via* stoichiometric palladation.[109]

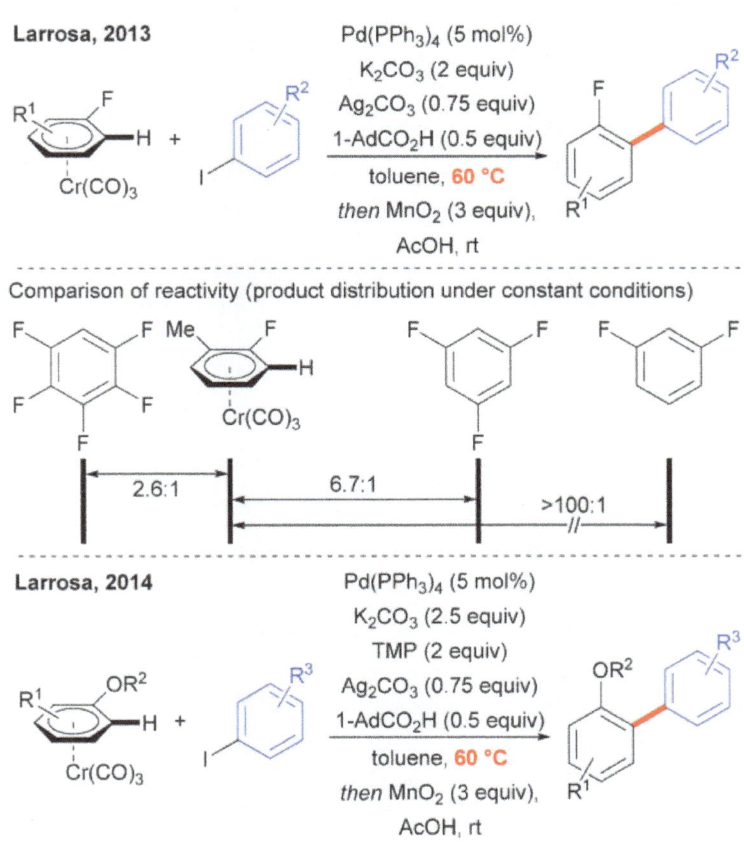

Scheme 46: Substrate activation by Cr^0 complexation in catalytic direct arylation.[110,111] TMP = 2,2,6,6-tetramethylpiperidine.

In a subsequent study, the same group extended the Cr^0-complexation strategy to the selective *ortho*-arylation of Cr^0-coordinated alkoxyarenes.[111] Under these conditions, the direct arylation of free anisole is both four orders of magnitude less efficient and less regioselective than the analogous process with the Cr^0 complexed substrate.[112] While the complexation enables the C–H activation of arenes to be carried out at lower temperatures, the conditions of the complexation and the oxidative cleavage of the chromium fragment can be a limitation to functional group tolerance.

REACTION PARTNER

Very mild transformations can result from the combination of an efficient catalyst system and very active reaction partners. With many of the established catalyst systems, C–H activation itself can occur readily at room temperature under certain circumstances as indicated by numerous deuteration studies in the literature. Despite this, most C–H activation methods possess significant kinetic isotope effects, indicating participation of C–H bond breaking in the rate determining step. Reasons for this apparent contradiction include the reversibility of deprotonation-type C–H activation favoring the starting material side of the equilibrium, or that interactions between a reaction partner and the catalyst can inhibit C–H activation, for example by outcompeting substrate binding. These factors are controlled by the specific catalyst structure, additives and reaction partners. When C–H activation is facile, other aspects of a transformation might necessitate higher reaction temperatures or otherwise less mild reaction conditions. Thus, if the reaction partners are highly active in combination with the catalyst system, mild transformations can be achieved. This can either be due to a ready functionalization of the organometallic intermediate or through modification of the catalyst itself by the reaction partner. These latter es make use of the catalyst activation strategy described earlier with the reaction partner playing a dual role as both reactant and activator.

This chapter on reactive reaction partners encompasses many transformations that do not easily fall into any of the other categories and for which the mildness of the transformation cannot be otherwise explained. In many methods described in other parts of this review, highly active reaction partners are also employed, but, in our opinion, another aspect of the reaction represents the concept of achieving mildness more accurately. The included examples are meant to represent the concepts and we do not seek a comprehensive collection.

Organometallic Reagents

Organo-main group metal compounds are highly reactive reagents in general. In the context of transition metal-catalyzed C–H functionalization, these compounds find use as catalyst modifiers (*vide supra*) and, especially, as reaction partners for direct coupling processes between the C–H and the C–metal constituents. Several catalyst systems, mostly based on first-row transition metals, have been reported that enable such couplings at or below room temperature. On the other hand, the use of (super)stoichiometric amounts of reactive organometallic reagents can be a limitation in terms of functional group tolerance and overall practical applicability.

The group of Nakamura presented a series of methods for direct C–H arylation with aryl Grignard or arylzinc reagents under iron catalysis, which proceed at 0 °C in very short reaction times (Scheme 47A).[113] The catalyst system consists of a Fe^{III} pre-catalyst such as $Fe(acac)_3$, the bipyridine ligand dtbpy and dichloroisobutane as an oxidant. Depending on the precise conditions, the Grignard reagents either act themselves as coupling partners or can be first transformed into arylzinc reagents *in situ*. Moreover, the organomagnesium derivatives themselves can be generated *in situ* from the corresponding aryl bromides and elemental magnesium. The latter conditions avoid the presence of stoichiometric amounts of the organometallic reagent, improving the mildness and substrate scope of the transformation.[113f] N-heterocycles, imines and amides have been used as directing groups to activate aryl and vinyl C–H bonds. Mechanistic investigations suggested that the C–H activation event occurs with aryliron species formed upon transmetalation from the aryl Grignard reagent.[113e] The C–C bond-forming reductive elimination from the metalacyclic diaryliron species was found to require the presence of the external oxidant in a stoichiometric experiment. A recent detailed computational study on the arylation with zinc compounds uncovered a unique two-state reactivity of the iron catalyst (Scheme 48).[114]

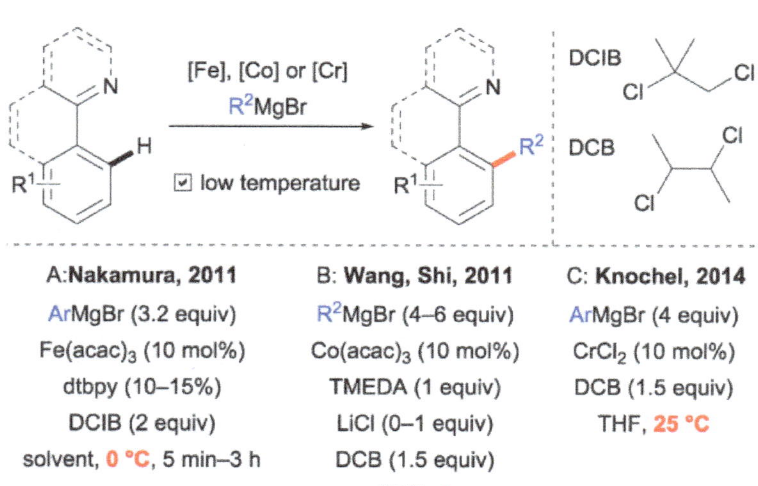

Scheme 47: Iron-, cobalt- and chromium-catalyzed directed C–H arylation/alkylation with Grignard reagents.[113,115,117]

In one likely scenario, the high-spin Fe^{II} reactant complex undergoes C–H cleavage after a spin transition to the low-spin singlet state, returning to the high-spin state after cyclometalation. The actual metalation proceeds *via* σ-bond metathesis, where the proton is transferred to a phenyl ligand on iron simultaneously to the formation of the Fe–C bond. Subsequent to a transmetalation, a one-electron oxidation to Fe^{III} triggers the C–C reductive elimination.

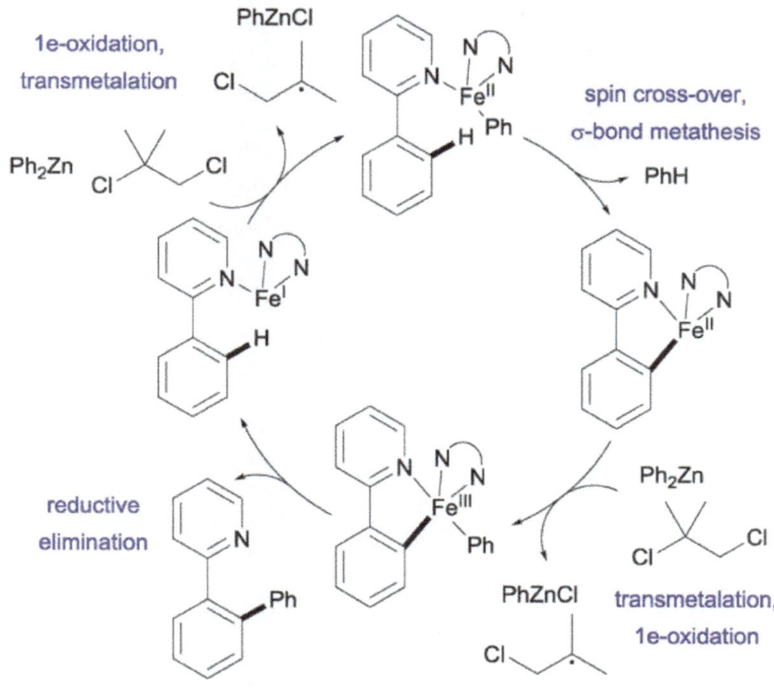

Scheme 48: Proposed mechanism of the Fe-catalyzed C–H arylation.[114]

Direct C–H arylation and even alkylation using Grignard reagents at low temperatures has also been achieved with related catalyst systems based on cobalt and chromium. Wang and Shi presented a method using a catalyst system consisting of $Co(acac)_3$ as pre-catalyst, TMEDA as ligand and 2,3-dichlorobutane (DCB) as oxidant for the arylation and alkylation of benzoquinolines and phenylpyridines at room temperature (Scheme 47B).[115] From mechanistic studies, the involvement of radical pathways was ruled out and a C–H activation pathway involving oxidative addition onto Co^{I}–aryl species was proposed, followed by a sequence of reductive elimination of a hydrocarbon, oxidation, transmetalation and reductive elimination of the product (Scheme 49). In light of the similarity of this reaction to the iron-catalyzed variant, the analogous σ-bond metathesis mechanism is plausible, too. It is interesting to note the differences between iron and cobalt in these related transformations, though. Kinetic isotope effects of 3.1 and 3.4 from intra- and intermolecular competition experiments measured for the iron-catalyzed process suggest a participation of C–H cleavage in the rate determining step.[113g] On the other hand, a KIE of 1.04 was observed in a parallel experiment using cobalt catalysis, suggesting C–H cleavage is not rate limiting.[115]

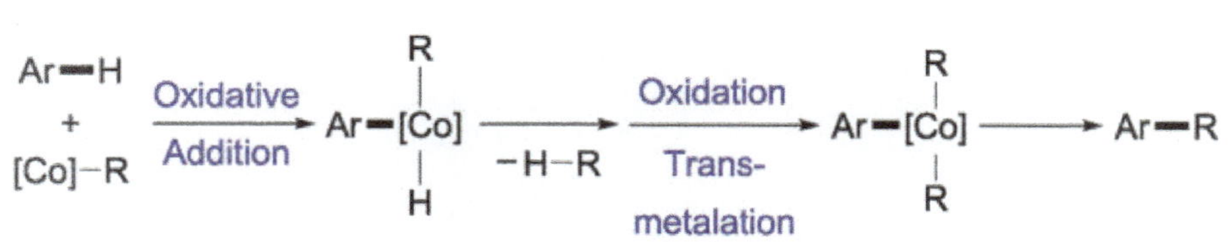

Scheme 49: Proposed mechanism of the cobalt-catalyzed direct coupling.[115]

Subsequently, the group of Nakamura published an alternative procedure for the alkylation of benzamides with alkyl Grignard reagents (Scheme 50).[116] The catalyst system employed in this room temperature transformation consisted of $Co(acac)_2$ and DMPU, while air was used as the sole oxidant. Furthermore, in a report by Knochel and co-workers, $CrCl_2$ was found to catalyze the direct arylation of arenes with N-heterocycle- or imine-based directing groups at room temperature without additional ligands using DCB as oxidant (Scheme 47C).[117]

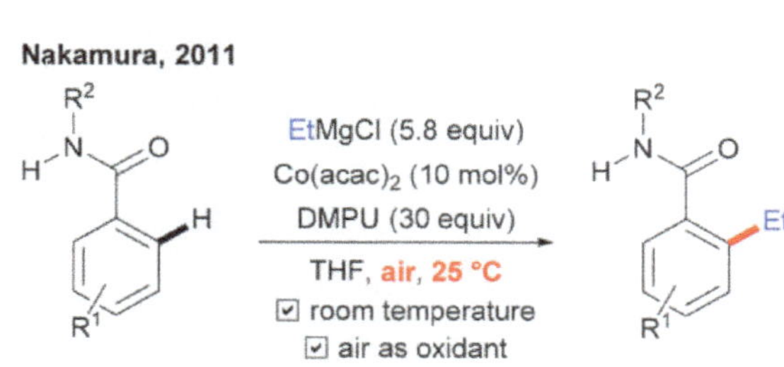

Scheme 50: Cobalt-catalyzed alkylation.[116]

Hypervalent Iodine Reagents

The use of high-valent iodine compounds is firmly established in organic synthesis due to their high and versatile reactivity. In C–H activation, I^{III} compounds have also proven to be highly active coupling partners enabling mild transformations. The high-valent iodine acts as a strongly electron withdrawing substituent, resulting in electrophilic organic residues that can react readily with organometallic nucleophiles generated upon C–H activation. The Rh^{III}-catalyzed coupling of arenes with high-valent iodine fragments can occur readily at room temperature in very short reaction times, generating diaryliodonium species from $PhI(OAc)_2$ (PIDA) or PhI(OH)(OTs) (Koser's reagent) *via* C–H activation (Scheme 51).[118] This transformation represents an umpolung of the substrate, as the nucleophilic organometallic intermediate is turned into an electrophile in the diaryliodonium product. A number of *in situ* transformations of the generated diaryliodonium species with nucleophiles such as azides, nitrites, halides and sulfonates have been developed (Scheme 52).[119] At first, the intermediacy of diaryliodonium species was not considered in these transformations and mechanisms involving Rh^{V} intermediates were proposed. However, the isolation of diaryliodonium products under similar reaction conditions[119b] and the propensity of these isolated products to react with the same nucleophiles without any catalyst highly suggests this order of events.

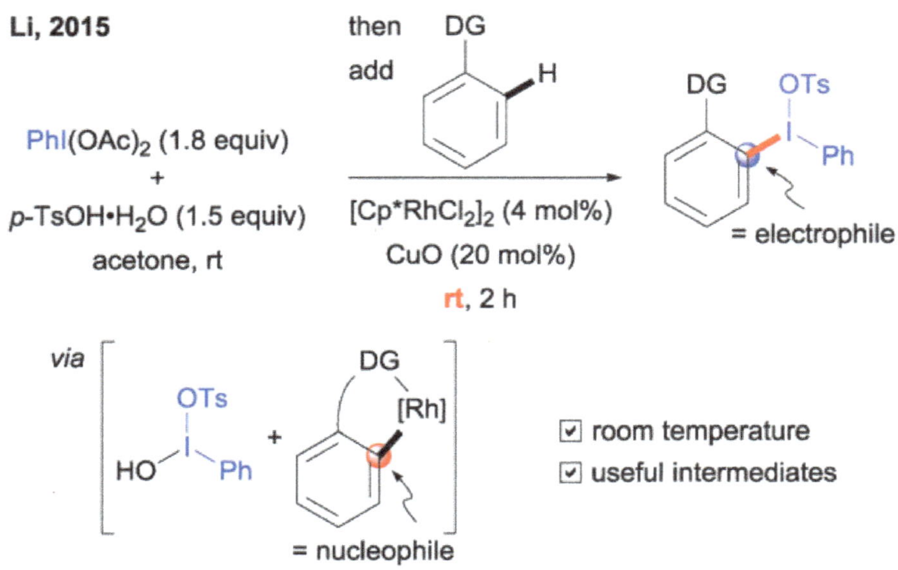

Scheme 51: Mild rhodium-catalyzed hyperiodination.[118]

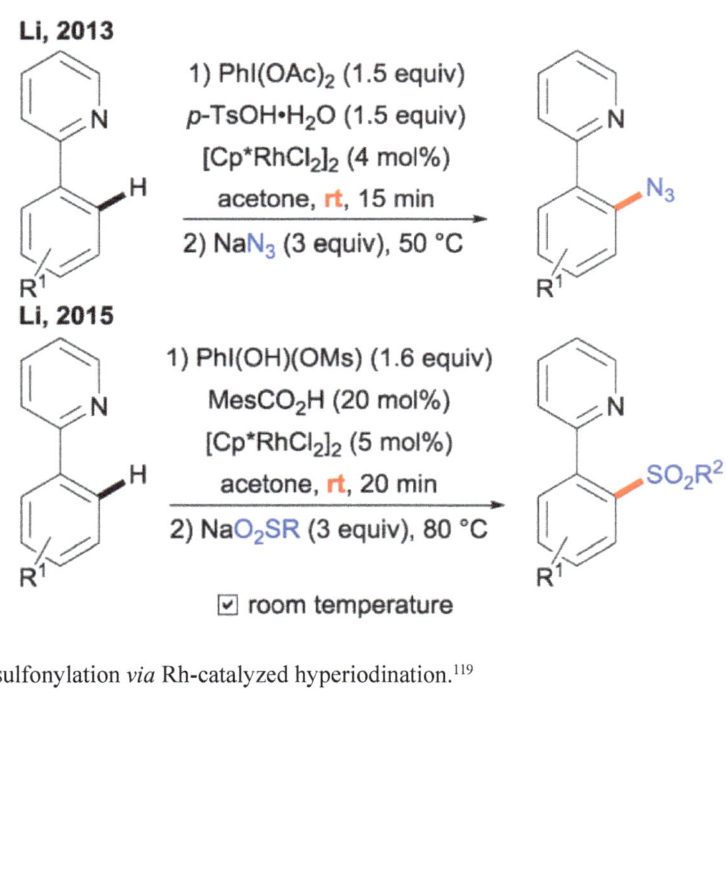

Scheme 52: C–H azidation and sulfonylation *via* Rh-catalyzed hyperiodination.[119]

In contrast to the C–H hyperiodination process discussed above, the majority of C–H functionalization reactions involving I^{III} reagents proceed through interaction of the metalated substrate with one of the residues on I^{III} rather than with the iodine itself. The C–H alkynylation reaction with alkynyl-substituted iodine(III) reagents (EBX derivatives) is a very general transformation proceeding under exceptionally mild conditions using a range of different catalytic systems. The Cp^*Rh^{III} and Cp^*Ir^{III}-catalyzed directed alkynylation of (hetero)arenes and olefins with TIPS-EBX was independently reported by the groups of Loh, Li and Glorius in 2014 (Scheme 53).[120] This method is compatible with a remarkably wide range of directing groups, demonstrates a broad substrate scope and can be carried out at temperatures between room temperature and 80 °C depending on the substrates.[23d,e] Several mechanisms have been proposed in the literature. Despite the considerable interest in this mild and versatile transformation, evidence regarding the mechanism is still rather scarce. One proposed scenario involves oxidative addition of the hypervalent iodine reagent into the metalacycle formed upon C–H activation followed by a product-releasing reductive elimination from a Rh^V intermediate (Scheme 54(1)). Another proposal invokes an alkyne insertion into the Rh–C bond of the metalacycle (Scheme 54(2)). α-Elimination of *o*-iodobenzoate could result in the formation of a Rh–vinylidene intermediate which should rearrange by 1,2-migration or *via* stepwise carbene insertion and elimination to afford the product.

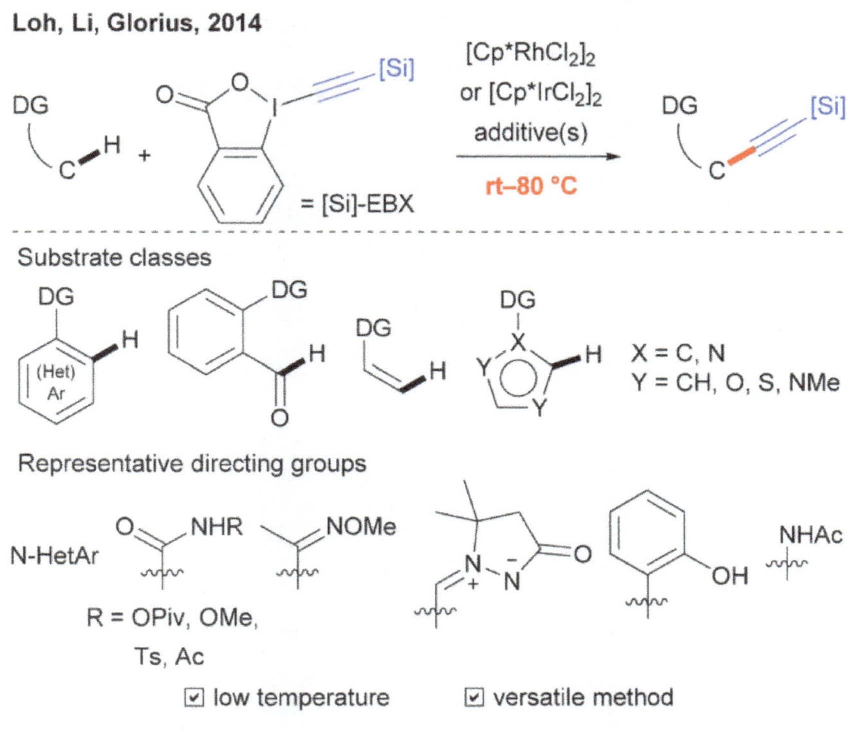

Scheme 53: Mild directed C–H alkynylation.[23d,e,120]

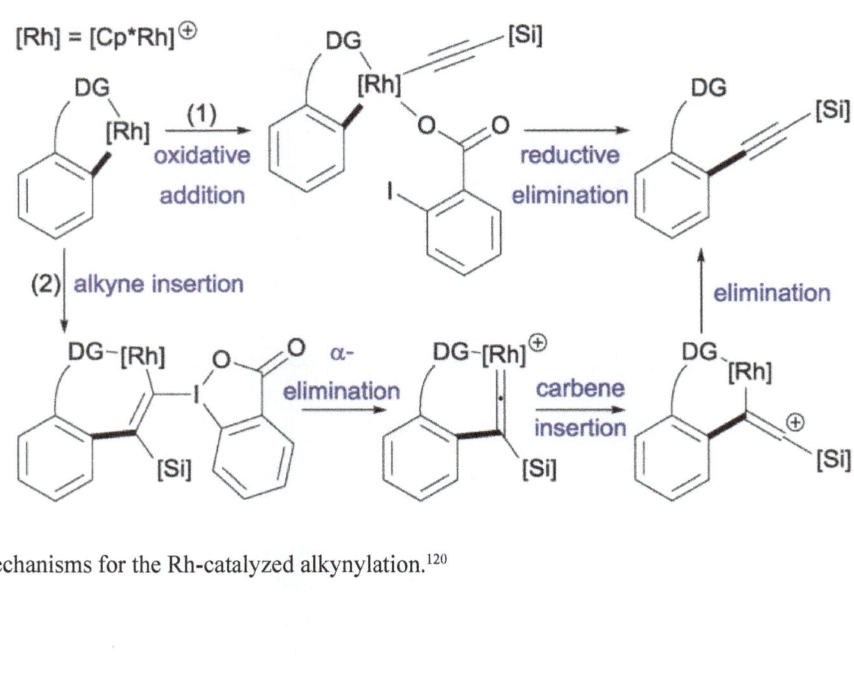

Scheme 54: Proposed mechanisms for the Rh-catalyzed alkynylation.[120]

Amination

Reliable methods for C–N coupling are of significant importance due to the synthetic utility of amine derivatives and their ubiquitous presence in bioactive compounds. While C–N bond-forming reductive elimination is generally recognized as a challenging step, mild transformations are required for the late stage introduction of nitrogen into complex molecules. In this context, many procedures have been developed for C–H amination or amidation using stoichiometric or *in situ*-generated electrophilic amine reagents or nitrene precursors as highly reactive nitrogen transfer reagents.[121] The reactive nature of these reagents can lead to mild reaction conditions with a wide range of substrates under Ir^{III}, Rh^{III}, Co^{III}, or Ru^{II} catalysis. When preformed amination reagents are employed, no additional oxidant is required, as the overall transformations are redox neutral. On the other hand, the pre-activation steps required to generate the reagents prior to C–H amination can reduce the attractiveness of the overall synthesis.

Electrophilic Amination Reagents

Mild C–N coupling reactions using electrophilic amine reagents were reported simultaneously by the groups of Yu and Glorius[122] using *N*-chloroamines under Cp^*Rh^{III} catalysis (Scheme 55) and later using *N*-aroyloxyamines under ruthenium- and iridium catalysis (Scheme 56).[65,123] The transformation proceeds at temperatures between room temperature and 40 °C and requires no external oxidant. The metalacycle formed upon C–H activation likely reacts as a nucleophile with the electrophilic amine derivatives in a formal nucleophilic substitution at nitrogen, though alternative mechanisms involving oxidative addition of the N–heteroatom bond to the catalyst cannot be completely ruled out.

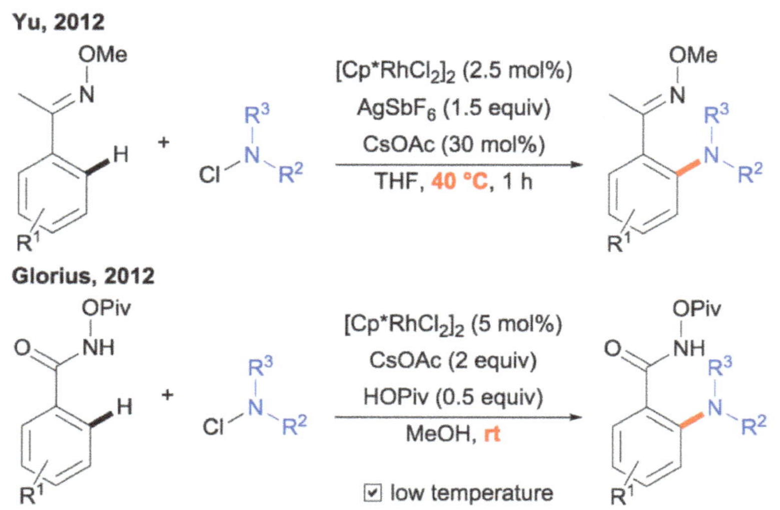

Scheme 55: Rh-catalyzed amination with *N*-chloroamines.[122]

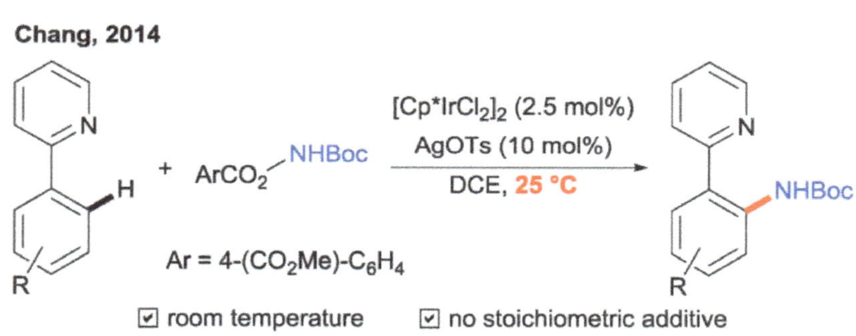

Scheme 56: Mild Ir-catalyzed amination with an electrophilic reagent.[123]

Nitrene Precursors as Nitrogen Transfer Reagents

Directed C–H amidation and amination reactions with organic azides as transfer reagents have attracted significant interest due to their broad applicability using a range of catalyst systems, azide derivatives and directing groups (Scheme 57).[121b,c] This development was sparked by the reports of additive-free Cp^*Rh^{III} catalyzed sulfonamidation and amination by the group of Chang in 2012 using tosyl azide and aryl azides, respectively.[124] Even milder conditions could be employed in a related amidation reaction using Cp^*Ir^{III} catalysts. In this system, the C–N coupling reaction with acyl azides could be carried out at 25 °C without any stoichiometric additives (Scheme 58).[125] Detailed investigations revealed a mechanism proceeding *via* C–H activation, nitrene formation at the metalacycle and subsequent nitrene insertion into the C–metal bond (Scheme 59).[126] Alternatively, a concerted amido insertion pathway might be relevant under specific conditions. The protodemetalation step, which results in the release of the product molecule, occurs upon C–H activation of another equivalent of the substrate. This process can be rate-limiting depending on the reaction partners, demonstrating the facile nature of the reaction between azides and metalacycles. The thermodynamic gain from the release of dinitrogen in the reaction probably contributes to the mildness of this transformation. Intramolecular hydrogen bonding between the newly-formed N–H moiety and the directing group is proposed to prevent product inhibition as well as difunctionalization.

A number of groups extended the amidation strategy to a broader range of transformations including a $C(sp^3)$–H amidation reaction, ruthenium- and cobalt-catalyzed protocols and a phosphoramidation process.[121] In many of these more challenging reactions, some aspects of mildness had to be compromised to attain reactivity, such as through increasing the reaction temperature or employing acidic or basic additives that influence various parts of the catalytic cycle.

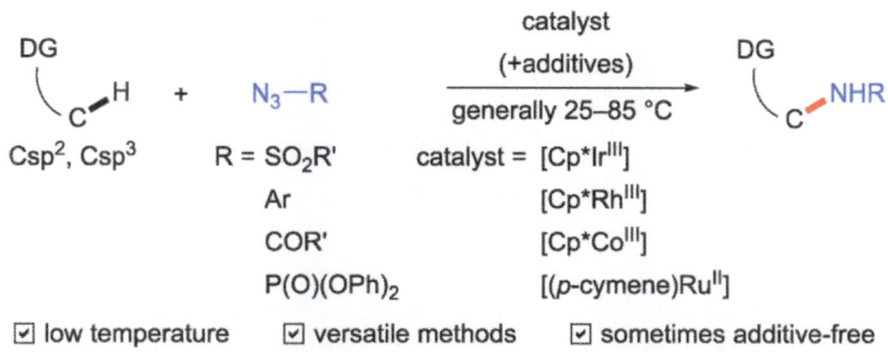

Scheme 57: Directed C–H amidation and amination with organic azides.[121]

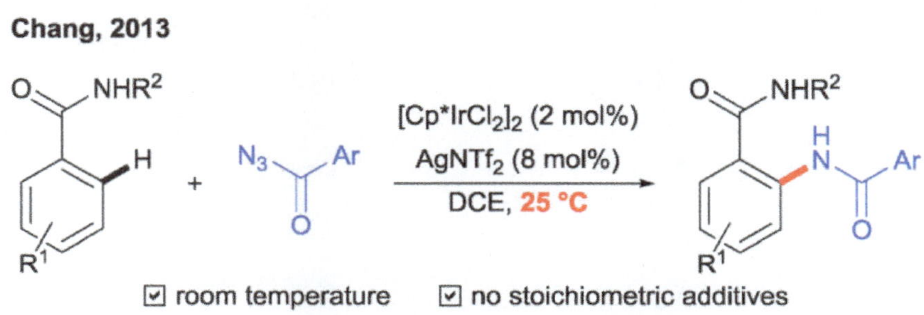

Scheme 58: Mild iridium-catalyzed C–H amidation.[125]

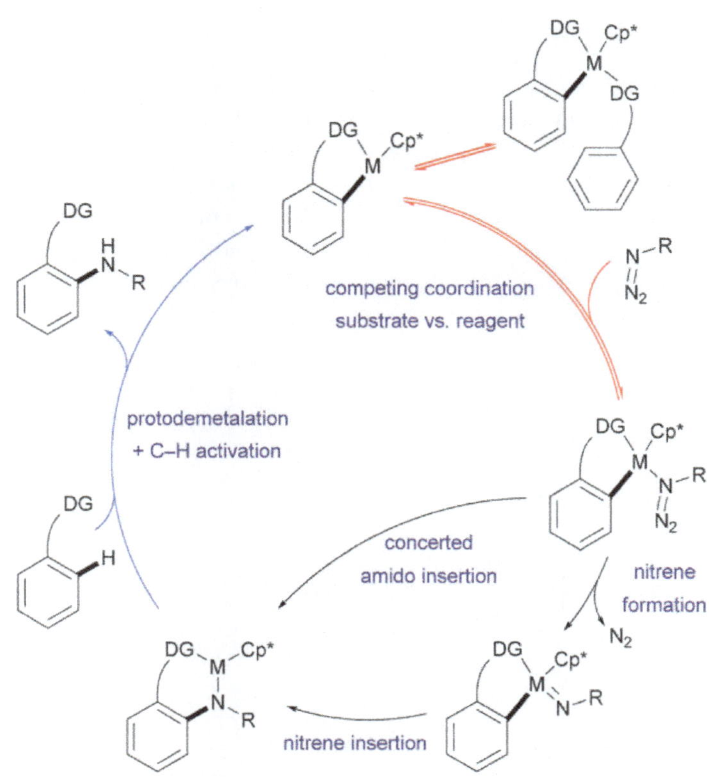

Scheme 59: Proposed mechanism for the directed amidation with organic azides.[126]

For example, the use of a combination of both basic and acidic additives was optimal for achieving mild Ir-catalyzed C–N coupling reactions using weak directing groups.[127] The initial hypothesis presumed that the binding efficiency of weakly coordinating substrates toward a metal center could be improved in the presence of both base and acid additives. Although basic species are recognized to assist CMD C–H activation, they also coordinate strongly to the metal catalyst and can consequently outcompete substrate binding when weakly coordinating directing groups are employed. Acidic additives, on the other hand, promote dissociation of such anionic ligands. After optimization, a well-orchestrated catalytic system combining Cp*$Ir^{III}$and catalytic amounts of acetic acid and lithium carbonate allowed for a room temperature ketone-directed C–H amidation reaction (Scheme 60).

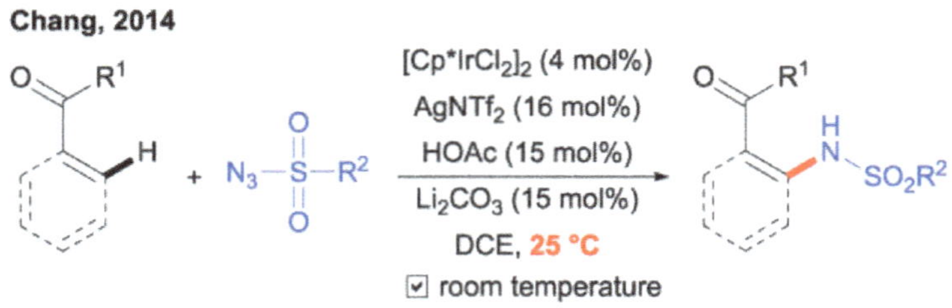

Scheme 60: Amidation of ketone derivatives.[127]

In search of an even milder and more general system for directed C–H amidation, the group of Chang studied the behavior of nitrene precursors in stoichiometric reactions with preformed rhodacycles. These studies revealed that product formation is inhibited by an excess of the substrate, phenylpyridine, due to preferential binding of the substrate to the rhodium over the benzoyl azide reaction partner (red in Scheme 59). Consequently, they found that aryl dioxazolones, which can also act as nitrene precursors upon extrusion of CO_2, can better compete for coordination to the rhodacycle. While this reagent is a more complex amide source, it can be synthesized easily, and its stability allows safe handling and indefinite storage. Using this insight, an additive-free rhodium-catalyzed amidation of aryl pyridine derivatives at 40 °C was developed with dioxazolones as new amidation reagents (Scheme 61).[128]

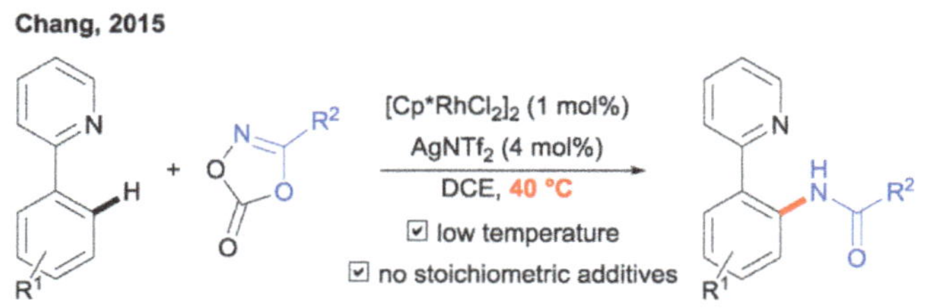

Scheme 61: Mild C–H amidation using dioxazolones.[128]

The general utility of dioxazolones as amidating agents was further demonstrated by the development of two room temperature amidation reactions: a cobalt-catalyzed process reported by Chang and co-workers (Scheme 62)[129a] and a rhodium-catalyzed directed C(sp^3)–H functionalization (Scheme 62).[129b] Both of these methods demonstrate the complementarity of the different catalyst metals for different aspects of C–H amidation. Under the conditions for the Co^{III}-catalyzed amidation of anilides, no product was observed with the corresponding Rh^{III}catalyst and lower yields were obtained with Ir^{III}. On the other hand, the rhodium-catalyzed C(sp^3)–H amidation proceeds with the same substrates at lower temperatures than the previously-described amidation with tosyl azides under iridium-catalysis.

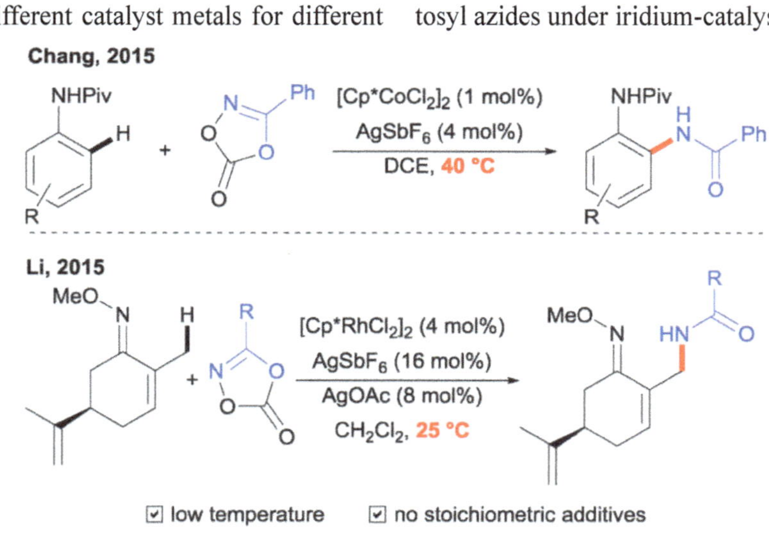

Scheme 62: Dioxazolones in mild cobalt-catalyzed amidation and C(sp^3)–H amidation.[129]

Amines and Oxidants

As an alternative to using pre-activated nitrogen coupling partners in C–H amination reactions, the combination of primary amines or amides with oxidants can be employed to generate the active N-transfer reagent *in situ*.

The group of Chang found that Cp*Ir^{III} catalyzes the directed amination of benzamides with anilines at room temperature (Scheme 63), and with primary alkyl amines at a slightly elevated temperature, in the presence of a silver salt oxidant.[130] In contrast to previously-reported palladium- and rhodium-catalyzed amidation methods using external oxidants, they observed no product formation upon using persulfate, [F$^+$], iodonium salts or any oxidants other than Ag^I either in a stoichiometric reaction with the cyclometalated amido complex or in catalytic reactions. Based on the reactivity of isolated intermediates of the amination reaction, the authors proposed that a nitrene species is generated at the iridacycle *via* stepwise oxidation (Scheme 63). Insertion of the nitrene into the Ir–C bond and protodemetalation then afford the product.

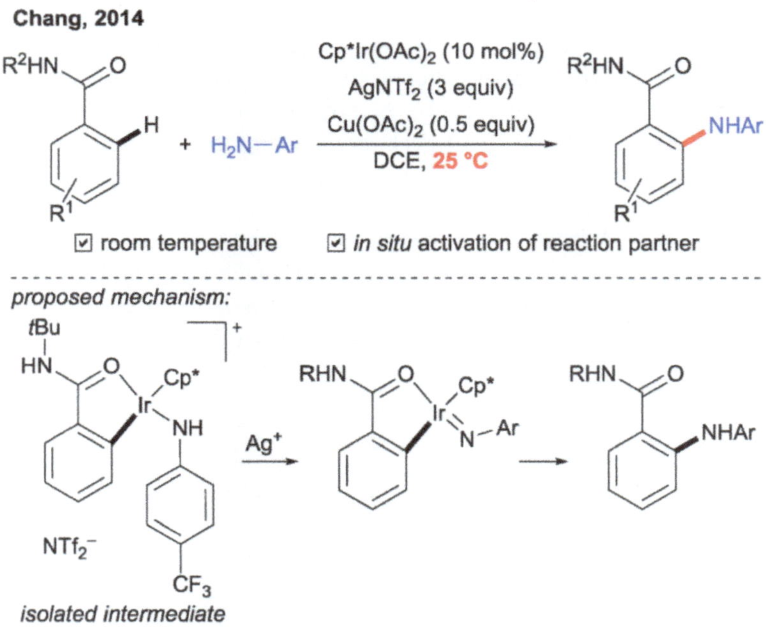

Scheme 63: Iridium-catalyzed C–H amination with anilines.[130]

Unsaturated Reaction Partners

The sequence of C–H activation *via* a deprotonation-type mechanism such as CMD, followed by insertion of an unsaturated moiety into the C–metal bond and protodemetalation can in principle proceed under additive-free conditions without the generation of side products, if the protons generated from C–H activation engage in the protodemetalation step. Accordingly, a number of reactions using unsaturated coupling partners have been presented that follow this reasoning. In some es, especially reactive coupling partners enable mild reactions that may be additive-free or run at room temperature.

The reaction of polar unsaturated moieties with organometallic species can be represented as a formal insertion into the C–metal bond or as a nucleophilic addition of the organometallic species onto the electrophilic part of the unsaturated system. The use of organo-transition metal nucleophiles generated upon C–H activation can result in more chemoselective and milder transformations than the use of more classical organo-main group metal nucleophiles. This has been demonstrated by the groups of Bergman and Ellman, and Shi in Cp*Rh^{III}-catalyzed additions of C–H metalated substrates to imines, aldehydes and isocyanates (*vide supra*) under additive-free conditions. Generally, these reactions proceed at slightly elevated temperatures. In addition, the hydroarylation of carbonyl derivatives achieved using these methods can be followed by various cyclization events to afford valuable heterocyclic products. This type of reaction has been already presented in recent reviews and is not discussed in detail here.[23a,131]

Cyclopropenones are highly strained molecules which exhibit an increased reactivity toward rhodacycles compared to ketones. This was demonstrated by the successful development of a low temperature method combining nucleophilic addition to a cyclopropenone, ring-opening and subsequent protodemetalation to generate chalcone derivatives (Scheme 64).[132] Notably, the strained cyclopropenone structure is responsible for the increased reactivity, as regular ketones are not known electrophiles in reactions with rhodacycles. In fact, the potential tertiary alcohol products of such ketone hydroarylation reactions undergo the reverse process affording the ketones and arenes *via* C–C bond cleavage.

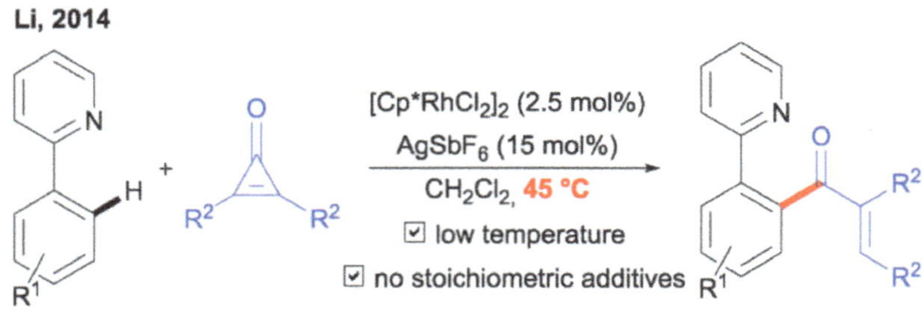

Scheme 64: Chalcone synthesis using cyclopropenones.[132]

Nitroso compounds have proven to be very reactive reaction partners for C–H nucleophilic addition reactions under mild conditions. A room temperature, Rh^{III}-catalyzed hydroarylation of nitrosobenzene was presented with phenyl pyridine derivatives by Zhou and Li in 2013 (Scheme 65).[133] A range of *N*,*N*-diarylhydroxylamines were obtained in this transformation, although electron-withdrawing substituents on the nitrosobenzene reaction partner were required, presumably due to their effect in increasing the electrophilicity of the nitrogen atom. Furthermore, copper-catalyzed oxidation of *N*-hydroxycarbamates can generate similarly reactive nitrosocarbonyl compounds *in situ*. This process was utilized in a dual Rh/Cu catalytic system for the C–H amination of *N*-methoxybenzamides leading to benzo[*c*]isoxazolones at room temperature.[134] Nitrosoformate derivatives, generated *in situ via* the copper-catalyzed cycle, participate as the electrophile in the hydroarylation process with rhodacycles formed upon *ortho*-C–H bond activation of the *N*-methoxybenzamide. This is followed by a nucleophilic substitution at the directing group, resulting in the formation of an N–O containing heterocycle (Scheme 66).

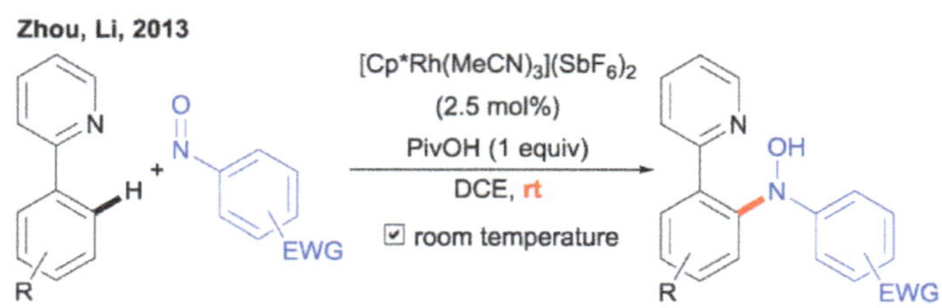

Scheme 65: Mild hydroarylation of nitrosobenzene.[133]

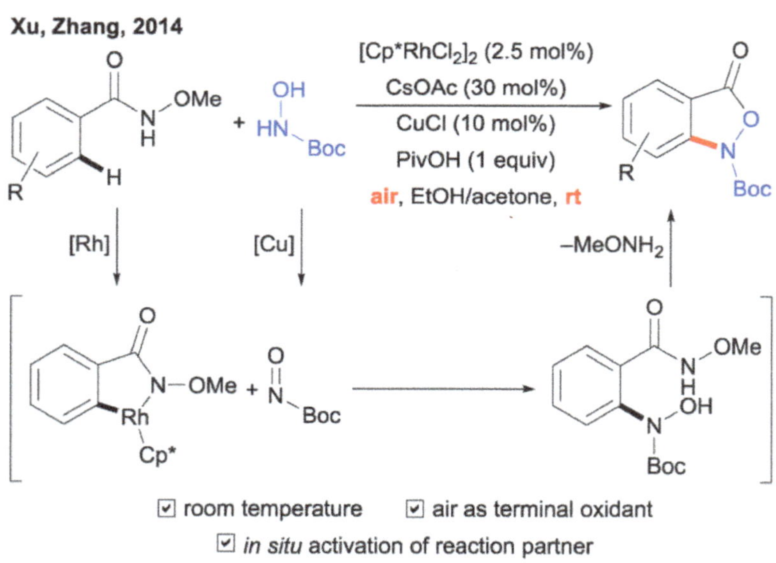

Scheme 66: Dual copper- and rhodium-catalyzed condensation.[134]

Olefins are classical coupling partners in C–H activation. Nevertheless, some carbon-based unsaturated moieties possess a higher reactivity toward organometallic intermediates, resulting in overall milder transformations. For example, allenes have been shown simultaneously by the groups of Glorius and Ma to undergo carboamination (*vide supra*) and hydroarylation reactions under exceptionally mild conditions under Cp^*Rh^{III} catalysis, demonstrating an increased reactivity compared to regular olefins.[93,135] The hydroarylation with *O*-methyl benzhydroxamates occurs readily at room temperature and even at −20 °C in what might be a low temperature record in C–H activation. The reaction proceeds with di- and trisubstituted allenes in a highly selective manner, resulting exclusively in the insertion of the lesser substituted double bond and formation of linear allylation products (Scheme 67). Also, axial chirality of the allene can be transferred to central chirality in the allyl product without loss of ee. KIE measurements suggest that C–H bond cleavage is rate determining, signifying the high reactivity of allenes toward rhodacycles. Cramer's group took advantage of the mildness and selectivity of this transformation in an enantioselective allylation reaction of benzamides using a Rh^{III} complex bearing a chiral Cp derivative as ligand. This process could be conducted smoothly at −20 °C, affording the corresponding allylated arenes with ee›s of up to 95% (Scheme 67).[136] In addition, an intramolecular hydroarylation process was developed using olefins tethered to the C–H activation substrate. Using another Cp derivative as a chiral ligand on Rh, high enantioselectivities were achieved at room temperature, demonstrating that the intramolecular alkene insertion is a more efficient process than the corresponding intermolecular variant.[137] Ma's group devised a transformation combining allene carborhodation and subsequent cyclization with a tethered olefin to prepare polycyclic products featuring eight-membered lactams.[138]

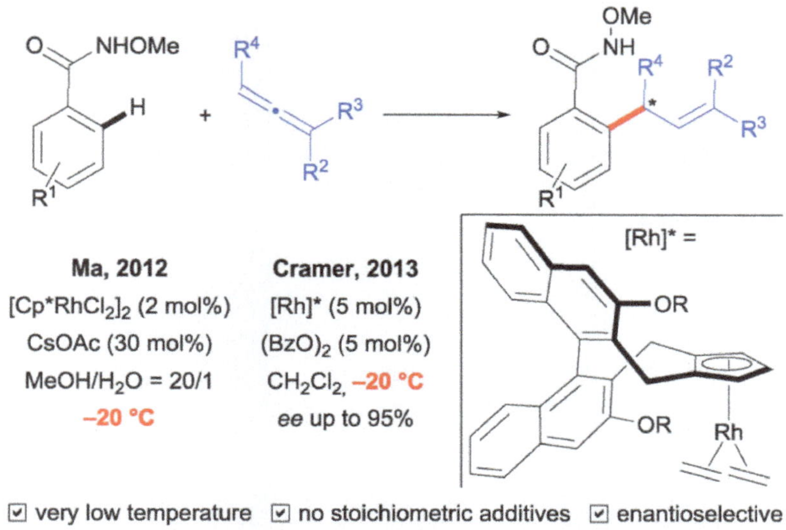

Scheme 67: Mild Rh-catalyzed allylation with allenes.[135,136]

Assisted Insertion

Direct C–H allylation using allyl transfer reagents carrying a leaving group has been shown to proceed under very mild conditions in a number of rhodium-, ruthenium-, and cobalt-catalyzed, formal S_N-type transformations. While these generally electron-rich olefins would not normally be expected to undergo facile insertion into a C–metal bond, the unifying feature of these reaction partners is the presence of additional donor moieties. Accordingly, insertion of the olefin is facilitated by an initial pre-coordination of the coupling partner to the metal species in a similar fashion to a directing group. Moreover, the resulting organometallic species generated upon insertion is stabilized by chelation (Scheme 68). The allyl moiety is subsequently formed upon β-elimination of the leaving group. The chelation furthermore largely prevents β-H-elimination by forming a rigid coordination environment where all hydrogen atoms are in an almost *trans* arrangement to the metal. If β-H-elimination occurs to some extent, the resulting metal hydride species can catalyze isomerization of the allyl products to conjugated styrene derivatives, in what is usually an undesired side-reaction.

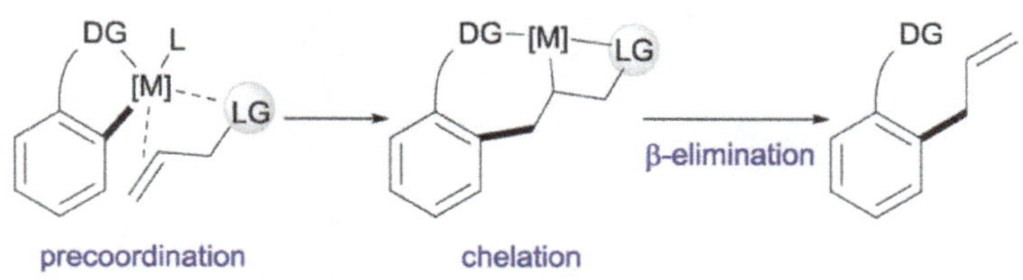

Scheme 68: Proposed general mechanistic steps of the coordination-assisted allylation.

Allyl carbonates have been used by the group of Glorius in the directed allylation reaction of various arenes with Cp^*Rh^{III} at 35 °C[139] and in an allylation of *N*-pyrimidyl (Pym) indoles using Cp^*Co^{III} at room temperature.[140] This latter process is successful at remarkably low catalyst loadings with turnover numbers of up to 2200 (Scheme 69). Similarly, allyl acetates were employed as reaction partners in a rare example of a Ru^{II}-catalyzed room temperature C–H functionalization process occurring without additives.[141] The group of Ma employed a similar approach to develop a very mild allenylation reaction of *N*-methoxybenzamides at 0 °C using propargyl carbonates as transfer reagents under Cp^*Rh^{III} catalysis (Scheme 70).[142] The authors proposed that the coordination of rhodium to the carbonate is essential in ensuring regioselective alkyne insertion.

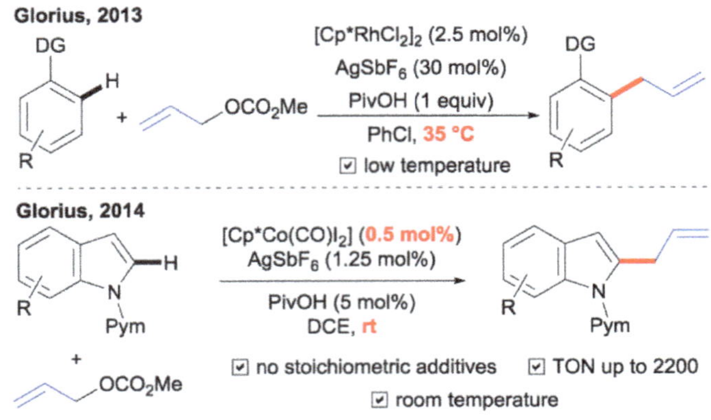

Scheme 69: Mild C–H allylation using allyl carbonates.[139,140]

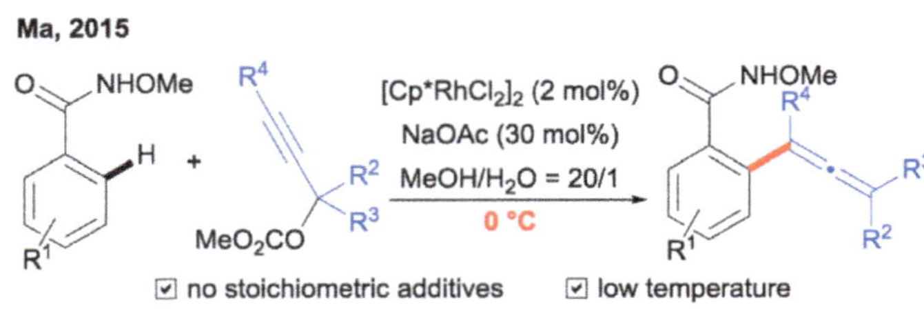

Scheme 70: Coordination-assisted allenylation.[142]

Complementary methods to introduce functionalized allyl moieties have been developed using cyclic reagents capable of undergoing ring opening either due to the presence of an intramolecular leaving group or in order to release ring strain. Thus, vinyloxirane was used as a reaction partner in the synthesis of allylic alcohols at room temperature with Cp*Rh^{III}.[143] Arenes and heteroarenes bearing a range of directing groups could be employed to generate the corresponding products as *E*/*Z* mixtures with selectivities between 1.5 : 1 and 6.1 : 1 (Scheme 71A). Similar products could also be obtained from vinyl 1,3-dioxolan-2-one using the same catalyst at 0 °C (Scheme 71B).[144] When no further substitution is present on the transfer reagent, the *E*-isomer is formed with >20 : 1 selectivity relative to the *Z* isomer. Geminally di-acceptor substituted vinyl cyclopropanes have been employed in directed allylation reactions at 30 °C using Cp*Rh^{III}.[145] In this e, product-releasing β-C elimination occurs after assisted olefin insertion into the rhodacycle.

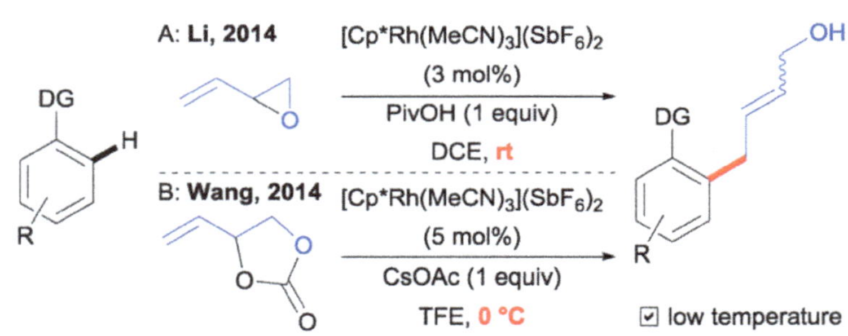

Scheme 71: Synthesis of allylic alcohols by C–H allylation.[143,144]

Radicals

The intrinsically high reactivity of radicals makes them interesting reaction partners in catalysis in general. In the context of directed C–H activation, unconventional selectivities may be observed using radicals as coupling partners. Due to the high reactivity of radicals, functionalization of the metalated species generated upon C–H activation is not kinetically relevant and other steps in the process determine the overall mildness. Indeed, besides the C–H activation step itself, radical generation and catalyst oxidation can be “mildness-limiting”. Thus, the controlled release of suitable radical species under mild conditions can enable overall mild C–H transformations.

One such approach involves the generation of aryl radicals from diaryliodonium or aryldiazonium salts. These compounds decompose thermally or upon reaction with radical initiators at elevated temperatures and C–H arylation reactions under palladium catalysis have accordingly been described at 100 °C. The group of Sanford devised a dual catalytic process where a photocatalyst is employed to both generate aryl radicals *in situ* and oxidize palladium during the subsequent arylation of an appropriate substrate. The authors proposed that the visible light-promoted radical generation should occur under milder conditions than the thermal or ionic pathways employed in previous methods. Combining Pd-catalyzed C–H activation with photocatalysis for the first time, the directed C–H arylation of various arene derivatives with aryldiazonium salts was achieved at room temperature using $Ru(bpy)_3Cl_2{\cdot}6H_2O$ as a photocatalyst and $Pd(OAc)_2$ as a C–H activation catalyst under irradiation with light from by a 26 W compact fluorescent lamp (CFL, Scheme 72A).[146] N-Heterocycles, amides and oximes were used as directing groups and generally no further additives, such as oxidants or bases, were required. Furthermore, a room temperature arylation process with easier-to-handle diaryliodonium salts was also developed using an Ir-based photocatalyst and $Pd(NO_3)_2$ (Scheme 72B).[147] The two protocols show complementary reactivity with some DGs, such as benzamides and oximes, but generally perform similarly with most demonstrated substrates. Experimental evidence suggests a radical mechanism in both methods. The authors proposed that a Pd^{III} species is generated upon radical addition to the metalacycle and that single electron oxidation to Pd^{IV} with concomitant regeneration the ground state photocatalyst precedes reductive elimination of the product (Scheme 73, grey pathway). A computational study by the group of Malacria later suggested a different order of events (Scheme 73, black pathway), where reductive elimination of the product occurs directly from the Pd^{III} intermediate generated upon radical addition and that subsequent oxidation of the resulting Pd^{I} species regenerates both the photo- and the C–H activation catalysts.[148]

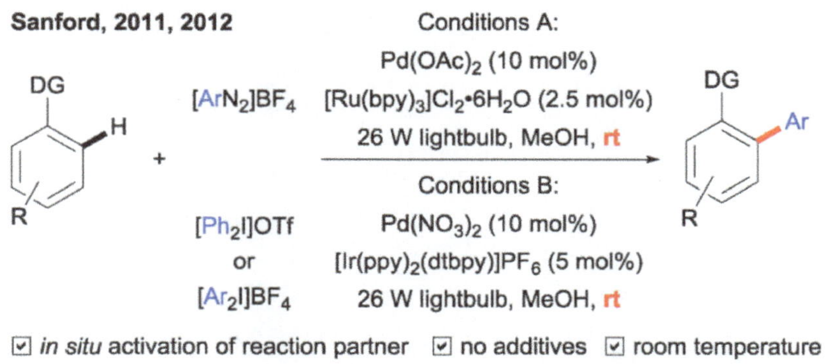

Scheme 72: Mild arylation *via* dual C–H activation and photocatalysis.[146,147]

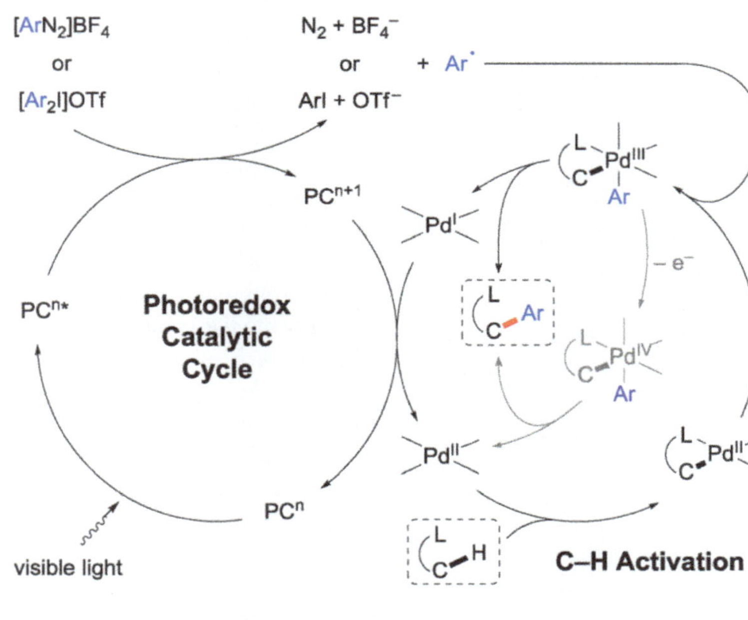

Scheme 73: Proposed mechanisms of the dual palladium- and photocatalytic C–H arylation.[146,148]

Palladium-catalyzed C–H carbonylation and carboxylation reactions have also been proposed to involve the generation of acyl radicals under oxidative conditions. While many such transformations require high reaction temperatures or strongly oxidizing conditions, the choice of suitable precursors and oxidants can result in much milder methods. Thus, the *ortho*-acylation of anilides at 40 °C was reported using *tert*-butyl hydroperoxide as oxidant and (hetero)aryl- or alkyl aldehydes as acylating reagents (Scheme 74).[149] Later, α-ketocarboxylic acids were similarly used as precursors for acyl radicals in Pd^{II}-catalyzed directed acylation at room temperature.[150] In this reaction, an organic photocatalyst (Eosin Y) was employed to facilitate visible light-promoted oxidative decarboxylation of the ketoacids in another example of dual photoredox/C–H activation catalysis. *ortho*-Carboxylation of anilides at room temperature was realized by the oxidative decomposition of diethyl azodicarboxylate (DEAD), generating ethoxyacyl radicals.[151] Ammonium peroxodisulfate and a catalytic amount of a Cu^{II} salt were found to be the optimal oxidants for this process. In each of these transformations, experimental evidence was presented to support the involvement of radical species as the coupling partner for the C–H palladated species.

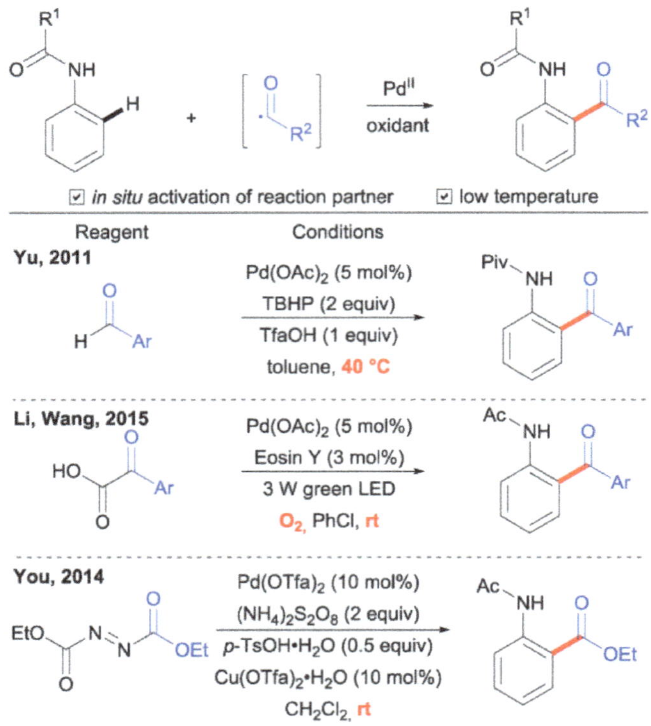

Scheme 74: Mild Pd-catalyzed C–H functionalization employing acyl radicals.[149–151]

OXIDANT

C–H activation reactions may be redox-neutral and hence not require the presence of an external oxidant to achieve catalytic turnover. However, in numerous transformations such as dehydrogenative coupling, reoxidation of the catalyst is necessary. The well-known Wacker process is an early example of redox co-catalysis, where a catalyst is re-oxidized to its active form after a reductive step in the mechanism by a redox catalyst. The redox catalyst in turn is then re-oxidized by a stoichiometric oxidant, preferably dioxygen from air. The feasibility of such a process is poorly rationalized in many es and depends on the specific reaction conditions and redox potentials of the species involved, which in turn are affected by any additives and reactants present. For this reason, a number of redox catalysts have been investigated in the literature and sometimes air or oxygen can even be used as oxidant without an additional mediator. In most es, however, oxidative C–H transformations still rely on the presence of (super) stoichiometric amounts of oxidants, which can be problematic in terms of functional group tolerance, waste generation and cost. In addition to numerous examples presented throughout the review where we commented on the use of air or oxygen as oxidant, we wish to demonstrate some recent examples employing redox co-catalysis or directly using air as the oxidant. For a more detailed discussion on this matter, the reader is referred to other recent reviews.[152]

The group of Sanford presented a method for the acyloxylation of C(sp^3)–H bonds using air as terminal oxidant which employed nitrate salts as redox co-catalysts (Scheme 75).[153] In using this system, stoichiometric amounts of the iodine(III) reagent PhI(OAc)$_2$, which is commonly used as the oxidant in such transformations, could be avoided. They proposed that a secondary NO$_2$/NO cycle mediates the oxidation of Pd^{II} to Pd^{IV} in the C–H activation cycle. Together with recent results from the Stahl group who employed fuming nitric acid as redox mediators in a benzene acetoxylation,[154] these results can also have interesting implications considering the common presence of nitrate impurities in commercial Pd(OAc)$_2$. Another approach to redox co-catalysis was taken by Jiao's group with *N*-hydroxyphthalimide (NHPI), which can initiate a sequence of steps resulting in the generation of hydroxyl radicals from oxygen (Scheme 75).[155] These hydroxyl radicals, in turn, then oxidize the C–H activation catalyst to a high-valent hydroxo species which then undergoes reductive elimination forming hydroxylated products. In this transformation, the redox-co-catalyst enables the use of oxygen as both a terminal oxidant and reactant.

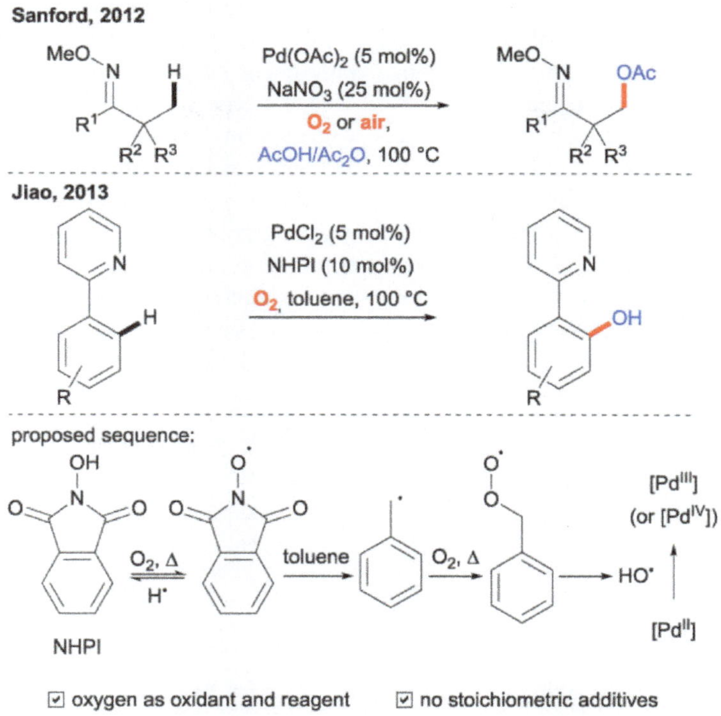

Scheme 75: Redox co-catalysts in Pd-catalyzed C–H activation.[153,155]

The group of Rueping demonstrated that photoredox catalysis can be employed to replace stoichiometric oxidant additives by air in C–H activation catalysis.[156] Ru and Ir based photocatalysts were employed to effect oxidation of the metal after a reductive step in Rh^{III}- and Ru^{II}-catalyzed oxidative olefinations as well as in the Pd^{II}-catalyzed indole synthesis originally reported by Glorius (Scheme 76).[157] As proposed by the authors, re-oxidation can either proceed directly with the excited photocatalyst (Scheme 77, red) or *via* superoxide anions generated upon single electron reduction of dioxygen by the reduced photocatalyst (Scheme 77, blue).[156]

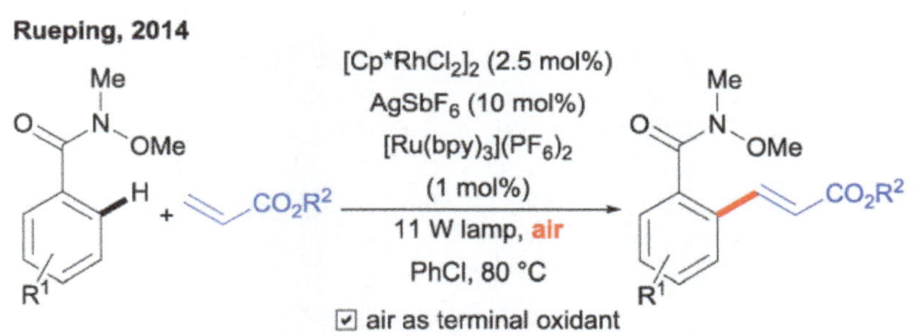

Scheme 76: Rh^{III}-catalyzed oxidative olefination using photoredox catalysis to replace stoichiometric amounts of oxidant with air.[156]

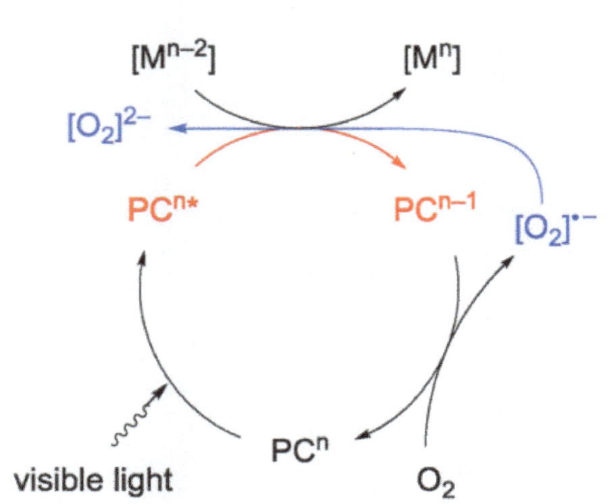

Scheme 77 Generalized scheme of the proposed metal reoxidation mechanisms under photoredox catalysis.[156]

A strategy for facilitating directed Pd-catalyzed C–H activation reactions conducted in the presence of heterocycles that relies on the relatively inefficient oxidation of Pd^0 by air was demonstrated by the group of Dai and Yu.[158] They argued that while heterocycles bind strongly to Pd^{II}, thus inhibiting transformations in their presence, a lower affinity for Pd^0 would allow exchange between the heterocycles and the *O*-methyl hydroxamate directing group present in their substrates. Furthermore, the hydroxamate should become an anionic ligand on deprotonation, thereby aiding the oxidation and binding strongly to Pd. Thus, by keeping the catalyst resting state in the lower oxidation state, catalyst saturation would be prevented because the oxidation to Pd^{II} should only occur once the metal is bound by the directing group. This concept was demonstrated in an annulation reaction with *tert*-butyl isonitrile affording iminoisoindolones (Scheme 78) which was proposed to proceed *via* oxidation to Pd^{II} and coordination to the directing group (1), isonitrile insertion (2), acyl migration (3), C–H activation (4) and, finally, reductive elimination. A very recent computational study confirmed this order of events and highlighted the crucial role of the *O*-methyl hydroxamate for the success of this transformation as it can be easily oxidized.[159]

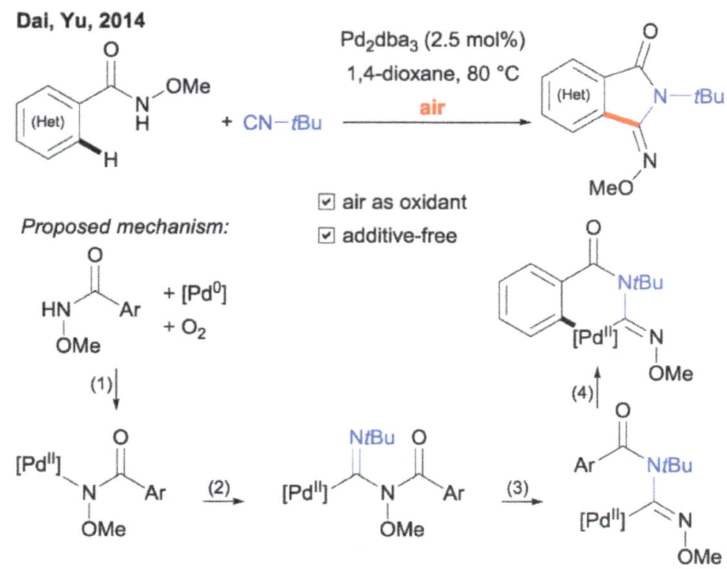

Scheme 78: Pd-catalyzed iminoisoindolinone synthesis under air.[158]

For several years, the group of Stahl has been working on technologies for the use of air and oxygen as terminal oxidants in C–H functionalization and oxidation catalysis, especially with palladium and copper.[152a] Building on insight gained from detailed mechanistic studies, they introduced 4,5-diazafluorenone (DAF) and related non-innocent redox active compounds as potent ligands facilitating oxidation of palladium. Among other transformations, they investigated an intramolecular Pd-catalyzed C–H amination of 2-aminobiphenyl derivatives affording carbazoles using oxygen in 1,4-dioxane (Scheme 79).[160] Noticing decomposition of the solvent under the reaction conditions, they observed that glycolic acid had an accelerating effect on the desired transformation. Glycolic acid can be formed by autoxidation of 1,4-dioxane and can itself be oxidized to peroxide species under the reaction conditions, thus aiding the catalyst re-oxidation.

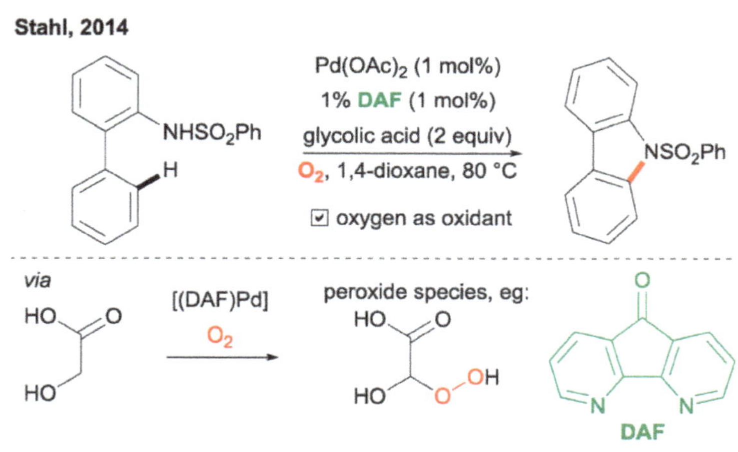

Scheme 79: Intramolecular C–H amination with oxygen as oxidant.[160]

In some es, oxidative C–H functionalization can also be carried out under air or oxygen without a redox co-catalyst system or special ligand. Often, the success of this process is difficult to predict and results from a judicious combination of the reaction conditions and reagents or products. One such example is an annulation reaction forming isocoumarins from benzoic acids and alkynes under Ru^{II}-catalysis reported by the group of Ackermann (Scheme 80).[161] This reaction proceeds at low temperatures using oxygen (or even air) as oxidant. Stoichiometric transformations with isolated ruthenacycles led to the observation and isolation of Ru^0–isocoumarin complexes, giving an opportunity to separately study the re-oxidation step. No reaction occurred in the presence of oxygen, $Cu(OAc)_2{\cdot}H_2O$ or NaOAc alone, and further studies revealed that acetic acid is required for re-oxidation to occur.

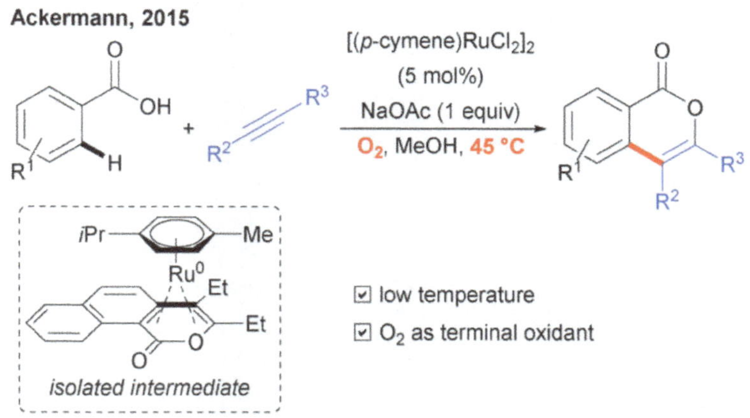

Scheme 80: Ru-catalyzed oxidative annulation reaction employing oxygen.[161]

CONCLUSION

As is evident from the many impressive examples discussed in the previous sections, the field of C–H activation has come a long way over the last few decades. Whereas harsh reaction conditions were once considered obligatory to achieve reasonable reactivity, nowadays efficient C–H transformations can often be conducted at or below ambient temperature, in the absence of acidic or basic additives and/or without strong external oxidants. In this review, we have sought to provide an update of this field by collecting together examples of mild C–H activation reactions disclosed since 2011. By categorizing the examples according to the conceptual strategies they have employed, we have also tried to demonstrate how greater insight into the mechanistic aspects of C–H activation has led to an improved understanding of how modifications to the substrate, reaction partners and catalyst can affect reactivity at each stage of the catalytic cycle.

For example, installing internal oxidants onto the substrate directing groups avoids the need for additional external oxidizing agents, while several room temperature reactions have been conducted employing olefinic reaction partners which bear carbonyl moieties capable of pre-coordinating the metal center. Furthermore, significant recent advances have been made using ancillary ligands to modulate catalyst reactivity. In Pd catalysis, MPAA ligands pioneered by the Yu group may assist C–H bond cleavage by acting as intramolecular proton shuttles, while σ-donating ligands can enhance oxidative addition-type C–H activation by increasing the electron richness of low-valent metal centers. The recent studies employing less well-established first-row transition metals such iron and cobalt as catalysts for C–H transformations have also opened up new promising avenues for investigation.

We hope that by providing an overview of the current state of the art and outlining the conceptual similarities between them, this review will act as a useful guide to the community and inspire future developments toward truly mild C–H transformations.

ACKNOWLEDGEMENTS

This work was supported by the Deutsche Fohungsgemeinschaft (Leibniz Award) and the Alexander von Humboldt Foundation (M. N. H.).

NOTES AND REFERENCES

1. For selected general reviews on C–H activation see: (*a*) T. W. Lyons and M. S. Sanford, *Chem. Rev.*, 2010, **110**, 1147–1169; (*b*) C. S. Yeung and V. M. Dong,*Chem. Rev.*, 2011, **111**, 1215–1292; (*c*) L. Ackermann, *Chem. Rev.*, 2011, **111**, 1315–1345; (*d*) D. Zhao, J. You and C. Hu, *Chem. – Eur. J.*, 2011, **17**, 5466–5492; (*e*) J. Wencel-Delord, T. Dröge, F. Liu and F. Glorius, *Chem. Soc. Rev.*, 2011, **40**, 4740–4761 ; (*f*) D. A. Colby, A. S. Tsai, R. G. Bergman and J. A. Ellman, *Acc. Chem. Res.*, 2012, **45**, 814–825; (*g*) P. B. Arockiam, C. Bruneau and P. H. Dixneuf, *Chem. Rev.*, 2012, **112**, 5879–5918; (*h*) N. Kuhl, M. N. Hopkinson, J. Wencel-Delord and F. Glorius, *Angew. Chem., Int. Ed.*, 2012, **51**, 10236–10254; (*i*) K. M. Engle and J.-Q. Yu, *J. Org. Chem.*, 2013, **78**, 8927–8955; (*j*) N. Kuhl, N. Schröder and F. Glorius, *Adv. Synth. Catal.*, 2014, **356**, 1443–1460 ; (*k*) F. Hu, Y. Xia, C. Ma, Y. Zhang and J. Wang, *Chem. Commun.*, 2015, **51**, 7986–7995 ; (*l*) C. Liu, J. Yuan, M. Gao, S. Tang, W. Li, R. Shi and A. Lei, *Chem. Rev.*, 2015, **115**, 12138–12204; (*m*) *C–H Bond Activation and Catalytic Functionalization I*, ed. P. H. Dixneuf and H. Doucet, Springer International Publishing, Cham, 2016, vol. 55; (*n*) *C–H Bond Activation and Catalytic Functionalization II*, ed. P. H. Dixneuf and H. Doucet, Springer International Publishing, Cham, 2016, vol. 56; (*o*) B. Zhao, Z. Shi and Y. Yuan, *Chem. Rec.*, 2016 DOI:10.1002/tcr.201500241.
2. For selected reviews concerning the application of the C–H activation see: (*a*) L. McMurray, F. O'Hara and M. J. Gaunt, *Chem. Soc. Rev.*, 2011, **40**, 1885–1898 ; (*b*) T. Newhouse and P. S. Baran, *Angew. Chem., Int. Ed.*, 2011, **50**, 3362–3374; (*c*) D. Y.-K. Chen and S.

W. Youn, *Chem. – Eur. J.*, 2012, **18**, 9452–9474; (*d*) J. Yamaguchi, A. D. Yamaguchi and K. Itami, *Angew. Chem., Int. Ed.*, 2012, **51**, 8960–9009; (*e*) J. Wencel-Delord and F. Glorius, *Nat. Chem.*, 2013, **5**, 369–375; (*f*) Y. Segawa, T. Maekawa and K. Itami, *Angew. Chem., Int. Ed.*, 2015, **54**, 66–81; (*g*) T. Cernak, K. D. Dykstra, S. Tyagarajan, P. Vachal and S. W. Krska, *Chem. Soc. Rev.*, 2016, **45**, 546–576 .

3. For selected publications on mechanisms of C–H activation, see: (*a*) D. L. Davies, S. M. A. Donald and S. A. Macgregor, *J. Am. Chem. Soc.*, 2005, **127**, 13754–13755; (*b*) S. I. Gorelsky, D. Lapointe and K. Fagnou, *J. Am. Chem. Soc.*, 2008, **130**, 10848–10849; (*c*) D. Balcells, E. Clot and O. Eisenstein, *Chem. Rev.*, 2010, **110**, 749–823; (*d*) D. Lapointe and K. Fagnou, *Chem. Lett.*, 2010, **39**, 1118–1126 ; (*e*) S. I. Gorelsky, *Coord. Chem. Rev.*, 2013, **257**, 153–164 ; (*f*) K. T. C. Carr, S. A. Macgregor and C. L. McMullin, *Top. Organomet. Chem.*, 2016, **55**, 53–76 ; (*g*) W. D. Jones, *Top. Organomet. Chem.*, 2016, **56**, 67–89 .

4. (*a*) F. Yang, F. Song, W. Li, J. Lan and J. You, *Adv.*, 2013, **3**, 9649–9652 ; (*b*) Z. She, Y. Shi, Y. Huang, Y. Cheng, F. Song and J. You, *Chem. Commun.*, 2014, **50**, 13914–13916 .

5. D. Munz, M. Webster-Gardiner, R. Fu, T. Strassner, W. A. Goddard III and T. B. Gunnoe, *ACS Catal.*, 2015, **5**, 769–775 .

6. P. Gandeepan and C.-H. Cheng, *J. Am. Chem. Soc.*, 2012, **134**, 5738–5741.

7. T.-S. Jiang and G.-W. Wang, *J. Org. Chem.*, 2012, **77**, 9504–9509.

8. For selected early examples see: (*a*) B.-F. Shi, N. Maugel, Y.-H. Zhang and J.-Q. Yu, *Angew. Chem., Int. Ed.*, 2008, **47**, 4882–4886; (*b*) D. H. Wang, K. M. Engle, B.-F. Shi and J.-Q. Yu, *Science*, 2010, **327**, 315–319; (*c*) B.-F. Shi, Y.-H. Zhang, J. K. Lam, D.-H. Wang and J.-Q. Yu, *J. Am. Chem. Soc.*, 2010, **132**, 460–461; (*d*) Y. Lu, D.-H. Wang, K. M. Engle and J.-Q. Yu, *J. Am. Chem. Soc.*, 2010, **132**, 5916–5921.

9. D. G. Musaev, A. Kaledin, B.-F. Shi and J.-Q. Yu, *J. Am. Chem. Soc.*, 2012, **134**, 1690–1698.

10. (*a*) K. M. Engle, D.-H. Wang and J.-Q. Yu, *J. Am. Chem. Soc.*, 2010, **132**, 14137–14151; (*b*) R. D. Baxter, D. Sale, K. M. Engle, J.-Q. Yu and D. G. Blackmond,*J. Am. Chem. Soc.*, 2012, **134**, 4600–4606.

11. For selected reviews on asymmetric C–H activation see: (*a*) R. Giri, B.-F. Shi, K. M. Engle, N. Maugel and J.-Q. Yu, *Chem. Soc. Rev.*, 2009, **38**, 3242–3272 ; (*b*) N. Cramer,*Chimia*, 2012, **66**, 869–872; (*c*) J. Wencel-Delord and F. Colobert, *Chem. – Eur. J.*, 2013, **19**, 14010–14017; (*d*) C. Zheng and S.-L. You, *Adv.*, 2014, **4**, 6173–6214 ; (*e*) J. Wencel-Delord and F. Colobert, *Synlett*, 2015, **26**, 2644–2658 .

12. (*a*) L. Chu, X.-C. Wang, C. E. Moore, A. L. Rheingold and J.-Q. Yu, *J. Am. Chem. Soc.*, 2013, **135**, 16344–16347; (*b*) L. Chu, K.-J. Xiao and J.-Q. Yu, *Science*, 2014, **346**, 451–455 ; For mechanistic studies, see: ; (*c*) M.-J. Zhou, T.-L. Yang and L. Dang, *J. Org. Chem.*, 2016, **81**, 1006–1020.

13. (*a*) Y. Izawa and S. S. Stahl, *Adv. Synth. Catal.*, 2010, **352**, 3223–3229; (*b*) D. Wang, Y. Izawa and S. S. Stahl, *J. Am. Chem. Soc.*, 2014, **136**, 9914–9917.

14. Y. Kuninobu, T. Iwanaga, T. Omura and K. Takai, *Angew. Chem., Int. Ed.*, 2013, **52**, 4431–4434.

15. Recent reviews on Pd^{IV} chemistry: (*a*) K. Muñiz, *Angew. Chem., Int. Ed.*, 2009, **48**, 9412–9423 ; (*b*) L.-M. Xu, B.-J. Li, Z. Yang and Z.-J. Shi, *Chem. Soc. Rev.*, 2010,**39**, 712–733 ; (*c*) P. Sehnal, R. J. K. Taylor and I. J. S. Fairlamb, *Chem. Rev.*, 2010, **110**, 824–889; (*d*) A. J. Hickman and M. S. Sanford, *Nature*, 2012, **484**, 177–185 . Pd^{III} in catalysis: ; (*e*) D. C. Powers and T. Ritter, *Top. Organomet. Chem.*, 2011, **35**, 129–156 ; (*f*) D. C. Powers and T. Ritter, *Acc. Chem. Res.*, 2012, **45**, 840–850.

16. J. J. Topczewski and M. S. Sanford, *Chem. Sci.*, 2015, **6**, 70–76 .

17. K. L. Hull, E. L. Lanni and M. S. Sanford, *J. Am. Chem. Soc.*, 2006, **128**, 14047–14049.

18. (*a*) J. M. Racowski, N. D. Ball and M. S. Sanford, *J. Am. Chem. Soc.*, 2011, **133**, 18022–18025; (*b*) A. Maleckis, J. W. Kampf and M. S. Sanford, *J. Am. Chem. Soc.*, 2013, **135**, 6618–6625.

19. (*a*) C. F. Rosewall, P. A. Sibbald, D. V. Liskin and F. E. Michael, *J. Am. Chem. Soc.*, 2009, **131**, 9488–9489; (*b*) P. A. Sibbald, C. F. Rosewall, R. D. Swartz and F. E. Michael, *J. Am. Chem. Soc.*, 2009, **131**, 15945–15951.

20. X. Wang, D. Leow and J.-Q. Yu, *J. Am. Chem. Soc.*, 2011, **133**, 13864–13867.

21. K. M. Engle, T.-S. Mei, X. Wang and J.-Q. Yu, *Angew. Chem., Int. Ed.*, 2011, **50**, 1478–1491.

22. R. Alam, L. T. Pilarski, E. Pershagen and K. J. Szabó, *J. Am. Chem. Soc.*, 2012, **134**, 8778–8781.

23. For selected reviews see: (*a*) D. A. Colby, R. G. Bergman and J. A. Ellman, *Chem. Rev.*, 2010, **110**, 624–655; (*b*) T. Satoh and M. Miura, *Chem. – Eur. J.*, 2010,**16**, 11212–11222; (*c*) F. W. Patureau, J. Wencel-Delord and F. Glorius, *Aldrichimica Acta*, 2012, **45**, 31–41 ; (*d*) G. Song and X. Li, *Acc. Chem. Res.*, 2015,**48**, 1007–1020; (*e*) J. Wencel-Delord, F. W. Patureau and F. Glorius, *Top. Organomet. Chem.*, 2016, **55**, 1–28 .

24. Y. Shibata and K. Tanaka, *Angew. Chem., Int. Ed.*, 2011, **50**, 10917–10921.

25. (*a*) Y. Hoshino, Y. Shibata and K. Tanaka, *Adv. Synth. Catal.*, 2014, **356**, 1577–1585 ; (*b*) M. Fukui, Y. Hoshino, T. Satoh, M. Miura and K. Tanaka, *Adv. Synth. Catal.*, 2014, **356**, 1638–1644 .

26. (*a*) Y. Takahama, Y. Shibata and K. Tanaka, *Chem. – Eur. J.*, 2015, **21**, 9053–9056; (*b*) F. W. Patureau and F. Glorius, *J. Am. Chem. Soc.*, 2010, **132**, 9982–9983.

27. (*a*) T. K. Hyster and T. Rovis, *Chem. Sci.*, 2011, **2**, 1606–1610 ; (*b*) T. K. Hyster and T. Rovis, *Chem. Commun.*, 2011, **47**, 11846–11848 ; (*c*) N. Umeda, K. Hirano, T. Satoh, N. Shibata, H. Sato and M. Miura, *J. Org. Chem.*, 2011, **76**, 13–24; (*d*) B. Ye and N. Cramer, *Acc. Chem. Res.*, 2015, **48**, 1308–1318.

28. S. Kawamorita, T. Miyazaki, H. Ohmiya, T. Iwai and M. Sawamura, *J. Am. Chem. Soc.*, 2011, **133**, 19310–19313.

29. E. C. Keske, B. D. Moore, O. V. Zenkina, R. Wang, G. Schatte and C. M. Crudden, *Chem. Commun.*, 2014, **50**, 9883–9886 .

30. (*a*) C. Cheng and J. F. Hartwig, *Science*, 2014, **343**, 853–857; (*b*) C. Cheng and J. F. Hartwig, *Chem. Rev.*, 2015, **115**, 8946–8975.

31. (*a*) T. Ishiyama, J. Takagi, K. Ishida, N. Miyaura, N. R. Anastasi and J. F. Hartwig, *J. Am. Chem. Soc.*, 2002, **124**, 390–391; (*b*) T. M. Boller, J. M. Murphy, M. Hapke, T. Ishiyama, N. Miyaura and J. F. Hartwig, *J. Am. Chem. Soc.*, 2005, **127**, 14263–14278.

32. T. Ishiyama, Y. Nobuta, J. F. Hartwig and N. Miyaura, *Chem. Commun.*, 2003, 2924–2925 .

33. S. M. Preshlock, B. Ghaffari, P. E. Maligres, S. W. Krska, R. E. Maleczka Jr. and M. R. Smith III, *J. Am. Chem. Soc.*, 2013, **135**, 7572–7582.

34. M. A. Larsen and J. F. Hartwig, *J. Am. Chem. Soc.*, 2014, **136**, 4287–4299.

35. S. A. Sadler, H. Tajuddin, I. A. I. Mkhalid, A. S. Batsanov, D. Albesa-Jove, M. Sing Cheung, A. C. Maxwell, L. Shukla, B. Roberts, D. C. Blakemore, Z. Lin, T. B. Marder and P. G. Steel, *Org. Biomol. Chem.*, 2014, **12**, 7318–7327 .

36. (*a*) S. Hwan Cho and J. F. Hartwig, *J. Am. Chem. Soc.*, 2013, **135**, 8157–8160 ; (*b*) C. W. Liskey and J. F. Hartwig, *J. Am. Chem. Soc.*, 2013, **135**, 3375–3378; (*c*) C. W. Liskey and J. F. Hartwig, *J. Am. Chem. Soc.*, 2012, **134**, 12422–12425.

37. S. Kawamorita, H. Ohmiya, K. Hara, A. Fukuoka and M. Sawamura, *J. Am. Chem. Soc.*, 2009, **131**, 5058–5059.

38. S. Kawamorita, R. Murakami, T. Iwai and M. Sawamura, *J. Am. Chem. Soc.*, 2013, **135**, 2947–2950.

39. Y. Kuninobu, H. Ida, M. Nishi and M. Kanai, *Nat. Chem.*, 2015, **7**, 712–717.

40. (*a*) L. Ackermann, *Acc. Chem. Res.*, 2014, **47**, 281–295; (*b*) C. Bruneau and P. H. Dixneuf, *Top. Organomet. Chem.*, 2016, **55**, 137–188 ; (*c*) S. Dana, M. R. Yadav and A. K. Sahoo, *Top. Organomet. Chem.*, 2016, **55**, 189–216 .

41. F. Kakiuchi, T. Kochi, E. Mizushima and S. Murai, *J. Am. Chem. Soc.*, 2010, **132**, 17741–17750.

42. J. Zhang, A. Ugrinov and P. Zhao, *Angew. Chem., Int. Ed.*, 2013, **52**, 6681–6684.

43. Selected reviews: (*a*) A. A. Kulkarni and O. Daugulis, *Synthesis*, 2009, **24**, 4087–4109; (*b*) J. J. Mousseau and A. B. Charette, *Acc. Chem. Res.*, 2013, **46**, 412–424; (*c*) T. K. Hyster, *Catal. Lett.*, 2015, **145**, 458–467 ; (*d*) Y. Liang, Y.-F. Liang and N. Jiao, *Org. Chem. Front.*, 2015, **2**, 403–415 ; (*e*) B. Su, Z.-C. Cao and Z.-J. Shi, *Acc. Chem. Res.*, 2015, **48**, 886–896 . Manganese: ; (*f*) C. Wang, *Synlett*, 2013, **24**, 1606–1613. Iron: ; (*g*) E. Nakamura and N. Yoshikai, *J. Org. Chem.*, 2010, **75**, 6061–6067; (*h*) I. Bauer and H.-J. Knölker, *Chem. Rev.*, 2015, **115**, 3170–3387; (*i*) L. Ilies and E. Nakamura, *Top. Organomet. Chem.*, 2016, **56**, 1–18 . Cobalt: ; (*j*) K. Gao and N. Yoshikai, *Acc. Chem. Res.*, 2014, **47**, 1208–1219; (*k*) L. Ackermann, *J. Org. Chem.*, 2014, **79**, 8948–8954; (*l*) N. Yoshikai, *ChemCatChem*, 2015, **7**, 732–734 ; (*m*) M. Moselage, J. Li and L. Ackermann, *ACS Catal.*, 2016, **6**, 498–525. Nickel: ; (*n*) S. Z. Tasker, E. A. Standley and T. F. Jamison, *Nature*, 2014, **509**, 299–309; (*o*) X.-H. Cai and B. Xie, *ARKIVOC*, 2015, 184–211 ; (*p*) N. Chatani, *Top. Organomet. Chem.*, 2016, **56**, 19–46 . Copper: ; (*q*) M. Zhang, *Appl. Organomet. Chem.*, 2010, **24**, 269–284 ; (*r*) A. E. Wendlandt, A. M. Suess and S. S. Stahl, *Angew. Chem., Int. Ed.*, 2011, **50**, 11062–11087; (*s*) R. T. Gephart III and T. H. Warren, *Organometallics*, 2012, **31**, 7728–7752 ; (*t*) K. Hirano and M. Miura, *Chem. Lett.*, 2015, **44**, 868–873 ; (*u*) K. Hirano and M. Miura, *Top. Organomet. Chem.*, 2016, **56**, 47–65 .

44. K. Gao, P.-S. Lee, T. Fujita and N. Yoshikai, *J. Am. Chem. Soc.*, 2010, **132**, 12249–12251.

45. P.-S. Lee, T. Fujita and N. Yoshikai, *J. Am. Chem. Soc.*, 2011, **133**, 17283–17295.

46. (*a*) Q. Chen, L. Ilies and E. Nakamura, *J. Am. Chem. Soc.*, 2011, **133**, 428–429; (*b*) Q. Chen, L. Ilies, N. Yoshikai and E. Nakamura, *Org. Lett.*, 2011, **13**, 3232–3234.

47. K. Gao, T. Yamakawa and N. Yoshikai, *Synthesis*, 2014, **46**, 2024–2039 .

48. (*a*) Z. Ding and N. Yoshikai, *Org. Lett.*, 2010, **12**, 4180–4183; (*b*) Z. Ding and N. Yoshikai, *Synthesis*, 2011, 2561–2566 ; (*c*) Z. Ding and N. Yoshikai,*Angew. Chem., Int. Ed.*, 2012, **51**, 4698–4701.

49. K. Gao and N. Yoshikai, *J. Am. Chem. Soc.*, 2011, **133**, 400–402.

50. (*a*) P.-S. Lee and N. Yoshikai, *Angew. Chem., Int. Ed.*, 2013, **52**, 1240–1244; (*b*) J. Dong, P.-S. Lee and N. Yoshikai, *Chem. Lett.*, 2013, **42**, 1140–1142 .

51. K. Gao and N. Yoshikai, *Angew. Chem., Int. Ed.*, 2011, **50**, 6888–6892.

52. K. Gao, P.-S. Lee, C. Long and N. Yoshikai, *Org. Lett.*, 2012, **14**, 4234–4237.

53. K. Gao and N. Yoshikai, *J. Am. Chem. Soc.*, 2013, **135**, 9279–9282.

54. K. Gao, R. Paira and N. Yoshikai, *Adv. Synth. Catal.*, 2014, **356**, 1486–1490 .

55. W. Song and L. Ackermann, *Angew. Chem., Int. Ed.*, 2012, **51**, 8251–8254.

56. B. Punji, W. Song, G. A. Shevchenko and L. Ackermann, *Chem. – Eur. J.*, 2013, **19**, 10605–10610.

57. M. Moselage, N. Sauermann, S. C. Richter and L. Ackermann, *Angew. Chem., Int. Ed.*, 2015, **54**, 6352–6355.

58. S. Asako, L. Ilies and E. Nakamura, *J. Am. Chem. Soc.*, 2013, **135**, 17755–17757.

59. S. Asako, J. Norinder, L. Ilies, N. Yoshikai and E. Nakamura, *Adv. Synth. Catal.*, 2014, **356**, 1481–1485 .

60. (*a*) E. R. Fruchey, B. M. Monks and S. P. Cook, *J. Am. Chem. Soc.*, 2014, **136**, 13130–13133; (*b*) B. M. Monks, E. R. Fruchey and S. P. Cook, *Angew. Chem., Int. Ed.*, 2014, **53**, 11065–11069.

61. (*a*) Q. Gu, H. H. Al Mamari, K. Graczyk, E. Diers and L. Ackermann, *Angew. Chem., Int. Ed.*, 2014, **53**, 3868–3871; (*b*) G. Cera, T. Haven and L. Ackermann,*Angew.*

Chem., Int. Ed., 2016, **55**, 1484–1488.

62. (*a*) C. Wang and Y. Huang, *Synlett*, 2012, **24**, 145–149 ; (*b*) M. Zhang, Y. Zhang, X. Jie, H. Zhao, G. Li and W. Su, *Org. Chem. Front.*, 2014, **1**, 843–895 ; (*c*) Z. Chen, B. Wang, J. Zhang, W. Yu, Z. Liu and Y. Zhang, *Org. Chem. Front.*, 2015, **2**, 1107–1295 .

63. Selected reviews: (*a*) J. Dupont, C. S. Consorti and J. Spencer, *Chem. Rev.*, 2005, **105**, 2527–2572; (*b*) M. Albrecht, in *C–H bond Activation: From Palladacycles: Synthesis, Characterisation and Applications*, ed. J. Dupont and M. Pfeffer, 2008, pp. 13–34; (*c*) M. Albrecht, *Chem. Rev.*, 2010, **110**, 576–623; (*d*) Y. F. Han and G.-X. Jin, *Chem. Soc. Rev.*, 2014, **43**, 2799–2823 .

64. (*a*) K. M. Engle, T.-S. Mei, M. Wasa and J.-Q. Yu, *Acc. Chem. Res.*, 2012, **45**, 788–802; (*b*) S. De Sarkar, W. Liu, S. I. Kozhushkov and L. Ackermann, *Adv. Synth. Catal.*, 2014, **356**, 1461–1479 .

65. M. Shang, S.-H. Zeng, S.-Z. Sun, H.-X. Dai and J.-Q. Yu, *Org. Lett.*, 2013, **15**, 5286–5289.

66. F. Péron, C. Fossey, J. Sopkova-de Oliveira Santos, T. Cailly and F. Fabis, *Chem. – Eur. J.*, 2014, **20**, 7507–7513 .

67. J. Guan, G.-J. Wu and F.-S. Han, *Chem. – Eur. J.*, 2014, **20**, 3301–3305.

68. Z.-J. Du, J. Guan, G.-J. Wu, P. Xu, L.-X. Gao and F.-S. Han, *J. Am. Chem. Soc.*, 2015, **137**, 632–635.

69. M. J. Tredwell, M. Gulias, N. Gaunt Bremeyer, C. C. C. Johansson, B. S. L. Collins and M. J. Gaunt, *Angew. Chem., Int. Ed.*, 2011, **50**, 1076–1079.

70. J. Wencel-Delord and F. Colobert, *Org. Chem. Front.*, 2016, **3**, 394–400 .

71. (*a*) Q. Dherbassy, G. Schwertz, M. Chessé, C. K. Hazra, J. Wencel-Delord and F. Colobert, *Chem. – Eur. J.*, 2016, **22**, 1735–1743 ; for pioneering work see: ; (*b*)T. Wesch, F. R. Leroux and F. Colobert, *Adv. Synth. Catal.*, 2013, **355**, 2139–2144 .

72. C. K. Hazra, Q. Dherbassy, J. Wencel-Delord and F. Colobert, *Angew. Chem., Int. Ed.*, 2014, **53**, 13871–13875.

73. C. Zhu, Y. Zhang, J. Kan, H. Zhao and W. Su, *Org. Lett.*, 2015, **17**, 3418–3421.

74. For a related example of direct functionalization of benzoic acid occuring preferentially in HFIP see: (*a*) G. Cheng, T.-J. Li and J.-Q. Yu, *J. Am. Chem. Soc.*, 2015, **137**, 10950–10953 . For a related example of mild dehydrogenative cross-coupling of arenes bearing carbonyl DGs occuring in HFIP see: ; (*b*) C. Zhang and Y. Rao, *Org. Lett.*, 2015, **17**, 4456–4459.

75. B. Liu, Y. Fang, Y. Gao, C. Sun, C. Xu and J. Zhu, *J. Am. Chem. Soc.*, 2013, **135**, 468–473.

76. (*a*) D. Zhao, S. Vásquez-Céspedes and F. Glorius, *Angew. Chem., Int. Ed.*, 2015, **54**, 1657–1661; (*b*) M. J. Ajitha and K. Huang, *Organometallics*, 2016, **35**, 450–455 .

77. (*a*) G. Rouquet and N. Chatani, *Angew. Chem., Int. Ed.*, 2013, **52**, 11726–11743; (*b*) L. C. M. tro and N. Chatani, *Chem. Lett.*, 2015, **44**, 410–421 .

78. H. Tang, X.-R. Huang, J. Yao and H. Chen, *J. Org. Chem.*, 2015, **80**, 4672–4682.

79. For selected reviews on C(sp^3)–H activation see: (*a*) C. Li, *Acc. Chem. Res.*, 2009, **42**, 335–344; (*b*) S. A. Girard, T. Knauber and C.-J. Li, *Angew. Chem., Int. Ed.*, 2014, **53**, 74–100; (*c*) G. Qiu and J. Wu, *Org. Chem. Front.*, 2015, **2**, 169–178 ; (*d*) O. Daugulis, J. Roane and L. D. Tran, *Acc. Chem. Res.*, 2015, **48**, 1053–1064; (*e*) R. K. Rit, M. R. Yadav, K. Ghosh and A. K. Sahoo, *Tetrahedron*, 2015, **71**, 4450–4459 ; (*f*) W. Zhang, N.-X. Wang and Y. Xing,*Synlett*, 2015, **26**, 2088–2098 .

80. R. K. Rit, M. R. Yadav and A. K. Sahoo, *Org. Lett.*, 2012, **14**, 3724–3727.

81. O. G. Mancheño and C. Bolm, *Chem. – Eur. J.*, 2007, **13**, 6674–6681 .

82. R. K. Rit, M. R. Yadav and A. K. Sahoo, *Org. Lett.*, 2014, **16**, 968–971.

83. H.-Y. Xiong, T. Besset, D. Cahard and X. Pannecoucke, *J. Org. Chem.*, 2015, **80**, 4204–4212.

84. L. Grigorjeva and O. Daugulis, *Angew. Chem., Int. Ed.*, 2014, **53**, 10209–10212.

85. (*a*) L. Grigorjeva and O. Daugulis, *Org. Lett.*, 2014, **16**, 4684–4687; (*b*) L. Grigorjeva and O. Daugulis, *Org. Lett.*, 2014, **16**, 4688–4690.

86. (*a*) F. W. Patureau and F. Glorius, *Angew. Chem., Int. Ed.*, 2011, **50**, 1977–1979; (*b*) H. Huang, X. Ji, W. Wu and H. Jiang, *Chem. Soc. Rev.*, 2015, **44**, 1155–1171 ; (*c*) J. Mo, L. Wang, Y. Liu and X. Cui, *Synthesis*, 2015, **47**, 439–459 .

87. (*a*) T. Gerfaud, L. Neuville and J. Zhu, *Angew. Chem., Int. Ed.*, 2009, **48**, 572–577; (*b*) J. Wu, X. Cui, L. Chen, G. Jiang and Y. Wu, *J. Am. Chem. Soc.*, 2009,**131**, 13888–13889; (*c*) Y. Tan and J. F. Hartwig, *J. Am. Chem. Soc.*, 2010, **132**, 3676–3677.

88. N. Guimond, C. Gouliaras and K. Fagnou, *J. Am. Chem. Soc.*, 2010, **132**, 6908–6909.

89. (*a*) S. Rakshit, C. Grohmann, T. Besset and F. Glorius, *J. Am. Chem. Soc.*, 2011, **133**, 2350–2353; (*b*) N. Guimond, S. I. Gorelsky and K. Fagnou, *J. Am. Chem. Soc.*, 2011, **133**, 6449–6457.

90. L. Xu, Q. Zhu, G. Huang, B. Cheng and Y. Xia, *J. Org. Chem.*, 2012, **77**, 3017–3024.

91. S. R. Neufeldt, G. Jiménez-Osés, J. R. Huckins, O. R. Thiel and K. N. Houk, *J. Am. Chem. Soc.*, 2015, **137**, 9843–9854.

92. (*a*) B. Li, H. Feng, S. Xu and B. Wang, *Chem. – Eur. J.*, 2011, **17**, 12573–12577; (*b*) B. Li, J. Ma, N. Wang, H. Feng, S. Xu and B. Wang, *Org. Lett.*, 2012, **14**, 736–739.

93. H. Wang and F. Glorius, *Angew. Chem., Int. Ed.*, 2012, **51**, 7318–7322.

94. (*a*) H. Wang, C. Grohmann, C. Nimphius and F. Glorius, *J. Am. Chem. Soc.*, 2012, **134**, 19592–19595; (*b*) M. Presset, D. Oehlrich, F. Rombouts and G. A. Molander, *Org. Lett.*, 2013, **15**, 1528–1531.

95. (*a*) T. K. Hyster, K. E. Ruhl and T. Rovis, *J. Am. Chem. Soc.*, 2013, **135**, 5364–5367; (*b*) S. Cui, Y. Zhang, D. Wang and Q. Wu, *Chem. Sci.*, 2013, **4**, 3912–3916 .

96. (*a*) T. K. Hyster, L. Knorr, T. R. Ward and T. Rovis, *Science*, 2012, **338**, 500–503; (*b*) B. Ye and N. Cramer, *Science*, 2012, **338**, 504–506; (*c*) B. Ye and N. Cramer, *Angew. Chem., Int. Ed.*, 2014, **53**, 7896–7899.

97. (*a*) G. Liu, Y. Shen, Z. Zhou and X. Lu, *Angew. Chem., Int. Ed.*, 2013, **52**, 6033–6037; (*b*) Z. Zhou, G. Liu, Y. Shen and X. Lu, *Org. Chem. Front.*, 2014, **1**, 1161–1165 ; (*c*) J. Zhou, J. Shi, Z. Qi, X. Li, H. E. Xu and W. Yi, *ACS Catal.*, 2015, **5**, 6999–7003 .

98. (*a*) T. Piou and T. Rovis, *J. Am. Chem. Soc.*, 2014, **136**, 11292–11295; (*b*) T. Piou and T. Rovis, *Nature*, 2015, **527**, 86–90.

99. (*a*) A. H. Sandtorv, *Adv. Synth. Catal.*, 2015, **357**, 2403–2435 ; (*b*) F. Bellina, *Top. Organomet. Chem.*, 2016, **55**, 77–102 ; (*c*) J.-F. Soulé and H. Doucet, *Top. Organomet. Chem.*, 2016, **55**, 103–118 .

100. (*a*) N. R. Deprez, D. Kalyani, A. Krause and M. S. Sanford, *J. Am. Chem. Soc.*, 2006, **128**, 4972–4973; (*b*) Y. Zhu, M. Bauer and L. Ackermann, *Chem. – Eur. J.*, 2015, **21**, 9980–9983; (*c*) A. J. Reay, T. J. Williams and I. J. S. Fairlamb, *Org. Biomol. Chem.*, 2015, **13**, 8298–8309 .

101. K. D. Hesp, R. G. Bergman and J. A. Ellman, *J. Am. Chem. Soc.*, 2011, **133**, 11430–11433.

102. (*a*) G. Liu and Y. Wu, *Top. Curr. Chem.*, 2010, **292**, 195–209; (*b*) C. J. Engelin and P. Fristrup, *Molecules*, 2011, **16**, 951–969.

103. E. M. Stang and M. C. White, *Nat. Chem.*, 2009, **1**, 547–551.

104. C. Engelin, T. Jensen, S. Rodriguez-Rodriguez and P. Fristrup, *ACS Catal.*, 2013, **3**, 294–302 .

105. A. J. Young and M. C. White, *Angew. Chem., Int. Ed.*, 2011, **50**, 6824–6827.

106. T. J. Osberger and M. C. White, *J. Am. Chem. Soc.*, 2014, **136**, 11176–11181.

107. A. Seoane, N. anova, N. Quiñones, J. L. Mascareñas and M. Gulías, *J. Am. Chem. Soc.*, 2014, **136**, 7607–7610.

108. F. Hu and M. Szostak, *ChemCatChem*, 2015, **7**, 1061–1063 .

109. M. J. Fuchter, D. K. Judge, M. Weimar and A. J. P. White, *Dalton Trans.*, 2013, **42**, 5615–5618 .

110. P. Ricci, K. Krämer, X. C. Cambeiro and I. Larrosa, *J. Am. Chem. Soc.*, 2013, **135**, 13258–13261.

111. P. Ricci, K. Krämer and I. Larrosa, *J. Am. Chem. Soc.*, 2014, **136**, 18082–18086.

112. M. Lafrance and K. Fagnou, *J. Am. Chem. Soc.*, 2006, **128**, 16496–16497.

113. (*a*) J. Norinder, A. Matsumoto, N. Yoshikai and E. Nakamura, *J. Am. Chem. Soc.*, 2008, **130**, 5858–5859; (*b*) N. Yoshikai, A. Matsumoto, J. Norinder and E. Nakamura, *Angew. Chem., Int. Ed.*, 2009, **48**, 2925–2928; (*c*) N. Yoshikai, A. Matsumoto, J. Norinder and E. Nakamura, *Synlett*, 2010, 313–316 ; (*d*) L. Ilies, S. Asako and E. Nakamura, *J. Am. Chem. Soc.*, 2011, **133**, 7672–7675; (*e*) N. Yoshikai, S. Asako, T. Yamakawa, L. Ilies and E. Nakamura, *Chem. – Asian J.*, 2011, **6**, 3059–3065; (*f*) L. Ilies, M. Kobayashi, A. Matsumoto, N. Yoshikai and E. Nakamura, *Adv. Synth. Catal.*, 2012, **354**, 593–596 ; (*g*) L. Ilies, E. Konno, Q. Chen and E. Nakamura, *Asian J. Org. Chem.*, 2012, **1**, 142–145 .

114. Y. Sun, H. Tang, K. Chen, L. Hu, J. Yao, S. Shaik and H. Chen, *J. Am. Chem. Soc.*, 2016, **138**, 3715–3730.

115. B. Li, Z.-H. Wu, Y.-F. Gu, C.-L. Sun, B.-Q. Wang and Z.-J. Shi, *Angew. Chem., Int. Ed.*, 2011, **50**, 1109–1113.

116. Q. Chen, L. Ilies, N. Yoshikai and E. Nakamura, *Org. Lett.*, 2011, **13**, 3232–3234.

117. O. M. Kuzmina and P. Knochel, *Org. Lett.*, 2014, **16**, 5208–5211.

118. F. Xie, Z. Zhang, X. Yu, G. Tang and X. Li, *Angew. Chem., Int. Ed.*, 2015, **54**, 7405–7409.

119. (*a*) F. Xie, Z. Qi and X. Li, *Angew. Chem., Int. Ed.*, 2013, **52**, 11862–11866; (*b*) F. Wang, X. Yu, Z. Qi and X. Li, *Chem. – Eur. J.*, 2016, **22**, 511–516.

120. (*a*) F. Xie, Z. Qi, S. Yu and X. Li, *J. Am. Chem. Soc.*, 2014, **136**, 4780–4787; (*b*) C. Feng and T.-P. Loh, *Angew. Chem., Int. Ed.*, 2014, **53**, 2722–2726; (*c*) K. D. Collins, F. Lied and F. Glorius, *Chem. Commun.*, 2014, **50**, 4459–4461 .

121. (*a*) M.-L. Louillat and F. W. Patureau, *Chem. Soc. Rev.*, 2014, **43**, 901–910 ; (*b*) K. Shin, H. Kim and S. Chang, *Acc. Chem. Res.*, 2015, **48**, 1040–1052; (*c*) J. G. Kim, K. Shin and S. Chang, *Top. Organomet. Chem.*, 2016, **55**, 29–51 .

122. (*a*) K.-H. Ng, Z. Zhou and W.-Y. Yu, *Org. Lett.*, 2012, **14**, 272–275; (*b*) C. Grohmann, H. Wang and F. Glorius, *Org. Lett.*, 2012, **14**, 656–659.

123. P. Patel and S. Chang, *Org. Lett.*, 2014, **16**, 3328–3331.

124. (*a*) J. Y. Kim, S. H. Park, J. Ryu, S. H. Cho, S. H. Kim and S. Chang, *J. Am. Chem. Soc.*, 2012, **134**, 9110–9113; (*b*) J. Ryu, K. Shin, S. H. Park, J. Y. Kim and S. Chang, *Angew. Chem., Int. Ed.*, 2012, **51**, 9904–9908.

125. J. Ryu, J. Kwak, K. Shin, D. Lee and S. Chang, *J. Am. Chem. Soc.*, 2013, **135**, 12861–12868.

126. (*a*) S. H. Park, J. Kwak, K. Shin, J. Ryu, Y. Park and S. Chang, *J. Am. Chem. Soc.*, 2014, **136**, 2492–2502; (*b*) T. M. Figg, S. Park, J. Park, S. Chang and D. G. Musaev, *Organometallics*, 2014, **33**, 4076–4085 .

127. J. Kim and S. Chang, *Angew. Chem., Int. Ed.*, 2014, **53**, 2203–2207.

128. Y. Park, K. T. Park, J. G. Kim and S. Chang, *J. Am. Chem. Soc.*, 2015, **137**, 4534–4542.

129. (*a*) J. Park and S. Chang, *Angew. Chem., Int. Ed.*, 2015, **54**, 14103–14107; (*b*) H. Wang, G. Tang and X. Li, *Angew. Chem., Int. Ed.*, 2015, **54**, 13049–13052.

130. (*a*) H. Kim, K. Shin and S. Chang, *J. Am. Chem. Soc.*, 2014, **136**, 5904–5907; (*b*) H. Kim and S. Chang, *ACS Catal.*, 2015, **5**, 6665–6669 .

131. X.-S. Zhang, K. Chen and Z.-J. Shi, *Chem. Sci.*, 2014, **5**, 2146–2159 .

132. S. Yu and X. Li, *Org. Lett.*, 2014, **16**, 1220–1223.

133. B. Zhou, J. Du, Y. Yang, H. Feng and Y. Li, *Org. Lett.*, 2013, **15**, 6302–6305.

134. W. Yang, J. Sun, X. Xu, Q. Zhang and Q. Liu, *Chem.*

Commun., 2014, **50**, 4420–4422 .

135. R. Zeng, C. Fu and S. Ma, *J. Am. Chem. Soc.*, 2012, **134**, 9597–9600.
136. B. Ye and N. Cramer, *J. Am. Chem. Soc.*, 2013, **135**, 636–639.
137. B. Ye, P. A. Donets and N. Cramer, *Angew. Chem., Int. Ed.*, 2014, **53**, 507–511.
138. S. Wu, R. Zeng, C. Fu, Y. Yu, X. Zhang and S. Ma, *Chem. Sci.*, 2015, **6**, 2275–2285 .
139. H. Wang, N. Schröder and F. Glorius, *Angew. Chem., Int. Ed.*, 2013, **52**, 5386–5389.
140. (*a*) D.-G. Yu, T. Gensch, F. de Azambuja, S. Vásquez-Céspedes and F. Glorius, *J. Am. Chem. Soc.*, 2014, **136**, 17722–17725; (*b*) T. Gensch, S. Vásquez-Céspedes, D.-G. Yu and F. Glorius, *Org. Lett.*, 2015, **17**, 3714–3717.
141. R. Manikandan, P. Madasamy and M. Jeganmohan, *Chem. – Eur. J.*, 2015, **21**, 13934–13938.
142. S. Wu, X. Huang, W. Wu, P. Li, C. Fu and S. Ma, *Nat. Commun.*, 2015, **6**, 7946 .
143. S. Yu and X. Li, *Org. Lett.*, 2014, **16**, 1200–1203.
144. S.-S. Zhang, J.-Q. Wu, Y.-X. Lao, X.-G. Liu, Y. Liu, W.-X. Lv, D.-H. Tan, Y.-F. Zeng and H. Wang, *Org. Lett.*, 2014, **16**, 6412–6415.
145. J.-Q. Wu, Z.-P. Qiu, S.-S. Zhang, J.-G. Liu, Y.-X. Lao, L.-Q. Gu, Z.-S. Huang, J. Li and H. Wang, *Chem. Commun.*, 2015, **51**, 77–80 .
146. D. Kalyani, K. B. McMurtrey, S. R. Neufeldt and M. S. Sanford, *J. Am. Chem. Soc.*, 2011, **133**, 18566–18569.
147. S. R. Neufeldt and M. S. Sanford, *Adv. Synth. Catal.*, 2012, **354**, 3517–3522.
148. G. Maestri, M. Malacria and E. Derat, *Chem. Commun.*, 2013, **49**, 10424–10426 .
149. C.-W. Chan, Z. Zhou and W.-Y. Yu, *Adv. Synth. Catal.*, 2011, **353**, 2999–3006 .
150. (*a*) C. Zhou, P. Li, X. Zhu and L. Wang, *Org. Lett.*, 2015, **17**, 6198–6201; (*b*) N. Xu, P. Li, Z. Xie and L. Wang, *Chem. – Eur. J.*, 2016, **22**, 2236–2242.
151. Y. Huang, G. Li, J. Huang and J. You, *Org. Chem. Front.*, 2014, **1**, 347–350 .
152. (*a*) A. N. Campbell and S. S. Stahl, *Acc. Chem. Res.*, 2012, **45**, 851–863; (*b*) W. Wu and H. Jiang, *Acc. Chem. Res.*, 2012, **45**, 1736–1748; (*c*) N. Gulzar, B. Schweitzer-Chaput and M. Klussmann, *Catal. Sci. Technol.*, 2014, **4**, 2778–2796 ; (*d*) O. Baslé, in *From C–H to C–C Bonds: Cross-Dehydrogenative-Coupling*, ed. C.-J. Li, Royal Society of Chemistry, 2015, pp. 197–218 .
153. K. J. Stowers, A. Kubota and M. S. Sanford, *Chem. Sci.*, 2012, **3**, 3192–3195 .
154. S. L. Zultanski and S. S. Stahl, *J. Organomet. Chem.*, 2015, **793**, 263–268.
155. Y. Yan, P. Feng, Q.-Z. Zheng, Y.-F. Liang, J.-F. Lu, Y. Cui and N. Jiao, *Angew. Chem., Int. Ed.*, 2013, **52**, 5827–5831.
156. (*a*) D. C. Fabry, J. Zoller, S. Raja and M. Rueping, *Angew. Chem., Int. Ed.*, 2014, **53**, 10228–10231; (*b*) J. Zoller, D. C. Fabry, M. A. Ronge and M. Rueping, *Angew. Chem., Int. Ed.*, 2014, **53**, 13264–13268; (*c*) D. C. Fabry, M. A. Ronge, J. Zoller and M. Rueping, *Angew. Chem., Int. Ed.*, 2015, **54**, 2801–2805.
157. S. Würtz, S. Rakshit, J. J. Neumann, T. Dröge and F. Glorius, *Angew. Chem., Int. Ed.*, 2008, **47**, 7230–7233 .
158. Y.-J. Liu, H. Xu, W.-J. Kong, M. Shang, H.-X. Dai and J.-Q. Yu, *Nature*, 2014, **515**, 389–393.
159. Y. Dang, X. Deng, J. Guo, C. Song, W. Hu and Z.-X. Wang, *J. Am. Chem. Soc.*, 2016, **138**, 2712–2723.
160. A. B. Weinstein and S. S. Stahl, *Catal. Sci. Technol.*, 2014, **4**, 4301–4307 .
161. S. Warratz, C. Kornhaaß, A. Cajaraville, B. Niepötter, D. Stalke and L. Ackermann, *Angew. Chem., Int. Ed.*, 2015, **54**, 5513–5517.

CHAPTER 18

SYNTHESIS OF FUNCTIONAL 'POLYOLEFINS'

Nicole M. G. Franssen, Joost N. H. Reek, and Bas de Bruin

FUNCTIONAL POLYMERS

Polymers bearing polar functionalized side groups are highly desired materials, due to their unique and rapidly expanding range of material properties. Compared to their non-functionalized analogues they exhibit beneficial properties with respect to adhesion, toughness, print/paintability, miscibility and rheological properties.[1] The wide variety of existing polar monomers and their ability to be copolymerized in various ratios with non-functionalized monomers such as ethene and propene allows the synthesis of a nearly infinite amount of different polymer formulations, each with completely different material properties. Nowadays, most functional polymers are synthesized *on demand*, *i.e.* the polymer composition and microstructure are tuned in such a way that they match the desired material properties for specialty applications (Fig. 1).[2]

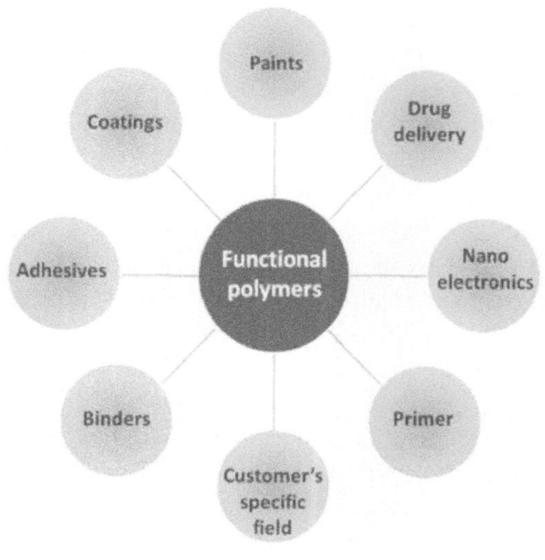

Figure 1: Examples of application areas of functional polymers.

Citation: Nicole M. G. Franssen, Joost N. H. Reek and Bas de Bruin, "Synthesis of functional 'polyolefins'," DOI: 10.1039/c3cs60032g.

FUNCTIONALIZED POLYOLEFINS

Due to the outstanding material properties of non-functionalized polyolefins (*i.e.* polyethene and polypropene), obtaining functional polymers based on polyolefins is of growing interest.[3]Introducing small amounts of polar functionalities into either polyethene or polypropene has a tremendous effect on the surface properties of the polymer.[2,4] In this way it is possible for polyolefins to enter a completely new area of applications, in which the material properties of polyolefins are combined with new beneficial surface properties imposed by the polar functionalities of the functionalized monomers.

Functionalized polyolefins can be classified into four categories, based on their structures (Fig. 2): randomly functionalized copolymers that exist as either branched (a) or linear structures (b), end-functionalized copolymers (c), block copolymers (d), and graft copolymers (e).[3,5] In the following sections, each class of functional polyolefins will be discussed separately. Although many polar comonomers have been used, giving rise to polymers with a wide range of different properties, we focus on copolymers of polyolefins and polar vinylmonomers (*i.e.* monomers in which the C=C double bond is directly substituted by the polar functionality), thereby excluding copolymers based on *e.g.* polyetheneglycol and poly-tetrahydrofuran (poly-THF). Most of the emphasis will lie on copolymers of acrylates, although some attention is given to co-monomers bearing other functionalities. Amounts of polar monomers incorporated will be listed in mol% if available, unless stated otherwise. These types of copolymers have been widely covered in the patent literature of the last few decades. However, since we want many readers to have access to the examples covered in this review, we mainly focused on articles published in scientific journals. Where appropriate, a reference to the corresponding patent literature is given as suggestion for further reading. Scientific publications that appeared in the literature in the last decade are the main focus of this review, although appropriate older examples are mentioned as well to give a complete overview.

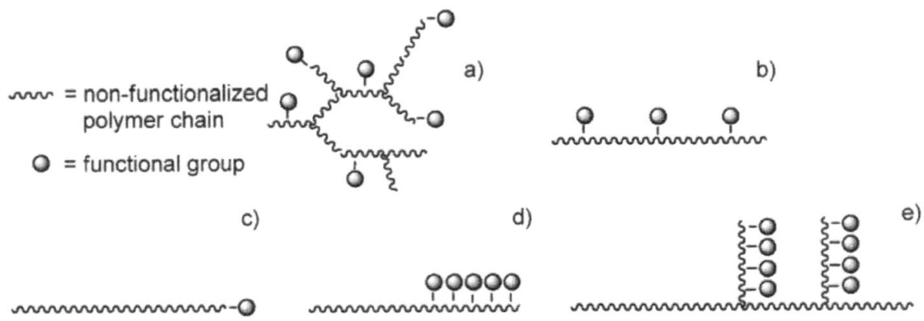

Figure 2: Functionalized polyolefins divided into 4 classes based on structural differences: randomly functionalized copolymers (a and b), chain-end functionalized polymers (c), block copolymers (d), and graft copolymers (e).

RANDOMLY FUNCTIONALIZED COPOLYMERS

Randomly functionalized polyolefins (Fig. 2a and b) offer a way to drastically alter the properties of the corresponding homo-polyolefins by applying a minimum of changes. The beneficial effect of randomly placed polar functionalities stems from breaking-up the crystal packing of the otherwise non-functionalized polyolefins, thereby broadening their application window. The resulting copolymers exhibit completely different properties compared to more ordered structures (Fig. 2c–e) due to the random distribution of monomers along the backbone. Randomly functionalized polyolefins exist in many variations, differing not only in the type of functional groups but also in the amount of functionalization and the polymer microstructure, *i.e.* the placement of functionalities along the non-functionalized chain (*cf.*Fig. 2a *vs.* b). Minor changes in the polymer structure often lead to major changes in the copolymer properties, making this class of copolymers an excellent starting point for the development of new materials. Functionalized polyolefins exhibiting linear structures as depicted in Fig. 2b are of particular interest since they can be regarded as a functionalized form of HDPE, thus benefiting from the highly developed HDPE chemistry, including properties and applications.[5,6]

Random copolymers can be synthesized in a variety of ways as depicted in Scheme 1, and each pathway will be discussed separately in the following sections.

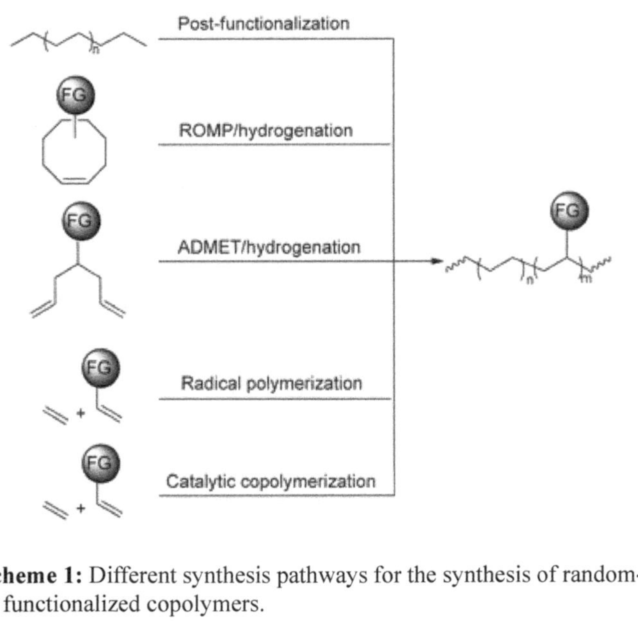

Scheme 1: Different synthesis pathways for the synthesis of randomly functionalized copolymers.

Polymer Post-Functionalization

Polymer post-functionalization is currently the most-used approach for the commercial synthesis of functionalized polyolefins, since this technique benefits most from the well-developed synthesis of non-functionalized polyolefins (for a review on the most recent advances in catalytic polyolefin synthesis, see Guan[7] and Qiao[8]). Polymer post-functionalization of existing polymeric structures can be carried out in the bulk as well as on the surface, thereby offering ample opportunities for obtaining desired material properties.[2] The amount of functionalization can be tuned by changes in stoichiometry and in principle any modification that can be carried out on a small molecule is useful for polymer modifications.[9] One of the most utilized approaches comprises reaction of free radicals with the C–H bonds of the polymer backbone, leading to H abstraction and the formation of a polymeric radical. These free radicals can be generated in various ways, with peroxide initiation being most common. In most cases maleic anhydride is added to this polymeric radical, giving rise to anhydride-functionalized polymer chains. These anhydride functionalities can in turn be converted to several other groups upon reaction with nucleophiles such as amines. A different post-functionalization approach deals with insertions of carbenes or nitrenes into the C–H bonds of the polymer backbone (Scheme 2). This method allows for somewhat more selectivity and control over the reaction in contrast to the above-described free radical approach. From an industrial perspective, surface modification techniques based on plasma, corona and flame treatment are relevant as well.[6] However, since non-functionalized polyolefins contain no reactive groups, chemical modifications of the inert C–H bonds generally require harsh conditions (for an example, see Scheme 2).[6,9] Many undesirable side reactions (*e.g.* cross-linking or chain scission; for more details see ref. 6 or 9) may occur under these conditions, thereby severely altering the properties of the polymer, also in an undesired manner.[10] Aside from this, polymer post-functionalization is often hampered by solubility problems and cumbersome purification steps. Although catalytic methodologies are being developed, this technique is still far from ideal due to the above-mentioned drawbacks.[9]

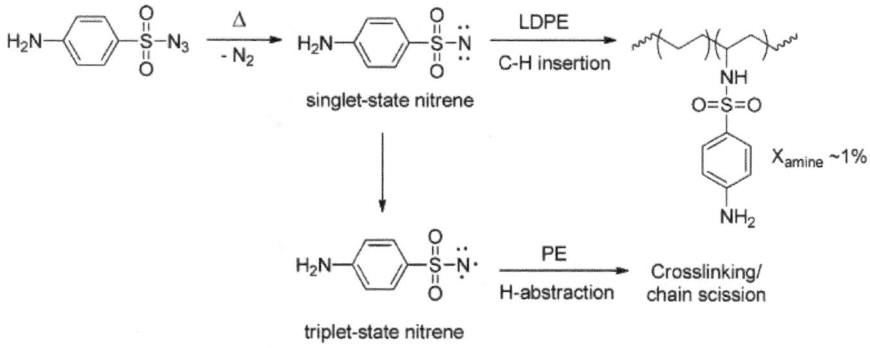

Scheme 2: Functionalization of polyethene *via* post-functionalization: C–H insertion of a nitrene generated from a tosylazide. The bottom pathway emphasizes the undesirable side reactions

Ring-Opening Metathesis Polymerization

Ring-opening metathesis polymerization (ROMP) of functionalized cyclic alkenes and subsequent hydrogenation of the resulting polyalkenamers is a viable alternative for obtaining functionalized polyolefins, although this technique requires specific monomers and occurs in a multi-step fashion. The production of linear, acyclic, functionalized polyolefins by homopolymerization of functionalized cyclooctenes (Scheme 1) is of particular interest. However, the number of reports describing ROMP of cyclooctenes followed by hydrogenation is rather limited, since they exhibit lower ring-strain and are therefore more reluctant towards polymerization than *e.g.* norbornene analogues. The presence of functional groups in cyclooctenes also hinders polymerization.[11] Due to the development of more efficient and polar-tolerant catalysts (mainly based on ruthenium)[12,13] the field is rapidly emerging and polymerizations of (functionalized) cyclooctenes are of growing importance.

Homopolymerization of cyclooctenes functionalized in the 5-position gave rise to polymers that resemble copolymers of non-functionalized and functionalized olefins, theoretically containing 25% of the corresponding polar vinyl monomers separated by either 6, 7 or 8 methylene units.[14–21] In this way, functional polyolefins bearing acrylate and ketone functionalities have been prepared (Scheme 3). Similar strategies have been applied to the synthesis of copolymers with highly regio- and stereoregular alcohol and amine functionalities (Scheme 4).[22–27]Due to their way of synthesis, these copolymers have precise sequence distributions that have a major effect on polymer crystallinity and physical properties.[28] More randomly distributedcopolymers with lower contents of polar monomers (<10%) are accessible *via* ROMP copolymerization of functionalized and non-functionalized cyclic monomers (Scheme 5).[11,28] The whole field of ROMP has recently been reviewed and summarized by Buchmeiser[29] and Slugovc.[30]

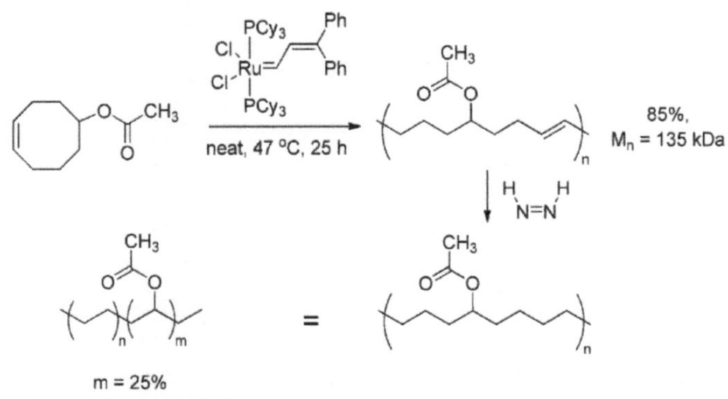

Scheme 3: Synthesis of functionalized polymers resembling copolymers of ethene and vinyl acetate *via* ROMP followed by hydrogenation.

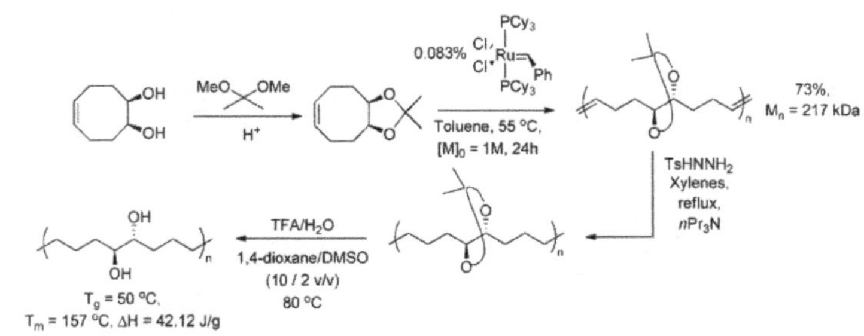

Scheme 4: Highly stereo- and regioregular analogues of hydroxy-functionalized PE obtained by ROMP of protected dihydroxy-functionalized cyclooctene.

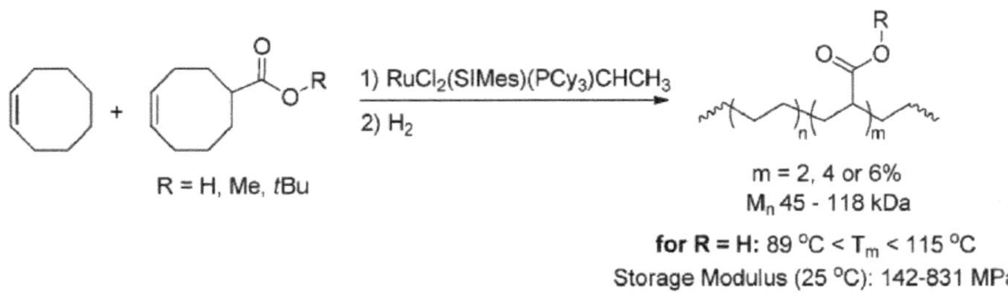

Scheme 5: Synthesis of analogues of ethene-acrylate copolymers obtained by ROMPcopolymerization of functionalized and non-functionalized cyclooctenes containing theoretically 2, 4 or 6% incorporation of polar vinyl monomers.

Acyclic Diene Metathesis Polycondensation

Acyclic diene metathesis polycondensation (ADMET), which is in fact based on the same elemental reaction as ROMP, followed by hydrogenation, is also a suitable approach for the synthesis of linear functional polyolefins, albeit with similar drawbacks as discussed above for ROMP. Instead of cyclic monomers, ADMET requires the synthesis of special symmetrically functionalized α,ω-diene monomers. The presence of polar functionalities in these monomers can lead to catalyst poisoning and often protection of the functional group prior topolymerization is required (for examples see *e.g.* Wagener *et al.*[31–33]). However, recent advances in the field of catalyst development have led to the discovery of certain metal complexes that do allow the synthesis of high-M_w polymers with polar functionalities. By applying these catalysts a wide variety of functional polymers was obtained, including those with alcohol,[34]acetate (Scheme 6),[35] ether,[36,37] ester,[28,38,39] amine/amide,[40] and halide (Scheme 7)[38,41–43] side groups. ADMET allows precise control over the polymer microstructure and as such this technique is used for the synthesis of model systems to study the effect of both the nature of the functionality and its distribution along the polymer backbone on the general material properties. The resulting structures are perfectly linear and therefore ADMET can function as a complementary technique to radical polymerization, which may yield similar copolymers but often with a high degree of branching.[44] More random distribution of the polar functionalities along the chain can be achieved by copolymerization of functionalized and non-functionalized α,ω-diene monomers, similar to ROMP (Scheme 8).[28] By carefully choosing the comonomers and tuning their relative rate of polymerization, alternating AB copolymers can be obtained with a degree of alternation between 90 and 99% (*i.e.* alternating diene metathesis polycondensation, ALTMET). The whole field of ADMET has recently been reviewed by Meier[44] and Wagener.[45]

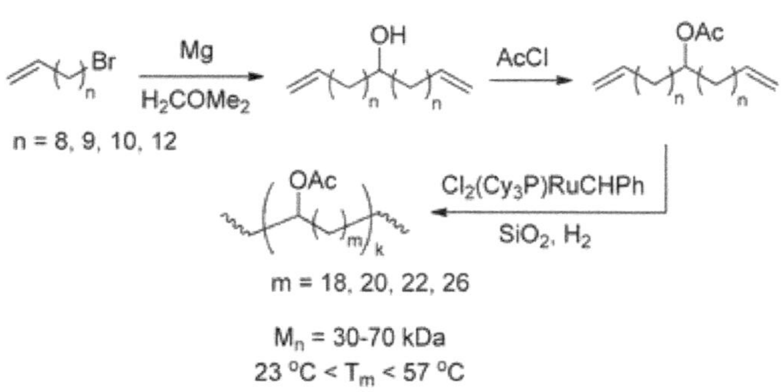

Scheme 6: Synthesis of acetate-functionalized monomers and subsequent ADMET polymerization.

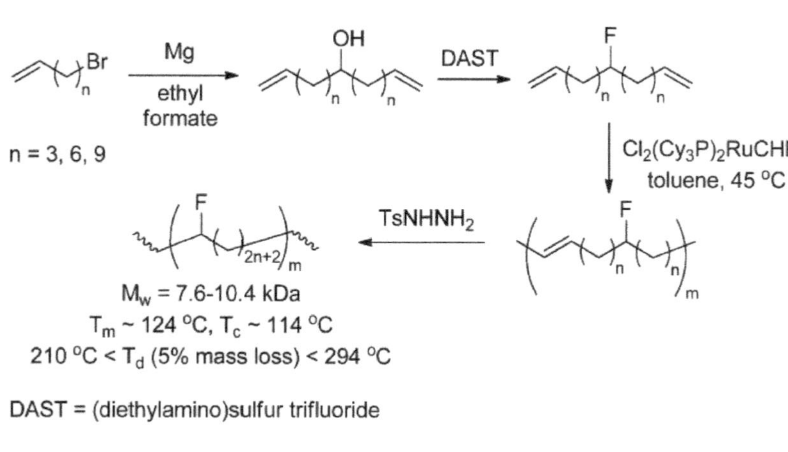

Scheme 7: Fluorinated polyethene obtained *via* ADMET polymerization of fluor-containing monomers.

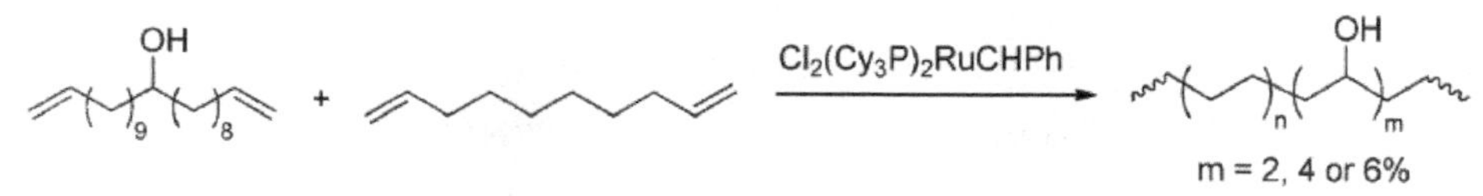

Scheme 8: Synthesis of analogues of ethene-vinyl alcohol copolymers containing theoretically 2, 4 or 6% incorporation of polar vinyl monomers obtained by ADMET copolymerization of functionalized monomers and 1,9-decadiene.

Direct Copolymerization of Polar and non-polar vinyl monomers

Despite some advantages of the above-mentioned techniques for obtaining functionalized polyolefins, direct copolymerization of polar vinyl monomers and non-polar olefins remains of particular interest.[46] Obviously, direct copolymerization circumvents the need for functionalization *via* often cumbersome polymer post-functionalization steps. By tuning the insertion efficiency of the functionalized monomer *vs.* the non-polar olefin, direct copolymerization allows quantitative control over the amount of functionalization. This difference in insertion efficiency, in combination with changes in the feed ratio of both monomers, also allows variations in copolymer architecture (*i.e.*, changing the arrangement of the monomers along the chain).[1] These features offer a broad window for tuning the copolymer properties, making direct copolymerization a powerful method for the development of a wide variety of new materials. However, despite being the most logical approach, direct copolymerization of polar and non-polar olefins is not that easy to achieve in a controlled way as is illustrated by the examples below. The two main copolymerization pathways to obtain such copolymers are based on radical polymerization or involve chain growth *via* catalytic processes, and each method will be discussed separately in the following sections.

Radical Pathways

Radical polymerization of olefins and polar vinyl monomers (*e.g.* acrylates, vinyl acetate and acrylic acid) is widely applied on a commercial scale, and the resulting copolymers find many applications in daily life.[47] The incorporated acetate or acid functionalities can be converted into their respective salts, giving rise to the formation of cross-linked ionomers, which find applications in areas that require extreme toughness (*e.g.* golf ball coatings) or enhanced sealing properties (*e.g.* bacon packaging).[48]

However, radical polymerization of non-polar olefins and polar vinyl monomers is not easily achieved, since non-polar olefins are poorly reactive monomers in radical polymerization. Due to the lack of stabilizing conjugation in non-polar olefins the stability of the generated radicals is too low to allow smooth polymerization, and to be successful quite harsh polymerization conditions (high temperatures and pressures) are required, resulting in the formation of highly branched polyolefins.[47,49] Since the reactivity of polar vinyl monomers towards radical polymerization is much higher, the resulting copolymers contain large amounts of polar functionalities.[5] Reports dealing with the synthesis of such copolymers under milder and more controlled conditions are rather scarce.

Earlier attempts deal with the use of boron-protected polar vinyl monomers, leading to the formation of highly electron-deficient monomers. In this way, alternating copolymers of ethene/propene and ethyl acrylate could be obtained by free-radical polymerizations under relatively mild conditions (25–50 °C, 6–20 atm. ethene), with considerable amounts of tacticity (Scheme 9).[50–52] At low ethene/propene pressures the acrylate content in the copolymers is higher than 50%.

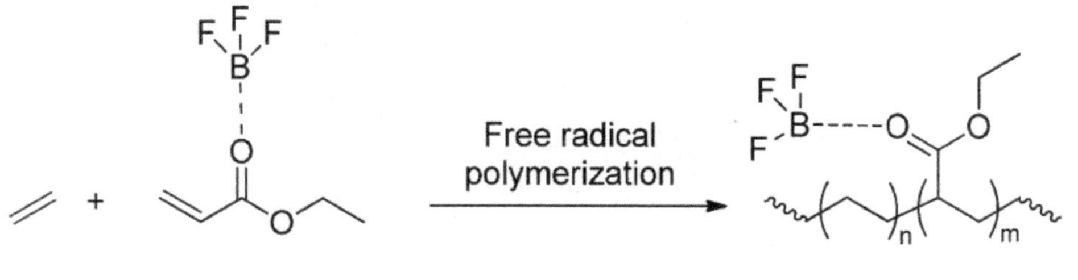

Scheme 9: Synthesis of alternating ethene-ethyl acrylate copolymers *via* free radical polymerization of ethene and boron-protected ethyl acrylate.

Copolymerizations using non-protected polar vinyl monomers have been developed as well, based on atom-transfer radical polymerization (ATRP) catalyzed by copper complexes (Scheme 10).[49,53] Copolymers of both ethene and propene with methyl acrylate were obtained containing up to 14 and 21% of non-functionalized olefin respectively (PDI < 1.5). Similar results were obtained in copolymerizations initiated by 2,2-azobis(isobutyronitrile) (AIBN), although those copolymers exhibited much broader molecular-weight distributions (PDI = 9 forethene).[54] Consecutive insertions of non-polar olefins were not observed, due to the reactivity difference between both monomers. The incorporation of olefin could be slightly tuned by varying the feed ratio of both monomers. A drawback of this metal-catalyzed process is that the copolymers often contain traces of metal impurities and are therefore colored.[54] Metal-free routes towards these copolymers have been developed based on nitroxide-mediated radical polymerization (NMRP).[55] This allowed the synthesis of copolymers of methyl acrylate and ethene/propene containing ~15% of non-functionalized olefin.

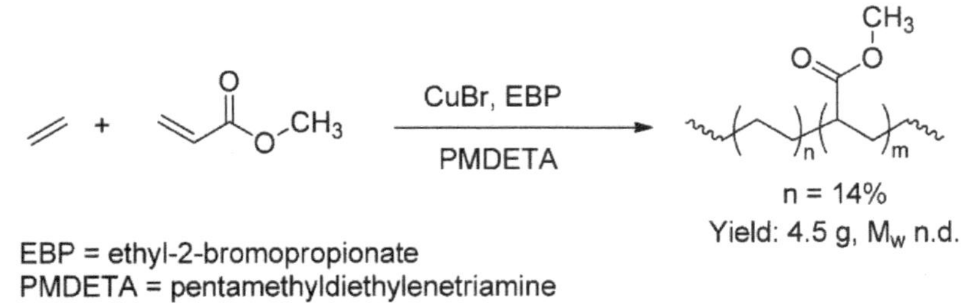

Scheme 10: Ethene-acrylate copolymers prepared *via* ATRP.

Reversible addition-fragmentation chain-transfer polymerization (RAFT) or degenerative-transfer (DT) polymerization of ethene and methylacrylate yields fully alternating copolymers, with a relatively narrow molecular-weight distribution (Scheme 11).[56]

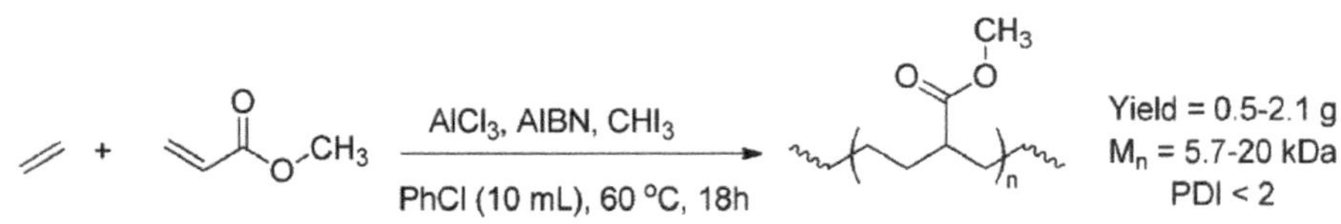

Scheme 11: Synthesis of alternating ethene-methyl acrylate copolymers *via* degenerative transfer polymerization.

Catalytic Routes

Transition-metal (TM) catalysed routes towards random copolymers with a linear structure are highly desired, and potentially allow the synthesis of functionalized copolymers with control over the amount of polar functionalities and their distribution along the polymer backbone. In that respect, TM catalysed processes are advantageous over radical polymerization reactions.[57] However, despite many years of research and recent advances in the field of coordination/insertion polymerization (*i.e.* migratory insertion of anolefin into a metal–alkyl bond) (Scheme 12), direct copolymerization of olefins with polar vinyl monomers still remains a challenge. In fact, achieving control over polymer composition and architecture, and development of catalysts that allow incorporation of a variety of functionalities during olefin polymerization are two of the main goals set by the Council for Chemical Research (CCR) to be achieved in 2020, driven by the future technology needs of the chemical industry in the area of catalysis.[54] Early transition-metal (ETM) catalysts are most active in the homopolymerization of ethene and propene, but they are easily poisoned by the functional groups in the co-monomers due to their high oxophilicity, and thus electronic or steric protecting strategies are required to prevent coordination to the metal center.[1,5] Recently, direct copolymerization has been successfully achieved by applying late transition-metal (LTM)catalysts (mainly based on palladium), since they are more tolerant to the presence of functional groups and allow their incorporation without the need for protection. However, due to the presence of the electron-withdrawing or -donating functionalities directly attached to the double bond, the electronic environment around the reactive C–C bonds is drastically changed compared to the non-functionalized analogues. This has a major effect on the relative rates of incorporation of both monomers, often leading to the formation of copolymers containing only a few per cent of polar monomers.[5] In principle, the effect of the polar functionality on the reactivity can be minimized by implementing a methylene spacer between the double bond and the functionality, but these monomers are less abundant and therefore more expensive.[1] Another problem that can hamper TM catalyzed copolymerization is β-heteroatom elimination of the functional group, which has been observed for vinyl acetate, vinyl halides and vinyl ethers.[1,5] Aside from this, after insertion of the polar monomer, the functional groups can coordinate to the metal center leading to the formation of stable chelates, breaking-up of which significantly slows down further insertions. Nevertheless, despite these possible problems, successful examples of direct copolymerization have been reported and the number of publications has rapidly grown in the past few years. The field of transition-metal catalyzed copolymerizations has been the subject of several reviews.[1,5,6,48,57–59] The following sections will give an overview of the current state-of-the-art with respect to controlled synthesis of functional copolymers *via* direct copolymerization of non-functionalized olefins and polar vinyl monomers.

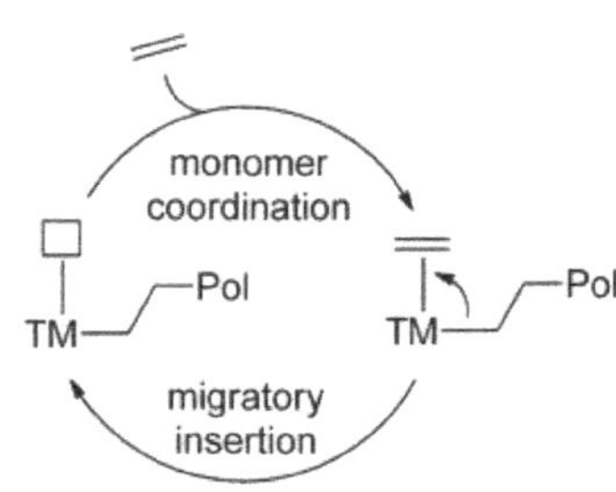

Scheme 12: Schematic representation of TM-catalyzed coordination/insertion polymerization.

Early Transition-Metal Catalysts

As stated above, ETM catalysts require the use of protecting strategies for incorporation of functional groups. Consequently, this approach requires post-polymerization reactions to obtain the desired functionalities. These protecting strategies mainly involve coordination of aluminium species to the polar groups, since aluminium is compatible with and often required for successful Ziegler/Natta or metallocene catalysis.[1] This approach has been used for the synthesis of polyethene and polypropene bearing amine,[60–62] ester (Scheme 13),[63–68] ether,[69,70] alcohol (Scheme 14),[61,62,64,65,70–78] thiol[62] and acid functionalities,[64,65,73,74] although in almost all cases the polymerization activity of the catalyst system dropped significantly compared to the homopolymerization of ethene/propene. Removal of the aluminium protection group is easily achieved by acid cleavage duringpolymer work-up. This approach has been protected in several patents (see *e.g.* Boffa[1] for examples).

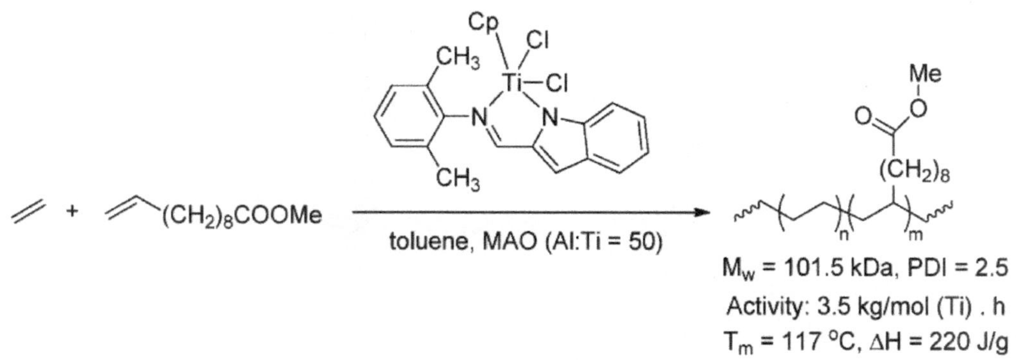

Scheme 13: ETM-catalyzed direct copolymerization of ethene and ester-functionalized vinylmonomers. The functional group is protected *in situ* with aluminium.

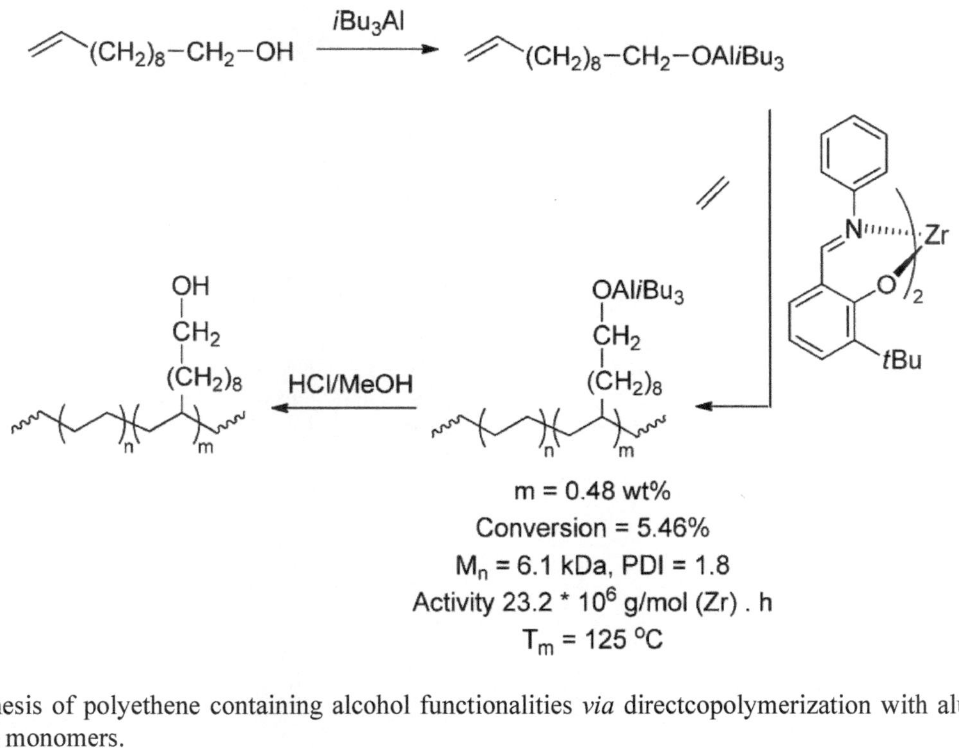

Scheme 14: Synthesis of polyethene containing alcohol functionalities *via* directcopolymerization with aluminium-protected OH-functionalized monomers.

Amine functionalities have also been protected with trimethylsilyl (TMS) groups (Scheme 15),[79,80] since monomers containing silyl functionalities can easily be copolymerized with ethene or propene using ETM catalysts (Scheme 16).[81–86] Different protection strategies have also been applied for copolymerization with acrylates, since acrylate monomers are the most difficult to be copolymerized with non-polar olefins due to the formation of inactive *O*-enolate complexes or π-allyl-η^3-complexes after insertion of the acrylate monomer into the M–C bond.[1,5] Formation of these inactive species can be circumvented by the use of acrylate salts as monomers, formed by complexation of the acrylate anions to metallocene cations (Scheme 17).[87] In this way, the synthesis of aminofunctional linear low density polyethene (LLDPE) could be realised.

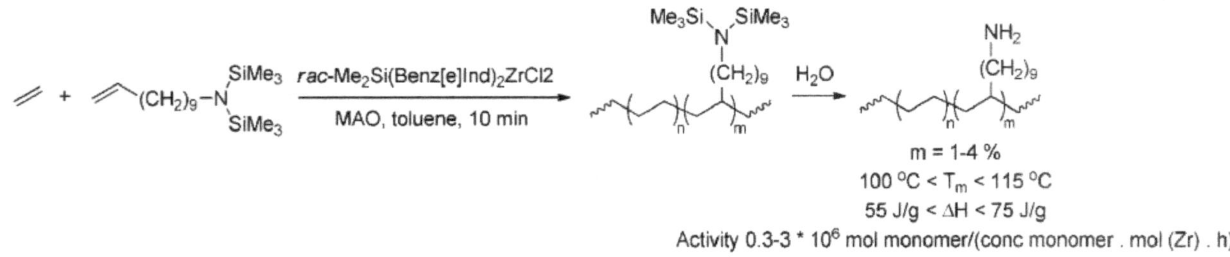

Scheme 15: ETM-catalyzed copolymerization of ethane and silyl-protected amine-functionalized monomers.

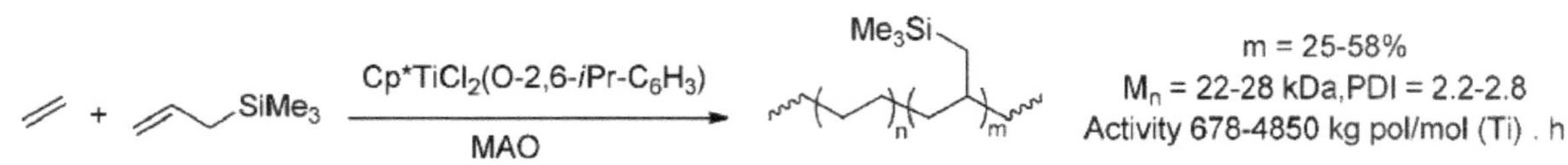

Scheme 16: Silyl-functionalized polyethene synthesized *via* direct copolymerization catalyzed by ETM complexes.

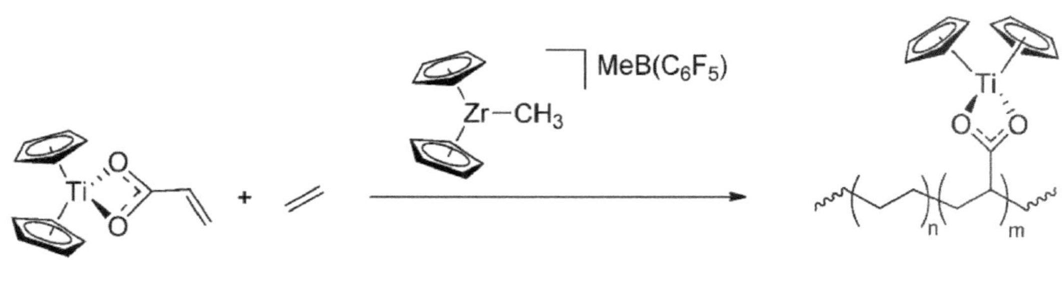

Scheme 17: Direct copolymerization of ethene and acrylate salts formed by complexation with metallocene cations.

Another approach that has been applied in combination with ETM catalysts is functionalization *via* borane monomers. Borane functionalities do not hamper ETM catalyzed polymerization, and since they are easily modified they form an excellent starting point for the introduction of various functionalities. Chung *et al.* reported formation of random copolymers of ethene and propene with vinyl monomers containing 9-borabicyclononane (9-BBN) functionalities.[88–91] These pendant borane functionalities could be converted to a variety of functional groups (Scheme 18), including alcohols, amines, halogens and radical species that can serve as initiators for the synthesis of graft copolymers (*vide infra*, Section 5.2).

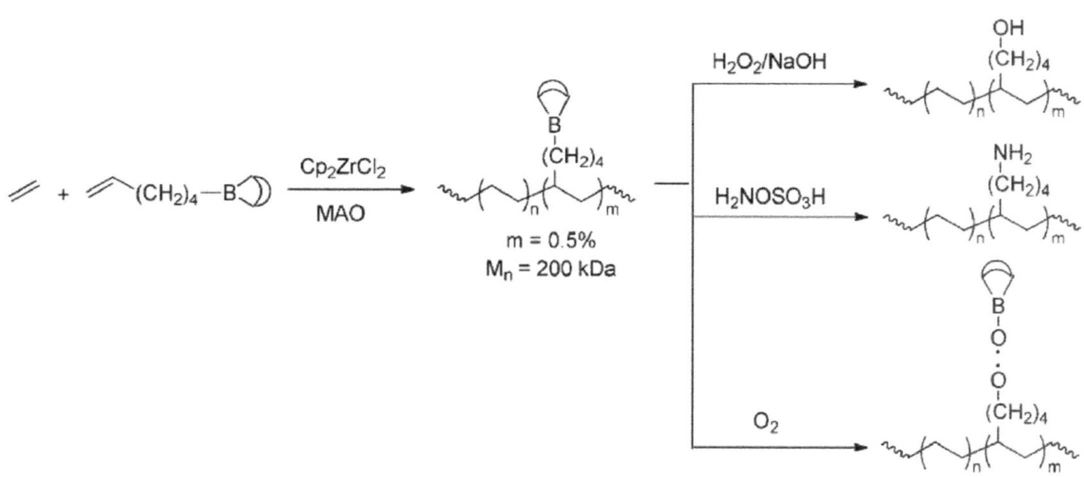

Scheme 18: Functionalization of polyethene *via* ETM-catalysed direct copolymerization with borane-containing monomers, followed by subsequent functionalization steps.

However, all the above-described examples deal with polar monomers containing several methylene groups as spacers between the C=C double bond and the polar group. Very few examples of direct copolymerization of polar vinyl monomers without this additional spacer have been reported for ETM catalysis, thereby limiting this approach to a selected class of monomers. Aside from this, the incorporation of polar monomers is rather low (in general less than 5%) due to diminished reactivity of the polar monomers compared to the non-functionalized analogues (similar to the lower reactivities of α-olefins), and the distribution along the polymer backbone is difficult to tune.

Late Transition-Metal Catalysts

Late transition-metal (LTM) catalysts are able to incorporate polar vinyl monomers that bear the polar functionality at the vinylic position, directly attached to the C=C double bond, *via* direct copolymerization with non-polar olefins, without the need for protection. The most successful examples have been reported for nickel and palladium catalysts.[5,92]

Palladium catalysts bearing α-diimine ligands are highly active in ethene homopolymerization and the scope of this reaction has been extended to copolymerizations of ethene/propenewith acrylates,[93–95] acrylic acid,[48] methyl vinyl ketone,[93] and monomers bearing acetyl functionalities (Scheme 19).[96] These systems yield highly branched copolymers (typically 100 branches per 1000 carbon atoms) due to rapid β-hydride elimination followed by reinsertion (*i.e.* chain walking), and the polar monomers are mainly located at the end of the branches. Incorporation of polar monomers up to 12% was achieved by varying the monomer feed ratio, although the polymerization efficiency decreased significantly upon incorporation of the polar monomers. The incorporation of polar monomers could be increased to more than 20% by slightly modifying the catalyst, *i.e.* by blocking the axial positions.[95] For none of these systems consecutive insertions of polar monomers were observed, and the system proved to be inactive towards homopolymerization of polar vinyl monomers. Copolymerization of olefins withmethyl methacrylate could not be achieved with these Pd-diimine catalysts and this feature was ascribed to the very low insertion rate of methacrylates.[97,98] Acrylonitrile and other nitrogen-containing monomers also hampered copolymerization, most likely *via* *N*-coordination of the monomers to the catalyst.[48] The catalyst system is more tolerant to functional groups if additional methylene spacers are present between the C=C double bond and the functionality similar to the ETM systems discussed above.[99] Although this approach is viable, it is not a unique feature of these Pd-diimine catalysts and therefore of minor importance.

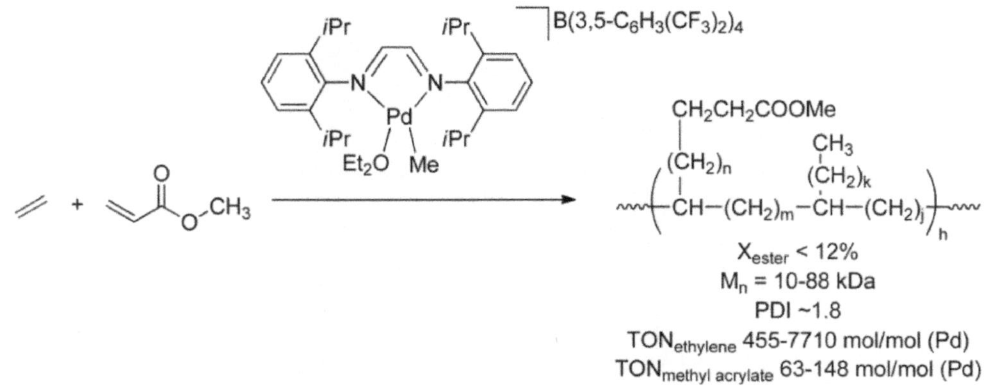

Scheme 19: Synthesis of highly branched ethene-acrylate copolymers *via* coordination/insertion polymerization catalyzed by a Pd-diimine complex.

The corresponding nickel-diimine catalysts showed far less activity in the copolymerization of olefins and polar vinyl monomers, due to strong coordination of the oxygen functionalities in the polar monomers to the metal center, thereby poisoning the catalyst.[100] This problem can be overcome by applying quite harsh polymerization conditions (high temperatures (120 °C) and pressures up to 340 bar), and in this way copolymers of ethene and methyl acrylate were obtained with around 1% incorporation of the polar monomers (Scheme 20).[101,102] In contrast to the above-mentioned Pd systems, these Ni catalysts give rise to copolymers with a wider range of microstructures, from moderately linear (*i.e.* 30 branches per 1000 carbon atoms) to highly branched, and the incorporation of acrylates takes place mainly in-chain. These Ni systems and their behavior in copolymerization reactions with polar monomers have mainly been discussed in the patent literature (for examples, see *e.g.* Nozaki[5] and Ittel[48]).

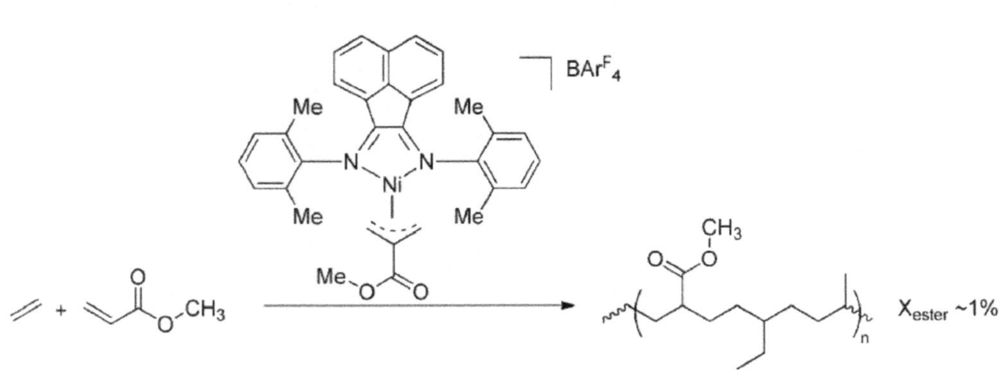

Scheme 20: Copolymerization of ethene and methylacrylate catalysed by a Ni-diiminecatalyst.

To overcome the problem of *O*-coordination to the oxophilic Ni center, several anionic ligands have been used to achieve copolymerization of ethene and methyl acrylate, mainly based on [P–O], [P–N] or [N–O] coordination to the metal center.

Application of these catalysts in combination with Lewis acids as cocatalysts resulted in the formation of moderately linear low-M_w copolymers (M_n ~ 1500 Da) with rather low amounts of polar functionalities (<1%), although the methyl acrylate units are incorporated in the polymer backbone rather than at the branches (Scheme 21).[103–105] The main advantage of these systems is that they allow copolymerization of olefins with methyl methacrylate, and copolymers with a MMA content up to 17% could be obtained (Scheme 22).[106–109] Papers dealing with MMA contents of 80% have been reported as well, but these high acrylate contents point in the direction of radical polymerization rather than sole coordination/insertion polymerization.[110]

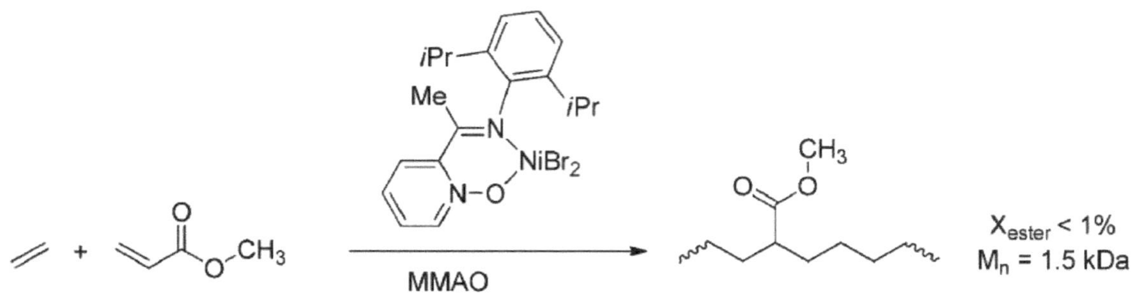

Scheme 21: Synthesis of low-M_w copolymers of ethene and methylacrylate obtained with a Ni catalyst bearing a [N–O] ligand.

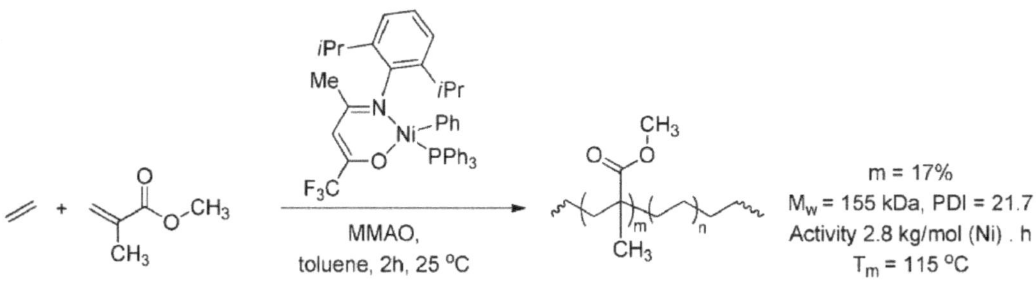

Scheme 22: Copolymerization of ethene and methyl methacrylate catalysed by a Ni[N–O] complex.

Along these lines, palladium complexes bearing phosphine–sulfonate [P–O] ligands have been applied in copolymerization reactions of ethene and several acrylates, giving rise to linearcopolymers with acrylate incorporation up to 17%.[57,111–114] The degree of branching in the copolymers is very low (1 branch per 1000 carbon atoms), giving rise to truly linear copolymerswith the polar monomers positioned along the polymer backbone. No consecutive acrylate insertions were observed for these systems Recently it was found that by slightly modifying the catalysts, the acrylate incorporation can be increased up to 52% giving rise to consecutive acrylate insertions (albeit up to 3 or 4) (Scheme 23).[115] Unfortunately, the M_w of the correspondingcopolymers decreased drastically (to as low as 2 kDa) upon incorporation of acrylate. This approach has been extended to the copolymerization of ethene and a variety of co-monomers, including acrylonitrile,[116] vinyl ether,[117] vinyl acetate,[118] acryl amides (Scheme 24),[119] acrylic acid,[120,121] vinyl sulfones[122] and allylic monomers containing functional groups.[123,124]

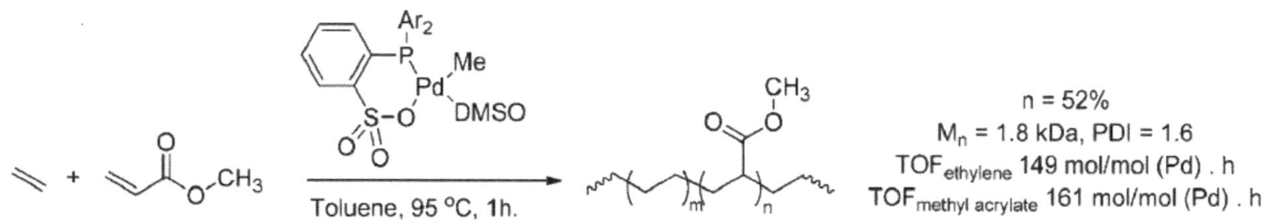

Scheme 23: Synthesis of ethene–methyl acrylate copolymers showing consecutive acrylate insertions, obtained *via* coordination/insertion polymerization using a Pd[P–O] catalyst.

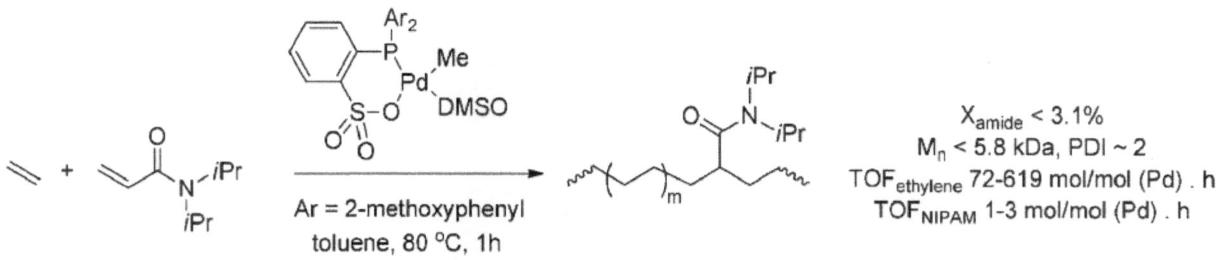

Scheme 24: Pd[P–O]-catalysed copolymerization of ethene and *N*-isopropylacrylamide(NIPAM).

The above-described examples emphasise the potential of this approach and the number of publications has rapidly grown in the past few years. The tolerance of these catalysts towards a large variety of functional groups makes them excellent candidates for the synthesis of copolymers of olefins and polar vinyl monomers. Up to now, Pd[P–O] systems have only been applied for copolymerizations with ethene. Another drawback is the limited occurrence of consecutive polar monomer insertions, which limits the potential of this approach for block copolymer synthesis. However, since the research in this field is still ongoing, future improvements of the catalysts and polymerization conditions are expected to broaden the scope of these catalyst systems further.

CHAIN-END FUNCTIONALIZED COPOLYMERS

Chain-end functionalization is a key step in polyolefin synthesis, since this is the starting point for constructing more complex macromolecular architectures such as block copolymers and graft copolymers from the otherwise unreactive polyolefin chains (*vide infra*). Polyolefins bearing a terminal functional group at either one end or both ends (*i.e.* telechelic polymers) are prepared by three general methods: controlled end-capping of living TM-catalyzed polymerization (a), *in situ* chain transfer reactions during TM catalyzed coordination polymerization (b), or modification of preformed unsaturated chain ends (c).[3,6,125,126] These approaches are discussed in the following paragraphs, with emphasis on those examples that have been exploited for the synthesis of more complex macrostructures.

End-Capping of Living Polymerizations

Although end-capping of living polymerization is the classical approach for introducing terminal functionalities, this approach is less suitable for end-functionalization of polyolefins with polar groups.[3] Aside from being metal-consuming and thus of limited practical value, this approach requires the use of very specific catalysts under certain polymerization conditions and, in fact, not many examples have been reported.[127] Living coordination polymerization of polypropene was achieved using a vanadium catalyst at a very low temperature (−65 °C) and syndiotactic polypropene end-capped with iodine was obtained by addition of I_2 at the end of the polymerization.[128–131] Aldehyde-functionalized polypropene was prepared in a similar process by reaction with carbon monoxide (Scheme 25).[132,133]

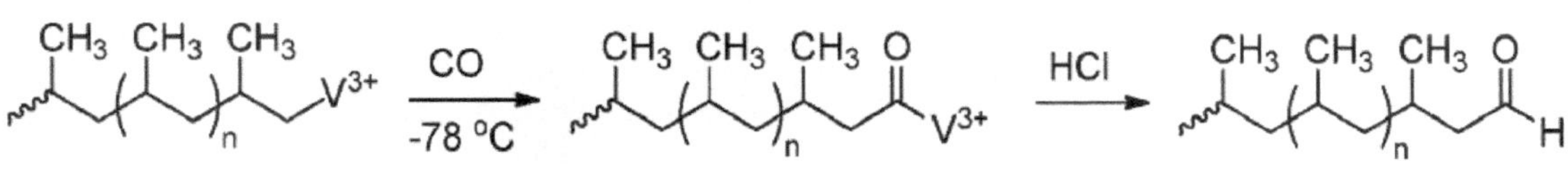

Scheme 25: Synthesis of aldehyde-functionalized polypropene by end-capping of living vanadium-catalyzed polymerization by carbon monoxide.

End-capped polyethene was obtained by using a cobalt catalyst bearing *para*-substituted phenyl groups.[134] These phenyl groups are able to initiate living coordination polymerization of ethene, thereby introducing the terminal functionality in the initiation step. More recently, this approach has been extended to the synthesis of acrylate-functionalized polyethenes *via* living coordination polymerization at selected palladium-diimine catalysts.[135,136] Pd complexes chelated by one acrylate unit were able to initiate living polymerization of ethene, thus allowing end-functionalization of polyethene in the initiation step. This approach has also been used for the synthesis of telechelic polyethenes, in which the second end functionality was introduced by insertion of an acrylate unit after the polymerization reaction (Scheme 26).[136]

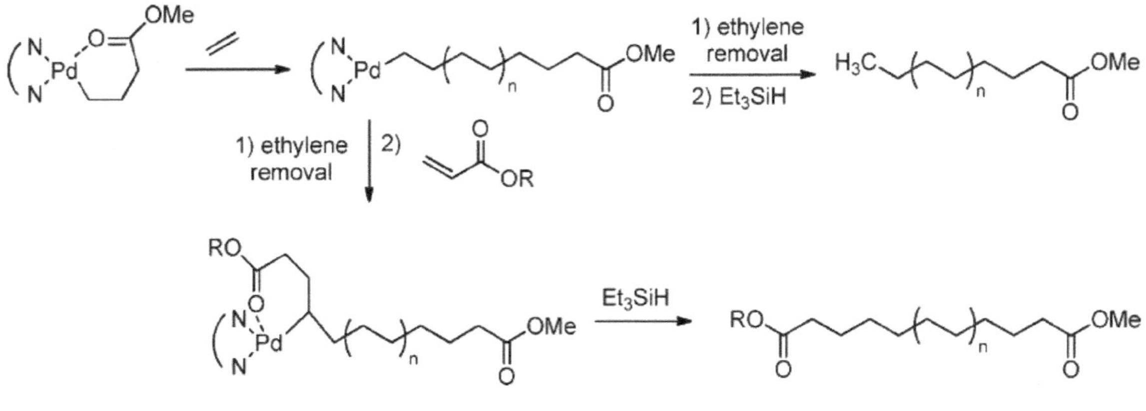

Scheme 26: Synthesis of polyethene functionalized with an acrylate unit at either one or both chain ends *via* end-capping of living polymerization catalysed by palladium.

Introduction of functional groups during the initiation step is not limited to the use of LTM catalysts. ETM catalysts bearing phenoxyimine (FI) ligands were used to obtain both polyethene and polypropene end-capped with aluminium or silyl functionalities, which can be converted to hydroxyl end groups (Scheme 27).[137] Similar to the LTM systems above, this method can be extended to the synthesis of telechelic polymers by insertion of functionalized monomers after polymerization of the non-functionalized monomers.

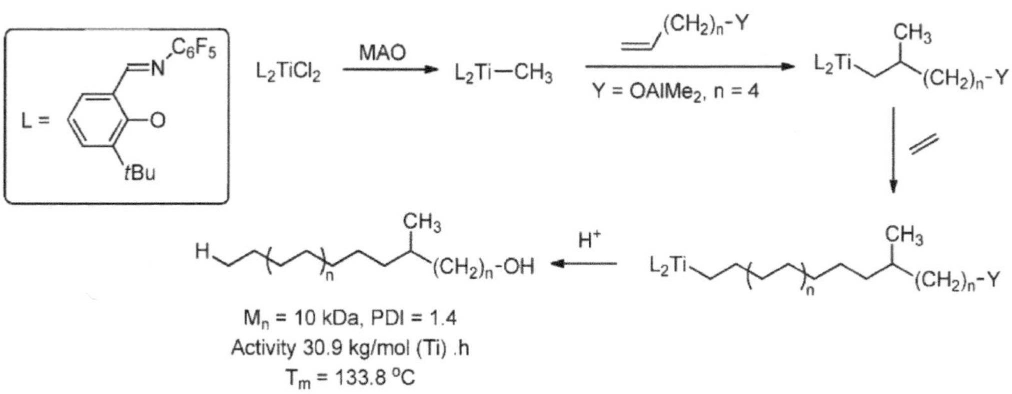

Scheme 27: Introduction of functional groups during the initiation of polyethene growth using a titanium catalyst bearing phenoxyimine (FI) ligands.

Chain-Transfer Reactions

End-functionalization *via* chain transfer is the most efficient approach for obtaining polyolefins bearing a terminal functionality. The process is very metal-efficient, since each catalytically active species produces multiple polymer chains, all bearing the desired end functionality. The overall polymerization efficiency is therefore retained, while the molecular weight of the resulting polymers can be tuned by varying the chain-transfer agent (CTA) to monomer ratio ([CTA]/[monomer]). Different varieties of chain-transfer reactions have been applied for this purpose and a general overview is given in Scheme 28. A well-known chain transfer reaction in polyolefin synthesis (often the cause of formation of low-M_w materials) is β-hydride elimination, yielding polyolefin chains with vinylic end groups that can be further modified through chemical transformation reactions. Chain transfer to a (main-group) metal, either catalytic or irreversible, is a different approach that is of growing importance due to the well-understood behavior of ETM/metal alkyl catalyst systems in polyolefin synthesis. By taking advantage of the reactivity of the newly formed metal–carbon bonds different functional groups can be introduced and examples of this approach will be provided in the following paragraphs. Another chain-transfer approach is direct introduction of functionalities using suitable (functionalized) chain-transfer agents (CTA) that terminate the growing polymer chain with a functional group while the catalyst is simultaneously re-activated for the formation of a new polyolefin chain.[3,6] The field of chain-transfer reactions in polyolefin synthesis has recently been reviewed by Marks.[138]

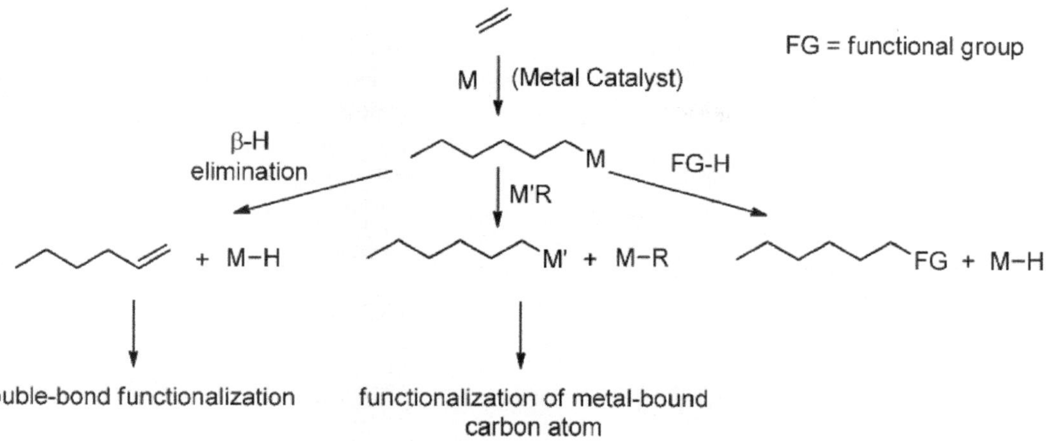

Scheme 28: General overview of chain-transfer reactions that can be used for end-functionalization of polyolefins.

Chain transfer of the growing polymer chain to a second (main-group) metal present in the polymerization mixture has already been known for a long time[139] and is mainly observed for ETM catalysts that operate in cooperation with zinc and main-group metal alkyl compounds. Under certain conditions this chain transfer occurs in a catalytic fashion, giving rise to a process called chain-shuttling or catalytic chain-transfer polymerization (CCTP).[140] This approach is very suitable for obtaining polyolefin chains that are fully end-capped with a (main-group) metal, although the non-catalytic version of this chain-transfer process has also been successfully used for end-functionalization of polyolefins. Chain transfer from ETM catalysts to zinc yielding Zn-alkyl terminated polypropene was first utilised to synthesise polypropenes with a variety of functional groups (hydroxyl,[141,142] carbonyl,[143] halogen,[143] vinyl[144,145] or amine groups[146]), making use of the high reactivity of Zn–C bonds towards many reagents (Scheme 29). *Catalytic* chain transfer from ETM catalysts to zinc has also been applied and led to the synthesis of polyolefins bearing alkyne or azide functionalities.[147] For a more general overview of the potential of catalytic chain transfer to zinc in polymerization, see *e.g.* Gibson,[148] Arriola[149] and Sita[150] and references therein. Transfer to zinc has also been used in combination with LTM catalysts, yielding hydroxyl-terminated polyethene.[151]

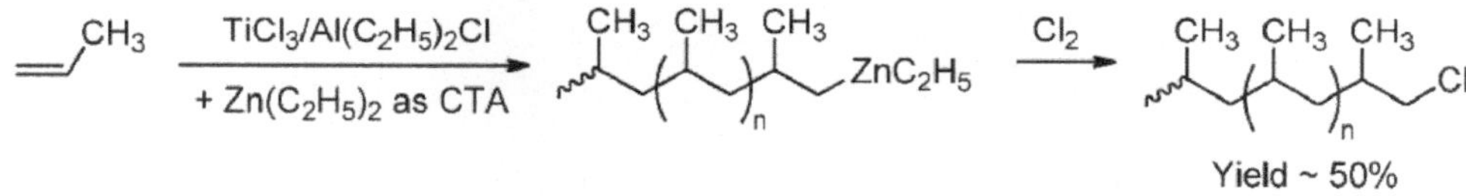

Scheme 29: Polypropene end-functionalized with halogen functionalities *via* intermediate chain transfer to zinc.

Similar results are obtained when ETM catalysed polymerizations are conducted in the presence of aluminium alkyls, leading to polyolefins bearing hydroxyl groups (Scheme 30).[152–158]

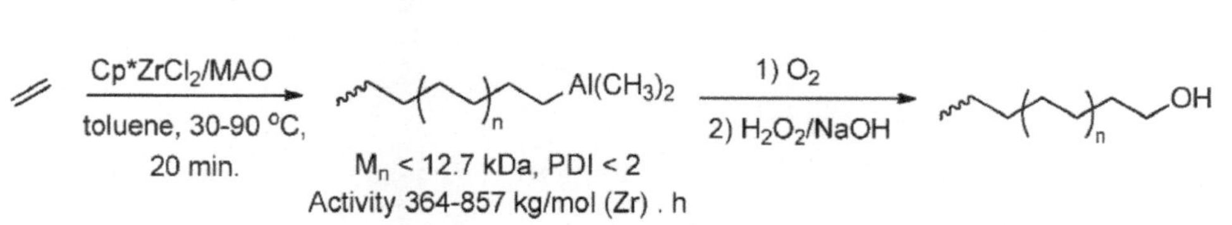

Scheme 30: Synthesis of hydroxy-functionalized polyethene *via* ETM catalysis in combination with chain transfer to aluminium.

Transfer to magnesium followed by subsequent functionalization has been exploited for lanthanide catalysts, giving rise to polymers with various functionalities,[159] such as[152] alkoxyamine-functionalized polymers,[160,161] polymers bearing thiocarbonyl functionalities[162] and azide-functionalized polyethene (Scheme 31).[163]

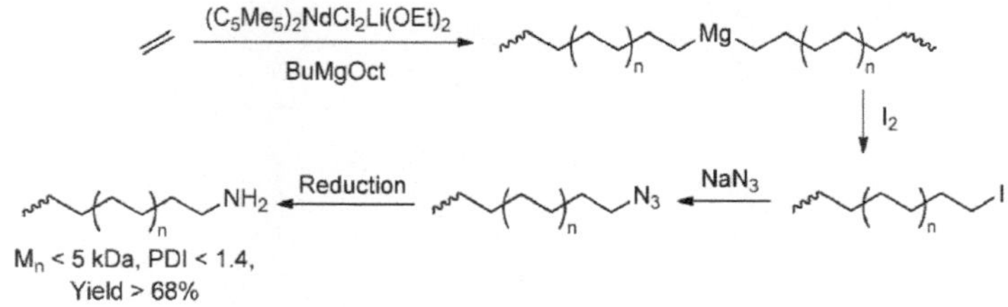

Scheme 31: End-functionalization of polyethene *via* lanthanide catalysis in combination with catalytic chain transfer to magnesium alkyls.

In the presence of various silanes, chain transfer of the growing polymer chain to the silyl functionality can occur from both ETM and lanthanide catalysts (Scheme 32).[164–167] This process is versatile, and both polyethene and polypropene bearing a wide range of silyl functionalities have been prepared. Although this should in principle be a good starting point for the synthesis of polyolefins with a variety of other functionalities,[168] additional modifications of the silyl functionalities were not reported.

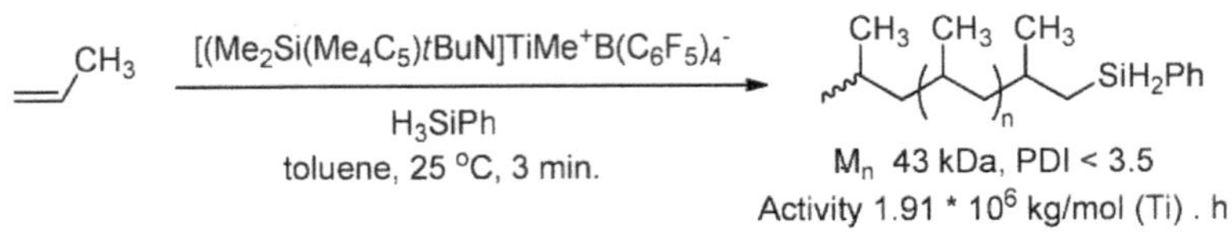

Scheme 32: Chain transfer to silanes as effective method for the synthesis of silyl end-capped polypropene.

Chain-transfer to organoboranes leads to the formation of polymers with a terminal borane functionality and a corresponding metal-hydride species, which is still active towardspolymerization. Chung and Xu reported formation of polyethene terminated with a 9-BBN moiety, obtained *via* chain-transfer from ETM catalysts to the 9-BBN dimer.[169,170] The boranefunctionalities can be oxidised to yield polyolefins with a terminal hydroxyl group,[169] but more interesting is their ability to be used as initiators in free radical polymerization *via* controlledoxidation (Scheme 33). This allows the formation of block copolymers *via* coordination/insertion polymerization in combination with subsequent free radical polymerization (*vide infra*).

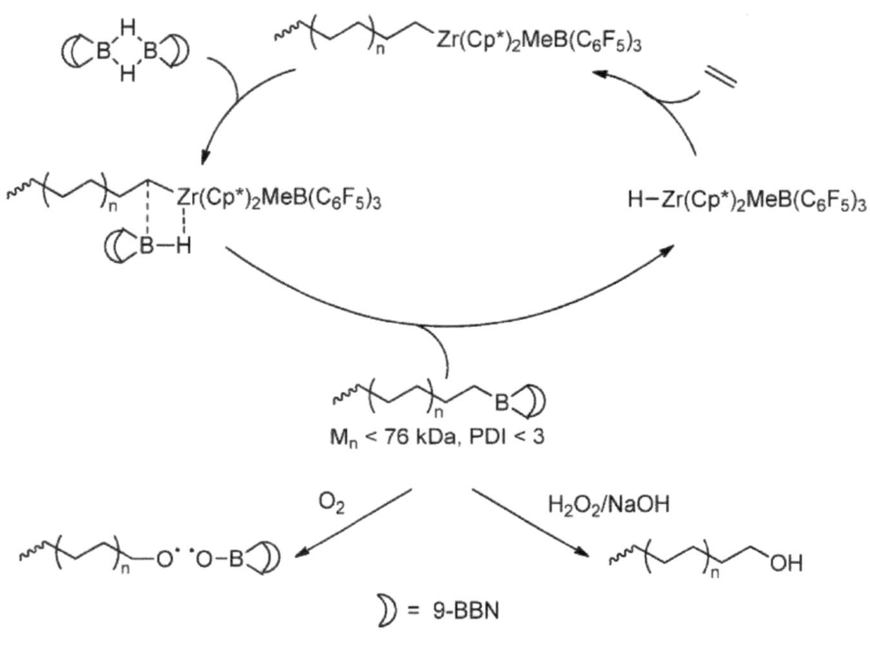

Scheme 33: Functionalization of polyethene with a terminal borane group and subsequent transformations into a hydroxyl functionality and a suitable initiator for radical polymerization.

Another chain-transfer reaction that is of interest is the transfer to styrene derivatives (in particular *p*-methylstyrene) that in turn can function as initiators for the growth of polar polymersegments (*vide infra*, Section 4.3). These styrene functionalities can be introduced by first inserting the styrene-derivative *via* a coordination/insertion mechanism, followed by chain transfer to hydrogen, yielding the styrene-terminated polymer chain and a metal-hydride catalyst (Scheme 34).[171–173] These systems benefit from the low activity of the ETM catalysts towards the incorporation of styrenes and, in fact, the system becomes inactive after insertion of the styrene-derivative. Chain-transfer to hydrogen is necessary to regenerate catalyst and allow the formation of additional polymer chains. This approach works for both polyethene and polypropene and several functionalized styrene-derivatives have been used, including styrenes bearing chloride and amine groups.[172]

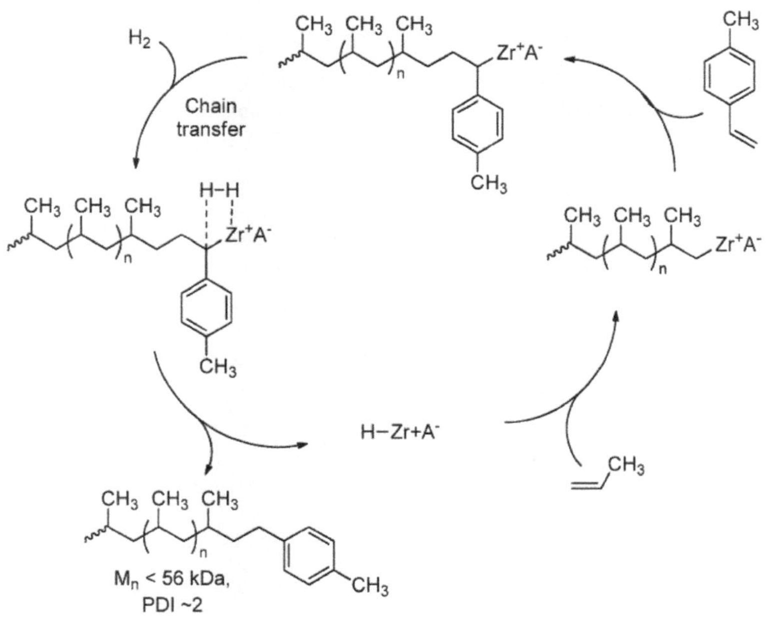

Scheme 34: Chain transfer of a growing polypropene chain to *p*-methylstyrene.

Functionalization of Unsaturated Chain Ends

Functionalization of polyolefins by chemical modification of preformed unsaturated chain-ends has been applied for the introduction of several functionalities, mainly *via* hydroboration, hydrosilylation and hydroalumination reactions.[3,126] However, this approach has several disadvantages, which limits its potential for the synthesis of end-functionalized polyolefins. The chain-end unsaturation of the preformed polyolefins needs to be nearly quantitative to ensure complete functionalization, although several ETM systems exist to date that mainly terminate *via* β-hydride elimination (*e.g.* $Et(Ind)_2ZrCl_2$ that has been applied by Chung[174]). The subsequent chemical transformations are hampered by the very low concentration of the double bonds in the polymer mixture and, even more importantly, by the poor solubility of these polymers in the reaction medium/solvent associated with the use of these reagents.[6,125]

Hydroalumination and subsequent oxidation have been used to functionalise polypropene with a hydroxyl group.[157,175] Later, this approach was used for hydroxy-functionalization of ethene–propene copolymers (EPR), followed by further conversion of the hydroxyl group to a methacryloyl functionality suitable for copolymerization reactions (Scheme 35).[175] Introduction of halogen functionalities at the chain end was also achieved *via* this method.[176]

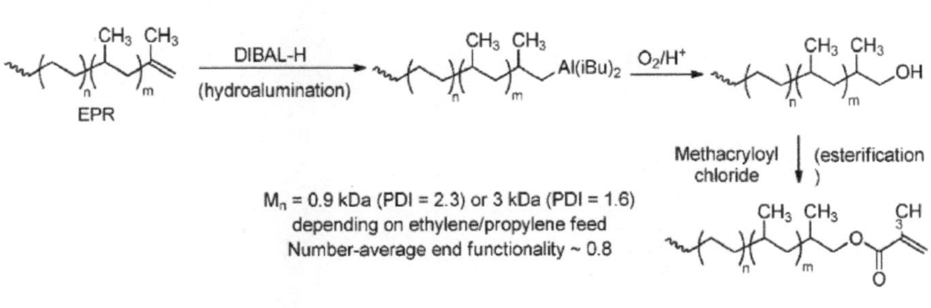

Scheme 35: Hydroalumination of an unsaturated ethene–propene copolymer used for the introduction of a methacryloyl functionality.

Hydrosilylation reactions have been applied for converting the unsaturated chain ends of polypropene into a suitable initiator for radical polymerization (Scheme 36),[177] in order to synthesize block copolymers (*vide infra*).

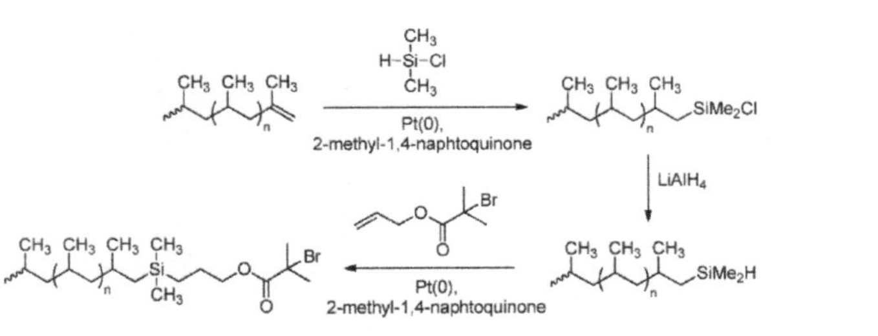

Scheme 36: Formation of a polypropene macro-initiator for radical polymerization byhydrosilylation of an unsaturated chain end.

Hydroboration is by far the most successful transformation reaction, which has been attributed to the advantageous reactivity and solubility of dialkylborane reagents in hydrocarbonmedia.[6] Polypropene or polyethene-*co*-hexadiene bearing an unsaturated chain end was transformed into borane-functionalized polypropene by a hydroboration reaction with alkylboranes.[174,178] Subsequent transformations of these newly formed polypropenes led to the formation of polyolefins end-functionalized with hydroxyl,[174,178] amine,[174] silyl[174] or maleic anhydride functionalities[179,180] (see Scheme 37).

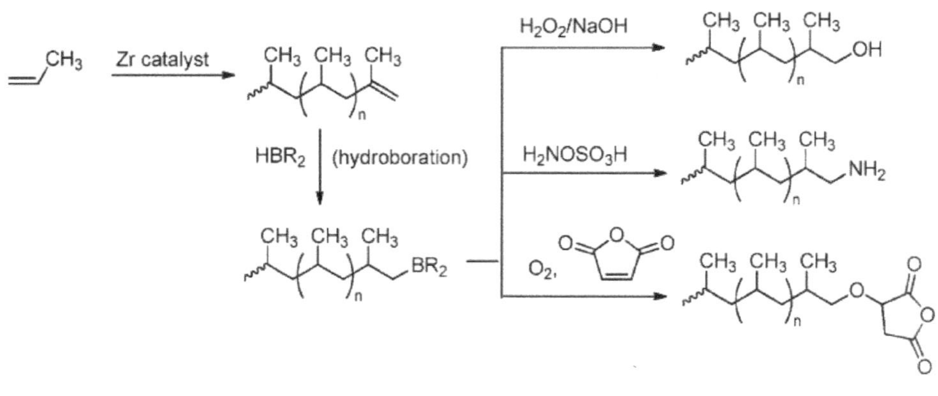

Scheme 37: End-functionalization of polypropene *via* hydroboration reactions converting the unsaturated chain ends.

SEGMENTED COPOLYMERS: BLOCK AND GRAFT COPOLYMERS

Segmented copolymers (*i.e.* block copolymers and graft copolymers) are highly desirable materials, since they offer a large quantity of functional groups while preserving the original properties of the non-functionalized polyolefin chain (*e.g.* crystallinity, melting point and hydrophobicity). Both block and graft copolymers are ideally suited to act as compatibilisers inpolymer blends, where the functionalized segment(s) provide(s) good adhesion to polar surfaces and the non-polar segment interpenetrates into polyolefin homopolymer domains.[6,181]Segmented copolymers may exhibit phase separation at the microscopic level, whether in the melt, solid state or solution (for an example, see Fig. 3).[181,182] Although phase separation of segmented copolymers is a spontaneous process driven by thermodynamics,[183] the ordering of the crystalline domains in thin films can be influenced in various ways depending on the desired properties and potential applications (for an overview see Hamley[184]). Due to the more simple molecular architecture of block copolymers compared to graft copolymers, the morphologies of (di)block copolymers exhibit a long-range lattice order while morphologies of graft copolymers often lack this long range order.[185] This long-range order makes block copolymers excellent materials for applications in microelectronics, *e.g.* as integrated circuits (for a review on the applications of block copolymers in electronics, see Kim *et al.*[186]) Another phenomenon mainly observed for block copolymers is self-assembly in solution, which is studied extensively as a mimic for protein folding, although its potential reaches much further than that.[187] However, most applications are based on functional polymers other than polyolefins and this field is therefore not discussed in detail.

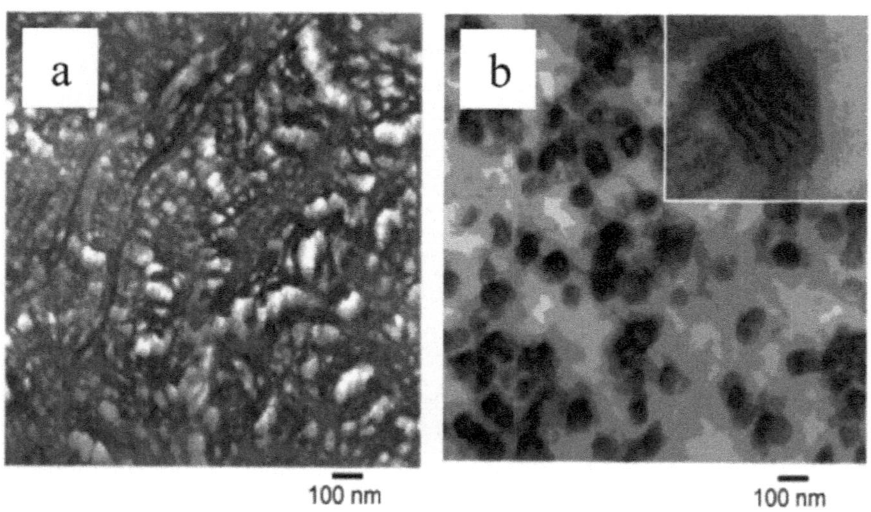

Figure 3: TEM pictures of polypropene-polymethylmethacrylate block-copolymers (PP-*b*-PMMA) showing phase separation at the microscopic level, with different morphologies for a PMMA content of 60% (a) and 75% (b).[181]

Synthesis of block copolymers

The synthesis of block copolymers based on polyolefins is rather difficult and often occurs in a multi-step fashion, involving two or more mechanistically distinct polymerizationtechniques.[188] Most techniques are not compatible with the large differences between polar and non-polar vinyl monomers, thereby hampering block copolymer synthesis *via* sequential monomer addition based on the same polymerization mechanism.[189] Radical polymerization by sequential monomer addition is hampered by the lack of reactivity towards non-polar alkenes(and especially ethene) compared to polar vinyl monomers.[49] TM catalysed block copolymer synthesis is limited by either the high oxophilicity of early TM catalysts or the low reactivity of late TMs towards polar monomers. Therefore, most successful synthesis pathways for block copolymers are based on transformation reactions, involving cross-over between distinctpolymerization mechanisms. A general overview of such transformation reactions is given in Fig. 4, including cross-over pathways that are not suitable for the synthesis of functionalized polyolefins.[188] Using this cross-over approach, first, monomer A is fed to the system, which is pre-polymerised *via* a suitable mechanism. The growing chain-end is converted into a suitable initiator for polymerization of monomer B *via* a different mechanism, and several different functionalities have been discussed in Section 4.2. Polymerization of B can occur (a) *via* the direct fresh feed of B into the prepolymer solution using the prepolymer directly as a macroinitiator, or (b) by isolation of the pre-polymer and applying that as macro-initiator in a second step (Scheme 38). The former method is convenient since it occurs in one pot, although this method can only be successful if the end-functionalization of the first pre-polymer occurs almost quantitatively and selectively to prevent fouling of the reaction mixture by homopolymers of A. The second approach yields exclusively block copolymers, although this requires the sometimes cumbersome isolation and purification of pre-polymers of A.[49,190]

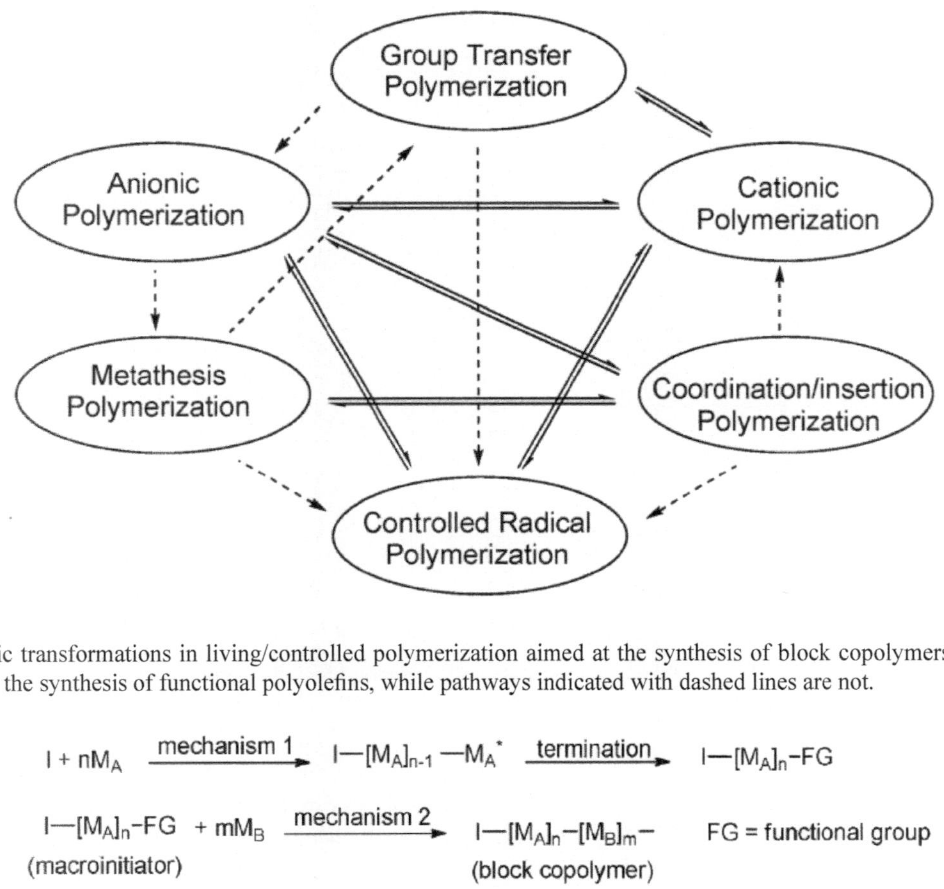

Figure 4: Mechanistic transformations in living/controlled polymerization aimed at the synthesis of block copolymers. Solid lines indicate pathways suitable for the synthesis of functional polyolefins, while pathways indicated with dashed lines are not.

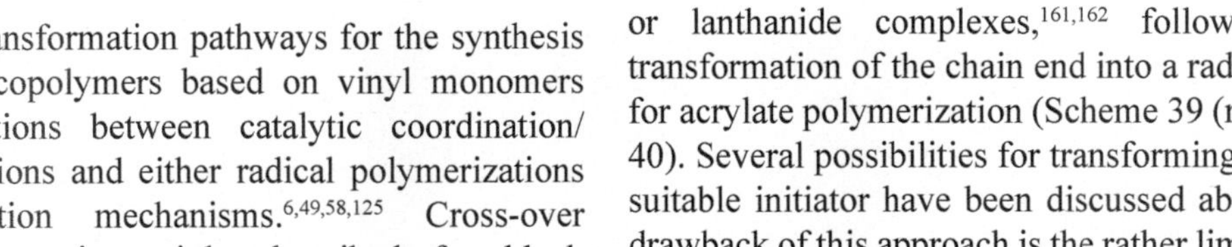

Scheme 38: Schematic representation of copolymer synthesis *via* mechanistic transformation reactions.

The two main transformation pathways for the synthesis of functional block copolymers based on vinyl monomers are cross-over reactions between catalytic coordination/ insertion polymerizations and either radical polymerizations or conjugate addition mechanisms.[6,49,58,125] Cross-over to radical mechanisms is mainly described for block copolymers of ethene/propeneand acrylates. The non-polar block is formed by catalytic coordination polymerization at either early[151,169,170,177,181,191–196]or late TM catalysts,[151,197–199] or lanthanide complexes,[161,162] followed by subsequent transformation of the chain end into a radical initiator suitable for acrylate polymerization (Scheme 39 (ref. 200) and Scheme 40). Several possibilities for transforming the chain end into a suitable initiator have been discussed above (Section 4.2). A drawback of this approach is the rather limited control over the polymerization in terms of stereoselectivity due to the use of radical polymerization.

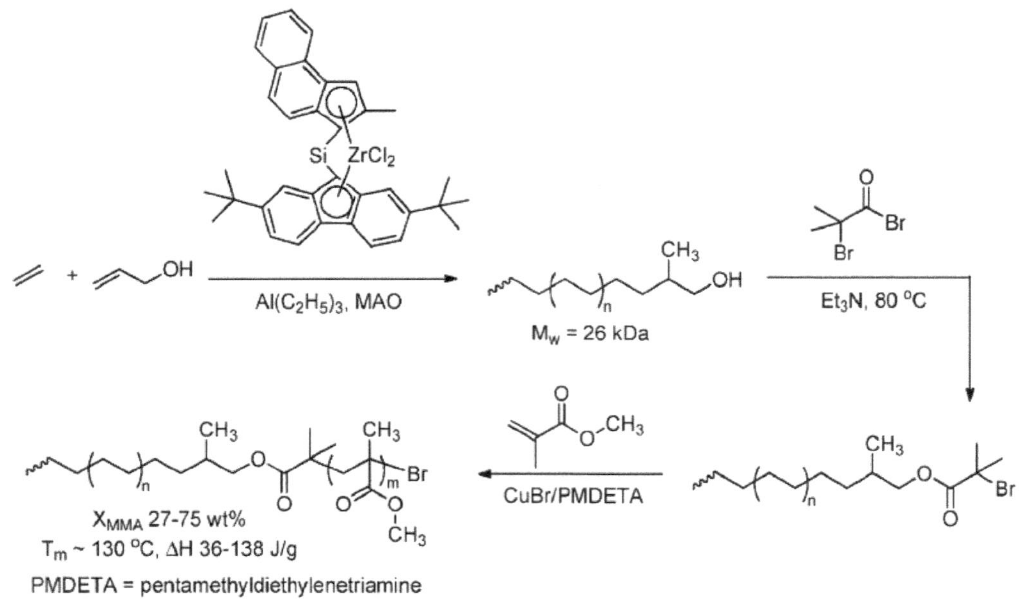

Scheme 39: PE-*b*-PMMA copolymer synthesized *via* ETM catalyzed coordination/insertion polymerization followed by cross-over to radical polymerization.

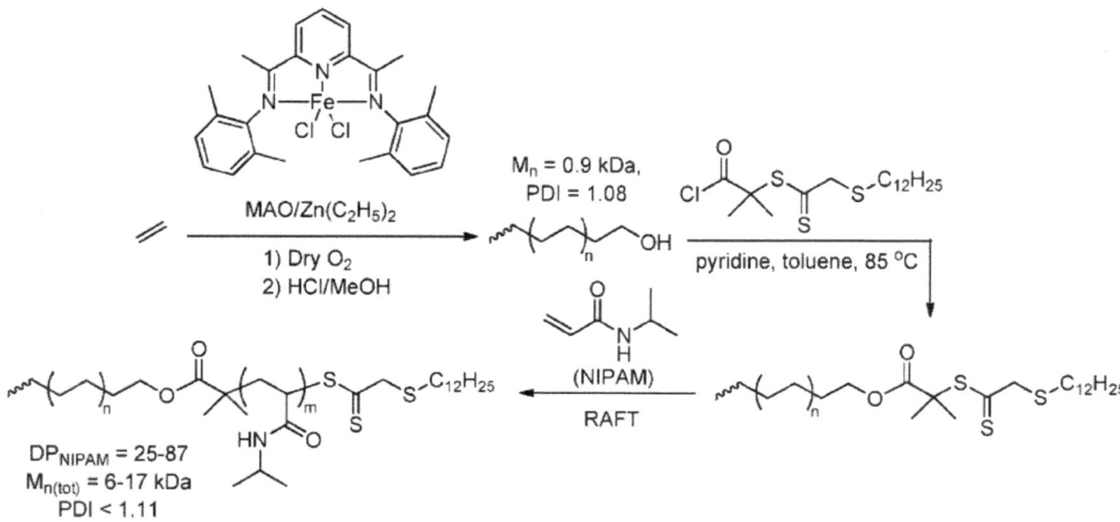

Scheme 40: Synthesis of PE-*b*-poly(NIPAM) copolymers *via* LTM catalyzed coordination/insertion polymerization and subsequent radical polymerization.

Very recently, Monteil and coworkers developed a pathway for the synthesis of multiblock copolymers from ethene and acrylates, based on a reversible shuttling process, going back and forth from catalytic insertion polymerization to radical polymerization (Scheme 41), occurring at the same metal (Ni) center.[201,202] This so-called dual radical/catalytic polymerization allows the synthesis of multiblock copolymers of various compositions that can be tuned by varying the feed ratio of both monomers and the absolute ethene pressure. As stated above, the stereoselectivity of the consecutive polar monomer insertions is moderate as a result of the underlying radical mechanism.

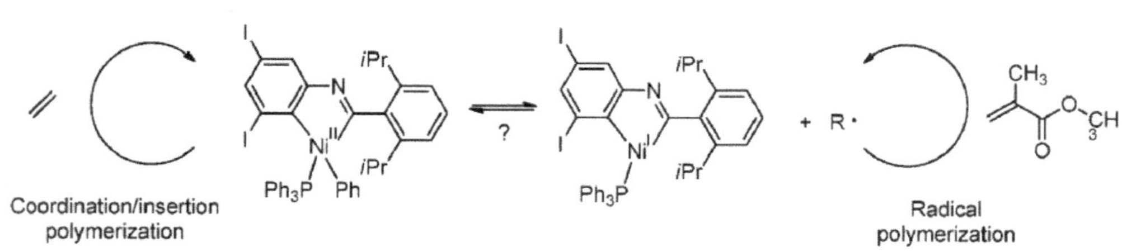

Scheme 41: Reversible shuttling process between a Ni^{II} species responsible for coordination/insertion polymerization of ethene and a Ni^{I} species active in controlled radical polymerization of acrylates. Multiple shuttling cycles result in the formation of ethene-acrylate multiblock copolymers.

Transformation of coordination/insertion polymerization to addition polymerization of acrylates (*i.e.* metal-catalysed anionic polymerization) is limited to the use of lanthanide (metallocene)[203–209] or early-TM metallocene[210,211] catalysts for the synthesis of the non-polar block (for an example, see Scheme 42). Polymerizations of the polar monomers, to some extent, take place stereoselectively, allowing the synthesis of isotactic and syndiotactic (rich) blocks.[210,212] However, this approach is only unidirectional: due to mechanistic incompatibilities block copolymers can only be obtained starting with the non-polar block. Copolymerization attempts starting with acrylate polymerization only lead to the formation of homopolymers.[58]This also implies that this cross-over approach is not suitable for the synthesis of statistical, random copolymers, which is a major drawback.

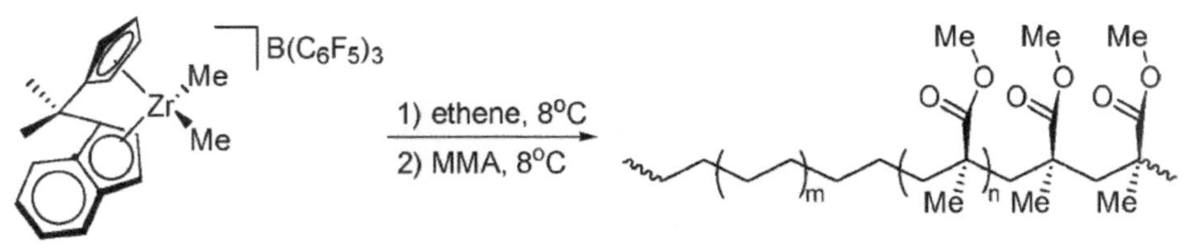

Scheme 42: Ethene–methylmethacrylate (MMA) copolymerization produced by mechanistic cross-over from coordination polymerization to addition polymerization using an ETMmetallocene catalyst.[210]

A different approach to the synthesis of block copolymers is linking two preformed polymer segments together *via* a coupling reaction. Despite the fact that coupling reactions of any kind are widely available and can nowadays be applied very efficiently for coupling of polymers (for a review see *e.g.* Schubert *et al.*[213]), this approach is not often used for linking non-polar polyolefin segments and segments functionalized with polar groups. This is due to the very limited abundance of polyolefin segments terminated with a functionality suitable for coupling reactions, since the field of end-functionalization of polyolefins has only recently started to bloom (see Section 4.2). Nevertheless, this approach will likely gain importance in the near future and the first successful example of coupling polyethene segments to polyacrylates *via* (hetero) Diels–Alder chemistry has been reported (Scheme 43).[214]

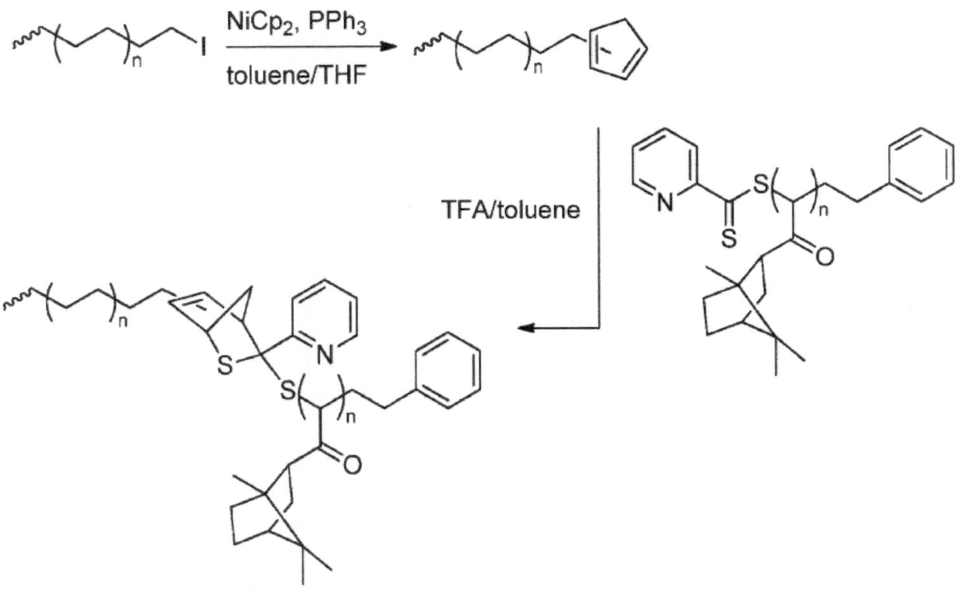

Scheme 43: Formation of a block copolymer by coupling a preformed non-polar and a preformed polar segment *via* a hetero Diels–Alder reaction.

The above-mentioned examples clearly emphasise the need for the development of alternative synthesis pathways for polar/non-polar block copolymers that allow more control over the synthesis and preferably do not require multiple consecutive reaction steps. Despite recent progress there is clearly room for optimisation in block copolymer synthesis.

Synthesis of Graft Copolymers

For similar reasons as discussed above for the synthesis of block copolymers, graft copolymers are generally prepared *via* transformation reactions involving two mechanistically distinctpolymerization mechanisms. Whereas this is rather straight-forward in the case of block copolymers (*i.e.* involving modification of one chain end), the synthesis of graft copolymers is more challenging since it requires modifications on multiple sites along the polymer backbone. Nevertheless, graft copolymers based on polyolefins have been described in the literature and the section below gives an overview of their synthesis pathways.[125]

Three general methods exist for the synthesis of graft copolymers: grafting onto (a), grafting from (b) and grafting through (c) approaches (Fig. 5) (for a more general review on graft copolymers see *e.g.* Hadjichristidis[215] and Huang[216]).

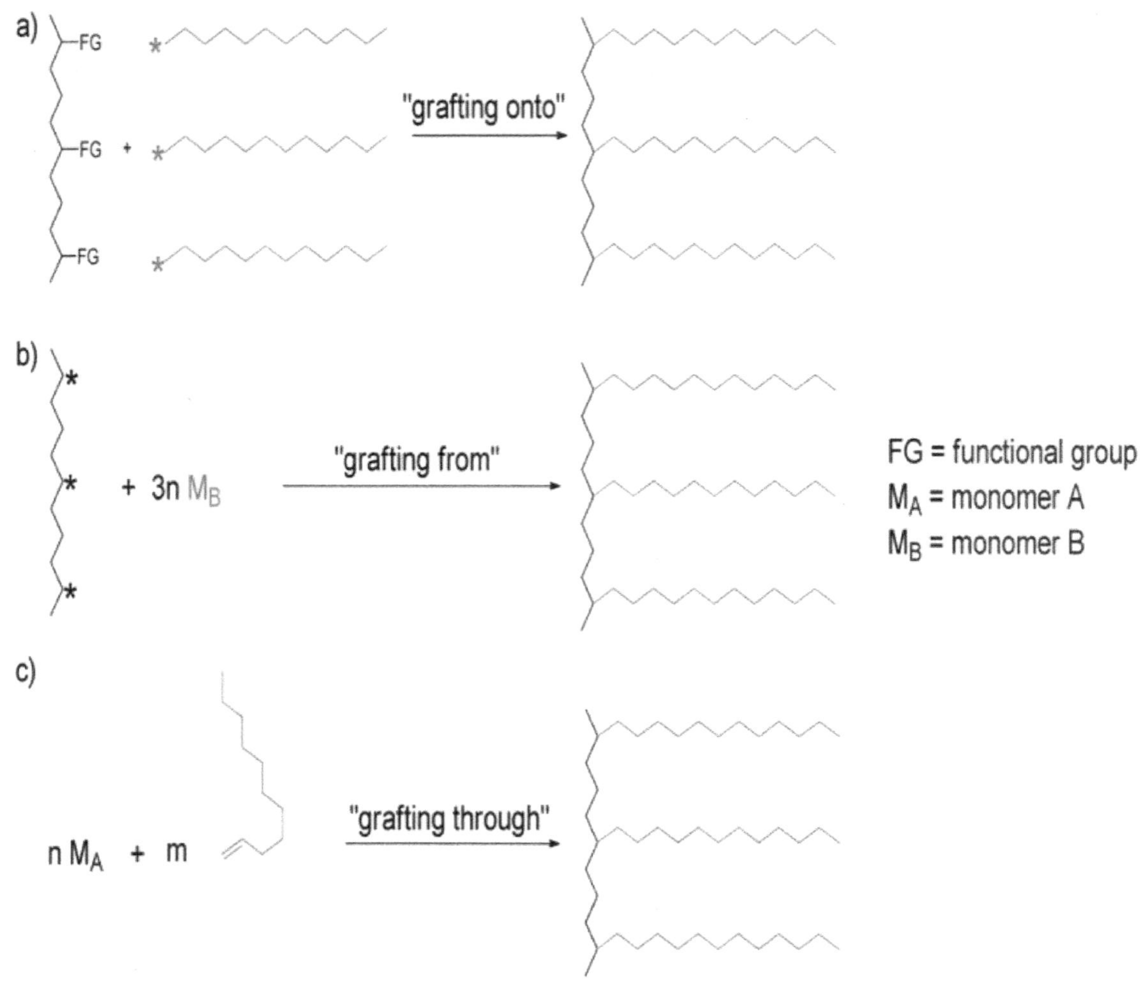

Figure 5: General methods for the synthesis of graft copolymers.

Grafting Onto

The "grafting onto" approach requires the synthesis of a polymer backbone with randomly placed functionalities, as well as the formation of end-functionalizedpolymers bearing functional groups with a complementary reactivity to those of the main chain. Coupling of the two functionalities in a separate step leads to the synthesis of graft copolymers in which the end-functionalized polymers become branches of the main chain. In principle, this technique is very suitable for obtaining graft copolymers based on polyolefins, since it allows the synthesis of the polymer backbone and the branches in separate steps (and thus by different polymerization mechanisms). Although this approach has been used to couple polyolefins to several other polymer segments (*e.g.* Li *et al.*[217]), to our knowledge, no examples of this approach exist in the literature covering the synthesis of polyolefin-*graft*-poly(polarvinyl monomer) copolymers.[216]

Grafting From

The second approach, "grafting from", involves the synthesis of a polymer backbone bearing randomly placed functionalities, which can serve as an initiator for the polymerization of a second monomer *via* a different mechanism. Similar to block copolymer synthesis, the most successful transformations are those based on catalytic coordination/insertion polymerization of the polyolefin backbone, combined with radical polymerization to form the functionalized branches.[3,125] The polymer backbone is synthesized *via* ETM-catalyzed insertion polymerization and the functionalities are incorporated *via* direct copolymerization of selected functional monomers with either ethene or propene (or both). The nature of the co-monomer depends highly on the desired follow-up polymerization reaction to form the polymeric branches. Common functionalities are *p*-methylaryl (for an example, see Chung *et al.*[218]) 9-borabicyclononane (9-BBN, see Scheme 44),[89,91] alkoxyamine,[219] hydroxyl (Scheme 45)[220,221] and vinylic groups,[174,222] some of which require additional transformations in order to serve as initiator in subsequent radical polymerization of the polar monomers. Although some of these examples are formally outside the scope of this review, they contribute to a better understanding of the possible ways to functionalize polyolefins and are therefore included in this section. Various forms of radical polymerization have been applied, including peroxyborane-initiated radical polymerization (Scheme 44),[90,223] nitroxide-mediated radical polymerization (NMRP)[219] and atom-transfer radical polymerization (ATRP) (Scheme 45),[220–222]leading to the formation of polyolefin-*graft*-polyacrylate copolymers.

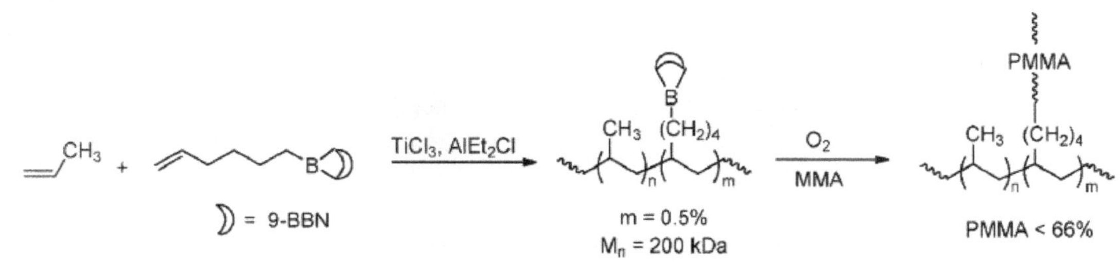

Scheme 44: Synthesis of PP-*g*-PMMA copolymers *via* ETM catalysed coordination/insertionpolymerization followed by peroxyborane-initiated radical polymerization. The boraneinitiation sites were incorporated in the polymer backbone *via* direct copolymerization of borane-functionalized monomers with propene.

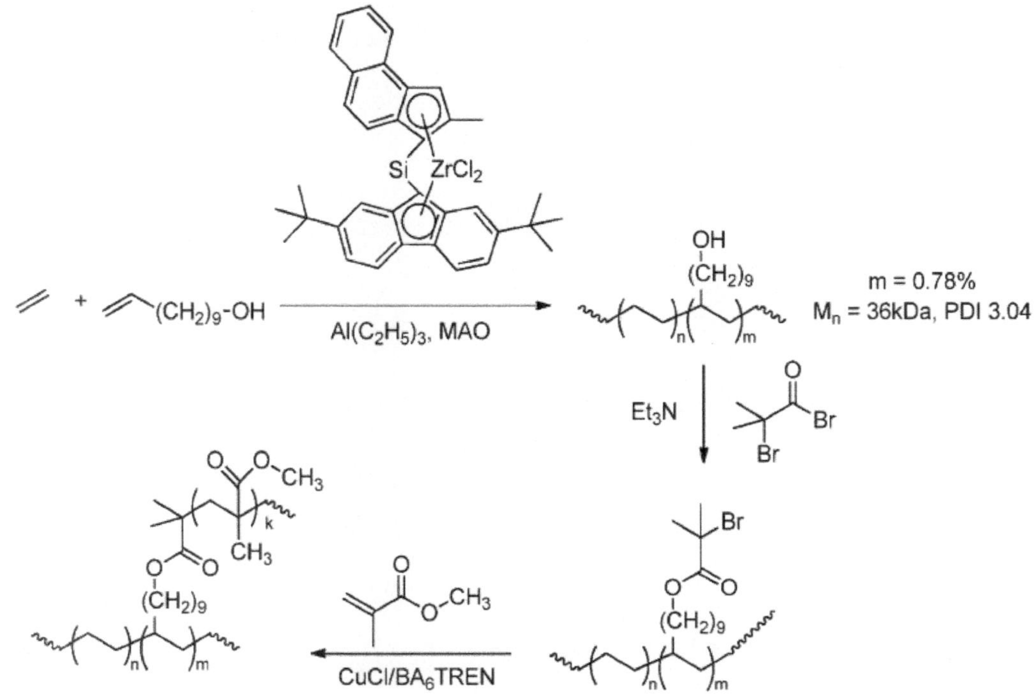

Scheme 45: ETM-catalyzed coordination/insertion polymerization followed by ATRP for the preparation of PE-*g*-PMMA copolymers.

The polar segments can also be introduced by cross-over to anionic polymerization from *e.g. p*-methylstyrene functionalities,[218] although examples in this field are scarce due to the special conditions required for incorporation of polar monomers.[224] Although the above-mentioned examples indicate that the "grafting from" approach has proven to be successful for the synthesis of graft copolymers, a major drawback of this method is that it requires multiple reaction steps.

Grafting through

In the "grafting through" or macro-monomer approach a polymer–oligomer chain with a polymerisable end group is formed, that can be copolymerized with other monomers to yield graft copolymers. The polyolefin macro-monomer is synthesized *via* coordination polymerization using either ETM (Scheme 46)[151,175] or LTM[135] catalysts, and the end group is transformed into a methacryloyl functionality that can be copolymerized with other acrylate monomers *via* ATRP. The polyolefin segment can also be prepared *via* catalytic chain shuttling polymerization using lanthanide/magnesium catalysis.[163] After introduction of a suitable functionality *via* "click" chemistry, these macro-monomers were copolymerized with other acrylates *via* radical polymerization (Scheme 47).

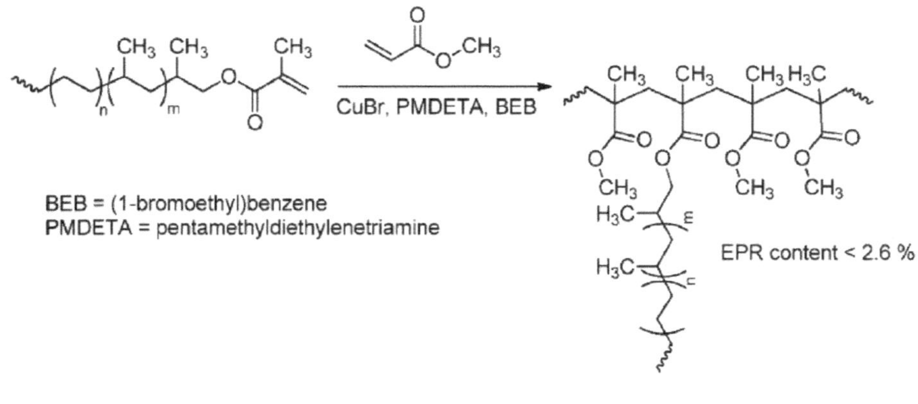

Scheme 46: PMMA-*g*-PE/PP copolymers obtained by copolymerization of MMA with PE/PP macro-monomers *via* the "grafting through" approach. For the synthesis of the macro-monomer, see Scheme 35.

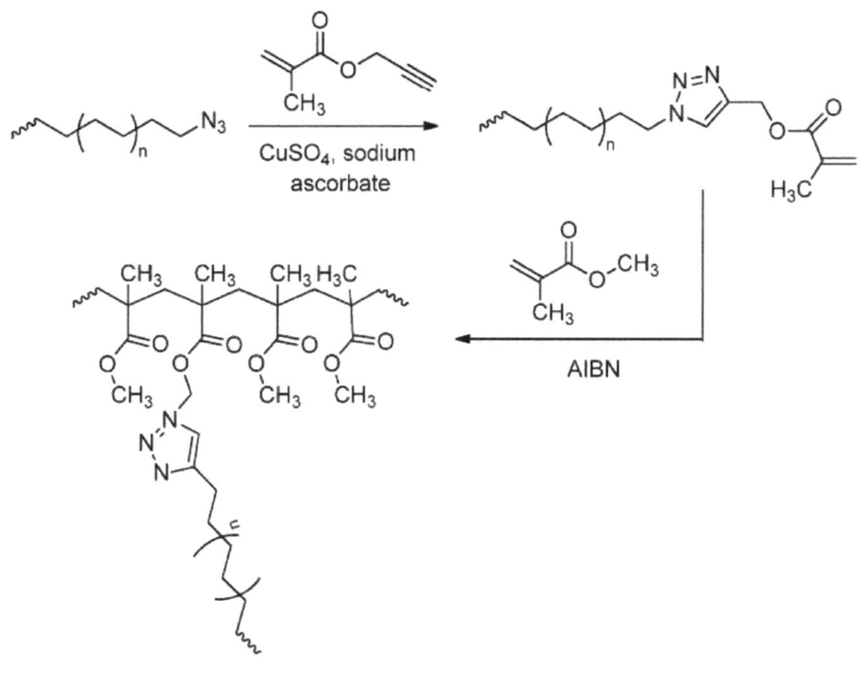

Scheme 47: PE macro-initiator prepared *via* catalytic chain-shuttling (see also Scheme 31) and subsequent 'click' reaction with propargyl methacrylate. The obtained macro-monomer was used in a radical copolymerization with methylmethacrylate to obtain PMMA-*g*-PE copolymers.

The "grafting through" approach yields polyacrylate-*graft*-polyolefin copolymers, thus being complementary to the graft copolymers obtained *via* the "grafting from" method. The length of the grafts and the grafting density can easily be tuned by changing the degree of polymerization of the side chains and the polymer backbone, respectively, allowing the synthesis of graft copolymers in a controlled way.[216] However, this approach has several drawbacks. Due to the incompatibility of the monomers and the high viscosity of the polymerization medium the conversions of the ATRP reaction are generally low, leading to the presence of unreacted macro-monomers in the copolymer mixture that need to be removed in additional steps. Similar to the systems discussed above, the "grafting through" process occurs in a multi-step fashion, although the sensitivity of the catalytic systems to the functionalities necessary for ATRP (*i.e.*halogens) often requires several additional transformations, making this approach somewhat more cumbersome.[125]

ALTERNATIVE POLAR MONOMERS

The above-described examples illustrate that the most successful routes towards functionalized polyolefins involve, at some point, polymerization *via* a coordination–insertion mechanism, catalyzed by transition metals. Transition-metal catalysis offers the advantages of more controlled polymerizations, *e.g.* in terms of polymer microstructure and stereoregularity. However, attempts to obtain functionalized polyolefins solely by transition-metal catalyzed coordination–insertion processes have met only limited success (*vide supra*). This has to do with the large differences of non-polar olefins and polar vinyl monomers in their reactivity towards different metals. Therefore, the development of alternative pathways to introduce functionalities directly into the growing polymer chain *via* a transition metal catalyzed process, based on monomers other than polar vinyl compounds is of growing importance. One such approach is the use of so-called C1 monomers that build up a polymer chain with one backbone-carbon unit at a time, as an alternative to traditional olefinic monomers or C2 monomers that extend the growing polymer chain by two carbon units at each insertion step.

The most widely used C1 monomer is carbon monoxide (CO) and ample examples exist in the literature describing *co-* and *ter*-polymerizations of CO and olefins formed *via* migratory-insertion processes, leading to successful introduction of ketone functionalities in the polymer chain (Scheme 48) (for a review on this topic see *e.g.* Drent,[225] Bianchini,[226] Sen[227] and Nozaki[5]). However, this approach is obviously not suitable for the direct synthesis of a wider variety of functionalized polymers without the need for any further modifications.

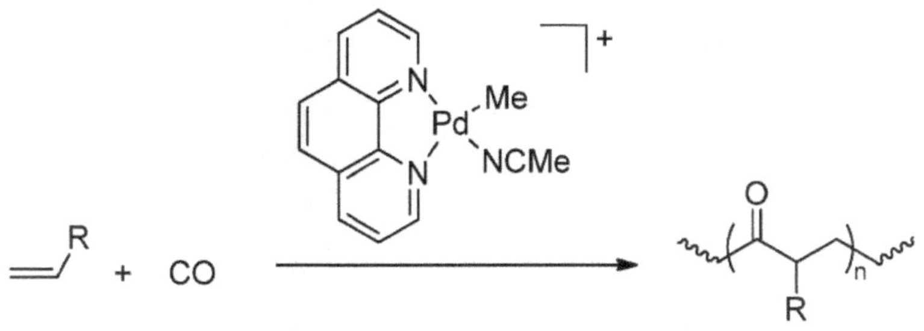

Scheme 48: Synthesis of alternating CO-olefin copolymers *via* a coordination/insertion mechanism catalyzed by Pd catalysts.

Carbenes (*i.e.* CRR' units) can be regarded as a more versatile class of C1 monomers since they allow incorporation of a wider variety of functionalities by varying the substituents. Carbenes are readily generated from diazo compounds[228] and other carbene precursors (*e.g.* ylides of sulfoxides[229] and α,α-dihalocarbons). Diazo compounds are excellent and widely applied carbene precursors due to their ease of preparation and their relative 'green' character: the only byproduct upon carbene generation is dinitrogen, while other carbene precursors give rise to formation of byproducts that are not so easily removed from the reaction mixture (*e.g.* dmso).[230,231] The *large scale applicability* of carbene precursors is somewhat limited in substrate scope due to some potential safety hazards associated with some diazo compounds. Many diazo compounds are toxic, and some of them are even potentially explosive under certain conditions.[232] Especially aliphatic diazoalkanes are very unstable and require handling at low temperatures and low concentrations. Diazocarbonyl compounds (diazoketones or diazoesters), on the other hand, are much more stable and quite safe to work with. Hence they are frequently applied as carbene precursors in several organic transformations, even in large scale industrial reactions. They are excellent substrates for the introduction of highly-functionalized carbon units, since a wide variety of substituted diazocarbonyl compounds is readily available *via* well-defined synthesis routes developed in the past decades.[233] Because the handling of aliphatic carbene precursors is more problematic, much recent effort has been put into the development of safer alternative methods that allow either *in situ* generation of the inherently unstable *aliphatic* diazo compounds (for a review see Fulton *et al.*[234]) or avoid the use ofdiazo compounds entirely (for a review see *e.g.* Müller[235]). The availability of these alternative methods to generate (aliphatic) carbenes will certainly stimulate further research into the use of carbenes in (catalytic) organic transformations in general.

In the presence of transition metals, carbenes can undergo migratory insertion into TM–C bonds, similar to insertions of alkenes and CO[236,237] (Scheme 49) and this elementary step has received quite some recent interest as a tool for the introduction of functionalities. While some main-group metals and several Lewis acids are suitable for the generation of carbenes and catalysis of the subsequent migratory insertion reactions, the most successful results were obtained with late transition-metal (LTM) catalysts and at these metal centres migratory insertions occur both in a stoichiometric and in a catalytic fashion.[238]

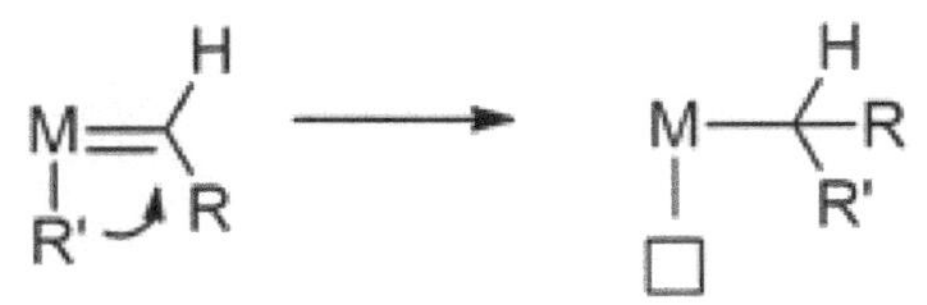

Scheme 49: Migratory insertion of carbenes.

Carbene Polymerization

If the reactions are carried out using an excess of carbene precursor with respect to the metal catalyst, multiple carbene insertions can occur resulting in the formation of oligomers orpolymers (provided that the rate constant for carbene insertion is larger than that for elimination–termination processes or follow-up reactions). As such, metal-catalysed insertions ofcarbenes provide a viable pathway for the synthesis of densely functionalized polymers that (so far) cannot be synthesised in any other way. The following sections give an overview of the development of TM catalysed coordination/insertion polymerization based on carbenes. The whole field of so-called 'C1 polymerization' including catalysis by non-LTM systems has recently been reviewed by de Bruin *et al.*[239] and Ihara.[240]

Polymers from Diazo-Alkanes and Phenyl Diazomethane

Reports dealing with the polymerization of diazoalkanes date back to the early 1900's and although many different metals have been attempted as catalysts for this reaction, most of the earlier reports deal with the activity of copper[241,242] and gold[243–245] (Scheme 50). With both metals high molecular-weight polymers were obtained (*i.e.* M_w up to 50 kDa for polymers of diazomethane) and this topic has been reviewed in the 1970's.[242,246] The activity of other metals is less extensively studied, but there are some reports mentioning the activity of nickel,[247] iridium[248] and palladium[249] towards polymerization of diazoalkanes. The behavior of the nickel system is rather unique, since it produces exclusively high-M_w polymers without the formation of byproducts (*i.e.* dimers and oligomers) and the catalyst remains active after all monomers have been converted.[247] The use of palladium catalysts, on the other hand, leads to formation of small amounts of low-M_w oligomers containing large amounts of azo-groups, indicating that the reaction is not fully selective towards carbene insertion (Scheme 51).

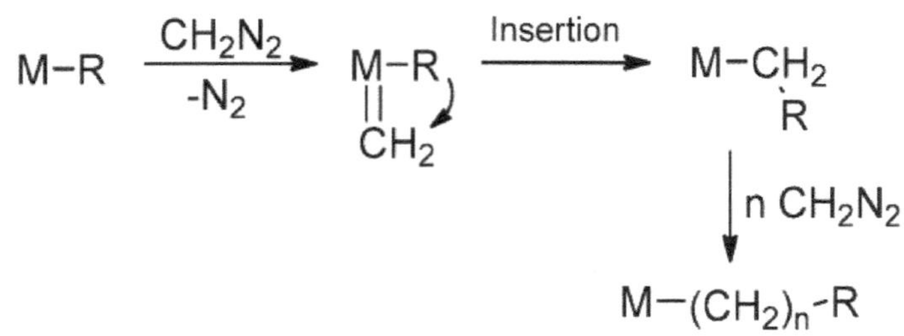

Scheme 50: General mechanism for polymerization of diazomethane by TM catalysts.

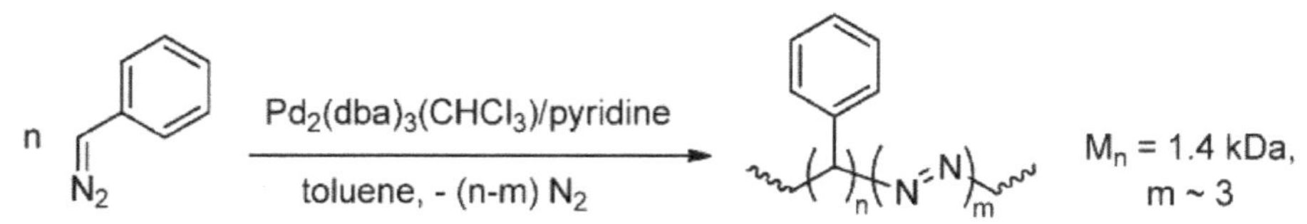

Scheme 51: Polymerization of phenyl diazomethane by $Pd_2(dba)_3(CHCl_3)$.

Aside from the above-described homogeneous catalysts also some heterogeneous metal surfaces catalyze the polymerization of carbenes (Au, Cu, Ti, Fe, Mg, W, Ni, V, Mn, Ta, Pt, Co, Zn, Cd, Cr, Al, Mo; listed in order of decreasing yield), with thin films of gold being the most active.[243,250–252] In a way, these polymerization reactions, involving multiple migratory carbeneinsertion steps, also model key steps of hydrocarbon formation in Fischer–Tropsch synthesis.[253,254]

Polymers from Carbenes Bearing Polar Functionalities

C1 polymerization of carbenes bearing polar functionalities is much more challenging due to the increased stability of the corresponding carbene precursors and most metal catalysts form only the corresponding dimers from these precursors.[247,255,256]

From 2003 on Ihara and co-workers investigated the polymerization of diazo compounds bearing polar functionalities catalysed by a variety of Pd^{II} catalysts. They successfully subjected diazoesters,[249,256–259] diazoketones[249,260–262] and diazoacetamides[263] to so-called 'poly(substituted methylene) synthesis', which allowed them to obtain various highly-functionalized oligomers/polymers in either homo- or co-polymerization processes, yielding low M_w atactic material. These results indicate that the catalyst system is not sensitive to the presence of functional groups in the monomer, emphasizing the wide range of applicability of this technique towards the synthesis of functional materials. Recently the authors have shown that higher-M_w polymers of diazoesters (up to 24 kDa) could be obtained in reasonable yields by applying a $Pd^0(NHC)/BPh_4^-$ system as a catalyst.[258,259] This approach has been extended to the synthesis of poly(ether ketone)s[264] and poly(ester ether ketone)s[265] by reaction of diazoketones with aromatic diols/cyclic ethers and diacids/cyclic ethers respectively (Fig. 6).

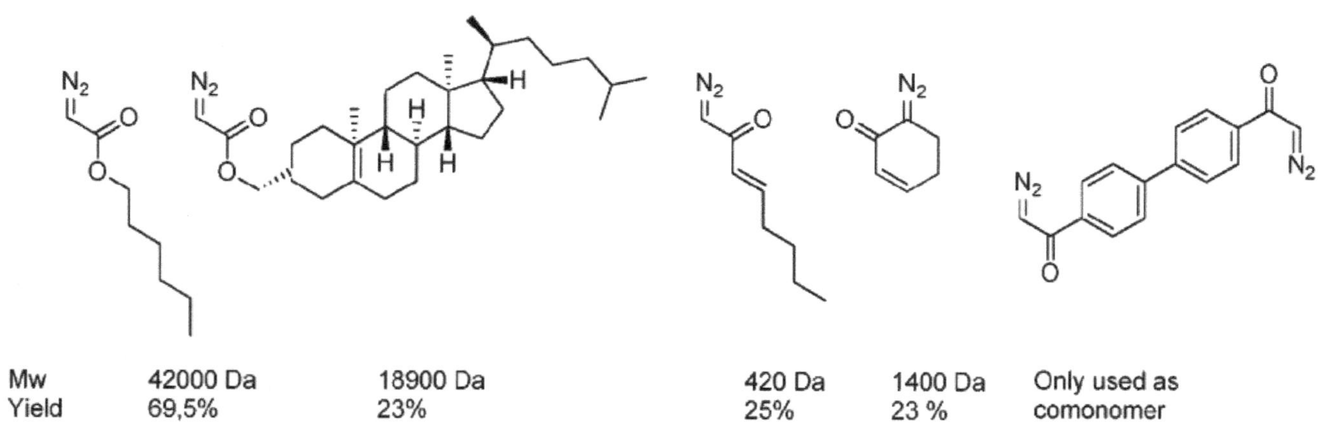

Figure 6: Structure of some diazo compounds attempted as monomers in the Pd catalysedcarbene polymerization.

Stereoselective polymerization of diazoesters was achieved by applying rhodium(diene) catalysis, giving rise to high-M_w, highly syndiotactic (co)polymers in high yields (up to 95%) (Scheme 52). The remaining products are atactic oligomers ($M_w \sim 1200$ Da) and dimers.[266–270] Using this approach, a variety of functionalized carbene precursors could be polymerised in a stereoselective way, such as *n*-butyl diazoacetate, 3-butenyl diazoacetate,[270] benzyl diazoacetate and ^{t}Bu-diazoacetate, and the reaction allows the synthesis of random and [homo-A]-[random-B > A]-type block copolymers (Scheme 53).[268] The resulting syndiotactic polymers are highly crystalline materials that reveal both thermotropic and lyotropic liquid crystalline properties.[268,270] The M_w of the polymers can be tuned by varying the Rh catalyst and the polymerization conditions.[268] Recently, it was shown that these Rh catalysts are also able to copolymerise functionalized carbenes with non-functionalized carbenes generated from either diazomethane[271] or sulfoxonium ylides[272] (Scheme 54). The resulting copolymers exhibit a blocky microstructure and the functional-group content can be tuned in a large window by varying the feed ratio of the two monomers. Copolymerization of functionalized carbenes withethene is also possible, albeit leading to quite inhomogeneous polymer mixtures.[273] These results emphasise the potential of this new technique for the development of new materials that are up to now inaccessible *via* traditional olefin polymerization.

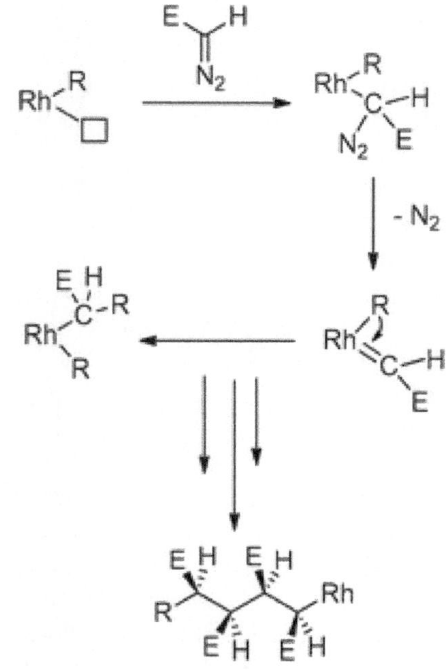

Scheme 52: Rh-mediated carbene polymerization: formation of syndiotactic polymers from diazo-esters.

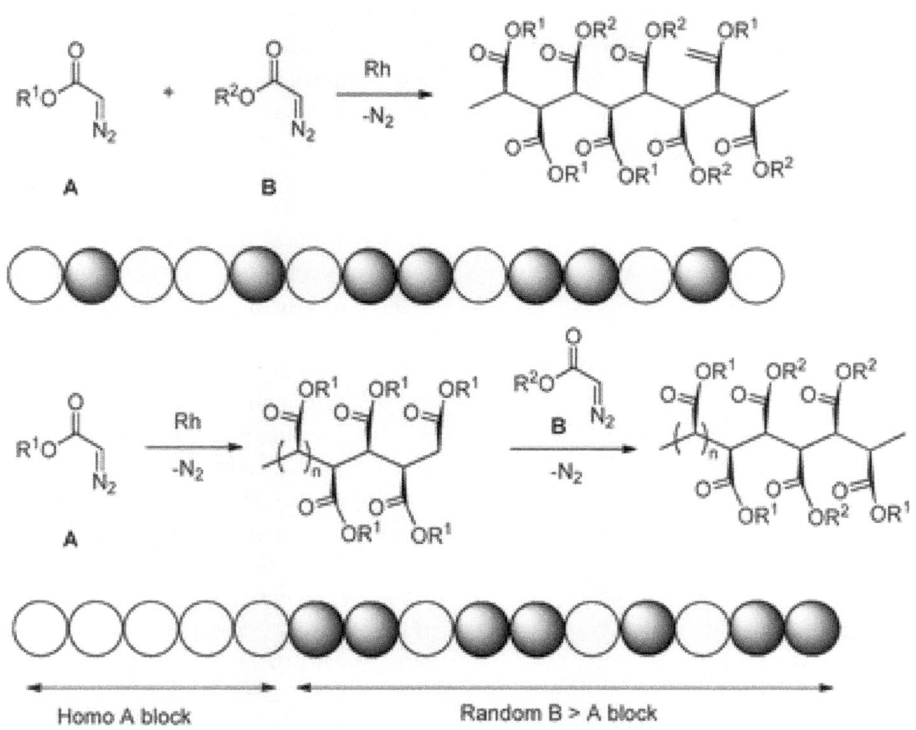

Scheme 53: Random and [homo-A]-[random-B > A] block copolymers synthesized *via* Rh-mediated carbene polymerization.

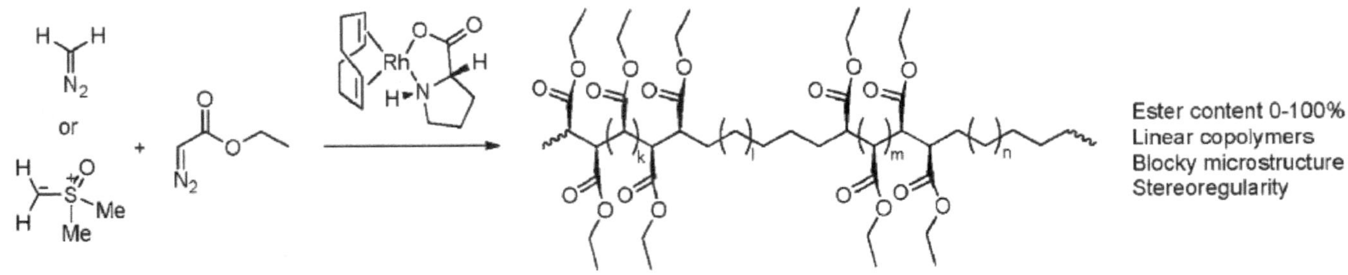

Scheme 54: Rh-catalyzed copolymerization of functionalized carbenes with eitherdiazomethane or dimethylsulfoxonium ylide, yielding copolymers with a blocky microstructure.

This Rh-mediated carbene polymerization has the advantage of giving stereoregular (syndiotactic) polymers with high molecular weights from polar C1 monomers,[239] whereas other C1polymerization techniques mainly yield atactic and/ or low-M_w material. This approach seems to work exclusively with Rh catalysts. Most of the corresponding iridium analogues did not show polymerization activity towards carbenes, with only one exception.[266,274] No other metals able to polymerise carbenes in a stereoselective manner have been reported. The stereoregularity of the polymers obtained by Rh catalysis is, according to DFT calculations, most likely a result of chain-end control during the propagation steps, since the Rh catalysts used in these reactions are non-chiral.[266] The reaction proceeds *via* a migratory insertion mechanism (Scheme 55)[266] and no or little chain transfer seems to occur.[275–278] Most recently, detailed mechanistic studies revealed that the active species responsible for this polymerization is a Rh^{III}(allyl)(alkyl) species (Fig. 7) that can be formed by modifications of rather simple Rh^{I}(diene) precursors (*e.g.* 1,5-cyclooctadiene (cod) derivatives) under the applied reaction conditions.[267,277,279] These findings are expected to lead to future improvements of this approach in terms of yields, initiation efficiencies of the catalysts and in the synthesis of stereoregular polymers with different tacticities. This clearly indicates that this new reaction has not yet been exploited at its full potential.

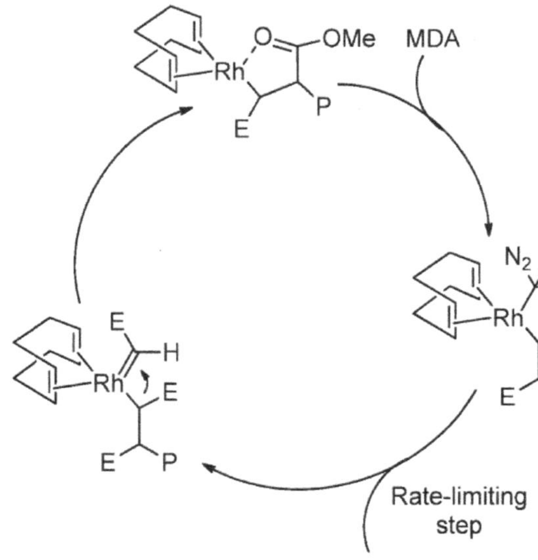

Scheme 55: Proposed mechanism for the Rh-mediated polymerization of polar functionalized carbenes, proceeding *via* a migratory insertion pathway (P = growing polymer chain , E = COOMe).

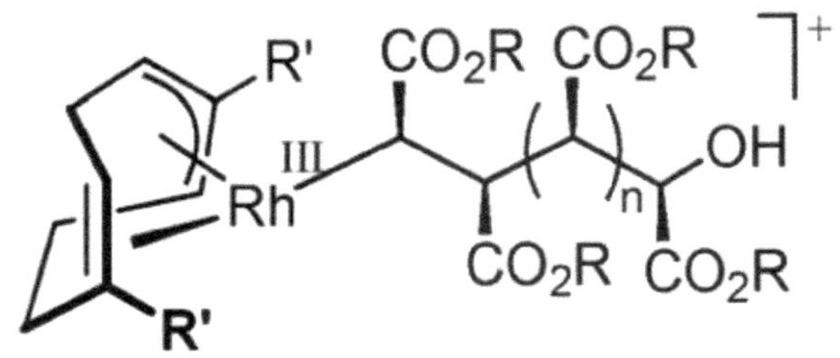

Figure 7: Rh^{III}(allyl)(alkyl) species proposed as active species for the Rh-catalyzed stereoselective polymerization of diazoesters.

CONCLUDING REMARKS

The research area reviewed here has seen a remarkable amount of activity over the past decade, resulting in the development of a variety of improved methods that allow the synthesis of functional polyolefins in a more controlled way. Each pathway described in this review has its own advantages and limitations and its potential for the synthesis of functional polyolefins has been described in detail.

Despite many recent advances in this field, the synthesis of functional polyolefins in a controlled way with tuneable amount of functionalities still remains a challenge. Randomly functionalized copolymers are still the most difficult to prepare in a controlled way, although advances in the field of late transition-metal catalyzed polymerization allow nowadays the synthesis of these materials *via* a direct copolymerization reaction of non-functionalized olefins and functional polar monomers. This approach is somewhat limited with respect to incorporation of large amounts of polar functionalities, since consecutive insertions of polar monomers take place only up to three or four times. Attempts to improve the catalysts used in this process are still ongoing, so we expect this technique to gain more potential in the future. Recent advances in the field of metathesis polymerization allowed the synthesis of analogues of these randomly functionalized polyolefins with very precise control over the distribution of functionalities along the polymer backbone. The resulting polymers have shown to be excellent structures for studying the effect of functional-group distributions on the crystal packing and physical properties. Chain-end functionalized polyolefins are readily available *via* a variety of pathways, although this approach benefits most from the recent advances in the field of chain-transfer reactions, which allow almost quantitative chain-end functionalization. The resulting materials are excellent starting points for the synthesis of block and graft copolymers in which the functional fragments are synthesized in a sequential reaction step *via* a different reaction mechanism. So far, these materials are not available *via* a single reaction mechanism due to the intrinsic reactivity differences between polar vinyl monomers and non-functionalized olefins. However, as stated above, advances in the field of late transition-metal catalysis might lead to the development of pathways for the synthesis of block and graft copolymers *via* a coordination/insertion mechanism in the (near) future.

To overcome some of these challenges, alternative pathways that lead to analogues of functional polyolefins are continuously being developed. One such promising alternative is the use of functionalized carbenes (C1 monomers) instead of the traditional polar vinyl monomers for the insertion of the polar functionalities. These carbenes can be polymerized to high molecular weight, highly stereoregular and densely functionalized polymers in a transition-metal catalysed process.

ACKNOWLEDGEMENTS

This work is part of the research programme of the Dutch Polymer Institute DPI, PO Box 902, 5600 AX Eindhoven, The Netherlands, project #646/647. We further kindly thank the European Research Council (ERC, EU, 7th framework program, grant agreement 202886-CatCIR) and the University of Amsterdam (UvA) for financial support.

NOTES AND REFERENCES

1. L. S. Boffa and B. M. Novak, *Chem. Rev.*, 2000, **100**, 1479 .
2. A. O. Patil, *Chem. Innovation*, 2000, **30**, 19 CAS.
3. J.-Y. Dong and Y. Hu, *Coord. Chem. Rev.*, 2006, **250**, 47 .
4. I. Novák, E. Borsig, L. Hrčková, A. Fiedlerová, A. Kleinová and V. Pollák, *Polym. Eng. Sci.*, 2007, **47**, 1207.
5. A. Nakamura, S. Ito and K. Nozaki, *Chem. Rev.*, 2009, **109**, 5215 .
6. T. C. Chung, *Functionalization of polyolefins*, Academic Press, London, 2002 .
7. *Metal Catalysts in Olefin Polymerization*, ed. Z. Guan, Springer, 2009, vol. 26, p. 255 .
8. J. Qiao, M. Guo, L. Wang, D. Liu, X. Zhang, L. Yu, W. Song and Y. Liu, *Polym. Chem.*, 2011, **2**, 1611 .
9. N. K. Boaen and M. A. Hillmeyer, *Chem. Soc. Rev.*, 2005, **34**, 267 .
10. A. Peacock, *J. Macromol. Sci., Part C: Polym. Rev.*, 2001, **41**, 285 .
11. J. Alonso-Villanueva, J. L. Vilas, I. Moreno, J. M. Laza, M. Rodríguez and L. M. León, *J. Macromol. Sci., Part A: Pure Appl. Chem.*, 2011, **48**, 211 .
12. M. R. Buchmeiser, *Chem. Rev.*, 2000, **100**, 1565 .
13. C. W. Bielawski and R. H. Grubbs, *Prog. Polym. Sci.*, 2007, **32**, 1 .
14. M. A. Hillmyer, W. R. Laredo and R. H. Grubbs, *Macromolecules*, 1995, **28**, 6311 .
15. S. J. McLain, E. F. McCord, S. D. Arthur, E. Hauptman, J. Feldman and W. A. Nugent, *Polym. Mater.: Sci. Eng.*, 1997, **76**, 246 CAS.
16. D. A. Bansleben, T. C. Huynh-Tran, R. L. Blanski, P. A. Hughes, W. P. Roberts, R. H. Grubbs and G. R. Hartfield, *U.S. Pat.* 6,203,293, 2001 .
17. D. A. Bansleben, T. C. Huynh-Tran, R. L. Blanski, P. A. Hughes, W. P. Roberts, R. H. Grubbs and G. R. Hartfield, *PCT Int. Pat. Appl.* WO00/18579, 2000 .
18. S. Y. Cho and U. H. Cho, *Korean Pat. Appl.* KR349626, 2002 .
19. S. Y. Cho and U. H. Cho, *Korean Pat. Appl.* KR2001036073A, 2001 .
20. H. Yang, M. Islam, C. Budde and A. E. Rowan, *J. Polym. Sci., Part A: Polym. Chem.*, 2003, **41**, 2107 .
21. C. H. Stephens, H. Yang, M. Islam, S. P. Chum and S. J. Rowan, *J. Polym. Sci., Part B: Polym. Phys.*, 2003, **41**, 2062 .
22. D. A. Bansleben, T. C. Huynh-Tran, R. L. Blanski, P. A. Hughes, W. P. Roberts, R. H. Grubbs and G. R. Hartfield, *PCT Int. Pat. Appl.* WO99/50331, 1999 .
23. D. A. Bansleben, T. C. Huynh-Tran, R. L. Blanski, P. A. Hughes, W. P. Roberts, R. H. Grubbs and G. R. Hartfield, *U.S. Pat.*, 6,153,714, 2000 .
24. D. A. Bansleben, T. C. Huynh-Tran, R. L. Blanski, P. A. Hughes, W. P. Roberts, R. H. Grubbs and G. R. Hartfield, *U.S. Pat.*, 6,506,860, 2003 .
25. O. A. Scherman, R. Walker and R. H. Grubbs, *Macromolecules*, 2002, **35**, 5366 .
26. O. A. Scherman, R. Walker and R. H. Grubbs, *Macromolecules*, 2005, **38**, 9009 .
27. J. P. Jordan, O. A. Scherman and R. H. Grubbs, *Polym. Prepr. (Am. Chem. Soc., Div. Polym. Chem.)*, 2003, **44**, 841 .
28. S. E. Lehman Jr., K. B. Wagener, L. Saunders Baugh, S. P. Rucker, D. N. Schulz, M. Varma-Nair and E. Berluche, *Macromolecules*, 2007, **40**, 2643 .
29. M. R. Buchmeiser, in *Handbook of Ring-Opening Polymerization*, ed. P. Dubois, O. Coulembier and J.-M. Raquez, Wiley-VCH Verlag GMBH, 2009, p. 197 .
30. A. Leitgeb, J. Wappel and C. Slugovc, *Polymer*, 2010, **51**, 2927 .
31. T. W. Baughman, C. D. Chan, K. I. Winey and K. B. Wagener, *Macromolecules*, 2007, **40**, 6564 .
32. K. L. Opper, B. Fassbender, G. Brunklaus, H. W. Spiess and K. B. Wagener, *Macromolecules*, 2009, **42**, 4407 .
33. K. L. Opper, D. Markova, M. Klapper, K. Müllen and K. B. Wagener, *Macromolecules*, 2010, **43**, 3690 .
34. D. J. Valenti and K. B. Wagener, *Macromolecules*, 1998, **31**, 2764 .
35. M. D. Watson and K. B. Wagener, *Macromolecules*, 2000, **33**, 5411 .
36. T. W. Baughman and E. van der Aa, *Macromolecules*, 2005, **38**, 2550 .
37. T. W. Baughman, E. van der Aa and K. B. Wagener, *Macromolecules*, 2006, **39**, 7015 .
38. M. D. Watson and K. B. Wagener, *Macromolecules*, 2000, **33**, 8963 .
39. K. L. Opper and K. B. Wagener, *Ma*, 2009, **30**, 915 .

40. J. K. Leonard, Y. Wei and K. B. Wagener, *Macromolecules*, 2012, **45**, 671 .
41. E. Boz, A. J. Nemeth, K. B. Wagener, K. Jeon, R. Smith, F. Nazirov and M. R. Bockstaller, *Macromolecules*, 2008, **41**, 1647 .
42. E. Boz, A. J. Nemeth, I. Ghiviriga, K. Jeon, R. G. Alamo and K. B. Wagener, *Macromolecules*, 2007, **40**, 6545 .
43. E. Boz, A. J. Nemeth, R. G. Alamo and K. B. Wagener, *Adv. Synth. Catal.*, 2007, **349**, 137 .
44. H. Mutlu, L. Montero de Espinosa and M. Meier, *Chem. Soc. Rev.*, 2011, **40**, 1404 .
45. K. L. Opper and K. B. Wagener, *J. Polym. Sci., Part A: Polym. Chem.*, 2011, **49**, 821 .
46. A. R. Padwa, *Prog. Polym. Sci.*, 1989, **14**, 811 .
47. *Encyclopedia of Polymer Science and Engineering*, ed. H. F. Mark, N. M. Bikales, C. G. Overberger and G. Mendes, Wiley, New York, 1986, vol. 13 .
48. S. D. Ittel, L. K. Johnson and M. Brookhart, *Chem. Rev.*, 2000, **100**, 1169 .
49. M. Ouchi, T. Terashima and M. Sawamoto, *Chem. Rev.*, 2009, **109**, 4963 .
50. A. L. Logothetis and J. M. McKenna, *J. Polym. Sci., Part A: Polym. Chem.*, 1977, **15**, 1441 .
51. A. L. Logothetis and J. M. McKenna, *J. Polym. Sci., Part A: Polym. Chem.*, 1977, **15**, 1431 .
52. A. L. Logothetis and J. M. McKenna, *J. Polym. Sci., Part A: Polym. Chem.*, 1978, **16**, 2797 .
53. S. Liu, S. Elyashiv and A. Sen, *J. Am. Chem. Soc.*, 2001, **123**, 12738 .
54. S. Liu and A. Sen, *J. Polym. Sci., Part A: Polym. Chem.*, 2004, **42**, 6175 .
55. B. Gu, S. Liu, J. D. Leber and A. Sen, *Macromolecules*, 2004, **37**, 5142 .
56. S. Liu, B. Gu, H. A. Rowlands and A. Sen, *Macromolecules*, 2004, **37**, 7924 .
57. A. Berkefeld and S. Mecking, *Angew. Chem., Int. Ed.*, 2008, **47**, 2538 .
58. E. Y.-X. Chen, *Chem. Rev.*, 2009, **109**, 5157 .
59. D. Takeuchi, *Dalton Trans.*, 2010, **39**, 311 .
60. A. W. Langer and R. R. Haynes, *U.S. Pat.*, 3,775,279, 1973 .
61. C.-E. Wilén, M. Auer and J. H. Näsman, *J. Polym. Sci., Part A: Polym. Chem.*, 1992, **30**, 1163 .
62. H. Hagihara, T. Ishihara, H. T. Ban and T. Shiono, *J. Polym. Sci., Part A: Polym. Chem.*, 2008, **46**, 1738 .
63. M. D. Purgett and O. Vogl, *J. Polym. Sci., Part A: Polym. Chem.*, 1989, **27**, 2051 .
64. K. Hakala, B. Löfgren and T. Helaja, *Eur. Polym. J.*, 1998, **34**, 1093 .
65. P. Aaltonen and B. Löfgren, *Eur. Polym. J.*, 1997, **33**, 1187 .
66. M. D. Purgett and O. Vogl, *J. Macromol. Sci., Part A: Pure Appl. Chem.*, 1987, **24**, 1465 .
67. W. Zuo, M. Zhang and W.-H. Sun, *J. Polym. Sci., Part A: Polym. Chem.*, 2009, **47**, 357 .
68. H. Terao, S. Ishii, M. Mitani, H. Tanaka and T. Fujita, *J. Am. Chem. Soc.*, 2008, **130**, 17636 .
69. M. Fernandes and W. Kaminsky, *Macromol. Chem. Phys.*, 2009, **210**, 585 .
70. K. Hakala, T. Helaja and B. Löfgren, *J. Polym. Sci., Part A: Polym. Chem.*, 2000, **38**, 1966 .
71. X. Zhang, S. Chen, H. Li, Z. Zhang, Y. Lu, C. Wu and Y. Hu, *J. Polym. Sci., Part A: Polym. Chem.*, 2005, **43**, 5944 .
72. H. Hagihara, K. Tsuchihara, J. Sugiyama, K. Takeuchi and T. Shiono, *J. Polym. Sci., Part A: Polym. Chem.*, 2004, **42**, 5600 .
73. M. M. Marques, S. G. Correira, J. R. Ascenso, A. F. G. Ribeiro, P. T. Gomes, A. R. Dias, P. Foster, M. D. Rausch and J. C. W. Chien, *J. Polym. Sci., Part A: Polym. Chem.*, 1999, **37**, 2457 .
74. X. Zhang, S. Chen, H. Li, Z. Zhang, Y. Lu, C. Wu and Y. Hu, *J. Polym. Sci., Part A: Polym. Chem.*, 2007, **45**, 59 .
75. P. Aaltonen and B. Löfgren, *Macromolecules*, 1995, **28**, 5333 .
76. P. Aaltonen, G. Fink, B. Löfgren and J. Seppälä, *Macromolecules*, 1996, **29**, 5255 .
77. N. Kawahara, S. Kojoh, S. Matsuo, H. Kaneko, T. Matsugi and N. Kashiwa, *J. Mol. Catal. A: Chem.*, 2005, **241**, 156 .
78. Y. Huang, K. Yang and J.-Y. Dong, *Macromol. Rapid Commun.*, 2006, **27**, 1278 .
79. M. J. Schneider, R. Schäfer and R. Mülhaupt, *Polymer*, 1997, **38**, 2455 .
80. V. Kotzabasakis, N. Petzetakis, M. Pitsikalis and N. Hadjichristidis, *J. Polym. Sci., Part A: Polym. Chem.*, 2009, **47**, 876 .
81. S. A. Amin and T. J. Marks, *J. Am. Chem. Soc.*, 2006, **128**, 4506 .
82. S. A. Amin and T. J. Marks, *J. Am. Chem. Soc.*, 2007, **129**, 2938 .
83. J. Liu and K. Nomura, *Macromolecules*, 2008, **41**, 1070 .
84. K. Nomura, K. Kakinuki, M. Fujiki and K. Itagaki, *Macromolecules*, 2008, **41**, 8974 .
85. D.-J. Byun, S.-M. Chin, C. J. Han and S. Y. Kim, *Polym. Bull.*, 1999, **43**, 333 .
86. S. H. Lipponen and J. Seppälä, *Organometallics*, 2011, **30**, 528 .
87. B. M. Novak and H. Tanaka, *Polym. Mater.: Sci. Eng.*, 1999, **80**, 45 CAS.
88. S. Ramakrishnan, E. Berluche and T. C. Chung, *Macromolecules*, 1990, **23**, 378 .
89. T. C. Chung and D. Rhubright, *Macromolecules*, 1991, **24**, 970 .
90. T. C. Chung and D. Rhubright, *Macromolecules*, 1993, **26**, 3467 .
91. T. C. Chung, L. Lu and C. L. Li, *Polym. Int.*, 1995, **37**, 197 .
92. S. Ito and K. Nozaki, *Chem. Rec.*, 2010, **10**, 315 .

93. L. K. Johnson, S. Mecking and M. Brookhart, *J. Am. Chem. Soc.*, 1996, **118**, 267 .

94. S. Mecking, L. K. Johnson, L. Wang and M. Brookhart, *J. Am. Chem. Soc.*, 1998, **120**, 888 .

95. C. S. Popeney, D. Camacho and Z. Guan, *J. Am. Chem. Soc.*, 2007, **129**, 10062 .

96. W. Li, X. Zhang, A. Meetsma and B. Hessen, *J. Am. Chem. Soc.*, 2004, **126**, 12246 .

97. A. Sen and S. Borkar, *J. Organomet. Chem.*, 2007, **692**, 3291 .

98. S. Borkar, H. Yennawar and A. Sen, *Organometallics*, 2007, **26**, 4711 .

99. G. Chen, X. S. Ma and Z. Guan, *J. Am. Chem. Soc.*, 2003, **125**, 6697 .

100. A. Michalak and T. Ziegler, *Organometallics*, 2003, **22**, 2660 .

101. S. J. McLain, K. J. Sweetman and L. K. Johnson, *PMSE prepr.*, 2002, **86**, 320

102. L. K. Johnson, A. Bennet, K. Dobbs, E. Hauptman, A. Ionkin, S. D. Ittel, E. F. McCord, S. J. McLain, C. Radzewich, Z. Yin, L. Wang, Y. Wang and M. Brookhart, *PMSE prepr.*, 2002,**86**, 319 .

103. M. Brasse, J. Cámpora, P. Palma, E. Álvarez, V. Cruz, J. Ramos and M. L. Reyes, *Organometallics*, 2008, **27**, 4711 .

104. L. Wang, E. Hauptman, L. K. Johnson, E. F. McCord, Y. Wang and S. D. Ittel, *Pat.*, WO 0,192,342, 2001 .

105. L. Wang, L. K. Johnson and A. S. Ionkin, *Pat.*, WO 0,192,348, 2001 .

106. C. Carlini, V. de Luise, M. Martinelli, A. M. R. Galetti and G. Sbrana, *J. Polym. Sci., Part A: Polym. Chem.*, 2006, **44**, 620 .

107. A. M. R. Galetti, C. Carlini, S. Giaiacopi, M. Martinelli and G. Sbrana, *J. Polym. Sci., Part A: Polym. Chem.*, 2007, **45**, 1134 .

108. V. C. Gibson and A. Tomov, *Chem. Commun.*, 2001, 1964 .

109. X.-F. Li, Y.-G. Li, Y.-S. Li, Y.-X. Chen and N.-H. Hu, *Organometallics*, 2005, **24**, 2502 .

110. C. Carlini, M. Martinelli, A. M. R. Galetti and G. Sbrana, *Macromol. Chem. Phys.*, 2002, **303**, 1606 .

111. E. Drent, R. van Dijk, R. van Ginkel, B. van Oort and R. I. Pugh, *Chem. Commun.*, 2002, 744 .

112. T. Kochi, K. Yoshimura and K. Nozaki, *Dalton Trans.*, 2006, 25 .

113. K. M. Skupov, P. R. Marella, M. Simard, G. P. A. Yap, N. Allen, D. Conner, B. L. Goodall and J. P. Claverie, *Macromol. Rapid. Commun.*, 2007, **28**, 2033 .

114. D. Guironnet, L. Caporaso, B. Neuwald and I. Göttker-Schnetmann, *J. Am. Chem. Soc.*, 2010, **132**, 4418 .

115. D. Guironnet, P. Roesle, T. Rünzi, I. Göttker-Schnetmann and S. Mecking, *J. Am. Chem. Soc.*, 2009, **131**, 422 .

116. T. Kochi, S. Noda, K. Yoshimura and K. Nozaki, *J. Am. Chem. Soc.*, 2007, **129**, 8948 .

117. S. Luo, J. Vela, G. R. Lief and R. F. Jordan, *J. Am. Chem. Soc.*, 2007, **129**, 8946 .

118. S. Ito, K. Munakata, A. Nakamura and K. Nozaki, *J. Am. Chem. Soc.*, 2009, **131**, 14606 .

119. T. Friedberger, P. Wucher and S. Mecking, *J. Am. Chem. Soc.*, 2012, **134**, 1010 .

120. J. C. Daigle, L. Piche and J. P. Claverie, *Macromolecules*, 2011, **44**, 1760 .

121. T. Rünzi, D. Fröhlich and S. Mecking, *J. Am. Chem. Soc.*, 2010, **132**, 17690 .

122. C. Bouilhac, T. Rünzi and S. Mecking, *Macromolecules*, 2010, **43**, 3583 .

123. J.-C. Daigle, L. Piche, A. Arnold and J. P. Claverie, *ACS Macro Lett.*, 2012, **1**, 343 .

124. S. Ito, M. Kanazawa, K. Munakata, N. Yoshikuni, J.-I. Kuroda, Y. Okimura and K. Nozaki, *J. Am. Chem. Soc.*, 2011, **133**, 1232 .

125. R. G. Lopez, F. D'Agosto and C. Boisson, *Prog. Polym. Sci.*, 2007, **32**, 419 .

126. M. J. Yanjarappa and M. Sivaram, *Prog. Polym. Sci.*, 2002, **27**, 1347 .

127. G. J. Domski, J. M. Rose, G. W. Coates and A. D. Bolig, *Prog. Polym. Sci.*, 2007, **32**, 30 .

128. Y. Doi, S. Ueki and K. Soga, *Makromol. Chem.*, 1979, **180**, 1359 .

129. Y. Doi, S. Ueki and K. Soga, *Macromolecules*, 1979, **12**, 814 .

130. Y. Doi, S. Suzuki and K. Soga, *Macromolecules*, 1986, **19**, 2896 .

131. Y. Doi and T. Keii, *Adv. Polym. Sci.*, 1986, **73/74**, 201 .

132. Y. Doi and M. Murata, *Makromol. Chem., Rapid Commun.*, 1984, **5**, 811 .

133. Y. Fukui and M. Murata, *Macromol. Chem. Phys.*, 2001, **202**, 1430 .

134. M. Brookhart, J. M. DeSimone, B. E. Grant and M. J. Tanner, *Macromolecules*, 1995, **28**, 5378 .

135. S. C. Hong, S. Jia, M. Teodorescu, T. Kowalewski, K. Matyaszewsi, A. C. Gottfried and M. Brookhart, *J. Polym. Sci., Part A: Polym. Chem.*, 2002, **40**, 2736 .

136. A. C. Gottfried and M. Brookhart, *Macromolecules*, 2003, **36**, 3085 .

137. H. Makio and T. Fujita, *Macromol. Rapid. Commun.*, 2007, **28**, 698 .

138. S. B. Amin and T. J. Marks, *Angew. Chem., Int. Ed.*, 2008, **47**, 2006 .

139. K. Ziegler, E. Holzkamp and H. Breil, *Angew. Chem.*, 1955, **67**, 541 .

140. J.-F. Pelletier, A. Mortreux and X. Olonde, *Angew. Chem., Int. Ed.*, 1996, **35**, 1854 .

141. T. Shiono, K. Yoshida and K. Soga, *Makromol. Chem., Rapid Commun.*, 1990, **11**, 285 .

142. D. R. Burfield, *Polymer*, 1984, **25**, 1817 .

143. T. Shiono, H. Kurosawa and K. Soga, *Makromol. Chem.*, 1992, **193**, 2751 .

144. T. Shiono, H. Kurosawa and K. Soga, *Macromolecules*, 1995, **28**, 437 .

145. H. Kurosawa, T. Shiono and K. Soga, *Macromol. Chem.*

Phys., 1994, **195**, 1381 .

146. T. Shiono, H. Kurosawa and K. Soga, *Macromolecules*, 1994, **27**, 2635 .
147. C. Zhang, H. Niu and J.-Y. Dong, *Appl. Organomet. Chem.*, 2011, **25**, 632 .
148. G. J. P. Britovsek, S. A. Cohen, V. C. Gibson and M. van Meurs, *J. Am. Chem. Soc.*, 2004, **126**, 10701 .
149. D. J. Arriola, E. M. Carnahan, P. D. Hustad, R. L. Kuhlman and T. T. Wenzel, *Science*, 2006, **312**, 714 .
150. L. Sita, *Angew. Chem., Int. Ed.*, 2009, **48**, 2464 .
151. H. Kaneyoshi, K. Inoue and K. Matyaszewski, *Macromolecules*, 2005, **38**, 5425 .
152. D.-J. Byun and S. Y. Kim, *Macromolecules*, 2000, **33**, 1921 .
153. C. J. Han, M. S. Lee, D.-J. Byun and S. Y. Kim, *Macromolecules*, 2002, **35**, 8923 .
154. J. Imuta, N. Kashiwa and Y. Toda, *J. Am. Chem. Soc.*, 2002, **124**, 1176 .
155. S. Tsubaki, J. Jin, T. Sano, T. Uozumi and K. Soga, *Macromol. Chem. Phys.*, 2001, **202**, 1757 .
156. S. Kojoh, T. Tsutsui, M. Kioka and N. Kashiwa, *Polym. J.*, 1999, **31**, 332 .
157. K. K. Kang, T. Shiono and T. Ikeda, *Macromolecules*, 1997, **30**, 1231 .
158. T. Shiono, K. K. Kang, H. Hagihara and T. Ikeda, *Macromolecules*, 1997, **30**, 5997 .
159. J. Mazzolini, E. Espinosa, F. D'Agosto and C. Boisson, *Polym. Chem.*, 2010, **1**, 793 .
160. R. Godoy Lopez, C. Boisson, F. D'Agosto, R. Spitz, F. Boisson and D. Bertin, *Macromolecules*, 2004, **37**, 3540 .
161. R. Godoy Lopez, C. Boisson, F. D'Agosto, R. Spits, F. Boisson, D. Gigmes and D. Bertin, *J. Polym. Sci., Part A: Polym. Chem.*, 2007, **45**, 2705 .
162. R. Godoy Lopez, C. Boisson, F. D'Agosto, R. Spitz, F. Boisson, D. Gigmes and D. Bertin, *Macromol. Rapid. Commun.*, 2006, **27**, 173 .
163. R. Briquel, J. Mazzolini, T. Le Bris, O. Boyron, F. Boisson, F. Delolme, F. D'Agosto, C. Boisson and R. Spitz, *Angew. Chem., Int. Ed.*, 2008, **47**, 9311 .
164. P.-F. Fu and T. Marks, *J. Am. Chem. Soc.*, 1995, **117**, 10474 .
165. K. Koo, P.-F. Fu and T. J. Marks, *Macromolecules*, 1999, **32**, 981 .
166. K. Koo and T. J. Marks, *J. Am. Chem. Soc.*, 1999, **121**, 8791 .
167. K. Koo and T. J. Marks, *J. Am. Chem. Soc.*, 1998, **120**, 4019 .
168. S. E. Thomas, *Organic Synthesis: The Roles of Boron and Silicon*, Oxford Chemistry Primers, New York, 1991, p. 47 .
169. G. Xu and T. C. Chung, *J. Am. Chem. Soc.*, 1999, **121**, 6763 .
170. T. C. Chung and G. Xu, *Macromolecules*, 2001, **34**, 8040 .
171. T. C. Chung and J.-Y. Dong, *J. Am. Chem. Soc.*, 2001, **123**, 4871 .
172. J.-Y. Dong, Z. M. Wang, H. Hon and T. C. Chung, *Macromolecules*, 2002, **35**, 9352 .
173. J.-Y. Dong and T. C. Chung, *Macromolecules*, 2002, **35**, 1622 .
174. T. C. Chung, H. L. Lu and C. L. Li, *Macromolecules*, 1994, **27**, 7533 .
175. H. Kaneko, S. Kojoh, N. Kawahara, S. Matsuo, T. Matsugi and N. Kashiwa, *J. Polym. Sci., Part A: Polym. Chem.*, 2005, **43**, 5103 .
176. T. Shiono and K. Soga, *Macromol. Rapid. Commun.*, 1992, **13**, 371 .
177. K. Matyaszewski, J. Saget, J. Pyun, M. Schlögl and B. Rieger, *J. Macromol. Sci., Part A: Pure Appl. Chem.*, 2002, **A39**, 901 .
178. T. C. Chung and D. Rhubright, *J. Polym. Sci., Part A: Polym. Chem.*, 1993, **31**, 2759 .
179. B. Lu and T. C. Chung, *Macromolecules*, 1998, **31**, 5943 .
180. B. Lu and T. C. Chung, *Macromolecules*, 1999, **32**, 2525 .
181. T. Matsugi, S. Kojoh, N. Kawahara, S. Matsuo and H. Kaneko, *J. Polym. Sci., Part A: Polym. Chem.*, 2003, **41**, 3965 .
182. H. Kaneko, J. Saito, N. Kawahara, S. Matsuo and T. Matsugi, *J. Polym. Sci., Part A: Polym. Chem.*, 2009, **47**, 812 .
183. S. Förster and T. Plantenberg, *Angew. Chem., Int. Ed.*, 2002, **41**, 688 .
184. I. W. Hamley, *Prog. Polym. Sci.*, 2009, **34**, 1161 .
185. M. Xenidou and N. Hadjichristidis, *Macromolecules*, 1998, **31**, 5690 .
186. H.-C. Kim, S. Park and W. D. Hinsberg, *Chem. Rev.*, 2010, **110**, 146 .
187. J. Rodríguez-Hernández, F. Chécot, Y. Gnanou and S. Lecommandoux, *Prog. Polym. Sci.*, 2005, **30**, 691 .
188. Y. Yagci and M. A. Tasdelen, *Prog. Polym. Sci.*, 2006, **31**, 1133 .
189. T. C. Chung, *Prog. Polym. Sci.*, 2002, **27**, 39 .
190. K. V. Bernaerts and F. E. Du Prez, *Prog. Polym. Sci.*, 2006, **31**, 671 .
191. Y. Doi and T. Koyama, *Makromol. Chem.*, 1985, **186**, 11 .
192. K. Inoue and K. Matyaszewski, *J. Polym. Sci., Part A: Polym. Chem.*, 2004, **42**, 496 .
193. A. Dix and S. Ptacek, *Macromol. Symp.*, 2006, **236**, 186 .
194. Q. Yi, G. Fan, X. Wen, J.-Y. Dong and C. C. Han, *Macromol. React. Eng.*, 2009, **3**, 91 .
195. T. C. Chung, H. L. Lu and W. Janvikul, *Polymer*, 1997, **138**, 1495 .
196. T. C. Chung and H. L. Lu, *J. Mol. Catal. A: Chem.*, 1997, **115**, 115 .
197. K. Zhang, Z. Ye and R. Subramanian, *Macromolecules*, 2008, **41**, 640 .

198. W.-J. Wang, P. Liu, B.-G. Li and S. Zhu, *J. Polym. Sci., Part A: Polym. Chem.*, 2010, **48**, 3024 .

199. Y. Zhao, X. Shi, H. Gao, L. Zhang, F. Zhu and Q. Wu, *J. Mater. Chem.*, 2012, **22**, 5737 .

200. Selective end-functionalization by the hydroxy-functionalized monomer is only observed for allyl alcohol. In case of longer methylene spacers the polar monomer is incorporated in the polymer chain as well.

201. A. Leblanc, J.-P. Broyer, C. Boisson, R. Spitz and V. Monteil, *Pure Appl. Chem.*, 2012, **84**(10), 2113 .

202. A. Leblanc, E. Grau, J.-P. Broyer, C. Boisson, R. Spitz and V. Monteil, *Macomolecules*, 2011, **44**, 3293 .

203. H. Yasuda, F. Furo, H. Yamamoto and A. Nakamura, *Macromolecules*, 1992, **25**, 5115 .

204. J. Gromada, T. Chenal, A. Mortreux and F. Leising, *J. Mol. Catal. A: Chem.*, 2002, **182/183**, 523 .

205. K. Tanaka, M. Furo and E. Ihara, *J. Polym. Sci., Part A: Polym. Chem.*, 2001, **39**, 1382 .

206. G. Desurmont, Y. Li and H. Yasuda, *Organometallics*, 2000, **19**, 1811 .

207. G. Desurmont, T. Tokimitsu and H. Yasuda, *Macromolecules*, 2000, **33**, 7679 .

208. M. D. K. Ingal, S. J. Joray, S. J. Duffy, D. P. Long and P. A. Biaconi, *J. Am. Chem. Soc.*, 2000, **122**, 7845 .

209. G. Desurmont, M. Tanaka, Y. Li, H. Yasuda, T. Tokimitsu, S. Stone and A. Yanagase, *J. Polym. Sci., Part A: Polym. Chem.*, 2000, **38**, 4095 .

210. H. Frauenrath, S. Balk, H. Keul and H. Höcker, *Macromol. Rapid. Commun.*, 2001, **22**, 1147 .

211. S. Collins and D. G. Ward, *J. Am. Chem. Soc.*, 1992, **114**, 5460 .

212. V. Kotzabasakis, S. Mourmouris, M. Pitsikalis, N. Hadjichristidis and D. J. Lohse, *Macromolecules*, 2011, **44**, 1952 .

213. K. Kempe, A. Krieg, C. R. Becer and U. S. Schubert, *Chem. Soc. Rev.*, 2012, **41**, 171 .

214. E. Espinosa, M. Glassner, C. Boisson, C. Barner-Kowollik and F. D'Agosto, *Macromol. Rapid. Commun.*, 2011, **32**, 1447 .

215. N. Hadjichristidis, H. Iatrou, M. Pitsikalis and J. Mays, *Prog. Polym. Sci.*, 2006, **31**, 1068 .

216. C. Feng, Y. Li, D. Yang, J. Hu, X. Zhang and X. Huang, *Chem. Soc. Rev.*, 2011, **40**, 1282 .

217. M. Hong, J.-Y. Liu, B.-X. Li and Y.-S. Li, *Macromolecules*, 2011, **44**, 5659 .

218. H. L. Lu, S. Hong and T. C. Chung, *J. Polym. Sci., Part A: Polym. Chem.*, 1999, **37**, 2795 .

219. N. B. Bowden, M. Dankova, W. Wiyatno, C. J. Hawker and R. M. Waymouth, *Macromolecules*, 2002, **35**, 9246 .

220. Y. Inoue, T. Matsugi, N. Kashiwa and K. Matyaszewski, *Macromolecules*, 2004, **37**, 3651 .

221. J. Imuta, Y. Toda and N. Kashiwa, *Chem. Lett.*, 2001, 710 .

222. C. Cao, J. Zou, J.-Y. Dong, Y. Hu and T. C. Chung, *J. Polym. Sci., Part A: Polym. Chem.*, 2005, **43**, 429 .

223. T. C. Chung and W. Janvikul, *J. Organomet. Chem.*, 1999, **581**, 176 .

224. D. Baskaran and A. H. E. Müller, in *Controlled and Living Polymerizations*, ed. A. H. E. Müller and K. Matyaszewski, Wiley-VCH Verlag GmbH & Co., Weinheim, 2009, p. 1 .

225. E. Drent and P. H. M. Budzelaar, *Chem. Rev.*, 1996, **96**, 663 .

226. C. Bianchini and A. Meli, *Coord. Chem. Rev.*, 2002, **225**, 35 .

227. A. Sen, *Synthesis of alkene-carbon monoxide copolymers and cooligomers*, Kluwer Academic Publishing, Dordrecht, 2003 .

228. M. P. Doyle, M. A. McKervey and T. Ye, *Modern Catalytic Methods for Organic Synthesis with Diazo Compounds*, John Wiley & Sons, New York, 2000 .

229. X. Zhou and K. Shea, *J. Am. Chem. Soc.*, 2000, **122**, 11515 .

230. P. T. Anastas and J. C. Warner, *Green Chemistry: Theory and Practice*, Oxford University Press, New York, 1998 .

231. Of course, if we include the energy load and the byproducts generated in their synthesis, diazocompounds are less 'green' in terms of atom and energy efficiency.

232. M. Regitz and G. Maas, *Diazocompounds – Properties and Synthesis*, Academic Press, Orlando, 1986 .

233. Z. Zhang and J. Wang, *Tetrahedron*, 2008, **64**, 6577 .

234. J. R. Fulton, V. K. Aggarwal and J. de Vicente, *Eur. J. Org. Chem.*, 2005, 1479 .

235. P. Muller, *Acc. Chem. Res.*, 2004, **37**, 243 .

236. P. W. N. M. van Leeuwen, *Homogeneous Catalysis: Understanding the Art*, Kluwer Academic Publishers, Dordrecht, 2004 .

237. P. W. N. M. van Leeuwen, K. Morokuma and J. H. van Lenthe, *Theoretical aspects of homogeneous catalysis*, Kluwer Academic Publishers, Dordrecht, 1995 .

238. N. M. G. Franssen, A. J. C. Walters, J. N. H. Reek and B. de Bruin, *Catal. Sci. Technol.*, 2011, **1**, 153 CAS.

239. E. Jellema, A. L. Jongerius, J. N. H. Reek and B. de Bruin, *Chem. Soc. Rev.*, 2010, **39**, 1706 .

240. E. Ihara, *Adv. Polym. Sci.*, 2010, **231**, 191 .

241. A. Loose, *J. Prakt. Chem.*, 1909, **124**, 185 .

242. N. Imoto and T. Nakaya, *J. Macromol. Sci., Rev. Macromol. Chem.*, 1972, **C7**(1), 1 .

243. A. G. Nasini, *Makromol. Chem.*, 1961, **44**, 550 .

244. A. G. Nasini and G. T. L. Saini, *Chimia*, 1962, 127 CAS.

245. A. G. Nasini, G. Saini and L. Trossarelli, *Pure Appl. Chem.*, 1962, **4**, 255 .

246. G. W. Cowell and A. Ledwith, *Q. Rev., Chem. Soc.*, 1970, **24**, 119 .

247. H. Werner and J. H. Richards, *J. Am. Chem. Soc.*, 1968, **90**, 4976 .

248. F. D. Mango and I. Dvouretzky, *J. Am. Chem. Soc.*, 1966, **88**, 1654 .

249. E. Ihara, M. Kida, M. Fujioka, N. Haida, T. Itoh and K. Inoue, *J. Polym. Sci., Part A: Polym. Chem.*, 2007, **45**, 1536 .

250. K. Sheshadri, S. V. Atre, Y.-T. Tao, M.-T. Lee and D. L. Allara, *J. Am. Chem. Soc.*, 1997, **119**, 4698 .

251. W. Guo and K. Jennings, *Langmuir*, 2002, **18**, 3123 .

252. D. Bai and K. Jennings, *J. Am. Chem. Soc.*, 2006, **127**, 3048 .

253. J. Lin, C. Chang, C. J. Jenks, M. X. Yank, T. H. Wentzlaff and B. E. Bent, *J. Catal.*, 1994, **147**, 250 .

254. J. P. Hindermann, G. J. Hutchings and A. Kiennemann, *Catal. Rev. Sci. Eng.*, 1993, **35**, 1 .

255. G. D. Buckley, L. H. Cross and N. H. Ray, *J. Chem. Soc.*, 1950, 2714 .

256. K. Lorey, *J. Prakt. Chem.*, 1929, **124**, 185 .

257. E. Ihara, N. Haida, M. Iio and K. Inoue, *Macromolecules*, 2003, **36**, 36 .

258. E. Ihara, Y. Ishiguro, N. Yoshida, T. Hiraren, T. Itoh and K. Inoue, *Macromolecules*, 2009, **42**, 8608 .

259. E. Ihara, H. Takahashi, M. Akazawa, T. Itoh and K. Inoue, *Macromolecules*, 2011, **44**, 3287 .

260. E. Ihara, M. Fujioka, N. Haida, T. Itoh and K. Inoue, *Macromolecules*, 2005, **38**, 2101 .

261. E. Ihara, Y. Goto, T. Itoh and K. Inoue, *Polym. J.*, 2009, **41**, 1117 .

262. E. Ihara, T. Hiraren, T. Itoh and K. Inoue, *J. Polym. Sci., Part A: Polym. Chem.*, 2008, **46**, 1638 .

263. E. Ihara, T. Hiraren, T. Itoh and K. Inoue, *Polym. J. (Tokyo)*, 2008, **40**, 1094 .

264. E. Ihara, K. Saiki, Y. Goto, T. Itoh and K. Inoue, *Macromolecules*, 2010, **43**, 4589 .

265. E. Ihara, Y. Hara, T. Itoh and K. Inoue, *Macromolecules*, 2011, **44**, 5955 .

266. E. Jellema, P. H. M. Budzelaar, J. N. H. Reek and B. de Bruin, *J. Am. Chem. Soc.*, 2007, **129**, 11631 .

267. E. Jellema, A. L. Jongerius, A. J. C. Walters, J. M. M. Smits, J. N. H. Reek and B. de Bruin, *Organometallics*, 2010, **29**, 2823 .

268. E. Jellema, A. L. Jongerius, G. Alberda van Ekenstein, S. D. Mookhoek, T. J. Dingemans, E. M. Reingruber, A. Chojnacka, P. J. Schoenmakers, R. Sprenkels, E. R. H. van Eck, J. N. H. Reek and B. de Bruin, *Macromolecules*, 2010, **43**, 8892 .

269. M. Rubio, E. Jellema, M. A. Siegler, A. L. Spek, J. N. H. Reek and B. de Bruin, *Dalton Trans.*, 2009, 8970 .

270. D. G. H. Hetteheid, C. Hendriksen, W. I. Dzik, J. M. M. Smits, E. R. H. van Eck, A. E. Rowan, V. Busico, M. Vacatello, V. Van Axel Castelli, A. Segre, E. Jellema, T. G. Bloemberg and B. de Bruin, *J. Am. Chem. Soc.*, 2006, **128**, 9746 .

271. N. M. G. Franssen, K. Remerie, T. Macko, J. N. H. Reek and B. de Bruin, *Macromolecules*, 2012, **45**, 3711.

272. A. I. Olivos-Suarez, M. P. del Río, K. Remerie, J. N. H. Reek and B. de Bruin, *ACS Catal.*, 2012, **2**, 2046 .

273. N. M. G. Franssen, J. N. H. Reek and B. de Bruin, *Dalton Trans.*, 2013 **10.1039/c3dt32941k** , accepted.

274. B. Bantu, K. Wurst and M. R. Buchmeiser, *J. Organomet. Chem.*, 2007, **692**, 5272 .

275. A. J. C. Walters, E. Jellema, M. Finger, P. Aarnoutse, J. M. M. Smits, J. N. H. Reek and B. de Bruin, *ACS Catal.*, 2012, 246 .

276. M. Finger, J. N. H. Reek and B. de Bruin, *Organometallics*, 2011, **30**, 1094 .

277. M. Finger, M. Lutz, J. N. H. Reek and B. de Bruin, *Eur. J. Inorg. Chem.*, 2012, 1437 .

278. A. J. C. Walters, O. Troeppner, I. Ivanović-Burmazović, C. Tejel, M. P. del Río, J. N. H. Reek and B. de Bruin, *Angew. Chem., Int. Ed.*, 2012, **51**, 5157 .

279. N. M. G. Franssen, M. Finger, J. N. H. Reek and B. de Bruin, *Dalton Trans.*, 2013, **42**, 8970 .

CHAPTER 19

CHEMO-, REGIO-, AND STEREOSELECTIVE IRON-CATALYSED HYDROBORATION OF ALKENES AND ALKYNES

Mark D. Greenhalgh and Stephen P. Thomas

Boronic acid derivatives have become ubiquitous in chemical synthesis. The facile stereospecific transformation of these diversely functionalized building blocks into a wide variety of functional groups has made them key intermediates in organic syntheses.[1] Alkyl boronic esters are generally easy to isolate, purify and store, and can be used in a wide variety of transformations including Suzuki–Miyaura cross-coupling reactions for the generation sp^3–sp^2 C–C bonds (Figure 1).[2] Alkyl boronic esters are commonly prepared by reaction of alkyllithium and magnesium reagents with a boron source;[1a] however these methods are limited by poor functional-group tolerance and atom economy. Transition-metal-catalysed processes have the potential to overcome these problems. Direct borylation of alkanes using Rh, Ir, Ru, and Re catalysts under photochemical or thermal conditions has been reported,[3] but these methods can suffer from forcing reaction conditions. Rhodium and iridium complexes are known to catalyse the addition of catechol- and pinacolborane to olefins under mild conditions, and with good functional group tolerance.[4] Many regio- and enantioselective examples have been reported,[4,5] however competitive dehydroboration,[6] and the relative instability of catecholborane[5b,7] can effect synthetic utility. The copper-catalysed synthesis of alkyl boronic esters from primary and secondary alkyl halides has been reported with good functional group tolerance,[2e] however long reaction times, excess B_2pin_2, and relatively high catalyst loadings were required.

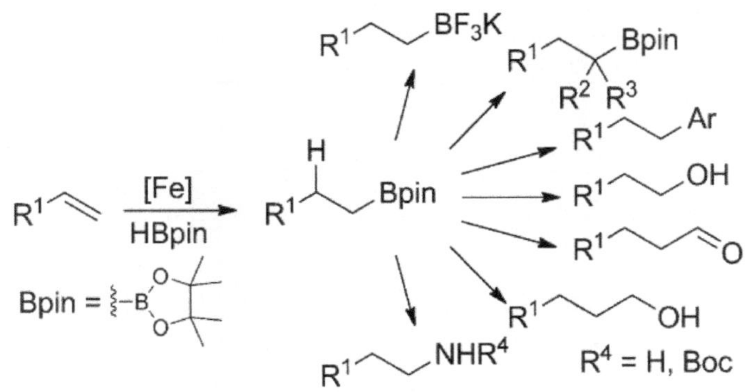

Figure 1: Selected synthetic transformations of alkylboronic esters.[1,2]

Iron offer significant advantages as a catalyst due to its low toxicity, low cost, natural abundance and sustainable long-term commercial availability.[8] Ritter reported the 1,4-hydroboration of terminal 1,3-dienes using an iron(II) iminopyridine complex, which was reduced to an active catalyst *in situ* using elemental magnesium.[9] Good to excellent regioselectivity and excellent stereoselectively were demonstrated. Enthaler has shown that $Fe_2(CO)_9$ can catalyse the hydroboration of terminal and internal alkynes with pinacolborane to give vinyl boronic esters in up to 99 : 1 dr.[10] Recently, Huang[11] and Chirik[12] have reported the iron-catalysed hydroboration of alkenes using pinacolborane. Huang found that a bipyridyl phosphine iron(II) complex activated with sodium triethylborohydride produced a highly active catalyst for the hydroboration of terminal, and 1,1-disubstituted alkenes. Chirik reported that

Citation: Mark D. Greenhalgh and Stephen P. Thomas, "Chemo-, regio-, and stereoselective iron-catalysed hydroboration of alkenes and alkynes," DOI: 10.1039/c3cc46727a.

bis(imimo)pyridine iron(0) bis(dinitrogen) complexes[13] would catalyse the addition of pinacolborane to terminal-, 1,1- and 1,2-disubstituted alkenes. Functional group tolerance has been demonstrated for tertiary amine, silyl, ether, acetal and tosyl-protected alcohol substrates; however both methods suffer from the use of highly air- and moisture sensitive pre-catalysts.

Herein we report the iron-catalysed hydroboration of alkenes and alkynes using a bench stable iron(II) pre-catalyst and pinacolborane to give alkyl and vinyl boronic esters directly. Iron(II) salts are reduced to highly active, low-valent species by reaction with a Grignard reagent,[14] which we speculated may provide simple access to catalysts for hydroboration. Using 4-phenylbutene as a model substrate, initial studies focussed on hydroboration using pinacolborane (HBpin), bis(imino)pyridine iron(II) complex [**1**-$FeCl_2$][15] (5 mol%), and tolylmagnesium bromide (TolMgBr) as activating agent in tetrahydrofuran. Using 5 mol% TolMgBr gave only a low yield of the linear boronic ester **3a** (Table 1, entry 1); however 10 and 15 mol% TolMgBr gave the linear boronic ester **3a** directly in excellent yield and complete regioselectivity (entries 2 and 3). Use of 25 mol% TolMgBr, led to a decreased yield of **3a** (entry 4). The system was equally active using 1 mol% pre-catalyst, which could be prepared *in situ* by simple combination of $FeCl_2$ (1 mol%) and free bis(imino) pyridine ligand **1** (1 mol%) prior to the addition of substrate and activator. Activation with EtMgBr (3 mol%) gave equal results to that using TolMgBr (entries 5 and 6). To demonstrate increased industrial applicability, the hydroboration was developed to operate in both toluene and under 'solvent-free' conditions.[16] A suspension of iron complex [**1**-$FeCl_2$] (1 mol%) in either toluene, or neat alkene,[17] could be activated for alkene hydroboration by the addition of *n*-BuLi (3 mol%) (entry 7). Using 'solvent-free' reaction conditions, the gram-scale hydroboration of 4-phenylbutene **2a** with pinacolborane was complete within 1 minute using just 0.2 mol% catalyst, corresponding to a catalyst turn-over frequency of 30 000 mol h^{-1} (entry 8). To the best of our knowledge this represents the most efficient iron catalyst reported to date for the hydroboration of olefins. Activation (reduction) of the iron(II) pre-catalyst [**1**-$FeCl_2$] using TolMgBr allowed the average oxidation state of iron to be calculated by quantifying the formation of bitolyl. Maximum catalytic activity corresponded to an average oxidation state of iron(I).[18,19]

Table 1: Optimization of iron-catalyzed hydroboration: solvent and activating agent^a

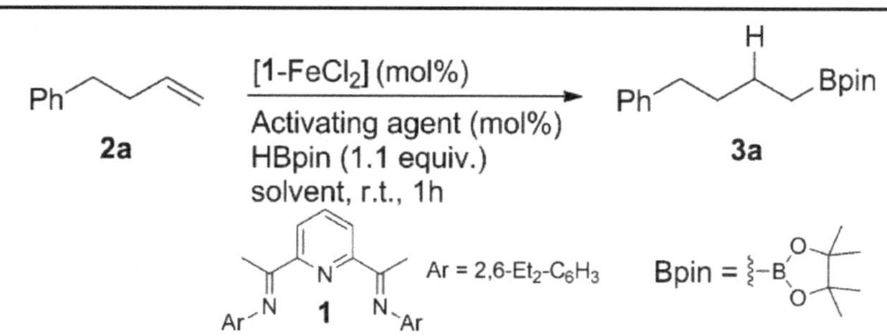

Entry	[Fe] (mol%)	Activating agent (mol%)	Solvent	Yield^b (%)
1	5	TolMgBr (5)	THF	18
2	5	TolMgBr (10)	THF	92
3	5	TolMgBr (15)	THF	91
4	5	TolMgBr (25)	THF	52
5^c	1	EtMgBr (3)	THF	90
6^c	1	TolMgBr (3)	THF	89
7^d	1	*n*-BuLi (3)	Toluene	92
8^e	0.2	*n*-BuLi (0.6)	'Solvent-free'	94^f

^{a}Conditions: 4-phenylbutene (0.7 mmol), [1-$FeCl_2$] (5 mol%), activating agent (*x* **mol%), HBpin (1.1 equiv.), solvent (0.25 M), 1 h, r.t.**

^{b} **Yield determined by** ^{1}H NMR of the crude reaction mixture using 1,3,5-trimethoxybenzene as internal standard.

^{c} **$FeCl_2$ (1 mol%) and 1 (1 mol%) complexed** ***in situ*** **in place of [1-$FeCl_2$].**

^{d} **[1-$FeCl_2$]** (1 mol%). *e* **Conditions: 4-phenylbutene (7.33 mmol), [1-$FeCl_2$]** (0.2 mol%), *n*-BuLi (0.6 mol%), HBpin (1.1 equiv.), 1 min, r.t. *f* **Isolated yield (1.80 g).**

The functional group tolerance of the hydroboration was then investigated using the *in situ* complexation of $FeCl_2$ (1 mol%) and **1** (1 mol%) in tetrahydrofuran, and activation with EtMgBr (3 mol%) (Table 2). Aryl fluoride, chloride and bromide substituted alkenes **2b–d** were tolerated under the reaction conditions giving linear boronic esters **3b–d** in excellent yield and regioselectivity, with no cleavage of the aryl-halide bond observed.[14b,20] Unprotected amine **2e** and alcohol **2f** were successfully hydroborated, and gave linear boronic esters **3e** and **3f**with complete control of regiochemistry and in excellent and moderate yield, respectively. Substrates containing more than one unsaturated group were then investigated. Ester substituted alkene **2g** was chemoselectively hydroborated in excellent yield and regioselectivity, with no observed C–O bond cleavage,[21] or ester reduction. Secondary amide **2h** reacted to give only a moderate yield of the linear boronic ester **3h**, along with unreacted starting material, suggesting catalyst deactivation. Aldimine substituted alkene **2i** was chemoselectively hydroborated at the alkene, with less than 10% aldimine reduction observed. 4-Vinylcyclohexene **2j** was also chemoselectively hydroborated at the terminal alkene giving linear boronic ester **3j** in excellent yield and regioselectivity, with the internal alkene intact. Using 'solvent-free' reaction conditions, 1,1- and 1,2-disubstituted alkenes were also suitable substrates for hydroboration. Boronic esters **3k–m** were isolated in good to excellent yield, and in the e of **3k** and **3l** with perfect regioselectivity for the linear boronic ester product. The developed methodology was also applied to the hydroboration of alkynes. (*Z*)-Vinyl boronic esters **5a** and **5b** were stereoselectively synthesised in excellent yield within one hour, with no observed *anti*-addition of pinacolborane, representing the most active and stereoselective iron catalyst reported for the hydroboration of alkynes.

Table 2: Iron-catalyzed hydroboration of olefins: scope and functional group tolerance[a]

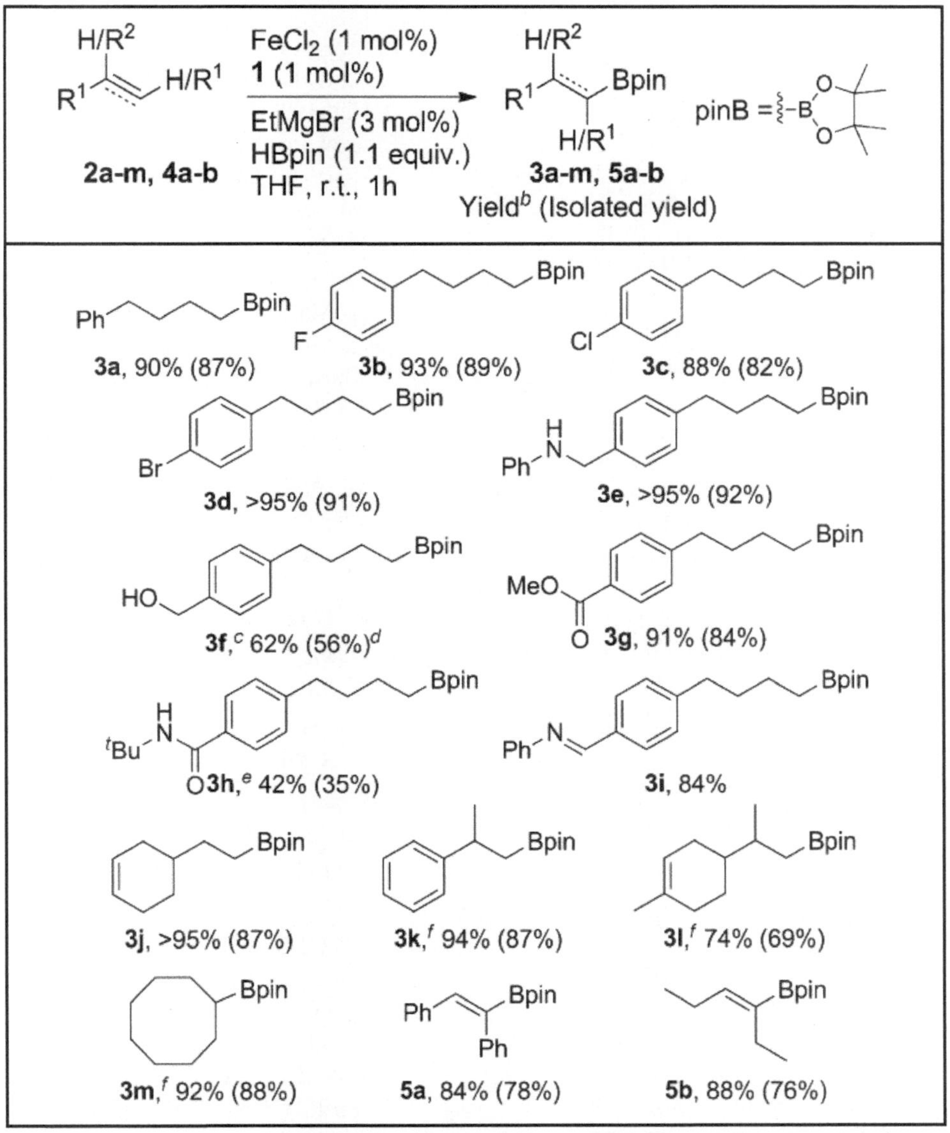

[a] Conditions: olefin (0.7 mmol), $FeCl_2$ **(1 mol%), 1 (1 mol%), EtMgBr (3 mol%), HBpin (1.1 equiv.), THF (0.25 M), 1 h, r.t.**

[b] **Yield determined by** ^1H NMR of the crude reaction mixture using 1,3,5-trimethoxybenzene as internal standard, isolated yield given in parentheses.

[c] TolMgBr (105 mol%) used.

[d] Product isolated as the diol following oxidation.

[e] **$FeCl_2$ (5 mol%), 1 (5 mol%), EtMgBr (15 mol%) used.**

[f] **Conditions: olefin (0.7 mmol), [1-$FeCl_2$]** (1 mol%), *n*-BuLi (3 mol%), HBpin (1.1 equiv.), 1 h, r.t.

Finally, the developed methodology was applied to the hydrogermylation of styrene using commercially available triethylgermanium hydride, giving the linear hydrogermylation product6 in 86% isolated yield and with complete control of regiochemistry (Scheme 1). To the best of our knowledge, this is the first example of an iron-catalysed alkene hydrogermylation,[22] but more significantly illustrates the generality of this iron catalyst in the activation of small molecules, and indicates the potential for further synthetic applications.

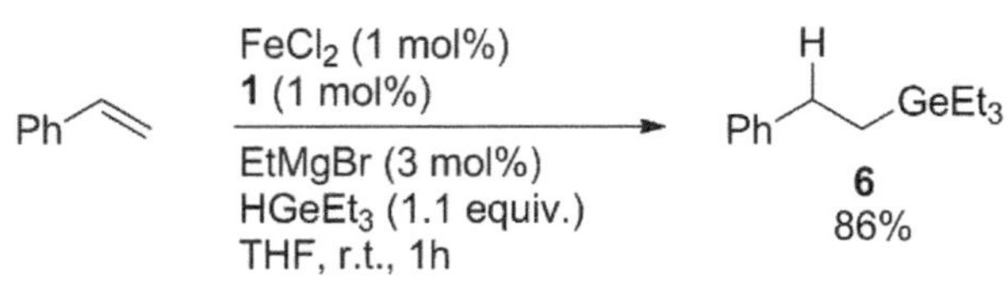

Scheme 1: Iron-catalysed hydrogermylation of styrene.

In summary, we have reported a highly functional group tolerant, operationally simple, chemo-, regio- and stereoselective iron-catalyzed hydroboration of alkenes and alkynes, which uses just 1 mol% iron catalyst [$FeCl_2$ (1 mol%), ligand **1** (1 mol%)] and 1.1 equivalents of pinacolborane at room temperature. All reagents used were commercially available, easy to handle and store, and the active iron catalyst was generated *in situ*. Terminal, 1,1- and 1,2-disubstituted aryl and alkyl alkenes and alkynes bearing an unprecedented diversity of functional groups were successfully hydroborated with excellent control of all aspects of selectivity (chemo, regio and stereochemistry). The methodology was shown to operate under 'solvent-free' conditions and on gram-scale, improving industrial applicability and the ease of product isolation. Preliminary mechanistic experiments suggest that an iron(I) catalyst may be formed under the reaction conditions. The use of an *in situ* generated iron catalyst greatly simplifies practical requirements, and should allow the non-expert to fully utilize this synthetic methodology.

M.D.G. thanks the University of Edinburgh for the provision of a studentship. S.P.T. thanks the Royal Society for generous funding.

NOTES AND REFERENCES

1. (*a*) H. C. Brown, *Organic Synthesis via Boranes*, Wiley, New York, 1975; (*b*) H. C. Brown and B. Singaram, *Pure Appl. Chem.*, 1987, **59**, 879 ; (*c*) *Science of Synthesis: Vol. 6, Boron Compounds*, ed. D. E. Kaufmann and D. S. Matteson, Georg Thieme Verlag, Stuttgart-New York, 2004; (*d*) *Boronic Acids: Preparation and Applications in Organic Synthesis and Medicine*, ed. D. G. Hall, Wiley-VCH, Weinheim, 2005; (*e*) S. N. Mlynarski, A. S. Karns and J. P. Morken, *J. Am. Chem. Soc.*, 2012, **134**, 16449 ; (*f*) J. L. Stymiest, V. Bagutski, R. M. French and V. K. Aggarwal, *Nature*, 2008, **456**, 778 ; (*g*) V. Bagutski, R. M. French and V. K. Aggarwal, *Angew. Chem., Int. Ed.*, 2010, **49**, 5142 .
2. For reviews see: (*a*) H. Doucet, *Eur. J. Org. Chem.*, 2008, 2013 ; (*b*) R. Jana, T. P. Pathak and M. S. Sigman, *Chem. Rev.*, 2011, **111**, 1417 ; For selected examples see: ; (*c*) M. Sato, N. Miyaura and A. Suzuki, *Chem. Lett.*, 1989, 1405 ; (*d*) S. D. Dreher, S.-E. Lim, D. L. Sandrock and G. A. Molander, *J. Org. Chem.*, 2009, **74**, 3626 ; (*e*) C.-T. Yang, Z.-Q. Zhang, H. Tajuddin, C.-C. Wu, J. Liang, J.-H. Liu, Y. Fu, P. G. Steel, T. B. Marder and L. Liu, *Angew. Chem., Int. Ed.*, 2012, **51**, 528 .
3. (*a*) H. Chen, S. Schlecht, T. C. Semple and J. F. Hartwig, *Science*, 2000, **287**, 1995 ; (*b*) I. A. I. Mkhalid, J. H. Barnard, T. B. Marder, J. M. Murphy and J. F. Hartwig, *Chem. Rev.*, 2010, **110**, 890 .
4. (*a*) K. Burgess and M. J. Ohlmeyer, *Chem. Rev.*, 1991, **91**, 1179 ; (*b*) I. Beletskaya and A. Pelter, *Tetrahedron*, 1997, **53**, 4957 ; (*c*) N. Miyaura, in *Catalytic Heterofunctionalization*, ed. A. Togni and H. Grützmacher, Wiley-VCH, Weinheim, 2001, ch. 1, pp. 1–45.
5. (*a*) T. Hayashi, in *Comprehensive Asymmetric Catalysis*, ed. E. C. Jacobsen, A. Pfaltz and H. Yamamoto, Springer, Berlin, 1999, ch. 9, pp. 351–367; (*b*) C. M. Crudden and D. Edwards, *Eur. J. Org. Chem.*, 2003, 4695 ; (*c*) J. M. Brown, in *Modern Rhodium-Catalyzed Organic Reactoins*, ed. P. A. Evans, Wiley-VCH, Weinheim, 2005, ch. 2. pp. 33–54.
6. (*a*) K. Burgess, W. A. van der Donk, S. A. Westcott, T. B. Marder, R. T. Baker and J. C. Calabrese, *J. Am. Chem. Soc.*, 1992, **114**, 9350 ; (*b*) S. A. Westcott and T. B. Marder, *Organometallics*, 1993, **12**, 975 ; (*c*) R. T. Baker, J. C. Calabrese, S. A. Westcott, P. Nguyen and T. B. Marder, *J. Am. Chem. Soc.*, 1993, **115**, 4367 .
7. S. A. Westcott, H. P. Blom and T. B. Marder, *Inorg. Chem.*, 1993, **32**, 2175 .
8. (*a*) S. Enthaler, K. Junge and M. Beller, *Angew. Chem., Int. Ed.*, 2008, **47**, 3317 ; (*b*) BGS (2012a), Risk list 2012, Nottingham, U.K., British Geological Survey.
9. J. Y. Wu, B. Moreau and T. Ritter, *J. Am. Chem. Soc.*, 2009, **131**, 12915 .
10. M. Haberberger and S. Enthaler, *Chem.–Asian J.*, 2013, **8**, 50 .
11. L. Zhang, D. Peng, X. Leng and Z. Huang, *Angew. Chem., Int. Ed.*, 2013, **52**, 3676 .
12. J. V. Obligacion and P. J. Chirik, *Org. Lett.*, 2013, **15**, 2680 .
13. (*a*) S. C. Bart, E. Lobkovsky and P. J. Chirik, *J. Am. Chem. Soc.*, 2004, **126**, 13794 ; (*b*) S. K. Russell, J. M. Darmon, E. Lobkovsky and P. J. Chirik, *Inorg. Chem.*, 2010, **49**, 2782 ; (*c*) A. M. Tondreau, C. C. H. Atienza, K. J. Weller, S. A. Nye, K. M. Lewis, J. G. P. Delis and P. J. Chirik, *Science*, 2012, **335**, 567 .
14. Grignard reagents are known to reduce Fe(II) salts to low-valent, highly reactive species, see: (*a*) B. Bogdanovic and M. Schwickardi, *Angew. Chem., Int. Ed.*, 2000, **39**, 4610 ; (*b*) A. Fürstner, R. Martin, H. Krause, G. Seidel, R. Goddard and C. W. Lehmann, *J. Am. Chem. Soc.*, 2008, **130**, 8773 .
15. For Fe(II) complex see: (*a*) M. S. Brookhart and B. L. Small, *Wo. Pat.*, 1998030612 A1, 1998 ; (*b*) G. J. P. Britovsek, V. C. Gibson and S. K. Spitzmesser, *Wo. Pat.*, 2000015646 A1, 2000; (*c*) R. Schmidt, M. B. Welch, R. D. Knudsen, S. Gottfried and H. G. Alt, *J. Mol. Catal. A: Chem.*, 2004, **222**, 9 . For Fe(0) analogue see ref. 13*b*.

16. (*a*) R. K. Henderson, C. Jimenez-Gonzalez, D. J. C. Constable, S. R. Alston, G. G. A. Inglis, G. Fisher, J. Sherwood, S. P. Binks and A. D. Curzons, *Green Chem.*, 2011, **13**, 854 .

17. *In situ* complexation of $FeCl_2$ and **1** in toluene or neat alkene not possible due to the low solubility of the Fe(II) complex.

18. See ESI‡ for further details.

19. For recent examples of Fe(I) in cross-coupling reactions see: (*a*) J. Kleimark, A. Hedström, P.-F. Larsson, C. Johansson and P.-O. Norrby, *ChemCatChem*, 2009, **1**, 152 ; (*b*) A. Hedström, U. Bollmann, J. Bravidor and P.-O. Norrby, *Chem.–Eur. J.*, 2011, **17**, 11991 ; (*c*) J. Kleimark, P.-F. Larsson, P. Emany, A. Hedström and P.-O. Norrby, *Adv. Synth. Catal.*, 2011, **354**, 448 ; (*d*) C. J. Adams, *et al.* , *J. Am. Chem. Soc.*, 2012, **134**, 10333 ; (*e*) R. B. Bedford, E. Carter, P. M. Cogswell, N. J. Gower, M. F. Haddow, J. N. Harvey, D. M. Murphy, E. C. Neeve and J. Nunn, *Angew. Chem., Int. Ed.*, 2013, **52**, 1285 ; (*f*) M. Guisán-Ceinos, F. Tato, E. Buñuel, P. Calle and D. J. Cárdenas, *Chem. Sci.*, 2013, **4**, 1098 .

20. (*a*) T. Hatakeyama, Y. Kondo, Y. Fujiwara, H. Rakaya, S. Ito, E. Nakamura and M. Nakamura, *Chem. Commun.*, 2009, 1216 ; (*b*) W. M. Czaplik, S. Grupe, M. Mayer and A. Jacobi von Wangelin, *Chem. Commun.*, 2010, **46**, 6350 .

21. R. J. Trovitch, E. Lobkovsky, M. W. Bouwkamp and P. J. Chirik, *Organometallics*, 2008, **27**, 6264 .

22. For Fe-catalysed hydrogermylation of alkynes see: M. Itazaki, M. Kamitani and H. Nakazawa, *Chem. Commun.*, 2011, **47**, 7854 .

CHAPTER 20 SPECIFIC AND QUANTITATIVE LABELING OF BIOMOLECULES USING CLICK CHEMISTRY

Kenichi Horisawa

Keywords: Click chemistry, fluorescent labeling, metabolic labeling, bioorthogonal reaction

Specific and highly efficient fluorescent labeling techniques for biomolecules, especially for proteins, are required for the quantitative analyses of bio-phenomena and for subsequent systems biology. Although expression of exogenous proteins fused with fluorescent tags, such as green fluorescent protein, is the most widely used method for quantitative bio-analysis, the following problems need to be considered carefully: (1) precise stoichiometric control in living cells is difficult, and (2) the bulkiness of the fluorescent tags restricts analysis of the inherent physical and biological properties of the proteins. Therefore, novel techniques to specifically and stoichiometrically label intrinsic proteins or other biomolecules in living cells should be developed. Click chemistry reactions (e.g., Huisgen cycloaddition and Staudinger ligation) are the most promising approaches for this purpose, because these chemical reactions have following advantages: (1) bioorthogonal reactions; (2) mild reaction conditions suitable for fragile biomolecules, cells, and tissues; (3) extremely high reaction ratio; (4) small size of the functional groups for the cross-coupling reactions; (5) stable covalent bonding; and (6) simple metabolic labeling procedures in living cells, using various biomolecular analogs. Diverse quantitative biological studies have been carried out using this technology (e.g., quantification of novel synthesized proteins and observation of post-translational modifications). In this review, I explain the basics of chemical probing with click chemistry, and discuss its recent applications in the field of quantitative biology. Furthermore, I discuss the capability, significance, and future of the chemical probing of proteins, with an emphasis on the use of click chemistry in the field of the quantitative biology.

QUANTITATIVE FLUORESCENCE IMAGING OF BIOMOLECULES IN LIVING CELLS

Fluorescent imaging tools such as fluorescent microscopy are one of the most efficient and widely used modern techniques in the life sciences for analyzing the quantitative behavior of biomolecules in living cells, tissues, and organisms (Stephens and Allan, 2003). Although a lot of alternative imaging technologies exist, such as electron microscopy, autoradiography, and immunochemistry, the fluorescent labeling of biomolecules and their subsequent observation with various optical instruments shows greater advantages, especially in the area of high temporal resolution, as this is one of the most important factors that needs to be analyzed for understanding biomolecular dynamics. Around the end of the last century and at the beginning of this century, the application of green fluorescent protein (Shimomura et al., 1962; Xue et al., 1993; Chalfie et al., 1994; Heim and Tsien, 1996) and its derivatives (Shaner et al., 2005) drastically improved the available fluorescence imaging methods for live samples. The huge impact of genetic fluorescent labeling in living cells can be seen from the wide variety of available fluorescent proteins. Indeed, the Nobel prize was awarded to Drs. Shimomura, Chalfie, and Tsien, in 2008 for the discovery and application of green fluorescent protein.

An alternative to fluorescent protein tagging is the use of fluorescent chemical compounds. These chemicals have long been employed in the field of bioscience. Table 1 shows the difference in the properties of fluorescent proteins and chemicals. As live imaging tools, fluorescent proteins are more predominantly used as compared to their chemical counterparts, because the genetic labeling procedure is very easy and reliable. However, fluorescent proteins also show certain disadvantages, especially during quantitative analysis (Table 1). The biggest issue of the protein-tagging method is the difficulty in controlling the stoichiometry of the target proteins inside the cell. Basically, the number of the proteins is controlled at the transcriptional level in living cells. Although the transcriptional activity can be roughly regulated through an appropriate choice of promoters or other artificial molecular systems, strict stoichiometric control is beyond the capability of the current molecular biological technology. The second problem pertains to the adverse effects of the tagged

Citation: Horisawa K (2014) Specific and quantitative labeling of biomolecules using click chemistry. Front. Physiol. 5:457. doi: 10.3389/fphys.2014.00457.

fluorescent proteins. The molecular size of the fluorescent proteins is larger than that of the fluorescent chemicals and other widely used conventional protein tags (e.g., FLAG, HA, V5, T7, and Myc tags). The bulkiness of the fluorescent protein tag is likely to affect the behavior, stability, and function of the target proteins.

Table 1. Comparison of fluorescent proteins and chemicals

Property	Fluorescent proteins	Fluorescent chemicals
Molecule size	Large	Small
Timing for labeling	During translation	Anytime
Transgene expression	Essential	Nonessential
Labeling position	Basically C/N-terminus	Reactive side-chains or C/N-terminus
Variation	Relatively narrow	Wide
Labeling efficiency	High	Case by case

For the reasons stated above, caution must be exercised when using fluorescent proteins for quantitative analyses. On the other hand, some fluorescent chemicals exhibit properties appropriate for the quantitative fluorescent observation of biomolecules, including small molecule size, various labeling positions, nonessentiality of transgene expression, and wide variation of their optical spectrum (Jung et al., 2013). However, the efficient and specific conjugation of fluorescent chemicals to the target proteins poses a major problem. For example, the NHS ester and isothiocyanate coupling reactions are the most general methods for the fluorescent labeling of proteins, although their target functional (amino) group is present not only in the target protein but also in other biomolecules such as DNA and RNA. Therefore, this makes it impossible to label target molecules specifically if presented with a mixture of biomolecules.

Click chemistry reactions have been recently developed for chemical coupling. This chemical method has the potential to be a breakthrough in the field of live fluorescence imaging. In the following sections, I introduce some of the applications of click chemistry for cell and tissue imaging, and discuss the various applications of this technology in the field of quantitative biology.

CLICK CHEMISTRY IN BIOLOGICAL STUDIES

Click chemistry does not correspond to one particular chemical reaction. This concept is related to the use of novel chemical reactions, as proposed by Prof. K. Barry Sharpless of the Scripps Research Institute (Kolb et al., 2001). A click chemistry reaction shows the following properties: (1) it uses a solvent that is benign or easily removable, such as water; (2) it only generates inoffensive byproducts; (3) it gives very high chemical yields; (4) it does not need extremely high temperature or pressure; and (5) the products from the reactions are physiologically stable. All of the characteristics above are regarded to be suitable for the chemical labeling of biomolecules, because almost all biomolecules, especially biopolymers, are fragile in the extreme conditions necessary for carrying out the standard chemical reactions. Thus, the number of applications focusing on the use of click chemistry in the life sciences is increasing every year (Best, 2009; Lang and Chin, 2014a).

Among the various click chemistry reactions, the azide-alkyne Huisgen cycloaddition is the most widely employed reaction in biological studies. This is a coupling reaction between the azide and alkyne groups, which form a very stable triazole ring as a linker (Figure 1A) (Huisgen, 1963). For the progression of this reaction, no additional factors are needed, such as heating and high pressure; the only requirement is the use of a monovalent copper ion as a catalyst (Rostovtsev et al., 2002; Tornøe et al., 2002). This reaction is termed the copper-catalyzed azide-alkyne cycloaddition (CuAAC). The most important characteristic of the CuAAC is that the azide and the alkyne do not react with any other molecules inside the cells and tissues. This property enables highly bioorthogonal fluorescent labeling in living and complex samples (Best, 2009; Lang and Chin, 2014a).

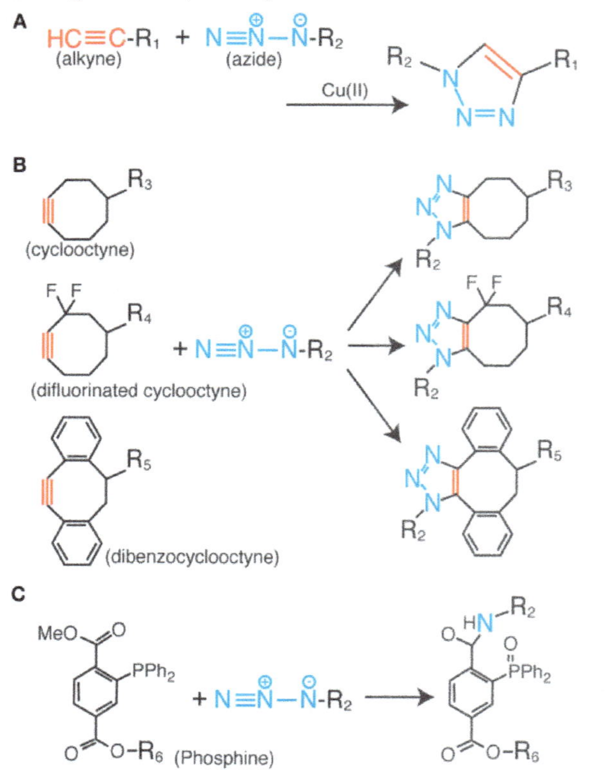

Figure 1. Huisgen cycloaddition and Staudinger ligation. (A) Copper-catalyzed azide-alkyne cycloaddition (CuAAC). (B) Copper-free azide-alkyne cycloaddition reactions using cyclic alkynes. (C) Staudinger-Bertozzi ligation.

Although the fluorescent labeling of various biomolecules such as proteins, peptides, sugar chains, DNAs, RNAs, and lipids via CuAAC has proven to be successful (Lahann, 2009; Lang and Chin, 2014a), several problems still remain to be solved. The most serious issue for CuAAC is the cytotoxicity of the copper ion (Boyce and Bertozzi, 2011). High concentrations of this metal ion catalyst make it impossible to fluorescently label living cells and organisms. As a solution to the problem, copper-free click reactions have been developed in recent years. In these reactions, cyclic derivatives of the alkynyl group [e.g., cyclooctyne (Agard et al., 2004), difluorinated cyclooctyne (Baskin et al., 2007), and dibenzocyclooctyne (Sletten et al., 2010)] are employed as reactive partners for the azide group (Figure 1B). These kinds of cyclic alkyne derivatives harness the intrinsic energy from their distorted structures; therefore, no additional energy or catalysis is needed. Some of these derivatives are already commercially available as fluorescence labeling reagents for living cells, and have been successfully employed in live cell analyses. However, the relatively bulky size of the cyclic alkyl groups affects the membrane permeability of the labeling reagents; therefore, almost all of the previous examples are limited to extracellular labeling studies.

The Staudinger-Bertozzi ligation is regarded as an effective alternative to CuAAC. This chemical reaction also involves a coupling reaction between an azide and a phosphine (Figure 1C) (Saxon and Bertozzi, 2000). Similar to the azide and the alkyne, the phosphine group also does not react with any of the functional groups on the biomolecules. However, the Staudinger-Bertozzi ligation has some useful characteristics for the fluorescent labeling of biomolecules in living cells. Unlike CuAAC, no catalyst is necessary for the coupling reaction between the azide and the phosphine groups. Indeed, several studies showed successful live cell fluorescent imaging with this technology (Chang et al., 2007; Hangauer and Bertozzi, 2008). Although this technology has a potential to be a universal tool for the live cell imaging, the membrane permeability of phosphine-dyes is a remaining issue to be solved.

More recently, novel bioorthogonal uses of "click" reactions have been published (Kodama et al., 2007; Song et al., 2008; Tong et al., 2009; Nguyen et al., 2011). Among them, the most promising reaction for live cell imaging study is the inverse-electron-demand Diels-Alder cycloaddition, which is a coupling reaction between strained alkenes and tetrazines (Blackman et al., 2008). A lot of studies for the live cell imaging by using the reaction have been published in recent years (Selvaraj et al., 2011; Lang et al., 2012; Plass et al., 2012; Liu et al., 2012a). This technology might become a primary tool for live cell imaging studies, because this reaction has not only harmlessnesss for living cells but also a very fast rate constant as compared with other reactions (Lang and Chin, 2014b), whish is an important parameter for the quantitative labeling.

INDUCTION METHODS FOR THE ANCHOR GROUPS OF BIOMOLECULES IN LIVING SAMPLES

The application of click chemistry in bioscience is increasing every year. To date, a wide variety of "clickable" reagents for the fluorescent labeling of biomolecules have been identified, many of which are also commercially available. However, the more important and challenging issue pertaining to the fluorescent labeling of living samples with click chemistry is the induction of the anchor groups (i.e., the azide, alkyne, and phosphine groups) into the biomolecules inside living samples. The most widely used induction method for the anchor groups in living cells is through metabolic incorporation, which uses monomer analogs for biopolymers, such as nucleotides, sugars, and amino acids (Figure 2A). These molecular anchors are of a relatively smaller size, thereby making it easier to enzymatically incorporate the derivatives into the biopolymers.

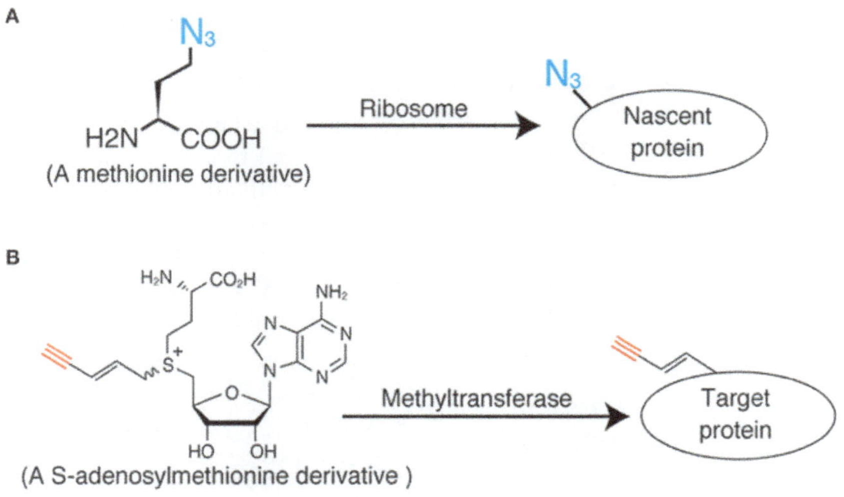

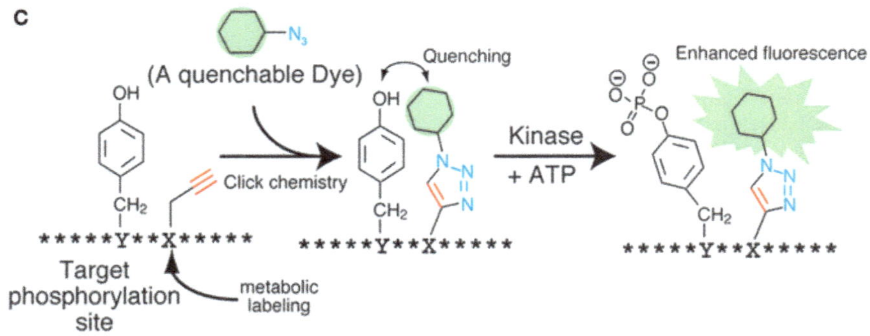

Figure 2. Applications of click chemistry for biomolecular labeling. (A) An example of the metabolic labeling methods. (B) An example of the enzymatic labeling methods. (C) A fluorescent chemosensor for detecting phosphorylation.

Among the metabolic labeling targets identified for click labeling, the sugar chains on the membrane proteins hold the biggest promise. Although the sugar chains on the membrane proteins have important roles in various physiological phenomena, including cell-cell adhesion, signal transduction, and immunologic response, their specific characteristics such as localization, amount, half-life, function, and the difference between cell types continue to remain elusive. In order to visualize the sugar chains containing particular types of sugars, different types of unnatural monosaccharide residues containing an azide group (azide sugars), i.e., N-azidoacetylmannosamine (Saxon and Bertozzi, 2000), N-azidoacetylglucosamine (Vocadlo et al., 2003), N-azidoacetylgalactosamine (Hang et al., 2003), and 6-azidofucose (Sawa et al., 2006), were produced and used for metabolic labeling studies. These derivatives were successfully incorporated into the sugar chains on the cell membrane through the intrinsic metabolic machinery, and were clearly visualized by following the click ligation with fluorescent dyes. The live imaging of the sugar chains uses this technique, and such experiments have already been done, not only in cultured cells but also in the tissues of living organisms. The copper-free click reagents such as difluorinated cyclooctyne (Figure 1B) have made this possible (Baskin et al., 2007). Therefore, the sugar chains could be regarded as a relatively easier target as compared to the other intracellular biopolymers.

Proteins have also been a primary target for analyses during metabolic labeling and click ligation. Several labeling reagents for newly synthesized proteins have been developed in recent years. Methionine derivatives bearing azido or alkyne groups, termed L-azidehomoalanine and L-homopropagylglycine, respectively (Wang et al., 2008), are the most widely used and commercially available reagents for the metabolic labeling of nascent proteins in living cells (Figure 2A). These are counterparts of the ^{35}S-labeled methionine used in radiographic analyses, and are able to label the nascent proteins synthesized during a particular period or under specific signals. Although these methionine analogs are useful and reliable, pretreatment of cells with a methionine-free medium is necessary (Soundrarajan et al., 2012), and this can affect the cells physiologically. Hence, alternatives with different labeling machineries were developed. The derivatives of puromycin, which is an inhibitor of protein biosynthesis, are one of the most promising reagents for the labeling of nascent proteins. The labeling method using these reagents needs no culture condition change, unlike the methionine analogs. Liu et al. reported an alkyne-conjugated puromycin derivative for nascent protein labeling in cultured cells and tissues (Liu et al., 2012b). Beatty and colleagues attempted the live cell imaging of nascent proteins in live fibroblast cells by using the L-azidehomoalanine- and a BODIPY®-conjugated cyclooctyne (Beatty et al., 2011). The fluorescent signals inside the cells were confirmed using confocal microscopy and protein fractionation. However, these data also revealed the preferential click labeling of membrane proteins, which might have resulted from the low cell penetrability of BODIPY®-conjugated cyclooctyne. In order to develop a reliable protein-labeling method for quantitative analyses, the development of membrane-penetrable click reagents seems essential. Genetic code reprograming technologies, which change the usage of codons, are also powerful method to introduce unnatural amino acids into proteins. Several teams have reported a lot of successful result of protein labeling by using the genetic code reprograming (Zhang et al., 2002; Chin et al., 2003).

The biosynthesis of RNA and DNA has also been analyzed in living cells using metabolic labeling followed by click ligation. The data collected from these kinds of analyses can be used to understand DNA replication, DNA repair, and transcriptional control. For example, the use of 5-ethynyl-2′deoxyuridine as a DNA precursor or the use of 5-ethynyluridine as an RNA precursor during microscopic or flow cytometric analyses enables quantitative analyses of DNA replication and whole transcription (Figure 2A) (Darzynkiewicz et al., 2011).

Metabolic labeling coupled with click labeling is a highly effective tool to analyze biomolecular transitions during the cell cycle, cell differentiation, signal responses, and apoptosis. However, the metabolic labeling methods for the anchor groups are not suitable for the analyses of particular species of biomolecules. Therefore, more specific labeling methods will be needed in the future.

NOVEL LABELING METHODS AND THE USE OF CLICK CHEMISTRY IN QUANTITATIVE BIOLOGY

Several technologies are currently under development for the molecular species-specific labeling of the anchor chemical groups present in biomolecules. The enzymatic labeling method for substrates is one such successful example. Many kinds of biomolecules such as proteins and nucleic acids are modified by methyl groups following neogenesis. These serve as important intrinsic functional markers, and are controlled by various methyltransferases. Prof. E. Weinhold and his colleagues developed analogs of S-adenosylmethionine, a methyl-group donor for almost all of the methyltransferases (Figure 2B) (Klimasauskas and Weinhold, 2007). These analogs work well *in vitro* and *in vivo*, and can modify the substrates with the anchor groups instead of the methyl groups in an enzyme-specific manner. By using this technology, DNAs (Schmidt et al., 2008), RNAs (Motorin et al., 2010), and proteins (Peters et al., 2010) have been successfully labeled with an alkyne or an azide. Similar enzymatic labeling methods for protein substrates have also been studied for other post-translational modifications, including phosphorylation (Lee et al., 2009) and acetylation (Yang et al., 2010). Although these technologies could potentially be used for substrate-specific labeling in living cells, an improvement of the molecular architecture of the methyl-donor analogs is necessary, because the degree of specificity and efficiency is still not up to the mark.

Other novel ideas using click chemistry for biological quantification have also been proposed. For example, Kamaruddin et al. devised a fluorescent chemosensor for detecting tyrosine phosphorylation, by metabolically labeling the residues near the phosphorylation sites with a quenchable dye (Figure 2C) (Kamaruddin et al., 2011). The strategy might enable quantitative phosphorylation analysis in an enzyme-specific manner in living cells.

Novel fluorescent labeling techniques have been continually devised in the past decade. Many more innovative fluorescence labeling methods that use click chemistry are expected to be developed in the near future to enhance research in the field of quantitative biology.

ACKNOWLEDGMENT

This work was supported by JSPS and MEXT KAKENHI Grant Numbers 24500447, 24116523.

REFERENCES

1. Agard, N. J., Prescher, J. A., and Bertozzi, C. R. (2004). A strain-promoted [3 + 2] azide-alkyne cycloaddition for covalent modification of biomolecules in living systems. *J. Am. Chem. Soc.* 126, 15046–15047. doi: 10.1021/ja044996f
2. Baskin, J. M., Prescher, J. A., Laughlin, S. T., Agard, N. J., Chang, P. V., Miller, I. A., et al. (2007). Copper-free click chemistry for dynamic *in vivo*imaging. *Proc. Natl. Acad. Sci. U.S.A.* 104, 16793–16797. doi: 10.1073/pnas.0707090104
3. Beatty, K. E., Szychowski, J., Fisk, J. D., and Tirrell, D. A. (2011). A BODIPY-cyclooctyne for protein imaging in live cells. *Chembiochem* 12, 2137–2139. doi: 10.1002/cbic.201100277
4. Best, M. D. (2009). Click chemistry and bioorthogonal reactions: unprecedented selectivity in the labeling of biological molecules. *Biochemistry*48, 6571–6584. doi: 10.1021/bi9007726
5. Blackman, M. L., Royzen, M., and Fox, J. M. (2008). Tetrazine ligation: fast bioconjugation based on inverse-electron-demand Diels-Alder reactivity. *J. Am. Chem. Soc.* 130, 13518–13519. doi: 10.1021/ja8053805
6. Boyce, M., and Bertozzi, C. R. (2011). Bringing chemistry to life. *Nat. Methods* 8, 638–642. doi: 10.1038/nmeth.1657
7. Chalfie, M., Tu, Y., Euskirchen, G., Ward, W. W., and Prasher, D. C. (1994). Green fluorescent protein as a marker for gene expression. *Science*263, 802–805.
8. Chang, P. V., Prescher, J. A., Hangauer, M. J., and Bertozzi, C. R. (2007). Imaging cell surface glycans with bioorthogonal chemical reporters. *J. Am. Chem. Soc.* 129, 8400–8401. doi: 10.1021/ja070238o
9. Chin, J. W., Cropp, T. A., Anderson, J. C., Mukherji, M., Zhang, Z., and Schultz, P. G. (2003). An expanded eukaryotic genetic code. *Science* 301, 964–967. doi: 10.1126/science.1084772
10. Darzynkiewicz, Z., Traganos, F., Zhao, H., Halicka, H. D., and Li, J. (2011). Cytometry of DNA replication and RNA synthesis: historical perspective and recent advances based on "click chemistry." *Cytometry A* 79, 328–337. doi: 10.1002/cyto.a.21048
11. Hang, H. C., Yu, C., Kato, D. L., and Bertozzi, C. R. (2003). A metabolic labeling approach toward proteomic analysis of mucin-type O-linked glycosylation. *Proc. Natl. Acad. Sci. U.S.A.* 100, 14846–14851. doi: 10.1073/pnas.2335201100
12. Hangauer, M. J., and Bertozzi, C. R. (2008). A FRET-based fluorogenic phosphine for live-cell imaging with the Staudinger ligation. *Angew. Chem. Int. Ed. Engl.* 47, 2394–2397. doi: 10.1002/anie.200704847
13. Heim, R., and Tsien, R. Y. (1996). Engineering green fluorescent protein for improved brightness, longer wavelengths and fluorescence resonance energy transfer. *Curr. Biol.* 6, 178–182.
14. Huisgen, R. (1963). 1,3-Dipolar cycloadditions. Past and future. *Angew. Chem. Int. Ed. Engl.* 2, 565–598. doi: 10.1002/anie.196305651
15. Jung, D., Min, K., Jung, J., Jang, W., and Kwon, Y. (2013). Chemical biology-based approaches on fluorescent labeling of proteins in live cells.*Mol. Biosyst.* 9, 862–872. doi: 10.1039/c2mb25422k
16. Kamaruddin, M. A., Ung, P., Hossain, M. I., Jarasrassamee, B., O'Malley, W., Thompson, P., et al. (2011). A facile, click chemistry-based approach to assembling fluorescent chemosensors for protein tyrosine kinases. *Bioorg. Med. Chem. Lett.* 21, 329–331. doi: 10.1016/j.bmcl.2010.11.005
17. Klimasauskas, S., and Weinhold, E. (2007). A new tool for biotechnology: AdoMet-dependent methyltransferases. *Trends Biotechnol.* 25, 99–104. doi: 10.1016/j.tibtech.2007.01.006
18. Kodama, K., Fukuzawa, S., Nakayama, H., Sakamoto, K., Kigawa, T., Yabuki, T., et al. (2007). Site-specific functionalization of proteins by organopalladium reactions. *Chembiochem* 8, 232–238. doi: 10.1002/cbic.200600432
19. Kolb, H. C., Finn, M. G., and Sharpless, K. B. (2001). Click chemistry: diverse chemical function from a few good reactions. *Angew. Chem. Int. Ed. Engl.* 40, 2004–2021.

doi: 10.1002/1521-3773(20010601)40:11<2004::AID-ANIE2004>3.0.CO;2-5

20. Lahann, J. (ed.). (2009). *Click Chemistry for Biotechnology and Materials Science*. Chichester: John Wiley & Sons, Ltd.

21. Lang, K., and Chin, J. W. (2014a). Cellular incorporation of unnatural amino acids and bioorthogonal labeling of proteins. *Chem. Rev.* 114, 4764–4806. doi: 10.1021/cr400355w

22. Lang, K., and Chin, J. W. (2014b). Bioorthogonal reactions for labeling proteins. *ACS Chem. Biol.* 9, 16–20. doi: 10.1021/cb4009292

23. Lang, K., Davis, L., Wallace, S., Mahesh, M., Cox, D. J., Blackman, M. L., et al. (2012). Genetic Encoding of bicyclononynes and trans-cyclooctenes for site-specific protein labeling *in vitro* and in live mammalian cells via rapid fluorogenic Diels-Alder reactions. *J. Am. Chem. Soc.* 134, 10317–10320. doi: 10.1021/ja302832g

24. Lee, S. E., Elphick, L. M., Anderson, A. A., Bonnac, L., Child, E. S., Mann, D. J., et al. (2009). Synthesis and reactivity of novel gamma-phosphate modified ATP analogues. *Bioorg. Med. Chem. Lett.* 19, 3804–3807. doi: 10.1016/j.bmcl.2009.04.028

25. Liu, D. S., Tangpeerachaikul, A., Selvaraj, R., Taylor, M. T., Fox, J. M., and Ting, A. Y. (2012a). Diels-Alder cycloaddition for fluorophore targeting to specific proteins inside living cells. *J. Am. Chem. Soc.* 134, 792–795. doi: 10.1021/ja209325n

26. Liu, J., Xu, Y., Stoleru, D., and Salic, A. (2012b). Imaging protein synthesis in cells and tissues with an alkyne analog of puromycin. *Proc. Natl. Acad. Sci. U.S.A.* 109, 413–418. doi: 10.1073/pnas.1111561108

27. Motorin, Y., Teimer, R., Koynov, K., Willnow, S., Weinhold, E., and Helm, M. (2010). Expanding the chemical scope of RNA: methyltransferases to site-specific alkynylation of RNA for click labeling. *Nucleic. Acids. Res.* 39, 1943–1952. doi: 10.1093/nar/gkq825

28. Nguyen, D. P., Elliott, T., Holt, M., Muir, T. W., and Chin, J. W. (2011). Genetically encoded 1,2-aminothiols facilitate rapid and site-specific protein labeling via a bio-orthogonal cyanobenzothiazole condensation. *J. Am. Chem. Soc.* 133, 11418–11421. doi: 10.1021/ja203111c

29. Peters, W., Willnow, S., Duisken, M., Kleine, H., Macherey, T., Duncan, K. E., et al. (2010). Enzymatic site-specific functionalization of protein methyltransferase substrates with alkynes for click labeling. *Angew. Chem. Int. Ed. Engl.* 49, 5170–5173. doi: 10.1002/anie.201001240

30. Plass, T., Milles, S., Koehler, C., Szymański, J., Mueller, R., Wiessler, M., et al. (2012). Amino acids for Diels-Alder reactions in living cells.*Angew. Chem. Int. Ed. Engl.* 51, 4166–4170. doi: 10.1002/anie.201108231

31. Rostovtsev, V. V., Green, L. G., Fokin, V. V., and Sharpless, K. B. (2002). A stepwise huisgen cycloaddition process: copper(I)-catalyzed regioselective "ligation" of azides and terminal alkynes. *Angew. Chem. Int. Ed. Engl.* 41, 2596–2599. doi: 10.1002/1521-3773(20020715)41:14<2596::AID-ANIE2596>3.0.CO;2-4

32. Sawa, M., Hsu, T. L., Itoh, T., Sugiyama, M., Hanson, S. R., Vogt, P. K., et al. (2006). Glycoproteomic probes for fluorescent imaging of fucosylated glycans *in vivo*. *Proc. Natl. Acad. Sci. U.S.A.* 103, 12371–12376. doi: 10.1073/pnas.0605418103

33. Saxon, E., and Bertozzi, C. R. (2000). Cell surface engineering by a modified Staudinger reaction. *Science* 287, 2007–2010. doi: 10.1126/science.287.5460.2007

34. Schmidt, F. H. G., Hüben, M., Gider, B., Renault, F., Teulade-Fichou, M. P., and Weinhold, E. (2008). Sequence-specific Methyltransferase-Induced Labelling (SMILing) of plasmid DNA for studying cell transfection. *Bioorg. Med. Chem.* 16, 40–48. doi: 10.1016/j.bmc.2007.04.054

35. Selvaraj, R., Liu, S., Hassink, M., Huang, C. W., Yap, L. P., Park, R., et al. (2011). Tetrazine-trans-cyclooctene ligation for the rapid construction of integrin $\alpha v\beta_3$ targeted PET tracer based on a cyclic RGD peptide. *Bioorg. Med. Chem. Lett.* 21, 5011–5014. doi: 10.1016/j.bmcl.2011.04.116

36. Shaner, N. C., Steinbach, P. A., and Tsien, R. Y. (2005). A guide to choosing fluorescent proteins. *Nat. Methods* 2, 905–909. doi: 10.1038/nmeth819

37. Shimomura, O., Johnson, F. H., and Saiga, Y. (1962). Extraction, purification and properties of aequorin, a bioluminescent protein from the luminous hydromedusan, Aequorea. *J. Cell. Comp. Physiol.* 59, 223–239.

38. Sletten, E. M., Nakamura, H., Jewett, J. C., and Bertozzi, C. R. (2010). Difluorobenzocyclooctyne: synthesis, reactivity, and stabilization by beta-cyclodextrin. *J. Am. Chem. Soc.* 132, 11799–11805. doi: 10.1021/ja105005t

39. Song, W., Wang, Y., Qu, J., and Lin, Q. (2008). Selective functionalization of a genetically encoded alkene-containing protein via "photoclick chemistry" in bacterial cells. *J. Am. Chem. Soc.* 130, 9654–9655. doi: 10.1021/ja803598e

40. Soundrarajan, N., Sokalingam, S., Raghunathan, G., Budisa, N., Paik, H. J., Yoo, T. H., et al. (2012). Conjugation of proteins by installing BIO-orthogonally reactive groups at their N-termini. *PLoS ONE* 7:e46741. doi: 10.1371/journal.pone.0046741

41. Stephens, D. J., and Allan, V. J. (2003). Light microscopy techniques for live cell imaging. *Science* 300, 82–86. doi: 10.1126/science.1082160

42. Tong, G. J., Hsiao, S. C., Carrico, Z. M., and Francis, M. B. (2009). Viral capsid DNA aptamer conjugates as multivalent cell-targeting vehicles. *J. Am. Chem. Soc.* 131, 11174–11178. doi: 10.1021/ja903857f

43. Tornøe, C. W., Christensen, C., and Meldal, M. (2002). Peptidotriazoles on solid phase: [1,2,3]-triazoles by regiospecific copper(I)-catalyzed 1,3-dipolar cycloadditions of terminal alkynes to azides. *J. Org. Chem.* 67, 3057–3064. doi: 10.1021/jo011148j

44. Vocadlo, D. J., Hang, H. C., Kim, E. J., Hanover, J. A., and Bertozzi, C. R. (2003). A chemical approach for identifying O-GlcNAc-modified proteins in cells. *Proc. Natl. Acad. Sci. U.S.A.* 100, 9116–9121. doi: 10.1073/pnas.1632821100

45. Wang, A., Winblade-Nairn, N., Johnson, R. S., Tirrell, D. A., and Grabstein, K. (2008). Processing of N-terminal unnatural amino acids in recombinant human interferon-beta in *Escherichia coli*. *Chembiochem* 9, 324–330. doi: 10.1002/cbic.200700379

46. Xue, D., Tu, Y., and Chalfie, M. (1993). Cooperative interactions between the Caenorhabditis elegans homeoproteins UNC-86 and MEC-3. *Science*261, 1324–1328.

47. Yang, Y.-Y., Ascano, J. M., and Hang, H. C. (2010). Bioorthogonal chemical reporters for monitoring protein acetylation. *J. Am. Chem. Soc.* 132, 3640–3641. doi: 10.1021/ja908871t

48. Zhang, Z., Wang, L., Brock, A., and Schultz, P. G. (2002). The selective incorporation of alkenes into proteins in *Escherichia coli*. *Angew. Chem. Int. Ed. Engl.* 41, 2840–2842. doi: 10.1002/1521-3773(20020802)41:15<2840::AID-ANIE2840>3.0.CO;2-#

INDEX

Q

R

S

T

U

W